Statistical Inference
Theory of Estimation

Manoj Kumar Srivastava
Associate Professor
Department of Statistics
Institute of Social Sciences
Dr. Bhim Rao Ambedkar University
Agra

Abdul Hamid Khan
Professor and Ex-Chairman
Department of Statistics and Operations Research
Aligarh Muslim University
Aligarh

Namita Srivastava
Associate Professor
Department of Statistics
St. John's College
Agra

PHI Learning Private Limited

Delhi 110092
2014

₹695.00

STATISTICAL INFERENCE: Theory of Estimation
Manoj Kumar Srivastava, Abdul Hamid Khan, and Namita Srivastava

ISBN-978-81-203-4930-8

The export rights of this book are vested solely with the publisher.

Published by Asoke K. Ghosh, PHI Learning Private Limited, Rimjhim House, 111, Patparganj Industrial Estate, Delhi-110092 and Printed by Rekha Printers Private Limited, New Delhi-110020.

Contents

Preface *vii*

1. INTRODUCTION **1–37**

1.1 General Theory of Estimation *1*
- 1.1.1 Introduction *1*
- 1.1.2 Statistical Model *2*
- 1.1.3 Sample Information and Estimators *3*
- 1.1.4 Loss Function *3*
- 1.1.5 Optimal Estimators and Their Criteria *4*
- 1.1.6 Examples on Point Estimation *6*

1.2 Probability Distributions *8*
- 1.2.1 Group Family of Distributions *8*
- 1.2.2 Group of Transformations *9*
- 1.2.3 Exponential Family of Distributions *11*
- 1.2.4 Non-regular (Pitman) Family of Distributions *12*
- 1.2.5 Discrete Distributions *13*
- 1.2.6 Continuous Distributions *14*
- 1.2.7 Multivariate Distributions *19*
- 1.2.8 Exact Sampling Distributions *21*

1.3 Calculus and Analysis *22*

1.4 Small '$\mathbf{o}(\cdot)$' and Big '$\mathbf{O}(\cdot)$' Notations *27*

1.5 Some Results Useful for Consistency *28*

2. DATA SUMMARIZATION AND PRINCIPLE OF SUFFICIENCY **38–109**

2.1 Introduction *38*
- 2.1.1 Problem of Point Estimation *40*
- 2.1.2 Existence of Uniformly Minimum Risk Estimator and Unbiasedness *40*

2.2 Data Summarization—Sufficient Statistics 41
2.2.1 Reconstruction of the Original Sample by a Sufficient Statistic *T* 42
2.3 Construction of a Sufficient Statistic 43
2.4 Minimal Sufficient Statistic 46
2.5 Data Summarization through Ancillary Statistics 50
2.5.1 Independence of Complete Sufficient Statistic with Ancillary Statistic 54
2.6 Convex Functions in Estimation Problems 58
2.7 Improved Estimators through Sufficiency 60
2.8 Solved Examples 61
Exercises 102

3. UNBIASED ESTIMATION **110–183**

3.1 Introduction 110
3.2 UMVU Estimation 112
3.3 Sufficiency and UMVU Estimation 116
3.4 Solved Examples 117
Exercises 171

4. INFORMATION INEQUALITY **184–264**

4.1 Introduction 184
4.2 Regular Family of Distributions, Score Function, and Fisher Information 185
4.3 Lower Bounds for Variance of Unbiased Estimators 190
4.3.1 Most Efficient, UMVU Estimators, and Attainment of CR Lower Bound 193
4.3.2 Family for which CR Lower Bound is Attained (Most Efficient Estimators are UMVUEs) 194
4.3.3 Bhattacharyya Lower Bound 202
4.3.4 Chapman, Robbin, and Kiefer Lower Bound 206
4.3.5 Relationship between CRLB and CRKLB 207
4.4 Solved Examples 208
Exercises 256

5. ASYMPTOTIC THEORY AND CONSISTENCY **265–336**

5.1 Introduction 265
5.2 Consistency of an Estimator 266
5.3 Method for Checking Consistency 269
5.4 Invariance Principle of Consistency under Continuous Functions 274
5.5 Rate of Consistency 276
5.6 Methods of Constructing Consistent estimators 278
5.6.1 Method of Moments (MoM) 278
5.6.2 Method of Percentiles (MoP) 280
5.7 Optimality and CAN Estimators 282
5.7.1 Optimality among Consistent Estimators 282

5.7.2 Asymptotic Normality of Consistent Estimators (CAN) 283
5.7.3 Principle of Invariance of CAN Estimators 285
5.7.4 Asymptotic Efficiency of MLE 288
5.8 Methods of Finding CAN Estimators and their Properties 289
5.8.1 Method of Moments (MoM) Estimation 290
5.8.2 Method of Percentiles (MoP) 297
5.9 Solved Examples 298
Exercises 332

6. METHODS OF ESTIMATION 337–444

6.1 Introduction 337
6.2 Method of Moments (MoM) 337
6.3 Method of Minimum Chi–Square (MoMCS) Estimation 340
6.4 Method of Modified Minimum Chi-Square (MoMMCS) Estimation 343
6.5 Method of Least Squares (MoLS) Estimation 344
6.6 Method of Maximum Likelihood (MoML) Estimation 345
6.6.1 Calculus of Finding MLE 347
6.6.2 MLE under a Transformation 349
6.6.3 Modified MoME and MLE in Exponential Families 351
6.6.4 General Properties of MLE 354
6.6.5 Invariance Property of MLE 356
6.6.6 Large Sample Properties of MLEs for Regular Models 358
6.6.7 Superefficiency 367
6.6.8 Variance Stabilization 368
6.6.9 Maximum Likelihood Estimators Using Fisher's Scoring Method 369
6.7 Solved Examples 373
Exercises 423

7. PRINCIPLE OF EQUIVARIANCE 445–493

7.1 Introduction 445
7.2 Principle of Equivariance 446
7.2.1 Basic Concepts and Definitions 446
7.2.2 Examples on Principle of Equivariance 447
7.2.3 Formal Structure 449
7.3 Minimum Risk Equivariant Estimator for Location Family 451
7.3.1 Pitman Estimator 452
7.4 Pitman Estimator for Scale Families 456
7.5 Pitman Estimator for Location-Scale Families 460
7.6 Solved Examples 464
Exercises 488

8. BAYES AND MINIMAX ESTIMATION 494–676

8.1 Introduction 494
8.1.1 Elements of Decision Theory 495
8.1.2 Bayes and Minimax Estimation—Elementary Concepts 499
8.1.3 Bayes and Minimax Ordering 499

8.2 Bayes Estimation *502*
8.2.1 Limit of Bayes Estimators *512*
8.2.2 Generalized Bayes Estimator *512*
8.2.3 Empirical Bayes Estimator *512*
8.2.4 Hierarchical Bayes Estimators *513*
8.2.5 Bayes Estimate and Admissibility *514*
8.2.6 Bayesian Inference Agrees to the Likelihood Principle *517*
8.3 Natural Conjugate Prior Distributions *518*
8.4 Duality between Loss Function and Prior Distribution *520*
8.5 Noninformative Priors *522*
8.5.1 Jeffreys Invariance Principle *527*
8.5.2 Invariance of Jeffreys Prior under Reparameterization *528*
8.5.3 Jeffreys Prior in Exponential Distribution *530*
8.5.4 Jeffreys Prior Violates the Likelihood Principle *532*
8.6 Minimax Estimation *533*
8.6.1 Minimax Estimator *533*
8.7 Solved Examples *541*
Exercises *658*

9. CONFIDENCE INTERVAL ESTIMATION 677–790

9.1 Introduction *677*
9.2 Basic Notations and Definitions *678*
9.3 Methods of Constructing Confidence Intervals *679*
9.3.1 Confidence Interval Based on a Pivotal Quantity *679*
9.3.2 Confidence Interval by Inverting Acceptance Region of a Test *694*
9.3.3 Confidence Intervals Based on Posterior Distribution *696*
9.3.4 Confidence Intervals Based on Large Samples *699*
9.3.5 Confidence Intervals Based on Chebyshev's Inequality *704*
9.4 Optimality of Confidence Interval Estimators *705*
9.4.1 Shortest-Length Confidence Interval *705*
9.4.2 Minimum Probability of False Coverage Confidence Intervals—Unbiased Confidence Intervals *712*
9.4.3 Minimum Expected Length Confidence Interval *713*
9.5 Equivariant Confidence Intervals *715*
9.6 Solved Examples *715*
Exercises *775*

Index **791–794**

Preface

During our teaching at Master's level, we realized the need of a book for core papers on Statistical Inference namely *Theory of Estimation* and *Testing of Hypotheses*, which could cover most of the syllabi of undergraduate and postgraduate courses prescribed by different universities and competitive examinations. We felt that the students require books solely dedicated to the *Theory of Estimation* and *Testing of Hypotheses*, emphasizing on concept building, detailing of proofs of main theorems, their real life applications in different statistical models and critical and analytical remarks to explain and develop more insight of the subject. In order to serve these academic requirements of the students, we have written a book on *Testing of Hypotheses*, which was published by PHI Learning in the year 2010. In the same sequel, the present book is being released on the *Theory of Estimation* that includes both point and interval estimation. The two books are independent of each other, therefore, can be read or taught in either order.

The text is a full semester course on Theory of Point and Interval Estimation covering the syllabi of Master's level of various Indian universities and of different competitive examinations such as I.A.S., I.S.S., UGC/CSIR-NET etc. This book discusses the Theory of Estimation by using classical approach. Bayesian approach to the Theory of Estimation is also discussed that includes sections on Empirical Bayes, Hierarchical Bayes and Equivariant estimators. The book deals with small sample theory of parametric estimation where optimal estimators and their statistical properties are discussed by using the criteria of unbiasedness, equivariance and minimaxity. The large-sample approach to the theory of point estimation leading to asymptotic optimality theory is also discussed.

The material in each chapter is self-contained and supplemented by numerous solved problems and exercises of varied nature, so selected and framed at different levels of difficulty that it suits the requirements of the students at that level. In addition to it, each chapter provides not only practice problems for students but also many additional results as complementary material to the main text. Thus, the solved problems and exercises that illustrate the applications of different theorems and results discussed form an important part of the book and will prove beneficial for the students.

The book is organized in nine chapters:

Chapter 1 of the book introduces problem of estimation in a statistical model along with certain real life examples and briefly lists basics of mathematical statistics, calculus of integrals and differentiation and fundamentals of large sample theory which are needed to grasp the concepts used in the book.

Chapter 2 introduces the problem of estimation in Decision Theoretic set up as a starting point for a course on Theory of Point Estimation where the data summarization through the principle of sufficiency, Halmos and Savage (1949) factorization theorem to characterize a sufficient statistic and the minimal sufficient statistic that results into greatest reduction in data summarization by introducing levels of data summarization and by partitioning over the sample space are discussed. The chapter also introduces the Basu (1955) theorem that states that complete sufficient statistic is independent of every ancillary statistic. Basu theorem discusses as to how it results into simplification of conditional calculations on its applications. The chapter discusses the Rao (1949) and Blackwell (1947) theorem under the convex loss function that gives an improved estimator by using a sufficient statistic. Numerous examples illustrating the theory are given in the form of solved examples and exercises at the end of the chapter.

Chapter 3 introduces unbiasedness as an impartiality principle of an estimator and shows that how the problem of estimation, i.e. of finding uniformly minimum variance unbiased estimator (UMVUE) is simplified by adopting this principle and the principle of sufficiency. Further, in this chapter two methods of finding UMVUE are discussed, based on the Cramer–Rao lower bound (1946) for variance of an unbiased estimator and by using the property of an UMVUE that it is uncorrelated with every zero estimator. The role of a complete sufficient statistic in finding an UMVUE, known as Lehmann–Scheffe theorem (1950) has been discussed. To illustrate the applications of the theorems and methods of finding UMVUE in various statistical situations, numerous examples and exercises are given at the end of the chapter.

In Chapter 4, the Fisher information (1922) has been defined as a measure of information contained in a sample about a parameter of a statistical model and subsequently it is used under certain regularity conditions to obtain lower bound for the variance of an unbiased estimator which is known as Cramer–Rao (1946) lower bound. Further, relaxing on these regularity conditions, a less sharp lower bound is introduced known as Chapman, Robbins and Kiefer lower bound (1951). Bhattacharya (1946) series of lower bounds has been introduced that approaches to the minimum variance of an unbiased estimator under certain regularity conditions. Further, these lower bounds are obtained. Numerous analytical questions relating to them are explained.

Large sample properties of an estimator such as consistency, consistent asymptotically normality (CAN), best asymptotically normality (BAN), and methods to construct such estimators are discussed in Chapter 5. Also, large sample optimality criterion due to Rao (1963), has also been discussed for judging between such estimators. Numerous examples and exercises, illustrating the applications of the theorems and results of the chapter in various statistical models, are given.

Chapter 6 deals with the conventional methods of estimation and discusses the importance of these methods. In particular, discussion on method of maximum likelihood estimation (MLE) is given in much detail. Fisher's scoring method of finding these estimators and large sample properties of MLE's are the main attraction of this chapter. Numerous illustrations of finding estimators are given at the end of the chapter.

In Chapter 7, the principle of equivariance is discussed as an impartiality criterion by restricting only to such estimators which satisfy certain symmetry requirements corresponding to such symmetry structure present in the statistical model. Pitman estimators for location, scale and location-scale models are derived and are calculated for various statistical models with numerous exercises at the end of the chapter.

Chapter 8 discusses Bayes and minimax estimation under the Decision Theoretic settings as central core of Bayesian estimation. Different types of loss functions and their applicability in different situations are discussed. A number of popular natural conjugate prior distribution are listed for different sampling distributions. A detailed discussion is given on Jeffreys (1961) noninformative priors. Limit of Bayes, generalized Bayes, extended Bayes, empirical Bayes and hierarchical Bayes estimation have also been discussed. A large number of examples and exercises on Bayes and minimax estimation are given at the end of the chapter.

Chapter 9 deals with confidence interval estimation by pivoting the cumulative distribution function. It also explains as to how these are obtained by inverting the acceptance region of a testing problem. This chapter also discusses the construction of credible intervals. Optimality of confidence intervals by considering the criteria of shortest-length, the criterion of minimum probability of false coverage among unbiased confidence interval and the Pratt (1961) and Guenther (1971) minimum expected length criterion are discussed. Numerous examples are given for obtaining optimal confidence interval estimators.

We are thankful to Prof. R.K. Singh, University of Lucknow, for critical remarks that have led to the improvement of the presentation.

Inspite of our best efforts, there might be still some errors and misprints in the presentation. We owe these mistakes and request the readers to kindly bring these to our notice with their comments.

Authors

Chapter 6 deals with the conventional methods of estimation and discusses the importance of these methods. In particular, discussion on method of maximum likelihood estimation (MLE) is given in much detail. Fisher's scoring method of finding these estimators and large sample properties of MLE's are the main attraction of this chapter. Numerous illustrations of finding estimators are given at the end of the chapter.

In Chapter 7, the principle of equivariance is discussed as an impartiality criterion by restricting only to such estimators which satisfy certain symmetry requirements corresponding to such symmetry structure present in the statistical model. Pitman estimators for location, scale and location-scale models are derived and are calculated for various statistical models with numerous exercises at the end of the chapter.

Chapter 8 discusses Bayes and minimax estimation under the Decision Theoretic settings as central core of Bayesian estimation. Different types of loss functions and their applicability in different situations are discussed. A number of popular natural conjugate prior distribution are listed for different sampling distributions. A detailed discussion is given on Jeffreys (1961) noninformative priors. Limit of Bayes, generalized Bayes, extended Bayes, empirical Bayes and hierarchical Bayes estimation have also been discussed. A large number of examples and exercises on Bayes and minimax estimation are given at the end of the chapter.

Chapter 9 deals with confidence interval estimation by pivoting the cumulative distribution function. It also explains as to how these are obtained by inverting the acceptance region of a testing problem. This chapter also discusses the construction of credible intervals. Optimality of confidence intervals by considering the criteria of shortest length, the criterion of minimum probability of false coverage among unbiased confidence interval and the Pratt (1961) and Guenther (1971) minimum expected length criterion are discussed. Numerous examples are given for obtaining optimal confidence interval estimators.

We are thankful to Prof. R.K. Singh, University of Lucknow, for critical remarks that have led to the improvement of the presentation.

Inspite of our best efforts, there might be still some errors and misprints in the presentation. We owe these mistakes and request the readers to kindly bring these to our notice with their comments.

Authors

1

Introduction

1.1 GENERAL THEORY OF ESTIMATION

1.1.1 Introduction

In an estimation problem, one needs to obtain a point or an interval estimate of the value of some unknown characteristic of the population, which is known as a parameter. The parameter describes the nature of the population. One can make statistical inference about the population by knowing its parameter. The experimenter collects observations on the nature of the population. He, then, uses these observations, to estimate the parameter. By an interval estimate, we mean a measure of accuracy of the point estimate by prescribing an interval that contains the plausible values of the parameter.

During the period 1920–30, R.A. Fisher and others have laid down the foundations of the *theory of estimation*. Specially, Fisher (1922), in his paper entitled "*On Mathematical Foundation of Theoretical Statistics*", describes the problem of estimation as the combination of the following three statistical procedures:

1. Specification of the *statistical model*, which is known as the *distribution of the population.*
2. *Estimation of the parameter* θ that indexes the population distribution, on the basis of a sample of observations drawn from the model in (1).
3. Problem of obtaining *sampling distribution of the estimators* of the θ in (2), for the purpose of comparing their performances and finally to arrive at one estimator of θ, using these sampling distributions, which is the most optimal in some desirable sense.

We formulate the problems of estimation in this section and discuss them in subsequent chapters. We also briefly list some elementary results from the general theory of statistics in this section, which are useful for development of theory of estimation in subsequent chapters. Section 1.2 discusses all kinds of distributions, which are relevant to the theory of statistical inference. Section 1.3 discusses some useful results from calculus and analysis. Some notations and results are given in Sections 1.4 and 1.5, which are useful for discussion on consistency and asymptotic theory of estimation in Chapter 5.

1.1.2 Statistical Model

An *experiment* ξ is called a *statistical* (*random*) *experiment* if it can be repeated under identical conditions infinitely, where all possible outcomes are known but the outcome of a particular trial is not known before it is conducted. The set of all possible outcomes is denoted by Ω and a σ–field generated by it is denoted by $\mathcal{S}$. The pair $(\Omega, \mathcal{S})$ is called a sample space of the statistical experiment ξ.

A single point set in $\mathcal{S}$ is called an *elementary event* and every subset in $\mathcal{S}$ is called an *event.* A sample space $(\Omega, \mathcal{S})$ is called *finite* if Ω is finite; *discrete* if Ω is countable; and *continuous* if $\Omega = \mathbb{R}^1$ or Ω is an interval in $\mathbb{R}^1$. In case $\Omega = \mathbb{R}^1$, the σ–field $\mathcal{S}$ contains all single points and all closed, open, and semi closed intervals as events. The σ–field $\mathcal{S}$ is then called the *Borel field* and is denoted by $\mathcal{B}$. The corresponding sample space in this case is denoted by $(\Omega, \mathcal{B})$.

A set function defined on a *sample space* $(\Omega, \mathcal{S})$ is called a *probability measure* if it satisfies the following three conditions:

1. $P(A) \geq 0 \;\forall A \in \mathcal{S}$
2. $P(\Omega) = 1$
3. For any sequence of disjoint sets $\{A_j\}$, $A_j \in \mathcal{S}$,

$$P\left(\bigcup_{j=1}^{\infty} A_j\right) = \sum_{j=1}^{\infty} P(A_j)$$

The property (3) is called the property of countable additivity. The triplet $(\Omega, \mathcal{S}, P)$ is called the *probability space.*

A finite real valued function defined on $\mathcal{S}$ i.e. $X: \mathcal{S} \to \chi \subseteq \mathbb{R}^1$, is called a *random variable* (*rv*) if its inverse image is an event i.e.,

$$X^{-1}(B) = \{w: X(w) \in B\} \in \mathcal{S} \;\forall\; B \in \mathcal{B}$$

where B is the Borel set in $\mathcal{B}$. The set of values taken by the random variable X is denoted by χ. $\mathcal{B}$ is the Borel field generated by the class of all semiclosed intervals B in $\mathbb{R}^1$. B may be $(-\infty, x]$, $x \in \mathbb{R}^1$.

A random variable X is defined on the probability space $(\Omega, \mathcal{S}, P)$, which induces a probability space $(\mathbb{R}^1, \mathcal{B}, F)$, where the function $F(\cdot)$ is defined as

$$F(B) = P\{w: X(w) \in B\} = P\{X^{-1}(B)\} \;\forall\; B \in \mathcal{B}$$

F is called the cumulative distribution function (*cdf*) of X. If $B = (-\infty, x]$, then $F(B)$ simplifies to

$$F(x) = F(-\infty, x] = P\{w: X(w) \leq x\} \;\forall\; x \in \mathbb{R}^1$$

An important property of a distribution function is that it is nondecreasing and right continuous.

A random variable X on the probability space $(\Omega, \mathcal{S}, P)$ is called *discrete* if it takes on at most countable values. If X takes on atmost countable values $x_1, x_2, \ldots$ with probabilities $P\{w: X(w) = x_i\} = p_i$, $i = 1, 2, \ldots, p_i \geq 0$ so that $\Sigma p_i = 1$, then the set of numbers $\{p_i\}$ is called the *probability mass function* (*pmf*) of the random variable X. Further, a random variable (*rv*) X is called *continuous* if F is absolutely continuous. In case F is continuous, there exists a function f such that

$$F(x) = \int_{-\infty}^{x} f(u)du$$

The function f is called the *probability density function* (*pdf*) of the *rv* X. If F is absolutely continuous and f is continuous, then

$$\frac{\partial}{\partial x} F(x) = f(x)$$

Consider a random experiment ξ that results into an *rv* X, describing the stochastic behaviour of some characteristics of the population. The distribution $f(\cdot, \theta)$ of an *rv* X is known as *statistical model* where θ is a measure of some physical quantity, relating to the population, known as parameter which takes on values from $\Theta \subseteq \mathbb{R}^1$, which is known as parameter space. The statistical model describes the stochastic behavior of the characteristic X in the experiment. The model $f(\cdot, \theta)$ is referred to as *pmf* if X is discrete and *pdf* if X is continuous.

If the parameter θ is known, then the *pdf* $f(\cdot, \theta)$ would have been completely specified and, hence, the stochastic behavior of the characteristic under study would have been completely known. Then, in this case, we could have made any statistical inference about θ. But, practically, the functional form f of the model is known; however, the parameter θ is not known. The problem of estimation is to estimate the parameter θ based on information $X_1, X_2, \ldots, X_n$ collected on the behavior of random variable X and based on the loss incurred to the experimenter due to incorrect estimation. This is the subject matter of the theory of estimation, which we will discuss throughout the book under different statistical settings. Hence, the justification of the theory of point and interval estimation.

1.1.3 Sample Information and Estimators

After specifying the statistical model $f(\cdot, \theta)$, the purpose of estimation theory is to arrive at some reasonable estimator $\delta(\mathbf{X})$ of the unknown parameter θ as its guess value based on the random observations $X_1, \ldots, X_n$ drawn from the model. We expect that the guessed estimate is very close to the parameter θ. The value taken on by the estimator $\delta(\mathbf{X})$ corresponding to the sample $(X_1, \ldots, X_n) = (x_1, \ldots, x_n)$ is called the estimate and is denoted by $\delta(\mathbf{x})$.

1.1.4 Loss Function

The fixed but unknown quantity θ is estimated by the estimator $\delta(\mathbf{X})$. This choice of the statistician $\delta(\mathbf{X})$ as a guess of θ results into a loss to the statistician. This loss is a real-valued function of the difference between $\delta(\mathbf{X})$ and θ and is known as the *loss function* and is denoted by $L(\theta,\delta(\mathbf{X}))$. If it is low, the loss incurred to the statistician is also low. However, if the difference is high, the loss incurred to the statistician is also very high.

If the estimate $\delta(\mathbf{x}) = \theta$, then there is no loss to the experimenter or the statistician. However, if $\delta(\mathbf{x}) \neq \theta$, the statistician has to bear a positive loss. More formally, estimating θ by δ, the consequences met by the statistician are measured by the loss function $L(\theta,\delta)$ so that

$$L(\theta,\delta) \geq 0 \quad \text{whatever be } \theta \text{ and } \delta$$

and
$$L(\theta,\theta) = 0 \quad \text{whatever be } \theta \tag{1.1.1}$$

where $L(\theta, \delta)$ is an *rv* since δ is an *rv*.

For a suitable loss function the risk of an estimator $\delta(\mathbf{x})$ is defined as

$$R(\theta,\delta) = E_\theta L(\theta, \delta(\mathbf{x})) \tag{1.1.2}$$

where the expectation is considered with respect to the assumed model. Note that the risk in Eq. (1.1.2) is an average loss and it is assumed that the expectation is defined and $R(\theta, \delta)$ is finite, i.e., $R(\theta, \delta) < \infty \ \forall \ \theta$. In the problem of estimation, the risk is considered as a measure of accuracy of an estimator. Usually, a well defined class of estimators are considered by applying some suitable impartiality principle such as unbiasedness or equivariance. Such classes contains reasonable estimators in them and are usually small in dimension so that the search for an *optimal estimator* become easier. The statistician then tries to obtain an optimal estimator for θ in such classes for which the risk is minimum.

Some popular and mathematically convenient loss functions are listed as follows:

1. $L(\theta,\delta) = (\delta - \theta)^2$
2. $L(\theta,\delta) = w(\theta)(\delta - \theta)^2$, where $w(\theta)$ is some function of θ (1.1.3)
3. $L(\theta,\delta) = |\delta - \theta|^p$ for some $p > 0$

Detailed description of different types of loss functions is given in Chapter 8.

1.1.5 Optimal Estimators and Their Criteria

Since the risk of an estimator δ in Eq. (1.1.2) is the measure of its accuracy, one would like to choose such an estimator which minimizes this risk. We will see in the following chapters that there does not exist such an estimator in the class of all estimators of θ and, therefore, we require to reduce the class of all estimators by imposing some impartiality principle such as *unbiasedness* or *equivariance*. In such classes, we generally search for an estimator for which the risk is uniformly minimum over other estimators. However, this criterion of finding a best estimator is very strong and in many cases, it does not exist except in some commonly occurring situations. Therefore, the use of this optimality criterion limits its applicability.

The other two criteria of measuring the performance of an estimator are *Bayes risk* and *minimax risk*. The Bayes risk of an estimator δ with respect to a given prior distribution $\pi(\theta)$ is defined by

$$r(\pi, \delta) = E^\theta R(\theta, \delta)$$

$$= \begin{cases} \int R(\theta,\delta)\pi(\theta)d\theta, & \text{when } \pi(\theta) \text{ is continuous} \\ \sum R(\theta_i,\delta)\pi(\theta_i), & \text{when } \pi(\theta) \text{ is discrete} \end{cases} \tag{1.1.4}$$

An estimator that minimizes Eq. (1.1.4) is called the *Bayes estimator* of θ with respect to a given *prior distribution* $\pi(\theta)$. Next, the maximum risk of an estimators is defined by

$$\sup_{\theta} R(\theta,\delta) \tag{1.1.5}$$

An estimator δ that minimizes Eq. (1.1.5) is called the *minimax estimator*.

Therefore, as we see, there can be different measures of accuracy of estimators leading to different optimal estimators. The choice of optimality criterion actually depends on the experimental circumstances and the statistical problem in hand.

The measure of accuracy of an estimator depends on the choice of an appropriate loss function among several available loss functions. Gauss (1821) proposed the use of squared error loss function as a measure of the accuracy of an estimator and justified its use on the ground of its mathematical simplicity and convenience.

The general theory of estimation is developed for a large class of loss functions where not many restrictions are placed on them. Only one restriction of convexity needs to be imposed on the loss function $L(\theta,\delta)$. The loss functions (1) and (2) in Eq. (1.1.3) are convex and (3) is also convex for $p \geq 1$. Squared error loss function is convex, results in simple calculations, and yields closed form (explicit) estimators. In this set–up of estimation problem, we try summarizing the data $(X_1,\ldots,X_n)$ to some statistic $T(X_1,\ldots,X_n)$ which takes on values from a space of smaller dimension before we attempt to estimate θ, so that there is no loss of information in doing so. In other words, $(X_1,\ldots,X_n)$ and $T(X_1,\ldots,X_n)$ must contain the same amount of information about θ. Such statistics are known as *sufficient statistics*, a complete discussion of which is given in Chapter 2. Two immediate advantages of this summarization are as follows:

1. We now deal with a smaller dimension mostly 1 or 2 of T as compared to the dimension n of sample space χ depending on θ and the model $f(\cdot,\theta)$.
2. Given any estimator of θ, we get another estimator possibly a randomized estimator as a function of sufficient statistic that has the same risk as of the considered one.

In addition to this, if the loss function $L(\theta,\delta)$ is, for a fixed θ, a convex function of δ, then the problem of estimation further simplifies in a number of ways. One basic property of the convex loss function defined on an open interval is that there exists a line passing through a point on it, so that it keeps all the points on it, above the line. This property proves the famous *Jensen's inequality*. This inequality can be used to show that given any estimator, there exists an estimator depending only on a sufficient statistic which is better than the considered one, provided the loss function $L(\theta,\delta)$ is strictly convex in δ. This result is known as the famous *Rao-Blackwell theorem*. Thus, when the loss function is convex, the problem of finding an optimal estimator from a big class simplifies by restricting only to such estimators in the class which depend on a sufficient statistic. Further, the consideration of convex loss function has an additional advantage. Given any randomized estimator of the parameter, there exists a nonrandomized estimator, which is uniformly better if the loss function is strictly convex. However, if the loss function is only convex, then there exists a nonrandomized estimator, which is as good as a given randomized estimator. Thus, the problem of estimation now reduces in restricting the class of estimators to such a class where the estimators are the function of a sufficient statistic and they are nonrandomized.

The Fisher (1922) specification of the model gives rise to different families of distributions. The exponential and group families of distributions (defined in Section 1.2) are most popular, since both of them result in great simplification of data. In exponential families, data can be reduced to a statistic of fixed dimension, irrespective of the sample size. The distributions in group families have some symmetry structures. Recognizing these structures and imposing these symmetry structures on the estimators reduces the dimensionality of the class of estimators. Therefore, it finally results in the simplification of finding an optimal estimator. The location family of distributions is, in particular, a group family. In this case, the problem of estimation is to estimate the location parameter θ under the loss function of the type $L(\theta,\delta) = \rho(\delta - \theta)$, where ρ is a function of $(\delta - \theta)$. One chooses such an estimator $\delta = \delta_0 - a$ that satisfies some symmetry requirement (or equivariance), similar to the symmetry structure of the model, such that the risk

$$R(0,\delta) = E_0\,\rho(\delta_0(X) - \delta) = R(0, \delta)$$

attains a minimum for some value of a. If ρ in the loss function $L(\theta,\delta) = \rho(\delta - \theta)$ is convex and not monotone, then $E_0\,\rho(\delta_0(X) - a)$ takes on its minimum in a closed interval. This minimum is unique if ρ is strictly convex. The estimator $\delta = \delta_0 - a_0$, where a_0 is the minimizing value of $E_0\,\rho(\delta_0(X) - a)$, is called the *minimum risk equivariant estimator* of the location parameter θ. Thus, an optimal estimator for a location parameter exists for convex loss functions. It is for this reason that estimation theory discusses the problem of estimating the parameter θ under convex loss functions, more particularly for squared error loss functions.

The second part of estimation theory relates to the estimation of a measure of accuracy of the point estimate by prescribing an interval that contains the plausible values of the parameter. Thus, the confidence interval serves as interval estimates at a certain level say at 95% level. Such an interval is called 95% *confidence interval*. This means that if sampling is repeated under identical conditions, the confidence interval will contain 95% of times the true value of the parameter. Thus, 95% is the level of confidence that the interval estimate contains the true value of the parameter.

1.1.6 Examples on Point Estimation

We will discuss here some examples which will demonstrate the problem of estimation.

It may be of interest to estimate the proportion of population that is in favour of an effective anticorruption bill to be passed by the parliament. To estimate this unknown proportion, one considers a random sample of opinions from the population and uses it to estimate the percentage of the population that favors the bill to be passed.

Another example is where a country's defence system uses radar to locate the enemy's plane or submarine entering into a restricted region. Based on the echo received at radar system, the estimator of the position of the aeroplane is ascertained. These echoes are technically called the noisy signals that contain all the information on the position of the aircraft, along with the uncertainty component. This forms a statistical model or probabilistic model indexed by some parameter relating to the exact position of the aircraft.

***Example* 1.1** (Pearson (1920) and Kale (1999)) One may be interested in the probability $g(\theta)$ that atleast 25 seeds of a particular type will germinate in a batch of 30 seeds, $g(\theta)=\sum_{j=25}^{30}\binom{30}{j}\theta^j(1-\theta)^{30-j}$. This probability can be estimated on the basis of an experiment in which 200 seeds are sown, one in each plot, and the number of germinated seeds are recorded. If we define an *rv* X_i which takes on values 1 or 0 according as the seed in the *i*th plot germinates or not, we have the data $(x_1,\ldots,x_{200})$ regarded as a realization on *iid rvs* $(X_1,\ldots,X_{200})$ with the probability that the seed germinates $P(X_i = 1) = \theta$ and that it does not $P(X_i = 0) = 1 - \theta$. The purpose of estimation theory is to estimate θ on the basis of data $(x_1,\ldots,x_{200})$ and, finally, estimate the probability of germinating at least 25 seeds in a batch of 30 seeds, i.e., $g(\theta)$, for future use. In this problem, one may estimate the probability that the seed germinates, θ, on the basis of sample observations $(x_1,\ldots,x_{200})$ corresponding to the Bernoulli trials repeated 200 times. Intuitively, the proportion of sample seeds that germinates, i.e., $\delta(x) = \Sigma x_i/n = \bar{x}$ will serve as an unbiased estimator of θ. Thus, the estimate $\delta(x_1,\ldots,x_n) = \bar{x}$ is the value taken by the random variables $\delta(X_1,\ldots,X_n) = \bar{X}$ corresponding to the observed sample $(x_1,\ldots,x_n)$ on the random vector $(X_1,\ldots,X_n)$. Thus, we can obtain the sampling distribution of the estimator δ. By using the technique of transformation of random variable, the sampling distribution of $n\delta = \Sigma X_i$ is $b(n, \theta)$. One can have different estimators of θ having their different sampling distributions. Based on these sampling distributions, the mean squared error, and some characteristics, such as mean and variance, the performance of an estimator is judged. Now, having obtained some optimal estimator of θ, one can obtain the optimal estimator of $g(\theta)$ by plugging-in estimator δ in $g(\theta)$, i.e., $\hat{g}(\theta)=\sum_{j=25}^{30}\binom{30}{j}\delta^j(1-\delta)^{30-j}$.

***Example* 1.2** It is well known that the number of times a particular event occurs in a unit time interval can be modelled by Poisson distribution, e.g., the number of alpha particles hitting the Geiger counter, the number of accidents per hour, the number of calls received per unit of time in a telephone exchange, the number of malarial parasites found in 1 cc of blood, etc.

Consider car movements on a national highway during a three-day vacation, which is expected to be high in such vacations. The number of car accidents per hour $(X_1,\ldots,X_{72})$ were recorded and the number of intervals in which exactly k accidents had occurred were counted (Rohatgi, 2006).

Number of car accidents occurred per hour: k	0 or 1	2	3	4	5	6	7	≥ 8
Number of intervals(hrs.): N_k	4	10	15	12	12	6	6	7

The number of accidents per hour can be modelled by a Poisson *rv*. The observations $(X_1,\ldots,X_{72})$ are *iid* Poisson *rvs* with mean parameter θ, where X_i is the number of accidents in the *i*th hour. One can suggest that the average number of accidents per hour $\delta(X_1, X_2,\ldots,X_{72})$ = $\bar{X}=\sum_{i=1}^{72}X_i/72=\sum kN_k/\sum N_k$ as a reasonable estimator of θ, so that its sampling distribution can be derived by noting that ΣX_i is $P(72\theta)$. Similarly, we can suggest several other estimators of θ and can assess their performances based on their sampling distributions respectively.

***Example* 1.3** We may consider another example of measuring some physical quantity, θ, like distance, temperature, or the gravity of earth. Lehmann (1983) has named this estimation problem as the problem of measurement. In this, n measurements $x_1,\ldots,x_n$ of this physical quantity are taken to estimate the parameter θ. These measurements involve errors due to measurement, skill of the experimenter, etc., and, thus, are subject to variations. We can model this problem as follows:

$$X_i = \theta + e_i$$

where e_is are random errors, which are distributed according to a distribution F, which is symmetric about 0. Thus, the distribution of X_i is given by

$$P(X_i \le x) = P(\theta + e_i \le x) = P(e_i \le x - \theta) = F(x - \theta)$$

so that the distribution of X_i is symmetric about θ. More specifically, one may consider that the errors e_is are distributed normally with mean 0 and variance σ^2, i.e., F is $N(0, \sigma^2)$. We may be interested in estimating the true unknown value of the physical quantity $g_1(\theta, \sigma^2) = \theta$ or in assessing the ability of the experimenter by estimating $g_2(\theta, \sigma^2) = \sigma^2$, where (θ, σ^2) are the parameters of the measurement model. We may consider $\delta_1(X_1,\ldots,X_n) = \bar{X}$ as an estimator of θ and $\delta_2(X_1,\ldots,X_n) = (1/n)\ \Sigma(X_i - \bar{X})^2$ as an estimator of σ^2 with their sampling distributions, respectively, $\delta_1 \sim N(\theta, \sigma^2/n)$ and $n\delta_2/\sigma^2 \sim \chi^2_{n-1}$. There can be several such estimators of θ and σ^2 alongwith their corresponding sampling distributions. Thus, in the problem of estimation, one will be interested in estimating the parameter θ by such an estimator which is optimal in some reasonable sense, say, for example, in terms of mean squared error or bias, etc., by utilizing its sampling distribution.

We will discuss some families of distributions which cover most of the statistical models arising in commonly occurred statistical situations.

1.2 PROBABILITY DISTRIBUTIONS

The group and the exponential family of distributions are such families of distributions which cover most of the statistical models and have the advantage that they result in great reduction of data. The parameter θ involved in the model $f(\cdot,\theta)$ plays an important role in determining the shape of the distribution. For example, in group families, the scale parameter σ stretches or contracts the shape of the *pdf* f while the location parameter locates the graph at μ. Another type of parameter is the shape parameter that determines the shape of the distribution. For example, in $P(x;\lambda) = (x!)^{-1} \exp(-\lambda)\lambda^x$ and $f(x;\alpha) = [\Gamma(\alpha)]^{-1} x^{\alpha-1} \exp(-x)$, $x > 0$, λ and α determine the shape of the distributions, respectively.

1.2.1 Group Family of Distributions

Consider an *rv* Y with a fixed *df* F and define a new *rv* X by adding some constant a to Y.

$$X = Y + a \tag{1.2.1}$$

The class of transformations in Eq. (1.2.1) is denoted by

$$\mathcal{G} = \{g: g(x) = y + a, \forall\, a \in \mathbb{R}^1\}$$

The distribution of X is given by

$$F_X(x) = P(X \le x) = P(Y + a \le x) = P(Y \le x - a) = F(x - a) \tag{1.2.2}$$

The family of distributions corresponding to the transformations in $\mathcal{G}$

$$\mathcal{F} = \{F: F_X(x) = F(x - a), a \in \mathbb{R}^1\} \tag{1.2.3}$$

is called a location family of distributions. The type of transformation in Eq. (1.2.1) is known as location transformation. Next, if we define a new *rv* by

$$X = bY,\ b > 0 \tag{1.2.4}$$

We can write the corresponding class of transformations by

$$\mathcal{G} = \{g: g(x) = by,\ b > 0\}$$

so that the distribution of X is given by

$$F_X(x) = P(X \le x) = P(bY \le x)$$

$$= P\left(Y \le \frac{x}{b}\right) = F\left(\frac{x}{b}\right), \quad \text{since } b > 0 \tag{1.2.5}$$

The family of distributions corresponding to the above transformations in $\mathcal{G}$

$$\mathcal{F} = \left\{F: F_X(x) = F\left(\frac{x}{b}\right), b > 0\right\} \tag{1.2.6}$$

is called scale family of distributions. The transformation in Eq. (1.2.4) is known as scale transformation. On combining the transformations in Eqs. (1.2.1) and (1.2.4), we get location and scale transformations

$$X = a + bY,\ a \in \mathbb{R}^1, b > 0 \tag{1.2.7}$$

The corresponding class of transformations is given by

$$\mathcal{G} = \{g: g(x) = a + by, a \in \mathbb{R}^1, b > 0\}$$

The family of distribution of X,

$$\mathcal{F} = \left\{F: F_X(x) = F\left(\frac{x-a}{b}\right), a \in \mathbb{R}^1, b > 0\right\} \tag{1.2.8}$$

is called location-scale family of distributions.

1.2.2 Group of Transformations

Let X be an *rv* taking on its values from the sample space χ. Consider a class of one–to–one transformations

$$\mathcal{G} = \{g: g: \chi \to \chi\}$$

A class of transformations $\mathcal{G}$ is said to be a *group* if it is closed under composition and inversion.

Let $g_1 \in \mathcal{G}$ and $g_2 \in \mathcal{G}$. If the composition of g_1 with g_2, $g_2 \circ g_1 \in \mathcal{G}$, then the class of transformations $\mathcal{G}$ is said to be *closed under composition*. If a transformation $g \in \mathcal{G}$ implies that its inverse $g^{-1} \in \mathcal{G}$, then the class of transformations $\mathcal{G}$ is said to be *closed under inversion*. A transformation e in $\mathcal{G}$ is said to be an *identity transformation* if $e(x) = x$. If $\mathcal{G}$ is a group transformations, $g \in \mathcal{G}$ implies $g^{-1} \in \mathcal{G}$, then $g^{-1} \circ g = e$ also belongs to $\mathcal{G}$. This implies that $\mathcal{G}$ is a group and, therefore, closed under composition. Thus, identity transformation is a member of the group of transformations $\mathcal{G}$. Consider the inverse $(g^{-1})^{-1}$ of g^{-1}; both belong to $\mathcal{G}$. The composition of g^{-1} with $(g^{-1})^{-1}$, i.e., $(g^{-1})^{-1} \circ g^{-1} = g \circ g^{-1} = e$, which is the identity transformation, belongs to $\mathcal{G}$.

A group of transformations $\mathcal{G}$ is said to be commutative if for every $g_1, g_2 \in \mathcal{G}$,

$$g_1 \circ g_2 = g_2 \circ g_1$$

We can easily see that the classes of location, scale, and location-scale transformations, given in Eqs. (1.2.1), (1.2.4) and (1.2.7), are closed under composition and inversion. Therefore, they form a group. Further, note that the groups defined in Eqs. (1.2.1) and (1.2.4) are commutative. However, Eq. (1.2.7) is not commutative. It is for these reasons that the families of distributions in Eqs. (1.2.3), (1.2.6) and (1.2.8) are called group families.

The distributions in the group families involve some kind of symmetry, which in turn is used to make a corresponding symmetry requirement for the estimators. This leads us to a great reduction of data and simplification of the problem. It is one of the main reasons why group families along with exponential families play a key role in statistical inference.

We list here some distributions that belong to a group family of distributions:

1. Normal distribution, $N(\mu,\sigma^2)$, with $\mu \in \mathbb{R}^1$ as location and $\sigma > 0$ as scale parameter;
2. Double exponential, $DE(\mu,\sigma)$, with $\mu \in \mathbb{R}^1$ as location and $\sigma > 0$ as scale parameter;
3. Cauchy, $C(\alpha,\beta)$, with $\beta \in \mathbb{R}^1$ as location and $\alpha > 0$ as scale parameter;
4. Logistic, $L(\mu,\sigma)$, with $\mu \in \mathbb{R}^1$ as location and $\sigma > 0$ as scale parameter;
5. Exponential, $E(\mu,\sigma)$, with $\mu \in \mathbb{R}^1$ as location and $\sigma > 0$ as scale parameter;
6. Uniform, $U(\theta - (c/2),\theta + (c/2))$, with $\theta > 0$ as location parameter.

In the location-scale family of distributions with *pdf*

$$\frac{1}{\sigma} f\left(\frac{x-\mu}{\sigma}\right), \sigma > 0,$$

the location parameter μ shifts the graph of the *pdf f* and locates at μ without changing its shape and the scale parameter σ stretches the graph of f if $\sigma > 1$ and contracts if $\sigma < 1$. Similarly, the location-scale parameter first stretches or contracts the shape of f and then places the graph at μ.

1.2.3 Exponential Family of Distributions

Most of the commonly used distributions belong to a very general family of distributions, which is known as exponential family. A family of distributions $\{f(x;\theta),\ \theta \in \Theta\}$ is said to constitute a k parameter exponential family of distributions if *pdf* $f(x;\theta)$ is of the form

$$f(x;\theta) = \exp\left\{\sum_{i=1}^{k} \eta_i(\theta)T_i(x) - D(\theta)\right\}h(x) \tag{1.2.9}$$

where η_i and $D(\theta)$ are real-valued functions of the parameter θ and T_i's and h are real-valued statistics. This family of distributions is also called the "Koopman–Pitman–Darmois" (1937–1938) family of distributions, since these authors had independently studied the properties of these distributions. The function $D(\theta)$ in Eq. (1.2.9) is obtained so that

$$\exp(D(\theta)) = \sum_x h(x)\exp\left\{\sum_{i=1}^{n} \eta_i(\theta)T_i(x)\right\}$$

when X is discrete, and

$$\exp(D(\theta)) = \int_x h(x)\exp\left\{\sum_{i=1}^{n} \eta_i(\theta)T_i(x)\right\}dx$$

when X is continuous. One important property of an exponential family is that the support of the distribution does not depend on the parameter θ. One can easily see that $N(\mu, \sigma^2)$ is a member of exponential family whereas $U(\theta_1, \theta_2)$ is not.

If it is more convenient to use the parameterization η_i, the density in Eq. (1.2.9) of the form

$$f(x; \boldsymbol{\eta}) = \exp\left\{\sum_{i=1}^{k} \eta_i T_i(x) - A(\boldsymbol{\eta})\right\}h(x) \tag{1.2.10}$$

is called the *exponential density in canonical form.* The parameters $\boldsymbol{\eta} = (\eta_1,\ldots,\eta_k)'$ are called the natural parameters and the corresponding space from which η_is take on values is called the natural parameter space. If the statistic $\mathbf{T} = (T_1,\ldots,T_k)$ or the parameters $\boldsymbol{\eta} = (\eta_1,\ldots,\eta_k)$ satisfy some linear constraints, that is, they are not independent, then the parameters η_i are statistically meaningless and called unidentifiable. $\boldsymbol{\eta}$ can be reduced to such η_is so that they are independent and the parameter space Ξ becomes a linear subspace of dimension less than k, wherein the former parameter space, the one before reduction, lies in this linear subspace. Thus, we can assume, without loss of generality, that all $\eta_1,\ldots,\eta_k$ are independent if the dimension of Ξ is k. Further, if Ξ contains k-dimensional rectangle, we say that the family given by Eq. (1.2.10) is of full rank.

Let $X_1,\ldots,X_n$ be a random sample drawn from the exponential family with density given in Eq. (1.2.9). The joint density of $X_1,\ldots,X_n$ is given by

$$f(\mathbf{x}; \boldsymbol{\theta}) = \exp\left\{\sum_{i=1}^{k} \eta_i(\boldsymbol{\theta})\sum_{j=1}^{n} T_i(x_j) - nD(\boldsymbol{\theta})\right\}\prod_{j=1}^{n} h(x_j) \tag{1.2.11}$$

which is also an exponential family. One important property of the exponential family of distributions is that there exists a sufficient statistic $(t_1,\ldots,t_k)$ of fixed dimension for the parameter $\boldsymbol{\eta} = (\eta_1,\ldots,\eta_k)$, irrespective of the size of the sample drawn from this distribution. Consider a random sample $X_1,\ldots,X_n$ from $N(\mu,\sigma^2)$, which is an exponential family. For any sample size, the two–dimensional statistic $(\Sigma X_i, \Sigma X_i^2)$ is sufficient for (μ,σ^2).

The exponential families of distributions are also called regular families since they satisfy certain mild regularity conditions apart from the property that its support does not depend on the parameter θ.

We have an important result

$$E_\eta\left(\sum_{i=1}^{n} T_j(X_i)\right) = n\frac{\partial A(\eta)}{\partial \eta_j} \tag{1.2.12}$$

and

$$\text{cov}\left(\sum_{i=1}^{n} T_j(x_i), \sum_{i=1}^{n} T_k(x_i)\right) = n\frac{\partial^2 A(\eta)}{\partial \eta_j \partial \eta_k} \tag{1.2.13}$$

The statistic **T** in Eq. (1.2.10) contains all the information about $\boldsymbol{\eta}$ or $\boldsymbol{\theta}$ which is contained in the data. It is due to this reason, we are interested in the family of distributions of $\mathbf{T} = (T_1,\ldots,T_k)$.

Theorem 1.2.1 If the *rv* X is distributed according to an exponential family with density Eq. (1.2.10), then $\mathbf{T} = (T_1,\ldots,T_k)'$ is distributed according to an exponential family with density

$$f(\mathbf{T};\boldsymbol{\eta}) = \exp\left\{\sum_{i=1}^{k} \eta_i T_i - A(\eta)\right\} k(\mathbf{T}) \tag{1.2.14}$$

that is, the distribution of **T** is again an exponential family.

Binomial distribution, Poisson distribution, negative binomial distribution, normal distribution, gamma distribution, inverse gamma distribution, inverse chi-square distribution, scaled inverse chi-square distribution and beta distribution are some examples of exponential family of distributions.

1.2.4 Non-regular (Pitman) Family of Distributions

A family of distributions for which the support of the distributions depends on the parameter θ and like regular families, there exists a sufficient statistics of fixed dimension, is called the non-regular (*Pitman*) family of distributions.

The *pdf* of the distributions in non-regular families is of the form

$$f(x;\theta) = c(\theta)\, h(x) I_{(\eta_1(\theta),\eta_2(\theta))}(x) \tag{1.2.15}$$

where $h(x) > 0$, η_1 and η_2 are extended real-valued functions of θ, $\eta_1 < \eta_2$, η_1 may be $-\infty$ and η_2 may be $+\infty$, and

$$[c(\theta)]^{-1} = \int_{\eta_1(\theta)}^{\eta_2(\theta)} h(x)\,dx < \infty$$

If it is convenient to treat $\eta_i(\theta)$s as numbers rather than functions of θ, then the form of Eq. (1.2.15) with this natural parameterization can be written as

$$f(x;\boldsymbol{\eta}) = c(\boldsymbol{\eta})h(x)\, I_{(\eta_1,\eta_2)}(x) \tag{1.2.16}$$

where, $-\infty \le \eta_1 < \eta_2 \le +\infty$. If we take $c(\boldsymbol{\eta}) = 1/(\eta_2 - \eta_1)$ and $h(x) \equiv 1$, then Eq. (1.2.16) becomes the *pdf* of uniform distribution $U(\eta_1, \eta_2)$.

The exponential distribution $E(\mu,\sigma)$ with *pdf*

$$f(x; \mu, \sigma) = \frac{1}{\sigma}\exp\left(-\frac{x-\mu}{\sigma}\right)I_{(\mu,\infty)}(x)$$

with location parameter $\mu \in \mathbb{R}^1$ and scale parameter $\sigma > 0$ belongs to the Pitman family of distributions.

1.2.5 Discrete Distributions

Discrete Uniform Distribution

A random variable X is said to have a discrete uniform distribution on n points $\{x_1, x_2,\ldots,x_n\}$ with *pmf* $P\{X = x_i\} = 1/n$, $i = 1, 2,\ldots,n$. The mean and variance are, respectively, $(n + 1)/2$ and $(n^2 - 1)/12$.

Bernoulli Distribution

Consider a random variable X that takes on values on two points 1 and 0 say with probabilities $P(X = 1) = p$, $P(X = 0) = 1 - p$, $0 < p < 1$. This distribution is called Bernoulli distribution with *pmf* $P(X = x) = p^x (1 - p)^{1-x}$, $0 < p < 1$, $x = 0, 1$. The mean and variance are, respectively, p and $p(1 - p)$.

Binomial Distribution

Binomial distribution can be considered as the distribution of sum of n independent and identically distributed Bernoulli random variables with parameter p. The *pmf* of Binomial distribution is given by

$$P(X = k) = \binom{n}{k} p^k (1-p)^{n-k}, 0 \le p \le 1,\ k = 0, 1, 2,\ldots,n$$

It is denoted by $b(n,p)$. The mean and variance are, respectively, given by np an $np(1 - p) = npq$, where $q = (1 - p)$.

Negative Binomial Distribution

A random variable X is said to have negative binomial distribution if its *pmf* for a fixed positive integer $r \ge 1$ is given by

$$P(X = x) = \binom{x+r-1}{x} p^r (1-p)^x, x = 0, 1, 2,\ldots,0 < p < 1,\ r \ge 1$$

It is denoted by $NB(r; p)$. The mean and variance are, respectively, $r(1 - p)/p = rq/p$ and $r(1 - p)/p^2 = rq/p^2$.

Poisson Distribution

A random variable X is said to have Poisson distribution with parameter $\lambda > 0$ if its *pmf* is given by

$$P(X = x) = \frac{e^{-\lambda}\lambda^x}{x!}, x = 0, 1, 2,\ldots, \lambda > 0$$

It is denoted by $P(\lambda)$. The mean and variance are, respectively, given by λ and λ

Geometric Distribution

A random variable X is said to have geometric distribution if its *pmf* is given by

$$P(X = x) = p(1 - p)^x, x = 0, 1, 2,\ldots; 0 < p < 1$$

It is denoted by $Geo(p)$. Note that geometric distribution is negative binomial $NB(1, p)$. The mean and variance are, respectively, given by $(1-p)/p = q/p$ and $(1-p)/p^2 = q/p^2$

Hypergeometric Distribution

A random variable X is said to have hypergeometric distribution if its *pmf* is given by

$$P(X = x) = \frac{\binom{A}{x}\binom{B}{n-x}}{\binom{A+B}{n}}, x = 0, 1,\ldots, n, x \in \{\max(0, n-B), \min(n, A)\}$$

It is denoted by $Hyp(A, B, n)$. The mean and variance are, respectively, given by $nA/(A+B)$ and $(nAB/(A+B)^2)((A+B-n)/(A+B-1))$.

Multinomial Distribution

Consider a random experiment where its outcome belongs to one of the k-categories. Let the probability of outcome belongs to the ith category be denoted by θ_i, $0 < \theta_i < 1$ and $\sum_{i=1}^{k}\theta_i = 1$. Suppose the experiment is conducted n times independently and the frequencies X_i that the outcomes belong to the ith category are recorded such that $\Sigma X_i = n$. Then the random variable $\mathbf{X} = (X_1,\ldots,X_n)'$ is said to have multinomial distribution if its *pmf* is given by

$$f(\mathbf{x}|n, \theta) = \frac{n!}{x_1!x_2!\ldots\left(n - \sum_{i=1}^{k-1} x_i\right)!}\theta_1^{x_1}\theta_2^{x_2}\cdots\left(1 - \sum_{i=1}^{k-1}\theta_i\right)^{n-\sum_{i=1}^{k-1}x_i}, \quad \sum_{i=1}^{k-1} x_i \le n, \; \sum_{i=1}^{k}\theta_i = 1$$

The mean, variance, and covariance, are respectively, given by $E(X_i) = n\theta_i$; $V(X_i) = n\theta_i(1 - \theta_i)$, $i = 1, 2,\ldots,(k - 1)$, and $\text{cov}(X_i, X_j) = -n\theta_i\theta_j$, $i, j, = 1, 2,\ldots,(k - 1)$, $i \neq j$.

1.2.6 Continuous Distributions

Uniform Distribution

A random variable X is said to have uniform distribution if its *pdf* is given by

$$f(x; a, b) = \frac{1}{b-a}, a \le x \le b$$

The mean and variance are, respectively, given by $(a + b)/2$ and $(b - a)^2/12$. It is denoted by $U[a, b]$.

Note. Let X be a continuous *rv* with *cdf* F. Then $F(X) \sim U[0, 1]$. This result is known as the *probability integral transform*. The following result is helpful in generating a sample from a particular distribution with the help of the uniform distribution in case F is discrete or continuous *cdf*:

Let F be a *cdf* and $X \sim U[0,1]$. Then there exists a function $h(X)$ such that

$$P\{h(X) \le x\} = F(x), \ \forall x \in \mathbb{R}^1$$

That is, $h(X)$ has *cdf* F. We take

$$h(x) = F^{-1}(x)$$

where
$$F^{-1}(y) = \inf\{x: F(x) \ge y\}$$

Thus,
$$P\{F^{-1}(X) \le x\} = P\{X \le F(x)\} = F(x), \ \forall x$$

For example, if one is to draw a random sample of size n from a distribution with *cdf* $F(x) = 1 - e^{-x}$, $x > 0$, one calculates $h(x) = F^{-1}(x) = -\log(1 - x)$ so that $h(X)$ has the required *cdf* F. For this, one draws a number, say, x_1 from $U[0, 1]$. It gives the corresponding observation from F by $-\log(1 - x_1)$. Similarly, one can draw the required n observations from F.

Normal Distribution

A random variable X is said to have normal distribution with parameters μ and σ^2 if its *pdf* is given by

$$f(x; \mu, \sigma^2) = \frac{1}{\sqrt{2\pi}\,\sigma} \exp\left(-\frac{1}{2\sigma^2}(x-\mu)^2\right), x \in \mathbb{R}^1, \mu \in \mathbb{R}^1, \sigma > 0$$

It is denoted by $N(\mu, \sigma^2)$. The mean and variance of the normal distribution are μ and σ^2 respectively.

Gamma Distribution

A random variable X is said to have gamma distribution with parameters α and β if its *pdf* is given by

$$f(x; \alpha, \beta) = \frac{1}{\Gamma(\alpha)\beta^\alpha} x^{\alpha-1} \exp\left(-\frac{x}{\beta}\right), 0 < x < \infty, \alpha > 0, \beta > 0$$

It is denoted by $G_1(\alpha, \beta)$, with mean $\alpha\beta$ and variance $\alpha\beta^2$.

The gamma distribution of second kind, $G_2(\alpha, \beta)$, is defined with *pdf*

$$f(x; \alpha, \beta) = \frac{\beta^\alpha}{\Gamma(\alpha)} x^{\alpha-1} \exp(-\beta x), 0 < x < \infty, \alpha > 0, \beta > 0$$

with mean α/β and variance α/β^2.

Notes.

(i) $G_1(1, \beta)$ is called exponential distribution with *pdf*

$$f(x;\beta)=\frac{1}{\beta}\exp\left(-\frac{x}{\beta}\right),\ x>0$$

(ii) $G_2(\alpha, 1/\beta) \equiv G_1(\alpha, \beta)$

(iii) If $X \sim P(\lambda)$, then the *cdf* of X,

$$P[X \le k-1] = P(X<k)$$

$$=\frac{1}{(k-1)!}\int_{\lambda}^{\infty} x^{k-1}\exp(-x)\,dx \qquad (1.2.17)$$

is the *incomplete gamma function.*

(iv) $G_1((n/2),2)$ is χ_n^2 (Chi-square distribution with n degrees of freedom). The *pdf* of χ_n^2 is

$$f(x;n)=\frac{1}{\Gamma(n/2)2^{n/2}}x^{(n/2)-1}\exp\left(-\frac{x}{2}\right),\ 0<x<\infty$$

with mean n and variance $2n$.

Inverted–Gamma Distribution

The distribution of $X = 1/Y$ is said to have inverted gamma distribution where Y is $G_2(\alpha, \beta)$. The *pdf* of X is then given by

$$f(x;\alpha,\beta)=\frac{\beta^{\alpha}}{\Gamma(\alpha)}x^{-(\alpha+1)}\exp\left(-\frac{\beta}{x}\right),\ 0<x<\infty,\quad \alpha,\beta>0$$

with mean $\beta/(\alpha-1)$ and variance $\beta^2/((\alpha-1)^2(\alpha-2))$, provided $\alpha > 2$. It is denoted by $IG_2(\alpha, \beta)$. The *pdf* of inverted gamma, $IG_1(\alpha, \beta)$, is given by

$$f(x;\alpha,\beta)=\frac{1}{\beta^{\alpha}\Gamma(\alpha)}x^{-(\alpha+1)}\exp\left(-\frac{1}{\beta x}\right),\ 0<x<\infty,\quad \alpha,\beta>0$$

where Y is $G_1(\alpha, \beta)$, with mean $1/\beta(\alpha-1)$ and variance $1/\{\beta^2(\alpha-1)^2(\alpha-2)\}$, provided $\alpha > 2$.

Inverse Chi-square Distribution

The distribution of $X = 1/Y$, $Y \sim \chi_n^2$ is said to have inverse Chi-square distribution if its *pdf* is given by

$$f(x;n)=\frac{2^{-n/2}}{\Gamma(n/2)}x^{-((n/2)+1)}\exp\left(-\frac{1}{2x}\right),\ 0<x<\infty$$

with mean $1/(n-2)$ and variance $2/((n-2)^2(n-4))$, provided $n > 4$. It is denoted by $\chi^{-2} \equiv IG_1(n/2,1/2)$.

Scaled–inverse Chi-square Distribution

A random variable X is said to have scaled-inverse chi-square distribution if its *pdf* is given by

$$f(x; n, \sigma_0^2) = \frac{(\sigma_0^2/2)^{n/2}}{\Gamma(n/2)} x^{-((n/2)+1)} \exp\left(-\frac{\sigma_0^2}{2x}\right), 0 < x < \infty$$

It is denoted by $\chi^{-2}(n, \sigma_0^2)$ or $SI\text{–}\chi^2(n, \sigma_0^2)$. If $\sigma_0 = 1$, then scaled-inverse Chi-square distribution reduces to inverse Chi-square distribution.

Beta Distribution

A random variable X is said to have beta distribution if its *pdf* is given by

$$f(x; \alpha, \beta) = \frac{1}{B(\alpha, \beta)} x^{\alpha-1}(1-x)^{\beta-1}, 0 < x < 1, \alpha > 0, \beta > 0$$

It is denoted by $X \sim Be(\alpha, \beta)$. The mean and variance of beta distribution are $\alpha/(\alpha + \beta)$ and $\alpha\beta/\{(\alpha + \beta)^2 (\alpha + \beta + 1)\}$, respectively. Note that if $X \sim b(n,p)$, then

$$\sum_{x=0}^{k-1} \binom{n}{x} p^x q^{n-x} = \frac{1}{B(k, n-k+1)} \int_p^1 t^{k-1}(1-t)^{n-k} dt$$

Pareto Distribution

A random variable X is said to have Pareto distribution if its *pdf* is given by

$$f(x; \theta, \alpha) = \frac{\alpha\theta^\alpha}{(x+\theta)^{\alpha+1}}, x > 0, \theta > 0, \alpha > 0$$

where θ is a scale parameter and α is a shape parameter. The mean and variance, respectively, are $\theta/(\alpha - 1)$, $\alpha > 1$ and $\alpha\theta^2/(\alpha - 1)^2 (\alpha - 2)$, $\alpha > 2$. This is denoted by $Pa(\theta, \alpha)$.

Consider the transformation $Y = X + \theta$ to get another form of the Pareto distribution

$$f(y; \alpha, \theta) = \frac{\alpha\theta^\alpha}{y^{\alpha+1}}, y > \theta$$

The mean and variance are, respectively, given by $\alpha\theta/(\alpha - 1)$, $\alpha > 1$ and $\alpha\theta^2/(\alpha - 1)^2 (\alpha - 2)$, $\alpha > 2$.

Note. If we consider the transformation $X = \log(Y/\theta)$, the *pdf* of X is given by

$$f(x; \alpha) = \alpha \exp(-\alpha x), x > 0$$

which is $G_1(1, 1/\alpha)$. Thus, the properties of *exponential distribution*, which are preserved under monotonic transformations, hold for the Pareto distribution too.

Logistic Distribution

Logistic distribution is obtained from the Pareto distribution by using the transformation $Y = \log(X/\theta)$ where $X \sim Pa(\theta, \alpha)$. The *pdf* and *cdf* of logistic distribution are given by

$$f(y;\alpha)=\frac{\alpha e^{y}}{(1+e^{y})^{\alpha+1}},\ -\infty<y<\infty$$

$$F(y;\alpha)=1-(1+e^{y})^{-\alpha},\ -\infty<y<\infty$$

Another form of logistic distribution with location parameter μ and scale parameter σ is obtained from the transformation $X=\mu+\sigma Y$ and $\alpha=1$. The *pdf* and *cdf* of logistic distribution in this case are, respectively, given by

$$f(x;\mu,\sigma)=\frac{1}{\sigma}\frac{\exp[(x-\mu)/\sigma]}{\{1+\exp[(x-\mu)/\sigma]\}^{2}},\ -\infty<x<\infty,\ \mu\in\mathbb{R}^{1},\ \sigma>0$$

$$F(x;\mu,\sigma)=\exp\left(\frac{x-\mu}{\sigma}\right)\left[1+\exp\left(\frac{x-\mu}{\sigma}\right)\right]^{-1}$$

The mean and variance are given by μ and $\pi^2\sigma^2/3$, respectively.

Weibull Distribution, $W(\alpha,\beta)$

The Weibull distribution is obtained from the transformation $Y = X^{1/\alpha}$, $\alpha > 0$ where $X \sim G_1(1, \beta)$. The *rv* Y is said to have Weibull distribution with *pdf* and *cdf* given by

$$f(y;\alpha,\beta)=\frac{\alpha}{\beta}y^{\alpha-1}\exp\left(-\frac{y^{\alpha}}{\beta}\right),\ y>0$$

$$F(y;\alpha,\beta)=1-\exp\left(-\frac{y^{\alpha}}{\beta}\right),\ y>0$$

respectively.

The r-th moment, mean, and variance are, respectively, given by $E(Y^r) = \beta^{r/\alpha}\,\Gamma\{1 + (r/\alpha)\}$, $E(Y) = \beta^{1/\beta}\,\Gamma(1 + (1/\alpha))$, and $V(Y) = \beta^{2/\alpha}[\Gamma(1 + (2/\alpha)) - \{\Gamma(1 + (1/\alpha))\}^2]$.

Rayleigh Distribution

The Weibull distribution $W(2,\theta)$ is known as the Rayleigh distribution. Its *pdf* is given by

$$f(x;\theta)=\frac{2}{\theta}x\exp\left(-\frac{x^{2}}{\theta}\right),\ x>0$$

Lognormal Distribution

A random variable X is said to have lognormal distribution if its *pdf* is given by

$$f(x;\mu,\sigma)=\frac{1}{x\sigma\sqrt{2\pi}}\exp\left(-\frac{(\log x-\mu)^{2}}{2\sigma^{2}}\right),\ x\ge 0,\ \mu\in\mathbb{R}^{1},\ \sigma>0$$

The *cdf* is given by

$$F(X\le x)=\Phi\left(\frac{\log x-\mu}{\sigma}\right),\ \text{where } \Phi \text{ is the } cdf \text{ of an } N(0,1)\ rv.$$

The rth moment, mean, and variance are, respectively, given by $E(X^r) = \exp\{r\mu + (r^2\sigma^2/2)\}$, $E(X) = \exp\{\mu + (\sigma^2/2)\}$, and $V(X) = \exp(2\mu + 2\sigma^2) - \exp(2\mu + \sigma^2)$.

Note. log X follows normal distribution.

Laplace or Double Exponential Distribution

A random variable X is said to have double exponential distribution if its *pdf* is given by

$$f(x;\mu,\alpha) = \frac{1}{2\sigma}\exp\left(-\frac{|x-\mu|}{\sigma}\right), x \in \mathbb{R}, \mu \in \mathbb{R}, \sigma > 0$$

where μ is called the location parameter and σ is called the scale parameter. The mean and variance are, respectively, given by μ and $2\sigma^2$.

Cauchy Distribution

A random variable X is said to have Cauchy distribution if its *pdf* is given by

$$f(x;\mu,\theta) = \frac{\mu}{\pi}\frac{1}{\mu^2 + (x-\theta)^2},\ x \in \mathbb{R}^1, \theta \in \mathbb{R}^1, \mu > 0$$

It is denoted by $C(\mu,\theta)$. The moments of order < 1 exist, but the moments of order ≥ 1 do not exist. That is why the mean and variance of Cauchy *rv* X do not exist.

1.2.7 Multivariate Distributions

Bivariate Normal Distribution

A bivariate *rv* (X, Y) is said to have bivariate normal distribution if the joint *pdf* is of the form

$$f(x,y) = \frac{1}{2\pi\sigma_1\sigma_2\sqrt{1-\rho^2}}\exp\left\{-\frac{1}{2}\frac{1}{(1-\rho)^2}\left[\left(\frac{x-\mu_1}{\sigma_1}\right)^2 - 2\rho\left(\frac{x-\mu_1}{\sigma_1}\right)\left(\frac{y-\mu_2}{\sigma_2}\right) + \left(\frac{y-\mu_2}{\sigma_2}\right)^2\right]\right\},$$

$$x \in \mathbb{R}^1, y \in \mathbb{R}^1, \mu_1 \in \mathbb{R}^1, \mu_2 \in \mathbb{R}^1, \sigma_1 > 0, \sigma_2 > 0, |\rho| < 1$$

Some of its important properties are listed as follows:

1. If $(X, Y) \sim BN\left((\mu_1,\mu_2), \begin{pmatrix}\sigma_1^2 & \rho\sigma_1\sigma_2 \\ \rho\sigma_1\sigma_2 & \sigma_2^2\end{pmatrix}\right)$, then marginally, $X \sim N(\mu_1,\sigma_1^2)$ and $Y \sim N(\mu_2,\sigma_2^2)$, and the correlation coefficient between X and Y is ρ.
2. Conditional on $X = x$, the distribution of Y is $N(\mu_x, \sigma_x^2)$, where

$$E[Y|x] = \mu_x = \mu_2 + \rho\frac{\sigma_2}{\sigma_1}(x-\mu_1)$$

and

$$V[Y|x] = \sigma_x^2 = \sigma_2^2(1-\rho^2)$$

Multivariate Normal Distribution

A k-dimensional random vector $\mathbf{X} = (X_1, X_2, \ldots, X_k)'$ has a non-singular multivariate normal distribution with mean vector μ and dispersion matrix Σ if $\mathbf{X}$ has a *pdf*

$$f(\mathbf{x}; \boldsymbol{\mu}, \Sigma) = (2\pi)^{-k/2} |\Sigma|^{-1/2} \exp\left[-\frac{1}{2}(\mathbf{x}-\boldsymbol{\mu})' \Sigma^{-1} (\mathbf{x}-\boldsymbol{\mu})\right], \mathbf{x} \in \mathbb{R}^n$$

where Σ is a $k \times k$ symmetric positive definite matrix, $\Sigma = (\sigma_{ij})$, $i, j = 1, \ldots, k$. We will denote it by $\mathbf{X} \sim N_k(\boldsymbol{\mu}, \Sigma)$. The *moment generating function* (*mgf*) of $\mathbf{X}$ is given by

$$M_{\mathbf{X}}(\mathbf{t}) = \exp\left(\mathbf{t}'\boldsymbol{\mu} + \frac{\mathbf{t}'\Sigma\mathbf{t}}{2}\right)$$

Properties

1. Suppose $X \sim N_k(\boldsymbol{\mu}, \Sigma)$ and A is a given $m \times k$ matrix. Consider the linear transformation $\mathbf{Y} = A\mathbf{X} + \mathbf{a}$. Then

$$\mathbf{Y} \sim N_m(A\boldsymbol{\mu} + \mathbf{a}, A\Sigma A')$$

2. Suppose that the k-dimensional random vectors $\mathbf{X}$ and $\boldsymbol{\mu}$ and covariance matrix Σ are partitioned as

$$\mathbf{X} = \begin{bmatrix} \mathbf{X}_1 \\ \mathbf{X}_2 \end{bmatrix}, \quad \boldsymbol{\mu} = \begin{bmatrix} \boldsymbol{\mu}_1 \\ \boldsymbol{\mu}_2 \end{bmatrix}, \quad \Sigma = \begin{bmatrix} \Sigma_{11} & \Sigma_{12} \\ \Sigma_{21} & \Sigma_{22} \end{bmatrix}$$

where $\mathbf{X}_1 = (X_1, \ldots, X_{k_1})'$, $\mathbf{X}_2 = (X_{k_1+1}, \ldots, X_{k_1+k_2})'$, $k = k_1 + k_2$.

$\boldsymbol{\mu}_1$ and $\boldsymbol{\mu}_2$ are corresponding mean vectors. The marginal *pdf* of $\mathbf{X}_1$ is $N_{k_1}(\boldsymbol{\mu}_1, \Sigma_{11})$ and the marginal *pdf* of $\mathbf{X}_2$ is $N_{k-k_1}(\boldsymbol{\mu}_2, \Sigma_{22})$.

The conditional distribution of $\mathbf{X}_1|\mathbf{X}_2$ is

$$\mathbf{X}_1|\mathbf{X}_2 \sim N_{k_1}(\boldsymbol{\nu}_1, \Sigma_{11} - \Sigma_{12}\Sigma_{22}^{-1}\Sigma_{21})$$

where, $\boldsymbol{\nu}_1 = \boldsymbol{\mu}_1 + \Sigma_{12}\Sigma_{22}^{-1}(\mathbf{X}_2 - \boldsymbol{\mu}_2)$.

Dirichlet Distribution

A random variable $\mathbf{X} = (X_1, \ldots, X_k)'$ follows Dirichlet distribution with parameter vector $\boldsymbol{\alpha} = (\alpha_1, \alpha_2, \ldots, \alpha_k)'$ if its *pdf* is given by

$$f(\mathbf{x}; \boldsymbol{\alpha}) = \frac{\Gamma(\alpha_0)}{\Gamma(\alpha_1)\ldots\Gamma(\alpha_k)} x_1^{\alpha_1-1} \ldots x_k^{\alpha_k-1} = \frac{1}{D(\alpha_1, \ldots, \alpha_k)} x_1^{\alpha_1-1} \ldots x_k^{\alpha_k-1},$$

$$x_i > 0, \Sigma_{i=1}^k x_i = 1, \; \alpha_i > 0, \; \alpha_0 = \Sigma_{i=1}^k \alpha_i$$

It is denoted by $D(\alpha_1, \ldots, \alpha_k)$. The mean, variance, and covariance are, respectively, given by $E(X_i) = \alpha_i/\alpha_0$, $V(X_i) = \{\alpha_i(\alpha_0 - \alpha_i)\}/\{\alpha_0^2(\alpha_0 + 1)\}$ and $\text{cov}(X_i, X_j) = -\alpha_i\alpha_j/\{\alpha_0^2(\alpha_0 + 1)\}$.

Note. For $k = 2$, the Dirichlet distribution reduces to beta distribution.

1.2.8 Exact Sampling Distributions

Chi-square (χ^2) Distribution

A random variable X is said to have chi-square distribution if its *pdf* is given by

$$f(x; n) = \frac{1}{\Gamma(n/2)\, 2^{n/2}} \exp(-x/2)\, x^{(n/2)-1}, 0 < x < \infty$$

It is $G_1(n/2, 2)$ and is denoted by χ^2-distribution with n degrees of freedom (*df*). The rth moment, mean and variance are, respectively, given by $E(X^r) = 2^r\Gamma((n/2) + r)/\Gamma(n/2)$, $E(X) = n$, and $V(X) = 2n$.

***t*-Distribution**

Let $X \sim N(\mu, \sigma^2)$ and $Y/\sigma^2 \sim \chi^2_n$ and X and Y be independent. Then the *rv*

$$T = \frac{X}{\sqrt{Y/n}}$$

is said to have a *non-central t-distribution* with parameter (also called non-centrality parameter) $\delta = \mu/\sigma$ and degrees of freedom n. The *pdf* of non-central t-distribution is given by

$$f_n(t; \delta) = \frac{n^{n/2} e^{-(\delta^2/2)}}{\sqrt{\pi}\, \Gamma(n/2)(n+t^2)^{(n+1)/2}} \sum_{s=0}^{\infty} \Gamma\left(\frac{n+s+1}{2}\right)\left(\frac{\delta^s}{s!}\right)\left(\frac{2t^2}{n+t^2}\right)^{s/2}$$

Putting $\delta = 0$, we get the *pdf* $f_n(t)$ of a central t-distribution

$$f_n(t) = \frac{\Gamma\left(\frac{n+1}{2}\right)}{\Gamma\left(\frac{n}{2}\right)\sqrt{n\pi}}\left(1+\frac{t^2}{n}\right)^{-(n+1)/2}, -\infty < t < \infty$$

The mean and variance are, respectively, given by

$$E(T) = \frac{\delta\, \Gamma\left(\frac{n-1}{2}\right)}{\Gamma\left(\frac{n}{2}\right)}\sqrt{\left(\frac{n}{2}\right)}, \quad n > 1$$

and

$$V(T) = \frac{n(1+\delta^2)}{n-2} - \frac{\delta^2 n}{2}\left(\frac{\Gamma\left(\frac{n-1}{2}\right)}{\Gamma\left(\frac{n}{2}\right)}\right)^2, \quad n > 2$$

***F*-Distribution**

Let X and Y be independent χ^2 *rvs* with m and n degree of freedom, respectively. The *rv*

$$F = \frac{X/m}{Y/n}$$

is said to have an F-distribution with (m,n) *df* and we say $F \sim F(m, n)$

$$g(f) = \frac{\Gamma\left(\frac{m+n}{2}\right)}{\Gamma\left(\frac{m}{2}\right)\Gamma\left(\frac{n}{2}\right)}\left(\frac{m}{n}\right)\left(\frac{m}{n}f\right)^{\left(\frac{m}{2}\right)-1}\left(1+\frac{m}{n}f\right)^{-\left(\frac{m+n}{2}\right)}, f > 0$$

We now briefly list the results on calculus and analysis which may be required as prerequisite to discuss the theory of estimation in different chapters.

1.3 CALCULUS AND ANALYSIS

Mean Value Theorem

Let f be a continuous function in the interval $[a, b]$. Then there exists a point x_0 in $[a,b]$ such that

$$\int_a^b f(x)dx = (b-a)f(x_0)$$

Leibnitz Rule

The Leibnitz differentiation under the integral sign is defined as

$$\frac{\partial}{\partial\theta}\int_{\alpha(\theta)}^{\beta(\theta)} f(x,\theta)\,dx = \int_{\alpha(\theta)}^{\beta(\theta)} \frac{\partial}{\partial\theta} f(x,\theta)dx + f\left(\beta(\theta),\theta\right)\frac{\partial\beta(\theta)}{\partial\theta} - f\left(\alpha(\theta),\theta\right)\frac{\partial\alpha(\theta)}{\partial\theta}$$

L'Hospital's Rule

Let f and g be two functions, such that $f(a) = 0 = g(a)$ at some point a. Let the first derivatives of these functions f' and g' exist and are continuous. If $\lim_{x\to a} f'(x)/g'(x)$ exists, then

$$\lim_{x\to a}\frac{f(x)}{g(x)} = \lim_{x\to a}\frac{f'(x)}{g'(x)}$$

The same result can be extended to the situation when $f'(a) = 0 = g'(a)$ and f'' and g'' exist and are continuous to yield

$$\lim_{x \to a} \frac{f'(x)}{g'(x)} = \lim_{x \to a} \frac{f''(x)}{g''(x)}$$

Similarly, this result can be extended to any number of times.

Support of the *pdf f*

The set $S(\theta) = \{x: f(x;\theta) > 0\}$ is called the support of the *pdf f*.

Relative and Absolute (Global) Extrema

Consider a function f defined on a set $E \subseteq \mathbb{R}^1$. The function f has a relative maximum at some point x_0 in E such that

$$f(x_0) \geq f(x) \ \forall \ x \in E \cap (a, b)$$

for some interval (a, b) containing x_0. If

$$f(x_0) \geq f(x) \ \forall \ x \in E$$

then, we say that f has absolute (global) maximum at x_0. Similar definitions may be given for relative and absolute minimum. Relative maximum or minimum is called relative extremum and absolute maximum or minimum is called absolute extremum. There is no direct method of finding absolute extrema. The following results are helpful in finding relative extrema.

If the function f has a relative extremum at some point x_0 in D, which is an interior point of E, and $x_0 \in (a, b) \subseteq E$ and $(\partial/\partial x)f(x)$ is defined for $x \in (a, b) \subseteq E$, then

$$\left.\frac{\partial}{\partial x} f(x)\right|_{x=x_0} = 0, x_0 \in (a, b)$$

Further, if

$$\left.\frac{\partial^2}{\partial x^2} f(x)\right|_{x=x_0} < 0$$

then f has a relative maximum at x_0 and if

$$\left.\frac{\partial^2}{\partial x^2} f(x)\right|_{x=x_0} > 0$$

then f has a relative minimum at x_0. Further, if the function f is continuous and

$$\frac{\partial}{\partial x} f(x) = \begin{cases} > 0 \ \forall \ x \in (a, x_0) \\ < 0 \ \forall \ x \in (x_0, b) \end{cases}$$

then x_0 is a relative maximum of f. If

$$\frac{\partial}{\partial x} f(x) = \begin{cases} < 0 \ \forall \ x \in (a, x_0) \\ > 0 \ \forall \ x \in (x_0, b) \end{cases}$$

then x_0 is a relative minimum of f. If

$$\frac{\partial}{\partial x} f(x) > 0 \quad \forall\, x \in (a, x_0) \cup (x_0, b)$$

or

$$\frac{\partial}{\partial x} f(x) < 0 \quad \forall\, x \in (a, x_0) \cup (x_0, b)$$

then x_0 is not a relative extremum at x_0.

Taylor's Expansion

Let a function f be such that its $(n + 1)$st derivative $f^{(n+1)}(x)$ is continuous in $|x - a| \le h$. Then the function can be expanded about a and it is known as *Taylor's expansion*

$$f(x) = f(a) + \sum_{i=1}^{n} f^{(i)}(a)\frac{(x-a)^i}{i!} + R_n$$

where the remainder R_n is given by

$$R_n = \frac{1}{n!}\int_a^x (x-t)^n\, f^{(n+1)}(t)\,dt$$

Power Series

A series of the form

$$c_0 + c_1 x + c_2 x^2 + \cdots + c_n x^n + \cdots = \sum_{i=0}^{\infty} c_i\, x_i \tag{1.3.1}$$

is known as *power series*. If the series in [Eq. (1.3.1)] converges for $x = x_0 \neq 0$, it converges absolutely for x in the interval $(-|x_0|, |x_0|)$ and uniformly in every closed interval $[-\varepsilon, \varepsilon]$, where $0 < \varepsilon < |x_0|$.

A function $f = \Sigma_{i=1}^{\infty} c_i x^i$ which converges in a nonzero radius is uniquely represented by

$$f(x) = \sum_{i=0}^{\infty} f^{(i)}(0)\, \frac{x^i}{i!} \tag{1.3.2}$$

where $f^{(0)}(0) = f(0)$.

The expansion in Eq. (1.3.2) is known as *Maclaurin's expansion*. One important expansion of the function $f(x) = (1 + x)^n$ is into a power series with $|x| < 1$

$$f(x) = (1+x)^n = \binom{n}{0}x^0 + \binom{n}{1}x^1 + \cdots + \binom{n}{n}x^n$$

which is known as *binomial series*.

Stirling's Approximation

Stirling's approximation, for large n, is given by

$$\frac{n!}{\sqrt{2\pi}\, n^{n+(1/2)}\, \exp(-n)} \longrightarrow 1$$

Monotonic Function

A function f defined on $E \subseteq \mathbb{R}^1$ is said to be nondecreasing, if

$$f(x) \le f(y) \ \forall \ x, y \in E, x < y$$

and strictly increasing, if $\quad f(x) < f(y) \ \forall \ x, y \in E, x < y$

Similarly, a function f is said to be nonincreasing, if

$$f(x) \ge f(y) \ \forall \ x, y \in E, x < y$$

and strictly decreasing, if $\quad f(x) > f(y) \ \forall \ x, y \in E, x < y$

A function f is said to be monotonic if it is either increasing or decreasing.

Let the function $y = f(x)$ be differentiable at every point $x \in \mathbb{R}^1$. Then f is increasing if $f' > 0$ and decreasing if $f' < 0$. If the function is monotone, the inverse of the function $x = f^{-1}(y)$ exists. This is continuous and monotone in (α, β). $x = f^{-1}(y)$ is differentiable and satisfies

$$\frac{dx}{dy}\frac{dy}{dx} = 1$$

$$\Rightarrow \qquad \frac{df^{-1}(y)}{dy} = \frac{dx}{dy} = \frac{1}{dy/dx} = \left(\frac{df(x)}{dx}\bigg|_{x=f^{-1}(y)} \right)^{-1}$$

Laplace Transform

If the integral $M(t) = \int_0^\infty \exp(-xt)\, f(x)\, dx$ converges for some value of t, $M(t)$ is called *unilateral Laplace transform* of the function f. If $M(t)$ converges at $t = t_0$, it converges for all $t > t_0$. In this case, we can differentiate the integral $\int_0^\infty \exp(-xt) f(x)\, dx$ any number of times. Thus, we can expand $M(t)$ in the Maclaurin series

$$M(t) = \sum_{i=1}^{\infty} M^{(i)}(0) \frac{t^i}{i!}$$

in the interval of convergence, where

$$M^{(i)}(t) = \int_0^\infty (-x)^i \exp(-xt)\, f(x)\, dx, \quad t > t_0, \quad i = 1, 2, \ldots$$

If the function f is continuous, then the Laplace transform of f, $M(t)$, is unique. The integral $M(t) = \int_{-\infty}^{\infty} \exp(-xt) f(x)\, dx$ is called the bilateral Laplace transform of f. We can express the bilateral Laplace transform of f by

$$\int_{-\infty}^{\infty} \exp(-xt)\, f(x)\, dx = \int_{-\infty}^{0} \exp(-xt)\, f(x)\, dx + \int_{0}^{\infty} \exp(-xt)\, f(x)\, dx$$

$$= \int_{0}^{\infty} \exp(xt)\, f(-x)\, dx + \int_{0}^{\infty} \exp(-xt)\, f(x)\, dx$$

Thus, the bilateral Laplace transform breaks into two unilateral transforms and the results for unilateral transform also extend to bilateral Laplace transform.

Absolute Continuous Function

A real-valued function f defined on $E = [a, b] \subseteq \mathbb{R}^1$ is said to be absolutely continuous on E if for every $\varepsilon > 0$ there exists a $\delta > 0$ such that for every set of disjoint subintervals $\{(x_i, y_i)\}$, $\Sigma_{i=1}^{n} |f(y_i) - f(x_i)| < \varepsilon$ whenever $\Sigma_{i=1}^{n} |y_i - x_i| < \delta$. Thus, every absolutely continuous function is continuous. Moreover, if the function f is absolutely continuous, it is differentiable almost everywhere and, it is the indefinite integral of its derivative.

Convexity of Loss Function

A real-valued function f on $\mathbb{R}^n$ is convex if

$$f\{\alpha \mathbf{x} + (1 - \alpha)\mathbf{y}\} \leq \alpha f(\mathbf{x}) + (1 - \alpha) f(\mathbf{y}) \tag{1.3.3}$$

$\forall$ $\mathbf{x}, \mathbf{y} \in \mathbb{R}^n$, $0 < \alpha < 1$. If the inequality is strict for $\mathbf{x} \neq \mathbf{y}$, then f is said to be strictly convex. If the function $-g$ is convex, then g is concave. The convex loss functions are bowl-shaped while the concave loss functions are upsidedown bowl-shaped.

Checking a function for convexity by using its definition given in Eq. (1.3.3) is difficult. The following result can be used for this purpose:

Consider a function f on $\mathbb{R}^n$ for which all second-order derivatives exist and are finite. The function f is convex if and only if

$$\mathbf{D} = \left\| \frac{\partial^2}{\partial x_i\, \partial x_j} f(\mathbf{x}) \right\|$$

is nonnegative definite, or $\mathbf{y}'\mathbf{D}\mathbf{y} \geq 0$ for all $\mathbf{y} \in \mathbb{R}^n$ and $\mathbf{x} \in \mathbb{R}^n$. Similarly, a function f is concave if $-\mathbf{D}$ is nonnegative definite. Further, if $\mathbf{D}$ is positive definite, then the function f is strictly convex and if it is negative definite, then f is strictly concave.

Jensen's Inequality

Let $f(\mathbf{x})$ be a convex function defined on $\mathbb{R}^n$ so that $E|\mathbf{X}| < \infty$. Then

$$f(E(\mathbf{X})) \leq E(f(\mathbf{X}))$$

with strict inequality if f is strictly convex and $\mathbf{X}$ is not concentrated at one point. The first application of convex loss functions using Jensen's inequality is, that one should restrict attention only to nonrandomized estimators. The second application of convex loss functions, when a sufficient statistic T for θ exists, is that given any estimator of θ, there exists a nonrandomized estimator which is a function of the sufficient statistic T, which is better than the considered estimator. Thus, nonrandomized estimators, which are the function of the sufficient statistics, need only be considered. We will discuss these results based on Jensen's inequality in much detail in Chapter 2, where we observe that convex loss functions are helpful in reducing the estimation problem to a great extent.

The results of the following two sections are to be read along with Chapter 5 where large sample properties of different estimators of parameter θ have been discussed.

1.4 SMALL 'o(·)' AND BIG 'O(·)' NOTATIONS

We introduce here little o and big O notations indicating the size of a quantity when n is tending to infinity.

Definition 1.4.1 (Small 'oh': $o(1)$) A non-random sequence $\{a_n\}$ is $o(1)$ if

$$a_n \to 0 \text{ as } n \to \infty$$

Definition 1.4.2 ($o_P(1)$) A random sequence $\{a_n\}$ is

(i) $o_P(1)$ if

$$a_n \xrightarrow{P} 0 \text{ as } n \to \infty$$

(ii) $o_P(b_n)$ if for some monotonic sequence $\{b_n\}$,

$$\frac{a_n}{b_n} \xrightarrow{P} 0 \text{ as } n \to \infty$$

Note that $b_n o_p(1) = o_P(b_n)$ implies $\sqrt{n}\, o_P(1/\sqrt{n}) = o_P(1)$. Further, note that $o_P(1) \cdot o_P(1) = o_P(1)$.

Definition 1.4.3 (Big 'oh': $O(1)$) A non–random sequence $\{a_n\}$ is $O(1)$ if there exists some constant M so that

$$a_n < M \quad \text{as} \quad n \to \infty$$

Definition 1.4.4 ($O_P(1)$) A random sequence $\{a_n\}$ is

(i) $O_P(1)$ if for every $\varepsilon > 0$ there exists some constant M, such that

$$P(|a_n| < M) \geq 1 - \varepsilon \text{ as } n \to \infty,$$

(ii) $O_P(b_n)$ if for some monotonic sequence $\{b_n\}$, $\{a_n/b_n\}$ is $O_P(1)$.

Note, that $O_P(1) \cdot O_P(1) = O_P(1)$ and $o_P(1) \cdot O_P(1) = o_P(1)$. Further, if $\sqrt{n}\bar{X}_n \xrightarrow{d} Z \sim N(0,1)$, then $\sqrt{n}\bar{X}_n$ is $O_P(1)$, i.e., $\bar{X}_n = O_P\left(n^{-1/2}\right)$.

Generally with $b_n = \sqrt{n}, n, 1/\sqrt{n}$ we interpret $X_n = o_P(b_n)$ by saying that X_n converges to 0 faster than the rate b_n and $X_n = O_P(b_n)$ by saying that X_n converges to 0 no slower than the rate b_n. However, if $X_n = O_P(n)$, it does not imply that X_n has a size n for large n.

Further, if we have two sequences of positive real numbers $\{b_n\}$ and $\{c_n\}$ and if $X_n = O_P(b_n)$ and $Y_n = O_P(c_n)$, then $X_n + Y_n = o_P(b_n + c_n)$ and $X_n \cdot Y_n = o_P(b_n \cdot c_n)$ holds; however, $X_n - Y_n = o_P(b_n - c_n)$ or $X_n/Y_n = O_P(b_n/c_n)$ does not necessarily hold.

1.5 SOME RESULTS USEFUL FOR CONSISTENCY

We state some important results which are useful for consistency evaluations discussed in Chapter 6:

Convergence in Probability

A sequence of estimators $\{\delta_n\}_1^\infty$ converges in probability to the true value of the parameter θ if

$$\lim_{n\to\infty} P\left(\left|\delta_n - \theta\right| > \varepsilon\right) = 0 \quad \text{for all } \varepsilon > 0$$

It is denoted by $\delta_n \xrightarrow{P} \theta$.

Chebyshev's Weak Law of Large Numbers (WLLNs)

Consider a random sample $X_1, X_2,\ldots,X_n$ from a population having *pdf* $f(x;\theta)$ with finite mean and variance, namely θ and σ^2. Then

$$\overline{X}_n \xrightarrow{P} \theta = E(X)$$

or

$$P\left(\left|\overline{X}_n - E(X)\right| > \varepsilon\right) \to 0$$

as $n \to \infty$ for any arbitrary small quantity ε. In other words, sample average converges in probability to the population mean $E(X)$.

Kolmogorov's Strong Law of Large Numbers (SLLNs)

The standard Strong Law of Large Numbers, established by Kolmogorov, for *iid rvs* holds as long as the population mean exists. In statistical view, the sample mean will always converge to the population mean as long as the population mean exists, without any further condition on moments. In fact, the sample mean converges to a finite limit, if and only if, the population mean is finite, in which case, the limit is the population mean.

It states that if $X_1, X_2,\ldots,X_n$ are *iid rvs* with $E(|X_i|) < \infty$ and $E(X_i) = \theta$, then

$$\overline{X}_n \xrightarrow{a.s.} \theta = E(X_i)$$

Conversely, if $\overline{X}_n \xrightarrow{a.s.} \theta$ which is finite, then $E(X_i) = \theta$.

Markov's Weak Law of Large Numbers (WLLNs)

Let $X_1, X_2,\ldots,X_n$ be uncorrelated random variables with finite means $E(X_i) = \theta_i < \infty$ and bounded variances $V(X_i) = \sigma_i^2 \le B < \infty$, $i = 1,2,\ldots,n$. Then

$$\frac{1}{n}\sum_1^n (X_i - \theta_i) \xrightarrow{P} 0 \text{ or } \overline{X} \xrightarrow{P} \lim_{n\to\infty} \frac{1}{n}\sum_1^n \theta_i$$

This law of large numbers is due to Markov and is useful in estimating the parameters in linear models.

Some Useful Results on Probability Limits

The following results are due to Slutsky. Let a and b be two constants.

(i) If $T_n \xrightarrow{P} b$, then $a \cdot T_n \xrightarrow{P} a \cdot b$.

(ii) If $S_n \xrightarrow{P} a$ and $T_n \xrightarrow{P} b$, then $S_n + T_n \xrightarrow{P} a + b$.

(iii) If $S_n \xrightarrow{P} a$ and $T_n \xrightarrow{P} b$, then (a) $S_n/T_n \xrightarrow{P} a/b$ provided $b \neq 0$, and (b) $S_n \cdot T_n \xrightarrow{P} a \cdot b$.

(iv) If $T_n \xrightarrow{P} a$ and h is some continuous function, then $h(T_n) \xrightarrow{P} h(a)$.

Convergence in Case of Random Vectors

Let $\mathbf{X}_1, \mathbf{X}_2, \ldots, \mathbf{X}_n$ be n random vectors each in $\mathbb{R}^k$. Then for some vector of constants $\mathbf{c}$ in $\mathbb{R}^k$,

$$\mathbf{X}_n \xrightarrow{P} \mathbf{c}$$

if

$$X_{n_i} \xrightarrow{P} c_i, \; i = 1, 2, \ldots, k$$

where X_{n_i} is the ith component of $\mathbf{X}_n$ and c_i is the ith component of $\mathbf{c}$.

We demonstrate the above results, by an example, under certain assumptions on population moments, that sample variance is a consistent estimator of population variance and so is sample standard deviation for population standard deviation. Let $X_1, X_2, \ldots, X_n$ be an *iid* sample with $E(X_i) = \theta$ and $V(X_i) = \sigma^2 < \infty$. Denote the sample mean and sample variance by $(1/n)\Sigma_1^n X_i = \bar{X}$ and $(1/n)\Sigma_1^n (X_i - \bar{X})^2 = S_n^2$, respectively. By Chebyshev's LLNs, $(1/n)\Sigma_1^n X_i = \bar{X} \xrightarrow{P} E(X_i) = \theta$ and by Komogorov's LLNs, $(1/n)\Sigma_1^n X_i^2 \xrightarrow{P} E(X_i^2) = V(X_i) + \{E(X_i)\}^2 = \sigma^2 + \theta^2$, since $E(|X_i^2|) < \infty$. By Slutsky's theorem, $[(1/n)\Sigma_1^n X_i]^2 \xrightarrow{P} \theta^2$. On putting together these results and using the additive property of convergence in probability, we get $S_n^2 = (1/n)\Sigma_1^n X_i^2 - [(1/n)\Sigma_1^n X_i]^2 \xrightarrow{P} (\sigma^2 + \theta^2) - \theta^2 = \sigma^2$. Given that transformation, the use of Slutsky's theorem gives $(S_n^2)^{1/2} \xrightarrow{P} (\sigma^2)^{1/2}$, i.e., $S \xrightarrow{P} \sigma$. Also, note that $\mathbf{X}_n = \begin{pmatrix} \bar{X} \\ S_n^2 \end{pmatrix} \xrightarrow{P} \begin{pmatrix} \theta \\ \sigma^2 \end{pmatrix} = \mathbf{c}$.

Convergence in Distribution

Consider a random sample $X_1, X_2, \ldots, X_n$ drawn from a population $f(x;\theta)$ with $E(X_i) = g(\theta)$ and $V(X_i) = \sigma^2(\theta) < \infty$. Define some sequence of statistics $Z_1, Z_2, \ldots, Z_n$ based on these sample observations. For example, Z_n may be defined as $Z_n = (\bar{X}_n - g(\theta))/(\sigma(\theta)/\sqrt{n})$. We say the sequence of statistics Z_n converges in distribution to a random variable Y, if for all y in $\mathbb{R}^1$,

$$P(Z_n \leq y) \rightarrow P(Y \leq y) \text{ as } n \rightarrow \infty$$

where $P(Z_n \leq y)$ is the *cdf* of Z_n and $P(Y \leq y)$ is the *cdf* of Y. We denote it by $Z_n \xrightarrow{d} Y$.

Lindeberg–Levy Central Limit Theorem (CLT)

Let $X_1, X_2,\ldots,X_n$ be an *iid* sample with $E(X_i) = g(\theta)$ and $V(X_i) = \sigma^2(\theta) < \infty$. Define

$$Z_n = \frac{\overline{X}_n - g(\theta)}{\sigma(\theta)/\sqrt{n}} = \sqrt{n}\left(\frac{\overline{X}_n - g(\theta)}{\sigma(\theta)}\right)$$

which is known as the *Z*-statistic. The approximating distribution of Z_n is standard normal distribution

$$Z_n = \sqrt{n}\left(\frac{\overline{X}_n - g(\theta)}{\sigma(\theta)}\right) \xrightarrow{d} Z \sim N(0,1)$$

that is, for all z in $\mathbb{R}^1$

$$P(Z_n \le z) \to \Phi(z) \quad \text{as } n \to \infty$$

where Φ is the *cdf* of standard normal distribution. Another notation of the above convergence is $Z_n \overset{A}{\sim} N(0, 1)$ or $\overline{X}_n \overset{A}{\sim} N(g(\theta), \sigma^2(\theta)/n)$, where *A* stands for asymptotic distribution.

Lindeberg–Feller Central Limit Theorem (CLT)

Let $X_1, X_2,\ldots,X_n$ be an *iid* sample with $E(X_i) = \theta_i$ and $V(X_i) = \sigma_i^2 < \infty$. Define $\theta_n = (1/n)\Sigma_1^n\theta_i$ and $\overline{\sigma}_n^2 = (1/n)\Sigma_1^n\sigma_i^2$. If the following two conditions hold

(i) $\lim\limits_{n\to\infty} \max\limits_i \dfrac{\sigma_i}{n\overline{\sigma}_n^2} = 0$

(ii) $\lim\limits_{n\to\infty} \overline{\sigma}_n^2 = \overline{\sigma}^2 < \infty$

then

$$T_n = \sqrt{n}\left(\frac{\overline{X}_n - \overline{\theta}_n}{\overline{\sigma}_n}\right) \xrightarrow{d} Z \sim N(0,1)$$

or

$$\sqrt{n}(\overline{X}_n - \overline{\theta}_n) \xrightarrow{d} Z \sim N(0, \overline{\sigma}^2)$$

Note that condition (i) ensures that every individual variance is bounded in limits and condition (ii) ensures that the average of variances is finite. The condition for Lindeberg–Feller CLT to hold is that the limiting variance of $\sqrt{n}\overline{X}_n$ is finite.

Slutsky's Theorem

(i) Consider $\{T_n\}$ as a sequence of random variables. Given that $T_n \xrightarrow{d} T$ and g is continuous, the asymptotic behavior of $g(T_n)$ is

$$g(T_n) \xrightarrow{d} g(T)$$

(ii) Let $\{T_n\}$ and $\{S_n\}$ be two sequences.

(a) If $T_n \xrightarrow{d} T$ and $S_n \xrightarrow{d} b$ for some constant b, then

$$\begin{pmatrix} T_n \\ S_n \end{pmatrix} \xrightarrow{d} \begin{pmatrix} T \\ b \end{pmatrix}$$

(b) If $T_n \xrightarrow{d} T$ and $S_n \xrightarrow{P} b$, then

$$T_n + S_n \xrightarrow{d} T + b$$

$$T_n S_n \xrightarrow{d} bT \quad (T_n S_n \xrightarrow{P} 0 \text{ if } S_n \xrightarrow{P} 0)$$

$$\frac{T_n}{S_n} \xrightarrow{d} \frac{T}{b} \quad \text{if } b \neq 0$$

(c) If $T_n \xrightarrow{d} T$ and $T_n - S_n \xrightarrow{P} 0$, then $S_n \xrightarrow{d} T$

(iii) *Corollary of Slutsky's theorem.* Let $T_n \in \mathbb{R}^q$, $S_n \in \mathbb{R}^r$, $T_n \xrightarrow{d} T$, $S_n \xrightarrow{d} c$, and f be a function f: $\mathbb{R}^{q+r} \to \mathbb{R}^k$, such that $\begin{pmatrix} T \\ c \end{pmatrix}$ is a point among the points of continuity under f. Then

$$f(T_n, S_n) \xrightarrow{d} f(T, c)$$

Notes.

1. $(X_n)^2 \xrightarrow{d} (X)^2 \sim \chi_1^2$ whenever $X_n \xrightarrow{d} X \sim N(0, 1)$
2. Let $T_n = T \sim U(0, 1)$. Clearly, $T_n \xrightarrow{d} T$. Define

$$S_n = \begin{cases} T & \text{if } n \text{ is odd} \\ 1 - T & \text{if } n \text{ is even} \end{cases}$$

Here, $S_n \xrightarrow{d} T$. The result

$$\begin{pmatrix} T_n \\ S_n \end{pmatrix} \xrightarrow{d} \begin{pmatrix} T \\ T \end{pmatrix}$$

does not hold since the condition $S_n \xrightarrow{d} b$ in ii(a) is not satisfied.

3. Let $X_1, X_2, \ldots, X_n$ be an *iid* sample with $E(X_i) = \theta$ and $V(X_i) = \sigma^2 > 0$. Consider the following statistics: $\bar{X}_n = (1/n)\, \Sigma_1^n X_i$ and $S_n^2 = (1/n)\, \Sigma_1^n X_i^2 - (\bar{X}_n)^2$. We have already shown that for estimating σ^2 from the data, $S_n^2 \xrightarrow{P} \sigma^2$ and $S_n \xrightarrow{P} \sigma$. From, Lindeberg–Levy central limit theorem, we have

$$\sqrt{n}\left(\bar{X}_n - \theta\right) \xrightarrow{d} N(0, \sigma^2)$$

By weak law of large numbers, we have

$$\bar{X}_n \xrightarrow{d} \theta$$

$$\left(\frac{1}{n}\right) \sum_1^n X_i^2 \xrightarrow{d} E(X^2)$$

Using these convergence results in the above corollary, we have

$$\frac{1}{n}\sum_1^n X_i^2 - \left(\frac{1}{n}\sum_1^n X_i\right)^2 = S_n^2 \xrightarrow{d} E(X^2) - \theta^2 = \sigma^2$$

Again, on using Slutsky's theorem, we have

$$\sqrt{n}\frac{(\bar{X}_n - \theta)}{S_n} \xrightarrow{d} N(0,1)$$

or
$$t = \sqrt{n-1}\frac{(\bar{X}_n - \theta)}{S_n} \xrightarrow{d} N(0,1)$$

We may alternatively proceed to show the convergence of statistic $(\bar{X}_n - \theta)/(S_n/\sqrt{n})$ to a standard normal variate $Z \sim N(0, 1)$ in distribution differently by expressing the statistic as follows:

$$\frac{\bar{X}_n - \theta}{S_n/\sqrt{n}} = \left(\frac{\bar{X}_n - \theta}{\sigma/\sqrt{n}}\right)\left(\frac{\sigma}{S_n}\right) = T_n \cdot U_n$$

where, $U_n = \sigma/S_n$. We have already shown that $S_n^2 \xrightarrow{P} \sigma^2$. Further, using LLNs and Slutsky's theorem on this convergence, we get $S_n \xrightarrow{P} \sigma$, which then implies $1/S_n \xrightarrow{P} 1/\sigma$ or, $U_n = (\sigma/S_n) \xrightarrow{P} (\sigma/\sigma) = 1$. Therefore,

$$T_n \cdot U_n \xrightarrow{d} Z \cdot 1$$

So,
$$\sqrt{n}\frac{(\bar{X}_n - \theta)}{S_n} \xrightarrow{d} N(0,1)$$

We can denote this result by $\sqrt{n}\ (\bar{X}_n - \theta) \overset{A}{\sim} N(0, S_n^2)$ or $\bar{X} \overset{A}{\sim} N(\theta, S_n^2/n)$. This result shows that S_n^2 is a consistent estimator of asymptotic variance of $\sqrt{n}(\bar{X}_n - \theta)$, i.e., σ^2, or $S_n^2 \xrightarrow{P} \sigma^2$ or that S_n^2/n is a consistent estimator of asymptotic variance of the statistic $\bar{X}$.

Note that the above asymptotic result requires only the first two moments and not the entire probabilistic structure. Therefore, we can call this asymptotic result as distribution free.

These examples prepare a base to discuss consistency in Chapter 6.

4. If $a_n \to a$, $b_n \to b$, and $T_n \xrightarrow{P} x$, then $a_n + b_nT_n \xrightarrow{P} a + bx$

First-Order Delta Method (Cramer)

In many cases, the calculations of mean and variance of an estimator are not easy. However, if this estimator can be expressed as a function of one or more averages, so that their means and variances can be calculated exactly, then in such cases, Cramer suggested simple approximation to mean and variance of the estimator; it is known as *delta method*.

Suppose the estimator δ can be expressed as some function of one average S, i.e., $\delta = g(S)$, so that its mean $E(S) = \theta$ and its variance $V(S) = v(\theta)$ can be calculated. The use of first-order Taylor's approximation gives

$$\delta = g(S) \approx g(\theta) + (S-\theta)g'(S)|_{S=\theta}$$

This gives $$E(\delta) = g(\theta) \quad \text{and} \quad V(\delta) = [g'(\theta)]^2\, v(\theta)$$

If we use second-order Taylor's approximation, it gives

$$\delta = g(S) \approx g(\theta) + (S-\theta)g'(S)|_{S=\theta} + \frac{1}{2!}(S-\theta)^2 g''(S)|_{S=\theta}$$

$$\Rightarrow \qquad E(\delta) = g(\theta) + \frac{1}{2}v(\theta)\,g''(\theta)$$

and $$V(\delta) = v(\theta)g'^2 + \frac{1}{4}\mu_4(\theta)g''^2 + \mu_3(\theta)g'g''$$

where $\mu_3(\theta)$ and $\mu_4(\theta)$ are the third and fourth–order central moments. Whenever θ and $v(\theta)$ are unknown, we generally plug–in their quick estimates in the above expression.

In case $\boldsymbol{\theta}$ and **S** are vectors and T is scalar, similar calculations as above, gives

$$E(\delta) = \mathbf{g}(\boldsymbol{\theta}) + \frac{1}{2}\,\text{trace}[\mathbf{g}''(\boldsymbol{\theta})\,\mathbf{v}(\boldsymbol{\theta})]$$

and $$V(\delta) = (\mathbf{g}'(\boldsymbol{\theta}))'\,\mathbf{v}(\boldsymbol{\theta})(\mathbf{g}'(\boldsymbol{\theta}))$$

where $\mathbf{g}'(\boldsymbol{\theta})$ and $\mathbf{g}''(\boldsymbol{\theta})$ are the vector and matrix of derivatives respectively.

If certain assumptions are satisfied and the sample size is large, so that central limit theorem holds for S, then we have

$$E(\delta - g(\theta)) \xrightarrow{d} N(0, [g'(\theta)]^2\, v(\theta))$$

Note that if one is to stabilize the variance of δ by way of choosing g, one has merely to solve

$$[g'(\theta)]^2\, v(\theta) = \text{constant}$$

for g. By stabilizing variance, we mean $V(\delta)$ is independent of θ. We will explain the procedure of stabilizing this variance by an example. Let $X_1, X_2, \ldots, X_n$ be *iid* $G_1(1,\theta)$. Consider the estimator $\delta = g(\bar{X}) = g(S)$ for estimating θ, where $E(\bar{X}) = \theta$ and $V(\bar{X}) = \theta^2/n = v(\theta)$. We choose such a g so that the variance of δ is stabilized. We solve

$$[g'(\theta)]^2\, v(\theta) = c \text{ (some constant)}$$

$$g'(\theta)\theta = c\sqrt{n}$$

$$\int\left(\frac{d}{d\theta}g(\theta)\right)d\theta = \sqrt{n}\,c\int\left(\frac{d}{d\theta}\log\theta\right)d\theta$$

This gives g $$g(\bar{X}) \propto \log \bar{X}$$

Therefore, for this choice of g i.e. $\delta = g(\bar{X}) = \log(\bar{X})$, the variance of δ.

$$V(\delta) = [g'(\theta)]^2\, v(\theta) = \frac{1}{\theta^2}\frac{\theta^2}{n} = \frac{1}{n}$$

stabilizes since it becomes independent of θ.

In order to formalize the above concepts, consider $\{\delta_n\}$ as a sequence of random variables. Cramer provided the following method to study the asymptotic behavior of the sequence $\{g(\delta_n)\}$.

Delta Method (Version 1). Consider a sequence $\{\delta_n\}$ so that

(i) $\sqrt{n}(\delta_n - \theta) \xrightarrow{d} N(0,\sigma^2(\theta))$ and

(ii) g is a given function, $g(\delta_n) \in \mathbb{R}^1$, so that g' exists and $g'(\theta) \neq 0$ at every θ.

Then $$\sqrt{n}(g(\delta_n) - g(\theta)) \xrightarrow{d} N(0, [g'(\theta)]^2\sigma^2(\theta))$$

or $$g(\delta_n) \text{ is } AN(g(\theta), [g'(\theta)]^2\sigma^2(\theta)/n).$$

Delta Method (Version 2). Consider a sequence $\{\delta_n\}$, $\delta_n \in \mathbb{R}^q$, so that

(i) $\sqrt{n}(\delta_n - \theta) \xrightarrow{d} X$ and

(ii) let g be a given function, $g(\delta_n) \in \mathbb{R}^r$, so that the Jacobian

$$\nabla g = \left(\frac{\partial g_i}{\partial X_j}\right)$$

is continuous in the neighbourhood of $E(X) = \theta$.

Then $$\sqrt{n}(g(\delta_n) - g(\theta)) \xrightarrow{d} N(0, \nabla g(\theta)X)$$

Delta Method (Version 3). In the setup of version 2, if $X \sim N(0, \Sigma)$, then

$$\sqrt{n}(g(\delta_n) - g(\theta)) \xrightarrow{d} N(0, \nabla g(\theta)\, \Sigma[\nabla g(\theta)]')$$

Next, consider the random variables X and Y so that $\mu(x) = E(X)$ and $\mu(y) = E(Y)$. Then

$$f(x, y) \approx f(\mu(x), \mu(y)) + \frac{\partial}{\partial x} f(\mu(x), \mu(y))(x - \mu(x)) + \frac{\partial}{\partial y} f(\mu(x), \mu(y))(y - \mu(y))$$

$$f(x, y) \sim N(f(\mu(x), \mu(y)), V(f(x,y)))$$

We consider here a few illustrative examples on first-order delta method to study the asymptotic behavior of some function $g(\delta_n)$. These examples also illustrate the difficulty that arises due to degenerate limit law and second-order delta method as a solution to such problems. Let $X_1, X_2,\ldots,X_n$ be *iid* with $E(X_i) = \theta$ and $V(X_i) = \sigma^2$. By central limit theorem, we have

$$\sqrt{n}(\bar{X}_n - \theta) \xrightarrow{d} N(0, \sigma^2)$$

For estimating the parametric function $g(\theta) = \theta^2$, we set $g(X_i) = X_i^2$. Note that this function is continuous everywhere. We have

$$\nabla g(\theta) = \frac{\partial}{\partial \theta} g(\theta) = \frac{\partial \theta^2}{\partial \theta} = 2\theta$$

Using delta method, we have

$$\sqrt{n}(\bar{X}_n^2 - \theta^2) \xrightarrow{d} N(0, (\nabla g(\theta))^2 \sigma^2) = N(0, 4\theta^2\sigma^2)$$

Notice, from above, that the square of sample mean as an estimator of θ^2 has one problem in its asymptotic behavior: its asymptotic distribution becomes degenerate limit law when $\theta = 0$. Direct use of delta method in this case is a problem. To circumvent this problem, consider the case when $\theta = 0$

$$\sqrt{n}\bar{X}_n \xrightarrow{d} N(0, \sigma^2)$$

By Slutsky's theorem, we have

$$n\bar{X}_n^2 \xrightarrow{d} \sigma^2 \chi_1^2$$

where χ_1^2 is chi-squared distribution with one degree of freedom.

Consider another example where $X_1, X_2, \ldots, X_n$ are *iid* $b(1, \theta)$. By central limit theorem, we have

$$\sqrt{n}(\bar{X}_n - \theta) \xrightarrow{d} N(0, \theta(1 - \theta))$$

Further, for estimating $g(\theta) = \theta(1 - \theta)$, $g(\bar{X}_n) = \bar{X}_n(1 - \bar{X}_n)$ is an estimator, for which

$$g'(\theta) = \frac{\partial}{\partial \theta}\theta(1-\theta) = 1 - 2\theta$$

Thus, by delta method, we have

$$\sqrt{n}(\bar{X}_n (1 - \bar{X}_n) - \theta(1 - \theta)) \xrightarrow{d} N(0, (1 - 2\theta)^2\, \theta(1 - \theta))$$

It is clear from the distribution of the estimator that $\bar{X}_n (1 - \bar{X}_n)$ converges into a degenerate limit law when $\theta = 1/2$. In fact, this degenerate law arises for such θ for which $g'(\theta) = 0$. To deal with this situation, we consider the second-order Taylor's expansion of the function $g(\delta_n)$ about $g(\theta)$

$$g(\delta_n) - g(\theta) = g'(\theta)(\delta_n - \theta) + \frac{1}{2} g''(\theta + \xi(\delta_n - \theta))(\delta_n - \theta)^2$$

Notice, that the first term vanishes and the second term is meaningful in studying the asymptotic behavior of $g(\delta_n)$ under the situation $g'(\theta) = 0$. This result is known as *second-order delta method* and is summarized as below.

Second-Order Delta Method

Let $\{\delta_n\}$ be a sequence of random variables so that

(i) $\sqrt{n}(\delta_n - \theta) \xrightarrow{d} N(0, \sigma^2(\theta))$ and

(ii) g is a given function so that g' and g'' exist and $g'(\theta) = 0$, $g''(\theta) \neq 0$ at every θ.

Then,
$$\sqrt{n}(g(\delta_n) - g(\theta)) \xrightarrow{d} \sigma^2(\theta)\frac{g''(\theta)}{2}\chi_1^2$$

The application of second-order delta method in the above example immediately gives the asymptotic distribution of the estimator $g(\delta_n) = \bar{X}_n(1 - \bar{X}_n)$

$$\sqrt{n}(\bar{X}_n(1-\bar{X}_n)-\theta(1-\theta)) \xrightarrow{d} \frac{1}{8}(-2)\chi_1^2 = -\frac{1}{4}\chi_1^2$$

Approximation for Mean and Variance

Let X follows some *cdf* F with $E(X) = \theta$ and $V(X) = \sigma^2 < \infty$. Let g be a given transformation, and one is interested in estimating $E[g(X)]$ and $V[g(X)]$. For this, Taylor's first-order expansion is considered

$$g(X) = g(X)|_{X=\theta} + g'(X)|_{X=\theta}(X-\theta)$$

Using this expansion, the quantities $E[g(X)]$ and $V[g(X)]$ are approximated by $g(\theta)$ and $[g'(\theta)]^2\sigma^2$, respectively.

Asymptotic Distribution of Median

Let $\mathbf{X} = (X_1, X_2,\ldots,X_n)$ be a random sample from a population with *pdf* belonging to some family of distributions $\{f(x; \theta), \theta \in \Theta\}$. A natural choice for estimating population median $\tilde{\theta}$ particularly when we require a robust estimator, is sample median

$$\tilde{X} = \text{Median}\,(X_1, X_2,\ldots,X_n) = \begin{cases} X_{(n+1)/2}, & \text{if } n \text{ is odd} \\ \dfrac{X_{(n/2)} + X_{((n/2)+1)}}{2}, & \text{if } n \text{ is even} \end{cases}$$

A standard result regarding the sample median is as follows:

$$E_\theta(\tilde{X}) = \tilde{\theta} + O\left(\frac{1}{n}\right) \to \tilde{\theta} \text{ and } V_\theta(\tilde{X}) = \frac{1}{4nf_1^2} + O\left(\frac{1}{n}\right) \to 0$$

Indeed,

$$\tilde{X} \sim AN\left(\tilde{\theta}, \frac{1}{4n\,f_1^2}\right)$$

where f_1 is the ordinate corresponding to the population median. This result is true whenever the distribution is continuous and $f_1 > 0$. Hence, by the sufficient condition of consistency, sample median $\tilde{X}$ is consistent for population median $\tilde{\theta}$. We consider an example where asymptotic distribution of sample median is obtained. Let a random sample $X_1,\ldots,X_n$ has been drawn from normal distribution $N(\theta,\sigma^2)$. The asymptotic distribution of sample median is given by

$$\tilde{X} \sim AN(\tilde{\theta}, v)$$

where $\tilde{\theta} = \theta$ and $v = \dfrac{1}{4nf_1^2}$.

Here,

$$f_1 = \frac{1}{\sigma\sqrt{2\pi}} \text{ and } v = \frac{2\pi\sigma^2}{4n} = \frac{\pi}{2}\frac{\sigma^2}{n}$$

This gives

$$\tilde{X} \sim AN\left(\theta, \frac{\pi}{2}\frac{\sigma^2}{n}\right)$$

Note that $\bar{X}$ is MLE for θ and by central limit theorem, $\sqrt{n}(\bar{X} - \theta) \xrightarrow{d} N(0, \sigma^2/n)$. Further, note that the asymptotic distributions of $\bar{X}$ and $\tilde{X}$ have the same location parameter since the population distribution is symmetric; however, their variances are different. Both $\bar{X}$ and $\tilde{X}$ are CAN estimators of θ, and $\bar{X}$ is most efficient with $V(\bar{X}) = \sigma^2/n = 1/nI(\theta)$. The asymptotic efficiency of sample median $\tilde{X}$ with respect to sample mean $\bar{X}$ is given by

$$e(\tilde{X}, \bar{X}) = \frac{1/nI(\theta)}{AV(\tilde{X})} = \frac{\sigma^2/n}{(\pi/2)(\sigma^2/n)} = \frac{2}{\pi} = 0.64$$

Since $e(\tilde{X}, \bar{X}) < 1$, median is an inefficient estimator of θ as compared to mean. The interpretation of ARE is that if we require n sample size by median $\tilde{X}$, to estimate population mean θ, to reach to a certain level of accuracy, we require only $0.64n$ sample size for mean $\bar{X}$ to estimate θ at the same level of accuracy. It is, however, noted that in case of dirty data, where sample observations are subjected to measurement errors, the use of sample median $\tilde{X}$ is preferred over sample mean $\bar{X}$, since in the former estimator, not all magnitudes of the sample observations are used in its computation.

Another example: Let $X_1, X_2, \ldots, X_n$ be a random sample from a Cauchy population with *pdf*

$$f(x;\theta) = \left(\frac{1}{\pi}\right)\frac{1}{1+(x-\theta)^2}, -\infty < x < \infty$$

The asymptotic distribution of sample median is given by

$$\tilde{X} \sim AN\left(\vartheta, \frac{1}{4nf_1^2}\right)$$

where, $f_1^2 = \dfrac{1}{\pi}$ and $\vartheta = \dfrac{\pi^2}{4n}$.

2 Data Summarization and Principle of Sufficiency

2.1 INTRODUCTION

Definition 2.1.1 (Game) A *game* is represented by a triplet $(\Theta, \mathcal{A}, L)$, where Θ is a set of all possible states of *nature*, referred to as a *parameter space*, $\mathcal{A}$ is a set of actions available to the statistician, $L(\theta, a)$: $\Theta \times \mathcal{A} \to \mathbb{R}^1$ is the *loss* incurred to the statistician as a result of joint choices of nature and statistician, and (θ, a) is an element in the cartesian product of Θ and $\mathcal{A}$ defined by $\Theta \times \mathcal{A} = \{(\theta, a): \theta \in \Theta, a \in \mathcal{A}\}$.

Definition 2.1.2 (Statistical decision problem) Consider a random experiment, corresponding to which there is a random variable X, defined on a probability space $(\Theta, \mathcal{B}, P)$, where $\mathcal{B}$ is the σ-field (Borel field in case of $\mathbb{R}^1$) generated by the subsets of the sample space χ, χ being a set of all possible values of random observable X; P is some probability measure defined on $\mathcal{B}$, with F_θ as the distribution function of the random variable (*rv*) X that depends on an unknown quantity θ. *Statistical decision problem* is a game between nature and statistician, combined with a random variable X so that the probability distribution of X is F_θ, $\theta \in \Theta$.

Usually, in statistical inference, which is a particular case of statistical decision problem, it is assumed that the functional form of the distribution function F_θ is known, except possibly its parameter θ. Therefore, the statistical decision problem is defined as a game denoted by a triplet $(\Theta, \mathcal{A}, L)$, combined with X on $(\Theta, \mathcal{B}, P)$, so that $X \sim F_\theta$, $\theta \in \Theta$, where F_θ is some distribution in a family of distributions $\mathcal{F}$. In this, the statistician collects information on X in the form of a sample $\mathbf{X} = (X_1, X_2,\ldots,X_n)'$ from the population having its distribution function P_θ, $\theta \in \Theta$, and based on these observations $\mathbf{X}$, he makes some inference about F_θ.

Definition 2.1.3 (Parameter) The quantity θ that indexes the distribution function F_θ of an *rv* X is known as a *parameter*. It is such a characterizing quantity of the population that if known, tells everything about the population.

Definition 2.1.4 (Statistic) Based on the observations $\mathbf{X} = \mathbf{x}$, the statistician chooses an action $\delta(\mathbf{x}) \in \mathcal{A}$ by way of defining a mapping δ on the sample space χ into action space $\mathcal{A}$ denoted by $\delta: \chi \rightarrow \mathcal{A}$ (or $\mathbf{x} \rightarrow \delta(\mathbf{x})$), $\delta(\mathbf{x})$ is known as a *statistic*. More specifically, a statistic is a function of observations $X_1, X_2,\ldots,X_n$, $T = T(X_1, X_2,\ldots,X_n)$. For example, $(1/n)\ \Sigma X_i = \bar{X}$, mean of observations; $\max\{X_i\} = X_{(n)}$, maximum of the observations; and median $\{X_i\}$, median of the observations, are all statistics. A statistic is itself a random variable since it is a function of random observations. It must have a distribution function explaining its stochastic behavior. This distribution function is induced from the distribution function of the random observations.

Definition 2.1.5 (Estimator) The action $\delta(\mathbf{x})$ in $\mathcal{A}$, when used for estimating θ, is known as an *estimator*.

Definition 2.1.6 (Risk of an estimator) The *risk* of choosing δ is the average loss, which is defined as a consequence of choosing an estimator $\delta(\mathbf{x})$ as a guess value of θ when the true value of the parameter is θ. Formally, the risk of an estimator δ is defined as

$$R(\theta, \delta) = E_\theta L(\theta,\delta(\mathbf{X})) = \int L(\theta,\delta(\mathbf{x}))dF_\theta(\mathbf{x})$$

where the statistician incurs a loss $L(\theta,\delta(\mathbf{x}))$ and F_θ is the joint *cdf* of the observations $X_1,\ldots,X_n$ that depend on $\theta \in \Theta$.

In the present Section, the formal introduction of the problem of point estimation is given. In Section 2.2, the concepts of data summarization and sufficient statistics are introduced. Section 2.3 discusses the Fisher–Neyman criterion of constructing sufficient statistic. In Section 2.4, the concepts of maximum summarization of data and minimal sufficient statistic are introduced and two procedures have been explained to construct minimal sufficient statistic. In Section 2.5, the degree of data summarization is characterized through the amount of ancillary information it carries. It also defines complete sufficient statistic and explains how it results into maximum summarization of data. Section 2.6 defines convex loss functions and discusses their role in estimation problems. In Section 2.7, given an estimator, it is shown as to how it can be further improved through a sufficient statistic; the result is known as the famous Rao–Blackwell theorem.

Definition 2.1.7 (Partition of the sample space induced by T) Consider a statistic T. Denote the set of all possible values of T by $I(T)$. $I(T)$ is known as the range set of T. Corresponding to each t in $I(T)$, we get a set

$$A_t = [\mathbf{x}\colon T(\mathbf{x}) = t]$$

The collection of all such sets A_t, $t \in I(T)$, is known as the *partition set of the sample space χ, induced by the statistic T.*

Under the joint probability distribution of $\mathbf{X}$, the probability of some partition A_t is

$$P(A_t) = P_\theta(T(\mathbf{X}) = t)$$

$$= \begin{cases} \displaystyle\int_{\mathbf{x}\in A_t} f(\mathbf{x};\theta)d\mathbf{x}, & \text{if } \mathbf{x} \text{ is continuous} \\ \displaystyle\sum_{\mathbf{x}\in A_t} P_\theta(\mathbf{X} = \mathbf{x}), & \text{if } \mathbf{x} \text{ is discrete} \end{cases}$$

2.1.1 Problem of Point Estimation

Consider the sample observations $(X_1, X_2,\ldots,X_n)$ drawn on a random variable X from some population with distribution F_θ, which is a member of a family of probability distributions $\mathcal{F}$. Our objective in statistical inference or, more particularly, in point estimation is to determine one $F_\theta \in \mathcal{F}$ that could likely have generated this sample.

To formalize this problem in a mathematical setup, consider Θ and $\mathcal{A}$ as real lines $\mathbb{R}^1$; define the loss function by

$$L(\theta, \delta) = c\rho(\theta - \delta) \tag{2.1.1}$$

where $c > 0$ and ρ is some even, differentiable, and convex function of $(\theta - \delta)$. Usually, the parameter θ is unknown. A function δ defined on χ into Θ is known as an estimator of θ. Among several such δs usually, a statistician chooses one such δ for which

$$R(\theta, \delta) = cE_\theta\rho(\theta - \delta(\mathbf{X}))$$

is minimum for every $\theta \in \Theta$. In case

$$\rho(\theta - \delta(\mathbf{x})) = (\theta - \delta(\mathbf{x}))^2$$

$L(\theta,\delta)$ in Eq. (2.1.1) is known as squared error loss function of an estimator $\delta(\mathbf{X})$. The statistical decision problem $(\Theta, \mathcal{A}, L)$ with $X \sim F_\theta$, $\theta \in \Theta$, $\mathcal{A} \subseteq \mathbb{R}^1$, $\Theta \subseteq \mathbb{R}^1$, and $L(\theta, \delta)$ as in Eq. (2.1.1) is known as the *problem of point estimation*. For example, consider $X \sim N(\theta, \sigma^2)$, $\Theta = \{(\theta, \sigma^2)$: $\theta \in \mathbb{R}^1$, $\sigma^2 > 0\}$, and the problem of point estimation is to estimate the unknown parameter vector $(\theta,\sigma^2)'$ on the basis of a sample observation $\mathbf{X} = (X_1, X_2,\ldots,X_n)'$ from $N(\theta,\sigma^2)$. In this, the statistician tries to construct such a statistic $\mathbf{T} = (T_1,T_2)$ for estimating (θ,σ^2), which is a function defined on χ into Θ, so that it is best in the sense of minimum mean square error of $\mathbf{T}$.

2.1.2 Existence of Uniformly Minimum Risk Estimator and Unbiasedness

Consider a family of distributions $\mathcal{F} = \{F_\theta, \theta \in \Theta\}$, where F_θ is the distribution function of an *rv* X. The cumulative distribution function (*cdf*) F_θ is referred to as a statistical model. Our interest is to estimate θ on the basis of a random sample $(X_1, X_2,\ldots,X_n)$. In other words, our interest is to identify a specific distribution function F_θ in $\mathcal{F}$ on the basis of these random observations which have generated them. Thus, formally, one is interested in such functions δ defined on χ into Θ, δ: $\chi \to \Theta$, which are reasonable estimators of θ in some sense.

The accuracy of such a function δ is measured by the loss L, so that $L(\theta, \delta) \geq 0\ \forall\ \theta$ and $\delta > 0$. L attains zero when $\delta \equiv \theta\ \forall\ \theta \in \Theta$. The accuracy of δ is measured by its risk, $R(\theta, \delta) = E[L(\theta, \delta(X))]$. The risk, therefore, becomes a criterion of judging between the estimators. In other words, the estimator δ_0 is preferred over all other estimators if its risk is the minimum over all θ. Note that for every fixed value $\theta = \theta_0$, we have an estimator with

$$\delta(x) \equiv \theta_0 \quad \text{for all values of } x \tag{2.1.2}$$

with $R(\theta_0, \delta) = 0$. The loss L may not be zero at other values of θ. The minimum risk estimator has to have zero risk at every θ, which is not possible by any such estimator δ in Eq. (2.1.2), unless $R(\theta, \delta) = 0\ \forall \theta$, which is practically not possible. Therefore, uniformly minimum risk estimator does not exist in the class of all estimators. To circumvent this problem, a desirable condition of unbiasedness of an estimator or some other suitable criterion needs to be introduced that technically eliminates all such estimators from consideration which favour very strongly one or more values of θ at the cost of neglecting others. This condition is based on the principle of balancing under and over estimation of errors obtained from $\delta(x) - \theta$, known as the unbiasedness property of an estimator. If one restricts to a class of all unbiased estimators, the dimension of the class reduces, the class becomes smaller, and the search for an optimal estimator becomes simplified. The optimality criterion that an estimator must satisfy and other details and discussions on unbiasedness have been discussed in Chapter 3. In the present chapter, we are interested in such a data summarizing statistic which reduces the sample data $\mathbf{X} = (X_1, X_2, \ldots, X_n)'$ to a statistic of much smaller dimension than of the sample space χ, which is n.

2.2 DATA SUMMARIZATION—SUFFICIENT STATISTICS

Usually, the data collected on the behaviour of the parameter θ in the form of a sample $X_1, X_2, \ldots, X_n$ (i) is not only voluminous but costly to maintain, (ii) there remains a possibility of loss of data in storage or transit, (iii) is not scientific to retain a voluminous data, and (iv) the sample $X_1, X_2, \ldots, X_n$ contains some additional information which are not relevant for estimating θ. That is why the statistician tries to reduce the dimensionality of sample space by considering a summary statistic $T(\mathbf{X})$, to replace the data $(X_1, \ldots, X_n)$ which discards all such information in the data that is not relevant for estimating θ. Thus, a sufficient statistic T does not use irrelevant information and, thus, extracts all the useful information which is only relevant for estimating θ. Also, it has simpler data structure and distribution than of the original sample $X_1, \ldots, X_n$. The sufficiency principle, therefore, talks of such inferential procedure which depends on data only through sufficient statistics. The dimension of a sufficient statistic is, therefore, generally smaller than the sample size n. This process of getting a desirable summary statistic $(T(\mathbf{X}))$ which replaces the entire sample observations is known as *data summarization*. This concept was first given by Sir Ronald A. Fisher in 1925.

The formal definition of a sufficient statistic T, under the assumption that the model is correct and the parameter θ is the only unknown value, is given as follows:

Definition 2.2.1 (Sufficiency) A statistic T is said to be sufficient for a family of distributions $\mathcal{F} = \{F_\theta, \theta \in \Theta\}$ if the conditional distribution of $\mathbf{X}|T = t$ is independent of θ for all t.

The intuitive basis for this definition is that once we know the value of T, the distribution of data does not depend upon θ, and thus, it contains no information about θ. More clearly, one may understand by a sufficient statistic T that if the original sample $\mathbf{X}$ is discarded after computing T, one can reconstruct the original sample, any time one wishes, by just knowing the value of T. Note that the conditional distribution of $\mathbf{X}|t$ is a completely known distribution since it is independent of unknown θ. Hence, the corresponding random quantity can be easily generated from this conditional distribution $\mathbf{X}|t$ by a random number generator.

The above definition of sufficiency means that there is enough information in T by which one can reconstruct the distribution of $\mathbf{X}$ without having referred back to the data.

Definition 2.2.2 (Principle of sufficiency) Formally, the sufficiency principle states that if T is a sufficient statistic for θ and if x_1 and x_2 are any two different values of the *rv* X, so that T takes on the same value on these values i.e.

$$T(x_1) = T(x_2)$$

then, the inferences drawn on θ from observing $X = x_1$ is the same as from observing $X = x_2$.

2.2.1 Reconstruction of the Original Sample by a Sufficient Statistic *T*

By the sufficiency of a statistic T, we mean the potential in the statistic $T = t$ towards reconstructing the original sample. Basically, one draws a random sample $Y_1, Y_2, \ldots, Y_n$ from the conditional distribution of $\mathbf{X}|t$. The sample $Y_1, Y_2, \ldots, Y_n$ is easily obtained with the help of random number table or by a similar mechanism, since the conditional distribution of $X|t$ is independent of θ. The conditional distribution of $\mathbf{Y}|t$ is same as the conditional distribution of $\mathbf{X}|t$; therefore, their unconditional distributions are also the same.

Consider a discrete variable case and assume that T is sufficient for θ. Given a value t of T, we can define a conditional distribution of $\mathbf{X}|T = t$ on the restricted sample space A_t. We, then, generate a pseudo sample $Y_1, Y_2, \ldots, Y_n$ from A_t by the random number generator.

We may note that the probability distribution

$$P(\mathbf{Y} = \mathbf{x}|T(\mathbf{X}) = T(\mathbf{x})) = P(\mathbf{X} = \mathbf{x}|T(\mathbf{X}) = T(\mathbf{x}))$$

is defined on the set $A_{T(\mathbf{x})}$. Now, the events $\{\mathbf{Y} = \mathbf{x}\}$ and $\{\mathbf{X} = \mathbf{x}\}$ are the subsets of $\{T(\mathbf{X}) = T(\mathbf{x})\}$. It can easily be seen, now, that the unconditional distributions of $\mathbf{Y}$ and $\mathbf{X}$ are the same, i.e.,

$$P_\theta(\mathbf{Y} = \mathbf{x}) = P_\theta(\mathbf{X} = \mathbf{x}) \ \forall \mathbf{x} \text{ and } \forall \theta$$

Therefore, the new sample $Y_1, Y_2, \ldots, Y_n$ and the original data $X_1, X_2, \ldots, X_n$ carry equal amount of probabilistic information about θ since $Y_1, Y_2, \ldots, Y_n$ can be regarded as another sample from the same population from where the original data $X_1, X_2, \ldots, X_n$ is drawn. Therefore, we can "recover data" if we discard $X_1, X_2, \ldots, X_n$ and retain T. It is in this sense we mean that T is "sufficient." Note that there may be several statistics which are sufficient in the above sense.

We have noted that the construction of original sample $\mathbf{X}$ by getting observations on Y depends on T as well as on some random mechanism. We, therefore, noticed that the estimator $T(\mathbf{Y})$ not only depends on T but also on the drawing mechanism of $\mathbf{Y}$; and that the estimator of θ is a random estimator which for each fixed value of $\mathbf{X} = \mathbf{x}$ is a random variable $T(\mathbf{Y})|_{\mathbf{x}}$ with a known distribution. We denote this randomized estimation $T(\mathbf{Y})|_{\mathbf{x}}$ by $T_{\mathbf{x}}$. The risk of the randomized estimator $T(\mathbf{Y})$ is, therefore, given by

$$R(\theta,T) = E_\theta^{\mathbf{X}} E^{\mathbf{Y}_{\mathbf{X}}} L(\theta, T_{\mathbf{X}})$$

Result. Let T be a sufficient statistic for a family of distributions $\mathcal{F} = \{F_\theta, \theta \in \Theta\}$ and $\delta(\mathbf{X})$ be an estimator of θ. Then, there exists an estimator of θ which is a function of T whose risk function is same as that of $\delta(\mathbf{X})$.

The examples relating the above concepts are given in Section 2.8.

2.3 CONSTRUCTION OF A SUFFICIENT STATISTIC

We can check that a statistic T is sufficient (i) if, the conditional distribution of $\mathbf{X}|T = t$ is free of θ or (ii) if, the distribution of data $\mathbf{X}$ and some statistic T are $p(\mathbf{x}; \theta)$ and $q(t; \theta)$ respectively, then the statistic T is sufficient if the tth conditional

$$\frac{p(\mathbf{x}; \theta)}{q(t; \theta)}$$

on A_t is free of θ (but may depend on $\mathbf{x}$), for all t and for all θ. This procedure for checking a statistic to be sufficient requires (a) a guess of a statistic as the candidate of a sufficient statistic and (b) to figure out the distribution of T. Meeting these requirements is somewhat ad hoc and sometimes inconvenient. Halmos and Savage (1949) developed a simple result as a solution to this problem that characterizes a sufficient statistic by considering the joint distribution of sample observations. This is a recast of the Fisher-Neyman (1925) concept of sufficiency. This is stated in the following theorem where we assume that the distribution function F of an *rv* X has a density with respect to a common measure μ, i.e., F is dominated with respect to μ.

Theorem 2.3.1 (Fisher–Neyman Factorization Theorem) Let the family of distributions $\mathcal{F} = \{F_\theta, \theta \in \Theta\}$ be dominated with respect to a σ-finite measure μ. A statistic T is said to be sufficient for $\mathcal{F}$, if and only if there exist non-negative functions $g(\theta, T(\mathbf{x}))$ and $h(\mathbf{x})$ such that the joint density can be expressed as

$$P_\theta(\mathbf{X} = \mathbf{x}) = h(\mathbf{x})\, g(\theta, T(\mathbf{x})) \quad \text{(a.e. } \mu) \tag{2.3.1}$$

where the first factor h depends on sample observation $\mathbf{x}$ and is free of θ and the second factor depends on θ and $T(\mathbf{x})$ only through $\mathbf{x}$.

Proof. If part (Sufficient Part): We assume here that the *rv* X is discrete random variable with *pmf* $P(x; \theta)$. If T is sufficient for F_θ, then by definition, $P\{\mathbf{X}|T = t\}$ is independent of θ. We have

$$P_\theta\{\mathbf{X} = \mathbf{x}\} = P_\theta\{\mathbf{X} = \mathbf{x}, T(\mathbf{X}) = T(\mathbf{x})\}$$

$$= P_\theta\{T(\mathbf{X}) = T(\mathbf{x})\}\, P\{\mathbf{X} = \mathbf{x}|T(\mathbf{X}) = T(\mathbf{x})\}$$

with $h(\mathbf{x}) = P\{\mathbf{X} = \mathbf{x}|T(\mathbf{X}) = T(\mathbf{x})\}$ for $\mu_\theta\ (\mathbf{X} = \mathbf{x}) > 0$ for some θ and $h(\mathbf{x}) = 0$ for $\mu_\theta(\mathbf{x}) = 0$ for all θ; $g(\theta, T(\mathbf{x})) = P_\theta\{T = T(\mathbf{x})\}$; and Eq. (2.3.1) follows.

Only if (Necessary Part): Suppose (2.3.1) holds. Then for $T = t_0$, we have

$$P_\theta\{T = t_0\} = \sum_{\{\mathbf{x}: T(\mathbf{x}) = t_o\}} P_\theta\{\mathbf{X} = \mathbf{x}\}$$

$$= g(\theta, t_0) \sum_{\{\mathbf{x}: T(\mathbf{x}) = t_0\}} h(\mathbf{x}) \tag{2.3.2}$$

For $T = t_0$, define

$$P_\theta\{(\mathbf{X} = \mathbf{x})| T = t_0\} = \frac{P_\theta\{\mathbf{X} = \mathbf{x}, T(\mathbf{x}) = t_0\}}{P_\theta\{T = t_0\}} \tag{2.3.3}$$

The conditional probability in Eq. (2.3.3) is well defined if we assume $P_\theta\{T = t_0\} > 0$ for some $\theta > 0$. On putting the value of $P_\theta\{T = t_0\}$ given by Eq. (2.3.2) in Eq. (2.3.3), we get

$$P_\theta\{(\mathbf{X} = \mathbf{x}) \mid T = t_0\} = \frac{g(\theta, t_0)\, h(\mathbf{x})}{g(\theta, t_0) \displaystyle\sum_{\{\mathbf{x}:\, T(\mathbf{x}) = t_0\}} h(\mathbf{x})}$$

$$= \frac{h(\mathbf{x})}{\displaystyle\sum_{\{\mathbf{x}:\, T(\mathbf{x}) = t_0\}} h(\mathbf{x})}$$

which is independent of θ. Therefore, by definition of sufficiency, T is sufficient for θ. ▣

***Corollary* 2.3.1** A necessary and sufficient condition for a statistic T to be sufficient for a family $\mathcal{F}$ of distributions of X is that for any two densities p_1 and p_2 in $\mathcal{F}$, the ratio $p_1(\mathbf{x})/p_2(\mathbf{x})$ is a function only of $T(\mathbf{x})$.

Invariance Property of a Sufficient Statistic

If T is a sufficient statistic for θ, then $h(T)$ is sufficient for $h(\theta)$ whenever h is a one-to-one function.

We finally state a theorem which helps in showing that a statistic is not sufficient:

Theorem 2.3.2 Let $T(\mathbf{X})$ be a statistic such that for some $\theta_1, \theta_2 \in \Theta$ and $\mathbf{x}, \mathbf{y}$ in the support of the distribution $f(\cdot, \theta)$ $\theta \in \Theta$, we have

1. $T(\mathbf{x}) = T(\mathbf{y})$
2. $f(\mathbf{x}; \theta_1)\, f(\mathbf{y}; \theta_2) \neq f(\mathbf{x}; \theta_2)\, f(\mathbf{y}; \theta_1)$

Then the statistic T is not sufficient for θ.

Proof. Let us denote the support of $f(\cdot;\, \theta_1)$ by $S(\theta_1)$ and of $f(\cdot;\, \theta_2)$ by $S(\theta_2)$, so that $S(\theta_1) = \{\mathbf{x}: f(\mathbf{x};\, \theta_1) > 0\}$ and $S(\theta_2) = \{\mathbf{x}: f(\mathbf{x};\, \theta_2) > 0\}$. Consider (2) with a situation in which one of its side is zero and other is non-zero. This can happen when either $\mathbf{x}$ or $\mathbf{y}$ is in $S(\theta_1)$ and not in $S(\theta_2)$. Further, if T were sufficient, this along with (1) implies that both $\mathbf{x}$ and $\mathbf{y}$ are in $S(\theta_1)$ and $S(\theta_2)$. Therefore, T is not sufficient.

Next, suppose that both the sides of (2) are positive, i.e., the points $\mathbf{x}$ and $\mathbf{y}$ belong to $S(\theta_1)$ and $S(\theta_2)$ and that if $T(\mathbf{X})$ were sufficient, we have

$$\frac{f(\mathbf{x}; \theta_1)}{f(\mathbf{x}; \theta_2)} = \frac{f(\mathbf{x} \mid T(\mathbf{x}), \theta_1)\, f(T(\mathbf{x}) \mid \theta_1)}{f(\mathbf{x} \mid T(\mathbf{x}), \theta_2)\, f(T(\mathbf{x}) \mid \theta_2)} \tag{2.3.4}$$

On using the fact that $T(\mathbf{X})$ is sufficient, i.e., given T, the distribution of $\mathbf{X}$, T is independent of θ and that if (1) holds, Eq. (2.3.4) gives

$$\frac{f(\mathbf{x}; \theta_1)}{f(\mathbf{x}; \theta_2)} = \frac{f(\mathbf{y} \mid T(\mathbf{y}), \theta_1)\, f(T(\mathbf{y}) \mid \theta_1)}{f(\mathbf{y} \mid T(\mathbf{y}), \theta_2)\, f(T(\mathbf{y}) \mid \theta_2)} = \frac{f(\mathbf{y}; \theta_1)}{f(\mathbf{y}; \theta_2)}$$

This gives $f(\mathbf{x};\, \theta_1)\, f(\mathbf{y};\, \theta_2) = f(\mathbf{x};\, \theta_2)\, f(\mathbf{y};\, \theta_1)$, which shows that (2) does not hold. Therefore, (1) and (2) show that $T(\mathbf{X})$ is not sufficient. ▣

Notes. *Features of sufficient statistics.*

1. If there are two distinct sufficient statistics, then each of them can be expressed as a function of the other; but not necessarily every function of a sufficient statistic is also sufficient.
2. If T is sufficient and $T = h(U)$ for some function h, then U is also sufficient. Further, if U is sufficient and h is one-to-one, then T is also sufficient and it is said to be equivalent to U. In this case, both the statistics T and U carry exactly the same amount of information about θ. If h is not one-to-one, then T shows greater reduction of data. Therefore, T may or may not be sufficient. If $\Sigma_{i=1}^{n} X_i$ is sufficient, then $(\Sigma_{i=1}^{m} X_i, \Sigma_{i=m+1}^{n} X_i)$ is sufficient; $\Sigma_{i=1}^{n} X_i$ is a greater reduction of data.
   If T is sufficient for θ, then any linear combination of T is also sufficient for θ. For example, consider a random sample $X_1, X_2,\ldots,X_n$ from $N(\theta, \sigma^2)$, σ^2 being known. If $T_1 = \Sigma X_i$ is sufficient for θ, $T_2 = a(\Sigma X_i) + b$ is also sufficient for θ. Note that $T_2 = T_1$ for $a = 1$, $b = 0$.
3. Let T be a sufficient statistic for θ. Then $\phi(T)$ is sufficient for θ and T is a sufficient statistic for $\psi(\theta)$. Then, the functions ϕ and ψ are 1-1 and onto.
4. By data summarization, we generally mean the reduction in dimension of the range space of a sufficient statistic. i.e., a sufficient statistic $T(\mathbf{X})$ takes values from its range space which is usually of a dimension (k, say) much smaller than the dimension of the sample space $\chi \subseteq \mathbb{R}^n$, which is n, i.e., $k < n$ without any loss of information on θ.
5. Let $X_1, X_2,\ldots,X_n$ be *iid* with density f.
   (i) the original data $X_1, X_2,\ldots,X_n$ is always sufficient for θ. This is a trivial statistic since it is the original data and there is no data reduction at this stage. Note that
   $$f(\mathbf{x}; \theta) = 1 \cdot f(\mathbf{x}; \theta) = h(\mathbf{x})\, g(T(\mathbf{x}), \theta)$$
   which implies $h(\mathbf{x}) = 1$, $T(\mathbf{x}) = \mathbf{x}$.
   (ii) Order statistic $T(\mathbf{X}) = (X_{(1)}, X_{(2)},\ldots,X_{(n)})$ is always sufficient for θ. Here, the dimension of the order statistic is n, which is the same as the dimension of data. However, it is non-trivial since $n!$ data points reduce to one value of T (see solved Example 2.11).
6. Apart from the other implications of factorization theorem, one implication is that on the basis of likelihood function, the inference about θ remains the same for all values of data $\mathbf{x}$ for which $T(\mathbf{x})$ is same.
7. If $T(\mathbf{X})$ is a sufficient statistic and $\hat{\theta}(\mathbf{X})$ is a maximum likelihood estimator of θ, then $\hat{\theta}$ is a function of T. Moreover, if $\hat{\theta}$ is unique and sufficient, then it is minimal sufficient for θ. Note that for any other sufficient statistic U if T can be expressed as a function of U, then T is minimal sufficient. The formal definition of minimal sufficient is given in Section 2.4.
8. Let T be a sufficient statistic for θ and U be a Bayes estimator of θ. U is a function of T.
9. Sufficient statistic can be used to improve any estimator (biased or unbiased) of θ (through the Rao–Blackwell theorem 2.7.1).
10. Theorem 2.3.1 holds also for $\boldsymbol{\theta}$ and $\mathbf{T}$ when they are vectors. Here, $\mathbf{T}$ is said to be jointly sufficient for $\boldsymbol{\theta}$. However, jth component of $\mathbf{T}$ is not necessarily sufficient for jth components of $\boldsymbol{\theta}$, whereas the converse holds under mild conditions.

11. If $\mathbf{U}$ and $\mathbf{V}$ are equivalent statistics and $\mathbf{U}$ is sufficient for $\boldsymbol{\theta}$, then $\mathbf{V}$ is also sufficient for $\boldsymbol{\theta}$.
12. Suppose the distribution of X belongs to a k-parameter exponential family of distributions. Then the natural statistic $\mathbf{T}(\mathbf{X})$ is sufficient for $\boldsymbol{\theta} \in \Theta \subseteq \mathbb{R}^k$.
13. $X_1, X_2, \ldots, X_n$ is a random sample from
    (i) $P(\theta)$, $\theta > 0 \cdot \Sigma_{i=1}^n X_i$ is sufficient for θ;
    (ii) $N(\mu, \sigma^2)$. Then $(\Sigma X_i, \Sigma X_i^2)$ is jointly sufficient for $N(\mu, \sigma^2)$. Note that $(\bar{X}, S^2)$ is also jointly sufficient for (μ, σ^2) [use equivalence of statistics];
    (iii) $G_1(\alpha, \beta)$, $\alpha > 0$, $\beta > 0$, α is a shape parameter and β is a scale parameter, $(\Pi_{i=1}^n X_i, \Sigma_{i=1}^n X_i)$ is jointly sufficient for (α, β);
    (iv) Be(α, β), $\alpha > 0, \beta > 0$. $(\Pi_{i=1}^n X_i, \Pi_{i=1}^n (1 - X_i))$ is jointly sufficient for (α, β);
    (v) BV $N(0, 0, 1, 1, \rho)$. $(\Sigma_{i=1}^n X_i^2, \Sigma_{i=1}^n Y_i^2, \Sigma_{i=1}^n X_i Y_i)$ is sufficient statistic for ρ;
    (vi) $U(0, \theta)$, $\theta > 0$. $X_{(n)}$ is sufficient for θ.

2.4 MINIMAL SUFFICIENT STATISTIC

We have made three observations so far:

1. The whole sample data is a sufficient statistic: *no summarization of data.*
2. One-to-one function of a sufficient statistic is also sufficient: *same degree of summarization.*
3. A sufficient statistic is obtained by many other sufficient statistics as their functions: *greater reduction, better summarization.*

There can be several sufficient statistics and each of them may represent different levels of data summarization, but the amount of information they carry about θ is the same, which is equal to the amount of information carried by the entire sample data itself. Thus, in the presence of several sufficient statistics for a model, one is interested in identifying that statistic which provides the maximum summarization of data while still retaining the entire information equivalent to the information carried by the sample data.

In this regard, we may state that a statistic T summarizes the data with maximum reduction without any loss in information on θ, if and only if,

1. T is sufficient for θ, and
2. if S is any other sufficient statistic for θ, then T is a function of S, i.e., there exists a function h so that $T = h(S)$ a.e. μ.

Such a statistic T is called minimal sufficient statistic. To clarify this, if one is at some level of data summarization in terms a sufficient statistic S, then maximum possible data summarization can be achieved by defining summarizing functions h of increasing levels so that $T = h(S)$ as long as T remains sufficient. In other words, one starts with some sufficient statistic T and summarizes it further and further by way of taking its functions, so that irrelevant information are discarded till such an extent when further reduction results into a statistic which is not sufficient.

This maximum summarization of data can also be explained in terms of defining partitions on the sample space χ that is induced by a statistic T. Consider two statistics T and S. Let the image spaces of χ under T and S be given by $I(T) = \{t: t = T(\mathbf{x}) \text{ for some } \mathbf{x} \in \chi\}$ and $I(S) = \{s: s = S(\mathbf{x}) \text{ for some } \mathbf{x} \in \chi\}$, respectively. The statistics T and S then induce collections of disjoint partition sets, namely $\{A_t: t \in I(T)\}$ and $\{B_s, s \in I(S)\}$, respectively. Now, saying that T is a function of S means that $T(\mathbf{x}) = T(\mathbf{y})$ whenever $S(\mathbf{x}) = S(\mathbf{y})$ or each B_s is a subset of some A_t. Therefore, a statistic that induces a **coarsest partition** on the sample space is known as minimal sufficient statistic. It, therefore, results into the greatest possible summarization of data while retaining all the relevant information on θ.

Normally, we expect the dimension of a minimal sufficient statistic to be the smallest among all sufficient statistics, but in some cases, the dimensions of a minimal sufficient statistic and of a sufficient statistic are both the same. For example, single data X from $N(0, \sigma^2)$ or $X_1, X_2,\ldots,X_n$ *iid* from a location family, say, a Cauchy distribution, where the order statistic $T(\mathbf{X}) = (X_{(1)}, X_{(2)},\ldots,X_{(n)})$ is minimal sufficient. Most often, it is equal to the number of free parameters, i.e., the dimension of the parameter space, but in some cases, it is larger than the number of free parameters, e.g., $X \sim U(\theta, \theta + 1)$. In Example 2.8.23, $(X_{(1)}, X_{(n)})$ is minimal sufficient for θ. Further, note that the minimal sufficient statistic is not unique; if U is minimal sufficient and $T = f(U)$, for some f which is one-to-one, then T is also minimal sufficient.

Suppose T is a minimal sufficient statistic and that $\mathbf{x}$ and $\mathbf{y}$ are observed data from some experiment. Then by sufficiency principle, identical inference is drawn if $T(\mathbf{x}) = T(\mathbf{y})$.

A minimal sufficient statistic, by using the above definition, is not easy to obtain. A simpler test is sought, instead. There are two commonly adopted techniques for constructing it: (1) Lehmann and Scheffe (1950) result, based on likelihood equivalence relation, and (2) Lehmann's procedure.

(1) Likelihood Equivalence Procedure

Theorem 2.4.1 Let $X_1, X_2,\ldots,X_n$ be an *iid* from $f(\mathbf{x}; \theta)$ (*pdf* or *pmf*), where $f(x; \theta)$ belong to some family of distributions $\mathcal{F}$. Suppose there exists a statistic T so that for any two points $\mathbf{x}$ and $\mathbf{y}$ in the sample space, the ratio $f(\mathbf{x}; \theta)/f(\mathbf{y}; \theta)$ is free of θ if and only if $\mathbf{x}$ and $\mathbf{y}$ are on the same orbits induced by T. The orbits are made of points such that for any two points in the same orbit we must have $\{T(\mathbf{x}) = T(\mathbf{y})\}$. Then $T(\mathbf{x})$ is a minimal sufficient for θ.

Proof. Consider the statistic T, denote the image of χ under T by $I(T) = \{t: T(\mathbf{x}) \text{ for some } \mathbf{x} \in \chi\}$ and partition sets by $A_t = \{\mathbf{x}: T(\mathbf{x}) = t\}$, $t \in I(T)$ induced by T. Choose an arbitrary point $\mathbf{x} \in \chi$ so that it falls on some A_t with $T(\mathbf{x}) = t$. Choose any point $\mathbf{x}(t)$ from A_t. Naturally, $T(\mathbf{x}) = t = T[\mathbf{x}(t)]$. By the assumption in the theorem, $f(\mathbf{x};\theta)/f(\mathbf{x}(t);\theta)$ is independent of θ. Set this quantity equal to $h(\mathbf{x})$. Define a function g on $I(T)$ so that $g(t; \theta) = f(\mathbf{x}(t); \theta)$. We have

$$f(\mathbf{x};\theta) = f(\mathbf{x}(t);\theta)\frac{f(\mathbf{x};\theta)}{f(\mathbf{x}(t);\theta)} = g(t;\theta)\,h(\mathbf{x})$$

Hence, by factorization theorem, $T(\mathbf{x})$ is sufficient for θ.

Consider, $U(\mathbf{X})$ as some other sufficient statistic. Let the corresponding image space be denoted by $I(U)$ and partition sets by B_u for $u \in I(U)$. Similarly as above, we define a function g' on $I(U)$ so that $g'(u;\theta) = f(\mathbf{x}(u);\theta)$.

Let **x** and **y** be two data points on the same orbit or partition B_u, i.e. $U(\mathbf{x}) = U(\mathbf{y})$. Then the ratio

$$\frac{f(\mathbf{x};\theta)}{f(\mathbf{y};\theta)} = \frac{g'\{u(\mathbf{x});\theta\}\,h'(\mathbf{x})}{g'\{u(\mathbf{y});\theta\}\,h'(\mathbf{y})} = \frac{h'(\mathbf{x})}{h'(\mathbf{y})}$$

is independent of θ. The assumption in the theorem implies that $T(\mathbf{x}) = T(\mathbf{y})$, i.e., $B_u \subset A_t$ or $T(\mathbf{x})$ is a function of $U(\mathbf{x})$. Thus, $T(\mathbf{x})$ is minimal.

A safer version of the theorem, which avoids the possibility of the density to be zero, is that

$$f(\mathbf{x};\theta) = h(\mathbf{x},\mathbf{y})f(\mathbf{y};\theta) \text{ for all } \theta$$

implies $T(\mathbf{x}) = T(\mathbf{y})$, where h is some function that does not depend on θ. Then $T(\mathbf{X})$ is minimal sufficient.

(2) Lehmann's Procedure

Lehmann, by utilizing the above theorem, gave another procedure to construct a minimal sufficient statistic. This procedure is much useful when considered for non-exponential families. The procedure assumes:

1. the family of densities, $\mathcal{L} = \{f_0(\mathbf{x}),\ldots,f_k(\mathbf{x})\}$, has finite members and each member is dominated;
2. the corresponding measurable space is Euclidean;
3. the family of densities $\mathcal{L}$ has the same support.

Conditions (1) and (2) imply the existence of minimal sufficient statistic. Under the assumptions (1) to (3) and for fixed θ and θ_0 in Θ, a statistic U is sufficient if and only if the ratio $f(x|\theta)/f(x|\theta_0)$, by factorization theorem, $f(x;\theta)/f(x;\theta_0) = \{g(U,\theta)\cdot h(x)\}/\{g(U,\theta_0)\cdot h(x)\} = k(U)$, is a function of $U(\mathbf{x})$ only. Define a statistic T as a collection of likelihood ratios

$$T(\mathbf{x}) = \left(\frac{f_1(\mathbf{x})}{f_0(\mathbf{x})}, \frac{f_2(\mathbf{x})}{f_0(\mathbf{x})}, \ldots, \frac{f_k(\mathbf{x})}{f_0(\mathbf{x})}\right)$$

This implies that U is sufficient, if and only if, T is a function of U. Thus, by definition, the constructed statistic T is minimal sufficient.

Note that the above procedure of obtaining minimal sufficient statistic has been developed for a finite number of densities in $\mathcal{L}$. It can also be extended to the case when there are at most countable number of densities in $\mathcal{L}$. Further, if $\mathcal{L}$ is uncountable which contains finite $\mathcal{L}_0$, $\mathcal{L} \supset \mathcal{L}_0$, U is sufficient for $\mathcal{L}_0$ whenever it is sufficient for $\mathcal{L}$. Since T is minimal sufficient for $\mathcal{L}_0$, T is a function of U and we conclude that T is minimal sufficient for $\mathcal{L}$. That is how a minimal sufficient statistic for uncountable family of distributions can be obtained.

Notes.

1. **Regular exponential family of distributions with one parameter:** The statistical models for various practical situations are the most commonly used distributions such as normal, exponential, binomial, Poisson, etc., which belong to the exponential family.

Main results on sufficiency, completeness, best unbiased estimators, etc., hold for the exponential family of distributions. To check these results for a given distribution, it is, thus, enough to check that the distribution belongs to the exponential family. Complete and sufficient statistic exists for exponential family, as shown in the following results: Let there be an *iid* sample $X_1, X_2, \ldots, X_n$ drawn from the exponential density

$$f(x; \theta) = c(\theta)h(x)\exp\{p(\theta)T(x)\}$$

where the support $\{x: f(x;\theta) > 0\}$ is independent of θ. The joint density of sample observations is

$$f(\mathbf{x};\theta) = [c(\theta)]^n \exp\left\{p(\theta)\cdot\sum_1^n T(x_i)\right\}\prod_{i=1}^n h(x_i) \qquad (2.4.1)$$

Here,

$$h(\mathbf{x}) = \prod_{i=1}^n h(x_i),\ g\left(\sum_1^n T(x_i),\ p(\theta)\right) = [c(\theta)]^n \cdot \exp\left\{p(\theta)\sum_1^n T(x_i)\right\}$$

Therefore, $T(\mathbf{X}) = \Sigma_1^n T(X_i)$ is sufficient for θ by Theorem 2.3.1 and also complete. The property of completeness of a statistic is discussed in Section 2.5.

2. **Regular exponential family of distributions with two or more parameters**

(a) Let $X_1, X_2, \ldots, X_n$ be *iid* $f(x; \theta_1, \theta_2)$ which is a *pdf* in two parameter exponential family distributions defined by

$$f(x; \theta_1, \theta_2) = c(\theta_1, \theta_2)\, h(x) \exp\{p_1(\theta_1, \theta_2)\, T_1(x) + p_2(\theta_1, \theta_2)\, T_2(x)\}$$

where support S is independent of (θ_1, θ_2). The joint density of sample observations is

$$f(\mathbf{x}; \theta_1, \theta_2) = \prod_{i=1}^n h(x_i)\exp\left\{p_1(\theta_1,\theta_2)\sum_1^n T_1(x_i) + p_2(\theta_1,\theta_2)\sum_1^n T_2(x_i)\right\}$$

By Neyman factorization theorem, the statistic $\{\Sigma_1^n T_1(X_i),\ \Sigma_1^n T_2(X_i)\}$ is jointly sufficient for (θ_1, θ_2).

(b) Let $X_1, X_2, \ldots, X_n$ be a random sample drawn from the exponential probability density of the form

$$f(x;\boldsymbol{\theta}) = c(\boldsymbol{\theta})h(x)\exp\left\{\sum_{j=1}^k p_j(\boldsymbol{\theta})T_j(x)\right\} \qquad (2.4.2)$$

where $\boldsymbol{\theta} = (\theta_1, \theta_2, \ldots, \theta_d)' \in \Theta \subseteq \mathbb{E}^d$, and $p_1, \ldots, p_k$ are linearly independent. The joint density of sample observations is again exponential with the density

$$f(\mathbf{x}; \theta) = [c(\boldsymbol{\theta})]^n \left(\prod_{i=1}^n h(x_i)\right)\exp\left\{\sum_{j=1}^k p_j(\boldsymbol{\theta})\sum_{i=1}^n T_j(x_i)\right\} \qquad (2.4.3)$$

where $T(\mathbf{X}) = (\Sigma_{i=1}^n T_1(X_i), \ldots, \Sigma_{i=1}^n T_k(X_i))$, d is the dimension of the vector $\boldsymbol{\theta}$, and k is the dimension of the vector $\mathbf{p}$.

(i) If p_j's are linearly independent or $d \le k$, then $\mathbf{T} = (\mathbf{T}_1(\mathbf{X}),\ldots,\mathbf{T}_k(\mathbf{X}))$ is minimal sufficient. Further, if $d < k$, then the family of distributions with the densities of the form Eq. (2.4.2) is known as curved exponential family. However, $\mathbf{T}$ is not complete if $d < k$. For example, $(\Sigma X_i, \Sigma X_i^2)$ is sufficient for θ in $N(\theta, \theta^2)$ but not complete.

(ii) if p_j's are functionally independent or $d = k$, then the family is known as full-rank exponential family or regular exponential family. $\mathbf{T}$ is always minimal sufficient statistic for θ; it is also complete sufficient statistic (discussed in Section 2.5).

3. Consider a k-parameter exponential density $f(x;\theta) = \exp[\boldsymbol{\pi}'(\theta)\mathbf{T}(x) - D(\theta)]\, h(x)$, where $\boldsymbol{\pi}$ and $\mathbf{T}$ are k-dimensional. Suppose there exist $\{\theta_0, \theta_1,\ldots,\theta_k\} \subset \Theta$ so that the vectors $\boldsymbol{\pi}(\theta_i) - \boldsymbol{\pi}(\theta_0)$, $0 \le i \le k$, are linearly independent in $\mathbb{R}^k$; then $\mathbf{T}$ is minimal sufficient for θ. If the exponential family is of full rank, then $\mathbf{T}$ is also complete sufficient.

2.5 DATA SUMMARIZATION THROUGH ANCILLARY STATISTICS

We have seen that there can be several sufficient statistics with varied degrees of data summarization while discussing several models under different statistical situations. The degree of data summarization by a sufficient statistic is characterized by the amount of ancillary information it carries. We say a sufficient statistic has resulted into maximal reduction if it contains no ancillary information.

A statistic V is said to be *ancillary* if its distribution does not depend on θ, and first order ancillary if $E_\theta[V(X)]$ is independent of θ. As opposed to a sufficient statistic, an ancillary statistic contains no information on θ.

Fisher argued that the inference must be conditional on the value of an ancillary statistic. His contention is demonstrated through a simple example. Let $X_1,\ldots,X_n$ be an *iid* $U(\theta - 1, \theta + 1)$. A reasonable estimator of θ is $T_1(\mathbf{X}) = (X_{(1)} + X_{(n)})/2$, known as mid-range. An ancillary statistic is $T_2(\mathbf{X}) = X_{(n)} - X_{(1)}$; T_2 takes value in (0, 2). Given the value of $T_2 = 2$ implies $X_{(n)} = \theta + 1$, and $X_{(1)} = \theta - 1$ gives $T_1(\mathbf{X}) = \theta$ which is the true value of θ. This indicates higher accuracy of $\mathbf{T}_1$. Moreover, the ancillary statistic when added to some statistic forms minimal sufficient. Although one may think that adding it to some statistic does not change the information content on θ. Indeed this is not true. It is shown by an example. Let $X_1,\ldots,X_n$ be an *iid* $U(\theta, \theta + 1)$, $T_1(\mathbf{X}) = (X_{(1)} + X_{(n)})/2$, $T_2(\mathbf{X}) = X_{(n)} - X_{(1)}$. $T_2(\mathbf{X})$ is ancillary since $T_2 \sim Be(n - 1, 2)$. Note that in the minimal sufficient statistic (T_1, T_2), an ancillary statistic T_2 is a part of it. Another example we may consider: $(X, Y) \sim BVN(0, 0, 1, 1, \rho)$, (X^2, Y^2, XY) is minimal sufficient for ρ, though X and Y are ancillary since $X \sim N(0, 1)$ and $Y \sim N(0, 1)$.

Note that a minimal sufficient statistic may contain much ancillary information, although ancillary statistic by itself contains no information on θ. More formally, we say a sufficient statistic T has no ancillary material resulting into maximum summarization of data if

$$E_\theta[h(T)] = c\ \forall\ \theta \Rightarrow h(t) = c \quad a.e.\ \ \mathcal{F}$$

or

$$E_\theta[h(T)] = 0\ \forall\ \theta \Rightarrow h(t) = 0 \quad a.e.\ \ \mathcal{F}$$

i.e., no non-constant function of T can have constant expectation, where $\mathcal{F} = \{F_\theta,\ \theta \in \Theta\}$. A sufficient statistic T that brings out the maximum simplification of the statistical problem is called *complete*.

Definition 2.5.1 (Complete statistic and completeness of family of distributions) A statistic T or the induced family $f_T(t;\theta)$, $\theta \in \Theta$ is said to be *complete* for $\mathcal{F}$ if for any Borel measurable function h,

$$E_\theta[h(T)] = 0 \;\forall\, \theta \text{ implies } h(t) = 0 \;\; a.e.\; \mathcal{F} \tag{2.5.1}$$

The property of completeness is useful for investigating whether there exists only one unbiased estimator for estimating θ. Suppose T is some statistic. $\hat{\theta}_1(T)$, an estimator which is a function of T, is unbiased for θ so that $E_\theta\,[\hat{\theta}_1(T)] = \theta$ for all $\theta \in \Theta$. Consider another unbiased estimator $\hat{\theta}_2(T)$ different from $\hat{\theta}_1(T)$ so that $E_\theta[\hat{\theta}_2(T)] = \theta$ for all $\theta \in \Theta$. If T is complete, then

$$E_\theta[\hat{\theta}_1(T) - \hat{\theta}_2(T)] = 0 \quad \text{for all } \theta$$

$$\Rightarrow \qquad \hat{\theta}_1(T) = \hat{\theta}_2(T) \quad a.e.\; t$$

Hence, the property of completeness helps in identifying a situation where an unbiased estimator, based on a complete statistic T, is unique. In fact, such an estimator is a minimum risk estimator (shown in Chapter 3).

The complete sufficiency is often easy to obtain in exponential families and in families in which the range of the distributions depends on the parameter θ, e.g., $U(\theta_1, \theta_2)$. However, in general, establishing minimality or completeness is usually difficult, while establishing sufficiency remains easy.

The order statistic obtained from a random sample drawn from a continuous distribution is shown to be complete in the following theorem.

Theorem 2.5.1 Let $\mathcal{F}$ be a class of absolutely continuous distribution functions F so that $\mathcal{F}$ is convex. Also, $\mathcal{F}$ contains all uniform densities. Let $X_1,\ldots,X_n$ be *iid* $F \in \mathcal{F}$. Show that the order statistic $T(\mathbf{X}) = (X_{(1)},\ldots,X_{(n)})$ is complete.

Proof. An estimator T_0 is a function of T, $T_0 = h(T)$ if and only if

$$T_0\,(\mathbf{x}_{\boldsymbol{\pi}}) = T_0\,(\mathbf{x}) \;\forall\, \boldsymbol{\pi}'s$$

where $\mathbf{x}_{\boldsymbol{\pi}} = (x_{\pi_1},\ldots,x_{\pi_n})'$ and $(\pi_1,\ldots,\pi_n)$ is one of the $n!$ permutations of numbers $(1, 2,\ldots,n)$. Consider *cdf* $F_1,\ldots,F_n$ from $\mathcal{F}$ with corresponding densities $f_1,\ldots,f_n$. For real positive numbers $\alpha_1,\ldots,\alpha_n$, there is some F in $\mathcal{F}$ so that its density is $f(x) = \Sigma_{i=1}^n \alpha_i f_i(x)/\Sigma_{i=1}^n \alpha_i$. Consider now the expectation of $h(T)$ at f

$$\int \cdots \int T_0(x_1,\ldots,x_n) \prod_{j=1}^{n} f(x_j)\, d\mathbf{x} = 0$$

$$\int \cdots \int T_0(x_1,\ldots,x_n) \prod_{j=1}^{n} \frac{\sum_{i=1}^{n} \alpha_i f_i(x_j)}{\sum_{i=1}^{n} \alpha_i}\, d\mathbf{x} = 0 \tag{2.5.2}$$

The left-hand side of Eq. (2.5.2) is a polynomial in $\alpha_1, \alpha_2,\ldots,\alpha_n$. This polynomial is identically equal to zero, which implies that the corresponding coefficients are also zero, i.e.,

$$\sum_{\pi\in\Pi}\int \dots \int T_0(x_1,\dots,x_n)\prod_{j=1}^{n} f_i(x_{\pi_j})\,d\mathbf{x} = 0$$

$$\sum_{\pi\in\Pi}\int \dots \int T_0(x_\pi)\prod_{j=1}^{n} f_i(x_j)\,d\mathbf{x} = 0$$

$$n!\int \dots \int h(T(\mathbf{x}))\prod_{j=1}^{n} f_i(x_j)\,d\mathbf{x} = 0$$

We have $f_i(x) = \dfrac{1}{b_i - a_i}, \quad a_i < x < b_i$

$$\int_{a_1}^{b_1} \dots \int_{a_n}^{b_n} h(T(\mathbf{x}))\prod_{j=1}^{n}\frac{1}{b_i - a_i}\,d\mathbf{x} = 0$$

$$\int_{a_1}^{b_1} \dots \int_{a_n}^{b_n} h(T(\mathbf{x}))\,d\mathbf{x} = 0$$

i.e., the integral of $h(T)$ over any n-dimensional rectangle is 0. This implies

$$P_F\{h(T)\} = 0 \;\forall\; F \in \mathcal{F}$$

This proves the theorem. ▣

The property of completeness of a statistic depends on the associated parameter space, see Exercises 2.12 and 2.15. Further, note that if the family $\mathcal{F}_0$ is complete and that $\mathcal{F}_0 \subset \mathcal{F}$, then $\mathcal{F}$ is also complete. It can be observed that the family should have a large number of distributions for the family to be complete. See Example 2.8.28 (i). It is important to note that if a nonconstant function of T is ancillary or first-order ancillary, then T is not complete.

Notes.

1. It is important to note that the property of completeness is a property of a family of distributions $\mathcal{F}$ which is induced by some statistic T as θ varies.
2. If U is complete and $T = k(U)$, then T is also complete.
3. If T_1 and T_2 are equivalent statistics and T_1 is complete, then T_2 is also complete.
4. Two results frequently used for showing a statistic T to be complete for θ are
   (i) If the infinite series $\Sigma_{i=0}^{\infty} a_i x^i$ converges to zero for all x in some interval I, then $a_0 = a_1 = \cdots = 0$.
   (ii) (a) If the integral $\int_0^{\infty} \exp(-xt)s(x)dx$ converges to some value of t, then

$$H(t) = \int_0^{\infty} \exp(-xt)\,s(x)\,dx$$

is called the *unilateral Laplace transform* of s.

(b) If the integral $\int_{-\infty}^{\infty} \exp(-xt)s(x)\,dx$ converges to some value of t, then

$$H_1(t) = \int_{-\infty}^{\infty} \exp(-xt)\,s(x)\,dx$$

is called the *bilateral transform* of s. If the function s is continuous, then H or H_1 is unique.

5. In order to prove that a given statistic T is not complete (i) show that T is nonconstant and ancillary or first-order ancillary or (ii) show that $E(T)$ does not depend on θ.
6. If a complete sufficient statistic T exists, it is minimal sufficient. But the converse is not necessarily true. It is for this reason that one checks whether a minimal sufficient statistic is complete, since if a minimal sufficient statistic is not complete, then there does not exist any complete sufficient statistic for that family.
7. One importance of completeness is that if a statistic is unbiased and is a function of a complete sufficient statistic, then it is best for estimating θ (in terms of variance) in the class of all unbiased estimators of θ for convex loss functions. This result is stated in Chapter 3 in the form of Theorem 3.3.1 which is popularly known as the Lehmann–Scheffe theorem. It is further important to mention that the application of the Rao–Blackwell theorem in this case results in the same estimator each time it is invoked for the improvement for any choice of initial unbiased estimator. This shows that further improvement is not possible and that unbiased estimator, which is a function of complete sufficient statistic, is best (see Note 2 to Rao–Blackwell theorem). Indeed, if a complete sufficient statistic exists, there is one and only one unbiased estimator, which is a function of a complete sufficient statistic.
8. The families of binomial, Poisson, normal and gamma distributions are all complete families except the normal $N(\theta_0,\sigma^2)$, with θ_0 known. The Normal family of distributions $N(\theta_0,\sigma^2)$, is not complete since we have a function $g(X) = X - \theta_0$ so that $E[g(X)] = E[X - \theta_0] = 0 \nRightarrow X - \theta_0 = 0$.

   If the exponential family in Eq. (2.4.3) is a full exponential family, i.e., $d = k$, then

$$T(\mathbf{X}) = \left(\sum_{i=1}^{n} T_1(X_i),\ldots,\sum_{i=1}^{n} T_k(X_i)\right)$$

   is complete if the parameter space

$$(p_1(\boldsymbol{\theta}),\ldots,p_k(\boldsymbol{\theta})) : \boldsymbol{\theta} \in \Theta$$

   contains an open set in $\mathbb{R}^k$. This condition is necessary but not sufficient.
9. Let $X_1, X_2,\ldots,X_n$ be a random sample from the Bernoulli distribution $b(1, \theta)$, $\theta \in (0, 1)$. For some Borel measurable function h and $T(\mathbf{X}) = \Sigma X_i$, $E_\theta\,[h(T)]$ is a polynomial in $\theta/(1 - \theta)$ and $E_\theta[h(T)] = 0\ \forall \theta$ implies the coefficients are zero, i.e., $h(t) = 0$ *a.e.* t. This shows that $T(\mathbf{X}) = \Sigma_1^n X_i$ is complete.
10. Let $X_1, X_2,\ldots,X_n$ be a random sample from Poisson population with parameter $\theta > 0$. On taking $T(\mathbf{X}) = \Sigma_1^n X_i$, $E_\theta[h(T(\mathbf{X}))]$ can be expressed as a power series in

θ. The condition $E_\theta h(T) = 0\ \forall\theta > 0$ implies that the coefficient must be zero, i.e., $h(t) = 0$ a.e. t. This shows that $\Sigma_1^n X_i$ is complete.

11. Let $X_1, X_2,\ldots,X_n$ be a random sample from exponential family with parameter $\theta > 0$. Consider a statistic $T(\mathbf{X}) = \Sigma_1^n X_i$. Note, that $E_\theta[h(T(\mathbf{X}))]$ is a Laplace function. By uniqueness property of the Laplace transform and $E_\theta[h(T(\mathbf{X}))] = 0\ \forall\ \theta$, we have $h(t) = 0\ \forall\ t$. The result generalizes to k-parameter exponential families where sufficient statistic $\mathbf{T} = \{T_1(\mathbf{X}),\ldots,T_k(\mathbf{X})\}$ is also complete.
12. Let $X_1,\ldots,X_n$ be *iid* $N(\theta, 1)$, $S_{n-1}^2 = (n-1)^{-1}\Sigma_{i=1}^n (X_i - \bar{X})^2$, $(n-1)\ S_{n-1}^2 \sim \chi_{(n-1)}^2$, $(n-1)S_{n-1}^2$ is ancillary; Let $X_1,\ldots,X_n$ be *iid* $f(x;\ \theta) = f(x - \theta)$, location density, $T(\mathbf{X}) = X_{(n)} - X_{(1)}$ is ancillary; Let $X_1,\ldots,X_n$ be *iid* from $f(x;\ \sigma) = (1/\sigma)f(x/\sigma)$, the scale density, $T(\mathbf{X}) = (X_1/X_n,\ldots,X_{n-1}/X_n)$ is ancillary.
X_1, X_2 *iid* from $N(0,1)$, X_1/X_2 is $C(0,1)$; $X_1,\ldots,X_n$ *iid* $N(0,\sigma^2)$, $T(\mathbf{X}) = ((X_1 - \bar{X})/S_{n-1},\ldots,(X_n - \bar{X})/S_{n-1})$ is ancillary.
13. Let $X_1,\ldots,X_n$ be *iid* $U(\theta, \theta + 1)$, the joint density of $(X_{(1)},X_{(n)})$ is $g_{1n}\ (x_1,x_n) = n(n-1)\ [x_n - x_1]^{n-2}$, $\theta < x_1 < x_n < \theta + 1$, the density of $T(\mathbf{X}) = X_{(n)} - X_{(1)}$ is $g_{T(t)} = n(n-1)\ t^{n-2}\ (1 - t)$, $0 < t < 1$, T is ancillary.

Heuristically, for the purpose of estimating θ, a sufficient statistic contains as much information on θ, as is contained in the data. Whereas a complete sufficient statistic retains only that information which is relevant for estimating θ by discarding all irrelevant information present in the sample observations. Possibly, a complete sufficient statistic may not contain any information on θ in case the data itself has no information. By taking the intersection of these two groups, if they exist, the complete sufficient statistic is a statistic that contains as much information as do the sample data and no irrelevant information.

A statistic T is said to be boundedly complete if the condition in Eq. (2.5.1) holds for all bounded functions h.

Note that completeness implies bounded completeness but the converse is not necessarily true (Ferguson, 1967). Note also that the property of completeness alone, without sufficiency, has very little significance. Therefore, throughout the book, sufficiency would remain in the background whenever we discuss completeness. Since completeness implies maximum reduction, the complete sufficient statistics are always minimal sufficient but the converse is not necessarily true. However, complete sufficient statistic may not exist whereas minimal sufficient statistic always exists.

2.5.1 Independence of Complete Sufficient Statistic with Ancillary Statistic

Debabrata Basu's theorem (1955) states that a complete sufficient statistic is independent of every ancillary statistic. This theorem is greatly useful as a tool to prove independence of two statistics by showing that one is complete and the other is ancillary. This situation can also be viewed as to what happens to an ancillary statistic V whenever a minimal sufficient statistic T is complete. The answer to this question is stated in a form of a theorem due to Basu stating that "such T and V are independent" or a complete sufficient statistic is independent of all ancillary statistics.

Theorem 2.5.2 (Basu, 1955) If T is a bounded complete sufficient statistic for the family $\mathcal{F} = \{F_\theta, \theta \in \Theta\}$ and V is any ancillary statistic, then T and V are independently distributed (conditionally) on every $\theta \in \Theta$.

Proof By the definition of ancillary statistic, we have for V, $P_V(A) = P(V \in A)$ is independent of θ. Consider a function of $T = t$, $f(t) = P(V \in A|T = t)$, and $E_\theta^T\ [f(T)] = P_\theta^T P^{V|T}\ (V \in A|T)$ $= P(V \in A) = P_V(A)$ (independent of θ since V is ancillary). Therefore, $E_\theta^T\ [f(T) - P_V(A)] = 0$ $\forall\ \theta$ implies $f(T) = P_V(A)\ \forall\,\theta$ (since T is bounded complete) or $P(V \in A|T = t) = P(V \in A)$ $\forall\, t$ a.e. $\mathcal{F}$. This shows that V and T are independent. ▣

The converse of Basu's theorem, that if $T(\mathbf{X})$ is independent of every ancillary statistic, then $T(\mathbf{X})$ is complete, is not necessarily true. Lehmann (1981) argued that the reason why the converse fails to hold is that the ancillary is the property of distribution of a statistic, while completeness is the property of expectations under a family of distributions. Lehmann then considered first-order ancillary in place of ancillary and proved the following result:

Theorem 2.5.3 T is complete if and only if the correlation coefficient

$$\rho_\theta(V, g(T)) = 0$$

for all θ, for every bounded first-order ancillary V and every bounded real-valued function of T.

The word "conditionally" in the theorem was inserted by Basu (1982) to distinguish the frequentist framework from that of Bayesian. Lehmann (1981) pointed out that the properties of minimality and completeness of a sufficient statistic are of different nature. Virtually, in almost all statistical problems, the minimal sufficient statistic exists. However, this does not guarantee that there does not exist any function of T which is ancillary, (Boos and Hughes Liver (1998)). Basu's theorem tells us that if T is complete in addition to being sufficient, no ancillary statistic other than the constants can be computed from T. Thus, by Basu's theorem, the informative part of the given data is taken out and the no–information part is separated through the characterization by complete sufficient statistic.

We will now discuss the ancillary statistics for location, scale, and location–scale models. The approach that we follow here of finding ancillary statistics is due to Hogg and Craig (1978). This approach is useful since it is simple to use. Further, we use Basu's theorem to show that a complete sufficient statistic is independent of every ancillary statistic obtained by this approach for each of the above models. The results are summarized in Corollaries 2.5.1. to 2.5.3.

***Corollary* 2.5.1** Let $X_1, X_2, \ldots, X_n$ be a random sample drawn from a distribution having a *pdf* of the form

$$f(x; \theta) = f(x - \theta) \tag{2.5.3}$$

for every real θ. Here, θ is called the location parameter and this *pdf* is called location density. Let $T = u_1(X_1, X_2, \ldots, X_n)$ be a complete sufficient statistic for θ. Then the statistic T is stochastically independent of every statistic $V = u(X_1, X_2, \ldots, X_n)$ that satisfies

$$u(x_1 + d, x_2 + d, \ldots, x_n + d) = u(x_1, x_2, \ldots, x_n) \tag{2.5.4}$$

for all real d. The statistic V that satisfies the condition given in Eq. (2.5.4) is an ancillary statistic.

Proof. Let us define a one-to-one transformation on $X_1, X_2,\ldots,X_n$ defined by $Z_i = X_i - \theta$, $i = 1, 2,\ldots,n$ so that it requires that the joint *pdf* of $Z_1, Z_2,\ldots,Z_n$ is given by

$$f(z_1, z_2,\ldots,z_n) = f(z_1)f(z_2)\ldots f(z_n)$$

is independent of θ. On using the functional nature of u in Eq. (2.5.4), we have

$$V = u(Z_1 + \theta, Z_2 + \theta,\ldots,Z_n + \theta) = u(Z_1, Z_2,\ldots,Z_n) \tag{2.5.5}$$

We see that V is a function of $Z_1, Z_2,\ldots,Z_n$ alone and not of θ. Therefore, the statistic V must have a distribution that does not depend on θ. This shows that V is an ancillary statistic. By Basu's theorem, the complete sufficient statistic T for θ and every ancillary statistic V are stochastically independent.

Thus, for the location parameter θ, any statistic V that satisfies the condition given in Eq. (2.5.4) is ancillary and by Basu's theorem, it is stochastically independent of the complete sufficient statistic T for θ. ▣

***Corollary* 2.5.2** Let $X_1, X_2,\ldots,X_n$ be a random sample drawn from a distribution having a *pdf* of the form

$$f(x;\theta) = \frac{1}{\theta} f\left(\frac{x}{\theta}\right) \tag{2.5.6}$$

for every real θ, where θ is a scale parameter, and this *pdf* is called scale density. Let $T = u_1(X_1, X_2,\ldots,X_n)$ be a complete sufficient statistic for θ. Then the statistic T is stochastically independent of every such statistic $V = u(X_1, X_2,\ldots,X_n)$ that satisfies

$$u(cx_1, cx_2,\ldots,cx_n) = u(x_1, x_2,\ldots,x_n) \tag{2.5.7}$$

for all $c > 0$. The statistic V that satisfies Eq. (2.5.7) is an ancillary statistic.

Proof. Let us define a one-to-one transformation on $X_1, X_2,\ldots,X_n$ by $Z_i = X_i/\theta$, $i = 1, 2,\ldots,n$, so that it requires that the joint *pdf* of $Z_1, Z_2,\ldots,Z_n$ given by

$$f(z_1, z_2,\ldots,z_n) = f(z_1)f(z_2)\ldots f(z_n)$$

does not depend on θ. On using the functional nature of u in Eq. (2.5.7), we have

$$V = u(\theta Z_1, \theta Z_2,\ldots,\theta Z_n) = u(Z_1, Z_2,\ldots,Z_n)$$

which is a function of $Z_1, Z_2,\ldots,Z_n$ alone and not of θ. Therefore, the statistic V must have a distribution that does not depend on θ. This shows that V is an ancillary statistic. Further, by Basu's theorem, the complete sufficient statistic T for θ is stochastically independent with every such ancillary statistic V.

Thus, for the scale parameter θ, any statistic V that satisfies the condition given in Eq. (2.5.7) is ancillary and by Basu's theorem, it is stochastically independent of the complete sufficient statistic T for θ. ▣

***Corollary* 2.5.3** Let $X_1, X_2,\ldots,X_n$ be a random sample drawn from a distribution having a *pdf* of the form

$$f(x;\theta_1,\theta_2)=\frac{1}{\theta_2}f\left(\frac{x-\theta_1}{\theta_2}\right) \tag{2.5.8}$$

for $-\infty < \theta_1 < \infty$, $\theta_2 > 0$, where θ_1 is a location parameter and θ_2 is a scale parameter, and this *pdf* is known as location-scale density. Let $T = u_1(X_1, X_2,\ldots,X_n)$ be a joint complete sufficient statistic for the parameters θ_1 and θ_2. Then the statistic T is stochastically independent of every such statistic $V = u(X_1, X_2,\ldots,X_n)$ that satisfies

$$u(cx_1 + d, cx_2 + d,\ldots,cx_n + d) = u(x_1, x_2,\ldots,x_n) \tag{2.5.9}$$

for all $-\infty < d < \infty$, $c > 0$. The statistic V that satisfies Eq. (2.5.9) is an ancillary statistic.

Notes.

1. The existence of a complete sufficient statistic is a prerequisite of Basu's theorem. In Basu's theorem the complete sufficient statistic T carries all the information on θ whereas the ancillary statistic V carries no information on θ.
2. Let $X_1, X_2,\ldots,X_n$ be *iid* sample from $G_1(\alpha,\beta)$, shape parameter $\alpha > 0$ and scale parameter $\beta > 0$. Define $U = \bar{X}$, V = geometric mean. U is complete sufficient statistic and U/V is ancillary for β. Therefore, U and U/V are independent.
3. Let $X_1,\ldots,X_n$ be *iid* $U(\theta - 1/2, \theta + 1/2)$ and $T(\mathbf{X}) = (X_{(1)}, X_{(n)})$ be minimal sufficient. The joint density of $(X_{(1)}, X_{(n)})$ is $g_{1n}(x_{(1)}, x_{(n)}) = n(n-1)[x_{(n)} - x_{(1)}]^{n-2}$, $\theta - 1/2 < x_{(1)} < x_{(n)} < \theta + 1/2$, and the marginal density of $U(\mathbf{X}) = X_{(n)} - X_{(1)}$ is $g(u;\theta) = n(n-1)\, u^{n-2}.(1-u)$, $0 < u < 1$, which is $Be(n-1, 2)$. This shows that U is ancillary. T and U are not independent since U is a function of T. Note that the statistic T is not complete. The importance of completeness in Basu's theorem is that if T were complete, then T and U would have been independent.
4. If T and V are equivalent statistics (results into same summarization of data) and V is ancillary for θ, then T is also ancillary for θ.
5. The existence and use of boundedly complete sufficient statistic in Basu's theorem has a special significance for finding UMP unbiased tests having Neyman's structure in multiparameter testing problem.
6. Usually, an ancillary statistic is seen as a component of a minimal sufficient statistic, if the dimension of the minimal sufficient statistic is larger than the dimension of θ (see Example 2.8.26).
7. Let $X_1,\ldots,X_n$ be a random sample drawn from a location density of the form $f(x;\theta) = f(x-\theta)$, where θ is a location parameter. A statistic $T(\mathbf{X})$ is said to be location invariant if and only if $T(\mathbf{X} + a) = T(\mathbf{X})\ \forall\ a$. Then T is ancillary, for example $S_{n-1} = \sqrt{(1/(n-1))\Sigma(X_i - \bar{X})^2}$ is an ancillary statistic.
8. Let $X_1,\ldots,X_n$ be a random sample drawn from a scale density of form $f(x;\sigma) = (1/\sigma) f(x/\sigma)$, where $\sigma > 0$ is a scale parameter. A statistic $T(\mathbf{X})$ is said to be scale invariant if and only if $T(a\mathbf{X}) = T(\mathbf{X})\ \forall\ a > 0$ and scale equivalent if and only if $T(a\mathbf{X}) = aT(\mathbf{X})\ \forall\ a > 0$. If T is scale invariant then T is ancillary. The statistic $V(\mathbf{X}) = f((X_1/X_n),\ldots,(X_{n-1}/X_n))$ for some function f an ancillary statistic for σ.
9. Let $X_1,\ldots,X_n$ be a random sample drawn from a location-scale density of the form $f(x;\theta,\sigma) = (1/\sigma)f((x-\theta)/\sigma)$, where $\theta \in \mathbb{R}^1$ is a location parameter and $\sigma > 0$ is a scale parameter. A statistic $T(\mathbf{X})$ is said to be location-scale invariant if and only if

$T(a\mathbf{x} + b) = T(\mathbf{X})\ \forall\ a > 0$ and $b \in \mathbb{R}^1$. If T is location-scale invariant, then T is ancillary. In this case, if T_1 and T_2 are two statistic such that $T_1(a\mathbf{x} + b) = aT_1$ and $T_2(a\mathbf{x} + b) = aT_2$, then T_1/T_2 is ancillary.

10. For some families like (i) location families $\{U(\theta, \theta + 1), \theta \in \mathbb{R}^1\}$, $\{C(\theta, 1), \theta \in \mathbb{R}^1\}$, and (ii) curved exponential families $\{N(\theta, \theta^2), \theta \in \mathbb{R}^1\}$, $\{BVN(0, 0, 1, 1, \rho): -1 \le \rho \le 1\}$, there do not exist complete sufficient statistics.

2.6 CONVEX FUNCTIONS IN ESTIMATION PROBLEMS

We have seen that the problem of estimating θ is to identify F_θ from $\mathcal{F}$ which has generated X. For this, we collect data on X and utilize them to get such an estimate $T(x)$ of θ that falls close to θ. Consequently, by choosing an estimator T, the loss incurred to the statistician is measured by the real quality

$$L(\theta, T) \ge 0\ \forall\ \theta, T$$

so that

$$L(\theta, \theta) = 0\ \forall\ \theta \tag{2.6.1}$$

i.e., the loss is zero when θ is estimated correctly. The risk in taking T is measured by

$$R(\theta, T) = E_\theta L(\theta, T(X))$$

This is an average loss over all the values of X on the sample space χ. In any estimation situation, one is interested in such an estimator T for which $R(\theta, T)$ is minimum for every value of θ. We, thus, choose L in Eq. (2.6.1) so that it

(i) does not assign abnormally high cost to high deviations of T from θ;
(ii) provides mathematical solution to the problem of minimization of $R(\theta, T)$ over a space of estimators T for every $\theta \in \Theta$; and
(iii) provides explicit estimators.

More particularly, we characterize a property of a loss function under which it brings about a better estimator through sufficiency and results into an explicit T, so that $E\rho(T - \theta)$ is minimum, with ρ even. This property is known as convexity of loss function $L(\theta, T)$ in T. We discuss these notions in more detail.

Definition 2.6.1 A real valued function $L: I \to \mathbb{R}^1$, $I = (a, b)$, is said to be convex, if for every $x, y \in I$, $x < y$, and any $0 < \alpha < 1$

$$L[\alpha x + (1 - \alpha)y]\ \le \alpha L(x) + (1 - \alpha)L(y) \tag{2.6.2}$$

and, is said to be strictly convex if the inequality in Eq. (2.6.2) is strict.

Convex functions are continuous and their derivatives exist at every point in I. The converse also holds and is helpful in examining whether a given function is convex. If L is defined on I and is differentiable throughout, it is convex if and only if $L'(x) \le L'(y)\ \forall\ x, y \in I, x < y$. L is strictly convex if and only if $L'(x) < L'(y)\ \forall\ x, y \in I, x < y$, where $L'(x)$ is the derivative of L evaluated at x and $L'(y)$ is the derivative of L evaluated y. If the second derivative of L exists, these necessary and sufficient conditions may equivalently be written as $L''(x) \ge 0\ \forall\ x \in I$, and L is strictly convex if and only if $L''(x) > 0\ \forall\ x \in I$.

A set S

$$S = \{\mathbf{z}: \mathbf{z} = \alpha\mathbf{x} + (1-\alpha)\mathbf{y},\quad 0 < \alpha < 1,\quad \mathbf{x}, \mathbf{y} \in \mathbb{E}^k\}$$

is said to be a convex set in $\mathbb{E}^k$ if it contains the line made by joining any two points in S. Consider S_0 as an open convex subset of S in $\mathbb{E}^k$. A real-valued function L on S_0 is said to be convex if

$$L[\alpha\mathbf{x} + (1-\alpha)\mathbf{y}] \le \alpha L(\mathbf{x}) + (1-\alpha)L(\mathbf{y}) \quad \forall\, \mathbf{x}, \mathbf{y} \in S_0$$

$$x_i < y_i,\ \forall\ i = 1, 2,\ldots,k$$

L is strictly convex if the inequality is strict for all $\mathbf{x}$ and $\mathbf{y}$. Let L be defined over an open convex set S in $\mathbb{E}^k$ and that it is twice differentiable in S. Define a matrix of second order partial derivatives of L by

$$H = \left\| \frac{\partial^2}{\partial x_i y_i} L(\mathbf{x}) \right\|$$

L is convex, if and only if, the matrix H is positive semidefinite, i.e., $\mathbf{z}'\mathrm{H}\mathbf{z} \ge 0\ \forall\ \mathbf{z} \in \mathbb{E}^k$ and $\forall\ \mathbf{x} \in S$. L is strictly convex, if H is positive definite. The matrix H is called *Hessian matrix*.

Another property of convex functions is that there exists a point on its curve so that the points on the line passing through it fall below the curve. Formally stated, if $L(x)$ is convex in I, then there exists a point $[t, L(t)]$ on the curve of L and a line passing through

$$P(x) = c(x - t) + L(t)$$

so that $\qquad P(x) \le L(x) \quad \forall\ x \in I$

Further, we consider X as a random variable taking on its values from I according to the probability distribution F. Defining

$$E[L(X)] = \int L(x)\,dF(x)$$

Further, $E[P(X)] = CE(X - t) + L(t) = L(t)$ if $t = E(X)$.

This gives

$$E(P(X)) = L(E(x))$$

$$= L\left(\int x\,dF(x)\right)$$

$$= L(\alpha x + (1-\alpha)y)$$

for $F(x) = \alpha$ and $F(y) = 1 - \alpha$. Thus,

$E(P(X)) \le E[L(X)]$ gives the inequality in Eq. (2.6.2), i.e.

$$L(\alpha x + (1-\alpha)y) \le \alpha L(x) + (1-\alpha)L(y)$$

Using these notions, a more general result is stated which is known as Jensen's inequality.

Theorem 2.6.1 (Jensen's Inequality) If L is a convex function defined on I and X is a random variable taking on its values from I, then

$$L[E(X)] \le E[L(X)] \tag{2.6.3}$$

The inequality in Eq. (2.6.3) is strict when L is strictly convex.

Proof. Let the line $P(x) = c(x - t) + L(t)$ pass through the point $[t, L(t)]$ on the curve L. If $t = E(X)$, we have $P(t) = L(t)$. Consider

$$\begin{aligned} E[L(X)] \geq E[P(X)] &= E[c(X - t) + L(t)] \\ &= (cE(X) - t) + L(t) \\ &= P(E(X)) = P(t) = L(t) \\ &= L[E(X)] \end{aligned}$$

that proves the theorem, and the inequality in Eq. (2.6.3) holds. Further, the inequality is strict for L strictly convex.

As we have seen earlier in Section 2.2.1 while discussing sufficient statistics that the estimators based on the entire sample and the one which is based on some sufficient statistic have same risk functions. There we did not impose any restriction on the loss function.

2.7 IMPROVED ESTIMATORS THROUGH SUFFICIENCY

However, now, if we impose a restriction on the action space and on the loss function that (i) action space is a convex set and (ii) loss function is convex in $a \in \mathcal{A}$ for each θ, then we get a superior estimator through a sufficient statistic. The result is stated in the form of a theorem due to Rao and Blackwell which provides us a way to reduce the variance of an unbiased estimator while keeping the property of unbiasedness intact.

Theorem 2.7.1 (Rao–Blackwell Theorem) Let X be a random variable with its distribution F_θ in $\mathcal{F} = \{F_\theta\colon \theta \in \Theta\}$, and T be a sufficient statistic for $\mathcal{F}$. Let h be an estimator of θ based on the sample data X and $L(\theta, h)$ be a convex function in h. The risk in choosing h is defined as $R(\theta, h) = EL(\theta, h(X))$, so that it is finite. Then there exists an estimator δ defined by

$$\delta(t) = E[h(X)|T = t]$$

so that

$$R(\theta, \delta) < R(\theta, h) \ \forall \ \theta \in \Theta$$

Proof. Consider

$$\begin{aligned} R(\theta, \delta) &= E_\theta L(\theta, \delta(T)) = E_\theta^T L(\theta, E(h(X)|T)) \\ &< E_\theta^T E^{X|T} L(\theta, h(X)) \quad \text{(by Jensen's inequality)} \\ &= E_\theta^X L(\theta, h(X)) = R(\theta, h) \ \forall \ \theta \end{aligned}$$

▣

Notes.

1. Given any estimator based on the sample $\mathbf{X}$, one gets an improved estimator which depends on a sufficient statistic T if the loss function is strictly convex, and one gets a no worse estimator if the loss function is convex. Note that the theorem holds under the assumption that the loss function is strictly convex in δ.
2. The theorem holds for unbiased and biased estimators where both can be considered for improvement. The improved estimator is unbiased if and only if the original estimator is unbiased. There may be several such improved estimators. Therefore, the theorem may be viewed as weak on this count; however, it is suggestive of considering only

such estimators which are the function of sufficient statistics for estimating θ. Further, if the sufficient statistic in the theorem is also complete, i.e., it admits no unbiased estimator of zero, the Rao–Blackwell theorem gives back the same improved estimator $\delta(t) = E[H(x)|T = t]$ which was put in for improvement. The estimator $\delta(t)$ is known as uniformly minimum variance unbiased estimator of θ and is unique. This estimator is obtained from Lehmann and Scheffe Theorem 3.3.1.

3. The better estimator $\delta(T)$ is the weighted average of estimator $h(X)$ on the partition of sample space χ on which $T(X) = t$, where the weights are conditional probability of X given $T(X) = t$. Note that these weights are independent of θ since T is sufficient. Therefore, $\delta(t)$ is a function of X only through $T(X) = t$.
4. Another advantage of using convex loss functions is that given any randomized estimator, there exists a nonrandomized estimator which is as good as the randomized estimator and better if the loss function is strictly convex. Indeed, nonrandomized estimators are simple to handle as compared to randomized estimators. The Rao–Blackwell theorem suggests us to consider a nonrandomized estimator for estimating θ as a function of a sufficient statistic.
5. The deviation $(\delta - \theta)$ is an error in estimating θ by $\delta(x)$. This error, $(\delta - \theta)$, or some even function, $\rho(\delta - \theta)$, is usually considered as a measure of inaccuracy in the estimator δ at $X = x$. This measure is expressed by a loss function $L(\theta, \delta) = \rho(\delta - \theta)$ at $X = x$. The risk in choosing an estimator δ at θ is expressed by $R(\theta, \delta) = E_\theta L(\theta, \delta(X)) = E_\theta \rho(\delta - \theta)$. The function $\rho(\theta, \cdot)$ is a convex function in δ which takes its values from $\mathbb{R}^1$. Therefore, $R(\theta, \cdot)$ is also convex. $(\delta - \theta) \xrightarrow{P} \pm\infty$ as $\delta \to \pm\infty$ and $\rho(\delta - \theta) \to +\infty$ as $\delta \to \pm\infty$. This shows that $R(\theta, \cdot)$ is not monotone and bounded from below. Further, $R(\theta, \cdot)$ takes on its minimum value since R is continuous. The function R is continuous since R is convex. The set on which the minimum value is attained is an interval denoted by $I(R)$ since $R(\theta, \cdot)$ is convex, and $I(R)$ is closed since R is continuous. Note that $I(R)$ is just a single point when R is strictly convex.

In the problem of estimation, one is generally interested in characterizing such an estimator δ that minimizes the risk $R(\theta, \delta)$ uniformly in θ in a class of reasonable estimators. To ensure the existence of such estimator, one considers convex loss function or strictly convex loss function, if possible. This simplifies the estimation problem to a great extent.

2.8 SOLVED EXAMPLES

***Example* 2.1** Consider $X \sim f(x;\theta)$, $x = 1, 2, 3$; $\theta = \theta_1, \theta_2, \theta_3$ with probability function

x	θ_1	θ_2	θ_3
1	0.1	0.2	0.3
2	0.7	0.4	0.1
3	0.2	0.4	0.6

Show that the statistic

$$T = \begin{cases} 0, & \text{if } x \text{ is odd } (x = 1, 3) \\ 1, & \text{if } x \text{ is even } (x = 2) \end{cases}$$

is sufficient for θ.

Solution. The distribution of T is given by $f(t;\theta)$

T	θ_1	θ_2	θ_3
0	0.3	0.6	0.9
1	0.7	0.4	0.1

The conditional probability function of $\mathbf{x}|t$ is given by:

When $t = 0$

x	θ_1	θ_2	θ_3
1	1/3	1/3	1/3
2	0	0	0
3	2/3	2/3	2/3

When $t = 1$

x	θ_1	θ_2	θ_3
1	0	0	0
2	1	1	1
3	0	0	0

Since the distribution of $\mathbf{X}|t$, $f(\mathbf{x}|t)$ does not depend on θ, T is sufficient for θ.

***Example* 2.2** Three coins have probability θ of getting a head as 1/2, 1/3, 1/4, respectively. Let each coin be flipped twice and X and Y be the numbers recorded as 1 and 0 for H and T, respectively. Define the statistics by S_1: $(X + Y)$ (sum of numbers on first and second flips) and S_2: $(X \times Y)$ (product of these numbers). Show that S_1 is sufficient for θ but S_2 is not.

Solution. The statistic S_1 may take values 0, 1, 2 and S_2 as 0, 1. The joint distribution of XY, $f((x, y)|\theta)$, is given by

(X, Y)	$\theta = 1/2$	$\theta = 1/3$	$\theta = 1/4$
(0, 0)	1/4	4/9	9/16
(0, 1)	1/4	2/9	3/16
(1, 0)	1/4	2/9	3/16
(1, 1)	1/4	1/9	1/16

The distributions of S_1 and S_2 are given by

S_1	$\theta = 1/2$	$\theta = 1/3$	$\theta = 1/4$
0	1/4	4/9	9/16
1	2/4	4/9	6/16
2	1/4	1/9	1/16

S_2	$\theta = 1/2$	$\theta = 1/3$	$\theta = 1/4$
0	3/4	8/9	15/16
1	1/4	1/9	1/16

The conditional distribution of $X, Y|S_1$ is

$P[X, Y|S_1 = 0]$

(X, Y) \ θ	1/4	1/3	1/4
(0, 0)	1	1	1
(0, 1)	0	0	0
(1, 0)	0	0	0
(1, 1)	0	0	0

$P[X, Y|S_1 = 1]$

(X, Y) \ θ	1/4	1/3	1/4
(0, 0)	0	0	0
(0, 1)	1/2	1/2	1/2
(1, 0)	1/2	1/2	1/2
(1, 1)	0	0	0

$P[X, Y|S_1 = 2]$

(X, Y) \ θ	1/2	1/3	1/4
(0, 0)	0	0	0
(0, 1)	0	0	0
(1, 0)	0	0	0
(1, 1)	1	1	1

We see that for each value of S_1, the conditional distribution of $X, Y|S_1$ is independent of θ. Therefore, S_1 is sufficient.

Next, consider the conditional distribution of $X, Y|S_2$, $P[X, Y|S_2 = 0]$.

(X, Y) \ θ	1/2	1/3	1/4
(0, 0)	1/3	4/8	9/15
(0, 1)	1/3	2/8	3/15
(1, 0)	1/3	2/8	3/15
(1, 1)	0	0	0

S_2 is not sufficient since there is at least one value of $S_2 = 0$ for which the conditional distribution of $(X, Y)|S_2$ depends on θ.

***Example* 2.3** Let X be a discrete random variable with the following *pmf* $f(x;\ \theta)$.

x	1	2	3	4	5	6
θ_1	1/30	2/30	3/30	8/30	4/30	12/30
θ_2	1/60	2/60	3/60	20/60	4/60	30/60

where $\chi = \{1, 2, 3, 4, 5, 6\}$, $\Theta = \{\theta_1, \theta_2\}$. Show that the minimal sufficient statistic is

$$T(x) = \begin{cases} a, & \text{if } x \in A_1 \\ b, & \text{if } x \in A_2 \end{cases}$$

where the partitions are $A_1 = \{1, 2, 3, 5\}$ and $A_2 = \{4, 6\}$.

Solution. Adopting the likelihood equivalence principle of constructing minimal sufficient statistic T, the ratios $f(x;\theta)/f(y;\theta)\ \forall\ x, y \in \chi$ are considered, and we search for partitions on which it is constant over θ. Note a statistic T so that

$$\frac{f(x;\theta)}{f(y;\theta)} \text{ is free of } \theta \Leftrightarrow T(x) = T(y)$$

$\dfrac{f(x;\theta)}{f(y;\theta)}$	$\dfrac{f(1;\theta)}{f(2;\theta)}$	$\dfrac{f(1;\theta)}{f(3;\theta)}$	$\dfrac{f(1;\theta)}{f(4;\theta)}$	$\dfrac{f(1;\theta)}{f(5,\theta)}$	$\dfrac{f(1;\theta)}{f(6;\theta)}$	$\dfrac{f(2;\theta)}{f(3;\theta)}$	$\dfrac{f(2;\theta)}{f(4;\theta)}$
θ_1	1/2	1/3	1/8	1/4	1/12	2/3	2/8
θ_2	1/2	1/3	1/20	1/4	1/30	2/3	2/20

$\dfrac{f(2;\theta)}{f(5;\theta)}$	$\dfrac{f(2;\theta)}{f(6;\theta)}$	$\dfrac{f(3;\theta)}{f(4;\theta)}$	$\dfrac{f(3;\theta)}{f(5;\theta)}$	$\dfrac{f(3;\theta)}{f(6;\theta)}$	$\dfrac{f(4;\theta)}{f(5;\theta)}$	$\dfrac{f(4;\theta)}{f(6;\theta)}$	$\dfrac{f(5;\theta)}{f(6;\theta)}$
2/4	2/12	3/8	3/4	3/12	8/4	8/12	4/12
2/4	2/30	3/20	3/4	3/30	20/4	20/30	4/30

The table suggests that the ratio $f(x;\theta)/f(y;\theta)$ is independent of θ on the partitions $A_1 = \{1, 2, 3, 5\}$ and $A_2 = \{4, 6\}$. Therefore, the minimal sufficient statistic is given by

$$T(x) = \begin{cases} a, & \text{if } x \in A_1 \\ b, & \text{if } x \in A_2 \end{cases}$$

The partitions A_{t_1} and A_{t_2} are the coarsest partitions and are known as minimal sufficient partitions.

***Example* 2.4** Consider the following *pmf* of a random variable of X:

x	1	2	3	4	5	6
$f(x;\theta_1)$	1/14	2/14	3/14	3/14	4/14	1/14
$f(x;\theta_2)$	1/18	2/18	5/18	5/18	4/18	1/18

Find a minimal sufficient statistic for (θ_1, θ_2).

Solution. Consider the ratios

$\frac{f(x;\theta)}{f(y;\theta)}$	$\frac{f(1;\theta)}{f(2;\theta)}$	$\frac{f(1;\theta)}{f(3;\theta)}$	$\frac{f(1;\theta)}{f(4;\theta)}$	$\frac{f(1;\theta)}{f(5;\theta)}$	$\frac{f(1;\theta)}{f(6;\theta)}$	$\frac{f(2;\theta)}{f(3;\theta)}$	$\frac{f(2;\theta)}{f(4;\theta)}$
θ_1	1/2	1/3	1/3	1/4	1	2/3	2/3
θ_2	1/2	1/3	1/5	1/4	1	2/5	2/5

$\frac{f(2;\theta)}{f(5;\theta)}$	$\frac{f(2;\theta)}{f(6;\theta)}$	$\frac{f(3;\theta)}{f(4;\theta)}$	$\frac{f(3;\theta)}{f(5;\theta)}$	$\frac{f(3;\theta)}{f(6;\theta)}$	$\frac{f(4;\theta)}{f(5;\theta)}$	$\frac{f(4;\theta)}{f(6;\theta)}$	$\frac{f(5;\theta)}{f(6;\theta)}$
2/4	2	1	3/4	3	3/4	3	4
2/4	2	1	5/4	5	5/4	5	4

The table suggests the minimal partitions as $A_1 = \{1, 2, 5, 6\}$ and $A_2 = \{3, 4\}$, and the minimal sufficient statistic is given by

$$T(x) = \begin{cases} a, & \text{if } x \in A_1 \\ b, & \text{if } x \in A_2 \end{cases}$$

***Example* 2.5** Consider an *rv* X with *pmf* $f(x;\theta)$ given by

	θ_1	θ_2	θ_3
x_1	1/4	0	0
x_2	1/4	1/4	0
x_3	1/4	1/4	1/4
x_4	1/4	1/4	1/4
x_5	0	1/4	1/4
x_6	0	0	1/4

The corresponding minimal partitions are given by

$$A_1 = \{x_3, x_4\};\ A_2 = \{x_1\};\ A_3 = \{x_2\};\ A_4 = \{x_5\};\ A_5 = \{x_6\}$$

Note. These partitions report a little summarization of data.

***Example* 2.6** Let T_1 and T_2 be the two statistics defined on the same space χ. Then T_1 is a function of T_2, if and only if,

$$T_1(x) = T_1(y) \Rightarrow T_2(x) = T_2(y) \text{ for all } x, y \in \chi$$

Solution. Suppose T_1 is a function of T_2, i.e., there exists a function h such that $T_1 = h(T_2)$. Then at any two points x and y in χ, we have $T_1(x) = h[T_2(x)]$ and $T_1(y) = h[T_2(y)]$. If $T_1(x) = T_1(y)$, it implies $h[T_2(x)] = h[T_2(y)]$ or $T_2(x) = T_2(y)$. Conversely, suppose for any two points

x and y in χ, we have $T_1(x) = T_1(y) \Rightarrow T_2(x) = T_2(y)$. Then there exists some function h such that $T_1 = h(T_2)$.

***Example* 2.7 (Exponential families with one- and two-parameters).** Let there be an *iid* sample $X_1, X_2, \ldots, X_n$ from one of the following *full-rank* exponential families. Find a sufficient statistic for θ in each of the following cases:

(i) binomial $B(n, \theta)$: $f(x;\ \theta) = \binom{n}{x} \theta^x(1-\theta)^{n-x},\ x = 0, 1, 2, \ldots, n$

(ii) Poisson $P(\theta)$: $f(x;\ \theta) = \dfrac{\exp(-\theta)\theta^x}{x!},\ x = 0, 1, 2, \ldots$

(iii) Geometric (θ): $p(x;\ \theta) = (1-\theta)^x\theta,\ x = 0, 1, 2, \ldots$

(iv) Normal $N(\theta, 1)$: $f(x;\ \theta) = \dfrac{1}{\sqrt{2\pi}} \exp\left\{-\dfrac{1}{2}(x-\theta)^2\right\},\ -\infty < x < \infty$

(v) Normal $N(0, \sigma^2)$: $f(x;\ \sigma^2) = \dfrac{1}{\sigma\sqrt{2\pi}} \exp\left\{-\dfrac{x^2}{2\sigma^2}\right\},\ -\infty < x < \infty$

(vi) Normal $N(\theta, \sigma^2)$: $f(x;\ \theta, \sigma^2) = \dfrac{1}{\sigma\sqrt{2\pi}} \exp\left\{-\dfrac{1}{2\sigma^2}(x-\theta)^2\right\},\ -\infty < x < \infty$

(vii) Exponential: $f(x,\ \theta) = \dfrac{1}{\theta} \exp\left\{-\dfrac{x}{\theta}\right\},\ x \ge 0$

(viii) Gamma $G_2(\alpha, \beta)$: $f(x;\ \alpha, \beta) = \dfrac{\beta^\alpha}{\Gamma(\alpha)} x^{\alpha-1} \exp\{-\beta x\},\ x > 0$

(ix) Beta (α, β):

$$f(x;\ \alpha, \beta) = \frac{1}{\beta(\alpha, \beta)} x^{\alpha-1}(1-x)^{\beta-1},\ 0 < x < 1;\ \alpha, \beta > 0$$

(x) Trinomial distribution:

$$f(x, y;\ \theta_1, \theta_2) = \frac{n!}{x!\,y!(n-x-y)!} \theta_1^x\, \theta_2^y (1-\theta_1-\theta_2)^{n-x-y},\ x + y \le n$$

(xi) Bivariate normal distribution:

$$f(x, y;\ \mu_1, \mu_2, \sigma_1^2, \sigma_2^2, \rho) = BVN(\mu_1, \mu_2, \sigma_1^2, \sigma_2^2, \rho)$$

Solution. Refer to Eq. (2.4.2), where the exponential family of distributions is defined by the *pdf*

$$f(x;\theta) = c(\theta)h(x) \exp\left\{\sum_{j=1}^{k} p_j(\theta)T_j(x)\right\}$$

(i) If $X \sim b(n, \theta)$, then

$$f(x;\theta) = \binom{n}{x} \theta^x(1-\theta)^{n-x};\ x = 0, 1, 2, \ldots, n$$

$$= (1-\theta)^n \binom{n}{x} \left(\frac{\theta}{1-\theta}\right)^x = (1-\theta)^n \binom{n}{x} \exp\left\{x \log\left(\frac{\theta}{1-\theta}\right)\right\}$$

$$c(\theta) = (1-\theta)^n, \; h(x) = \binom{n}{x}$$

$$p(\theta) = \log\left(\frac{\theta}{1-\theta}\right), \; T(x) = x$$

This shows that ΣX_i is sufficient for θ.

(ii) If $X \sim P(\theta)$, then

$$f(x;\theta) = \frac{\exp(-\theta)\theta^x}{x!}, \; x = 0, 1, 2, \ldots$$

$$= \exp(-\theta)\frac{1}{x!}\exp(x \log\theta)$$

$c(\theta) = \exp(-\theta)$, $h(x) = (1/x!)$, $p(\theta) = \log\theta$, $T(x) = x$. This shows that ΣX_i is sufficient for θ.

(iii) If $X \sim$ geometric(θ), then

$$p(x;\, \theta) = (1-\theta)^x\theta, \; x = 0, 1, 2, \ldots$$

$$p(x;\, \theta) = \theta\exp\{x \log(1-\theta)\}$$

Here, $h(x) = 1$, $c(\theta) = \theta$, $p(\theta) = \log(1-\theta)$, and $T(x) = x$. This is a Full-rank exponential family. We have, $E_p(X_1) = (1-\theta)/\theta$.

(iv) If $X \sim N(\theta, 1)$, then

$$f(x;\theta) = \frac{1}{\sqrt{2\pi}}\exp\left[-\frac{1}{2}(x-\theta)^2\right], \; -\infty < x < \infty$$

$$= \frac{1}{\sqrt{2\pi}}\exp\left\{-\frac{x^2}{2}\right\}\exp\left\{-\frac{\theta^2}{2}\right\}\exp\{\theta x\}$$

$$h(x) = \frac{1}{\sqrt{2\pi}}\exp\left\{-\frac{x^2}{2}\right\}$$

$$c(\theta) = \exp\left\{-\frac{\theta^2}{2}\right\}$$

$$p(\theta) = \theta, \; T(x) = x$$

Therefore, $\Sigma_1^n X_i$ is sufficient for θ.

(v) If $X \sim N(0\ ,\ \sigma^2)$, then

$$f(x;\sigma^2) = \frac{1}{\sigma\sqrt{2\pi}}\exp\left\{-\frac{x^2}{2\sigma^2}\right\}, -\infty < x < \infty$$

$$h(x) = \frac{1}{\sqrt{2\pi}},\quad c(\theta) = \frac{1}{\sigma}$$

$$p(\theta) = -\frac{1}{2\sigma^2},\quad T(x) = x^2$$

Therefore, $T = \Sigma X_i^2$ is sufficient for σ^2.

(vi) If $X \sim N(\theta,\ \sigma^2)$, then

$$f(x;\theta,\sigma^2) = \frac{1}{\sqrt{2\pi}\sigma}\exp\left\{-\frac{1}{2\sigma^2}(x-\theta)^2\right\}$$

$$= \exp\left\{-\frac{\theta^2}{2\sigma^2}\right\}\frac{1}{\sqrt{2\pi}\sigma}\exp\left\{\frac{\theta}{\sigma^2}x - \frac{1}{2\sigma^2}x^2\right\}$$

$$c(\theta,\sigma^2) = \frac{1}{\sigma}\exp\left\{-\frac{\theta^2}{2\sigma^2}\right\}, h(x) = \frac{1}{\sqrt{2\pi}}, p_1(\theta,\sigma^2) = \frac{\theta}{\sigma^2}, T_1(x) = x,$$

$$p_2(\theta,\sigma^2) = -\frac{1}{2\sigma^2}, T_2(x) = x^2$$

Therefore, $(\Sigma_1^n X_i^2,\ \Sigma X_i)$ is jointly sufficient for $(\theta,\ \sigma^2)$.

(vii) If $X \sim E(\theta)$, then

$$f(x;\theta) = \frac{1}{\theta}\exp\left\{-\frac{x}{\theta}\right\}, x \ge 0$$

$$h(x) = 1,\quad c(\theta) = \frac{1}{\theta}$$

$$p(\theta) = -\frac{1}{\theta},\quad T(x) = x$$

ΣX_i is sufficient for θ.

(viii) If $X \sim G_2(\alpha,\ \beta)$, then

$$f(x;\alpha,\beta) = \frac{\beta^\alpha}{\Gamma(\alpha)}x^{\alpha-1}\exp\{-\beta x\}, x > 0$$

$$= \frac{\beta^\alpha}{\Gamma(\alpha)}\frac{1}{x}\exp\{\alpha\ln x - \beta x\}$$

$$c(\alpha, \beta) = \frac{\beta^\alpha}{\Gamma(\alpha)}, \quad h(x) = \frac{1}{x}, \quad p_1(\alpha, \beta) = \alpha, \quad T_1(x) = \ln x$$

$$p_2(\alpha, \beta) = -\beta, \quad T_2(x) = x$$

This implies that $(\Sigma_1^n \ln X_i, \Sigma X_i) = (\ln \Pi_1^n X_i, \Sigma_1^n X_i)$ or $(\Pi_1^n X_i, \Sigma_1^n X_i)$ is jointly sufficient for (α, β).

(ix) If $X \sim Be(\alpha, \beta)$, then

$$f(x; \alpha, \beta) = \frac{1}{B(\alpha, \beta)} x^{\alpha-1}(1-x)^{\beta-1}, \qquad 0 < x < 1; \alpha, \beta > 0$$

$$= \frac{1}{B(\alpha, \beta)} \frac{1}{x(1-x)} \exp\{\alpha \ln x + \beta \ln(1-x)\}$$

where $c(\alpha, \beta) = 1/[\beta(\alpha, \beta)]$, $h(x) = 1/[x(1-x)]$, $p_1(\alpha, \beta) = \alpha$, $T_1(x) = \ln x$, $p_2(\alpha, \beta) = \beta$, $T_2(x) = \ln(1-x)$.

This shows that $(\Sigma_1^n \ln X_i, \Sigma_1^n \ln(1-X_i))$ or $(\Pi_1^n X_i, \Pi_1^n(1-X_i))$ is jointly sufficient for (α, β).

(x) Consider the trinomial distribution

$$f(x, y; \theta_1, \theta_2) = \frac{n!}{x!\, y!(n-x-y)!} \left(\frac{\theta_1}{1-\theta_1-\theta_2}\right)^x \left(\frac{\theta_2}{1-\theta_1-\theta_2}\right)^y (1-\theta_1-\theta_2)^n$$

$$= (1-\theta_1-\theta_2)^n \frac{n!}{x!y!(n-x-y)!} \exp\left\{x \ln\left(\frac{\theta_1}{1-\theta_1-\theta_2}\right) + y \ln\left(\frac{\theta_2}{1-\theta_1-\theta_2}\right)\right\}$$

$$c(\theta_1, \theta_2) = (1-\theta_1-\theta_2)^n, h(x, y) = \frac{n!}{x!y!(n-x-y)!}$$

$$p_1(\theta_1, \theta_2) = \ln\left(\frac{\theta_1}{1-\theta_1-\theta_2}\right), T_1(x) = x, p_2(\theta_1, \theta_2) = \ln\left(\frac{\theta_2}{1-\theta_1-\theta_2}\right), T_2(y) = y$$

Therefore, $(\Sigma X_i, \Sigma Y_i)$ is jointly sufficient for (θ_1, θ_2).

***Example* 2.8** Show that $U(0, \theta)$, $0 < x < \theta$, does not belong to an exponential family of distributions.

Solution. The uniform density of $U(0, \theta)$ is given by

$$f(x; \theta) = \frac{1}{\theta}, \quad 0 < x < \theta$$

If this density belongs to the exponential family, its form must have been as given in Eq. 2.4.2, in which the density is 0 if and only if $\exp(T(x)) = 0$ or $h(x) = 0$ for every θ. Whereas, the uniform density is 0 if and only if $x \notin (0, \theta)$, which is a contradiction. Therefore, uniform density does not belong to an exponential family.

***Example* 2.9**

(i) Let X be a random variable with *pmf*

$$P(X=-1;\,\theta)=\theta \quad \text{and} \quad P(X=x;\,\theta)=(1-\theta)^2\theta^x$$

x = 0, 1, 2,…,where $0 < \theta < 1$. Find a sufficient and a minimal sufficient statistic for θ. Also, show that X is not complete but bounded for the complete family $\{P_\theta;\ 0 < \theta < 1\}$.

(ii) Let $X_1, X_2,\ldots,X_n$ be a random sample of size n from the following distribution:

$$P(X=x_1;\,\theta)=\frac{1-\theta}{2},\quad P(X=x_2;\,\theta)=\frac{1}{2},\quad P(X=x_3;\,\theta)=\frac{\theta}{2}$$

and 0 elsewhere; $\Theta = \{0 < \theta < 1\}$. Find a sufficient statistic for θ.

(iii) Let $X_1, X_2,\ldots,X_n$ be a random sample of size n from a population with *pdf*

$$f(x;\,\theta)=\begin{cases}\theta(1+x)^{-(1+\theta)}, & x>0, \theta>0\\ 0, & \text{otherwise}\end{cases}$$

Find a sufficient statistic for θ.

(iv) Let $X_1, X_2,\ldots,X_n$ be *iid* from

$$f(x;\,\theta)=\frac{x}{\theta}\exp\left(-\frac{x^2}{2\theta}\right),\quad x>0,\quad \theta>0$$

Show that the statistic $T = \Sigma X_i^2$ is minimal sufficient for θ. Also show that $S = \Sigma X_i$ is not sufficient.

(v) Let $X_1, X_2,\ldots,X_n$ be *iid* from the inverse Gaussian distribution

$$f(x;\,\theta,\eta)=\left\{\frac{\eta}{2\pi}\right\}^{1/2} x^{-3/2}\exp\left\{-\frac{\eta(x-\theta)^2}{2\theta^2 x}\right\}, x>0$$

where θ is mean and η is precision parameter. Let $\tau = \eta/\theta^2$. Find the sufficient statistic for $(\tau/2,\ \eta/2)$.

Solution.

(i) We can write $P(X = x;\,\theta) = \theta^x\,[\theta^2 I_{[-1]}(x) + (1-\theta)^2 I_A(x)]$, where $A = \{x: x = 0,1,2,\ldots\}$. Then by factorization Theorem 2.3.1, $g(\theta, T(x)) = \theta^x$. Therefore, $T(X) = X$ is sufficient for θ. Next, consider the two points x and y and the ratio

$$\frac{f(x;\,\theta)}{f(y;\,\theta)}=\frac{\theta^x[\theta^2 I_{[-1]}(x)+(1-\theta)^2 I_A(x)]}{\theta^y[\theta^2 I_{[-1]}(y)+(1-\theta)^2 I_A(y)]}$$

where,

$$I_A(x)=\begin{cases}1, & \text{if } x\in A\\ 0, & \text{otherwise}\end{cases}$$

This ratio is independent of θ, if and only if, $x = y$ for if not, either x or y is -1 and other is 0, 1, 2,…,the ratio is either $\theta/(\theta^y(1-\theta)^2)$ or $\theta/(\theta^x(1-\theta)^2)$, which is not independent of θ. Therefore, $T(X) = X$ is a minimal sufficient statistic for θ.

For some function $h(X)$,

$$E_\theta[h(X)] = 0, \ \forall\ 0 < \theta < 1$$

implies

$$\theta h(-1) + \sum_{x=0}^{\infty} h(x)(1-\theta)^2 \theta^x = 0$$

$$\theta(1-\theta)^{-2} h(-1) + \sum_{x=0}^{\infty} h(x)\theta^x = 0$$

$$\sum_{x=0}^{\infty} x\, \theta^x h(-1) + \sum_{x=0}^{\infty} h(x)\, \theta^x = 0$$

$$\sum_{x=0}^{\infty} \{xh(-1) + h(x)\}\theta^x = 0$$

This implies $xh(-1) + h(x) = 0$ for $x = 0, 1, \ldots, \infty$, or $h(x) = -xh(-1) \neq 0$ if $h(-1) \neq 0$ *a.e. x*.

Hence, X is not complete for $\{P(x;\theta), 0 < \theta < 1\}$. Further, consider h to be bounded. $h(-1) \neq 0$ implies that $h(x)$ is unbounded. Hence, $h(x) = 0$ for $x = -1, 0, 1, \ldots$. Thus, X is boundedly complete for $\{P(x;\theta), 0 < \theta < 1\}$.

(ii) The joint density of sample observations is

$$P(\mathbf{x};\theta) = \left(\frac{1-\theta}{2}\right)^{n(x_1)} \left(\frac{1}{2}\right)^{n(x_2)} \left(\frac{\theta}{2}\right)^{n(x_3)} = 2^{-n}(1-\theta)^{n(x_1)}\theta^{n(x_3)}$$

where $n(x_1)$, the number of observations in $X_1, X_2, \ldots, X_n$, take value x_1,
$n(x_2)$, the number of observations in $X_1, X_2, \ldots, X_n$, take value x_2, and
$n(x_3)$, the number of observations in $X_1, X_2, \ldots, X_n$ take value x_3.

By factorization theorem, $(n(x_1), n(x_3))$ is sufficient for θ.

(iii) The joint density of sample observations is

$$f(\mathbf{x};\theta) = \theta^n \exp\left\{-(1-\theta)\sum_{i=1}^{n} \log(1+x_i)\right\}$$

$$= \theta^n \exp\left\{-\sum_{i=1}^{n} \log(1+x_i)\right\} \exp\left\{\theta \sum_{i=1}^{n} \log(1+x_i)\right\}$$

Therefore, by factorization theorem, $T(\mathbf{X}) = \Sigma_{i=1}^{n} \log(1 + X_i)$ is sufficient for θ. Note that this distribution belongs to an exponential family.

(iv) Consider two points $\mathbf{x}$ and $\mathbf{y}$.

$$\frac{f(\mathbf{x};\theta)}{f(\mathbf{y};\theta)} = \frac{\dfrac{1}{\theta^n}\prod x_i \exp\left(\dfrac{-\sum x_i^2}{2\theta}\right)}{\dfrac{1}{\theta^n}\prod y_i \exp\left(\dfrac{-\sum y_i^2}{2\theta}\right)} = \prod\left(\frac{y_i}{x_i}\right)\exp\left[-\frac{1}{2\theta}\left(\sum x_i^2 - \sum y_i^2\right)\right]$$

Therefore, the statistic $T = \Sigma X_i^2$ is minimal sufficient for θ.

Consider now a point $\mathbf{x}$ on the circle obtained by intersecting a sphere $\Sigma x_i^2 = b_1$ by a given plane $\Sigma x_i = a_0$ and another point $\mathbf{y}$ on the other circle obtained by intersecting a sphere $\Sigma x_i^2 = b_2$ by the same plane $\Sigma x_i = a_0$. We then have $S(\mathbf{x}) = \Sigma x_i = \Sigma y_i = S(\mathbf{y})$, and for any two different θ_1 and θ_2,

$$f(\mathbf{x};\theta_1)f(\mathbf{y};\theta_2) = \frac{\prod x_i \prod y_i}{\theta_1^n\,\theta_2^n}\exp\left(-\frac{1}{2\theta_1}\sum x_i^2 - \frac{1}{2\theta_2}\sum y_i^2\right)$$

$$f(\mathbf{x};\theta_2)f(\mathbf{y};\theta_1) = \frac{\prod x_i \prod y_i}{\theta_1^n\,\theta_2^n}\exp\left(-\frac{1}{2\theta_2}\sum x_i^2 - \frac{1}{2\theta_1}\sum y_i^2\right)$$

Note that $f(\mathbf{x};\theta_1)f(\mathbf{y};\theta_2) \neq f(\mathbf{x};\theta_2)f(\mathbf{y};\theta_1)$. Therefore, by Theorem 2.3.2, the statistic $S = \Sigma x_i$ is not sufficient.

(v) The reparameterization $\tau = \eta/\theta^2$ yields

$$f(x;\tau,\eta) = \{2\pi\}^{-1/2}x^{-3/2}\exp\left\{(\tau\eta)^{1/2} + \frac{1}{2}\log\eta - \frac{1}{2}\tau x - \frac{\eta}{2}x^{-1}\right\}$$

The joint density of $X_1, X_2, \ldots, X_n$ is

$$f(\mathbf{x};\tau,\eta) = \{2\pi\}^{-n/2}\prod_{i=1}^{n} x_i^{-3/2}\exp\left\{n(\tau\eta)^{1/2} + \frac{n}{2}\log\eta - \frac{\tau}{2}\sum x_i - \frac{\eta}{2}\sum\frac{1}{x_i}\right\}$$

This shows that $(\Sigma X_i, \Sigma 1/X_i)$ is jointly sufficient for $(\tau/2, \eta/2)$.

***Example* 2.10** Let $(X_1, X_2, \ldots, X_n)$ be a random sample from some discrete distribution $f(x;\theta)$. Show that the statistic $T(\mathbf{X}) = (X_1, X_2, \ldots, X_{n-1})$ is not sufficient.

Solution. The conditional distribution

$$P_\theta\{\mathbf{X} = \mathbf{x} | X_1 = x_1, \ldots, X_{n-1} = x_{n-1}\} = P_\theta(X_n = x_n)$$

is not independent of the parameter θ, which shows that the statistic $T(\mathbf{X}) = (X_1, X_2, \ldots, X_{n-1})$ is not sufficient. In fact, any statistic that is based on only a part of the sample data violates the principle of sufficiency.

***Example* 2.11 (Nonparametric estimation).** Let $X_1, \ldots, X_n$ be independently distributed according to an unknown distribution f where the corresponding distribution function F belongs to a family of distributions $\mathcal{F} = \{F$: F is some continuous distribution function$\}$. This is a situation of nonparametric estimation where we cannot specify any more information on f other than that f is continuous. Show that the order statistics $T(\mathbf{X}) = (X_{(1)}, \ldots, X_{(n)})$ is sufficient for f.

Solution. The observations $X_1, \ldots, X_n$ are distinct since f is continuous. The joint distribution of the sample observations is

$$f_{\mathbf{x}}(\mathbf{x}) = \prod_{i=1}^{n} f_X(x_i)$$

Consider the order statistics $T(\mathbf{X}) = (X_{(1)}, X_{(2)},\ldots,X_{(n)})$ so that $X_{(1)} < X_{(2)},\ldots,< X_{(n)}$. The joint distribution of $T(\mathbf{X})$ is

$$f_T(x_{(1)}, x_{(2)},\ldots, x_{(n)}) = n!\prod_{i=1}^{n} f_X(x_{(i)}) \text{ where } x_{(1)} < x_{(2)} < \cdots < x_{(n)}$$

Therefore, the conditional distribution of $\mathbf{X}|[T(\mathbf{X}) = (x_{(1)},\ldots,x_{(n)})]$, $x_{(1)} < x_{(2)} < \cdots < x_{(n)}$, is given by

$$f(\mathbf{x}|(x_{(1)}, x_{(2)},\ldots, x_{(n)})) = \frac{1}{n!}$$

which is independent of f. So, the order statistic $T(\mathbf{X})$ is sufficient. After having discarded the original sample $\mathbf{X}$, the reconstructed sample $\mathbf{Y}$ given $T = t$ would be any one sample among $n!$ samples obtained by assigning random labelling to $(X_{(1)},\ldots,X_{(n)})$. Such a sample $\mathbf{Y}$ and the original sample $\mathbf{X}$ have the same distribution. This shows too that the order statistic T is sufficient statistics for F. Therefore, there always exists a sufficient statistic for continuous densities.

The order statistic $T(\mathbf{X})$ gains a little summarization over the sample since only weak information on the density, that it is continuous, was available. Mostly, in all nonregular families of distributions, we hardly get sufficient statistics of a smaller dimension and, therefore, order statistics, though of higher dimension, is the only best summarization possible in such cases.

***Example* 2.12**

(i) We have seen that the order statistic $T(\mathbf{X}) = (X_{(1)}, X_{(2)},\ldots,X_{(n)})$, when the sample observations are independently drawn from some model f, is sufficient for f. The statistic T though sufficient reports little summarization due to the lack of information on f in terms of θ. Investigate whether by imposing a structural form on f such as $f(x;\,\theta) = f(x - \theta)$, location model, further summarizing statistics exist.

(ii) Show that the statistics $V_1(\mathbf{X}) = k_1[T(\mathbf{X})] = (X_{(n)} - X_{(1)},\ldots,X_{(n)} - X_{(n-1)})$ and $V_2(\mathbf{X}) = k_2\,[T(\mathbf{X})] = (X_{(2)} - X_{(1)},\ldots,X_{(n)} - X_{(1)})$ are ancillary for θ, and thus, T is not complete. Also, show that the statistics $R = X_{(n)} - X_{(1)}$ and $S_{n-1}^2 = [1/(n-1)]\Sigma(X_i - \bar{X})^2$ are ancillary.

Solution.

(i) On choosing a structure on f so that f is a normal, exponential, and uniform distribution contributes extensive reduction of data (as shown further in the solution), where the dimension of minimal sufficient statistic is as low as one or two; whereas contrary to it, by choosing a structure on f such as Cauchy, double exponential, and logistic distribution, we achieve no reduction and, thus, the order statistic $T(\mathbf{X})$ is minimal sufficient.

Consider a location family $\mathcal{L} = \{N(\theta, 1){:}\,\theta \in \mathbb{R}^1\}$ and $\mathcal{L}_0 = \{N(\theta_0, 1), N(\theta_1, 1)\}$. Appealing to Lehmann's procedure, the minimal sufficient statistic for $\mathcal{L}_0$ is given by

$$\frac{f(\mathbf{x};\theta_1)}{f(\mathbf{x};\theta_0)} = \exp\left\{-n(\theta_0 - \theta_1)\bar{x} - \frac{n}{2}(\theta_1^2 - \theta_0^2)\right\}$$

or $T(\mathbf{X}) = \bar{X}$. Since $\bar{X}$ is sufficient for $\mathcal{L}$ and $\mathcal{L}_0 \subset \mathcal{L}$, $T(\mathbf{X}) = \bar{X}$ is minimal sufficient for $\mathcal{L}$. In this example, we have seen that making a structural assumption on F in $\mathcal{L}$, results in data reduction from order statistic to $\bar{X}$, i.e., from sufficient to minimal sufficient. Similarly, if f in $\mathcal{F}$ is given by exponential density $f(x;\ \theta) = \exp\{-(x - \theta)\}$, then $x \geq \theta$. Let $\mathcal{L}_0 = \{f(x;\ \theta_0), f(x;\ \theta_1)\}$. Appealing to Lehmann's procedure, the minimal sufficient statistic is given by

$$\frac{f(\mathbf{x};\theta_1)}{f(\mathbf{x};\theta_0)} = \frac{\exp\left\{-\sum_{i=1}^{n}(x_i - \theta_1)\right\}\prod_{i=1}^{n}k(\theta_1, x_i)}{\exp\left\{-\sum_{i=1}^{n}(x_i - \theta_0)\right\}\prod_{i=1}^{n}k(\theta_0, x_i)}$$

$$= \exp\{n(\theta_1 - \theta_0)\}\frac{k(\theta_1, x_{(1)})}{k(\theta_0 - x_{(1)})}$$

i.e., $T(\mathbf{X}) = X_{(1)}$. On using factorization theorem, one can easily see that $X_{(1)}$ is sufficient for $\mathcal{L}$. Further, $\mathcal{L} \supset \mathcal{L}_0$ gives $X_{(1)}$ is minimal sufficient for $\mathcal{L}$. Proceeding similarly, one can easily show that $T(\mathbf{X}) = (X_{(1)}, X_{(2)})$ is minimal sufficient for $\mathcal{L} = \{U[\theta - (1/2), \theta + (1/2)],\ \theta > 0\}$.

However, if we consider double exponential, logistic, and Cauchy distributions, knowing their forms, apart from knowing that they are members of a location family, do not contribute further. In these cases, the order statistic $T(\mathbf{X})$ is the only best summarization and it is minimal sufficient statistic.

We may state that imposing a structural restriction on f does not necessarily result into better summarization.

(ii) The joint density of data is

$$f(\mathbf{x};\theta) = \prod_{i=1}^{n} f(x_i - \theta)$$

Consider the transformation $Z_i = X_i - \theta$. The joint density of $\mathbf{Z}$ is $f_{\mathbf{Z}}(\mathbf{z}) = \Pi_{i=1}^{n} f(z_i)$, which is independent of θ. Therefore, the statistics $R = X_{(n)} - X_{(1)}$ and $S_{n-1}^2 = [1/(n-1)]\ \Sigma(X_i - \bar{X})^2$ are ancillary. Further, the distributions of the statistics V_1 and V_2 do not depend on θ. Therefore, V_1 and V_2 are ancillary for θ. T is not complete since it is a function of ancillary statistics V_1 and V_2.

***Example* 2.13**

(i) **Location family.** Let $X_1, X_2, \ldots, X_n$ be a random sample from each of the following location families:

(a) $X \sim N(\theta, 1)$, $f(x;\ \theta) = (1/\sqrt{2\pi})\exp[-(x - \theta)^2/2]$

(b) $X \sim E(\theta)$, $f(x;\ \theta) = \exp[-(x - \theta)]$, $x > \theta$

(c) $X \sim \text{Logistic}(\theta)$, $f(x;\ \theta) = \dfrac{\exp[-(x-\theta)]}{\{1 + \exp[-(x-\theta)]\}^2}$

(d) $X \sim C(1, \theta)$, $f(x;\theta) = 1/\{\pi[1 + (x - \theta)^2]\}$

(e) $X \sim DE(\theta)$, $f(x;\theta) = (1/2)\exp(-|x - \theta|)$

For each of these families, let $(X_{(1)}, X_{(2)},\ldots,X_{(n)})$ be the ordered sample. Show that the statistic $(Y_1, Y_2,\ldots,Y_{n-1}) = (X_{(n)}\ X_{(1)}, X_{(n)} - X_{(2)},\ldots,X_{(n)} - X_{(n-1)})$ is ancillary. Also, investigate whether the ancillary statistic $(Y_1, Y_2,\ldots,Y_{n-1})$ is independent of the *minimal sufficient statistic* in each case.

(ii) **Location and scale family.** Let $X_1, X_2,\ldots,X_n$ be *iid* from $N(\theta, \sigma^2)$. Show that the statistics $(X_{(n)} - X_{(1)})/S_{n-1}$, $(X_{(n)} - \bar{X})/S_{n-1}$ and $\Sigma_{i=1}^{n-1}(X_{i+1} - X_i)^2/S_{n-1}^2$ are ancillary. Use Basu's theorem to show that $(\bar{X}, S^2)$ is independent of each of these ancillary statistics.

Solution.

(i) The distributions in (a) to (e) belong to the location family of distributions having their *pdf*s of the form as given in Eq. (2.5.3). Denoting the statistics $(Y_1, Y_2,\ldots,Y_{n-1})$ by V, we have

$$\begin{aligned}V(X_1 + d, X_2 + d,\ldots,X_n + d) &= \{((X_{(n)} + d) - (X_{(1)} + d)),\\ &\quad ((X_{(n)} + d) - (X_{(2)} + d)),\ldots,((X_{(n)} + d) - (X_{(n-1)} + d))\}\\ &= V(X_1, X_2,\ldots,X_n)\end{aligned}$$

This shows that the statistic V satisfies the condition given in Eq. (2.5.4). Thus, the statistic V is ancillary. In (a), $\bar{X}$ is a complete sufficient statistic for θ; therefore, by Basu's theorem, $(Y_1, Y_2,\ldots,Y_{n-1})$ is independent of the complete sufficient (*minimal sufficient*) statistic $\bar{X}$. In (b), $T(\mathbf{X}) = X_{(1)}$ is sufficient. It is minimal sufficient since no further reduction in $X_{(1)}$ is possible. Setting $Y_n = X_{(1)}$, the joint *pdf* of $(Y_1, Y_2,\ldots, Y_{n-1}, Y_n)$ is $f(y_1, y_2,\ldots,y_n) = n!\ \exp[-n(y_1 - \theta)]\ \exp(-ny_n)\ \Pi_{i=2}^{n-1}\exp(y_i)$, $\theta < y_{n-1} < y_{n-2} < \cdots < y_1$, $\theta < y_n < \infty$. This shows that the ancillary statistic $(Y_1, Y_2,\ldots,Y_{n-1})$ is independent of the minimal sufficient statistic $X_{(1)}$. In (c), (d), and (e), the order statistics $T(\mathbf{X}) = (X_{(1)}, X_{(2)},\ldots,X_{(n)})$ is minimal sufficient and the ancillary statistic $(Y_1, Y_2,\ldots,Y_{n-1})$ is a function of these order statistics. Therefore, the ancillary statistic $(Y_1, Y_2,\ldots,Y_{n-1})$ is not independent of the minimal sufficient statistic $T(\mathbf{X})$.

(ii) Each of the statistics $(X_{(n)} - X_{(1)})/S_{n-1}$, $(X_{(n)} - \bar{X})/S_{n-1}$, and $\Sigma_{i=1}^{n-1}(X_{i+1} - X_i)^2/S_{n-1}^2$ satisfies the condition given in Eq. (2.5.9). Therefore, these statistics are ancillary. Further, the statistic $(\bar{X}, S_{n-1}^2)$ is complete since the distribution $N(\theta, \sigma^2)$ is a fullrank exponential family of distributions. Now, by invoking Basu's theorem, the statistic $(\bar{X}, S_{n-1}^2)$ and the ancillary statistic $(X_{(n)} - X_{(1)})/S_{n-1}$ are stochastically independent. Similar arguments show the independence of $(\bar{X}, S_{n-1}^2)$ with ancillary statistics $(X_{(n)} - \bar{X})/S_{n-1}$ and $\Sigma_{i=1}^{n-1}(X_{i+1} - X_i)^2/S_{n-1}^2$.

***Example* 2.14 (Ancillary statistics for scale family).** (i) Let $X_1, X_2,\ldots,X_n$ be *iid* observations from scale family with density

$$f(x;\sigma) = \frac{1}{\sigma} f\left(\frac{x}{\sigma}\right), \quad \sigma > 0$$

Show that any statistic that depends on the observations $X_1, X_2, \ldots, X_n$ only through $(X_1/X_n, \ldots, X_{n-1}/X_n)$ is ancillary.

(ii) [Hogg and Craig (1978)] Let X and Y be two random variables so that $E(X^k)$ and $E(Y^k) \neq 0$ exist for $k = 1, 2, \ldots$. If the ratio $R = X/Y$ and its denominator Y are stochastically independent, prove that $E[(X/Y)^k] = E(X^k)/E(Y^k)$.

(iii) Let $X_1, X_2, \ldots, X_n$ be *iid* exponential distribution

$$f(x;\theta) = \frac{1}{\theta}\exp\left\{-\frac{x}{\theta}\right\}, x > 0, \theta > 0$$

Define $U = \Sigma_1^n X_i$ and $\mathbf{V} = (V_1, V_2, \ldots, V_n)' = (X_1/U, X_2/U, \ldots, X_n/U)'$. Show that U and $\mathbf{V}$ are independent. Compute $E[X_1/U]$. Also, show that the statistic $T(\mathbf{X}) = \bar{X}/\left[\sqrt{(n-1)^{-1}\Sigma(X_i - \bar{X})^2}\right]$ is ancillary.

(iv) Let X_1, X_2 be *iid* $N(0, \sigma^2)$. Find an ancillary statistic for σ^2.

Solution.

(i) Since the statistic$(X_1/X_n, \ldots, X_{n-1}/X_n)$ satisfies the condition given in Eq. (2.5.2), it is ancillary. Further, any function of the statistic $(X_1/X_n, \ldots, X_{n-1}/X_n)$, e.g., $X_n/\Sigma X_i$, is also ancillary.

(ii) Expressing

$$X^k = Y^k\left(\frac{X}{Y}\right)^k$$

and on taking expectation, we get

$$E(X^k) = E\left[Y^k\left(\frac{X}{Y}\right)^k\right] = E(Y^k)\,E\left(\frac{X}{Y}\right)^k$$

since X/Y and Y are independent. This gives the required result

$$E\left(\frac{X}{Y}\right)^k = \frac{E(X^k)}{E(Y^k)}$$

(iii) Since the exponential distribution belongs to an exponential family, the statistic $U = \Sigma_1^n X_i$ is complete. The statistic $\mathbf{V}$ satisfies the condition given in Eq. (2.5.7). Therefore, $\mathbf{V}$ is ancillary. Further, by Basu's theorem, U and $\mathbf{V}$ are independent. Using the result in (ii), we get $E[X_1/U] = E(X_1)/E(U) = 1/n$.

The statistic $T(\mathbf{X}) = \bar{X}/\left[\sqrt{(n-1)^{-1}\Sigma(X_i - \bar{X})^2}\right]$ is ancillary since it satisfies the condition in Eq. (2.5.7). It is because of this reason that the t-test based on the exponential data is of no use.

(iv) The statistic X_1/X_2 is ancillary since the distribution of $(X_1/\sigma)/(X_2/\sigma)$, the ratio of two $N(0,1)$ variables, is $C(0,1)$, which is independent of σ^2.

***Example* 2.15** Let T be a minimal sufficient statistic. Show that a necessary condition for a sufficient statistic U to be complete is that U is minimal.

Solution. Suppose the statistic U is sufficient and complete for the family $\mathcal{F} = \{F(x;\ \theta)$: $\theta \in \Theta\}$, $\mathcal{F}(\theta; \theta \in \Theta)$, so that $E_\theta(U^2) < \infty$. Let V be some other sufficient statistic which is unbiased for its expectation $E_\theta V$. The statistic $g(U) = E_\theta(V|U)$, by the Lehmann Scheffe theorem, is UMVUE of $E_\theta V$. Let T be another sufficient statistic for θ. If $g(U)$ were not the function of T, then $h(T) = E_\theta(g(U)|T)$ would have been unbiased for $E_\theta(V)$ and by the Rao–Blackwell theorem, $V_\theta[h(T)] \le V_\theta[g(U)]$. This is a contradiction to the fact that $g(U)$ is UMVUE for $E_\theta(V)$.Therefore, $g(U)$ is a function of T. The arbitrary choices of g and T show that U must be a function of every sufficient statistic T. Hence, by definition, U is minimal sufficient. Therefore, a necessary condition for a sufficient statistic U to be complete is that U must be minimal.

***Example* 2.16**

(i) Let $X_1, X_2,\ldots,X_n$ be *iid* bernoulli, $b(1, \theta)$, variate for estimating the unknown probability of success θ, with *pmf* $P(x;\theta) = \theta^x(1-\theta)^{1-x}$, $x = 0,\ 1$. Show that $T = \Sigma X_i$ is a (a) sufficient, (b) minimal sufficient, and (c) complete sufficient statistic for θ. After recording T, construct the sample observations that carry the same information on θ as were contained in the original sample $X_1,\ldots,X_n$.

(ii) Let $X_1,\ X_2,\ldots,X_n$ be *iid* from $b(m,\ \theta)$. Find a sufficient statistic for θ. Show that $T = \Sigma X_i$ is complete for θ.

Solution.

(i) The sample $\mathbf{X}$ contains information on n trials by noting '1' if a trial results into a success or '0' if it results into a failure, and the total number of successes in n trials is recorded by $T = \Sigma X_i$. We note that the trials which result into failure do not contribute in estimating θ. We, therefore, easily ignore them by considering the statistic $T(\mathbf{X}) = \Sigma X_i$. Given $T = t$, the conditional distribution of $\mathbf{X}$ becomes $P(\mathbf{X} = \mathbf{x}|t) = 1/\binom{n}{t}$. In other words, all possible t positions of success in the sample give equal information for estimating θ. After the original sample $\mathbf{X}$ is discarded, an equivalent sample $\mathbf{Y}$, from the distribution $P(\mathbf{X} = \mathbf{x}|t) = 1/\binom{n}{t}$, using random number tables, can be constructed by randomly assigning t 1's and $(n - t)$ 0's to n coordinates of the sample. Notice that the value t of T is sufficient towards rebuilding the original sample $\mathbf{X}$, and thus, the reconstruction of sample observations is possible, without any loss in information, since the unconditional distributions of $\mathbf{Y}$ and $\mathbf{X}$ are the same.

Consider the ratio of *pmfs*

$$\frac{f(\mathbf{x};\ p)}{f(\mathbf{y};\ p)} = \frac{\theta^{\sum_1^n x_i}(1-\theta)^{n-\sum_1^n x_i}}{\theta^{\sum_1^n y_i}(1-\theta)^{n-\sum_1^n y_i}}$$

$$= \theta^{\sum_1^n x_i - \sum_1^n y_i}(1-\theta)^{-\sum_1^n x_i + \sum_1^n y_i}$$

This ratio is independent of θ, if and only if, $T(\mathbf{x}) = \Sigma_1^n x_i = \Sigma_1^n y_i = T(\mathbf{y})$. Therefore, $T(\mathbf{X}) = \Sigma_1^n X_i$ is minimal sufficient for θ.

Now, to show that $T = \Sigma X_i$ is complete, consider some function g

$$\sum_{t=1}^{n} g(t)\binom{n}{t}\theta^t(1-\theta)^{n-t} = 0 \quad \forall\theta \quad (\text{since } T \sim b(n,\theta))$$

$$(1-\theta)^n \sum_{t=1}^{n} g(t)\binom{n}{t}\left(\frac{\theta}{1-\theta}\right)^t = 0 \quad \forall\theta$$

The left-hand side expression is a polynomial in $[\theta/(1-\theta)]$. If this vanishes, then the coefficients also vanish. We have

$$g(t)\binom{n}{t} = 0, \quad \text{a.e. } t$$

or
$$g(t) = 0, \quad \text{a.e. } t$$

Therefore, $T = \Sigma X_i$ is complete.

(ii) The joint density, $f(\mathbf{x};\theta) = \prod_1^n \binom{m}{x_i}\theta^{\sum_1^n x_i}(1-\theta)^{nm-\sum_1^n x_i}$

We have

$$h(\mathbf{x}) = \prod_{i=1}^{n}\binom{m}{x_i}; \quad g(t,\theta) = \left(\frac{\theta}{1-\theta}\right)^{\sum_1^n x_i}(1-\theta)^{nm}$$

Therefore, $T = \Sigma_1^n X_i$ is sufficient for θ. Now, consider some function g

$$(1-\theta)^{mn}\sum_{t=1}^{mn} g(t)\binom{mn}{t}\left(\frac{\theta}{1-\theta}\right)^t = 0 \quad \forall\theta \quad (\text{since } T \sim b(mn,\theta))$$

The left-hand side expression is a polynomial in $[\theta/(1-\theta)]$. If this vanishes, then the coefficients also vanish.

$$g(t)\binom{mn}{t} = 0 \quad \text{a.e. } t$$

$$g(t) = 0 \quad \text{a.e. } t$$

Therefore, $T = \Sigma X_i$ is complete.

***Example* 2.17**

(i) Consider X_1 and X_2 as independent observations from the Poisson distribution, $P(\theta)$, $\theta > 0$. Define a statistic $T(\mathbf{X}) = X_1 + X_2$. Prove that $T(\mathbf{X})$ is a sufficient statistic for θ, by reconstructing the observations by knowing $T = t$, so that they provide the same information on θ that the original sample observations initially carried. Also, show that $X_1 - X_2$ is not sufficient.

(ii) Consider $X_1, X_2,\ldots,X_n$ as *iid* from $P(\theta)$, $\theta > 0$. Find a sufficient statistic for θ. Show that $T = \Sigma X_i$ is complete for θ.

(iii) Show that the statistic $W = (T, S_{n-1}^2)$ where $T = \bar{X}$ and $S_{n-1}^2 = [1/(n-1)]\Sigma(X_i - \bar{X})^2$ is not complete.

(iv) Let $X_1, X_2,\ldots,X_n$ be *iid* $P(\theta)$, $\theta > 0$. Show that the order statistics $T(\mathbf{X}) = (X_{(1)},\ldots,X_{(n)})$ is not complete.

Solution.

(i) The joint distribution of sample observations is given by

$$P(\mathbf{X} = \mathbf{x}; \theta) = \frac{\theta^{x_1+x_2}\exp(-2\theta)}{x_1!\,x_2!}$$

The conditional distribution of $\mathbf{X}|T = t$ is given by

$$P(\mathbf{X}|t) = \frac{\dfrac{\theta^t \exp(-2\theta)}{x_1!(t-x_1)!}}{\dfrac{(2\theta)^t\exp(-2\theta)}{t!}} = \binom{t}{x_1}\left(\frac{1}{2}\right)^{x_1}\left(\frac{1}{2}\right)^{t-x_1} \tag{2.8.1}$$

which is independent of θ. Therefore, $T(\mathbf{X})$ is verified as a sufficient statistic for θ. Further, the conditional distribution in Eq. (2.8.1) is $b(t, 1/2)$. Therefore, Y_1 is the number of heads in tossing a fair coin t times and $Y_2 = t - Y_1$ is the number of tails. Note that the conditional distribution of $(Y_1, Y_2)|t$ is the same as the conditional distribution of $(X_1, X_2)|t$. Therefore, their unconditional distributions are also same. This illustrates as to how the sample observations Y_1 and Y_2 can be reconstructed by recording $T = t$ from the original sample observations X_1 and X_2 before they are discarded and that Y_1 and Y_2 contain as much information on θ as was contained in the original sample.

(ii) The joint density is

$$f(\mathbf{x}; \theta) = \frac{1}{\prod_{i=1}^{n} x_i!}\exp(-n\theta)\,\theta^{\sum x_i} = \frac{\exp(-n\theta)\exp(\sum x_i \log\theta)}{\prod_{i=1}^{n} x_i!}$$

Using factorization theorem, $T = \Sigma X_i$ is sufficient for θ. Note that the joint density can be written as

$$f(\mathbf{x}; \theta) = \frac{1}{\prod_{i=1}^{n} x_i!}\frac{t!}{n^t}\frac{(n\theta)^t\exp(-n\theta)}{t!}$$

$$= f_{X|T}(\mathbf{x}|T = t)\, f_T(t|\theta) \qquad (\text{since } T = \sum X_i \sim P(n\theta))$$

Further, for some function g,

$$\sum g(t)\frac{\exp(-n\theta)(n\theta)^t}{t!} = 0 \qquad (\text{since } \sum X_i \sim P(n\theta))$$

$$\exp(-n\theta)\sum g(t)\frac{(n\theta)^t}{t!} = 0$$

The above equation is a polynomial in $(n\theta)$. Hence the coefficient must vanish. So, $g(t) = 0$ a.e. t, which proves that T is complete.

(iii) Consider a function g of W

$$g(W) = T - S_{n-1}^2$$

Then $E_\theta[g(W)] = 0$ (since T and S_{n-1}^2 are unbiased estimators of θ) does not imply $g(W) = T - S_{n-1}^2 = 0$. This shows that W is not complete. Another way to show that the statistic W is not complete is to note that W is sufficient, since $W = (T, S_{n-1}^2)$ and T is sufficient. By definition, W is not minimal sufficient and, therefore, W is not complete. Moreover, the necessary and sufficient condition for a sufficient statistic to be complete in exponential families is that the dimension of the statistic must be the same as the dimension of the parameter. Here, $\dim(W) = 2$ and $\dim(\theta) = 1$. Therefore, W is not complete.

(iv) The distribution of the order statistics $T(\mathbf{X}) = (X_{(1)},\ldots,X_{(n)})$ is given by

$$P(x_{(1)}, x_{(2)},\ldots, x_{(n)};\, \theta) = n!\prod_1^n f(x_{(i)}) \text{ where } x_{(1)} < x_{(2)} < \cdots < x_{(n)}$$

$$E[g(X_{(1)}, X_{(2)},\ldots, X_{(n)})] = n! \sum_{x_{(1)},\ldots,x_{(n)}} g(x_{(1)},\ldots, x_{(n)})\frac{\exp(-n\theta)\,\theta^{\sum x_{(i)}}}{\prod(x_{(i)})!} = 0$$

$$= n!\exp(-n\theta)\sum_{t=0}^{\infty}\left[\sum_{\substack{x_{(1)},\ldots,x_{(n)} \\ \text{s.t.}\,\Sigma x_{(i)}=t}} \frac{g(x_{(1)},\ldots, x_{(n)})}{\Pi(x_{(i)})!}\right]\theta^t = 0$$

$\not\Rightarrow$ $g(x_{(1)},\ldots,x_{(n)}) = 0$ for $x_{(1)},\ldots,x_{(n)}$ such that $\Sigma x_{(i)} = t$

Therefore, the order statistics $T(\mathbf{X}) = (X_{(1)},\ldots,X_{(n)})$ for the Poisson distribution is not complete.

***Example* 2.18** Consider a class of distribution functions $\mathcal{F} = \{F_{\theta_1}, F_{\theta_2}\}$ where F_{θ_1} is $N(0,1)$ and F_{θ_2} is $C(1,0)$. Find a sufficient statistic for $\mathcal{F}$.

Solution. Since

$$\frac{f_{\theta_1}}{f_{\theta_2}} = \frac{\dfrac{1}{\sqrt{2\pi}}\exp\left(-\dfrac{1}{2}x^2\right)}{\dfrac{1}{\pi}\dfrac{1}{1+x^2}}$$

is a function of x^2, $T(X) = X^2$ is a sufficient statistic for $\mathcal{F}$.

***Example* 2.19**

(i) Let $X_1, X_2,\ldots,X_n$ be a random sample of size n drawn from population with *pmf*

$$f(x;\theta) = P_\theta\{X = x\} = c(\theta)\,2^{-x/\theta},\ \ x = \theta, \theta+1,\ldots,\theta>0$$

and $c(\theta) = 2^{[1-(1/\theta)]}\,(2^{1/\theta}-1)$. Find a sufficient statistic for θ.

(ii) Suppose $X_1, X_2,\ldots,X_n$ are *iid* with density

$$f(x;\theta) = \frac{\theta}{x^2},\ \theta < x, \theta > 0$$

Show that $T(\mathbf{X}) = X_{(1)}$ is a sufficient statistic for θ.

(iii) Let $X_1, X_2,\ldots,X_n$ be a random sample of size n drawn from the population

$$P(X = x;\theta) = \begin{cases} \theta, & \text{if } x = 1 \\ \dfrac{1-\theta}{\lambda-1}, & \text{if } x = 2, 3,\ldots,\lambda \\ 0, & \text{otherwise} \end{cases}$$

where θ and λ are unknown. Find a sufficient statistic for θ.

Solution.

(i) The joint density is

$$f(\mathbf{x};\theta) = [c(\theta)]^n\, 2^{-(1/\theta)\Sigma x_i}\cdot I(\theta, x_{(1)})$$

where,

$$I(a,b) = \begin{cases} 1, & \text{if } a \le b \\ 0, & \text{otherwise} \end{cases}$$

Therefore, by factorization theorem, $(\Sigma X_i, X_{(1)})$ is sufficient for θ.

(ii) The joint density of sample observations is

$$f(\mathbf{x};\theta) = \theta^n \prod\left(\frac{1}{x_i^2}\right) I(\theta, x_{(1)})$$

By factorization theorem, $T(\mathbf{X}) = X_{(1)}$ is sufficient for θ.

(iii) The joint *pmf* of sample observations is

$$P(\mathbf{x};\theta) = \theta^{n(1)}\left\{\frac{(1-\theta)}{(\lambda-1)}\right\}^{n(2)}$$

where $n(1)$ is the number of times the observations $X_1, X_2,\ldots,X_n$ take value 1 and $n(2)$ is the number of times they take values from 2 to λ. Therefore, by factorization theorem, $(n(1), n(2))$ is jointly sufficient for (θ, λ).

***Example* 2.20**

(i) Let $X_1, X_2,\ldots,X_n$ be *iid rvs* with *pdf*

$$f(x;\theta) = \frac{2}{\theta^2}(\theta - x),\quad 0 < x < \theta$$

Find a minimal sufficient statistic for θ.

(ii) Let $X_1, X_2, \ldots, X_n$ be *iid rvs* from the shifted exponential distribution with *pdf*

$$f(x;\theta,\sigma)=\frac{1}{\sigma}e^{-\frac{1}{\sigma}(x-\theta)}, \quad x\ge\theta$$

Find a sufficient and a minimal sufficient statistic for (θ, σ). Also, show that $(X_{(1)}, \Sigma_{i=1}^n(X_i - X_{(1)}))$ is a complete sufficient statistic for (θ, σ).

(iii) Let $X_1, X_2, \ldots, X_n$ be *iid* from $f(x;\theta) = \exp[-(x-\theta)]$, $x \ge \theta$. Show that the first-order statistic $T = X_{(1)}$ is complete and sufficient for θ. Also, show that the statistic $T = X_{(1)}$ is independent of S_n^2.

Solution.

(i) Consider the ratio of densities at points **x** and **y**

$$\frac{f(\mathbf{x};\theta)}{f(\mathbf{y};\theta)}=\frac{\left(\dfrac{2}{\theta^2}\right)^{n/2}\prod(\theta-x_i)I(x_{(n)},\theta)}{\left(\dfrac{2}{\theta^2}\right)^{n/2}\prod(\theta-y_i)I(y_{(n)},\theta)}=\frac{\prod(\theta-x_i)I(x_{(n)},\theta)}{\prod(\theta-y_i)I(y_{(n)},\theta)}$$

This ratio is independent of θ, if and only if, $x_{(1)} = y_{(1)}, x_{(2)} = y_{(2)}, \ldots, x_{(n)} = y_{(n)}$. Therefore, the statistic $T(\mathbf{X}) = (X_{(1)}, \ldots, X_{(n)})$ is minimal sufficient for θ.

(ii) Shifted exponential distribution(θ, σ) is not an exponential family of distribution, since its support depends on θ, which is not allowed for exponential families. The joint density of sample observations is given by

$$f(\mathbf{x};\theta,\sigma)=\prod_{i=1}^{n}\frac{1}{\sigma}\exp\left\{-\frac{1}{\sigma}(x_i-\theta)\right\}I(\theta,x_i)$$

$$=\left(\frac{1}{\sigma}\right)^n\exp\left\{-\frac{1}{\sigma}\sum_1^n(x_i-\theta)\right\}I(\theta,x_{(1)})$$

where $h(\mathbf{x})=1,\ g\left(\sum_1^n x_i, x_{(1)};\theta,\sigma\right)=\exp\left\{-\frac{1}{\sigma}\sum_1^n(x_i-\theta)\right\}I(\theta,x_{(1)})$.

This shows that $(X_{(1)}, \Sigma_1^n X_i)$ is jointly sufficient for (θ, σ). Also, note that $(X_{(1)}, \Sigma_{i=1}^n (X_i - X_{(1)}))$ is sufficient for (θ, σ).

Consider the ratio of *pdfs*

$$\frac{f(\mathbf{x};\theta,\sigma)}{f(\mathbf{y};\theta,\sigma)}=\frac{\exp\left(-\frac{1}{\sigma}\sum(x_i-\theta)\right)I(\theta,x_{(1)})}{\exp\left(-\frac{1}{\sigma}\sum(y_i-\theta)\right)I(\theta,y_{(1)})}$$

$$=\exp\left[-\frac{1}{\sigma}\left(\sum x_i-\sum y_i\right)\right]\frac{I(\theta,x_{(1)})}{I(\theta,y_{(1)})}$$

This ratio is independent of the parameters (θ, σ), if and only if, $\Sigma x_i = \Sigma y_i$ and $x_{(1)} = y_{(1)}$. Therefore, $(x_{(1)}, \Sigma x_i)$ is minimal sufficient for (θ, σ).

(iii) The joint density is

$$f(\mathbf{x}; \theta) = \exp(-\Sigma x_i) \exp(n\theta) I(\theta, x_{(1)})$$

Therefore, by factorization theorem, $T = X_{(1)}$ is a sufficient statistic for θ. Further, the *pdf* of T is

$$f_T(t; \theta) = n[1 - F(t; \theta)]^{n-1} f(t; \theta)$$
$$= n \exp[-n(t - \theta)]$$

Now, for some function g,

$$E[g(T)] = 0 \tag{2.8.2}$$

$$\int_{\theta}^{\infty} g(t) \exp[-n(t-\theta)] dt = 0$$

By differentiating Eq. (2.8.2) with respect to θ, we get

$$g(\theta) = 0 \ \forall \ \theta$$

Thus, $T = X_{(1)}$ is complete.

Consider the transformation $Z = X - \theta$, $f(z;0) = \exp(-z)$. The joint distribution of $(X_1,\ldots,X_n)$ is the same as the joint distribution of $(Z_1 + \theta,\ldots,Z_n + \theta)$. Since $S_n^2 = (1/n)\ \Sigma(X_i - \bar{X})^2 = (1/n)\ \Sigma(Z_i + \theta - \bar{Z} - \theta)^2 = (1/n)\ \Sigma(Z_i - \bar{Z})^2$ is a function of Z_i's and they are independent of the parameter θ, S_n^2 is, therefore, ancillary. Since $X_{(1)}$ is complete, by Basu's theorem, it is independent of S_n^2.

***Example* 2.21**

(i) Consider a random sample $X_1, X_2,\ldots,X_n$ drawn from $U(0, \theta)$, $0 < x < \theta$. Let $T(\mathbf{X}) = X_{(n)}$, the nth order statistic. Show that $T = X_{(n)}$ is sufficient and complete.

(ii) Let $X_1, X_2,\ldots,X_n$ be *iid* from $U(\theta_1, \theta_2)$, $\theta_1, \theta_2 > 0$, $\theta_1 < \theta_2$. Find a sufficient statistic for (θ_1, θ_2).

(iii) Consider a random sample $X_1, X_2,\ldots,X_n$ from $f(x; \theta)$ given by

$$f(x; \theta) = c(\theta)m(x), \quad 0 < x < \theta$$

Find a sufficient statistic for θ.

(iv) Let $X_1, X_2,\ldots,X_n$ be *iid rvs* with *pdf* $f(x; \theta_1, \theta_2)$.

$$f(x; \theta_1, \theta_2) = c(\theta_1, \theta_2)\, m(x), \quad \theta_1 < x < \theta_2$$

Find a sufficient statistic for (θ_1, θ_2).

Solution.

(i) If a random sample $X_1, X_2,\ldots,X_n$ is drawn from an unknown population F and the i-th order statistic $X_{(i)}$ is fixed at a, then each $(i - 1)$-order statistic to the left to $X_{(i)}$

is independently distributed as $f(x)/F(a)$ and each $(n - i)$-order statistic to the right of $X_{(i)}$ is independently distributed as $f(x)/(1 - F(a))$. Further, given $X_{(i)} = a$, each of the two sets, to the left and right of $X_{(i)} = a$, are also independent. If F is $U(0, \theta)$, then $(n - 1)$-order statistics to the left of fixed $T(\mathbf{X}) = X_{(n)} = t$ are independently distributed as

$$\frac{f(x)}{F(t)} = \frac{1/\theta}{t/\theta} = \frac{1}{t}$$

Notice that the above ratio is independent of θ. The reconstruction of the sample is obtained by drawing $n - 1$ random observations from $U(0, t)$ and the last observation as t given $T = X_{(n)} = t$. This shows that T is a sufficient statistic for θ.
Alternatively, consider the distribution of T

$$F_T(t;\theta) = P(X_{(n)} \le t;\ \theta) = P(\text{all } X_i \le t)$$

$$= P(X_1 \le t \text{ and } X_2 \le t \text{ and } \ldots X_n \le t) = [P(X_i \le t)]^n = \left(\frac{t}{\theta}\right)^n$$

So, the *pdf* of T is

$$f_T(t;\ \theta) = \frac{n}{\theta^n} t^{n-1} I_{(0,\,\theta)}(t)$$

where, $I_{(a,\,b)}(x)$ is an indicator function defined as

$$I_{(a,b)}(x) = \begin{cases} 1, & \text{if } a < x < b \\ 0, & \text{otherwise} \end{cases}$$

or it can be directly written as

$$f_T(t;\ \theta) = n\left(\frac{t}{\theta}\right)^{n-1} \frac{1}{\theta}\ I_{(0,\,\theta)}(t)$$

$$= \frac{n}{\theta^n} t^{n-1}\ I_{(0,\,\theta)}(t)$$

It clarifies that the conditional distribution of $\mathbf{X}$ given $T = t$ is

$$f(\mathbf{x};\ t) = \frac{f(\mathbf{x}, t;\ \theta)}{f_T(t;\ \theta)}$$

$$= \frac{(1/\theta^n)\ I_{(0,\,\theta)}(t)}{(n/\theta^n) t^{n-1} I_{(0,\,\theta)}(t)} = \frac{1}{n\, t^{n-1}}$$

It is independent of θ. Therefore, $T = X_{(n)}$ is sufficient.
Alternatively, consider the joint density of sample observations

$$f(\mathbf{x};\ \theta) = \frac{1}{\theta^n} \prod_{i=1}^{n} I(0,\ x_i) \prod_{i=1}^{n} I(x_i,\ \theta)$$

$$= \frac{1}{\theta^n} I(0,\ x_{(1)})\ I(x_{(n)},\ \theta)$$

Here, $I(a, b) = 1$ if $a \le b$ and 0 otherwise.

$$h(x) = k(0, x_{(1)}), \quad g(x_{(n)}, \theta) = \frac{1}{\theta^n} g(x_{(n)}, \theta)$$

Hence, by factorization theorem, $T = X_{(n)}$ is sufficient statistic for θ.

(ii) The joint density of sample observations is

$$f(\mathbf{x}; \theta_1, \theta_2) = \prod_{i=1}^{n} \frac{1}{(\theta_2 - \theta_1)} I(\theta_1, x_i)\, I(x_i, \theta_2)$$

$$= \frac{I(\theta_1, x_{(1)})\, I(x_{(n)}, \theta_2)}{(\theta_2 - \theta_1)^n} = h(\mathbf{x})\, g(x_{(1)}, x_{(n)}; \theta_1, \theta_2)$$

where $h(\mathbf{x}) = 1$ and $g(x_{(1)}, x_{(n)}; \theta_1, \theta_2) = I(\theta_1, x_{(1)})\, I(x_{(n)}, \theta_2)/(\theta_2 - \theta_1)^n$. Therefore, $(X_{(1)}, X_{(n)})$ is jointly sufficient for (θ_1, θ_2).

(iii) Consider the joint density

$$f(\mathbf{x}; \theta) = \prod_{i=1}^{n} I(0, x_i) I(x_i, \theta) c(\theta) m(x_i)$$

$$= [c(\theta)]^n\, I(x_{(n)}, \theta)\, I(0, x_{(1)}) \prod_{i=1}^{n} m(x_i)$$

$$= h(\mathbf{x})\, g(t, \theta)$$

where $h(\mathbf{x}) = I(0, x_{(1)}) \prod_{i=1}^{n} m(x_i)$ and $g(t, \theta) = [c(\theta)]^n\, I(x_{(n)}, \theta)$. This shows that $T = X_{(n)}$ is sufficient for θ. Alternatively,

$$\int_0^{\theta} f(x; \theta) dx = c(\theta) \int_0^{\theta} m(x) dx = 1$$

implies

$$\int_0^{\theta} m(x)\, dx = \frac{1}{c(\theta)}$$

The distribution of X is given by

$$F_X(x) = \int_0^{x} f(y) dy = c(\theta) \int_0^{x} m(y) dy = \frac{c(\theta)}{c(x)}$$

The density of $T = X_{(n)}$ is given by

$$g(t; \theta) = n[F_X(t)]^{n-1} f(t)\, I(t, \theta)$$

$$= n \left[\frac{c(\theta)}{c(t)}\right]^{n-1} c(\theta)\, m(t)\, I(t, \theta)$$

$$= \frac{n}{[c(t)]^{n-1}}[c(\theta)]^n\, m(t)\, I(t, \theta)$$

The conditional distribution of **X** given $T = t$ is given by

$$f(\mathbf{x}|t) = \frac{f(\mathbf{x}; \theta)}{g(t; \theta)}$$

$$= \frac{[c(t)]^{n-1}\prod m(x_i)}{nm(t)}$$

which is independent of θ, which shows that $T = X_{(n)}$ is sufficient for θ.
Note that by setting $c(\theta) = 1/\theta$ and $m(x) = 1$, the above density becomes $U(0, \theta)$.

(iv) $\int_{\theta_1}^{\theta_2} f(x; \theta_1, \theta_2)dx = 1$ gives $\int_{\theta_1}^{\theta_2} m(x)dx = \frac{1}{c(\theta_1, \theta_2)}$

and
$$F_X(x; \theta_1, \theta_2) = \int_{\theta_1}^{x} f(y; \theta_1, \theta_2)dy = \frac{c(\theta_1, \theta_2)}{c(\theta_1, x)}$$

and
$$F_X(y; \theta_1, \theta_2) - F_X(x; \theta_1, \theta_2) = \int_{x}^{y} f(z; \theta_1, \theta_2)dz = \frac{c(\theta_1, \theta_2)}{c(x, y)}$$

If $T_1 = X_{(1)}$ and $T_2 = X_{(n)}$, the joint density of $X_{(1)}$ and $X_{(n)}$ is given by

$$f_{T_1, T_2}(t_1, t_2; \theta_1, \theta_2) = n(n-1)\left[\frac{c(\theta_1, \theta_2)}{c(t_1, t_2)}\right]^{n-2} c(\theta_1, \theta_2)m(t_1)c(\theta_1, \theta_2)m(t_2)I(\theta_1, t_1)I(t_2, \theta_2)$$

Consider now

$$f(\mathbf{x}|t_1, t_2) = \frac{f(\mathbf{x}; \theta_1, \theta_2)}{f_{T_1, T_2}(t_1, t_2; \theta_1, \theta_2)}$$

$$= \frac{[c(\theta_1, \theta_2)]^n \prod_1^n m(x_i) I(\theta_1, t_1) I(\theta_2, t_2)}{n(n-1)[c(\theta_1, \theta_2)]^n \left[\frac{1}{c(t_1, t_2)}\right]^{n-2} m(t_1)m(t_2)I(\theta_1, t_1)I(t_2, \theta_2)}$$

$$= \frac{[c(t_1, t_2)]^{n-2}}{n(n-1)} \frac{\prod_1^n m(x_i)}{m(t_1)m(t_2)}$$

which is independent of θ_1 and θ_2. Therefore, $(T_1, T_2) = (X_{(1)}, X_{(n)})$ are jointly sufficient for (θ_1, θ_2). The same can, alternatively, be shown by considering the joint density of sample observations

$$f(x;\theta_1,\theta_2)=[c(\theta_1,\theta_2)]^n\prod_1^n m(x_i)I(\theta_1,x_i)I(x_i,\theta_2)$$

$$=\prod_1^n m(x_i)[c(\theta_1,\theta_2)]^n I(\theta_1,x_{(1)})I(x_{(n)},\theta_2)$$

This gives $h(\mathbf{x}) = \prod_1^n m(x_i)$

and $\quad g(x_{(1)}, x_{(2)};\theta_1,\theta_2) = [c(\theta_1,\theta_2)]^n I(\theta_1, x_{(1)}) I(x_{(n)},\theta_2)$

Therefore, $(T_1, T_2) = (X_{(1)}, X_{(n)})$ is sufficient for (θ_1, θ_2).

***Example* 2.22** Let $X_1, X_2,\ldots,X_n$ be *iid* $U(0, \theta)$. Show that the statistic $T(\mathbf{X}) = X_{(n)}/X_{(1)}$ is ancillary.

Solution. Since $Y_i = X_i/\theta \sim U(0, 1)$ and $T(\mathbf{X}) = X_{(n)}/X_{(1)} = (X_{(n)}/\theta)/(X_{(1)}/\theta) = Y_{(n)}/Y_{(1)}$, the distribution of $T(\mathbf{X})$ does not depend on θ. Therefore, $T(\mathbf{X})$ is ancillary. One can, alternatively, show that $T(\mathbf{X})$ is an ancillary statistic since it satisfies the condition given in Eq. (2.5.7).

***Example* 2.23**

(i) Let $X \sim U(\theta, \theta + 1)$, $\theta \in \mathbb{R}^1$. Show that the statistic $T(X) = X$ is not complete.

(ii) Let $X_1, X_2,\ldots,X_n$ be *iid* $U(\theta, \theta + 1)$, $\theta \in \mathbb{R}^1$ *rvs*. Show that the statistic $T(\mathbf{X}) = (X_{(1)}, X_{(n)})$ is minimal sufficient for θ. Also, show that the statistic $V(\mathbf{X}) = (X_{(n)} - X_{(1)})$ is ancillary and that $T(\mathbf{X})$ is not complete.

Solution.

(i) If we choose a function $h(x)$ so that $h(x) = x$, set $E_\theta(X) = \int_\theta^{\theta+1} x\,dx = 0\ \forall\ \theta$. It does not imply $x = 0$ since $\theta < x < \theta + 1$. Alternatively, on differentiating $E_\theta(X) = \int_\theta^{\theta+1} x\,dx$ with respect to θ, we get $(\theta + 1) - (\theta) = 1 \neq 0$. Therefore, T is not complete for θ.

(ii) The joint *pdf* of sample observations is

$$f(\mathbf{x};\theta)=\prod_{i=1}^n I(\theta,x_i)\,I(x_i,\theta+1)$$

$$=I(\theta,x_{(1)})I(x_{(n)},\theta+1)$$

The ratio of densities is

$$\frac{f(\mathbf{x};\theta)}{f(\mathbf{y};\theta)}=\frac{I(\theta,x_{(1)})I(x_{(n)},\theta+1)}{I(\theta,y_{(1)})I(y_{(n)},\theta+1)}$$

Note that the numerator and denominator in the above ratio depends on the parameter θ. These are positive for the same values of θ, if and only if, $T_1(\mathbf{x}) = x_{(1)} = y_{(1)} = T_1(\mathbf{y})$ and $T_2(\mathbf{x}) = x_{(n)} = y_{(n)} = T_2(\mathbf{y})$. In this case, the ratio reduces to 1 and becomes independent of θ. This shows that $(X_{(1)}, X_{(n)})$ is minimal sufficient for θ. Note again that a one-to-one function of $(X_{(1)}, X_{(n)})$, namely $(X_{(n)} - X_{(1)}, X_{(1)} + X_{(n)}/2)$, is also minimal sufficient.

Consider the transformations $U_i = X_i - \theta$. Then $U_1, U_2,\ldots,U_n$ are *iid* $U(0, 1)$. Also, consider that the distribution of $V = (X_{(n)} - X_{(1)})$ is the same when $\theta = 0$ or any other value. So the distribution of V does not depend on θ and, therefore, V is ancillary. Since $T = (X_{(1)}, X_{(n)})$ has a joint distribution in $\mathbb{R}^2$, $V(\mathbf{X}) = (X_{(n)} - X_{(1)})$ has a nondegenerate distribution in $[0, 1]$. So $E_\theta[V(T)]$ exists and

$$E_\theta[V(T)] = c \qquad \text{(independent of } \theta)$$

Since V is ancillary,

$$E_\theta[V(T) - c] = 0 \quad \not\Rightarrow \quad V(t) = c \quad \text{a.e. } t$$

and since V, which is a function of T, has nondegenerate distribution, this proves that $T(\mathbf{X}) = (X_{(1)}, X_{(n)})$ is not complete. Alternatively, we can show that $(X_{(1)}, X_{(n)})$ is not complete as follows:

We have

$$f(x; \theta) = 1, \quad \theta < x < \theta + 1$$

$$F(x) = \int_\theta^x dt = x - \theta$$

$$F(y) - F(x) = y - x$$

The joint *pdf* of $X_{(1)}$ and $X_{(n)}$ is

$$f_{1n}(x,y) = n(n-1)[F(y) - F(x)]^{n-1} f(x) f(y), \; x < y$$

$$= n(n-1)(y - x)^{n-2}$$

$$E[X_{(n)} - X_{(1)}] = n(n-1)\int\int (y - x)^{n-1} dx\, dy$$

$$= \frac{n-1}{n+1} \text{ as } f_{1,n+1}(x, y) = (n+1)n.(y - x)^{n-2}$$

is joint *pdf* of $X_{(1)}$ and $X_{(n+1)}$.

This can also be obtained as

$$E(X_{(n)}) = \int_\theta^{\theta+1} y\, n\, F(y)^{n-1} f(y)\, dy = n \int_\theta^{\theta+1} y(y - \theta)^{n-1} dy$$

$$= n\int_0^1 u^{n-1}(u + \theta) du = \frac{n}{n+1} + \theta$$

$$E(X_{(1)}) = \int_\theta^{\theta+1} x n[1 - F(x)]^{n-1} f(x) dx = n \int_\theta^{\theta+1} x(\theta + 1 - x)^{n-1} dx$$

$$= n\int_0^1 (1 + \theta - v)v^{n-1} dv = (1 + \theta) - \frac{n}{n+1}$$

$$E(X_{(n)} - X_{(1)}) = \frac{n-1}{n+1} \Rightarrow X_{(n)} - X_{(1)} - \frac{n-1}{n+1} = 0$$

This is an example where the dimension of a sufficient statistic is more than the dimension of the parameter space, which are in this case two and one, respectively.

Note. Under the situation where support involves the parameter θ, care must be taken that for densities in the likelihood, the ratio must be positive for the same value of θ while using the likelihood equivalence principle to arrive at a minimal sufficient statistic.

Example 2.24 Let $X \sim U(-1/2, 1/2)$. Show that the statistic $T(X) = X$ is not complete.

Solution. On choosing a function $h(x) = x$ and setting $E_\theta(X) = \int_{-1/2}^{1/2} x\,dx = 0$, it does not imply $x = 0$ as $-1/2 < x < 1/2$. Therefore, T is not complete for θ.

Example 2.25

(i) Show that the family $U(-\theta, \theta)$ is not complete.

(ii) Let $X_1, X_2, \ldots, X_n$ be *iid* $U(-\theta, \theta)$, $\theta \in \mathbb{R}^1$. Show that the statistic $T(\mathbf{X}) = (X_{(1)}, X_{(n)})$ is not complete. Also, show that the statistic $U(\mathbf{X}) = \max(-X_{(1)}, X_{(n)})$ is also a sufficient statistic for θ.

(iii) Let $X_1, \ldots, X_n$ be *iid rvs* from the *pdf*

$$f(x; \theta) = \frac{1}{2\theta}, \ |x| < \theta, \ \theta > 0$$

Show that the statistic $T(\mathbf{X}) = \max\{-X_{(1)}, X_{(n)}\}$ is minimal and complete sufficient for θ.

Solution.

(i) $X \sim U(-\theta, \theta)$. Consider a function $g(x) = x$, $E[g(X)] = E(X) = \int_{-\theta}^{\theta} x \frac{1}{2\theta} dx = 0 \not\Rightarrow$ $X = 0$. Therefore, the family $U(-\theta, \theta)$ is not complete.

(ii) The joint density of data is

$$f(\mathbf{x}; \theta) = \left(\frac{1}{2\theta}\right)^n I(-\theta, x_{(1)})\, I(x_{(n)}, \theta)$$

By factorization theorem, $T(\mathbf{X}) = (X_{(1)}, X_{(n)})$ is sufficient. It is also clear that T is also minimal sufficient. Further, we have

$$F(x) = P(X_i \le x) = \int_{-\theta}^{x} \frac{1}{2\theta} dx = \frac{x}{2\theta} + \frac{1}{2};$$

$$1 - F(x) = \frac{1}{2} - \frac{x}{2\theta}$$

$$E(X_{(1)}) = \frac{n}{2\theta} \int_{-\theta}^{\theta} x \left(\frac{1}{2} - \frac{x}{2\theta}\right)^{n-1} dx$$

$$= -\frac{n-1}{n+1}\theta$$

and

$$E(X_{(n)}) = \frac{n}{2\theta} \int_{-\theta}^{\theta} x \left(\frac{1}{2} + \frac{x}{2\theta}\right)^{n-1} dx$$

$$= \frac{n-1}{n+1}\theta$$

So, if we consider a function $g(T) = X_{(1)} + X_{(n)}$, we have $E[g(T)] = 0$. But it does not imply that $g(T) = X_{(1)} + X_{(n)} = 0$. Therefore, T is not complete.

Note. The *pdfs* of $X_{(1)}$ and $X_{(n)}$ can, alternatively, be obtained by using the expressions

$$P[X_{(1)} < x] = 1 - P(\text{all } x_i > x) = 1 - \{1 - P(X_i < x)\}^n$$

$$= 1 - \left(\frac{1}{2} - \frac{x}{2\theta}\right)^n$$

and $$P[X_{(n)} < x] = P(\text{all } X_i < x) = \{P(X_i < x)\}^n = \left(\frac{x}{2\theta} + \frac{1}{2}\right)^n$$

Consider the joint density function of sample observations

$$f(\mathbf{x}; \theta) = \left(\frac{1}{2\theta}\right)^n I(-x_{(1)}, \theta)\, I(x_{(n)}, \theta)$$

$$= \left(\frac{1}{2\theta}\right)^n I[\max(-x_{(1)}, x_{(n)}), \theta]$$

which by factorization theorem shows that $U(\mathbf{X}) = \max(-X_{(1)}, X_{(n)})$ is a sufficient statistic for θ.

(iii) $T(\mathbf{X})$ is sufficient for θ can be easily seen by factorization theorem. Consider the joint density of sample observation at points $\mathbf{x}$ and $\mathbf{y}$

$$\frac{f(\mathbf{x}; \theta)}{f(\mathbf{y}; \theta)} = \frac{\left(\frac{1}{2\theta}\right)^n I(-x_{(1)}, \theta) I(x_{(n)}, \theta)}{\left(\frac{1}{2\theta}\right)^n I(-y_{(1)}, \theta) I(y_{(n)}, \theta)} = \frac{I(\max\{-x_{(1)}, x_{(n)}\}, \theta)}{I(\max\{-y_{(1)}, y_{(n)}\}, \theta)}$$

The above ratio is independent of θ, if and only if

$$T(\mathbf{x}) = \max\{-x_{(1)}, x_{(n)}\} = \max\{-y_{(1)}, y_{(n)}\} = T(\mathbf{y})$$

This shows that $T(\mathbf{X}) = \max\{-X_{(1)}, X_{(n)}\}$ is minimal sufficient statistic.
The statistic T takes values from $(0, \infty)$ and its *cdf* is given by

$$F(t; \theta) = \begin{cases} P(T \le t) = 0, & t \le 0 \\ \int_{-t}^{t} \int_{-t}^{x_{(n)}} n(n-1)(x_{(n)} - x_{(1)})^{n-2} \left(\frac{1}{2\theta}\right)^n dx_{(1)} dx_{(n)}, & 0 < t \le \theta \\ 1, & t > \theta \end{cases}$$

Since $T(\mathbf{X}) \le t$, if and only if, $-t < x_{(1)} < x_{(n)} < t$, the *cdf* of T for $0 < t \le \theta$ reduces to

$$F(t; \theta) = \frac{t^n}{\theta^n}$$

Consider a statistic $Z = |X|$. The *cdf* of Z is $G(z,\theta) = P(Z \le z) = P(|X| \le z) = P(-z < X < z) = \int_{-z}^{z}(1/2\theta)dx = z/\theta$, $z \in (0, \theta)$. This shows that Z is $U(0, \theta)$ whenever X is $U(-\theta, \theta)$. Therefore, sample observations $X_1,\ldots,X_n$ *iid* $U(-\theta, \theta)$ are the same as the observations *iid* $U(0, \theta)$. We can easily observe that $T(\mathbf{X}) = \max\{-X_{(1)}, X_{(n)}\} = \max\{|X_1|,\ldots,|X_n|\} = \max\{Z_1,\ldots,Z_n\} = Z_{(n)}$. Thus, the statistics $T(\mathbf{x}) = \max\{-x_{(1)}, x_{(n)}\}$and $Z_{(n)}$ are equivalent. It is already shown that $Z_{(n)}$ is complete. Therefore, $T(\mathbf{X})$ is also complete.

***Example* 2.26** Let $X_1, X_2,\ldots,X_n$ be *iid* $U(\theta - 1/2, \theta + 1/2)$, $\theta \in \mathbb{R}^1$. Show that the statistic $T(\mathbf{X}) = (X_{(1)}, X_{(n)})$ is minimal sufficient but not complete.

Solution. The *cdf* of X is

$$F(x;\theta) = \begin{cases} 0, & x \le \theta - 1/2 \\ (x - \theta + 1)/2, & \theta - 1/2 < x < \theta + 1/2 \\ 1, & \theta + 1/2 \le x \end{cases}$$

The joint density of $X_{(1)}$ and $X_{(2)}$ is

$$f_{1,n}(x, y) = n(n-1)[F(y;\theta) - F(x;\theta)]^{n-2} f(x) f(y), x < y$$

$$= n(n-1)(y-x)^{n-2}, \quad \theta - \frac{1}{2} < x < y < \theta + \frac{1}{2}$$

Consider $T_1(\mathbf{X}) = (X_{(n)} - X_{(1)})$, $T_2(\mathbf{X}) = (X_{(1)} + X_{(n)})/2$, $|J| = 1$. The joint *pdf* of T_1 and T_2 is

$$g(t_1, t_2;\theta) = n(n-1)t_1^{n-2}, \quad 0 < t_1 < 1$$

and
$$\theta - \frac{1}{2} + \frac{t_1}{2} < t_2 < \theta + \frac{1}{2} - \frac{t_1}{2}$$

Therefore, the marginal density of T_1 is

$$g(t_1;\theta) = \int_{\theta - \frac{1}{2} + \frac{t_1}{2}}^{\theta + \frac{1}{2} - \frac{t_1}{2}} n(n-1)t_1^{n-2}\, dt_2$$

$$= n(n-1)t_1^{n-2}(1 - t_1), \qquad 0 < t_1 < 1$$

So, $T_1 \sim Be(n-1, 2)$. The statistic T_1, which is a function $T(\mathbf{X}) = (X_{(1)}, X_{(n)})$, is ancillary, and hence, $T(\mathbf{X})$ is not complete.

Alternatively,

$$E(X_{(n)} - X_{(1)}) = n(n-1)\iint (y-x)^{n-1}\, dx\, dy = \frac{n-1}{n+1}$$

$$E\left(X_{(n)} - X_{(1)} - \frac{n-1}{n+1}\right) = 0 \Rightarrow X_{(n)} - X_{(1)} = \frac{n-1}{n+1}$$

Therefore, $T(X) = (X_{(1)}, X_{(n)})$ is not complete.

This implies that no further reduction is possible in $T(\mathbf{X})$, so that the reduced statistic is sufficient. Therefore, $T(\mathbf{X})$ is minimal sufficient. This is an example where it shows that a minimal sufficient statistic may not be complete.

Since the dimension of $T(\mathbf{X})$ (two) is greater than the dimension of θ (one), consider a statistic $(U, V) = [(1/2)\,(X_{(1)} + X_{(n)}),\ X_{(n)} - X_{(1)}]$, which is equivalent to T. The statistic (U, V) involves an ancillary statistic V as its component. Thus, T is not complete. Moreover, it measures the accuracy of U.

Example 2.27

(i) Suppose $X_1, X_2,\ldots,X_n$ are *iid* $N(\theta, \sigma^2)$. Find sufficient statistic for (θ, σ^2).

(ii) Find a minimal sufficient statistic for (θ, σ^2).

(iii) Let $\bar{X}$ and S_{n-1}^2 be the sample mean and sample variance, respectively. Then show that $k\bar{X}/S_n$ estimates θ/σ.

(iv) Let $\mathbf{X}_1,\ldots,\mathbf{X}_n$ be *iid* $N_k(\mathbf{0},\Sigma)$. Show that $(\Sigma\mathbf{X}_i,\ \Sigma_i\mathbf{X}_i\mathbf{X}_i')$ or $(\Sigma X_i,\ (1/n),\ \Sigma_i\mathbf{X}_i\mathbf{X}_i' - \bar{\mathbf{X}}\bar{\mathbf{X}}')$ is sufficient for $(\boldsymbol{\theta}, \Sigma)$.

Solution.

(i) The joint density of sample observations is

$$f(\mathbf{x};\theta,\sigma^2) = \left(\frac{1}{2\pi\sigma^2}\right)^{n/2} \exp\left\{-\frac{1}{2\sigma^2}\sum_1^n (x_i-\theta)^2\right\}$$

$$= (2\pi\sigma^2)^{-n/2} \exp\left\{-\frac{n\theta}{2\sigma^2}\right\} \exp\left\{\left(-\frac{1}{2\sigma^2}\right)\sum x_i^2 + \left(\frac{\theta}{\sigma^2}\right)\sum x_i\right\}$$

which is an exponential family.

(a) Consider the case when σ^2 is known. We first use the definition of sufficiency to show that $T = \bar{X}$ is sufficient. The joint density by adding and subtracting $\bar{x}$ can be written as

$$f_{\mathbf{X}}(\mathbf{x};\theta,\sigma^2) = \left(\frac{1}{2\pi\sigma^2}\right)^{n/2} \exp\left[-\sum_1^n (x_i-\bar{x})^2 - \frac{n(\bar{x}-\theta)^2}{2\sigma^2}\right]$$

Note that $T \sim N(\theta,\sigma^2/n)$. Therefore, the conditional distribution of $\mathbf{X}|T(\mathbf{X}) = t$ is

$$f_{\mathbf{X}}(\mathbf{x}|t) = \frac{(1/2\pi\sigma^2)^{n/2} \exp\left[-\sum_1^n (x_1-\bar{x})^2 - n(\bar{x}-\theta)^2/(2\sigma^2)\right]}{(n/2\pi\sigma^2)^{1/2} \exp[-n(\bar{x}-\theta)^2/(2\sigma^2)]}$$

$$\propto \exp\left[-\sum_1^n (x_i-\bar{x})^2/(2\sigma^2)\right]$$

which is independent of θ. Thus, $\bar{X}$ is sufficient for θ.

Another way of showing $\bar{X}$ as sufficient is to use factorization theorem. Consider the joint density of sample observations

$$f(\mathbf{x};\theta)=\left(\frac{1}{2\pi\sigma^2}\right)^{n/2}\exp\left\{-\frac{1}{2\sigma^2}\sum_1^n(x_i-\bar{x})^2\right\}\cdot\exp\left\{-\frac{n}{2\sigma^2}(\bar{x}-\theta)^2\right\}$$

Appealing to the factorization theorem, $h(\mathbf{x})$ in the above expression is the second term and $g(\theta, t_1)$ is the third term. Therefore, $T_1 = \bar{X}$ is sufficient for θ.

(b) If θ is known, then $h(x) = 1$ and $g(\sigma, t_2) = \exp\{-1/(2\sigma^2)\Sigma_1^n(x_i - \theta)^2\}$. This gives that $T_2 = \Sigma_1^n(x_i - \theta)^2$ is sufficient for σ^2.

(c) If (θ, σ^2) both are unknown, the joint density function is

$$f(\mathbf{x};\theta,\sigma^2)=\left(\frac{1}{2\pi}\right)^{n/2}\frac{1}{\sigma^n}\exp\left\{-\frac{\sum_1^n(x_i-\bar{x})^2}{2\sigma^2}-\frac{n(\bar{x}-\theta)^2}{2\sigma^2}\right\}$$

$$h(\mathbf{x})=\left(\frac{1}{2\pi}\right)^{n/2};\ g(\theta,\sigma^2,t_1,t_2)=\frac{1}{\sigma^n}\exp\left\{-\frac{\sum_1^n(x_i-\bar{x})^2}{2\sigma^2}-\frac{n(\bar{x}-\theta)^2}{2\sigma^2}\right\}$$

showing that $(\bar{X}, \Sigma(X_i - \bar{X})^2)$ is jointly sufficient for (θ, σ^2). Observe that $(\bar{X}, \Sigma(X_i - \bar{X})^2)$ is different from (T_1, T_2). However, statistics are equivalent statistics in the sense they carry same amount of information for (θ, σ^2). It is important to mention that the joint sufficiency does not imply coordinate to coordinate sufficiency.

(ii) The joint density of sample observations may be written as a function of sample mean $\bar{X} = (1/n)\Sigma_1^n X_i$ and sample variance

$$S_{n-1}^2=\frac{1}{n-1}\sum_1^n(X_i-\bar{X})^2$$

$$f(\mathbf{x};\theta,\sigma^2)=\left(\frac{1}{2\pi\sigma^2}\right)^{n/2}\exp\left\{-\frac{n(\bar{x}-\theta)^2+\sum_1^n(x_i-\bar{x})^2}{2\sigma^2}\right\}$$

Appealing to the likelihood equivalence method of constructing minimal sufficient statistic, consider

$$\frac{f(\mathbf{x};\theta,\sigma^2)}{f(\mathbf{y};\theta,\sigma^2)}=\frac{\left(\frac{1}{2\pi\sigma^2}\right)^{n/2}\exp\left\{-\frac{n(\bar{x}-\theta)^2+\sum_1^n(x_i-\bar{x})^2}{2\sigma^2}\right\}}{\left(\frac{1}{2\pi\sigma^2}\right)^{n/2}\exp\left\{-\frac{n(\bar{y}-\theta)^2+\sum_1^n(y_i-\bar{y})^2}{2\sigma^2}\right\}}$$

$$=\exp\left\{-\frac{n(\bar{x}^2-\bar{y}^2)-2n\theta(\bar{x}-\bar{y})+(n-1)(s_{\mathbf{x}}^2-s_{\mathbf{y}}^2)}{2\sigma^2}\right\}$$

This ratio is independent of θ, if and only if, $T_1(\mathbf{x})=\bar{x}=\bar{y}=T_1(\mathbf{y})$ and $T_2(\mathbf{x})=s_{n-1}^2(\mathbf{x})=s_{n-1}^2(\mathbf{y})=T_2(\mathbf{y})$. Therefore, $(T_1, T_2)=(\bar{X}, S_{n-1}^2)$ is jointly minimal sufficient for (θ, σ^2). $(\Sigma X_i, \Sigma X_i^2)$ is also minimal sufficient for (θ, σ^2), since it is a one-to-one function of $(\bar{X}, S_{n-1}^2)$. In fact every one-to-one function of minimal sufficient statistic is also minimal sufficient.

(iii) Note that $\bar{X}$ and $(X_1-\bar{X}, X_2-\bar{X},\ldots,X_n-\bar{X})$ or $\bar{X}$ and S_{n-1}^2 are independent. Note that $(n-1)\,S_{n-1}^2/\sigma^2$ is $\chi^2(n-1)$ and $\sqrt{n}(\bar{X}-\theta)/\sigma$ is $t(n-1)$. This gives

$$E\left[\frac{\sqrt{n}(\bar{X}-\theta)}{S_{n-1}}\right]=0$$

$$E\left[\frac{\bar{X}}{S_{n-1}}\right]=\frac{\theta}{\sigma}\sqrt{n-1}\,E\left[\frac{1}{\sqrt{n-1}\,S_{n-1}/\sigma}\right]$$

$$=\frac{\theta}{\sigma}\sqrt{n-1}\int_0^\infty x^{-1/2}G_1\left(\frac{n-1}{2},2\right)dx$$

$$=\frac{\theta}{\sigma}\sqrt{n-1}\int_0^\infty x^{-1/2}\frac{1}{\Gamma\left(\frac{n-1}{2}\right)2^{(n-1)/2}}x^{((n-1)/2)-1}\exp\left(-\frac{x}{2}\right)dx$$

$$=\frac{\theta}{\sigma}\sqrt{n-1}\frac{1}{\Gamma\left(\frac{n-1}{2}\right)2^{\frac{n-1}{2}}}\int_0^\infty x^{\frac{n-2}{2}-1}\exp\left(-\frac{x}{2}\right)dx$$

$$=\frac{\theta}{\sigma}\sqrt{n-1}\frac{\Gamma\left(\frac{n-2}{2}\right)2^{\frac{n-2}{2}}}{\Gamma\left(\frac{n-1}{2}\right)2^{\frac{n-1}{2}}}=2^{-1/2}\sqrt{n-1}\frac{\Gamma\left(\frac{n-2}{2}\right)}{\Gamma\left(\frac{n-1}{2}\right)}\left(\frac{\theta}{\sigma}\right)$$

$$E\left[2^{1/2}\frac{\Gamma\left(\dfrac{n-1}{2}\right)}{\sqrt{n-1}\,\Gamma\left(\dfrac{n-2}{2}\right)}\frac{\bar{X}}{S_{n-1}}\right]=\left(\frac{\theta}{\sigma}\right)$$

$$E\left[k\frac{\bar{X}}{S_{n-1}}\right]=\left(\frac{\theta}{\sigma}\right)$$

where $k=2^{1/2}\dfrac{\Gamma[(n-1)/2]}{\sqrt{n-1}\cdot\Gamma[(n-2)/2]}$. Thus, $k(\bar{X}/S_{n-1})$ is an unbiased estimator of θ/σ.

In fact, it is UMVUE of θ/σ by the Lehmann–Scheffe theorem since it is a function of complete sufficient statistic $(\bar{X}, S_{n-1}^2)$.

(iv) Use factorization theorem.

Example 2.28

(i) Let X_1 and X_2 be *iid* $N(\theta, 1)$. Show that the statistic $T(\mathbf{X}) = X_1 + X_2$ is complete. Is the statistic $U(\mathbf{X}) = X_1 - X_2$ also complete?

(ii) Let $X_1,\ldots,X_n$ be *iid* $N(\theta,1)$, $\theta \in \mathbb{R}^1$. Show that the family of normal distributions $N(\theta, 1)$ is complete and that the statistic $V(\mathbf{X}) = (X_1 - \bar{X},\ldots,X_n - \bar{X})'$ is ancillary.

(iii) Show that the necessary and sufficient condition for $\Sigma a_i X_i$ and ΣX_i to be independent is $\Sigma a_i = 0$ in (ii).

(iv) Show in (ii) that $T(\mathbf{X}) = (\Sigma X_i)^2$ is not sufficient for θ.

Solution.

(i) $T \sim N(2\theta, 2)$. For any Borel measurable function $h(T)$ and for every θ, set

$$E_\theta h(T) = 0$$

$$\int_{-\infty}^{\infty} h(t)\frac{1}{\sqrt{4\pi}}\exp\left\{-\frac{1}{4}(t-2\theta)^2\right\}dt = 0$$

$$\exp(-\theta^2)\int_{-\infty}^{\infty} h(t)\exp\left(\frac{-t^2}{4}\right)\exp(t\theta)dt = 0 \tag{2.8.3}$$

Note that $\exp(-\theta^2) > 0$ and Eq. (2.8.3) is a bilateral Laplace transformation in θ. This implies $h(t) = 0$ a.e. t. So, there is no nonzero function of t such that its expectation is zero for every θ. This shows that T is complete. Further, the distribution of the statistic $U(\mathbf{X}) = X_1 - X_2$ is $N(0, 2)$, which is independent of the parameter θ. Therefore, $U(\mathbf{X})$ is ancillary. (Similarly, one can easily see that $X_1 + X_2 - 2X_3$ and, in general, $\Sigma a_i X_i$, where $\Sigma a_i = 0$, is ancillary.) Since $U(\mathbf{X})$ contains the ancillary material, it cannot be complete. Another way of showing that the statistic $U(\mathbf{X}) = X_1 - X_2$ is not complete is that $E_\theta(X_1 - X_2) = 0\ \forall\, \theta$ does not imply $X_1 - X_2 = 0$.

(ii) Since $V(\mathbf{X}) = (X_1 - \bar{X},\ldots,X_n - \bar{X})' \sim N_n(\mathbf{0}, I - n^{-1}\mathbf{11}')$, $V(\mathbf{X})$ is ancillary. The statistics $\bar{X}$ and $(\bar{X}, V(\mathbf{X}))$ are equivalent.

(iii) Using similar arguments as in (i), ΣX_i is complete. $\Sigma a_i X_i$ is ancillary, if and only if, $\Sigma a_i = 0$. Therefore, by Basu's theorem, ΣX_i and $\Sigma a_i X_i$ are independent, if and only if, $\Sigma a_i = 0$.

(iv) Consider two points **x** and **y** in $\mathbb{R}^n$ so that $\Sigma x_i = -\Sigma y_i$. In this case, the statistic $T(\mathbf{X}) = (\Sigma X_i)^2$ will have the same value at **x** and **y**, i.e., $T(\mathbf{x}) = T(\mathbf{y})$ but $f(\mathbf{x}; \theta_1)f(\mathbf{y}; \theta_2) \neq f(\mathbf{x}; \theta_2)f(\mathbf{y}; \theta_1)$, since $\exp(\theta_1\Sigma x_i + \theta_2\Sigma y_i) \neq \exp(\theta_2\Sigma x_i + \theta_1\Sigma y_i)$. Therefore, by Theorem 2.3.2, $T(\mathbf{X}) = (\Sigma X_i)^2$ is not sufficient.

***Example* 2.29** Let $X_1, X_2, \ldots, X_n$ be *iid* from $N(\theta, \sigma^2)$, σ^2 fixed. Use Basu's theorem to show that the sample mean $T(\mathbf{X}) = (1/n)\ \Sigma_1^n X_i$ and sample standard deviation $S_{n-1}(\mathbf{X}) = \sqrt{(n-1)^{-1}\Sigma_1^n (X_i - \overline{X})^2}$ are independent.

Solution. Note that $T(\mathbf{X})$ is a complete sufficient statistic for θ and $S(\mathbf{X})$ is ancillary since $(n-1)S^2/\sigma^2 \sim \chi^2\ (n-1)$. Another way to show that $S(\mathbf{X})$ is ancillary is to note that $Y_i = X_i - \theta \sim N(0, 1)$, $\overline{Y} = \overline{X} - \theta$, $X_i - \overline{X} = Y_i - \overline{Y} \Rightarrow S_{n-1}^2 = (1/(n-1))^{-1}\ \Sigma_1^n (Y_i - \overline{Y})^2$. Appealing to Basu's theorem, T and S_{n-1} are independent.

***Example* 2.30**

(i) Suppose a single observation X has been drawn from $N(0, \sigma^2)$, σ^2 being unknown. Find sufficient statistic for σ^2.

(ii) Let $X_1, \ldots, X_n$ be *iid* $N(0, \sigma^2)$. Show that $U = \Sigma X_i$ is not sufficient. Find a sufficient statistic for σ^2 and reconstruct the sample by recording the value of the sufficient statistic $T = \Sigma X_i^2 = t$. Also, show that $T = \Sigma X_i^2$ is minimal sufficient.

Solution.

(i) Let us decide to lose information on the sign of X by defining the statistic $T(X) = |X|$. Given $T = t$, $P(X = t|T = t) = 1/2 = P(X = -t|T = t)$ since $N(0, \sigma^2)$ is symmetric about 0. Therefore, the statistic $T = |X|$ is sufficient for σ^2. Further, that $P(|X| = t) = P(X = t) + P(X = -t)$. This allows the reconstruction of observation X by taking $X = t$ if toss of a fair coin turns up head and $X = -t$ if it turns up tail. Note that

$$P(Y = t|T = t) = \frac{1}{2} = P(Y = -t|T = t)$$

$$P(X = t|T = t) = \frac{1}{2} = P(X = -t|T = t)$$

Thus, the unconditional distribution of X is the same as that of Y.

(ii) The joint density of sample observations is

$$f(\mathbf{x}; \sigma^2) = \left(\frac{1}{2\pi\sigma^2}\right)^{n/2} \exp\left(-\frac{1}{2\sigma^2}\sum x_i^2\right)$$

The density of $U = \Sigma X_i \sim N(0, n\sigma^2)$ is

$$f(U; \sigma^2) = \left(\frac{1}{2\pi n\sigma^2}\right)^{1/2} \exp\left(-\frac{1}{2n\sigma^2}u^2\right)$$

The dependence of the conditional distribution of $X|U$ is

$$f(\mathbf{x}|u) = \frac{(2\pi n\sigma^2)^{1/2} \exp\left(-\frac{1}{2\sigma^2}\sum x_i^2\right)}{(2\pi\sigma^2)^{1/2} \exp\left(-\frac{1}{2n\sigma^2}u^2\right)}$$

$$= c\sigma^{1-n} \exp\left[-\frac{1}{2\sigma^2}\left(\sum x_i^2 - \frac{u^2}{n}\right)\right]$$

on σ^2, it verifies that $U = \Sigma X_i$ is not sufficient.

The statistic $T(\mathbf{X}) = \Sigma X_i^2$ is an intuitive choice for estimating σ^2. We would see how to reconstruct the sample on removing the original observations after recording $T(\mathbf{X}) = \Sigma X_i^2 = t$. Note that the points of the sample space are the points on the surface of the sphere $T = t$, with its origin $(0,\ldots,0)$ and with radius $\sqrt{t}$. The distribution of T is given by $G((n/2), 2\sigma^2)$. This gives the distribution of the points falling on the surface $T = t$

$$P(\mathbf{X} = \mathbf{x}|T = t) = \frac{\Gamma(n/2)2^{n/2}}{(2\pi)^{n/2}} \frac{1}{t^{(n/2)-1}}$$

which is independent of σ^2. Select any point $\mathbf{Y} = (X_1,\ldots,X_n)$ on the sphere $T = \Sigma X_i^2 = t$. The reconstructed sample $\mathbf{Y}$ carries the same information as the original sample $\mathbf{X}$ contained on σ^2. We say $T = \Sigma X_i^2$ is a sufficient statistic for σ^2.

Consider now the ratio of densities at points $\mathbf{x}$ and $\mathbf{y}$

$$\frac{f(\mathbf{x};\sigma^2)}{f(\mathbf{y};\sigma^2)} = \frac{\left(\frac{1}{2\pi\sigma^2}\right)^{n/2} \exp\left(-\frac{1}{2\sigma^2}\sum x_i^2\right)}{\left(\frac{1}{2\pi\sigma^2}\right)^{n/2} \exp\left(-\frac{1}{2\sigma^2}\sum y_i^2\right)} = \exp\left[-\frac{1}{2\sigma^2}\left(\sum x_i^2 - \sum y_i^2\right)\right]$$

This shows that $T = \Sigma X_i^2$ is minimal sufficient statistic.

***Example* 2.31** $\mathbf{X} = (X_1, X_2,\ldots,X_m)$ be *iid* $N(\theta_1, \sigma_1^2)$, $\mathbf{Y} = (Y_1, Y_2,\ldots,Y_n)$ be *iid* $N(\theta_2, \sigma_2^2)$, independent of X_is. Find complete sufficient statistic for $(\theta_1, \theta_2, \sigma_1^2, \sigma_2^2)$ (i) if $\theta_1 \neq \theta_2$ and (ii) if $\theta_1 = \theta_2$.

Solution.

(i) Write the joint density of data $(\mathbf{X}, \mathbf{Y})$, and use factorization theorem to show that $T(\mathbf{X}, \mathbf{Y}) = (\bar{X}, \bar{Y}, S_X^2, S_Y^2)$ is jointly sufficient for $(\theta_1, \theta_2, \sigma_1^2, \sigma_2^2)$. By Note 2 of Section 2.4, the sufficient statistic $T(\mathbf{X}, \mathbf{Y})$ is also complete.

(ii) If $\theta_1 = \theta_2 = \theta$, then $T(\mathbf{X}, \mathbf{Y}) = (\bar{X}, \bar{Y}, S_X^2, S_Y^2)$ is still sufficient but not complete since $(\bar{X} - \bar{Y})$ is first-order ancillary. $E(\bar{X} - \bar{Y}) = 0$ does not imply $\bar{X} - \bar{Y} = 0$ and note that $\bar{X} - \bar{Y} \sim N(0, (\sigma_1^2/m) + (\sigma_2^2/n))$. This is the reason why an UMVUE for θ does not exist and as an alternative, an MLE is obtained in Chapter 6 with some large sample properties.

***Example* 2.32 (Curved exponential family).** Let $X_1, X_2, \ldots, X_n$ be a random sample from $N(a\theta, \theta^2)$, where a is a known real number and $\theta > 0$.

(i) Show that the statistic $T(\mathbf{X}) = (\bar{X}, S^2)$ is sufficient for θ.
(ii) Show that the parameter space does not contain a two-dimensional open set.
(iii) Show that the family of distributions of $T(\mathbf{X})$ is not complete.

Solution.

(i) The joint *pdf* of sample observations is

$$f(\mathbf{x};\theta) = \left(\frac{1}{2\pi\theta^2}\right)^{n/2} \exp\left(-\frac{1}{2\theta^2}\sum (x_i - a\theta)^2\right)$$

$$= \left(\frac{1}{2\pi\theta^2}\right)^{n/2} \exp\left(-\frac{a^2}{2}\right) \exp\left(-\frac{1}{2\theta^2}\sum x_i^2 + \frac{a}{\theta}\sum x_i\right)$$

By factorization theorem, $T(\mathbf{X}) = (\Sigma X_i, \Sigma X_i^2) = (\bar{X}, S^2)$ is sufficient for θ.

(ii) The parameter space $\Theta = \{(a\theta, \theta^2): \theta > 0\}$ is just a line and, therefore, does not contain a two-dimensional open set.

(iii) We have $E(S_{n-1}^2) = \theta^2$ since $\Sigma(X_i - \bar{X})^2/\theta^2 \sim \chi^2_{n-1}$. Further, $E(\bar{X}^2) = V(\bar{X}) + \{E(\bar{X})\}^2 = (\theta^2/n) + a^2\,\theta^2$. Consider a function $g(\bar{X}, S_{n-1}^2) = [\{n/(1 + na^2)\}\,\bar{X}^2 - S_{n-1}^2]$, which has zero expectation for every θ, but the function is not identically equal to zero. Therefore, $T(\mathbf{X}) = (\bar{X}, S_{n-1}^2)$ is not complete.

***Example* 2.33** BSNL has recently floated 3G mobile services and the customer care is receiving complaints of poor functioning of some of the 3G features. Engineers have made some technical changes and now want to assess the chances of no complaints in the next one minute. They choose a crude estimator which estimates $\exp(-\theta)$ by 1 if there is no complaint in the first minute and zero otherwise

$$Y_i = \begin{cases} 1, & \text{if } X_i = 0 \\ 0, & \text{otherwise} \end{cases}$$

(i) They consider two estimators: One is maximum likelihood estimator

$$T_{1n} = \exp(-\bar{X}), \text{ where } \bar{X} = \frac{1}{n}\sum X_i$$

and the second is

$$T_{2n} = \bar{Y}_n = \frac{1}{n}\sum y_i$$

From asymptotic considerations, which estimator should the Engineers use?

(ii) Find a sufficient statistic in this case and use the Rao–Blackwell theorem to arrive at a better estimator.

Solution. We can model the complaints received by the Poisson process at an average rate of θ per minute. The number of such complaints on n successive one-minute periods have been recorded as $X_1, X_2, \ldots, X_n$. The probability of no complaints in a single minute is $\exp(-\theta)$.

(i) T_{2n} is unbiased for $\exp(-\theta)$, since Y_i's are *iid* and $E_\theta(Y_i) = P_\theta(X_i = 0) = \exp(-\theta)$.

(ii) Y_1 is an unbiased estimator of $\exp(-\theta)$ and $U(\mathbf{X}) = \Sigma_1^n X_i$ is a sufficient statistic of θ. Therefore, the Rao–Blackwell estimator is

$$\delta(\mathbf{X}) = E[Y_1|u] = \left(1 - \frac{1}{n}\right)^u$$

$\delta(\mathbf{X})$ is also unbiased. The average number of complaints, in the first n minutes, is $n\theta$. The estimator behaves nice since it estimates $\exp(-\theta)$ very closely $[1 - (1/n)]^{n\theta} \to e^{-\theta}$ when n is large. This shows that the improved estimator obtained by Rao–Blackwellization is much sensible, instead of taking originally a rough estimator. In fact, $\delta(\mathbf{X})$ is uniformly minimum variance unbiased estimator by the Lehmann–Scheffe theorem (discussed in Chapter 3).

***Example* 2.34** Let $X_1, X_2, \ldots, X_n$ be *iid rvs* from the *pmf*

$$P_N(x) = \frac{1}{N}, \quad x = 1, 2, \ldots, N$$

where N is some positive integer. Show that $T(\mathbf{X}) = X_{(n)}$ is a complete sufficient statistic for N.

Solution. The *cdf* of $T = X_{(n)}$ is

$$P(X_{(n)} \le t) = P(X_1 \le t, \ldots, X_n \le t) = \left(\frac{t}{N}\right)^n$$

The *pmf* of T is

$$P(X_{(n)} = t) = P(X_{(n)} \le t) - P(X_{(n)} \le t - 1) = \left(\frac{t}{N}\right)^n - \left(\frac{t-1}{N}\right)^n$$

The conditional distribution of $\mathbf{X}|T = t$ calculates to

$$P(\mathbf{X} = \mathbf{x}|T = t) = \frac{P(X_1 = x_1, \ldots, X_n = x_n \cap X_{(n)} = t)}{P(X_{(n)} = t)}$$

$$= \frac{\frac{1}{N^n}}{\left(\frac{t}{N}\right)^n - \left(\frac{t-1}{N}\right)^n} = \frac{1}{t^n - (t-1)^n}$$

which is independent of N. This shows that $T = X_{(n)}$ is sufficient for N.

For some function h,

$$E_N[h(T)] = \frac{1}{N^n}\sum_{t=1}^{N} h(t)[t^n - (t-1)^n] = 0 \quad \text{for } N = 1, 2, \ldots$$

It implies that if $N = 1$, we get $h(1) = 0$ or if $N = 2$, we get $h(2) = 0$ and so on. Therefore, $h(t) = 0$, for $N = 1, 2, \ldots$ proves that T is complete sufficient statistic.

***Example* 2.35** Consider the family of probability curves $\{P(x;\theta),\ \theta \in \Theta\}$ parameterized by θ in three-dimensional probability simplex with *pmf*

$$P(x;\theta) = \begin{cases} \theta^2, & x = 0 \\ 2\theta, & x = 1 \\ 1-\theta^2-2\theta, & x = 2 \end{cases}$$

where $\theta \in \Theta = [0, \sqrt{2} - 1]$. Show that the family $\{P(x;\theta),\ \theta \in \Theta\}$ is complete.

Solution. For any function,

$$\begin{aligned} E_\theta[h(X)] &= \theta^2 h(0) + 2\theta h(1) + (1 - \theta^2 - 2\theta)h(2) \\ &= \theta^2\,[h(0) - h(2)] + 2\theta[h(1) - 2\,h(2)] + h(2) \end{aligned}$$

This expression is a polynomial in θ of order 2. $E_\theta[h(X)] = 0 \Rightarrow$ all the three coefficients vanish, i.e., $[h(0) - h(2)] = 0$, $[h(1) - 2h(2)] = 0$, $h(2) = 0$. This implies that $h(x) = 0\ \forall\ x = 0, 1, 2$. Therefore, the family $\{P(x;\theta),\ \theta \in \Theta\}$ is complete.

***Example* 2.36** Let $X_1, X_2,\ldots,X_n$ be a random sample from double exponential distribution

$$f(x;\theta) = \frac{1}{2}\exp\{-|x-\theta|\};\ x \in \mathbb{R}^1$$

Examine whether sufficient and minimal sufficient statistics exist for θ.

Solution. The joint density is written as

$$\begin{aligned} f(\mathbf{x};\theta) &= \frac{1}{2^n}\exp\left\{-\sum|x_i-\theta|\right\},\quad \mathbf{x} \in \mathbb{R}^n \\ &= \frac{1}{2^n}\exp\left\{-\sum|x_{(i)}-\theta|\right\} \end{aligned}$$

By factorization theorem, $T(\mathbf{X}) = (X_{(1)},\ldots,X_{(n)})$ is sufficient statistic. Consider

$$\frac{f(\mathbf{x};\theta)}{f(\mathbf{y};\theta)} = \frac{\dfrac{1}{2^n}\exp\left\{-\sum|x_i-\theta|\right\}}{\dfrac{1}{2^n}\exp\left\{-\sum|y_i-\theta|\right\}}$$

This ratio is independent of θ, if and only if, $T(\mathbf{X}) = (X_{(1)},\ldots,X_{(n)}) = (Y_{(1)},\ldots,Y_{(n)}) = T(\mathbf{Y})$. Therefore, $T(\mathbf{X}) = (X_{(1)},\ldots,X_{(n)})$ is minimal sufficient for θ.

***Example* 2.37** Let $X_1,\ldots,X_n$ be a random sample from $C(0, \theta)$ given by its *pdf*

$$f(x;\theta) = \frac{1}{\pi}\frac{1}{1+(x-\theta)^2};\ -\infty < x < \infty,\ -\infty < \theta < \infty$$

Find a sufficient statistic for θ.

Solution. The joint distribution of sample observations is

$$f(\mathbf{x};\theta)=\frac{1}{\pi^n}\prod_{i=1}^{n}\frac{1}{1+(x_i-\theta)^2}$$

Therefore, by factorization theorem, the second term of the joint density is

$$g(x_{(1)},x_{(2)},\ldots,x_{(n)},\theta)$$

which gives that the ordered statistics

$$T(\mathbf{X})=(X_{(1)},X_{(2)},\ldots,X_{(n)})$$

is sufficient.

***Example* 2.38** Consider a random sample $X_1,\ldots,X_n$ from $N(0,\theta)$. Show that $T_1 = X_1$ is not a complete statistic for θ but $T_2 = X_1^2$ is complete for θ.

Solution. Consider some Borel measurable function $h(T_1) = h(X_1)$ so that

$$E[h(X_1)] = 0 \quad \forall\, \theta \in \Theta$$

The above equality holds only when h is an odd function, i.e., $h(x) = -h(-x)\ \forall\, x$

or $$h(x) + h(-x) = 0 \quad \forall\, x$$

This does not imply that

$$h(x) = 0 \quad \forall\, x$$

Therefore, $T_1 = X_1$ is not complete. Next, consider the statistic $T_2 = X_1^2$. If

$$E[h(X_1^2)] = 0$$

then $$\frac{1}{\sqrt{2\pi\theta}}\int_{-\infty}^{\infty} h(x^2)\exp\left(-\frac{x^2}{2\theta}\right)dx = 0 \quad \forall\, \theta$$

Consider, $x^2 = y$; $2x\,dx = dy$

$$\int_{-\infty}^{\infty} h(y)\frac{1}{2y^{1/2}}\exp\left(-\frac{y}{2\theta}\right)dy = 0$$

$$\int_{-\infty}^{\infty} \frac{h(y)}{\sqrt{y}}\exp\left(-\frac{y}{2\theta}\right)dy = 0$$

The above integral is a bilateral Laplace transformation in $(1/2\theta)$. This gives

$$\frac{h(y)}{\sqrt{y}} = 0 \qquad \text{a.e. } y$$

$$h(y) = 0 \qquad \text{a.e. } y$$

Thus, the statistic $T_2 = X_1^2$ is a complete statistic for θ.

EXERCISES

1. Let T be a sufficient statistic and U be some other statistic so that

$$T = g(U)$$

Is U also sufficient?

2. If T is sufficient and a statistic U is

$$U = g(T)$$

where g is single-valued. Is U also sufficient? Show. If g is one-to-one function on $\mathbb{R}^1$, show that T is sufficient for $g(\theta)$.

3. Let S be the support of $f(x;\theta)$, $\theta \in \Theta$, and let T be a statistic such that for some θ_1, $\theta_2 \in \Theta$ and $x, y \in S$, $x \neq y$, $T(x) = T(y)$ but $f(x;\theta_1)\cdot f(y;\theta_2) \neq f(x;\theta_2)\cdot f(y;\theta_1)$. Show that T is not a sufficient statistic for θ.

4. Let $\mathbf{X} = (X_1,\ldots,X_n)$ be a random sample from some discrete *pmf* $P(x;\theta)$. Show that $T(\mathbf{X}) = (X_1,\ldots,X_{n-1})$ is not a sufficient statistic.

5. Let $f(x; \theta)$, $\theta \in \Theta$, be a family of densities each of which is defined on the sample space χ. Let T_1 and T_2 be two statistics which have the following properties: $T_1(x) = T_1(y)$ if and only if $T_2(x) = T_2(y)$. Show that T_1 is a sufficient statistic for the specified family if and only if, T_2 is sufficient for that family.

6. Let T_1 and T_2 be two distinct sufficient statistics for θ. Show that T_1 is a function of T_2. Does it follow that every function of a sufficient statistic is itself sufficient?

7. If T is sufficient for $\theta \in \Theta$, then show that it is also sufficient for $\theta \in w \subseteq \Theta$.

8. Let $\mathcal{F} = \{f(x;\theta), \theta \in \Theta\}$ be a family of distributions so that the *pdf* $f(x;\theta)$ is symmetric about θ and $E_\theta[T(X)] < \infty\ \forall\ \theta$ and is an odd function of x. Show that such a family of distribution $\mathcal{F}$ is not complete.

9. Discrete distributions. Let (X, Y) be a bivariate random variable with joint distribution given as follows:

$P[Y = 0|X = 0] = 1/2$; $P[Y = 0|X = 1] = 1 - \theta$; $P[Y = 1|X = 0] = 1/2$; $P[Y = 1|X = 1] = \theta$; $P[X = 0] = 1 - \theta$; $P[X = 1] = \theta$

or, the joint distribution of (X, Y) is given by

$$P(x;\theta) = \begin{cases} \dfrac{1-\theta}{2}, & (x, y) = (0, 0), (0, 1) \\ \theta(1-\theta), & (x, y) = (1, 0) \\ \theta^2, & (x, y) = (1, 1) \end{cases}$$

Find a minimal sufficient statistic for θ.

10. Let X be an *rv* with *pmf* $P(X = -1; \theta) = \theta$ and $P(X = x; \theta) = (1 - \theta)^2\theta^x$, $x = 0, 1, 2,\ldots,0 < \theta < 1$. Show that X is sufficient and minimal sufficient for θ.

11. Suppose X is $b(2, \theta)$ random variable, where θ can take only two values 1/2 and 1/4. Show that this family is not complete by constructing a nonzero function $g(X)$ whose expectation is zero for both $\theta = 1/2$ and 1/4.

12. Show that the family
(i) $\{b(2, p), p = 1/2, p = 1/4\}$ is not complete.
(ii) $\{b(2, p), p = 1/2, p = 1/3, p = 1/4\}$ is complete.
(iii) $\{b(2, p), 0 < p < 1\}$ is complete.

13. Location-scale family of distributions. Based on a random sample $X_1, X_2, \ldots, X_n$ of size n from the following location-scale densities, examine if there exist sufficient, minimal sufficient, and complete sufficient statistics in each of the following cases:

(i) Discrete uniform distribution

$$X \sim P_{N_1,N_2}, P_{N_1,N_2}(x) = \frac{1}{N_2 - N_1}, \quad x = N_1, \ldots, N_2$$

N_1, N_2 are integers, (a) N_1 is known, N_2 is unknown; (b) N_2 is known, N_1 is unknown; (c) both N_1 and N_2 are unknown.

(ii) $X \sim f(x;\theta)$ (Displaced exponential)

$$f(x;\theta) = \frac{1}{\sigma}\exp\left[-\frac{(x-\theta)}{\sigma}\right], \theta < x < \infty$$

Hint. (T_1, T_2) is jointly sufficient for (θ, σ), where $T_1(\mathbf{X}) = X_{(1)}$ and $T_2(\mathbf{X}) = \Sigma_{i=1}^{n} X_i$ and $(T_1, T_2) = (X_{(1)}, \Sigma_{i=1}^{n}(X_i - X_{(1)}))$ is jointly minimal sufficient.

(iii) $X \sim f(x;\theta, p)$, $f(x;\theta, p) = (1-p)\,p^{x-\theta}$, $x = \theta, \theta+1, \ldots, 0 < p < 1$
(a) p is known, θ is unknown
(b) p is unknown, θ is known
(c) both p and θ are unknown

(iv) $X \sim f(x;\theta)$, $f(x;\theta) = (2/\theta^2)(\theta - x)$, $\quad 0 < x < \theta, \theta > 0$

(v) $X \sim DE(\theta, \sigma)$ (Double exponential or displaced Laplace)

$$f(x;\theta,\sigma) = \frac{1}{2\sigma}\exp\left\{-\frac{|x-\theta|}{\sigma}\right\}$$

(vi) $X \sim$ Three-parameter family with *pdf*

$$f(x;\theta,\sigma,p) = \frac{1}{\Gamma(p)}\frac{1}{\sigma}\left(\frac{x-\theta}{\sigma}\right)^{p-1}\exp\left\{-\left(\frac{x-\theta}{\sigma}\right)\right\}; \quad \sigma > 0, \quad \theta < x < \infty$$

Hint. Find sufficient statistic for (*i*) p if θ and σ are known, σ if θ and p are known, (p, σ) if θ is known; and (ii) θ if σ is known and $p = 1$.

(vii) $X \sim LN(\theta, \sigma)$ (Lognormal)

$$f(x;\theta,\sigma) = \frac{1}{x\sigma\sqrt{2\pi}}\exp\left\{-\frac{1}{2\sigma^2}(\log x - \theta^2)\right\}, \quad x > 0, \sigma > 0$$

(viii) $X \sim N(\theta, \sigma^2)$ (Normal)

$$f(x;\theta,\sigma) = \frac{1}{\sigma\sqrt{2\pi}}\exp\left\{-\frac{1}{2\sigma^2}(x-\theta)^2\right\}, x \in \mathbb{R}^1, \sigma > 0$$

Show also that the necessary and sufficient condition for $\Sigma_{i=1}^n a_i X_i$ and $\Sigma_{i=1}^n X_i$ to be independent is $\Sigma_{i=1}^n a_i = 0$. Show that $(\bar{X}, \Sigma X_i^2)$ is independent of $(X_{(n)} - X_{(1)})/S$, $(X_{(n)} - \bar{X})/S$, and $\Sigma_{i=1}^{n-1}(X_{i+1} - X_i)^2/S^2$.

Hint. $\Sigma_{i=1}^n X_i$ is complete sufficient for θ if σ^2 is known. $\Sigma_{i=1}^n (X_i - \theta)^2$ is complete sufficient for σ^2 if θ is known. $(\Sigma X_i, \Sigma X_i^2)$ is minimal sufficient for the family $N(\theta, \sigma^2)$. $T(\mathbf{X}) = (\Sigma_{i=1}^n X_i, \Sigma_{i=1}^n X_i^2)$ or $T(\mathbf{X}) = (\sqrt{n}\ \bar{X}, \Sigma_{i=1}^n (X_i - \bar{X})^2)$ is jointly complete sufficient for (θ, σ^2), when both θ and σ^2 are unknown.

(ix) $X \sim N(\theta, 1)$. Show also that the statistic $T_1(\mathbf{X}) = X_1$ is complete for θ, so is $T_2(\mathbf{X}) = \Sigma X_i$. Determine the conditional distribution of $X_1 | \bar{X}$. For $n = 2$, if $T = X_1 + X_2$ and $U = X_2 - X_1$, show that the continuous distribution of U given $T = t$ does depend on θ. Interpret this result in the light of sufficiency concept. If $X \sim N(\theta, \sigma^2)$, σ^2 known, show that $(\Sigma X_i)^2$ is not sufficient for θ.

(x) $X \sim N(0, \sigma^2)$. If $n = 1$, show that $T(X) = X_1^2$ is minimal sufficient for σ^2.

(xi) Let $X_1,\ldots,X_n$ be a random sample from $N(0, \theta)$, prove that $T_1(\mathbf{X}) = X_1$ is not a complete statistic for θ but $T_2(\mathbf{X}) = X_1^2$ is complete for θ.

(xii) $X \sim N(0, \sigma^2)$. If X^2 is a sufficient statistic for σ^2, show that X is not complete.

(xiii) Is $|X|$ a sufficient statistic for σ^2?

(xiv) If X^2 is a sufficient statistic for σ^2, show that X^2 is complete for σ^2. If $n > 1$, find a sufficient statistic for σ^2. Show also that ΣX_i^2 is a minimal sufficient statistic for σ^2 but ΣX_i is not sufficient. For $X \sim N(\theta, \sigma^2)$, θ known, show that $\sqrt{(1/n)\Sigma_{i=1}^n (X_i - \theta)^2}$ is sufficient for σ.

(xv) Let $(X_1,\ldots,X_n)$ be a random sample from $N(\mu, \sigma^2)$, where μ and σ^2 are both unknown. Prove also that $\mathbf{T} = (T_1, T_2)$ is a sufficient statistic for $\theta = (\mu, \sigma^2)$, where $T_1 = \sqrt{n}\,\bar{X}$ and $T_2 = \Sigma_i (X_i - \bar{X})^2$.

(xvi) Let $(X_1,\ldots,X_n)$ be a random sample from $N(\theta, \theta^2)$. Find a minimal sufficient statistic for θ.

(xvii) $X \sim N(b\theta, \theta^2)$, b is a real known number. Show that the statistic $T(\mathbf{X}) = (\Sigma_{i=1}^n X_i, \Sigma_{i=1}^n X_i^2)$ is sufficient for θ but the family of distribution of $T(\mathbf{X})$ is not complete.

(xviii) Let $(X_{11},\ldots,X_{1n_1})$ and $(X_{21},\ldots,X_{2n_2})$ be random samples drawn from $N(\theta_1, \sigma_1^2)$ and $N(\theta_2, \sigma_2^2)$, respectively, of sizes n_1 and n_2. Show that the family of distributions of the statistic $T(\mathbf{X}) = (\mathbf{1}'\mathbf{X}_1, \mathbf{1}'\mathbf{X}_2, \mathbf{X}_1'\mathbf{X}_1, \mathbf{X}_2'\mathbf{X}_2)$ is complete.

(xix) Based on a bivariate sample $(X_1, Y_1),\ldots,(X_n, Y_n)$ of size n from

$$f((x,y);\mu,\Sigma) = \frac{1}{2\pi\sigma_1\sigma_2\sqrt{1-\rho^2}}\exp\left\{-\frac{1}{2(1-\rho^2)}\left[\left(\frac{x-\mu_1}{\sigma_1}\right)^2 -2\rho\left(\frac{x-\mu_1}{\sigma_1}\right)\left(\frac{y-\mu_2}{\sigma_2}\right)+\left(\frac{y-\mu_2}{\sigma_2}\right)^2\right]\right\}$$

for $(x, y) \in \mathbb{R}^2$, $\boldsymbol{\theta} = (\mu_1, \mu_2, \sigma_1, \sigma_2, \rho)$, find sufficient statistic for $\boldsymbol{\theta}$.

(xx) Find also sufficient statistics for ρ in the following distribution:

$$f(x, y; \rho) = \frac{1}{2\pi\sqrt{1-\rho^2}} \exp\left\{-\frac{1}{2(1-\rho^2)}(x^2 - 2\rho xy + y^2)\right\}$$

(xxi) Let $X_1,\ldots,X_n$ be a random sample from N_k ($\boldsymbol{\theta}$, Σ) with *pdf*

$$f(\mathbf{x}; \boldsymbol{\theta}, \Sigma) = \frac{1}{(2\pi)^{k/2}\sqrt{|\Sigma|}} \exp\left\{-\frac{1}{2}(\mathbf{x}-\boldsymbol{\theta})'\Sigma^{-1}(\mathbf{x}-\boldsymbol{\theta})\right\}$$

where $\mathbf{x} = (x_1,\ldots,x_n)$, $\boldsymbol{\theta} = (\theta_1,\ldots,\theta_k)'$, $\mathbf{x} \in \mathbb{R}^k$, $\boldsymbol{\theta} \in \mathbb{R}^k$. $\Sigma = ((\sigma_{ij}), i, j = 1,\ldots,k)$ is the variance–covariance matrix of multivariate normal distribution. Also, discuss the case when $\Sigma = \text{diag } (\sigma_{ii}, i = 1,\ldots,k)$

(xxii) $X \sim C(\mu, \theta)$ (Cauchy)

$$f(x; \mu, \theta) = \frac{\mu}{\pi} \frac{1}{\mu^2 + (x-\theta)^2}, x \in \mathbb{R}^1, \theta \in \mathbb{R}^1, \mu > 0$$

Show that no non-trivial sufficient statistic exists. Also, show that there does not exist a single sufficient statistic for either parameter if the other is known or both are unknown.

(xxiii) $X \sim L(\theta, \sigma)$ (Logistic)

$$f(x; \theta, \sigma) = \frac{1}{\sigma} \frac{\exp\{-(x-\theta)/\sigma\}}{\{1+\exp[-(x-\theta)/\sigma]\}^2}$$

14. Exponential family of distributions. For a random sample $(X_1,\ldots,X_n)$ from a regular exponential family of distributions, obtain sufficient, minimal sufficient, and complete sufficient statistics in each of the following cases, if they exist:

(i) A biased coin is tossed repeatedly until the first head appears. Let X be the number of tosses required. The probability function is

$$P(x; \theta) = \theta(1-\theta)^{x-1}, x = 1, 2, \ldots, \theta \in \Theta = [0, 1]$$

where θ is the true probability of head in a single toss.
Hint. X is complete.

(ii) $X \sim b(1, \theta)$ (Bernoulli population) with *pmf*

$$P(x; 1, \theta) = \theta^x (1-\theta)^{1-x}$$

Hint. Show that the conditional distribution of the sample observations $\mathbf{X}$ given $\mathbf{X}|T(\mathbf{X}) = r$ is independent of θ. Interpret this result in the light of sufficiency concept. $T(\mathbf{X}) = \Sigma_{i=1}^n X_i$ is a minimal sufficient and complete sufficient statistic for θ. Sometimes the same question may be asked to show that the family $\{P(x; \theta), \theta \in (0, 1)\}$ with *pmf* $P(x; \theta) = \binom{2}{x} \theta^x (1-\theta)^{2-x}$, $x = 0, 1, 2$, is complete.

(iii) Bernoulli X_1 and X_2 are *iid rvs* with $X_i \sim b(1, p_i)$, $0 < p_i < 1$, $i = 1, 2$.
Hint. (X_1, X_2) is sufficient for (p_1, p_2).

(iv) Let X_1 and X_2 be independent *rvs* from $b(n_i, p_i)$, $i = 1, 2$, respectively.
Hint. The family of distributions of the statistic $T(\mathbf{X}) = (X_1, X_2)$ is complete.

(v) Let $X_1 \sim b(1, \theta)$ and $X_2 \sim b(1, 5\theta)$, $\theta \in [0, 1/5]$.
Hint. Consider the statistics $T_1(X) = X_1 + X_2$ and $T_2(X) = X_1 + 3X_2$ and check for sufficiency.

(vi) Let (X_1, X_2) be a random sample from the Bernoulli distribution $b(1, \theta)$, $\theta \in [0, 1]$. Show that the statistics $T_1(\mathbf{X}) = X_1 + X_2$ and $T_2(\mathbf{X}) = X_1 + 5X_2$ are sufficient for θ. Also, answer which of the two statistics are preferred?

(vii) $X \sim NB(m, \theta)$ with *pmf* $P(x; \theta) = \binom{m+x-1}{m-1}\theta^m (1-\theta)^x$, $x = 0, 1, \ldots$
Hint. $T(\mathbf{X}) = \Sigma_{i=1}^n X_i$ is complete.

(viii) $X \sim$ Geometric(θ) with *pmf* $p(X = k) = \theta(1-\theta)^k$, $k = 0, 1, 2, \ldots$; $0 < \theta < 1$.

(ix) $f(x; \theta) = q\, p^{x-\theta}$, $q = 1-p$, $0 < p < 1$, $x = \theta, \theta + 1, \ldots, \theta > 0$

(x) $X \sim P(\theta)$

$$P(x;\theta) = \frac{\exp\{-\theta\}\theta^x}{x!}$$

(a) Show that for $n = 2$, the statistic $T(\mathbf{X}) = X_1 + \alpha X_2$, for $\alpha > 1$, an integer, is not sufficient.

(b) Is the family of distribution of $T(\mathbf{X}) = (X_{(1)}, \ldots, X_{(n)})$ complete or boundedly complete?

Hint. $T(\mathbf{X}) = \Sigma_{i=1}^n X_i$ is minimal and complete sufficient statistic for θ.

(xi) Let X_1, X_2 be *iid* with $X_i \sim P(\theta_i)$, $\theta_i > 0$, $i = 1, 2$. Show that $(X_{(1)}, X_{(2)})$ is sufficient for (θ_1, θ_2).

(xii) Let $X \sim P(\theta)$, $\theta \geq 0$. Show that no non-trivial sufficient statistic exists.

(xiii) $X \sim G_1(\alpha, \beta)$ with *pdf* $f(x; \alpha, \beta) = \{1/[\Gamma(\alpha)\beta^\alpha]\} \exp\{-x/\beta\}x^{\alpha-1}$, $0 < x < \infty$.
Hint. $\Sigma_{i=1}^n X_i$ is complete sufficient for β if α is known. $\Pi_{i=1}^n X_i$ is complete sufficient for α if β is known. $(\Sigma_{i=1}^n X_i, \Pi_{i=1}^n X_i)$ is complete sufficient for (α, β).

(xiv) $f(x;\theta) = (1/\theta)\exp(-x/\theta)$, $x > 0$, $\theta > 0$. Obtain sufficient statistic for θ.

(xv) $X \sim \chi_d^2$ [or $G_1(d/2, 2)$] with *pdf*

$$f(x;d) = \frac{1}{\Gamma(d/2)2^{d/2}}\exp\left\{-\frac{x}{2}\right\}x^{[(d/2)-1]}, \quad x > 0$$

(xvi) $X \sim Be(\alpha, \beta)$ with *pdf* $f(x; \alpha, \beta) = [\Gamma(\alpha + \beta)/\{\Gamma(\alpha)\,\Gamma(\beta)\}]\, x^{\alpha-1}(1-x)^{\beta-1}$, $0 < x < 1$.

(xvii) $X \sim f(x; \theta)$ with *pmf* $P(x; \theta) = C(\theta)\, 2^{-x/\theta}$, $x = \theta, \theta + 1, \ldots$; $\theta > 0$ and $C(\theta) = 2^{\{1-(1/\theta)\}}[2^{(1/\theta)} - 1]$.

(xviii) Let $(X_1, Y_1), (X_2, Y_2), \ldots, (X_n, Y_n)$ be a random sample from the distribution from $P(x, y; \theta_1, \theta_2)$ (Trinomial distribution)

$$P(x, y; \theta_1, \theta_2) = \frac{(x+y+z)!}{(x)!\,(y)!\,(z)!}\theta_1^x\,\theta_2^y(1-\theta_1-\theta_2)^z$$

where $x, y = 0, 1, \ldots$; $z = n - x - y = 1, 2, \ldots, \theta_1 + \theta_2 < 1$.

(xix) Family of distributions $\mathcal{F}$ with *pdf* $f(x; N)$

$$f(x; N) = (N+1)x^N, \ 0 \le x \le 1, N \ge 0$$

Hint. The family $\mathcal{F}$ is complete.

(xx) $f(x;\theta) = \theta x^{\theta-1}$, $0 < x < 1$, $\theta > 0$

(xxi) $f(x;\theta) = [1/(|\theta|\sqrt{2\pi})] \exp[-(x-\theta)^2/2\theta^2]$, $x \in \mathbb{R}^1$, $\theta \in \mathbb{R}^1 - \{0\}$

(xxii) $f(x;\theta) = (1+\theta)x^\theta$, $0 < x < 1$, $\theta > 0$

(xxiii) $f(x;\theta) = Q(\theta) \exp\{-(x/\theta)\}$, $x = \theta, \theta+1, \ldots, \theta > 0$

(xxiv) $f(x;\theta) = \exp\{-x^{2n} + \theta_1 x + \theta_2 x^2 + \cdots + \theta_n x^n\}$
where $\boldsymbol{\theta} = (\theta_1, \ldots, \theta_n)' \in \Theta \subseteq \mathbb{R}^n$.

Hint. $T(\mathbf{X}) = (X_{(1)}, \ldots, X_{(n)})$ is a sufficient statistics for θ.

15. Let $X_1, \ldots, X_n$ be *iid* $N(\theta, 1)$, $\theta = 1, 2$. Show that the family of distributions $\{N(\theta, 1), \theta = 1, 2\}$ is not complete.

16. Consider the estimation of σ^2 based on the *iid* $N(0, \sigma^2)$ *rvs* $X_1, X_2, \ldots, X_n$. Which of the following sufficient statistics provides the maximum summarization of data?

$T_1(\mathbf{X}) = (X_1, X_2, \ldots, X_n)$

$T_2(\mathbf{X}) = (X_1^2, X_2^2, \ldots, X_n^2)$

$T_3(\mathbf{X}) = (X_1^2, X_2^2, \ldots, X_m^2, X_{m+1}^2, X_{m+2}^2, \ldots, X_n^2)$

$T_4(\mathbf{X}) = (X_1^2 + X_2^2 + \cdots + X_n^2)$

(i) T_1 (ii) T_3 (iii) T_2 (iv) T_4

17. Koopman–Pitman family of distributions. Let $\mathcal{F} = \{f(x; \theta), \theta \in \Theta\}$ be a family of *pdf*s of an *rv* X so that

$$f(x; \theta) = \exp\{Q(\theta)T(x) + D(\theta) + S(x)\}$$

where $Q(\theta)$ and $D(\theta)$ are some real-valued functions on Θ and $T(x)$ and $S(x)$ are the Borel measurable functions on $\mathbb{R}^1$, $[\partial Q(\theta)/\partial\theta] \neq 0 \ \forall \ \theta \in \Theta$. The support $S = \{x: f(x;\theta) > 0\}$ is independent of θ and $\{T(x), 1\}$ are linearly independent over the support S. Find $E_\theta[T(X)]$ and $V_\theta[T(X)]$.

Hint. Once differentiate $\int_{\mathbb{R}^1} \exp\{Q(\theta)T(x) + D(\theta) + S(x)\}dx = 1$ with respect to θ to calculate $E_\theta[T(X)]$ and twice differentiate with respect to θ to calculate $V_\theta[T(X)]$.

18. Some other continuous distributions. Find a sufficient statistic, minimal sufficient statistic, complete sufficient statistic (if they exist) based on a random sample of size n in each of the following cases:

(i) The Weibull distribution is derived from gamma distribution by a simple transformation, $X = Y^{1/\theta}$, $Y \sim G_1(1, \sigma)$, where the probability density of $G_1(\alpha, \beta)$ is of the form $f(x; \alpha, \beta) = cx^{\alpha-1}\exp(-x/\beta)$

$$f(x; \theta, \sigma) = \left(\frac{\theta}{\sigma}\right) x^{\theta-1} \exp\left(-\frac{x^\theta}{\sigma}\right), x > 0$$

(ii) Pareto distribution

Form 1: $$f(x; \sigma, \alpha) = \frac{\alpha\sigma^\alpha}{(x+\sigma)^{\alpha+1}}, \ x > 0, \sigma > 0, \alpha > 0$$

Form 2: $f(x;\sigma,\alpha)=\dfrac{\alpha\sigma^{\alpha}}{x^{\alpha+1}},\ x>\sigma,\sigma>0,\alpha>0$

(Form 2 can easily be derived from 1 by a simple transformation. Here, σ is known as scale parameter and α is known as shape parameter. Note that $Y = \ln(X/\sigma)$ is an exponential distribution.)

(iii) $f(x;\theta)=\dfrac{1+\theta}{(x+\theta)^2},\ 1\le x\le\infty$

(iv) $f(x;\theta)=\dfrac{\theta}{(1+x)^{1+\theta}},\ 0<x<\infty,\theta>0$

Hint. $\Pi_{i=1}^{n}(1 + X_i)$ is minimal sufficient statistics for θ.

19. Uniform Distribution. Consider $(X_1,\ldots,X_n)$ to be a random sample of size n from
  (i) a family of distribution $\mathcal{F} = \{U(0, \theta);\ \theta > 0\}$. Show that the family is complete. Show that $T(\mathbf{X}) = X_{(n)}$ is a complete sufficient statistic for θ.
  (ii) a family $\mathcal{F} = \{U(-\theta, \theta); -\infty < \theta < \infty\}$, known as Pitman's family of distributions. Show that $(X_{(1)}, X_{(n)})$ and $X_{(n)}$ are both sufficient statistics for θ. Which would you prefer and why? Show that the family of distributions is not complete. Prove that the statistic $T(\mathbf{X}) = \max(-X_{(1)}, X_{(n)})$ is a sufficient statistic for θ. Find other sufficient statistics for θ.
  (iii) $U(\theta_1, \theta_2)$, $\theta_1 < \theta_2$. Is the family of distributions of $T(\mathbf{X}) = (X_{(n)} - X_{(1)})$ complete or boundedly complete?
  (iv) Consider a family of distribution $\mathcal{F} = \{U(\theta, \theta + 1): -\infty < \theta < \infty\}$. Show that $\mathcal{F}$ is not complete.
  (v) $U[\theta - (1/2), \theta + (1/2)]$, $\theta \in \mathbb{R}^1$. Show that the statistics $T(\mathbf{X}) = (X_{(1)}, X_{(n)})$ is sufficient for θ but not complete.
  (vi) $f(x;\theta) = (1/\theta)$, $k\theta \le x \le (k + 1)\theta$, where k is an integer. Is the complete sufficient statistic exists?
  (vii) $f(x;\theta) = Q(\theta)T(x)$, $0 < x < \theta$, $0 < \theta < \infty$, where $T(x)$ is an increasing function of θ. Show that the family of *pdf* of $X_{(n)}$ is complete.

20. Let $(X_1,\ldots,X_n)$ be a random sample of size n from $U(\theta, \theta + 1)$, $-\infty < \theta < \infty$. Then show that the statistic $\mathbf{X} = (X_{(1)}, X_{(n)})$ is
  (i) minimal but not complete
  (ii) minimal and complete,
  (iii) complete but not minimal, and
  (iv) neither minimal nor complete.

21. Let $X_1, X_2,\ldots,X_n$ be a random sample of size n ($n \ge 4$) from $U(0, \theta)$ distribution. Which of the following statistics is not an ancillary statistic?
  (i) $X_{(n)}/X_{(1)}$,
  (ii) X_n/X_1,
  (iii) $(X_4 - X_1)/(X_3 - X_2)$,
  (iv) $X_{(n)} - X_{(1)}$

22. Suppose $X_1 \sim U(0, \theta)$ and $X_2 \sim U(0, 1 + \theta)$, and X_1, and X_2 are independent. Then
  (i) $X_{(1)}$ is sufficient for θ.

(ii) $X_{(2)}$ is sufficient for θ.
(iii) $\max\{X_1, X_2 - 1\}$ is sufficient for θ.
(iv) $\max\{X_1 + 1, X_2\}$ is sufficient for θ.

23. Suppose $X_1, X_2, \ldots, X_n$ are *iid* with density function

$$f(x; \theta) = \frac{\theta}{x^2}; \theta < x, \theta > 0$$

Then (i) $\Sigma_{i=1}^n 1/X_i^2$ is sufficient for θ, (ii) $\min_{(1 \le i \le n)} X_i$ is sufficient for θ, (iii) $\Pi_{i=1}^n 1/X_i^2$ is sufficient for θ, (iv) $(\max_{(1 \le i \le n)} X_i, \min_{(1 \le i \le n)} X_i)$ is not sufficient for θ, (v) $\max_{(1 \le i \le n)} X_i$ is the maximum likelihood estimator of θ, and (vi) $\min_{(1 \le i \le n)} X_i$ is the maximum likelihood estimator of θ.

24. $X_1, X_2, \ldots, X_n$ are n random observations from the population with density function

$$f(x; \theta) = \frac{\theta}{x^2} \exp\left(-\frac{\theta}{x}\right); \ \theta > 0, x > 0$$

Which of the following is sufficient for the above family of distributions?
(i) $\bar{X}$
(ii) Geometric mean of the observations
(iii) Harmonic mean of the observations
(iv) No sufficient statistic exists

25. $X_1, X_2, \ldots, X_n$ are n random samples from a distribution with the *pdf*

$$f(x; \sigma) = \frac{1}{2\sigma} \exp\left(-\frac{|x-3|}{\sigma}\right); \ \sigma > 0$$

For the above family, which one of the following is correct?
(i) $\Sigma_{i=1}^n |X_i - 3|$ is complete sufficient statistic for σ.
(ii) $n^{-1} \Sigma X_i$ is an unbiased estimator of σ.
(iii) $f(x;\sigma)$ does not belong to one parameter exponential family.
(iv) MLE of σ does not exist.

26. Let $X_1, X_2, \ldots, X_n$ be independent *rvs* with common probability distribution function

$$P(X \le x; \alpha, \beta) = \begin{cases} 0, & \text{if } x < 0 \\ (x/\beta)^\alpha, & \text{if } 0 < x \le \beta \\ 1, & \text{if } x \ge \beta \end{cases}$$

where $\alpha, \beta > 0$.
(i) Find a two-dimensional sufficient statistic for (α, β).
(ii) Find an unbiased estimator of $(1 + \alpha)^{-1}$ when $\beta = 1$.

27. Let $X_1, X_2, \ldots, X_n$ be *iid* from $G_1(\alpha, \beta)$. Consider $U = \Sigma_1^n X_i/n$ and $V = (\Pi_{i=1}^n X_i)^{1/n}$. U/V is ancillary for β; U and U/V are independent; and $E(V/U)^k = E(V^k)/E(U^k)$. Show.

3

Unbiased Estimation

3.1 INTRODUCTION

We have formulated the problem of estimating parameter θ by an estimator δ in Chapter 2, where the problem simplified through the data reduction principle known as principle of sufficiency. Having achieved this simplification, we will now discuss in the present chapter as to how one should choose between estimators to get an optimal estimator on the basis of their risks. Generally, we mean that an optimal estimator is the one that minimizes the risk (expected loss) at every value of θ, i.e., uniformly in θ. The (risk of an estimator) δ is defined by

$$R(\theta, \delta) = E_\theta L(\theta, \delta(\mathbf{X})) \tag{3.1.1}$$

as a function of θ. The loss function $L(\theta, \delta)$ is convex in δ and satisfies $L(\theta, \delta) \geq 0 \ \forall \ \theta, \delta$ and $L(\theta, \theta) = 0 \ \forall \ \theta$. In terms of risk, an estimator δ is treated as the most preferred over all other estimators of θ, which minimizes the risk $R(\theta,\delta)$ over all θ, among all δ. There can be several other criteria for defining an optimal estimator, but in this chapter, we consider the *risk* of an estimator as the criterion of judging between the estimators. We will see in the following arguments that a minimum risk estimator does not exist in the class of all estimators of θ, since the risk depends on the unknown parameter θ.

In the present section, we will discuss the impartiality principle of estimators which is known as unbiasedness and that simplifies the estimation problem to a great extent. The risk of an estimator is taken as a criterion of judging between estimators. In Section 3.2, we discuss three methods of finding uniformly minimum variance unbiased estimators (UMVUE). In Section 3.3, we discuss the role of complete sufficient statistic in minimum variance unbiased estimation, and the result is based on the famous Lehmann–Scheffe theorem.

Given any estimator δ of θ, there exist estimators $\delta'(x) = \theta \ \forall x$ for each fixed θ, so that $L(\theta, \delta'(x)) = 0 \ \forall x \Rightarrow R(\theta, \delta') = 0$. In other words, at every point θ, there exists an estimator δ' so that δ is not superior to δ'. This shows that in the class of all estimators of θ, denoted by

$C(\theta)$, there does not exist any estimator δ that can have minimum risk over all θ. The estimators in $C(\theta)$ of the type $\delta'(x) = \theta_0$, for some fixed θ_0, are the ones which favour some value of θ and neglect all other values. Such estimators are not reasonable estimators and need to be excluded from the class of all estimators. Note that the systematic errors of these estimators are, say, for $\theta = \theta_0$, $E_\theta[\delta'(\mathbf{X}) \equiv \theta_0] - \theta = \theta_0 - \theta \neq 0 \; \forall \; \theta \neq \theta_0$. Therefore, such estimators must be eliminated from $C(\theta)$ to get a reduced class of estimators by imposing a condition that ensures zero-systematic-errors (also called bias), i.e. $E_\theta \delta(\mathbf{X}) = \theta \; \forall \; \theta$. This restriction gives rise to an impartiality principle known as *unbiasedness* which relates to a parameter but that does not depend on it.

Definition 3.1.1 The bias of an estimator δ of θ is defined as

$$\text{bias}_\theta \, \delta(X) = E_\theta \delta(\mathbf{X}) - \theta \tag{3.1.2}$$

Definition 3.1.2 An estimator δ is said to be *unbiased* for θ if its bias is equal to zero identically in θ, i.e.,

$$\text{bias}_\theta \, \delta(\mathbf{X}) = 0 \; \forall \; \theta$$

or

$$E_\theta[\delta(\mathbf{X})] = \theta \; \forall \; \theta \tag{3.1.3}$$

Definition 3.1.3 A class of unbiased estimators of θ is denoted by

$$\mathcal{U}(\theta) = \{\delta\text{: } E_\theta \delta(\mathbf{X}) = \theta \; \forall \; \theta\} \tag{3.1.4}$$

Definition 3.1.4 An estimator $U(X)$ is said to be a *zero estimator* if $E_\theta U(X) = 0 \; \forall \; \theta \in \Theta$; and the class of zero estimators is denoted by $\mathcal{U}(0)$.

$$\mathcal{U}(0) = \{U\text{: } E_\theta U = 0 \quad \forall \; \theta \quad \text{and} \quad E_\theta U^2 < \infty, \forall \; \theta\} \tag{3.1.5}$$

Here, Θ is the parameter space.

Definition 3.1.5 If the loss function is squared error, i.e., $L(\theta, \delta) = (\delta - \theta)^2$, then the mean squared error of an estimator is defined as

$$\begin{aligned}\text{MSE}_\theta(\delta) &= E_\theta(\delta - \theta)^2 = V_\theta(\delta) + \{\text{bias}_\theta(\delta)\}^2 \\ &= \begin{pmatrix}\text{Variability of} \\ \text{the estimator}\end{pmatrix} + \begin{pmatrix}\text{Bias in the} \\ \text{estimator}\end{pmatrix}^2 \\ &= \begin{pmatrix}\text{Precision of the} \\ \text{estimator } \delta\end{pmatrix} + \begin{pmatrix}\text{Accuracy of the} \\ \text{estimator } \delta\end{pmatrix}^2\end{aligned} \tag{3.1.6}$$

An estimator is preferred over others if its MSE is small as compared to that of others, which is achieved by small variance and small bias both together. Controlling over bias does not necessarily result into low MSE. Sometimes, bearing a small amount of bias combined with substantial decrease in variance finally results into a high decrease in MSE. Small MSE of an estimator results into high probability that the estimator is too close to true θ by Chebyshev's inequality $P_\theta(|\delta - \theta| > \varepsilon) \leq (1/\varepsilon^2) \; E_\theta(\delta - \theta)^2 = (1/\varepsilon^2)\text{MSE}_\theta(\delta)$.

Frequently, the MSEs of two estimators, as a function of θ, cross each other, i.e., on one portion of the parameter space, one estimator behaves better and the situation reverses on other portions. This causes a situation where there does not exist a best MSE estimator. This situation arises since the class of estimators is too large a class to search for the best MSE estimator. However, in many situations, a best estimator exists when we restrict our search only to unbiased estimators. Hence, in the present section, we will consider only those estimators that control their bias so that it equals zero and attain maximum reduction in MSE, which is equal to variance in such cases.

In statistical inference, we make an inference on the population parameter θ or equivalently characterize such a population F_θ in $\mathcal{F}$ that could likely have generated the sample **X**. Unbiasedness is a property that holds uniformly for every member F_θ of the family $\mathcal{F}$. Reducing the class of all estimators $C(\theta)$ into a class of all unbiased estimators $\mathcal{U}(\theta)$, in fact, does not guarantee the existence of a uniformly minimum risk estimator. It is surely a better situation than of dealing with $C(\theta)$, where uniformly minimum risk estimator does not exist at all. It is important to mention, at this stage, that if one can bear some bias, then there can exist a biased estimator with much smaller risk as compared to a minimum risk unbiased estimator. Further, in this chapter, we will discuss the possibilities of getting a uniformly minimum risk estimator in $\mathcal{U}(\theta)$ for convex loss functions and particularly uniformly minimum variance estimator for squared error loss functions.

Next, we define locally and uniformly minimum variance unbiased estimators of θ for squared error loss functions, $L(\theta, \delta) = (\delta - \theta)^2$.

Definition 3.1.6 An estimator δ_0 is said to be *locally minimum variance unbiased estimator* (LMVUE) of θ_0 in the class of unbiased estimator of θ_0, $\mathcal{U}(\theta_0) = \{\delta : E_{\theta_0}\delta(X) = \theta_0, E_{\theta_0}\delta^2(X) < \infty\}$, if

$$R(\theta_0, \delta_0) \le R(\theta_0, \delta)$$

or, equivalently, $$V_{\theta_0}(\delta_0) \le V_{\theta_0}(\delta) \ \forall \ \delta \in \mathcal{U}(\theta_0)$$

Definition 3.1.7 An estimator δ_0 is said to be *uniformly minimum variance unbiased estimator* (UMVUE) of θ in the class of all unbiased estimators of θ, $\mathcal{U}(\theta) = \{\delta : E_\theta[\delta(X)] = \theta \ \forall \theta$ and $E_\theta\delta^2 < \infty\}$, if

$$V_\theta(\delta_0) \le V_\theta(\delta) \quad \forall \ \theta \text{ and } \forall \ \delta \in \mathcal{U}(\theta)$$

Note that, here, *uniformly* means that the minimum variance property of the estimator δ_0 is independent of the parameter θ.

Definition 3.1.8 (Efficiency of an unbiased estimator) Let δ_0 be the UMVUE of θ in $\mathcal{U}(\theta)$. The (efficiency of an estimator) δ in $\mathcal{U}(\theta)$ is defined as

$$e_\theta(\delta) = \frac{V_\theta(\delta_0)}{V_\theta(\delta)} \tag{3.1.7}$$

Note that $0 < e_\theta \le 1$. If $e_\theta = 1$, δ is UMVUE; δ is also called "most efficient."

3.2 UMVU ESTIMATION

Any estimator δ in $\mathcal{U}(\theta)$ can be improved by Rao–Blackwell theorem according to which, if a sufficient statistic T exists and the improved estimator $\phi(T) = E(\delta|T)$ too belongs to $\mathcal{U}(\theta)$, then the variance of $\phi(T)$ will be smaller than the variance of δ uniformly in θ. Therefore, the search for a best unbiased estimator of θ in $\mathcal{U}(\theta)$ may be restricted to the class of all unbiased estimators which are the functions of a sufficient statistic. Denote this restricted class by $\mathcal{U}(\theta|T)$. However, there is no certainty of existence of an UMVUE in $\mathcal{U}(\theta)$. We state few commonly used methods of finding an UMVUE, if it exists.

***Method* 1.** This method defines a lower bound on the variance of an unbiased estimator of θ, known as the Cramer–Rao lower bound. This lower bound holds under certain regularity conditions on the underlying *pmfs* or *pdfs* as is the *rv* discrete or continuous. If the variance of an unbiased estimator in $\mathcal{U}(\theta)$ attains this lower bound, such an estimator is known as UMVUE. Detailed discussion of this method is given in Chapter 4 where we call such estimators as minimum variance bound (MVB) estimators; such estimators are also called efficient estimators [$V(T)$ = CRLB, T being UMVUE]. One drawback of this method is that it does not give an UMVUE when CR variance lower bound is not sharp, that is, CR bound is smaller than uniformly minimum variance (UMVUE $\nRightarrow$ efficient estimator, efficient estimator $\Rightarrow$ UMVUE).

***Method* 2.** This method is based on zero estimators. Defining the class of unbiased estimators in terms of some unbiased estimator δ' and zero estimators by

$$\mathcal{U}(\theta) = \{\delta\text{: } \delta(X) = \delta'(X) - U(X)\ \forall\ U \in \mathcal{U}(0) \text{ and } \delta' \text{ is some unbiased estimator}\} \tag{3.2.1}$$

$$V_\theta[\delta(X)] = V_\theta[\delta'(X) - U(X)] = E_\theta(\delta' - U)^2 - \theta^2 \tag{3.2.2}$$

where $\mathcal{U}(0)$ is the class of zero estimators and $\mathcal{U}(\theta)$ is the class of all unbiased estimators of θ. This shows that the problem of minimization of variance $V_\theta\,(\cdot)$, with respect to δ in $U(\theta)$, translates into minimization of $E_\theta(\delta' - U)^2$ with respect to U in $\mathcal{U}(0)$. So, the UMVUE of θ is given by $\delta_0 = \delta' - U_0$, where U_0 is a zero estimator that minimizes $E_\theta(\delta' - U)^2$ for every θ. In fact, such an UMVUE is characterized by δ_0 in $\mathcal{U}(\theta)$ when it is uncorrelated with every zero estimator uniformly in θ. This result is stated in Theorem 3.2.1, that gives a necessary and sufficient condition for UMVUE to exist.

The method suffers from some drawbacks. Firstly, one needs to guess for a probable candidate δ_0 in $\mathcal{U}(\theta)$ that could be UMVUE, and secondly, it is difficult to characterize all zero estimators. However, one advantage is that the method does not require any condition on the *pdfs* and they can be general in nature. The price that we pay for this generality is in the difficulty which arises in verifying the existence of a best unbiased estimator. Sometimes, this method is helpful in showing that a particular unbiased estimator is not the best.

***Method* 3.** This method is based on the specification of the *pdfs* so that the class of zero estimators is empty, i.e., $\mathcal{U}(0) = \{\phi\}$ or characterizing such *pdfs* so that there does not exist any unbiased estimator of zero other than zero itself. Then, for any estimator δ_0 in $\mathcal{U}(\theta)$,

$$\text{cov}_\theta\,(\delta_0, 0) = 0 \qquad \forall\ \theta \in \Theta \tag{3.2.3}$$

which by method 2 implies that δ_0 is UMVUE. This situation is met when a complete sufficient statistic exists, where δ_0 is the function of the complete sufficient statistic, and it is an unbiased estimator of θ. Note that a UMVUE may not be an efficient estimator.

The procedures for this method are as follows:

Procedure 1: If T is a complete sufficient statistic, then estimator obtained by solving the set of equations

$$E_\theta[\delta(T)] = \theta \qquad \forall\, \theta \in \Theta \tag{3.2.4}$$

is UMVUE of θ. The solution to this set of equations is unique by Theorem 3.3.1.

Procedure 2: If T is a complete sufficient statistic and $\delta(X)$ is any estimator in $\mathcal{U}(\theta)$, then the conditional expectation of $\delta(X)$ given T, i.e., $\delta(T) = E_\theta[\delta(X)|T]$ is UMVUE of θ. One gets the same $\delta(T)$ as UMVUE irrespective of the choice of $\delta(X)$ from $\mathcal{U}(\theta)$, as mentioned in Theorem 3.3.1. Therefore, one should choose such a $\delta(X)$ in $\mathcal{U}(\theta)$ for which the computation of the conditional expectation simplifies.

The following theorem provides with the necessary and sufficient condition for anunbiased estimator δ to be UMVUE of θ. The theorem, by characterizing all unbiased estimators of zero, provides to check whether an given unbiased estimator δ of θ is uncorrelated with all zero–estimators; if so, δ is UMVUE, otherwise not. If no unbiased estimator of zero exists, then clearly δ is UMVUE. This situation relates to the concept of completeness.

Theorem 3.2.1 Let $\mathcal{U}(\theta)$ and $\mathcal{U}(0)$ be the class of all unbiased estimators of θ and 0, respectively. A necessary and sufficient condition for an estimator δ_0 to be UMVUE is that δ_0 is uncorrelated with every zero estimator U in $\mathcal{U}(0)$, or

$$\text{cov}_\theta(\delta_0, U) = 0 \quad \forall\, U \in \mathcal{U}(0),\ \forall\, \theta \tag{3.2.5}$$

Proof (*Necessary condition*). Given that δ_0 is UMVUE of θ in $\mathcal{U}(\theta)$, fix $U = U_0 \in \mathcal{U}(0)$ and $\theta = \theta_0 \in \Theta$. The estimator $\delta' = \delta_0 + \alpha U_0$ belongs to $\mathcal{U}(\theta)$ for all real α. Clearly, $V_{\theta_0}(\delta') \ge V_{\theta_0}(\delta_0)$ since δ_0 is UMVUE. This gives

$$V_{\theta_0}(\delta_0 + \alpha U_0) - V_{\theta_0}(\delta_0) \ge 0$$

$$V_{\theta_0}(\delta_0) + \alpha^2 V_{\theta_0}(U_0) + 2\alpha\,\text{cov}_{\theta_0}(\delta_0, U_0) - V_{\theta_0}(\delta_0) \ge 0$$

$$\phi(\alpha) = \alpha^2\, V_{\theta_0}(U_0) + 2\alpha\, \text{cov}_{\theta_0}(\delta_0, U_0) \ge 0$$

$\phi(\alpha) = 0$ gives two real roots

$$\alpha = 0 \text{ and } \alpha = -\frac{2\,\text{cov}_{\theta_0}(\delta_0, U_0)}{V_{\theta_0}(U_0)}$$

If α is chosen negative, it gives $\phi(\alpha) < 0$, which is a contradiction to the fact that $\phi(\alpha) \ge 0$. Therefore, α will be 0, which implies $\text{cov}_{\theta_0}(\delta_0, U_0) = 0$. Since U_0 and θ_0 were chosen arbitrarily, it proves $\text{cov}_\theta(\delta_0, U) = 0\ \forall\, \theta \in \Theta$ and $\forall\, U \in \mathcal{U}(0)$.

(*Sufficient condition*). Suppose $E_\theta(\delta_0 U) = 0\ \forall\, \theta,\ \forall\, U \in \mathcal{U}(0)$, holds for some δ_0 in $\mathcal{U}(\theta)$. Consider another estimator δ' from $\mathcal{U}(\theta)$. Then clearly, $(\delta_0 - \delta')$ is a zero estimator, $(\delta_0 - \delta') \in \mathcal{U}(0)$. This gives

$$E_\theta[\delta_0(\delta_0 - \delta')] = 0$$

or

$$E_\theta(\delta_0^2) = E_\theta(\delta_0\, \delta')$$

or

$$V_\theta(\delta_0) = \text{cov}_\theta(\delta_0, \delta') \qquad (\text{since } \delta_0, \delta' \in \mathcal{U}(\theta))$$

By Cauchy–Schwartz inequality $\text{cov}_\theta^2(\delta_0, \delta') \le V_\theta(\delta_0)\cdot V_\theta(\delta')$, we get

$$V_\theta(\delta_0) \le \sqrt{V_\theta(\delta_0)}\cdot\sqrt{V_\theta(\delta')}$$

$$V_\theta(\delta_0) \le V_\theta(\delta')\ \forall\ \theta$$

This proves that δ_0 is UMVUE of θ, since δ' was chosen arbitrary.

Note. One important remark on this theorem is that if one can identify a θ_0 from Θ and a zero estimator U from $\mathcal{U}(0)$ so that $\text{cov}_{\theta_0}(\delta_0, U) \ne 0$, then δ_0 cannot be the UMVUE of θ. This is so because one can construct an estimator $\delta_\alpha = \delta + \alpha U$ by choosing some α in the interval $(-2\text{cov}_{\theta_0}(\delta_0, U)/V_{\theta_0}(U), 0)$ so that $V_{\theta_0}(\delta_\alpha) \le V_{\theta_0}(\delta_0)$.

We will now show that if UMVUE exists, it is unique.

Theorem 3.2.2 If UMVUE exists, it is unique in $\mathcal{U}(\theta)$.

Proof. Let δ_1 and δ_2 be two UMVU estimators of θ in $\mathcal{U}(\theta)$. Since $\delta_1 - \delta_2 \in \mathcal{U}(0)$, we have

$$E_\theta[\delta_1(\delta_1 - \delta_2)] = 0\ \forall\ \theta$$

(since δ_1 and δ_2 are unbiased estimators of θ)

$$\text{cov}_\theta(\delta_1, \delta_2) = V_\theta(\delta_1)$$

$$\text{cov}_\theta(\delta_1, \delta_2) = \sqrt{V_\theta(\delta_1)}\ \sqrt{V_\theta(\delta_2)}$$

(since δ_1 and δ_2 are UMVUE of θ so that $V_\theta(\delta_1) = V_\theta(\delta_2)$)

The above condition is the equality condition of Cauchy–Schwartz inequality, which implies that δ_1 and δ_2 are correlated with probability one for all θ, i.e., $\delta_2 = a\delta_1 + b \Rightarrow a = 1, b = 0$, as δ_1 and δ_2 are unbiased estimators of $\theta \Rightarrow \delta_1 = \delta_2$ with probability one for all θ. Therefore, a UMVUE is unique if it exists.

We will now show that UMVUE is a symmetric function of sample observation $X_1,\ldots,X_n$ whenever they are *iid* from F_θ, $\theta \in \Theta$.

Theorem 3.2.3 Let $X_1,\ldots,X_n$ be *iid* from F_θ, $\theta \in \Theta$. The UMVUE of θ, if it exists, must necessarily be a symmetric function of observations $X_1,\ldots,X_n$.

Proof. An estimator δ is said to be a symmetric function of $X_1, X_2,\ldots,X_n$ if $\delta(X_{i_1}, X_{i_2},\ldots,X_{i_n}) = \delta(X_1, X_2,\ldots,X_n)$ for all $n!$ permutations $X_{i_1}, X_{i_2},\ldots,X_{i_n}$. Let $\delta(X_1, X_2,\ldots,X_n)$ be an unbiased estimator, not necessarily symmetric, of θ with $V_\theta[\delta(\mathbf{X})] = \sigma^2$. We denote the estimator δ by δ_π, based on the permutation statistic $(X_{i_1}, X_{i_2},\ldots,X_{i_n})$, where $(i_1,\ldots,i_n)$ is some permutation of $(1, 2,\ldots,n)$. Note, that $E_\theta(\delta_\pi) = \theta$, $V_\theta(\delta_\pi) = \sigma^2\ \forall\ \theta$ since the joint distributions of $X_1, X_2,\ldots,X_n$ and $X_{i_1}, X_{i_2},\ldots,X_{i_n}$ are same, as $X_1, X_2,\ldots,X_n$ are *iids* from $F_\theta, \theta \in \Theta$. Let us define an estimator

$$\delta' = \frac{1}{n!}\sum_\pi \delta_\pi$$

where Σ_π is the sum over all permutations π of $(1,2,\ldots,n)$. We have

$$E(\delta') = \frac{1}{n!}\sum_\pi E(\delta_\pi) = \frac{1}{n!}\sum_\pi \theta = \theta$$

This implies $\delta' \in \mathcal{U}(\theta)$. Further,

$$V_\theta(\delta') = \frac{1}{(n!)^2}\left[\sum_\pi V(\delta_\pi) + \sum_{\pi \neq \pi'}\sum \text{cov}(\delta_\pi, \delta_{\pi'})\right]$$

$$= \frac{1}{(n!)^2}\left[\sum_\pi V(\delta_\pi) + \sum_{\pi \neq \pi'}\sum r(\delta_\pi, \delta_{\pi'})\right]\sqrt{V(\delta_\pi)}\sqrt{V(\delta_{\pi'})}$$

(since $\mathbf{X}_\pi$ and $\mathbf{X}_{\pi'}$ have the same distribution)

$$\leq \frac{1}{(n!)^2}[n!\sigma^2 + n!(n!-1)\sigma^2] \quad (\text{since } \rho \leq 1)$$

$$= \sigma^2 = V_\theta(\delta) \ \forall \ \theta$$

This shows that corresponding to any estimator δ in $\mathcal{U}(\theta)$, there exists an estimator δ' in $\mathcal{U}(\theta)$, which is a symmetric function of $X_1,\ldots,X_n$ and δ' is superior to δ. Hence, the theorem is proved.

3.3 SUFFICIENCY AND UMVU ESTIMATION

We have already discussed sufficiency as a data reduction principle in Chapter 2 and have shown, for convex loss functions, that any estimator $U(\mathbf{X})$ of θ can be improved by considering an estimator $E_\theta[U(\mathbf{X})|T]$, assuming the existence of a sufficient statistic T for the family $\{F_\theta, \theta \in \Theta\}$. Note that this result holds even for those estimators which are not unbiased and stated in a form of a theorem, popularly known as Rao and Blackwell theorem. Since squared error loss function is convex in T, therefore this theorem holds for the setup of this chapter too. The relevance of the Rao and Blackwell theorem in unbiased estimation is that it restricts $\mathcal{U}(\theta)$ to a class of all such unbiased estimators which are functions of a sufficient statistic T. This restricted class of estimators is denoted by $\mathcal{U}(\theta|T)$. In fact, this results into a great reduction of the estimation problem, since $\mathcal{U}(\theta|T)$ is a much smaller class, and search for a UMVUE becomes simple.

Suppose, now, that there is only one estimator which is a function of sufficient statistic T, $E[U(\mathbf{X})|T]$, i.e., $\mathcal{U}(\theta|T)$ has only one member. Clearly, in this case, $E(U(\mathbf{X})|T)$ is the UMVUE. However, this fortunate situation arises only when the family of distributions $\mathcal{F} = \{F_\theta, \theta \in \Theta\}$ is complete or the statistic T is sufficient and complete. The Lehmann and Scheffe theorem summarizes this result.

Theorem 3.3.1 (Lehmann–Scheffe Theorem) Let T be a complete sufficient statistic for the family $\mathcal{F}$. If there exists an unbiased estimator of θ, there is one and only one unbiased estimator as function of T, $E[U(\mathbf{X})|T]$, in $\mathcal{U}(\theta|T)$. Clearly, $E[U(\mathbf{X})|T]$ is UMVUE of θ.

Proof. Let $U_1, U_2 \in \mathcal{U}(\theta)$. Then $E(U_1|T)$ and $E(U_2|T) \in \mathcal{U}(\theta|T)$. By completeness of T,

$$E_\theta^T\,[E(U_1|T)–E(U_2|T)] = 0 \ \forall \ \theta$$

implies $E[(U_1|T) - E(U_2|T)] = 0$ a.e. T. Note that $E(U_1|T) - E(U_2|T)$ is a function of T. Therefore, given an unbiased estimator U of θ in $\mathcal{U}(\theta)$ and on the application of Rao–Blackwell theorem, $E(U|T)$ is a unique UMVUE of θ in $\mathcal{U}(\theta|T)$.

UMVUE exists for exponential families, as stated in the following theorem.

Theorem 3.3.2 If $\mathcal{F}$ is an exponential family with full rank (that all parameters are distinct), then **T** is a unique UMVUE of $\boldsymbol{\theta}$.

Proof. Easily follows from the Lehmann–Scheffe theorem.

We will now discuss the relationship between a complete sufficient statistic and a minimal sufficient statistic by using the Rao–Blackwell theorem and the Lehmann–Scheffe theorem.

Theorem 3.3.3 If a complete sufficient statistic exists, it is also minimal sufficient.

The proof of this theorem is already given in Chapter 2. However, sufficiency does not imply completeness.

We will now illustrate the UMVU estimation for different families of distributions, both continuous and discrete.

3.4 SOLVED EXAMPLES

***Example* 3.1 (Non parametric estimation).** Let $X_1, X_2,\ldots,X_n$ be a random sample from an unknown population distribution function F, where F belongs to a family of distributions $\mathcal{F} = \{F_\theta : F_\theta$ is some continuous distribution function$\}$. It is shown in Example 2.11 that the order statistics $T(\mathbf{X}) = (X_{(1)}, X_{(2)},\ldots,X_{(n)})$ is sufficient for F. Show that the sample distribution function S_n based on $T(\mathbf{X})$ is unbiased for F_θ.

Solution. The empirical distribution function is defined as

$$S_n(x) = \begin{cases} 0, & X_{(1)} < x \\ m/n, & X_{(m)} \le x < X_{(m+1)} \\ 1, & X_{(n)} \ge x \end{cases}$$

On taking expectation of S_n,

$$E_F(S_n) = \sum_{i=1}^{n} \frac{i}{n} P(X_{(i)} \le x < X_{(i+1)})$$

$$= \sum_{i=1}^{n} \frac{i}{n}[P(X_{(i)} \le x) - P(X_{(i+1)} \le x)] \qquad [\text{since } P[X_{(r)} < a < X_{(s)}] = P[X_{(r)} < a] - P[X_{(s)} < a]$$

$$= \sum_{i=1}^{n} \frac{i}{n}\binom{n}{i}[F(x)]^i[1-F(x)]^{n-i}$$

$$= \left(\sum_{i=1}^{n}\binom{n-1}{i-1}\{[F(x)]^{i-1}[1-F(x)]^{n-i}\}\right)F(x)$$

$$= [F(x) + 1 - F(x)]^{n-1}F(x) = F(x)$$

Proceeding similarly,

$$E_F(S_n^2) = \sum \left(\frac{i}{n}\right)^2 \binom{n}{i} [F(x)]^i [1-F(x)]^{n-i}$$

$$= \frac{1}{n}\left\{\sum i \binom{n-1}{i-1} [F(x)]^{i-1} [1-F(x)]^{n-i}\right\} F(x)$$

$$= \frac{1}{n}[(n-1)F(x)+1]F(x)$$

$$= [F(x)]^2 + \frac{F(x)}{n}[1-F(x)]$$

$$\therefore \quad E_F(nS_n) = nF(x) \quad \text{and} \quad V_F(nS_n) = nF(x)[1-F(x)]$$

That is, $nS_n \sim b(n, F(x))$ where $F(x) = P(X \le x)$. Further, for large n, the sequence $\{[S_n - E_F(S_n)]/\sqrt{V_F(S_n)}\} = \{[\sqrt{n}(S_n - F(X))]/[\sqrt{F(X)(1-F(X))}]\}$, by the central limit theorem, converges to $N(0, 1)$ in distribution $\forall\, X$. These results also hold even when F is discrete.

***Example* 3.2** Let $X_1, X_2, \ldots, X_n$ be an independent sample drawn from (i) $f(x;\, \theta, \sigma^2)$ so that $E(X_i) = \theta\ \forall\, i$ (common mean θ) and $V(X_i) = \sigma^2\ \forall\, i$ (common variance σ^2). Show that $T(\mathbf{X}) = \bar{X}$ is the best linear unbiased estimator (BLUE) of θ, (ii) $f(x;\, \theta, \sigma^2)$ so that $E(X_i) = \theta\ \forall\, i$ and $V(X_i) = \sigma_i^2\ \forall\, i$ (X_i's are not identically distributed). Show that $T = \Sigma_1^n(X_i/\sigma_i^2)/\Sigma_1^n(1/\sigma_i^2)$ is the BLUE of θ with minimum variance $1/[\Sigma_1^n(1/\sigma_i^2)]$, the harmonic mean of $\sigma_1^2, \ldots, \sigma_n^2$.

Solution.

(i) A linear estimator of θ is given by

$$T(\mathbf{X}) = \Sigma_1^n a_i X_i$$

for some real constants a_i from $\mathbb{R}^1$. For T to be unbiased, $\Sigma_1^n a_i$ must be equal to 1. Minimum variance estimator T is determined by choosing a_i such that $V(T)$ is minimum while T remains unbiased ($\Sigma_1^n a_i = 1$)

$$V(T) = \sigma^2 \Sigma_1^n a_i^2$$

$$= \sigma^2[a_1^2 + \cdots + a_i^2 + \cdots + (1 - a_1 - \cdots - a_{n-1})^2]$$

This gives

$$\frac{\partial V(T)}{\partial a_i} = \sigma^2[2a_i - 2(1 - a_1 - \cdots - a_{n-1})] = 0$$

or

$$2[a_i - a_n] = 0 \Rightarrow a_i = a_n = \frac{1}{n}\ \forall\, i$$

This gives that $T(\mathbf{X}) = (1/n)\Sigma_1^n X_i$ is the BLUE of θ.

(ii) In this case, $V(T) = \Sigma_1^n a_i^2 \sigma_i^2$. We choose a_is so that $V(T)$ is minimum subject to the condition $\Sigma_1^n a_i = 1$.

$$V(T) = a_1^2\sigma_1^2 + a_2^2\sigma_2^2 + \cdots + (1 - a_1 - a_2 - \cdots - a_{n-1})^2\sigma_n^2$$

$$\frac{\partial V(T)}{\partial a_i} = 0,\ i = 1, 2, \ldots, (n-1)$$

$$2a_i\sigma_i^2 - 2(1 - a_1 - a_2 - \cdots - a_{n-1})\,\sigma_n^2 = 0$$

$$a_i\sigma_i^2 = a_n\sigma_n^2, \quad i = 1, 2, \ldots, (n-1)$$

$$\Rightarrow \qquad a_i \propto \frac{1}{\sigma_i^2} \Rightarrow a_i = \frac{1}{\sigma_i^2 \sum_1^n (1/\sigma_i^2)} \qquad (\text{since } \sum_1^n a_i = 1)$$

Therefore, the BLUE is given by

$$T(\mathbf{X}) = \frac{\sum_1^n \left(X_i/\sigma_i^2\right)}{\sum_1^n \left(1/\sigma_i^2\right)}$$

Further, the minimum variance corresponding to the BLUE $T(\mathbf{X})$ is

$$V(T(\mathbf{X})) = \frac{1}{\left[\sum_1^n \left(1/\sigma_i^2\right)\right]^2} \sum_1^n \left(\frac{1}{\sigma_i^2}\right)^2 \sigma_i^2$$

$$= \frac{1}{\sum_1^n \left(1/\sigma_i^2\right)} : \text{Harmonic mean of } \sigma_i^2$$

***Example* 3.3** Let X be an *rv* with *pmf*

$$P_\theta(X = -1) = \theta \text{ and } P_\theta(X = x) = (1-\theta)^2\theta^x, \quad x = 0, 1, 2, \ldots; \quad 0 < \theta < 1$$

Find the UMVUE for $g(\theta) = P_\theta(X = 0) = (1 - \theta)^2$.

Solution. The Lehmann–Scheffe theorem cannot be used, since we noted in Chapter 2 that $\mathbf{X}$ is though sufficient and minimal sufficient, but not complete. We will now use Theorem 3.2.1 to find UMVUE for $(1 - \theta)^2$. Let U be a zero estimator

$$E_\theta[U(X)] = 0 = U(-1)\theta + \sum_{x=0}^{\infty} U(x)(1-\theta)^2\theta^x$$

$$\Rightarrow \qquad U(-1)\theta(1-\theta)^{-2} + \sum_{x=0}^{\infty} U(x)\theta^x = 0$$

$$U(-1)\sum_{x=0}^{\infty} x\theta^x + \sum_{x=0}^{\infty} U(x)\theta^x = 0$$

$$(\text{since } \Sigma\theta^x = (1-\theta)^{-1},\ (1-\theta)^{-2} = \Sigma x\theta^{x-1} = \theta^{-1}\Sigma x\theta^x)$$

Now, $$\sum_{x=0}^{\infty}[xU(-1)+U(x)]\theta^x = 0$$

implies $$U(x) = -xU(-1) \text{ for } x = 0, 1, 2,\ldots$$
$$= cx \text{ for } x = 0, 1, 2,\ldots$$

A necessary and sufficient condition for δ to be UMVUE for $(1-\theta)^2$ is $\text{cov}(\delta, U) = 0$,

i.e., $$E_\theta[\delta(X)U] = 0 \ \ \forall\ \theta$$

$$E_\theta[X\delta(X)] = -\delta(-1)\theta + \sum_{x=0}^{\infty} x\delta(x)\theta^x(1-\theta)^2$$

$$-\delta(-1)\theta(1-\theta)^{-2} + \sum_{x=0}^{\infty} x\delta(x)\theta^x = 0$$

$$-\delta(-1)\sum x\theta^x + \sum_{x=0}^{\infty} x\theta^x\delta(x) = 0$$

$$\sum_{x=0}^{\infty}[-x\delta(-1)+x\delta(x)]\theta^x = 0$$

$$x\delta(x) = x\delta(-1) \text{ for } x = 0, 1, 2,\ldots$$

$$\delta(x) = \delta(-1) = c \text{ for } x = 1, 2,\ldots$$

and $\delta(0) = d$ where d is an arbitrary value. The expected value of δ is

$$E_\theta[\delta(X)] = \delta(-1)\theta + \delta(0)(1-\theta)^2 + \sum_{x=1}^{\infty}\delta(x)(1-\theta)^2\theta^x$$

$$= c\theta + d(1-\theta)^2 + c(1-\theta)^2\frac{\theta}{1-\theta}$$

$$= c[1-(1-\theta)^2] + d(1-\theta)^2 = c + (d-c)(1-\theta)^2$$

The estimation of $g(\theta) = (1-\theta)^2$ corresponds to $c = 0$ and $d = 1$. This gives that

$$\delta(X) = \begin{cases} 1, & \text{when } X = 0 \\ 0, & \text{otherwise} \end{cases}$$

is the UMVUE of $g(\theta) = (1-\theta)^2$.

***Example* 3.4** Let $X_1, X_2,\ldots,X_n$ be a random sample from the *pmf*

$$P(x; N) = \frac{1}{N}, \quad x = 1, 2,\ldots, N$$

Let $\psi(N)$ be a function of N. Find the UMVUE for $\psi(N)$.

Solution. Consider an unbiased estimator $\delta(X_1)$ of $\psi(N)$

$$E_N[\delta(X_1)] = \psi(N); \ \sum_{x=1}^{N} \delta(x)\frac{1}{N} = \psi(N); \ \sum_{x=1}^{N} \{\delta(x) - \psi(N)\} = 0$$

It gives that for any estimator $\delta(X_1)$ to be unbiased for $\psi(N)$, it must satisfy $\Sigma_{x=1}^{N}\delta(x) = N\psi(N)$. The estimator $\delta(x)$ with $\delta(x) = \psi(N)\ \forall\ x = 1, 2,\ldots,N$ satisfies this condition. We have already shown in Chapter 2 that the statistic $T(\mathbf{X}) = X_{(n)}$ is a complete sufficient statistic. It follows from Theorem 3.3.1 that $E_N[\delta(X_1)|T]$ is UMVUE of $\psi(N)$. We have

$$P\{X_1 = x_1|T = t\} = \begin{cases} \dfrac{t^{n-1} - (t-1)^{n-1}}{t^n - (t-1)^n}, & \text{if } x_1 = 1, 2, \ldots, t-1 \\[2ex] \dfrac{t^{n-1}}{t^n - (t-1)^n}, & \text{if } x_1 = t \end{cases}$$

$\therefore$

$$E_N[\delta(X_1)|T = t] = \frac{[t^{n-1} - (t-1)^{n-1}]\sum_{1}^{t-1}\delta(x) + t^{n-1}\delta(t)}{t^n - (t-1)^n}$$

$$= \frac{t^{n-1}\sum_{1}^{t}\delta(x) - (t-1)^{n-1}\sum_{1}^{t-1}\delta(x)}{t^n - (t-1)^n}$$

$$= \frac{t^n\psi(t) - (t-1)^n\psi(t-1)}{t^n - (t-1)^n}, \quad t \ge 1$$

is the UMVUE of $\psi(N)$.

***Example* 3.5** Let $X_1, X_2,\ldots,X_n$ be *iid rvs* with common *pmf*

$$P(x;\theta) = \frac{1}{2\theta}, \ |x| < \theta$$

Find the UMVUE for θ^r.

Solution. We have already shown that the sufficient statistic $T(\mathbf{X}) = \max\{-X_{(1)}, X_{(n)}\}$ is minimal and complete for θ with *pdf*

$$f(t;\theta) = \frac{nt^{n-1}}{\theta^n}, \ 0 < t < \theta$$

which is also the *pdf* of $Y_{(n)}$ when $Y_1, Y_2, \ldots, Y_n$ are *iid* $U(0,\theta)$. Consider

$$E_\theta(T^r) = \int_0^\theta t^r f(t;\theta)dt = \left(\frac{n}{\theta^n}\right)\int_0^\theta t^{r+n-1}dt$$

$$= \left(\frac{n}{\theta^n}\right)\cdot\frac{\theta^{n+r}}{n+r}$$

$$\Rightarrow \qquad E\left(\frac{n+r}{n}T^r\right) = \theta^r$$

This gives $\delta(T) = [(n + r)/n]T^r$. Note that for $r = 1$, $[(n + 1)/n]T$ is the UMVUE of θ.

***Example* 3.6** Let $X_1, X_2, \ldots, X_n$ be *iid* from $b(1, \theta)$, $\theta \in (0, 1)$.

(i) Given that T/n is an unbiased estimator of θ, $T = \Sigma_1^n X_i$, show that $[T/n]^2$ is not unbiased for θ^2 (note that unbiased estimators are not invariant).

(ii) Find an unbiased estimator of θ^2.

Solution.

(i) $$E_\theta\left(\frac{T}{n}\right)^2 = V_\theta\left(\frac{T}{n}\right) + \left\{E_\theta\left(\frac{T}{n}\right)\right\}^2 = \frac{\theta(1-\theta)}{n} + \theta^2 \neq \theta^2$$

(ii) $$E_\theta T^2 = V_\theta(T) + \{E_\theta(T)\}^2$$

$$= n\theta(1-\theta) + n^2\theta^2 = n\theta + n(n-1)\theta^2$$

$$E_\theta(T^2 - T) = n(n-1)\theta^2$$

$$E_\theta\left[\frac{T(T-1)}{n(n-1)}\right] = \theta^2$$

This shows that $\delta(T) = [T(T - 1)/n(n - 1)]$ is an unbiased estimator of θ^2. In fact, $\delta(T)$ is a UMVUE of θ^2 since it is based on the complete sufficient statistic T.

***Example* 3.7** Let X follows hypergeometric distribution with *pmf*

$$P_M(x) = \binom{N}{n}^{-1}\binom{M}{x}\binom{N-M}{n-x}$$

where $\max(0, M + n - N) \le x \le \min(M, n)$.

(i) Find the UMVUE for M when N is known and show that an unbiased estimator of N does not exist when M is known.

(ii) Find an unbiased estimator of M^2 when n, N are known.

Solution.

(i) The family of hypergeometric distribution is complete. $E_M(X) = (n/N)M$, $E_M(NX/n) = M$ gives that $T(X) = NX/n$ is a UMVUE of M. Further, if M is known and N is unknown, an unbiased estimator of N must satisfy

$$\sum_{x=0}^{\min(n,M)} T(x)\binom{N}{n}^{-1}\binom{M}{x}\binom{N-M}{n-x} = N$$

$$\sum_{x=0}^{\min(n,M)} T(x)\binom{M}{x}\binom{N-M}{n-x} = N\binom{N}{n} = \sum_{x=0}^{\min(n,M)} N\binom{M}{x}\binom{N-M}{n-x}$$

$T(x)$ must satisfy $T(x) = N \;\forall\; x = 0,1,\ldots,\min\{n, M\}$, which is not possible since N is unknown. Therefore, there does not exist any unbiased estimator of N when M is known.

(ii)
$$E[X(X-1)] = M(M-1)\cdot\frac{n(n-1)}{N(N-1)} = M(M-1)C$$

$$E\left[\frac{X(X-1)}{C}\right] = M^2 - M$$

$$E\left[\frac{X(X-1)}{C} + \frac{NX}{n}\right] = M^2$$

$$E\left[\frac{X(X-1)N(N-1)}{n(n-1)} + \frac{NX}{n}\right] = M^2$$

Therefore, $[X(X-1)N(N-1)/n(n-1)] + (NX/n)$ is an unbiased estimator of M^2.

***Example* 3.8 (Power series distribution) [Roy and Mitra (1957)].** Let $X_1, X_2,\ldots,X_n$ be a random sample from power series distribution with *pmf*

$$P_\theta\{X = x\} = \frac{a(x)\theta^x}{c(\theta)}; \quad x = 0, 1, 2,\ldots \tag{3.4.1}$$

where $\theta > 0$, $a(x) > 0$, and $c(\theta)$ is given by

$$c(\theta) = \sum_{x=0}^{\infty} a(x)\theta^x$$

Show that $T = \Sigma_1^n X_i$ is a complete sufficient statistic for θ and UMVUE of θ^r, where $r > 0$ is an integer, is

$$\delta(t) = \begin{cases} 0, & \text{if } t < r \\ \dfrac{A(t-r, n)}{A(t, n)}, & \text{if } t \ge r \end{cases}$$

where $A(t, n)$ is the coefficient of θ^t in the expansion of $[c(\theta)]^n$ and

$$A(t, n) = \sum_{x_1, x_2,\ldots,x_n} \prod_{i=1}^{n} a(x_i)$$

so that $\Sigma_1^n x_i = t$. Also, find the UMVUE of the *pmf* $P_\theta\{X = x\}$.

Solution. The distribution of the statistic $T = \Sigma_1^n X_i$ is given by

$$P_\theta(T=t) = \sum_{\{x\in\mathbb{R}^1: T(\mathbf{x})=t\}} \frac{\prod_{i=1}^n a(x_i)}{[c(\theta)]^n} \cdot \theta^t$$

Note that $\Sigma_{[\mathbf{x}:T(\mathbf{x})=t]} \Pi_{i=1}^n a(x_i)$ is the coefficient of θ^t in the expansion of $[c(\theta)]^n = [\Sigma_{x=0}^\infty a(x)\theta^x]^n$. We denote this term by A(t, n). Thus, the distribution of T is given by

$$P_\theta(T=t) = \frac{A(t,n)\theta^t}{[c(\theta)]^n}, \quad t = 0, 1, \ldots \tag{3.4.2}$$

Let $\theta = e^\eta$, $[c(\theta)]^n = e^{\beta(\eta)}$ and $A(t, n) = a(t)$. We have

$$P_\eta(T=t) = \exp(\eta t - \beta(\eta))a(t)$$

The distribution of T belongs to one-parameter exponential family, where T is a complete sufficient statistic for θ. The UMVUE of θ^r, by using T, is obtained by solving the equation for $\delta(t)$

$$E[\delta(T)] = \theta^r$$

$$\sum_{t=0}^\infty \delta(t)\frac{A(t,n)}{[c(\theta)]^n}\theta^t = \theta^r$$

$$\sum_{t=0}^\infty \delta(t)A(t,n)\theta^t = [c(\theta)]^n\theta^r$$

$$= \sum_{t=0}^\infty A(t,n)\theta^{t+r} = \sum_{y=r}^\infty A(y-r,n)\theta^y$$

$$= \sum_{y=0}^{r-1} 0\cdot\theta^y + \sum_{y=r}^\infty A(y-r,n)\theta^y$$

as $[c(\theta)]^n = \Sigma_t A(t, n)\theta^t$. On comparing the coefficients of θ^t on both sides of the above equation, we have

$$\delta(t) = \begin{cases} 0, & t = 0, 1, \ldots, r-1 \\ \dfrac{A(t-r,n)}{A(t,n)}, & t \ge r \end{cases}$$

The estimator $\delta(t)$ is unbiased for θ^r since

$$E[\delta(T)] = \sum_{t\ge r} \frac{A(t-r,n)}{A(t,n)}\frac{A(t,n)\theta^t}{[c(\theta)]^n}$$

$$= \theta^r \sum_{t-r\ge 0} \theta^{t-r}\frac{A(t-r,n)}{[c(\theta)]^n} = \theta^r$$

Now consider the problem of estimating the *pmf* $P_\theta(X = x)$ on the basis of sample observations. The estimator

$$U(X_1) = \begin{cases} 1, & \text{if } X_1 = x \\ 0, & \text{otherwise} \end{cases}$$

is an unbiased estimator of $P_\theta(X = x)$, since $E[U(X_1)] = P_\theta(X_1 = x) = [a(x)\theta^x]/[c(\theta)]$.

Since $T = \Sigma_1^n X_i$ is a complete sufficient statistic for θ, the UMVUE of $P_\theta(X_1 = x)$ is

$$\delta(t) = E[U(X_1)|T = t] = \frac{P_\theta\left(X_1 = x, \sum_1^n X_i = t\right)}{P\left(\sum_1^n X_i = t\right)}$$

$$= \frac{P_\theta(X_1 = x) \cdot P_\theta\left(\sum_2^n X_i = t - x\right)}{P\left(\sum_1^n X_i = t\right)}$$

$$= \frac{\dfrac{a(x)\theta^x}{c(\theta)} \dfrac{A(t-x, n-1)\theta^{t-x}}{[c(\theta)]^{n-1}}}{\dfrac{A(t, n)\theta^t}{[c(\theta)]^n}}$$

$$= a(x)\frac{A(t-x, n-1)}{A(t, n)}, \qquad n > 1, 0 \le x \le t$$

***Example* 3.9 (Negative binomial distribution).** Let $X_1, X_2, \ldots, X_n$ be *iid* from negative binomial distribution, $NB(m, \theta)$ distribution. Find the UMVUE of

$$P_\theta(X = x) = \binom{m + x - 1}{m - 1}\theta^m (1 - \theta)^x, \quad x = 0, 1 \ldots$$

Solution. $P_\theta(X = x)$ is a power series distribution with θ^x as $(1 - \theta)^x$ and $a(x) = \binom{m + x - 1}{m - 1}$. The statistic $T = \Sigma_1^n X_i$ is complete and sufficient for θ and it follows NB(nm, θ) with density

$$P_\theta(T = t) = \binom{nm + t - 1}{nm - 1}\theta^{nm}(1 - \theta)^t, \quad t = 0, 1, 2, \ldots$$

where $A(t, nm) = \binom{nm + t - 1}{nm - 1}$. Therefore, by using the result of Example 3.8, the UMVUE of $P_\theta(X = x)$ is given by

$$\delta(t) = \frac{a(x)A[t-x,(n-1)m]}{A(t, nm)}$$

$$= \frac{\binom{m+x-1}{m-1}\binom{(n-1)m+t-x-1}{(n-1)m-1}}{\binom{nm+t-1}{nm-1}}$$

An alternative method of finding the UMVUE of $P_\theta(T = t)$ is to consider an unbiased estimator of $P_\theta(T = t)$

$$U(X) = \begin{cases} 1, & \text{if } X_1 = x \\ 0, & \text{otherwise} \end{cases}$$

Note that $U(X)$ is unbiased since $E[U(X)] = \binom{m+x-1}{m-1}\theta^m(1-\theta)^x$. Using the fact that $T = \Sigma X_i$ is complete and sufficient statistic for θ, the UMVUE of $P_\theta(T = t)$ is given by

$$E[U(X)|T = t] = P\left[X_1 = x \Big| \sum_1^n X_i = t\right] = \frac{P[X_1 = x]\cdot\left[\sum_2^n X_i = t - x\right]}{P\left[\sum_1^n X_i = t\right]}$$

$$= \frac{\binom{m+x-1}{m-1}\binom{(n-1)m+t-x-1}{(n-1)m-1}}{\binom{nm+t-1}{nm-1}}$$

***Example* 3.10 (Geometric distribution).** Let $X_1, X_2,\ldots,X_n$ be *iid* from the geometric distribution with *pmf*

$$P_\theta(X = x) = \theta(1-\theta)^x,\ x = 0, 1,\ldots \tag{3.4.3}$$

(i) Find the UMVUE for (a) θ^r and (b) $(1 - \theta)^r$.

(ii) Show that the distribution of X belongs to a full-rank exponential family and $T(\mathbf{X}) = \bar{X} + 1$ is the UMVUE of $1/\theta$.

Solution.

(i) (a) Consider an estimator

$$U(\mathbf{X}) = \begin{cases} 1, & \text{if } X_1 = 0, X_2 = 0,\ldots, X_r = 0 \\ 0, & \text{otherwise} \end{cases}$$

We have

$$E_\theta[U(\mathbf{X})] = 1\cdot[P(X_1 = 0)\cdots P(X_r = 0)] + 0 = \theta^r$$

Therefore, $U(\mathbf{X})$ is an unbiased estimator of θ^r. Further, $T = \Sigma_1^n X_i \sim \text{NB}(n, \theta)$ with *pmf*

$$P_\theta(T=t) = \binom{t+n-1}{n-1}\theta^n(1-\theta)^t, t = 0, 1, 2, \ldots$$

Using the fact that $T = \Sigma_1^n X_i$ is a complete and sufficient statistic for θ, the UMVUE of θ^r is

$$\delta(t) = E[U(\mathbf{X})|T=t] = \frac{P_\theta(X_1 = 0, \ldots, X_r = 0)\, P_\theta\left(\sum_{r+1}^{n} X_i = t\right)}{P_\theta\left(\sum_{1}^{n} X_i = t\right)}$$

$$= \frac{\theta^r \binom{n-r+t-1}{n-r-1}\theta^{n-r}(1-\theta)^t}{\binom{t+n-1}{n-1}\theta^n(1-\theta)^t} = \frac{\binom{n-r+t-1}{n-r-1}}{\binom{n+t-1}{n-1}}$$

$$= \frac{(n-r+t-1)!}{t!(n-r-1)!} \cdot \frac{(n-1)!t!}{(n+t-1)!}$$

(b) By Eq. (3.4.2) in Example 3.8, the distribution of T is a power series distribution with $(1-\theta)^t$ at the place of θ^t; $[C(\theta)]^n = (1/\theta^n)$;

$$A(t, n) = \binom{n+t-1}{n-1} \tag{3.4.4}$$

Therefore, by Example 3.8, the UMVUE of $(1-\theta)^r$ is given by $[A(t-r, n)]/[A(t, n)]$, $t \geq r$,

$$\delta(t) = \frac{\binom{n+t-r-1}{n-1}}{\binom{n+t-1}{n-1}}$$

One may check $\delta(T)$ for its unbiasedness

$$E_\theta[\delta(T)] = \sum_{t=r}^{\infty} \frac{\binom{n+t-r-1}{n-1}}{\binom{n+t-1}{n-1}} \cdot \binom{n+t-1}{n-1} \cdot \theta^n(1-\theta)^t$$

$$= \theta^n(1-\theta)^r \sum_{t=r}^{\infty} \binom{n+t-r-1}{n-1}(1-\theta)^{t-r}$$

Let $t - r = y$. We have

$$E_\theta[\delta(T)] = \theta^n (1-\theta)^r \sum_{t=0}^{\infty} \binom{n+y-1}{n-1} (1-\theta)^y$$

$$= \theta^n (1-\theta)^r \theta^{-n} = (1-\theta)^r$$

(ii) The geometric distribution belongs to the one-parameter exponential family given in Eq. (2.4.2) with $T(x) = x$, $p(\theta) = \log(1 - \theta)$ and $c(\theta) = \theta$. In the joint *pdf* of X_1, $X_2,\ldots,X_n$, the statistic $T(\mathbf{X}) = \Sigma X_i$ does not satisfy any linear constraint. $p(\theta)$ takes values from $(-\infty, 0)$ since θ ranges over (0, 1). This family is of full rank since the space of $p(\theta)$ contains a non-empty interior. We have

$$E_\theta(X) = \frac{1-\theta}{\theta}$$

$$\Rightarrow \qquad E_\theta(\bar{X}) = \frac{1}{\theta} - 1$$

$T(\mathbf{X}) = \bar{X} + 1$ is the UMVUE of $1/\theta$ since $\bar{X}$ is complete and sufficient statistic θ.

***Example* 3.11** Let $X \sim P(\theta)$. Comment on the existence of an unbiased estimator of θ^{-1}.

Solution. If $\delta(X)$ is an unbiased estimator of θ^{-1}, then

$$E_\theta[\delta(X)] = \frac{1}{\theta}$$

$$\sum_{x=0}^{\infty} \delta(x) \frac{e^{-\theta}\theta^x}{x!} = \frac{1}{\theta}$$

If $\theta \to 0$, then the right-hand side of the above expression tends to ∞ whereas the left-hand side tends to $[\delta(0)\cdot 0]/0! = 0$. Therefore, there does not exist any unbiased estimator of θ.

***Example* 3.12** Consider the distribution of the number of telephone calls in a fixed time period t on a PBX box of an office which is a Poisson distribution $P(\theta)$ (a limiting form of binomial distribution, $b(n, \theta)$, when θ is small and $n \to \infty$). Based on the sample observations $X_1, X_2,\ldots,X_n$, find the UMVUE of

(i) θ

(ii) θ^r by using the results for power series distribution due to Roy and Mitra

(iii) a parametric function $g(\theta)$ where g is some smooth function such that $g(\theta) = \Sigma_{j=0}^{\infty} a_j \theta_j$

(iv) θ^r, r is an integer greater than 1, using the result of (iii)

(v) $g(\theta) = \dfrac{1}{1-\theta}$

(vi) $P_\theta(X = k) = \dfrac{\exp(-\theta)\cdot\theta^k}{k!}$

(vii) $g(\theta) = P(X = 0) = \exp(-\theta)$

(viii) $g(\theta) = \Sigma_{x \le k} P_\theta(X = x) = P_\theta(0) + P_\theta(1) + \cdots + P_\theta(k)$

(ix) $g(\theta) = (1 + \theta)\exp(-\theta) = P_\theta(X = 0 \text{ or } 1)$. Let UMVUE and MLE (maximum likelihood estimator) of $g(\theta)$ be denoted by δ_1 and δ_2, respectively, where $\delta_2 = (1 + \bar{X})\exp(-\bar{X})$. Show that δ_2 is biased. ML estimation has been discussed in Chapter 5.

Solution.

(i) $T = \Sigma_1^n X_i$ is a complete sufficient statistic for θ; it follows the Poisson distribution $P(n\theta)$. Since $E(T) = n\theta$, (T/n) is the UMVUE of θ.

(ii) $P(\theta)$ is a power series distribution with

$$a(x) = \frac{1}{x!};\ C(\theta) = \exp(\theta); [C(\theta)]^n = \exp(n\theta) = 1 + n\theta + \frac{(n\theta)^2}{2!} + \cdots$$

The coefficient of θ^t is

$$A(t, n) = \frac{n^t}{t!}$$

Therefore, UMVUE θ^r is

$$\delta = \frac{A(t-r, n)}{A(t, n)} = \frac{n^{t-r}/(t-r)!}{n^t/t!}$$

$$= \frac{t!}{(t-r)!} \cdot \frac{1}{n^r}$$

Note that $\delta(t)$ is valid for all r since for $t = 0, 1, \ldots, r - 1$, $\delta(t) \equiv 0$.

(iii) An estimator $\delta(T)$ is unbiased for $g(\theta)$ if

$$E_\theta[\delta(T)] = g(\theta)$$

$$\sum_{t=0}^{\infty} \delta(t) \cdot \frac{\exp(-n\theta)(n\theta)^t}{t!} = g(\theta)$$

$$\sum_{t=0}^{\infty} \delta(t) \cdot \frac{n^t}{t!}\theta^t = g(\theta)\exp(n\theta) = \sum_{i=0}^{\infty} \frac{(n\theta)^i}{i!} \sum_{j=0}^{\infty} a_j \theta^j$$

$$= \sum_{i=0}^{\infty} \frac{n^i \theta^i}{i!} \sum_{j=0}^{\infty} a_j \theta^j = \sum_{t=0}^{\infty} \left(\sum_{\substack{i,\, j \text{ such that} \\ i+j=t}}^{\infty} \frac{n^i}{i!} a_j \right) \theta^t$$

On comparing the coefficients of θ^t on both sides for any θ,

$$\delta(t) \cdot \frac{n^t}{t!} = \sum_{\substack{i,j \text{ such that} \\ i+j=t}}^{\infty} \frac{n^i}{i!} a_j$$

$$\delta(T) = \frac{T!}{n^T} \sum_{\substack{i,j \text{ such that} \\ i+j=T}}^{\infty} \frac{n^i}{i!} a_j$$

This is the UMVUE of $g(\theta)$.

(iv) Similarly, as in (iii), an estimator $\delta(T)$ is unbiased for θ^r if

$$E_\theta[\delta(T)] = \sum_{t=0}^{\infty} \delta(t) \frac{\exp(-n\theta)(n\theta)^t}{t!} = \theta^r$$

$$\sum_{t=0}^{\infty} \delta(t) \frac{(n\theta)^t}{t!} = \theta^r \exp(n\theta) = \theta^r \sum_{x=0}^{\infty} \frac{(n\theta)^x}{x!} = \sum_{x=0}^{\infty} \frac{n^x}{x!} \theta^{x+r}$$

$$= \sum_{t=r}^{\infty} \frac{n^{t-r}}{(t-r)!} \theta^t \qquad \text{(by assuming } x + r = t\text{)}$$

$$\Rightarrow \qquad \delta(t) = \frac{t!}{(t-r)!} \frac{n^{t-r}}{n^t}, \quad t \geq r$$

Proceeding as a particular case of (iii), we have $g(\theta) = \Sigma_{j=0}^{\infty} a_j \theta^j = \theta^r$, $r \geq 1$; $a_r = 1$, $a_j = 0$ for $j \neq r$. The UMVUE of θ^r is

$$\delta(T) = \begin{cases} \dfrac{T!}{n^T} \cdot \dfrac{n^{T-r}}{(T-r)!}, & \text{if } T \geq r \\ 0, & \text{if } T < r \end{cases}$$

(v) *Method* 1: If $g(\theta) = \Sigma_{j=0}^{\infty} a_j \theta^j = (1 - \theta)^{-1} = 1 + \theta + \theta^2 + \theta^3 + \cdots + \theta^r + \ldots$, then $a_j = 1$ for all j. By using (iii), the UMVUE of $(1 - \theta)^{-1}$ becomes

$$\delta(T) = \sum_{j=0}^{T} \frac{T!}{j!} n^{j-T}$$

Method 2: $g(\theta) = (1 - \theta)^{-1} = \theta^0 + \theta^1 + \cdots + \theta^t + \theta^{t+1} + \cdots$

$$= \sum_{j=0}^{t} \theta^j + \sum_{j=t+1}^{\infty} \theta^j = \sum_{j=0}^{t} c_j \theta^j \text{ with } c_j = 1 \quad \forall\, j = 0, 1, \ldots$$

By using (iv), the UMVUE of θ^j is $(1/n^j) \cdot T!/(T - j)!$ for $j \leq t$ and 0 for $j > t$. The UMVUE of $1/(1 - \theta)$ is given by

$$\delta(T) = \sum_{j=0}^{T} \frac{1}{n^j} \cdot \frac{T!}{(T-j)!} = \sum_{j=0}^{T} \frac{T!}{j!} n^{j-T}$$

(vi) Using the Roy and Mitra result for power series distributions, the UMVUE of $P_\theta(X = k)$ is given by

$$\delta(t) = \frac{a(k)\cdot A(t-k, n-1)}{A(t, n)} = \frac{1}{k!}\cdot\frac{(n-1)^{t-k}}{(t-k)!}\cdot\frac{t!}{n^t}$$

$$= \frac{t!}{k!(t-k)!}\left(\frac{n-1}{n}\right)^{t-k}\left(\frac{1}{n}\right)^k$$

$$= \binom{t}{k}\left(\frac{1}{n}\right)^k\left(1-\frac{1}{n}\right)^{t-k} \quad \text{for } k = 0, 1, 2, \ldots, t$$

We illustrate another way to obtain the UMVUE of $P_\theta(x = k)$. Define an unbiased estimator of $P_\theta(X = k)$ by

$$U(\mathbf{X}) = \begin{cases} 1, & \text{if } X_1 = k \\ 0, & \text{otherwise} \end{cases}$$

Here, $T = \Sigma_1^n X_i \sim P(n\theta)$ is a complete sufficient statistic for θ. Therefore, the UMVUE of $P_\theta(k)$ is given by

$$\delta(t) = E[U(\mathbf{X})|T = t] = \frac{P(X_1 = k)P\left(\sum_2^n X_i = t-k\right)}{P\left(\sum_1^n X_i = t\right)}$$

$$= \frac{\dfrac{\exp(-\theta)\theta^k}{k!}\,\dfrac{\exp(-(n-1)\theta)[(n-1)\theta]^{t-k}}{(t-k)!}}{\dfrac{\exp(-n\theta)(n\theta)^t}{t!}}$$

$$= \binom{t}{k}\frac{(n-1)^{t-k}}{n^t} = \binom{t}{k}\left(\frac{1}{n}\right)^k\left(1-\frac{1}{n}\right)^{t-k}$$

Note that the variance of $\delta(T)$ is at least as small as the variance of $U(\mathbf{X})$.

(vii) On taking $k = 0$, the UMVUE of $P_\theta(X = 0) = \exp(-\theta)$,

$$\delta(T) = \left(1-\frac{1}{n}\right)^T$$

(viii) By (vi), the UMVUE of $\Sigma_{x\le k}P_\theta\,(X = x)$ is

$$\delta(T) = \sum_{x=0}^{k}\binom{T}{x}\left(\frac{1}{n}\right)^x\left(1-\frac{1}{n}\right)^{T-x}$$

(ix) By (vi), the UMVUE of $P_\theta(X = 0 \text{ or } 1) = \exp(-\theta) + \exp(-\theta)\theta$ is given by

$$\delta_1 = \binom{T}{0}\left(\frac{1}{n}\right)^0\left(1-\frac{1}{n}\right)^{T-0} + \binom{T}{1}\left(\frac{1}{n}\right)^1\left(1-\frac{1}{n}\right)^{T-1}$$

$$= \left(1-\frac{1}{n}\right)^T + \frac{T}{n}\left(1-\frac{1}{n}\right)^{T-1} = \left(1-\frac{1}{n}\right)^{T-1}\left(1-\frac{1}{n}+\frac{T}{n}\right)$$

$$= \left(1-\frac{1}{n}\right)^{T-1}\left(1+\frac{T-1}{n}\right)$$

The estimator δ_1 is unbiased. Assume that δ_2 is also unbiased for $g(\theta)$. This along with the fact that they are the functions of the complete sufficient statistic T gives $E_\theta(\delta_1 - \delta_2) = 0 \Rightarrow \delta_1 = \delta_2$ a.s. (almost surely); whereas the two estimators do not coincide. Thus, δ_2 cannot be unbiased.

***Example* 3.13** X follows a truncated Poisson distribution (at $X = 0$). Find a UMVUE of $\theta = (1/2)(1 - \exp(-\lambda)) = P_\lambda$ (X is even).

Solution. A Poisson distribution is called truncated at $x = 0$, if the Poisson distribution takes all values except 0, i.e., $x = 1, 2, 3, \ldots$. If $P(x;\lambda)$ is the *pmf* of the Poisson distribution, then

$$\sum_{x=1}^{\infty} P(x;\lambda) = 1 - P(0;\lambda) = 1 - \exp(-\lambda)$$

The *pmf* of the truncated Poisson distribution (at $x = 0$) is

$$P_\lambda^0(x) = \frac{1}{1-\exp(-\lambda)}\frac{\exp(-\lambda)\lambda^x}{x!}, \quad x = 1, 2, 3, \ldots$$

Let $U(X)$ be an unbiased estimator of $(1 - e^{-\lambda})$.

$$\sum_{x=1}^{\infty} U(x)\cdot\frac{1}{1-\exp(-\lambda)}\frac{\exp(-\lambda)\lambda^x}{x!} = \frac{1}{2}(1-\exp(-\lambda)) \quad \text{(Given)}$$

$$\sum_{x=1}^{\infty} U(x)\frac{\lambda^x}{x!} = \frac{1}{2}\exp(\lambda)[1-\exp(-\lambda)]^2 = \frac{1}{2}(\exp(\lambda)+\exp(-\lambda)-2) = \sum_{1}^{\infty}\frac{\lambda^{2x}}{(2x)!}$$

This gives

$$U(X) = \begin{cases} 1, & \text{if } X \text{ is even} \\ 0, & \text{if } X \text{ is odd} \end{cases}$$

Since X is a function of complete sufficient statistic and $U(X)$ is unbiased for $(1/2)(1 - \exp(-\lambda))$, $U(X)$ is UMVUE of $(1/2)(1 - \exp(-\lambda))$. Although $U(X)$ is UMVUE but it estimates the probability P_λ (X is even) by two extreme values: either 0 or 1. Therefore, $U(X)$ is not a reasonable estimator.

***Example* 3.14 (UMVUEs are not always sensible).** Assume that the customers on a cash counter are arriving, as a Poisson process, at an unknown rate of arrival θ per 5 minutes. The clerk wishes to leave the counter for 10 min for a cup of tea. Estimate, $\exp(-2\theta)$, the probability that no customer will arrive in the next 10 min.

Solution. Let X be the number of customers that arrive in the first 5 min. $X \sim P(\theta)$, X is a complete sufficient statistic for θ. The UMVUE of θ is obtained by solving the equation

$$E_\theta T(X) = \exp(-2\theta)$$

$$\sum_{x=0}^{\infty} T(x)\frac{\exp(-\theta)\theta^x}{x!} = \exp(-2\theta)$$

Multiplying both the sides by $\exp(\theta)$, we get

$$\sum_{x=0}^{\infty} T(x)\frac{\theta^x}{x!} = \exp(-\theta) = \sum_{x=0}^{\infty} (-1)^x \frac{\theta^x}{x!}$$

Left- and right-side series are the power series. They are convergent; therefore, their coefficients are equal

$$T(x) = (-1)^x$$

This gives that $T(X) = (-1)^X$ is the UMVUE of the probability of no arrival on the counter in the next 10 min. Although $T(X)$ is UMVUE, it lacks on the merit that it does not utilize the quantity as how many customers have arrived in the first 10 min and that it estimates the probability $\exp(-2\theta)$ by –1 when X is odd. Therefore, such estimators are avoided for their use.

***Example* 3.15**

(i) Let T_1 and T_2 be two estimators of θ in $\mathcal{U}(\theta)$ with their efficiencies e_1 and e_2, respectively. If $\rho = \rho_\theta$ is the correlation coefficient between T_1 and T_2, then show that

$$\rho_\theta \in \left[\sqrt{e_1 e_2} \mp \sqrt{(1-e_1)(1-e_2)}\right]$$

(ii) Show that the correlation coefficient between a most efficient unbiased estimator and any other unbiased estimator with efficiency e is $\sqrt{e}$.

(iii) Also, show that if T_1 and T_2 are in $\mathcal{U}(\theta)$ and have the same variance, i.e., have the same efficiency e, then

$$\rho_\theta(T_1, T_2) \geq 2e - 1$$

Solution.

(i) Let T be a UMVUE of θ in $\mathcal{U}(\theta)$ with minimum variance V_θ^0, say. Consider an estimator defined by

$$T_3 = a_1 T_1 + a_2 T_2$$

which is unbiased [$T_3 \in \mathcal{U}(\theta)$], if and only if

$$E_\theta(T_3) = a_1\theta + a_2\theta = \theta$$

i.e., $a_1 + a_2 = 1$.
Further, we must have

$$V_\theta(T_3) \geq V_\theta^0 \tag{3.4.5}$$

since V_θ^0 is the minimum variance corresponding to the UMVUE of T. Eq. (3.4.5) gives

$$a_1^2 V_\theta(T_1) + a_2^2 V_\theta(T_2) + 2a_1 a_2 \operatorname{cov}_\theta(T_1, T_2) \geq V_\theta^0$$

$$a_1^2 \frac{V_\theta^0}{e_1} + a_2^2 \frac{V_\theta^0}{e_2} + 2a_1 a_2 \operatorname{cov}_\theta(T_1, T_2) \geq V_\theta^0$$

$$\frac{a_1^2}{e_1} + \frac{a_2^2}{e_2} + 2\frac{a_1 a_2 \rho}{\sqrt{e_1 e_2}} \geq 1 = (a_1 + a_2)^2$$

since e_1 and e_2 are the efficiencies of T_1 and T_2, respectively. This gives

$$\left(\frac{1}{e_1} - 1\right)a_1^2 + \left(\frac{1}{e_2} - 1\right)a_2^2 = 2a_1 a_2\left(-\frac{\rho}{\sqrt{e_1 e_2}} + 1\right) \geq 0$$

or
$$\left(\frac{1}{e_1} - 1\right)\left(\frac{a_1}{a_2}\right)^2 + 2\left(-\frac{\rho}{\sqrt{e_1 e_2}} + 1\right)\left(\frac{a_1}{a_2}\right) + \left(\frac{1}{e_2} - 1\right) \geq 0$$

Let $y = a_1/a_2$. The above inequality reduces to

$$\left(\frac{1}{e_1} - 1\right)y^2 + 2\left(\frac{\rho}{\sqrt{e_1 e_2}} + 1\right)y + \left(\frac{1}{e_2} - 1\right) \geq 0$$

where $e_1 < 1 \Rightarrow (1/e_1) > 1 \Rightarrow [(1/e_1) - 1] > 0$. Similarly, $[(1/e_2) - 1] > 0$. This inequality holds, if and only if

$$4\left(-\frac{\rho}{\sqrt{e_1 e_2}} + 1\right)^2 - 4\left(\frac{1}{e_1} - 1\right)\left(\frac{1}{e_2} - 1\right) \leq 0 \tag{3.4.6}$$

Note that $ax^2 + bx + c \geq 0$, $\forall\ x$, $a > 0$, $c > 0$, if and only if $b^2 - 4ac \leq 0$. Eq. (3.4.6) simplifies to

$$(-\rho + \sqrt{e_1 e_2})^2 - (1 - e_1)(1 - e_2) \leq 0$$

$$\rho^2 - 2\sqrt{e_1 e_2}\,\rho + (e_1 + e_2 - 1) \leq 0$$

This implies that ρ lies between the roots of the equation

$$\rho^2 - 2\sqrt{e_1 e_2}\,\rho + (e_1 + e_2 - 1) = 0$$

which is given by

$$\frac{2\sqrt{e_1 e_2} \mp \sqrt{4e_1 e_2 - 4(e_1 + e_2 - 1)}}{2} = \sqrt{e_1 e_2} \mp \sqrt{e_1 e_2 - (e_1 + e_2 - 1)}$$

$$\therefore \qquad \rho_\theta \in \left[\sqrt{e_1 e_2} \mp \sqrt{(1 - e_1)(1 - e_2)}\right]$$

Note that if we consider $e_1 = 1$ and $e_2 = e$, then

$$\rho = \sqrt{e}$$

[See Example 3.17(ii).]

(ii) Let T_0 be UMVUE and T_1 be some estimator of θ in $\mathcal{U}(\theta)$ with efficiency $e_\theta(T_1) = e$. We have to show that

$$\rho_\theta(T_0, T_1) = \sqrt{e}$$

where ρ_θ is the correlation coefficient between T_0 and T_1. In the present case, $e(T_0) = 1$, $e(T_1) = e$. On putting these values in the result of (i), we get

$$\left[\sqrt{e_1 e_2} \mp \sqrt{(1-e_1)(1-e_2)}\right] = \sqrt{e}$$

$$\therefore \qquad \rho_\theta(T_0, T_1) = \sqrt{e}$$

Alternatively, we can prove it differently by defining an estimator T as a linear combination of T_0 and T_1

$$T = aT_0 + (1-a)T_1$$

The variance of T is

$$V_\theta(T) = a^2 V_\theta(T_0) + (1-a)^2 V_\theta(T_1) + 2a(1-a)\,\text{cov}_\theta(T_0, T_1)$$

On minimizing $V_\theta(T)$ with respect to a, we get

$$\frac{\partial V_\theta}{\partial a} = 2aV_0 - 2(1-a)V_1 + 2(1-2a)\,\text{cov}(T_0, T_1) = 0$$

On putting $V_\theta(T_1) = V_1 = V_\theta(T_0)/e = V_0/e$ and $\text{cov}(T_0, T_1) = \rho\sqrt{V_\theta(T_0)V_\theta(T_1)} = \rho V_0/e$ in the above equation, we get

$$aV_0 - (1-a)\frac{V_0}{e} + (1-2a)\frac{\rho V_0}{\sqrt{e}} = 0$$

$$a\left(1 + \frac{1}{e} - 2\frac{\rho}{\sqrt{e}}\right) = \frac{1}{e} - \frac{\rho}{\sqrt{e}}$$

$$a = \frac{1-\rho\sqrt{e}}{1-2\rho\sqrt{e}+e}$$

After some algebraic simplifications, we have

$$V_\theta(T) = \frac{(1-\rho^2)V_0}{1+e-2\rho\sqrt{e}}$$

$$= \frac{(1-\rho^2)V_0}{1+e-2\rho\sqrt{e}+\rho^2-\rho^2}$$

$$= \frac{(1-\rho^2)V_0}{(1-\rho^2)+(\sqrt{e}-\rho)^2}$$

$$V_\theta(T) \le V_\theta(T_0) \qquad \text{since } [(1-\rho^2)V_0/\{(1-\rho^2)+(\sqrt{e}-\rho)^2\}] < 1$$

This is a contradiction that T_0 is UMVUE. Therefore, $V_\theta(T) = V_\theta(T_0)$, if and only if, $\rho_\theta = \sqrt{e}$.

(iii) On putting $e(T_1) = e(T_2) = e$ in the result of (i), we get

$$\rho \in [e \mp (1-e)] = [2e-1, 1]$$

or
$$\rho \geq 2e - 1$$

***Example* 3.16** Let T_0 be a UMVUE of θ and T_1 be some estimator of θ in $\mathcal{U}(\theta)$ with efficiency $e_\theta < 1$. Using the result $\rho_\theta = \sqrt{e_\theta}$, ρ_θ is a correlation coefficient between T_0 and T_1. Show that no unbiased estimator of θ in $\mathcal{U}(\theta)$, as a linear combination of T_0 and T_1, can be a UMVUE of θ.

Solution. Consider an estimator as a convex combination of T_0 and T_1

$$T = a_0T_0 + a_1T_1;\ a_0 + a_1 = 1;\ a_0 \neq 0$$

and
$$E_\theta(T) = (a_0 + a_1)\theta = \theta\ \forall\ \theta$$

Clearly, $T \in \mathcal{U}(\theta)$. The variance of T is

$$V_\theta(T) = a_0^2V_\theta(T_0) + a_1^2V_\theta(T_1) + 2a_0a_1\text{cov}_\theta(T_0, T_1)$$

Now,
$$V_\theta(T_1) = \frac{V_\theta(T_0)}{e_\theta},\ \rho_\theta = \sqrt{e_\theta}\ \text{ gives}$$

$$V_\theta(T) = a_0^2V_\theta(T_0) + a_1^2\frac{V_\theta(T_0)}{e_\theta} + 2a_0a_1\sqrt{e_\theta}\frac{V_\theta(T_0)}{\sqrt{e_\theta}}$$

$$= V_\theta(T_0)\left(a_0^2 + \frac{a_1^2}{e_\theta} + 2a_0a_1\right)$$

Since $e_\theta < 1$, we have $(a_1^2/e_\theta) > a_1^2 \Rightarrow a_0^2 + (a_1^2/e_\theta) + 2a_0a_1 > a_0^2 + a_1^2 + 2a_0a_1$

$$a_0^2 + \frac{a_1^2}{e_\theta} + 2a_0a_1 > (a_0 + a_1)^2 = 1 \text{ gives } V_\theta(T) \geq V_\theta(T_0)\ \forall\ \theta$$

This shows that there cannot be an estimator T of θ in $\mathcal{U}(\theta)$, as a convex combination of T_0 and T_1, which is UMVUE of θ.

***Example* 3.17**

(i) Let T_1 and T_2 be two unbiased estimators of θ with $V(T_1) = \sigma_1^2$, $V(T_2) = \sigma_2^2$, $\rho(T_1, T_2) = \rho$. Define an unbiased estimator

$$T = a_1T_1 + a_2T_2$$

so that $a_1 + a_2 = 1$. For what values of a_1 and a_2, is $V(T)$ the least? Also, find the least variance.

(ii) If in (i) T_1 has minimum variance

$$V(T_1) = \sigma^2$$

and $$V(T_2) = \frac{\sigma^2}{e}$$

then show that $$\rho(T_1, T_2) = \sqrt{e}$$

that is, show that the correlation coefficient between a most efficient estimator and any other estimator with efficiency e is $\sqrt{e}$.

Solution.

(i) T is unbiased.

$$V(T) = a_1^2\sigma_1^2 + a_2^2\sigma_2{}^2 + 2\rho a_1 a_2 \sigma_1 \sigma_2 \tag{3.4.7}$$

Define a Lagranges function

$$\phi(a_1, a_2) = V(T) - \lambda(a_1 + a_2 - 1)$$

On solving the equations

$$\frac{\partial \phi}{\partial a_1} = 0 = \frac{\partial \phi}{\partial a_2}$$

$$2a_1\sigma_1^2 + 2\rho a_2 \sigma_1 \sigma_2 - \lambda = 0 \tag{3.4.8}$$

and $$2a_2\sigma_2^2 + 2\rho a_1 \sigma_1 \sigma_2 - \lambda = 0 \tag{3.4.9}$$

We can write Eqs. (3.4.8) and (3.4.9) as

$$2a_1^2\,\sigma_1^2 + 2\rho a_1 a_2 \sigma_1 \sigma_2 = a_1\lambda$$

and $$2a_2^2\,\sigma_2^2 + 2\rho a_1 a_2 \sigma_1 \sigma_2 = a_2\lambda$$

On adding them, we get

$$2a_1^2\,\sigma_1^2 + 2a_2^2\,\sigma_2^2 + 4\rho a_1 a_2 \sigma_1 \sigma_2 = \lambda$$

On putting the value of λ in Eqs. (3.4.8) and (3.4.9) and solving, we get

$$a_1 = \frac{\sigma_2^2 - \rho\sigma_1\sigma_2}{\sigma_1^2 + \sigma_2^2 - 2\rho\sigma_1\sigma_2} \tag{3.4.10}$$

and $$a_2 = \frac{\sigma_1^2 - \rho\sigma_1\sigma_2}{\sigma_1^2 + \sigma_2^2 - 2\rho\sigma_1\sigma_2} \tag{3.4.11}$$

On putting these values in Eq. (3.4.7), we get the minimum variance.

(ii) Consider an estimator

$$T = a_1 T_1 + a_2 T_2$$

for such a_1 and a_2 so that T is unbiased. Given that T_1 is the minimum variance or the most efficient, by Example 3.16, we cannot have any T which can be the most efficient other than T_1 itself. So,

$$a_1 = 1 \quad \text{and} \quad a_2 = 0$$

On using Eq. (3.4.8), we have

$$\frac{(\sigma^2/e) - \rho\sigma(\sigma/\sqrt{e})}{\sigma^2 + (\sigma^2/e) - 2\rho\sigma(\sigma/\sqrt{e})} = 1$$

$$\Rightarrow \qquad \frac{1/e - (\rho/\sqrt{e})}{\{1 + (1/e)\} - 2(\rho/\sqrt{6e})} = 1$$

which on solving gives

$$\rho = \sqrt{e}$$

***Example* 3.18** Let $X_1, X_2, \ldots, X_n$ be a random sample with $E(X_i) = \theta$, $V(X_i) = \sigma_i^2$ for $i = 1, \ldots, n$. Let a linear estimator $\delta(\mathbf{X}) = \Sigma_{i=1}^n \alpha_i X_i$ be unbiased for θ and have minimum variance. Suppose δ' be another linear unbiased estimator of θ. Show that

$$\text{cov}_\theta(\delta, \delta') = V_\theta(\delta)$$

Solution. We have seen in Example 3.2 that minimum variance linear unbiased estimator is $\delta(\mathbf{X}) = \Sigma_{i=1}^n \alpha_i X_i$, with $\alpha_i = (1/\sigma_i^2)/\Sigma_{i=1}^n(1/\sigma_i^2)$ and minimum variance is given by

$$V[\delta(\mathbf{X})] = \frac{1}{\sum_{i=1}^{n}\left(\frac{1}{\sigma_i^2}\right)}; \quad \sum \alpha_i = 1$$

Suppose δ' is any other linear unbiased estimator

$$\delta' = \sum_i \beta_i X_i; \ \sum_i \beta_i = 1$$

Note that $\Sigma_i \beta_i = 1$ comes from the unbiasedness of T'.

$$\text{cov}_\theta(\delta, \delta') = \text{cov}_\theta\left(\sum \alpha_i X_i, \sum \beta_i X_i\right) = \sum \alpha_i \beta_i V_\theta(X_i) + \sum\sum_{i \neq j} \alpha_i \beta_i \text{cov}_\theta(X_i, X_j) = \sum \alpha_i \beta_i V_\theta(X_i)$$

Since $\text{cov}_\theta(X_i, X_j) = 0 \ \forall \ i \neq j$, X_i's are independent. On putting the values of α_i and $V(X_i)$, we get

$$\text{cov}_\theta(\delta, \delta') = \frac{1}{\sum_{i=1}^{n}\left(\frac{1}{\sigma_i^2}\right)} \sum_i \left(\frac{1}{\sigma_i^2}\right)\beta_i \sigma_i^2 = \frac{1}{\sum_{1}^{n}\left(\frac{1}{\sigma_i^2}\right)} = V_\theta(\delta) \qquad \left(\text{Since} \sum \beta_i = 1\right)$$

***Example* 3.19** Suppose δ_1 and δ_2 are two unbiased estimators of θ with $V_\theta[\delta_1(X)] = \alpha\sigma^2 = V_\theta[\delta_2(X)]$, $\alpha > 1$, σ^2 is the variance of UMVUE. Show that the correlation coefficient between δ_1 and δ_2 is

$$\rho(\delta_1, \delta_2) \geq \frac{(2 - \alpha)}{\alpha}$$

Solution. Consider an unbiased estimator of θ

$$\delta_{1/2} = \frac{1}{2}(\delta_1 + \delta_2)$$

and let T be UMVUE of θ. This gives

$$V_\theta(T) \le V_\theta(\delta_{1/2})$$

$$\sigma^2 \le \frac{1}{4}[V_\theta(\delta_1) + V_\theta(\delta_2) + 2\text{cov}_\theta(\delta_1, \delta_2)]$$

$$= \frac{1}{2}(\alpha\sigma^2 + \rho\alpha\sigma^2);\ \rho \ge \frac{2-\alpha}{\alpha}$$

***Example* 3.20** Let X follows negative binomial distribution with *pmf*

$$P_\theta(X = x) = \binom{x+r-1}{r-1}\theta^r(1-\theta)^x,\quad x = 0, 1, \ldots$$

Find an unbiased estimation of $1/\theta$.

Solution. Note that $E(X) = r(1 - \theta)/\theta$. [if $Y_i \sim$ geometric distribution with parameter θ, then $X = \Sigma_1^r Y_i \sim \text{NB}(r, \theta)$ and $E(Y) = (1 - \theta)/\theta$]. Hence,

$$E\left(\frac{X}{r}\right) = \frac{1-\theta}{\theta};\quad E\left(\frac{X}{r}+1\right) = \frac{1-\theta}{\theta}+1 = \frac{1}{\theta}$$

Therefore, $(X + r)/r$ is unbiased for $1/\theta$. However, an unbiased estimator for $1/(1 - \theta)$ does not exist in the negative binomial distribution.

***Example* 3.21** Let $X \sim \text{NB}(1,\theta)$ (geometric distribution), and $X_1, X_2,\ldots,X_n$ be a random sample from this population. Find the UMVUE of $P_\theta(X = 0)$.

Solution. The *rv* X follows geometric distribution with parameter θ with *pmf*

$$P_\theta(x) = \theta(1 - \theta)^x,\ x = 0, 1, 2,\ldots$$

The joint density of sample observations is

$$P_\theta(\mathbf{x}) = \theta^n(1-\theta)^{\sum x_i} = \theta^n \exp\{\sum x_i \log(1-\theta)\}$$

which is an exponential family. Thus, $T = \Sigma X_i$ is complete sufficient statistic for θ having distribution $\text{NB}(n,\theta)$. Further, an unbiased estimator of $P_\theta(X = 0)$ is

$$U(\mathbf{X}) = \begin{cases} 1, & \text{if } X_1 = 0 \\ 0, & \text{otherwise} \end{cases}$$

Consider the Rao–Blackwellized estimator

$$\delta(t) = E_\theta[U(\mathbf{X})|\sum X_i = t] = P_\theta\left(X_1 = 0|\sum X_i = t\right) = \frac{P_\theta(X_1 = 0, \sum_{i=2}^n X_i = t)}{P_\theta\left(\sum X_i = t\right)} = \frac{P_\theta(X_1 = 0)P_\theta\left(\sum_{i=2}^n X_i = t\right)}{P_\theta\left(\sum X_i = t\right)}$$

(since sample observations are independent)

$$= \frac{\theta\binom{t+(n-1)-1}{t}\theta^{n-1}(1-\theta)^t}{\binom{t+n-1}{t}\theta^n(1-\theta)^t} = \frac{\binom{t+n-2}{t}}{\binom{t+n-1}{t}}$$

Therefore, $\delta(T) = \binom{T+n-2}{T}\Big/\binom{T+n-1}{T}$ is a UMVUE for $P_\theta(X = 0)$, since T is complete.

***Example* 3.22** Let X be a single observation from

(i) Bernoulli, $b(1, \theta)$. Show that the unbiased estimator of θ^2 does not exist.
(ii) Bernoulli, $b(1, \theta^2)$. Show that the unbiased estimator of θ does not exist.

Solution.

(i) If h is an unbiased estimator of θ, it must satisfy

$$E_\theta h(X) = \theta^2 \ \forall\ \theta \in (0, 1)$$

$$\sum_{x=0}^{1} h(x)\theta^x(1-\theta)^{1-x} = \theta^2$$

$$h(1)\theta + h(0)(1 - \theta) = \theta^2$$

$$\theta^2 - [h(1) - h(0)]\theta - h(0) = 0$$

We make use of a standard result: if a power series vanishes in an open interval, then each of the coefficients must be zero. Note that the coefficient of θ^2 is not zero. Therefore, $E_\theta\, h(X) = \theta^2$ cannot hold. Thus, there does not exist any h which is unbiased for θ^2.

(ii) Similarly, in this case, we get

$$h(1)P_\theta(X = 1) + h(0)P_\theta(X = 0) = \theta$$

$$h(1)\theta^2 + h(0)(1 - \theta^2) = \theta$$

$$[h(1) - h(0)]\theta^2 - \theta + h(0) = 0$$

Note that the coefficient of θ is not zero. Therefore, $E_\theta h(X) = \theta$ cannot hold. Thus, there does not exist any h which is unbiased for θ.

***Example* 3.23 (Unbiased estimators do not always exist).** Let X follows binomial distribution $b(n, \theta)$; $\theta \in (0,1)$. Show that the unbiased estimators of $1/\theta$ and $\theta/(1 - \theta)$ do not exist.

Solution. Let $h(X)$ be an unbiased estimator of $1/\theta$

$$\sum_{x=0}^{n} h(x)\binom{n}{x}\theta^x(1-\theta)^{n-x} = \frac{1}{\theta} \tag{3.4.12}$$

Note that the left-hand side of the above equation is a polynomial in θ. We say $1/\theta$ is unbiased estimable if there exists some polynomial in θ on the left side of the equation that is equal to $1/\theta$ for all $\theta \in (0, 1)$. If we allow $\theta \to 0$, i.e., the probability of getting success approaching 0, then $X \to 0$. Therefore, the left-hand side of Eq. (3.4.12) will approach $h(0) < \infty$. Whereas, the right-hand side of the equation will approach ∞. This shows that there does not exist a polynomial in θ on the left side of Eq. (3.4.12) which is equal to $1/\theta$ for $\theta \to 0$; and thus, we state that there does not exist an unbiased estimator of $1/\theta$.

Similarly, $\theta/(1 - \theta)$ is unbiased estimable if for some T,

$$\sum_{x=0}^{n} h(x)\binom{n}{x}\theta^x (1-\theta)^{n-x} = \frac{\theta}{1-\theta}$$

If $\theta \to 1$, the left-hand side approaches $h(1) < \infty$ and the right-hand side approaches ∞, and we also conclude that no unbiased estimator $h(X)$ of $\theta/(1 - \theta)$ exists.

Another way of checking for the existence of an unbiased estimator for $(1/\theta)$ is, by assuming that T is an unbiased estimator

$$E_\theta[h(X)] = \frac{1}{\theta}$$

Assume $\theta/(1 - \theta) = u$. We have $\theta = u/(1 + u), 1 - \theta = 1 - [u/(1 + u)] = 1/(1 + u)$, and

$$\sum_{x=0}^{n} T(x)\binom{n}{x}\left(\frac{\theta}{1-\theta}\right)^x (1-\theta)^n = \frac{1}{\theta}$$

$$\sum_{x=0}^{n} T(x)\binom{n}{x}u^x = \frac{1}{u}(1+u)^{n+1} = \sum_{x=1}^{n+1}\binom{n+1}{x}u^{x-1} = \sum_{x=-1}^{n}\binom{n+1}{x+1}u^x$$

Since the power of u on the right-hand side starts from u^{-1}, we cannot compare it with the left-hand side. Therefore, the unbiased estimator of $1/\theta$ is not possible.

***Example* 3.24 (Lehmann).** Let $X \sim b(1,\theta)$. Find an UMVUE of (i) $g(\theta) = \theta(1 - \theta)$ and (ii) $g(\theta) = \theta^2$.

Solution.

(i) $T = \Sigma X_i$ is a complete sufficient statistics for θ. The unbiasedness condition of an estimator $h(X)$ for estimating $\theta(1 - \theta)$ gives

$$E_\theta h(T) = \theta(1 - \theta)$$

$$\sum_{t=0}^{n} h(t)\binom{n}{t}\theta^x (1-\theta)^{n-t} = \theta(1-\theta)$$

$$\sum_{t=0}^{n} h(t)\binom{n}{t}\left(\frac{\theta}{1-\theta}\right)^t = \frac{\theta}{(1-\theta)^{n-1}}$$

Assume $u = \theta/(1 - \theta)$. We have

$$\sum_{t=0}^{n} h(t)\binom{n}{t}u^t = u(1+u)^{n-2} = \sum_{t=1}^{n-1}\binom{n-2}{t-1}u^t \text{ for } u \in (0, \infty)$$

Comparing the powers of u^t on both sides of the equation, we have

$$h(T) = \begin{cases} \dfrac{\binom{n-2}{T-1}}{\binom{n}{T}} = \dfrac{T(n-T)}{n(n-1)}, & T = 1, \ldots, n-1 \\ 0, & \text{otherwise} \end{cases}$$

which is a UMVUE of $\theta(1 - \theta)$.

Another way to proceed is to consider an unbiased estimator of $\theta(1 - \theta)$

$$U(\mathbf{X}) = \begin{cases} 1, & X_1 = 1, X_2 = 0 \\ 0, & \text{otherwise} \end{cases}$$

$E[U(\mathbf{X})] = \theta(1 - \theta)$. Therefore,

$$E[U(\mathbf{X}) | T] = P\left[X_1 = 0, X_2 = 0 \middle| \sum_{i=1}^{n} X_i = t\right]$$

$$= \frac{P\left[X_1 = 1, X_2 = 0, \sum_{i=3}^{n} X_i = t-1\right]}{P\left[\sum_{i=1}^{n} X_i = t\right]}$$

$$= \frac{\theta(1-\theta)\binom{n-2}{t-1}\theta^{t-1}(1-\theta)^{n-t-1}}{\binom{n}{t}\theta^t(1-\theta)^{n-t}} = \frac{\binom{n-2}{t-1}}{\binom{n}{t}}$$

It is the UMVUE of $\theta(1 - \theta)$.

(ii) Since

$$h(T) = \frac{T(n-T)}{n(n-1)} = \frac{T}{n} - \frac{T(T-1)}{n(n-1)}$$

is the UMVUE of $\theta(1 - \theta)$,

$$h(T) = \frac{T(T-1)}{n(n-1)}$$

is the UMVUE of θ^2. Another way to proceed is to consider an unbiased estimator of θ^2

$$U(\mathbf{X}) = \begin{cases} 1, & X_1 = 1, X_2 = 1 \\ 0, & \text{otherwise} \end{cases}$$

so that $E[U(\mathbf{X})] = \theta^2$. Therefore, $E[U(\mathbf{X})|T] = T(T-1)/[n(n-1)]$ is the UMVUE of θ^2.

***Example* 3.25** Let $X_1, X_2,\ldots,X_n$ be *iid* from $b(1,\theta)$, where θ is the probability of success. Find the UMVUE of (i) θ, (ii) θ^r, $r \le n$, where r is an integer, (iii) $\theta^r + (1-\theta)^{n-r}$, and (iv) $\theta^r(1-\theta)^s$. (v) Derive the one-step jackknife estimator of θ^2 based on the biased estimator $\delta(\mathbf{X}) = [(1/n)\Sigma_1^n X_i]^2$ to reduce its bias. Show that this jackknife estimator has reduced the bias; in this case, it reduces the bias to zero. Show that this jackknife estimator is a UMVUE of θ^2.

Solution.

(i) The joint *pmf* of sample observations is given by

$$P(x_1, x_2,\ldots, x_n; \theta) = \theta^{\sum x_i}(1-\theta)^{n-\sum x_i}$$

The statistic $T = \Sigma X_i$ is a complete sufficient statistic for θ since the *pmf* belongs to the exponential family of distributions. Thus, T/n is a UMVUE of θ. An alternative way of showing that T/n is the UMVUE of θ is to consider a poor unbiased estimator such as X_1 and to improve it by the Lehmann–Scheffe theorem through a complete sufficient statistic $T = \Sigma X_i$

$$\phi(t) = E_\theta[X_1|T = t]$$

Since $E_\theta\,[X_1|T = t] = E_\theta\,[X_i|T = t]\ \forall\ i = 1, 2,\ldots,n$, $\phi(t)$ simplifies to

$$\phi(t) = \frac{1}{n}\sum_{i=1}^{n} E_\theta[X_i|T = t] = \frac{1}{n}E_\theta\left[\sum_{i=1}^{n} X_i\Big|T = t\right] = \frac{t}{n}$$

In this example, the estimator $\phi(T)$ is the only function of the sufficient statistic T that is unbiased estimator of θ; therefore, $\phi(T)$ is a UMVUE. This has happened, since T is a complete sufficient statistic for θ. The performance of a Rao–Blackwellized estimator depends on the choice of the initial estimator. However, if we consider the trivial sufficient statistic $T(\mathbf{X}) = \mathbf{X}$, the Rao–Blackwell theorem pushes back the initial estimator as the Rao–Blackwellized estimator since

$$\phi(\mathbf{x}) = E_\theta[\delta(\mathbf{X})|\mathbf{X} = \mathbf{x}] = \delta(\mathbf{x})$$

for some initial estimator $\delta(\mathbf{X})$.

(ii) We have seen in the previous example that an unbiased estimator of some function of θ exists if and only if it can be expressed as a polynomial of degree not higher than n. Consider the estimation of θ^r, $r \le n$, where

$$\theta^r = P(X_1 = 1 = X_2 = \cdots = X_r)$$

Its unbiased estimator is given by

$$U(\mathbf{X}) = \begin{cases} 1, & \text{if } X_1 = 1 = X_2 = \cdots = X_r \\ 0, & \text{otherwise} \end{cases}$$

Since $T = \Sigma X_i$ is a complete sufficient statistic, the UMVUE of θ^r is given by

$$E[U(\mathbf{X})|T = t] = \begin{cases} > 0, & \text{if } r \le t \le n \\ = 0, & \text{if } r > t \end{cases}$$

In fact, $E[U(\mathbf{X})|T = t]$ is the ratio of probability of getting r successes in the first r trials and $(t - r)$ successes in the rest $(n - r)$ trials to the probability that $T = t$. We have

$$\delta_r(t) = E[U(\mathbf{X})|T = t] = \begin{cases} \dfrac{\binom{n-r}{t-r}}{\binom{n}{t}}, & 0 \le r < t \le n \\ 1\Big/\binom{n}{t}, & r = t \end{cases}$$

Therefore, the estimator $\delta_r\,(T)$ is a UMVUE of θ^r. We conclude that θ^r, $r \le n$, is unbiased-estimable if and only if it is a polynomial of degree $\le n$. Note that $T(T - 1)/n(n - 1)$ is a UMVUE of θ^2.

(iii) Consider the estimation of$(1 - \theta)^{n-r}$ where

$$(1 - \theta)^{n-r} = P(X_1 = 0 = X_2 = \cdots = X_{n-r})$$

Its unbiased estimator is given by

$$W(\mathbf{X}) = \begin{cases} 1, & \text{if } X_1 = 0 = X_2 = \cdots = X_{n-r} \\ 0, & \text{otherwise} \end{cases}$$

and $$\delta_{n-r}(t) = E[W(\mathbf{X})|T = t] = \frac{(1-\theta)^{n-r}\binom{r}{t}\theta^t(1-\theta)^{r-t}}{\binom{n}{t}\theta^t(1-\theta)^{n-t}}$$

$$= \begin{cases} \binom{r}{t}\Big/\binom{n}{t}, & 0 \le t < r \le n \\ 1\Big/\binom{n}{t}, & r = t \end{cases}$$

Therefore, $\delta_0 = \delta_r(T) + \delta_{n-r}(T)$

$$\delta_0(T) = \begin{cases} \binom{r}{t}\Big/\binom{n}{t}, & 0 \le t < r \\ 2\Big/\binom{n}{t}, & r = t \\ \binom{n-r}{t-r}\Big/\binom{n}{t}, & r+1 \le t \le n \end{cases}$$

is a UMVUE of $\theta^r + (1 - \theta)^{n-r}$.

(iv) Consider, now, the estimation of $\theta^r(1-\theta)^s$, where

$$\theta^r(1-\theta)^s = P(X_1 = 1, X_2 = 1, \ldots, X_r = 1, X_{r+1} = 0, \ldots, X_{r+s} = 0)$$

Its unbiased estimator is given by

$$U(\mathbf{X}) = \begin{cases} 1, & \text{if } X_1 = 1, X_2 = 1, \ldots, X_r = 1, X_{r+1} = 0, \ldots, X_{r+s} = 0 \\ 0, & \text{otherwise} \end{cases}$$

Since $T = \Sigma X_i$ is complete sufficient statistic, the UMVUE of $\theta^r(1-\theta)^s$ is given by

$$E[U(\mathbf{X})|T = t] = \frac{P\left[X_1 = 1, X_2 = 1, \ldots, X_r = 1, X_{r+1} = 0, \ldots, X_{r+s} = 0, \ \sum_{i=r+s+1}^{n} X_i = t - r\right]}{P\left[\sum_1^n X_i = t\right]}$$

$$= \frac{\theta^r(1-\theta)^s \binom{n-r-s}{t-r} \theta^{t-r}(1-\theta)^{n-s-t}}{\binom{n}{t}\theta^t(1-\theta)^{n-t}}$$

$$= \begin{cases} \binom{n-r-s}{t-r} \Big/ \binom{n}{t}, & 0 \le t < r \le n \\ 1 \Big/ \binom{n}{t}, & r = t \end{cases}$$

(v) In fact, jackknife is a technique of bias reduction in an estimator which was initially proposed by Quenoulle (1956). This technique is described as follows:

Let $X_1, X_2, \ldots, X_n$ be a random sample from a given distribution $f(x;\theta)$. Based on these observations, let $\delta_n(\mathbf{X})$ be a biased estimator of θ. Let $\delta_n^{(i)}$ be the estimator obtained from δ_n based on $(n-1)$ observations after removing the ith observation from the sample. The one-step jackknife estimator of θ is defined as

$$jk(\delta_n) = n\delta_n - \frac{n-1}{n}\sum_{i=1}^{n} \delta_n^{(i)}$$

The bias in $jk(\delta_n)$ is smaller than in δ_n [see Casella and Berger (2001)]. Let $T = \Sigma_{i=1}^n X_i$. Then the given biased estimator is $\delta_n = (T/n)^2$. We have $\delta_n^{(i)} = \Sigma_{j\neq i}^n X_j/(n-1)^2$

$$\delta_n^{(i)} = \begin{cases} \dfrac{(T-1)^2}{(n-1)^2}, & \text{if the dropped } X_i \text{ is equal to } 1 \\ \dfrac{T^2}{(n-1)^2}, & \text{if the dropped } X_i \text{ is equal to } 0 \end{cases}$$

$$\sum_{i=1}^{n}\delta_n^{(i)}=\sum_{i=1}^{n}\frac{\left(\sum_{j\neq 1}^{n}X_j\right)^2}{(n-1)^2}=T\frac{(T-1)^2}{(n-1)^2}+(n-T)\frac{T^2}{(n-1)^2}$$

$$jk(\delta_n)=n\frac{T^2}{n^2}-\frac{1}{n(n-1)}\{T(T-1)^2+(n-T)T^2\}$$

$$=\frac{1}{n(n-1)}[T^2(n-1)-T(T-1)^2-(n-T)T^2]$$

$$=\frac{T(T-1)}{n(n-1)}$$

To assess the bias in the jackknife estimator $jk(\delta_n)$, we calculate

$$E(\delta_n)=E\left(\frac{T}{n}\right)^2=\frac{1}{n^2}[V(T)+\{E(T)\}^2]=\frac{1}{n^2}[n\theta(1-\theta)+n^2\theta^2]$$

$$=\frac{\theta}{n}+\left(1-\frac{1}{n}\right)\theta^2 \qquad \left(\text{since } T=\sum_{i=1}^{n}X_i\sim b(n,\theta)\right)$$

$$E(\delta_n^{(i)})=\frac{\theta}{n-1}+\left(1-\frac{1}{n-1}\right)\theta^2$$

$$E[jk(\delta_n)]=\theta+(n-1)\theta^2-[\theta+(n-2)\,\theta^2]=\theta^2$$

Generally, jackknife estimators yield small bias, but in the present case, the one-step jackknife technique has reduced the bias to zero. The jackknife estimator in the present case is UMVUE of θ^2 since it is the function of the complete sufficient statistic $T=\Sigma X_i$ and is an unbiased estimator of θ^2.

***Example* 3.26** Let $X_1, X_2,\ldots,X_n$ be *iid* from $b(k, \theta)$, where θ is the probability of success. Find the UMVUE of

(i) $g(\theta)=\theta$,

(ii) $P_\theta(X=1)=k\theta(1-\theta)^{k-1}$.

Solution.

(i) $T_n=\Sigma X_i$ is a complete sufficient statistic for θ and $E_\theta(T_n)=nk\theta$. Thus, T_n/nk is a UMVUE of θ.

(ii) If one considers a crude unbiased estimator of $P_\theta(X=1)$,

$$U(X_1)=\begin{cases}1, & \text{if } X_1=1\\ 0, & \text{otherwise}\end{cases}$$

We call it crude since it utilizes only one observation X_1 when n observations are available. It is unbiased since $E_\theta(U)=1\cdot P(X_1=1)=k\theta(1-\theta)^{k-1}$. Utilizing the estimator $U(X_1)$ and the complete sufficient statistic T_n in the Lehmann–Scheffe theorem, the UMVUE of $P_\theta(X = 1)$ is given by

$$\delta(T_n) = E(U|T_n = t) \qquad \text{(since } T_n \text{ is sufficient, } E \text{ is free of } \theta)$$

$$= P\left[X_1 = 1 \middle| \sum_{i=1}^{n} X_i = t\right] = \frac{P_\theta(X_1 = 1)P_\theta\left(\sum_{i=2}^{n} X_i = t-1\right)}{P_\theta\left(\sum_{i=1}^{n} X_i = t\right)}$$

(since X_1 is independent of $X_2,\ldots,X_n$)

$$= \frac{k\theta(1-\theta)^{k-1}\binom{k(n-1)}{t-1}\theta^{t-1}(1-\theta)^{k(n-1)-(t-1)}}{\binom{kn}{t}\theta^t(1-\theta)^{kn-t}}$$

$$= k\frac{\binom{k(n-1)}{t-1}}{\binom{kn}{t}}$$

Thus, the UMVUE of $P_\theta(X = 1) = k\theta(1 - \theta)^{k-1}$ is given by

$$\delta(T_n) = k\frac{\binom{k(n-1)}{T_n-1}}{\binom{kn}{T_n}}$$

***Example* 3.27** Let $X_1, X_2,\ldots,X_n$ be *iid* from $\mathcal{U}(0, \theta)$, $\theta > 0$. Find the UMVUE of (i) θ, and (ii) θ^r, (iii) $g(\theta)$, a differentiable function on $(0, \theta)$, and (iv) θ^2 and $1/\theta$ using the result of (iii).

Solution.

(i) The distribution of $T = X_{(n)}$, a complete sufficient statistic for θ, is given by

$$f_{X_{(n)}}(x;\theta) = \frac{nx^{n-1}}{\theta^n} I_{(0,\theta)}(x)$$

Therefore, $E(X_{(n)}) = [n/(n + 1)]\theta$ gives $E_\theta(T) = [n/(n + 1)]\theta$. Similarly, we have $E_\theta(T^2) = [n/(n + 2)]\theta^2$. We, thus, see that T is not an unbiased estimator of θ, though asymptotically unbiased. (We will see this property for large n in Chapter 5). Note that $T_u = [(n + 1)/n]T$ is a UMVUE of θ and its variance is given by $V_\theta(T_u) = [\theta^2/n(n + 2)]$. Further, consider a family of estimators $T_\alpha = \alpha T$, $\alpha \in \mathbb{R}^1$. This class contains biased and unbiased estimators. If we choose $\alpha = [(n + 1)/n]$, then $T_\alpha = T_u$. The mean square estimator of T_α, for some α in $\mathbb{R}^1$, is given by

$$\begin{aligned}\text{MSE}(T_\alpha) &= E_\theta(\alpha T - \theta)^2 \\ &= \alpha^2 E_\theta(T^2) - 2\alpha E_\theta(T\theta) + \theta^2\end{aligned}$$

On minimizing MSE(T_α) for variations in α, we get a choice

$$\alpha_0 = \frac{\theta E_\theta(T)}{E_\theta(T^2)} = \frac{\dfrac{n}{n+1}\theta^2}{\dfrac{n}{n+2}\theta^2} = \frac{n+2}{n+1}$$

The minimum mean square estimator in this class is given by

$$T_0 = \alpha_0 T = \frac{n+2}{n+1} X_{(n)}$$

with minimum mean square error

$$V_\theta(T_0) = \frac{(n+2)^2}{(n+1)^2}\frac{n}{(n+2)(n+1)^2}\theta^2 = \frac{n(n+2)}{(n+1)^4}\theta^2 < \frac{(n+1)^2}{(n+1)^4}\theta^2 < \frac{\theta^2}{n^2+2n} = V_\theta(T_u)$$

Note that the estimator T_0 is not unbiased since

$$E(T_0) = \frac{n+2}{n+1}\frac{n}{n+1}\theta$$

This example shows that sometimes biased estimators are superior to unbiased estimators. In this example, the biased estimator T_0 is superior to T_u, which is a UMVUE of θ.

(ii) We demonstrate the two methods of finding UMVUE of θ^r as follows:

Method 1: If T is a complete sufficient statistic for θ and if $E[\delta(T)] = \theta^r$, then $\delta(T)$ is UMVUE of θ^r. We have $f(x) = (1/\theta)$, $0 < x < \theta$, $F(x) = (x/\theta)$, $T = X_{(n)}$, the nth order statistic. Its distribution is

$$f_n(t) = n\frac{t^{n-1}}{\theta^n}, \quad 0 < t < \theta$$

Now consider $\quad E(T^r) = \int_0^\theta t^r f_n(t)dt = \frac{n}{n+r}\theta^r$

$$E\left(\frac{n+r}{n}T^r\right) = \theta^r$$

This gives $\delta(T) = [(n + r)/n]T^r$. Note that for $r = 1$, $[(n + 1)/n]T$ is the UMVUE of θ as shown in (*i*).

Method 2: Let $U(X)$ be an unbiased estimator of θ^r and let T be a complete sufficient statistic for θ. Then the UMVUE of θ^r is $E(U(X)|T)$. We have

$$E(X^r) = \int_0^\theta x^r \frac{1}{\theta}\,dx = \frac{\theta^r}{r+1}$$

This shows that $U(X) = (r + 1)X^r$ is unbiased for θ^r. Further, note the result

$$E(X^k | X_{(s)} = x) = \frac{x^k}{n} + \frac{s-1}{n} E[X^k | X \le x] + \frac{n-s}{n} E[X^k | X \ge x]$$

Since $T = X_{(n)}$ is a complete sufficient statistic for θ, the UMVUE of θ^r is given by

$$E[U(X)|T = t] = (r + 1)E[X^r|X_{(n)} = t]$$

$$= (r+1)\left[\frac{t^r}{n} + \frac{n-1}{n} E[X^r | X \le t]\right]$$

Now, $$E[X^r | X \le t] = \int_0^t x^r \frac{f(x)}{F(t)} dx$$

(since the distribution of X is truncated from the left)

$$= \frac{\theta}{t} \frac{1}{\theta} \int_0^t x^r dx = \frac{t^r}{r+1}$$

This gives $$\delta(t) = E[U(X) | T = t] = (r+1)\left[\frac{t^r}{n} + \frac{n-1}{n} \cdot \frac{t^r}{r+1}\right]$$

$$= (r+1) \cdot \frac{t^r}{n}\left[1 + \frac{n-1}{r+1}\right] = \left(\frac{n+r}{n}\right) t^r$$

Therefore, $\delta(T) = [(n + r)/n]\, X^r_{(n)}$ is the UMVUE of θ^r.

(iii) Let $T(X_{(n)})$ be an unbiased estimator of $g(\theta)$

$$E_\theta[T(X_{(n)})] = g(\theta) = \int_0^\theta T(x) n \theta^{-n} x^{n-1} dx$$

$$\theta^n g(\theta) = \int_0^\theta nT(x) x^{n-1} dx \;\; \forall\, \theta > 0$$

Differentiating both the sides with respect to θ, we get

$$T(\theta) = g(\theta) + n^{-1} \theta\, g'(\theta) \;\; \forall\, \theta$$

Hence, the UMVUE of $g(\theta)$ is given by

$$T(X_{(n)}) = g(X_{(n)}) + n^{-1} X_{(n)}\, g'(X_{(n)}), \quad 0 < X_{(n)} < \theta$$

(iv) Using the result of (iii), the UMVUE of $g(\theta) = \theta$ is

$$T(X_{(n)}) = X_{(n)} + n^{-1} X_{(n)} = \left(\frac{n+1}{n}\right) X_{(n)}$$

The UMVUE of $g(\theta) = \theta^{-1}$ with $g'(\theta) = -\theta^{-2}$ is

$$T(X_{(n)}) = \frac{1}{X_{(n)}} + n^{-1}\frac{X_{(n)}}{X_{(n)}^2} = \frac{1}{X_{(n)}}\left(1-\frac{1}{n}\right) = \left(\frac{n-1}{n}\right)\frac{1}{X_{(n)}}$$

The UMVUE of $g(\theta) = \theta^2$ with $g'(\theta) = 2\theta$ is

$$T(X_{(n)}) = X_{(n)}^2 + n^{-1}X_{(n)} \cdot 2X_{(n)} = \left(\frac{2+n}{n}\right)X_{(n)}^2$$

***Example* 3.28** Let $X_1, X_2,\ldots,X_n$ be *iid* from $\mathcal{U}(\theta, \theta+1)$, $\theta > 0$. Find an unbiased estimator of θ.

Solution. Let $Y_i = X_i - \theta$. $Y_1, Y_2,\ldots,Y_n$ are *iid* from $U(0,1)$. The distribution of Y_i's is independent of θ, so is any function of these observations. Therefore, the statistic $V = X_{(n)} - X_{(1)} = Y_{(n)} - Y_{(1)}$ is ancillary and the statistic $T = (X_{(1)}, X_{(n)})$ is not complete though sufficient. We have

$$f(x; \theta) = 1, \ \theta < x < \theta + 1$$

$$F(x) = \int_\theta^x dx = (x-\theta), \quad 1 - F(x) = (1+\theta-x)$$

The *pdf* of $X_{(r)}$ is given by

$$f_{X_{(r)}}(x;\theta) = \frac{n!}{(r-1)!(n-r)!}[F(x)]^{r-1}[1-F(x)]^{n-r} f(x)$$

From Example 2.23, we have

$$\therefore \qquad E(X_{(1)} + X_{(n)}) = 2\theta + 1$$

$$E_\theta\left[\frac{X_{(1)} + X_{(n)}}{2}\right] = \theta + \frac{1}{2}$$

Therefore, $(X_{(1)} + X_{(n)} - 1)/2$ is an unbiased estimator of θ.

***Example* 3.29** Let $X_1, X_2,\ldots,X_n$ be *iid* $G_2(1, \theta)$, $\theta > 0$, (exponential with parameter θ). The *pdf* of $G_2(1, \theta)$ is given by $f(x; \theta) = \theta \exp(-x\theta)$.

(i) Find the UMVUE for θ.
(ii) Find the UMVUE for $g(\theta) = F_\theta(t_0)$ where $F_\theta(t_0) = P_\theta(X_1 \le t_0)$, where t_0 is a fixed real number.
(iii) Find the UMVUE for $g(\theta) = \exp(-\theta)$.

Solution.

(i) $T = \Sigma_1^n X_i$ is a complete sufficient statistic with $T \sim G_2(n, \theta)$. The estimator $(n-1)T^{-1}$ is UMVUE for θ since $E_\theta(T^{-1}) = [\theta/(n-1)]$.
(ii) We now define an unbiased estimator of $P_\theta(X_1 \le t_0)$

$$U(\mathbf{X}) = \begin{cases} 1, & \text{if } X_1 \le t_0 \\ 0, & \text{otherwise} \end{cases}$$

so that $E_\theta[U(\mathbf{X})] = P_\theta(X_1 \le t_0) = 1 - \exp(-\theta t)$.
Therefore, the UMVUE of $P_\theta(X_1 \le t_0)$ is

$$\delta(T) = E_\theta\,[U(\mathbf{X})|T] = P_\theta(X_1 \le t_0\,|T)$$

$$\delta(t) = P_\theta\left(\frac{X_1}{T} \le \frac{t_0}{t}\middle| T = t\right) \text{ when } T = t$$

Since $X_1 \sim G_2(1, \theta)$ and $\Sigma_2^n X_i \sim G_2(n-1, \theta)$, $W = X_1/T \sim Be(1, n-1)$. Note that if $X \sim G_2(p_1, \theta)$, $Y \sim G_2(p_2, \theta)$ and if X and Y are independent, then $X/(X+Y) \sim Be(p_1, p_2)$. This shows that X_1/T is ancillary. The following result can, alternatively, be used to show that the statistic X_1/T is ancillary:
For scale family, if $U(cx_1,\ldots,cx_2) = U(x_1,\ldots,x_n)$, then $U(x_1,\ldots,x_n)$ is ancillary.
Further, it is known that $T = \Sigma_1^n X_i$ is a complete sufficient statistic. Therefore, by Basu's theorem, X_1/T and T are independent. Thus, the UMVUE of $P_\theta(X_1 \le t_0)$ is given by

$$\delta_1(T) = P_\theta\left(\frac{X_1}{T} \le \frac{t_0}{t}\right)$$

$$= (n-1)\int_0^{t_0/t} (1-w)^{n-2}\,dw$$

$$= 1 - \left(1 - \frac{t_0}{t}\right)^{n-1}$$

Therefore

$$\delta_1(T) = \begin{cases} 1 - \left(1 - \dfrac{t_0}{T}\right)^{n-1}, & \text{if } T > t_0 \\ 1, & \text{if } T \le t_0 \end{cases}$$

Thus, $\delta_1(T) = 1 - [1 - \{\min(t_0, T)\}/T]^{n-1}$ is the UMVUE of $P_\theta(X_1 \le t_0)$.

(iii) We may write $\exp(\theta t_0) = P_\theta(X_1 > t_0) = 1 - P_\theta(X_1 \le t_0)$. Its value at $t_0 = 1$ is $\exp(-\theta) = P_\theta(X_1 > 1) = 1 - P_\theta(X_1 \le 1)$. The UMVUE of $\exp(-\theta)$, using the result of (ii), is

$$\delta(T) = \begin{cases} (1 - T^{-1})^{n-1}, & \text{if } T > 1 \\ 1, & \text{if } T \le 1 \end{cases}$$

***Example* 3.30** Let $X_1, X_2,\ldots,X_n$ be a random sample of size n from a life distribution $[E(\theta, 1)]$ with its *pdf*

$$f(x;\theta) = \exp\{-(x - \theta)\}, \quad x > \theta$$

Find the UMVUE of the reliability function

$$R(t, \theta) = P[x \ge t;\, \theta] = \int_t^\infty f(x;\theta)\,dx$$

where $f(x;\theta)$ is *pdf* of the above life distribution.

Solution. The reliability function is a parametric function

$$R(t,\theta) = \int_t^{\infty} \exp\{-(x-\theta)\}dx$$

$$= \exp\{-(t-\theta)\}$$

consider an unbiased estimator of $R(t, \theta)$

$$U(\mathbf{X}) = \begin{cases} 1, & \text{if } X_1 > 1 \\ 0, & \text{otherwise} \end{cases}$$

The statistic $X_{(1)}$ is a complete sufficient statistic. Thus, by Lehmann–Scheffe theorem, the UMVUE of $R(t; \theta)$ is given by

$$E[U(\mathbf{X})|X_{(1)} = x] = P[X_1 > t|X_{(1)} = x]$$

By using the result

$$P[X_1 < t|X_{(1)} = x] = \begin{cases} \dfrac{1-1}{n}\dfrac{F(t)}{F(x)}, & \text{if } t < x \\ \dfrac{1}{n} + \dfrac{n-1}{n}\dfrac{F(t)-F(x)}{1-F(x)}, & \text{if } t \geq x \end{cases}$$

due to Nagaraja and Nevzorov (1997, pp. 271–284), Feller (1965, p. 163) and Galambos and Kotz (1978, p. 62), we have

$$P[X_1 < t|X_{(1)} = x] = \frac{1}{n} + \frac{n-1}{n}\left[1 - \frac{1-F(t)}{1-F(x)}\right], \quad t \geq x$$

$$= 1 - \frac{n-1}{n}\frac{R(t,\theta)}{R(x,\theta)}$$

$$= 1 - \frac{n-1}{n} \cdot \frac{\exp[-(t-\theta)]}{\exp[-(x-\theta)]}$$

$$= 1 - \frac{n-1}{n}\exp[-(t-x)]$$

This gives the UMVUE of $R(t,\theta)$ by

$$E[U(\mathbf{X})|X_{(1)} = x] = \left(1 - \frac{1}{n}\right)\exp[-(t-x)]$$

We will discuss another technique of finding UMVUE of $R(t,\theta)$ due to Sathe and Varde (1969). Consider the transformation

$$Y_i = (n-i+1)(X_{(i)} - X_{(i-1)}),\ X_{(0)} = 0,\ i = 1, 2, \ldots, n$$

where $X_{(i)} = X_{i:n}$ is the ith order statistic Y_is are then mutually independent. The *pdf* of $X_{(1)}$ is given by

$$f(x_{(1)};\theta) = n\exp[-n(x_{(1)} - \theta)],\ x_{(1)} \geq \theta$$

This gives the *pdf* of $Y_1 = n\ X_{(1)}$ by

$$f(y_1;\theta) = \exp[-(y_1 - n\theta)],\ \ y_1 > n\theta$$

To obtain the *pdf*s of Y_i, $i = 2,\ldots,n$, we consider the joint *pdf* of $X_{(i-1)}$, $X_{(i)}$

$$f(x, y;\ \theta) = \frac{n!}{(i-2)!\,0!\,(n-i)!} \qquad [F(x)]^{i-2}[1-F(y)]^{n-i}\ f(x;\theta)f(y;\theta), y \geq x \geq \theta$$

$$= \frac{n!}{(i-2)!(n-i)!}\{1-\exp[-(x-\theta)]\}^{i-2}\cdot\{\exp[-(y-\theta)]\}^{n-i}\cdot\exp[-(x-\theta)]\cdot\exp[-(y-\theta)]$$

Let

$$y_i = (n - i + 1)(y - x)$$

$$v = x$$

We have, $x = v$, $\ y = \dfrac{y_i}{n-i+1} + v$ and

$$|J| = \frac{1}{n-i+1}$$

Thus, the joint *pdf* of y_i, v is given by

$$f(y_i, v; \theta) = \frac{n!}{(i-2)!(n-i)!}\exp[-(v-\theta)]$$

$$\cdot\{1-\exp[-(v-\theta)]\}^{i-2}\left\{\exp\left[-\frac{y_i}{n-i+1}+v-\theta\right]\right\}^{n-i+1}\cdot\frac{1}{n-i+1}$$

This gives the marginal distribution of y_i

$$f(y_i;\theta) = \exp(-y_i)\frac{n!}{(i-2)!(n-i+1)!}\int_{\theta_1}^{\infty}\{\exp[-v-\theta]\}^{n-i+2}\{1-\exp[-(v-\theta)]\}^{i-2}\,dv$$

Let $\exp[-(v - \theta)] = z$. We have

$$f(y_i;\theta) = \exp(-y_i)\frac{n!}{(i-2)!(n-i+1)!}\int_0^1 z^{n-i+1}(1-z)^{i-z}\,dz$$

$$= \exp(-y_i),\ y_i \geq 0$$

Thus, $Y_i \sim G_1(1,\ 1)$, $i = 2,\ldots,n$ and Y_i are mutually independent.

In this technique, we obtain UMVUEs of $R(t,\theta) - 1/n[R(t,\theta)]^n$ and $1/n[R(t,\theta)]^n$ and then add to get the UMVUE of $R(t,\theta)$. Consider, first an unbiased estimator

$$U_1(\mathbf{X}) = \begin{cases} 1 - \dfrac{1}{n}, & \text{if } y_2 + \dfrac{y_1}{n} > t \\ 0, & \text{otherwise} \end{cases}$$

of $R(t, \theta) - (1/n)\,[R(t, \theta)]^n$. The estimator $U_1(\mathbf{X})$ is an unbiased estimator.

This is shown below.

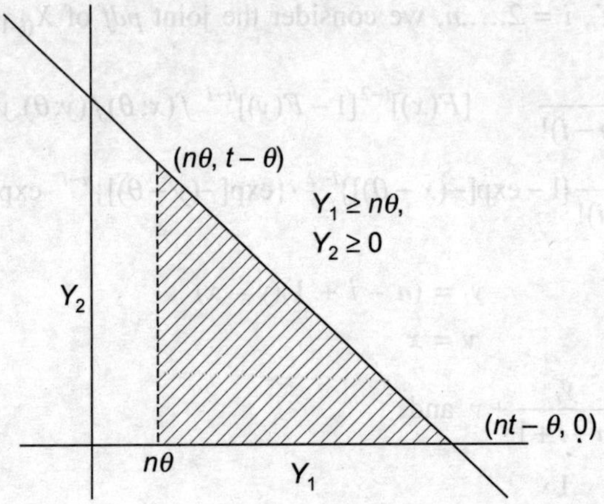

Now, we have

$$P\left(Y_2 + \frac{Y_1}{n} \le t\right) = \int_{n\theta}^{nt} \left(\int_0^{t - \frac{y_1}{n}} \exp(-y_2)\,dy_2 \right) \exp[-(y_1 - n\theta)]\,dy_1$$

$$= \int_{n\theta}^{nt} \left\{ 1 - \exp\left[-\left(t - \frac{y_1}{n} \right) \right] \right\} \exp[-(y_1 - n\theta)]\,dy_1$$

$$= \int_{n\theta}^{nt} \exp[-(y_1 - n\theta)]\,dy_1 - \int_{n\theta}^{nt} \exp\left\{ -\left[\left(1 - \frac{1}{n} \right) y_1 + (t - n\theta) \right] \right\} dy_1$$

$$= 1 - \exp[-n(t-\theta)] + \frac{n}{n-1} \exp\left\{ -\left[\left(1 - \frac{1}{n} \right) nt + (t - n\theta) \right] \right\}$$

$$- \frac{n}{n-1} \exp\left\{ -\left[\left(1 - \frac{1}{n} \right) n\theta + (t - n\theta) \right] \right\}$$

$$= 1 - \exp[-n(t-\theta)] + \frac{n}{n-1} \exp[-n(t-\theta)] - \frac{n}{n-1} \exp[-(t-\theta)]$$

$$= 1 + \frac{1}{n-1} \exp[-n(t-\theta)] - \frac{n}{n-1} \exp[-(t-\theta)]$$

This gives $E[U_1(\mathbf{X})] = Y_2 + (Y_1/n) \le t$, $y_1 \ge n\theta$, $y_2 \ge 0$ gives the range of y_1 and y_2, respectively, $(n\theta, nt)$ and $(0, t - (y_1/n))$. Thus,

$$E[U_1(\mathbf{X})] = \left(1 - \frac{1}{n}\right) P(Y_2 - Y_{1/n} \ge t)$$

$$= \left(1 - \frac{1}{n}\right)\left\{\frac{n}{n-1}\exp[-(t-\theta)] - \frac{1}{n-1}\exp[-n(t-\theta)]\right\}$$

$$= \exp[-(t-\theta)] - \frac{1}{n}\exp[-n(t-\theta)]$$

$$= R(t,\theta) - \frac{1}{n}[R(t,\theta)]^n$$

The first order statistic $X_{(1)}$ is a complete sufficient statistic for θ. Thus, UMVUE of $R(t, \theta) - (1/n)[R(t, \theta)]^n$ is given by

$$E[U_1(\mathbf{X})| X_{(1)}] = \left(1 - \frac{1}{n}\right) P\left[Y_2 + \frac{Y_1}{n} > t \,\middle|\, X_{(1)} = x_{(1)}\right]$$

$$= \left(1 - \frac{1}{n}\right) P[Y_2 > t - x_{(1)}]$$

where, $Y_2 \sim G_1(1, 1)$ is ancillary statistic. Thus,

$$E[U_1(\mathbf{X})| X_{(1)}] = \left(1 - \frac{1}{n}\right) \int_{t-x_{(1)}}^{\infty} \exp(-y_2)dy_2$$

$$= \begin{cases} \left(1 - \frac{1}{n}\right)\exp[-(t - x_{(1)})], & \text{if } x_{(1)} \le t \\ \left(1 - \frac{1}{n}\right), & \text{if } x_{(1)} \ge t \end{cases}$$

is the UMVUE of $R(t, \theta) - (1/n)^{[R(t, \theta)]}$.

Similarly, the UMVUE of $(1/n)[R(t, \theta)]^n$ is given by

$$U_2(\mathbf{X}) = \begin{cases} 1/n, & \text{if } X_{(1)} > t \\ 0, & \text{otherwise} \end{cases}$$

Note that

$$E[U_2(\mathbf{X})] = \frac{1}{n} P[X_{(1)} > t]$$

$$= \frac{1}{n}\int_t^{\infty} \exp[-n(x-\theta)]\,dx$$

$$= \frac{1}{n}\exp[-n(t-\theta)]$$

$$= \frac{1}{n}[R(t,\theta)]^n$$

Thus, the UMVUE of the reliability function $R(t, \theta)$ is given by

$$R^*(t, \hat{\theta}) = \begin{cases} \left(1 - \dfrac{1}{n}\right)\exp[-(t - X_{(n)})], & \text{if } X_{(1)} \le t \\ 1 - \dfrac{1}{n} + \dfrac{1}{n} = 1, & \text{if } X_{(1)} > t \end{cases}$$

***Example* 3.31** Let $X_1, X_2, \ldots, X_n$ be a random sample drawn from the two-parameter exponential-life distribution, $E(\theta, \sigma)$, with *pdf*

$$f(x; \theta, \sigma) = \frac{1}{\sigma}\exp\left\{-\frac{(x-\theta)}{\sigma}\right\}, \qquad x > \theta, \sigma > 0$$

(i) Find the UMVUE of θ when σ is known and σ is unknown.
(ii) Find the UMVUE of σ when θ is known and θ is unknown.
(iii) Find the UMVUE of $g(\theta, \sigma) = E_{\theta,\sigma}X_1$.
(iv) Show that the UMVUE of the reliability function $R(t, \theta) = P(X \ge t; \theta, \sigma)$ is given by

$$R^*(t; \hat{\theta}, \hat{\sigma}) = \begin{cases} 1, & \text{if } t < X_{(1)} \\ \left(1 - \dfrac{1}{n}\right)\left(1 - \dfrac{t - X_{(1)}}{\sum_1^n (X_i - X_{(1)})}\right)^{n-2}, & \text{if } X_{(1)} < t < X_{(1)} + \sum_1^n (X_i - X_{(1)}) \\ 0 & \text{if } t > X_{(1)} + \sum_1^n (X_i - X_{(1)}) \end{cases}$$

Solution. The density $f(x; \theta, \sigma)$ does not belong to an exponential family of distributions, since its support depends on the parameters θ, i.e., it changes with θ, even though it is generally referred to as exponential family with θ as a location parameter and σ as a scale parameter. In fact, this distribution belongs to the location–scale family of distributions.

The distribution of the sufficient statistic $T = X_{(1)}$ is

$$f(t; \theta, \sigma) = \frac{n}{\sigma}\exp\left(-\frac{n}{\sigma}(t - \theta)\right), \quad t > \theta$$

Consider some Borel–measurable function $h(T)$ so that

$$E[h(T)] = \frac{n}{\sigma}\exp\left(\frac{n\theta}{\sigma}\right)\int_\theta^\infty h(t)\exp\left(-\frac{nt}{\sigma}\right)dt = 0$$

This implies

$$\int_\theta^\infty h(t)\exp\left(-\frac{nt}{\sigma}\right)dt = 0$$

Differentiating both the sides with respect to θ gives

$$h(\theta)\exp\left(-\frac{n\theta}{\sigma}\right) = 0 \qquad \forall\, \theta \in (0, \infty)$$

which implies $$h(t)\exp\left(-\frac{nt}{\sigma}\right)=0$$

or $$h(t) = 0$$

Thus, if σ is known, then $T = X_{(1)}$ is complete sufficient statistic for a location parameter θ in the location family $E(\theta, \sigma)$.
Let $Y = (X - \theta)/\sigma$. Then the *pdf* of Y is

$$f(y; \theta, \sigma) = \exp(-y), y > 0$$

i.e., $Y \sim G_1(1, 1)$. Therefore, $Q = \Sigma_1^n(X_i - \theta)/\sigma \sim G_1(n, 1)$.

Now, $$Q=\sum\left(\frac{X_i-\theta}{\sigma}\right)=\sum\frac{(X_i-X_{(1)})}{\sigma}+\frac{n(X_{(1)}-\theta)}{\sigma}$$
$$=Q_2+Q_1$$

$Q_1 \sim G_1(1, 1)$ since the distribution of the complete sufficient statistic $X_{(1)}$ is

$$f(x_{(1)};\theta,\sigma)=\frac{n}{\sigma}\exp\left(-\frac{n}{\sigma}(x_{(1)}-\theta)\right), x_{(1)}>\theta$$

For a location family, a statistic $U(\mathbf{X})$ is ancillary if

$$U(X_1, X_2,\ldots,X_n) = U(X_1 + c, X_2 + c,\ldots,X_n + c), c \in \mathbb{R}^1$$

Clearly, the statistic $\Sigma(X_i - X_{(1)})$ is ancillary. By Basu's theorem, the statistics Q_1 and Q_2 are independent. Thus, $Q_2 \sim G_1(n - 1, 1)$ since $Q \sim G_1(n, 1)$.

(i) When σ is known, $X_{(1)}$ is complete and sufficient for θ and

$$E(Q_1)=E\left[\frac{n(X_{(1)}-\theta)}{\sigma}\right]=1$$

Thus, $X_{(1)} - (\sigma/n)$ is the UMVUE of θ.
When σ is not known, the UMVUE of θ is

$$X_{(1)}-\frac{1}{n(n-1)}\sum(X_i-X_{(1)})$$

as $[1/(n - 1)]\Sigma_1^n(X_i - X_{(1)})$ is the UMVUE of σ shown in (ii).

(ii) When θ is known, $\Sigma(X_i - \theta)$ is a complete sufficient statistic for σ

$$E(Q)=E\left[\frac{\sum(X_i-\theta)}{\sigma}\right]=n$$

$\Sigma(X_i - \theta)/n$ is a UMVUE of σ. When θ is unknown, $E(Q_2) = n - 1$. This implies $\Sigma(X_i - X_{(1)})/(n - 1)$ is a UMVUE of σ.

Note that when (θ, σ) is unknown, we have seen that $X_{(1)}$ and $\Sigma_1^n(X_i - X_{(1)})$ are complete and sufficient for (θ, σ); further, $n(X_{(1)} - \theta)/\sigma$ and $\Sigma_1^n(X_i - X_{(1)})/\sigma$ are independently distributed as $G_1(1, 1)$ and $G_1(n - 1, 1)$.

Therefore, $[1/(n - 1)]\ \Sigma_1^n(X_i - X_{(1)})$ is UMVUE for σ and $X_{(1)} - [1/n(n - 1)]\Sigma_1^n(X_i - X_{(1)})$ is UMVUE for θ.

(iii) $g(\theta, \sigma) = E_{\theta,\sigma}(X_1) = \int_\theta^\infty (x/\sigma) \exp\{-(x - \theta)/\sigma\}\, dx = \theta + \sigma$. Therefore, the UMVUE of $g(\theta, \sigma) = \theta + \sigma$ is given by

$$X_{(1)} - \frac{1}{n(n-1)}\sum_1^n (X_i - X_{(1)}) + \frac{1}{(n-1)}\sum_1^n (X_i - X_{(1)})$$

$$= X_{(1)} + \frac{1}{n}\sum_1^n (X_i - X_{(1)})$$

(iv) Proceeding similarly as in Example 3.30, consider the transformation

$$Y_i = (n - i + 1)(X_{(i)} - X_{(i-1)}),\ X_{(0)} = 0, i = 1, 2, \ldots, n$$

where $X_{(i)}$ is the ith order statistic. Y_is are then mutually independent the *pdf* of $X_{(1)}$ is given by

$$f(x; \theta, \sigma) = \frac{n}{\sigma}\exp\left[\frac{-n(x_{(1)} - \theta)}{\sigma}\right] \quad x_{(1)} \geq \theta$$

The *pdf* of y_1 is given by

$$f(y_1; \theta, \sigma) = \frac{1}{\sigma}\exp\left[\frac{-(y_1 - n\theta)}{\sigma}\right] \quad y_1 \geq n\theta$$

and Y_i are $G_1(1, \sigma)$, $i = 2, \ldots, n$.

Consider an estimator

$$U_1(\mathbf{X}) = \begin{cases} 1 - \dfrac{1}{n}, & \text{if } Y_2 + \dfrac{Y_1}{n} > t \\ 0, & \text{otherwise} \end{cases}$$

The estimator $U_1(\mathbf{X})$ can be shown to be unbiased for $R(t, \theta, \sigma) - (1/n)[R(t, \theta, \sigma)]^n$ as follows. Also consider

$$E[U_1(\mathbf{X})] = \left(1 - \frac{1}{n}\right) P\left(Y_2 + \frac{Y_1}{n} > t\right)$$

Like as in Example 3.30, we may obtain

$$P\left(Y_2 + \frac{Y_1}{n} \leq t\right) = 1 + \frac{1}{n-1}\exp\left[\frac{-n(t - \theta)}{\sigma}\right] - \frac{n}{n-1}\exp\left[\frac{-(t - \theta)}{\sigma}\right]$$

This gives

$$E[U_1(\mathbf{X})] = R(t, \theta, \sigma) - \frac{1}{n}[R(t, \theta, \sigma)]^n.$$

Since $(X_{(1)})$, $\Sigma(X_i - X_{(1)})$ is jointly complete sufficient statistic for (θ, σ) the UMVUE of $R(t, \theta, \sigma) - (1/n)[R(t, \theta, \sigma)]^n$ is given by

$$\begin{aligned} & E[U_1(\mathbf{X}) | X_{(1)}, \sum_1^n (X_i - X_{(1)})] \\ &= \left(1 - \frac{1}{n}\right) P\left[Y_2 + \frac{Y_1}{n} > t \,\middle|\, X_{(1)} = x_{(1)}, \sum_1^n (X_i - X_{(1)}) = \sum_1^n (x_i - x_{(1)}) \right] \\ &= \left(1 - \frac{1}{n}\right) P\left[\frac{Y_2}{\sum_1^n (X_i - X_{(1)})} > \frac{t - x_{(1)}}{\sum_1^n (x_i - x_{(1)})} \,\middle|\, x_{(1)}, \sum_1^n (x_i - x_{(1)}) \right] \end{aligned}$$

We may express

$$\begin{aligned} \sum_{i=2}^n Y_i &= Y_n + \cdots + Y_2 = (X_{(n)} - X_{(n-1)}) + 2(X_{(n-1)} - X_{(n-2)}) + \cdots \\ &\quad + (n-1)(X_{(2)} - X_{(1)}) - (X_{(1)} - X_{(1)}) \\ &= \sum_1^n X_{(i)} - n X_{(1)} = \sum_1^n (X_{(i)} - X_{(1)}) \end{aligned}$$

$Y_2/\Sigma_2^n Y_i \sim b(1, n-2)$ since $Y_2 \sim G_1(1, \sigma)$, $\Sigma_2^n Y_i \sim G_1(n-1, \sigma)$ and Y_is are independent.

Clearly, $U^* = Y_2/\Sigma_2^n$ is ancillary. Thus, by Basu's theorem, we have

$$\begin{aligned} & E\left[U_1(X) | X_{(1)}, \sum_1^n (x_i - x_{(1)}) \right] \\ &= \left(1 - \frac{1}{n}\right) P[U^* > u^*] \\ &= \left(1 - \frac{1}{n}\right)(n-2) \int_{u^*}^1 (1-x)^{n-3}\, dx \\ &= \left(1 - \frac{1}{n}\right)(1 - u^*)^{n-2} \\ &= \begin{cases} 1 - \dfrac{1}{n}, & \text{if } t < x_{(1)} \\ \left(1 - \dfrac{1}{n}\right)\left(1 - \dfrac{t - x_{(1)}}{\sum (x_i - x_{(1)})}\right)^{n-2}, & \text{if } x_{(1)} < t < x_{(1)} + \sum (x_i - x_{(1)}) \\ 0, & \text{if } t > x_{(1)} + \sum (x_i - x_{(1)}) \end{cases} \end{aligned}$$

Consider an estimator

$$U_2(\mathbf{X}) = \begin{cases} 1/n, & \text{if } \dfrac{Y_1}{n} > t \\ 0, & \text{otherwise} \end{cases}$$

The estimator U_2 is unbiased estimator of $(1/n)[R(t, \theta, \sigma)]^n$ since

$$E[U_2(\mathbf{X})] = \frac{1}{n}P\left[\frac{Y_1}{n} > t\right] = \frac{1}{n}P[X_{(1)} > t]$$

$$= \frac{1}{n}\exp\left[\frac{-n(t-\theta)}{\sigma}\right]$$

$$= \frac{1}{n}[R(t, \theta, \sigma)]^n$$

U_2 is UMVUE of $(1/n)[R(t, \theta, \sigma)]^n$ since it is a function of complete sufficient statistic $(X_{(1)}, \Sigma_1^n(X_i - X_{(1)}))$ and is unbiased.

Thus, by adding the UMVUEs of $R(t, \theta, \sigma) - (1/n)[R(t, \theta, \sigma)]^n$ and $(1/n)[R(t, \theta, \sigma)]^n$, UMVUE of $R(t, \theta, \sigma)$ is given by

$$R^*(t; \hat{\theta}, \hat{\sigma}) = \begin{cases} 1 - \dfrac{1}{n} + \dfrac{1}{n} = 1, & \text{if } t < X_{(1)} \\ \left(1 - \dfrac{1}{n}\right)\left(1 - \dfrac{t - X_{(1)}}{\sum(X_i - X_{(1)})}\right)^{n-2}, & \text{if } X_{(1)} < t < X_{(1)} + \sum_1^n (X_i - X_{(1)}) \\ 0, & \text{if } t > X_{(1)} + \sum_1^n (X_i - X_{(1)}) \end{cases}$$

***Example* 3.32** Let $X_1, X_2, \ldots, X_n$ be *iid* $f(x;\theta)$, where $f(x;\theta)$ is some density for which the mean(θ) and variance (σ^2) exist. Show that

$$E(S_n^2) \neq \sigma^2; \quad \text{where } S_n^2 = \frac{1}{n}\sum_1^n (X_i - \bar{X})^2$$

Suggest an unbiased estimator of σ^2 and θ^2.

Solution. Let $E(X_i) = \theta \ \forall \ i$

$$E_\theta(S_n^2) = \frac{1}{n}E_\theta\left[\sum_1^n (X_i - \bar{X})^2\right]$$

$$= \frac{1}{n}E_\theta\left[\sum_1^n (X_i - \theta)^2 - n(\bar{X} - \theta)^2\right]$$

$$= \frac{1}{n}\left(n\sigma^2 - n\frac{\sigma^2}{n}\right) = \sigma^2\left(1 - \frac{1}{n}\right) \neq \sigma^2$$

Thus, $[n/(n-1)]\, S_n^2$ is unbiased for σ^2. Using

$$V(\bar{X}) = E(\bar{X}^2) - [E(\bar{X})]^2$$

$$E(\bar{X}^2) = \frac{\sigma^2}{n} + \theta^2$$

and

$$E\left(\frac{1}{n-1} S_n^2\right) = \frac{\sigma^2}{n}$$

Thus

$$\bar{X}^2 - \frac{1}{n-1} S_n^2$$

or

$$\bar{X}^2 - \frac{1}{n} S_{n-1}^2$$

is an unbiased estimator of θ^2.

***Example* 3.33** Let $X_1, X_2, \ldots, X_n$ be *iid* $N(\theta, \sigma^2)$, θ known. Find an unbiased estimator of σ based on the statistic $\Sigma_1^n |X_i - \theta|$.

Solution. $$E_{\sigma^2}\left(\sum_1^n |X_i - \theta|\right) = \sum_1^n E|X_i - \theta| \quad \text{(since } X_i\text{s are independent)}$$

$$= nE|X_i - \theta| \quad \text{(since } X_i\text{s are identical)}$$

$$= n\sigma\sqrt{2/\pi}$$

since $E|X_i - \mu|$ is the mean deviation of $N(\theta, \sigma^2)$. Thus, $\sqrt{\pi/2}\,(1/n)\Sigma_1^n |X_i - \mu|$ is an unbiased estimator of σ.

***Example* 3.34** Let $X_1, X_2, \ldots, X_n$ be *iid* $N(\theta, \sigma^2)$.

(i) Find the UMVUE of θ^r, $r = 1, 2, 3$ when σ is known and σ is unknown.
(ii) Find the UMVUE of σ^r when θ is known and θ is unknown.
(iii) Find the UMVUE of θ/σ (when one measures θ in terms of σ units).
(iv) Find the UMVUE of θ/σ^2.
(v) Find the UMVUE of x_p so that $P(X \le x_p) = p \in (0,1)$ (when p is known).
(vi) Find the UMVUE of probability $P_\theta\,(X \le x)$ for some fixed x in $\mathbb{R}^1$ when $\sigma = 1$.
(vii) Denote the *pdf* of $N(\theta,\sigma^2)$ by $f(x;\theta,\sigma^2)$. Show that the UMVUE of $g(\theta) = f(x;\theta,1)$ when $\sigma = 1$ is $f(x;t/n,(n-1)/n)$ where $T(X) = \Sigma_{i=1}^n X_i$.
(viii) Find the UMVUE of the probability $P_\theta(X \le x)$ for some fixed x in $\mathbb{R}^1$ when σ is unknown.

Solution. We know that $(\bar{X}, S^2)$ is jointly complete sufficient statistic for (θ,σ^2).

(i) When σ is known, the statistic $\bar{X}$ is complete and sufficient for θ. Therefore, by the Lehmann–Scheffe theorem, $\bar{X}$ is the UMVUE of θ. Let $Y = \bar{X} - \theta \sim N(0, \sigma^2/n)$. For calculating $E(Y^2)$, note that $E(Y^{2m+1}) = E(\bar{X} - \theta)^{2m+1} = 0$, $m = 0, 1, 2,\ldots$, and $E(Y^{2m}) = 1{\cdot}3\ldots(2m-1)(\sigma^2/n)$, $m = 1, 2,\ldots$. Therefore, $E(Y) = E(\bar{X} - \theta) = 0$ implies that $\bar{X}$ is a UMVUE for θ.

Further, $E(Y^2) = 1\cdot(\sigma^2/n)$; $V(\bar{X}) = E(\bar{X}^2) - \theta^2 = \sigma^2/n$. That is, $(\bar{X}^2 - \sigma^2/n)$ is a UMVUE of θ^2.

Next, $E(\bar{X} - \theta)^3 = 0$ gives $[\bar{X}^3 - 3\bar{X}(\sigma^2/n)]$ as a UMVUE of θ^3.

If σ^2 is not known, letting $V^2 = \Sigma_1^n(X_i - \bar{X})^2$, $V^2/\sigma^2 \sim \chi^2_{n-1}$. So, $E(V^2/\sigma^2) = (n-1)$ gives $E[V^2/(n-1)] = \sigma^2$. Since $(\bar{X}, \Sigma(X_i - \bar{X})^2)$ is jointly complete sufficient statistic for (θ, σ^2), $A^2/(n-1)$ is UMVUE of σ^2. UMVUE of θ^3 is

$$\bar{X}^3 - 3\bar{X}\frac{V^2}{n(n-1)} = \bar{X}^3 - 3\bar{X}\frac{1}{n(n-1)}\sum_1^n (X_i - \bar{X})^2$$

by the independence of $\bar{X}$ and $\Sigma(X_i - \bar{X})^2$.

Similarly, UMVUE of θ^2 is given by

$$\bar{X}^2 - \frac{1}{n(n-1)}\sum (X_i - \bar{X})^2$$

(ii) When θ is known, $U = \Sigma_{i=1}^n(X_i - \theta)^2$ is a complete sufficient statistic for σ^2. For estimating σ, $r = 1$, $A(n, 1) = [\Gamma(n/2)]/[2^{(1/2)}\Gamma((n+1)/2)]$; $A(n,1)\ (\Sigma_{i=1}^n(X_i - \theta)^2)^{1/2}$ is a UMVUE of σ.

Since $U/\sigma^2 \sim \chi_n^2$ for every $\sigma^2 > 0$, we have

$$E\left(\frac{U}{\sigma^2}\right)^{r/2} = \frac{2^{r/2}\Gamma\left(\dfrac{n+r}{2}\right)}{\Gamma\left(\dfrac{n}{2}\right)} = \frac{1}{A(n, r)} \quad \text{(say)}$$

$$E[A(n, r)\, U^{r/2}] = \sigma^r \quad \text{for } r > -n$$

Therefore, $A(n,r)\, U^{r/2} = A(n,r)\, [\Sigma_{i=1}^n(X_i - \theta)^2]^{r/2}$ is a UMVUE of σ^r.

If (θ, σ^2) is unknown, $\bar{X}$ and $V = \Sigma(X_i - \bar{X})^2$ are jointly complete sufficient statistic for (θ, σ^2). Note that $\bar{X}$ is a UMVUE of θ and $(V/\sigma^2) \sim \chi^2_{n-1}$. So,

$$E\left(\frac{V}{\sigma^2}\right)^{r/2} = \frac{1}{A(n-1, r)}$$

$$E[A(n-1, r)V^{r/2}] = \sigma^r$$

$A(n-1, r)\, V^{r/2} = A(n-1, r)\, [\Sigma_1^n(X_i - \bar{X})^2]^{r/2}$ is a UMVUE of σ^r. On taking $r = 2$, we get $A(n-1, 2) = 1/(n-1)$. This gives that

$$A(n-1, 2)V = \frac{1}{n-1}\sum_1^n (X_i - \bar{X})^2 = S^2_{n-1}$$

is a UMVUE of σ^2.

(iii) Note that $\bar{X}$ is UMVUE of θ and

$$A(n-1, -1)\, V^{-1/2} = A(n-1, -1)\, [(n-1)\, S^2]^{-1/2}$$
$$= A(n-1, -1)\, (n-1)^{-1/2}\, S^{-1}$$

is a UMVUE of σ^{-1} provided $-1 > (1-n)$ or $n > 2$.

Since the statistic $A(n-1, -1)\,(n-1)^{-1/2}\,\bar{X}/S$ is an unbiased estimator of (θ/σ) and function of the complete sufficient statistic $(\bar{X}, S)$, it is a UMVUE of (θ/σ) provided $n > 2$.

(iv) $A(n-1, -2)V^{-2/2} = [(n-3)/(n-1)]S^{-2}$ is a UMVUE of σ^{-2}, provided $n > 3$. The estimator $[(n-3)/(n-1)]\,(\bar{X}/S^2)$ is a UMVUE for (θ/σ^2) since it is unbiased for (θ/σ^2) and is the function of the complete sufficient statistic $(\bar{X}, S)$.

(v) Suppose we have scores X_i of n candidates in an IAS preliminary examination and we wish to estimate the cut-off marks which $(1-p)100\%$ candidates must qualify, $X \geq x_p$, so that $P(X \geq x_p) = 1 - p$, for some preassigned probability $(1-p)$. In other words, we wish to estimate x so that

$$P(X \leq x_p) = p = \Phi\left(\frac{x_p - \theta}{\sigma}\right)$$

On solving this for x_p, we get

$$x_p = \theta + \Phi^{-1}(p)\cdot\sigma = \theta + z_p\cdot\sigma$$

On combining the UMVUE $\bar{X}$ of θ and $A(n-1,1)\sqrt{n-1}S$ of σ, we get $\bar{X}+z_p\cdot A(n-1,1)\sqrt{n-1}S$ as a UMVUE of x_p.

(vi) Consider a situation where a statistician working in a cigarette manufacturing company wishes to estimate the probability that tobacco processed by some new method has nicotine percentage below a prescribed level, say 2%. This problem in statistics is known as the problem of reliability of some procedure, as in the present example it is tobacco processing. For simplicity, consider $\sigma = 1$. For a given x, one wishes to estimate $g(\theta) = P_\theta(X \leq x)$.

Let us define an estimator

$$U(\mathbf{X}) = \begin{cases} 1, & \text{if } X_1 \leq x \\ 0, & \text{otherwise} \end{cases}$$

$$E[U(\mathbf{X})] = P_\theta\,[X_1 \leq x]$$

This shows that $U(\mathbf{X})$ is an unbiased estimator of $g(\theta) = P_\theta(X \leq x)$. On utilizing the fact that $T/n = \bar{X}$ is a complete sufficient statistic, UMVUE of $g(\theta)$ is given by

$$\begin{aligned} E[U(\mathbf{X})|T] &= P[X_1 \leq x|T] \\ &= P[X_1 - \bar{X} \leq x - \bar{x}|T = t] \quad \text{for } T = t \end{aligned}$$

$X_1 - T/n$ is ancillary since $X_1 - (T/n) \sim N[0, 1-(1/n)]$. The statistic $X_1 - T/n$ may alternatively be shown as ancillary since it satisfies $U(X_1,\ldots,X_n) = U(c + X_1,\ldots,c + X_n)$. Further, we know that the statistic T/n is complete. Therefore, by Basu's theorem, $X_1 - T/n$ and T/n are independent. Thus,

$$\begin{aligned} E[U(\mathbf{X})|T] &= P\left[X_1 - \frac{T}{n} \leq x - \frac{t}{n}\right] \\ &= \Phi\left[\sqrt{\frac{n}{n-1}}\left(x - \frac{t}{n}\right)\right] \end{aligned}$$

is a UMVUE of $g(\theta) = P_\theta(X \leq x)$ where Φ is the *cdf* of $N(0, 1)$.

(vii) On differentiating the UMVUE $\Phi\left[\sqrt{[n/(n-1)]}(x-(t/n))\right]$ of $g(\theta) = P_\theta(X \le x)$ in (vi) with respect to x, we get $f(x; (t/n), 1 - (1/n))$ as UMVUE of $g(\theta) = f(x;\theta,1)$.

(viii) If σ^2 is unknown, then $U(\mathbf{X})$ in (vi) continues to be an unbiased estimator of the probability $p = P(X_1 \le x)$. The statistic $(\bar{X}, S^2)$, where $\bar{X} = (1/n)\Sigma X_i$, $(n-1)S^2 = \Sigma(X_i - \bar{X})^2$, is complete sufficient for (μ, σ^2). Therefore, UMVUE of $p = P(X_1 \le x)$ is given by

$$E[U(\mathbf{X})|\bar{X}, S] = P[X_1 \le x|\bar{X}, S]$$

$$= P\left[\frac{X_1 - \bar{X}}{S} \le \frac{x_1 - \bar{x}}{s}\middle|\bar{X} = \bar{x}, S = s\right]$$

Since $(X_1 - \bar{X})/S$ is ancillary, appealing to Basu's theorem, $(X_1 - \bar{X})/S$ is independent of $(\bar{X}, S)$. Thus,

$$E[U(\mathbf{X})|\bar{X}, S] = P\left[\frac{X_1 - \bar{X}}{S} \le \frac{x_1 - \bar{x}}{s}\right] = P\left[T \le \sqrt{\frac{n}{n-1}}\frac{x_1 - \bar{x}}{s}\right]$$

is a UMVUE of p.

We have

$$X_1 - \bar{X} \sim N\left(0, \frac{n-1}{n}\sigma^2\right)$$

or

$$\sqrt{\frac{n}{n-1}}\frac{X_1 - \bar{X}}{\sigma} \sim N(0,1)$$

Further

$$\frac{\sum(X_i - \bar{X})^2}{\sigma^2} = \frac{(n-1)S^2}{\sigma^2} \sim \chi^2_{n-1}$$

Therefore, the distribution of T is given by

$$T = \frac{\sqrt{\dfrac{n}{n-1}}\dfrac{X_1 - \bar{X}}{\sigma}}{\sqrt{\dfrac{(n-1)S^2}{(n-1)\sigma^2}}} = \sqrt{\frac{n}{n-1}}\frac{X_1 - \bar{X}}{S} \sim t_{n-1}$$

Therefore,

$$p = P\left(T \le \sqrt{\frac{n}{n-1}}\frac{x_1 - \bar{x}}{s}\right)$$

***Example* 3.35** Let $X_1, X_2, \ldots, X_n$ be *iid* $N(0, \sigma^2)$. Show that the UMVUE $T/n = (1/n)\Sigma_1^n X_i^2$ of σ^2 is not admissible.

Solution. $T = \Sigma_1^n X_i^2$ is a complete sufficient statistic of σ^2; $(T/\sigma^2) \sim \chi^2_n$, $E(T) = n\sigma^2$, $V(T) = 2n\sigma^4$. Since

$$E_{\sigma^2}\left(\frac{T}{n}\right) = \sigma^2$$

T/n is a UMVUE of σ^2. Consider next a class of estimators of the type $T_c = c{\cdot}T$. The risk of T_c is

$$
\begin{aligned}
R(\sigma^2, T_c) &= E_\sigma(\sigma^2 - cT)^2 \qquad \text{(for squared error loss)}\\
&= E\{c(T - n\sigma^2) + (nc-1)\sigma^2\}^2\\
&= c^2 2n\sigma^4 + (nc-1)^2\sigma^4\\
&= \sigma^4\{1 - 2nc + n(n+2)c^2\} \qquad (3.4.13)
\end{aligned}
$$

Minimizing $R(\sigma^2, T_c)$ with respect to c, we have

$$\frac{\partial}{\partial c}R(\sigma^2, T_c) = \sigma^4\{-2n + 2n(n+2)c\} = 0$$

$$c = \frac{1}{n+2}$$

The minimum squared error estimator $T_{[1/(n+2)]} = [T/(n+2)]$. The risk of $T_{[1/(n+2)]}$ is given by

$$R(\sigma^2, T_{1/(n+2)}) = \frac{2\sigma^4}{(n+2)}$$

Further, the risk of T/n is obtained by putting $c = 1/n$ in Eq. (3.4.13)

$$R\left(\sigma^2, \frac{T}{n}\right) = V\left(\frac{T}{n}\right) = \frac{2\sigma^4}{n} > R(\sigma^2, T_{1/(n+2)})$$

This shows that the UMVUE of σ^2, (T/n), is not admissible, though $T_{1/(n+2)}$ is admissible.

***Example* 3.36** Let $X_1, X_2, \ldots, X_m$ be *iid* $N(\theta, \sigma^2)$ and $Y_1, Y_2, \ldots, Y_n$ be *iid* $N(\theta, \tau^2)$.

(i) If θ, σ^2 and τ^2 are all unknown, show that a UMVUE of θ does not exist.
(ii) If σ^2 and τ^2 are known, find the UMVUE of θ.
(iii) Suggest a reasonable estimator of θ when σ^2 and τ^2 are not known.

Solution.

(i) The joint density of sample observations is

$$
\begin{aligned}
f(\mathbf{x}, \mathbf{y}; \theta, \sigma^2, \tau^2) &= \left(\frac{1}{2\pi}\right)^{\frac{m+n}{2}} \frac{1}{\sigma^m \tau^n} \exp\left\{-\frac{1}{2\sigma^2}\sum(x_i - \theta)^2 - \frac{1}{2\tau^2}\sum(y_i - \theta)^2\right\}\\
&= c\exp\left\{-\frac{1}{2\sigma^2}\sum x_i^2 - \frac{1}{2\tau^2}\sum y_i^2 + \frac{\theta}{\sigma^2}\sum x_i + \frac{\theta}{\tau^2}\sum y_i\right\}\\
&= c\exp\left[-\frac{1}{2\tau^2}\left(\frac{\tau^2}{\sigma^2}\sum x_i^2 + \sum y_i^2\right) + \frac{\theta}{\tau^2}\left(\frac{\tau^2}{\sigma^2}\sum x_i + \sum y_i\right)\right]
\end{aligned}
$$

Let us assume that $\alpha = \tau^2/\sigma^2$ is known, say equal to α_0. The joint density is

$$f(\mathbf{x}, \mathbf{y}; \theta, \tau^2, \alpha_0) = c \exp\left[-\frac{1}{2\tau^2}\left(\alpha_0 \sum x_i^2 + \sum y_i^2\right) + \frac{\theta}{\tau^2}\left(\alpha_0 \sum x_i + \sum y_i\right)\right]$$

which is an exponential family with joint complete sufficient statistic $(\alpha_0 \Sigma X_i^2 + \Sigma Y_i^2, \alpha_0 \Sigma X_i + \Sigma Y_i)$ for $(-1/\tau^2, \theta/\tau^2)$ when τ^2/σ^2 is known. We consider an estimator $\delta_{\alpha_0}(\mathbf{X},\mathbf{Y})$ as a function of this complete sufficient statistic which is unbiased for θ

$$E\left(\alpha_0 \sum X_i + \sum Y_i\right) = \theta(\alpha_0 m + n)$$

so that $\delta_{\alpha_0}(X, Y) = (\alpha_0 \Sigma X_i + \Sigma Y_i)/(\alpha_0 m + n)$ is UMVUE of θ. If $\alpha = \tau^2/\sigma^2$ was not known so that $\alpha > 0$, the estimator $\delta_\alpha(\mathbf{X},\mathbf{Y}) = (\alpha \Sigma X_i + \Sigma Y_i)/(\alpha m + n)$ would have continued to be unbiased. Suppose, in this case, $\delta_0(\mathbf{X}, \mathbf{Y})$ is UMVUE of θ. This implies

$$V(\delta_0; \theta, \alpha) \le V(\delta_\alpha; \theta, \alpha) \quad \forall\ \theta, \alpha > 0$$

Therefore, it also holds at $\alpha = \alpha_0$

$$V(\delta_0; \theta, \alpha_0) \le V(\delta_{\alpha_0}; \theta, \alpha_0) \quad \forall\ \theta$$

This along with the fact that the UMVUE δ_{α_0} is a unique estimator of θ implies that

$$\delta_0(\mathbf{x}, \mathbf{y}) = \delta_{\alpha_0}(\mathbf{x}, \mathbf{y}) \text{ a.e. } \mathbf{x}, \mathbf{y}$$

This is a contradiction since α was chosen arbitrarily and δ_0 cannot be equal to several estimators δ_α simultaneously.

(ii) Given that σ^2 and τ^2 are known, an unbiased estimator of θ is given by

$$\beta\bar{X} + (1 - \beta)\,\bar{Y}$$

so that
$$E_\theta(\beta\bar{X} + (1 - \beta)\,\bar{Y}) = \theta$$
$$V_\theta(\beta\bar{X} + (1 - \beta)\bar{Y}) = \beta^2\, V(\bar{X}) + (1 - \beta)^2\, V(\bar{Y})$$
$$= \beta^2 \frac{\sigma^2}{m} + (1 - \beta)^2 \frac{\tau^2}{n}$$

On minimizing $V(\cdot)$ for β, we have

$$\frac{\partial}{\partial\beta} V(\cdot) = 2\beta\frac{\sigma^2}{m} + 2(1 - \beta)(-1)\frac{\tau^2}{n} = 0$$

$$\beta = \frac{\tau^2/n}{(\sigma^2/m) + (\tau^2/n)}$$

Therefore, $(\tau^2/n)/\{(\sigma^2/m) + (\tau^2/n)\}\bar{X} + (\sigma^2/m)/\{(\sigma^2/m) + (\tau^2/n)\}\bar{Y}$ is UMVUE of θ when σ^2 and τ^2 are known.

(iii) A reasonable estimator of θ and when σ^2 and τ^2 are not known is given by

$$\frac{\hat{\tau}^2/n}{\hat{\sigma}^2/m + \hat{\tau}^2/n}\bar{X} + \frac{\hat{\sigma}^2/m}{\hat{\sigma}^2/m + \hat{\tau}^2/n}\bar{Y}$$

where, $\hat{\sigma}^2$ and $\hat{\tau}^2$ are some reasonable estimators of σ^2 and τ^2, respectively.

***Example* 3.37** Let $(X_1,Y_1), (X_2,Y_2),\ldots,(X_n,Y_n)$ be a sample from a bivariate normal population with parameters θ_1, θ_2, σ_1^2, σ_2^2, and ρ. Consider the problem of estimating θ when $\theta_1 = \theta_2 = \theta$. Note that a complete sufficient statistic does not exist in this case. Consider the class of all unbiased estimators

$$T_\alpha = \alpha\bar{X} + (1-\alpha)\bar{Y}$$

(i) Find the variance of T_α.

(ii) Choose $\alpha = \alpha_0$ to minimize $V(T_\alpha)$, and consider the estimator

$$T_{\alpha_0} = \alpha_0\bar{X} + (1-\alpha_0)\bar{Y}$$

Compute $V(T_{\alpha_0})$. If $\sigma_1 = \sigma_2 = \sigma$, the BLUE of θ (in the sense of minimum variance) is

$$T_1 = \frac{\bar{X}+\bar{Y}}{2}$$

irrespective of whether σ and ρ are known or unknown.

(iii) If $\sigma_1 \neq \sigma_2$ and ρ, σ_1, σ_2 are unknown, replace these values in α_0 by their corresponding estimators

$$\hat{\alpha} = \frac{S_2^2 - S_{11}}{S_1^2 + S_2^2 - 2S_{11}}$$

Show that $$T_2 = \bar{Y} + (\bar{X} - \bar{Y})\hat{\alpha}$$

is an unbiased estimator of θ.

Solution. The sample $(X_1,Y_1), (X_2,Y_2),\ldots,(X_n,Y_n)$ is drawn from a bivariate normal population $N(\boldsymbol{\theta}, \Sigma)$, where $\boldsymbol{\theta} = (\theta_1, \theta_2)'$ and variance covariance matrix

$$\Sigma = \begin{pmatrix} \sigma_1^2 & \rho\sigma_1\sigma_2 \\ \rho\sigma_1\sigma_2 & \sigma_2^2 \end{pmatrix}$$

where ρ is the correlation between X and Y. Let the class of estimators be defined as

$$C(\alpha) = \{T_\alpha : T_\alpha = \alpha\bar{X} + (1-\alpha)\bar{Y},\ 0 \le \alpha \le 1\}$$

(i) $V(T_\alpha) = \alpha^2 V(\bar{X}) + (1-\alpha)^2\, V(\bar{Y}) + 2\alpha(1-\alpha)\operatorname{cov}(\bar{X}, \bar{Y})$

$$= \alpha^2\frac{\sigma_1^2}{n} + (1-\alpha)^2\frac{\sigma_2^2}{n} + 2\alpha(1-\alpha)\cdot\rho\sigma_1\sigma_2$$

(ii) For minimizing $V(T_\alpha)$ with respect to α, we set

$$\frac{\partial}{\partial\alpha}V(T_\alpha) = 0$$

$$2\alpha\cdot\frac{\sigma_1^2}{n} - 2(1-\alpha)\frac{\sigma_2^2}{n} + 2(1-2\alpha)\rho\sigma_1\sigma_2 = 0$$

$$\alpha\left(\frac{\sigma_1^2}{n} + \frac{\sigma_2^2}{n} - 2\rho\sigma_1\sigma_2\right) = \frac{\sigma_2^2}{n} - \rho\sigma_1\sigma_2$$

We have $$\alpha_0 = \frac{\sigma_2^2 - n\rho\sigma_1\sigma_2}{\sigma_1^2 + \sigma_2^2 - 2n\rho\sigma_1\sigma_2}$$

The minimum variance estimator is then given by

$$T_{\alpha_0} = \alpha_0\bar{X} + (1 - \alpha_0)\bar{Y}$$

$$V(T_{\alpha_0}) = \frac{\alpha_0^2}{n}(\sigma_1^2 + \sigma_2^2 - 2n\rho\sigma_1\sigma_2) - 2\frac{\alpha_0}{n}(\sigma_2^2 - 2n\rho\sigma_1\sigma_2) + \frac{\sigma_2^2}{n}$$

$$= \frac{1}{n}\left[\frac{(\sigma_2^2 - n\rho\sigma_1\sigma_2)^2}{(\sigma_1^2 + \sigma_2^2 - 2n\rho\sigma_1\sigma_2)} - 2\frac{(\sigma_2^2 - n\rho\sigma_1\sigma_2)^2}{(\sigma_1^2 + \sigma_2^2 - 2n\rho\sigma_1\sigma_2)} + \sigma_2^2\right]$$

$$= \frac{\sigma_1^2\sigma_2^2(1 - n^2\rho^2)}{n(\sigma_1^2 + \sigma_2^2 - 2n\rho\sigma_1\sigma_2)}$$

If $\sigma_1 = \sigma_2 = \sigma$, we have

$$\alpha_0 = \frac{(1 - n\rho)\sigma^2}{2(1 - n\rho)\sigma^2} = \frac{1}{2}$$

The minimum variance estimator is then given by

$$T_1 = \frac{\bar{X} + \bar{Y}}{2}\left(\text{at } \alpha_0 = \frac{1}{2}\right)$$

$$V(T_1) = \frac{\sigma^2}{2n}(1 + n\rho)$$

Note that T_1 does not depend on σ^2 and ρ whereas its variance does.

(iii) Consider again α_0 when $\sigma_1 \neq \sigma_2$ and σ_1, σ_2 and ρ are unknown quantities. Let S_1^2, S_2^2, and S_{11} are estimators of σ_1^2/n, σ_2^2/n, and $\rho\sigma_1\sigma_2$ respectively. By putting these estimators in α, we get an estimator

$$\hat{\alpha} = \frac{S_2^2 - S_{11}}{S_1^2 + S_2^2 - 2S_{11}}$$

Here, $\hat{\alpha}$ is a function of sample variance–covariance matrix $\begin{pmatrix} S_1^2 & S_{12} \\ S_{12} & S_2^2 \end{pmatrix}$ which is independent of sample mean vector $(\bar{X}, \bar{Y})$. Consider the expectation of the estimator

$$T_2 = \bar{Y} + (\bar{X} - \bar{Y})\hat{\alpha}$$
$$= \hat{\alpha}\bar{X} + (1 - \hat{\alpha})\bar{Y}$$

Thus, T_2 is unbiased since it is a convex combination of two unbiased estimators.

***Example* 3.38** Let $X_1, X_2,...,X_n$ and $Y_1, Y_2,...,Y_m$ be independent samples from $N(\mu,\sigma_1^2)$ and $N(\mu,\sigma_2^2)$, respectively, where μ, σ_1^2 and σ_2^2 are unknown. Let $\rho = \sigma_2^2/\sigma_1^2$ and $\theta = m/n$. Consider the problem of unbiased estimation of μ.

(i) If ρ is known, show that

$$\delta_0 = \alpha\bar{X} + (1-\alpha)\bar{Y}$$

where $\alpha = \rho/(\rho + \theta)$ is BLUE of μ. Also, compute $V(\delta_0)$.

(ii) If ρ is unknown, the unbiased estimator

$$\delta_1 = \frac{\bar{X}+\theta\bar{Y}}{1+\theta}$$

is optimum in the neighbourhood of $\rho = 1$. Find the variance of δ_1.

(iii) Compute the efficiency of δ_1 relative to δ_0.

(iv) Another unbiased estimator of μ is

$$\delta_2 = \frac{[\rho F(m-1)/(n-1)]\bar{X}+\theta\bar{Y}}{\theta+\rho F(m-1)/(n-1)}$$

where $F = S_2^2/\rho S_1^2$ is $F(m-1, n-1)$ *rv*. Show.

Solution.

(i) δ_0 is unbiased since

$$E(\delta_0) = \theta$$

$$V(\delta_0) = \alpha^2\frac{\sigma_1^2}{n} + (1-\alpha)^2\frac{\sigma_2^2}{m}$$

since X and Y observations are drawn independently. Further,

$$\frac{\partial}{\partial\alpha}V(\delta_0) = 0$$

gives $$\alpha\left(\frac{\sigma_1^2}{n}+\frac{\sigma_2^2}{m}\right) = \frac{\sigma_2^2}{m}$$

$$\Rightarrow \qquad \alpha_0 = \frac{\sigma_2^2/m}{\sigma_1^2/n+\sigma_2^2/m} = \frac{\sigma_2^2/\sigma_1^2}{m/n+\sigma_2^2/\sigma_1^2} = \frac{\rho}{\theta+\rho}$$

So, $$V(\delta_0)\big|_{\alpha=\alpha_0} = \frac{1}{(\theta+\rho)^2}\left[\rho^2\frac{\sigma_1^2}{n}+\theta^2\frac{\sigma_2^2}{m}\right]$$

$$= \frac{1}{(\theta+\rho)^2}\frac{\sigma_1^2}{n}\left[\rho^2+\theta^2\frac{\sigma_2^2/\sigma_1^2}{m/n}\right] = \frac{1}{(\theta+\rho)^2}\frac{\sigma_1^2}{n}\left(\rho^2+\theta^2\frac{\rho}{\theta}\right)$$

$$= \frac{1}{(\theta+\rho)^2}\frac{\sigma_1^2}{n}(\rho+\theta)\rho = \frac{\rho}{\theta+\rho}\frac{\sigma_1^2}{n} \qquad (3.4.14)$$

(ii) Consider α when approximately $\sigma_1 = \sigma_2$, i.e., nearly $\rho = 1$. Then,

$$\alpha = \frac{1}{1+\theta}$$

This gives a new unbiased estimator

$$\delta_1 = \delta_0\Big|_{\alpha=\frac{1}{1+\theta}} = \frac{1}{1+\theta}\bar{X} + \left(1 - \frac{1}{1+\theta}\right)\bar{Y} = \frac{\bar{X}+\theta\bar{Y}}{1+\theta}$$

Using Eq. (3.4.14), we get

$$V(\delta_1) = \frac{1}{1+\theta}\frac{\sigma_1^2}{n}$$

(iii) $$e(\delta_0|\delta_1) = \frac{V(\delta_1)}{V(\delta_0)}$$

$$= \frac{\dfrac{1}{1+\theta}\dfrac{\sigma_1^2}{n}}{\dfrac{\rho}{\theta+\rho}\dfrac{\sigma_1^2}{n}} = \frac{(\theta+\rho)}{\rho(1+\theta)}$$

(iv) $$\frac{S_1^2}{\sigma_1^2} \sim \chi^2(n-1), \frac{S_2^2}{\sigma_2^2} \sim \chi^2(m-1)$$

$$F = \frac{S_2^2/\sigma_2^2(m-1)}{S_1^2/\sigma_1^2(n-1)} = \frac{S_2^2/S_1^2}{(\sigma_2^2/\sigma_1^2)[(m-1)/(n-1)]}$$

$$= \frac{S_2^2/S_1^2}{\rho[(m-1)/(n-1)]} = F \sim F(m-1, n-1)$$

We may consider $S_2^2/S_1^2 = \rho[(m-1)/(n-1)]F$ as an estimator of $\dfrac{\sigma_2^2}{\sigma_1^2}$ in

$$\alpha = \frac{\sigma_2^2/\sigma_1^2}{(m/n) + \sigma_2^2/\sigma_1^2}$$

$$\Rightarrow \qquad \hat{\alpha} = \frac{\rho F(m-1)/(n-1)}{\theta + \rho F(m-1)/(n-1)}$$

Thus, we have

$$\delta_2 = \delta_0\Big|_{\hat{\alpha}} = \frac{\rho F(m-1)/(n-1)}{\theta + \rho F(m-1)/(n-1)}\bar{X} + \frac{\theta}{\theta + \rho F(m-1)/(n-1)}\bar{Y}$$

$$= \frac{[\rho F(m-1)/(n-1)]\bar{X} + \theta\bar{Y}}{\theta + \rho F(m-1)/(n-1)}$$

Note. Unbiased estimators are not always sensible. For example, if $X \sim P(\theta)$, $\theta > 0$, the estimator $T(X) = [-(k-1)]^X$ is unbiased for $\exp(-k\theta)$, k is an integer and $k > 1$; the estimator T may take negative values whereas the estimand $\exp(-k\theta)$ is positive.

The units of measurement of unbiased estimators may be different. For example, $X_1, X_2,\ldots,X_n$ are *iid* from $P(\theta)$, $\theta > 0$. $\bar{X}$, $S^2 = (n-1)^{-1}\Sigma_1^n(X_i - \bar{X})^2$, $\alpha\bar{X} + (1-\alpha)S^2$, $0 < \alpha < 1$, are all unbiased estimators of θ but have different units of measurement.

EXERCISES

1. Let T be the minimum variance unbiased estimator (UMVUE) of θ. Then prove that T^k(k is positive integer) is the MVUE for $E(T^k)$ provided $E(T^{2k}) < \infty$.

2. Let $X_1, X_2,\ldots,X_n$ be a random sample from a population with mean μ and variance σ^2.

(i) Find an unbiased estimator of μ^2.

(ii) Show that $S_1^2 = (1/n)\Sigma_{i=1}^n(X_i - \bar{X})^2$ is a biased estimator of σ^2, however, $S_n^2 = (1/n)\Sigma_{i=1}^n(X_i - \mu)^2$ is unbiased when μ is known, and $S_{n-1}^2 = [1/(n-1)]\,\Sigma_{i=1}^n(X_i - \bar{X})^2$ is unbiased for σ^2 if μ is unknown.

3. Let T_0 be some unbiased estimator of θ so that its product with zero estimator is also a zero estimator. Then, which of the following is true?

(i) T_0 is not UMVUE

(ii) T_0 is UMVUE

(iii) $T_0\ U$ is UMVUE, where U is some zero estimator

(iv) None

4. Let $X_1, X_2,\ldots,X_n$ be a random sample drawn from some distribution F_θ, $\theta \in \Theta$. If T is a complete sufficient statistic for θ so that $E[T(\mathbf{X})] = \theta$, then for any other unbiased estimator T' of θ, show that (i) $E(T'|T) = T$ and (ii) $V(T) < V(T')$.

Hint. $E_\theta(T) = \theta = E_\theta(T') = E_\theta^T\ E[T'|T]$ (since T, T' and the Rao–Blackwellized estimators are all unbiased). Thus, $E_\theta^T\ [E(T'|T) - T] = 0$ implies $g(T) = E[T'|T] - T = 0$ since T is complete. Further, $V(T') = E^T V(T'|T) + V^T E(T'|T) = E^T V(T'|T) + V^T(T)$ gives $V(T) < V(T')$.

5. Consider a regression model

$$Y_i = \theta_0 + \theta_1 x_i + \varepsilon_i$$

where

$$x_i = \begin{cases} 1, & \text{if } i = 1,\ldots,n_1 \\ 0, & \text{if } i = n_1+1,\ldots,n \end{cases}$$

and ε_i are uncorrelated random variables with mean 0 and common variance σ^2. Let T_1 and T_2 be two estimators of θ_1 given by $T_1 = Y_1 - Y_n$ and $T_2 = \bar{Y}_1 - Y_n$ where $\bar{Y}_1 = n_1^{-1}\Sigma Y_i$.

(i) Verify whether T_1 and T_2 are unbiased and find their variances.

(ii) If possible, propose an unbiased estimator of θ_1 which has variance smaller than T_1 and T_2 with justification.

6. An inspector samples at random from a large lot of a particular item. He measures certain characteristics of each item drawn and stops inspection when he has obtained $d(>1)$ defective items, by which time his total sample size is X. Find a UMVUE of θ, the proportion of defective items in the lot.

7. Assume the number of accidents at some intersection follows the Poisson distribution with unknown rate of accidents per week, θ. The district administration wishes to estimate θ to decide whether a traffic light should be installed in order to reduce the number of accidents at this intersection. Suppose the number of accidents X have been recorded in n successive weeks.

(i) Find the UMVUE of θ^2, θ^r, $r \le n$.

(ii) Given an unbiased estimator of $\exp(-\theta)$ that there were no accidents in the week

$$\delta(X) = \begin{cases} 1, & \text{if } X_1 = 0 \\ 0, & \text{otherwise} \end{cases}$$

and $T = \Sigma X_i$ is complete sufficient, find the UMVUE of $\exp(-\theta)$.

(iii) Show that both $\bar{X}$ and S^2 are unbiased for θ. Improve S^2 through the complete sufficient statistic $T = \Sigma X_i$ and conclude that $V(\bar{X}) \le V(S^2)$. Can you use the results of Exercise 3.4 to show this inequality?

(iv) Find the UMVUE of $g(\theta) = (1 + \theta)\exp(-\theta)$.

(v) Find the UMVUE of $g(\theta) = P(X_1 = 0, X_2 = 0) = \exp(-2\theta)$. Discuss the demerits of the estimator obtained. Show that for large n, it is equivalent to the MLE.

Hint. (i) T is complete sufficient statistic for θ. Therefore, the estimator $T^{(r)}/n^r = T(T-1)\ldots(T-r+1)/n^r$ is UMVUE of θ^r since

$$E(T^{(r)}) = \sum \frac{t(t-1)\ldots(t-r+1)(n\theta)^t \exp(-n\theta)}{t!}$$

$$= (n\theta)^r \sum \frac{(n\theta)^{t-r} \exp(-n\theta)}{(t-r)!} = (n\theta)^r$$

Thus, for $r = 2$, $\dfrac{T^{(2)}}{n^2} = \dfrac{T(T-1)}{n^2}$ is UMVUE of θ^2.

Another way to proceed is to consider $E_\theta(X_1X_2) = \theta^2$ since the observations are independent.

$$\phi(T) = E_\theta(X_1X_2 \mid T)$$

$$= E_\theta^{X_2|T}\left[X_2 E_\theta^{(X_1|X_2,T)}(X_1 \mid X_2, T)\right]$$

$$= E_\theta^{X_2|T}\left[\frac{X_2(T-X_2)}{(n-1)} \middle| T\right]$$

$$= \left[T\frac{T}{n} - T\frac{T}{n}\left(1-\frac{1}{n}\right) - \left(\frac{T}{n}\right)^2\right]\frac{1}{(n-1)} = \frac{T(t-1)}{n^2}$$

since $\quad X_i|T \sim b\left(T, \frac{1}{n}\right), X_1|T_1, X_2 \sim b\left(T - X_2, \frac{1}{n-1}\right)$

$$\text{(iii)}\quad \varphi(T) = E(S^2|T) = \frac{1}{(n-1)}E\left[\left(\sum X_i^2 - n\bar{X}^2\right)\Big|T\right]$$

$$= \frac{1}{(n-1)}\left[E\left(\sum X_i^2|T\right) - n\left(\frac{T}{n}\right)^2\right]$$

$$= \frac{1}{(n-1)}\left[\left\{T\frac{1}{n}\left(1-\frac{1}{n}\right) - n\left(\frac{T}{n}\right)^2\right\} - n\left(\frac{T}{n}\right)^2\right] = \frac{T}{n} = \bar{X}$$

Alternatively, one can proceed with the fact that T is a complete sufficient statistic. Therefore, T/n is UMVUE of θ and it is unique. The estimator $\varphi(T) = E(S^2|T)$ is unbiased for θ and it is also the unique UMVUE of θ. Therefore, $\varphi(T) = E(S^2|T) = T/n = \bar{X}$. (iv) Find $P(X_1 \leq 1)$.

8. (Example on an absurd unbiased estimator). Let X follows a truncated Poisson distribution with *pmf*

$$f(x;\theta) = \frac{\exp(-\theta)\theta^x}{\{1-\exp(-\theta)\}x!}; x = 1, 2, 3, \ldots$$

Show that the estimate based on X

$$T(X) = \begin{cases} 0, & \text{if } x \text{ is odd} \\ 2, & \text{if } x \text{ is even} \end{cases}$$

is the only unbiased estimator of $\{1 - \exp(-\theta)\}$.

Hint. $\exp(y) = 1 + y + y^2/2! + \cdots$ and $\exp(-y) = 1 - y + y^2/2! - \cdots$ gives $[(\exp(\theta) + \exp(-\theta)]/2 - 1 = 1 + y^2/2! + y^4/4! + \cdots = \Sigma_{x \text{ even}}\theta^x/x!$. The expectation of $T(X)$ is $E[T(X)] = 2\{\exp(-\theta)/[1 - \exp(-\theta)]\}\{[\exp(\theta) + \exp(-\theta) - 2]/2\} = 1 - \exp(-\theta)$.

9. A random variable X has the truncated binomial distribution with mass function

$$P(x;\theta) = \frac{\binom{n}{x}\theta^x(1-\theta)^{n-x}}{1-(1-\theta)^n}, \quad x = 1, 2, \ldots, n$$

Show that X is complete and sufficient for θ. Determine the minimum variance unbiased estimator of $\theta/[1 - (1 - \theta)^n]$.

10. $X_1, X_2,\ldots,X_n$ are n independent random observations identically distributed as $P(X_i = 1) = 1 - P(X_i = 0) = p$. Some functions of p are given below:

(i) $p^{n-1}(1 - p)$

(ii) $(1 + p^3)/(1 + p)$

(iii) $(p^2 + 1)/(1 + p)$

Which of the above functions have a unique UMVU estimator?

(a) (i) and (ii) (b) (i) and (iii) only
(c) (ii) and (iii) only (d) (i), (ii), and (iii)

11. Let $X_1, X_2, \ldots, X_n$ be a random sample from a Bernoulli distribution with parameter $\theta (0 < \theta < 1)$. Obtain UMVU estimator of $g(\theta) = \theta^2 (1 - \theta)$.

12. Let $(X_1, X_2) \sim$ multinomial(n, θ_1, θ_2). Find the UMVUE of $g(\theta) = \theta_1 \theta_2$.

13. Let $X_1, X_2, \ldots, X_n$ be a random sample from NB(k, θ). Find the UMVUE of θ.

Hint. $T = \Sigma_1^n X_i$ is a complete sufficient statistic for θ and $T \sim$ NB(nk, θ).

14. Consider an *rv* with *pmf*

$$f(x;\, \theta) = Bf_1(x;\, \theta)\, f_2^2(x;\, \theta),\ x = 0, 1, \ldots, m$$

where $f_1(x; \theta)$ and $f_2(x; \theta)$ are binomial $b(m, \theta)$ and geometric (m, θ) *pmfs*, respectively, i.e.

$$f_1(x;\, \theta) = \binom{m}{x} \theta^x (1-\theta)^{m-x},\ x = 0, 1, 2, \ldots, m$$

and $$f_2(x;\, \theta) = \theta(1 - \theta)^x,\ x = 0, 1, \ldots, m$$

(i) Show that $B = \theta^{-2}(1 - \theta)^{-m} [1 + \theta(1 - \theta)]^{-m}$ and the distribution belongs to the one-parameter exponential family. Find $E(X)$.

(ii) Based on a sample $X_1, X_2, \ldots, X_n$ from $f(x; \theta)$, find a one-dimensional sufficient statistic for θ. Is it complete? Find the UMVUE of $[1 + \theta(1 - \theta)]^{-1}$ when $m = 1$.

Hint. $1 = \Sigma_{x=0}^m f(x;\theta) = B\theta^2(1 - \theta)^m \, \Sigma_{x=0}^m \binom{m}{x} \theta^x(1 - \theta)^x$. This gives $B = \theta^{-2}(1 - \theta)^{-m}$ $[1 + \theta(1 - \theta)]^{-m}$. The *pdf* is given by $f(x;\theta) = \theta^{-2}(1 - \theta)^{-m}[1 + \theta(1 - \theta)]^{-m}\binom{m}{x}$ $\theta^x(1-\theta)^{m-x}\theta^2(1-\theta)^{2x} = [1 + \theta(1 - \theta)]^{-m}\binom{m}{x}\theta^x(1-\theta)^x$. This gives $E(X) = [1 + \theta(1-\theta)]^{-m} m\theta(1 - \theta)$. Since $T = \Sigma_{i=1}^n X_i$ is the complete sufficient statistic for θ and $1 - (T/n)$ is unbiased for $[1 + \theta(1 - \theta)]^{-1}$, $1 - (T/n)$ is UMVUE for $[1 + \theta(1 - \theta)]^{-1}$.

15. Let X be an *rv* having *pdf*

$$f(x;\, \theta) = \left(\frac{\theta}{x}\right)^{|x|} (1-\theta)^{1-|x|},\ x = -1, 0, +1,\ \ 0 \le \theta \le 1$$

so that $E(|X|) = \theta$, $V(X) = \theta(1 - \theta)$

(i) Let X be a single observation drawn from the above distribution. Define an unbiased estimator by

$$\delta(x) = \begin{cases} 2, & \text{if } x = 1 \\ 0, & \text{otherwise} \end{cases}$$

and find a still better estimator and show that it is better.

(ii) Based on n independent observations $X_1, X_2,...,X_n$ drawn from the above distribution, find the UMVUE of θ.

16. Let $X_1, X_2,...,X_n$ be *iid* from a geometric distribution with *pmf*

$$f(x;\,\theta) = \theta(1-\theta)^{x-1},\ \ x = 1, 2,...,0 < \theta < 1$$

Find the UMVUE of θ^{-1}.

17. Let a random sample $X_1, X_2,...,X_n$ of size n ($n \geq 3$) is drawn from a bernoulli population $b(1,\theta)$, $0 < \theta < 1$. Given that $\delta(\mathbf{X}) = X_1 \cdot X_2 \cdot X_3$ is an unbiased estimator of θ^3, find the UMVUE of θ^3.

Hint. δ is unbiased since the observations are independent. $T = \Sigma_{i=1}^{n} X_i$ is a complete sufficient statistic with $T \sim b(n,\,\theta)$. UMVUE of θ is

$$\varphi(t) = E(X_1 \cdot X_2 \cdot X_3 | t) = \frac{P(X_1 = X_2 = X_3 = 1, \sum_{i=4}^{n} X_i = t-3)}{P(T=t)} = \begin{cases} \binom{n-3}{t-3} \Big/ \binom{n}{t} & t \geq 3 \\ 0 & t < 3 \end{cases}$$

18. Let $X_1, X_2,...,X_n$ be n independent random variables from Rayleigh distribution

$$f(x;\,\theta) = \frac{2x}{\theta}\exp\left\{-\frac{x^2}{\theta}\right\},\ x \geq 0$$

where $E(X) = \sqrt{\pi\theta/2}$ $V(X) = [1 - (\pi/4)]\theta$. Show that $T(\mathbf{X}) = (1/n)\ \Sigma_{i=1}^{n} X_i^2$ is UMVUE of θ.

19. Let $X_1, X_2,...,X_n$ be a sample taken from the distribution with *pdf*

$$f(x;\,\theta) = \frac{1}{2}\theta^3 x^2 \exp(-\theta x),\ x > 0,\ \theta > 0$$

Find sufficient statistic for θ. Is it also complete? Find the UMVUE of $g_1(\theta) = \theta$; $g_2(\theta) = (1/\theta)$; and $g_3(\theta) = \log\theta$. Which of these admits an efficient estimator?

20. Let $X_1, X_2,...,X_n$ be *iid* from a Laplace distribution

$$f(x;\,\theta) = \frac{1}{2}\exp\{-|x-\theta|\},\ -\infty < x < \infty,\ -\infty < \theta < \infty$$

Discuss UMVU estimation of θ.

21. Let the time to failure of a piece of equipment be modelled by $G_2(1,\,\beta)$. A sample $X_1, X_2,...,X_n$ is drawn to estimate the mean failure time, i.e., $1/\beta$. Two estimators are available

$$T_1(\mathbf{X}) = \overline{X},\ V(T_1) = \frac{1}{n\beta^2},\ T_2(\mathbf{X}) = nX_{(1)},\ V(T_2) = \frac{1}{\beta^2}$$

Which estimate would you prefer as a reasonable estimator?

(i) T_1 since it is UMVUE
(ii) T_1 and T_2 are equally reasonable
(iii) T_2
(iv) Neither T_1 nor T_2.

Hint. $E(X) = \int_0^\infty x\beta\exp(-x\beta)dx = \beta\Gamma(2)/\beta^2 = 1/\beta$, $E(X^2) = \beta\Gamma(3)/\beta^3 = 2/\beta^2$, $V(X) = 1/\beta^2$, $V(\bar{X}) = 1/(n\beta^2)$, $E(nX_{(1)}) = \beta n^2\int x\exp(-nx\beta)dx = \beta n^2\Gamma(2)/(n\beta)^2 = 1/\beta$, $E(nX_{(1)})^2 = \beta n^3\int x^2\exp(-nx\beta)dx = \beta n^3\,\Gamma(3)/(n\beta)^3 = 2/\beta^2$, $V(nX_{(1)}) = 1/\beta^2$

22. Let $X_1, X_2,\ldots,X_n$ be a random sample drawn from $G_2(\alpha, \beta)$ with α known. Find the UMVUE of β^{-1}.

23. Let $X_1, X_2,\ldots,X_n$ be a random sample from $G_2(2, \theta)$ distribution. Find the UMVUE of $g_1(\theta) = (1/\theta)$ and $g_2(\theta) = P_\theta(X_1 > c)$, c is some constant, $c > 0$.

24. Let $X_1, X_2,\ldots,X_n$ be a random sample drawn from the inverse Gaussian distribution with *pdf*

$$f(x;\theta) = \frac{1}{\sqrt{2\pi x^3}}\exp\left\{-\frac{(x-\theta)^2}{2x\theta^2}\right\}, x > 0 \text{ where } \theta > 0$$

Find the UMVUE of θ and $1/\theta$.

Hint. ΣX_i is complete sufficient statistic for θ, $\Sigma X_i/n$ is unbiased for θ. $(n/\Sigma X_i - 1/n)$ is UMVUE of $1/\theta$.

25. $X_1, X_2,\ldots,X_n$ is a random sample drawn from lognormal distribution with *pdf*

$$f(x;\mu,\sigma) = \frac{1}{\sigma\sqrt{2\pi}x}\exp\left(-\frac{1}{2\sigma^2}(\log x-\mu)^2\right);\quad x > 0, \mu \in \mathbb{R}^1\ \sigma > 0$$

For the above family, which one of the following statements is correct?

(i) $E(\bar{X}) = \mu$ (ii) $E\left[\log\left(\prod X_i\right)^{1/n}\right] = \mu$

(iii) $V(\bar{X}) = \dfrac{\sigma^2}{n}$ (iv) $E(\log\bar{X}) = \mu$

26. Let $X_1, X_2,\ldots,X_n$ be a random sample drawn from the Weibull (θ) distribution

$$f(x;\theta) = \frac{2x}{\theta}\exp\left(-\frac{x^2}{\theta}\right), x > 0 \text{ where } \theta > 0$$

(i) Find a minimal sufficient statistic for θ.
(ii) By $E(X^2) = \theta$, find an unbiased estimator of θ based on the statistic in (i).

27. Let $X_1, X_2,\ldots,X_n$ be a random sample drawn from the distribution with *pdf*

$$f(x;\theta) = \theta x^{\theta-1}, 0 < x < 1, \theta > 0$$

(i) Show that the geometric mean of the sample, $T(\mathbf{X}) = (\Pi_{i=1}^n X_i)^{1/n}$, is complete sufficient statistic and find the UMVUE of θ. Calculate the distribution of X^θ and find the UMVUE of θ based on a single observation.

(ii) If $Z = -\Sigma \log X_i$, show that $(n-1)/Z$ is an unbiased estimator for θ and its efficiency is $(n-2)/n$.

Hint. (i) $Y = -\log X \sim E(1/\theta)$ with *pdf* $f(y;\theta) = \theta \exp(-\theta y)$ which is $G_2(1, \theta)$. Thus, from the distribution of a function of an *rv*, $X^\theta \sim U(0, 1)$. (ii) $Z = -\Sigma \log X_i \sim G_2(n, \theta)$, $E(Z^{-1}) = \int_0^\infty z^{-1}[\theta^n/\Gamma(n)] \exp(-\theta z) z^{n-1}\, dz = \theta/(n-1)$. This gives $E[(n-1)\, Z^{-1}] = \theta$. Further, $E[(n-1)\, Z^{-1}]^2 = (n-1)(n-2)\theta^2$; $V[(n-1)Z^{-1}] = E[(n-1)Z^{-1}]^2 - [E\{(n-1)\, Z^{-1}\}]^2 = (n-2)^{-1}\theta^2$.

28. Let $X_1, X_2, \ldots, X_n$ be *iid* from a gamma distribution $G_1(3, \theta)$ with *pdf*

$$f(x;\theta) = \frac{1}{2\theta^3} x^2 \exp\left(-\frac{x}{\theta}\right), \quad x \ge 0, \theta > 0$$

Find the UMVUE of θ.

29. Let $X_1, X_2, \ldots, X_n$ be *iid* from a beta distribution $Be(\theta, 4\theta)$ with *pdf*

$$f(x;\theta) = \frac{\Gamma(5\theta)}{\Gamma(\theta)\Gamma(4\theta)} x^{\theta-1}(1-x)^{4\theta-1}, 0 < x < 1, \theta > 0$$

(i) Show that $T(\mathbf{X}) = \Pi_{i=1}^n X_i\,(1 - X_i)^4$, $T_k(\mathbf{X}) = [T(\mathbf{X})]^k$; $k = 2, 3, \ldots$, are all sufficient statistics for θ.

(ii) Find the UMVUE of θ.

30. Let $X_1, \ldots, X_n$ be a random sample from a rectangular distribution with *pdf*

$$f(x;\theta) = \frac{1}{\theta}; \; 0 < x < \theta$$

where $0 < \theta < \infty$.

(i) Show that the estimators $[(n+1)/n]X_{(n)}$ and $(2/n)\Sigma X_i$ are unbiased estimators of θ, where $X_{(n)}$ is the largest order statistic. Find the variance and, hence, the efficiency of these estimators. Show that $[(n+1)/n]X_{(n)}$ is a minimum variance unbiased (MVU) estimator of θ.

(ii) Find c so that $T = cS$ where (a) $S = \bar{X}$; (b) $S = X_{(n)}$; (c) $S = X_{(n)} - X_{(1)}$ is an unbiased estimator of θ. Which of these unbiased estimators is the best?

31. If $X_1, \ldots, X_n$ is a random sample from the uniform population with its density

$$f(x;\theta) = 1 \text{ if } \theta < x < \theta + 1$$

show that $\bar{X} - (1/2)$ is an unbiased estimator of θ. Comment on the existence of UMVUE of θ.

Hint. $E(X) = \int_\theta^{\theta+1} x\,dx = \theta + (1/2)$. The statistic $(X_{(1)}, X_{(n)})$ is sufficient for θ, however, complete sufficient statistic does not exist.

32. Let $X_1, X_2, \ldots, X_n$ be a random sample from $U(\theta, 2\theta)$ distribution where $\theta > 0$. There are two usual estimators of θ, one obtained by the method of moments $\delta_1 = \bar{X}/3$, and the other is $\delta_2 = [(n+1)/(2n+1)]X_{(n)}$ which is based on the maximum likelihood estimator $X_{(n)}/2$. Which of these two estimators is better?

33. Let X_1, X_2, X_3 be a sample of size 3 from $U(0, \theta)$, $\theta > 0$ is unknown.

(i) Show that $X_{(1)}/\theta$ is distributed as Be(1, 3).

(ii) Compute $E[X_{(1)}]$. Construct an unbiased estimator of θ using $X_{(1)}$.

(iii) Show that $X_{(3)}$ is a sufficient statistic.

(iv) Find the conditional *pdf* of $X_{(1)}$ given $X_{(3)}$.

(v) Compute $W = E[4X_{(1)}|X_{(3)}]$. Show that W is unbiased for θ.

(vi) Which of estimators W and the one constructed in (ii) is better for estimating θ? Show details to justify your answer.

34. Let n random observations $X_1, X_2, \ldots, X_n$ have been drawn from the exponential distribution with mean θ.

(i) Construct an unbiased estimator of θ based on first-order statistic $X_{(1)}$.

(ii) Find the UMVUE of θ using the estimator in (i).

Hint.
$$f_{X_1}(X_{(1)}; \theta) = \frac{n}{\theta}\exp\left\{-\frac{n x_{(1)}}{\theta}\right\}$$

$X_{(1)} \sim$ exponential (θ/n), $E[X_{(1)}] = (\theta/n)$. $T = \Sigma X_i$ is a complete sufficient statistic and it follows exponential $(n\theta)$. Given an unbiased estimator $\delta_1 = nX_{(1)}$ of θ, we have $\varphi(t) = E[\delta_1|t]$. Completeness of T, $E_\theta^T[\varphi(T) - (T/n)] = 0$ implies that $\varphi(T) = (T/n)$ is the UMVUE of θ.

35. Obtain an unbiased estimator of σ from $N(\theta, \sigma^2)$ if θ is known.

36. $X_1, X_2, \ldots, X_n$ are random observations from the two-parameter family of distributions $N(\mu, \sigma^2)$. Let $\bar{X} = n^{-1}\Sigma X_i$, $S_{n-1}^2 = (n-1)^{-1}\Sigma(X_i - \bar{X})^2$.

Assertion (A): $T = (\bar{X}^2 - S_{n-1}^2/n)$ is the unique UMVUE of μ^2.

Reason (R): $(\bar{X}, S_{n-1}^2)$ is sufficient for the two-parameter exponential family of distributions $N(\mu, \sigma^2)$ and $E(T) = \mu^2$.

(i) Both A and R are individually true and R is the correct explanation of A.

(ii) Both A and R are individually true but R is not the correct explanation of A.

(iii) A is true but R is false.

(iv) A is false but R is true.

37. State and prove the Rao–Blackwell theorem. What is the implication of the result. When will the estimation in the theorem be the best?

38. Let X be a single observation from normal population $N(0, \theta^2)$; $\theta^2 > 0$. Show that X^2 is a complete sufficient statistic for θ^2. Examine if X is also a complete sufficient for θ^2.

39. Let $X_1, X_2, \ldots, X_n$ be a random sample from $N(\theta, \sigma^2)$.

(i) Find the UMVUE of θ^2 if σ^2 is known.

(ii) Check whether
$$T = \frac{1}{n}\sum_{i=1}^{n}|X_i - \theta|$$

is unbiased estimator of σ if θ is known. If not, find an unbiased estimator of σ.

Hint. $E(T) = E\left[\left(\frac{1}{n}\right)\sum_{i=1}^{n}|X_i - \theta|\right] = \sqrt{\frac{2}{\pi}}\,\sigma$

40. Consider a random sample $X_1, X_2, \ldots, X_n$ from $N(\theta, \sigma^2)$. The UMVUE of the critical value u for given p

$$P(X_1 \le u) = \Phi\left(\frac{u - \theta}{\sigma}\right) = p$$

is

(i) $\bar{X} + A(n-1, 1)S_{n-1}\Phi^{-1}(p)$ (ii) $S_{n-1}\Phi^{-1}(p)$

(iii) $\bar{X} - S_{n-1}\Phi^{-1}(p)$ (iv) $\Phi^{-1}\left[\sqrt{n/(n-1)}(p - \bar{X})\right]$

where $A(n, r)$ is defined in the Example 3.8 and $S_{n-1}^2 = (n-1)^{-1}\Sigma_{i=1}^{n}(X_i - \bar{X})^2$.

41. Obtain unbiased estimator of σ from $N(\theta, \sigma^2)$ if θ is known. Also find the UMVUE of σ^r where r is a known positive constant, r is not necessarily an integer.

42. Let $X_1, X_2, \ldots, X_n$ be a random sample from a normal distribution $N(\theta, \theta)$, $\theta > 0$ and both $\bar{X}$ and S^2 are unbiased for θ.

(i) Which estimator would you prefer?

(ii) Does the complete sufficient statistic exists in this case?

(iii) Does there exists the UMVUE of θ. Is one of $\bar{X}$ or S^2 UMVUE of θ.

43. Let $X_1, X_2, \ldots, X_n$ be *iid* random variables from

(*i*) $N(\theta, \theta^2)$ (ii) $N(\theta, \theta)$

Discuss optimal unbiased estimators of θ. Does an UMVUE of θ exists in each case?

Hint. Complete sufficient statistic for θ does not exist.

44. Let $X_1, X_2, \ldots, X_n$ be random sample from the $N(0, \theta)$ population. Find the UMVUE of $g(\theta) = \theta^2$.

45. Let $X_1, X_2, \ldots, X_n$ be random sample from the $N(\theta, 1)$ distribution. Find the UMVUE of $\Phi(c - \theta) = P(X_i \le c)$ for some constant c where Φ is the standard normal *cdf*.

46. The measurement of radius of a circle is modelled by

$$X = \theta + \varepsilon$$

where θ is the correct radius and ε is the error that follows $N(0, \sigma^2)$, σ^2 is unknown. Let n independent observations $X_1, X_2, \ldots, X_n$ on the radius of the circle have been made. Find the UMVUE of the area of the circle.

Hint. $\bar{X}$ is the complete sufficient statistic so that $\bar{X} \sim N(\theta, (\sigma^2/n))$, $(\sigma^2/n) = E\bar{X}^2 - \theta^2$, $\pi\theta^2 = E\pi[\bar{X}^2 - (S^2/n)]$.

47. Let $X_1, X_2, \ldots, X_n$ be a random sample from $N(\theta, \theta^2)$, $\theta > 0$. The two unbiased estimators of θ are $\bar{X}$ and kS, $k = \dfrac{\sqrt{n-1}\,\Gamma[(n-1)/2]}{\sqrt{2}\,\Gamma(n/2)}$

(i) Show that the convex combination of these estimators $\delta_\alpha = \alpha\bar{X} + (1 - \alpha)kS$ is unbiased for θ for every α, $0 < \alpha < 1$.

(ii) Calculate the variance of δ_α and obtain α at which it is minimum.

(iii) Show that the two-dimensional statistic $(\bar{X}, S^2)$ is a sufficient statistic but not complete.

Hint. $V(T_\alpha) = \alpha^2 V(\bar{X}) + (1 - \alpha)^2 V(kS)$, $(\partial/\partial\alpha)\, V(T_\alpha) = 0$

$$\Rightarrow \quad 2\alpha V(\bar{X}) - 2(1 - \alpha)V(kS) = 0$$

$$\Rightarrow \quad \alpha = \frac{V(kS)}{V(\bar{X}) + V(kS)} = \frac{k^2E(S^2) - \theta^2}{\theta^2/n + \{k^2E(S^2) - \theta^2\}}$$

$$= \frac{k^2\theta^2 - \theta^2}{(\theta^2/n) + \{k^2\theta^2 - \theta^2\}} = \frac{k^2 - 1}{(1/n) + k^2 - 1}$$

$(\bar{X}, S^2)$ is sufficient by the factorization theorem. Since $\bar{X}$ and kS are unbiased for θ, it gives $E_\theta(\bar{X} - kS) = 0$. However, it does not imply that $\bar{X} = kS$. Therefore, $(\bar{X}, S^2)$ is not complete.

48. Let $X_1, X_2, \ldots, X_n$ be a random sample from the distribution with *pdf*

$$f(x; \theta) = \frac{3x^2}{\theta^3} I_{(0,\theta)}(x)$$

where $\theta > 0$ is unknown. Find the UMVUE of θ.

49. Let $X_1, X_2, \ldots, X_n$ be a random sample from the population with *pdf*

$$f(x; \theta) = (\theta + 1)x^\theta;\ x \in (0, 1);\ \theta > -1$$

Show that $-\{(n - 1)/\Sigma \log X_i\} - 1$ is an unbiased estimator of θ. Show that it is UMVUE.

Hint. Consider the transformation $Y = -\log X$, $0 < Y < \infty$. Then $f_Y(y; \theta) = (\theta + 1) \exp[-(\theta + 1)y]$, which is $G_2(1, (\theta + 1))$. Therefore, $U = \Sigma Y_i = -\Sigma \log X_i \sim G_2(n, (\theta + 1))$. This gives $E(1/U) = [(\theta + 1)^n/\Gamma(n)] \int_0^\infty u^{-1}\, u^{n-1} \exp[-(\theta + 1)u]du = (\theta + 1)/(n - 1)$. This implies $E[-\{(n - 1)/(\Sigma \log X_i)\} - 1] = \theta$

50. Let $X_1, X_2, \ldots, X_n$ be a random sample from the distribution with *pdf*

$$f(x; \theta) = \theta(1 + x)^{-(1+\theta)},\ 0 < x < \infty,\ \theta > 0$$

Find the UMVUE of θ.

Hint. $f(\mathbf{x};\theta) = \theta^n \exp\{-(1 + \theta)\Sigma_{i=1}^n \log(1 + x_i)\}$ is one-parameter exponential family of distributions. $T(\mathbf{X}) = \Sigma_{i=1}^n \log(1 + X_i)$ is complete sufficient statistic. Let $y = \log(1 + x)$; $f(y;\theta) = \theta\exp(-\theta y)$. So, $Y \sim G_2(1,\theta)$, $T = \Sigma Y_i \sim G_2(n,\theta)$, $E(T) = n/\theta$. So, $E(1/T) = [\theta^n/\Gamma(n)] \int_0^\infty t^{-1} t^{n-1} \exp(-\theta t)dt = [\theta^n/\Gamma(n)][\Gamma(n - 1)/\theta^{n-1}] = [\theta/(n - 1)]$; $E[(n - 1)/T] = \theta$; so, $\{(n - 1)/[\Sigma\log(1 + X_i)]\}$ is UMVUE of θ.

51. Let $X_1, X_2, \ldots, X_n$ be the survival times (in months) of n patients suffering from AIDS. Assume that X_is are independent and follow a common exponential distribution with mean $\theta > 0$.

(i) Show that the unbiased estimators of θ are given by $\delta_1(\mathbf{X}) = \overline{X}$ and $\delta_2(\mathbf{X}) = nX_{(1)}$. Which of these estimators is UMVUE?

(ii) Comment on the applicability of the estimators in (i).

52. Let $X_1, X_2,\ldots,X_n$ be a sample from gamma distribution $G_2(a, b)$ with *pdf*

$$f(x; a, b) = \frac{a^b}{\Gamma(b)} x^{b-1} \exp(-ax), x > 0$$

where b is known, $n > (1/b)$, and $a > 0$ is unknown. Find the UMVUE of a.

Hint. Find a suitable choice of an unbiased estimation of a as some multiple of $1/X_1$. Use that $T = \Sigma X_i$ is a complete sufficient statistic and $T \sim G_2(nb,a)$. Observe that X_1/T is Be$[b, (n-1)b]$, i.e., it is ancillary. Use Basu's theorem to observe that X_1/T and T are independent.

53. Define sufficiency and completeness of a statistic and comment on their importance. Show that if a sufficient statistic is complete, then it is minimal sufficient.

54. Show that if T_1 and T_2 are UMVUEs of their expectations, then T_1T_2 is UMVUE of $E(T_1T_2)$.

Hint. Let $E_\theta(T_1) = g_1(\theta)$ and $E_\theta(T_2) = g_2(\theta)$ $\forall$ θ and U is some zero estimator. $\text{cov}(T_1T_2, U) = E^{T_1}[\text{cov}^{T_2|T_1}(T_1T_2, U)] + \text{cov}^{T_1}[E^{T_2|T_1}(T_1 T_2), U] = E^{T_1}[T_1\text{cov}(T_2, U)] + \text{cov}^{T_1}[T_1g_2(\theta), U] = 0 + g_2(\theta)\,\text{cov}^{T_1}(T_1, U) = 0$

55. Show that if T_is are UMVUEs of $g_i(\theta)$, $i = 1,\ldots,m$, then $\Sigma_{i=1}^m a_iT_i$ is UMVUE of $\Sigma_{i=1}^m a_ig_i(\theta)$.

56. Let $\{T_n\}$ be a sequence of UMVUEs and T be a statistic with $E_\theta T^2 < \infty$ such that $E\{T_n - T\}^2 \to 0$ as $n \to 0$ for θ. Then T is also a UMVUE. Prove it.

57. Let T_0 be a UMVUE of $g(\theta)$ and T_1 be an unbiased estimator with efficiency $e_\theta < 1$. Then can any unbiased linear function of T_0 and T_1 be a UMVUE of $g(\theta)$?

58. Define sufficient, minimal sufficient, and complete statistics and bring out their relevance in obtaining UMVU estimators.

59. Give an example of a sufficient statistic which is not complete.

60. Let T_1 and T_2 be UMVUEs of a parameter θ. Show that no unbiased combination of T_1 and T_2 can be a UMVUE of θ.

61. Let $X_1, X_2,\ldots,X_n$ be independent random variables where X_i has *pdf*

$$f(x_i; \theta_i) = \frac{1}{\theta_i} \exp\left\{-\frac{x_i}{\theta_i}\right\}, x_i > 0$$

where $\theta_i = (\alpha + i\beta)^{-1}$ and α and β are unknown parameters. Find a set of minimal sufficient statistics for α and β.

62. Let $X_1, X_2,\ldots,X_m$ and $Y_1, Y_2,\ldots,Y_n$ be independent samples from normal populations $N(2\theta, 1)$ and $N(\theta, 1)$, respectively. Starting with the unbiased estimator $X_1 - Y_1$ for θ, use the Rao–Blackwell theorem to find the unbiased estimator Z which is a function of the sufficient statistic. Compare Z with $\overline{Y} = (1/n)\,\Sigma_{i=1}^n Y_i$ for efficiency as an estimator of θ.

63. Suppose $X_1, X_2, \ldots, X_n$ are *iid* with density function

$$f(x;\theta) = \frac{\theta}{x^2},\ \theta < x,\ \theta > 0$$

Then find the sufficient statistic for θ. Also, find the UMVUE of θ.

Hint. $f_{X_{(r)}}(x) = \{n!/[(n-r)!r!]\}\ [1 - F(x)]^{n-r}\ [F(x)]^{r-1} f(x);\ F(x) = \int_\theta^x (\theta/x^2)dx = \theta[x^{-1}/{-1}]_\theta^x = 1 - (\theta/x);\ 1 - F(x) = (\theta/x);\ f_{X_{(1)}}(x_{(1)}:\theta) = (n/\theta)(\theta/(x_{(1)})^{n+1}\ i(\theta, x_{(1)})$: $f_{\mathbf{X}}(\mathbf{x};\theta) = \theta^n \prod_{i=1}^n [1/(x_i^2)]$

$$f_{\mathbf{X}|X_{(1)}}(\mathbf{x};\theta) = \frac{1}{n} x_{(1)}^{n+1} \prod_{i=2}^n \frac{1}{x_{(i)}^2}$$

64. (Pitman family of distributions). Let $X_1, X_2, \ldots, X_n$ be a random sample from the Pitman family of distribution given by the *pdf*

$$f(x;\,\theta) = c(\theta)\,u(x),$$

(i) $a < x < \theta$ (ii) $\theta < x < b$

Show that the UMVUE of a parametric function $g(\theta)$ is

$$\delta(X_{(n)}) = g(X_{(n)}) + \frac{g'(X_{(n)})}{nu(X_{(n)})c(X_{(n)})}$$

and
$$\delta(X_{(1)}) = g(X_{(1)}) - \frac{g'(X_{(1)})}{nu(X_{(1)})c(X_{(1)})}$$

for the cases (i) and (ii), respectively.

Hint. Consider the case (i). $[c(\theta)]^{-1} = \int_a^\theta u(x)dx$, $F(x) = [c(\theta)]/[c(x)]$. $X_{(n)}$ is a complete sufficient statistic with *pdf*

$$f_{X_{(n)}}(x;\theta) = nu(x)\frac{[c(\theta)]^n}{[c(x)]^{n-1}},\ a < x < \theta$$

Therefore, an estimator δ that is some function of $X_{(n)}$ and unbiased for θ will be UMVUE of θ. We have

$$\int_a^\theta f(x;\theta)dx = 1 \Rightarrow c(\theta)\int_a^\theta u(x)dx = 1 \Rightarrow \int_a^\theta u(x)dx = \frac{1}{c(\theta)}$$

Differentiating both the sides with respect to θ, we get

$$u(\theta) = -\frac{c'(\theta)}{[c(\theta)]^2}$$

Further, we have $E[\delta(X_{(n)})] = g(\theta)$

$$\int_a^\theta \delta(x)\,nu(x)\frac{[c(\theta)]^n}{[c(x)]^{n-1}}dx = g(\theta)$$

$$\int_a^\theta \delta(x)\,nu(x)\frac{1}{[c(x)]^{n-1}}dx = \frac{g(\theta)}{[c(\theta)]^n}$$

Differentiating both the sides with respect to θ, we get

$$\delta(\theta)\frac{nu(\theta)}{[c(\theta)]^{n-1}} = \frac{[c(\theta)]^n[g'(\theta) - ng(\theta)][c(\theta)]^{n-1}c'(\theta)}{[c(\theta)]^{2n}}$$

$$\delta(\theta) = -\frac{ng(\theta)c'(\theta)}{nu(\theta)[c(\theta)]^2} + \frac{g'(\theta)}{nu(\theta)c(\theta)} = g(\theta) + \frac{g'(\theta)}{nu(\theta)c(\theta)}$$

By replacing θ by the variable $X_{(n)}$,

$$\delta(X_{(n)}) = g(X_{(n)}) - \frac{g'(X_{(n)})}{nu(X_{(n)})c(X_{(n)})}$$

is UMVUE for $g(\theta)$ for $a < x < \theta$. Proceeding similarly for case (ii), we get

$$\delta(X_{(1)}) = g(X_{(1)}) - \frac{g'(X_{(1)})}{nu(X_{(1)})c(X_{(1)})}$$

is UMVUE for $g(\theta)$ for $\theta < x < b$.

65. Use the required results to show the UMVUE of θ^r based on a random sample X_1, $X_2,\ldots,X_n$ from each of the location families with *pdf*s

(i) $f(x;\theta) = \exp\{-(x - \theta)\}$, $x > \theta$
(ii) $f(x;\theta) = \theta^{-1}, 0 < x < \theta$, $n > r$
(iii) $f(x;\theta) = \exp(-x)/[\exp(-\theta) - \exp(-b)]$, $\theta < x < b$, b is known,

are $\delta(X_{(1)}) = X_{(1)}^r - \dfrac{rX_{(1)}^{r-1}}{n}$, $\delta(X_{(n)}) = \dfrac{(n+r)}{n}X_{(n)}^r$,

$$\delta(X_{(1)}) = X_{(1)}^r - \frac{rX_{(1)}^{r-1}[1 - \exp(-(b - X_{(1)}))]}{n}, \text{ respectively.}$$

Hint. Here, $g(\theta) = \theta^r$, $g'(\theta) = r\theta^{r-1}$

(i) $c(\theta) = \exp(\theta)$, $u(x) = \exp(-x)$. The UMVUE of θ^r is

$$\delta(X_{(1)}) = X_{(1)}^r - \frac{rX_{(1)}^{r-1}}{n\exp(-X_{(1)})\exp(X_{(1)})} = X_{(1)}^r - \frac{rX_{(1)}^{r-1}}{n}.$$

(ii) $c(\theta) = \theta^{-1}$, $u(x) = 1$. The UMVUE of θ^r is

$$\delta(X_{(n)}) = X_{(n)}^r + \frac{rX_{(n)}^{r-1}}{nX_n^{(-1)}} = \frac{(n+r)}{n}X_{(n)}^r.$$

(iii) $c(\theta) = [\exp(\theta) - \exp(-b)]^{-1}$, $u(x) = \exp(-x)$. The UMVUE of θ^r is

$$\delta(X_{(1)}) = X_{(1)}^r - \frac{rX_{(1)}^{r-1}}{n\exp(-X_{(1)})}[\exp(-X_{(1)}) - \exp(-b)] = X_{(1)}^r - \frac{rX_{(1)}^{r-1}[1 - \exp(-(b - X_{(1)}))]}{n}.$$

4

Information Inequality

4.1 INTRODUCTION

We have discussed the problem of estimation of $g(\theta)$ in Chapter 3 through the concepts of unbiasedness, consistency, sufficiency, and completeness. Problem arises when UMVUE does not exist or even if it exists, it cannot be obtained by the methods discussed in Chapter 3. The performance of an unbiased estimator δ at some point $\theta = \theta_0$ can then be measured by the fact that how close is the variance of δ, $V_\theta(\delta)$, to the minimum variance that an unbiased estimator can achieve at θ_0, $V_L(\theta_0)$. Here, L stands for *local*, with the meaning of $V_L(\theta_0)$ that the minimum variance at the point $\theta = \theta_0$. If $V_\theta(\delta)$ is close to $V_L(\theta)$ for all $\theta \in \Theta$, we may say that δ is a reasonable estimator but we are not sure that δ is the best. Moreover, the calculations of $V_L(\theta)$ are usually lengthy and involved. These difficulties demand a more comprehensive approach for finding a best unbiased estimator of $g(\theta)$. In this chapter, the lower bounds $B(\theta)$ for the variance, i.e., smallest variance, that can be attained by an unbiased estimator of $g(\theta)$ are derived. Generally, these lower bounds are simple to calculate. The performance of an unbiased estimator is judged by the closeness of its variance to the lower bound $B(\theta)$. Indeed, these lower bounds are not sharp, i.e., close to $V_L(\theta)$, for all $\theta \in \Theta$. Some sharper lower bounds like Bhattacharyya bound are also discussed.

In Section 4.2, score function and Fisher information is defined as a measure of information contained in the sample about the parameter θ. On utilizing Fisher information, Rao and Cramer, Bhattacharyya lower bounds for variance of an unbiased estimator for regular models and Chapman, Robbins, and Kiefer lower bounds for non-regular models are discussed in Section 4.3.

4.2 REGULAR FAMILY OF DISTRIBUTIONS, SCORE FUNCTION, AND FISHER INFORMATION

We mention here some regularity conditions under which results on the lower bound for the variance of an estimator is discussed in the following sections:

Regularity conditions

Consider a random sample $X_1, X_2,\ldots,X_n$ drawn from a distribution $f(x;\theta)$, where the distribution $f(x;\theta)$ belongs to a family of probability density functions (*pdfs*), $\{f(x;\theta),\ \theta \in \Theta\}$. Let the joint density of sample observations be denoted by $f(\mathbf{x};\theta)$. We state that the family of *pdfs* $\{f(x;\theta),\ \theta \in \Theta\}$ is regular, if it satisfies the following regularity conditions.

Single parameter case

(i) $\Theta \subseteq \mathbb{R}$, Θ is an open interval;

(ii) The support of the distribution $f(x;\theta)$, $S(\theta) = \{x: f(x;\theta) > 0\}$ is independent of θ. In other words, the family $\{f(x;\theta),\ \theta \in \Theta\}$ has a common support;

(iii) For every $\mathbf{x}$ in $S(\theta)$ and θ in Θ, the derivatives $(\partial/\partial\theta)\, f(\mathbf{x};\theta)$ and $[\partial^2/(\partial\theta^2)]\, f(\mathbf{x};\theta)$ exist and are finite;

(iv) Let $\delta(\mathbf{X})$ be some statistic for estimating θ, so that $E_\theta|\delta(\mathbf{X})| < \infty\ \forall\ \theta$, $E_\theta[\delta(\mathbf{X})] = g(\theta) < \infty\ \forall\ \theta$ and $V_\theta[\delta(\mathbf{X})] < \infty\ \forall\ \theta$. Then, differentiating

$$E_\theta[\delta(\mathbf{X})] = \int \delta(\mathbf{x}) f(\mathbf{x};\theta)\, dx = g(\theta)$$

with respect to θ, we have

$$\frac{\partial}{\partial\theta} E_\theta[\delta(\mathbf{X})] = \int \delta(\mathbf{X}) \frac{f'(\mathbf{x};\theta)}{f(\mathbf{x};\theta)} f(\mathbf{x};\theta) d\mathbf{x} = g'(\theta)$$

$$\Rightarrow \qquad \int \delta(\mathbf{x}) \left[\frac{\partial}{\partial\theta} \log f(\mathbf{x};\theta)\right] f(\mathbf{x};\theta) d\mathbf{x} = g'(\theta)$$

or

$$E[\delta(\mathbf{X}) S(\mathbf{X},\theta)] = g'(\theta)$$

where

$$S(\mathbf{X},\theta) = \frac{\partial}{\partial\theta} \log f(\mathbf{X};\theta)$$

In other words, differentiation and integration operations are interchangeable.

It is important to make distinction between the notations $S(\theta)$ and $S(\mathbf{X},\theta)$. $S(\theta)$ denotes the support of the distribution with probability density function $f(x;\theta)$ and $S(\mathbf{X},\theta)$ denotes the score function of the sample.

Multi parameter case

(i) $\Theta \subseteq \mathbb{R}^k$, Θ is non-degenerate open interval in $\mathbb{R}^k$;

(ii) $S(\boldsymbol{\theta})$ is independent of $\boldsymbol{\theta}$, i.e., the range of $\mathbf{X}$ does not depend on $\boldsymbol{\theta}$;

(iii) For every $\mathbf{x}$ in S and $\boldsymbol{\theta}$ in Θ, the derivatives for all $i = 1, 2,\ldots,k$, $[\partial/(\partial\theta_i)]\, f(\mathbf{x};\boldsymbol{\theta})$, $[\partial^2/(\partial\theta_i^2)]\, f(\mathbf{x};\boldsymbol{\theta})$ exist and are finite;

(iv) For any unbiased estimator $\delta(\mathbf{X})$ of $g(\boldsymbol{\theta})$, $E_\theta[\delta(\mathbf{X})] = g(\boldsymbol{\theta})$ with $E_\theta(\delta^2) < \infty\ \forall\ \boldsymbol{\theta} \in \Theta$

$$\frac{\partial}{\partial \theta_i} E_\theta[\delta(\mathbf{X})] = \frac{\partial}{\partial \theta_i} \int \delta(\mathbf{x}) f(\mathbf{x}; \boldsymbol{\theta})\, d\mathbf{x}$$

$$= \int \delta(\mathbf{x}) \left[\frac{\partial}{\partial \theta_i} \log f(\mathbf{x}; \theta) \right] f(\mathbf{x}; \theta) d\mathbf{x}$$

$$= \frac{\partial}{\partial \theta_i} g(\boldsymbol{\theta})\ \forall\, i = 1, 2, \ldots, k$$

$$\Rightarrow \quad E[\delta(\mathbf{X}) S_i(\mathbf{X}, \theta)] = \frac{\partial}{\partial \theta_i} g(\boldsymbol{\theta})$$

where

$$S_i(\mathbf{X}, \boldsymbol{\theta}) = \frac{\partial}{\partial \theta_i} \log f(\mathbf{X}; \boldsymbol{\theta})$$

Score Function and Fisher Information: Let us consider a random sample $X_1, X_2, \ldots, X_n$ from a density $f(x;\theta)$. Let us denote the joint density of sample observations by $f(\mathbf{x};\theta)$. Consider two points θ and $\theta + \delta$ in Θ. The rate of change in the density at $\mathbf{x}$ due to small changes of the value of θ is given by

$$\frac{\partial}{\partial \theta} f(\mathbf{x}; \theta) = \underset{\delta \to 0}{\text{limit}} \frac{f(\mathbf{x}; \theta + \delta) - f(\mathbf{x}; \theta)}{\delta}$$

Further, the relative rate at which the density changes at $\mathbf{x}$ is given by

$$S(\mathbf{x}, \theta) = \frac{1}{f(\mathbf{x}; \theta)} \underset{\delta \to 0}{\text{limit}} \frac{f(\mathbf{x}; \theta + \delta) - f(\mathbf{x}; \theta)}{\delta}$$

$$= \frac{1}{f(\mathbf{x}; \theta)} \frac{\partial}{\partial \theta} f(\mathbf{x}; \theta)$$

$$= \frac{\partial}{\partial \theta} \log f(\mathbf{x}; \theta) \tag{4.2.1}$$

The first derivative of log-likelihood in Eq. (4.2.1) is called *score* of the sample; we denote it by $S(\mathbf{x},\theta)$. It measures the sensitivity of log–likelihood to small changes in the value of θ. For example, a small value of score for a given value of θ indicates that the log-likelihood is unaffected by small changes in θ indicating the sample contains little information about θ. For a given value of θ, the score is a function of sample observations. Therefore, it is a random variable. Under the assumption of interchangeability of integration and differentiation operations, the random variable $S(\mathbf{X}, \theta)$ is centered about 0, since

$$E_\theta[S(\mathbf{X}, \theta)] = E_\theta \left(\frac{\partial}{\partial \theta} \log f(\mathbf{X}; \theta) \right)$$

$$= \int \frac{\partial}{\partial \theta} \log f(\mathbf{x}; \theta) f(\mathbf{x}; \theta) d\mathbf{x}$$

$$= \int \frac{1}{f(\mathbf{x};\theta)} \left(\frac{\partial}{\partial\theta} f(\mathbf{x};\theta) \right) f(\mathbf{x};\theta) d\mathbf{x}$$

$$= \frac{\partial}{\partial\theta} \int f(\mathbf{x};\theta) d\mathbf{x} \qquad \text{[by assumption (iv)]}$$

$$= \frac{\partial}{\partial\theta}(1) = 0 \tag{4.2.2}$$

The variance of a score function measures the strength of information contained in the sample observations about θ. Small variance for a given value of θ indicates that all the samples have their score near 0 and, thus, all the samples contain little information about the true value of θ. Therefore, the variance of the score becomes the natural measure of information that the sample contains about θ

$$I_{\mathbf{X}}(\theta) = E_\theta[S^2(\mathbf{X}, \theta)] = V_\theta[S(\mathbf{X}, \theta)] \qquad \text{[from Eq. (4.2.2)]} \tag{4.2.3}$$

Therefore, the squared average of the relative rate of change in the density, $E_\theta^X[(\partial/\partial\theta) \log f(\mathbf{X};\theta)]^2$, at some point θ measures the strength by which the value of θ can be distinguished from its neighboring values. This quantity is denoted by $I(\theta)$. A high value of $I(\theta)$ indicates that θ can be more accurately estimated by the sample observations $\mathbf{X}$, and we expect that we will get an unbiased estimator for estimating θ with smaller variance. So, $I(\theta_0)$ measures the information that $\mathbf{X}$ contains about the parameter θ. This is known as *Fisher information*. We notice a reverse relationship between Fisher information (variance of the score) and the variance of any unbiased estimator. This becomes the intuitive basis for the Cramer–Rao lower bound for the variance of an unbiased estimator, which holds under regularity conditions (i) to (iv).

Score function and Fisher information are additive in a random sample: If $X_1, X_2, \ldots, X_n$ is a random sample drawn from $f(x;\theta)$, then the score function for the entire sample is defined as

$$S(\mathbf{X}, \theta) = \frac{\partial}{\partial\theta} \log f(\mathbf{X};\theta) = \frac{\partial}{\partial\theta} \log \prod_{i=1}^{n} f(X_i;\theta)$$

$$= \frac{\partial}{\partial\theta} \sum_{i=1}^{n} \log f(X_i;\theta) = \sum_{i=1}^{n} \frac{\partial}{\partial\theta} \log f(X_i;\theta)$$

$$= \sum_{i=1}^{n} S(X_i, \theta)$$

Next,

$$I_{\mathbf{X}}(\theta) = E_\theta \left[\frac{\partial}{\partial\theta} \log f(\mathbf{X};\theta) \right]^2 = E_\theta \left[\sum_{i=1}^{n} S(X_i, \theta) \right]^2$$

$$= \left[\sum_{i=1}^{n} E_\theta S(X_i, \theta)^2 + \sum_{i \neq j}^{n} E_\theta S(X_i, \theta) \cdot E_\theta S(X_j, \theta) \right]$$

since the observations are independent. By Eqs. (4.2.2) and (4.2.3),

$$I_{\mathbf{X}}(\theta) = \sum_{i=1}^{n} I_{X_i}(\theta) = nI_X(\theta) \tag{4.2.4}$$

since the *rvs* are identically distributed. The higher the value of Fisher information number $I_{\mathbf{X}}(\theta)$, the better information we have in the sample about θ, and the better estimation of θ is possible.

Computationally convenient form of $I_X(\theta)$: Assume that the second derivative of $f(x;\theta)$ with respect to θ exists, is finite, and the regularity condition (iv) holds. On differentiating

$$\int f(x;\theta)dx = 1$$

with respect to x, we have

$$\int f'(x;\theta)dx = \int \frac{\partial}{\partial\theta} \log f(x;\theta)\, f(x;\theta)dx = 0$$

On differentiating it again with respect to θ, we get

$$\int \left[\frac{\partial}{\partial\theta} \log f(x;\theta)\right]^2 f(x;\theta)dx + \int \left[\frac{\partial^2}{\partial\theta^2} \log f(x;\theta)\right] f(x;\theta)dx = 0$$

This gives

$$E\left[\frac{\partial}{\partial\theta} \log f(X;\theta)\right]^2 = E\left[-\frac{\partial^2}{\partial\theta^2} \log f(X;\theta)\right]$$

or

$$I_X(\theta) = E\left[-\frac{\partial^2}{\partial\theta^2} \log f(X;\theta)\right] \tag{4.2.5}$$

The alternative expression in Eq. (4.2.5) is computationally convenient. Note that the cost that one pays to achieve this computational ease is the regularity condition that we assume to hold.

$I(\theta)$ changes under reparameterization: Consider the Fisher information in X, $I_X(\theta)$, when parameterization is θ. If the parameterization is chosen by $\theta = g(\eta)$, then the information that X contains on η is shown by

$$I_X^0(\eta) = I_{\mathbf{X}}[g(\eta)]\, [g'(\eta)]^2$$

assuming g is differentiable. So, $I_X(\theta)$ depends on what parameterization is chosen.

Note. It is important to mention that the information $I_{\mathbf{X}}(\theta)$ increases with the increase in n. In other words, for large n, we can estimate θ with more ease and efficiency since the sample observations carry more information about θ.

Fisher information when $\boldsymbol{\theta}$ is a vector: Let $X_1, X_2, \ldots, X_n$ be a random sample from the density $f(x; \boldsymbol{\theta})$, $\boldsymbol{\theta} = (\theta_1, \ldots, \theta_k)'$, $\boldsymbol{\theta} \in \Theta \subseteq \mathbb{R}^k$; and the regularity conditions (i) to (iv) hold. In this case, the information contained in the sample $X_1, X_2, \ldots, X_n$ about the parameter $\boldsymbol{\theta}$ is defined by the following $k \times k$ matrix which is known as Fisher information matrix

$$I_{\mathbf{X}}(\boldsymbol{\theta}) = \|I_{ij}(\theta)\|$$

where
$$I_{ij}(\boldsymbol{\theta}) = E_{\boldsymbol{\theta}}[S_i(\mathbf{X}, \boldsymbol{\theta})\, S_j(\mathbf{X}, \boldsymbol{\theta})]$$

$$= E_{\boldsymbol{\theta}}\left[\frac{\partial}{\partial \theta_i}\log f(\mathbf{X}; \boldsymbol{\theta}) \cdot \frac{\partial}{\partial \theta_j}\log f(\mathbf{X}; \boldsymbol{\theta})\right]$$

$$= \text{cov}_{\boldsymbol{\theta}}[S_i(\mathbf{X}, \boldsymbol{\theta}), S_j(\mathbf{X}, \boldsymbol{\theta})]$$

since $E[S_i(\mathbf{X},\boldsymbol{\theta})] = 0$ for all i by assumption (iv). Note that $I_{\mathbf{X}}(\boldsymbol{\theta})$ is the variance-covariance matrix of the random quantities $S_i(\mathbf{X},\boldsymbol{\theta})$, $i = 1,\ldots,k$. It is positive semi-definite.

Note. If X follows one–parameter exponential distribution, (refer Eq. 2.4.1)

$$f(x;\, \theta) = c(\theta)h(x)\exp\{p(\theta)T(x)\}$$

On reparameterization $\pi = p(\theta)$ and $T(x) = x$, the density changes to

$$f(x;\pi) = c(\pi)h(x)\exp(\pi x)$$

This form of density is known as canonical form. On differentiating

$$\int c(\pi)h(x)\exp(\pi x)dx = 1$$

with respect to π, we get

$$\int [c'(\pi)h(x)\exp(\pi x) + xc(\pi)h(x)\exp(\pi x)]dx = 0 \tag{4.2.6}$$

$$\Rightarrow \qquad \frac{c'(\pi)}{c(\pi)} + E(X) = 0$$

or
$$E(X) = -\frac{c'(\pi)}{c(\pi)}$$

Differentiating Eq. (4.2.6) again with respect to π, we get

$$\int [c''(\pi)\,h(x)\,\exp(\pi x) + xc'(\pi)h(x)\,\exp(\pi x) + xc'(\pi)h(x)\,\exp(\pi x) + x^2c(\pi)h(x)\,\exp(\pi x)]dx = 0$$

$$\Rightarrow \qquad \frac{c''(\pi)}{c(\pi)} + 2E(X)\frac{c'(\pi)}{c(\pi)} + E(X^2) = 0$$

$$E(X^2) = -\frac{c''(\pi)}{c(\pi)} - 2E(X)\frac{c'(\pi)}{c(\pi)} = -\frac{c''(\pi)}{c(\pi)} + 2[E(X)]^2$$

or
$$V(X) = -\frac{c''(\pi)}{c(\pi)} + \left\{\frac{c'(\pi)}{c(\pi)}\right\}^2$$

Further, consider

$$\log f = \log c(\pi) + \log h(x) + \pi x$$

$$\frac{\partial}{\partial \pi}\log f = \frac{c'(\pi)}{c(\pi)} + x$$

$$\frac{\partial^2}{\partial \pi^2}\log f = \frac{c''(\pi)}{c(\pi)} - \left\{\frac{c'(\pi)}{c(\pi)}\right\}^2$$

The Fisher information about π is then given by

$$I_X(\pi) = E\left[-\frac{\partial^2}{\partial \pi^2}\log f\right] = -\frac{c''(\pi)}{c(\pi)} + \left\{\frac{c'(\pi)}{c(p)}\right\}^2$$

The Fisher information about θ is given by

$$I_X(\theta) = I_X[p(\theta)]\,[p'(\theta)]^2$$

Further, if $X_1, X_2, \ldots, X_n$ are *iid* $f(x;\theta)$, the Fisher information about θ is

$$I_{\mathbf{X}}(\theta) = nI_X(\theta)$$

4.3 LOWER BOUNDS FOR VARIANCE OF UNBIASED ESTIMATORS

Uniformly minimum variance unbiased estimation has been the criterion for judging the estimators as best in Chapter 3, whereas we have seen that they exist in case of exponential families. For other families, such estimators often do not exist. In such cases, what an unbiased estimator can achieve the minimum variance locally at point $\theta = \theta_0$ becomes the criterion for judging the estimators. We state an estimator as reasonably optimal if its variance is close to the local minimum variances at all points of θ. One problem for such judgments arises that these local minimum variances are difficult to calculate. Statisticians like Frechet, Rao, Cramer, Chapman, and Robbins worked for easy solutions to this problem and came out with simple solutions by calculating lower bounds for the variances of unbiased estimators. Though calculating these lower bounds is much simple, usually these bounds are not sharp. In other words, the calculation of these bounds gives their values below the minimum variance of an unbiased estimator at some local point $\theta = \theta_0$. We state two such lower bounds first due to Rao and Cramer, and second due to Chapman and Robbins in Theorems 4.3.1 and 4.3.7, respectively.

Theorem 4.3.1 (Rao and Cramer Lower Bound) Suppose a family of *pdfs* $\mathcal{F} = \{f(x;\,\theta),\ \theta \in \Theta\}$ satisfies the regularity conditions (i) to (iv). Let a random sample $X_1, X_2, \ldots, X_n$ is drawn from a population with *pdf* $f(x;\,\theta)$ in $\mathcal{F}$, where θ is not known. Let $\delta(\mathbf{X})$ be an unbiased estimator of $g(\theta)$, so that its second moment exists and is finite. Then

$$V_\theta[\delta(\mathbf{X})] \geq \frac{[g'(\theta)]^2}{I_{\mathbf{X}}(\theta)} \tag{4.3.1}$$

Proof. On differentiating

$$\int f(\mathbf{x};\theta)d\mathbf{x}=1$$

with respect to θ, we get

$$\int \frac{f'(\mathbf{x};\theta)}{f(\mathbf{x};\theta)} f(x;\theta)d\mathbf{x}=0$$

or

$$\int \frac{1}{f(\mathbf{x};\theta)}\frac{\partial}{\partial\theta} f(\mathbf{x};\theta)\, f(\mathbf{x};\theta)d\mathbf{x}=0$$

$$\int S(\mathbf{x},\theta)\, f(\mathbf{x};\theta)d\mathbf{x}=0$$

$$E[S(\mathbf{X},\theta)]=0$$

Further,

$$V[S(\mathbf{X},\theta)]=E[S(\mathbf{X},\theta)]^2-\{E[S(\mathbf{X},\theta)]\}^2$$

$$=E[S(\mathbf{X},\theta)]^2 \quad \text{[by Eq. (4.2.2)]}$$

$$=E\left[\frac{\partial}{\partial\theta}\log f(\mathbf{X};\theta)\right]^2=I_{\mathbf{X}}(\theta) \tag{4.3.2}$$

The estimator $\delta(\mathbf{X})$ is an unbiased estimator of θ

$$E[\delta(\mathbf{X})]=\int \delta(\mathbf{x}) f(\mathbf{x};\theta)d\mathbf{x}=g(\theta) \tag{4.3.3}$$

On differentiating Eq. (4.3.3) with respect to θ, we get

$$\int \delta(\mathbf{x})\frac{f'(\mathbf{x};\theta)}{f(\mathbf{x};\theta)} f(x;\theta)d\mathbf{x}=g'(\theta)$$

$$\int \delta(\mathbf{x})\frac{1}{f(\mathbf{x};\theta)}\frac{\partial}{\partial\theta} f(\mathbf{x};\theta)\, f(\mathbf{x};\theta)d\mathbf{x}=g'(\theta)$$

$$E[\delta(\mathbf{X})S(\mathbf{X},\theta)]=g'(\theta)$$

$$\therefore \qquad \operatorname{cov}[\delta(\mathbf{X}),S(\mathbf{X},\theta)]=g'(\theta) \tag{4.3.4}$$

By the Cauchy–Schwarz inequality,

$$[\operatorname{cov}(\delta(\mathbf{X}),S(\mathbf{X},\theta))]^2\le V[\delta(\mathbf{X})]\;V[S(\mathbf{X},\theta)]$$

We obtain, by Eqs. (4.3.2) and (4.3.4),

$$V_\theta[\delta(\mathbf{X})]\ge\frac{[g'(\theta)]^2}{I_{\mathbf{X}}(\theta)}$$

Hence, the theorem is proved. ▣

Comments on CR lower bound

1. Regularity condition (iv) holds for an exponential family but not necessarily true for a non-exponential family.

2. CRLB, $B(\theta)$, depends only on the parametric function $g(\theta)$ and the joint density $f(\mathbf{x};\theta)$. This lower bound is uniform for any unbiased estimator.
3. The computation of the Cramer–Rao lower bound (CRLB) in Eq. (4.3.1) could be made much simpler by using an alternative expression for $I_{\mathbf{X}}(\theta)$ given in Eq. (4.2.5).
4. CRLB in *iid* case: If $X_1, X_2,\ldots,X_n$ are *iid* from $f(x;\theta)$, then by Eq. (4.2.4), $I_{\mathbf{X}}(\theta) = nI_{X_1}(\theta)$, where $I_{X_1}(\theta) = E_\theta[(\partial/\partial\theta)\log f(X_1;\theta)]^2$. In this case, the CRLB $B(\theta)$ is given by

$$V_\theta[\delta(\mathbf{X})] \geq \frac{[g'(\theta)]^2}{nI_{X_1}(\theta)} = B(\theta) \tag{4.3.5}$$

5. Fisher's information contained in the sample $X_1, X_2,\ldots,X_n$ on the parameter θ, $I_{\mathbf{X}}(\theta)$, increases with the increase in sample size n. Consequently, we have a smaller lower bound on the variance of an unbiased estimator of $g(\theta)$.
6. The regularity condition (ii) states that the support $S(\theta)$ of the density $f(x;\ \theta)$ does not depend on the parameter θ. We have noted, while proving Theorem 4.3.1 on lower bound for the variance of an unbiased estimator, that assumption (ii) was never used in the proof. In fact, for all such families for which support $S(\theta)$ depends on θ, either (iv) does not hold or if (iv) holds, then inequality Eq. (4.3.1) does not hold. So, in brief, assumption (ii) excludes all such families $f(x;\ \theta)$ for which assumption (iv) or CRLB in Eq. (4.3.1) does not hold. In addition to it, if assumptions (ii) and (iii) hold, then (iv) holds if for all $\delta(\mathbf{X})$ with $E_\theta|\delta(\mathbf{X})| < \infty\ \forall\ \theta \in \Theta$, $E_\theta[\delta(\mathbf{X})(\partial/\partial\theta)\log f(\mathbf{X};\ \theta)]$ and $E_\theta[\delta(\mathbf{X})(\partial/\partial\theta)\, f(\mathbf{X};\ \theta)]$ are continuous functions of θ. We must remember that assumptions (i) to (iv) hold when the density $f(\mathbf{x};\ \theta)$ belongs to one-parameter exponential family.
7. In some cases, when regularity conditions are satisfied and UMVUEs exist, the CRLB $B(\theta)$ is not sharp. In other words, in these cases, the variance of the best estimator fails to reach the CRLB; and UMVUEs are not most efficient. This may be considered as a drawback of defining an estimator as the most efficient corresponding to the CRLB. However, under such situations, one fails to decide whether one should continue the search for an estimator that could attain the CRLB or just no estimator can attain it. These questions cannot be resolved unless we propose some other method that can provide a sharper lower bound (i.e., greater than the CRLB). One such sharp lower bound is due to Chapman and Robbins discussed in Theorem 4.3.7.

   Let $X_1, X_2,\ldots,X_n$ be independent random variables from $N(\theta^{1/3},\ 1)$. The CR lower bound for estimating θ, $1/I(\theta) = 9\theta^4/n$ is smaller than the greatest lower bound to the variance of unbiased estimators of θ, $9\theta^4/n + 18\theta^2/n^2 + 6/n^3$. It is clear from the above example that the CR lower bound is not always sharp. So, if the equality in the CR lower bound is not attained, it must not mean that the minimum variance estimator is not an optimal estimator. Such an estimator may be the only unbiased estimator.

   We have seen that the underlying regularity conditions for the Carmer and Rao lower bound ruled out mainly Pitman's families of distributions. Moreover, these bounds were not sharp.

8. In cases where the regularity conditions are not satisfied, we cannot talk of the CRLB, even though UMVUEs may still exist. This is sometimes seen as a criticism of defining most efficient estimators, since they may not exist at all, whereas contrary to it UMVUEs exist.

We now state a few definitions for the regular family of *pdf*s $\{f(x;\theta),\ \theta \in \Theta\}$.

Definition 4.3.1 (Most efficient estimator) An unbiased estimator δ_0 is said to be the most efficient estimator for a regular family of distributions $\{f(x;\theta),\ \theta \in \Theta\}$ if

$$V_\theta(\delta_0) = \text{CRLB} = \frac{[g'(\theta)]^2}{I_{\mathbf{X}}(\theta)}$$

i.e., δ_0 is the best estimator of $g(\theta)$ in the sense that it achieves the minimum value for the average squared deviation $E_\theta[\delta_0 - g(\theta)]^2$ for all θ.

Definition 4.3.2 (Efficiency of an estimator) The efficiency of an estimator δ, when δ_0 is given to be the most efficient estimator for a regular family $\{f(x;\theta)\ ,\ \theta \in \Theta\}$, is defined by

$$e(\delta, \theta) = \frac{\text{CRLB}}{V_\theta(\delta)} = \frac{[I_{\mathbf{X}}(\theta)]^{-1}}{V_\theta(\delta)}$$

The estimators become better and better with increase in their efficiencies; generally, the efficiency of an estimator $e < 1$ and when it attains 1, the corresponding estimator is said to be the most efficient.

Definition 4.3.3 (Asymptotically most efficient estimator) An estimator δ is said to be asymptotically most efficient if

$$\lim_{n\to\infty} e(\delta, \theta) = 1$$

δ is asymptotically unbiased if

$$\lim_{n\to\infty} E_\theta[\delta(\mathbf{X})] = \theta$$

We discuss next the relationship between unbiased estimators, most efficient estimators, and UMVUEs in the following subsection.

4.3.1 Most Efficient, UMVU Estimators, and Attainment of CR Lower Bound

We assume here that the observations $X_1, X_2, \ldots, X_n$ in the sample are *iid* from a regular density $f(x;\theta)$, where $\delta(\mathbf{X})$ is a function of a sufficient statistic and is an unbiased estimator of $g(\theta)$. Then δ is the most efficient (clearly UMVUE) if and only if the equality in Eq. (4.3.1) is attained. This happens when and only when $\delta(\mathbf{X})$ and $S(\mathbf{X}, \theta)$ are linearly related, i.e., the correlation between $\delta(\mathbf{X})$ and the score function $S(\mathbf{X}, \theta)$,

$$\rho[\delta(\mathbf{X}), S(\mathbf{X}, \theta)] = 1$$

$$S(\mathbf{X}, \theta) = a(\theta) + \delta(\mathbf{X})\, c(\theta) \tag{4.3.6}$$

Note that only δ is a statistic, S is not a statistic—it is a function of both $\mathbf{X}$ and θ. Therefore, $a(\theta)$ and $c(\theta)$ are functions of θ. Equations (4.2.2) and (4.3.6) give

$$a(\theta) + c(\theta)g(\theta) = 0$$

$$a(\theta) = -c(\theta)g(\theta)$$

On putting the value of $a(\theta)$ in Eq. (4.3.6), we get

$$S(\mathbf{X}, \theta) = \frac{\partial}{\partial \theta} \log f(\mathbf{X}; \theta) = c(\theta)[\delta(\mathbf{X}) - g(\theta)] \tag{4.3.7}$$

Condition (4.3.7) is the condition of linearity between the score and the unbiased estimator of $g(\theta)$. If the condition is satisfied, then $\delta(\mathbf{X})$ is not only UMVUE but also the most efficient (attains CR lower bound) for estimating $g(\theta)$.

Condition Eq. (4.3.7) is a necessary and sufficient condition for the existence of a most efficient estimator δ of $g(\theta)$. It is further clear from Eq. (4.3.7) that there is at most one function $g(\theta)$ which can be efficiently estimated. In other words, if Eq. (4.3.7) holds, then $g(\theta)$ is the only function of θ for which the UMVU estimator exists. Note that if $g(\theta)$ is not an identity function, then there does not exist an unbiased estimator of θ that is the most efficient. The variance of such $\delta(\mathbf{X})$ is given by [from Eq. (4.3.10)]

$$V_\theta(\delta) = \frac{g'(\theta)}{c(\theta)}$$

4.3.2 Family for Which CR Lower Bound is Attained (Most Efficient Estimators are UMVUEs)

One step ahead, to characterize the family $\{f(x; \theta), \theta \in \Theta\}$ for which UMVUEs are also most efficient, we integrate Eq. (4.3.7) with respect to θ.

$$\int \frac{\partial}{\partial \theta} \log f(\mathbf{x}; \theta) d\theta = \int c(\theta)[\delta(\mathbf{x}) - g(\theta)] d\theta$$

$$\log f(\mathbf{x}; \theta) = \int c(\theta)[\delta(\mathbf{x}) - g(\theta)] d\theta + S(\mathbf{x})$$

$$= \delta(\mathbf{x}) \int c(\theta) d\theta - \int c(\theta) g(\theta) d\theta + S(\mathbf{x})$$

$$= Q(\theta)\delta(\mathbf{x}) - D(\theta) + A(\mathbf{x}) \tag{4.3.8}$$

where $\int c(\theta)d\theta = Q(\theta)$ and $\int c(\theta)g(\theta)d\theta = D(\theta)$. We then have

$$f(\mathbf{x}; \theta) = \exp\{Q(\theta)\delta(\mathbf{x}) - D(\theta) + S(\mathbf{x})\}$$

Hence, if the CRLB is attained, the corresponding family $\{f(x; \theta), \theta \in \Theta\}$ must be a one-parameter exponential family; $\delta(\mathbf{X})$ is the most efficient (CRLB) estimator of $g(\theta) = D'(\theta)/Q'(\theta)$ and it must be a sufficient statistic. Wherein, if Eq. (4.3.8) is not satisfied, the variance of a UMVUE would be larger than the CR lower bound.

If we consider the *pdf* in one one-parameter exponential family of distributions [see Eq. (2.4.2)]

$$f(x;\,\theta) = c(\theta)h(x)\exp\{p(\theta)T(x)\}$$

$$\log f(x;\,\theta) = \log c(\theta) + \log h(x) + p(\theta)T(x)$$

$$\frac{\partial}{\partial\theta}\log f(x;\theta) = \frac{c'(\theta)}{c(\theta)} + p'(\theta)T(x)$$

$$T(x) = \frac{1}{p'(\theta)}\frac{\partial}{\partial\theta}\log f(x;\theta) - \frac{c'(\theta)}{p'(\theta)c(\theta)}$$

$$\delta(\mathbf{X}) = \frac{1}{n}\sum_1^n T(X_i) = \frac{1}{np'(\theta)}\sum^n \frac{\partial}{\partial\theta}\log f(X_i;\theta) - \frac{c'(\theta)}{p'(\theta)c(\theta)}$$

$$E_\theta[\delta(\mathbf{X})] = -\frac{c'(\theta)}{p'(\theta)c(\theta)} = g(\theta)$$

$\delta(\mathbf{X}) = (1/n)\,\Sigma_1^n T(X_i)$ is the most efficient estimator of $g(\theta)$.

The constant $c(\theta)$ may be evaluated by assuming that Eq. (4.3.7) holds

$$S(\mathbf{X},\,\theta) = c(\theta)[\delta(\mathbf{X}) - g(\theta)]$$

This gives
$$V(S) = I(\theta) = c^2(\theta)\,V\delta(\mathbf{X}) \tag{4.3.9}$$

Further, in this case, that is when the CRLB is attained,

$$V[\delta(\mathbf{X})] = \frac{[g'(\theta)]^2}{I_{\mathbf{X}}(\theta)}$$

On putting the value of $V(\delta)$ in Eq. (4.3.9), we get

$$I(\theta) = [c(\theta)]^2\,\frac{[g'(\theta)]^2}{I_{\mathbf{X}}(\theta)}$$

or
$$c(\theta) = \frac{I_{\mathbf{X}}(\theta)}{g'(\theta)} \tag{4.3.10}$$

To sum up, we have shown that the CRLB estimator is (a) sufficient; (b) the corresponding density belongs to a one-parameter exponential family; and (c) condition (4.3.7) holds.

Conversely, if T is sufficient and Eq. (4.3.7) holds, from the Eq. (4.3.9),

$$V[\delta(\mathbf{X})] = \frac{I_{\mathbf{X}}(\theta)}{[c(\theta)]^2} \tag{4.3.11}$$

On putting the value of $c(\theta)$ in Eq. (4.3.11) from Eq. (4.3.10), we get

$$V[\delta(\mathbf{X})] = \frac{[g'(\theta)]^2}{I_{\mathbf{X}}(\theta)} = \text{CRLB}$$

This shows that the variance of $\delta(\mathbf{X})$ attains CRLB or $\delta(\mathbf{X})$ is the most efficient estimator. Further, on putting the value of $I_{\mathbf{X}}(\theta)$ in Eq. (4.3.11) from Eq. (4.3.10), we get

$$V_\theta[\delta(X)] = \left|\frac{g'(\theta)}{c(\theta)}\right| \tag{4.3.12}$$

It, thus, proves the following theorem.

Theorem 4.3.2 A necessary and sufficient condition for an estimator to be the CRLB estimator (most efficient) is that it is sufficient and Eq. (4.3.7) holds.

Therefore, for one-parameter exponential family, one can find an estimator so that its variance attains the CR lower bound, it is UMVUE, and most efficient. But for non–exponential families, the CR bounds may not be sharp or attainable. In this situation, there is no guarantee that the variance of a UMVUE may attain the CR lower bound.

Further, if the regularity conditions for the CR inequality are not satisfied, the bound may still be calculated, but it remains meaningless, as one may get an unbiased estimator whose variance may still be smaller than this bound (see solved Example 4.9).

We see that if there exists some function $g(\theta)$ for which $\delta(\mathbf{X})$ is the most efficient estimator, then $\delta(\mathbf{X})$ is a sufficient statistic for θ. Whereas, if θ admits a sufficient statistic, there may not be any function $g(\theta)$ for which the most efficient estimator, as a function of sufficient statistic, exists.

Like UMVUEs, the most efficient estimators are also unique. However, maximum likelihood estimators (MLEs) may not be unique (see Chapter 6). For example, consider $X \sim U(\theta - (1/2), \theta + (1/2))$ with corresponding likelihood function $L(\theta) = f(x;\, \theta) = 1$ if $[\theta - (1/2)] < x < \theta + (1/2)$. The MLE of θ is $\alpha[X_{(1)} + (1/2)] + (1-\alpha)[X_{(n)} - (1/2)]$ for $\theta < X_{(1)} + (1/2)$, $\theta > X_{(n)} - (1/2)$.

If an unbiased estimator attains the CRLB, i.e., it is the most efficient, then it is MLE, but the converse is not necessarily true. However, MLEs are asymptotically CRLB estimators (most efficient). In fact, we prove a more general result in Chapter 6 that for a sequence of MLEs $\{\delta_n\}$,

$$\sqrt{n}(\delta_n - \theta) \xrightarrow{d} N(0, I_{\mathbf{X}}^{-1}(\theta))$$

and, therefore, MLE is not only consistent and asymptotically normal but also asymptotically most efficient.

The application of Theorem 4.3.2 is as follows: Let $X_1, X_2, \ldots, X_n$ be a random sample from one-parameter exponential family with *pdf*

$$f(x;\, \theta) = c(\theta)h(x)\exp\{p(\theta)T(x)\}$$

If $E_\theta[T(X)] = g(\theta)$, then the estimator $\delta(\mathbf{X}) = (1/n)\Sigma_{i=1}^{n} T(X_i)$ is unbiased for $g(\theta)$. The estimator $\delta(\mathbf{X})$ is the CRLB estimator of $g(\theta)$

$$V_\theta\left(\frac{1}{n}\sum_{i=1}^{n} T(X_i)\right) = \frac{[g'(\theta)]^2}{I_{\mathbf{X}}(\theta)}$$

***Corollary* 4.3.1** A more general form of the CR lower bound is considered when $[\delta(\mathbf{X})]$ is considered as a biased estimator of $g(\theta)$, with bias $b(\theta)$ so that

$$E_\theta[\delta(\mathbf{X})] = g(\theta) + b(\theta)$$

$$\frac{\partial}{\partial\theta}E_\theta[\delta(\mathbf{X})] = g'(\theta) + b'(\theta)$$

and
$$\text{MSE}_\theta[\delta(\mathbf{X})] = V_\theta(\delta) + \text{Bias}^2(\theta)$$

The CR lower bound for the mean square error of a biased estimator is then given by

$$V_\theta[\delta(\mathbf{X})] \geq b^2(\theta) + \frac{[g'(\theta) + b'(\theta)]^2}{I_X(\theta)}$$

Theorem 4.3.3 Let $X_1, X_2, \ldots, X_n$ be a random sample from some *pdf* $f(x;\theta)$, $\theta \in \Theta$, and let $\delta(\mathbf{X})$ be a sufficient statistic and also unbiased for $g(\theta)$. Further, if the *pdf* of $\delta(\mathbf{X})$ is a member of exponential family of distributions, then δ is the CRLB estimator, i.e.,

$$V_\theta[\delta(\mathbf{X})] = \frac{[g'(\theta)]^2}{I_{\mathbf{X}}(\theta)}$$

but the converse may not be true.

Proof. Let the *pdf* of $T = \delta(\mathbf{X})$ be of the form

$$f(t;\theta) = \exp\{t\,Q(\theta) - D(\theta) + S(t)\} \tag{4.3.13}$$

$$\int f(t;\theta)dt = 1$$

The gives
$$\int \exp\{tQ(\theta) - D(\theta) + S(t)\}dt = 1 \tag{4.3.14}$$

On differentiating both sides of Eq. (4.3.14) with respect to θ, we get

$$\int f(t;\theta)\{Q'(\theta)t - D'(\theta)\}dt = 0 \tag{4.3.15}$$

$$E_\theta[\delta(\mathbf{X})] = \frac{D'(\theta)}{Q'(\theta)} = g(\theta) \tag{4.3.16}$$

On differentiating Eq. (4.3.15) with respect to θ again, we get

$$\int[\{Q'(\theta)t - D'(\theta)\}^2 f(t;\theta) + \{Q''(\theta)t - D''(\theta)\}f(t;\theta)]dt = 0$$

$$E_\theta[Q'(\theta)\delta(\mathbf{X}) - D'(\theta)]^2 + E_\theta[Q''(\theta)\delta(\mathbf{X}) - D''(\theta)] = 0$$

$$E_\theta(Q'^2\delta^2(\mathbf{X}) + D'^2 - 2Q'D'\delta(\mathbf{X})) + E_\theta(Q''\delta(\mathbf{X}) - D'') = 0$$

$$E_\theta(Q'^2\delta^2(\mathbf{X})) = E_\theta(2Q'D'\delta(\mathbf{X}) - Q''\delta(\mathbf{X}) + D'' - D'^2)$$

On putting the value of $E_\theta[\delta(\mathbf{X})]$ from Eq. (4.3.16), we get

$$Q'^2 E_\theta(\delta(\mathbf{X})^2) = 2Q'D' \cdot \frac{D'}{Q'} - Q'' \cdot \frac{D'}{Q'} + D'' - D'^2$$

$$= D'^2 - \frac{D'}{Q'} Q'' + D''$$

$$E_\theta(\delta(\mathbf{X}))^2 = \frac{1}{Q'^3}[Q'D'^2 - D'Q'' + Q'D''] \tag{4.3.17}$$

$V_\theta[\delta(\mathbf{X})]$ is calculated by Eqs. (4.3.16) and (4.3.17)

$$V_\theta[\delta(\mathbf{X})] = E_\theta(\delta(\mathbf{X}))^2 - [E_\theta(\delta(\mathbf{X}))]^2$$

$$= \frac{1}{Q'^3}(Q'D'^2 - D'Q'' + Q'D'') - \frac{D'^2}{Q'^2}$$

$$= \frac{1}{Q'^3}(Q'D'^2 - D'Q'' + Q'D'' - Q'D'^2)$$

$$= \frac{1}{Q'^3}(Q'D'' - Q''D')$$

Further, on taking the logarithm of Eq. (4.3.13), we get

$$\log f(t;\, \theta) = tQ(\theta) - D(\theta) + S(t)$$

and

$$\frac{\partial}{\partial\theta} \log f(t;\, \theta) = tQ'(\theta) - D'(\theta)$$

Fisher's information is given by

$$I_{\delta(\mathbf{X})}(\theta) = E_\theta\left[\frac{\partial}{\partial\theta} \log f(\delta(\mathbf{X});\, \theta)\right]^2$$

$$= E_\theta[\delta(\mathbf{X}) \cdot Q'(\theta) - D'(\theta)]^2 \quad \text{[from Eq. (4.3.16)]}$$

$$= Q'^2 E_\theta \delta^2(\mathbf{X}) + D'^2 - 2Q'D' E_\theta \delta(\mathbf{X})$$

From Eqs. (4.3.16) and (4.3.17), we have

$$I_{\delta(\mathbf{X})}(\theta) = \frac{1}{Q'}(Q'D'^2 - Q''D' + Q'D'') + D'^2 - 2Q'D'\frac{D'}{Q'}$$

$$= \frac{1}{Q'}(Q'D'' - Q''D') \tag{4.3.18}$$

Further,

$$g'(\theta) = \frac{1}{Q'^2}(Q'D'' - Q''D') \tag{4.3.19}$$

Eqs. (4.3.18) and (4.3.19) gives the CR lower bound

$$\text{CRLB} = \frac{[g'(\theta)]^2}{I_\delta(\theta)} = \frac{\frac{1}{Q'^4}(Q'D'' - Q''D')^2}{\frac{1}{Q'}(Q'D'' - Q''D')}$$

$$= \frac{1}{Q'^3}(Q'D'' - Q''D')$$

Note that this is equal to $V_\theta(\delta)$. Therefore, δ is the CRLB estimator of $g(\theta)$.

Theorem 4.3.4 (Multivariate Extension of CR Inequality) Suppose the regularity conditions (i) to (iv) are satisfied and the information matrix $\mathbf{I_X}(\boldsymbol{\theta})$ is positive definite. Then for any unbiased estimator $\boldsymbol{\delta}(\mathbf{X})$ of $\mathbf{g}(\boldsymbol{\theta})$ with $E_{\boldsymbol{\theta}}\boldsymbol{\delta}^2 < \infty \ \forall \ \boldsymbol{\theta} \in \Theta$,

$$V_{\boldsymbol{\theta}}(\boldsymbol{\delta}) \geq \nabla\mathbf{g}(\boldsymbol{\theta})\mathbf{I_X}^{-1}(\boldsymbol{\theta})\nabla'\mathbf{g}(\boldsymbol{\theta}) \tag{4.3.20}$$

where ∇ is known as the gradient matrix

$$\nabla\mathbf{g}(\boldsymbol{\theta}) = \frac{\partial}{\partial\boldsymbol{\theta}}E_{\boldsymbol{\theta}}[\boldsymbol{\delta}(\mathbf{X})] = \frac{\partial}{\partial\boldsymbol{\theta}}\mathbf{g}(\boldsymbol{\theta})$$

Proof. Consider r, the correlation between $\mathbf{a}'\boldsymbol{\delta}(\mathbf{X})$ and $\mathbf{b}'\mathbf{S}(\mathbf{X}, \boldsymbol{\theta})$, as

$$r = \frac{\text{cov}_{\boldsymbol{q}}[\mathbf{a}'\boldsymbol{\delta}(\mathbf{X}), \mathbf{b}'\mathbf{S}(\mathbf{X}, \boldsymbol{\theta})]}{\sqrt{V_{\boldsymbol{q}}[\mathbf{a}'\boldsymbol{\delta}(\mathbf{X})]\,V_{\boldsymbol{q}}[\mathbf{b}'\mathbf{S}(\mathbf{X}, \boldsymbol{\theta})]}} \tag{4.3.21}$$

where

$$\begin{aligned}\text{cov}_{\boldsymbol{\theta}}[\mathbf{a}'\boldsymbol{\delta}(\mathbf{X}), \mathbf{b}'\mathbf{S}(\mathbf{X}, \boldsymbol{\theta})] &= \mathbf{a}'\text{cov}_{\boldsymbol{\theta}}(\boldsymbol{\delta}(\mathbf{X}), \mathbf{S}(\mathbf{X}, \boldsymbol{\theta}))\mathbf{b} \\ &= \mathbf{a}'E_{\boldsymbol{\theta}}[\boldsymbol{\delta}(\mathbf{X})\mathbf{S}'(\mathbf{X}, \boldsymbol{\theta})]\mathbf{b} \quad (\text{since } E_{\boldsymbol{\theta}}\mathbf{S} = \mathbf{0})\end{aligned}$$

Further,

$$E_{\boldsymbol{\theta}}[\boldsymbol{\delta}(\mathbf{X})] = \int\boldsymbol{\delta}(\mathbf{x})f(\mathbf{x}; \boldsymbol{\theta})d\mathbf{x} = \mathbf{g}(\boldsymbol{\theta})$$

$$\begin{aligned}\frac{\partial}{\partial\boldsymbol{\theta}}E_{\boldsymbol{\theta}}[\boldsymbol{\delta}(\mathbf{X})] = \nabla\mathbf{g}(\boldsymbol{\theta}) &= \int\boldsymbol{\delta}(\mathbf{x})\mathbf{S}'(\mathbf{x}, \boldsymbol{\theta})f(\mathbf{x}; \boldsymbol{\theta})d\mathbf{x} \quad [\text{by assumption (iv)}] \\ &= E_{\boldsymbol{\theta}}(\boldsymbol{\delta}(\mathbf{X})\mathbf{S}'(\mathbf{X}, \boldsymbol{\theta}))\end{aligned}$$

This gives

$$\text{cov}_{\boldsymbol{\theta}}[\mathbf{a}'\ \boldsymbol{\delta}(\mathbf{X}), \mathbf{b}'\mathbf{S}(\mathbf{X}, \boldsymbol{\theta})] = \mathbf{a}'\nabla\mathbf{b} \tag{4.3.22}$$

and

$$\begin{aligned}V_{\boldsymbol{\theta}}[\mathbf{b}'\mathbf{S}(\mathbf{X}, \boldsymbol{\theta})] &= \mathbf{b}'V_{\boldsymbol{\theta}}[\mathbf{S}(\mathbf{X}, \boldsymbol{\theta})]\mathbf{b} \\ &= \mathbf{b}'E_{\boldsymbol{\theta}}[\mathbf{S}(\mathbf{X}, \theta)\mathbf{S}'(\mathbf{X}, \boldsymbol{\theta})]\mathbf{b} \\ &= \mathbf{b}'\mathbf{I_X}(\boldsymbol{\theta})\mathbf{b}\end{aligned} \tag{4.3.23}$$

Equations (4.3.21), (4.3.22), and (4.3.23) give

$$r = \frac{\mathbf{a}'\ \nabla\mathbf{b}}{\sqrt{\mathbf{b}'\mathbf{I_X}(\boldsymbol{\theta})\mathbf{b}}\sqrt{\mathbf{a}'V_{\boldsymbol{\theta}}[\boldsymbol{\delta}(\mathbf{X})]\mathbf{a}}} \tag{4.3.24}$$

Note that $\mathbf{b}'\mathbf{I_X}(\boldsymbol{\theta})\mathbf{b} > 0$, since $\mathbf{I_X}(\boldsymbol{\theta})$ has been shown to be positive definite. We know from the results of the theory of multivariate analysis that the maximum of the right-hand side expression of Eq. (4.3.24) over **b** is the multiple correlation coefficient between $\boldsymbol{\delta}(\mathbf{X})$ and $\mathbf{S}(\mathbf{X},\boldsymbol{\theta})$.

Since r is invariant to the scale transformations, so a_is maximizing the right-hand side of Eq. (4.3.24) are not unique. We may, therefore, without loss in generality, impose a constraint

$$V_{\boldsymbol{\theta}}[\mathbf{b}'\mathbf{S}(\mathbf{X}, \boldsymbol{\theta})] = \mathbf{b}'\mathbf{I_X}(\boldsymbol{\theta})\mathbf{b} = 1 \qquad (4.3.25)$$

(since $\mathbf{I_X}(\boldsymbol{\theta})$ is positive definite)

Therefore, the problem reduces to the maximization of $\mathbf{b}'\nabla'\mathbf{a}$ subject to Eq. (4.3.25). By Lagranges' multipliers method, we maximize

$$H(\mathbf{b}, \lambda) = \mathbf{b}'\,\nabla'\mathbf{a} - \frac{1}{2}\,\lambda[\mathbf{b}'\mathbf{I_X}(\boldsymbol{\theta})\mathbf{b} - 1]$$

with respect to **b** subject to Eq. (4.3.25). We then have

$$\frac{\partial}{\partial \mathbf{b}} H(\mathbf{b}, \lambda) = 0$$

$$\nabla'\mathbf{a} - \lambda\mathbf{I_X}(\boldsymbol{\theta})\mathbf{b} = 0$$

$$\mathbf{b} = \frac{1}{\lambda}\mathbf{I_X^{-1}}(\boldsymbol{\theta})\nabla'\mathbf{a} \qquad (4.3.26)$$

Equations (4.3.25) and (4.3.26) give

$$\frac{1}{\lambda}\mathbf{a}'\nabla\mathbf{I_X^{-1}}(\boldsymbol{\theta})\mathbf{I_X}(\boldsymbol{\theta})\frac{1}{\lambda}\mathbf{I_X^{-1}}(\boldsymbol{\theta})\nabla'\mathbf{a} = 1$$

$$\lambda^2 = \mathbf{a}'\nabla\mathbf{I_X^{-1}}(\boldsymbol{\theta})\nabla'\mathbf{a}$$

$$\lambda = \pm\sqrt{\mathbf{a}'\nabla\mathbf{I_X^{-1}}(\boldsymbol{\theta})\nabla'\mathbf{a}} \qquad (4.3.27)$$

From Eqs. (4.3.26) and (4.3.27), we get

$$\mathbf{b} = \pm\frac{\mathbf{I_X^{-1}}(\boldsymbol{\theta})\nabla'\mathbf{a}}{\sqrt{\mathbf{a}'\nabla\mathbf{I_X^{-1}}(\boldsymbol{\theta})\nabla'\mathbf{a}}} \qquad (4.3.28)$$

The maximum value of r (the multiple correlation coefficient) from Eqs. (4.3.24) and (4.3.28) is

$$r_0 = \max_{\mathbf{b}}\frac{\mathbf{a}'\nabla\mathbf{b}}{\sqrt{\mathbf{a}'V_{\boldsymbol{\theta}}[\boldsymbol{\delta}(\mathbf{X})]\mathbf{a}}}$$

$$= \frac{\mathbf{a}'\nabla\mathbf{I_X^{-1}}(\boldsymbol{\theta})\nabla'\mathbf{a}}{\sqrt{\mathbf{a}'\nabla\mathbf{I_X^{-1}}(\boldsymbol{\theta})\nabla'\mathbf{a}}\sqrt{\mathbf{a}'V_{\boldsymbol{\theta}}[\boldsymbol{\delta}(\mathbf{X})]\mathbf{a}}} \qquad (4.3.29)$$

$$r_0^2 = \frac{\mathbf{a}'\nabla\mathbf{I_X^{-1}}(\boldsymbol{\theta})\nabla'\mathbf{a}}{\mathbf{a}'V_{\boldsymbol{\theta}}[\boldsymbol{\delta}(\mathbf{X})]\mathbf{a}} \le 1$$

Always, $0 \le r_0 \le 1$, and $r_0^2 \le 1$. We have

$$\mathbf{a}'V_{\boldsymbol{\theta}}[\boldsymbol{\delta}(\mathbf{X})]\mathbf{a} - \mathbf{a}'\nabla\mathbf{I}_{\mathbf{X}}^{-1}(\boldsymbol{\theta})\nabla'\mathbf{a} \ge 0$$

$$\mathbf{a}'[V_{\theta}[\boldsymbol{\delta}(\mathbf{X})] - \nabla\mathbf{I}_{\mathbf{X}}^{-1}(\theta)\nabla']\mathbf{a} \ge 0$$

$$E_{\boldsymbol{\theta}}[\boldsymbol{\delta}(\mathbf{X}) - \boldsymbol{\theta}]\,[\boldsymbol{\delta}(\mathbf{X}) - \boldsymbol{\theta}]' = V_{\boldsymbol{\theta}}[\boldsymbol{\delta}(\mathbf{X})] \ge \nabla\mathbf{I}_{\mathbf{X}}^{-1}(\boldsymbol{\theta})\nabla' \;\forall\; \mathbf{a}$$

This proves the theorem. ▣

***Corollary* 4.3.4** Under the same conditions as in Theorem 4.3.4 with $g(\boldsymbol{\theta})$ as real-valued, the lower bound to the variance of any unbiased estimator $\delta(\mathbf{X})$ of $g(\boldsymbol{\theta})$ is given by Eq. (4.3.20), where $\nabla g(\boldsymbol{\theta})$ is a column vector.

The equality in CRLB in Eq. (4.3.20) is achieved if $\boldsymbol{\delta}(\mathbf{X})$ and $\mathbf{S}(\mathbf{X}, \boldsymbol{\theta})$ are linearly related, that is,

$$\mathbf{S}(\mathbf{X}, \boldsymbol{\theta}) = c(\boldsymbol{\theta})[\boldsymbol{\delta}(\mathbf{X}) - \mathbf{g}(\boldsymbol{\theta})]$$

We now state a few definitions for the regular family of distributions $\{f(x;\, \boldsymbol{\theta}),\, \boldsymbol{\theta} \in \Theta\}$.

Definition 4.3.4 (Most efficient estimator) An unbiased estimator $\boldsymbol{\delta}_0(\mathbf{X})$ is said to be the most efficient estimator for a regular family $\{f(x;\, \boldsymbol{\theta}),\, \boldsymbol{\theta} \in \Theta\}$ if

$$E_{\boldsymbol{\theta}}[\boldsymbol{\delta}_0(\mathbf{X}) - \mathbf{g}(\boldsymbol{\theta})][\boldsymbol{\delta}_0(\mathbf{X}) - \mathbf{g}(\boldsymbol{\theta})]' = \nabla\mathbf{I}_{\mathbf{X}}^{-1}(\boldsymbol{\theta})\nabla'$$

Theorem 4.3.5 An estimator $\boldsymbol{\delta}_0(\mathbf{X})$ is the most efficient if and only if

$$\mathbf{S}(\mathbf{X}, \boldsymbol{\theta}) = \mathbf{I}_{\mathbf{X}}(\boldsymbol{\theta})\nabla^{-1}[\boldsymbol{\delta}_0(\mathbf{X}) - \mathbf{g}(\boldsymbol{\theta})] \tag{4.3.30}$$

Furthermore, any unbiased most efficient estimator is an MLE.

Proof. Let Eq. (4.3.30) holds. By the definition of Fisher's information matrix.

$$\begin{aligned}\mathbf{I}_{\mathbf{X}}(\boldsymbol{\theta}) &= E_{\boldsymbol{\theta}}\mathbf{S}(\mathbf{X}, \boldsymbol{\theta})\mathbf{S}'(\mathbf{X}, \boldsymbol{\theta})\\ &= \mathbf{I}_{\mathbf{X}}(\boldsymbol{\theta})\nabla^{-1}E_{\boldsymbol{\theta}}[\boldsymbol{\delta}_0(\mathbf{X}) - \mathbf{g}(\boldsymbol{\theta})]\,[\boldsymbol{\delta}_0(\mathbf{X}) - \mathbf{g}(\boldsymbol{\theta})]'\,(\nabla^{-1})'\,\mathbf{I}_{\mathbf{X}}'(\boldsymbol{\theta})\end{aligned}$$

This gives

$$E_{\boldsymbol{\theta}}[\boldsymbol{\delta}_0(\mathbf{X}) - \mathbf{g}(\boldsymbol{\theta})][\boldsymbol{\delta}_0(\mathbf{X}) - \mathbf{g}(\boldsymbol{\theta})]' = \nabla\mathbf{I}_{\mathbf{X}}^{-1}(\boldsymbol{\theta})\nabla'$$

which proves that $\boldsymbol{\delta}_0(\mathbf{X})$ is the most efficient.

Conversely, suppose $\boldsymbol{\delta}_0(\mathbf{X})$ is the most efficient, that is,

$$E_{\boldsymbol{\theta}}[\boldsymbol{\delta}_0(\mathbf{X}) - \mathbf{g}(\boldsymbol{\theta})]\,[\boldsymbol{\delta}_0(\mathbf{X}) - \mathbf{g}(\boldsymbol{\theta})]' = \nabla I_X^{-1}(\boldsymbol{\theta})\nabla'$$

Since $\boldsymbol{\delta}_0(\mathbf{X})$ is unbiased,

$$E_{\boldsymbol{\theta}}[\boldsymbol{\delta}_0'(\mathbf{X})] = \mathbf{g}'(\boldsymbol{\theta})$$

$$\nabla' = \frac{\partial}{\partial\boldsymbol{\theta}}E_{\boldsymbol{\theta}}[\boldsymbol{\delta}_0'(\mathbf{X})] = \frac{\partial}{\partial\boldsymbol{\theta}}\mathbf{g}'(\boldsymbol{\theta}) = E_{\boldsymbol{\theta}}[\mathbf{S}(\mathbf{X}, \boldsymbol{\theta})\boldsymbol{\delta}_0'(\mathbf{X})]$$

gives

$$E_{\boldsymbol{\theta}}\mathbf{S}(\mathbf{X}, \boldsymbol{\theta})\,[\boldsymbol{\delta}_0(\mathbf{X}) - \mathbf{g}(\boldsymbol{\theta})]' = E_{\boldsymbol{\theta}}[\mathbf{S}(\mathbf{X}, \boldsymbol{\theta})\boldsymbol{\delta}_0'(\mathbf{X})] = \nabla'$$

By the Cauchy–Schwarz inequality,

$$[E_{\boldsymbol{\theta}}\mathbf{S}(\mathbf{X}, \boldsymbol{\theta})\, (\boldsymbol{\delta}_0(\mathbf{X}) - \mathbf{g}(\boldsymbol{\theta}))']^2 \le E_{\boldsymbol{\theta}}\mathbf{S}(\mathbf{X}, \boldsymbol{\theta})\mathbf{S}'(\mathbf{X}, \boldsymbol{\theta}) E_{\boldsymbol{\theta}}[\boldsymbol{\delta}_0(\mathbf{X}) - \mathbf{g}(\boldsymbol{\theta})]\, [\boldsymbol{\delta}_0(\mathbf{X}) - \mathbf{g}(\boldsymbol{\theta})]'$$
$$\le \nabla \mathbf{I}_{\mathbf{X}}(\boldsymbol{\theta})\mathbf{I}_{\mathbf{X}}^{-1}(\boldsymbol{\theta})\nabla' = \nabla\nabla'$$

Equality holds if and only if

$$\mathbf{S}(\mathbf{X}, \boldsymbol{\theta}) = c(\boldsymbol{\theta})[\boldsymbol{\delta}_0(\mathbf{X}) - \mathbf{g}(\boldsymbol{\theta})] \tag{4.3.31}$$

Multiplying both the sides of Eq. (4.3.31) by $(\boldsymbol{\delta}_0(\mathbf{X}) - \mathbf{g}(\boldsymbol{\theta}))'$ and taking expectation, we get

$$E_{\boldsymbol{\theta}}\mathbf{S}(\mathbf{X}, \boldsymbol{\theta})\, (\boldsymbol{\delta}_0(\mathbf{X}) - \mathbf{g}(\boldsymbol{\theta}))' = c(\boldsymbol{\theta})E_{\boldsymbol{\theta}}[\boldsymbol{\delta}_0(\mathbf{X}) - \mathbf{g}(\boldsymbol{\theta})]\, [\boldsymbol{\delta}_0(\mathbf{X}) - \mathbf{g}(\boldsymbol{\theta})]'$$
$$\nabla' = c(\boldsymbol{\theta})\nabla \mathbf{I}_{\mathbf{X}}^{-1}(\boldsymbol{\theta})\nabla'$$
$$\nabla'(\nabla \mathbf{I}_{\mathbf{X}}^{-1}(\boldsymbol{\theta})\nabla')^{-1} = c(\boldsymbol{\theta})$$
$$\nabla'(\nabla')^{-1}\mathbf{I}_{\mathbf{X}}(\boldsymbol{\theta})\nabla^{-1} = c(\boldsymbol{\theta})$$
$$\mathbf{I}_{\mathbf{X}}(\boldsymbol{\theta})\nabla^{-1} = c(\boldsymbol{\theta}) \tag{4.3.32}$$

Equations (4.3.31) and (4.3.32) give

$$\mathbf{S}(\mathbf{X}, \boldsymbol{\theta}) = \mathbf{I}_{\mathbf{X}}(\boldsymbol{\theta})\nabla^{-1}[\boldsymbol{\delta}_0(\mathbf{X}) - \mathbf{g}(\boldsymbol{\theta})]$$

Given that $\boldsymbol{\delta}_0(\mathbf{X})$ is unbiased and most efficient, Eq. (4.3.32) holds. Let $\boldsymbol{\delta}_{\text{ml}}(\mathbf{X})$ be the maximum likelihood estimator of $\boldsymbol{\theta}$. Then Eq. (4.3.30) at $\boldsymbol{\theta} = \boldsymbol{\delta}_{\text{ml}}(\mathbf{X})$ is

$$\mathbf{S}(\mathbf{X}, \boldsymbol{\delta}_{\text{ml}}(\mathbf{X})) = \mathbf{I}_{\mathbf{X}}(\boldsymbol{\theta})\nabla^{-1}[\boldsymbol{\delta}_0(\mathbf{X}) - \mathbf{g}(\boldsymbol{\delta}_{\text{ml}}(\mathbf{X}))]$$

By the definition of maximum likelihood estimator, $\boldsymbol{\delta}_{\text{ml}}(\mathbf{X})$ is obtained by setting $\mathbf{S}(\mathbf{X}, \boldsymbol{\theta}) = \mathbf{0}$. This gives

$$\mathbf{I}_{\mathbf{X}}(\boldsymbol{\theta})\nabla^{-1}[\boldsymbol{\delta}_0(\mathbf{X}) - \mathbf{g}(\boldsymbol{\delta}_{\text{ml}}(\mathbf{X}))] = \mathbf{0}$$
$$\boldsymbol{\delta}_0(\mathbf{X}) = \mathbf{g}(\boldsymbol{\delta}_{\text{ml}}(\mathbf{X}))$$

Consequently, $\boldsymbol{\delta}_0(\mathbf{X})$ is the maximum likelihood estimator. This proves the theorem. ▣

4.3.3 Bhattacharyya Lower Bound

There arises a situation when the UMVUE exists but the corresponding minimum variance does not reach CRLB. In other words, CRLB is not sharp. As one solution to this problem, Bhattacharyya (1946), in addition to other regularity conditions (i) to (iv) of Section 4.2 for CR inequality, assumes that

$$\int \delta(\mathbf{X}) f(x_1,\ldots,x_n;\, \theta)dx_1\ldots dx_n = g(\theta) \tag{4.3.33}$$

is differentiable m times with respect to θ under the integral sign and defines m variables by

$$S_i = \frac{f^{(i)}(x_1,\ldots, x_n;\, \theta)}{f(x_1,\ldots, x_n;\, \theta)},\ i = 1,\ldots, m \tag{4.3.34}$$

Bhattacharyya has obtained a series of lower bounds. Each of these lower bounds is sharper as compared to the previous ones and approach the minimum variance that corresponds to the UMVUE as this series advances.

Theorem 4.3.6 Let the regularity conditions (i) to (iv) of Section 4.2 and the condition (4.3.34) hold. Consider $\delta(\mathbf{X})$ as an unbiased estimator of $g(\theta)$. Then the lower bound for the variance of an unbiased estimator of $g(\theta)$ is given by

$$V_\theta[\delta(\mathbf{X})] \geq \boldsymbol{v}' A^{-1} \boldsymbol{v} = L_m$$

where
$$A = \begin{pmatrix} E(S_1^2) & \cdots & E(S_1 S_m) \\ \vdots & \ddots & \vdots \\ E(S_m S_1) & \cdots & E(S_m^2) \end{pmatrix}$$

$\boldsymbol{v} = [\partial g(\theta)/\partial\theta, \ldots, \partial^m g(\theta)/\partial\theta^m]'$ and λ^{11} is the leading element of A^{-1}. Further,

$$L_1 \leq L_2 \leq L_3 \leq \cdots \leq L_m$$

These lower bounds approach minimum variance as $m \to \infty$.

Proof Let $X \sim f(x;\theta)$, $\theta \in \Theta$, θ be the single parameter. The CR lower bound for the variance of an unbiased estimator $g(\theta)$, under certain regularity conditions, is given by

$$V[\delta(\mathbf{X})] \geq \frac{1}{E(S_1^2)} = \frac{1}{I_{\mathbf{X}}(\theta)}$$

where
$$S_1 = \frac{\partial}{\partial\theta} \log f(x_1, \ldots, x_n; \theta) = \frac{f'(x_1, \ldots, x_n; \theta)}{f(x_1, \ldots, x_n; \theta)}$$

and
$$E(S_1^2) = V(S_1) = I_{\mathbf{X}}(\theta)$$

In CR inequality, it was assumed that Eq. (4.3.33) is differentiable once with respect to θ under the integral sign. Then using the following inequality

$$\text{corr}^2(\delta, S_1) = \frac{\text{cov}(\delta, S_1)}{V(\delta)\,V(S_1)} \leq 1$$

CR inequality was obtained. Similar to this, Bhattacharyya defined m variables $(S_1, \ldots, S_m)$ by assuming that Eq. (4.3.33) is differentiable m times with respect to θ under the integral sign in addition to other regularity conditions.

Consider a set of p variables $(X_1, X_2, \ldots, X_p) = (X_1, \mathbf{X}_2')$ with variance covariance matrix

$$D(\mathbf{X}) = \Sigma \left(\begin{array}{c|ccc} \sigma_{11} & \sigma_{12} & \cdots & \sigma_{1p} \\ \hline \sigma_{21} & \sigma_{22} & \cdots & \sigma_{2p} \\ \vdots & \vdots & \vdots & \vdots \\ \sigma_{p1} & \sigma_{p2} & \cdots & \sigma_{pp} \end{array}\right) = \begin{pmatrix} \sigma_{11} & \boldsymbol{\sigma}' \\ \boldsymbol{\sigma} & \Sigma_{22} \end{pmatrix}$$

The multiple correlation coefficient between X_1 and a linear combination of $X_2,\ldots,X_p$ is given by

$$\rho^2 = \frac{\boldsymbol{\sigma}'\Sigma_{22}^{-1}\boldsymbol{\sigma}}{\sigma_{11}} \le 1$$

or

$$\sigma_{11} \ge \boldsymbol{\sigma}'\Sigma_{22}^{-1}\boldsymbol{\sigma}$$

We use this result for the set of $(m + 1)$ variables $\delta^* = (\delta, S_1,\ldots,S_m)$ and find the multiple correlation between δ and a linear combination of $S_1,\ldots,S_m$.

Consider
$$\int f(x_1,\ldots, x_n; \theta)dx_1 \ldots dx_n = 1$$

$$\int \frac{f'(x_1,\ldots, x_n; \theta)}{f(x_1,\ldots, x_n; \theta)} f(x_1,\ldots, x_n; \theta)dx_1 \ldots dx_n = 0$$

$$\Rightarrow \qquad E(S_i) = 0, \quad i = 1,\ldots,m \tag{4.3.35}$$

and
$$\int \delta f(x_1,\ldots,x_n; \theta)dx_1\ldots dx_n = g(\theta)$$

or
$$\int \delta(\mathbf{X}) \frac{f'(x_1,\ldots, x_n; \theta)}{f(x_1,\ldots, x_n; \theta)} f(x_1,\ldots, x_n; \theta)dx_1 \ldots dx_n = \frac{\partial g(\theta)}{\partial \theta}$$

$$\Rightarrow \qquad \begin{cases} E(\delta(\mathbf{X})S_1) = (\partial/\partial\theta)g(\theta) \\ E(\delta(\mathbf{X})S_i) = (\partial^i/\partial\theta^i)g(\theta),\ i = 2,\ldots,m \end{cases}$$

or
$$\operatorname{cov}(\delta(\mathbf{X}), S_i) = \begin{cases} (\partial/\partial\theta)\, g(\theta), & i = 1 \\ (\partial^i/\partial\theta^i)g(\theta), & i = 2,\ldots, m, \end{cases} \tag{4.3.36}$$

On using the results given in Eqs. (4.3.35) and (4.3.36), the variance–covariance matrix of δ^* is given by

$$D[\delta^*(\mathbf{X})] = \left(\begin{array}{c|ccc} V(\delta(\mathbf{X})) & \operatorname{cov}(\delta(\mathbf{X}), S_1) & \cdots & \operatorname{cov}(\delta(\mathbf{X}), S_m) \\ \hline \operatorname{cov}(S_1, \delta(\mathbf{X})) & V(S_1) & \cdots & \operatorname{cov}(S_1, S_m) \\ \vdots & \vdots & \vdots & \vdots \\ \operatorname{cov}(S_m, \delta(\mathbf{X})) & \operatorname{cov}(S_m, S_1) & \cdots & V(S_m) \end{array}\right)$$

$$= \begin{pmatrix} V(\delta(\mathbf{X})) & (\partial/\partial\theta)g(\theta) & \cdots & (\partial^m/\partial\theta^m)g(\theta) \\ (\partial/\partial\theta)g(\theta) & V(S_1) & \cdots & \operatorname{cov}(S_1, S_m) \\ \vdots & \vdots & \vdots & \vdots \\ (\partial^m/\partial\theta^m)g(\theta) & \operatorname{cov}(S_m, S_1) & \cdots & V(S_m) \end{pmatrix}$$

$$= \left(\begin{array}{c|ccc} V[\delta(\mathbf{X})] & (\partial/\partial\theta)g(\theta) & \cdots & (\partial^m/\partial\theta^m)g(\theta) \\ \hline (\partial/\partial\theta)g(\theta) & E(S_1^2) & \cdots & E(S_1S_m) \\ \vdots & \vdots & & \vdots \\ (\partial^m/\partial\theta^m)g(\theta) & E(S_mS_1) & \cdots & E(S_m^2) \end{array}\right)$$

$$= \left(\begin{array}{c|c} V[\delta(\mathbf{X})] & (\partial/\partial\theta)g(\theta) \quad \cdots \quad (\partial^m/\partial\theta^m)g(\theta) \\ \hline (\partial/\partial\theta)g(\theta) & \\ \vdots & A \\ (\partial^m/\partial\theta^m)g(\theta) & \end{array}\right)$$

Thus, the multiple correlation coefficient between $\delta(\mathbf{X})$ and the linear combination of $S_1,\ldots,S_m$ is

$$\rho^2_{\delta(S_1, S_2,\ldots, S_m)} = \frac{(\partial/\partial\theta)g(\theta) \quad \cdots \quad (\partial^m/\partial\theta^m)g(\theta)A^{-1}\begin{pmatrix}(\partial/\partial\theta)g(\theta)\\ \vdots \\ (\partial^m/\partial\theta^m)g(\theta)\end{pmatrix}}{V[\delta(\mathbf{X})]} \le 1$$

This proves that $V_\theta[\delta(\mathbf{X})] \ge \boldsymbol{\nu}' A^{-1}\ \boldsymbol{\nu} = L_m$.
At $m = 1$, we have the lower bound

$$\lambda^{11} = \frac{1}{E(S_1^2)} = \frac{1}{I_{\mathbf{X}}(\theta)}$$

for the estimation of θ. Note that at $m = 1$, the Bhattacharyya lower bound equals the CR lower bound. Further, at m,

$$A^{-1} = \begin{pmatrix}\lambda^{11} & \cdots & \lambda^{1m}\\ \vdots & \ddots & \vdots \\ \lambda^{m1} & \cdots & \lambda^{mm}\end{pmatrix}$$

It may be seen that $L_i \le L_j$, $i < j$, by noting that

$$L_m = V[\delta(\mathbf{X})]\rho^2_{\delta(S_1, S_2,\ldots,S_m)}$$

The multiple correlation coefficient increases as m increases. This gives

$$V[\delta(\mathbf{X})]\rho^2_{\delta(S_1, S_2,\ldots,S_i)} \le V[\delta(\mathbf{X})]\rho^2_{\delta(S_1, S_2,\ldots,S_j)}$$

$\Rightarrow$

$$L_i \le L_j$$

This finally implies that

$$L_1 \le L_2 \le L_3 \le \cdots \le L_m$$

Therefore, if L_1 is not sharp (i.e., CRLB is not attained), we calculate L_2, L_3,…,etc., to get sharper bounds. These lower bounds approach minimum variance as $m \to \infty$. This completes the proof. ▣

Examples 4.4.21 to 4.4.23 illustrate that how the Bhattacharyya bounds are obtained and how do they sharpen the CRLB.

We next discuss a variance lower bound for the variance of an estimator due to Chapman, Robbins, and Kiefer, which does not require any regularity condition and gives even a sharper lower bound. In addition to this, the Chapman and Robbins lower bound can be used even when the parameter space Θ is discrete, as opposed to CRLB, which necessarily requires Θ to be open.

4.3.4 Chapman, Robbin, and Kiefer Lower Bound

Theorem 4.3.7 (Chapman, Robbin and Kiefer Lower Bound) Let $X_1, X_2, \ldots, X_n$ be a random sample from $f(x;\theta)$, so that $f(x;\theta)$ is some density in the family $\{f(x;\theta);\ \theta \in \Theta\}$, $\Theta \subset \mathbb{R}^1$ and Θ is arbitrary (open or discrete). Consider g to be some function defined on Θ and let $\delta(\mathbf{X})$ be some unbiased estimator of $g(\theta)$ so that its second moment exists $E_\theta[\delta^2(X)] < \infty\ \forall\ \theta \in \Theta$). Assume that for $\theta_1\ (\neq \theta) \in \Theta$, the densities $f(\cdot;\ \theta_1)$ and $f(\cdot;\ \theta)$ are different *a.e.* with respect to a common Lebesgue measure μ

$$\mu[\mathbf{x}|f(\mathbf{x};\ \theta_1) = f(\mathbf{x};\ \theta)] = 0$$

and $S(\theta_1) \subset S(\theta)$, where $S(\theta)$ is the support of the distribution $f(x;\theta)$. Then

$$V_\theta[\delta(\mathbf{X})] \geq \sup_{[\theta_1 : S(\theta_1) \subset S(\theta),\ \theta_1 \neq \theta]} \frac{[g(\theta_1) - g(\theta)]^2}{V_\theta\left[\dfrac{f(\mathbf{X};\ \theta_1)}{f(\mathbf{X};\ \theta)}\right]} \quad \forall\ \theta \in \Theta$$

Proof. Define

$$l(\mathbf{x}, \theta) = \begin{cases} \dfrac{f(\mathbf{x};\ \theta_1) - f(\mathbf{x};\ \theta)}{f(\mathbf{x};\ \theta)}, & \text{if } f(\mathbf{x};\ \theta) > 0 \\ 0, & \text{otherwise} \end{cases} \tag{4.3.37}$$

so that

$$\begin{aligned} E_\theta[l(\mathbf{X}, \theta)] &= \int_{S(\theta)} \frac{f(\mathbf{x};\ \theta_1) - f(\mathbf{x};\ \theta)}{f(\mathbf{x};\ \theta)} f(\mathbf{x};\ \theta) d\mathbf{x} \\ &= \int_{S(\theta)} f(\mathbf{x};\ \theta_1) d\mathbf{x} - \int_{S(\theta)} f(\mathbf{x};\ \theta) d\mathbf{x} \\ &= \int_{S(\theta_1)} f(\mathbf{x};\ \theta_1) d\mathbf{x} - \int_{S(\theta)} f(\mathbf{x};\ \theta) d\mathbf{x} = 0 \end{aligned} \tag{4.3.38}$$

[since $S(\theta_1) \supset S(\theta)$]

$$\begin{aligned} \text{cov}_\theta(\delta(\mathbf{X}), l(\mathbf{X}, \theta)) &= E_\theta[\delta(\mathbf{X}) \cdot l(\mathbf{X}, \theta)] \quad \text{[by Eq. (4.3.37)]} \\ &= \int_{s(\theta)} \delta(\mathbf{x}) \frac{f(\mathbf{x};\ \theta_1) - f(\mathbf{x};\ \theta)}{f(\mathbf{x};\ \theta)} f(\mathbf{x};\ \theta) d\mathbf{x} \end{aligned}$$

$$= \int_{s(\theta_1)} \delta(\mathbf{x}) f(\mathbf{x}; \theta_1) d\mathbf{x} - \int_{s(\theta)} \delta(\mathbf{x}) f(\mathbf{x}; \theta) d\mathbf{x} \qquad \text{(since } S(\theta_1) \supset S(\theta))$$

$$= g(\theta_1) - g(\theta) \qquad \text{(since } \delta(\mathbf{X}) \text{ is unbiased for } g(\theta))$$

By the Cauchy–Schwarz inequality,

$$\text{cov}_\theta^2[\delta(\mathbf{X}), l(\mathbf{X}, \theta)] \le V_\theta[\delta(\mathbf{X})]\, V_\theta[l(\mathbf{X}, \theta)]$$

Using Eqs. (4.3.37) and (4.3.38), we get

$$V_\theta[\delta(\mathbf{X})] \ge \frac{[g(\theta_1) - g(\theta)]^2}{E_\theta\left[\dfrac{f(\mathbf{X}; \theta_1)}{f(\mathbf{X}; \theta)} - 1\right]^2} \quad \forall\, \theta_1 \ne \theta$$

This gives

$$V_\theta[\delta(\mathbf{X})] \ge \sup_{\{\theta_1 : \theta_1 \ne \theta \text{ and } S(\theta_1) \subset S(\theta)\}} \frac{[g(\theta_1) - g(\theta)]^2}{E_\theta\left[\dfrac{f(\mathbf{X}; \theta_1)}{f(\mathbf{X}; \theta)} - 1\right]^2}$$

An alternative form of CRKLB

Assume $g(\theta) = \theta$, $\theta_1 = \theta + h$, $h \ne 0$, so that $S(\theta_1) \subset S(\theta)$. We have $[g(\theta_1) - g(\theta)]^2 = [\theta + h - \theta]^2 = h^2$ and $E_\theta[\{f(\mathbf{X}; \theta + h)/f(\mathbf{X}; \theta)\} - 1\}^2 = E_\theta[f(\mathbf{X}; \theta + h)/f(\mathbf{X}; \theta)]^2 - 1$. Defining $M(\theta, h) = (1/h^2)[\{f(\mathbf{X}; \theta + h)/f(\mathbf{X}; \theta)\}^2 - 1]$, the CRKLB may be expressed as

$$V_\theta[\delta(\mathbf{X})] \ge \sup_{h \ne 0} \frac{h^2}{E_\theta\left[\dfrac{f(\mathbf{X}; \theta + h)}{f(\mathbf{X}; \theta)} - 1\right]^2}$$

$$= \sup_{h \ne 0} \frac{1}{E_\theta \dfrac{1}{h^2}\left[\left\{\dfrac{f(\mathbf{X}; \theta + h)}{f(\mathbf{X}; \theta)}\right\}^2 - 1\right]}$$

$$= \frac{1}{\inf_{h \ne 0} E_\theta M(\theta, h)}$$

4.3.5 Relationship between CRLB and CRKLB

The CRLB does not hold in case the distribution of X belongs to Pitman's family; as opposed to it, the CRKLB does not require any regularity condition for it to hold. We show here that both these lower bounds are identical when the regularity conditions are satisfied. Consider the quantity

$$\text{CRKLB} = \frac{[g(\theta_1) - g(\theta)]^2}{V_\theta\left[\dfrac{f(\mathbf{X};\theta_1)}{f(\mathbf{X};\theta)}\right]}$$

$$= \frac{\left[\dfrac{g(\theta_1) - g(\theta)}{(\theta_1 - \theta)}\right]^2}{E_\theta\left[\dfrac{1}{(\theta_1 - \theta)}\dfrac{f(\mathbf{X};\theta_1) - f(\mathbf{X};\theta)}{f(\mathbf{X};\theta)}\right]^2}$$

Consider the limit $\theta_1 \to \theta$ on both the sides

$$\text{CRKLB} = \frac{\left[\lim\limits_{\theta_1\to\theta}\dfrac{g(\theta_1) - g(\theta)}{(\theta_1 - \theta)}\right]^2}{E_\theta\left[\dfrac{1}{f(\mathbf{X};\theta)}\lim\limits_{\theta_1\to\theta}\dfrac{f(\mathbf{X};\theta_1) - f(\mathbf{X};\theta)}{(\theta_1 - \theta)}\right]^2}$$

$$= \frac{\left[\dfrac{\partial}{\partial\theta}g(\theta)\right]^2}{E_\theta\left[\dfrac{1}{f(\mathbf{X};\theta)}\dfrac{\partial}{\partial\theta}\log f(\mathbf{X};\theta)\right]^2}$$

$$= \frac{[g'(\theta)]^2}{E_\theta\left[\dfrac{\partial}{\partial\theta}\log f(\mathbf{X};\theta)\right]^2} = \text{CRLB}$$

Thus, both the lower bounds are identical whenever the regularity conditions are satisfied.

4.4 SOLVED EXAMPLES

***Example* 4.1** Let $X_1, X_2, \ldots, X_n$ be *iid*

(i) location density $f(x;\mu) = f(x - \mu)$, for example, $N(\mu, 1)$, $C(1, \mu)$, $\text{DE}(\mu, 1)$,
(ii) scale density $f(x;\sigma) = (1/\sigma)f(x/\sigma)$, for example, gamma, exponential, Cauchy, and logistic.

Here, the form of the density f is known but the location parameter μ in (i) and the scale parameter σ in (ii) are unknown. Assume that the regularity conditions for the CR inequality to hold are satisfied. Find the CR bounds in both the cases.

Solution.

(i) Log likelihood for single observation is given by

$$\log f(x;\mu) = \log f(x - \mu)$$

$$\frac{\partial}{\partial\mu}\log f(x;\mu) = -\frac{f'(x-\mu)}{f(x-\mu)}$$

$$I_X(\mu) = E_\mu\left[\frac{\partial}{\partial\mu}\log f(X;\mu)\right]^2$$

$$= \int\left[\frac{f'(x-\mu)}{f(x-\mu)}\right]^2 f(x-\mu)dx$$

$$= \int\frac{\{f'(x-\mu)\}^2}{f(x-\mu)}dx$$

$$= \int\frac{\{f'(y)\}^2}{f(y)}dy$$

Fisher's information in the sample about the location parameter μ is

$$I_{\mathbf{X}}(\mu) = nI_X(\mu)$$

Thus, for any unbiased estimator δ of μ, we have

$$V_\mu[\delta(\mathbf{X})] \geq \frac{1}{nI_{X_1}(\mu)}$$

$$V_\mu\left[\sqrt{n}(\delta(\mathbf{X})-\mu)\right] \geq \frac{1}{I_X(\mu)}$$

(ii) Log likelihood in this case is given by

$$\log f(x;\sigma) = -\log\sigma + \log f\left(\frac{x}{\sigma}\right)$$

$$\frac{\partial}{\partial\sigma}\log f(x;\sigma) = -\frac{1}{\sigma} - \frac{x}{\sigma^2}\frac{f'(x/\sigma)}{f(x/\sigma)} = -\frac{1}{\sigma}\left[1+\frac{x}{\sigma}\frac{f'(x/\sigma)}{f(x/\sigma)}\right]$$

$$I_X(\sigma) = E_\theta\left\{\frac{\partial}{\partial\sigma}\log f(x;\sigma)\right\}^2 = \frac{1}{\sigma^2}\int\left[1+\frac{x}{\sigma}\frac{f'(x/\sigma)}{f(x/\sigma)}\right]^2\frac{1}{\sigma}f\left(\frac{x}{\sigma}\right)dx$$

$$= \frac{1}{\sigma^2}\int\left[1+y\frac{f'(y)}{f(y)}\right]^2 f(y)dy$$

This gives

$$I_{\mathbf{X}}(\sigma) = nI_X(\sigma)$$

Thus, for any unbiased estimator $\delta(\mathbf{X})$ of scale parameter θ,

$$V_\theta[\delta(\mathbf{X})] \geq \frac{1}{nI_{X_1}(\sigma)}$$

$$V_\sigma\left[\sqrt{n}(\delta(\mathbf{X})-\sigma)\right] \geq \frac{1}{I_{X_1}(\sigma)}$$

***Example* 4.2** Let $X_1, X_2, \ldots, X_n$ be a random sample drawn from $b(1,\theta)$. Find the lower bound of an unbiased estimator of θ based on these observations. Find the UMVUE of θ and show that it attains the CR bound.

Solution. The regularity conditions (i) to (iv) for the CR bound to hold can be easily seen to be satisfied. Fisher's information can be computed as

$$I_{\mathbf{X}}(\theta) = nI_{X_1}(\theta)$$

$$= nE_\theta\left[\frac{\partial}{\partial\theta}\log f(X_1;\theta)\right]^2$$

$$= \frac{n}{\{\theta(1-\theta)\}^2}E_\theta(X_1-\theta)^2 = \frac{n}{\theta(1-\theta)}$$

Therefore, the CRLB is given by

$$V_\theta[\delta(\mathbf{X})] \geq \frac{1}{I_{\mathbf{X}}(\theta)} = \frac{\theta(1-\theta)}{n}$$

We are interested in estimating $g(\theta) = \theta$. Consider

$$\log f(\mathbf{x};\theta) = \left(\sum x_i\right)\log\theta + \left(n - \sum x_i\right)\log(1-\theta)$$

$$\frac{\partial}{\partial\theta}\log f(\mathbf{x};\theta) = \frac{\sum x_i}{\theta} - \frac{(n-\sum x_i)}{(1-\theta)} = \frac{\sum x_i - n\theta}{\theta(1-\theta)}$$

$$= \frac{\left(\sum x_i/n\right)-\theta}{[\theta(1-\theta)]/n} \tag{4.4.1}$$

On comparing Eq. (4.4.1) with Eq. (4.3.7), we conclude that the CRLB holds with

$$c(\theta) = \frac{n}{\theta(1-\theta)},\quad \delta(\mathbf{X}) = \frac{\sum X_i}{n} \tag{4.4.2}$$

and that $\delta(\mathbf{X})$ is the most efficient estimator with minimum variance

$$V_\theta[\delta(\mathbf{X})] = \left|\frac{g'(\theta)}{c(\theta)}\right| = \frac{\theta(1-\theta)}{n} \qquad [\text{here, } g'(\theta) = 1]$$

Note that the variance of the estimator $\delta(\mathbf{X}) = \Sigma X_i/n$ in Eq. (4.4.2) achieves CRLB(θ). Therefore, it is the CR lower bound estimator (most efficient). We will discuss maximum likelihood (ML) estimation in Chapter 6 where we will show that $\bar{X}$ is the MLE and that the MLE is asymptotically normal

$$\sqrt{n}(\bar{X}-\theta) \rightarrow N(0, \theta(1-\theta))$$

Note that the asymptotic variance of the MLE $\bar{X}$ is equal to CRLB(θ).

***Example* 4.3** Let $X_1, X_2, \ldots, X_n$ be a random sample from the Poisson distribution $P(\theta)$ with parameter θ. Obtain the CRLBs and examine whether there exist such estimators that their variances attain these lower bounds for estimating the parametric functions $g(\theta)$:

(i) θ (ii) $\exp(-\theta)$ (iii) θ^2

Solution.

(i) The joint density of sample observations is

$$f(\mathbf{x}; \theta) = \frac{\exp(-n\theta)\theta^{\Sigma x_i}}{\prod_{i=1}^{n}(x_i!)}$$

$$\log f(\mathbf{x}; \theta) = \left(\sum x_i\right)\log\theta - n\theta - \sum_{1}^{n}\log(x_i!)$$

$$S(\mathbf{x}, \theta) = \frac{\partial}{\partial\theta}\log f(\mathbf{x}; \theta) = \frac{n\bar{x}}{\theta} - n = \frac{n}{\theta}(\bar{x}-\theta) \tag{4.4.4}$$

Since this is in the form of Eq. (4.3.7) (score and estimator are linearly related), the CRLB is attained. Therefore, $\delta(\mathbf{X}) = \bar{X}$ is the CRLB estimator (most efficient) with $c(\theta) = (n/\theta)$, $g(\theta) = \theta$. By Eq. (4.3.12), the minimum variance of $\delta(\mathbf{X})$ is

$$V_\theta[\delta(\mathbf{X})] = \left|\frac{g'(\theta)}{c(\theta)}\right| = \frac{\theta}{n}$$

However, Fisher's information in the sample about θ, $I_{\mathbf{X}}(\theta)$, is

$$I_{\mathbf{X}}(\theta) = E_\theta\left[\frac{\partial}{\partial\theta}\log f(\mathbf{X}; \theta)\right]^2$$

$$= \frac{1}{\theta^2}E_\theta\left(\sum X_i - n\theta\right)^2$$

$$= \frac{1}{\theta^2}V_\theta\left(\sum X_i\right) = \frac{n\theta}{\theta^2} = \frac{n}{\theta}, \qquad [\text{since } \Sigma X_i \sim P(n, \theta)]$$

Thus, the CRLB, $g'(\theta) = 1$

$$V_\theta[\delta(\mathbf{X})] \geq \frac{[g'(\theta)]^2}{I_{\mathbf{X}}(\theta)} = \frac{1}{n/\theta} = \frac{\theta}{n}$$

Therefore, we see that $\delta(\mathbf{X}) = \bar{X}$ is the most efficient estimator (UMVUE).

(ii) The CRLB for estimating $g(\theta) = \exp(-\theta)$ is

$$V[\delta(\mathbf{X})] \geq \frac{[g'(\theta)]^2}{I_{\mathbf{X}}(\theta)} = \frac{\theta \exp(-2\theta)}{n} = \text{CRLB}[g(\theta)] \tag{4.4.5}$$

It is clear from Eq. (4.4.4) that there does not exist any unbiased estimator of $\exp(-\theta)$ the variance of which attains CRLB.

Consider an unbiased estimator of $\exp(-\theta)$

$$\delta_1(\mathbf{X}) = \frac{1}{n}\sum_{i=1}^{n} h(x_i) \tag{4.4.6}$$

where

$$h(X_i) = \begin{cases} 1, & \text{if } X_i = 0 \\ 0, & \text{otherwise} \end{cases}$$

We have

$$E[h(X_1)] = \exp(-\theta);\ E[h^2(X_1)] = \exp(-\theta)$$

$$V[h(X_1)] = \exp(-\theta)\ [1 - \exp(-\theta)]$$

$$V(\delta_1) = \frac{\exp(-\theta)[1 - \exp(-\theta)]}{n}$$

Since $V[\delta_1(\mathbf{X})] > \text{CRLB}[\exp(-\theta)]$, $\delta_1(\mathbf{X})$ is not the CRLB estimator. In case of $n = 1$, $\delta_1(\mathbf{X})$ is the only unbiased estimator. Therefore, the UMVUE and its variance achieve CRLB. However, if $n > 1$, there can be several unbiased estimators satisfying Eq. (4.4.6). One such estimator which is a function of the complete sufficient statistic ΣX_i is given by

$$\delta_2(\mathbf{X}) = \left(1 - \frac{1}{n}\right)^{\sum X_i}$$

For this estimator, we have

$$E[\delta_2(\mathbf{X})] = \sum\left(1 - \frac{1}{n}\right)^t \frac{\exp(-n\theta)(n\theta)^t}{t!} = \sum \frac{\exp(-n\theta)\left[n\theta\left(1 - \frac{1}{n}\right)\right]^t}{t!}$$

$$= \exp(-n\theta)\exp[(n-1)\theta] = \exp(-\theta)$$

$$E[\delta_2^2(\mathbf{X})] = \exp(-n\theta)\sum \frac{\left[n\theta\left(1-\frac{1}{n}\right)^2\right]^t}{t!} = \exp(-n\theta)\exp\left[n\theta\left(1-\frac{1}{n}\right)^2\right]$$

$$= \exp\left[-2\theta + \frac{\theta}{n}\right]$$

This gives

$$V[\delta_2(\mathbf{X})] = \exp(-2\theta)\left[\exp\left(\frac{\theta}{n}\right) - 1\right]$$

Clearly, $\delta_2(\mathbf{X})$ is a UMVUE with its variance $V_\theta[\delta_2(\mathbf{X})] = \exp(-2\theta)\ [\exp(\theta/n) - 1]$. However, $V_\theta[\delta_2(\mathbf{X})] > \text{CRLB}[\exp(-\theta)]\ \forall\, \theta$. Therefore, it is not the CRLB estimator.

(iii) From Eq. (4.4.4), it is clear that $(\partial/\partial\theta) \log f(\mathbf{x};\, \theta)$ cannot be expressed in the form of $c(\theta)(\delta - \theta^2)$. Therefore, there does not exist any unbiased estimator of θ^2, the variance of which attains the CRLB. The CR lower bound for estimating θ^2 is

$$V[\delta(\mathbf{X})] \ge \frac{(2\theta)^2}{n/\theta} = \frac{4\theta^3}{n} = \text{CRLB}(\theta^2)$$

We have seen that this lower bound cannot be achieved by any estimator. However, $\Sigma_1^n X_i$ is complete, $\delta(\mathbf{X}) = \bar{X}^{\,2} - (\bar{X}/n)$ is unbiased for θ^2. Hence, $\delta(\mathbf{X})$ is the UMVUE of θ^2 with its variance $(4\theta^3/n) + (2\theta^2/n^2) > \text{CRLB}(\theta^2) = 4\theta^3/n$.

***Example* 4.4** Let $X_1, X_2, \ldots, X_n$ be a random sample drawn from $C(1, \theta)$ with *pdf*

$$f(x;\theta) = \frac{1}{\pi}\frac{1}{1+(x-\theta)^2},\ -\infty < x < \infty$$

(i) Show that the CRLB for the variance of an unbiased estimator of θ is $2/n$.

(ii) Show that the CRLB is not achieved by any unbiased estimator of θ.

Solution.

(i) We have

$$\log(f(x;\theta)) = -\log \pi - \log[1 + (x-\theta)^2]$$

This gives

$$\frac{\partial}{\partial\theta}\log f(x;\theta) = \frac{2(x-\theta)}{1+(x-\theta)^2}$$

$$E\left(\frac{\partial}{\partial\theta}\log f(X;\theta)\right)^2 = \frac{1}{\pi}\int_{-\infty}^{\infty}\frac{4(x-\theta)^2}{\{1+(x-\theta)^2\}^3}dx$$

Let $x - \theta = \tan u$. We have $dx = \sec^2 u\, du$. This gives

$$E\left(\frac{\partial}{\partial\theta}\log f(X;\theta)\right)^2 = \frac{2}{\pi}\int_0^{\pi/2}\frac{4\tan^2 u}{\sec^6 u}\sec^2 u\, du$$

$$= \frac{2}{\pi}\int_0^{\pi/2} 4\sin^2 u\cos^2 u\, du$$

On using the result

$$2\int_0^{\pi/2}\sin^{2m-1}\theta\cos^{2n-1}\theta\, d\theta = B(m, n)$$

and setting $2m - 1 = 2$, $m = 3/2$, $n = 3/2$, we have

$$E\left(\frac{\partial}{\partial\theta}\log f(X;\theta)\right)^2 = \frac{1}{\pi}\frac{[\Gamma(3/2)]^2}{\Gamma(3)} = \frac{4}{\pi}\frac{(1/4)\pi}{2} = \frac{1}{2}$$

Thus, $$I_{\mathbf{X}}(\theta) = nI_{X_1}(\theta) = \frac{n}{2}$$

and the CRLB for the variance of an unbiased estimator of θ is given by

$$V_\theta[\delta(\mathbf{X})] \ge [I_{\mathbf{X}}(\theta)]^{-1} = \frac{2}{n}$$

(ii) The likelihood function is given by

$$f(\mathbf{x};\theta) = \left(\frac{1}{\pi}\right)^n\prod_{i=1}^{n}\left\{\frac{1}{1+(x_i-\theta)^2}\right\}$$

or $$\log f(\mathbf{x};\theta) = -n\log\pi - \sum_{i=1}^{n}\log\{1+(x_i-\theta)^2\}$$

This implies

$$\frac{\partial}{\partial\theta}\log f(\mathbf{x};\theta) = 2\sum_{i=1}^{n}\frac{(x_i-\theta)}{1+(x_i-\theta)^2}$$

This expression cannot be expressed in the form Eq. (4.3.7). Therefore, in case of sampling from the Cauchy population, the CRLB is not attainable by any unbiased estimator of θ.

***Example* 4.5** Let $X_1, X_2, \ldots, X_n$ be a random sample from $N(\theta, \sigma^2)$, where θ is unknown but σ^2 is known. Find the UMVUE of (i) $g(\theta) = \theta$ and (ii) $g(\theta) = \theta^2$. Show the UMVUE of θ in (i) attains the CR lower bound.

Solution.

(*i*) $g'(\theta) = 1$

$$f(\mathbf{x};\theta,\sigma^2)=\left(\frac{1}{2\pi\sigma^2}\right)^{n/2}\exp\left\{-\frac{1}{2}\sum\left(\frac{x_i-\theta}{\sigma}\right)^2\right\}$$

$$\log f(\mathbf{x};\theta,\sigma^2)=-\frac{n}{2}\log(2\pi\sigma^2)-\frac{1}{2}\sum\left(\frac{x_i-\theta}{\sigma}\right)^2$$

$$\frac{\partial}{\partial\theta}\log f(\mathbf{x};\theta,\sigma^2)=\frac{2}{2\sigma^2}\sum(x_i-\theta)=\frac{n(\bar{x}-\theta)}{\sigma^2}$$

This is the form of Eq. (4.3.7) with

$$c(\theta)=\frac{n}{\sigma^2},\quad \delta=\bar{X},\quad g(\theta)=0$$

Therefore, the CRLB is attained by the estimator $\delta = \bar{X}$ with minimum variance

$$V_\theta(\bar{X})=\left|\frac{g'(\theta)}{c(\theta)}\right|=\frac{\sigma^2}{n}$$

Also, $\delta = \bar{X}$ is the UMVUE.

(ii) $g(\theta) = \theta^2$, $g'(\theta) = 2\theta$. The CRLB is given by

$$V_\theta(T)\geq\frac{[g'(\theta)]^2}{I_{\mathbf{X}}(\theta)}=\frac{4\theta^2}{n/\sigma^2}$$

$$=\frac{4\theta^2\sigma^2}{n}=\text{CRLB}(\theta^2)$$

Since $(\partial/\partial\theta)\log f(\mathbf{x};\ \theta,\ \sigma^2)$ cannot be expressed in the form of $c(\theta)[\delta(\mathbf{X}) - \theta^2]$, CRLB($\theta^2$) cannot be attained by any unbiased estimator. However, we have a UMVUE of θ^2, namely $\delta_1 = \bar{X}^2 - (\sigma^2/n)$. We now calculate the variance of δ_1. Note that $Y = \sqrt{n}(\bar{X} - \theta)/\sigma \sim N(0, 1)$ since $X \sim N(\theta, \sigma^2)$. Thus, $Y^2 \sim \chi_1^2$. This gives $V(Y^2) = V[(n/\sigma^2)(\bar{X} - \theta)^2] = (n^2/\sigma^4)[V(\bar{X}^2) - 2\theta V(\bar{X})] = 2$. This implies $V(\bar{X}^2) - 2\theta V(\bar{X}) = 2\sigma^4/n^2$. Thus, the minimum variance of δ_1 is $V(\delta_1) = V(\bar{X}^2) = (4\theta^2\sigma^2/n) + (2\sigma^4/n^2)$. This shows that δ_1 is the UMVUE of θ^2 but not the CRLB estimator.

***Example* 4.6** Let $X_1, X_2,\ldots,X_n$ be a random sample from $N(\theta, \sigma^2)$, where θ is known. Discuss the estimation of $g(\sigma^2)$ on the basis of the Cramer–Rao lower bound:

(i) $g(\sigma^2) = \sigma^2$ (ii) $g(\sigma^2) = \sigma$

Solution.

(i) $$\frac{\partial}{\partial\sigma^2}g(\sigma^2)=g'(\sigma^2)=1$$

$$\log f(x;\theta,\sigma^2)=-\log\left(\sqrt{2\pi}\right)-\frac{1}{2}\log\sigma^2-\frac{1}{2}\left(\frac{x-\theta}{\sigma}\right)^2$$

$$\frac{\partial}{\partial\sigma^2}\log f(x;\theta,\sigma^2) = -\frac{1}{2\sigma^2} + \frac{1}{2}\frac{(x-\theta)^2}{\sigma^4}$$

This gives
$$S(\mathbf{x},\sigma^2) = -\frac{n}{2\sigma^2} + \frac{1}{2}\frac{\sum(x_i-\theta)^2}{\sigma^4}$$

$$= \frac{n}{2\sigma^4}\left[\frac{1}{n}\sum(x_i-\theta)^2 - \sigma^2\right] \tag{4.4.7}$$

We see that $(\partial/\partial\sigma^2)\, f(\mathbf{x};\ \theta,\ \sigma^2)$ can be expressed in the form of Eq. (4.3.7) with $g(\sigma^2) = \sigma^2$, $c(\sigma^2) = (n/2\sigma^4)$, $\delta(\mathbf{X}) = (1/n)\Sigma(X_i - \theta)^2$. Therefore, the CRLB is attained by the variance of the estimator δ and by $V_{\sigma^2}(\delta) = |(g'(\sigma^2))/c(\sigma^2)| = (2\sigma^4)/n$. Note that the estimator δ is unbiased for σ^2 and its variance is given by $2\sigma^4/n$. This variance expression can also be obtained as follows:

Since $X \sim N(\theta, \sigma^2)$, $[(X - \theta)/\sigma] \sim N(0, 1)$, it implies

$$U = \sum\frac{(X_i-\theta)^2}{\sigma^2} \sim \chi^2_{(n)} \tag{4.4.8}$$

$$E\left[\sum\frac{(X_i-\theta)^2}{\sigma^2}\right] = n$$

$$E\left[\frac{1}{n}\sum(X_i-\theta)^2\right] = \sigma^2$$

$$E[\delta] = \sigma^2$$

This shows that δ is unbiased for σ^2. Moreover,

$$V\left[\sum\frac{(X_i-\theta)^2}{\sigma^2}\right] = 2n$$

$$V(\delta) = V\left[\frac{1}{n}\sum(X_i-\theta)^2\right] = \frac{2\sigma^4}{n} = \frac{1}{I(\sigma^2)}$$

This shows that the variance of δ is equal to the CRLB. Therefore, δ is the CRLB estimator.

(ii) For estimating $g(\sigma^2) = \sigma$, it is clear from Eq. (4.3.7) that the CRLB estimator of σ does not exist. However, the CR lower bound for the variance of an unbiased estimator of σ is given by calculating

$$g'(\sigma^2) = \frac{\partial}{\partial\sigma^2}g(\sigma^2) = \frac{\partial}{\partial\sigma^2}\sqrt{\sigma^2} = \frac{1}{2\sqrt{\sigma^2}} = \frac{1}{2\sigma}$$

$$V_{\sigma^2}(\delta) \ge \frac{[g'(\sigma^2)]^2}{I_{\mathbf{X}}(\sigma^2)} = \frac{1/4\sigma^2}{n/2\sigma^4} = \frac{\sigma^2}{2n}$$

We will check whether $\delta = \sqrt{(1/n)\Sigma(X_i - \theta)^2}$ is an unbiased estimator of σ. Using the Eq. (4.4.8), we calculate

$$E(U^{r/2}) = \int_0^\infty u^{r/2} G_1\left(\frac{n}{2}, 2\right) du$$

$$= \int_0^\infty u^{r/2} \frac{1}{\Gamma(n/2)2^{n/2}} u^{(n/2)-1} \exp(-u/2)\, du$$

$$= \frac{\Gamma[(n+r)/2]}{\Gamma(n/2)} 2^{r/2}$$

This gives

$$E(\delta) = E\left(\frac{\sigma\sqrt{U}}{\sqrt{n}}\right) = \frac{\sigma}{\sqrt{n}} E(\sqrt{U})$$

$$= \frac{\Gamma[(n+1)/2]}{\Gamma(n/2)} \sqrt{2/n}\,\sigma \neq \sigma \tag{4.4.9}$$

This shows that the estimator δ is not an unbiased estimator of σ. The variance of δ is given by

$$V(\delta) = V\left(\frac{\sigma\sqrt{U}}{\sqrt{n}}\right) = \frac{\sigma^2}{n} V\left(\sqrt{U}\right)$$

$$= \frac{\sigma^2}{n}\left\{E(U) - \left[E\left(\sqrt{U}\right)\right]^2\right\}$$

$$= \frac{\sigma^2}{n}\left\{n - \left[\frac{\Gamma(n+1)/2}{\Gamma(n/2)}\right]^2 2\right\}$$

An unbiased estimator of σ can be obtained from Eq. (4.4.9)

$$\delta_1 = \sqrt{\frac{n}{2}} \frac{\Gamma(n/2)}{\Gamma[(n+1)/2]} \delta = \sqrt{\frac{n}{2}} \frac{\Gamma(n/2)}{\Gamma[(n+1)/2]} \frac{\sigma}{\sqrt{n}} \sqrt{U}$$

$$= \frac{\Gamma(n/2)}{\Gamma[(n+1)/2]} \frac{\sigma}{\sqrt{2}} \sqrt{U}$$

and

$$V(\delta_1) = \sigma^2 \left\{\frac{n}{2}\left[\frac{\Gamma(n/2)}{\Gamma[(n+1)/2]}\right]^2 - 1\right\}$$

The efficiency of δ_1 is given by

$$e(\delta_1, \sigma) = \frac{\text{CRLB}}{\text{V}(\delta_1)} = \frac{\sigma^2/2n}{\sigma^2\left\{\frac{n}{2}\left[\frac{\Gamma(n/2)}{\Gamma[(n+1)/2]}\right]^2 - 1\right\}}$$

Note that $e(\delta_1, \sigma) \to 1$ as $n \to \infty$. Therefore, δ_1 is asymptotically the most efficient.

We next consider another estimator of σ on using the result when θ is not known that $\Sigma(X_i - \bar{X})^2/\sigma^2 = U' \sim \chi^2_{(n-1)} = G_1[(n-1)/2, 2]$. Further,

$$E\left[\sqrt{(1/(n-1))\sum(X_i-\bar{X})^2}\right] = \frac{\sigma}{\sqrt{n-1}}E[\sqrt{U'}]$$

$$= \sigma\sqrt{[2/(n-1)]}\frac{\Gamma(n/2)}{\Gamma[(n-1)/2]}$$

This gives that

$$\delta_2 = \sqrt{(n-1)/2}\frac{\Gamma[(n-1)/2]}{\Gamma(n/2)}\sqrt{[1/(n-1)]\sum(X_i-\bar{X})^2}$$

is an unbiased estimator of σ. Further, the variance of δ_2 is given by

$$V(\delta_2) = \sigma^2\left\{\frac{n-1}{2}\left[\frac{\Gamma[(n-1)/2]}{\Gamma(n/2)}\right]^2 - 1\right\}$$

Since $V(\delta_2) > V(\delta_1)$, δ_1 is more efficient. We next use the result $E(|Y|^r) = E(|X-\theta|^r) = 2^{r/2}\sigma^r\Gamma[(r+1)/2]/\sqrt{\pi}$ to construct an unbiased estimator of σ, namely

$$\delta_3 = \sqrt{\frac{\pi}{2}}\frac{1}{n}\sum|X_i - \theta|$$

Since $E(|X-\theta|) = \sqrt{2/\pi}\sigma$ and $E(|X-\theta|^2) = 2\sigma^2\Gamma(3/2)/\sqrt{\pi} = \sigma^2$, the variance of δ_3 is given by

$$V(\delta_3) = \frac{\pi}{2}\frac{1}{n}\sigma^2\frac{(\pi-2)}{\pi}$$

$$= \frac{\sigma^2}{2n}(\pi-2)$$

The efficiency of δ_3 is given by

$$e(\delta_3, \sigma) = \frac{\sigma^2/2n}{\sigma^2(\pi-2)/2n} = 0.88$$

which is independent of n.

***Example* 4.7** Let δ_1 and δ_2 be two unbiased estimators of θ with $V_\theta(\delta_1) = V_\theta(\delta_2)$. Then show that

$$\text{corr}(\delta_1, \delta_2) \geq 2e - 1$$

where $e = e(\delta_1) = e(\delta_2)$.

Solution. Let V be the CRLB and

$$V_\theta(\delta_1) = V_\theta(\delta_2) = V_1 \text{ (say)}$$

Then, the efficiency of δ_1 is

$$e(\delta_1) = \frac{V}{V_1} = e \text{ or } V_1 = \frac{V}{e}$$

Define an unbiased estimator as a function of δ_1 and δ_2

$$\delta_3 = \frac{1}{2}(\delta_1 + \delta_2)$$

Clearly, $E_\theta(\delta_3) = \theta$, and

$$V_\theta(\delta_3) = \frac{1}{4}[V_\theta(\delta_1) + V_\theta(\delta_2) + 2\text{cov}_\theta(\delta_1, \delta_2)]$$

$$= \frac{1}{4}[V_1 + V_1 + 2\text{cov}_\theta(\delta_1, \delta_2)] \geq V$$

(V is the variance of the most efficient estimator)

$$V_1 + \text{cov}_\theta(\delta_1, \delta_2) \geq 2\text{V}$$

$$\frac{V}{e} + \frac{V}{e}\rho(\delta_1, \delta_2) \geq 2V$$

Therefore, $\text{corr}(\delta_1, \delta_2) \geq 2e - 1$.

***Example* 4.8** Let δ be the most efficient estimator of θ and δ_1 be some other unbiased estimator with efficiency e. Then show that

$$\rho(\delta, \delta_1) = \sqrt{e}$$

Solution. Equation (4.3.7) holds since δ is the most efficient.

$$\frac{\partial}{\partial \theta}\log f = c(\theta)(\delta - \theta)$$

The variance of δ is

$$V_\theta(\delta) = \frac{g'(\theta)}{c(\theta)} = \frac{1}{c(\theta)} = V(\text{say})$$

This gives $c(\theta) = (1/V)$ so that

$$\frac{\partial}{\partial \theta}\log f = \frac{1}{V}(\delta - \theta) \tag{4.4.8}$$

It is given that δ_1 is an unbiased estimator of θ with efficiency e. Thus,

$$e = \frac{V_\theta(\delta)}{V_\theta(\delta_1)} = \frac{V}{V_1}$$

We have

$$\text{cov}_\theta(\delta, \delta_1) = E_\theta[(\delta - \theta)(\delta_1 - \theta)]$$

$$= E_\theta\left[V\frac{\partial}{\partial\theta}(\delta_1 - \theta)\log f\right] \qquad \text{(From Eq. (4.4.8))}$$

$$= V \cdot E_\theta\left[\delta_1 \frac{\partial}{\partial\theta}\log f\right] \quad \left(\text{since } E_\theta\left(\frac{\partial}{\partial\theta}\log f\right) = 0\right) \qquad (4.4.9)$$

It is given that δ_1 is an unbiased estimator of θ. Thus,

$$\int \delta_1 f(\mathbf{x}; \theta) d\mathbf{x} = \theta$$

Differentiating both the sides with respect to θ, we get

$$\int \delta_1 \frac{\partial}{\partial\theta} f(\mathbf{x}; \theta) d\mathbf{x} = 1$$

or

$$E_\theta\left[\delta_1 \frac{\partial}{\partial\theta}\log f(\mathbf{x}; \theta)\right] = 1 \qquad (4.4.10)$$

Equations (4.4.9) and (4.4.10) give

$$\text{cov}_\theta(\delta, \delta_1) = V$$

Therefore,

$$\text{corr}(\delta, \delta_1) = \frac{\text{cov}_\theta(\delta, \delta_1)}{\sqrt{V_\theta(\delta)}\sqrt{V_\theta(\delta_1)}}$$

$$= \frac{V}{\sqrt{V}\sqrt{V/e}} = \sqrt{e}$$

***Example* 4.9** Let $X_1, X_2, \ldots, X_n$ be a random sample from $U(0, \theta)$. Examine the application of the CRLB theorem for the estimation of θ.

Solution. The support of $U(0, \theta)$, that is $S(\theta) = \{x: 0 < x < \theta\}$, depends on the parameter θ. This violates the regularity conditions (ii) and (iv) and the CRLB theorem does not produce meaningful result. We see this by the following calculations:

$$\log f(x; \theta) = -\log\theta$$

$$\frac{\partial}{\partial\theta}\log f(x; \theta) = -\frac{1}{\theta}$$

$$I_{\mathbf{X}}(\theta) = nI_{X_1}(\theta) = nE_\theta\left[\frac{\partial}{\partial\theta}\log f(X; \theta)\right]^2 = \frac{n}{\theta^2}$$

Let T be an unbiased estimator of θ. CRLB(θ) is given by

$$V_\theta(T) \geq \frac{1}{I_{\mathbf{X}}(\theta)} = \frac{\theta^2}{n}$$

Note that

$$E_\theta(X) = \frac{\theta}{2}$$

$$E(2X) = \theta$$

Based on the sample observations $X_1, X_2, \ldots, X_n$, $2\bar{X}$ is an unbiased estimator of θ. Further,

$$E(X^2) = \frac{1}{\theta}\int_0^\theta x^2 dx = \frac{\theta^2}{3}$$

$$V(X) = \frac{\theta^2}{12}$$

This gives $V(\bar{X}) = (\theta^2/12n)$ or $V(2\bar{X}) = (\theta^2/3n) < (\theta^2/n) = \text{CRLB}(\theta)$.

Therefore, we have encountered one unbiased estimator of θ whose variance is even smaller than CRLB(θ). This contradicts that CRLB(θ) is the lower bound of the variance of the unbiased estimator of θ.

Another example that we may consider is

$$T = \frac{n+1}{n} X_{(n)}$$

$$E(T) = \theta$$

$$V_\theta(T) = \frac{\theta^2}{n(n+2)} < \frac{\theta^2}{n}$$

where $X_{(n)} = \max(X_1, X_2, \ldots, X_n)$ is the C–S (Complete sufficient statistic) for θ. Hence, the CR inequality does not hold.

***Example* 4.10** Let $X_1, X_2, \ldots, X_n$ be *iid* from $f(x; \theta)$,

$$f(x;\theta) = \begin{cases} \exp[-(x-\theta)], & \text{if } \theta < x < \infty, -\infty < \theta < \infty \\ 0, & \text{otherwise} \end{cases}$$

Is the above family of distributions regular in the sense of CRLB? Examine this case for unbiased estimation of θ.

Solution. Since the support of this distribution is not independent of the parameter θ, the regularity conditions (ii) and (iv) are not satisfied. Thereby, the CRLB is not valid in this case. This we can see by the following calculations.

The density of X_1 is

$$f(x_1; \theta) = \exp[-(x_1 - \theta)]$$

$$\log f(x_1; \theta) = -(x_1 - \theta)$$

$$\frac{\partial}{\partial\theta}\log f(x_1;\theta)=1$$

$$I_{X_1}(\theta)=E_\theta\left[\frac{\partial}{\partial\theta}\log f(X_1;\theta)\right]^2=1$$

$$I_{\mathbf{X}}(\theta)=nI_{X_1}(\theta)=n$$

Therefore,

$$\text{CRLB}(\theta)=\frac{1}{I_{\mathbf{X}}(\theta)}=\frac{1}{n}$$

Further, consider the first-order statistic $X_{(1)}$. Its *pdf* is given by

$$g(x_{(1)};\theta)=n\exp\{-n(x_{(1)}-\theta)\}$$

$$E(X_{(1)}-\theta)^r=n\int_\theta^\infty (x_{(1)}-\theta)^r\exp\{-n(x_{(1)}-\theta)\}\,dx_{(1)}$$

$$=n\int_\theta^\infty y^r\exp(-ny)\,dy=\frac{n\Gamma(r+1)}{n^{r+1}}$$

This gives

$$E(X_{(1)}-\theta)=\frac{1}{n}$$

$$E(X_{(1)}-\theta)^2=\frac{2}{n^2}$$

$$E(X_{(1)})=\theta+\frac{1}{n}$$

$$V(X_{(1)})=V(X_{(1)}-\theta)=\frac{1}{n^2}=V(\delta)$$

where $\delta=[X_{(1)}-(1/n)]$ is an unbiased estimator of θ. Note that $V(\delta)=1/n^2<1/n=\text{CRLB}(\theta)$. Therefore, the CRLB is not valid.

***Example* 4.11** Let $X_1, X_2,\ldots,X_n$ be a random sample from $G_2(p,\theta)$, $p>0$, $\theta>0$.

(i) Find the CRLB for the variance of an unbiased estimator of $g(\theta)=\theta$, when p is known. Calculate the variance of the UMVUE of θ. Does it achieve the CR bound? Is $\bar{X}/p$ the most efficient for $1/\theta$?

(ii) Find the CRLB for the variance of an unbiased estimator of $g(p)=(\partial/\partial p)\log\Gamma(p)$ when θ is known.

Solution.

(i) Consider

$$f(\mathbf{x};\theta)=\frac{\theta^{np}}{[\Gamma(p)]^n}\left(\prod_{i=1}^{n}x_i\right)^{p-1}\exp(-\theta\sum x_i)$$

$$\log f(\mathbf{x};\theta)=(p-1)\log\prod_{i=1}^{n}x_i-n\log\Gamma(p)+np\log\theta-\theta\sum x_i$$

$$\frac{\partial}{\partial\theta}\log f(\mathbf{x};\theta)=S(\mathbf{x};\theta)=\frac{np}{\theta}-\sum x_i \tag{4.4.11}$$

$$\frac{\partial^2}{\partial\theta^2}\log f(\mathbf{x};\theta)=-\frac{np}{\theta^2}$$

$$I_{\mathbf{X}}(\theta)=\frac{np}{\theta^2}$$

$$\text{CRLB}(\theta)=\frac{\theta^2}{np}$$

On writing Eq. (4.4.11) in the form of Eq. (4.3.7), we have

$$S(\mathbf{x},\theta)=\frac{\partial}{\partial\theta}\log f(\mathbf{x};\theta)=\frac{\sum x_i}{\theta}\left(\frac{np}{\sum x_i}-\theta\right) \tag{4.4.12}$$

Since $np/\Sigma X_i$ is not an unbiased estimator of θ, we conclude, from Eq. (4.4.12), that the CR bound for estimating θ cannot be attained. We have $Y=\Sigma X_i\sim G_2(np,\theta)$. This gives

$$E(Y^r)=\frac{\theta^{np}}{\Gamma(np)}\frac{\Gamma(np+r)}{\theta^{np+r}}=\frac{\Gamma(np+r)}{\Gamma(np)}\frac{1}{\theta^r}$$

$$E(Y^{-r})=\frac{\Gamma(np-r)}{\Gamma(np)}\theta^r$$

$$E(Y)=\frac{np}{\theta},\ E\left(\frac{1}{Y}\right)=\frac{\theta}{np-1},\ E(Y^2)=\frac{(np+1)np}{\theta^2},\ V(Y)=\frac{np}{\theta^2}$$

Thus,
$$\frac{np-1}{Y}=\frac{np-1}{\sum X_i}$$

is an unbiased estimator of θ. It is a UMVUE since it is a function of the sufficient statistic ΣX_i. The corresponding minimum variance of this estimator is given by

$$V\left(\frac{np-1}{\sum X_i}\right)=(np-1)^2[EY^{-2}-(EY^{-1})^2]$$

$$= (np-1)^2\left[\frac{1}{(np-1)(np-2)} - \frac{1}{(np-1)^2}\right]\theta^2$$

$$= \frac{1}{np-2}\theta^2$$

which is greater than the CR bound θ^2/np. Clearly, this UMVUE is not the most efficient.

However, it is clear from Eq. (4.4.12) that $\bar{X}/p$ is the CR bound estimator (most efficient) of $1/\theta$ since Eq. (4.4.12) can be expressed as $S(\mathbf{X},\theta) = -np[(\bar{X}/p) - (1/\theta)]$. The quantities $g(\theta) = 1/\theta$, $g'(\theta) = -1/\theta^2$, $I_{\mathbf{X}}(\theta) = np/\theta^2$ give the CRLB,

$$\text{CRLB} = \frac{1/\theta^4}{np/\theta^2} = \frac{1}{np\theta^2}$$

The variance of $\bar{X}/p = Y/np$ is given by

$$V\left(\frac{\bar{X}}{p}\right) = \left(\frac{1}{np}\right)^2 V(Y) = \frac{1}{np\theta^2}$$

which is equal to the CRLB.

(ii) $\log f(\mathbf{x};p) = np\log\theta - n\log[\Gamma(p)] + (p-1)\Sigma\log x_i - \theta\Sigma x_i$

$$\frac{\partial}{\partial p}\log(\mathbf{x};p) = n\log\theta - n\frac{\partial}{\partial p}\log[\Gamma(p)] + \sum\log x_i$$

$$= n\left(\frac{1}{n}\sum\log x_i + \log\theta - \frac{\partial}{\partial p}\log\Gamma(p)\right) \qquad (4.4.13)$$

Further,
$$\frac{\partial^2}{\partial p^2}\log f(\mathbf{x};p) = -n\frac{\partial^2}{\partial p^2}\log[\Gamma(p)] \qquad (4.4.14)$$

The Fisher information about p in the sample is, thus, given by

$$I_{\mathbf{X}}(p) = n\frac{\partial^2}{\partial p^2}\log[\Gamma(p)]$$

Therefore, the CR lower bound for the variance of an unbiased estimator of $g(p) = (\partial/\partial p)\log\Gamma(p)$ is given by

$$V_p[\delta(\mathbf{X})] \ge \frac{1}{n}\frac{\partial^2}{\partial p^2}\log[\Gamma(p)] \quad \left[\text{since } g'(p) = \frac{\partial^2}{\partial p^2}\log\Gamma(p)\right] \qquad (4.4.15)$$

Further, on comparing Eq. (4.4.13) with Eq. (4.3.14), we get $\delta(\mathbf{X}) = \log\theta + (1/n)\Sigma\log X_i$, which is an unbiased estimator of $(\partial/\partial p)\log\Gamma(p)$, with $g(p) = (\partial/\partial p)\log\Gamma(p)$, $c(p) = n$, and $\delta(\mathbf{X})$ is a CRLB estimator. The variance of δ is given by

$$V_p(\delta) = \frac{g'(p)}{c(p)} = \frac{1}{n}\frac{\partial^2}{\partial p^2}\log\Gamma(p)$$

which is equal to the CRLB given in Eq. (4.4.15).

The unbiasedness of the estimator $\delta(\mathbf{X}) = \log\theta + (1/n)\Sigma\log X_i$ for $(\partial/\partial p)\log\Gamma(p)$ can be shown differently. Consider the differentiation of the quantity

$$\int_0^\infty x^{p-1}\exp(-\theta x)dx = \frac{\Gamma(p)}{\theta^p}$$

with respect to p. Differentiation of $x^{p-1} = \exp[(p-1)\log x]$ with respect to p yields $(\partial/\partial p)x^{p-1} = x^{p-1}\log x$. This gives

$$\int_0^\infty (\log x)x^{p-1}\exp(-\theta x)dx = \frac{\Gamma'(p)\theta^p - \theta^p\log\theta\,\Gamma(p)}{\theta^{2p}}$$

$$= \frac{\Gamma(p)}{\theta^p}\left[\frac{\Gamma'(p)}{\Gamma(p)} - \log\theta\right]$$

or

$$\frac{\theta^p}{\Gamma(p)}\int_0^\infty (\log x)x^{p-1}\exp(-\theta x)dx = \frac{\partial}{\partial p}\log\Gamma(p) - \log\theta$$

Therefore, $\log\theta + (1/n)\Sigma\log X_i$ is an unbiased estimator of $(\partial/\partial p)\log\Gamma(p)$.

***Example* 4.12** Let $X_1, X_2, \ldots, X_n$ be *iid rv*s having negative binomial distribution NB(r, θ)

$$f(x; r, \theta) = \binom{x+r-1}{r-1}\theta^r(1-\theta)^x, \quad x = 0, 1, \ldots$$

What is the CR lower bound for the variance of an unbiased estimator of θ.

Solution. On writing the density for single random variable X,

$$f(x; r, \theta) = \binom{x+r-1}{r-1}\theta^r(1-\theta)^x$$

$$\log f(x; r, \theta) = \log\binom{x+r-1}{r-1} + r\log\theta + x\log(1-\theta)$$

$$\frac{\partial}{\partial\theta}\log f(x; r, \theta) = \frac{r}{\theta} - \frac{x}{1-\theta} = -\frac{1}{1-\theta}\left[x - \frac{r(1-\theta)}{\theta}\right] \tag{4.4.16}$$

Differentiating

$$\Sigma\binom{x+r-1}{r-1}(1-\theta)^x = \theta^{-r}$$

with respect to θ, we get

$$\sum x \binom{x+r-1}{r-1}(1-\theta)^{x-1} = r\theta^{-r-1}$$

$$\frac{1}{1-\theta}\sum x \binom{x+r-1}{r-1}\theta^r(1-\theta)^x = \frac{r}{\theta}$$

This gives

$$E(X) = \frac{r(1-\theta)}{\theta}$$

Again differentiating with respect to θ, we get

$$\sum x(x-1)\binom{x+r-1}{r-1}(1-\theta)^{x-2} = r(r+1)\theta^{-r-2}$$

$$E[X(X-1)] = \frac{r(r+1)(1-\theta)^2}{\theta^2}$$

$$V(X) = E[X(X-1)] + E(X) - [E(X)]^2 = \frac{r(1-\theta)}{\theta^2}$$

The Fisher information in a single observation from Eq. (4.4.16) is given by

$$I_X(\theta) = \frac{1}{(1-\theta)^2}E\left[X - \frac{r(1-\theta)}{\theta}\right]^2 = \frac{1}{(1-\theta)^2}\frac{r(1-\theta)}{\theta^2}$$

$$= \frac{r}{\theta^2(1-\theta)}$$

Thus,

$$I_{\mathbf{X}}(\theta) = \frac{nr}{\theta^2(1-\theta)}$$

Here, $g(\theta) = \theta$ and $g'(\theta) = 1$. Therefore, the CRLB for the variance of an unbiased estimator of θ is

$$\text{CRLB}(\theta) = \frac{\theta^2(1-\theta)}{nr}$$

***Example* 4.13** Let $X_1, X_2, \ldots, X_n$ be a random sample from an exponential distribution with *pdf* $f(x;\ \theta) = \theta \exp(-\theta x)$, $x \geq 0$. Derive the CRLB to the variance of an unbiased estimator of $\exp(-\theta)$.

Solution. We have

$$f(\mathbf{x};\theta) = \theta^n \exp(-\theta \Sigma x_i)$$

$$\log f(\mathbf{x};\ \theta) = n \log\theta - \theta\sum x_i$$

$$\frac{\partial}{\partial\theta}\log f = \frac{n}{\theta} - \sum x_i$$

$$\frac{\partial^2}{\partial\theta^2}\log f = -\frac{n}{\theta^2}$$

$$I_{\mathbf{X}}(\theta) = E\left[-\frac{\partial^2}{\partial\theta^2}\log f\right] = \frac{n}{\theta^2}$$

Here, $g(\theta) = \exp(-\theta)$ and $g'(\theta) = -\exp(-\theta)$. Therefore, the CRLB is given by

$$\text{CRLB} = \frac{\{-\exp(-\theta)\}^2}{n/\theta^2} = \frac{\exp(-2\theta)\theta^2}{n}$$

***Example* 4.14** Let $X_1, X_2, \ldots, X_n$ be a random sample from the distribution with *pdf*

$$f(x;\,\theta) = \theta x^{\theta-1},\ \ 0 < x < 1$$

Find the CRLB for the variance of an unbiased estimator for θ.

Solution. Note $Y = -\log X \sim G_2\,(1,\theta)$. This gives $Z = \Sigma Y_i \sim G_2\,(n,\theta)$. The *pdf* of Z is

$$f(z;\theta) = \frac{\theta^n}{\Gamma(n)} z^{n-1}\exp(-\theta z)$$

$$\log f = \log\frac{z^{n-1}}{\Gamma(n)} + n\log\theta - \theta z$$

$$\frac{\partial}{\partial\theta}\log f = \frac{n}{\theta} - z$$

$$\frac{\partial^2}{\partial\theta^2}\log f = -\frac{n}{\theta^2}$$

Fisher information in the sample is

$$I_{\mathbf{X}}(\theta) = I_Z(\theta) = \frac{n}{\theta^2}$$

Therefore, the CRLB for θ is given by

$$\text{CRLB}(\theta) = [I_Z(\theta)]^{-1} = \frac{\theta^2}{n}$$

Consider

$$E\left(\frac{1}{Z}\right) = \frac{\theta^n}{\Gamma(n)}\int_0^\infty z^{n-2}\exp(-\theta z)\,dz$$

$$= \frac{\theta^n}{\Gamma(n)}\cdot\frac{\Gamma(n-1)}{\theta^{n-1}} = \frac{\theta}{n-1}$$

which gives that $\delta = (n - 1)/Z$ is an unbiased estimator of θ. Further,

$$E\left(\frac{1}{Z^2}\right) = \frac{\theta^n}{\Gamma(n)}\int_0^{\infty} z^{n-3}\exp(-\theta z)dz$$

$$= \frac{\theta^n}{\Gamma(n)}\frac{\Gamma(n-2)}{\theta^{n-2}} = \frac{\theta^2}{(n-1)(n-2)}$$

$$V\left(\frac{1}{Z}\right) = \frac{\theta^2}{(n-1)^2(n-2)}$$

$$V\left(\frac{n-1}{Z}\right) = \frac{\theta^2}{n-2}$$

Since $V(\delta) > \text{CRLB}(\theta)$, the CRLB is not attained by the variance of an unbiased estimator

$$\delta = \frac{n-1}{Z} = \frac{n-1}{\sum Y_i} = -\frac{(n-1)}{\sum_1^n \log X_i}$$

However, the efficiency of δ is

$$e(\delta,\theta) = \frac{\text{CRLB}(\theta)}{V(\delta)} = \frac{\theta^2/n}{\theta^2/(n-2)} = \frac{n-2}{n}$$

which is less than 1 and $e(\delta, \theta) \to 1$ as $n \to \infty$. Therefore, though δ is not the most efficient, but it is asymptotically the most efficient.

***Example* 4.15** Show that if the CRLB estimator exists, it is unique.

Solution. Let δ_1 and δ_2 be two unbiased estimators of θ, which are also the CRLB, i.e. $E(\delta_1) = \theta = E(\delta_2)$ and $V(\delta_1) = V(\delta_2) = V_0$ (say). We then have another unbiased estimator $\delta_0 = [(\delta_1 + \delta_2)/2]$, $E(\delta_0) = \theta$, and

$$V(\delta_0) = \frac{1}{4}[V(\delta_1) + V(\delta_2) + 2\,\text{cov}(\delta_1, \delta_2)]$$

$$= \frac{1}{4}(2V_0 + 2\rho V_0), \quad \text{where } \rho = \text{corr}(\delta_1, \delta_2)$$

$$= \frac{V_0}{2}(1 + \rho) \geq V_0 \qquad \text{(since } V_0 \text{ is CRLB)}$$

This implies $\rho \geq 1$. Therefore, $\rho = 1$. So, δ_1 and δ_2 are linearly related. We have

$$\delta_1 = a\delta_2 + b$$

$$E(\delta_1) = aE(\delta_2) + b$$

$$\Rightarrow \qquad b = 0,\ a = 1$$

Thus, $\delta_1 = \delta_2$ a.e. This shows that the CRLB estimator is unique.

***Example* 4.16** Let $X_1, X_2, \ldots, X_n$ be a random sample from binomial $b(m, \theta)$, $0 < \theta < 1$, m is known. Obtain the CRLB for the unbiased estimator of θ and $\theta(1 - \theta)$. Also, check whether there exist estimators attaining these lower bounds.

Solution. The *pdf* of X is given by

$$f(x;\theta) = \binom{m}{x}\theta^x(1-\theta)^{m-x}$$

$$\log f = \log\binom{m}{x} + x\log\theta + (m-x)\log(1-\theta)$$

$$\frac{\partial}{\partial\theta}\log f = \frac{x}{\theta} - \frac{m-x}{1-\theta} = \frac{x - m\theta}{\theta(1-\theta)}$$

Thus, Fisher information in X_1 is

$$I_{X_1}(\theta) = E_\theta\left[\frac{\partial}{\partial\theta}\log f\right]^2 = \frac{m}{\theta(1-\theta)}$$

and
$$I_{\mathbf{X}}(\theta) = nI_X(\theta) = \frac{nm}{\theta(1-\theta)}$$

The CRLB for estimating $g(\theta) = \theta$ is given by

$$\text{CRLB}(\theta) = \frac{1}{I_{\mathbf{X}}(\theta)} = \frac{\theta(1-\theta)}{nm}$$

and for estimating $g(\theta) = \theta(1 - \theta)$ is

$$\text{CRLB}[\theta(1-\theta)] = \frac{\theta(1-\theta)(1-2\theta)^2}{nm}$$

However, $(\partial/\partial\theta)\log f$ can not be expressed as $c(\theta)[T - \theta(1 - \theta)]$. Therefore, there does not exist an unbiased estimator of $\theta(1 - \theta)$ whose variance can attain $\text{CRLB}[\theta(1 - \theta)] = \theta(1 - \theta)(1 - 2\theta)^2/nm$.

***Example* 4.17** Let $X_1, X_2, \ldots, X_n$ be a random sample from the exponential population with *pdf*

$$f(x;\theta) = \begin{cases} \dfrac{1}{\theta}\exp\left(\dfrac{-x}{\theta}\right), & x > 0 \\ 0, & \text{otherwise} \end{cases}$$

Find the CRLBs for the variance of unbiased estimators of θ and θ^2. Check whether there exist estimators attaining this bound.

Solution. We have

$$\log f(x;\theta) = -\log\theta - \frac{x}{\theta}$$

$$\frac{\partial}{\partial \theta} \log f(x;\theta) = -\frac{1}{\theta} + \frac{x}{\theta^2}$$

$$\frac{\partial^2}{\partial \theta^2} \log f(x;\theta) = \frac{1}{\theta^2} - \frac{2x}{\theta^3}$$

$$I_X(\theta) = -E\left[\frac{\partial^2}{\partial \theta^2} \log f\right]$$

$$= -\frac{1}{\theta^2} + \frac{2\theta}{\theta^3} = \frac{1}{\theta^2}$$

Fisher information in the sample $X_1, X_2, \ldots, X_n$ is

$$I_{\mathbf{X}}(\theta) = n \cdot I_{X_1}(\theta) = \frac{n}{\theta^2}$$

Therefore, the CRLB for the variance of an unbiased estimator of $g(\theta) = \theta$ is given by

$$\text{CRLB}(\theta) = \frac{1}{n/\theta^2} = \frac{\theta^2}{n} \tag{4.4.17}$$

and for $g(\theta) = \theta^2$ is given by

$$\text{CRLB}(\theta^2) = \frac{(2\theta)^2}{n/\theta^2} = \frac{4\theta^4}{n} \tag{4.4.18}$$

Further,

$$\frac{\partial}{\partial \theta} \log f(\mathbf{x};\theta) = \sum_{i=1}^{n} \frac{\partial}{\partial \theta} \log f(x_i;\theta) \tag{4.4.19}$$

$$= -\frac{n}{\theta} + \frac{\sum x_i}{\theta^2}$$

$$= \frac{n}{\theta^2}(\bar{x} - \theta)$$

Comparing Eq. (4.4.19) with Eq. (4.3.7), we conclude that the CRLB is attained for estimating $g(\theta) = \theta$ by the estimator $\delta(\mathbf{X}) = \bar{X}$ with minimum variance

$$V_\theta(T) = \frac{1}{n^2} \sum V(X_i) = \frac{\theta^2}{n} = \text{CRLB}(\theta) \qquad \text{[from Eq. (4.4.17)]}$$

However, $(\partial/\partial\theta) \log f(\mathbf{x};\theta)$ cannot be expressed in the form of Eq. (4.3.7) for estimating θ^2, i.e., $c(\theta)[\delta(\mathbf{X}) - \theta^2]$. Therefore, there does not exist an unbiased estimator of θ^2 whose variance can attain $\text{CRLB}(\theta^2) = 4\theta^4/n$ given in Eq. (4.4.18).

***Example* 4.18** Let $O_1, O_2, \ldots, O_n$ be a random sample drawn from a trinomial distribution with parameters θ_1, θ_2; $\theta_1, \theta_2 > 0$. Find the CRLB for $\theta = (\theta_1, \theta_2)'$.

Solution. Let A_1, A_2, A_3 be the three categories into which sample observations fall with frequencies, $n(A_1) = x_1$, $n(A_2) = x_2$, $n(A_3) = x_3$, respectively. The joint distribution is

$$f(x_1, x_2, x_3; \theta_1, \theta_2) = \frac{n!}{x_1!x_2!x_3!}\theta_1^{x_1} \cdot \theta_2^{x_2} \cdot (1-\theta_1-\theta_2)^{x_3}$$

where $x_1 + x_2 + x_3 = n$; $\theta_1, \theta_2 > 0$, $\theta_1 + \theta_2 < 1$, $\theta_3 = 1 - \theta_1 - \theta_2$. Also, $E(X_i) = n\theta_i$, $i = 1, 2, 3$. Further, $\nabla \mathbf{g}(\boldsymbol{\theta})$ in Theorem 4.3.4 is $(1,1,1)'$. We next compute $I_{\mathbf{X}}(\boldsymbol{\theta})$. Consider

$$\log f(x_1, x_2, x_3 ; \theta_1, \theta_2) = \log\frac{n!}{x_1!x_2!x_3} + x_1 \log\theta_1 + x_2 \log\theta_2 + x_3 \log(1-\theta_1-\theta_2)$$

$$\frac{\partial}{\partial\theta_1}\log f = \frac{x_1}{\theta_1} - \frac{x_3}{(1-\theta_1-\theta_2)}$$

$$\frac{\partial^2}{\partial\theta_1^2}\log f = -\frac{x_1}{\theta_1^2} - \frac{x_3}{(1-\theta_1-\theta_2)^2}$$

This gives
$$I_{11}(\boldsymbol{\theta}) = E_\theta\left[-\frac{\partial^2}{\partial\theta_1^2}\log f\right]$$

$$= n\left(\frac{1}{\theta} + \frac{1}{1-\theta_1-\theta_2}\right) = \frac{n(1-\theta_2)}{\theta_1(1-\theta_1-\theta_2)}$$

Similarly,
$$I_{22}(\boldsymbol{\theta}) = E_\theta\left[-\frac{\partial^2}{\partial\theta_2^2}\log f\right]$$

$$= \frac{n(1-\theta_1)}{\theta_2(1-\theta_1-\theta_2)}$$

Further,
$$\frac{\partial}{\partial\theta_2}\log f = \frac{x_2}{\theta_2} - \frac{x_3}{(1-\theta_1-\theta_2)}$$

$$\frac{\partial^2}{\partial\theta_1\partial\theta_2}\log f = -\frac{x_3}{(1-\theta_1-\theta_2)^2}$$

$$I_{12}(\boldsymbol{\theta}) = E_\theta\left[-\frac{\partial^2}{\partial\theta_1\partial\theta_2}\log f\right]$$

$$= \frac{n}{(1-\theta_1-\theta_2)}$$

Therefore, the information matrix is given by

$$\mathbf{I}_{\mathbf{X}}(\boldsymbol{\theta})=\begin{pmatrix} I_{11}(\boldsymbol{\theta}) & I_{12}(\boldsymbol{\theta}) \\ I_{21}(\boldsymbol{\theta}) & I_{22}(\boldsymbol{\theta}) \end{pmatrix}$$

$$=n^2\begin{pmatrix} \dfrac{(1-\theta_2)}{\theta_1(1-\theta_1-\theta_2)} & \dfrac{1}{(1-\theta_1-\theta_2)} \\ \dfrac{1}{(1-\theta_1-\theta_2)} & \dfrac{(1-\theta_1)}{\theta_2(1-\theta_1-\theta_2)} \end{pmatrix}$$

$$=\frac{n^2}{(1-\theta_1-\theta_2)^2}\begin{pmatrix} \dfrac{(1-\theta_2)}{\theta_1} & 1 \\ 1 & \dfrac{(1-\theta_1)}{\theta_2} \end{pmatrix}$$

$$|\mathbf{I}_{\mathbf{X}}(\boldsymbol{\theta})|=\frac{n^2}{(1-\theta_1-\theta_2)^2}\left(\frac{(1-\theta_1)(1-\theta_2)}{\theta_1\theta_2}-1\right)$$

$$=\frac{n^2}{(1-\theta_1-\theta_2)^2}\cdot\frac{(1-\theta_1-\theta_2)}{\theta_1\theta_2}$$

$$=\frac{n^2}{\theta_1\theta_2(1-\theta_1-\theta_2)}$$

$$\text{Adjoint matrix}=\begin{pmatrix} \dfrac{n(1-\theta_1)}{\theta_2(1-\theta_1-\theta_2)} & \dfrac{-n}{(1-\theta_1-\theta_2)} \\ \dfrac{-n}{(1-\theta_1-\theta_2)} & \dfrac{n(1-\theta_2)}{\theta_1(1-\theta_1-\theta_2)} \end{pmatrix}$$

This gives

$$[\mathbf{I}_{\mathbf{X}}(\boldsymbol{\theta})]^{-1}=\frac{Adj[\mathbf{I}_{\mathbf{X}}(\boldsymbol{\theta})]}{|\mathbf{I}_{\mathbf{X}}(\boldsymbol{\theta})|}$$

$$=\begin{pmatrix} \dfrac{\theta_1(1-\theta_1)}{n} & -\dfrac{\theta_1\theta_2}{n} \\ -\dfrac{\theta_1\theta_2}{n} & \dfrac{\theta_2(1-\theta_2)}{n} \end{pmatrix}$$

Therefore, the CRLB for the variance of any unbiased estimator of $\boldsymbol{\theta}$ is given by

$$V_{\boldsymbol{\theta}}(\boldsymbol{\delta})=\begin{pmatrix} V_{\boldsymbol{\theta}}(\delta_1) & \mathrm{cov}_{\boldsymbol{\theta}}(\delta_1,\delta_2) \\ \mathrm{cov}_{\boldsymbol{\theta}}(\delta_1,\delta_2) & V_{\boldsymbol{\theta}}(\delta_2) \end{pmatrix}$$

$$\geq \begin{pmatrix} \frac{\theta_1(1-\theta_1)}{n} & -\frac{\theta_1\theta_2}{n} \\ -\frac{\theta_1\theta_2}{n} & \frac{\theta_2(1-\theta_2)}{n} \end{pmatrix}$$

If we consider $\delta_i = (X_i/n)$ as an unbiased estimator of θ_i, then $\boldsymbol{\delta} = (\delta_1, \delta_2)'$ is the most efficient and also the UMVUE for $\boldsymbol{\theta}$ since $V_{\boldsymbol{\theta}}(\delta_i) = [\theta_i(1-\theta_i)/n] = \text{CRLB}(\theta_i)$, $i = 1, 2$.

***Example* 4.19** Let $X_1, X_2, \ldots, X_n$ be a random sample from $N(\theta_1, \theta_2)$, where $\theta_1 \in \mathbb{R}^1$ and $\theta_2 > 0$, both θ_1 and θ_2 are unknown parameters. Find the CR lower bound for the variance of an unbiased estimator of θ_2^2. Comment whether $\bar{X}$ and $(n-1)^{-1}\Sigma_1^n(X_i - \bar{X})^2$ are the most efficient estimators of θ_1 and θ_2, respectively.

Solution. Let us consider a single *rv* X_1. Its *pdf* is given by

$$f(x; \theta_1, \theta_2) = \frac{1}{\sqrt{2\pi\theta_2}} \exp\left\{-\frac{1}{2\theta_2}(x-\theta_1)^2\right\}$$

$$\log f(x; \theta_1, \theta_2) = -\frac{1}{2}\log(2\pi) - \frac{1}{2}\log\theta_2 - \frac{1}{2\theta_2}(x-\theta_1)^2$$

$$\frac{\partial}{\partial\theta_1}\log f = \frac{(x-\theta_1)}{\theta_2}$$

$$\frac{\partial^2}{\partial\theta_1^2}\log f = -\frac{1}{\theta_2}$$

$$I_{11}(\boldsymbol{\theta}) = nE_{\theta}\left[-\frac{\partial^2}{\partial\theta_1^2}\log f\right] = \frac{n}{\theta_2}$$

Further,
$$\frac{\partial}{\partial\theta_2}\log f = -\frac{1}{2\theta_2} + \frac{(x-\theta_1)^2}{2\theta_2^2}$$

$$\frac{\partial^2}{\partial\theta_1\partial\theta_2}\log f = -\frac{(x-\theta_1)}{\theta_2^2}$$

$$I_{12}(\boldsymbol{\theta}) = nE_{\theta}\left[-\frac{\partial^2}{\partial\theta_1\partial\theta_2}\log f\right] = 0$$

Consider
$$\frac{\partial^2}{\partial\theta_2^2}\log f = \frac{1}{2\theta_2^2} - \frac{(x-\theta_1)^2}{\theta_2^3}$$

This gives
$$I_{22}(\boldsymbol{\theta}) = nE_{\boldsymbol{\theta}}\left[-\frac{\partial^2}{\partial\theta_2^2}\log f\right]$$
$$= n\left[\frac{1}{\theta_2^3}E_{\boldsymbol{\theta}}(X-\theta_1)^2 - \frac{1}{2\theta_2^2}\right] = \frac{n}{2\theta_2^2}$$

We then have
$$\mathbf{I_X}(\boldsymbol{\theta}) = \begin{pmatrix} I_{11} & I_{12} \\ I_{21} & I_{22} \end{pmatrix}$$
$$= \begin{pmatrix} \frac{n}{\theta_2} & 0 \\ 0 & \frac{n}{2\theta_2^2} \end{pmatrix}$$
$$|\mathbf{I_X}(\boldsymbol{\theta})| = \frac{n^2}{2\theta_2^3}$$

$[\mathbf{I_X}(\boldsymbol{\theta})]^{-1}$ is computed by
$$[\mathbf{I_X}(\boldsymbol{\theta})]^{-1} = \frac{\text{Adj}[\mathbf{I_X}(\boldsymbol{\theta})]}{|\mathbf{I_X}(\boldsymbol{\theta})|}$$
$$= \begin{pmatrix} \frac{\theta_2}{n} & 0 \\ 0 & \frac{2\theta_2^2}{n} \end{pmatrix}$$

Here also $\nabla\mathbf{g}(\boldsymbol{\theta}) = (1, 1)'$, since δ_1 and δ_2 in $\boldsymbol{\delta} = (\delta_1, \delta_2)'$ are unbiased for θ_1 and θ_2, respectively. Therefore, by Theorem 4.3.4, the CRLB for $\boldsymbol{\theta} = (\theta_1, \theta_2)'$ is given by
$$V_{\boldsymbol{\theta}}(\boldsymbol{\delta}) \geq \begin{pmatrix} \frac{\theta_2}{n} & 0 \\ 0 & \frac{2\theta_2^2}{n} \end{pmatrix}$$

which gives
$$V_{\boldsymbol{\theta}}(\delta_1) \geq \frac{\theta_2}{n} \text{ and } V_{\boldsymbol{\theta}}(\delta_2) \geq \frac{2\theta_2^2}{n}$$

Since $\bar{X}$ is an unbiased estimator for θ_1 with variance $V_{\boldsymbol{\theta}}(\bar{X}) = (\theta_2/n) = \text{CRLB}(\theta_1)$, it is the most efficient. Further, $[1/(n-1)]\ \Sigma_1^n(X_i - \bar{X})^2$ is unbiased for θ_2 with
$$V_{\boldsymbol{\theta}}\left[\frac{1}{(n-1)}\sum_1^n (X_i - \bar{X})^2\right] = \frac{2\theta_2^2}{(n-1)} > \frac{2\theta_2^2}{n} = \text{CRLB}(\theta_2)$$

which shows that $[1/(n-1)]\ \Sigma_1^n(X_i-\bar{X})^2$ is not the most efficient. However,

$$e\left(\frac{1}{(n-1)}\sum_1^n (X_i-\bar{X})^2;\boldsymbol{\theta}\right)=\frac{\text{CRLB}(\theta_2)}{V_\theta\left[\frac{1}{(n-1)}\sum_1^n (X_i-\bar{X})^2\right]}$$

$$=\frac{2\theta_2^2/n}{2\theta_2^2/(n-1)}=\frac{(n-1)}{n}\to 1$$

as $n\to\infty$. Therefore, $[1/(n-1)]\ \Sigma_1^n(X_i-\bar{X})^2$ is not the most efficient but it is asymptotically the most efficient.

***Example* 4.20 (Weibull distribution).** Let $X_1, X_2,\ldots,X_n$ be *iid* $f(x;\boldsymbol{\theta})$

$$f(x;\boldsymbol{\theta})=\frac{\beta}{\alpha}\left(\frac{x}{\alpha}\right)^{\beta-1}\exp\left\{-\left(\frac{x}{\alpha}\right)^{\beta}\right\}$$

$$\boldsymbol{\theta}=(\alpha,\beta)'\in(0,\infty)$$

where $f(x;\boldsymbol{\theta})$ is the density of the Weibull distribution. Find the CR lower bound for the variance of an unbiased estimator of $g(\boldsymbol{\theta})=E_\theta(X)=\int_0^\infty xf(x;\boldsymbol{\theta})dx=\alpha\Gamma[1+(1/\beta)]$. Comment on the estimation of $g(\boldsymbol{\theta})=E_\theta(X)$ when β is known, i.e., $\beta=\beta_0$.

Solution. We have

$$\log f(x;\boldsymbol{\theta})=\log\left(\frac{\beta}{\alpha}\right)+(\beta-1)\log\left(\frac{x}{\alpha}\right)-\left(\frac{x}{\alpha}\right)^{\beta}$$

Scores are given by

$$S(x,\alpha)=\frac{\partial}{\partial\alpha}\log f(x;\boldsymbol{\theta})=\frac{\beta}{\alpha}\left\{\left(\frac{x}{\alpha}\right)^{\beta}-1\right\}$$

$$S(x,\beta)=\frac{\partial}{\partial\beta}\log f(x;\boldsymbol{\theta})=\frac{1}{\beta}-\frac{1}{\beta}\left\{\left(\frac{x}{\alpha}\right)^{\beta}-1\right\}\log\left(\frac{x}{\alpha}\right)^{\beta}$$

Now,
$$I_{11}(\boldsymbol{\theta})=E_\theta[S^2(X,\alpha)]=\left(\frac{\beta}{\alpha}\right)^2E_\theta\left[\left(\frac{X}{\alpha}\right)^{\beta}-1\right]^2$$

Let $Y=(X/\alpha)^\beta$. The distribution of Y is given by

$$f(y)=\exp(-y)$$

Then $I_{11}(\boldsymbol{\theta})$ simplifies to

$$I_{11}(\boldsymbol{\theta})=\left(\frac{\beta}{\alpha}\right)^2E_\theta(Y-1)^2=\frac{\beta^2}{\alpha^2}$$

Consider the differentiation of $S(x,\beta)$ with respect to β to get

$$\frac{\partial^2}{\partial\beta^2}\log f(x;\boldsymbol{\theta}) = -\frac{1}{\beta^2} - \log\left(\frac{x}{\alpha}\right)\left[\left(\frac{x}{\alpha}\right)^{\beta}\log\left(\frac{x}{\alpha}\right)\right]$$

$$I_{22}(\boldsymbol{\theta}) = \frac{1}{\beta^2}E\left\{1+\left(\frac{X}{\alpha}\right)^{\beta}\left[\log\left(\frac{X}{\alpha}\right)^{\beta}\right]^2\right\} = \frac{1}{\beta^2}E[1+Y(\log Y)^2]$$

We have $\qquad E[Y(\log Y)^2] = \int y(\log y)^2 \exp(-y)dy$

and $\qquad \Gamma(p) = \int y^{p-1}\exp(-y)dy$

On differentiating $\Gamma(p)$ with respect to p and using

$$y^{p-1} = \exp[(p-1)\log y]$$

$$\frac{\partial}{\partial p}y^{p-1} = y^{p-1}\log y$$

we get

$$\Gamma'(p) = \int (\log y) y^{p-1}\exp(-y)dy$$

$$\Gamma''(p) = \int (\log y)^2 y^{p-1}\exp(-y)dy$$

$$E[Y(\log Y)^2] = \Gamma''(2) = \int (\log y)^2 y\exp(-y)dy$$

$$E[Y(\log Y)^2] = \frac{\Gamma''(2)}{\Gamma(2)} = \frac{1}{\Gamma(2)}\int (\log y)^2 y\exp(-y)dy$$

This gives

$$I_{22}(\boldsymbol{\theta}) = \frac{1}{\beta^2}E[1+Y(\log Y)^2] = \frac{1}{\beta^2}\left(1+\frac{\Gamma''(2)}{\Gamma(2)}\right) = \frac{[1+\Gamma''(2)]}{\beta^2}$$

Now, we will calculate $I_{12}(\boldsymbol{\theta})$. Consider the differentiation of

$$S(x,\alpha) = \frac{\partial}{\partial\alpha}\log f(x;\boldsymbol{\theta}) = \frac{\beta}{\alpha}\left\{\left(\frac{x}{\alpha}\right)^{\beta} - 1\right\}$$

with respect to β

$$\frac{\partial^2}{\partial\alpha\,\partial\beta}\log f(X;\boldsymbol{\theta}) = \frac{1}{\alpha}\left\{\left(\frac{X}{\alpha}\right)^{\beta} - 1\right\} + \frac{\beta}{\alpha}\left(\frac{X}{\alpha}\right)^{\beta}\log\left(\frac{X}{\alpha}\right)$$

$$= \frac{1}{\alpha}[(Y-1) + Y\log Y]$$

$$I_{12}(\boldsymbol{\theta}) = E_\theta[S(X,\alpha)S(X,\beta)]$$
$$= \frac{1}{\alpha}E_\theta\{(Y-1)+Y\log Y\}$$
$$= \frac{1}{\alpha}\left[0+\frac{\Gamma'(2)}{\Gamma(2)}\right] = \frac{\Gamma'(2)}{\alpha}$$

Therefore, Fisher information matrix based on the sample $X_1, X_2, \ldots, X_n$, where X_is *iid* $f(x;\boldsymbol{\theta})$ is

$$\mathbf{I_X}(\boldsymbol{\theta}) = \begin{pmatrix} \dfrac{n\beta^2}{\alpha^2} & \dfrac{n\Gamma'(2)}{\alpha} \\ \dfrac{n\Gamma'(2)}{\alpha} & \dfrac{n[1+\Gamma''(2)]}{\beta^2} \end{pmatrix}$$

$$|\mathbf{I_X}(\boldsymbol{\theta})| = \frac{n^2}{\alpha^2}\{1+\Gamma''(2)-[\Gamma'(2)]^2\}$$

By using Jensen's inequality,

$$[E(Y(\log Y))]^2 \le E[(Y(\log Y)^2]$$

or
$$[\Gamma'(2)]^2 \le \Gamma''(2)$$

Thus,
$$|\mathbf{I_X}(\boldsymbol{\theta})| \ge 0$$

gives that the information matrix is nonsingular.

$$\text{Adjoint matrix} = \begin{pmatrix} \dfrac{n[1+\Gamma''(2)]}{\beta^2} & -\dfrac{n\Gamma'(2)}{\alpha} \\ -\dfrac{n\Gamma'(2)}{\alpha} & \dfrac{n\beta^2}{\alpha^2} \end{pmatrix}$$

This gives

$$[\mathbf{I_X}(\boldsymbol{\theta})]^{-1} = \frac{Adj[\mathbf{I_X}(\boldsymbol{\theta})]}{|\mathbf{I_X}(\boldsymbol{\theta})|}$$

$$= \begin{pmatrix} \dfrac{\alpha^2}{\beta^2}\dfrac{1+\Gamma''(2)}{n\{1+\Gamma''(2)-[\Gamma'(2)]^2\}} & -\dfrac{\alpha\Gamma'(2)}{n\{1+\Gamma''(2)-[\Gamma'(2)]^2\}} \\ -\dfrac{\alpha\Gamma'(2)}{n\{1+\Gamma''(2)-[\Gamma'(2)]^2\}} & \dfrac{\beta^2}{n\{1+\Gamma''(2)-[\Gamma'(2)]^2\}} \end{pmatrix}$$

We are interested in estimating a real-valued function $g(\boldsymbol{\theta}) = \alpha\Gamma[1 + (1/\beta)]$. We have $\nabla g(\boldsymbol{\theta}) = (\Gamma[1 + (1/\beta)], -\alpha\Gamma'[1 + (1/\beta)]/\beta^2)' = \Gamma[1 + (1/\beta)](1, -\alpha \text{ dgm}[1 + (1/\beta)]/\beta^2)'$ where

$\text{dgm}[1 + (1/\beta)] = \Gamma'[1 + (1/\beta)]/\Gamma[1 + (1/\beta)]$ is a digamma function. Therefore, the CR lower bound for the variance of unbiased estimators $\delta(\mathbf{X})$ of $g(\boldsymbol{\theta})$

$$V_{\boldsymbol{\theta}}(\delta) \geq \nabla' g(\boldsymbol{\theta}) \mathbf{I}_{\mathbf{X}}^{-1}(\theta) \nabla g(\boldsymbol{\theta})$$

$$= \frac{\Gamma^2\left(1+\frac{1}{\beta}\right)}{n\{1+\Gamma''(2)-[\Gamma'(2)]^2\}} \frac{\alpha^2}{\beta^2}\left[1+\Gamma''(2)+2\Gamma'(2)\,\text{dgm}\left(1+\frac{1}{\beta}\right)+\text{dgm}^2\left(1+\frac{1}{\beta}\right)\right]$$

If $\beta = \beta_0$ is known, $\boldsymbol{\theta} = (\alpha, \beta_0)'$, $g(\boldsymbol{\theta}) = E_{\boldsymbol{\theta}}(X) = \{\alpha\Gamma[1 + (1/\beta_0)]\}$, $\nabla g(\boldsymbol{\theta}) = \Gamma[1 + (1/\beta_0)]$, $I_{11} = (n\beta_0^2/\alpha^2)$, $I_{11}^{-1} = (\alpha^2/n\beta_0^2)$, $\Gamma^2 g(\theta)\, I_{11}^{-1} = (\alpha^2/n\beta_0^2)\, \Gamma^2[1 + (1/\beta_0)]$. The CR lower bound in this case when β is known for the variance of unbiased estimators of $g(\boldsymbol{\theta})$

$$V_\theta(\delta) \geq \left(\frac{\alpha^2}{n\beta_0^2}\right)\Gamma^2\left(1+\frac{1}{\beta_0}\right)$$

Note that this lower bound is less than or equal to the bound when β is unknown. This is because we have better information for estimating $g(\boldsymbol{\theta}) = E_{\boldsymbol{\theta}}(X)$ when β is known as compared to when β is unknown.

***Example* 4.21 (Negative binomial).** Let $X \sim NB(1, p)$ be a geometric discrete random variable with *pmf*

$$f(x;\theta) = \theta(1-\theta)^x,\; x = 0, 1, 2, \ldots$$

where x is the number of failures preceeding the first success; $1 + x$ be the number of independent Bernoulli trials needed to get a success. Show that the CRLB is not sharp and that Bhattacharyya suggests a sequence of lower bounds approaching a minimum variance that can be attained by some unbiased estimator of θ.

Solution. Let us calculate S_1 and S_2 in this case

$$f(x;\theta) = (1-\theta)^x\,\theta$$

$$f'(x;\theta) = (1-\theta)^x - x(1-\theta)^{x-1}\theta = \frac{f(x;\theta)}{\theta} - \frac{x}{(1-\theta)} f(x;\theta)$$

$$= f(x;\theta)\left(\frac{1}{\theta} - \frac{x}{1-\theta}\right)$$

$$S_1 = \frac{f'(x;\theta)}{f(x;\theta)} = \frac{1}{\theta} - \frac{x}{1-\theta}$$

Next, consider

$$f''(x;\theta) = f'(x;\theta)\left(\frac{1}{\theta} - \frac{x}{1-\theta}\right) + f(x;\theta)\left[-\frac{1}{\theta^2} - \frac{x}{(1-\theta)^2}\right]$$

$$= f(x;\theta)\left(\frac{1}{\theta}-\frac{x}{1-\theta}\right)^2 + f(x;\theta)\left[-\frac{1}{\theta^2}-\frac{x}{(1-\theta)^2}\right]$$

$$= f(x;\theta)\left[\left(\frac{x}{1-\theta}\right)^2 - \frac{x(2-\theta)}{\theta(1-\theta)^2}\right]$$

$$S_2 = \frac{f''(x;\theta)}{f(x;\theta)} = \frac{x^2}{(1-\theta)^2} - \frac{x(2-\theta)}{\theta(1-\theta)^2}$$

For $m = 1$, $A = [a_{11}]$, $\boldsymbol{\nu} = (\nu_1) = (\partial/\partial\theta)g(\theta) = (1)$, and

$$\frac{\partial^2}{\partial\theta^2}\log f(x;\theta) = -\frac{1}{\theta^2} - \frac{x}{(1-\theta)^2}$$

gives

$$a_{11} = E(S_1^2) = V(S_1) = I_X(\theta) = E_\theta\left[-\frac{\partial^2}{\partial\theta^2}\log f(X;\theta)\right]$$

$$= \frac{1}{\theta^2} + \frac{1}{(1-\theta)^2}E_\theta(X)$$

$$= \frac{1}{\theta^2} + \frac{1}{(1-\theta)^2}\frac{(1-\theta)}{\theta}$$

Since $E_\theta(X) = (1 - \theta)/\theta$, we get

$$a_{11} = \frac{1}{\theta}\left[\frac{1}{\theta} + \frac{1}{(1-\theta)}\right] = \frac{1}{\theta^2(1-\theta)}$$

Thus, the Bhattacharyya lower bound for $m = 1$ (CRLB)

$$V_\theta(\delta) \geq \boldsymbol{\nu}'A\boldsymbol{\nu} = a_{11}^{-1} = \theta^2(1-\theta) = L_1$$

For single observation, let

$$\delta(X) = \begin{cases} 1, & X = 0 \\ 0, & X > 0 \end{cases}$$

Then δ is the UMVUE and its variance is $\theta(1 - \theta)$. Further, the CRLB (Bhattacharyya lower bound, $m = 1$) is not sharp since CRLB(θ, $m = 1$) $= \theta^2(1 - \theta) < \theta(1 - \theta)$ = MV (minimum variance). We, next, calculate the second Bhattacharyya bound, $m = 2$, for which we require the first four moments of the negative binomial distribution. We have

$$\boldsymbol{\nu} = (\nu_1, \nu_2)' = \left(\frac{\partial}{\partial\theta}g(\theta), \frac{\partial^2}{\partial\theta^2}g(\theta)\right)' = (1, 0)'$$

Note that $\theta^{-1} = \Sigma_{x=0}^{\infty}(1-\theta)^x$. Then then $\theta^{-2} = \Sigma x(1-\theta)^{x-1} = E(X)/[\theta(1-\theta)]$

$$E(X) = \frac{1-\theta}{\theta}$$

and $2\theta^{-3} = \Sigma x(x-1)(1-\theta)^{x-2} = E[X(X-1)]/[\theta(1-\theta)^2]$

$$E[X(X-1)] = 2\frac{\theta(1-\theta)^2}{\theta^3} = 2\left(\frac{1-\theta}{\theta}\right)^2$$

Thus, in general, the factorial moments of X from the negative binomial distribution are computed by

$$E[X^{(m)}] = E[X(X-1)\ldots(X-\overline{m-1})] = (m)!\left(\frac{1-\theta}{\theta}\right)^m$$

or $$E[X] = E[X^{(1)}] = \frac{1-\theta}{\theta}$$

$$E(X^2) = E[X(X-1)+X] = E[X^{(2)}] + E[X^{(1)}]$$

$$= \frac{2(1-\theta)^2}{\theta^2} + \frac{1-\theta}{\theta} = \frac{1-\theta}{\theta^2}[2(1-\theta)+\theta]$$

$$= \frac{1-\theta}{\theta^2}[1+(1-\theta)] = \frac{(1-\theta)+(1-\theta)^2}{\theta^2}$$

$$E(X^3) = E[X(X-1)(X-2)+3X(X-1)+X]$$

$$= E[X^{(3)}] + 3E[X^{(2)}] + E[X^{(1)}]$$

$$= \frac{6(1-\theta)^3}{\theta^3} + \frac{6(1-\theta)^2}{\theta^2} + \frac{(1-\theta)}{\theta}$$

$$= \frac{(1-\theta)+4(1-\theta)^2+(1-\theta)^3}{\theta^3}$$

$$E(X^4) = E[X(X-1)(X-2)(X-3)+6X(X-1)(X-2)+7X(X-1)+X]$$

$$= \frac{(1-\theta)+11(1-\theta)^2+11(1-\theta)^3+(1-\theta)^4}{\theta^4}$$

We have $$a_{11} = E(S_1^2) = V(S_1) = E\left(\frac{1}{\theta} - \frac{X}{1-\theta}\right)^2$$

$$= \frac{1}{\theta^2} - \frac{2E(X)}{\theta(1-\theta)} + \frac{E(X^2)}{(1-\theta)^2} = \frac{1}{(1-\theta)\theta^2}$$

Similarly, we may calculate

$$a_{12} = E(S_1S_2) = \text{cov}(S_1, S_2) = -\frac{2}{\theta^3(1-\theta)}$$

$$a_{22} = E(S_2^2) = V(S_2) = \frac{4(2-\theta)}{\theta^4(1-\theta)^2}$$

This gives the variance and covariance matrix of S_1 and S_2 by

$$A = \|a_{ij}\|$$

$$= \begin{bmatrix} \dfrac{1}{\theta^2(1-\theta)} & -\dfrac{2}{\theta^3(1-\theta)} \\ \dfrac{2}{\theta^3(1-\theta)} & \dfrac{4(2-\theta)}{\theta^4(1-\theta)^2} \end{bmatrix}$$

$$|A| = \frac{4}{\theta^6(1-\theta)^3}$$

$$A^{-1} = \frac{\text{Adj}(A)}{\det(A)} = \frac{\begin{bmatrix} a_{22} & -a_{12} \\ -a_{12} & a_{11} \end{bmatrix}}{\det(A)}$$

The second-order Bhattacharyya lower bound is given by

$$L_2 = (\boldsymbol{\nu}'A^{-1}\boldsymbol{\nu})$$

$$= \frac{1}{\det(A)}(1 \quad 0)\begin{bmatrix} a_{22} & -a_{12} \\ -a_{12} & a_{11} \end{bmatrix}\begin{pmatrix} 1 \\ 0 \end{pmatrix}$$

$$= \theta^2(1-\theta)[1+(1-\theta)] > L_1$$

Thus, L_2 is a sharper lower bound as compared to L_1. We can easily show that

$$L_m = \theta^2(1-\theta)[1 + (1-\theta) + (1-\theta)^2 + \cdots + (1-\theta)^{m-1}]$$

so that $$L_1 < L_2 < L_3 \cdots < L_m < \ldots;\ \theta \in (0, 1)$$

and $$\lim_{m\to\infty} L_m = \theta^2(1-\theta)\frac{1}{1-(1-\theta)} = \theta(1-\theta)$$

= Minimum variance among all unbiased estimators (or variance of UMVUE)

Thus, in this example we see that, though the CRLB is not sharp, the Bhattacharyya bounds give a series of lower bounds that incrementally converge to the minimum variance that an unbiased estimator can attain.

***Example* 4.22** Let $X_1, X_2,\ldots,X_n$ be a random sample from the Bernoulli distribution $b(1, \theta)$ with *pmf*

$$f(x;\theta) = \theta^x(1-\theta)^{1-x},\ x = 0, 1;\ 0 < \theta < 1$$

Show that there does not exist an estimator of θ^2 attaining the CRLB. Obtain the Bhattacharyya lower bound of order two for θ^2.

Solution. We have

$$\log f(x;\theta) = x\log\theta + (1-x)\log(1-\theta)$$

$$S_1 = \frac{f'(x;\theta)}{f(x;\theta)} = \frac{\partial}{\partial\theta}\log f(x;\theta) = \frac{x}{\theta} - \frac{(1-x)}{(1-\theta)}$$

$$= \frac{x-\theta}{\theta(1-\theta)}$$

and
$$E(S_1^2) = V(S_1) = nI_X(\theta) = \frac{n}{\theta(1-\theta)}$$

This gives
$$L_1 = \frac{[g'(\theta)]^2}{V(S_1)} = \frac{4\theta^3(1-\theta)}{n} = \text{CRLB}(\theta^2)$$

Writing for the sample $X_1, X_2,\ldots,X_n$,

$$\frac{\partial}{\partial\theta}\log(\mathbf{x};\theta) = \sum_{i=1}^{n}\frac{\partial}{\partial\theta}\log f(x_i;\theta)$$

$$= \sum_{i=1}^{n}\frac{(x_i-\theta)}{\theta(1-\theta)}$$

$$= \frac{n}{\theta(1-\theta)}(\overline{x}-\theta)$$

This is not in the form of Eq. (4.3.7), i.e. $c(\theta)[\delta - \theta^2]$. Therefore, there does not exist an unbiased estimator of θ^2 whose variance can attain the CRLB.

Next, we calculate the Bhattacharyya lower bound of order 2

$$L_2 = [v_1 \quad v_2]\begin{bmatrix} a_{11} & a_{12} \\ a_{21} & a_{22} \end{bmatrix}^{-1}\begin{bmatrix} v_1 \\ v_2 \end{bmatrix}$$

Here, $g(\theta) = \theta^2$; $v_1 = (\partial/\partial\theta)g(\theta) = 2\theta$, $v_1 = (\partial^2/\partial\theta^2)g(\theta) = 2$

$$a_{11}(n) = na_{11}(1)$$

$$a_{12}(n) = na_{12}(1)$$

$$a_{22}(n) = na_{22}(1) + 2n(n-1)\,[a_{11}(1)]^2$$

$$S_1 = \frac{f'(x;\theta)}{f(x;\theta)} = \frac{1}{f(x;\theta)}\frac{\partial}{\partial\theta}f(x;\theta)$$

$$S_2 = \frac{f''(x;\theta)}{f(x;\theta)} = \frac{1}{f(x;\theta)}\frac{\partial^2}{\partial\theta^2}f(x;\theta)$$

$$a_{11}(1) = E_\theta\left[\frac{1}{f}\frac{\partial}{\partial\theta}f\right]^2 = E_\theta\left[\frac{(X-\theta)^2}{\theta^2(1-\theta)^2}\right] = \frac{\theta(1-\theta)}{\theta^2(1-\theta)^2} = \frac{1}{\theta(1-\theta)}$$

$$a_{12}(1) = E_\theta[S_1S_2] = E_\theta\left[\frac{1}{f}\frac{\partial}{\partial\theta}f\frac{1}{f}\frac{\partial^2}{\partial\theta^2}f\right]$$

where $$\frac{1}{f}\frac{\partial^2}{\partial\theta^2}f = \frac{1}{f}\frac{\partial}{\partial\theta}\frac{\partial}{\partial\theta}f = \frac{1}{f}\frac{\partial}{\partial\theta}f\left[\frac{x}{\theta} - \frac{(1-x)}{(1-\theta)}\right]$$

$$= \frac{1}{f}\left[\left(\frac{x}{\theta} - \frac{(1-x)}{(1-\theta)}\right)\frac{\partial}{\partial\theta}f + f\left(-\frac{x}{\theta^2} - \frac{(1-x)}{(1-\theta)^2}\right)\right]$$

$$= \frac{1}{f}\left[f\left(\frac{x}{\theta} - \frac{(1-x)}{(1-\theta)}\right)^2 - f\left(\frac{(1-\theta)^2x + \theta^2 - \theta^2x}{\theta^2(1-\theta)^2}\right)\right]$$

$$= \frac{(x-\theta)^2}{\theta^2(1-\theta)^2} + \frac{\theta^2x - \theta^2 - (1-\theta)^2x}{\theta^2(1-\theta)^2}$$

$$= \frac{x^2 - 2\theta x + \theta^2 + \theta^2x - \theta^2 - x + 2\theta x - \theta^2x}{\theta^2(1-\theta)^2}$$

$$= \frac{x^2 - x}{\theta^2(1-\theta)^2} = \frac{x(x-1)}{\theta^2(1-\theta)^2}$$

This gives $$a_{12}(1) = E_\theta\left[\frac{(X-\theta)}{\theta(1-\theta)}\frac{X(X-1)}{\theta^2(1-\theta)^2}\right]$$

$$= \frac{1}{\theta^3(1-\theta)^2}E_\theta[X^3 - \theta X^2 - X^2 + \theta X]$$

$$= \frac{1}{\theta^3(1-\theta)^2}(\theta - \theta^2 - \theta + \theta^2) = 0$$

Since $$\mu'_{r+1} = \theta\mu'_r + \left(\frac{\partial\mu'_r}{\partial\theta}\right)\theta(1-\theta)$$

$\Rightarrow$ $$\mu'_2 = \theta^2 + \theta(1-\theta) = \theta = \mu'_3$$

$\Rightarrow$ $$E_\theta(X^k) = \theta \;\forall\; k$$

Similarly,

$$a_{22}(1) = E_\theta\left[\frac{1}{f}\frac{\partial^2}{\partial\theta^2}f\frac{1}{f}\frac{\partial^2}{\partial\theta^2}f\right]$$

$$= E_\theta\left[\frac{X(X-1)}{\theta^2(1-\theta)^2}\right]^2$$

$$= \frac{1}{\theta^4(1-\theta)^4}E_\theta[X^2(X-1)^2]$$

$$= \frac{1}{\theta^4(1-\theta)^4}E_\theta[X^4-2X^3+X^2] = 0$$

Therefore,

$$a_{11}(n) = n\,a_{11}(1) = \frac{n}{\theta(1-\theta)}$$

$$a_{12}(n) = n\,a_{12}(1) = 0$$

$$a_{22}(n) = n\,a_{22}(1) + 2n(n-1)\,[a_{11}(1)]^2$$

$$= 0 + 2n(n-1)\frac{1}{\theta^2(1-\theta)^2}$$

$$= \frac{2n(n-1)}{\theta^2(1-\theta)^2}$$

This gives the variance–covariance matrix of S_1, S_2

$$A = \|a_{ij}(n)\| = \begin{bmatrix} \dfrac{n}{\theta(1-\theta)} & 0 \\ 0 & \dfrac{2n(n-1)}{\theta^2(1-\theta)^2} \end{bmatrix}$$

$$A^{-1} = \frac{\text{adj}(A)}{\det(A)} = \frac{1}{\dfrac{2n^2(n-1)}{\theta^3(1-\theta)^3}}\begin{bmatrix} \dfrac{2n(n-1)}{\theta^2(1-\theta)^2} & 0 \\ 0 & \dfrac{n}{\theta(1-\theta)} \end{bmatrix}$$

$$= \begin{bmatrix} \dfrac{\theta(1-\theta)}{n} & 0 \\ 0 & \dfrac{\theta^2(1-\theta)^2}{2n(n-1)} \end{bmatrix}$$

Therefore, the Bhattacharyya lower bound of order 2 is given by

$$L_2 = \mathbf{v}'\Delta^{-1}\mathbf{v} = [2\theta \quad 2]\begin{bmatrix} \dfrac{\theta(1-\theta)}{n} & 0 \\ 0 & \dfrac{\theta^2(1-\theta)^2}{2n(n-1)} \end{bmatrix}\begin{bmatrix} 2\theta \\ 2 \end{bmatrix}$$

$$= \frac{\theta(1-\theta)}{n}4\theta^2 + \frac{\theta^2(1-\theta)^2}{2n(n-1)}4$$

$$= \frac{4\theta^3(1-\theta)}{n} + \frac{2\theta^2(1-\theta)^2}{n(n-1)}$$

It can easily be seen that the CRLB ($m = 1$)

$$L_1 = \frac{[g'(\theta)]^2}{V(S_1)} = \frac{4\theta^3(1-\theta)}{n} = \text{CRLB}(\theta^2)$$

$$= \frac{[g'(\theta)]^2}{a_{11}(n)} = \frac{(2\theta)^2}{[n/\theta(1-\theta)]} = \frac{4\theta^3(1-\theta)}{n}$$

This gives
$$L_2 = \text{CRLB} + O\left(\frac{1}{n}\right) = L_2 + O\left(\frac{1}{n}\right)$$

This shows that the Bhattacharyya lower bound increases with the increase in its order and one gets sharper lower bounds.

***Example* 4.23** Let $X_1, X_2, \ldots, X_n$ be a random sample from $N(\theta, 1)$ for estimating θ and θ^2. Find the second Bhattacharyya lower bound for the variance of an unbiased estimator of θ^2.

Solution. Define

$$S_1 = \frac{f'(\mathbf{x};\theta)}{f(\mathbf{x};\theta)} = \frac{\partial}{\partial\theta}\log f(\mathbf{x};\theta)$$

$$= \frac{\partial}{\partial\theta}\left[-\log\sqrt{2\pi} - \frac{1}{2}\sum(x_i-\theta)^2\right] = n(\bar{x}-\theta)$$

$$S_2 = \frac{f''(\mathbf{x};\theta)}{f(\mathbf{x};\theta)} = \frac{1}{f(\mathbf{x};\theta)}\frac{\partial^2}{\partial\theta^2}f(\mathbf{x};\theta)$$

$$= \frac{1}{f}\frac{\partial}{\partial\theta}(S_1 f) = \frac{1}{f}\frac{\partial}{\partial\theta}[f\cdot n(\bar{x}-\theta)]$$

$$= n^2(\bar{x}-\theta)^2 - n$$

$$v_1 = \frac{\partial}{\partial\theta} g(\theta) = \frac{\partial}{\partial\theta}\theta^2 = 2\theta;\ v_2 = \frac{\partial^2}{\partial\theta^2}\theta^2 = 2$$

$$a_{11}(n) = E_\theta[S_1^2] = n^2 E_\theta(\bar{X} - \theta)^2 = n^2\,(1/n) = n$$

$$a_{12}(n) = E_\theta[S_1 S_2] = E_\theta\{[n(\bar{X} - \theta)][n^2(\bar{X} - \theta)^2 - n]\}$$

$$= E_\theta[n^3(\bar{X} - \theta)^3 - n^2\,(\bar{X} - \theta)] = 0$$

$$a_{22}(n) = E_\theta[S_2^2] = E_\theta[n^4(\bar{X} - \theta)^4 - 2n^3(\bar{X} - \theta)^2 + n^2]$$

$$= n^4\left(\frac{3V(\bar{X})}{n}\right) - 2n^3\frac{1}{n} + n^2 = n^4\frac{3}{n^2} - n^2 = 2n^2$$

Therefore, the variance–covariance matrix of S_1 and S_2

$$A = \begin{bmatrix} a_{11}(n) & a_{12}(n) \\ a_{21}(n) & a_{22}(n) \end{bmatrix}$$

$$= \begin{bmatrix} n & 0 \\ 0 & 2n^2 \end{bmatrix}$$

$$\text{Adj}(A) = \begin{bmatrix} 2n^2 & 0 \\ 0 & n \end{bmatrix}$$

$$\det(A) = 2n^3$$

$$A^{-1} = \frac{\text{Adj}(A)}{\det(A)} = \begin{bmatrix} 1/n & 0 \\ 0 & 1/2n^2 \end{bmatrix}$$

The Bhattacharyya lower bound of order one for estimating θ is

$$V(\delta) \ge L_1 = \frac{1}{n}$$

and that of order two is

$$V(\delta) \ge L_2 = \frac{1}{n}$$

It can be seen in this case that

$$L_1 = L_2 = \cdots = L_m = \frac{1}{n}$$

Hence, no improvement is possible beyond the CRLB, where $\bar{X}$ is the UMVUE of θ. The Bhattacharyya lower bound of order two for estimating θ^2 is given by

$$\begin{aligned} L_2 &= \mathbf{v}'A^{-1}\mathbf{v} \\ &= [2\theta \quad 2]\begin{bmatrix} 1/n & 0 \\ 0 & 1/2n^2 \end{bmatrix}\begin{bmatrix} 2\theta \\ n \end{bmatrix} \\ &= \frac{4\theta^2}{n} + \frac{2}{n} = \frac{[g'(\theta)]^2}{a_{11}(n)} + \frac{2}{n^2} \\ &= L_1(\theta^2) + \frac{2}{n^2} \end{aligned}$$

which shows that $L_2 > L_1$. Proceeding similarly, we get sharper and sharper lower bounds by considering higher-order Bhattacharyya bounds.

***Example* 4.24** Let $X_1, X_2, \ldots, X_n$ be a random sample from the distribution

$$f(x;\theta) = \theta \exp(-\theta x), \quad x > 0, \quad \theta > 0$$

Find the Bhattacharyya bound of order 2 for the variance of an unbiased estimator of θ.

Solution. $g(\theta) = \theta$, $v_1 = 1$, $v_2 = 0$, $\mathbf{v} = (1, 0)'$

For single observation, $f = \theta \exp(-\theta x)$

$$\begin{aligned} \frac{\partial}{\partial\theta} f &= \exp(-\theta x) - \theta x \exp(-\theta x) \\ &= \theta \exp(-\theta x)\left(\frac{1}{\theta} - x\right) = f\left(\frac{1}{\theta} - x\right) \end{aligned}$$

or

$$S_1 = \frac{1}{f}\frac{\partial}{\partial\theta} f = \left(\frac{1}{\theta} - x\right)$$

Further,

$$\begin{aligned} \frac{\partial^2}{\partial\theta^2} f &= f\left(-\frac{1}{\theta^2}\right) + \left(\frac{1}{\theta} - x\right)\frac{\partial}{\partial\theta} f \\ &= f\left(-\frac{1}{\theta^2}\right) + f\left(\frac{1}{\theta} - x\right)^2 \end{aligned}$$

$$\begin{aligned} S_2 &= \frac{1}{f}\frac{\partial^2}{\partial\theta^2} f = \left(\frac{1}{\theta} - x\right)^2 - \frac{1}{\theta^2} \\ &= \frac{1}{\theta^2} - \frac{2x}{\theta} + x^2 - \frac{1}{\theta^2} = x^2 - \frac{2x}{\theta} \end{aligned}$$

$$a_{11}(1) = E_\theta[S_1^2] = E_\theta\left(X - \frac{1}{\theta}\right)^2 = V_\theta(X) = \frac{1}{\theta^2}$$

Similarly, $$a_{12}(1) = E_\theta[S_1 S_2] = E_\theta\left[\frac{1}{f}\frac{\partial}{\partial\theta}f\frac{1}{f}\frac{\partial^2}{\partial\theta^2}f\right]$$

$$= E_\theta\left[\left(\frac{1}{\theta} - X\right)\left(X^2 - \frac{2X}{\theta}\right)\right]$$

$$= E_\theta\left[\frac{3X^2}{\theta} - \frac{2X}{\theta^2} - X^3\right] = \left[\frac{3(2/\theta^2)}{\theta} - \frac{2(1/\theta)}{\theta^2} - \frac{6}{\theta^3}\right]$$

since $$E_\theta(X^K) = \frac{\Gamma(K+1)}{\theta^K}$$

This gives

$$a_{12}(1) = -\frac{2}{\theta^3}$$

$$a_{22}(1) = E_\theta[\delta_2^2]$$

$$= E_\theta\left(\frac{1}{f}\frac{\partial^2}{\partial\theta^2}f\right)^2 = E_\theta\left(X^2 - \frac{2X}{\theta}\right)^2$$

$$= E_\theta\left(X^4 - \frac{4X^3}{\theta} + \frac{4X^2}{\theta^2}\right)$$

$$= \frac{1}{\theta^4}4\cdot 3\cdot 2\cdot 1 - \frac{4}{\theta}\cdot\frac{1}{\theta^3}3\cdot 2\cdot 1 + \frac{4}{\theta^2}\cdot\frac{1}{\theta^2}\cdot 2\cdot 1$$

$$= \frac{24}{\theta^4} - \frac{24}{\theta^4} + \frac{8}{\theta^4} = \frac{8}{\theta^4}$$

we get $$a_{11}(n) = n\,a_{11}(1) = \frac{n}{\theta^2}$$

$$a_{12}(n) = n\,a_{12}(1) = -\frac{2n}{\theta^3}$$

$$a_{22}(n) = n\,a_{22}(1) + 2n(n-1)[a_{11}(1)]^2$$

$$= \frac{8n}{\theta^4} + \frac{2n(n-1)}{\theta^4} = \frac{8n + 2n^2 - 2n}{\theta^4}$$

$$= \frac{2n^2 + 6n}{\theta^4}$$

Therefore, the variance–covariance matrix of S_1 and S_2 is given by

$$A = \begin{bmatrix} \dfrac{n}{\theta^2} & -\dfrac{2n}{\theta^3} \\ -\dfrac{2n}{\theta^3} & \dfrac{(2n^2+6n)}{\theta^4} \end{bmatrix}$$

$$\det(A) = \frac{1}{\theta^6}(2n^3 + 6n^2 - 4n^2)$$

$$= \frac{2n^2(n+1)}{\theta^6}$$

$$\text{Adj}(A) = \begin{bmatrix} \dfrac{(2n^2+6n)}{\theta^4} & \dfrac{2n}{\theta^3} \\ \dfrac{2n}{\theta^3} & \dfrac{n}{\theta^2} \end{bmatrix}$$

We then have $$A^{-1} = \frac{\text{Adj}(A)}{\det(A)} = \frac{1}{\dfrac{2n^2(n+1)}{\theta^6}} \begin{bmatrix} \dfrac{2n^2+6n}{\theta^4} & \dfrac{2n}{\theta^3} \\ \dfrac{2n}{\theta^3} & \dfrac{n}{\theta^2} \end{bmatrix}$$

$$= \begin{bmatrix} \dfrac{\theta^2(n+3)}{n(n+1)} & \dfrac{\theta^3}{n(n+1)} \\ \dfrac{\theta^3}{n(n+1)} & \dfrac{\theta^4}{2n(n+1)} \end{bmatrix}$$

The Bhattacharyya lower bound of order 2 is then given by

$$L_2 = \mathbf{v}'A^{-1}\mathbf{v}$$

$$= [1 \quad 0] \begin{bmatrix} \dfrac{\theta^2(n+3)}{n(n+1)} & \dfrac{\theta^3}{n(n+1)} \\ \dfrac{\theta^3}{n(n+1)} & \dfrac{\theta^4}{2n(n+1)} \end{bmatrix} \begin{bmatrix} 1 \\ 0 \end{bmatrix}$$

$$= \frac{\theta^2(n+3)}{n(n+1)} = \frac{[g'(\theta)]^2}{\delta_{11}(n)} \cdot \frac{n+3}{n+1}$$

$$= \frac{(n+3)}{(n+1)} L_1 > L_1 \qquad \left[\text{since } \frac{(n+3)}{(n+1)} > 1\right]$$

This shows that the Bhattacharyya lower bound becomes shaper with the increase in the order of which it is calculated at.

***Example* 4.25** Let $X \sim U(0, \theta)$, $\theta > 0$. Find the CRKLB in cases of (i) single observation (no data problem) and (ii) on the basis of a random sample (data problem).

Solution.

(i) $g(\theta) = \theta$, $S(\theta_1) \subset S(\theta)$ implies $\theta_1 < \theta$.

$$\text{CRKLB} = \sup_{\theta_1<\theta} \frac{[g(\theta_1) - g(\theta)]^2}{V_\theta\left[\dfrac{f(X;\theta_1)}{f(X;\theta)}\right]}$$

$$= \sup_{\theta_1<\theta} \frac{(\theta_1 - \theta)^2}{E_\theta\left[\dfrac{f(X;\theta_1)}{f(X;\theta)}\right]^2 - 1}$$

$$= \sup_{\theta_1<\theta} \frac{(\theta_1 - \theta)^2}{E_\theta\left[\dfrac{(1/\theta_1)I_{(0,\theta_1)}(X)}{(1/\theta)I_{(0,\theta)}(X)}\right]^2 - 1}$$

$$= \sup_{\theta_1<\theta} \frac{(\theta_1 - \theta)^2}{\displaystyle\int_0^{\theta_1}\left(\frac{\theta}{\theta_1}\right)^2 \frac{1}{\theta}dx - 1}$$

$$= \sup_{\theta_1<\theta} \frac{(\theta_1 - \theta)^2}{\left(\dfrac{\theta}{\theta_1} - 1\right)} = \sup_{\theta_1<\theta} \frac{\theta_1(\theta - \theta_1)^2}{(\theta_1 - \theta)}$$

$$= \sup_{\theta_1<\theta} \theta_1(\theta - \theta_1)$$

Let $\varphi(\theta_1) = \theta_1(\theta - \theta_1)$; $(\partial/\partial\theta_1)\varphi(\theta_1) = \theta - 2\theta_1 = 0$, $\theta_1 = (\theta/2)$. This gives

$$\text{CRKLB} = \sup_{\theta_1<\theta} \varphi(\theta_1) = \frac{\theta}{2}\left(\theta - \frac{\theta}{2}\right) = \frac{\theta^4}{4}$$

(ii) $$E_\theta\left[\frac{f(\mathbf{X};\theta_1)}{f(\mathbf{X};\theta)}\right]^2 - 1 = \left(\frac{\theta}{\theta_1}\right)^{2n} \int_0^{\theta_1}\dots\int_0^{\theta_1} f(\mathbf{x};\theta)\,d\mathbf{x} - 1$$

$$= \left(\frac{\theta}{\theta_1}\right)^{2n} \prod_1^n \int_0^{\theta_1} \frac{1}{\theta}dx - 1$$

$$= \left(\frac{\theta}{\theta_1}\right)^{2n} \left(\frac{\theta_1}{\theta}\right)^n - 1 = \left(\frac{\theta}{\theta_1}\right)^n - 1$$

Therefore, the CRKLB is given by

$$\text{CRKLB} = \sup_{\theta_1<\theta} \frac{(\theta-\theta_1)^2}{\left(\dfrac{\theta}{\theta_1}\right)^n - 1}$$

Let
$$\varphi(\theta_1) = \frac{(\theta-\theta_1)^2\theta_1^n}{\theta^n - \theta_1^n} \tag{4.4.20}$$

$$\frac{\partial}{\partial\theta_1}\varphi(\theta_1) = \frac{(n\theta_1^{n-1}(\theta-\theta_1)^2 - 2(\theta-\theta_1)\theta_1^n)(\theta^n-\theta_1^n) + n\theta_1^{n-1}\theta_1^n(\theta-\theta_1)^2}{(\theta^n-\theta_1^n)^2}$$

$$[n\theta_1^{n-1}(\theta-\theta_1)^2 - 2(\theta-\theta_1)\theta_1^n](\theta^n-\theta_1^n) + n\theta_1^{2n-1}(\theta-\theta_1)^2 = 0$$

$$[n(\theta-\theta_1) - 2\theta_1]\cdot(\theta^n-\theta_1^n) + n\theta_1^n(\theta-\theta_1) = 0$$

$$n(\theta-\theta_1)\theta^n - 2\theta_1\theta^n - n(\theta-\theta_1)\theta_1^n + 2\theta_1^{n+1} + n\theta_1^n\theta - n\theta_1^{n+1} = 0$$

$$n\theta^{n+1} - n\theta_1\theta^n - 2\theta_1\theta^n = 0$$

$$2\theta_1^{n+1} - (2+n)\theta_1\cdot\theta^n + n\theta^{n+1} = 0 \tag{4.4.21}$$

Solving this equation for θ_1 so that $\varphi(\theta_1)$ attains maximum, $\theta_1 = \theta^{(0)}$, the CRKLB is then given by

$$\text{CRKLB} = \frac{(\theta-\theta_1^{(0)})^2(\theta_1^{(0)})^n}{\theta^n - (\theta_1^{(0)})^n} \tag{4.4.22}$$

For example, when $n = 1$, Eq. (4.4.21) reduces to

$$2\theta_1^2 - 3\theta_1\cdot\theta + \theta^2 = 0$$

$$(2\theta_1 - \theta)(\theta_1 - \theta) = 0$$

$$\Rightarrow \qquad \theta_1 = \theta,\ \theta_1 = \frac{\theta}{2}$$

$$\varphi(\theta_1)\big|_{\theta_1=\theta} = 0;\ \varphi(\theta_1)\big|_{\theta_1=\frac{\theta}{2}} = \frac{\theta^4}{4}$$

This gives CKRLB for $n = 1$

$$\text{CRKLB} = \frac{\theta^4}{4}, \text{ as shown earlier}$$

For sufficient statistic $Y = X_{(n)}$ for θ with *pdf*

$$g(y) = n[F(y)]^{n-1} f(y)$$

$$E_\theta(Y) = \frac{n}{n+1}\theta$$

This gives that $\delta = [(n+1)/n]Y$ is an unbiased estimator for θ. Further,

$$E_\theta(Y^2) = \frac{n}{n+2}\theta^2$$

$$V_\theta(Y) = E_\theta(Y^2) - [E_\theta(Y)]^2 = \left[\frac{n}{n+2} - \left(\frac{n}{n+1}\right)^2\right]\theta^2$$

$$= n\theta^2\left[\frac{1}{(n+2)} - \frac{n}{(n+1)^2}\right]$$

$$= n\theta^2\left[\frac{n^2+2n+1-n^2-2n}{(n+1)^2(n+2)}\right]$$

$$= \frac{n\theta^2}{(n+1)^2(n+2)}$$

This gives $$V_\theta(\delta) = \frac{(n+1)^2}{n^2}\frac{n\theta^2}{(n+1)^2(n+2)} = \frac{\theta^2}{n(n+2)}$$

At $n = 1$,

$$V_\theta(\delta) = \frac{\theta^2}{3} > \frac{\theta^4}{4} = \text{CRKLB}$$

Note that δ is the UMVUE. Therefore, the CRKLB is not attainable.

***Example* 4.26** Let $X_1, X_2, \ldots, X_n$ constitute a random sample from the density

$$f(x;\theta) = \frac{x}{\theta^2}\exp\left(-\frac{x^2}{2\theta^2}\right), x > 0, \theta > 0$$

Find the CR lower bound to the variance of an unbiased estimator of θ^2.

Solution. We have

$$\frac{\partial}{\partial\theta}\log f(x;\theta) = \frac{\partial}{\partial\theta}\left(\log x - \log\theta^2 - \frac{x^2}{2\theta^2}\right) = \frac{x^2}{\theta^3} - \frac{2}{\theta}$$

$$\frac{\partial^2}{\partial\theta^2}\log f(x;\theta) = -\frac{3x^2}{\theta^4} + \frac{2}{\theta^2}$$

$$I_X(\theta) = E_\theta\left[-\frac{\partial^2}{\partial\theta^2}\log f\right] = -\frac{2}{\theta^2} + \frac{3}{\theta^4}E_\theta(X^2)$$

$$E_\theta(X^2) = \int\left(\frac{x^3}{\theta^2}\right)\exp\left(\frac{-x^2}{2\theta^2}\right)dx$$

Let $(x^2/2\theta^2) = y$. This implies $xdx = \theta dy$

$$E_\theta(X^2) = 2\theta^2\int_0^\infty y^{2-1}\exp(-y)dy = 2\theta^2\Gamma(2) = 2\theta^2$$

$$I_{\mathbf{X}}(\theta) = nI_X(\theta) = n\left(-\frac{2}{\theta^2} + \frac{3}{\theta^4}2\theta^2\right) = \frac{4n}{\theta^2}$$

Therefore,

$$\text{CRLB} = \frac{[g'(\theta)]^2}{I_{\mathbf{X}}(\theta)} = \frac{4\theta^2}{4n/\theta^2} = \frac{\theta^4}{n}$$

***Example* 4.27** Let $X_1, X_2, \ldots, X_n$ be *iid rv*s with mean μ and finite variance v. Calculate the efficiency of the consistent estimator $\delta_1 = [2/\{n(n+1)\}]\ \Sigma_1^n iX_i$ relative to $\bar{X}$.

Solution. The efficiency of δ_1 relative to δ_2 is given by

$$\text{eff}(\delta_1, \delta_2) = \frac{V(\delta_1)}{V(\delta_2)}$$

where
$$V(\delta_1) = \frac{4}{n^2(n+1)^2}\sum_1^n i^2 V(X_i)$$

$$= \frac{4}{n^2(n+1)^2}v\sum_1^n i^2 = \frac{4}{n^2(n+1)^2}\frac{n(n+1)(2n+1)}{6}v$$

$$= \frac{2(2n+1)}{3n(n+1)}v$$

Further,
$$V(\delta_2) = V(\bar{X}) = \frac{1}{n^2}\sum_1^n v = \frac{1}{n}v$$

Therefore,
$$\text{eff}(\delta_1|\delta_2) = \frac{2(2n+1)}{3(n+1)} = \frac{2\left(2+\frac{1}{n}\right)}{3\left(1+\frac{1}{n}\right)}$$

$$\rightarrow \frac{4}{3} > 1 \text{ as } n \rightarrow \infty$$

This shows that δ_2 is asymptotically more efficient as compared to δ_1.

***Example* 4.28** Let $X_1, X_2, \ldots, X_n$ be a random sample from a three-point distribution; $P[X = y_1] = [(1 - \theta)/2]$, $P[X = y_2] = (1/2)$, $P[X = y_3] = (\theta/2)$, where $0 < \theta < 1$. Does the CR inequality apply in this case? If so, find out the lower bound for the variance of an unbiased estimator of θ.

Solution. Note that the support $S(\theta) = \{y_1, y_2, y_3\}$ does not involve θ.

$$\sum_x \frac{\partial}{\partial\theta} f(\mathbf{x}; \theta) = \sum_{n_1}\sum_{n_2}\sum_{n_3} \frac{\partial}{\partial\theta}\left(\frac{1-\theta}{2}\right)^{n_1}\left(\frac{1}{2}\right)^{n_2}\left(\frac{\theta}{2}\right)^{n_3} \quad \text{so that } n_1 + n_2 + n_3 = n$$

$$= \sum\sum\sum\left[n_1\left(\frac{1-\theta}{2}\right)^{n_1-1}\left(-\frac{1}{2}\right)\left(\frac{1}{2}\right)^{n_2}\left(\frac{\theta}{2}\right)^{n_3} + \left(\frac{1-\theta}{2}\right)^{n_1}\left(\frac{1}{2}\right)^{n_2}\frac{n_3}{2}\left(\frac{\theta}{2}\right)^{n_3-1}\right]$$

$$= \sum\sum\sum\left[-\frac{n_1}{2}\frac{2}{1-\theta} + \frac{n_3}{3}\frac{2}{\theta}\right]\left(\frac{1-\theta}{2}\right)^{n_1}\left(\frac{1}{2}\right)^{n_2}\left(\frac{\theta}{2}\right)^{n_3}$$

$$= -\frac{1}{(1-\theta)}E(n_1) + \frac{1}{\theta}E(n_3)$$

$$= -\frac{1}{(1-\theta)}n\frac{(1-\theta)}{2} + \frac{1}{\theta}E(n_3)$$

$$= -\frac{1}{(1-\theta)}n\frac{(1-\theta)}{2} + \frac{1}{\theta}n\frac{\theta}{2} = 0$$

$$= \frac{\partial}{\partial\theta}\sum_x f(\mathbf{x}; \theta) = \frac{\partial}{\partial\theta}1 = 0$$

Thus, we easily see that the regularity conditions, for the CR inequality to apply, are clearly satisfied.

Fisher information in the sample is computed by

$$I_{\mathbf{X}}(\theta) = nI_{X_1}(\theta)$$

where
$$\log L = n_1 \log\left(\frac{1-\theta}{2}\right) + n_2 \log\left(\frac{1}{2}\right) + n_3 \log\left(\frac{\theta}{2}\right)$$

$$\frac{\partial}{\partial\theta}\log L = n_1\frac{2}{1-\theta}\left(-\frac{1}{2}\right) + 0 + n_3\frac{1}{\theta} = -\frac{n_1}{1-\theta} + \frac{n_3}{\theta}$$

$$\frac{\partial^2}{\partial\theta^2}\log L = -\frac{n_1}{(1-\theta)^2} - \frac{n_3}{\theta^2}$$

This gives
$$I_{\mathbf{X}}(\theta) = nI_{X_1}(\theta) = nE\left(-\frac{\partial}{\partial\theta^2}\log L\right) = \frac{n}{2\theta(1-\theta)}$$

Therefore, $$\text{CRLB} = \frac{[g'(\theta)]^2}{I_{\mathbf{X}}(\theta)} = \frac{2\theta(1-\theta)}{n}$$

***Example* 4.29** Let $X_1, X_2, \ldots, X_n$ be a random sample from

$$f(x;\theta) = \exp[-(x-\theta)];\ x \ge \theta$$

Find the CRKLB for the variance of an unbiased estimator of θ.

Solution. $g(\theta) = \theta$, $g(\theta_1) = \theta_1$. Let

$$A(\theta,\theta_1) = \frac{f(\mathbf{x};\theta_1)}{f(\mathbf{x};\theta)} = \frac{\exp\left(-\sum(x_i-\theta_1)\right)\prod_1^n I_{(\theta_1,\infty)}(x_i)}{\exp\left(-\sum(x_i-\theta)\right)\prod_1^n I_{(\theta,\infty)}(x_i)}$$

$$E_\theta[A(\theta,\theta_1)] = \exp(n(\theta_1-\theta))\int_{\theta_1}^{\infty}\ldots\int_{\theta_1}^{\infty}\exp\left(-\sum(x_i-\theta)\right)d\mathbf{x}$$

$$= 1 \qquad [\text{since } S(\theta_1) \subset S(\theta) \Rightarrow \theta_1 \ge \theta]$$

Further, $$E_\theta[A^2] = \exp[2n(\theta_1-\theta) - n(\theta_1-\theta)]$$

$$= \exp[n(\theta_1-\theta)]$$

$$V_\theta(Z) = \exp[n(\theta_1-\theta)] - 1$$

Therefore, the CRKLB is given by

$$\text{CRKLB} = \sup_{\theta_1 \ge \theta}\frac{(\theta_1-\theta)^2}{\exp[n(\theta_1-\theta)]-1} \cong \frac{0.4715}{n^2}$$

Note that $X_{(1)}$ is sufficient for θ and $E_\theta(X_{(1)}) = (1/n)$. Therefore, $\delta = X_{(1)} - (1/n)$ is unbiased for θ with variance, $V_\theta = (1/n^2) > (0.4715/n^2)$. This shows that the CRKLB is not sharp in this case.

***Example* 4.30** Let $X_1, X_2, \ldots, X_n$ be *iid* $G_2(1, \theta)$ *rvs*.

(i) Show that the estimator $\delta(\mathbf{X}) = [(n-1)/(n\bar{X})]$ is a UMVUE for θ with variance $[\theta^2/(n-2)]$.

(ii) Show that the minimum variance from the FCR inequality is (θ^2/n).

Solution.

(i) $X \sim G_2(1,\theta)$; $\Sigma X_i \sim G_2(n,\theta)$. The joint *pdf* of sample observations is given by

$$f(\mathbf{x};\theta) = \theta^n \exp(-\theta\Sigma x_i)$$

By factorization theorem, $Y = \Sigma X_i$ is sufficient for θ. Further,

$$E_\theta\left[\frac{n-1}{n\bar{X}}\right] = E_\theta\left[\frac{n-1}{Y}\right]$$

$$= \frac{(n-1)\theta^n}{\Gamma(n)}\int_0^\infty y^{-1}y^{n-1}\exp(-\theta y)dy = \theta$$

By the Lehmann–Scheffe theorem, since $n\bar{X}$ is complete and sufficient and that $[(n-1)/n\bar{X}]$ is unbiased for θ, $T = [(n-1)/(n\bar{X})]$ is UMVUE for θ.

$$V_\theta\left(\frac{n-1}{n\bar{X}}\right) = \frac{\theta^2}{(n-2)}$$

(ii) The Fisher information is given by

$$I_{\mathbf{X}}(\theta) = nI_{X_1}(\theta) = \frac{n}{\theta^2}$$

Therefore, $$\text{CRLB} = \frac{1}{I_{\mathbf{X}}(\theta)} = \frac{\theta^2}{n}$$

Note that CRLB is not sharp in this case since it is smaller than the variance of a UMVUE, i.e., $\theta^2/(n-2)$.

EXERCISES

1. Show by examples the following situations for regular families:
 (i) There does not exist an unbiased estimator of $g(\theta)$.
 (ii) An unbiased estimator of $g(\theta)$ exists but there does not exist a UMVUE.
 (iii) The CR lower bound is not sharp, i.e., a UMVUE of $g(\theta)$ exists but its variance is greater than the CR lower bound.
 (iv) A UMVUE of $g(\theta)$ exists and its variance is equal to the CR lower bound.

2. Let $X_1,\ldots,X_n$ be *iid* $f(x;\theta)$ *rvs* with common mean μ and variance σ^2. Find the efficiency of the unbiased estimate $(2\Sigma_{i=1}^n iX_i)/[n(n+1)]$ relative to $\bar{X}_n$.

3. Show that if $X_1,\ldots,X_n$ are *iid* with *pdf* $f(x;\theta) = \exp\{A(\theta)B(x) + C(\theta) + D(x)\}$, then $T = \Sigma_{i=1}^n[B(X_i)/n]$ attains the Rao–Cramer lower bound for an unbiased estimator of $E(T)$.

Hint. Show that if $T(x_1,\ldots,x_n)$ is a linear function of $(\partial/\partial\theta)\log f(\mathbf{x};\theta)$, equality holds in the RC–inequality.

4. If X is a random variable assuming values 0, 1, 2, 3 with probabilities θ, θ^2, θ^3, $1-\theta-\theta^2-\theta^3$, respectively. Find the RC–bound to the variance of an unbiased estimator of θ.

5. Let $X_1,\ldots,X_n$ be *iid* Bernoulli $b(1, p)$ variables. Show that $T_n = S_n(n - S_n)(n - 1 - S_n)$, where $S_n = \Sigma_{i=1}^{n} X_i$, is an unbiased estimator of pq^2 and find the RC–lower bound to the variance of this estimator.

6. Let $X_1,\ldots,X_n$ be independent *rvs* with common *pdf* $P(X_i = 1) = p$, $P(X_i = 0) = 1 - p$, $0 \le p \le 1$, $i = 1,\ldots,n$. Let $T(X_1,\ldots,X_n)$ be an unbiased estimator of p. Find a lower bound to the variance of T and check the regularity conditions.

7. Compute the CRK lower bound for the variance of an estimator of θ based on X with $P_\theta[X = 0] = 1 - \theta$, $P_\theta[X = 1] = \theta$, $0 < \theta < 1$.

8. Suppose that $X_1,\ldots,X_n$ are *iid* $b(1,\theta)$, where $n \ge 2$ and $0 < \theta < 1$ is the unknown parameter.
(i) Derive the UMVUE of $g(\theta)$, where $g(\theta) = \theta(1 - \theta)$.
(ii) Find the CRLB for estimating $g(\theta) = \theta(1 - \theta)$.

9. Let $X_1,\ldots,X_n$ be *iid* $b(1,\theta)$ random variables.
(i) Find the Fisher information $I_X(\theta)$.
(ii) Find the CRLB for unbiased estimators of $g(\theta) = \theta$.
(iii) The MLE for θ is $\bar{X}$. Find $V(\bar{X})$.
(iv) Does the MLE achieve the CR lower bound? Is this surprising? Explain.

10. Let $X_1, X_2,\ldots,X_n$ be *iid* $P(\theta)$. Choose the correct option
(i) The estimator $T_0 = [(n - 1)/n]^{\Sigma X_i}$ of $\exp(-\theta)$ is
(a) CRLB estimator (b) neither UMVUE nor CRLB
(c) UMVUE and most efficient (d) UMVUE but not most efficient
(ii) Which of the following classes of estimators correctly characterize the family of MVB estimators?
(a) $\bar{X}$
(b) $A\bar{X} + B$, where A and B are constants
(c) $f(\bar{X})$, where f is uni-valued function of $\bar{X}$
(d) All polynomials in $\bar{X}$, where $\bar{X} = n^{-1}\Sigma_{i=1}^{n} X_i$

11. Let $X_1,\ldots,X_n$ be *iid* $P(\theta)$ random variables.
(i) Find the UMVUE for θ.
(ii) Find the Fisher information $I_X(\theta)$.
(iii) Find the CRLB for the variance of an unbiased estimator of $g(\theta) = \theta$.
(iv) The MLE for θ in $\bar{X}$. Find $V(\bar{X})$.
(v) Does the MLE achieve the CRLB? Is this surprising? Explain.

12. Let X follows a Poisson distribution with parameter θ and suppose it is desired to estimate $P(X = 0) = \exp(-\theta)$ on the basis of a sample of size one. If the estimator T is defined as

$$T = \begin{cases} 1, & \text{if } X = 0 \\ 0, & \text{otherwise} \end{cases}$$

then which one of the following statements about T is correct?
(i) T is biased for $\exp(-\theta)$.
(ii) $T/2$ is unbiased for $\exp(-\theta)$ and its variance attains the CR lower bound.

(iii) The variance of T attains the CR lower bound and, hence, T is the minimum variance estimator of $\exp(-\theta)$.

(iv) T is only an unbiased estimator of $\exp(-\theta)$ and, hence, is the uniformly minimum variance estimator of $\exp(-\theta)$.

13. Let $X_1,\ldots,X_n$ be independent samples from $P(\theta)$. We want to estimate the function $\exp(-\theta)$.

(i) Define the function

$$U(X) = \begin{cases} 1, & \text{if } X = 0 \\ 0, & \text{otherwise} \end{cases}$$

Find the UMVUE of $\exp(-\theta)$ by using the Rao–Blackwell theorem.

Hint. Use Rao–Blackwellization with respect to the complete sufficient statistic $\Sigma_{i=1}^{n} X_i$.

(ii) Compute the CR lower bound for the variance of unbiased estimators of $\exp(-\theta)$. Compute the variance of UMVUE in (i) and check whether it achieves this bound. (Notice that the variance of UMVUE, though it is best, does not achieve the CR lower bound. This is so because UMVUE is not a function of sufficient statistic in the Poisson density when it is written in the form of an exponential density. This is the example where it shows that CR lower bound is not sharp and one can obtain sharper lower bounds, say due to Bhattacharyya.)

(iii) Find the MLE of $\exp(-\theta)$ and using the delta method, show that the approximate large sample variance of the MLE of $\exp(-\theta)$ is the same as the CR lower bound given in (ii).

14. Let $X_1,\ldots,X_n$ be a random sample from $P(\theta)$ population.

(i) Find CRLB(θ^2) for the variance of an unbiased estimator of $g(\theta) = \theta^2$.

(ii) The UMVUE of θ^2 is $\delta(\mathbf{X}) = \bar{X}^2 - (\bar{X}/n)$. Is $V_\theta(\delta)$ = FCRLB(θ)?

15. A sample of size one is drawn from a geometric population having *pmf* $f(x;\theta) = \theta(1-\theta)^x$, $x = 0, 1, 2,\ldots, 0 < \theta < 1$. Consider the following estimator of θ:

$$T_1(X) = \begin{cases} 1, & \text{if } X = 0 \\ 0, & \text{otherwise} \end{cases}$$

$$T_2(X) = \frac{1}{1+X}$$

(i) Which one is the best statement among the following with regard to the above estimators of θ?

(a) $T_2(X)$ is MLE and is unbiased.

(b) $T_1(X)$ is UMVUE.

(c) $T_1(X)$ attains the CRLB.

(d) T_2 is MVBE (minimum variance bound estimator).

(ii) Show that there exists a parametric function $g(\theta)$ such that there exists an MVB unbiased estimator δ of $g(\theta)$. Find $g(\theta)$, δ, and $V(\delta)$.

16. Let $X_1,\ldots,X_n$ be n independent random variables such that for any i,

$$P(X_i = r) = p(1-p)^r,\ r = 0, 1,\ldots, 0 < p < 1$$

Find a sufficient statistics for p and, hence, obtain an unbiased estimator of $1/p$. Also obtain the variance of the estimator and compare it with the Rao–Cramer lower bound for the unbiased estimator of $1/p$.

17. Let $X_1,\ldots,X_n$ be *iid* $N(\theta,1)$. Show that the best unbiased estimator of θ^2 is $\bar{X}^2 - (1/n)$. Calculate its variance and show that it is greater than the CRLB.

18. Let $X_1,\ldots,X_n$ be n independent observations from a normal $N(\theta, 1)$ population. Show that $\bar{X}$ is complete sufficient for θ, $-\infty < \theta < \infty$. It is desired to estimate the proportion p of the distribution $N(\theta, 1)$ which lies above a given number. A statistic T_1 is defined as follows:

$$T_1 = \begin{cases} 0, & \text{if } X_1 \le c \\ 1, & \text{if } X_1 > c \end{cases}$$

where X_1 is the first observation.

(i) Show that T_1 is an unbiased estimator of p.

(ii) Deduce the expression for the statistic T defined by $T = E(T_1|\bar{X})$ and show that the variance of an unbiased estimator of θ^k attain the $(k-1)$th, Bhattacharyya lower bound if it is a polynomial of degree k in $\bar{X}$.

19. Let $X_1, X_2,\ldots,X_n$ be *iid* $N(\theta,\sigma^2)$ where (θ,σ^2) are both unknown. Calculate the CR lower bound for estimating σ^2. Is that attainable?

20. Assume $X_i \sim N(\theta,\sigma^2)$, where (θ,σ^2) are both unknown. Which of the following estimators are unbiased for θ?

(i) $T_1 = \dfrac{X_1 + X_2 + X_3 + X_4}{4}$

(ii) $T_2 = \dfrac{2(X_1 + X_2)}{6} + \dfrac{X_3 + X_4}{6}$

(iii) $T_3 = \dfrac{X_1 - X_2 + X_3 - X_4}{4}$

Which of the above estimators is the most efficient? Which is the most efficient estimator among all unbiased estimators?

21. Let $X_1,\ldots,X_n$ be *iid* $N(\theta, \sigma^2)$ random variable where θ is known and $\sigma^2 > 0$ is unknown. Consider the estimation of $g(\sigma^2) = \sigma^{2r}$, $r > 0$. The estimator

$$\delta(\mathbf{X}) = \frac{\Gamma(n/2)S^k}{2^k\Gamma(r + n/2)}$$

is the UMVUE of σ^{2r}, when $S = \Sigma_{i=1}^n (X_i - \theta)^2$, a complete sufficient statistic and $S \sim \sigma^2\chi_n^2$. Note that if $Y \sim \chi_n^2$, then

$$E(Y^r) = \frac{\Gamma[r+(n/2)]2^r}{\Gamma(n/2)}$$

(i) Show that $$V_{\sigma^2}[\delta(\mathbf{X})] = \delta^{4r}\left[\frac{\Gamma(n/2)\Gamma[2r+(n/2)]}{\Gamma[r+(n/2)]\Gamma[r+(n/2)]} - 1\right]$$

$$= c(r)\sigma^{4r}$$

(ii) For $r = 2$, calculate CRLB(σ^2) and show that

$$V_{\sigma^2}[\delta(\mathbf{X})] - \text{CRLB}(\sigma^4) > 0$$

22. (i) Let $X_1, X_2, \ldots, X_n$ be independent sample from $N(0, \sigma^2)$ where $\sigma > 0$ is unknown. Let $T = n^{-1}\Sigma_{i=1}^n X_i^2$ and $\bar{X} = n^{-1}\Sigma_{i=1}^n X_i$. Is T UMVUE for σ^2? Obtain the CR lower bound for the variance of unbiased estimators of σ^2. Is this bound attainable? Also, determine the CR lower bound for estimating σ. Find the UMVUE for σ and calculate its efficiency.

(ii) Let $X_1, X_2, \ldots, X_n$ be a random sample from $N(0, \sigma)$. Determine whether the variance of the UMVUE of σ achieves the CR lower bound and also find its efficiency.

23. If $X_1, \ldots, X_n$ are n *iid rvs* each having an exponential density $f(x;\theta) = (1/\theta)\exp(-x/\theta)$, $x > 0$.

(i) Find the total score $S(\mathbf{X}, \theta) = (\partial/\partial\theta)\log f(\mathbf{x};\theta)$ for the entire sample and the information contained in the sample about θ, $E[(\partial/\partial\theta)S(\mathbf{X},\theta)]$.

(ii) Find the variance of the most efficient unbiased estimator of θ by means of the CR lower bound. Prove that the sample mean is an efficient estimator of θ.

24. Show that $\bar{X} = n^{-1}\Sigma X_i$ in random sampling from

$$f(x;\theta) = \frac{1}{\theta}\exp\left(-\frac{x}{\theta}\right), \qquad 0 < x < \infty$$

where $0 < \theta < \infty$, is a minimum variance bound (MVB) estimator of θ and has variance (θ^2/n). Show that Fisher information number in $I_{\mathbf{X}}(\theta)$ is (n/θ^2).

25. Let $X_1, \ldots, X_n$ be a random sample from a population with *pdf*

$$f(x;\theta) = \theta\exp(-\theta x); \; x > 0, \; \theta > 0$$

Find the CRLB for the variance of an unbiased estimator of θ.

26. Let $X_1, \ldots, X_n$ be *iid* $E(\theta)$ ($G_1(1,\theta)$) random variables.

(i) Find $I_{\mathbf{X}}(\theta)$.

(ii) Find the CRLB for estimating $g(\theta) = \theta^2$.

(iii) If $T = \Sigma_{i=1}^n X_i$, it can be shown that the UMVUE of θ^2 is

$$W = \frac{\Gamma(n)}{\Gamma(n+2)}T^2$$

Is $V_\theta(W)$ equal to the CRLB in part (ii)?

Hint. $E(T^r) = [1/(\Gamma(n)\theta^r)]\int_0^\infty t^r t^{n-1}\exp(-t/\theta)dt = [\Gamma(r+n)/\Gamma(n)]\theta^2$, $E(T^2) = [\Gamma(n+2)/\Gamma(n)]\theta^2$

27. Based on a random sample of size n from the distribution with *pdf*

$$f(x;\alpha) = \frac{1}{\Gamma(\alpha)} \exp(-x)\, x^{\alpha-1}, \qquad x \geq 0,\ \alpha > 0$$

obtain an estimator of $(\partial/\partial\alpha)\log\Gamma(\alpha)$ which attains the RC lower bound.

28. If $f(x;\theta) = [\theta^p/\Gamma(p)]\exp(-\theta x)x^{p-1}$, where p is known and θ is unknown. Show that there exists an unbiased estimate of $(1/\theta)$ based on a sample of size one whose variance is equal to the CR lower bound.

29. The *rv* X follows gamma distribution $G_1(p, \theta)$ with *pdf*

$$f(x;\theta) = \frac{1}{\Gamma(p)\theta^p} x^{p-1} \exp\left(-\frac{x}{\theta}\right); 0 < x < \infty, \theta > 0, p \text{ is known}$$

Find the CRLB for estimating θ^2 so that

$$V(\delta) \geq [nI_X(\theta^2)]^{-1}$$

Use X^2 to construct an unbiased estimator δ of θ^2. Find $V(\delta)$ and show that it satisfies the above inequality.

30. Let $X_1,\ldots,X_n$ be a random sample from a gamma distribution

$$f(x;\alpha,p) = \begin{cases} \dfrac{1}{p^{\alpha+1}\Gamma(\alpha+1)} x^{\alpha} \exp(-x/p), & 0 \leq x \leq \infty \\ 0, & \text{otherwise} \end{cases}$$

Let T_{1n} be an unbiased estimator α, where p is known, and T_{2n} be an unbiased estimator of p, where α is known. Find the Rao–Cramer lower bound for the variance of these estimators.

31. Let $X_1,\ldots,X_n$ be a random sample from the distribution with *pdf*

$$f(x;\theta) = \theta x^{\theta-1}, 0 < x < 1, \theta > 0$$

Show that the variance of the UMVUE does not achieve the CR lower bound. Comment on the efficiency of UMVUE.

32. Let $X_1,\ldots,X_n$ be *iid* from a distribution with *pdf*

$$f(x;\theta) = \theta x^{\theta-1},\quad 0 < x < 1,\quad \theta > 0$$

(i) Find the MLE of θ.
(ii) What is the MLE of θ^{-1}? Explain.
(iii) Find the Fisher information $I_X(\theta)$. You may use the fact that $\log(X) \sim E(1/\theta) \equiv G_2(1,\theta)$
(iv) Find the CRLB for the unbiased estimators of $g(\theta) = 1/\theta^2$.

33. Let $X_1,\ldots,X_n$ be a random sample of size n from a *pdf* $f(x;\theta) = \theta/(1+x)^{1+\theta}$, if $x > 0$ (= 0, otherwise), for some $\theta > 0$. Find the CR lower bound to the variance of unbiased estimator of (i) $1/\theta$, (ii) $\exp(\theta)$, and (iii) θ^2.

34. Let $X_1,\ldots,X_n$ be *iid* $f(x;\theta) = [(\log\theta)/(\theta-1)]\,\theta^x I_{(0,1)}(x)$, $\theta > 0$. Find a function of θ for which there exists an unbiased estimator whose variance coincides with the CR lower bound.

35. (Casella) Let $X_1,\ldots,X_n$ be a random sample drawn from the following distributions. Is there a function g of θ, $g(\theta)$, for which there exists an unbiased estimator whose variance attains CRLB? If so, find it. If not, show why not:

(i) $f(x;\theta) = \theta x^{\theta-1}$, $0 < x < 1$

(ii) $f(x;\theta) = [(\log\theta)/(\theta-1]\theta^x$, $0 < x < 1$, $\theta > 1$

36. Let $X_1,\ldots,X_n$ be *iid rvs* from a distribution $f(x;\theta) = (1-\theta) + (\theta/2\sqrt{x})$, $0 < x < 1$, $0 < \theta < 1$. Find the lower bound for $E_\theta(X_n - \theta)^2$. (Note that X_n is a biased estimator of θ).

37. Let $X_1,\ldots,X_n$ be *iid* with common *pdf* $\exp[-(x-\theta) - \exp(x-\theta)]$. Find the CR lower bound for the variance of an unbiased estimator of (i) θ, (ii) θ^2, (iii) $\theta^2 + \theta$, and (iv) $\exp(\theta)$.

38. Let $X_1,\ldots,X_n$ be *iid rvs* with *pdf*

$$f(x;\theta_1,\theta_2) = \frac{1}{\sqrt{2\pi\theta_2}}\exp\left[-\frac{(x-\theta_1)^2}{2\theta_2}\right], -\infty < x < \infty; -\infty < \theta_1 < \infty; 0 < \theta_2 < \infty$$

Let $T(X_1,\ldots,X_n)$ be an unbiased estimator of θ_2. Find the Rao–Cramer lower bound to $V(T)$.

39. Find the Chapman–Robbins–Kiefer lower bound of variance of an unbiased estimator of θ based on a sample of size n from the distribution

(i) $f(x;\theta) = \begin{cases} \exp(-(x-\theta)), & \theta < x < \infty \\ 0, & \text{otherwise} \end{cases}$

(ii) $N(\mu, 1)$

40. For the *pdf*

$$f(x;\theta) = \frac{1}{\pi\{1+(x-\theta)^2\}}, \quad -\infty < x < \infty$$

derive the Rao–Cramer lower bound of variance of an unbiased estimator of θ.

41. Let $X_1,\ldots,X_n$ be independent identically distributed Rayleigh random variables with *pdf*

$$f(x;\theta) = \begin{cases} \dfrac{2x}{\theta}\exp\left\{-\dfrac{x^2}{\theta}\right\}, & x > 0 \\ 0, & x \le 0 \end{cases}$$

$\theta > 0$ and is unknown. $E(X) = \sqrt{(\pi/2)}\,\theta$ and $V(X) = [(4-\pi)/2]\theta$.

(i) Show that $f(x;\theta)$ belongs to the one-parameter exponential family, and $\Sigma_{i=1}^n X_i^2$ is a complete sufficient statistic.

(ii) Show that X_1^2 is an unbiased estimator of θ.

Hint. Use the substitution $W = X^2$ and find the *pdf* of W or use μ substitution with $\mu = (x^2/\theta)$.

(iii) Show that the CRLB for estimating θ is (θ^2/n).
(iv) Find the UMVUE of θ.
(v) Also, show that $\delta(\mathbf{X}) = (1/2n)\Sigma_{i=1}^{n} X_i^2$ is the UMVUE and the CRLB estimator of θ.

42. Let $X_1,\ldots,X_n$ be a random sample from the distribution with *pdf*

$$f(x;\theta) = \frac{1}{2}\theta^3 x^2 \exp\{-\theta x\},\ x > 0,\ \theta > 0$$

Which of the following three functions of θ:

$$g_1(\theta) = \theta,\ g_2(\theta) = \frac{1}{\theta},\ g_3(\theta) = \ln\theta$$

admits an efficient estimator? Why? Find this estimator.

43. Suppose $X_1,\ldots,X_n$ are n independent observations from a probability distribution with density

$$f(x;\theta) = \begin{cases} 3x^2\theta\exp(-\theta x^3), & \text{for } x > 0 \\ 0, & \text{otherwise} \end{cases}$$

where $\theta > 0$ is an unknown parameter.

(i) Find $h(\theta) = E(X_1^n)$.
(ii) Find uniformly minimum variance unbiased estimator (UMVUE) of $h(\theta)$.
(iii) Calculate the CR bound for the variance of an unbiased estimator of $h(\theta)$. Does the UMVUE of $h(\theta)$ attain this bound?

44. Let $X_1,\ldots,X_n$ be a random sample drawn from the population with *pdf*

$$f(x;\theta) = \begin{cases} \dfrac{\theta}{(1+x)^{1+\theta}}, & x > 0, \theta > 0 \\ 0, & \text{otherwise} \end{cases}$$

Find the Cramer–Rao lower bound to the variance of the unbiased estimator of θ^{-1}, $\exp(\theta)$ and θ^2.

45. Let $X_1,\ldots,X_n$ be independent observations from $U(0,\theta)$ and $g(\theta) = \theta^2$.

(i) Find $E_\theta[(\partial/\partial\theta)\log(\mathbf{X};\theta)$ where $\log f(\mathbf{X};\theta)$ is the log likelihood of θ for the data.
(ii) Find the MLE of θ and hence of $g(\theta)$.
(iii) Find $E[\hat{g}(\theta)]$ and hence derive an unbiased estimator of $g(\theta)$ of the form $k\hat{g}(\theta) = t(x_1, x_2,\ldots,x_n) = t$(say) for suitable constant k.
(iv) Find $V[\hat{g}(\theta)]$.
(v) Find $\{g'(\theta)\}^2/\{E_\theta^{[(\partial/\partial\theta)\log f(\mathbf{x};\theta)]}\}^2$
(vi) Explain why the values in (iv) and (v) are in apparent contradiction to the result of Cramer–Rao theorem.

46. (i) Show that Cramer–Rao inequality holds good for a distribution $f(x;\theta)$ whose range (a,b) depends on θ, provided that $E_\theta[(\partial/\partial\theta)\log f(\mathbf{x};\theta)] = 0$, where $f(\mathbf{x};\theta)$ holds. Show that this is so iff $f(a;\theta) = f(b;\theta) = 0$. If in addition $[(\partial/\partial\theta)f(\mathbf{x};\theta)|_{\mathbf{x}=(a,\ldots,a)} = (\partial/\partial\theta)f(\mathbf{x};\theta)|_{\mathbf{x}=(b,\ldots,b)} = 0]$, then $E[(\partial/\partial\theta)\log f(\mathbf{x};\theta)]^2 = -E_\theta[(\partial^2/\partial\theta^2)\log f(\mathbf{x};\theta)]$ and the MVB for $g(\theta)$ can be written in the form

$$V_\theta(T) \geq -\frac{\{g'(\theta)\}^2}{E\left(\dfrac{\partial^2 \log f(\mathbf{X};\theta)}{\partial\theta^2}\right)}$$

(ii) Apply this result to show that the MVB holds for the estimation of θ in

$$dF(x) = \frac{1}{\sqrt{p}}\,(x-\theta)^{p-1}\exp\{-(x-\theta)\}dx,\ x \geq \theta;\ p > 2$$

and is equal $[(p-2)/n]$, but is not attainable since there is no single sufficient statistic for θ.

5 Asymptotic Theory and Consistency

5.1 INTRODUCTION

In the previous chapters, we have discussed small sample properties for evaluating estimators, where it has been noted that in some statistical models it is difficult to find an efficient estimator for some parameter θ. In such cases, the change of evaluation criterion is almost essential and we look at such properties of an estimator that it has in the presence of moderate data, *i.e.*, when we let the sample size to increase larger and larger without bounds; we expect that we eventually know the truth about the parameter θ as we get more and more data. The phenomenon of increasing sample size is known as *asymptopia* and such criterion is known as *asymptotic properties* of estimators. The related theory is called *asymptotic theory*. There are two main concepts related to asymptotic theory: one, the *consistency* and second, *the asymptotic normality*. Consistency provides us the answer as to how the estimator works, i.e., does it provide us with the correct answer if we give it enough data? Further, asymptotic normality provides us the answer whether an estimator is efficient in some asymptotic sense?

We will discuss these properties of an estimator in the present chapter. If an estimator converges or gets closer and closer to the true value of the parameter, when n gets larger and larger, it is known as consistency. Consistency talks of limiting behaviour of an estimator as the sample size increases without bounds. It does not mean that the observed value of the estimator is necessarily close to the true value of the parameter based on some specific sample of size n. If only a small sample is available, then for an estimator based on these sample observations, it does not matter whether this estimator is consistent or not. However, if the estimator is not consistent, then there is no immediate advantage of increasing the sample size to base our estimator on this sample.

Further, consider the distributional behaviour of the average of random variables with increasing number of variables to be averaged. If these averages behave normally, this phenomenon is known as *asymptotic normality*. We will be using two statistical tools: one the law of large numbers for proving consistency results and second, the *central limit theorem* (CLT) for proving asymptotic normality results.

Let us consider a random sample $X_1,\ldots,X_n$ of size n from the family of distributions $\{f(x;\theta), \theta \in \Theta\}$. Before we proceed to define consistency, we recall the definition of convergence of a sequence of random variables $\{X_n\}_{n=1}^{\infty}$ to some constant a in probability.

Definition 5.1.1 (Convergence in probability) A sequence of random variables $\{X_n\}_{n=1}^{\infty}$ is said to *converge in probability* to a constant a if for every $\varepsilon > 0$,

$$\lim_{n\to\infty} P_\theta\left(|X_n - a| < \varepsilon\right) = 1$$

This convergence is denoted by $X_n \xrightarrow{P} a$ as $n \to \infty$. Now, define a sequence of estimators $\{\delta_n\}_{n=1}^{\infty}$

$$\begin{aligned} \delta_1 &= g(X_1), \\ \delta_2 &= g(X_1, X_2), \\ \vdots \quad & \quad \vdots \\ \delta_n &= g(X_1,\ldots,X_n) \end{aligned}$$

Note that every term in this sequence is a random variable. The *asymptotic unbiasedness* of estimators δ_n is defined as follows:

Definition 5.1.2 (Asymptotic unbiasedness) An estimator δ_n is said to be asymptotically unbiased for θ if its bias vanishes as the sample size increases to infinity, i.e.,

$$B(\delta_n) = E(\delta_n) - \theta \to 0 \text{ as } n \to \infty$$

Having introduced some elementary concepts related to the theory of asymptotic studies of estimators in Sections 1.5, 1.6, and in the present section, we introduce the concept of consistency of an estimator in Section 5.2. In Section 5.3, we discuss the method for checking the consistency of an estimator. Section 5.4 discusses the invariance principle of consistency under continuous functions. In Section 5.5, we define the concept of rate of consistency. Section 5.6 discusses the methods of constructing consistent estimators. Section 5.7 defines optimality among consistent estimators and also discusses the asymptotic normality of consistent asymptotically normal (CAN) estimators. In Section 5.8, the methods of finding CAN estimators and their properties are discussed.

5.2 CONSISTENCY OF AN ESTIMATOR

We consider here a sequence of estimators $\{\delta_n\}_{n=1}^{\infty}$ in order to study the behavior of $\delta_n(\mathbf{X})$ as sample size n increases to infinity. The large sample property of a sequence of estimators is defined here and is known as *consistency*. The property of consistency not only considers the behaviour of bias in the estimator as the sample size grows large, but it also looks at its variance. See Theorem 5.3.1.

Definition 5.2.1 (Consistency) A sequence of estimators $\{\delta_n\}$ is said to be consistent for a parameter θ if for every $\varepsilon > 0$ howsoever small and $\theta \in \Theta$,

$$\lim_{n\to\infty} P_\theta\left(|\delta_n - \theta| < \varepsilon\right) = 1$$

Symbolically, it is denoted by $\delta_n \xrightarrow{P} \theta$ as $n \to \infty$ for every $\theta \in \Theta$. Here, $\{|\delta_n - \theta|\}$ is $o_p(1)$, i.e., $\delta_n - \theta = o_p(1)$ or $\delta_n = \theta + o_p(1)$.

Let $X_1, X_2, \ldots, X_n$ be *iid* $P(\theta)$ with mean θ. Then, $\exp(\bar{X}_n) = \exp(\theta) + O_P(1)$ and $\exp(\bar{X}_n) = \exp(\theta) + O_P(n^{-1/2})$. See also Slutsky's theorem in Chapter 1.

In the definition of consistency $\delta_n \xrightarrow{P} \theta$, it does not imply that for every given $\varepsilon > 0$, we can find a stage N such that $|\delta_n - \theta| < \varepsilon$, $\forall\, n > N$; rather it speaks only of the convergence of the sequence of probabilities $\{P(|\delta_n - \theta| < \varepsilon)\}$ to unity, i.e., for given $\varepsilon > 0$, $\eta > 0$, howsoever small ε and η, there exists a stage $n_0(\varepsilon, \eta, \theta)$ such that

$$P_n(\delta_n, \varepsilon, \theta) = P_\theta|\,|\delta_n - \theta| \le \varepsilon] \ge 1 - \eta\ \forall\, n \ge n_0\ (\varepsilon, \eta, \theta) \text{ and } \forall\, \theta \in \Theta$$

This property is known as *weak consistency* or *simple consistency*. In fact, the quantity $P_n(\delta_n, \varepsilon, \theta)$ is used as a criterion for choosing between estimators among all the consistent estimators of θ. The faster $P_n(\delta_n, \varepsilon, \theta)$ approaches 1 as n approaches ∞, the better is the estimator δ_n. So, the level of accuracy achieved by the estimator δ_n [i.e., $P_n(\delta_n, \varepsilon, \theta) > 1 - \eta$ $\forall\, \theta$] for some value of ε and η is measured by knowing the stage, minimum $n_0(\varepsilon, \eta, \theta)$, at which $P_n(\delta_n, \varepsilon, \theta) > 1 - \eta\ \forall\, n \ge n_0\ (\varepsilon, \eta, \theta)\ \forall\, \theta$. Indeed, if δ_1 is an estimator that achieves a level of accuracy specified by (ε, η) at smaller sample size as compared to the sample size required by some other estimator δ_2 that also achieves the same level of accuracy for every θ, we say that δ_1 is preferred over δ_2. Sometimes, this phenomenon is referred to as the rate of convergence of an estimator to the true value of θ. If δ_1 converges to true θ faster than δ_2 in probability as n approaches infinity, then δ_1 is preferred over δ_2.

Next, we discuss the consistency of an estimator when $\boldsymbol{\theta}$ is a vector. Suppose $\boldsymbol{\theta} = (\theta_1, \ldots, \theta_k)' \in \Theta \subseteq \mathbb{R}^k$ and $\boldsymbol{\delta}_n = (\delta_n^{(1)}, \delta_n^{(2)}, \ldots, \delta_n^{(k)})$, is a vector statistic for estimating $\boldsymbol{\theta}$.

Definition 5.2.2 (Marginal consistency) $\boldsymbol{\delta}_n$ is said to be marginal consistent for $\boldsymbol{\theta}$ if

$$P_\theta|\,|\delta_n^{(i)} - \theta_i| < \varepsilon] \xrightarrow{P} 1\ \forall\, \varepsilon > 0, \forall\, \boldsymbol{\theta} \in \Theta, \forall\, i = 1, \ldots, k \text{ as } n \to \infty$$

Thus, the vector-valued statistic $\boldsymbol{\delta}_n$ is said to be consistent for the vector-valued parameter $\boldsymbol{\theta}$ if $\boldsymbol{\delta}_n$ is component-wise consistent for $\boldsymbol{\theta}$.

Let us define a distance function between $\boldsymbol{\delta}_n$ and $\boldsymbol{\theta}$ by the Euclidean distance

$$d(\boldsymbol{\delta}_n, \boldsymbol{\theta}) = \|\boldsymbol{\delta}_n - \boldsymbol{\theta}\| = \{(\boldsymbol{\delta}_n - \boldsymbol{\theta})' (\boldsymbol{\delta}_n - \boldsymbol{\theta})\}^{1/2}$$

$$= \left\{\sum_{i=1}^{k} (\delta_n^{(i)} - \theta_i)^2\right\}^{1/2}$$

This is called the distance norm and is equivalent to $\max_i |\delta_n^{(i)} - \theta_i|$. Using this distance norm, we can frame another definition of consistency.

Definition 5.2.3 (Joint consistency) $\boldsymbol{\delta}_n$ is said to be jointly consistent for $\boldsymbol{\theta}$ if

$$P_{\theta}[\|\boldsymbol{\delta}_n - \boldsymbol{\theta}\| < \varepsilon] \xrightarrow{P} 1 \ \forall\, \varepsilon > 0,\ \forall\, \boldsymbol{\theta} \in \Theta \quad \text{as } n \to \infty$$

We state the following result without proof.

Theorem 5.2.1 $\boldsymbol{\delta}_n$ is marginally consistent for $\boldsymbol{\theta}$ if and only if $\boldsymbol{\delta}_n$ is jointly consistent for $\boldsymbol{\theta}$.

The mean square error of an estimator plays an important role in proving the consistency of an estimator. It is discussed in the next section. The mean square error of an estimator $\boldsymbol{\delta}_n$ is expressed by

$$\text{MSE}_{\theta}(\delta_n) = E_{\theta}(\delta_n - \theta)^2 = E_{\theta}[\delta_n - E(\delta_n)]^2 + \{E_{\theta}(\delta_n - \theta)\}^2$$
$$= V_{\theta}(\delta_n) + \{\text{Bias}_{\theta}(\delta_n)\}^2$$

We have seen in the previous chapters that getting a small MSE of δ_n often involves a trade-off between variance and bias, and by not insisting δ_n to be unbiased, the variance of δ_n can be substantially reduced. This establishes the reason why biased estimators are preferred over unbiased estimators. Note that no trade-off is possible in case of unbiased estimators, since MSE and variance are both equal.

Choosing between estimators may also be restated using Chebyshev's inequality in terms of the rate at which $\text{MSE}_{\theta}(\delta_n)$ approaches zero as n approaches infinity. Therefore, estimator δ_1 is preferred over δ_2 if for $\forall\ n \geq n_0$,

$$\text{MSE}_{\theta}(\delta_1) \leq \text{MSE}_{\theta}(\delta_2)\ \forall\ \theta$$

and $$\text{MSE}_{\theta}(\delta_1) < \text{MSE}_{\theta}(\delta_2) \text{ for at least one } \theta$$

A detailed account of the rate of convergence of probability $P_n(\delta_n, \varepsilon, \theta)$ to 1 as $n \to \infty$ as a criterion of choosing between estimators has been elaborated in Section 5.5. In fact, the quantities $P(\delta_n, \varepsilon, \theta)$ and η may be considered as those which determine the accuracy of an estimator δ_n at some n, specified by ε and η. We expect accuracy as a function of ε, and η increases as n increases. An estimator that reaches some specified level of accuracy with small n is treated as superior. This idea has been illustrated through examples.

The drawback of the property of consistency is that it is a limiting behaviour of an estimator. It does not tell us about the behaviour of an estimator for finite sample sizes. Informally, the definition says that as the sample size grows larger and larger without bounds, the estimator δ_n converges to the true value of the parameter θ with a very high probability. The notion of consistency refers to the behaviour of a sequence of estimators $\{\delta_n\}$ that as the data accumulates, δ_n gets closer and closer to the true value of θ. It is not regarded as a good property, since in practice, we deal with finite sample sizes, not growing without bounds. Consistency is not a relevant behaviour of a sequence of estimators $\{\delta_n\}$. However, if an estimator is not consistent, then there is no obvious advantage in increasing the sample size without bounds.

Note that consistency is a property of a sequence of estimators $\{\delta_n\}$ rather than that of a point estimator. More precisely, for each value of θ, the probability structure associated with the sequence $\{\delta_n\}$ in the family of distributions $\{f(x; \theta),\ \theta \in \Theta\}$ is such that the sequence $\{\delta_n\}$ converges to θ. We see that the property of consistency deals with an entire family of distributions $\{f(x; \theta),\ \theta \in \Theta\}$, in contrast to the usual definition of the convergence of random variables, where it deals with just one probability structure.

5.3 METHOD FOR CHECKING CONSISTENCY

Basically, there are two principles for proving the consistency of an estimator δ_n for estimating θ: one by definition and another by the use of a particular law of large numbers (LLN). By the law of large numbers, we mean a set of conditions under which a sample mean of random variables converges to a population expectation. There are two laws of large numbers: first is Chebyshev's weak LLN and second is Kolmogorov's weakest LLN. The former uses Chebyshev's inequality in proving its result. Therefore, its name is Chebyshev's WLLN. The results on these LLN are briefly summarized in Section 1.5 for their ready reference. Checking for consistency by using its definition can be intractable. However, the use of Chebyshev's LLN is relatively easy. Chebyshev's inequality is given by

$$P_\theta\left(|\delta_n - \theta| \geq \varepsilon\right) \leq \frac{E_\theta(\delta_n - \theta)^2}{\varepsilon^2}$$

On setting $E_\theta(\delta_n - \theta)^2 = 0$ as $n \to \infty$ in the above inequality for every $\theta \in \Theta$, we get

$$\lim_{n\to\infty} P_\theta(|\delta_n - \theta| \geq \varepsilon) = 0$$

This shows that the sequence of estimators $\{\delta_n\}$ is consistent. So, we have another definition of consistency:

Definition 5.3.1 (Mean squared error consistency) A sequence of estimators $\{\delta_n\}$ is said to be consistent for θ if the mean squared estimation error tends to zero as n becomes large, i.e.,

$$\lim_{n\to\infty} E_\theta(\delta_n - \theta)^2 = 0 \tag{5.3.1}$$

The condition (5.3.1) is known as mean squared error consistency.

Definition 5.3.2 (Fisher consistency) A sequence of estimators $\{\delta_n\}$ is said to be Fisher consistent, if it is the same functional of empirical (sample) distribution function as is the parameter of the population distribution function:

$$\delta_n = g(F_n),\ \theta = g(F(\theta))$$

where F_n and $F(\theta)$ are sample and population distribution functions defined by $F_n(x) = (1/n)$ $\Sigma_1^n I\{X_i \leq x\}$and $F(x;\ \theta) = P_\theta\{X \leq x\}$, respectively. We cannot say much about the relationship between the Fisher consistency and usual consistency. However, for some g sufficiently smooth,

$$\delta_n = g(F_n) \to g(F(\theta))$$

since $$F_n(x) \to F(x;\ \theta) \text{ as } n \to \infty$$

at every continuous point x of $F(\theta)$.

Consider one example illustrating the Fisher consistency. Let $X_1, X_2, \ldots, X_n$ be *iid* with $E(X_i) = \theta$ and $V(X_i) = \sigma^2 < \infty$. Sample mean $\bar{X}$ and sample variance S_n^2 are Fisher consistent for θ and σ^2, respectively, whereas sample variance S_{n-1}^2(with divisor $n - 1$) is not Fisher consistent.

Note from Eq. (5.3.1) that if $V_\theta(\delta_n) + \{\text{Bias}_\theta(\delta_n)\}^2 \to 0$ as $n \to \infty$, it implies that $\text{bias}_\theta\, \delta_n \to 0$ as $n \to \infty$. Thus, consistent estimators are asymptotically unbiased. The above details give rise to the following theorem, where conditions (i) and (ii) are sufficient conditions for an estimator δ_n to be consistent for θ.

Theorem 5.3.1 If $\{\delta_n\}$ is a sequence of estimators satisfying

(i) $\lim\limits_{n\to\infty} V_\theta(\delta_n) = 0$ and

(ii) $\lim\limits_{n\to\infty} \text{Bias}_\theta(\delta_n) = 0$ (asymptotically unbiased)

for every $\theta \in \Theta$, then δ_n is consistent for θ.

Proof. Consider the Chebyshev's inequality

$$P_\theta\left(\left|\delta_n - \theta\right| \ge \varepsilon\right) \le \frac{E_\theta(\delta_n - \theta)^2}{\varepsilon^2} = \frac{V_\theta(\delta_n) + \text{Bias}_\theta^2(\delta_n)}{\varepsilon^2}$$

On taking limits and by the conditions given in the theorem, we have

$$\lim_{n\to\infty} P_0\left(\left|\delta_n - \theta\right| \ge \varepsilon\right) \le \frac{1}{\varepsilon^2}\left(\lim_{n\to\infty} V_\theta(\delta_n) + \lim_{n\to\infty} \text{Bias}_\theta^2(\delta_n)\right) = 0$$

$$\lim_{n\to\infty} P_\theta(\left|\delta_n - \theta\right| \ge \varepsilon) = 0$$

This proves the theorem. ▣

The sufficient conditions (i) and (ii) in Theorem 5.3.1 may be restated as $E_\theta(\delta_n - \theta)^2 \to 0$ as $n \to \infty$, $\forall\ \theta \in \Theta$. We may proceed to its proof, alternatively, as follows: Let $G_n(t)$ be the distribution function of the statistic δ_n. Then

$$E(\delta_n - \theta)^2 = \int_{-\infty}^{\infty} (t-\theta)^2\, dG_n(t) \ge \int_{|t-\theta|\ge\varepsilon} (t-\theta)^2\, dG_n(t)$$

$$\ge \varepsilon^2 \int_{|t-\theta|\ge\varepsilon} dG_n(t) = \varepsilon^2 P[\left|\delta_n - \theta\right| \ge \varepsilon] \tag{5.3.2}$$

Note that it is a regular Chebyshev's inequality. Equation (5.3.2) gives

$$P_\theta\left[\left|\delta_n - \theta\right| \ge \varepsilon\right] \le \frac{E_\theta(\delta_n - \theta)^2}{\varepsilon^2}$$

$$\Rightarrow \qquad \lim_{n\to\infty} P_\theta\left[\left|\delta_n - \theta\right| \ge \varepsilon\right] = 0\ \forall\ \varepsilon > 0 \text{ and } \forall\ \theta \in \Theta \tag{5.3.3}$$

(since $E_\theta(\delta_n - \theta)^2 \to 0$ as $n \to \infty$, $\forall\ \theta \in \Theta$)

Note that if conditions (i) and (ii) of Theorem 5.3.1 are satisfied, then the sequence of estimators is mean squared error consistent for θ, which immediately implies that it is consistent for θ; however, the converse is not necessarily true. Further, note in particular that if an estimator δ_n is unbiased, then it is consistent for θ if $V(\delta_n) \to 0$ as $n \to \infty\ \forall\ \theta \in \Theta$.

Consider the estimation of the parametric function $P(X > 0)$. We define

$$\delta_i(\mathbf{X}) = \begin{cases} 1, & \text{if } X_i > 0 \\ 0, & \text{otherwise} \end{cases}$$

We have $$E[\delta_i(\mathbf{X})] = 1 \cdot P(X_i > 0) + 0 \cdot P(X_i \le 0) = P(X_i > 0)$$

Consider $$E(\delta_i^2(\mathbf{X})) = P(X_i > 0)$$

$$V[\delta(\mathbf{X})] = \frac{1}{n} P(X > 0)[1 - P(X > 0)] \to 0 \text{ as } n \to \infty$$

This shows that $\delta(\mathbf{X})$ is an unbiased and consistent estimator for $P(X > 0)$.

Consider, next, a random sample $X_1,\ldots,X_n$ from $N(\theta, \sigma^2)$ where both θ and σ^2 are unknown. The estimator $\bar{X}$ is consistent for θ by Theorem 5.3.1, since $\bar{X} \sim N(\theta, (\sigma^2/n))$, $E(\bar{X}) = \theta$, $V(\bar{X}) = (\sigma^2/n) \to 0$ as $n \to \infty$. Note that the rate of convergence is n^{-1}. Alternatively,

$$P\left[|\bar{X} - \theta| > \varepsilon\right] = \int_{\{\bar{x}:|\bar{x}-\theta|>\varepsilon\}} f(\bar{x}, \theta)\, d\bar{x}$$

$$= \int_{\{\bar{x}:|\bar{x}-\theta|>\varepsilon\}} \frac{\sqrt{n}}{\sqrt{2\pi}\sigma} \exp\left[-\frac{n}{2\sigma^2}(\bar{x} - \theta)^2\right] d\bar{x}$$

$$= 1 - \int_{\{\bar{x}:|\bar{x}-\theta|\le\varepsilon\}} \frac{\sqrt{n}}{\sqrt{2\pi}\sigma} \exp\left[-\frac{n}{2\sigma^2}(\bar{x} - \theta)^2\right] d\bar{x}$$

Let $(\bar{x} - \theta)/(\sigma/\sqrt{n}) = z$; $d\bar{x} = (\sigma/\sqrt{n})dz$. We have

$$P\left[|\bar{X} - \theta| > \varepsilon\right] = 1 - \int_{|z|\le(\varepsilon\sqrt{n}/\sigma)} \frac{1}{\sqrt{2\pi}} \exp\left(-\frac{z^2}{2}\right) dz$$

$$= 1 - \left[\Phi\left(\frac{\varepsilon\sqrt{n}}{\sigma}\right) - \Phi\left(-\frac{\varepsilon\sqrt{n}}{\sigma}\right)\right]$$

Now, as $n \to \infty$, we have

$$\lim_{n\to\infty} P\left[|\bar{X} - \theta| > \varepsilon\right] = 1 - \lim_{n\to\infty} \Phi\left(\frac{\varepsilon\sqrt{n}}{\sigma}\right) + \lim_{n\to\infty} \Phi\left(-\frac{\varepsilon\sqrt{n}}{\sigma}\right)$$

$$= 1 - 1 + 0 = 0$$

This is an example of an estimator where it is unbiased and consistent for θ. However, if we consider an estimator $\delta_n = X_n$, we have $E_\theta(X_n) = \theta \ \forall \ \theta \in \Theta$. Consider

$$P_\theta\left[|\delta_n - \theta| < \varepsilon\right] = P_\theta\left[\left|\frac{\delta_n - \theta}{\sigma}\right| < \frac{\varepsilon}{\sigma}\right]$$

$$= P\left[|Z| < \frac{\varepsilon}{\sigma}\right] = \Phi\left(\frac{\varepsilon}{\sigma}\right) - \Phi\left(-\frac{\varepsilon}{\sigma}\right)$$

where $Z \sim N(0,1)$

$$\lim_{n\to\infty} P_\theta[|\delta_n - \theta| < \varepsilon] = \Phi\left(\frac{\varepsilon}{\sigma}\right) - \Phi\left(-\frac{\varepsilon}{\sigma}\right) \neq 1 \ \forall\ \varepsilon > 0$$

Therefore, the estimator X_n is unbiased but not consistent for θ.

Further, $U = \Sigma_1^n(X_i - \bar{X})^2/\sigma^2 \sim \chi^2_{n-1}$. Let $S_n^2 = (1/n)\Sigma(X_i - \bar{X})^2$ and $S_{n-1}^2 = [1/(n-1)]$ $\Sigma(X_i - \bar{X})^2$, $E(S_n^2) = [(n-1)/n]\sigma^2$, $V(S_n^2) = [2(n-1)/n^2]\sigma^4$. This gives $E(S_n^2) \to \sigma^2$ as $n \to \infty$ and $V(S_n^2) \to 0$ as $n \to \infty$. Therefore, by Theorem 5.3.1 S_n^2 is consistent but not unbiased for σ^2, since $B(S_n^2) = E(S_n^2) - \sigma^2 = [(n-1)/n]\sigma^2 - \sigma^2 = -\sigma^2/n \to 0$ as $n \to \infty$. This part of the example shows that a biased estimator may be a consistent estimator for σ^2. However, consistent estimator S_n is unbiased in the limiting sense ($n \to \infty$).

Further, $U = (n-1)\ S_{n-1}^2/\sigma^2 \sim \chi^2_{n-1}$, $E(S_{n-1}^2) = \sigma^2$. This shows that S_{n-1}^2 is an unbiased estimator of σ^2. $V(S_{n-1}^2) = [2/(n-1)]\sigma^4$, $V(S_{n-1}^2) \to 0$ as $n \to \infty$. Therefore, by Theorem 5.3.1, S_{n-1}^2 is also a consistent estimator of σ^2. This is an example which shows that consistent estimators are not unique. Note that the estimators shown in this example are consistent and they converge at the same rate of n^{-1}. More formally, it is stated in Theorem 5.4.1.

Moreover, it is important to note that unbiasedness does not imply consistency. For example, consider a random sample $X_1,\ldots,X_n$ from $N(\theta, \sigma^2)$ and the estimator $\delta(\mathbf{X}) = X_1$, so that $E(X_1) = \theta$ and $V(X_1) = \sigma^2$. Note, that X_1 is unbiased but not consistent for θ.

Differently, we can construct examples so that $B_\theta(\delta_n) \to \infty$ as $n \to \infty$ and δ_n is consistent for θ. Let $\mathbf{X} = (X_1,\ldots,X_n)$ be a random sample from $N(\theta, 1)$, $-\infty < \theta < \infty$.
Define

$$\delta_n(\mathbf{X}) = \begin{cases} \bar{X}, & \text{with probability} \left(1 - \dfrac{1}{n}\right) \\ n^2, & \text{with probability} \dfrac{1}{n} \end{cases}$$

Note that

$$E(\delta_n) = \left(1 - \frac{1}{n}\right)\theta + n^2\frac{1}{n} = \left(1 - \frac{1}{n}\right)\theta + n \to \infty \text{ as } n \to \infty$$

since $\bar{X} \sim N(\theta, (1/n))$. Thus, the bias

$$B_\theta(\delta_n) = \left(n - \frac{\theta}{n}\right) \to \infty \text{ as } n \to \infty$$

$$V(\delta_n) = EV(\delta_n) = E\left[\left(\bar{X}^2 \cdot \left(1 - \frac{1}{n}\right) + n^4 \cdot \frac{1}{n}\right) - \left(\bar{X} \cdot \left(1 - \frac{1}{n}\right) + n^2 \cdot \frac{1}{n}\right)^2\right]$$

$$= \left(1 - \frac{1}{n}\right)\left[\frac{\theta^2}{n} + \frac{1}{n} + n(n^2 - 2\theta)\right] \to \infty \text{ as } n \to \infty$$

Though

$$\lim_{n\to\infty} P[|\delta_n - \theta| < \varepsilon] = \lim_{n\to\infty} P[|\bar{X} - \theta| < \varepsilon] = 1 \ \forall \ \theta \in \Theta \text{ and } \varepsilon > 0$$

This is an example which shows that even if the (sufficient) conditions (i) and (ii) in Theorem 5.3.1 are not satisfied, δ_n is consistent for θ. This example shows that the conditions (i) and (ii) in Theorem 5.3.1 are sufficient but not necessary.

It is, thus, important to note that these conditions are sufficient but are not necessary, as one can construct examples of a sequence of estimators which is consistent and with very heavy tails, so that most of the mass is concentrated around the true value θ. In this case, $\text{Bias}_\theta\, \delta_n \to 0$ as $n \to \infty$ and $V_\theta(\delta_n) \to \infty$ as $n \to \infty$, but the estimator remains consistent. If a general distribution sample variance is a biased estimator of the variance of the distribution, the sample correlation coefficient is a biased estimator of population correlation coefficient for all values of n. In both these cases, the estimators are consistent, but biased for even larger and larger values of n. These are some simple examples which show that consistent estimators may be biased even if sample sizes grow larger and larger. Commonly, statisticians are not very keen for estimators to be unbiased; instead what is important to them is that on the average, how close are the estimators to the true value of the parameter θ, which is measured by mean squared error, $E_\theta[\delta_n - \theta]^2$. More clearly, statisticians prefer a little biased estimator δ_1 over an unbiased estimator δ_2 in a situation where a certain amount of bias is compensated for by a low variance or the variance plus bias squared is smaller than the variance of δ_2 for all values of θ, i.e.,

$$V_\theta(\delta_1) + \text{Bias}_\theta^2(\delta_1) \le V_\theta(\delta_2) \ \forall \ \theta \in \Theta$$

Not that in the above example,

$$\text{MSE}_{\mu,\sigma^2}\left[\frac{1}{n}\sum(X_i - \bar{X})^2\right] \le \text{MSE}_{\mu,\sigma^2}\left[\frac{1}{n-1}\sum(X_i - \bar{X})^2\right]$$

though $[1/(n-1)]\Sigma(X_i - \bar{X})^2$ is an unbiased estimator of population variance σ^2.

We have seen that both the estimators of σ^2, S_n^2, and S_{n-1}^2 are consistent estimators of σ^2. However, S_{n-1}^2 is unbiased and S_n^2 is biased. This gives

$$\text{MSE}(S_n^2) = \frac{2(n-1)}{n^2}\sigma^4$$

and

$$\text{MSE}(S_{n-1}^2) = V(S_{n-1}^2) = \frac{2\sigma^4}{n-1}$$

and notice that

$$\text{MSE}(S_n^2) < \text{MSE}(S_{n-1}^2)$$

Thus, we have seen that a little biased estimator can have smaller mean squared error. So, in search for a good estimator, one keeps the following two criteria in mind: (i) the bias of an estimator and (ii) the concentration of an estimator around the true value of the parameter.

Note. UMVUE and maximum likelihood estimators (MLEs) are consistent so are usually the method of moment estimators (MME).

Further, the property of consistency is preserved under continuous transformations. More formally, if δ_n is consistent for θ and g is a continuous function, then $g(\delta_n)$ is consistent for $g(\theta)$. This property is discussed in the next Section 5.4.

5.4 INVARIANCE PRINCIPLE OF CONSISTENCY UNDER CONTINUOUS FUNCTIONS

The property of consistency of a sequence of estimators is such a fundamental property of an estimator that if it is not satisfied, the worth of the estimator becomes doubtful. One reason for the above argument is that, possibly, a large number of estimators may be expressed as $\{a_n\delta_n + b_n\}$ as consistent estimators, whenever $\{\delta_n\}$ is consistent, as stated in Theorem 5.4.1. The basis of this argument is the invariance property, under continuous functions, of consistent estimators, stated in Theorem 5.4.2. Therefore, inconsistent estimators cannot be considered as reasonable estimators for estimating θ.

Theorem 5.4.1 Let $\{\delta_n\}$ be a sequence of consistent estimators of a parameter θ, and $\{a_n\}$ and $\{b_n\}$ be the sequences of constants satisfying

(i) $\lim_{n\to\infty} a_n = 1$ and

(ii) $\lim_{n\to\infty} b_n = 0$.

Then, the sequence of estimators $\{a_n\delta_n + b_n\}$ is a consistent sequence of estimators of the parameter θ.

Consider a random sample $X_1, X_2,\ldots,X_n$ from $N(\theta, \sigma^2)$

$$S_{n-1}^2 = \frac{n}{n-1}\frac{1}{n}\sum_{1}^{n}(X_i - \bar{X})^2$$

$$= \frac{n}{n-1}S_n^2 \xrightarrow{P} \sigma^2$$

Since $a_nY_n + b_n \xrightarrow{P} aY + b$ if $\{a_n\} \to a$ and $\{b_n\} \to b$. So, the estimator $\bar{X}$ is consistent for θ and the estimators S_n^2 and S_{n-1}^2 are consistent estimators for σ^2. We have already seen in Chapter 3 that $\bar{X}$ and S_{n-1}^2 are UMVUEs for θ and σ^2.

Next, consider an example that shows that for some continuous function g of θ, $g(\delta_n)$ is not necessarily unbiased for $g(\theta)$, whenever δ_n is unbiased for θ; but as opposed to it, $g(\delta_n)$ is consistent for $g(\theta)$ whenever δ_n is consistent for θ. Note that $\delta_n = X_n$, $E(X_n) = \theta$, $E(X_n^2) = V(X_n) + \{E(X_n)\}^2 = \sigma^2 + \theta^2 \neq \theta^2$; X_n^2 is not an unbiased estimator for θ^2. Similarly, $\bar{X}$ is an unbiased estimator of θ but $\bar{X}^2$ is not an unbiased estimator for θ^2 since $E(\bar{X}^2) = V(\bar{X}) + \{E(\bar{X})\}^2 = (\sigma^2/n) + \theta^2 \neq \theta^2$. However, for g continuous, $g(\theta) = \theta^2$ (say), so that for every $\varepsilon > 0$, there exists an $\eta > 0$ such that

$$|\delta_n^2 - \theta^2| < \varepsilon \text{ whenever } |\delta_n - \theta| < \eta$$

or
$$|\delta_n - \theta| < \eta \Rightarrow |\delta_n^2 - \theta^2| < \varepsilon$$

which implies
$$\{x: |\delta_n^2 - \theta^2| < \varepsilon\} \supseteq \{x: |\delta_n - \theta| < \eta\}$$

By the definition of probability,

$$P(|\delta_n{}^2 - \theta^2\,| < \varepsilon) \geq P(|\delta_n - \theta| < \eta)$$

$$\Rightarrow \quad \lim_{n\to\infty} P(|\delta_n{}^2 - \theta^2| < \varepsilon) \geq \lim_{n\to\infty} P(|\delta_n - \theta| < \eta) = 1 \;\forall\; \delta > 0$$

Hence, δ_n^2 is a consistent estimator of θ^2. This shows that $\bar{X}^2$ is a consistent estimator of θ^2, whenever $\bar{X}$ is consistent for θ.

A complete result relating to this property of invariance of consistency for continuous functions has been given in Theorem 5.4.2.

Theorem 5.4.2 Let a sequence of estimators $\{\delta_n\}$ be consistent for θ and g be some continuous function of δ_n. Then, $\{g(\delta_n)\}$ is consistent for $g(\theta)$.

Proof. The transformation g is continuous, which implies that for every $\varepsilon > 0$, there exists an $\eta > 0$ so that

$$|g(\delta_n) - g(\theta)| < \varepsilon$$

whenever

$$|\delta_n - \theta| < \eta$$

If we denote $\{x{:}|g(\delta_n) - g(\theta)| < \varepsilon\}$ and $\{x{:}\ |\delta_n - \theta| < \eta\}$, respectively, by A_ε and B_η, we have

$$B_\eta \subset A_\varepsilon$$

which by definition of probability becomes

$$P_\theta(B_\eta) \le P_\theta(A_\varepsilon)$$

$$\lim_{n\to\infty} P_\theta(B_\eta) \le \lim_{n\to\infty} P(A_\varepsilon) \qquad \text{(on taking limits on both sides)}$$

$$1 \le \lim_{n\to\infty} P(A_\varepsilon) \qquad \text{(since } \delta_n \text{ is consistent for } \theta\text{)}$$

Therefore, $\lim_{n\to\infty} P(A_\varepsilon) = 1\ \forall\, \theta \in \Theta$ and for all $\varepsilon > 0$ implies that $g(\delta_n)$ is consistent for $g(\theta)$.

More generally, consider a random sample $\mathbf{X} = (X_1, X_2,\ldots,X_n)$ from population with *pdf* $f(x;\theta)$ so that the population mean $E(X) = \theta$ exists and is finite. Then, by the weak law of large numbers (WLLN), sample mean $\bar{X}$ is a consistent estimator of population mean θ. By the invariance property of consistency, the consistent estimator of $\theta^{1+r} (r > 0)$, which is a continuous function of θ, is $\bar{X}^{1+r}$ (e.g., $\bar{X}^2$ is a consistent estimator of θ^2).

Similarly, the invariance property of consistency ensures that δ^2 is consistent fort θ^2, $\exp(-\delta)$ is a consistent estimator for $\exp(-\theta)$ whenever δ is consistent for θ. However, unbiased estimators lack the property of invariance.

Consider a random sample $X_1, X_2,\ldots,X_n$ from $N(\theta, \sigma^2)$. Further, the consistent estimator of positive square root of σ^2, i.e., the standard deviation σ is, by Theorem 5.4.2, $\sqrt{S_n^2}$. Here, this square root is a continuous transformation since $S_n^2 > 0$ and $\sqrt{S_n^2} > 0$. Thus, the property of consistency is retained under a continuous transformation.

Let $X \sim b(n, \theta)$. The estimator $\delta_n = (X/n)$ is unbiased for θ, since $E(X/n) = \theta$ and its variance $V(X/n) = (1/n^2)n\theta(1 - \theta) = [\theta(1 - \theta)/n] \to 0$ as $n \to \infty$. Therefore, by Theorem 5.4.2, the estimator $\delta_n = (X/n)$ is consistent for θ at the rate n^{-1}. Further, for estimating parametric functions $g(\theta) = \theta(1 - \theta)$, note that the function $\theta(1 - \theta)$ is continuous. Therefore, by Theorem 5.4.2, its consistent estimator is given by $(X/n)[1 - (X/n)]$; and for estimating $g(\theta) = 1/\theta$, which is also a continuous function of θ, by Theorem 5.4.2, its consistent estimator is given by (n/X).

Consider now a data problem. Let $X_1, X_2,\ldots,X_n$ be *iid* $b(m, \theta)$. The estimator $\delta_n = (1/m)\bar{X}$ is unbiased for θ. $V[(1/m)\bar{X}] = (1/mn)\theta(1-\theta) \to 0$ as $n \to \infty$. Therefore, the estimator $\delta_n = (1/m)\bar{X}$ is a consistent estimator for θ at the rate n^{-1}.

The multivariate version of Theorem 5.4.2 is stated without proof.

Theorem 5.4.3 Suppose $\boldsymbol{\delta}$ is jointly consistent for $\boldsymbol{\theta}$ and g is a continuous function so that $g\colon \Theta \to \mathbb{R}^s$, where $\Theta \subset \mathbb{R}^k$, $s \le k$. Then $g(\boldsymbol{\delta})$ is consistent for $g(\boldsymbol{\theta})$.

5.5 RATE OF CONSISTENCY

In the definition of consistency of an estimator, it does not give us an idea as how fast or slow $P_n(\varepsilon,\theta) = P_\theta(|\delta_n - \theta| < \varepsilon)$ converges to unity as n approaches infinity. In the present section, we consider the rate at which $P_n(\varepsilon, \theta)$ converges to unity as n approaches infinity as the criterion of choosing between consistent estimators.

Let $X_1, X_2,\ldots,X_n$ be *iid* $N(\theta, \sigma^2)$ random variables, where σ^2 is known. The sample mean $\bar{X}$ is a UMVU estimator of θ. By the WLLN, $\bar{X} \xrightarrow{P} \theta$ since $E(X_i) = \theta$. Thus, $\bar{X}$ is consistent for θ. Although $\bar{X}$ is consistent for θ, it does not give us an idea as how fast or slow $P_n(\varepsilon, \theta) = P_\theta[|\bar{X} - \theta| < \varepsilon]$ converges to unity as n approaches infinity. This can be explored by two methods: one by Chebyshev's inequality and the second by exact method.

Chebyshev's Inequality method

Using Chebyshev's inequality, we have

$$P_n(\varepsilon, \theta) = P_\theta[|\bar{X} - \theta| < \varepsilon] \ge 1 - \frac{(\sigma^2/n)}{\varepsilon^2} \quad \forall\, \theta \in \Theta \tag{5.5.1}$$

Thus, for weak consistency of an estimator defined in Eq. (5.1.1), the minimum sample size on which the estimator is based, n_0, so that $P_n(\varepsilon, \theta) \to 1$ is obtained such that

$$1 - \frac{\sigma^2}{n\varepsilon^2} \ge 1 - \eta$$

or

$$n_0 \ge \frac{\sigma^2}{\eta\varepsilon^2}$$

Note that n_0 does not depend on θ. Therefore, $P_n(\varepsilon, \theta) \to 1$ uniformly in θ. This shows that $\bar{X}$ is uniformly consistent for θ. Since sample size is an integer, $n \ge [\sigma^2/(\eta\varepsilon^2)] + 1$, where $[x]$ is the largest integer less than the real number x. Therefore, the smallest n is given by $n_0 = [\sigma^2/\eta\varepsilon^2] + 1$. We calculate n_0 for a combination of values of ε and η at fixed $\sigma^2 = 1$ for the purpose of illustration.

Exact method

We have

$$P_n(\varepsilon, \theta) = P_\theta[|\bar{X} - \theta| < \varepsilon] = P_\theta\left[-\frac{\varepsilon}{\sigma/\sqrt{n}} < \frac{\bar{X} - \theta}{\sigma/\sqrt{n}} < \frac{\varepsilon}{\sigma/\sqrt{n}}\right]$$

$$= \Phi\left(\frac{\sqrt{n}\varepsilon}{\sigma}\right) - \Phi\left(-\frac{\sqrt{n}\varepsilon}{\sigma}\right)$$

[since $\bar{X} \sim N(\theta, (\sigma^2/n))$, Φ is the *cdf* of $N(0, 1)$ distribution]

$$= \Phi\left(\frac{\sqrt{n}\varepsilon}{\sigma}\right) - \left[1 - \Phi\left(\frac{\sqrt{n}\varepsilon}{\sigma}\right)\right]$$

[since $N(0, 1)$ is symmetric about 0]

$$= 2\Phi\left(\frac{\sqrt{n}\varepsilon}{\sigma}\right) - 1 \tag{5.5.2}$$

Note that $\Phi[(\sqrt{n}\varepsilon)/\sigma] \to 1$ as $n \to \infty$. So, $P_n(\varepsilon,\theta) \to 1$ as $n \to \infty$ shows that $\bar{X}$ is consistent for θ. The smallest n_0 at which $\bar{X}$ achieves a certain accuracy level specified by ε and η is obtained by the condition

$$P_n(\varepsilon, \theta) \geq 1 - \eta$$

or

$$2\Phi\left(\frac{\sqrt{n}\varepsilon}{\sigma}\right) - 1 \geq 1 - \eta$$

$$\Phi\left(\frac{\sqrt{n}\varepsilon}{\sigma}\right) \geq 1 - \frac{\eta}{2}$$

$$n \geq \frac{\sigma^2}{\varepsilon^2}\left\{\Phi^{-1}\left(1 - \frac{\eta}{2}\right)\right\}^2$$

Therefore, the smallest n is given by

$$n_0 = \left[\frac{\sigma^2}{\varepsilon^2}\left[\Phi^{-1}\left(1 - \frac{\eta}{2}\right)\right]^2\right] + 1 \tag{5.5.3}$$

The entries in Tables 5.5.1(a) and (b) are the minimum sample sizes required by the estimator $\bar{X}$ to reach a certain level of accuracy as specified by ε and η. For example, the minimum sample size, as per the exact calculations, that requires for the estimator $\bar{X}$ of θ to be correct up to two decimal places ($\varepsilon = 10^{-2}$), with probability atleast 0.99 [$P_n(\varepsilon,\theta) \geq 1 - \eta = 1 - 0.01 = 0.99$], is given by 66349.

Table 5.5.1(a) Approximations by Chebyshev's inequality

ε \ η	0.5	0.1	0.05	0.01	0.001
0.5	8	40	80	400	4000
0.1	200	1000	2000	10^4	10^5
0.05	800	4000	8000	4×10^4	4×10^5
0.01	2×10^4	10^5	2×10^5	10^6	10^7
0.001	2×10^6	10^7	2×10^7	10^8	10^9

Table 5.5.1(b) Exact calculations

ε \ η	0.5	0.1	0.05	0.01	0.001
0.5	2	11	16	27	44
0.1	46	271	385	664	1083
0.05	182	1083	1537	2654	4332
0.01	4550	27056	38415	66349	108276
0.001	454937	2705544	3841460	6634897	10827566

Note that on comparing Tables 5.5.1(a) and 5.5.1(b), the rough calculations of minimum sample sizes by Chebyshev's inequality required by $\overline{X}$ to reach a given level of accuracy are much larger than actually required, as shown by exact calculations. We, clearly, see that $P_n(\varepsilon,\theta)$ is an increasing function of sample size.

Thus, the rate for an estimator δ_n at which $P_n(\varepsilon, \theta)$ converges to unity as n approaches infinity is taken as a criterion of choosing between estimators.

5.6 METHODS OF CONSTRUCTING CONSISTENT ESTIMATORS

Generally, the methods used in practice to generate consistent estimators are method of moments (MoM), method of maximum likelihood (MLE), and method of percentiles (MoP). These methods have been discussed in Chapter 6 in full details, but for the purpose of their use in the present context, we outline them and their large sample features in brief.

5.6.1 Method of Moments (MoM)

Consider $X \sim f(x;\theta)$, where $f(x;\theta)$ is in some family of distribution $\{f(x;\theta);\ \theta \in \Theta\}$. Let $X_1, X_2,\ldots,X_n$ be a random sample from $f(x;\theta)$, where the parameter θ is unknown. Based on these sample observations, we want to construct an estimator $\delta(X_1, X_2,\ldots,X_n)$ of θ.

Suppose for some function g, g: $\chi \to \mathbb{R}^1$, so that the function μ

$$\mu(\theta) = E_\theta[g(X)]$$

has a continuous inverse μ^{-1}. The solution of the moment equation

$$\overline{g}(\mathbf{X}) = \frac{1}{n}\sum_{1}^{n} g(X_i) = \mu(\theta)$$

i.e.,

$$\delta = \mu^{-1}[\overline{g}(X)]$$

is known as the method of moments estimator (MoM) of θ. The condition for μ^{-1} to exist and to be continuous is that $|\partial\mu/\partial\theta| \neq 0$. This condition is sufficient but not necessary. Since sample observations $X_1, X_2,\ldots,X_n$ are *iid* from $f(x;\ \theta)$, by WLLN, $\overline{g}(\mathbf{X})$ is consistent for $\mu(\theta)$. We have

$$\overline{g}(X) \xrightarrow{P} E_\theta(X_1) = \mu(\theta)$$

Under the assumption that μ^{-1} exists and is continuous and by using the invariance property of consistency under continuous transformation

$$\delta = \mu^{-1}(\bar{g}) \xrightarrow{P} \mu^{-1}[\mu(\theta)] = \theta$$

we have $\delta = \mu^{-1}(\bar{g})$ is consistent for θ. Usually, we consider $g(X) = X^k$, $k = 1, 2$, and the quantity $\mu_k(\theta) = E_\theta(X^k)$ is known as the kth moment of X.

Consider $X \sim \exp(\theta) \equiv G_2(1,\theta)$, where $f(x;\theta) = \theta\exp(-\theta x)$, $\theta > 0$, $E(X^r) = \theta\Gamma(r+1)/\theta^{r+1}$, $E(X) = (1/\theta) = \bar{X} = m_1'$, $E(X^2) = (2/\theta^2) = (1/n)\Sigma X_i^2 = \overline{X^2} = m_2'$. Here, $m_r' = (1/n)\Sigma X_i^r = \overline{X^r}$. The two MoM estimators, namely $\delta_1 = 1/\bar{X}$ and $\delta_2 = \sqrt{2/\overline{X^2}}$ are consistent for θ. Now the question is that which of the two estimators is better? The answer to this question is given on the basis of their asymptotic variances in Section 5.8.1 where it is shown that the MoM estimators are CAN.

In case $\boldsymbol{\theta}$ is vector-valued, the above method, holds for this situation too and also the obtained estimator (MoM) is consistent. Consider, for example, a random sample $X_1, X_2,\ldots,X_n$ from $N(\theta, \sigma^2)$, where (θ, σ^2) is a vector of unknown parameters. We have for $g(X) = X$,

$$m_1' = \bar{X} \xrightarrow{P} E(X) = \mu_1' = \theta$$

(by the weak law of large numbers)

Also,

$$m_1'^2 \xrightarrow{P} \theta^2$$

(by invariance property of consistency under continuous transformations)

and for $g(X) = X^2$,

$$m_2' = \frac{1}{n}\sum_1^n X_i^2 \xrightarrow{P} E(X^2) = \mu_2'$$

(by weak law of large numbers)

where

$$\mu_2' = V(X) + \{E(X)\}^2 = \sigma^2 + \theta^2$$

Therefore, by the definition of consistency,

$$(m_1'^2, m_2') \xrightarrow{P} (\theta^2, \sigma^2 + \theta^2)$$

This gives

$$m_2' - m_1'^2 = \frac{1}{n}\sum_1^n (X_i - \bar{X})^2 = S_n^2 \xrightarrow{P} \sigma^2$$

This procedure of obtaining the consistent estimators for θ and σ^2 is known as the *method of moments*.

We, therefore, formalize this procedure in case of vector parameters. Assume that a random sample $X_1,\ldots,X_n$ is drawn from a population with *pdf* or *pmf* $f(x;\boldsymbol{\theta})$, $\boldsymbol{\theta} = (\theta_1,\ldots,\theta_k)$, $\boldsymbol{\theta} \in \Theta$. Moment equations are

$$\begin{aligned} m_1' &= \mu_1'(\boldsymbol{\theta}) \\ m_2' &= \mu_2'(\boldsymbol{\theta}) \\ &\vdots \\ m_k' &= \mu_k'(\boldsymbol{\theta}) \end{aligned} \tag{5.6.1}$$

where m_j' and μ_j' are sample and population raw moments, respectively which are defined by

$$m_j' = \frac{1}{n}\sum_{i=1}^{n} X_i^j \text{ and } \mu_j'(\boldsymbol{\theta}) = E_\theta(X^j)$$

The method of moments estimator of $\boldsymbol{\theta}$ involves solving the system of Eq. (5.6.1) for $\boldsymbol{\theta}$ in terms of sample moments $m_1',\ldots,m_k'$.

By the WLLN,

$$m_j' \xrightarrow{P} \mu_j'(\boldsymbol{\theta}) \;\forall\, j = 1,\ldots,k$$

If μ_j' is invertable and continuous in $\boldsymbol{\theta}$, then by the result that consistency is preserved under a continuous transformation stated in Theorem 5.4.2, we have

$$\mu_j'^{-1}(m_j') \xrightarrow{P} \theta_j \;\forall\, j = 1,\ldots,k$$

Usually, whether μ_j' are invertible and continuous in $\boldsymbol{\theta}$, in practice, they are checked by a sufficient condition that $|\partial(\mu_1',\ldots,\mu_k')/\partial(\theta_1,\ldots,\theta_k)| \neq 0$. However, this condition is not necessary. Formally, this method has been discussed in Chapter 6.

5.6.2 Method of Percentiles (MoP)

Assume that a sample $X_1,\ldots,X_n$ on a continuous random variable X is drawn from a population with density $f(x;\theta)$, $\theta \in \Theta \subseteq \mathbb{R}^1$. We define $100p\%$ population percentile $x_p(\theta)$ as a solution to the equation

$$F(x_p;\theta) = \int_{-\infty}^{x_p(\theta)} f(x;\theta)dx = p, \quad 0 < p < 1$$

or

$$x_p(\theta) = F_\theta^{-1}(p)$$

where $x_p(\theta)$ is unique if $dF/d\theta \neq 0$ at $x_p(\theta)$.

Consider $r = [np] + 1$. Then, the rth order statistics $X_{(r)}$ of sample observations $X_1,\ldots,X_n$, is known as the $100p\%$ sample percentile. The equation in which we equate sample percentiles to population percentiles

$$X_{(r)} = x_p(\theta)$$

is known as the percentile equation. When we solve this equation for θ, the solution in terms of $X_{(r)}$ is known as the estimator obtained by the method of percentiles. The percentile equation admits a unique solution if $(\partial/\partial\theta)x_p(\theta) \neq 0$.

Next, we will prove a theorem where we show that $X_{(r)}$ is a consistent estimator of $x_p(\theta)$.

Theorem 5.6.1 Let $X_1,\ldots,X_n$ be a random sample of size n on a continuous *rv* X, having its *pdf* $f(x;\ \theta)$ and *df* $F(x;\theta)$ so that $dF/dx = f(x;\theta) > 0$. Then

$$X_{(r)} \xrightarrow{P} x_p(\theta)$$

Proof. The sample $X_1, X_2,\ldots,X_n$ has been drawn on a continuous *rv* X with *df* F. The corresponding o.s. (order statistics) are given by $(X_{(1)},\ldots,X_{(n)})$. From the result $F(X) \sim U(0,1)$,

$\{F(X_{(1)}),\ldots,F(X_{(n)})\}$ can be treated as order statistics $(V_{(1)},\ldots,V_{(n)})$ of sample $V_1,\ldots,V_n$ of size n from $U(0, 1)$. Thus,

$$V_{(r)} \sim \text{Be}(r, n-r+1)$$

and
$$E(V_{(r)}) = \frac{r}{n+1};\ E(V_{(r)}^2) = \frac{r(r+1)}{(n+1)(n+2)} \tag{5.6.2}$$

This gives
$$E(V_{(r)} - p)^2 = \frac{r(r+1)}{(n+1)(n+2)} - 2p\cdot\frac{r}{(n+1)} + p^2$$

We have taken
$$r = [np] + 1$$
$$np \le r \le np + 1$$
$$p \le \frac{r}{n} \le p + \frac{1}{n}$$
$$\frac{r}{n} \to p \text{ as } n \to \infty$$

This gives
$$E(V_{(r)} - p)^2 \xrightarrow{P} 0 \text{ as } n \to \infty$$

Using Eq. (5.6.2) in Chebyshev's inequality,

$$P\left(\left|V_{(r)} - p\right| < \varepsilon\right) \ge 1 - \frac{E(V_{(r)} - p)^2}{\varepsilon^2}\ \forall\, \varepsilon > 0$$

$$V_{(r)} = F(X_{(r)}) \xrightarrow{P} p$$

Further, F^{-1} exists and is continuous since $dF/dx = f(x) > 0$ at $x_p(\theta)$. Therefore, by the invariance of consistency under continuous transformations, Theorem 5.4.2 gives

$$F^{-1}[F(X_{(r)})] \xrightarrow{P} F^{-1}(p) = x_p(\theta)$$
$$X_{(r)} \xrightarrow{P} x_p(\theta)$$

This proves what is asserted in the theorem. ▣

Consider a random sample $X_1,\ldots,X_n$ from $N(\theta, 1)$. The 50-percentile equation is given by

$$X_{[n/2]+1} = x_{(1/2)}(\theta) = F_\theta^{-1}\left(\frac{1}{2}\right) = (100 \times 0.5)\% \text{ of population percentile} = \theta$$

This gives that sample median is a consistent estimator of θ.

Let us consider the estimation of the pth quantile (or $100p\%$ point) of $N(\theta, \sigma^2)$ so that

$$P(X \le x_p) = p$$

or
$$P\left(\frac{X-\theta}{\sigma} \le \frac{x_p - \theta}{\sigma}\right) = p$$

We then have
$$\frac{x_p - \theta}{\sqrt{\sigma^2}} = z_p$$
(where z_p is $100p\%$ point of $N(0, 1)$ distribution)
$$x_p = \theta + z_p\sqrt{\sigma^2}$$

where x_p is unknown since θ and $\sqrt{\sigma^2}$ are unknown. Further, $\theta + z_p\sqrt{\sigma^2}$ is a continuous function of vector of parameters $(\theta, \sqrt{\sigma^2})$. Therefore, $\bar{X} + z_p\sqrt{S_n^2}$ is a consistent estimator for $100p\%$ point of $N(\theta, \sigma^2)$, i.e., $x_p = \theta + z_p\sqrt{\sigma^2}$.

Next, consider the estimation of probability $P[X \le x_0]$ for given x_0. We may write
$$P[X \le x_0] = P\left[\frac{X-\theta}{\sqrt{\sigma^2}} \le \frac{x_0 - \theta}{\sqrt{\sigma^2}}\right] = \Phi\left(\frac{x_0-\theta}{\sqrt{\sigma^2}}\right)$$

Since Φ is a continuous transformation of $(\theta, \sqrt{\sigma^2})'$, $\Phi(x_0 - \bar{X}/\sqrt{S_n^2})$ is a consistent estimator of $P[X \le x_0]$.

5.7 OPTIMALITY AND CAN ESTIMATORS

5.7.1 Optimality among Consistent Estimators

We have seen that the consistency of an estimator indicates a high degree of its concentration around the true value of the parameter θ for large n. Intutively, the estimators that converge to the true value of the parameter are naturally preferred ones; and the one which converges to θ most rapidly is then usually treated as the most optimal. In other words, an estimator for which $P_n(\varepsilon, \theta)$ converges to unity at the fastest rate as n tends to infinity for all $\theta \in \Theta$ is the most optimal. For such an optimal estimator, by Chebyshev's inequality, MSE(δ) would approach zero at the fastest rate as n tends to infinity. This criterion of judging estimators is called rate of convergence criterion. However, there is no set principle of assessing the rate of convergence of an estimator of θ or the rate at which $P_n(\varepsilon, \theta)$ converges to unity as n tends to infinity. This may be done by approximating $P_n(\varepsilon, \theta)$ for large n. Consider a sequence of real positive numbers $\{a_n\}$, $a_n > 0$, so that
$$X_n = a_n(\delta - \theta) \xrightarrow{d} X$$
where X is a non-degenerate *rv* with continuous *df* H. The above convergence is called the convergence in distribution. Then,
$$\begin{aligned} P_n(\varepsilon, \theta) &= P[|\delta - \theta| < \varepsilon] = P[a_n|\delta - \theta| < a_n\varepsilon] \\ &\cong H(a_n\,\varepsilon) - H(-a_n\,\varepsilon) \end{aligned}$$

We consider a random sample of size n from some distribution function F and a consistent estimator δ with an appropriate blow-up factor a_n so that
$$a_n(\delta - \theta) \xrightarrow{d} N(0, \sigma_\delta^2(\theta))$$

or
$$\delta \sim AN\left[\theta, \frac{\sigma_\delta^2(\theta)}{a_n^2}\right]$$

Such estimators are known as *consistent asymptotically normal* (CAN) estimators. We then have

$$P_n(\varepsilon, \theta) = \Phi\left(\frac{a_n \varepsilon}{\sigma_\delta(\theta)}\right) - \Phi\left(-\frac{a_n \varepsilon}{\sigma_\delta(\theta)}\right) \tag{5.7.1}$$

Therefore, the rate at which $P_n(\varepsilon, \theta) \to 1$ can be studied by the rate at which $\Phi(a_n\varepsilon/\sigma_\delta(\theta)) \to 1$ and $\Phi(-a_n\varepsilon/\sigma_\delta(\theta)) \to 0$. Expressing $P_n(\varepsilon, \theta)$ in Eq. (5.7.1) as

$$P_n(\varepsilon, \theta) = 2\Phi\left(\frac{a_n \varepsilon}{\sigma_\delta(\theta)}\right) - 1$$

shows that the rate $P_n(\varepsilon, \theta)$ at which it approaches unity is inversely proportional to $\sigma_\delta^2(\theta)/a_n^2$. Therefore, the stochastic convergence criterion of judging between consistent estimators equivalently states that an estimator is preferred over another if the former has a smaller value of $\sigma_\delta^2(\theta)/a_n^2$. Consider two CAN estimators δ_1 and δ_2 with different blow-up factors a_n and b_n, respectively. The estimator δ_2 is preferred to δ_1 if $AV_\theta(\delta_2) \le AV_\theta(\delta_1)$, i.e., $\sigma_{\delta_2}^2(\theta)/b_n^2 \le \sigma_{\delta_1}^2(\theta)/a_n^2$, where AV stands for asymptotic variance. The limit of the ratios

$$\frac{AV_\theta(\delta_2)}{AV_\theta(\delta_1)} = \frac{\sigma_{\delta_2}^2(\theta)/b_n^2}{\sigma_{\delta_1}^2(\theta)/a_n^2}$$

is called *asymptotic relative efficiency* (*ARE*) of δ_1 with respect to δ_2, i.e.,

$$ARE_\theta(\delta_1, \delta_2) = \lim_{n\to\infty} \frac{\sigma_{\delta_2}^2(\theta)/b_n^2}{\sigma_{\delta_1}^2(\theta)/a_n^2}$$

If $ARE_\theta(\delta_1, \delta_2) < 1 \ \forall\ \theta \in \Theta$, we prefer δ_2 over δ_1.

In brief, two asymptotically unbiased and normally distributed estimators can be compared on the basis of their asymptotic variances. That is, if

$$\sqrt{n}\,[\delta_1 - g(\theta)] \xrightarrow{d} N(0, \sigma_{\delta_1}^2(\theta))$$

and
$$\sqrt{n}\,[\delta_2 - g(\theta)] \xrightarrow{d} N(0, \sigma_{\delta_2}^2(\theta))$$

then $ARE_\theta(\delta_1, \delta_2) < 1 \ \forall\ \theta$ implies that δ_2 is asymptotically more efficient than δ_1.

5.7.2 Asymptotic Normality of Consistent Estimators (CAN)

Definition 5.7.1 (Consistent asymptotically normal (CAN) estimator) A sequence of estimators $\{\delta_n\}$ is said to be consistent asymptotically normal for $g(\theta)$ if

(i) $\{\delta_n\}$ is consistent for $g(\theta)$ and

(ii) for some appropriate blow-up factor a_n so that δ_n is asymptotically normal,

$$a_n[\delta_n - g(\theta)] \xrightarrow{d} N(0, \sigma^2_{\delta_n}(\theta)) \ \forall \ \theta \in \Theta$$

where $g(\theta)$ is called the *asymptotic mean* and $\sigma^2_{\delta_n}(\theta)$ is called the *asymptotic variance*. Usually, the blow-up factor a_n is taken as $\sqrt{n}$. In this case δ_n is

$$AN\left[g(\theta), \frac{\sigma^2_{\delta_n}(\theta)}{n}\right] \ \forall \ \theta \in \Theta$$

Definition 5.7.2 (Best asymptotically normal (BAN) estimator) A sequence of estimators $\{\delta_n\}$ is said to be best *asymptotically normal* for θ if

(i) $\{\delta_n\}$ is consistent for θ and
(ii) $\delta_n \sim AN(\theta, [1/(nI_X(\theta))]$

where $I_X(\theta)$ is the Fisher information in a single observation X about θ.

The definition of CAN estimators, when $\boldsymbol{\theta}$ is a vector parameter, $\boldsymbol{\theta} \in \Theta \subseteq \mathbb{R}^k$, is given as follows:

Definition 5.7.3 (CAN) A consistent estimator $\boldsymbol{\delta}$ is said to be a CAN estimator for $\boldsymbol{\theta}$ if for a sequence $\{a_n\}$, $a_n > 0$, $a_n \to \infty$,

$$a_n(\boldsymbol{\delta} - \boldsymbol{\theta}) \xrightarrow{d} N_k(\mathbf{0}, \Sigma)$$

where N_k denotes a non-singular k-dimensional multivariate normal distribution with mean $\mathbf{0}$ and variance–covariance matrix Σ which is a symmetric positive definite. Usually, $a_n = \sqrt{n}$.

In the above definition,

$$a_n(\delta_i - \theta_i) \xrightarrow{d} N(0, \sigma_{ii})$$

i.e., δ_i is CAN for θ_i, $i = 1,\ldots,k$. On using this, $\mathbf{l}'\delta \xrightarrow{d} N_1(\mathbf{l}'\boldsymbol{\theta}, \mathbf{l}'\Sigma\mathbf{l})$, for some $\mathbf{l}$ in $\mathbb{R}^k$.

Analogous to the univariate central limit theorem, the multivariate central limit theorem can be a useful tool towards generating CAN estimators for the vector parameter $\boldsymbol{\theta}$. Let $(\mathbf{X}_1, \mathbf{X}_2,\ldots,\mathbf{X}_n)$ be *iid rvs* on $\mathbf{X}$, $E(\mathbf{X}) = \boldsymbol{\theta}$, and $V(\mathbf{X}) = \Sigma$, Σ is p.d., $\bar{\mathbf{X}} = (\bar{X}_1, \bar{X}_2,\ldots,\bar{X}_k)'$, $\bar{X}_j = (1/n)\ \Sigma^n_{i=1}X_{ji}$, $j = 1,\ldots,k$.

$$\sqrt{n}\,(\bar{\mathbf{X}} - \boldsymbol{\theta}) \xrightarrow{d} N_k(\mathbf{0}, \Sigma)$$

Definition 5.7.4 (Asymptotic relative efficiency) Let δ_1 and δ_2 be two CAN estimators of θ. Then the *ARE* of δ_1 relative to δ_2 is defined by

$$ARE_\theta(\delta_1, \delta_2) = \lim_{n\to\infty} \frac{\sigma^2_{\delta_2}(\theta)/b_n^2}{\sigma^2_{\delta_1}(\theta)/a_n^2}$$

where $a_n(\delta_1 - \theta) \xrightarrow{d} N(0, \sigma^2_{\delta_1}(\theta))$ and $b_n(\delta_2 - \theta) \xrightarrow{d} N(0, \sigma^2_{\delta_2}(\theta))$.

Rao (1963) has shown that the Cramer–Rao lower bound for the variance of an unbiased estimator under the following regularity conditions also holds as a lower bound for the asymptotic variances of CAN estimators.

Regularity conditions

Let $X_1, X_2, \ldots, X_n$ be a random sample drawn identically from $f(x;\theta)$. The parameter space is non-degenerate open interval. We define

$$S(X,\theta) = \frac{\partial}{\partial\theta}\log f(X;\theta) \text{ and } H(X,\theta,\theta_0) = \log\frac{f(X;\theta)}{f(X;\theta_0)}$$

and $$I_X(\theta) = E_\theta\left[\frac{\partial}{\partial\theta}\log f(X;\theta)\right]^2 = E_\theta[S(X;\theta)]^2$$

and assume that

(i) $I(\theta)$ is a continuous function of θ;

(ii) (a) $E_\theta[S(X,\theta)]|_{\theta=\theta_0} = (\theta - \theta_0)I_X(\theta_0) + O(\theta - \theta_0)$;

(b) $V_\theta[S(X,\theta))|_{\theta=\theta_0} = I_X(\theta_0) + O(1)$;

(c) $\text{cov}_\theta[S(X,\theta), S(X,\theta_0)] = I_X(\theta_0) + O(1)$;

(iii) $E_\theta|S(X,\theta)|^{2+\eta}$ exists for some $\eta > 0$ in compact interval of θ;

(iv) (a) $E_\theta[H(X,\theta,\theta_0)]|_{\theta=\theta_0} = -[(\theta - \theta_0)^2/2]\, I_X(\theta_0) + O(\theta - \theta_0)^2$;

(b) $V_\theta[H(X,\theta,\theta_0)]|_{\theta=\theta_0} = (\theta - \theta_0)^2 I_X(\theta_0) + O(\theta - \theta_0)^2$;

(c) $\text{cov}_\theta[S(X,\theta), H(X, \theta, \theta_0)]|_{\theta=\theta_0} = (\theta - \theta_0)I_X(\theta_0) + O(\theta - \theta_0)$.

The information inequality for CAN estimators proved by Rao is stated in the following theorem.

Theorem 5.7.1 (Asymptotically most efficient estimator) [Rao (1963)] Let $\{\delta_n\}$ be a sequence of CAN estimators so that

$$\sqrt{n}[\delta_n - g(\theta)] \xrightarrow{d} N(0, \sigma^2_{\delta n}(\theta)) \;\forall\; \theta \in S(\theta)$$

where $S(\theta)$ is a compact interval of θ and $\sigma^2_{\delta_n}(\theta)$ is bounded. Then under the assumptions (i) to (iv),

$$\sigma^2_{\delta_n}(\theta) \geq \frac{[g'(\theta)]^2}{I_X(\theta)} \;\forall\, \theta$$

An estimator δ_0 for which $\sigma^2_{\delta_0}(\theta) = [g'(\theta)]^2/I(\theta) \;\forall\; \theta$, i.e., δ_0 achieves the Cramer–Rao lower bound, is said to be an asymptotically most efficient estimator. The efficiency of an estimator δ_n is defined as

$$e_\theta(\delta_n) = \lim_{n\to\infty}\frac{\sigma^2_{\delta_0}(\theta)}{\sigma^2_{\delta_n}(\theta)} = \lim_{n\to\infty}\frac{1}{\sigma^2_{\delta_n}(\theta)I(\theta)}$$

5.7.3 Principle of Invariance of CAN Estimators

As for a consistent estimator δ, $g(\delta)$ is consistent for $g(\theta)$ for some continuous function g. Similarly, for a CAN estimator δ too, $g(\delta)$ is a CAN estimator for $g(\theta)$ for some continuous and differentiable function g. This property is stated in the following theorem.

Theorem 5.7.2 Suppose δ is CAN for θ so that $a_n(\delta - \theta) \xrightarrow{d} N(0, \sigma_\delta^2(\theta))$. Consider some g which is a differentiable function of θ, so that $(\partial/\partial\theta)g$ is continuous and nonvanishing $[(\partial/\partial\theta)g \neq 0$ at any $\theta]$. Then $g(\delta)$ is CAN for $g(\theta)$ and

$$a_n[g(\delta) - g(\theta)] \xrightarrow{d} N\left[0, \left(\frac{\partial}{\partial\theta}g\right)^2 \sigma_\delta^2(\theta)\right]$$

Proof. If g is differentiable implies that it is continuous. By Theorem 5.4.2, $g(\delta)$ is consistent for $g(\theta)$. On applying the mean value theorem, we get

$$g(\delta) - g(\theta) = (\delta - \theta)\frac{\partial g}{\partial\theta} + R_n \tag{5.7.2}$$

where $|R_n| \leq B|\delta - \theta|^{1+\eta}$ for some $\eta > 0$. On multiplying Eq. (5.7.2) by a_n, where a_n are real positive real numbers increasing with n, we have

$$a_n[g(\delta) - g(\theta)] = a_n(\delta - \theta)\frac{\partial g}{\partial\theta} + a_nR_n \tag{5.7.3}$$

Here, $$|a_nR_n| \leq B|a_n(\delta - \theta)|\,|\delta - \theta|^\eta,\ \eta > 0$$

Since (i) $a_n(\delta - \theta) \xrightarrow{d} N(0, \sigma_\delta^2(\theta))$ and (ii) $\delta \xrightarrow{P} \theta$ or $(\delta - \theta) \xrightarrow{P} 0$ imply together with Slutsky's theorem that

$$B|a_n(\delta - \theta)||\delta - \theta|^\eta \xrightarrow{P} 0$$

This shows $$|a_nR_n| \xrightarrow{P} 0$$

It is given that δ is CAN. So,

$$a_n(\delta - \theta) \xrightarrow{d} N(0, \sigma_\delta^2(\theta))$$

or $$a_n(\delta - \theta)\frac{\partial g}{\partial\theta} \xrightarrow{d} N\left(0, \sigma_\delta^2(\theta)\left(\frac{\partial g}{\partial\theta}\right)^2\right) \tag{5.7.4}$$

On combining Eqs. (5.7.3) and (5.7.4), we get

$$a_n[g(\delta) - g(\theta)] \xrightarrow{d} N\left(0, \sigma_\delta^2(\theta)\left(\frac{\partial g}{\partial\theta}\right)^2\right)$$

Theorem 5.7.3 Let $a_n(\boldsymbol{\delta} - \boldsymbol{\theta}) \xrightarrow{d} N_k(\mathbf{0}, \Sigma)$, $\boldsymbol{\theta} \in \Theta \subseteq \mathbb{R}^k$, $\mathbf{g}$: $\mathbb{R}^k \to \mathbb{R}^s$ so that $\mathbf{g}(\boldsymbol{\theta}) = (g_1(\boldsymbol{\theta}),\ldots,g_s(\boldsymbol{\theta}))'$ and $\mathbf{L}_{(s\times k)} = (\partial g_i/\partial\theta_j)$ such that $\mathbf{L}\Sigma\mathbf{L}'$ is pd (positive definite). Then,

$$a_n[\mathbf{g}(\boldsymbol{\delta}) - \mathbf{g}(\boldsymbol{\theta})] \xrightarrow{d} N_s(\mathbf{0}, \mathbf{L}\Sigma\mathbf{L}') \tag{5.7.5}$$

Proof. The transformation g is such that $\partial g_i/\partial\theta_j$ is continuous and nonvanishing for every $j = 1,\ldots,k$ and $i = 1,\ldots,s$. This implies that $g_i(\boldsymbol{\delta})$ is consistent for $g_i(\boldsymbol{\theta})$, $i = 1,\ldots,s$, or that $\mathbf{g}(\boldsymbol{\delta})$ is consistent for $\mathbf{g}(\boldsymbol{\theta})$. By mean value theorem, we have

$$\mathbf{g}(\boldsymbol{\delta}) - \mathbf{g}(\boldsymbol{\theta}) = \mathbf{L}(\boldsymbol{\delta} - \boldsymbol{\theta}) + R$$

or

$$a_n[\mathbf{g}(\boldsymbol{\delta}) - \mathbf{g}(\boldsymbol{\theta})] = a_n\mathbf{L}(\boldsymbol{\delta} - \boldsymbol{\theta}) + a_nR$$

where $a_nR \xrightarrow{P} 0$, $a_n(\boldsymbol{\delta} - \boldsymbol{\theta}) \xrightarrow{d} N_k(\mathbf{0}, \Sigma)$, $a_n\mathbf{L}(\boldsymbol{\delta} - \boldsymbol{\theta}) \xrightarrow{d} N_s(\mathbf{0}, \mathbf{L}\Sigma\mathbf{L}')$.

This gives

$$a_n[\mathbf{g}(\boldsymbol{\delta}) - \mathbf{g}(\boldsymbol{\theta})] \xrightarrow{d} N_s(\mathbf{0}, \mathbf{L\Sigma L'})$$

Theorem 5.7.2 does not hold when $(\partial/\partial\theta)g$ is 0 at some point $\theta = \theta_0$. In this case, the limiting distribution of CAN estimator δ is not necessarily normal with inflating factor a_n replaced by a_n^2. A generalized result is stated in the following theorem.

Theorem 5.7.4 Let δ be a CAN estimator for θ, so that $a_n(\delta - \theta) \xrightarrow{d} N(0, \sigma_\delta^2(\theta))$ and let g be a differentiable function such that $\partial g/\partial\theta$ is continuous. Further, at some point $\theta = \theta_0$, $\partial g/\partial\theta = 0$ and $\partial^2 g/\partial\theta^2 \neq 0$. Then

$$a_n^2[g(\delta) - g(\theta_0)] \xrightarrow{d} \frac{\sigma_\delta^2(\theta)}{2}\left(\frac{\partial^2 g}{\partial\theta^2}\right)_{\theta=\theta_0} \chi_1^2 \tag{5.7.6}$$

Proof. Expanding $g(\delta)$ by the Taylor series around $g(\theta_0)$, we have

$$g(\delta) = g(\theta_0) + (\delta - \theta_0)\frac{\partial g}{\partial\theta}\bigg|_{\theta=\theta_0} + \frac{1}{2!}(\delta - \theta_0)^2\frac{\partial^2 g}{\partial\theta^2}\bigg|_{\theta=\theta_0} + R_n$$

$$g(\delta) - g(\theta_0) = \frac{1}{2!}(\delta - \theta_0)^2\frac{\partial^2 g}{\partial\theta^2}\bigg|_{\theta=\theta_0} + R_n \tag{5.7.7}$$

since $\partial g/\partial\theta = 0$ at $\theta = \theta_0$. On multiplying both the sides of Eq. (5.7.7) by a_n^2, we have

$$a_n^2[g(\delta) - g(\theta_0)] = \frac{a_n^2(\delta - \theta_0)^2}{2!}\frac{\partial^2 g}{\partial\theta^2}\bigg|_{\theta=\theta_0} + a_n^2R_n \tag{5.7.8}$$

where $|R_n| \le B|\delta - \theta_0|^{2+\eta}$ for $\eta > 0$.

$$|a_n^2R_n| \le Ba_n^2(\delta - \theta_0)^2|g - \theta_0|^\eta \tag{5.7.9}$$

$\delta \xrightarrow{P} \theta_0$ or $\delta - \theta_0 \xrightarrow{P} 0$ or $|\delta - \theta_0|^\eta \xrightarrow{P} 0$ since δ is consistent for θ. This gives

$$Ba_n^2(\delta - \theta_0)^2\,|g - \theta_0|^\eta \xrightarrow{P} 0 \tag{5.7.10}$$

Equations (5.7.8) and (5.7.9) give

$$|a_n^2R_n| \xrightarrow{P} 0 \tag{5.7.11}$$

or

$$R_n \xrightarrow{P} 0$$

and

$$a_n(\delta - \theta_0) \xrightarrow{d} N(0, \sigma_\delta^2(\theta_0))$$

or
$$\frac{a_n(\delta-\theta_0)}{\sigma_\delta(\theta_0)} \xrightarrow{d} N(0,1)$$

or
$$\frac{a_n^2(\delta-\theta_0)^2}{\sigma_\delta^2(\theta_0)} \xrightarrow{d} \chi_1^2$$

or
$$\frac{\sigma_\delta^2(\theta_0)}{2}\frac{a_n^2(\delta-\theta_0)^2}{\sigma_\delta^2(\theta_0)} \xrightarrow{d} \frac{\sigma_\delta^2(\theta_0)}{2}\chi_1^2$$

or
$$\frac{a_n^2(\delta-\theta_0)^2}{2} \xrightarrow{d} \frac{\sigma_\delta^2(\theta_0)}{2}\chi_1^2$$

or
$$\frac{a_n^2(\delta-\theta_0)^2}{2}\left.\frac{\partial^2 g}{\partial\theta^2}\right|_{\theta=\theta_0} \xrightarrow{d} \frac{\sigma_\delta^2(\theta_0)}{2}\left(\frac{\partial^2 g}{\partial\theta^2}\right)_{\theta=\theta_0}\chi_1^2 \tag{5.7.12}$$

On using Eqs (5.7.11) and (5.7.12) in Eq. (5.7.8), we get

$$a_n^2[g(\delta)-g(\theta_0)] \xrightarrow{d} \frac{\sigma_\delta^2(\theta_0)}{2}\left(\frac{\partial^2 g}{\partial\theta^2}\right)_{\theta=\theta_0}\chi_1^2$$

5.7.4 Asymptotic Efficiency of MLE

ML estimator is consistent under some regularity conditions, which indicates that it is a reasonable estimator which performs well. However, under these regularity conditions, it performs no inferior than any other reasonable estimator, since it is asymptotically unbiased and its variance approaches to the CR lower bound of an unbiased estimator as $n \to \infty$. This result is stated in the following theorem and its proof is deferred to Chapter 6 for the purpose of brevity, where ML estimation has been discussed at length.

Theorem 5.7.5 Let $X_1, X_2,\ldots,X_n$ be *iid* from $f(x;\theta)$. The MLE δ_n of θ under certain regularity conditions is

$$\sqrt{n}(\delta_n-\theta) \xrightarrow{d} N\left(0, \frac{1}{I_X(\theta)}\right) \quad \forall\,\theta$$

or δ_n is $AN(\theta, [nI_X(\theta)]^{-1})$ for all θ. That is, the MLE δ_n is a consistent and an asymptotically efficient estimator of θ. Further, if g is continuous and differentiable in θ, then

$$\sqrt{n}[g(\delta_n)-g(\theta)] \xrightarrow{d} N\left(0, [g'(\theta)]^2\,\frac{1}{I_X(\theta)}\right) \quad \forall\,\theta$$

or $g(\delta_n)$ is $AN(g(\theta), [g'(\theta)]^2[nI_X(\theta)]^{-1})$ for all θ. That is, the MLE $g(\delta_n)$ is a consistent and an asymptotically efficient estimator of $g(\theta)$. The asymptotic variance of MLE $g(\delta_n)$ of $g(\theta)$ is

$$AV[g(\delta_n)] = \frac{[g'(\theta)]^2}{nI_X(\theta)}$$

For a finite sample size n, this variance can be approximated by plug-in estimator

$$AV[g(\delta_n)] \cong \frac{[g'(\delta_n)]^2}{-\frac{\partial^2}{\partial\theta^2}\log L(\theta;\mathbf{X})\Big|_{\theta=\delta_n}}$$

The denominator is called the observed Fisher information.

In case $\boldsymbol{\theta}$ is vector, $\boldsymbol{\theta} \in \mathbb{R}^k$, just as in case of θ scalar, the MLE $\boldsymbol{\delta}_n$ is consistent for $\boldsymbol{\theta}$, where each component of $\boldsymbol{\delta}_n$ converges in probability to the corresponding component of $\boldsymbol{\theta}$. Further, $\boldsymbol{\delta}_n$ is asymptotically unbiased and its variance approaches the CR lower bound of an unbiased estimator. The following theorem states this result.

Theorem 5.7.6 Let $X_1, X_2,\ldots,X_n$ be *iid* $f(x;\boldsymbol{\theta})$, $\boldsymbol{\theta} \in \mathbb{R}^k$. The MLE $\boldsymbol{\delta}_n$ of θ under certain regularity conditions is

$$\sqrt{n}(\boldsymbol{\delta}_n - \boldsymbol{\theta}) \xrightarrow{d} N_k\,[\mathbf{0}, \mathbf{I}_{\mathbf{X}}^{-1}(\boldsymbol{\theta})] \quad \forall\, \boldsymbol{\theta}$$

or $\boldsymbol{\delta}_n$ is $AN(\boldsymbol{\theta}, [n\mathbf{I}_{\mathbf{X}}(\boldsymbol{\theta})]^{-1})$ for all $\boldsymbol{\theta}$ where $\mathbf{I}_{\mathbf{X}}(\boldsymbol{\theta})$ is the Fisher information matrix for a sample of size n. We assume that $\mathbf{I}_{\mathbf{X}}(\boldsymbol{\theta})$ is a non-singular matrix. That is, the MLE $\boldsymbol{\delta}_n$ is a consistent and asymptotically efficient estimator of $\boldsymbol{\theta}$. Further, if $g(\boldsymbol{\theta})$ is a real-valued function, continuous and differentiable in $\boldsymbol{\theta}$, then

$$\sqrt{n}[g(\boldsymbol{\delta}_n) - g(\boldsymbol{\theta})] \xrightarrow{d} N_k\,(\mathbf{0}, \mathbf{D}'\,\mathbf{I}_{\mathbf{X}}^{-1}(\boldsymbol{\theta})\mathbf{D}) \quad \forall\, \boldsymbol{\theta}$$

or $g(\boldsymbol{\delta}_n)$ is $AN_k(g(\boldsymbol{\theta}), \mathbf{D}'[n\mathbf{I}_{\mathbf{X}}(\boldsymbol{\theta})]^{-1}\mathbf{D})$ for all $\boldsymbol{\theta}$ where, $\mathbf{D} = [(\partial g/\partial\theta_1),\ldots,(\partial g/\partial\theta_k)]'$. That is, the MLE $g(\boldsymbol{\delta}_n)$ is a consistent and asymptotically efficient estimator of $g(\boldsymbol{\theta})$. The asymptotic variance of the MLE $g(\boldsymbol{\delta}_n)$ of $g(\boldsymbol{\theta})$ is

$$AV[g(\boldsymbol{\delta}_n)] = \mathbf{D}'[n\mathbf{I}_{\mathbf{X}}(\boldsymbol{\theta})]^{-1}\mathbf{D}$$

In Theorem 5.7.5, it is assumed that $(\partial g/\partial\theta)$ is continuous and nonvanishing everywhere. In some cases, this condition is not satisfied. Consider a random sample $(X_1, X_2,\ldots,X_n)$ *iid* $N(\theta, 1)$. $\bar{X}$ is CAN for θ since $\bar{X} \sim N(\theta, (1/n))$. Consider the estimation of $g(\theta) = \theta^2$. In this case, $(\partial/\partial\theta)g(\theta) = 2\theta = 0$ for $\theta = 0$. Thus, the theorem fails in this case. The choice of norming constant $a_n = \sqrt{n}$ makes $a_n(\bar{X}^2 - \theta^2) = \sqrt{n}(\bar{X}^2 - \theta^2) \to 0$ in probability at $\theta = 0$, and therefore, the limiting distribution is not normal. If the condition of $\partial g/\partial\theta$ being nonvanishing is violated, the choice of considering sequences of norming constants $\{a_n^2\}$ at the place of $\{a_n\}$ saves us, but the cost that we pay is that the limiting distribution of $a_n^2\,(\bar{X}^2 - \theta^2)$ is non-normal. A general result, in this case, is stated in the next Section 5.8.

5.8 METHODS OF FINDING CAN ESTIMATORS AND THEIR PROPERTIES

We discuss here two methods of finding CAN estimators, namely method of moments and method of percentiles.

5.8.1 Method of Moments (MoM) Estimation

Let $X_1, X_2,\ldots,X_n$ be *iid* from a population with *pdf* $f(x;\theta)$ so that $E(X) = \mu(\theta)$ and $V(X) = \sigma^2(\theta)$. The moment equation is

$$\bar{X} = \mu(\theta)$$

The method of moments estimator $\hat{\theta} = \mu^{-1}(\bar{X})$ is CAN for θ. This is shown in the following theorem.

Theorem 5.8.1 Let $X_1, X_2,\ldots,X_n$ be *iid* from $f(x;\theta)$, $\theta \in \Theta \subseteq \mathbb{R}^1$, and for some function $g(X)$, $E[g(X)] = \mu(\theta)$, $V[g(X)] = \sigma^2(\theta)$. Suppose the quantity $[\partial\mu(\theta)/\partial\theta]^{-1} \neq 0$ and is continuous. Then $\mu^{-1}[\bar{g}(X)]$ is consistent for θ and

$$\sqrt{n}\{\mu^{-1}[\bar{g}(\mathbf{X})] - \theta\} \xrightarrow{d} N\left\{0, V[g(X_1)]\left(\frac{\partial\mu(\theta)}{\partial\theta}\right)^{-2}\right\}$$

i.e., $\mu^{-1}[\bar{g}(X)]$ is CAN for θ, where $\bar{g}(\mathbf{X}) = (1/n)\Sigma_1^n g(X_i)$.

Proof. By central limit theorem,

$$\sqrt{n}[\bar{g}(\mathbf{X}) - \mu(\theta)] \xrightarrow{d} N(0, V[g(X)]) = N(0, \sigma^2(\theta))$$

Consider the moment equation

$$\bar{g}(\mathbf{X}) = \mu(\theta)$$

On solving this equation, we get the method of moments (MoM) estimator $\hat{\theta} = \mu^{-1}[\bar{g}(\mathbf{X})]$. The condition that $\partial\mu(\theta)/\partial\theta$ is nonvanishing ($\neq 0$) and continuous is a sufficient condition for μ^{-1} to exist and to be continuous and differentiable. Then, the results of Section 5.6.1 imply that $\mu^{-1}[\bar{g}(\mathbf{X})]$ is consistent for θ.

Consider, now, the reparametrization $\eta = \mu(\theta)$ so that $\theta = \mu^{-1}(\eta)$. We have

$$\sqrt{n}[\bar{g}(\mathbf{X}) - \eta] \xrightarrow{d} N(0, \sigma^2[\mu^{-1}(\eta)])$$

and
$$\frac{\partial\mu^{-1}(\eta)}{\partial\eta} = \frac{\partial\theta}{\partial\eta} = \frac{1}{(\partial\eta/\partial\theta)} = \left(\frac{\partial\eta}{\partial\theta}\right)^{-1}$$

$$\left[\text{since}\left(\frac{\partial\eta}{\partial\theta}\right)\left(\frac{\partial\theta}{\partial\eta}\right) = 1\right]$$

Given that $(\partial\eta/\partial\theta)^{-1}$ exists and is nonvanishing and continuous implies that $\partial\mu^{-1}(\eta)/\partial\eta$ exists and is nonvanishing and continuous. Therefore, by Theorem 5.7.2,

$$\sqrt{n}\{\mu^{-1}[g(\mathbf{X})] - \mu^{-1}(\eta)\} \xrightarrow{d} N\left(0, \sigma^2[\mu^{-1}(\eta)]\left(\frac{\partial\mu^{-1}(\eta)}{\partial\eta}\right)^2\right)$$

This gives $\sqrt{n}\{\mu^{-1}[\overline{g}(\mathbf{X})]-\theta\} \xrightarrow{d} N\left(0, \sigma^2(\theta)\left(\frac{\partial\mu(\theta)}{\partial\theta}\right)^{-2}\right)$

$$= N\left(0, V[g(X)]\left(\frac{\partial\mu(\theta)}{\partial\theta}\right)^{-2}\right)$$

Alternatively, this theorem can be proved by expanding the function $\mu^{-1}(\overline{g})$ by the Taylor expansion at the point $\mu(\theta)$

$$\mu^{-1}(\overline{g}) = \mu^{-1}[\mu(\theta)] + [\mu^{-1}(\cdot)]'\big|_{\mu(\theta)}[\overline{g}-\mu(\theta)] + \frac{[\mu^{-1}(\cdot)]''\big|_b}{2!}[\overline{g}-\mu(\theta)]^2 \tag{5.8.1}$$

where $b \in [\mu(\theta), \overline{g}]$. This simplifies to

$$\mu^{-1}(\overline{g}) - \theta = [\mu^{-1}(\cdot)]'\big|_{\mu(\theta)}[\overline{g}-\mu(\theta)] + \frac{[\mu^{-1}(\cdot)]''\big|_b}{2!}[\overline{g}-\mu(\theta)]^2$$

since $\mu^{-1}[\mu(\theta)] = \theta$. We may rewrite

$$\sqrt{n}[\mu^{-1}(\overline{g}) - \theta] = [\mu^{-1}(\cdot)]'\big|_{\mu(\theta)}\sqrt{n}[\overline{g}-\mu(\theta)] + \frac{[\mu^{-1}(\cdot)]''\big|_b}{2!}\frac{1}{\sqrt{n}}\{\sqrt{n}[\overline{g}-\mu(\theta)]\}^2 \tag{5.8.2}$$

The last term in Eq. (5.8.2) converges to 0 because $\sqrt{n}$ is in the denominator. The factor $[\mu^{-1}(\cdot)]'|_{\mu(\theta)}$ in the first term reduces to

$$[\mu^{-1}(\cdot)]'\big|_{\mu(\theta)} = \frac{\partial}{\partial\mu(\theta)}\mu^{-1}[\mu(\theta)] = \frac{1}{\mu'(\theta)}$$

and by the central limit theorem, the quantity $\sqrt{n}[\overline{g} - \mu(\theta)]$ in the first term in Eq. (5.8.2) converges in distribution to

$$\sqrt{n}[\overline{g}-\mu(\theta)] \xrightarrow{d} N(0, V_\theta[g(X_1)])$$

where X_1 is a single observation. Therefore, the left side of Eq. (5.8.2) converges in distribution to

$$\sqrt{n}[\mu^{-1}(\overline{g})-\theta] \xrightarrow{d} \frac{1}{\mu'(\theta)}N(0, V_\theta[g(X_1)]) = N\left(0, V[g(X_1)]\left(\frac{\partial\mu(\theta)}{\partial\theta}\right)^{-2}\right)$$

This proves what is asserted in the theorem. ▣

One application of this theorem is that an MoM estimator is preferred if its asymptotic variance is small. Consider an example where $X \sim E(\theta)$, $f(x;\theta) = \theta\exp(-\theta x)$. $\delta_1 = 1/\overline{X}$ and $\delta_2 = \sqrt{2/\overline{X^2}}$ are two MoM estimators of θ. The result of Theorem 5.8.1 is used to answer the question as to which of the two is better. We may easily show that δ_1 is superior to δ_2 on the basis of their asymptotic variances.

The multivariate version of the theorem can be stated as follows: Consider a random sample $(X_1, X_2,\ldots,X_n)$ from the population with *pdf* $f(x;\boldsymbol{\theta})$, $\boldsymbol{\theta} \in \Theta \subset \mathbb{R}^k$. Define a k-dimensional statistic based on a single observation X by $\mathbf{U}(X) = (U_1(X), U_2(X),\ldots,U_k(X))'$. We have a random sample on $\mathbf{U}(X)$ namely $\{U_1(X_i), U_2(X_i),\ldots,U_k(X_i)\}_{i=1}^n$ corresponding to a random sample $(X_1, X_2,\ldots,X_n)$, so that $E[U_j(X)] = \pi_j(\boldsymbol{\theta})$, $j = 1, 2,\ldots,k$, and $\text{cov}(U_r, U_s) = \sigma_{rs}(\boldsymbol{\theta})$, where $\Sigma = [\sigma_{rs}(\theta)]_{\substack{r=1,\ldots,k\\ s=1,\ldots,k}}$ is positive definite. The moment equation, in this case, is given by

$$\bar{\mathbf{U}}(\mathbf{X}) = \boldsymbol{\pi}(\boldsymbol{\theta}) \tag{5.8.3}$$

$$\bar{U}_j(\mathbf{X}) = \frac{1}{n}\sum_1^n U_j(X_i) = \pi_j(\boldsymbol{\theta}),\ j = 1,\ldots,k$$

On assuming that the inverse function of π, namely ϕ, exists and is continuous, the solution of the moment Eq. (5.8.3) is given by

$$\boldsymbol{\delta} = (\delta_1, \delta_2,\ldots,\delta_k)' = [\phi_1(\bar{U}_1,\ldots,\bar{U}_k),\ldots,\phi_k(\bar{U}_1,\ldots,\bar{U}_k)] = \boldsymbol{\phi}(\bar{\mathbf{U}})$$

The estimator $\boldsymbol{\delta}$ is known as the *moments estimator* of $\boldsymbol{\theta}$. We will now show that $\boldsymbol{\delta}$ is a CAN estimator of $\boldsymbol{\theta}$ in the following theorem.

Theorem 5.8.2 Let $\mathbf{B} = (\partial\boldsymbol{\pi}/\partial\boldsymbol{\theta})$ be non-singular. Then, the moment estimator $\boldsymbol{\delta} = \boldsymbol{\phi}(\bar{\mathbf{U}})$ for $\boldsymbol{\theta}$ is CAN, i.e.,

$$\sqrt{n}(\boldsymbol{\delta} - \boldsymbol{\theta}) \xrightarrow{d} N\left(\mathbf{0}, \mathbf{B}^{-1}\Sigma(\mathbf{B}^{-1})'\right)$$

Proof. $E(\bar{\mathbf{U}})$ exists and is finite. $D(\bar{\mathbf{U}}) = (1/n)\Sigma$. Therefore,

$$\bar{\mathbf{U}} \xrightarrow{P} \boldsymbol{\pi}(\boldsymbol{\theta})$$

and
$$\sqrt{n}[\bar{\mathbf{U}} - \boldsymbol{\pi}(\boldsymbol{\theta})] \xrightarrow{d} N(\mathbf{0}, \Sigma)$$

i.e., $\bar{\mathbf{U}}$ is CAN for $\boldsymbol{\pi}(\boldsymbol{\theta})$.

Since $\mathbf{B} = (\partial\boldsymbol{\pi}/\partial\boldsymbol{\theta})$ is non-singular, i.e.,

$$\left|\frac{\partial(\pi_1(\boldsymbol{\theta}),\ldots,\pi_k(\boldsymbol{\theta}))}{\partial(\theta_1,\ldots,\theta_k)}\right| \neq 0$$

the moment equations $\bar{\mathbf{U}} = \boldsymbol{\pi}(\boldsymbol{\theta})$ would admit a unique solution, say $\boldsymbol{\delta} = \boldsymbol{\phi}(\bar{\mathbf{U}})$

In this case,

$$\mathbf{C} = \left(\frac{\partial(\phi_1,\ldots,\phi_k)}{\partial(\theta_1,\ldots,\theta_k)}\right) = \mathbf{B}^{-1}$$

Next, consider the reparametrization $\boldsymbol{\eta} = \boldsymbol{\pi}(\boldsymbol{\theta})$. We have, since $\mathbf{B}$ is nonsingular and $\boldsymbol{\pi}$ has a unique inverse $\boldsymbol{\phi}$ so that $\boldsymbol{\theta} = \boldsymbol{\phi}(\boldsymbol{\eta})$, with Jacobian of transformation $\mathbf{C}$, so that $\mathbf{C} = \mathbf{B}^{-1}$, $\mathbf{BC} = \mathbf{CB} = \mathbf{I}$. $\mathbf{C}$ is non-singular, $|\mathbf{C}| \neq 0$. By Theorem 5.7.3,

$$\sqrt{n}[\boldsymbol{\phi}(\bar{\mathbf{U}}) - \boldsymbol{\phi}(\boldsymbol{\eta})] \xrightarrow{d} N(\mathbf{0}, \mathbf{C}\Sigma\mathbf{C}')$$

$$\sqrt{n}(\boldsymbol{\delta} - \boldsymbol{\theta}) \xrightarrow{d} N\left(\mathbf{0}, \mathbf{B}^{-1}\Sigma(\mathbf{B}^{-1})'\right)$$

This proves the theorem. ▣

One can easily see that the moment estimator is a function of the sufficient statistic when the density $f(x;\theta)$ belongs to one-parameter exponential family of distributions. The following theorem, in this case, illustrates a relationship between asymptotic variance of a CAN estimator with CRLB where CAN estimator is obtained through moment equations. In fact, it is shown that in this case, the asymptotic variance of such a CAN estimator attains CRLB. However, in some situations, there may not exist an unbiased estimator of θ which attains CRLB. In these situations, the asymptotic variance of such a CAN estimator is equal to the variance of a UMVUE of θ and is larger than CRLB. An illustration of this result is given in Section 5.9 of solved examples.

Theorem 5.8.3 Consider a random sample $\mathbf{X} = (X_1, X_2,\ldots,X_n)$ *iid* with $f(x;\theta)$, so that $f(x;\theta)$ belongs to one-parameter exponential family. Then, the moment estimator $\hat{\theta}$ based on a sufficient statistic is CAN for θ

$$\sqrt{n}(\hat{\theta}-\theta) \xrightarrow{d} N\left(0, \frac{1}{I_X(\theta)}\right)$$

where $I_X(\theta)$ is Fisher information contained in a single observation X on the parameter θ.

Proof. The *pdf* of the distribution belonging to one-parameter exponential family is given by

$$f(x;\ \theta) = \exp\{Q(\theta)T(x) + D(\theta) + S(x)\}$$

where $Q(\theta)$ and $D(\theta)$ are real-valued functions on Θ and $T(x)$ and $S(x)$ are some Borel-measurable functions on sample space $\chi \subseteq \mathbb{R}^1$. We know that the statistic $T(X)$ is a complete sufficient statistic for $Q(\theta)$. Consider

$$\log f = Q(\theta)T(x) + D(\theta) + S(x)$$

$$\therefore \qquad \frac{\partial}{\partial\theta}\log f = Q'(\theta)T(x) + D'(\theta) \tag{5.8.4}$$

We have noted in subsection 4.3.2 that

$$E_\theta\left(\frac{\partial}{\partial\theta}\log f\right) = 0$$

which, here, implies that

$$Q'(\theta)\, E_\theta[T(X)] + D'(\theta) = 0$$

or

$$E_\theta[T(X)] = -\frac{D'(\theta)}{Q'(\theta)} = g(\theta) \ \ \text{(say)} \tag{5.8.5}$$

On using the result of Eq. (5.8.5), we get

$$\left(\frac{\partial}{\partial\theta}\log f\right)^2 = Q'^2(\theta)T^2(x) + 2Q'(\theta)D'(\theta)T(x) + D'^2(\theta)$$

$$E_\theta\left(\frac{\partial}{\partial\theta}\log f\right)^2 = Q'^2 E_\theta(T^2) - 2Q'D'\frac{D'}{Q'} + D'^2$$

$$= Q'^2 E_\theta(T^2) - D'^2 = Q'^2 \left[E_\theta(T^2) - \left(\frac{D'}{Q'} \right)^2 \right]$$

$$= Q'^2 [E_\theta(T^2) - (E_\theta T)^2] = Q'^2 V_\theta(T) \tag{5.8.6}$$

The identity $E_\theta[(\partial/\partial\theta) \log f]^2 = I_X(\theta)$ implies

$$V_\theta(T) = \frac{I_X(\theta)}{Q'^2} \tag{5.8.7}$$

On using Eqs. (5.8.5) and (5.8.7) in the central limit theorem, we get

$$\sqrt{n}[T_n - g(\theta)] \xrightarrow{d} N\left(0, \frac{I_X(\theta)}{Q'^2}\right)$$

where $T_n = (1/n) \Sigma_1^n T(X_i)$.
The moment equation is given by

$$T_n = \frac{1}{n} \sum_1^n T(X_i) = g(\theta)$$

To investigate whether g^{-1} exists and is continuous, we consider the differentiation of Eq. (5.8.4) with respect to θ

$$\frac{\partial^2}{\partial\theta^2} \log f = Q''T + D''$$

$$E_\theta\left(-\frac{\partial^2}{\partial\theta^2} \log f \right) = I_X(\theta) = -Q'' E_\theta(T) - D''$$

$$= Q'' \frac{D'}{Q'} - D'' = \frac{Q''D' - D''Q'}{Q'}$$

On using Eq. (5.8.5), we have

$$I_X(\theta) = \frac{Q''D' - D''Q'}{Q'} = \left[\frac{\partial}{\partial\theta} g(\theta) \right] Q'$$

This gives

$$\frac{\partial}{\partial\theta} g(\theta) = \frac{I_X(\theta)}{Q'}$$

This implies that $[(\partial/\partial\theta) g(\theta)]^{-1}$ is nonvanishing and continuous. Therefore, by Theorem 5.8.1, the moment estimator $\hat{\theta} = g^{-1}(T)$ is CAN for θ, so that

$$\sqrt{n}[g^{-1}(T) - \theta] \xrightarrow{d} N\left[0, \sigma_T^2(\theta) \left(\frac{\partial g}{\partial\theta} \right)^{-2} \right]$$

$$\sqrt{n}[g^{-1}(T)-\theta] \xrightarrow{d} N\left(0, \frac{I_X(\theta)}{Q'^2}\frac{Q'^2}{[I_X(\theta)]^2}\right)$$

or
$$\sqrt{n}[g^{-1}(T)-\theta] \xrightarrow{d} N\left(0, \frac{1}{I_X(\theta)}\right)$$

This completes the proof of the theorem. ▣

The multivariate version of the above result is stated in the following theorem.

Theorem 5.8.4 Let $(X_1, X_2,\ldots,X_n)$ be a random sample from $f(x;\ \boldsymbol{\theta})$, where $f(x;\ \boldsymbol{\theta})$ belongs to k-parameter exponential family $\{f(x;\boldsymbol{\theta});\ \boldsymbol{\theta} \in \Theta \subset \mathbb{R}^k\}$. Then, the moment estimator based on the minimal sufficient statistic is CAN for $\boldsymbol{\theta}$

$$\sqrt{n}(\hat{\boldsymbol{\theta}}-\boldsymbol{\theta}) \xrightarrow{d} N(0, \mathbf{I}_{\mathbf{X}}^{-1}(\boldsymbol{\theta}))$$

Proof. The *pdf* $f(x;\boldsymbol{\theta})$ belonging to the k-parameter exponential family is given by

$$f(x;\ \boldsymbol{\theta}) = \exp\{\mathbf{Q}'(\boldsymbol{\theta})\mathbf{T}(x) + D(\boldsymbol{\theta}) + S(x)\}$$

where $\mathbf{Q}'\mathbf{T} = \Sigma_{j=1}^{k} Q_j(\boldsymbol{\theta})\, T_j(x)$, $\boldsymbol{\theta} \in \Theta \subset \mathbb{R}^k$. Since $|d\mathbf{Q}/d\boldsymbol{\theta}| \neq 0$, for reparametrization of the problem $\boldsymbol{\eta} = \mathbf{Q}(\boldsymbol{\theta})$, there exists a unique inverse transformation $\mathbf{Q}^{-1}$ so that $\boldsymbol{\theta} = \mathbf{Q}^{-1}(\boldsymbol{\eta}) = \boldsymbol{\ell}(\boldsymbol{\eta})$. The density under this parametrization is

$$\log f(x;\ \boldsymbol{\ell}(\boldsymbol{\eta})) = \boldsymbol{\eta}'\mathbf{T}(x) + D_1(\boldsymbol{\eta}) + S(x)$$

where $D_1(\boldsymbol{\eta}) = D(\boldsymbol{\ell}(\boldsymbol{\eta}))$. Consider

$$\frac{\partial}{\partial \boldsymbol{\eta}}\log f(x;\ \boldsymbol{\ell}(\boldsymbol{\eta})) = \mathbf{T}(x) + \frac{\partial}{\partial \boldsymbol{\eta}}D_1(\boldsymbol{\eta})$$

$$E_\eta\left[\frac{\partial}{\partial \boldsymbol{\eta}}\log f(X;\ \boldsymbol{\ell}(\boldsymbol{\eta}))\right] = 0 = E[\mathbf{T}(X)] + \frac{\partial}{\partial \boldsymbol{\eta}}D_1(\boldsymbol{\eta})$$

$$E[\mathbf{T}(X)] = -\frac{\partial}{\partial \boldsymbol{\eta}}D_1(\boldsymbol{\eta}) = \mathbf{g}(\boldsymbol{\eta}) \tag{5.8.8}$$

This gives
$$\frac{\partial}{\partial \boldsymbol{\eta}}\log f = [\mathbf{T}(x) - \mathbf{g}(\boldsymbol{\eta})] \tag{5.8.9}$$

On differentiating Eq. (5.8.9) with respect to $\boldsymbol{\eta}$, we get

$$\begin{aligned}
E\left[-\frac{\partial^2}{\partial \boldsymbol{\eta}^2}\log f\right] &= E\left[\left(\frac{\partial}{\partial \boldsymbol{\eta}}\log f\right)\left(\frac{\partial}{\partial \boldsymbol{\eta}}\log f\right)'\right] = E\left(-\frac{\partial^2 \log f}{\partial \eta_i \partial \eta_j}\right) \\
&= \mathbf{I}_{\mathbf{X}}(\boldsymbol{\eta}) = \frac{\partial \mathbf{g}(\boldsymbol{\eta})}{\partial \boldsymbol{\eta}} = \mathbf{G}\ (\text{say}) \\
&= E\{[\mathbf{T}(X) - \mathbf{g}(\boldsymbol{\eta})][\mathbf{T}(X) - \mathbf{g}(\boldsymbol{\eta})]'\} \qquad (5.8.10) \\
&= D(\mathbf{T})
\end{aligned}$$

where $\mathbf{I}_{\mathbf{X}}(\boldsymbol{\eta})$ is Fisher information matrix of $\boldsymbol{\eta}$.

The moment equations

$$\bar{\mathbf{T}}(\mathbf{x}) = \mathbf{g}(\boldsymbol{\eta}) \tag{5.8.11}$$

admit a unique solution $\hat{\boldsymbol{\eta}} = (\hat{\eta}_1,\ldots,\hat{\eta}_k) = \mathbf{g}^{-1}[\bar{\mathbf{T}}(\mathbf{x})]$ since $|\mathbf{G}| = |\partial \mathbf{g}(\boldsymbol{\eta})/\partial\boldsymbol{\eta}| = |\partial(g_1,\ldots,g_k)/\partial(\eta_1,\ldots,\eta_k)| \neq 0$. Here, $\bar{\mathbf{T}}(\mathbf{x}) = [\bar{T}_1(\mathbf{x}),\ldots,\bar{T}_k(\mathbf{x})]'$, $\bar{T}_j(\mathbf{x}) = (1/n)\Sigma_{i=1}^n T_j(x_i)$, $j = 1,\ldots,k$, $\mathbf{g}(\boldsymbol{\eta}) = [g_1(\boldsymbol{\eta}),\ldots,g_k(\boldsymbol{\eta})]'$. By the multivariate central limit theorem,

$$\sqrt{n}[\bar{\mathbf{T}} - \mathbf{g}(\boldsymbol{\eta})] \xrightarrow{d} N(\mathbf{0}, \mathbf{I}_{\mathbf{X}}(\boldsymbol{\eta}))$$

so that $\bar{\mathbf{T}}$ is CAN for $\mathbf{g}(\boldsymbol{\eta})$ and that $|\partial\mathbf{g}/\partial\boldsymbol{\eta}| \neq 0$. Thus, the asymptotic distribution of the moment estimator $\hat{\boldsymbol{\eta}} = \mathbf{g}^{-1}[\bar{\mathbf{T}}(\mathbf{x})]$ is given by using Theorem 5.7.3

$$\sqrt{n}[\mathbf{g}^{-1}(\bar{\mathbf{T}}) - \boldsymbol{\eta}] \xrightarrow{d} N(0, \mathbf{G}^{-1}\mathbf{I}_{\mathbf{X}}(\mathbf{G}^{-1})') \tag{5.8.12}$$

where $\mathbf{G} = \mathbf{I}(\boldsymbol{\eta})$, which gives $\mathbf{G}^{-1}\mathbf{I}\mathbf{G}^{-1} = \mathbf{I}^{-1}(\boldsymbol{\eta})$. Therefore,

$$\sqrt{n}(\hat{\boldsymbol{\eta}} - \boldsymbol{\eta}) \xrightarrow{d} N(\mathbf{0}, \mathbf{I}_{\mathbf{X}}^{-1}(\boldsymbol{\eta}))$$

Using $\boldsymbol{\theta} = \mathbf{Q}^{-1}(\boldsymbol{\eta}) = \boldsymbol{\ell}(\boldsymbol{\eta})$ and the moment estimator $\hat{\boldsymbol{\eta}}$ give uniquely

$$\hat{\boldsymbol{\theta}} = \boldsymbol{\ell}(\hat{\boldsymbol{\eta}})$$

Since $\mathbf{L} = \partial\boldsymbol{\theta}/\partial\boldsymbol{\eta} = (\partial\mathbf{Q}(\boldsymbol{\eta})/\partial\boldsymbol{\eta})^{-1}$, $|\partial\mathbf{Q}/\partial\boldsymbol{\theta}| \neq 0$ implies that $|\mathbf{L}| \neq 0$. Thus, Theorem 5.7.3 gives

$$\sqrt{n}(\hat{\boldsymbol{\theta}} - \boldsymbol{\theta}) \xrightarrow{d} N[\mathbf{0}, \mathbf{L}\mathbf{I}_{\mathbf{X}}^{-1}(\boldsymbol{\eta})\mathbf{L}'] \tag{5.8.13}$$

Consider, now, Fisher information matrix for $\boldsymbol{\theta}$, $\mathbf{I}(\boldsymbol{\theta})$.

$$\mathbf{I}_{\mathbf{X}}(\boldsymbol{\theta}) = E\left[\left(\frac{\partial}{\partial\boldsymbol{\theta}}\log f\right)\left(\frac{\partial}{\partial\boldsymbol{\theta}}\log f\right)'\right]$$

where
$$\frac{\partial}{\partial\boldsymbol{\theta}}\log f = \frac{\partial\boldsymbol{\eta}}{\partial\boldsymbol{\theta}}\frac{\partial}{\partial\boldsymbol{\eta}}\log f = \mathbf{L}^{-1}\frac{\partial}{\partial\boldsymbol{\eta}}\log f$$

This gives
$$\mathbf{I}_{\mathbf{X}}(\boldsymbol{\theta}) = E\left[\mathbf{L}^{-1}\frac{\partial}{\partial\boldsymbol{\eta}}\log f\left(\frac{\partial}{\partial\boldsymbol{\eta}}\log f\right)'(\mathbf{L}^{-1})'\right]$$

$$= \mathbf{L}^{-1}E\left[\left(\frac{\partial}{\partial\boldsymbol{\eta}}\log f\right)\left(\frac{\partial}{\partial\boldsymbol{\eta}}\log f\right)'\right](\mathbf{L}^{-1})'$$

$$= \mathbf{L}^{-1}\mathbf{I}(\boldsymbol{\eta})(\mathbf{L}^{-1})'$$

This gives
$$\mathbf{L}\mathbf{I}(\boldsymbol{\eta})\mathbf{L}' = \mathbf{I}_{\mathbf{X}}^{-1}(\boldsymbol{\theta}) \tag{5.8.14}$$

Equations (5.8.13) and (5.8.14) finally give

$$\sqrt{n}(\hat{\boldsymbol{\theta}}-\boldsymbol{\theta}) \xrightarrow{d} N(\mathbf{0}, \mathbf{I}_{\mathbf{X}}^{-1}(\boldsymbol{\theta}))$$

5.8.2 Method of Percentiles (MoP)

If we denote the pth population percentile by $x_p(\theta)$ so that

$$\int_{-\infty}^{x_p(\theta)} f(x;\theta)dx = p$$

the percentile equation is given by

$$X_{([np]+1)} = x_p(\theta)$$

The solution of this equation, $\hat{\theta} = x_p^{-1}(X_{([np]+1)})$, is called the percentile estimator of θ. The following theorem asserts that the percentile estimator is also CAN for θ.

Theorem 5.8.5 (David, 1981) The pth sample percentile for $0 < p < 1$

$$\sqrt{n}[X_{([np]+1)} - x_p(\theta)] \xrightarrow{d} N\left(0, \frac{p(1-p)}{[f(x_p(\theta))]^2}\right), f(x_p(\theta)) > 0$$

or $X_{([np]+1)}$ is CAN for $x_p(\theta)$. If $[(\partial/\partial\theta)x_p(\theta)]^{-1}$ is nonvanishing and continuous, then

$$\sqrt{n}[x_p^{-1}(X_{([np]+1)}) - \theta] \xrightarrow{d} N\left(0, \frac{p(1-p)}{[f(x_p(\theta))]^2}\left(\frac{\partial x_p(\theta)}{\partial\theta}\right)^{-2}\right)$$ ▣

The multivariate version of generating CAN estimators through the method of percentiles has been stated in the following theorems.

Consider $X_1, X_2,\ldots,X_n$ as a random sample from $f(x;\boldsymbol{\theta})$, $\boldsymbol{\theta} \in \Theta \subseteq \mathbb{R}^k$. Let $\mathbf{X}_p = (X_{(i_1)},\ldots,X_{(i_k)})'$ be the sample percentiles, with $i_j = [np_j] + 1$, $j = 1,\ldots,k$, corresponding to the population k-percentile points $(x_{(p_1)}(\boldsymbol{\theta}),\ldots,x_{(p_k)}(\boldsymbol{\theta}))$ for some $p_1,\ldots,p_k$ so that $0 < p_1 < p_2 < \cdots < p_k < 1$.

Theorem 5.8.6 (David, 1981) Under the above setup,

$$\sqrt{n}[\mathbf{X}_p - \mathbf{x}_p(\theta)] \xrightarrow{d} N(\mathbf{0}, \Sigma)$$

with

$$\sigma_{jj} = V\left[\sqrt{n}\left(X_{(ij)} - x_{p_j}(\boldsymbol{\theta})\right)\right] = \frac{p_i(1-p_i)}{[f(x_{p_i}(\boldsymbol{\theta}))]^2}$$

$$\sigma_{jk} = \text{cov}\left[\sqrt{n}\left(X_{(ij)} - x_{p_j}(\boldsymbol{\theta})\right), \sqrt{n}\left(X_{(ik)} - x_{p_k}(\boldsymbol{\theta})\right)\right]$$

$$= \frac{p_i(1-p_j)}{f(x_{p_i}(\boldsymbol{\theta}))\, f(x_{p_j}(\boldsymbol{\theta}))}, \quad j < k$$

▣

The percentile equations

$$\mathbf{X}_p = \mathbf{x}_p(\boldsymbol{\theta})$$

admit a unique solution $\hat{\boldsymbol{\theta}} = \boldsymbol{\phi}(\mathbf{X}_p)$ for some $\boldsymbol{\phi}$ if $\mathbf{B} = (\partial \mathbf{x}_p/\partial \boldsymbol{\theta})$ is nonsingular or $|\mathbf{B}| \neq 0$. We have the following theorem that shows that $\hat{\boldsymbol{\theta}} = \boldsymbol{\phi}(\mathbf{X}_p)$ is CAN estimator for $\boldsymbol{\theta}$.

Theorem 5.8.7 If **B** is nonsingular, then the percentile estimator $\hat{\boldsymbol{\theta}} = \boldsymbol{\phi}(\mathbf{X}_p)$ is CAN

$$\hat{\boldsymbol{\theta}} - \boldsymbol{\theta}) \xrightarrow{d} N\left[\mathbf{0}, \mathbf{B}^{-1}\Sigma(\mathbf{B}^{-1})'\right]$$

▣

5.9 SOLVED EXAMPLES

***Example* 5.1** Let $X_1, X_2, \ldots, X_n$ be *iid* with common *pdf* $f(x;\theta)$, $\theta \in \Theta$, and mean $E(X) = \theta$ exists. Find consistent estimator for θ^2, $\theta^{1+r} (r > 0)$, θ^{-1}.

Solution. Here, $T_n = \bar{X}$ is consistent for $E(X) = \theta$, which implies that by the invariance property of consistent estimators, consistent estimator for θ^2 is $\bar{X}^2$; for θ^{1+r} is $\bar{X}^{1+r}$, $r > 0$; and for θ^{-1} is $\bar{X}^{-1}$.

***Example* 5.2** Consider a random sample $X_1, X_2, \ldots, X_n$ from the populations (i) $N(\theta, 1)$; (ii) $U(\theta - 1, \theta + 1)$. Obtain consistent estimators of θ by the method of moments and the method of percentiles.

Solution. In both the cases, the moment equation is

$$\bar{X} = \theta$$

which gives $\bar{X}$ as a consistent estimator of θ. Further, for $p = (1/2)$, we get the percentile equation

$$X_{([n/2]+1)} = x_{(1/2)}(\theta) = \theta$$

This gives that the sample median is a consistent estimator of population median θ.

***Example* 5.3** Let $X_1, X_2, \ldots, X_n$ be *iid* from a continuous distribution with *pdf* $f(x;\theta)$ so that the population median (Md) exists and $1/f(\text{Md}) < \infty$. Show that the sample median is a consistent estimator of Md.

Solution. Let the sample median be denoted by $\tilde{X}_n$. For this, we have

$$E(\tilde{X}_n) = \text{Md} + O\left(\frac{1}{n}\right)$$

$$V(\tilde{X}_n) = \frac{1}{4n[f(\text{Md})]^2} + O\left(\frac{1}{n}\right)$$

This shows that $E(\tilde{X}_n) \to \text{Md}$ and $V(\tilde{X}_n) \to 0$ as $n \to \infty \ \forall\ \theta$. Thus, by Theorem 5.3.1, $\tilde{X}_n$ is consistent for population median Md.

***Example* 5.4** Let $X_1, X_2, \ldots, X_n$ be *iid* from a distribution with $E(X_i) = \theta$ and $E|X_i|^2 < \infty$. Show that the estimator $\delta_n(\mathbf{X}) = 2[n(n+1)]^{-1}\Sigma iX_i$ is consistent for θ.

Solution. Consider

$$E(\delta_n) = 2[n(n+1)]^{-1}\theta\sum i = \theta$$

and

$$V(\delta_n) = 4[n(n+1)]^{-2}\sum i^2 V(X_i)$$

$$= 4[n(n+1)]^{-2}\left[\frac{n(n+1)(2n+1)}{6}\right]\sigma^2$$

Since $E|X_i|^2 < \infty$ implies $E(X_i - \theta)^2 = \sigma^2 < \infty$, $V(X_i)$ exists and is finite. We have

$$V(\delta_n) = \frac{2(2n+1)}{3n(n+1)}\sigma^2 \to 0 \text{ as } n \to \infty$$

Hence, δ_n is consistent for θ.

***Example* 5.5** Let $X_1, X_2, \ldots, X_n$ be a random sample from a population with *cdf* $F(x;\theta)$, $\theta \in \Theta$. Find an unbiased and consistent estimator of $F(x;\theta) = P_\theta(X \le x)$ for each fixed x.

Solution. Define

$$Y_i = \begin{cases} 1, & \text{if } X_i \le x \\ 0, & \text{otherwise} \end{cases}$$

Note that $P_\theta[Y_i = 1] = P_\theta[X_i \le x] = F(x;\theta)$; and $P_\theta\,[Y_i = 0] = P_\theta[X_i > x] = 1 - F(x;\theta)$.

Corresponding to the random sample $X_1, X_2, \ldots, X_n$ from a population $F(x;\theta)$, we have, for fixed x, $Y_1, \ldots, Y_n$ as *iid* from $b(1, F(x;\theta))$, with $E(Y_i) = F(x;\theta)$ and $V(Y_i) = F(x;\theta)[1 - F(x;\theta)]$.

If we consider the empirical distribution as the estimator of $F(x;\theta)$, then

$$F_n(X) = \frac{1}{n}\sum_1^n Y_i$$

We have $E_F[F_n(X)] = (1/n)\Sigma_1^n Y_i = F(x;\theta)$ which shows that $F_n(X)$ is an unbiased estimator of $F(x)$. Next

$$V[F_n(X)] = \frac{1}{n^2}\sum V(Y_i) = \frac{1}{n}F(x;\theta)[1 - F(x;\theta)] \to 0 \text{ as } n \to \infty$$

Therefore, by Theorem 5.3.1, $F_n(x)$ is an unbiased and consistent estimator of population distribution function $F(x;\theta)$ at each point x.

Extreme order statistics have an important role in the study of consistency when $X_1, X_2, \ldots, X_n$ are *iid* from the Pitman family of distributions, i.e., when the range of the *rvs* depends on the parameter. We state the following example which shows that extreme order statistics (o.s.) are consistent for the parameter.

***Example* 5.6** Let $X_1, X_2, \ldots, X_n$ be a random sample on *rv* X that follows a continuous distribution with *cdf* F and X takes on values from an interval. Then,

(i) $X_{(1)} \xrightarrow{P} \theta_1$ and $X_{(n)} \xrightarrow{P} \theta_2$ whenever $X \in (\theta_1, \theta_2)$

(ii) $X_{(n)} \xrightarrow{P} \theta$ and $(X_{(1)} \xrightarrow{P} -\infty)$ whenever $X \in (-\infty, \theta)$

(iii) $X_{(1)} \xrightarrow{P} \theta$ and $(X_{(n)} \xrightarrow{P} \infty)$ whenever $X \in (\theta, \infty)$

Solution. The distribution functions of $X_{(1)}$ and $X_{(n)}$ are given by $1 - [1 - F(x)]^n$ and $[F(x)]^n$.

(i) The *cdf* of $X_{(1)}$ when $x \in (\theta_1, \theta_2)$ is given by

$$G_1(x) = \begin{cases} 0, & \text{if } x < \theta_1 \\ 1-[1-F(x)]^n, & \text{if } \theta_1 \le x < \theta_2 \\ 1, & \text{if } x \ge \theta_2 \end{cases}$$

The *cdf* of $X_{(n)}$ is

$$G_n(x) = \begin{cases} 0, & \text{if } x < \theta_1 \\ [F(x)]^n, & \text{if } \theta_1 \le x < \theta_2 \\ 1, & \text{if } x \ge \theta_2 \end{cases}$$

On taking limit as $n \to \infty$

$$\lim_{n\to\infty} G_1(x) = \begin{cases} 0, & \text{if } x < \theta_1 \\ 1, & \text{if } x \ge \theta_1 \end{cases}$$

and

$$\lim_{n\to\infty} G_n(x) = \begin{cases} 0, & \text{if } x < \theta_2 \\ 1, & \text{if } x \ge \theta_2 \end{cases}$$

Note that $\lim_{n\to\infty} G_1(x)$ is the *cdf* of the *rv* which is degenerate at θ_1 and $\lim_{n\to\infty} G_n(x)$ is the *cdf* of the *rv* which is degenerate at θ_2. Consider

$$P[|X_{(1)} - \theta_1| < \varepsilon] = P[\theta_1 - \varepsilon < X_{(1)} < \theta_1 + \varepsilon]$$

$$= G_1(\theta_1 + \varepsilon) - G_1(\theta_1 - \varepsilon) \to 1 - 0 = 1$$

for any $\varepsilon > 0$. Therefore, we have

$$P[|X_{(1)} - \theta_1| < \varepsilon] \to 1 \text{ for every } \varepsilon > 0 \text{ as } n \to \infty$$

or

$$X_{(1)} \xrightarrow{P} \theta$$

Similarly, one can show that $X_{(n)} \xrightarrow{P} \theta_2$.

(ii) If the range of X is $(-\infty, \theta)$

$$G_1(x) = \begin{cases} 1-[1-F(x)]^n, & \text{if } x < \theta \\ 1, & \text{if } x \ge \theta \end{cases}$$

$$G_n(x) = \begin{cases} [F(x)]^n, & \text{if } x < \theta \\ 1, & \text{if } x \ge \theta \end{cases}$$

$$\lim_{n\to\infty} G_1(x) = \begin{cases} 1, & \text{if } x < \theta \\ 1, & \text{if } x \ge \theta \end{cases}$$

$$\lim_{n\to\infty} G_n(x) = \begin{cases} 0, & \text{if } x < \theta \\ 1, & \text{if } x \ge \theta \end{cases}$$

We have $\quad P[|X_{(n)} - \theta| < \varepsilon] = G_n(\theta + \varepsilon) - G_n(\theta - \varepsilon) \to 1 \text{ as } n \to \infty \qquad (5.9.1)$

$$P[X_{(1)} \ge x'] \to 0 \text{ as } n \to \infty \;\forall\; x' \in \mathbb{R}^1$$

which implies $\quad P[X_{(1)} < x'] \to 1 \text{ as } n \to \infty \;\forall\; x' \in \mathbb{R}^1$

$$X_{(1)} \to -\infty \qquad (5.9.2)$$

By Eqs. (5.9.1) and (5.9.2), $X_{(n)} \xrightarrow{P} \theta$ and $X_{(1)}$ diverges to $-\infty$ as $n \to \infty$.

(iii) Proceeding with similar arguments as in (ii), one can easily prove the results.

***Example* 5.7** Consider a random sample $X_1, X_2, \ldots, X_n$ from the Poisson population $P(\theta)$, $\theta > 0$. Find a consistent estimator of $P(X = 0) = \exp(-\theta)$ and compare it with UMVUE $\delta_0 = [(n-1)/n]^{\Sigma X_i}$. Compare the two estimators $(1/n)\Sigma_{i=1}^{n} I(X_i = 0)$ and MLE for estimating $P(X = 0) = \exp(-\theta)$ by calculating ARE. Investigate whether it is CAN. Also, show that the estimator $\bar{X}\exp(-\bar{X})$ for $P(X = 1) = \theta\exp(-\theta)$ may not be CAN at $\theta = 1$.

Solution. The moment equation is

$$\frac{1}{n}\sum_{1}^{n} X_i = m_1' = \mu_1'(\theta) = \theta$$

which gives $\hat{\theta} = \bar{X}$. This is a consistent estimator of θ. On using the central limit theorem, we get

$$\sqrt{n}(\bar{X} - \theta) \xrightarrow{d} N(0, \theta)$$

Thus, $\bar{X}$ is CAN for θ. Note that $\bar{X}$ is UMVUE too. Since $(\partial/\partial\theta)\exp(-\theta) = -\exp(-\theta) \neq 0$, $\exp(-\theta)$ is a continuous function of θ. By Theorem 5.4.2,

$$\exp(-\bar{X}) \xrightarrow{P} \exp(-\theta)$$

i.e., $\exp(-\bar{X})$ is consistent for $\exp(-\theta)$. Further,

$$E[\exp(-\bar{X})] = \sum_{x=0}^{\infty} \exp\left(-\frac{x}{n}\right)\frac{\exp(-n\theta)(n\theta)^x}{x!}$$

$$\left(\text{since } \sum X_i \sim P(n\theta)\right)$$

$$= \exp(-n\theta)\sum_{x=0}^{\infty}\frac{[n\theta\exp(-1/n)]^x}{x!}$$

$$= \exp(-n\theta)\exp\left[n\theta\exp\left(-\frac{1}{n}\right)\right]$$

$$= \exp\left\{n\theta\left[\exp\left(-\frac{1}{n}\right) - 1\right]\right\}$$

Similarly, $$E[\exp(-2\bar{X})] = \exp\left\{n\theta\left[\exp\left(-\frac{2}{n}\right)-1\right]\right\}$$

This gives

$$V[\exp(-\bar{X})] = E[\exp(-2\bar{X})] - \{E[\exp(-\bar{X})]\}^2$$

$$= \exp\left\{n\theta\left[\exp\left(-\frac{2}{n}\right)-1\right]\right\} - \exp\left\{2n\theta\left[\exp\left(-\frac{1}{n}\right)-1\right]\right\}$$

$$\simeq \exp\left[-2\theta+\left(\frac{2\theta}{n}\right)\right] - \exp\left[-2\theta+\left(\frac{\theta}{n}\right)\right]$$

$$= \left\{\exp\left(\frac{2\theta}{n}\right) - \exp\left(\frac{\theta}{n}\right)\right\}\exp(-2\theta)$$

$$= \exp\left(\frac{\theta}{n}\right)\left\{\exp\left(\frac{\theta}{n}\right)-1\right\}\exp(-2\theta)$$

By the invariance property of CAN estimators, since for $g(\theta) = \exp(-\theta)$, $(\partial/\partial\theta)g(\theta) = -\exp(-\theta) < 0$ for $\forall\ \theta > 0$ and is a continuous function of θ, we have

$$\sqrt{n}[\exp(-\bar{X}) - \exp(-\theta)] \xrightarrow{d} N[0,\ \theta\exp(-2\theta)]$$

Note that the asymptotic variance of CAN estimator, $\exp(-\bar{X})$, for estimating $\exp(-\theta)$ achieves the CR lower bound for any unbiased estimator of $\exp(-\theta)$, $\text{CRLB}(\theta) = \{[(\partial/\partial\theta)\exp(-\theta)]^2/nI_X(\theta)\} = \exp(-2\theta)/(n/\theta) = (\theta/n)\exp(-2\theta)$ (as noted later in this example), though the variance corresponding to its UMVUE, namely $\{(n-1)/n\}^{\Sigma X_i} = \delta_0(\bar{X})$ is greater than CRLB.

The variance of the UMVUE, $\delta_0 = [(n-1)/n]^T$, where $T = \Sigma_1^n X_i$, of $\exp(-\theta)$ is given by

$$V_\theta(\delta_0) = E_\theta(\delta_0^2) - [E_\theta(\delta_0)]^2$$

Let $a = [(n-1)/n]$. We have

$$E_\theta(\delta_0^2) = E(a^{2T}) = \sum_{t=0}^{\infty} a^{2t}\,\frac{\exp(-n\theta)(n\theta)^t}{t!}$$

$$= \exp(-n\theta)\sum_{t=0}^{\infty}\frac{(n\theta a^2)^t}{t!} = \exp(-n\theta)\exp(a^2 n\theta)$$

$$= \exp\left[\frac{(n-1)^2}{n^2}n - n\right]\theta = \exp\left[\frac{1}{n}(1-2n)\theta\right]$$

This gives $$V_\theta(\delta_0) = \exp\left[\frac{1}{n}(1-2n)\theta\right] - \exp(-2\theta)$$

$$= \exp\left(\frac{\theta}{n}\right)\exp(-2\theta) - \exp(-2\theta)$$

$$= \left[\exp\left(\frac{\theta}{n}\right) - 1\right]\exp(-2\theta)$$

Note that $V[\exp(-\bar{X})] > V(\delta_0)$. The variance of the estimator $\exp(-\bar{X})$ is higher than the minimum variance of UMVUE δ_0 by the factor $\exp(\theta/n)$. Further, note that $V(\delta_0) > \text{CRsLB}(\theta)$ $\forall\ \theta > 0$.

The MLE of θ is $\bar{X}$ and by using Zehna's theorem, $\exp(-\bar{X})$ is MLE of $\exp(-\theta)$. Further, $\exp(-\bar{X})$ is asymptotically unbiased, since $n[\exp(-1/n) - 1] = -1 + (1/n2!) - (1/n^2 3!) + \cdots \simeq -1$, and $\exp\{[\exp(-1/n) - 1]n\theta\} \to \exp(-\theta)$. We have already noted that the variance of $\exp(-\bar{X})$ is $\exp(\theta/n)[\exp(\theta/n) - 1]\exp(-2\theta)$. Next, consider the variance of the estimator $(1/n)\Sigma I(X_i = 0)$

$$V\left[\left(\frac{1}{n}\right)\sum I(X_i = 0)\right] = \frac{1}{n}\{[EI^2(X=0)] - [EI(X=0)]^2\}$$

$$= \exp(-\theta)[1 - \exp(-\theta)]/n$$

Thus, the ARE of the estimator $(1/n)\Sigma I(X_i = 0)$ with respect to the MLE $\exp(-\bar{X})$ is given by

$$\text{ARE}\left[\left(\frac{1}{n}\right)\sum I(X_i = 0), \exp(-\bar{X})\right] = \lim_{n\to\infty} \frac{\theta\exp(-2\theta)/n}{\exp(-\theta)[1-\exp(-\theta)]/n}$$

$$= \frac{\theta}{[\exp(\theta)-1]} = \frac{1}{(\theta/2!) + (\theta^2/3!) + \cdots} < 1 \text{ since } \theta > 0$$

This shows that the estimator $\exp(-\bar{X})$ is preferred over $(1/n)\Sigma I(X_i = 0)$.

Now, consider the problem of estimating $P(X = 1) = \theta\exp(-\theta) = g(\theta)$. Note in this case $(\partial/\partial\theta)g(\theta) = (1 - \theta)\exp(-\theta) \neq 0$ for $\theta \neq 1$. Therefore, by Theorem 5.7.2,

$$\sqrt{n}[\bar{X}\exp(-\bar{X}) - \theta\exp(-\theta)] \xrightarrow{d} N[0, \theta(1-\theta)^2\exp(-2\theta)]$$

for any $\theta \neq 1$. Thus, $\bar{X}\exp(-\bar{X})$ is CAN for $\theta\exp(-\theta)$ for any $\theta \neq 1$, but at $\theta = 1$, it may not be CAN. Note that at $\theta = 1$, $(\partial/\partial\theta)g(\theta) = 0$. Therefore, we use Theorem 5.7.4. Note that $(\partial^2/\partial\theta^2)g(\theta) = -(2-\theta)\exp(-\theta)$, $(\partial^2/\partial\theta^2)g(\theta)|_{\theta=1} = -\exp(-1) \neq 0$ and is continuous. Thus, the asymptotic distribution of $\bar{X}\exp(-\bar{X})$ is not normal but

$$n[\bar{X}\exp(-\bar{X}) - \theta\exp(-\theta)] \xrightarrow{d} \frac{1}{2}[-\exp(-1)]\chi_1^2$$

Note that the choice of norming constant is n instead of $\sqrt{n}$.

***Example* 5.8** Consider a random sample drawn from the truncated Poisson distribution with *pmf*

$$P[X = x] = \frac{\exp(-\theta)\theta^x}{x!}[1-\exp(-\theta)]^{-1}$$

$x = 1, 2, 3, \ldots$. Find a consistent estimator for θ.

Solution. Here,

$$E(X) = \theta\exp(-\theta)\sum_{1}^{\infty}\frac{\theta^{x-1}}{(x-1)!}[1-\exp(-\theta)]^{-1}$$
$$= \theta[1-\exp(-\theta)]^{-1}$$

The moment equation is

$$\bar{X} = m_1' = \mu_1'(\theta) = \theta[1-\exp(-\theta)]^{-1}$$

Since $(\partial/\partial\theta)\mu_1'(\theta) = (\partial/\partial\theta)\theta[1 - \exp(-\theta)]^{-1} = [1 - \exp(-\theta)]^{-1} - \theta \exp(-\theta)/[1 - \exp(-\theta)]^2 \neq 0$, $\mu_1'^{-1}$ exists and is a continuous function of θ. This gives the unique solution.

***Example* 5.9 (Estimator which is neither unbiased nor consistent).** Let $X_1, X_2, \ldots, X_n$ be a random sample from the Cauchy population with *pdf*

$$f(x;\theta) = \frac{1}{\pi[1+(x-\theta)^2]}, \ -\infty < x < \infty, -\infty < \theta < \infty$$

Show that $\bar{X}$ is neither unbiased nor consistent for population median θ. Find a consistent estimator for θ.

Solution. The characteristic function for the given Cauchy distribution is given by

$$\phi_{\Sigma X_i}(t) = \prod_{i=1}^{n}\phi_{X_i}(t) = [\phi_X(t)]^n$$

since $X_1, X_2, \ldots, X_n$ are random and drawn identically from $C(\theta, 1)$. Thus,

$$\phi_{\Sigma X_i}(t) = [\exp(\mu it - |t|)]^n = \exp(n\mu it - n|t|)$$

$$\therefore \quad \phi_{\bar{X}}(t) = \phi_{\Sigma X_i}\left(\frac{t}{n}\right) = \exp(\mu it - |t|) = \phi_X(t)$$

This shows that for the Cauchy population, X and $\bar{X}$ have the same distribution, whatever be the sample size. Hence, for the Cauchy population, $E(\bar{X})$ does not exist, which shows that $\bar{X}$ is not an unbiased estimator for θ.

Further, utilizing the above result, that X and $\bar{X}$ have the same distributions, we have

$$P_\theta[|\bar{X} - \theta| < \varepsilon] = P_\theta[|X - \theta| < \varepsilon] \ \forall \ \varepsilon > 0$$
$$= P_\theta[-\varepsilon < X - \theta < \varepsilon] = P_\theta[\theta - \varepsilon < X < \theta + \varepsilon]$$

$$= \frac{1}{\pi} \int_{\theta-\varepsilon}^{\theta+\varepsilon} \frac{1}{1+(x-\theta)^2} dx = \frac{1}{\pi} \int_{-\varepsilon}^{\varepsilon} \frac{1}{1+t^2} dt$$

(on putting $x - \theta = t$)

$$= \frac{2}{\pi} \int_0^{\varepsilon} \frac{1}{1+t^2} dt = \frac{2}{\pi} \tan^{-1}(\varepsilon)$$

This gives $$\lim_{n\to\infty} P_\theta\left[\left|\bar{X}-\theta\right| < \varepsilon\right] = \frac{2}{\pi} \tan^{-1}(\varepsilon) \neq 1 \ \forall\ \varepsilon > 0$$

Hence, $\bar{X}$ is not consistent for θ. Moreover, by the method of percentiles, for $p = 1/2$, the percentile equation is given by

$$X_{([n/2]+1)} = x_{(1/2)}(\theta) = \theta$$

which gives that sample median $X_{([n/2]+1)}$ is a consistent estimator for population median θ.

***Example* 5.10 (Estimator which is unbiased but not Consistent).** Let $X_1, X_2, \ldots, X_n$ be a random sample from $N(\theta, 1)$. Construct an estimator of θ which is unbiased but not consistent.

Solution. Consider

$$T(\mathbf{X}) = X_1,\ E(X_1) = \theta,\ \forall\ \theta \in \mathbb{R}^1$$

X_1 is unbiased for θ. However,

$$P_\theta[|T-\theta| < \varepsilon] = P_\theta[|X_1 - \theta| < \varepsilon]$$
$$= \Phi(\varepsilon) - \Phi(-\varepsilon) \neq 1 \ \forall\ \varepsilon > 0$$

since $X_1 - \theta \sim N(0, 1)$; Φ denotes the probability distribution function of standard normal variate. Therefore, $T = X_1$ is not consistent for θ.

***Example* 5.11** Let $X_1, X_2, \ldots, X_n$ be *iid* from Be$(\theta, 1)$. Find a consistent estimator for θ. Also find its CAN estimator.

Solution. Consider

$$E_\theta(X) = \theta \int_0^1 x x^{\theta-1} dx = \frac{\theta}{1+\theta} = \mu(\theta)$$

$$E_\theta(X^2) = \int_0^1 \theta x^{\theta+1} dx = \frac{\theta}{\theta+2}$$

$$V_\theta(X) = \frac{\theta}{\theta+2} - \left(\frac{\theta}{1+\theta}\right)^2$$

This gives $$E_\theta(\bar{X}) = \frac{1}{n}\sum_1^n \frac{\theta}{1+\theta} = \frac{\theta}{1+\theta}$$

$$V_\theta(\bar{X}) = \frac{1}{n^2}\sum_1^n V(X_i) = \frac{1}{n}\left[\frac{\theta}{\theta+2} - \left(\frac{\theta}{1+\theta}\right)^2\right] = \frac{\theta}{n(1+\theta)^2(2+\theta)} \to 0 \text{ as } n \to \infty$$

By Theorem 5.4.1, $\bar{X}$ is consistent for $\theta/(1 + \theta)$. Further, by the invariance property of consistent estimators and that $\theta/(1 + \theta)$ is a continuous function of θ, $\bar{X}/(1 + \bar{X})$ is a consistent estimator for θ.

On using the method of moments to generate CAN estimator for θ, we have

$$\bar{X} = \frac{\theta}{1+\theta} = \mu(\theta) \tag{5.9.3}$$

Here, $\partial\mu/\partial\theta = 1/(1 + \theta)^2$. $(\partial\mu/\partial\theta)^{-1} = (1 + \theta)^2 > 0$ and is continuous. Therefore, by Theorem 5.8.1, the solution of the moment Eq. (5.9.3)

$$\hat{\theta} = \frac{\bar{X}}{1+\bar{X}}$$

is CAN for θ, i.e.,

$$\sqrt{n}(\hat{\theta} - \theta) \xrightarrow{d} N\left[0, \frac{\theta(\theta+1)^2}{(\theta+2)}\right] \tag{5.9.4}$$

The problem of estimating θ may alternatively be handled by considering the transformation $U = -\log X$. We have

$$h(u, \theta) = \theta \exp(-\theta u), \ \theta > 0, y > 0$$

In this case, $E(U^r) = \theta\Gamma(r + 1)/\theta^{r+1}$ gives $E(U) = (1/\theta) = \mu(\theta)$, $V(U) = (1/\theta^2) = \sigma^2(\theta)$. Further, note that $[\partial\mu(\theta)/\partial\theta] = -(1/\theta^2)$, $[\partial\mu(\theta)/\partial\theta]^{-1} = -\theta^2 < 0$ and is continuous. Therefore, by Theorem 5.8.1, the solution of moment equation $\bar{u} = (1/\theta)$ i.e.,

$$\hat{\theta} = \frac{n}{-\sum \log X_i} \tag{5.9.5}$$

is CAN for θ so that

$$\sqrt{n}(\hat{\theta} - \theta) \xrightarrow{d} N(0, \theta^2) \tag{5.9.6}$$

By comparing Eqs. (5.9.4) and (5.9.6), we may easily note that

$$\theta^2 < \frac{\theta(\theta+1)^2}{(\theta+2)}$$

$$AV\left(\frac{n}{-\sum \log X_i}\right) < AV\left(\frac{\bar{X}}{1+\bar{X}}\right)$$

The *pdf* corresponding to the distribution Be(θ, 1)

$$f(x; \theta) = \theta x^{\theta-1} = \exp\{(\theta - 1)\log x + \log\theta\}$$

is one-parameter exponential family with complete sufficient statistic $T(x) = \log x$. The moment estimator, $\hat{\theta}$, given in Eq. (5.9.5) is CAN estimator for θ so that

$$\sqrt{n}(\hat{\theta} - \theta) \xrightarrow{d} N\left(0, \frac{1}{I_X(\theta)}\right)$$

where $I_X(\theta) = E\left[-\dfrac{\partial^2 \log f}{\partial \theta^2}\right] = -\dfrac{\partial}{\partial \theta}\left[\dfrac{1}{\theta} + \log x\right] = \dfrac{1}{\theta^2}$

$\therefore$ $$AV(\hat{\theta}) = \frac{1}{nI_X(\theta)} = \frac{\theta^2}{n} = \text{CRLB}$$

This shows that the asymptotic variance of moment estimator, in case $f(x;\theta)$ is a member of exponential family, which is the function of complete sufficient statistic, is equal to CRLB for estimating θ. Note that in this case, there does not exist any unbiased estimator which attains CRLB for estimating θ since

$$\frac{\partial}{\partial \theta}\log f = \frac{1}{\theta} + \log x \tag{5.9.7}$$

is not of the form in Eq. (4.3.7).
Moreover,

$$E\left(\frac{n-1}{-\sum \log X_i}\right) = (n-1)\int_0^{\infty} t^{-1}\frac{\theta^n}{\Gamma(n)}t^{n-1}\exp(-\theta t)dt$$

$$= \frac{(n-1)\theta^n}{\Gamma(n)}\frac{\Gamma(n-1)}{\theta^{n-1}} = \theta$$

since $T = \Sigma U_i = -\Sigma \log X_i \sim G_2(n, \theta)$. We see here that the estimator $(n-1)/(-\Sigma \log X_i)$ is a function of complete sufficient statistic $-\Sigma \log X_i$ which is unbiased for θ. Therefore, is UMVUE for θ. Its variance is given by

$$V\left(\frac{n-1}{-\sum \log X_i}\right) = \frac{\theta^2}{(n-2)}$$

$$> \frac{\theta^2}{n} = \frac{1}{nI_X(\theta)} = \text{CRLB}$$

We, therefore, observe that the variance of UMVUE for θ is larger than CRLB, though the moment estimator $\hat{\theta}$ in Eq. (5.9.5) attains its variance to CRLB. This is an example that shows that the variance of a biased estimator is smaller than CRLB.

***Example* 5.12** Consider a regression model with one regressor

$$y_i = X_i\beta + \varepsilon_i,\ i = 1,\ldots,n \tag{5.9.8}$$

where the regressor X_i is independent of each ε_i. The errors ε_i are *iid* with $E(\varepsilon_i) = 0$ and $V(\varepsilon_i) = \sigma^2 < \infty$. It is given that $\lim_{n\to\infty}(1/n)\Sigma_1^n X_i^2 = A$ is positive and finite. Obtain a consistent estimator of β and a consistent estimator of its asymptotic variance.

Solution. This linear model is given under the Gauss–Morkovian setup. Therefore, the least squared estimator of β is $\hat{\beta} = (\mathbf{X}'\mathbf{X})^{-1}\mathbf{X}'\mathbf{y} = (\Sigma_1^n X_i^2)^{-1}\Sigma_1^n X_i y_i = (\Sigma_1^n X_i^2)^{-1}\Sigma_1^n X_i(X_i\beta + \varepsilon_i) = \beta + (\Sigma_1^n X_i^2)^{-1}\Sigma_1^n X_i \varepsilon_i$. This gives

$$\hat{\beta} - \beta = \left(\sum_1^n X_i^2\right)^{-1} \sum_1^n X_i\varepsilon_i = \left(\frac{1}{n}\sum_1^n X_i^2\right)^{-1} \left(\frac{1}{n}\sum_1^n X_i\varepsilon_i\right)$$

$$= \left(\frac{1}{n}\sum_1^n X_i^2\right)^{-1} \left(\frac{1}{n}\sum_1^n U_i\right) \tag{5.9.9}$$

where $U_i = X_i \varepsilon_i$. The assumption in the example gives

$$\frac{1}{n}\sum_1^n X_i^2 \to A \text{ as } n \to \infty$$

Using Slutsky's theorem, since inversion is a continuous function, we have

$$\left(\frac{1}{n}\sum_1^n X_i^2\right)^{-1} \to A^{-1} \text{ as } n \to \infty \tag{5.9.10}$$

Consider, next, the term

$$\frac{1}{n}\sum_1^n U_i$$

Here, $E(U_i) = X_i E(\varepsilon_i) = 0$ and $V(U_i) = X_i^2 V(\varepsilon_i) = X_i^2\sigma^2$ which is uniformly bounded by the assumption in the example. Therefore, Markov's LLNs holds to give

$$\frac{1}{n}\sum_1^n U_i \xrightarrow{P} 0 \tag{5.9.11}$$

Using Eqs. (5.9.10) and (5.9.11) in Eq. (5.9.9), we get

$$\hat{\beta} - \beta \xrightarrow{P} 0$$

Hence, the LS estimator $\hat{\beta}$ is a consistent estimator for β.
Next, consider Eq. (5.9.9) and multiply both the sides by $\sqrt{n}$

$$\sqrt{n}(\hat{\beta} - \beta) = \left(\frac{1}{n}\sum_1^n X_i^2\right)^{-1} \sqrt{n}\left(\frac{1}{n}\sum_1^n U_i\right)$$

$$= \left(\frac{1}{n}\sum_1^n X_i^2\right)^{-1} \sqrt{n}\bar{U} \tag{5.9.12}$$

Now, consider the variance of the second term $\sqrt{n}\bar{U}$

$$V(\sqrt{n}\bar{U}) = \frac{1}{n}\sum_1^n V(X_i\varepsilon_i) = \frac{\sigma^2}{n}\sum_1^n X_i^2$$

On taking limits on both the sides, we get

$$\lim_{n\to\infty} V(\sqrt{n}\bar{U}) = \lim_{n\to\infty} \frac{\sigma^2}{n}\sum_{1}^{n} X_i^2 = \sigma^2 A < \infty \tag{5.9.13}$$

by the assumption in the question. This shows that the asymptotic variance of the statistic $\sqrt{n}\bar{U}$ is finite. Therefore, by the Lindeberg–Feller CLT,

$$\sqrt{n}\bar{U} \xrightarrow{d} N(0, \sigma^2 A) \tag{5.9.14}$$

By substituting Eqs. (5.9.13) and (5.9.14) in Eq. (5.9.12), we get

$$\sqrt{n}(\hat{\beta} - \beta) \xrightarrow{d} N(0, \sigma^2 A^{-1})$$

or

$$\hat{\beta} \xrightarrow{d} N\left(\beta, \frac{\sigma^2}{n} A^{-1}\right)$$

or

$$\hat{\beta} \overset{A}{\sim} N\left(\beta, \frac{\sigma^2}{n} A^{-1}\right)$$

Note that the asymptotic variance of the consistent estimator $\hat{\beta}$ is

$$AV(\hat{\beta}) = \frac{\sigma^2}{n} A^{-1}$$

This asymptotic variance cannot be computed, since it involves unknown quantities σ^2 and A. In order to estimate this asymptotic variance, we need to find a consistent estimator of $\sigma^2 A^{-1}/n$. For this, consider the asymptotic convergence of $(1/n)\Sigma_1^n \varepsilon_i^2$. It is easy to show that

$$\frac{1}{n}\sum_{1}^{n} \varepsilon_i^2 \xrightarrow{P} E(\varepsilon_i^2)$$

Here, $(1/n)\Sigma_1^n \varepsilon_i^2 = (1/n)\Sigma_1^n (y_i - \hat{\beta}X_i)^2 = \hat{\sigma}^2$ and $E(\varepsilon_i^2) = V(\varepsilon_i) + [E(\varepsilon_i)]^2 = \sigma^2$. Thus, $\hat{\sigma}^2$ is a consistent estimator of σ^2

$$\hat{\sigma}^2 \xrightarrow{P} \sigma^2 \tag{5.9.15}$$

By using the assumption in the theorem, we get

$$\left(\frac{1}{n}\sum_{1}^{n} X_i^2\right)^{-1} \xrightarrow{P} A^{-1} \text{ as } n \to \infty \tag{5.9.16}$$

Equations (5.9.15) and (5.9.16) combine to give

$$\hat{\sigma}^2 \left(\frac{1}{n}\sum_{1}^{n} X_i^2\right)^{-1} \xrightarrow{P} \sigma^2 A^{-1}$$

Therefore, the consistent estimator of $AV(\hat{\beta})$ is

$$\frac{1}{n}\hat{\sigma}^2\left(\frac{1}{n}\sum_1^n X_i^2\right)^{-1} = \hat{\sigma}^2\left(\sum_1^n X_i^2\right)^{-1}$$

Note that it is a classical LS estimate, $\hat{\sigma}^2(\mathbf{X}'\mathbf{X})^{-1}$, of the variance of $\hat{\boldsymbol{\beta}}$, where $\mathbf{X} = (X_1, X_2,\ldots,X_n)'$.

***Example* 5.13** Consider a random sample $X_1, X_2,\ldots,X_n$ drawn from $U(0,\theta)$, $\theta \in \Theta = (0, \infty)$. Show

(i) $X_{(n)} \xrightarrow{P} \theta$ and $X_{(n-1)} \xrightarrow{P} \theta$.

(ii) $Y_n = 2\overline{X}$ is consistent for θ.

(iii) Show that the asymptotic bias of the estimator $X_{(n)}$ is $(1 - (1/n))\theta$ and that it is not BAN.

(iv) The estimator $\delta_0 = (n + 2)X_{(n)}/(n + 1)$ is the least MSE estimator in the class of estimators of the type $\delta = cX_{(n)}$, $c > 0$.

(v) $X_{(1)}$ is not consistent for θ.

Solution.

(i) The density of $X_{(r)}$ is given by

$$f_{X_{(r)}}(x;\theta) = \frac{n!}{(r-1)!\,(n-r!}\left(\frac{x}{\theta}\right)^{r-1}\left(1-\frac{x}{\theta}\right)^{n-r}\frac{1}{\theta},\ 0<x<\theta$$

$$f_{X_{(n-1)}}(x;\theta) = \frac{n(n-1)x^{n-2}(\theta-x)}{\theta^n},\ 0<x<\theta$$

$$f_{X_{(n)}}(x;\theta) = \frac{nx^{n-1}}{\theta^n},\ 0<x<\theta$$

$$E(X_{(n-1)}^k) = \frac{n(n-1)}{\theta^n}\int_0^\theta x^{n+k-2}(\theta-x)dx$$

$$= \frac{n(n-1)}{\theta^n}\left[\frac{\theta^{n+k}}{n+k-1} - \frac{\theta^{n+k}}{n+k}\right]$$

$$= \frac{n(n-1)}{(n+k)(n+k-1)}\theta^k$$

$$E(X_{(n)}^k) = \frac{n}{\theta^n}\int_0^\theta x^{n+k-1}dx = \frac{n}{n+k}\theta^k$$

Using the above results, we get

$$E(X_{(n-1)}) = \frac{n-1}{n+1}\theta \to \theta \text{ as } n \to \infty$$

$$V(X_{(n-1)}) = E(X^2_{(n-1)}) - E^2(X_{(n-1)})$$

$$= \frac{n(n-1)}{(n+2)(n+1)}\theta^2 - \left(\frac{n-1}{n+1}\theta\right)^2$$

$$= \frac{2(n-1)}{(n+1)^2(n+2)}\theta^2 \to 0 \text{ as } n \to \infty$$

Therefore, $X_{(n-1)}$ is a consistent estimator for θ.

Similarly, $$E(X_{(n)}) = \frac{n}{n+1}\theta \to \theta \text{ as } n \to \infty$$

$$V(X_{(n)}) = E(X^2_{(n)}) - E^2(X_{(n)})$$

$$= \frac{n}{(n+1)^2(n+2)}\theta^2 \to 0 \text{ as } n \to \infty$$

Thus, $X_{(n)}$ and $X_{(n-1)}$ are both consistent estimators for θ but not unbiased. However, they are asymptotically unbiased. Also, note that $X_{(n)}$ is not UMVUE.

(ii) $E(X) = \theta/2$ and $E(X^2) = \theta^2/3$ gives $E(Y_n) = \theta$ and $V(Y_n) = (1/3n)\theta^2 \to 0$ as $n \to \infty$. So, by Theorem 5.3.1, $Y_n = 2\bar{X}$ is consistent for θ.

(iii) Let $Y_n = n(\theta - X_{(n)})$

$$F_{Y_n}(Y) = P(Y_n \le y) = P[n(\theta - X_{(n)}) \le y]$$

$$P\left(\theta - X_{(n)} \le \frac{y}{n}\right) = P\left(X_{(n)} \ge \theta - \frac{y}{n}\right)$$

$$= 1 - P\left(X_{(n)} \le \theta - \frac{y}{n}\right) = 1 - P\,(\text{all } X_i\text{s are} \le \theta - (y/n))$$

$$= 1 - \left[P\left(X \le \theta - \frac{y}{n}\right)\right]^n$$

$$= 1 - \left(1 - \frac{y}{n\theta}\right)^n$$

$$\lim_n F_{Y_n}(y) = 1 - \lim_{n\to\infty}\left(1 - \frac{y}{n\theta}\right)^n = 1 - \exp\left(-\frac{y}{\theta}\right) = G_1(1, \theta)$$

This shows that $n(\theta - X_{(n)}) \xrightarrow{L} G_1(1, \theta)$ and $E[n(\theta - X_{(n)})] \xrightarrow{P} \theta$. This implies that $E(X_{(n)}) \to (1 - (1/n))\theta$ as $n \to \infty$. Therefore, $X_{(n)}$ is asymptotically unbiased. Moreover,

$X_{(n)}$ is not asymptotically normal, though consistent. Hence, $X_{(n)}$ is not BAN. It is important to note that we cannot check whether asymptotic variance is equal to $1/I_X(\theta)$ since $U(0, \theta)$ is a member of the Pitman family.

(iv) Consider, next, the class of estimators

$$C = \{\delta_n: \delta_n = cX_{(n)}, c > 0\}$$

The MSE of an estimator T_n in C is

$$\text{MSE}(\delta_n) = E(cX_{(n)} - \theta)^2 = c^2E(X_{(n)}^2) - 2c\theta E(X_{(n)}) + \theta^2$$

$$= c^2 \frac{n}{n+2}\theta^2 - 2c\frac{n}{n+1}\theta^2 + \theta^2$$

$$\frac{\partial}{\partial c}\text{MSE}(\delta_n) = 2c\frac{n}{n+2}\theta^2 - 2\frac{n}{n+1}\theta^2 = 0$$

$$\Rightarrow \qquad c = \frac{n+2}{n+1}$$

$$\frac{\partial^2}{\partial c^2}\text{MSE}(\delta_n) = 2\frac{n}{n+2}\theta^2 > 0$$

Therefore, $\text{MSE}(\delta_n)$ is minimum at

$$c = \frac{n+2}{n+1}$$

and the corresponding estimator is

$$\delta_0 = \frac{n+2}{n+1}X_{(n)}$$

Note that $[(n + 1)/n]X_{(n)}$ is UMVUE of θ.

(v) The *pdf* of $X_{(1)}$ is given by

$$f_{X_{(1)}}(x; \theta) = \frac{n}{\theta^n}(\theta - x)^{n-1}; \quad 0 < x < \theta$$

We have

$$E(X_{(1)}^k) = \frac{n}{\theta^n}\int_0^{\theta} x^k(\theta - x)^{n-1}\,dx$$

$$= \frac{n}{\theta^n}\theta^{k+n}\int_0^1 y^k(1-y)^{n-1}\,dy \qquad \text{(by letting } x = \theta y\text{)}$$

$$= n\theta^k\frac{\Gamma(k+1)\Gamma(n)}{\Gamma(n+k+1)}$$

This gives $$E(X_{(1)}) = \frac{\theta}{n+1}; E(X_{(1)}^2) = \frac{2}{(n+1)(n+2)}\theta^2$$

and $$V(X_{(1)}) = \theta^2\left[\frac{2}{(n+1)(n+2)} - \left(\frac{1}{n+1}\right)^2\right] = \frac{n}{(n+1)^2(n+2)}\theta^2$$

Note that one of the sufficient conditions of Theorem 5.3.1 that $E(X_{(1)}) \to \theta$ as $n \to \infty$ is not satisfied. Therefore, we cannot use this theorem to check whether $X_{(1)}$ is consistent. We, thus, use the definition to check this. Calculate

$$P_\theta[|X_{(1)} - \theta| < \varepsilon] = P_\theta[\theta - \varepsilon < X_{(1)} < \theta + \varepsilon]$$

$$= \int_{\theta-\varepsilon}^{\theta+\varepsilon} n\left(1 - \frac{x}{\theta}\right)^{n-1} \frac{1}{\theta}dx$$

Let $1 - (x/\theta) = y$. $dx = -\theta dy$.

$$P_\theta[|X_{(1)} - \theta| < \varepsilon] = \frac{n}{\theta}\int_{-\varepsilon/\theta}^{\varepsilon/\theta} y^{n-1}\theta dy$$

$$= \left(\frac{\varepsilon}{\theta}\right)^n - \left(-\frac{\varepsilon}{\theta}\right)^n = \begin{cases} 2\left(\dfrac{\varepsilon}{\theta}\right)^n, & n \text{ odd} \\ 0, & n \text{ even} \end{cases}$$

This probability oscillates between 0 and $2(\varepsilon/\theta)^n$ as $n \to \infty$ when $\varepsilon > \theta$, though it approaches 0 for choices of $\varepsilon < \theta$. This shows that $X_{(1)}$ is not consistent for θ.

***Example* 5.14** Let $X_1, X_2, \ldots, X_n$ be a random sample drawn from $U(0,\theta)$. Show that the estimator $T_n(\mathbf{X}) = (\Pi_{i=1}^n X_i)^{1/n}$ is consistent for $\theta\exp(-1)$.

Solution. We have

$$f(x;\theta) = \frac{1}{\theta}, 0 < x < \theta$$

Consider the transformation $y = -\log(x/\theta)$ or $x = \theta\exp(-y)$. The *pdf* of y is

$$g(y) = \exp(-y), 0 < y < \infty$$

which is $G_2(1, 1)$. Let $V = \Sigma Y_i = -\Sigma\log(X_i/\theta) = -\log\Pi(X_i/\theta)$, $(\Pi X_i) = \theta^n\exp(-V)$. Thus, the distribution of V is $G_2(n, 1)$. Using this, we can calculate

$$E\left[\left(\prod X_i\right)^{1/n}\right] = \frac{\theta}{\Gamma(n)} \int_0^\infty v^{n-1} \exp(-v) \exp\left(\frac{-v}{n}\right) dv$$

$$= \frac{\theta}{\Gamma(n)} \int_0^\infty v^{n-1} \exp\left[-v\left(1+\frac{1}{n}\right)\right] dv = \frac{\theta}{\Gamma(n)} \frac{\Gamma(n)}{\left(1+\dfrac{1}{n}\right)^n}$$

$$= \theta\left(1+\frac{1}{n}\right)^{-n} \to \theta \exp(-1) \text{ as } n \to \infty$$

$$E\left[\left(\prod X_i\right)^{2/n}\right] = \frac{\theta^2}{\Gamma(n)} \int_0^\infty v^{n-1} \exp\left[-v\left(1+\frac{2}{n}\right)\right] dv$$

$$= \theta^2\left(1+\frac{2}{n}\right)^{-n}$$

$$V\left[\left(\prod X_i\right)^{1/n}\right] = \theta^2\left[\left(1+\frac{2}{n}\right)^{-n} - \left(1+\frac{1}{n}\right)^{-2n}\right]$$

$$\to \theta^2[\exp(-2) - \exp(-2)] = 0 \text{ as } n \to \infty$$

Hence, the sufficient conditions (i) and (ii) of Theorem 5.3.1 hold. Therefore, the estimator $T_n = (\Pi_{i=1}^n X_i)^{1/n}$ is consistent for $\theta\exp(-1)$.

***Example* 5.15** Let $X_1, X_2, \ldots, X_n$ be a random sample drawn from the population with density

$$f(x; \theta) = \begin{cases} 1, & \text{if } \theta < x < \theta + 1 \\ 0, & \text{otherwise} \end{cases}$$

Show that sample mean $\bar{X}$ is an unbiased and consistent estimator of $\theta + (1/2)$.

Solution. We have

$$E_\theta(X) = \frac{1}{2} \int_\theta^{\theta+1} x\, dx = \frac{(\theta+1)^2 - \theta^2}{2} = \theta + \frac{1}{2}$$

$$E_\theta(X^2) = \frac{(\theta+1)^3 - \theta^3}{3} = \frac{3\theta^2 + 3\theta + 1}{3} = \theta^2 + \theta + \frac{1}{3}$$

$$V(X) = \frac{1}{12}$$

$$E(\bar{X}) = \theta + \frac{1}{2}$$

$$V(\bar{X}) = \frac{1}{12n} \to 0 \text{ as } n \to \infty$$

This shows that $\overline{X}$ is an unbiased estimator of $\theta + (1/2)$ and by Theorem 5.3.1, it is a consistent estimator.

***Example* 5.16** Let $\mathbf{X} = (X_1, X_2,\ldots,X_n)$ be a random sample from $U(\theta - 1, \theta + 1)$. Show that $\overline{X}$ and sample median $\tilde{X}$ are consistent estimators of θ. Show that the estimators $\delta_1 = X_{(1)} + 1$, $\delta_2 = X_{(n)} - 1$, and $\delta_3 = (\delta_1 + \delta_2)/2$ are consistent for θ. Also, obtain the minimum sample size required by these estimators to achieve at an accuracy level as specified by ε and η.

Solution.

$$E(X) = \frac{1}{2}\int_{\theta-1}^{\theta+1} x dx = \frac{1}{4}[(\theta+1)^2 - (\theta-1)^2] = \theta$$

$$E(X^2) = \frac{1}{6}[(\theta+1)^3 - (\theta-1)^3] = \frac{3\theta^2+1}{3}$$

$$V(X) = \frac{1}{3}$$

This gives

$$E(\overline{X}) = \theta$$

$$V(\overline{X}) = \frac{1}{3n} \to 0 \text{ as } n \to \infty$$

Therefore, $\overline{X}$ is consistent for θ. The asymptotic distribution of $\overline{X}$, by choosing a norming constant $a_n = 1/\sqrt{\text{MSE}(\overline{X})} = \sqrt{3n}$ as a rule of thumb, is given by $\sqrt{3n}(\overline{X}-\theta) \xrightarrow{d} N(0,1)$. Further, for some $\varepsilon > 0$, $P[|\overline{X} - \theta| < \varepsilon] = P_n(\varepsilon, \theta) = P[-\sqrt{3n}\varepsilon < |\overline{X} - \theta|\sqrt{3n} < \sqrt{3n}\varepsilon] = \Phi(\sqrt{3n}\varepsilon) - \Phi(-\sqrt{3n}\varepsilon) = 2\Phi(\sqrt{3n}\varepsilon) - 1$. For given ε and η, the smallest sample size n_0 may be determined such that

$$P[|\overline{X}_n - \theta| < \varepsilon] \geq 1 - \eta$$

$$2\Phi(\sqrt{3n}\varepsilon) - 1 \geq 1 - \eta$$

$$\Phi(\sqrt{3n}\varepsilon) \geq 1 - \frac{\eta}{2}$$

$$n \geq \frac{1}{3\varepsilon^2}\left\{\Phi^{-1}\left(1-\frac{\eta}{2}\right)\right\}^2$$

$$n_0 = \left[\frac{1}{3\varepsilon^2}\left[\Phi^{-1}\left(1-\frac{\eta}{2}\right)\right]^2\right] + 1$$

Further,

$$E(\tilde{X}) = E(\overline{X}) = \theta$$

since the rectangular distribution is symmetric over $(\theta - 1, \theta + 1)$, and

$$V(\tilde{X}) = \frac{1}{4nf_1^2}$$

Since $f_1 = (1/2)$, we get

$$V(\tilde{X}) = \frac{1}{4n(1/4)} = \frac{1}{n} \to 0 \text{ as } n \to \infty$$

By the sufficient condition of consistency, the sample median $\tilde{X}$ is consistent for θ.

Let $Y = (X - \theta + 1)/2$. Then, $Y \sim U(0, 1)$. The *pdf* of $U = Y_{(1)}$ is given by

$$f_1(u; \theta) = n(1-u)^{n-1}, 0 < u < 1$$

and the *pdf* of $V = Y_{(n)}$ is given by

$$f_n(v; \theta) = nv^{n-1}, 0 < v < 1$$

Further, the joint *pdf* of $(U, V) = (Y_{(1)}, Y_{(n)})$ is given by

$$f_{1,n}(u, v; \theta) = n(n-1)(v-u)^{n-2}, 0 < u < v < 1$$

Using these distributions, we compute their moments

$$E(Y_{(1)}) = \frac{1}{n+1}; E(Y_{(n)}) = \frac{n}{n+1}$$

$$E(Y_{(1)}^2) = \frac{2}{(n+1)(n+2)}$$

$$E(Y_{(n)}^2) = \frac{n}{n+2}$$

$$V(Y_{(1)}) = V(Y_{(n)}) = \frac{n}{(n+1)^2(n+2)}$$

$$\text{cov}(Y_{(1)}, Y_{(n)}) = \frac{1}{(n+1)^2(n+2)}$$

Consider now the estimators $\delta_1 = X_{(1)} + 1$ and $\delta_2 = X_{(n)} - 1$ of θ. We have

$$E(\delta_1) = E(2Y_{(1)} + \theta - 1 + 1) = 2E(Y_{(1)}) + \theta$$

$$= \theta + \frac{2}{(n+1)}$$

$$V(\delta_1) = \frac{4n}{(n+1)^2(n+2)}$$

$$E(\delta_2) = \theta - \frac{2}{n+1}$$

$$V(\delta_2) = \frac{4n}{(n+1)^2(n+2)}$$

$$\text{cov}(\delta_1, \delta_2) = \frac{4}{(n+1)^2(n+2)}$$

$$\text{MSE}(\delta_1) = \frac{4n}{(n+1)^2(n+2)} + \left(\frac{2}{n+1}\right)^2 = \frac{8}{(n+1)(n+2)} = \text{MSE}(\delta_2)$$

$$= \frac{8}{n^2}\left(1+\frac{1}{n}\right)^{-1}\left(1+\frac{2}{n}\right)^{-1}$$

$$= \frac{8}{n^2}\left(1-\frac{1}{n}+\frac{1}{n^2 2!}-\frac{1}{n^3 3!}+\cdots\right)\left(1-\frac{2}{n}+\left(\frac{2}{n}\right)^2\frac{1}{2!}-\left(\frac{2}{n}\right)^3\frac{1}{3!}+\cdots\right)$$

$$= \frac{8}{n^2}\left[1-\frac{3}{n}+O\left(\frac{1}{n^2}\right)\right]$$

$$= \frac{8}{n^2}-\frac{24}{n^3}+O\left(\frac{1}{n^4}\right)$$

Consider the estimator $\delta_3 = (\delta_1 + \delta_2)/2$.

$$E(\delta_3) = \frac{1}{2}[E(\delta_1+\delta_2)] = \theta$$

$$V(\delta_3) = \frac{2}{(n+1)(n+2)} = \frac{2}{n^2}\left[\left(1+\frac{1}{n}\right)^{-1}\left(1+\frac{2}{n}\right)^{-1}\right]$$

$$= \frac{2}{n^2}-\frac{6}{n^3}+O\left(\frac{1}{n^4}\right)$$

It immediately follows that δ_3 is unbiased but δ_1 and δ_2 are asymptotically unbiased, and that δ_1, δ_2, and δ_3 are all consistent for θ since all of them converge to θ in quadratic mean, i.e., $E(\delta_i - \theta)^2 \xrightarrow{P} 0$, $i = 1, 2, 3$. If we compare these estimators in terms of MSE, note that the MSE of $\bar{X}$ converges to 0 at the rate $1/3n$, that of δ_1 and δ_2 converge to 0 at the rate $8/n^2$, and of δ_3 to 0 at the rate $2/n^2$. The rate of convergence of the MSE of δ_3 to 0 is faster as compared to other estimators. Therefore, δ_3 is the best estimator for θ.

We, next, derive the asymptotic distribution of δ_1, δ_2, and δ_3. We choose the norming constant so that the limiting distributions of δ_1 and δ_2 are nondegenerate as $a_n = n$ since $\text{MSE}(\delta_1)$ and $\text{MSE}(\delta_2)$ are of the order $1/n^2$. Consider

$$Z_n = n(\delta_1 - \theta),\ 0 < Z_n < 2n$$

$$= 2nY_{(1)}$$

The *cdf* of Z_n

$$P[Z_n \le z] = P\left[Y_{(1)} \le \frac{z}{2n}\right] = 1-\left(1-\frac{z}{2n}\right)^n \quad \text{if } 0 < z < 2n$$

$$\to 1-\exp\left(-\frac{z}{2}\right) \quad \text{as } n \to \infty$$

$$\therefore \qquad Z_n = n(\delta_1 - \theta) \xrightarrow{d} G_1(1, 2)$$

Thus,
$$P_n(\varepsilon, \theta) = P[|\delta_1 - \theta| < \varepsilon] = P[Z_n < n\varepsilon]$$
$$\simeq 1 - \exp\left(-\frac{n\varepsilon}{2}\right)$$

The minimum sample size n_0 required by δ_1 to reach the accuracy as specified by choosing ε and η is determined by the condition

$$1 - \exp\left(-\frac{n\varepsilon}{2}\right) = P_n(\varepsilon, \theta) \geq 1 - \eta; \exp\left(-\frac{n\varepsilon}{2}\right) \leq \eta$$

$$-\frac{n\varepsilon}{2} \leq \log \eta \text{ or } n \geq -\frac{2}{\varepsilon}\log \eta$$

$$n_0(\delta_1) = \left[-\frac{2}{\varepsilon}\log \eta\right] + 1$$

Next, we may define

$$Z_n = n(\theta - \delta_2) = 2n(1 - Y_{(n)})$$

The *cdf* of Z_n is given by

$$P[Z_n \leq z] = P\left[1 - Y_{(n)} \leq \frac{z}{2n}\right] = P\left[Y_{(n)} - 1 \geq -\frac{z}{2n}\right]$$

$$= P\left[Y_{(n)} \geq 1 - \frac{z}{2n}\right] = 1 - P\left[Y_{(n)} \leq 1 - \frac{z}{2n}\right]$$

$$= 1\left(1 - \frac{z}{2n}\right)^2 \to 1 - \exp\left(\frac{z}{2}\right) \text{ as } n \to \infty$$

This gives

$$P_n(\varepsilon, \theta) = P[(|\theta - \delta_2|) < \varepsilon] = P[Z_n < n\varepsilon] = 1 - \exp\left(-\frac{n\varepsilon}{2}\right)$$

Therefore, the minimum sample size n_0 required by δ_2 to reach the accuracy as specified by choosing ε and η is determined by the condition

$$n_0(\delta_2) = \left[-\frac{2}{\varepsilon}\log \eta\right] + 1 = n_0(\delta_1)$$

Similarly, we can show that the asymptotic distribution of

$$Z_n = n(\delta_3 - \theta) = n\left(\frac{\delta_1 + \delta_2}{2} - \theta\right) = \frac{n}{2}[(\delta_1 - \theta) - (\theta - \delta_2)]$$

$$P[n(\delta_3 - \theta) \leq z] = \frac{1}{2} P[n(\delta_1 - \theta) - n(\theta - \delta_2) \leq z]$$

$$\leq \frac{1}{2}\left\{P\left[(\delta_1 - \theta) \leq \frac{z}{2n}\right] + P\left[(\theta - \delta_2) \leq \frac{z}{2n}\right]\right\}$$

$$\to 1 - \exp\left(-\frac{z}{2}\right) \text{ as } n \to \infty$$

This gives, for large n,

$$P_n(\varepsilon, \theta) = P[|\delta_3 - \theta| < \varepsilon] = P[|n(\delta_3 - \theta)| < n\varepsilon] = 1 - \exp(-n\varepsilon)$$

and the minimum sample size required by δ_3 to achieve at an accuracy level as specified by ε and η is obtained by the condition

$$1 - \exp(-n\varepsilon) = P_n(\varepsilon, \theta) \geq 1 - \eta$$

$$\exp(-n\varepsilon) \leq \eta \quad \text{or} \quad -n\varepsilon \leq \log \eta$$

$$n \geq -\frac{1}{\varepsilon}\log \eta$$

$$n_0(\delta_3) = \left[-\frac{1}{\varepsilon}\log \eta\right] + 1$$

Next, we will show that the sample median statistic is consistent and MSE($\tilde{X}$) → 0 at the rate (1/n). Consider n as odd, i.e., $n = 2m + 1$. In this case, the sample median is given by $\tilde{X} = X_{([n/2]+1)} = X_{(m+1)}$. Its *pdf* is given by

$$f_{m+1}(x; \theta) = \frac{1}{B(m+1, m+1)}\left(\frac{x - \theta + 1}{2}\right)^m \left(1 - \frac{x - \theta + 1}{2}\right)^m \frac{1}{2}, \quad \theta - 1 < x < \theta + 1$$

Under the transformation $y = [(x - \theta + 1)/2]$, the distribution of the corresponding median statistic $Y_{(m+1)}$ is given by

$$g_{m+1}(y; \theta) = \frac{1}{B(m+1, m+1)} y^m (1 - y)^m, 0 < y < 1$$

$$E(Y_{(m+1)}) = \frac{1}{2}$$

This gives

$$\frac{1}{2} = E(Y_{(m+1)}) = E\left(\frac{X_{(m+1)} - \theta + 1}{2}\right)$$

or

$$E(X_{(m+1)}) = \theta$$

$$E(Y_{(m+1)}^2) = \frac{1}{B(m+1, n-m)} \int_0^1 y^{m+2} (1 - y)^m \, dy$$

$$= \frac{m+2}{2(2m+3)}$$

$$V(Y_{(m+1)}) = \frac{1}{4(2m+3)}$$

This gives
$$V(X_{(m+1)}) = 4V(Y_{(m+1)}) = \frac{1}{2m+3}$$

$$\text{MSE}(X_{(m+1)}) = \frac{1}{2m+3} \to 0 \quad \text{as} \quad n \to \infty$$

$\therefore$
$$X_{(m+1)} \xrightarrow{P} \theta$$

Further,
$$\text{MSE}(X_{(m+1)}) = \frac{1}{n+2} = \frac{1}{n\left(1+\frac{2}{n}\right)} = \frac{1}{n}\left(1+\frac{2}{n}\right)^{-1}$$

$$= \frac{1}{n}\left(1 - \frac{2}{n} + O\left(\frac{1}{n^2}\right)\right)$$

$$\text{MSE}(X_{(m+1)}) \to 0 \text{ at the rate } \frac{1}{n}$$

Consider the case when n is even, i.e., $n = 2m$. In this case, the sample median is $\tilde{X} = X_{([n/2]+1)} = X_{(m+1)}$. Its *pdf* is given by

$$f_{m+1}(x; \theta) = \frac{1}{B(m+1, m)}\left(\frac{x-\theta+1}{2}\right)^m \left(1 - \frac{x-\theta+1}{2}\right)^{m-1} \frac{1}{2}, \ \theta - 1 < x < \theta + 1$$

On using the transformation $y = (x - \theta + 1)/2$, the *pdf* of the corresponding median statistic $Y_{(m+1)}$ is given by

$$g_{m+1}(y; \theta) = \frac{1}{B(m+1, m)} y^m (1-y)^{m-1}, 0 < y < 1$$

We then have

$$EY_{(m+1)} = \frac{B(m+2, m)}{B(m+1, m)} = \frac{m+1}{2m+1}$$

$$V(Y_{(m+1)}) = \frac{m}{2(2m+1)^2}$$

$$E(\tilde{X}) = E(X_{(m+1)}) = 2E(Y_{(m+1)}) + \theta - 1$$

$$= \theta + \frac{1}{2m+1}$$

$$V(\tilde{X}) = V(X_{(m+1)}) = 4V(Y_{(m+1)}) = \frac{2m}{(2m+1)^2}$$

$$\text{MSE}(\tilde{X}) = \text{MSE}(X_{(m+1)}) = \frac{2m}{(2m+1)^2} + \left(\frac{1}{2m+1}\right)^2$$

$$= \frac{1}{2m+1} = \frac{1}{n+1} = \frac{1}{n}\left(1+\frac{1}{n}\right)^{-1} = \frac{1}{n}\left[1-\frac{1}{n}+O\left(\frac{1}{n^2}\right)\right]$$

$$\therefore \qquad X_{(m+1)} \xrightarrow{P} \theta$$

and MSE$(X_{(m+1)}) \to 0$ at the rate $1/n$. Further, the asymptotic distribution of $100p\%$ sample percentile, David (1981), so that $f[x_p(\theta)] \neq 0$, is

$$X_{([np]+1)} = X_{(m+1)} \xrightarrow{d} N\left(x_p(\theta), \frac{p(1-p)}{n[f(x_p)]^2}\right)$$

For $X \sim U(\theta - 1, \theta + 1)$, $F[x_p(\theta)] = (1/2)x_p(\theta)$, $p = 1/2$,

$$\int_{\theta-1}^{x_p(\theta)} \frac{1}{2}dx = \frac{1}{2}$$

$$x_p(\theta) - \theta + 1 = 1$$

$$x_p(\theta) = \theta$$

This gives

$$X_{(m+1)} \xrightarrow{d} N\left(\theta, \frac{1}{n}\right)$$

We then have

$$P_n(\varepsilon, \theta) = P[|\sqrt{n}(X_{(m+1)} - \theta)| < \sqrt{n}\varepsilon] = 2\Phi(\sqrt{n}\varepsilon) - 1 \ge 1 - \eta$$

Thus, $n_0(X_{(m+1)})$ is obtained so that

$$\Phi(\sqrt{n}\varepsilon) \ge 1 - \frac{\eta}{2}$$

$$n \ge \frac{1}{\varepsilon^2}\left\{\Phi^{-1}\left(1-\frac{\eta}{2}\right)\right\}^2$$

$$\therefore \qquad n_0(X_{(m+1)}) \ge \left[\frac{1}{\varepsilon^2}\left[\Phi^{-1}\left(1-\frac{\eta}{2}\right)\right]^2\right] + 1$$

***Example* 5.17** Consider a random sample $X_1, X_2, \ldots, X_n$ from $N(\theta, \sigma^2)$. Find CAN estimators for $(\theta, \sigma^2)'$.

Solution. Consider here

$$\mathbf{V}(X) = (V_1(X), V_2(X))' = (X, X^2)'$$

$$E[V_1(X)] = E(X) = \theta = \mu_1' = \pi_1(\theta, \sigma^2)$$

$$E[V_2(X)] = E(X^2) = \sigma^2 + \theta^2 = \mu_2' = \pi_2(\theta, \sigma^2)$$

$$V(V_1) = V(X) = EX^2 - \{E(X)\}^2 = \mu_2' - \mu_1'^2 = \sigma^2$$

$$V(V_2) = V(X^2) = \mu_4' - \mu_2'^2 = 2\sigma^4 + 4\sigma^2\theta^2$$

$$\text{cov}(V_1, V_2) = \mu_3' - \mu_1'\mu_2' = 2\sigma^2\theta$$

$$D(\mathbf{V}) = \Sigma = \begin{pmatrix} \sigma^2 & 2\sigma^2\theta \\ 2\sigma^2\theta & 2\sigma^4 + 4\sigma^2\theta^2 \end{pmatrix}$$

Therefore, by multivariate central limit theorem,

$$\sqrt{n}(\mathbf{V} - \boldsymbol{\pi}) \xrightarrow{d} n(\mathbf{0}, \Sigma)$$

where $E(\mathbf{V}) = (E(V_1), E(V_2))' = (\theta, \sigma^2 + \theta^2)' = \boldsymbol{\pi}$.

Consider the transformation $\mathbf{g}(\boldsymbol{\pi}) = (g_1(\boldsymbol{\pi}), g_2(\boldsymbol{\pi}))' = (\pi_1, \pi_2 - \pi_1^2)' = (\theta, \sigma^2)'$. Then,

$$\mathbf{L} = \left(\frac{\partial \mathbf{g}}{\partial \boldsymbol{\pi}}\right) = \begin{pmatrix} \dfrac{\partial g_1}{\partial \pi_1} & \dfrac{\partial g_1}{\partial \pi_2} \\ \dfrac{\partial g_2}{\partial \pi_1} & \dfrac{\partial g_2}{\partial \pi_2} \end{pmatrix} = \begin{pmatrix} 1 & 0 \\ -2\pi_1 & 1 \end{pmatrix}$$

$$= \begin{pmatrix} 1 & 0 \\ -2\theta & 1 \end{pmatrix}$$

$$|\mathbf{L}| \neq 0$$

$$\mathbf{g}(\bar{\mathbf{V}}) = (g_1(\bar{\mathbf{V}}), g_2(\bar{\mathbf{V}}))' = (\bar{V}_1, \bar{V}_2 - \bar{V}_1^2)' = \left(\frac{1}{n}\sum X_i, \frac{1}{n}\sum X_i^2 - \left(\frac{1}{n}\sum X_i\right)^2\right)'$$

$$= (\bar{X}, S^2)'$$

Therefore, by Theorem 5.7.3,

$$\sqrt{n}[g(\bar{\mathbf{V}}) - \mathbf{g}(\boldsymbol{\pi})] \xrightarrow{d} N_2(\mathbf{0}, \mathbf{L}\Sigma\mathbf{L}')$$

where

$$\mathbf{L}\Sigma\mathbf{L}' = \begin{pmatrix} \sigma^2 & 0 \\ 0 & 2\sigma^4 \end{pmatrix}$$

or

$$\sqrt{n}\left[\begin{pmatrix} \bar{X} \\ S^2 \end{pmatrix} - \begin{pmatrix} \theta \\ \sigma^2 \end{pmatrix}\right] \xrightarrow{d} N_2\left[\begin{pmatrix} 0 \\ 0 \end{pmatrix}, \begin{pmatrix} \sigma^2 & 0 \\ 0 & 2\sigma^4 \end{pmatrix}\right]$$

This shows that $(\bar{X}, S^2)'$ is CAN for $(\theta, \sigma^2)'$ and $\bar{X}$ and S^2 are asymptotically independent.

The above results can also be obtained differently by noting that the density of $N(\theta, \sigma^2)$ is a two-parameter exponential family

$$f(x; \theta, \sigma^2) = \exp\left\{\frac{\theta}{\sigma^2}x - \frac{1}{2\sigma^2}x^2 - \frac{1}{2}\left(\frac{\theta^2}{\sigma^2} + \log(2\pi\sigma^2)\right)\right\}$$

where $\mathbf{T}(X) = (T_1(X), T_2(X))' = (X, X^2)'$ is complete and sufficient statistic for (θ, σ^2). The moment equations, in this case, are

$$\bar{x} = m_1' = \theta$$

$$\frac{1}{n}\sum x_i^2 = m_2' = \theta^2 + \sigma^2$$

which give the solution $\hat{\boldsymbol{\theta}} = (\hat{\theta}, \hat{\sigma}^2)' = (\bar{X}, S^2)'$.
Fisher information matrix calculates to

$$I(\theta, \sigma^2) = \begin{pmatrix} 1/\sigma^2 & 0 \\ 0 & 1/2\sigma^4 \end{pmatrix}$$

$$I^{-1}(\theta, \sigma^2) = \begin{pmatrix} \sigma^2 & 0 \\ 0 & 2\sigma^4 \end{pmatrix}$$

Therefore, by Theorem 5.8.4,

$$\sqrt{n}\left[\begin{pmatrix} \bar{X} \\ S^2 \end{pmatrix} - \begin{pmatrix} \theta \\ \sigma^2 \end{pmatrix}\right] \xrightarrow{d} N_2\left[\begin{pmatrix} 0 \\ 0 \end{pmatrix}, \begin{pmatrix} \sigma^2 & 0 \\ 0 & 2\sigma^2 \end{pmatrix}\right]$$

***Example* 5.18** Let $X_1, X_2, \ldots, X_n$ be a random sample drawn from the exponential population with *pdf* $f(x;\theta) = \theta \exp\{-\theta x\}$, $\theta > 0$, $x > 0$, where $\theta = 1/(\text{mean lifetime of the } rv\ X)$ = failure rate. Which of the two MoM estimators $\hat{\theta}_1 = (1/\bar{X})$ and $\hat{\theta}_2 = \sqrt{2/[(1/n)\Sigma_1^n X_i^2]}$ is preferred? Find a consistent estimator for θ by using the percentile method.

Solution. Consider $g(X) = X$. We have

$$E(g(X)) = E(X) = \mu(\theta) = \frac{1}{\theta}$$

The moment equation is

$$\bar{g}(\mathbf{X}) = \bar{X} = \mu(\theta)$$

We get the moment estimator

$$\hat{\theta}_1 = \mu^{-1}(\bar{X}) = \frac{1}{\bar{X}}$$

Using $V[g(X)] = V(X) = 1/\theta^2$ and $\mu'(\theta) = -(1/\theta^2)$, the asymptotic variance of $\hat{\theta}_1$ is given by

$$V(\hat{\theta}_1) = V[g(X)]\left[\frac{\partial \mu(\theta)}{\partial \theta}\right]^{-2} = \left(\frac{1}{\theta^2}\right)\left(-\frac{1}{\theta^2}\right)^{-2} = \theta^2$$

Consider, next, $g(X) = X^2$, we have

$$E[g(X)] = E(X^2) = \mu(\theta) = \frac{2}{\theta^2}$$

The corresponding moment equation

$$\bar{g}(\mathbf{X}) = \frac{1}{n}\sum_1^n X_i^2 = \mu(\theta) = \frac{2}{\theta^2}$$

gives the moment estimator

$$\hat{\theta}_2 = \mu^{-1}\left(\frac{1}{n}\sum_1^n X_i^2\right) = \sqrt{\frac{2}{\left(\frac{1}{n}\right)\sum_1^n X_i^2}}$$

Since the calculation of $V[g(X)] = V(X^2)$ involves second- and forth-order moments of exponential distribution, the rth moment is given by

$$E_\theta(X^r) = \theta\int_0^\infty x^r \exp(-\theta x)dx = \theta\frac{\Gamma(r+1)}{\theta^{r+1}} = \frac{r!}{\theta^r}$$

Thus, $E_\theta(X^4) = 4!/\theta^4$, $E_\theta(X^2) = 2!/\theta^2$ give $V[g(X)] = 4!/\theta^4 - (2!/\theta^2)^2 = 20/\theta^4$ and $\mu'(\theta) = -4/\theta^3$. The asymptotic variance of $\hat{\theta}_2$ calculates to

$$V(\hat{\theta}_2) = \left(\frac{20}{\theta^4}\right)\left(-\frac{4}{\theta^3}\right)^{-2} = \frac{5}{4}\theta^2 > \theta^2 = V(\hat{\theta}_1)$$

Thus, the first estimator $\hat{\theta}_1$ is better than $\hat{\theta}_2$ since its asymptotic variance is small as compared to the latter.

Another consistent estimator may be obtained by considering the $100p\%$ percentile of X, $x_p(\theta)$, so that

$$\int_0^{x_p(\theta)} \theta\exp(-\theta x)dx = p$$

This gives

$$x_p(\theta) = -\frac{\log(1-p)}{\theta}$$

The percentile equation based on X is given by

$$X_{([np]+1)} = -\frac{\log(1-p)}{\theta}$$

which leads to the consistent estimator for θ as

$$\hat{\theta} = -\frac{\log(1-p)}{X_{([np]+1)}}$$

***Example* 5.19** Let $X_1, X_2, \ldots, X_n$ be *iid* from $G_1(\alpha,\beta)$ with *pdf* $f(x;\alpha,\beta) = [1/\Gamma(\alpha)\beta^\alpha] \exp(-x/\beta)x^{\alpha-1}$, $x > 0$, $\alpha > 0$, $\beta > 0$. Find a consistent estimators of $(\alpha\beta^2)$ and $(\alpha, \beta)'$.

Solution. We have

$$E(X_i) = \alpha\beta = \mu_1'$$

$$E(X_i^2) = \alpha\beta^2 + (\alpha\beta)^2 = \alpha\beta^2(1 + \alpha) = \mu_2'$$

$$\left|\frac{\partial(\mu_1'\mu_2')}{\partial(\alpha,\beta)}\right| = \begin{vmatrix} \frac{\partial\mu_1'}{\partial\alpha} & \frac{\partial\mu_1'}{\partial\beta} \\ \frac{\partial\mu_2'}{\partial\alpha} & \frac{\partial\mu_2'}{\partial\beta} \end{vmatrix} = \begin{vmatrix} \beta & \alpha \\ \beta^2(1+2\alpha) & 2\alpha\beta(1+\alpha) \end{vmatrix}$$

$$= \alpha\beta^2 > 0$$

Therefore, the moment equations give the unique solution for α and β

$$m_1' = \mu_1'(\alpha, \beta) = \alpha\beta$$
$$m_2' = \mu_2'(\alpha, \beta) = \alpha\beta^2(1 + \alpha)$$

This gives $$m_2' - m_1'^2 = \alpha\beta^2(1 + \alpha) - \alpha^2\beta^2 = \alpha\beta^2$$

$$\frac{m_2'}{m_2' - m_1'^2} = 1 + \alpha \quad \text{or} \quad \hat{\alpha} = \frac{m_1'^2}{m_2' - m_1'^2} = \frac{m_1'^2}{m_2}$$

$$\hat{\beta} = \frac{m_2' - m_1'^2}{m_1'} = \frac{m_2}{m_1'}$$

which are unique solutions for α and β and are consistent for $\theta = (\alpha, \beta)'$, by Theorem 5.8.1. Here, m_1' and m_2' are the raw sample moments and m_1 and m_2 are the central sample moments.

***Example* 5.20** Let $X_1, X_2, \ldots, X_n$ be *iid* from a Pareto distribution with *pdf* $f(x;\theta) = \theta/x^{\theta+1}$, $x \geq 1$, where $\theta > 0$ is known as shape parameter. This model is usually used as an income distribution model. Find a consistent estimator of θ.

Solution. The mean of the *rv* X exists, $E(X) = \mu(\theta) = \theta/(\theta - 1)$, only if $\theta > 1$. If we restrict Θ to $(1, \infty)$, by WLLN,

$$\bar{X} \xrightarrow{P} \frac{\theta}{\theta - 1}$$

and in this case, that $\bar{X}$ is consistent for $\theta/(\theta - 1)$. However, using the fact that $\theta/(\theta - 1)$ is a continuous transformation and the consistency is preserved under a continuous transformations, $\bar{X}/(\bar{X} - 1)$ is consistent for θ. Moreover, the corresponding moment equation is given by

$$\bar{X} = \mu(\theta) = \frac{\theta}{\theta - 1}$$

which gives a unique solution since $\partial\mu(\theta)/\partial\theta = -1/(\theta - 1)^2 \neq 0$ for θ in $\Theta = (1, \infty)$.

The method of moments fails if we consider Θ as $(0, \infty)$. In this situation, if we consider a transformation

$$Y = \log X$$

then the *pdf* of Y is given by

$$f(y;\, \theta) = \theta\exp(-\theta y),\ y \geq 0$$

which is a one-parameter exponential distribution.

The consistent estimator for θ, using the result of Example 5.9.11, is given by

$$\hat{\theta} = \frac{n-1}{\sum \log X_i}$$

In this case, $100p\%$ percentile of Y is

$$\int_0^{y_p(\theta)} \theta \exp(-\theta y)\, dy = p$$

This gives
$$y_p(\theta) = -\frac{\log(1-p)}{\theta}$$

The percentile equation based on Y is given by

$$Y_{([np]+1)} = y_p(\theta) = -\frac{\log(1-p)}{\theta}$$

or
$$\hat{\theta} = -\frac{\log(1-p)}{Y_{([np]+1)}} = -\frac{\log(1-p)}{\log X_{([np]+1)}}$$

Further, $100p\%$ population percentile is obtained by solving the equation

$$\int_1^{x_p(\theta)} \frac{\theta}{x^{\theta+1}}\, dx = p$$

We have
$$x_p(\theta) = (1-p)^{-1/\theta} = \exp\left\{-\frac{1}{\theta}\log(1-p)\right\}$$

The percentile equation is given by

$$X_{([np]+1)} = \exp\left\{-\frac{1}{\theta}\log(1-p)\right\}$$

Since $\partial x_p(\theta)/\partial\theta = -(1/\theta)\log(1-p)^{-(1/\theta)-1} \neq 0$, $x_p(\theta)$ is a continuous invertible function. Therefore, the consistent estimator of θ is given by

$$\hat{\theta} = -\frac{\log(1-p)}{\log X_{([np]+1)}}$$

***Example* 5.21** Consider $X_1, X_2, \ldots, X_n$ as a random sample from the Weibull distribution with *pdf* $f(x;\theta) = \theta x^{\theta-1}\exp(-x^{\theta})$, $x > 0$, $\theta > 0$. Find a MoM consistent estimator and the pth percentile CAN estimator for θ.

Solution. The Weibull distribution is generally used as a failure time distribution. We have

$$E(X) = \Gamma\left(\frac{1}{\theta}+1\right) = \mu_1'$$

and the moment equation is

$$m_1' = \Gamma\left(\frac{1}{\theta}+1\right) = \mu_1'(\theta)$$

where, though, $\Gamma[(1/\theta) + 1]$ is a monotone decreasing function from $(0, \infty)$ to $(\infty, 1)$, but it is non-invertible. This can be seen by assuming

$$\eta = \frac{1}{\theta}+1$$

so that $\Gamma[\eta(\theta)]$ is a function from $(0, \infty)$ to $(1, \infty)$. The function $\Gamma(\eta)$ is differentiable and continuous over $(1, \infty)$ so that

$$\Gamma(1) = 1 = \Gamma(2)$$

Therefore, by Rolle's theorem, there exists a point $\eta \in (1, 2)$ at which

$$\frac{\partial \Gamma(n)}{\partial \theta} = 0$$

Thus, $\mu_1'^{-1}$ does not exist and the method of moments fails to provide a consistent estimator for θ. To get around this problem, consider the log transformation and proceed with the observations $(\log X_1,\ldots,\log X_n)$. Consider the *mgf* of $\log X$

$$M_{\log X}(t) = E[\exp(t \log X)] = E(X^t) = \Gamma\left(\frac{t}{\theta}+1\right)$$

This gives $\quad \mu_1' = E(\log X) = M_{\log X}^{(1)}(t)\Big|_{t=0} = \frac{\partial}{\partial t} M_{\log X}(t)\Big|_{t=0}$

$$= \frac{1}{\theta}\frac{\partial}{\partial t}\Gamma\left(\frac{t}{\theta}+1\right)\Bigg|_{t=0} = \frac{1}{\theta}\Gamma'(1) = -\frac{\eta}{\theta}$$

where $-(\partial/\partial t)\Gamma(t)|_{t=1} = \eta \approx 0.577216$
Using this result, the moment equation is

$$\frac{1}{n}\sum_1^n \log X_i = \mu_1'(\theta) = -\frac{\eta}{\theta} \tag{5.9.17}$$

Solving Eq. (5.9.17) for θ, gives

$$\hat{\theta} = -\frac{n\eta}{\sum_1^n \log X_i} \tag{5.9.18}$$

By Theorem 5.8.1, $\hat{\theta}$ is consistent for θ. By the expression of the estimator $\hat{\theta}$ in Eq. (5.9.18), it appears that it can take negative values as an estimate of θ, where θ is always positive. To check this, let us consider the asymptotic behavior of $(1/n)\Sigma_1^n \log X_i$ based on the sequence $\{\log X_i\}_1^n$. We have $E[(1/n)\Sigma_1^n \log X_i] = -\eta/\theta$ and $V[(1/n)\Sigma_1^n \log X_i) = \pi^2/(6n\theta^2)$.

By the central limit theorem,

$$\frac{(1/n)\sum_{1}^{n}\log X_i - (-\eta/\theta)}{\sqrt{\pi^2/(6n\theta^2)}} \xrightarrow{d} N(0,1)$$

$$\frac{1}{n}\sum_{1}^{n}\log X_i \xrightarrow{d} N\left(-\frac{\eta}{\theta}, \frac{\pi^2}{6n\theta^2}\right)$$

Hence,
$$P\left[\frac{1}{n}\sum_{1}^{n}\log X_i > 0\right] = P\left[N(0,1) > \frac{\eta/\theta}{\sqrt{\pi^2/(6n\theta^2)}}\right]$$

$$= P\left[N(0,1) > \frac{\eta\sqrt{6n}}{\pi}\right] = 1 - \Phi\left(\frac{\eta\sqrt{6n}}{\pi}\right)$$

$$\to 0 \quad \text{as} \quad n \to \infty$$

Therefore, the consistent estimator $\hat{\theta}$ in Eq. (5.9.18), as a moment estimator based on $\{\log X_i\}_1^n$, behaves well for large n as far as the possibility that it may take negative values is concerned, since this possibility tends to zero as n tends to infinity.

Further, the distribution of $Y = X^\theta$ is

$$f(y;\theta) = \exp(-y), \quad y > 0$$

with $E(Y) = 1$.

The $100p\%$ population percentile based on $Y = X^\theta$ is found by solving the equation

$$\int_0^{y_p} \exp(-y)dy = p$$

$$1 - \exp(-y_p) = p$$

$$y_p = -\log(1-p) = x_p^\theta$$

The percentile equation is given by

$$Y_{([np]+1)} = X^\theta_{([np]+1)} = -\log(1-p)$$

The consistent estimator of θ by percentile method is, therefore,

$$\hat{\theta} = \frac{\log[-\log(1-p)]}{\log X_{([np]+1)}} = x_p^{-1}(X_{([np]+1)})$$

Further, we have

$$\log x_p = \frac{1}{\theta}\{\log[-\log(1-p)]\} = \frac{k(p)}{\theta}$$

$$\frac{1}{x_p(\theta)}\frac{\partial x_p(\theta)}{\partial\theta} = -\frac{k(p)}{\theta^2}$$

$$\left(\frac{\partial x_p(\theta)}{\partial \theta}\right)^{-1} = -\frac{\theta^2}{x_p(\theta)k(p)}$$

Since $[\partial x_p(\theta)/\partial\theta]^{-1}$ is nonvanishing and continuous, by Theorem 5.8.5, the pth percentile estimator $\hat{\theta} = k(p)/\log X_{([np]+1)}$ is CAN for θ, so that

$$\sqrt{n}(\hat{\theta}-\theta) \xrightarrow{d} N\left[0, \frac{p(1-p)}{[f(x_p(\theta))]^2}\frac{\theta^4}{x_p^2(\theta)k^2(p)}\right]$$

***Example* 5.22** Consider an *iid* sample $X_1, X_2,\ldots,X_n$ from the distribution with the *pdf* $f(x;\theta) = \exp\{-(x-\theta)\}$, $x > \theta$. (i) Show that $X_{(1)}$ is a consistent estimator of θ. (ii) Consider the class of estimators of the type $\delta = X_{(1)} + b$, $b \in \mathbb{R}^1$. Show that the smallest MSE estimator in this class is given by $\delta_0 = X_{(1)} - (1/n)$. Also, show that δ_0 is consistent for θ.

Solution. The corresponding distribution function is

$$F_X(x;\theta) = 1 - \exp\{-(x-\theta)\}$$

The *pdf* of $X_{(1)}$ is

$$f_{X_{(1)}}(x;\theta) = n\exp[-n(x-\theta)],\ x > \theta$$

(i) Consider

$$E(X_{(1)}-\theta) = n\int_{\theta}^{\infty}(x-\theta)\exp[-n(x-\theta)]dx$$

Letting $n(x-\theta) = y$, we get

$$E(X_{(1)}-\theta) = n\int_{0}^{\infty} y\exp(-y)dy = \frac{1}{n}$$

Thus, $X_{(1)}$ is asymptotically unbiased for θ since $E(X_{(1)}) \to \theta$ as $n \to \infty$. Next, consider

$$E(X_{(1)}-\theta)^2 = n\int_{\theta}^{\infty}(x-\theta)^2\exp[-n(x-\theta)]dx$$

$$= \frac{1}{n^2}\int_{0}^{\infty} y^2\exp(-y)dy = \frac{2}{n^2}$$

This gives $\quad V(X_{(1)}-\theta) = \frac{2}{n^2} - \frac{1}{n^2} = \frac{1}{n^2} = V(X_{(1)}) \to 0 \quad \text{as} \quad n\to\infty$

Thus, by Theorem 5.3.1, $X_{(1)}$ is consistent for θ.

(ii) The MSE of $\delta = X_{(1)} + b$ is given by

$$\begin{aligned}\text{MSE}(\delta) &= E(X_{(1)} + b - \theta)^2\\ &= E(X_{(1)}-\theta)^2 + 2bE(X_{(1)}-\theta) + b^2\\ &= \frac{2}{n^2} + \frac{2b}{n} + b^2\end{aligned}$$

The constant b for which MSE(δ) is the minimum is obtained by solving

$$\frac{\partial}{\partial b}\text{MSE}(\delta) = \frac{2}{n} + 2b = 0$$

$$b = -\frac{1}{n}$$

$$\frac{\partial^2}{\partial b^2}\text{MSE}(\delta) = 2 > 0$$

Therefore, the least MSE estimator, in the class of estimators of the type $\delta = X_{(1)} + b$, $b \in \mathbb{R}^1$, is $\delta_0 = X_{(1)} - (1/n)$. The estimator δ_0 is unbiased with

$$\text{MSE}(\delta_0) = V(\delta_0) = V(X_{(1)}) = \frac{1}{n^2}$$

So, The MSE of the least MSE estimator δ_0 is $1/n^2$. Clearly, δ_0 is consistent.

***Example* 5.23** Consider a random sample $(X_1, Y_1),\ldots,(X_n, Y_n)$ from *pmf* $P[(x, y); \theta, p]$

$$P[(x, y); \theta, p] = \binom{x}{y} p^y (1-p)^{x-y} \exp(-\theta)\frac{\theta^x}{x!}$$

$p \in (0, 1)$, $\theta > 0$, $y = 0, 1,\ldots,x$; $x = 0, 1, 2,\ldots$. Find CAN estimator for (θ, p).

Solution. We consider

$$\bar{\mathbf{V}} = (\bar{V}_1, V_2)' = (\bar{X}, \bar{Y})'$$

$$E(V_1) = E(X) = \theta = \pi_1$$

$$E(V_2) = E(Y) = E^X E^{Y|X}(Y) = E^X Xp = \theta p = \pi_2$$

$$V(V_1) = V(X) = \theta$$

$$V(V_2) = V(Y) = E^X V^{Y|X}(Y) + V^X E^{Y|X}(Y)$$

$$= E_1^X[Xp(1-p)] + V_1^X(Xp) = \theta p(1-p) + \theta p^2 = \theta p$$

$$\text{cov}(X, Y) = E(XY) - E(X)E(Y)$$

$$E(XY) = E^X X E^{Y|X} Y = E^X(X^2 p)$$

$$= p(\theta + \theta^2)$$

$$\text{cov}(X, Y) = \theta p + \theta^2 p - \theta^2 p = \theta p$$

$$D(\mathbf{V}) = \begin{pmatrix} \theta & \theta p \\ \theta p & \theta p \end{pmatrix} = \Sigma$$

Therefore, the multivariate central limit theorem gives

$$\sqrt{n}(\bar{\mathbf{V}} - \boldsymbol{\pi}) \xrightarrow{d} N_2(\mathbf{0}, \Sigma)$$

$$\sqrt{n}\left[\begin{pmatrix}\bar{X}\\ \bar{Y}\end{pmatrix}-\begin{pmatrix}\theta\\ \theta p\end{pmatrix}\right]\xrightarrow{d} N_2\left[\begin{pmatrix}0\\0\end{pmatrix},\begin{pmatrix}\theta & \theta p\\ \theta p & \theta p\end{pmatrix}\right]$$

$$\mathbf{g}(\boldsymbol{\pi})=(g_1,g_2)'=(\pi_1,(\pi_2/\pi_1))'=(\theta,p)'$$

$$\mathbf{L}=\left(\frac{\partial \mathbf{g}}{\partial \boldsymbol{\pi}}\right)=\begin{pmatrix}\dfrac{\partial g_1}{\partial \pi_1} & \dfrac{\partial g_1}{\partial \pi_2}\\ \dfrac{\partial g_2}{\partial \pi_1} & \dfrac{\partial g_2}{\partial \pi_2}\end{pmatrix}=\begin{pmatrix}1 & 0\\ -\dfrac{\pi_2}{\pi_1^2} & \dfrac{1}{\pi_1}\end{pmatrix}$$

$$=\begin{pmatrix}1 & 0\\ -\dfrac{\theta p}{\theta^2} & \dfrac{1}{\theta}\end{pmatrix}=\begin{pmatrix}1 & 0\\ -\dfrac{p}{\theta} & \dfrac{1}{\theta}\end{pmatrix}$$

$$\mathbf{L\Sigma L}'=\begin{pmatrix}\theta & \theta p\\ -p+p & -p^2+p\end{pmatrix}\begin{pmatrix}1 & -\dfrac{p}{\theta}\\ 0 & \dfrac{1}{\theta}\end{pmatrix}$$

$$=\begin{pmatrix}\theta & 0\\ 0 & \dfrac{p(1-p)}{\theta}\end{pmatrix}$$

$$\mathbf{g}(\bar{\mathbf{V}})=(g_1(\bar{\mathbf{V}}),g_2(\bar{\mathbf{V}}))'=(\bar{V}_1,(\bar{V}_2/\bar{V}_1))'=(\bar{X},(\bar{Y}/\bar{X}))'$$

Therefore, by Theorem 5.7.3,

$$\sqrt{n}\left[\begin{pmatrix}\bar{X}\\ \bar{Y}/\bar{X}\end{pmatrix}-\begin{pmatrix}\theta\\ p\end{pmatrix}\right]\xrightarrow{d} N_2\left[\begin{pmatrix}0\\0\end{pmatrix},\begin{pmatrix}\theta & 0\\ 0 & \dfrac{p(1-p)}{\theta}\end{pmatrix}\right]$$

This shows that $(\bar{X}, \bar{Y}/\bar{X})'$ is CAN for $(\theta, p)'$.

The same result may also be reached at by noting that the *pmf* $P[(x, y);\theta, p]$ belongs to two-parameter exponential family with $T_1(x, y) = y$ and $T_2(x, y) = x$. Here, Fisher information matrix for $(\theta, p)'$ is given by

$$I(\theta,p)=\begin{pmatrix}\dfrac{1}{\theta} & 0\\ 0 & \dfrac{\theta}{p(1-p)}\end{pmatrix}=\text{diag}\left(\frac{1}{\theta},\frac{\theta}{p(1-p)}\right)$$

The moment equations are

$$\bar{y}=E(Y)=\theta p$$

and

$$\bar{x}=E(X)=\theta$$

The moment estimator $(\bar{X}, \bar{Y}/\bar{X})'$of $(\theta, p)'$, which is a function of the complete sufficient statistics $(\Sigma_1^n X_i, \Sigma_1^n Y_i)$, and CAN by Theorem 5.7.3, so that

$$\sqrt{n}\left[\begin{pmatrix}\bar{X}\\ \bar{Y}/\bar{X}\end{pmatrix}-\begin{pmatrix}\theta\\ p\end{pmatrix}\right]\xrightarrow{d} N_2\left[\begin{pmatrix}0\\ 0\end{pmatrix}, \text{diag}\left(\theta, \frac{p(1-p)}{\theta}\right)\right]$$

since

$$I_X^{-1}(\theta, p) = \text{diag}\left(\theta, \frac{p(1-p)}{\theta}\right)$$

Consider a situation when $\bar{x} = 0$. This implies that $\bar{y} = 0$. The moment equations become

$$\theta = 0$$
$$\theta p = 0$$

which gives $\hat{\theta} = 0$, $\hat{p}$ = any value. Note that $\hat{\theta} = 0$ is not in the parameter space since $\theta > 0$, and $\hat{p}$ gives no unique solution. However, that situation $\bar{x} = 0$ will never happen at least for large n for sure since

$$P(\bar{X} = 0) = \exp(-n\theta) \to 0 \quad \text{as} \quad n \to \infty$$

for any $\theta > 0$. Therefore, the estimator $(\bar{X}, \bar{Y}/\bar{X})'$ is CAN, well-defined, and meaningful estimator for $(\theta, p)'$.

EXERCISES

1. Give an example of an unbiased estimator which is not consistent and an estimator which is consistent but biased.

2. Show that δ^2 is not necessarily unbiased estimator of θ^2 whenever δ is unbiased estimator of θ, but δ^2 is consistent for θ^2 whenever δ is consistent for θ.

3. (Dudewicz, 1976, pp 181–184) Consider a random sample from a distribution with *df* $F(x; \theta)$, $\theta \in \Theta$. Assume that $E_\theta(X) = \mu(\theta)$ and population median Md exist. Show that
   (i) $\bar{X}$ is unbiased and consistent for $\mu(\theta)$.
   (ii) sample median $\tilde{X}_n$, which is not generally unbiased, is a consistent estimator of population median Md.
   (iii) $\tilde{X}_n$ is an unbiased estimator of Md if the *pdf* $f(x;\theta)$ is symmetric about Md.

4. Let $X_1, X_2, \ldots, X_n$ be a random sample from a distribution with *cdf* $F(x)$. Show that the sample *cdf* $F_n(x)$ is a consistent estimator of $F(x)$. Also, show that F_n is CAN, i.e.,

$$\sqrt{n}[F_n(x) - F(x)] \xrightarrow{d} N(0, F(x)[1 - F(x)])$$

Hint. $nF_n(x) \sim b(n, F(x))$, $E(F_n) = F$, $V(F_n) = F(1 - F)/n \to 0$. It shows that $F_n(x) \xrightarrow{P} F(x)$. Further, direct application of SLLNs gives $F_n(x) \xrightarrow{a.s.} F(x)$. Further, by the Lindeberg–Levey CLT,

$$\sqrt{n}[F_n(x) - F(x)] \xrightarrow{d} N(0, F(x)[1 - F(x)]), \forall\, x \in \mathbb{R}^1$$

5. Let $\{\delta_n\}_1^\infty$ be a sequence of estimators for θ.

(i) If $E_\theta(\delta_n) \to \theta$ and $V(\delta_n) \to 0$ as $n \to \infty$, then show that δ_n is squared-error consistent for θ, $\delta_n \xrightarrow{2} \theta$.

(ii) If $\{\delta_n\}$ is consistent for θ and $|\delta_n - \theta| \le A < \infty$, $\forall\ \theta \in \Theta$ and $\forall\ \mathbf{x} \in \mathbb{R}^n$, then show that $\delta_n \xrightarrow{2} \theta$.

(iii) If, however, in (ii), $|\delta_n - \theta| \le A_n < \infty$, then prove that δ_n may not $\xrightarrow{2} \theta$.

Hint. (i) Consider

$$E|\delta_n - \theta|^2 = E[\delta_n - E(\delta_n) + [\delta_n - E(\delta_n)]]^2 = V(\delta_n) + \{B(\theta)\}^2$$

$$E|\delta_n - \theta|^2 \to 0 \quad \text{as} \quad n \to \infty$$

6. If δ_n is a consistent estimator of θ for some population with its *pdf* $f(x;\theta)$, then show that $\exp(\delta_n)$ is a consistent estimator of $\exp(\theta)$.

7. Let $X_1, X_2,\ldots,X_n$ be *iid* $b(1, \theta)$.

(i) Show that $\delta_n = \Sigma X_i/n$ is an unbiased and consistent estimator of θ but δ_n^2 is not unbiased but consistent estimator of θ^2.

(ii) Let

$$\delta_n = \sum_{i=1}^{n} X_i$$

(a) Examine the consistency of unbiased estimators $[\delta_n(\delta_n - 1)]/[n(n-1)]$ of θ^2 and $[\delta_n(\delta_n - 1)(\delta_n - 2)]/[n(n-1)(n-2)]$ of θ^3.

(b) Show that $(\delta_n/n)(1 - (\delta_n/n))^2$ is not unbiased for $\phi = \theta(1 - \theta)$ but its bias tends to zero as n tends to infinity.

(c) Show that $(\delta_n/n)(1 - (\delta_n/n))$ is not an unbiased estimator of ϕ and find its bias. Find c so that $S = cT$ is an unbiased estimator of ϕ.

(d) Let $\theta \ne 1/2$. Find the asymptotic mean and variance of $\bar{X}(1 - \bar{X})$.

(e) Show that the sample mean δ_n/n is an unbiased estimator of θ and find the variance of this estimator. Is this estimator consistent for θ? Is $(\delta_n/n)(1 - (\delta_n/n))$ consistent for ϕ?

(f) Calculate the variance of the MLE of θ.

Hint. Use approximation formula.

(g) Calculate the variance of the MLE of $\theta/(1 - \theta)$.

(iii) Show that $\bar{X}^{-1}$ is a consistent estimator of θ^{-1}.

8. Consider two experiments for estimating θ^2.

(i) Let $X_1, X_2,\ldots,X_n$ be *iid* $b(1,\theta^2)$.

(ii) Let $X_1, X_2,\ldots,X_n$ be *iid* $b(1,\theta)$.

Compare the estimators $\delta_a = \bar{X}$ and $\delta_b = \bar{X}^2$ of θ^2 in experiments (i) and (ii), respectively, on the basis of ARE.

Hint. Using central limit theorem and delta method, we have $\sqrt{n}(\delta_a - \theta^2) \xrightarrow{d} N[0, \theta^2(1 - \theta^2)]$ and $\sqrt{n}(\delta_b - \theta^2) \xrightarrow{d} N[0, 4\theta^3(1 - \theta)]$; ARE calculates to $(1 + \theta)/4\theta$; δ_b is better for $\theta < 1/3$ and δ_a is better for $\theta > 1/3$.

9. Let $X_1, X_2, \ldots, X_n$ be *iid* $P(\theta)$.

(i) Is $\bar{X}_n$ a mean squared error consistent sequence of estimators of θ?

(ii) Define the function

$$U(X) = \begin{cases} 1, & \text{if } X = 0 \\ 0, & \text{otherwise} \end{cases}$$

Using the above estimator and the Rao–Blackwell theorem, find the UMVUE of $\exp(-\theta)$. Find the MLE of $\exp(-\theta)$. Compute large sample variance of MLE using delta method and show that it is equal to the CR lower bound for an unbiased estimator of $\exp(-\theta)$. Consider two unbiased estimators $\delta_1 = \bar{X}$ and $\delta_2 = (X_1 + X_2)/2$ of θ. Find the relative efficiency of T_2 with respect to T_1.

10. Let $X_1, X_2, \ldots, X_n$ be a random sample from the distribution with *pdf* given by

$$f(x;\theta) = \begin{cases} \dfrac{2}{\theta^2}(\theta - x), & \text{for } x \in [0, \theta] \\ 0, & \text{otherwise} \end{cases}$$

Find an estimator of θ by the MoM method. Calculate its bias and variance. Check whether it is consistent.

11. Let $X_1, X_2, \ldots, X_n$ be *iid* Rayleigh distribution, $X \sim \mathcal{R}(\theta)$, with *pdf*

$$f(x;\theta) = \begin{cases} \dfrac{1}{\theta} x \exp\left\{-\dfrac{x^2}{2\theta}\right\}, & \text{for } x \in [0, \infty] \\ 0, & \text{otherwise} \end{cases}$$

where θ is an unknown positive constant with $E(X) = \sqrt{(\pi/2)\theta}$ and $V(X) = [(4 - \pi)/2]\theta$. Note that ΣX_i^2 is a complete sufficient statistic since $\mathcal{R}(\theta)$ belongs to the one-parameter exponential family. ML estimator is $\delta_n = (1/2n)\Sigma X_i^2$ and is unbiased. δ_n is UMVUE of θ, since it is the function of complete sufficient statistic for θ and is unbiased for θ. Show that δ_n is a consistent estimator of θ.

12. Let $X_1, X_2, \ldots, X_n$ be *iid* $N(\theta, \sigma^2)$, $\theta \neq 0$.

(i) Are the sequence of estimators $\bar{X}_n = (1/n)\Sigma_1^n X_i$ and $S_n^2 = (1/n)\Sigma_1^n (X_i - \bar{X})^2$, mean and mean squared error, consistent for θ and σ^2, respectively? Also, find the efficiency of $nS_n^2/(n-1)$.

(ii) Show that the MLE of θ^2 is $AN(a, b)$. Specify its asymptotic mean a and variance b.

13. Let $X_1, X_2, \ldots, X_n$ be *iid* $N(\theta, \theta)$, $\theta > 0$. Is $\bar{X} S_n^2$ consistent for θ^2?

Hint. $E(\bar{X}) = \theta$ and $E(S_n^2) = \theta$.

14. Suppose $X_1, X_2, \ldots, X_n$ is a random sample from the distribution of a random variable X when $X \sim N(0, \theta)$, where θ is population variance.

(i) Show that $\delta_c = c\Sigma_{i=1}^n |X_i|$ is consistent for θ for some c.

(ii) Given that $U = (n-1)S_{n-1}^2/\theta \sim \chi^2_{(n-1)}$, where S_{n-1}^2 is the adjusted sample variance, $S_{n-1}^2 = [1/(n-1)]\Sigma_{i=1}^n(X_i - \bar{X})^2$, show that S_{n-1}^2 is an unbiased and consistent estimator of θ. Also, calculate $V(S_{n-1}^2)$.

(iii) Let $\delta = (1/n)\Sigma_{i=1}^n X_i^2$. Set $U = n\delta/\theta = \Sigma_{i=1}^n Z_i^2$, where $Z_i = X_i/\sqrt{\theta}$. Obtain the distribution of $Z_1, Z_2, \ldots, Z_n$ and, hence, the distribution of U. Also, calculate $E(\delta)$ and $V(\delta)$. Show that δ is also unbiased, consistent estimator of θ and that it has smaller variance than S_{n-1}^2.

15. Let $X_1, X_2, \ldots, X_n$ be *iid* with *pdf* $f(x;\theta) = \theta\exp(-\theta x)$, $x > 0$, $\theta > 0$. Based on the biased estimator $1/\bar{X}$ of θ, suggest an unbiased estimator of θ and check whether it is consistent.

16. Let $X_1, X_2, \ldots, X_n$ be *iid* $G_1(\alpha, \beta)$, where α is known. Show that

(i) the estimator $\bar{X}/\alpha$ is unbiased and mean square consistent for β and

(ii) the estimator $[1/n\alpha(\alpha+1)]\Sigma_{i=1}^n X_i^2$ is unbiased and consistent for β^2.

17. Let $X_1, X_2, \ldots, X_n$ be a random sample from the Cauchy distribution with *pdf*

$$f(x;\theta) = \frac{1}{\pi[1+(x-\theta)^2]}, -\infty < x < \infty, \theta > 0$$

(i) Show that $\bar{X}_n$ is not a consistent estimator of θ.

(ii) Sample median $\tilde{X}$ is a consistent estimator of θ and its asymptotic efficiency is $8/\pi^2$.

Hint. Consider the sample mean, $\bar{X}_n$, as a reasonable estimator of θ. The distributions of $\bar{X}_n$ and of a single observation X_1, in case of the Cauchy distribution, are the same. This gives $V(\bar{X}_n) = V(X_1)$, which does not approaches 0 as $n \to \infty$. Thus, by Theorem 5.3.1, $\bar{X}_n$ is not consistent for θ. However, for large n, the sample median $\tilde{X}$

$$\tilde{X} \sim AN\left(\tilde{\theta}, \frac{1}{4n^2 f_1^2}\right)$$

where f_1 is the median or modal (due to symmetry) ordinate of the parent distribution. In case of the Cauchy population,

$$f_1 = [f(x;\theta)]_{x=\theta} = \frac{1}{\pi}$$

This gives
$$AV(\tilde{X}) = \frac{\pi^2}{4n} \to 0 \quad \text{as} \quad n \to \infty$$

Therefore, by Theorem 5.3.1 and results (13) and (14) of Section 1.5, the sample median $\tilde{X}$ is a consistent estimator for θ.

18. Let $X_1, X_2, \ldots, X_n$ be a random sample from the distribution with *pdf*

$$f(x;\theta) = \frac{1}{2}\theta^3 x^2 \exp(-\theta x), x > 0, \theta > 0$$

Which of the three parametric functions $g_1(\theta) = \theta$, $g_2(\theta) = 1/\theta$ and $g_3(\theta) = \ln\theta$ admits an efficient estimator? Find the MLE of $g_1(\theta)$, $g_2(\theta)$ and $g_3(\theta)$. Let $g(\theta)$ among g_1,

g_2, and g_3 be the function which admits an efficient estimator $\delta(X)$. Does there exist a consistent estimator of $g(\theta)$ whose variance is smaller than the variance of $\delta(X)$ for each?

Hint. We have $(\partial/\partial\theta)\log L(\theta, \mathbf{x}) = -3n[(1/3n)\Sigma_{i=1}^{n}X_i - 1/\theta]$. Therefore, $\delta_n(\mathbf{X}) = (1/3n)\Sigma_{i=1}^{n}X_i$ is efficient estimator of $1/\theta$. $3/\bar{X}$, $\bar{X}/3$, and $\ln(3/\bar{X})$ are MLEs of $g_1(\theta)$, $g_2(\theta)$ and $g_3(\theta)$, respectively, by using its invariance property. The estimator $\delta_n(\mathbf{X})$ is consistent for $1/\theta$ and so are the estimators of the type $a_n\delta_n(\mathbf{X})$ for $1/\theta$ so that $a_n \to 1$ as $n \to \infty$ or $a_n \to 1$ together with $|a_n| < 1$ as $n \to \infty$. The estimator $U_n(\mathbf{X}) = a_n\delta_n(\mathbf{X}) = [n/(n + 1)]\delta_n(\mathbf{X})$ is consistent so that $V(U_n) = a_n^2\, V(\delta_n) < V(\delta_n)$.

19. Assume readings on a voltage meter connected to test a circuit follow uniform distribution $U(\theta, \theta + 1)$. Let the random observations $X_1, X_2,\ldots,X_n$ be readings on the meter taking on its values in the interval $(\theta, \theta + 1)$, where θ is the true unknown value of the voltage of the circuit. Compute the bias and MSE of an estimator $\bar{X}$ used as an estimator of θ. Find its function that is an unbiased estimator of θ. Also, find a consistent estimator of θ.

(i) Show that $X_{(n)}$ is consistent and asymptotically unbiased for θ but not BAN.
(ii) Investigate whether $2\bar{X}$ is consistent for θ. Show that $\bar{X}$ is unbiased and consistent estimator of $\theta + (1/2)$.
(iii) Show that $(\Pi_{i=1}^{n}X_i)^{1/n}$ is consistent for θ/e.
(iv) Consider a class of estimators $C = \{\delta_a : \delta_a = aX_{(n)},\ a > 0\}$. Show that δ_a is least MSE estimator for $a = (n + 2)/(n + 1)$.

20. Consider a random sample $X_1, X_2,\ldots,X_n$ from the population with *pdf* $f(x;\theta) = (1/2)\exp[-(1/2)(x - \theta)]$, $x > \theta$, $\theta \in \mathbb{R}^1$. Obtain consistent estimator of θ by the method of moments and the method of percentiles.

21. Let $X_1, X_2,\ldots,X_n$ be *iid* $L(\theta)$ with *pdf*

$$f(x;\theta) \propto \exp\left\{-\frac{1}{\sigma}|x - \theta|\right\}$$

Show that sample median is better than sample mean in terms of ARE. Further, show that sample median is consistent for θ but not sample mean.

6 Methods of Estimation

6.1 INTRODUCTION

In statistical inference the characteristics of population are described by a random variable with probability density $f(x;\theta)$ with respect to a σ-finite measure. The probability density involves an unknown parameter (or a set of parameters), which characterizes the population completely. The parameters have meaningful physical interpretations about the population. The density is completely specified if the parameter is known, and any statistical inference about the population can easily be reached at in this case. However, the parameter θ is usually unknown. We then observe a random sample $X_1, X_2,...,X_n$ from the population $f(x;\theta)$ to estimate θ. We have discussed this estimation problem, in the previous chapters, based on these observations, using the principle of sufficiency and completeness and impartiality principle, such as unbiasedness, so that such estimators are optimal, say in the sense of mean squared error.

However, in the present chapter, we deal with the methods of estimating some parametric function $g(\theta)$ and suggest estimators $T(\mathbf{X})$ based on these sample observations. In Section 6.2, we introduce the method of moments (MoM) which is the oldest method of estimation given by Karl Pearson in the 18th century. In Sections 6.3 and 6.4, the method of minimum chi-square and the method of modified minimum chi-square (MoMMCS) are introduced. In Section 6.5, the method of least squares (MoLS) is discussed. In Section 6.6 the method of maximum likelihood estimation (MoML) and the large sample properties of such estimators are discussed.

6.2 METHOD OF MOMENTS (MoM)

The method of moments is the oldest method of estimation introduced by Karl Pearson in the late 18th century. This is a popular method of estimation because it is simple to use and always yields an estimator. We will now define this method formally.

Let $X_1, X_2, \ldots, X_n$ be a random sample drawn from a population with *pdf* (or *pmf*) $f(x;\boldsymbol{\theta})$, $\boldsymbol{\theta} \in \mathbb{R}^k$, and the parameters are expressible as

$$\theta_i = h_i\,(\mu_1', \mu_2', \ldots, \mu_k'),\ i = 1, 2, \ldots, k \tag{6.2.1}$$

for some known numerical functions $h_1, h_2, \ldots, h_k$, where population moments $\mu_j' = E(X^j)$, $j = 1, 2, \ldots, k$, are known to exist. The method of moments estimators $(T_1, T_2, \ldots, T_k)$ for estimating $(\theta_1, \theta_2, \ldots, \theta_k)$ is given by

$$T_1(X_1, X_2, \ldots, X_n) = h_1(m_1', m_2', \ldots, m_k')$$

$$\vdots \qquad \vdots \qquad \vdots$$

$$T_k(X_1, X_2, \ldots, X_n) = h_k(m_1', m_2', \ldots, m_k') \tag{6.2.2}$$

where $m_j' = (1/n)\Sigma_{i=1}^n X_i^j$ are sample moments. To be more specific, moment equations are given by

$$m_j' = E(X^j), \quad j = 1, \ldots, k$$

For example, let $X_1, X_2, \ldots, X_n$ be *iid rvs* with *pdf* or *pmf* $f(x;\theta)$. The MoM estimator of mean $E_\theta(X) = \mu_1'$ is $\hat{\mu}_1'(\theta) = \bar{X}$ and of variance $V_\theta(X) = \mu_2'(\theta) - [\mu_1'(\theta)]^2$ is $\hat{\sigma}^2 = \hat{\mu}_2' - \hat{\mu}_1'^2 = (1/n)\Sigma_1^n X_i^2 - \bar{X}^2 = (1/n)\Sigma_1^n (X_i - \bar{X})^2 = S^2$.

Some important results relating to the MoM estimators are listed as follows:

1. MoM estimators (MoMEs) are considered, mostly, as the first estimator to start with, since they are easy to compute. These estimators are not necessarily UMVUEs or most efficient, though they are generally consistent.
2. For estimating the joint moment $\mu_{XY} = E(XY)$, the MoME of μ_{XY} is given by $(1/n)\,\Sigma_1^n X_i Y_i$ by the above method.
3. For estimating population moment of order j, $\mu_j' = E(X^j)$, by the MoM the sample moment of order j, $m_j' = (1/n)\Sigma_{i=1}^n X^j$, is consistent and asymptotically unbiased, since $m_j' \xrightarrow{P} \mu_j'$ by WLLN.
4. Using the Lindeberg–Levy central limit theorem, assuming $E|X|^{2j} < \infty$, $m_j' = (1/n)\Sigma_{i=1}^n X_i^j \sim AN(EX^j, V(X^j)/n)$ where $\mu_j' = E(X^j)$. In other words, MoMEs are asymptotically normal.
5. However, the estimator $\mathbf{T} = (T_1, T_2, \ldots, T_k) = (h_1(\mathbf{m}'), \ldots, h_k(\mathbf{m}')] = \mathbf{h}(\mathbf{m}')$ in Eq. (6.2.2) is consistent for $\boldsymbol{\theta} = (\theta_1, \theta_2, \ldots, \theta_k) = (h_1(\boldsymbol{\mu}'), \ldots, h_k(\boldsymbol{\mu}'))$, $\boldsymbol{\mu}' = (\mu_1', \mu_2', \ldots, \mu_k')$ whenever h_i, $i = 1, 2, \ldots, k$, are continuous functions of $\boldsymbol{\theta}$ since $m_i' \xrightarrow{P} \mu_i'$, $i = 1, 2, \ldots, k$. Note that if $m_i' \xrightarrow{P} \mu_i'$, then
$$T_j = h_j(m_1', \ldots, m_k') \xrightarrow{P} h_j(\mu_1', \ldots, \mu_k') = \theta_j, \quad j = 1, 2, \ldots, k \tag{6.2.3}$$
whenever h_js are continuous.
6. Cramer has shown by imposing a few mild conditions on the functions h_j in Eq. (6.2.3) that $\mathbf{T}$ is asymptotically normal (see Chapter 5, Section 8.2). Suppose for some function $g(X)$, $E[g(X)] = \mu(\theta)$, $V[g(X)] = \sigma^2(\theta)$, and the quantity $[(\partial/\partial\theta)\mu(\theta)]^{-1} \neq 0$ and continuous. The moment equation is

$$\frac{g(X_1)+g(X_2)+\cdots+g(X_n)}{n}=\bar{g}(\mathbf{X})=\mu(\theta)=E[g(X)]$$

The MoME $T=\mu^{-1}[\bar{g}(\mathbf{X})]$ is CAN for θ, i.e., $\mu^{-1}[\bar{g}(\mathbf{X})]$ is consistent for θ and

$$\sqrt{n}\{\mu^{-1}[\bar{g}(\mathbf{X})]-\theta\}\xrightarrow{d}N\left\{0,\ V[g(X)]\left(\frac{\partial\mu(\theta)}{\partial\theta}\right)^{-2}\right\}$$

We can use the above result for comparing two MoMEs on the basis of their asymptotic variances.

7. MoM estimation fails when population moments do not exist, e.g., in case of the Cauchy distribution.
8. In case of single parameter θ, if $E(X_i)=\mu_1'(\theta)$ and $\{X_i\}_1^n$, is a sequence of *iid rvs*, then by WLLNs, $m_1'\xrightarrow{P}\mu_1'(\theta)$. In addition to it, if $\mu_1'^{-1}$ exists and is continuous, then $\mu_1'^{-1}(m_1')\xrightarrow{P}\mu_1'^{-1}\mu_1'(\theta)=\theta$. Therefore, $\mu_1'^{-1}(m_1')=\mu_1'^{-1}(\bar{X})$ is consistent for θ. Note that $\mu_1'^{-1}(\bar{X})$ is the solution of MoM equation $m_1'=\bar{X}=\mu_1'(\theta)$. Therefore, $\mu_1'^{-1}(\bar{X})$ is the MoME, which is consistent for θ provided $\mu_1'^{-1}$ exists and is continuous. A sufficient condition for this condition to hold is that $(\partial/\partial\theta)\mu\neq 0$.
9. In general, MoMEs are less efficient as compared to MLEs. However, if the likelihood equations are expressible as the linear function of sample moments, then

$$(\partial/\partial\theta_j)\log L=a_0+a_1m_1'+a_2m_2'+a_3m_3'+\cdots$$

or
$$f(x;\theta)=\exp\{b_0+b_1x+b_2x^2+b_3x^3+\cdots\} \qquad (6.2.4)$$

where *b*s are functions of $\boldsymbol{\theta}=(\theta_1,\theta_2,\ldots,\theta_k)$. In this case, the MoM and ML estimators are identical. In other words, if the *pmf* or *pdf* is of the general form, as in Eq. (6.2.4), then the moment equations and the maximum likelihood equations give the same solution, i.e., MoMEs and MLEs are the same. In this case, MoMEs are as efficient as MLEs. MoMs are not consistent in general, but if the *pdf* is of the form given by Eq. (6.2.4), then they are consistent.
10. The first occurrence of the use of the MoM was found when Karl Pearson addressed the problem of failure to fit the normal distribution to many real-life continuous variables. He has developed a generalized system of pdfs, popularly known as the Pearson distributions. *Pdf*s belonging to this family are determined by

$$\frac{\partial}{\partial x}\log f(x)=\frac{(x-a)}{b_0+b_1x+b_2x^2}$$

Pearson stated certain regularity conditions according to which, the first four moments $(\mu_1',\mu_2',\mu_3',\mu_4')$ exist. Kendall and Stuart, Vol. I (1958) have expressed these moments of the *pdf f* in terms of the constants a, b_0, b_1 and b_2. One can equivalently obtain $(\mu_1',\mu_2,\beta_3,\beta_4)$ from the four moments $(\mu_1',\mu_2',\mu_3',\mu_4')$ where $\mu_1'=$ mean, $\mu_2=$ variance, $\beta_1=$ skewness, and $\beta_2=$ kurtosis of the distribution. Pearson tabulated the values of (β_1,β_2) for the types of the curve to be guessed. Subsequently, the constants (a, b_0, b_1, b_2) can then be estimated as follows:

(i) Calculate sample moments (m_1', m_2', m_3', m_4'), then (m_1', m_2, b_1, b_2).
(ii) Use sample skewness b_1, sample kurtosis b_2 as given in Biometrika tables prepared by Pearson to guess the type of *pdf*.
(iii) Finally, estimate the parameters (a, b_0, b_1, b_2) by equating sample moments to the moments of the *pdf f* and finally solving the equations for a, b_0, b_1, b_2.

This is the standard procedure which is being used for a long time for the fitting of theoretical frequency curves to observed frequency distributions.

6.3 METHOD OF MINIMUM CHI–SQUARE (MoMCS) ESTIMATION

This method of estimation is appropriate in a situation when the sample $X_1, X_2, \ldots, X_n$ is drawn from a multinomial population with k mutually exclusive groups having probabilities $p_1, p_2, \ldots, p_k$, $\Sigma p_i = 1$. Suppose the observations are classified into these groups with observed frequencies $f_1, \ldots, f_k$, $\Sigma f_i = n$. The group probabilities p_is are the functions of the parameters $\boldsymbol{\theta} = (\theta_1, \ldots, \theta_s)'$ so that $p_i = p_i(\boldsymbol{\theta})$. The expected frequencies in k groups are given by $e_1 = np_1(\boldsymbol{\theta}), \ldots, e_k = np_k(\boldsymbol{\theta})$. Pearson proposed χ^2 as a measure of departure of observed frequencies from expected frequencies, which is defined by

$$\chi^2(\boldsymbol{\theta}) = \sum_{i=1}^{k} \frac{[f_i - e_i(\boldsymbol{\theta})]^2}{e_i(\boldsymbol{\theta})} = \sum \frac{f_i^2}{e_i(\boldsymbol{\theta})} - n \tag{6.3.1}$$

where $\Sigma_{i=1}^k e_i = n$ and $\Sigma_{i=1}^k f_i = n$.

Note that if e_is and f_is match, $\chi^2(\boldsymbol{\theta})$ attains 0 and we state that the sample $X_1, X_2, \ldots, X_n$ has come from population with such parameter $\boldsymbol{\theta}$ for which this situation occurs. The Method of minimum chi-square consists in obtaining such $\boldsymbol{\theta}$ for which this measure of discrepancy, i.e., $\chi^2(\boldsymbol{\theta})$ is minimum over $\boldsymbol{\theta} \in \Theta \subseteq \mathbb{R}^s$.

Using mathematical tools, MoMCSEs are obtained by solving k simultaneous equations

$$\frac{\partial}{\partial \theta_j} \chi^2(\boldsymbol{\theta}) = 0, \; j = 1, \ldots, k$$

so that
$$\frac{\partial^2}{\partial \theta_j^2} \chi^2(\boldsymbol{\theta}) > 0 \;\; \forall \;\; j = 1, \ldots, k$$

$$\Rightarrow \qquad \sum \frac{f_i^2}{e_i^2(\boldsymbol{\theta})} \frac{\partial e_i(\boldsymbol{\theta})}{\partial \theta_j} = 0 \tag{6.3.2}$$

Equations (6.3.2) are known as MCS equations.
However, if we write the likelihood

$$L(\mathbf{p}|\mathbf{f}) = \prod_{i=1}^{k} \frac{n!}{f_i!} p_i^{f_i}$$

we get
$$\log L = \text{const.} + \sum f_i \log p_i \tag{6.3.3}$$

Thus, the likelihood Eq. (6.3.3) are given by

$$\frac{\partial}{\partial\theta_j}L = \sum \frac{f_i}{p_i}\frac{\partial}{\partial\theta_j}p_i(\boldsymbol{\theta}) = 0$$

$$\sum \frac{f_i}{np_i}\frac{\partial}{\partial\theta_j}np_i(\boldsymbol{\theta}) = 0$$

or

$$\sum\left(\frac{f_i - e_i(\boldsymbol{\theta})}{e_i(\boldsymbol{\theta})}\right)\frac{\partial}{\partial\theta_j}e_i(\boldsymbol{\theta}) = 0 \tag{6.3.4}$$

(since $\Sigma e_i = 1\,\Sigma\partial e_i/\partial\theta_j = 0\ \forall\ j$)

Equation (6.3.2) can be written as

$$\sum \frac{(f_i^2 - e_i^2(\boldsymbol{\theta}))}{e_i^2(\boldsymbol{\theta})}\frac{\partial}{\partial\theta_j}e_i(\boldsymbol{\theta}) = 0$$

or

$$\sum \frac{(f_i - e_i)}{e_i}\frac{(f_i + e_i)}{e_i}\frac{\partial}{\partial\theta_j}e_i = 0$$

For large n, $(f_i + e_i)/e_i \xrightarrow{P} 2$, gives

$$\sum\left(\frac{f_i - e_i}{e_i}\right)\frac{\partial}{\partial\theta_j}e_i = 0 \tag{6.3.5}$$

This shows that likelihood Eq. (6.3.4) and minimum chi-square Eq. (6.3.5) are the same for large n and, therefore, MoMCSE and MLE become identical for large n. Therefore, MoMCSE share all asymptotic properties of MLE. MoMCSE are BAN estimators. Rao (1957), among others, has discussed large sample properties of MoMCSE independently for single parameter $\boldsymbol{\theta}$.

Theorem 6.3.1 MoMCS estimators are consistent.

Proof. We consider here just a single parameter θ for simplicity of proof. We will show that chi-square Eq. (6.3.5) has a solution with probability one, that converges to true value of $\theta = \theta_0$.

The left-hand expression of minimum chi-square Eq. (6.3.2) is

$$nH(\theta) = n\sum \frac{f_i^2}{e_i^2}\frac{\partial}{\partial\theta}e_i$$

and it is the function of θ. Expanding it by Taylor series about the true value θ_0, we get

$$nH(\theta) = nH(\theta)\big|_{\theta=\theta_0} + n(\theta - \theta_0)\frac{\partial}{\partial\theta}H(\theta)\big|_{\theta'} \tag{6.3.6}$$

$\theta' \in h\text{-nbd}(\theta_0)$ (i.e., h-neighborhood about θ_0) or $\theta' \in (\theta_0 - h,\ \theta_0 + h)$. The second term of Eq. (6.3.6) involves

$$\frac{\partial}{\partial\theta}H(\theta) = \frac{\partial}{\partial\theta}\sum\frac{f_i^2}{e_i^2}\frac{\partial}{\partial\theta}e_i$$

$$= -2\sum\frac{f_i^2}{e_i^3}\left(\frac{\partial}{\partial\theta}e_i\right)^2 + \sum\frac{f_i^2}{e_i^2}\frac{\partial^2}{\partial\theta^2}e_i$$

For large n,

$$\frac{f_i}{n} \xrightarrow{P} p_i \quad \text{or} \quad \frac{n^2(f_i/n)^2}{e_i^2} \xrightarrow{P} \frac{n^2 p_i^2}{e_i^2} = \frac{e_i^2}{e_i^2} = 1$$

implies $H(\theta)|_{\theta_0} \xrightarrow{P} \Sigma 1(\partial/\partial\theta)e_i = (\partial/\partial\theta)\Sigma e_i = 0$. By choosing proper continuity assumption on $(\partial/\partial\theta)H(\theta)|_{\theta'}$, $\theta' \in h\text{-nbd}(\theta_0)$, the quantity $(\partial/\partial\theta)H(\theta)|_{\theta'}$ can be made very close to $(\partial/\partial\theta)H(\theta)|_{\theta_0}$. Under these settings,

$$\frac{\partial}{\partial\theta}H(\theta)|_{\theta=\theta_0} \xrightarrow{P} -2\sum\frac{n^3 p_i^2}{(np_i)^3}\left(\frac{\partial}{\partial\theta}e_i\right)^2\Bigg|_{\theta_0}$$

$$= -2\sum\frac{1}{p_i}\left(\frac{\partial}{\partial\theta}e_i\right)^2\Bigg|_{\theta_0} \tag{6.3.7}$$

Hence, from Eqs. (6.3.6) and (6.3.7),

$$nH(\theta) \xrightarrow{P} -2n(\theta-\theta_0)\sum\left(\frac{1}{p_i}\right)\left(\frac{\partial}{\partial\theta}e_i\right)^2\Bigg|_{\theta_0}$$

We, therefore, see that the left-hand side of MoMCS-equation $nH(\theta) = 0$ converges to a positive quantity if $\theta < \theta_0$ and to a negative quantity if $\theta > \theta_0$. Hence, there exists one point in the h-nbd(θ_0) at which $nH(\theta) = 0$ with probability one, so that $\theta \xrightarrow{P} \theta_0$. This shows that MoMCS estimators are consistent.

Theorem 6.3.2 MoMCS estimators are unique.

Proof. We assume that the MoMCS equation $nH(\theta) = 0$ has two distinct solutions, T_1 and T_2. Clearly,

$$nH(\theta)|_{\theta=T_1} = 0,\ nH(\theta)|_{\theta=T_2} = 0$$

$$n\sum\frac{f_i^2}{e_i^2}\frac{\partial}{\partial\theta}e_i\Big|_{T_1} = 0 = n\sum\frac{f_i^2}{e_i^2}\frac{\partial}{\partial\theta}e_i\Big|_{T_2}$$

We assume that $H(\theta)$ is continuous and differentiable in some interval of θ, (T_1, T_2). Therefore, by Rolle's theorem, there exists one point θ' in (T_1, T_2) at which

$$n\frac{\partial}{\partial\theta}H(\theta)\Big|_{\theta'} = 0 \tag{6.3.8}$$

We have shown in Eq. (6.3.7) that $(\partial/\partial\theta)H(\theta)$ approaches a negative quantity, which is a contradiction to Eq. (6.3.8). Therefore, T_1 and T_2 cannot be different.

Theorem 6.3.3 Minimum Chi-square equation leads to a solution that minimizes χ^2.

Proof. Suppose an MoMCS estimator exists so that a solution to the MoMCS equation

$$\frac{\partial}{\partial \theta}\chi^2 = -nH(\theta) = 0 \tag{6.3.9}$$

exists, say T_0. By Theorem 6.3.1, $T_0 \xrightarrow{P} \theta_0$, where θ_0 is the true value of the parameter θ. Using Eq. (6.3.7),

$$H'(\theta)\big|_{\theta_0} = \frac{\partial}{\partial \theta}H(\theta)\big|_{\theta_0} \xrightarrow{P} (\text{–ve quantity})$$

In Eq. (6.3.9), we have

$$\frac{\partial^2}{\partial \theta^2}\chi^2 = -nH'(\theta) \xrightarrow{P} (\text{+ve quantity})$$

Therefore, the MoMCS estimator T_0 minimizes χ^2.

6.4 METHOD OF MODIFIED MINIMUM CHI-SQUARE (MoMMCS) ESTIMATION

The expression of chi-square in Eq. (6.3.1) involves parameter θ, both in the numerator and the denominator, and the MoMCS Eq. (6.3.2) often takes a complicated form. Modified MoMCS method introduces f_i in place of e_i in the denominator of the expression of chi-square. This takes the form

$$\chi'^2 = \sum_{i=1}^{k}\frac{(f_i - np_i)^2}{f_i} = \sum_i \frac{n^2 p_i^2}{f_i} - n \tag{6.4.1}$$

MoMMCS estimators are obtained by solving

$$\frac{\partial}{\partial \theta}\chi'^2 = 0 \quad \text{or} \quad 2n^2 \sum_i \frac{p_i}{f_i}\frac{\partial}{\partial \theta}p_i = 0 \tag{6.4.2}$$

Similarly, as the MoMCS and ML estimators are asymptotically equivalent, we will show here that the MoMMCS and ML estimators are equivalent. The likelihood in Eq. (6.3.3) can be written as

$$L(\mathbf{p}|\mathbf{f}) = \frac{n!}{\prod_i f_i!}\prod_i p_i^{f_i} = \frac{n!}{\prod_i f_i!}\prod\left(\frac{np_i}{f_i}\right)^{f_i}\prod\left(\frac{f_i}{n}\right)^{f_i}$$

$$\log L = \text{const.} + \sum f_i \log\left(\frac{np_i}{f_i}\right) \tag{6.4.3}$$

For large sample,

$$\frac{f_i}{n} \xrightarrow{P} p_i$$

or

$$\frac{f_i}{n} + \frac{b_i}{\sqrt{n}} \xrightarrow{P} p_i$$

where $\{b_i\}$ is a sequence of real numbers so that $\Sigma b_i = 0$. We can write, for large n,

$$p_i \approx \frac{f_i}{n} + \frac{b_i}{\sqrt{n}}$$

$$np_i = f_i + b_i\sqrt{n}$$

On putting the value of np_i in Eq. (6.4.3), we get

$$\log L = \text{const.} + \sum f_i \log\left(1 + \frac{b_i\sqrt{n}}{f_i}\right)$$

$$= \text{const.} + \sum b_i\sqrt{n} - \sum \frac{b_i^2 n}{2f_i} + \sum \frac{b_i^3 n^{3/2}}{3f_i^2} - \cdots$$

$$= \text{const.} - \frac{1}{2}\sum \frac{(np_i - f_i)^2}{f_i} + O(n^{-1/2}) \tag{6.4.4}$$

For large n, Eq. (6.4.4) reduces to

$$\log L = -\frac{1}{2}\sum \frac{(np_i - f_i)^2}{f_i} = -\frac{1}{2}\chi'^2 \tag{6.4.5}$$

Equation (6.4.5) shows that the maximization of $\log L$ amounts to the minimization of χ'^2. This shows that for large n, MoMMCS and ML estimators are asymptotically equivalent. However, Neyman (1948) has shown that MoMMCS estimators are the best asymptotically normal (BAN) estimators. We can sum up this section with the following remarks:

1. MoMCS, MoMMCS, ML estimators are asymptotically equivalent. They share same properties: They are (i) consistent, (ii) asymptotically normal, and (iii) efficient; and all of them are BAN estimators.
2. Practically, MoMCS or MoMMCS methods are preferred when ML equations take a complicated form. For doing so, there is no theoretical justification, except that of computational convenience. In such cases, MoMCS or MoMMCS methods generally provide a solution that serves as a starting solution to an iterative procedure that leads to the minimization of χ^2 or χ'^2. Very often, such procedures result in greater simplifications, as compared to solving ML equations by numerical methods.
3. One drawback of using the MoMCS or the MoMMCS method is that these classify data into certain classes, which becomes a drawback when it is available on a continuous variable. However, classifying such data into classes amounts to loss in information. Therefore, these estimation procedures suffer from not being utilizing some part of the data.

6.5 METHOD OF LEAST SQUARES (MoLS) ESTIMATION

Consider a statistical model

$$E(Y|x) = h(x;\mathbf{b}) \tag{6.5.1}$$

where $\mathbf{b} = (b_1, b_2, \ldots, b_s)'$, $\mathbf{b}$ is unknown, x is known; h is some known linear or nonlinear function, sometimes referred to as curve. The quantity $E(Y|x)$ is called the expected value of the *rv* Y depending on x. The value taken on by Y is known as the observed value. In experimental situations, $b_1, \ldots, b_s$ are known as fixed effects and the model in Eq. (6.5.1) is known as the fixed effect statistical model. Further,

$$e|x = y - E(Y|x) \tag{6.5.2}$$

is called residual or error. We assume $E(e|x) = 0$ and $V(e|x) = \sigma^2(x) = \sigma^2$. The unknown quantities $(\mathbf{b}, \sigma^2)$ are called the parameters of the model (6.5.1). To estimate these parameters, we draw n observations $(x_1, y_1), \ldots, (x_n, y_n)$ from the model in Eq. (6.5.1). The method of least square deals with estimating $(\mathbf{b}, \sigma^2)$ so that the residual sum of squares

$$D(\mathbf{b}, \sigma^2) = \sum_{i=1}^{n} \left(\frac{y_i - E(Y|x_i)}{\sigma} \right)^2 \tag{6.5.3}$$

is minimum.

Further, if $V(e_i|x_i) = \sigma_i^2(x_i) = \sigma_i^2$, then this method minimizes the quantity

$$D = \sum_{i=1}^{n} \left(\frac{1}{\sigma_i^2} \right) [y_i - E(Y|x_i)]^2 \tag{6.5.4}$$

The corresponding estimators are known as weighted least squares estimators. Indeed, MoLS can be used even when the underlying population distribution function is not known, but the experimental situations should be such that one can realistically believe that the observations are drawn independently. Further, if errors are *iid* with $E(e_i) = 0$, $\text{cov}(e_i, e_j) = 0$, and $V(e_i|x_i) = \sigma^2$, then the MoLS estimators of fixed effects are BLUEs. Moreover, if $e_i \sim N(0, \sigma_i^2)$, e_i are independent, then the MoLS and MLE estimators of fixed effects are identical and BLUE.

6.6 METHOD OF MAXIMUM LIKELIHOOD (MoML) ESTIMATION

The method of maximum likelihood estimation is one of the oldest and an important method of estimation. Gauss (1821) first conceptualized and used this method in discussing least square theory. Later, R.A. Fisher (1912) investigated several of its optimality properties and compared this with its competing methods.

To formally introduce this method, let us assume $X_1, X_2, \ldots, X_n$ to be a random sample drawn from a population with *pdf* (or *pmf*) $f(x;\boldsymbol{\theta})$, $\boldsymbol{\theta} \in \Theta \subseteq \mathbb{R}^k$. The joint distribution of sample observations, in this case, is given by

$$f(\mathbf{x};\boldsymbol{\theta}) = \begin{cases} \prod_1^n f(x_i;\boldsymbol{\theta}), & \text{if } X \text{ is discrete} \\ \prod_1^n P_\theta(x_i \le X_i \le x_i + dx_i), & \text{if } X \text{ is continuous} \end{cases} \tag{6.6.1}$$

$$= \prod_1^n f(x_i;\boldsymbol{\theta}) dx_i,$$

The joint density in Eq. (6.6.1) is a function of **x** for given $\boldsymbol{\theta}$, i.e., the joint density is a function of n observations. Whenever this is denoted as a function of $\boldsymbol{\theta}$ given **x**, we call it as likelihood function and denote it by $L(\boldsymbol{\theta}; \mathbf{x})$. The likelihood function is a function of parameters $(\theta_1,\ldots,\theta_k)$. Generally, k is either 1 or 2. Note the difference between likelihood function and joint density function. Though the likelihood function arises from a probability function, it is neither a probability density function nor relates to a probability attached to a model. However, it refers to a model or a distribution given the observations. Here, the quantity $L(\boldsymbol{\theta}; \mathbf{x})$ measures the likelihood that the observations **x** have come from the population with parameter $\boldsymbol{\theta}$. If for given two points **x** and **y**,

$$\frac{L(\boldsymbol{\theta}_1; \mathbf{x})}{L(\boldsymbol{\theta}_2; \mathbf{y})}$$

is the function of **x** and **y**, then the likelihood principle says that the inference about $\boldsymbol{\theta}$ based on **x** is the same as that based on **y**. Given a point **x**, if we have two points $\boldsymbol{\theta}_1$, $\boldsymbol{\theta}_2$ in Θ so that

$$L(\boldsymbol{\theta}_1; \mathbf{x}) > L(\boldsymbol{\theta}_2; \mathbf{x})$$

we, intuitively, say that it is more likely that the observation $\mathbf{X} = \mathbf{x}$ has come from a population with parameter $\boldsymbol{\theta}_1$ as compared to a population with parameter $\boldsymbol{\theta}_2$. This becomes the basis for defining the method of maximum likelihood estimation, which suggests to consider such $\boldsymbol{\theta}$ as a reasonable estimator $\hat{\boldsymbol{\theta}}(\mathbf{x})$ for which the observed sample **x** is most likely. In other words, the maximum likelihood estimator of $\boldsymbol{\theta}$ is that value of $\boldsymbol{\theta}$ in Θ at which the likelihood function $L(\boldsymbol{\theta}; \mathbf{x})$ as a function of $\boldsymbol{\theta}$ with **x** held fixed, attains its maximum.

Definition 6.6.1 A value of $\boldsymbol{\theta}$ in Θ, say $\boldsymbol{\theta}_0$, is said to be a maximum likelihood estimator (MLE) of $\boldsymbol{\theta}$, given a set of sample observations **x**, if

$$L(\boldsymbol{\theta}_0; \mathbf{x}) = \max_{\theta \in \Theta} L(\boldsymbol{\theta}; \mathbf{x}) \tag{6.6.2}$$

Here, the range of MLE $\boldsymbol{\theta}_0(\mathbf{x})$ is the same as it is for the parameters. $\boldsymbol{\theta}_0(\mathbf{x})$ must be in Θ; if not, then it is not the MLE of $\boldsymbol{\theta}$.

For the purpose of computational simplicity, definition 6.6.1 may equivalently be written as

$$\log L(\boldsymbol{\theta}_0; \mathbf{x}) = \max_{\theta \in \Theta} \log L(\boldsymbol{\theta}; \mathbf{x}) \tag{6.6.3}$$

since log is a monotonic function. The function log $L(\boldsymbol{\theta}; \mathbf{x})$ is called *log-likelihood function.*

Likelihood-based statistical inference uses the information as to how the likelihood or log-likelihood changes by varying $\boldsymbol{\theta}$. A function that represents this change

$$S(\mathbf{x}; \boldsymbol{\theta}) = \frac{\partial}{\partial \boldsymbol{\theta}} \log L(\boldsymbol{\theta}; \mathbf{x}) \tag{6.6.4}$$

is known as a *score function*. If log $L(\boldsymbol{\theta}; \mathbf{x})$ (i) is a function of $\boldsymbol{\theta}$ which is twice differentiable and (ii) the extrema of $L(\boldsymbol{\theta}; \mathbf{x})$ does not occur on the boundary, i.e., extrema is an interior point of the domain of the function $L(\cdot\,; \mathbf{x})$, i.e., Θ, then the MLEs of $\boldsymbol{\theta}$ are defined as solution of the equations

$$\frac{\partial}{\partial \theta_i} \log L(\boldsymbol{\theta}; \mathbf{x}) = 0;\ i = 1,\ldots, k \tag{6.6.5}$$

or
$$S(\mathbf{x},\boldsymbol{\theta}) = \left(\frac{\partial}{\partial\theta_1}\log L(\boldsymbol{\theta}),\ldots,\frac{\partial}{\partial\theta_k}\log L(\boldsymbol{\theta})\right)'$$
$$= (0,\ldots,0)' = \mathbf{0}$$

so that the $k \times k$ matrix

$$\left.\frac{\partial}{\partial\boldsymbol{\theta}}S(\mathbf{x},\boldsymbol{\theta})\right|_{\theta=\theta_0(\mathbf{x})} = \left\|\frac{\partial^2}{\partial\theta_i\partial\theta_j}\log L(\boldsymbol{\theta};\mathbf{x})\right\|_{\theta=\theta_0(\mathbf{x})} \tag{6.6.6}$$
$$= \left\|-J_{ij}(\boldsymbol{\theta}_0)\right\|$$

is negative definite or $\mathbf{J}(\boldsymbol{\theta}_0)$ is positive definite, where $\boldsymbol{\theta}_0(\mathbf{x})$ is the solution of Eq. (6.6.5). The matrix $\mathbf{J}(\boldsymbol{\theta}_0)$ is called the *observed information matrix*. It measures the amount of information about $\boldsymbol{\theta}$ available in the experiment. However, the matrix

$$\mathbf{I}(\boldsymbol{\theta}) = E_{\boldsymbol{\theta}}[S(\mathbf{X},\boldsymbol{\theta})\,S(\mathbf{X},\boldsymbol{\theta})']$$
$$= \left\|E_{\boldsymbol{\theta}}\left(\frac{\partial}{\partial\theta_i}\log L(\boldsymbol{\theta})\right)\left(\frac{\partial}{\partial\theta_j}\log L(\boldsymbol{\theta})\right)\right\|$$
$$= \left\|E_{\boldsymbol{\theta}}\left\{-\frac{\partial^2}{\partial\theta_i\partial\theta_j}\log L(\boldsymbol{\theta};\mathbf{x})\right\}\right\| = \left\|E_{\boldsymbol{\theta}}J_{ij}(\boldsymbol{\theta})\right\|$$
$$= E_{\boldsymbol{\theta}}[J(\boldsymbol{\theta})] \tag{6.6.7}$$

is called *Fisher information matrix*. A large value of $|\mathbf{I}(\boldsymbol{\theta}_0)|$ indicates that the likelihood function is more curved; thus, one can easily calculate the maximum of likelihood function over $\boldsymbol{\theta}$ to get the MLE.

The mean and variance-covariance matrix of score function for regular families are as follows:

$$E_{\boldsymbol{\theta}}[S(\mathbf{X},\boldsymbol{\theta})] = \mathbf{0} = (0,\ldots,0)'$$

and
$$V_{\boldsymbol{\theta}}[S(\mathbf{X},\boldsymbol{\theta})] = E[S(\boldsymbol{\theta};\mathbf{X})S(\boldsymbol{\theta};\mathbf{X})'] = I(\boldsymbol{\theta})$$

Note that in case of $X_1, X_2,\ldots,X_n$ *iid* from $f(x;\theta)$, θ scalar, the likelihood equation is given by

$$\sum_1^n S(X_i,\theta) = 0 \tag{6.6.8}$$

Equations (6.6.8) are referred to as *estimating equations* since their solution is an estimator.

6.6.1 Calculus of Finding MLE

It is important at this stage to make some comments on the calculation of MLE using calculus. Consider $\theta \in \Theta$ as scalar. Any point θ_0 in Θ is called a *local maxima*, if the first derivative of likelihood function at θ_0 is zero and the second derivative $(\partial^2/\partial\theta^2)L(\theta)|_{\theta_0} < 0$.

If the likelihood function $L(\theta)$ is strictly concave, i.e., $(\partial^2/\partial\theta^2)L(\theta) < 0$ for all $\theta \in \Theta$, then any local maxima is a *global maxima*.

If Θ is an interval with endpoints a and b not necessarily finite so that $a < b$ and if the likelihood function $L(\theta)$ is continuous on $[a, b]$, then both maxima and minima of L will exist.

If $L(\theta)$ is continuous on $[a, b]$ and differentiable on (a, b), then the maxima and minima of L will be among the extreme points, i.e., at which $L'(\theta) = 0$, points at which L' does not exist, and the endpoints a and b.

However, if the likelihood function $L(\theta)$ is continuous and differentiable on (a, b) and the likelihood equation

$$\frac{\partial}{\partial\theta}\log L(\theta) = 0 \tag{6.6.9}$$

has unique solution $\hat{\theta}$ so that

$$\left.\frac{\partial^2}{\partial\theta^2}\log L(\theta)\right|_{\hat{\theta}} < 0 \tag{6.6.10}$$

then, in this case, the local maxima $\hat{\theta}$ is the global maxima and, therefore, it is the MLE of θ. The reason why it is a global maxima, for if it is not, then there exists some other local maxima and, therefore, local maxima cannot be unique as in this case. To demonstrate the above case, Casella and Berger (2002) considered two functions

$$L_1(\theta) = -\theta^2 + 25$$

and

$$L_2(\theta) = \theta^3 - 1.5\theta^2 - 6\theta + 11$$

on the interval (–2, 4). We can easily see that on plotting L_1, local maxima is situated at $\theta = 0$ and it is unique. Therefore, it is a global maxima. If L_2 is plotted, then local maxima is situated at $\theta = -1$ but it is not unique. Therefore, it is not a global maxima.

We now formalize the above remarks. The first derivative being 0, from Eq. (6.6.9), alone, gives extreme points called extrema in the interior of the range of the function $L(\cdot\,; \mathbf{x})$ i.e., $R(L) = \Theta$. Here, $R(L)$ denotes the range of the function $L(\cdot\,; \mathbf{x})$. These extreme points can be local minima, local maxima, or points of inflexion. Therefore, one of these extreme points can be global maxima and that will be the MLE. To identify such global maxima, we check for a point $\hat{\theta}$ among the extreme points such that

$$\left.\frac{\partial^2}{\partial\theta^2}\log L(\theta; \mathbf{x})\right|_{\theta=\hat{\theta}} < 0 \tag{6.6.11}$$

This is known as local maxima. If the log-likelihood at this point, $\log L(\hat{\theta}; \mathbf{x})$, is larger than the log-likelihoods at all boundary points, then such local maxima is global maxima, i.e., $\hat{\theta}$ is MLE. In case likelihood equations give a unique solution and condition (6.6.11) holds, then this solution is MLE. However, if the extrema is located on the boundary of R, the first derivative may not be 0. Therefore, the first-derivative-zero condition, in Eq. (6.6.9), is only a necessary condition for a maxima in the interior, it is not a sufficient condition. Though finding an MLE reduces into a simple exercise in differential calculus, it sometimes poses problems and it gets messy for even those populations which are having simple pdfs. In these situations, we proceed with some general technique of direct maximization.

Note that if the support of the distribution involves parameter as in the case of the Pitman family of distributions, the first-derivative-zero condition cannot be used to find the MLE. In this case, some other technique of maximizing likelihood needs to be used.

There are some references made by Peressini and Uhl (1988), Sundaram (1996), and Bertsekas (1999) on optimization theory or nonlinear programming aimed to find a global minima or maxima of univariate and multivariate functions.

Consider here a few simple illustrations explaining the principle of maximum likelihood estimation. Suppose one is interested in estimating the proportion θ of good light bulbs manufactured by a company, where it is known that θ takes on only two possible values 0.6 and 0.9. If a single bulb is selected at random, tested, and found good, the likelihood function is $L(\theta;\, x = 1) = \theta^1\,(1 - \theta)^{1-1} = \theta$. The likelihood principle suggests of taking that possible value of θ which maximizes $L(\theta;\, x)$; in this case, $\hat{\theta} = 0.9$. Next, if two bulbs are selected at random and tested, found both as good, the likelihood function is $L(\theta;\, x = 2) = \binom{2}{2}\theta^2(1 - \theta)^{2-2} = \theta^2$. The likelihood principle suggests the same estimator $\hat{\theta} = 0.9$ as was found in the previous experiment ($n = 1$). Further, if two bulbs are selected at random, tested one of them as good, $L(\theta;x = 1) = \binom{2}{1}\theta^1(1 - \theta)^{2-1} = \theta(1 - \theta)$, $(\partial/\partial\theta)L(\theta;\, 1) = 1 - 2\theta = 0$ gives $\hat{\theta} = 1/2 = 0.5$ as an estimate under this principle. Next, consider a binomial experiment where one decides to stop after 10 trials and gets 3 successes; and also a negative binomial experiment where one decides to stop after 3 successes, and to achieve this, say, 10 trials were needed. In both these examples, the likelihood function is a function of θ and is maximum (the most plausible value of θ) at $\theta = 0.3$.

6.6.2 MLE under a Transformation

Let $X_1, X_2,\ldots,X_n$ be *iid* with $f(x;\boldsymbol{\theta})$ where $f(x;\boldsymbol{\theta})$ is the density of an *rv* X. Assume that $\hat{\boldsymbol{\theta}}(\mathbf{x})$ is the MLE of $\boldsymbol{\theta}$ based on the observed value $\mathbf{X} = \mathbf{x}$. Consider a transformation $Y = T(X)$. The *pdf* of Y is given by

$$f_Y(y;\boldsymbol{\theta}) = f_X[T^{-1}(y);\boldsymbol{\theta}]\left|\frac{\partial T^{-1}(y)}{\partial y}\right|$$

Let $Y_1, Y_2,\ldots,Y_n$ be *iid* from $f(y;\boldsymbol{\theta})$. The likelihood function is

$$L(\boldsymbol{\theta};\, \mathrm{y}) = \prod_{i=1}^{n} f_X[T^{-1}(y_i);\boldsymbol{\theta}]\prod_{i=1}^{n}\left|\frac{\partial T^{-1}(y_i)}{\partial y_i}\right| \tag{6.6.12}$$

and the log likelihood function is

$$\log L(\boldsymbol{\theta};\, \mathrm{y}) = \text{const.} + \sum_{i=1}^{n} \log f_X(T^{-1}(y_i);\boldsymbol{\theta})$$

$$= \text{const.} + \sum_{i=1}^{n} \log f_X(x_i; \boldsymbol{\theta})$$

$$= \text{const.} + \log L(\boldsymbol{\theta}; \mathbf{x}) \qquad (6.6.13)$$

Thus, MLEs in the original and the transformed problems are same, since their respective likelihoods are different only up to an additive constant. Hence,

$$\hat{\boldsymbol{\theta}}_Y(\mathbf{y}) = \hat{\boldsymbol{\theta}}_X(\mathbf{x}) = \hat{\boldsymbol{\theta}}_X(T^{-1}(y_i), \ldots, T^{-1}(y_n))$$

See the application of this result for lognormal distribution in Example (6.31).

In case of multi-parameter, it is difficult to obtain a solution of likelihood equation in closed form. So, it is difficult to show that a solution $\hat{\boldsymbol{\theta}}$ is global maxima of $\log L(\boldsymbol{\theta})$, since it involves a Hessian matrix which is usually difficult to calculate. Only in case of the k-parameter exponential family, the calculation of a Hessian matrix is avoided by showing the family as exponential family. However, in case of non-exponential families, where one is generally interested in only one or some of the parameters, as an alternative, we may proceed with one of the following methods.

(i) **Simultaneous maximization method:** In case of two parameters $\boldsymbol{\theta} = (\theta_1, \theta_2)'$, solve

$$\max_{\theta_1, \theta_2} L[(\theta_1, \theta_2); \mathbf{x}] \qquad (6.6.14)$$

The solution of Eq. (6.6.14) is called *extreme point*. It may be denoted by $\hat{\boldsymbol{\theta}} = (\hat{\theta}_1, \hat{\theta}_2)'$. If the Jacobian of the Hessian matrix

$$\mathbf{H} = \begin{pmatrix} \dfrac{\partial^2}{\partial \theta_1^2} L(\theta_1, \theta_2) & \dfrac{\partial^2}{\partial \theta_1 \partial \theta_2} L(\theta_1, \theta_2) \\ \dfrac{\partial^2}{\partial \theta_1 \partial \theta_2} L(\theta_1, \theta_2) & \dfrac{\partial^2}{\partial \theta_2^2} L(\theta_1, \theta_2) \end{pmatrix}_{(\hat{\theta}_1, \hat{\theta}_2)} \qquad (6.6.15)$$

is positive, then $\hat{\boldsymbol{\theta}} = (\hat{\theta}_1, \hat{\theta}_2)'$ is the MLE of $\boldsymbol{\theta}$.

(ii) **Casella method:**

Definition 6.1 (Profile likelihood) Consider $\boldsymbol{\theta} = (\theta_1, \theta_2, \ldots, \theta_k)' \in \Theta \subseteq \mathbb{R}^k$. If $\theta_2, \ldots, \theta_k$ are fixed, then the likelihood $L(\theta_1; \theta_2, \ldots, \theta_k, \mathbf{x})$ is called a profile likelihood of θ_1 for given $\theta_2, \ldots, \theta_k$ and $\mathbf{x}$. In case of two parameters $\boldsymbol{\theta} = (\theta_1, \theta_2)'$, we solve

$$\max_{\theta_1} \max_{\theta_2} L[(\theta_1, \theta_2); \mathbf{x}] \qquad (6.6.16)$$

Definition 6.2 (Conditional likelihood) If the pdfs in the likelihood function can be factorized in two factors, one factor as a function of $\theta_2, \ldots, \theta_k$ and $\mathbf{x}$, say, $S(\theta_2, \ldots, \theta_k, \mathbf{x})$, and the second factor involving θ_1 given $S(\theta_2, \ldots, \theta_k, \mathbf{x})$, i.e., $Q[\theta_1 | S(\theta_2, \ldots, \theta_k, \mathbf{x})]$, so that it is independent of $\theta_2, \ldots, \theta_k$, then Q is called the conditional likelihood of θ_1 given $S(\theta_2, \ldots, \theta_k, \mathbf{x})$.

However, Casella (2002), in this case, by considering L as a function of two variables θ_1 and θ_2 (say), gives an analytical solution to this situation, by outlining the following three conditions that must hold for the function $L(\theta_1, \theta_2)$ to have local maximum at $(\hat{\theta}_1, \hat{\theta}_2)$:

(i) The first-order partial derivatives with respect to θ_1 and θ_2 at $(\hat{\theta}_1, \hat{\theta}_2)$ are 0:

$$\frac{\partial}{\partial \theta_1} L(\theta_1, \theta_2)\Big|_{(\theta_1, \theta_2)=(\hat{\theta}_1, \hat{\theta}_2)} = 0$$

and
$$\frac{\partial}{\partial \theta_2} L(\theta_1, \theta_2)\Big|_{(\theta_1, \theta_2)=(\hat{\theta}_1, \hat{\theta}_2)} = 0 \tag{6.6.17}$$

(ii) At least one of the second-order partial derivatives is negative:

$$\frac{\partial^2}{\partial \theta_1^2} L(\theta_1, \theta_2)\Big|_{(\theta_1, \theta_2)=(\hat{\theta}_1, \hat{\theta}_2)} < 0$$

or
$$\frac{\partial^2}{\partial \theta_2^2} L(\theta_1, \theta_2)\Big|_{(\theta_1, \theta_2)=(\hat{\theta}_1, \hat{\theta}_2)} < 0 \tag{6.6.18}$$

(iii) the Jacobian of the second-order partial derivatives

$$|\mathbf{H}| = \begin{vmatrix} \dfrac{\partial^2}{\partial \theta_1^2} L(\theta_1, \theta_2) & \dfrac{\partial^2}{\partial \theta_1 \partial \theta_2} L(\theta_1, \theta_2) \\ \dfrac{\partial^2}{\partial \theta_1 \partial \theta_2} L(\theta_1, \theta_2) & \dfrac{\partial^2}{\partial \theta_2^2} L(\theta_1, \theta_2) \end{vmatrix}_{(\theta_1, \theta_2)=(\hat{\theta}_1, \hat{\theta}_2)} \tag{6.6.19}$$

is positive, $\mathbf{H} > 0$.

We may easily notice two problems at this stage:

(i) The calculations while checking for the above conditions, for interior maxima, are much involved.

(ii) We still need to check for maxima at boundaries, since the above conditions are only necessary conditions for maxima in the interiors of Θ. Casella (2002), among others, has discussed problems relating to it and has advised to avoid the use of second-derivative conditions and recommend some simpler technique, like the technique of successive maximization.

6.6.3 Modified MoME and MLE in Exponential Families

We have seen in Chapter 5 that MLEs are a function of sufficient statistics whereas MoMEs, in general, are not. The MoMEs can always be improved by conditioning on sufficient statistics. Such estimators are known as *modified MoMEs*. Such modified MoMEs and MLEs are usually the same. Davidson and Solomon (1974) have investigated the correspondence of MoMEs and MLEs in case of exponential families and concluded that both are identical for such families.

A family of *pdf* for *rv* X, namely $\{f(x;\boldsymbol{\theta}): \boldsymbol{\theta} = (\theta_1,\ldots,\theta_s)' \in \Theta_U\}$ with usual parameterization Θ_U is known as the k-parameter exponential family if

$$f(x;\boldsymbol{\theta}) = h(x)\, b(\boldsymbol{\theta}) \exp\left(\sum_{i=1}^{k} Q_i(\boldsymbol{\theta}) T_i(x)\right) \tag{6.6.20}$$

where $h(x)$ and $b(\boldsymbol{\theta})$ are real-valued functions, so that $h(x) \geq 0$ and $b(\boldsymbol{\theta}) \geq 0$. Here, U in the parameter space Θ_U refers to usual parametrization. Distributions like logistic, t, Cauchy, and uniform are not exponential families, since they cannot be put in the form as above. Further, if the support of a distribution depends on θ, the corresponding family cannot be an exponential family

If parameterization $\mathbf{Q}$ is used as natural parameter, so that $\mathbf{Q} \in \Theta_N$, then the form of the *pdf* for X, namely $\{f(x;\mathbf{Q}):\mathbf{Q} = (Q_1,\ldots,Q_k)' \in \Theta_N\}$ is called exponential family

$$f(x;\mathbf{Q}) = h(x) c(\mathbf{Q}) \exp\left(\sum_{i=1}^{k} Q_i T_i(x)\right) \tag{6.6.21}$$

where $h(x)$ and $c(\mathbf{Q})$ are real-valued functions so that $h(x) \geq 0$ and $c(\mathbf{Q}) \geq 0$. Here, N in Θ_U refers to the parameter space for natural parameters. The natural parameter space is the cross-product of the ranges of Q_i's, i.e., $\Theta_N = R(Q_1) \times \cdots \times R(Q_k)$. The function $h(x)\exp[\Sigma_{i=1}^{k} Q_i T_i(x)]$ is called the kernel function. Define a set $\bar{\Theta}_N$ of $\mathbf{Q}$ for which $[c(\mathbf{Q})]^{-1}$ is finite

$$\left\{\mathbf{Q} : [c(\mathbf{Q})]^{-1} = \int_{-\infty}^{\infty} h(x) \exp\left(\sum_{i=1}^{k} Q_i T_i(x)\right) dx < \infty\right\}$$

where $[c(\mathbf{Q})]^{-1}$ is the integral of the kernel function over its range. An important result for $\bar{\Theta}_N$ is that it is a convex set.

An exponential family in Eq. (6.6.21) is said to be a k-parameter regular exponential family if (i) $\Theta_N = \bar{\Theta}_N$, (ii) in the linear parameterization Q_i's and T_i's are all linearly independent, and (iii) Θ_N is a k-dimensional open set and is non-empty. Θ_N is the cross-product of k open intervals.

If we know that the *pdf* of X belongs to an exponential family and we now wish to verify whether the family is a k-parameter exponential family, we proceed with checking if $\dim(\Theta_U) = \dim(\Theta_N)$. If it is true, then $\Theta = \Theta_N$, i.e., condition (i) is satisfied. Then in this case, Θ_N is the cross-product of ranges of Q_is; so, condition (iii) is satisfied. Further, if Θ_N contains a k-dimensional rectangle, then Q_is and T_is, both are linearly independent; so, condition (ii) is satisfied. So, to verify if a given exponential family is a k-parameter regular exponential family, one needs to know whether $\dim(\Theta_U) = \dim(\Theta_N)$ and Θ_N contains a k-dimensional rectangle. Consider the family $N(a\theta, b\theta^2)$ which is a 2-dimensional exponential family with natural parameters $Q_1 = -1/2b\theta^2$ and $Q_2 = a/b\theta$ whereas the usual parameterization is determined by θ. Note, $\dim(\Theta_U) = k = 1 < 2 = k = \dim(\Theta_N)$. Further, note that $Q_1 = (-b/2a^2)Q_2^2$. Therefore, the natural parameter space is a parabola and it is not a 2-dimensional open set made up of the cross-product of the ranges of Q_1 and Q_2. Thus, the above family is not regular.

Suppose the *pdf* of an *rv* X comes from an exponential family given by Eq. (6.6.21), with natural parameterization $\mathbf{Q}$. Then the distribution of $\mathbf{T} = (T_1(X),\ldots,T_k(X))'$ is given by

$$f(\mathbf{T};\mathbf{Q}) = h_1(\mathbf{T}) c(\mathbf{Q}) \exp\left(\sum_{i=1}^{k} Q_i T_i\right). \tag{6.6.22}$$

which is again a k-parameter exponential family. The moments of $\mathbf{T}$ are

$$E_{\mathbf{Q}}[T_i(X)] = -\frac{\partial}{\partial Q_i}\log[c(\mathbf{Q})]$$

$$\Rightarrow \quad \operatorname{cov}_{\mathbf{Q}}[T_i(X), T_j(X)] = -\frac{\partial^2}{\partial Q_i \partial Q_j}\log[c(\mathbf{Q})] \tag{6.6.23}$$

Let the covariance matrix of $\mathbf{T}$ with elements as above be denoted by $\Sigma_{\mathbf{T}}$.

Consider a random sample $X_1, X_2, \ldots, X_n$ from the k-parameter exponential family in Eq. (6.6.21), with natural parameterization $\mathbf{Q}$, where the corresponding parameter space Θ_N is an open convex set.

Likelihood function is

$$L(\mathbf{Q}) = \left[\prod_1^n h(T_i)\right][c(\mathbf{Q})]^n \exp\left(Q_1 \sum_{j=1}^n T_1(X_j) + \cdots + Q_k \sum_{j=1}^n T_k(X_j)\right) \tag{6.6.24}$$

Davidson and Solomon gave the modified MoM moments equations as

$$\frac{1}{n}\sum_{j=1}^n T_i(x_j) = \frac{1}{n}E_\theta\left[\sum_{j=1}^n T_i(X_j)\right], i = 1,\ldots, k \tag{6.6.25}$$

Moment equations can be simplified by using the following results:

$$E_{\mathbf{Q}}[T_i(X_j)] = -\frac{\partial}{\partial Q_i}\log c(\mathbf{Q}), i = 1,\ldots, k;\ j = 1,\ldots, n \tag{6.6.26}$$

and
$$\operatorname{cov}_{\mathbf{Q}}[T_i(X_j), T_{i'}(X_j)] = -\frac{\partial^2}{\partial Q_i, Q_{i'}}\log c(\mathbf{Q}); i, i' = 1,\ldots, k \tag{6.6.27}$$

The log-likelihood function

$$\log L(\mathbf{Q}) = \sum_{i=1}^n \log h(T_i) + n\log c(\mathbf{Q}) + \sum_{j=1}^n Q_i T_i(X_j)$$

Denoting $\Sigma_{j=1}^n T_i(X_j)$ by T_i and differentiating $\log L(\mathbf{Q})$ with respect to Q_i, we get the likelihood equation

$$\frac{\partial}{\partial Q_i}\log L(\mathbf{Q}) = n\frac{\partial}{\partial Q_i}\log c(\mathbf{Q}) + T_i = -E[T_i(\mathbf{X})] + T_i(\mathbf{X}) = 0$$

$$\Rightarrow \quad E\sum_{j=1}^n T_i(X_j) = \sum_{j=1}^n T_i(X_j) \tag{6.6.28}$$

for $i = 1,\ldots,k$. The solutions of the above k-equations for $\mathbf{Q}$ are the candidates for MLE.

Further, the second derivative of log–likelihood is

$$\frac{\partial^2}{\partial Q_i Q_j}\log L(\mathbf{Q}) = n\frac{\partial^2}{\partial Q_i \partial Q_j}\log c(\mathbf{Q})$$

$$= -n\operatorname{cov}[T_i(X), T_j(X)] \tag{6.6.29}$$

Hence, the Hessian matrix is negative definite, since Σ_T is positive definite. This shows that the log-likelihood $\log L(\mathbf{Q})$ is strictly concave. Therefore, the solution of Eq. (6.6.28), $\hat{\mathbf{Q}}$, is unique and, consequently, is global maxima. Hence, $\hat{\mathbf{Q}}$ is MLE of $\mathbf{Q}$. Note that these MLEs are obtained by setting the sample moment of natural sufficient statistic T_i equal to their expected values. Therefore, MoMEs and MLEs of $Q_i(\boldsymbol{\theta})$ are identical in exponential families.

Hence, we finally conclude, that in case of k-parameter exponential family, one need not check the Hessian matrix. The solution of the likelihood equation will be unique. Therefore, this solution will be global maxima and, thus, it will be the MLE of $\mathbf{Q}$. One only needs to verify whether a given exponential family is regular, which is usually easier than to calculate the Hessian matrix. However, in case of one-parameter exponential family, verifying that the family is regular is somewhat difficult and, instead, the calculus involved in checking whether the extreme points are global maxima by showing the second-derivative of log-likelihood $\log L(\mathbf{Q})$ as negative is sometimes easy. Care must be taken in cases of obtaining MLE where the parameter space is not an open set, e.g., [0, 1], since here the family cannot be regular.

Further, for the problem of estimating a one-to-one function of $\eta_i = Q_i(\boldsymbol{\theta})$, namely $\pi(\eta_i)$, the MLEs of $\pi(\eta_1),\ldots,\pi(\eta_k)$ are obtained by $\pi(\hat{\eta}_1),\ldots,\pi(\hat{\eta}_k)$, where $\hat{\eta}_i = \hat{Q}_i(\boldsymbol{\theta})$.

6.6.4 General Properties of MLE

Levy (1985) and Moore (1971) have shown that if an MLE is unique, then it is a function of the minimal sufficient statistic. In case of exponential families, MLE is a function of the complete sufficient statistic.

We discuss here some useful properties of maximum likelihood estimators. We consider here the parameter, not as a vector but just as a single-valued parameter for the purpose of simplicity. Many of these properties can be analogously discussed for $\boldsymbol{\theta}$ as a vector.

Theorem 6.6.1 A consistent maximum likelihood estimator is unique.

Proof. Let us consider two distinct consistent MLEs of a parameter θ, namely $\hat{\theta}_1(\mathbf{x})$ and $\hat{\theta}_2(\mathbf{x})$ so that

$$\frac{\partial}{\partial\theta}\log L(\theta,\mathbf{x})\Big|_{\hat{\theta}_1} = 0 = \frac{\partial}{\partial\theta}\log L(\theta,\mathbf{x})\Big|_{\hat{\theta}_2} \tag{6.6.30}$$

Assume that $(\partial/\partial\theta)\log L(\theta;\mathbf{x})$ is continuous and differentiable in $(\hat{\theta}_1,\hat{\theta}_2)$. Then, by Rolle's theorem, there exists some value of θ^* in $(\hat{\theta}_1,\hat{\theta}_2)$ at which

$$\frac{\partial^2}{\partial\theta^2}\log L(\theta;\mathbf{x})\Big|_{\theta^*} = 0 \tag{6.6.31}$$

Further, θ^* is consistent, i.e., $\theta^* \xrightarrow{P} \theta$ since $\hat{\theta}_1$ and $\hat{\theta}_2$ are consistent estimators of θ. Therefore, by Theorem 6.6.7,

$$P\left\{\frac{\partial^2}{\partial\theta^2}\log L(\theta;\mathbf{x})\Big|_{\theta^*} < 0\right\} \to 1 \quad \text{as} \quad n\to\infty \tag{6.6.32}$$

This is a contradiction to the observation $(\partial^2/\partial\theta^2)\log L(\theta;\mathbf{x})|_{\theta^*} = 0$. Hence, it proves the theorem.

Theorem 6.6.2 MLEs, if they exist, are the functions of sufficient statistic.

Proof. We assume that the population shown by the family of distributions $\{f(x;\theta),\ \theta \in \Theta\}$ admits a sufficient statistic T. By factorization theorem,

$$L(\theta,\mathbf{x}) = f(\mathbf{x};\theta) = h(\mathbf{x})\cdot g(t,\theta)$$

$$\Rightarrow \qquad \log L = \log h(\mathbf{x}) + \log g(t,\theta)$$

$$\Rightarrow \qquad \frac{\partial}{\partial\theta}\log L = \frac{\partial}{\partial\theta}\log g(t,\theta) = 0$$

Therefore, any solution of likelihood equation $(\partial/\partial\theta)\log L = 0$ will be a function of t.
However, the converse of Theorem 6.6.2 is not true in general.

Theorem 6.6.3 If a CRLB estimator exists under the regularity conditions (i) to (iv) in Section 4.2, then it is a unique solution to the likelihood equation that maximizes the likelihood.

Proof. We assume that the regularity conditions (i) to (iv) stated in Section 4.2 are satisfied. Given that $\hat{\theta}$ is a CRLB estimator, the following condition must be satisfied:

$$\frac{\partial}{\partial\theta}\log f(\mathbf{x};\theta) = \frac{\partial}{\partial\theta}\log L(\theta;\mathbf{x}) = c(\theta)(\hat{\theta}-\theta) \tag{6.6.33}$$

Further, the likelihood equation

$$\frac{\partial}{\partial\theta}\log L(\theta;\mathbf{x}) = 0$$

gives

$$c(\theta)(\hat{\theta}-\theta) = 0 \tag{6.6.34}$$

Since $c(\theta)$ being the function of θ alone cannot be 0, $\theta = \hat{\theta}$ is the unique solution of the likelihood equation. Further, if we assume that $c(\theta)$ is differentiable with respect to θ and the second-order partial derivative of $(\partial/\partial\theta)\log L(\theta,\mathbf{x})$ exists, then from Eq. (6.6.33),

$$\frac{\partial^2}{\partial\theta^2}\log L(\theta;\mathbf{x}) = c'(\theta)(\hat{\theta}-\theta) - c(\theta)$$

The value of $(\partial^2/\partial\theta^2)\log L(\theta;\mathbf{x})\big|_{\theta=\hat{\theta}}$ is

$$\frac{\partial^2}{\partial\theta^2}\log L(\theta;\mathbf{x})\Big|_{\theta=\hat{\theta}} = -c(\theta) \tag{6.6.35}$$

The unique solution $\theta = \hat{\theta}$ maximizes the log likelihood if the second derivative of log-likelihood is negative and $c(\theta)$ in Eq. (6.6.35) is positive. We will now have to show that $c(\theta) > 0$. From the results on CR inequality, $\text{cov}[\delta(\mathbf{X}), S(\mathbf{X},\theta)] = g'(\theta) = (\partial/\partial\theta)\theta = 1$ or $E[(\delta(\mathbf{X}))(\partial/\partial\theta)\log L(\theta;\mathbf{x})] = 1$. In the case that CRLB is achieved,

$$[\delta(\mathbf{X}) - \theta] = c^{-1}(\theta)\frac{\partial}{\partial\theta}\log L(\theta;\mathbf{X})$$

This gives
$$c^{-1}(\theta)E\left[\frac{\partial}{\partial\theta}\log L(\theta;\mathbf{x})\right]^2 = 1$$

$$c(\theta) = E\left[\frac{\partial}{\partial\theta}\log L(\theta;\mathbf{x})\right]^2 > 0$$

This shows that the unique solution of likelihood equation, namely $\theta = \hat{\theta}$ maximizes the likelihood.

Note. Thus, the most efficient estimators are necessarily MLEs but the converse, that MLEs are the most efficient, is not always true. Example 6.21 illustrates that in case of normal distribution, the MLE $(1/n)\ \Sigma_{i=1}^n(X_i - \bar{X})^2$ of σ^2 is not the most efficient.

So far, we have considered the estimation of the parameter θ that indexes the distribution of the random variable X. However, very often, we are interested in estimating some function of θ, say $g(\theta)$. If the function is one-to-one, it is easy to see that the maximization of likelihood as a function of $g(\theta)$ amounts to the maximization of likelihood as a function of θ. In this case, if the MLE of θ is $\hat{\theta}$, then the MLE of $g(\theta)$ is the same function of $\hat{\theta}$ as $g(\theta)$ is of θ, i.e., $g(\hat{\theta})$.

However, difficulty arises when the function g is not one-to-one but many-to-one. In this case, if $\hat{\theta}$ is an MLE of θ, there could be some other point θ' at which $g(\hat{\theta}) = g(\theta')$. As a solution to this problem, we define an induced likelihood function for $\eta = g(\theta)$

$$L'(\eta;\mathbf{x}) = \underbrace{\sup}_{\{\theta:\, g(\theta)=\eta\}} L(\theta;\mathbf{x}) \tag{6.6.36}$$

and consider the problem of maximization of L' over η for obtaining the MLE of $\eta = g(\theta)$. The value η_0 maximizing $L'(\eta;\mathbf{x})$ is called the MLE of $\eta = g(\theta)$. Moreover, it is clear from Eq. (6.6.36) that the maxima of L' and L are the same.

6.6.5 Invariance Property of MLE

Zehna (1966) has investigated the problem of MLEs for $g(\theta)$ and has concluded that irrespective of g being one-to-one or many-to-one, the MLE of $g(\theta)$ is $g(\hat{\theta})$ whenever $\hat{\theta}$ is the MLE of θ. This is known as the invariance property of ML estimators. The results are summarized in the Theorems 6.6.4 and 6.6.5.

Theorem 6.6.4 [Zehna's Theorem (1966)] If $\hat{\theta}$ is the MLE of θ and $g(\theta)$ is a one-to-one function of θ with domain Θ, then $g(\hat{\theta})$ is the MLE of $g(\theta)$.

Proof. Since it is given that the mapping $\theta \to g(\theta)$ is one-to-one, we define the likelihood as a function of $\eta = g(\theta)$ by

$$L'(\eta;\mathbf{x}) = \prod_{i=1}^{n} f(x_i; g^{-1}(\eta))|J| = L(g^{-1}(\eta);\mathbf{x}) \tag{6.6.37}$$

If $\hat{\theta}$ is the MLE of θ so that

$$L(\hat{\theta}; \mathbf{x}) \geq L(\theta; \mathbf{x}) \; \forall \; \theta \in \Theta$$

then there are some $\hat{\eta}$ and η, so that $\hat{\theta} = g^{-1}(\hat{\eta})$, $\theta = g^{-1}(\eta)$

$$L(g^{-1}(\hat{\eta}), \mathbf{x}) \geq L(g^{-1}(\eta), \mathbf{x}) \; \forall \; \eta \in g(\Theta) = \Lambda$$

By Eq. (6.6.37),

$$L'(\hat{\eta}, \mathbf{x}) \geq L'(\eta, \mathbf{x}) \; \forall \; \eta \in g(\Theta) = \Lambda$$

Therefore, $\hat{\eta} = g(\hat{\theta})$ is the MLE of $g(\theta)$. This proves the theorem. ▣

Theorem 6.6.5 If $\hat{\theta}$ is the MLE of θ, g is some many-to-one function of θ, $g(\theta)$, then the MLE of $g(\theta)$ is $g(\hat{\theta})$.

Proof. We consider the likelihood function of $\eta = g(\theta)$ by $L'(\eta; \mathbf{x})$ as defined in Eq. (6.6.37). However, we have noted in the previous theorem that the maxima of functions L' and L are the same. Let us assume that $\hat{\eta}$ is the value that maximizes L', so that

$$L'(\hat{\eta}; \mathbf{x}) = \sup_{\eta} \sup_{\Theta_\eta = \{\theta: g(\theta) = \eta\}} L(\theta; \mathbf{x})$$

[by the definition of L' in Eq. (6.6.37)]

where Θ_η is the set of all values of θ which maps η under the function g. Note that the function $g(\theta)$ is not one-to-one. The likelihood $\sup_{\Theta_\eta = \{\theta: g(\theta) = \eta\}} L(\theta; \mathbf{x})$ is called the induced likelihood as a function of η.

$$\begin{aligned} L'(\hat{\eta}; \mathbf{x}) &= \sup_{\theta} L(\theta; \mathbf{x}) \\ &= L(\hat{\theta}; \mathbf{x}) \qquad \text{(since } \hat{\theta} \text{ is MLE of } \theta) \\ &= \sup_{\{\theta: g(\theta) = g(\hat{\theta})\}} L(\theta; \mathbf{x}) \\ &= L'(g(\hat{\theta}); \mathbf{x}) \end{aligned}$$

[by the definition of L' in Eq. (6.6.37)]

Hence, $\hat{\eta} = g(\hat{\theta})$ is the MLE of $g(\theta)$. ▣

It is important to note that the new parameterization η does not carry enough information that is needed to compute $f(\mathbf{x}; \eta)$ when the function g is not one-to-one. Then the question that arises is as to what is that $\hat{\eta} = g(\hat{\theta})$ which Zehna called the MLE of $g(\theta)$, when the actual likelihood function $M[g(\theta)]$ for reparameterization $\eta = g(\theta)$ is not computable? Indeed, $\hat{\eta} = g(\hat{\theta})$ maximizes the induced likelihood. Would it be correct to call this as the MLE of $g(\theta)$? Berk (1967) answered these questions. He considered a function

$$h(\theta) = (g(\theta), l(\theta)) = (\eta, \nu) = \xi$$

by specifying an additional parameter $\nu = l(\theta)$, so that h is one-to-one or invertible and has a well-defined likelihood $M(\xi) = L[h^{-1}(\xi)]$. Generally, simple identity function $l(\theta) = \theta$ gives $h(\theta)$ as one-to-one so that

$$h(\theta) = (g(\theta), \theta)$$

and $$\theta = h^{-1}(g(\theta), \theta) = h^{-1}(\xi)$$

where θ is obtained uniquely, but there can be several such functions h. Hence, the likelihood function $M(\xi) = M[h(\theta)]$ is well-defined and $h(\hat{\theta}) = [g(\hat{\theta}), l(\hat{\theta})]$ is the MLE of $\xi = h(\theta) = (g(\theta), l(\theta))$. Thus, $\hat{\eta} = g(\hat{\theta})$ is the MLE of $\eta = g(\theta)$.

MLEs do not possess small sample properties, in general, since, loosely speaking, they are like modes which maximize the *pdf* (*pmf*) (frequency) and mode as compared to mean and median does not possess good small sample properties. However, MLEs have many nice large sample properties. We will investigate them in Section 6.6.6.

6.6.6 Large Sample Properties of MLEs for Regular Models

We have noted that MLEs have poor small sample properties, since they are typically modes of the distribution $f(\mathbf{x};\theta)$. From large sample theory, the mode values approach mean and median. Therefore, MLEs have many attractive large sample properties. Thus, for large samples, MLE is an excellent estimator, since it is consistent, asymptotically efficient, approximately unbiased, and approximately UMVUE of θ. That is why MLE is the most widely used parametric estimation technique.

However, the large sample properties of MLEs hold under certain regularity conditions on a given statistical model $\{f(x;\theta);\ \theta \in \Theta\}$. These regularity conditions mainly relate to differentiability and the permission to interchange differentiation and integration operations. We state here the regularity conditions due to Cramer (1946) and Huzurbazar (1948), which were used by them to prove the results stated in Theorems 6.6.6 and 6.6.7. The family of distributions $\{f(x;\theta);\ \theta \in \Theta\}$ for which these regularity conditions are satisfied is called the *regular family of distributions* or *regular model*.

When θ is a scalar

(A1) The parameter space Θ is an open set in $\mathbb{R}^1$ and it contains an open set of which θ_0 is an interior point, for example $N_\delta(\theta_0) \in \Theta$. Further, the parameters are identifiable, that is, if $\theta_1 \neq \theta_2$, it implies $f(x;\theta_1) \neq f(x;\theta_2)$.

(A2) The support $S(\theta) = \{x\colon f(x;\theta) > 0\}$ does not depend on the parameter θ, i.e., $S(\theta) = S$.

(A3) The observations $X_1, X_2,\ldots,X_n$ are *iid* from $f(x;\theta)$.

(A4) $\log f(x;\theta)$ is differentiable in θ, $\theta \in N_\delta(\theta_0)$, i.e., $(\partial/\partial\theta)\log f$, $(\partial^2/\partial\theta^2)\log f$, and $(\partial^3/\partial\theta^3)\log f$ exist, for almost all values $x \in S$ and for every $\theta \in N_\delta(\theta_0)$ for some $\delta > 0$.

(A5) $(\partial^3/\partial\theta^3)\log f$ is continuos in θ. The quantity $\int f(x;\theta)dx = 1$ can be differentiated twice under the integral sign with respect to θ, so that

$$E_\theta\left(\frac{\partial}{\partial\theta}\log f\right) = 0$$

and $$E_\theta\left(\frac{\partial}{\partial\theta}\log f\right)^2 = E\left(-\frac{\partial^2}{\partial\theta^2}\log f\right) = I_X(\theta) > 0$$

i.e., the Fisher information is positive.

(A6) For every $\theta \in N_\delta(\delta_0)$, there exists a positive number (which may depend on θ_0) such that

$$\left|\frac{\partial^3}{\partial\theta^3}\log f(x;\theta)\right| < H(x)$$

with $E_{\theta_0}[H(X)] < k$ where k is independent of θ and is positive.

When θ is vector

(B1) Θ is an open subset in $\mathbb{R}^k$; it contains an open set of which $\boldsymbol{\theta}_0$ is an interior point.

(B2) All regularity conditions hold which are required for MLEs to be consistent.

(B3) $(\partial/\partial\theta_i)\log f$ and $(\partial^2/\partial\theta_i^2)\log f$ exist and are continuous for all x and all $\boldsymbol{\theta}$.

(B4) $\|J(\boldsymbol{\theta})\|$ exists and is positive definite, where

$$J_{ij}(\boldsymbol{\theta}) = E_{\boldsymbol{\theta}}\left[\frac{\partial}{\partial\theta_i}\log f(\mathbf{X};\boldsymbol{\theta})\frac{\partial}{\partial\theta_j}\log f(\mathbf{X};\boldsymbol{\theta})\right]$$

$$= E_{\boldsymbol{\theta}}\left[-\frac{\partial^2}{\partial\theta_i\partial\theta_j}\log f(\mathbf{X};\boldsymbol{\theta})\right] = nE_{\boldsymbol{\theta}}\left[-\frac{\partial^2}{\partial\theta_i\partial\theta_j}\log f(X_1;\theta)\right] \ \forall\ \boldsymbol{\theta}\in\Theta$$

and
$$E_{\boldsymbol{\theta}}\left[\frac{\partial}{\partial\theta_i}\log f(\mathbf{X};\boldsymbol{\theta})\right] = 0\ \forall\ \boldsymbol{\theta}\in\Theta$$

Theorem 6.6.6 [(Cramer (1946), Kale (1999)] Under the above regularity conditions stated in A1 to A6, $\hat{\theta}$, the maximum likelihood of θ, is a consistent estimator of θ, i.e., for every $\varepsilon > 0$ and every $\theta \in \Theta$,

$$\lim_{n\to\infty} P_\theta\left(\left|\hat{\theta}-\theta\right| \geq \varepsilon\right) = 0$$

Proof. We have assumed that sample observations $X_1,\ldots,X_n$ are *iid* from a distribution $f(x;\theta)$ (see assumption A3). The admissible values of θ constitute the parameter space Θ which contains an open set B so that $B \subset \Theta$ and the true value of the parameter $\theta = \theta_0$ is an interior point (see assumption A1). In other words, we assume that a δ-neighborhood of θ_0, $N_\delta(\theta_0)$, so that $N_\delta(\theta_0) \in \Theta$, θ_0, is an interior point of $N_\delta(\theta_0)$.

To prove that the given MLE, $\hat{\theta}$, is a consistent estimator of θ, we study the behaviour of the ratio of log-likelihoods in $N_\delta(\theta_0)$, i.e., $Y = L(\theta;\mathbf{x})/L(\theta_0;\mathbf{x})$, the ratio of likelihoods at θ and θ_0. We have $E_{\theta_0}(Y) = \int_{\mathbf{x}\in\mathbb{R}^n}[L(\theta;\mathbf{x})/L(\theta_0;\mathbf{x})]\,L(\theta_0;\mathbf{x})d\mathbf{x} = 1$. Consider $H(Y) = -\log Y$. $H(Y)$ is a strictly convex function since $H''(Y) = 1/Y^2 > 0$. We, therefore, have by Jensen's inequality, at $\theta = \theta_0$,

$$H[E_{\theta_0}(Y)] < E_{\theta_0}[H(Y)]$$

$$0 = -\log E_{\theta_0}(Y) < E_{\theta_0}(-\log Y)$$

$$E_{\theta_0}(-\log Y) > 0$$

$$\int_{\mathbf{x}\in\mathbb{R}^n} -\log\left[\frac{L(\theta;\mathbf{x})}{L(\theta_0;\mathbf{x})}\right] L(\theta_0;\mathbf{x})d\mathbf{x} > 0$$

$$\int_{\mathbf{x}\in\mathbb{R}^n} \log\prod_{i=1}^{n}\frac{f(x_i;\theta)}{f(x_i;\theta_0)}\prod_{i=1}^{n} f(x_i;\theta_0)d\mathbf{x} < 0$$

$$\int_{\mathbf{x}}\sum_{i=1}^{n}\log\left[\frac{f(x_i;\theta)}{f(x_i;\theta_0)}\right]\prod_{i=1}^{n} f(x_i;\theta_0)d\mathbf{x} < 0$$

$$\sum_{i=1}^{n}\int_{x_i}\log\left[\frac{f(x_i;\theta)}{f(x_i;\theta_0)}\right] f(x_i;\theta_0)dx_i < 0$$

$$\sum_{i=1}^{n} E_{\theta_0}\left[\log\frac{f(X_i,\theta)}{f(X_i,\theta_0)}\right] < 0$$

$$\sum_{i=1}^{n} -I_{X_i}(\theta,\theta_0) < 0$$

where
$$I_X(\theta,\theta_0) = -E_\theta\left[\log\frac{f(X;\theta)}{f(X;\theta_0)}\right] \tag{6.6.38}$$

is known as the *Kullback Leibler information* per unit of observation.

We, now, consider Taylor's expansion of $\log L(\theta;\mathbf{x})$ around θ_0 by

$$\log L(\theta;\mathbf{x}) = \log L(\theta_0;\mathbf{x}) + (\theta-\theta_0)\frac{\partial}{\partial\theta}\log L\Big|_{\theta=\theta_0} + \frac{(\theta-\theta_0)^2}{2}\frac{\partial^2}{\partial\theta^2}\log L\Big|_{\theta=\theta_0} + o(\theta-\theta_0)^2$$

$$\log L(\theta;\mathbf{x}) - \log L(\theta_0;\mathbf{x}) = \log\frac{L(\theta;\mathbf{x})}{L(\theta_0;\mathbf{x})}$$

$$= (\theta-\theta_0)\frac{\partial}{\partial\theta}\log L\Big|_{\theta=\theta_0} + \frac{(\theta-\theta_0)^2}{2}\frac{\partial^2}{\partial\theta^2}\log L\Big|_{\theta=\theta_0} + o(\theta-\theta_0)^2$$

We have

$$-\log\frac{L(\theta;\mathbf{x})}{L(\theta_0;\mathbf{x})} = -(\theta-\theta_0)\frac{\partial}{\partial\theta}\log L\Big|_{\theta=\theta_0} + \frac{(\theta-\theta_0)^2}{2}\left(-\frac{\partial^2}{\partial\theta^2}\log L\Big|_{\theta=\theta_0}\right) + o(\theta-\theta_0)^2 \tag{6.6.39}$$

Taking expectation on both sides of Eq. (6.6.39), we get

$$E_{\theta_0}\left(-\log\frac{L(\theta;\mathbf{X})}{L(\theta_0;\mathbf{X})}\right) = -(\theta-\theta_0)E_\theta\frac{\partial}{\partial\theta}\log L\Big|_{\theta=\theta_0}$$

$$+\frac{(\theta-\theta_0)^2}{2}E_\theta\left(-\frac{\partial^2}{\partial\theta^2}\log L\Big|_{\theta=\theta_0}\right) + o(\theta-\theta_0)^2$$

$$\sum_{i=1}^{n} I_{X_i}(\theta,\theta_0) = \frac{(\theta-\theta_0)^2}{2}I_X(\theta_0) + o(\theta-\theta_0)^2 > 0$$

Define $Y_i = \log[f(X_i, \theta)/f(X_i, \theta_0)]$ so that

$$E_{\theta_0}(Y_i) = \int \log \frac{f(x_i;\theta)}{f(x_i;\theta_0)} f(x_i;\theta)dx_i$$
$$= -I(\theta, \theta_0)$$

We, thus, have a sequence of *iid* random variables $\{Y_i\}$ with $E_\theta(Y_i) = -I(\theta,\theta_0)$. By the weak law of large numbers, at $\theta = \theta_0$,

$$\frac{1}{n}\sum_{i=1}^{n} Y_i \xrightarrow{P} E_{\theta_0}(Y_i) = -I(\theta, \theta_0)$$

$$\frac{1}{n}\sum_{i=1}^{n} \log \frac{f(x_i;\theta)}{f(x_i;\theta_0)} = \frac{1}{n}\log \frac{L(\theta;\mathbf{x})}{L(\theta_0;\mathbf{x})} \xrightarrow{P} = -I(\theta,\theta_0) < 0 \tag{6.6.40}$$

Therefore, when $\theta = \theta_0$,

$$P_{\theta_0}\left\{\log \frac{L(\theta;\mathbf{x})}{L(\theta_0;\mathbf{x})} < 0\right\} \to 1 \quad \text{as} \quad n \to \infty \ \forall\, \theta \in N_\delta(\theta_0) \tag{6.6.41}$$

Let us define two sets

$$A_n = \left\{\mathbf{x}\colon \log \frac{L(\theta';\mathbf{x})}{L(\theta_0;\mathbf{x})} < 0 \text{ for some } \theta' \text{ in } N_\delta(\theta_0) \text{ so that } \theta' < \theta_0\right\}$$

and
$$B_n = \left\{\mathbf{x}\colon \log \frac{L(\theta';\mathbf{x})}{L(\theta_0;\mathbf{x})} < 0 \text{ for some } \theta' \text{ in } N_\delta(\theta_0) \text{ so that } \theta' > \theta_0\right\}$$

By the Eq. (6.6.41), $P(A_n) \to 1$ and also $P(B_n) \to 1$ as $n \to \infty$. So, $P(A_n \cap B_n) \to 1$ as $n \to \infty$, where $A_n \cap B_n = \{\mathbf{x}\colon \log L(\theta';\mathbf{x}) < \log L(\theta_0;\mathbf{x})$ and $\log L(\theta'';\mathbf{x}) < \log L(\theta_0;\mathbf{x})$, $\theta' < \theta_0 < \theta''$, $(\theta', \theta'') \in N_\delta(\theta_0)\}$. This along with the fact that $\log L(\theta;\mathbf{x})$ is continuous shows that $\log L(\theta;\mathbf{x})$ first increases and then decreases as θ varies over $N_\delta(\theta_0)$. Therefore, $(\partial/\partial\theta)$ $\log L(\theta;\mathbf{x})$ is first positive and then negative. Thus, by regularity condition A5 that $(\partial/\partial\theta)$ $\log L(\theta;\mathbf{x})$ is continuous in θ, we have

$$P_{\theta_0}\left\{\exists \text{ a } \hat\theta \in N_\delta(\theta_0) \text{ s.t. } \frac{\partial}{\partial\theta}\log L(\theta;\mathbf{x})\bigg|_{\hat\theta} = 0\right\} \to 1 \quad \text{as} \quad n \to \infty$$

or
$$P_{\theta_0}\left\{\left|\hat\theta - \theta_0\right| \le \delta\right\} \ge P(A_n \cap B_n) \to 1 \quad \text{as} \quad n \to \infty$$

Therefore, $\hat\theta \xrightarrow{P} \theta_0$ as $n \to \infty$. This shows that with probability tending to 1 as $n \to \infty$, the likelihood equation admits a solution $\hat\theta$ which is a consistent estimator of θ.

Theorem 6.6.7 [Huzurbazar (1948)] With probability approaching 1 as $n \to \infty$, any consistent solution $\hat\theta$ of likelihood equation is a relative maxima of the likelihood, i.e.,

$$P_{\theta_0}\left[\frac{\partial^2}{\partial\theta^2}\log L(\theta;\mathbf{x})\Big|_{\theta=\hat\theta} < 0\right] \to 1 \quad \text{as} \quad n \to \infty$$

where, θ_0 is the true value of the parameter θ.

Proof. Let $\hat{\theta}$ be a consistent solution of the likelihood equation and θ_0 be the true value of the parameter θ. Since $(\partial^2/\partial\theta^2)\log L$ is continuous and differentiable in $(\theta_0, \hat{\theta})$ (by regularity conditions A4 and A5), by mean value theorem, there exists a point $\theta' \in (\theta_0, \hat{\theta})$, $(|\hat{\theta} - \theta'| \le |\hat{\theta} - \theta_0|)$ such that

$$\frac{1}{n}\frac{\partial^2}{\partial\theta^2}\log L\Big|_{\theta=\hat{\theta}} = \frac{1}{n}\frac{\partial^2}{\partial\theta^2}\log L\Big|_{\theta=\theta_0} + \frac{1}{n}(\hat{\theta}-\theta_0)\frac{\partial^3}{\partial\theta^3}\log L\Big|_{\theta=\theta'} + \cdots \tag{6.6.42}$$

Consider the first term on the right hand side of Eq. (6.6.42)

$$\frac{1}{n}\frac{\partial^2}{\partial\theta^2}\log L\Big|_{\theta_0} = \frac{1}{n}\sum_{i=1}^{n}\frac{\partial^2}{\partial\theta^2}\log f(X_i;\theta)\Big|_{\theta_0}$$

We have a sequence of *iid* random variables $\{(\partial^2/\partial\theta^2)\log f(X_i;\ \theta)|_{\theta_0}\}_1^n$ with $E_\theta\{(\partial^2/\partial\theta^2)\log f(X_i;\ \theta)\}|_{\theta_0} = -I_X(\theta_0)$. Therefore, by Khintchin's weak law of large numbers,

$$\frac{1}{n}\frac{\partial^2}{\partial\theta^2}\log L\Big|_{\theta_0} \xrightarrow{P} -I_X(\theta_0) \tag{6.6.43}$$

Consider now a sequence of *iid* random variables $\{|(\partial^3/\partial\theta^3)\log f(X_i;\ \theta)|\}_1^n$ so that $E_\theta|(\partial^3/\partial\theta^3)\log f(X_i;\ \theta)| < E_\theta[H(X)] < k$.
Therefore, by WLLN,

$$\frac{1}{n}\sum_{i=1}^{n}\left|\frac{\partial^3}{\partial\theta^3}\log f(X_i;\theta)\right| \xrightarrow{P} E_\theta\left|\frac{\partial^3}{\partial\theta^3}\log f(X_i;\theta)\right| < k \tag{6.6.44}$$

Further, at θ', so that $|\theta' - \theta_0| \le |\hat{\theta} - \theta_0|$ and $\hat{\theta} \xrightarrow{P} \theta_0$,

$$\frac{1}{n}\left|\frac{\partial^3}{\partial\theta^3}\log f(\mathbf{X};\theta)\right|_{\theta'} = \frac{1}{n}\left|\frac{\partial^3}{\partial\theta^3}\log L(\theta;\mathbf{X})\right|_{\theta'}$$

$$\le \frac{1}{n}\sum_{i=1}^{n}\left|\frac{\partial^3}{\partial\theta^3}\log f(X_i;\theta)\right|_{\theta'} \tag{6.6.45}$$

By combining Eqs. (6.6.44) and (6.6.45), we have

$$\frac{1}{n}\left|\frac{\partial^3}{\partial\theta^3}\log L(\theta;\mathbf{X})\right|_{\theta'} \xrightarrow{P} E_{\theta_0}\left|\frac{\partial^3}{\partial\theta^3}\log f(X_i;\theta_0)\right| < E_{\theta_0}H(X) < k \tag{6.6.46}$$

Therefore, by Eq. (6.6.46) and that $\hat{\theta}$ is a consistent estimator of θ, we have

$$P_{\theta_0}\left[\frac{1}{n}\left|\frac{\partial^3}{\partial\theta^3}\log L(\theta;\mathbf{X})\right|_{\theta'} < k\right] \to 1 \quad \text{as} \quad n \to \infty \tag{6.6.47}$$

and

$$\hat{\theta} \xrightarrow{P} \theta_0 \quad \text{as} \quad n \to \infty \tag{6.6.48}$$

Equations (6.6.47) and (6.6.48) together give the second term of Eq. (6.6.42)

$$\frac{1}{n}(\hat{\theta}-\theta_0)\frac{\partial^3}{\partial\theta^3}\log L\big|_{\theta'} \xrightarrow{P} 0 \quad \text{as} \quad n\to\infty \tag{6.6.49}$$

Hence, Eqs. (6.6.43) and (6.6.49) together with Eq. (6.6.42) give

$$\frac{1}{n}\frac{\partial^2}{\partial\theta^2}\log L\big|_{\hat{\theta}} \xrightarrow{P} -I_X(\theta_0)$$

$$\therefore \qquad P_{\theta_0}\left[\frac{\partial^2}{\partial\theta^2}\log L(\theta;\mathbf{X})\bigg|_{\hat{\theta}} < 0\right] \to 1 \text{ as } n\to\infty$$

This shows that the probability is tending to 1 as $n \to \infty$ and that any consistent solution of likelihood equation is a maxima of the likelihood function.

Theorem 6.6.8 [Huzurbazar (1948)] A consistent solution of likelihood equation is unique, or $P(\hat{\theta}_1$ and $\hat{\theta}_2$ are two distinct solutions of likelihood equation) $\to 0$ as $n \to \infty$.

Proof. Let $\hat{\theta}_1$ and $\hat{\theta}_2$ be two distinct consistent solutions of likelihood equation, where both $\hat{\theta}_1$ and $\hat{\theta}_2$ are the points in $N_\delta(\theta_0)$ and $(\partial/\partial\theta)\log L|_{\hat{\theta}_1} = 0 = (\partial/\partial\theta)\log L|_{\hat{\theta}_2}$. Consider the function $(\partial/\partial\theta)\log L$ which is given to be continuous and differentiable at every $\theta \in N_\delta(\theta_0)$ (see regularity condition A4). Then, by Rolle's theorem, there exists a point $\hat{\theta}_3 = \alpha\,\hat{\theta}_1 + (1-\alpha)\,\hat{\theta}_2$ for some α, $0 < \alpha < 1$, $\hat{\theta}_3 \in N_\delta(\theta_0)$, such that

$$\frac{1}{n}\frac{\partial^2}{\partial\theta^2}\log L\big|_{\hat{\theta}_3} = 0 \tag{6.6.50}$$

Since $(\partial^2/\partial\theta^2)\log L$ is continuous and differentiable at every $\theta \in N_\delta(\theta_0)$, by mean value theorem, similarly, as in Theorem 6.6.7, we expand $(\partial^2/\partial\theta^2)\log L|_{\hat{\theta}_3}$ around θ_0

$$\frac{1}{n}\frac{\partial^2}{\partial\theta^2}\log L\big|_{\hat{\theta}_3} = \frac{1}{n}\frac{\partial^2}{\partial\theta^2}\log L\big|_{\theta_0} + (\hat{\theta}_3-\theta_0)\cdot\frac{1}{n}\frac{\partial^3}{\partial\theta^3}\log L\big|_{\theta'} + \cdots \tag{6.6.51}$$

where $\theta' \in N_\delta(\theta_0)$

By similar arguments as were used in the proof of Theorem (6.6.7), we have

$$\frac{1}{n}\frac{\partial^2}{\partial\theta^2}\log L\big|_{\hat{\theta}_3} \xrightarrow{P} -I_X(\theta_0) \tag{6.6.52}$$

as $n \to \infty$, and

$$P_{\theta_0}\left[\frac{1}{n}\frac{\partial^2}{\partial\theta^2}\log L\big|_{\hat{\theta}_3} < 0\right] \to 1 \quad \text{as} \quad n\to\infty \tag{6.6.53}$$

This is a contradiction to the finding that $(1/n)(\partial^2/\partial\theta^2)\log L|_{\hat{\theta}_3} = 0$ in Eq. (6.6.50). Therefore, it proves, with probability approaching 0 as $n \to \infty$, that likelihood equation admits two distinct consistent roots in $N_\delta(\theta_0)$. This proves that the consistent solution of likelihood equation is unique.

Theorem 6.6.9 [Cramer (1946)] Under the regularity conditions A1 to A6, a consistent solution $\hat{\theta}$ of likelihood equation is asymptotically normal

$$\sqrt{n}(\hat{\theta}-\theta_0) \xrightarrow{d} N\left(0, \frac{1}{I(\theta)}\right) \tag{6.6.54}$$

where $1/(nI_X(\theta))$ is the Cramer–Rao lower bound for the variance of an unbiased estimator of θ, that is, $\hat{\theta}$ is a consistent and asymptotically most efficient estimator of θ.

Proof. Given that $\hat{\theta}$ is a consistent solution of the likelihood equation and θ_0 is the true value of θ, consider $N_\delta(\theta_0)$, $\delta = |\hat{\theta} - \theta_0|$. Taylor's expansion of the first derivative of log likelihood around the true value θ_0 gives

$$\frac{1}{n}\frac{\partial}{\partial\theta}\log L(\theta; \mathbf{X})\Big|_\theta$$

$$= \frac{1}{n}\frac{\partial}{\partial\theta}\log L(\theta; \mathbf{x})\Big|_{\theta_0} + (\theta-\theta_0)\frac{1}{n}\frac{\partial^2}{\partial\theta^2}\log L(\theta; \mathbf{x})\Big|_{\theta_0} + \frac{(\theta-\theta_0)^2}{2}\frac{1}{n}\frac{\partial^3}{\partial\theta^3}\log L(\theta; \mathbf{x})\Big|_{\theta'}, \tag{6.6.55}$$

$\forall\ \theta \in N_\delta(\theta_0)$, and for some $\theta' \in N_\delta(\theta_0)$. Now, substituting MLE $\hat{\theta}$ for θ and that $(1/n)(\partial/\partial\theta)\log L(\theta; \mathbf{x})|_{\hat{\theta}} = 0$, we get from Eq. (6.6.55),

$$\sqrt{n}(\hat{\theta}-\theta_0) = \frac{\dfrac{1}{\sqrt{n}}\dfrac{\partial}{\partial\theta}\log L(\theta; \mathbf{x})\Big|_{\theta_0}}{-\dfrac{1}{n}\dfrac{\partial^2}{\partial\theta^2}\log L(\theta; \mathbf{x})\Big|_{\theta_0} - \dfrac{(\theta-\theta_0)}{2}\dfrac{1}{n}\dfrac{\partial^2}{\partial\theta^2}\log L(\theta; \mathbf{x})\Big|_{\theta'}} \tag{6.6.56}$$

Note that $\{(\partial^2/\partial\theta^2)\log L(\theta; x_i)\}_1^n$ is a sequence of *iid* random variables with $E_{\theta_0}[(\partial^2/\partial\theta^2)\log L(\theta; X_i)] = -I_X(\theta_0)$. Therefore, by WLLN,

$$\frac{1}{n}\sum_1^n \frac{\partial^2}{\partial\theta^2}\log L(\theta; x_i)\Big|_{\theta_0} \xrightarrow{P} -I_X(\theta_0)$$

or

$$\frac{1}{n}\frac{\partial^2}{\partial\theta^2}\log L(\theta; \mathbf{x})\Big|_{\theta_0} \xrightarrow{P} -I_X(\theta_0) \tag{6.6.57}$$

Moreover, we note from Eq. (6.6.49) that

$$\frac{\theta-\theta_0}{2}\frac{1}{n}\frac{\partial^3}{\partial\theta^3}\log L(\theta; \mathbf{x})\Big|_{\theta'} \xrightarrow{P} 0 \quad \text{as} \quad n\to\infty \tag{6.6.58}$$

On combining Eqs. (6.6.57) and (6.6.58), we see that the denominator in Eq. (6.6.56)

$$-\frac{1}{n}\frac{\partial^2}{\partial\theta^2}\log L(\theta; \mathbf{x})\Big|_{\theta_0} - \frac{(\theta-\theta_0)}{2}\frac{1}{n}\frac{\partial^3}{\partial\theta^3}\log L(\theta; \mathbf{x})\Big|_{\theta'} \xrightarrow{P} I(\theta_0) \quad \text{as} \quad n\to\infty \tag{6.6.59}$$

Further, that $\{(\partial/\partial\theta)\log f(x_i;\theta)|_{\theta_0}\}$ is a sequence of *iid* random variables with $E[\partial/\partial\theta \log f(X_i;\theta)]_{\theta_0} = 0$ and $V[(\partial/\partial\theta)\log f(X_i;\ \theta)]_{\theta_0} = I_X(\theta_0)$. In view of this, by Lindeberg–Levy's central limit theorem, we have

$$U = \frac{1}{\sqrt{n}} \frac{\partial}{\partial \theta} \log L(\theta; \mathbf{X})\Big|_{\theta_0}$$

$$= \frac{1}{\sqrt{n}} \sum_{1}^{n} \frac{\partial}{\partial \theta} \log f(X_i; \theta) \xrightarrow{d} N[0, I_X(\theta_0)] \tag{6.6.60}$$

Therefore, if $U \sim N(0, I(\theta_0))$, then using Eq. (6.6.60) in Eq. (6.6.56), we get

$$\sqrt{n}(\hat{\theta} - \theta_0) \xrightarrow{d} \frac{U}{I_X(\theta_0)} \sim N\left[0, \frac{1}{I_X(\theta_0)}\right] \tag{6.6.61}$$

Further, note that the asymptotic variance of a maximum likelihood estimator is

$$AV(\hat{\theta}) = \frac{1}{nI_X(\theta_0)} = \text{CRLB} \tag{6.6.62}$$

Therefore, MLE is a consistent and asymptotically most efficient estimator of θ.

Note. Suppose that

$$Z_n = \sqrt{n}\left(\frac{X_n - \theta}{\sigma}\right) \xrightarrow{d} Z$$

where Z is $N(0, 1)$. Define

$$(X_n - \theta) = c_n \cdot Z_n = \left(\frac{\sigma}{\sqrt{n}}\right)\left(\sqrt{n}\frac{X_n - \theta}{\sigma}\right)$$

where $c_n \xrightarrow{P} 0$ and $Z_n \xrightarrow{d} Z$ as $n \to \infty$. By using Slutsky's theorem 1.5.3,

$$(X_n - \theta) = c_n \cdot Z_n \xrightarrow{d} 0 \cdot Z = 0 \quad \text{as} \quad n \to \infty$$

By the standard result, that convergence in distribution implies convergence in probability, we have

$$X_n - \theta \xrightarrow{P} 0 \text{ as } n \to \infty$$

So, if X_n is asymptotically normal, it is a consistent estimator of θ. This result shows that if $\hat{\theta}$ is asymptotically efficient, it is defined only when the estimator is asymptotically normal. This result by itself implies that $\hat{\theta}$ is consistent. Therefore, there is no need to discuss efficiency and consistency together since efficiency implies consistency [Casella (2002)].

Theorem 6.6.10 Under the assumptions (B1) to (B4), $\hat{\boldsymbol{\theta}}$, the MLE of $\boldsymbol{\theta}$, is asymptotically multivariate normal

$$\sqrt{n}(\hat{\boldsymbol{\theta}} - \boldsymbol{\theta}_0) \sim N_k[\mathbf{0}, \mathbf{I}_{\mathbf{X}}^{-1}(\boldsymbol{\theta}_0)] \tag{6.6.63}$$

so that $V(\hat{\boldsymbol{\theta}}) = (1/n)\mathbf{I}_{\mathbf{X}}^{-1}(\boldsymbol{\theta}_0)$, $\hat{\boldsymbol{\theta}}$ is most efficient. In other words, $\hat{\boldsymbol{\theta}}$ is the best estimator and distributed as asymptotically normal (BAN).

Proof. We assume that $\boldsymbol{\theta}_0$ is the true value of $\boldsymbol{\theta}$. Consider a point $\boldsymbol{\theta}_1$ in the sphere so that

$$|\boldsymbol{\theta}_1 - \boldsymbol{\theta}_0| < \eta \tag{6.6.64}$$

where $\eta = |\boldsymbol{\theta} - \boldsymbol{\theta}_0|$. We expand the function $(\partial/\partial\boldsymbol{\theta}) \log L$ about $\boldsymbol{\theta}_0$

$$\frac{1}{n}\frac{\partial}{\partial\boldsymbol{\theta}}\log L\bigg|_{\boldsymbol{\theta}} = \frac{1}{n}\frac{\partial}{\partial\boldsymbol{\theta}}\log L\bigg|_{\boldsymbol{\theta}_0} + A_n(\boldsymbol{\theta})\big|_{\boldsymbol{\theta}_1}(\boldsymbol{\theta} - \boldsymbol{\theta}_0) \tag{6.6.65}$$

where
$$A_n(\boldsymbol{\theta}) = \frac{1}{n}\left\| \frac{\partial^2}{\partial\theta_i \partial\theta_j}\log L \right\|$$

Let $\hat{\boldsymbol{\theta}}$ be the MLE of $\boldsymbol{\theta}$. Replacing $\boldsymbol{\theta}$ by $\hat{\boldsymbol{\theta}}$ in Eq. (6.6.65), we get

$$\frac{1}{n}\frac{\partial}{\partial\boldsymbol{\theta}}\log L\bigg|_{\hat{\boldsymbol{\theta}}} = \frac{1}{n}\frac{\partial}{\partial\boldsymbol{\theta}}\log L\bigg|_{\boldsymbol{\theta}_0} + A_n(\boldsymbol{\theta})\big|_{\boldsymbol{\theta}_1}(\hat{\boldsymbol{\theta}} - \boldsymbol{\theta}_0) = 0 \tag{6.6.66}$$

This implies
$$A_n(\boldsymbol{\theta})\big|_{\boldsymbol{\theta}_1}(\hat{\boldsymbol{\theta}} - \boldsymbol{\theta}_0) = -\frac{1}{n}\frac{\partial}{\partial\boldsymbol{\theta}}\log L\bigg|_{\boldsymbol{\theta}_0}$$

$$\sqrt{n}A_n(\boldsymbol{\theta})\big|_{\boldsymbol{\theta}_1}(\hat{\boldsymbol{\theta}} - \boldsymbol{\theta}_0) = -\sqrt{n}\left(\frac{1}{n}\frac{\partial}{\partial\theta_1}\log L, \ldots, \frac{1}{n}\frac{\partial}{\partial\theta_k}\log L\right)_{\boldsymbol{\theta}_0}$$
$$= -b_n(\boldsymbol{\theta}_0) \tag{6.6.67}$$

By the WLLNs,

$$\frac{1}{n}\frac{\partial^2}{\partial\theta_i\partial\theta_j}\log L\bigg|_{\boldsymbol{\theta}_0} \xrightarrow{P} E\left(\frac{\partial^2}{\partial\theta_i\partial\theta_j}\log L\right)_{\boldsymbol{\theta}_0} = -I_{ij}(\boldsymbol{\theta}_0) \tag{6.6.68}$$

We have
$$A_n(\boldsymbol{\theta}) \xrightarrow{P} -\mathbf{I}_{\mathbf{X}}(\boldsymbol{\theta}_0) \tag{6.6.69}$$

Since $\hat{\boldsymbol{\theta}}$ is MLE of $\boldsymbol{\theta}$, $\hat{\boldsymbol{\theta}} \xrightarrow{P} \boldsymbol{\theta}_0$, and $\boldsymbol{\theta}_1$ is such that

$$|\boldsymbol{\theta}_1 - \boldsymbol{\theta}_0| \le \eta$$

and
$$\eta = |\hat{\boldsymbol{\theta}} - \boldsymbol{\theta}_0|$$

Therefore, $\boldsymbol{\theta}_1 \xrightarrow{P} \boldsymbol{\theta}_0$. This gives

$$A_n(\boldsymbol{\theta}_1) \xrightarrow{P} -\mathbf{I}_{\mathbf{X}}(\boldsymbol{\theta}_0) \tag{6.6.70}$$

$\mathbf{I}_{\mathbf{X}}(\boldsymbol{\theta}_0)$ has been given to be positive definite. For large n, we may take $A_n(\boldsymbol{\theta}_1)$ as nonsingular, so that

$$\sqrt{n}(\hat{\boldsymbol{\theta}} - \boldsymbol{\theta}_0) = -A_n^{-1}(\boldsymbol{\theta}_1)b_n(\boldsymbol{\theta}_0) \tag{6.6.71}$$

From Eq. (6.6.65), we have

$$-A_n^{-1}(\boldsymbol{\theta}_1) \xrightarrow{P} \mathbf{I}_{\mathbf{X}}^{-1}(\boldsymbol{\theta}_0) \tag{6.6.72}$$

Moreover, $E_{\boldsymbol{\theta}}[b_n(\boldsymbol{\theta}_0)] = \mathbf{0}$ and the variance–covariance matrix of $b_n(\boldsymbol{\theta}_0)$ is $D[b_n(\boldsymbol{\theta}_0)] = \mathbf{I}_{\mathbf{X}}(\boldsymbol{\theta}_0)$. Hence, by central limit theorem,

$$\sqrt{n}[b_n(\boldsymbol{\theta}_0) - \mathbf{0}] \overset{d}{\sim} N_k[\mathbf{0}, \mathbf{I}_{\mathbf{X}}(\boldsymbol{\theta}_0)] \tag{6.6.73}$$

where N_k is a multivariate normal distribution of dimension k. Hence, on combining Eqs. (6.6.71), (6.6.72), and (6.6.73), we have

$$\sqrt{n}(\hat{\boldsymbol{\theta}} - \boldsymbol{\theta}_0) = -A_n^{-1}(\boldsymbol{\theta}_1)b_n(\boldsymbol{\theta}_0) \xrightarrow{d} N_k[\mathbf{0}, \mathbf{I}_{\mathbf{X}}^{-1}(\boldsymbol{\theta}_0)] \tag{6.6.74}$$

The asymptotic variance of $\hat{\boldsymbol{\theta}}$ is given by

$$V_{\boldsymbol{\theta}_0}(\hat{\boldsymbol{\theta}}) = \frac{1}{n}\mathbf{I}_{\mathbf{X}}^{-1}(\boldsymbol{\theta}_0) \tag{6.6.75}$$

which is equal to the CRLB. Therefore, $\hat{\boldsymbol{\theta}}$ is the most efficient. To sum up the results, we state that $\hat{\boldsymbol{\theta}}$ is BAN.

***Corollary* 6.6.1** Suppose one is interested in estimating the parametric function

$$\mathbf{g}(\boldsymbol{\theta}) = (g_1(\boldsymbol{\theta}),\ldots,g_r(\boldsymbol{\theta}))' \tag{6.6.76}$$

so that the matrix of first–order derivatives of g's exists

$$\mathbf{G}_{(r\times k)} = \left\| \frac{\partial}{\partial \theta_j} g_i(\boldsymbol{\theta}) \right\|_{\substack{i=1,\ldots,r \\ j=1,\ldots,k}}$$

Then,
$$\sqrt{n}[\mathbf{g}(\hat{\boldsymbol{\theta}}) - \mathbf{g}(\boldsymbol{\theta}_0)] \xrightarrow{d} N_r[\mathbf{0}, \mathbf{G}\mathbf{I}_{\mathbf{X}}^{-1}(\boldsymbol{\theta}_0)\mathbf{G}'] \tag{6.6.77}$$

and
$$AV[\mathbf{g}(\hat{\boldsymbol{\theta}})] = \frac{1}{n}\mathbf{G}\mathbf{I}_{\mathbf{X}}^{-1}(\boldsymbol{\theta}_0)\mathbf{G}' \tag{6.6.78}$$

Hence, $\mathbf{g}(\hat{\boldsymbol{\theta}})$ is also BAN.

6.6.7 Superefficiency

It was believed for a long time that if

$$\sqrt{n}(\hat{\theta} - \theta) \xrightarrow{d} N(0, v(\theta)) \tag{6.6.79}$$

then
$$v(\theta) \geq \frac{1}{I_X(\theta)} \quad \forall\ \theta$$

and that any estimator that achieves $1/I_X(\theta)$ is asymptotically most efficient until Hodges had given a counter-example by constructing one estimator for which its asymptotic variance is less than $1/I_X(\theta)$ at $\theta = 0$. This phenomenon is referred to as *superefficiency* and such an estimator is known as a *superefficient estimator*. The estimator proposed by Hodges is as follows:

$$\hat{\theta}' = \begin{cases} a\hat{\theta}, & \text{if } |\hat{\theta}| < n^{-1/4} \\ \hat{\theta}, & \text{otherwise} \end{cases} \tag{6.6.80}$$

Consider the probability

$$P\left[\sqrt{n}\,|\hat{\theta}' - \theta| < c\right] = P\left[\sqrt{n}\,|a\hat{\theta} - \theta| < c \text{ and } \sqrt{n}\,|\hat{\theta}| < n^{1/4}\right] + P\left[\sqrt{n}\,|\hat{\theta} - \theta| < c \text{ and } \sqrt{n}\,|\hat{\theta}| > n^{1/4}\right] \tag{6.6.81}$$

For $\theta = 0$, the second term in Eq. (6.6.81), i.e., $P[\sqrt{n}|\hat{\theta} - \theta| < c$ and $\sqrt{n}|\hat{\theta}| > n^{1/4}] \to 0$ as $n \to \infty$ and the first term, i.e., $P[\sqrt{n}|a\hat{\theta}| < c$ and $\sqrt{n}|\hat{\theta}| < n^{1/4}] \le P[\sqrt{n}|a\hat{\theta}| < c] \to P[N[0, a^2 v(0)] < c]\ \forall\ c$. This implies that for $\theta = 0$, $AV(\hat{\theta}') = v'(0) = a^2 v(0)$. Next, for $\theta \ne 0$, the second term dominates, i.e., $P[\sqrt{n}|\hat{\theta} - \theta| < c$ and $\sqrt{n}|\hat{\theta}| > n^{1/4}] < P[\sqrt{n}|\hat{\theta} - \theta| < c] \to P[N(0, v(\theta)) < c]\ \forall\ c$. This implies that for $\theta \ne 0$, $AV(\hat{\theta}') = v'(\theta) = v(\theta)$. Therefore

$$v'(\theta) = \begin{cases} v(\theta), & \text{for } \theta \ne 0 \\ a^2 v(0), & \text{for } \theta = 0 \end{cases} \tag{6.6.82}$$

If we had $\hat{\theta}$ for which $v(\theta) = 1/I_X(\theta)$, as in the case when $\hat{\theta}$ is the MLE of θ, the asymptotic variance of the estimator $\hat{\theta}'$ for $a < 1$

$$v'(\theta) = \frac{a^2}{I_X(\theta)} < \frac{1}{I_X(\theta)} \tag{6.6.83}$$

at $\theta = 0$. Therefore, Hodges makes a point, by his counter example, that it is not safe to call an estimator as asymptotically most efficient if its asymptotic variance equals $1/I_X(\theta)$ for there may exists an estimator whose asymptotic variance may even be less than $1/I_X(\theta)$ at $\theta = 0$. It was noted that superefficiency can be avoided by imposing additional restrictions on the class of estimators. However, *Le Cam* stated that superefficiency does not occur, except perhaps on a set with the Lebesgue measure zero. Superefficiency can be avoided by considering a class of estimators for which the asymptotic variance $v(\theta)$ is and continuous function of θ. Thus, in this case, one can safely use the optimality criterion of an estimator by stating that it is asymptotically most efficient if it achieves $1/I_X(\theta)$.

6.6.8 Variance Stabilization

The asymptotic variance of an estimator $\hat{\theta}$ so that

$$\sqrt{n}(\hat{\theta} - \theta) \xrightarrow{d} N[0, v(\theta)] \tag{6.6.84}$$

namely $v(\theta)$ can cause inconvenience, since it depends on the parameter θ. It is, thus, desirable to suggest such a transformation that makes the asymptotic variance independent of θ. For some smooth function g, delta method gives

$$\sqrt{n}[g(\hat{\theta}) - g(\theta)] \xrightarrow{d} N(0, [g'(\theta)]^2 v(\theta)) \tag{6.6.85}$$

If we choose such a function g that $g'(\theta) = c/\sqrt{v(\theta)}$ for some constant c, we have

$$\sqrt{n}[g(\hat{\theta}) - g(\theta)] \xrightarrow{d} N(0, c^2) \tag{6.6.86}$$

This is called *variance stabilization*, since after transformation, the variance becomes independent of θ. We will discuss one advantage of this result in constructing asymptotic confidence intervals in closed form in Chapter 9.

In case of MLEs, asymptotic variance, i.e., $1/I_X(\theta)$ too depends on the parameter θ and the variance stabilization may be approached at by a different method using the observed Fisher information. We have

$$I_X(\hat{\theta}) = E_\theta\left\{-\frac{\partial^2}{\partial\theta^2}\log L(\theta;\mathbf{X})\right\}\Bigg|_{\theta=\hat{\theta}} > 0$$

$$= -\frac{1}{n}\sum_1^n \frac{\partial^2}{\partial\theta^2}\log L(\theta; X_i)\Bigg|_{\theta=\hat{\theta}} + o_p(1) \tag{6.6.87}$$

The quantity $-(1/n)\Sigma_1^n(\partial^2/\partial\theta^2)\log L(\theta; X_i)\big|_{\theta=\hat{\theta}}$ is called the *observed Fisher information.* We have already shown that under certain regularity conditions, the sequence of MLEs $\{\hat{\theta}\}$

$$\hat{\theta} \xrightarrow{P} \theta \text{ and } \sqrt{n}(\hat{\theta}-\theta) \xrightarrow{d} N\left(0, \frac{1}{I_X(\theta)}\right) \tag{6.6.88}$$

Note that the asymptotic variance of MLE $\hat{\theta}$, $AV(\hat{\theta}) = 1/I_X(\theta)$, depends on θ. We now consider the following quantity for variance stabilization

$$\sqrt{nI_X(\hat{\theta})}(\hat{\theta}-\theta) = \sqrt{\frac{I_X(\hat{\theta})}{I_X(\theta)}}\sqrt{nI_X(\theta)}(\hat{\theta}-\theta)$$

The first term on the right-hand side $\sqrt{I_X(\hat{\theta})/I_X(\theta)} \xrightarrow{P} 1$ on assuming $I_X(\theta)$ is a continuous function of θ so that $I_X(\hat{\theta})/I_X(\theta) \xrightarrow{P} 1$ whenever $\hat{\theta} \xrightarrow{P} \theta$. The second term $\sqrt{nI_X(\theta)}\,(\hat{\theta} - \theta)$ $\xrightarrow{d} N(0, 1)$. Therefore, by Slutsky's theorem,

$$\sqrt{nI_X(\hat{\theta})}(\hat{\theta}-\theta) \xrightarrow{d} N(0,1) \tag{6.6.89}$$

where $I_X(\hat{\theta})$ is replaced by the observed Fisher information, i.e., $-(1/n)\Sigma_1^n(\partial^2/\partial\theta^2)\log L(\theta; X_i)\big|_{\theta=\hat{\theta}}$

6.6.9 Maximum Likelihood Estimators Using Fisher's Scoring Method

The maximum likelihood estimators are obtained by solving the likelihood equations $(\partial/\partial\theta)\log L(\theta;\, x) = 0$. Sometimes these likelihood equations are complicated and, subsequently, an analytic solution for θ is not possible. In such cases, we use some iterative procedure, starting with some trial value of θ, calculating correction factor at each iteration, and continuing the iteration process till corrections are negligible. Usually, the trial values are considered as some consistent estimators of θ such as the method of moment estimators. One such method is discussed here, which is one of the most popular methods of obtaining an MLE in a numerical manner, known as *Fisher's scoring method.* Let θ be a scalar parameter. Then, $S(\mathbf{x},\theta) = (\partial/\partial\theta)\log L(\theta;\mathbf{x})$ is known as *efficient score function*. The score function is defined in Section 4.2. It is a function of $\mathbf{x}$ and θ. However, for the purpose notational convenience it is denoted by $S(\theta)$ in the present section. Following this, the likelihood equation is given by $S(\theta) = 0$. Let $\hat{\theta}$ be the MLE of θ and θ_0 be the trial value of θ such that $\hat{\theta} = \theta_0 + h$, where h is so small that

$$S(\hat{\theta}) = 0 = S(\theta_0 + h)$$

But by Taylor's series expansion,

$$S(\hat{\theta}) = S(\theta_0) + hS'(\theta_0) + o(h)$$

$$0 = S(\theta_0) + hS'(\theta_0) \quad \text{when } o(h) \to 0$$

i.e.
$$S(\theta_0) + h^{(1)}S'(\theta_0) = 0 \tag{6.6.90}$$

However,
$$S'(\theta) = \frac{\partial^2}{\partial\theta^2}\log L$$

$$= \sum_1^n \frac{\partial^2}{\partial\theta^2}\log f(x_i;\theta) \xrightarrow{P} E\left[\frac{\partial^2}{\partial\theta^2}\log L\right] = -I_{\mathbf{X}}(\theta) \tag{6.6.91}$$

since the sum of n *iid* random variables converges in probability to its expectation. Therefore, for large n, Eq. (6.6.90) can be written as

$$S(\theta_0) - h^{(1)}I_{\mathbf{X}}(\theta_0) = 0$$

$$h^{(1)} = \frac{S(\theta_0)}{I_{\mathbf{X}}(\theta_0)} \tag{6.6.92}$$

so that the first approximation is $\hat{\theta}^{(1)} = \theta_0 + h^{(1)}$. Following this, the iteration process may be written as

$$\hat{\theta}^{(i+1)} = \hat{\theta}^{(i)} + \frac{S(\hat{\theta}^{(i)})}{I_{\mathbf{X}}(\hat{\theta}^{(i)})}; \quad i = 0, 1, 2,\ldots \tag{6.6.93}$$

The iteration is repeated by taking $\hat{\theta}^{(1)}$ as the next trial value and is continued till the corrected values $\hat{\theta}^{(i)}$ converge.

One can start with an easy-to-compute, $\sqrt{n}$-consistent, asymptotically normal but not the most efficient estimator so that

$$\sqrt{n}(\hat{\theta}^{(0)} - \theta) \xrightarrow{d} N[0, v(\theta)] \tag{6.6.94}$$

with $v(\theta) > 1/I_{\mathbf{X}}(\theta)$. Such estimators are known as cheap estimators. One remarkable point in using Fisher's scoring method is that just 1-step improvement

$$\hat{\theta}^{(1)} = \hat{\theta}^{(0)} + \frac{S(\hat{\theta}^{(0)})}{I_{\mathbf{X}}(\hat{\theta}^{(0)})} \tag{6.6.95}$$

generally results into the asymptotically most efficient estimator under certain regularity conditions. However, in practical situations, we need to run quite a few iterations to finally get the most efficient estimator.

Further,
$$I_{\mathbf{X}}(\theta) = E\left[-\frac{\partial^2}{\partial\theta^2}\log L\right] = nE\left(\frac{\partial}{\partial\theta}\log f\right)^2 \tag{6.6.96}$$

and
$$V(\hat{\theta}) \cong \frac{1}{I_{\mathbf{X}}(\theta)} \quad \text{for large } n \tag{6.6.97}$$

Analogously, in multiparameter case the iteration process is stated as

$$\hat{\boldsymbol{\theta}}^{(i+1)} = \hat{\boldsymbol{\theta}}^{(i)} + \left[\mathbf{I}_{\mathbf{X}}(\hat{\theta})^{(i)}\right]^{-1} \boldsymbol{S}(\hat{\boldsymbol{\theta}}^{(i)}),\ i = 0, 1, 2, \ldots \tag{6.6.98}$$

where $\boldsymbol{S}(\boldsymbol{\theta})$ is a vector, $\mathbf{I}_{\mathbf{X}}(\boldsymbol{\theta})$ is a matrix, and $[\mathbf{I}_{\mathbf{X}}(\hat{\theta})^{(i)}]^{-1}\ \boldsymbol{S}(\hat{\boldsymbol{\theta}}^{(i)})$ is the corrected vector. The process is repeated till consistent values of $\boldsymbol{\theta}$ are obtained.

Fisher scoring method for grouped data: Let $p_1, p_2, \ldots, p_k$ $(\Sigma_1^k p_i = 1)$ be the probabilities of k mutually exclusive classes and p_is be defined as functions of a single parameter θ. Let $f_1, f_2, \ldots, f_k$ $(\Sigma_1^k f_i = n)$ be the observed frequencies. Thus, we have an observation from a multinomial distribution with k classes and with cell probabilities $p_1(\theta), p_2(\theta), \ldots, p_k(\theta)$. So, the likelihood function for the observations $f_1, f_2, \ldots, f_k$ is given by

$$L = \frac{n!}{f_1!\, f_2! \cdots f_k!}\{p_1(\theta)\}^{f_1}\{p_2(\theta)\}^{f_2} \cdots \{p_k(\theta)\}^{f_k} \text{ for some } \theta \in \Theta$$

$$\Rightarrow \qquad \log L = \log c + \sum_1^k f_i \log p_i(\theta)$$

where $c = n!/(f_1!\, f_2! \ldots f_k!)$

$$S(\theta) = \frac{\partial}{\partial\theta}\log L = \sum_{i=1}^k \frac{f_i}{p_i(\theta)}\frac{\partial}{\partial\theta}p_i(\theta)$$

Further, $$S'(\theta) = \frac{\partial^2}{\partial\theta^2}\log L$$

$$= \sum_{i=1}^k \frac{f_i}{p_i(\theta)}\frac{\partial^2}{\partial\theta^2}p_i(\theta) - \sum_{i=1}^k \frac{f_i}{p_i^2(\theta)}\left(\frac{\partial}{\partial\theta}p_i(\theta)\right)^2 \tag{6.6.99}$$

On taking expectation in Eq. (6.6.99), we get

$$E\left[\frac{\partial^2}{\partial\theta^2}\log L\right] = \sum_{i=1}^k \frac{np_i(\theta)}{p_i(\theta)}\frac{\partial^2}{\partial\theta^2}p_i(\theta) - \sum_{i=1}^k \frac{np_i(\theta)}{p_i^2(\theta)}\left(\frac{\partial}{\partial\theta}p_i(\theta)\right)^2 \tag{6.6.100}$$

Further, on differentiating $\Sigma_1^k p_i = 1$ with respect to θ, we have

$$\sum_{i=1}^k \frac{\partial^2}{\partial\theta^2}p_i(\theta) = 0 \tag{6.6.101}$$

Equations (6.6.100) and (6.6.101) give

$$E\left[\frac{\partial^2}{\partial\theta^2}\log L\right] = -n\sum_1^k \frac{1}{p_i(\theta)}\left(\frac{\partial}{\partial\theta}p_i(\theta)\right)^2 \tag{6.6.102}$$

$$\therefore \qquad I_X(\theta) = n\sum_{i=1}^k \frac{1}{p_i(\theta)}\left(\frac{\partial}{\partial\theta}p_i(\theta)\right)^2 \tag{6.6.103}$$

Assuming θ_0 as the first trial value of θ, the correction factor becomes $h^{(1)} = S(\theta_0)/I(\theta_0)$. The iteration equations are given by

$$\theta^{(i+1)} = \theta^{(i)} + \frac{S(\theta^{(i)})}{I(\theta^{(i)})}, \quad i = 0, 1, 2,\ldots \tag{6.6.104}$$

We continue the iterations till a consistent value of θ is finally obtained.

For multivariate case, we have

$$S(\theta_i) = \sum_{j=1}^{k} \frac{f_j}{p_j} \frac{\partial}{\partial \theta_i} p_j, \; i = 1,\ldots,k$$

whenever $\boldsymbol{\theta} = (\theta_1,\ldots,\theta_k)' \in \Theta \subseteq \mathbb{R}^k$.

$$I_{ij} = n \sum_{\ell=1}^{k} \frac{1}{p_l(\theta)} \left(\frac{\partial}{\partial \theta_i} p_l(\theta) \frac{\partial}{\partial \theta_j} p_l(\theta) \right)$$

is the ijth element in the $k \times k$-dimensional information matrix, $\mathbf{I}(\boldsymbol{\theta})$. Corresponding iteration equation is given by

$$\boldsymbol{\theta}^{(i+1)} = \boldsymbol{\theta}^{(i)} + [\mathbf{I}(\boldsymbol{\theta}^{(i)})]^{-1} S(\boldsymbol{\theta}^{(i)}), \; i = 0, 1, 2\ldots \tag{6.6.105}$$

Notes.

1. If the density is not explicitly available, the maximum likelihood technique cannot be applied and in this case, one can safely use MoME.
2. In general, MLE is not unbiased for its estimand. However, it may have smaller mean square error than of an unbiased estimator. In case of normal distribution $N(\theta,\sigma^2)$, where both the parameters θ and σ^2 are unknown, S_n^2, the MLE of σ^2, is not unbiased for σ^2; but S_{n-1}^2, the sample variance, is unbiased for σ^2. Note that MSE(S_n^2) < MSE(S_{n-1}^2).
3. By definition, MLE attempts to find the mode of the distribution; it justifies as to why its small sample properties are poor. However, for large samples, mode tends to mean (if it exists) and median. Therefore, MLE has many good large sample properties.
4. MLE is not necessarily the sample mean of the distribution.
5. Likelihood equation may have several solutions, i.e., MLEs are not necessarily unique, although usually they are.
6. $\log L$ may not be differentiable everywhere in Θ or the solution of likelihood equation $(\partial/\partial\theta)\log L = 0$ may be a terminal value in Θ; in such a case, it is not the MLE.
7. MLE is numerically sensitive, i.e., non-robust.
8. Sometimes, likelihood equations are complicated and difficult to solve. In such cases, we resort to some numerical procedure to obtain the estimate.
9. MLE under certain regularity conditions is consistent. In case there are several MLEs, one of them is consistent. However, an MLE may not be unbiased, even if a unique MLE exists.
10. MLE is the function of a sufficient statistic. It can easily be shown by using factorization theorem.
11. MLE is invariant.
12. MLE is asymptotically efficient.

6.7 SOLVED EXAMPLES

***Example* 6.1** Consider $X_1, X_2,\ldots,X_n$ as a random sample from the population with *pmf*

$$P\{X=y_1\}=\frac{1-\theta}{2}, P\{X=y_2\}=\frac{1}{2}, P\{X=y_3\}=\frac{\theta}{2}, (0<\theta<1)$$

Find the MLE of θ.

Solution. Each observation X_i, $i = 1,\ldots,n$, takes one of the three values y_1, y_2, or y_3. Suppose y_1 occurs $n(y_1)$ times, y_2 occurs $n(y_2)$ times, and y_3 occurs $n(y_3)$ times in the sample so that $n(y_1) + n(y_2) + n(y_3) = n$. Thus, the likelihood function is given by

$$L(\theta; \mathbf{x})=\frac{n!}{n(y_1)!\,n(y_2)!\,n(y_3)!}\left(\frac{1-\theta}{2}\right)^{n(y_1)}\left(\frac{1}{2}\right)^{n(y_2)}\left(\frac{\theta}{2}\right)^{n(y_3)}$$

$$\log L=\log\frac{n!}{n(y_1)!\,n(y_2)!\,n(y_3)!}+n(y_1)\log\left(\frac{1-\theta}{2}\right)$$

$$+n(y_2)\log\left(\frac{1}{2}\right)+n(y_3)\log\left(\frac{\theta}{2}\right)$$

$$\frac{\partial}{\partial\theta}\log L=-n(y_1)\frac{2}{1-\theta}\frac{1}{2}+n(y_3)\frac{2}{\theta}\frac{1}{2}$$

$\frac{\partial}{\partial\theta}\log L=0$ gives

$$\frac{n(y_1)}{1-\theta}=\frac{n(y_3)}{\theta}$$

$$[n(y_1)+n(y_3)]\theta=n(y_3)$$

Thus, the MLE of θ given by

$$\hat{\theta}=\begin{cases}\dfrac{n(y_3)}{n(y_1)+n(y_3)}, & \text{if } n(y_1), n(y_2)>0\\ \text{any value between } (0,1), & \text{if } n(y_1)=0=n(y_3)\\ \text{no MLE}, & \text{if } n(y_1)=0 \text{ and } n(y_3)\neq 0\\ & \text{or } n(y_1)\neq 0, n(y_3)=0\end{cases}$$

***Example* 6.2** Consider the population made up of three different types of individuals occurring in the proportion θ^2, $2\theta(1-\theta)$, and $(1-\theta)^2$, respectively, where $0 < \theta < 1$. Let n_1, n_2, and n_3 denote the respective random sample sizes of the above three types of individuals. Find the MLE of θ.

Solution. The likelihood function is

$$L(\theta; \mathbf{x}) \propto \theta^{2n_1}[2\theta(1-\theta)]^{n_2}(1-\theta)^{2n_3}$$

$$\log L = \text{const.} + 2n_1\log\theta + n_2\log[2\theta(1-\theta)] + 2n_3\log(1-\theta)$$

$$\frac{\partial}{\partial\theta}\log L=\frac{2n_1}{\theta}+\frac{n_2}{2\theta(1-\theta)}2(1-2\theta)-2n_3\frac{1}{1-\theta}$$

$(\partial/\partial\theta)\log L = 0$ implies

$$2n_1(1-\theta) + n_2(1-2\theta) - 2n_3\theta = 0$$

$$(-2n_1 - 2n_2 - 2n_3)\theta + (2n_1 + n_2) = 0$$

$$\hat{\theta} = \frac{2n_1 + n_2}{2(n_1+n_2+n_3)} = \frac{2n_1+n_2}{2n}$$

Therefore, the MLE of θ is

$$f(x;\theta) = \begin{cases} \left(\dfrac{2n_1+n_2}{2n}\right), & \text{if } n_1, n_2 > 0 \\ \text{any value between } (0,1), & \text{if } n_1 = n_2 = n_3 = 0 \\ \text{no MLE}, & \text{if } n_2 = n_3 = 0 \text{ and } n_1 \neq 0 \\ & \text{or } n_1 = n_2 = 0,\ n_3 \neq 0 \end{cases}$$

***Example* 6.3** Let $X_1,\ldots,X_n$ be a random sample drawn from the population

$$f(x;\theta) = \begin{cases} \theta, & \text{if } x = 1 \\ \dfrac{1-\theta}{\lambda-1}, & \text{if } x = 2, 3, \ldots, \lambda \\ 0, & \text{otherwise} \end{cases}$$

where $\theta \in [0, 1]$, and $\lambda \in \{2, 3, \ldots\}$, are unknown. Find the MLE of θ and λ.

Solution. The likelihood function is given by

$$L(\theta, \lambda; x_1,\ldots,x_n) = \theta^{n(1)} \prod_{i=1}^{n} \left(\frac{1-\theta}{\lambda-1}\right) I_{\{2,3,\ldots,\lambda\}}(x_i)$$

$$\log L = n(1)\log\theta + [n - n(1)]\log\frac{1-\theta}{\lambda-1}$$

where $n(j)$ is the number of X_is which take value j for $j = 1, 2,\ldots,\lambda$ so that $n(1) + n(2) + \cdots + n(\lambda) = n$. Consider the case $n(1) < n$. We have

$$\frac{\partial \log L}{\partial\theta} = \frac{n(1)}{\theta} + \frac{[n-n(1)]}{1-\theta}(\lambda-1)\left(-\frac{1}{\lambda-1}\right) = 0$$

$$(1-\theta)n(1) - \theta[n - n(1)] = 0$$

Thus, the MLEs of θ and λ are given by

$$\hat{\theta} = \frac{n(1)}{n},\ \hat{\lambda} = X_{(n)}$$

Next, consider the case $n(1) = n$

$$L(\theta,\lambda; x_1,\ldots,x_n) = \theta^n = \text{const.}\ \forall\ \lambda$$

Therefore, the MLE of θ is $\hat{\theta} = 1$. Any value of λ in {2, 3,...} is the MLE of λ. Note that the MLE of λ is not unique.

***Example* 6.4** Let $X_1, X_2,\ldots,X_n$ be a random sample from $b(r, \theta)$ population with r known, $0 \le \theta \le 1$. Find the MLE of θ and $\theta(1 - \theta)$.

Solution. The likelihood function of θ is given by

$$L(\theta, \mathbf{x}) = \prod_{i=1}^{n} \binom{n}{x_i} \theta^{x_i} (1-\theta)^{r-x_i}, \ x_i = 0, 1, \ldots, r$$

$$= \sum \log \binom{n}{x_i} + \log\theta \sum x_i + \log(1-\theta) \sum (r - x_i)$$

$$\frac{\partial}{\partial\theta} \log L = \frac{\sum x_i}{\theta} - \frac{\sum (r - x_i)}{(1-\theta)} = 0$$

$$\hat{\theta} = \frac{\sum x_i}{nr} = \frac{\overline{x}}{r}$$

To examine whether $\hat{\theta}$ is the maxima of L, consider the second derivative

$$\frac{\partial^2}{\partial\theta^2} \log L\big|_{\hat{\theta}} = \left[-\frac{\sum x_i}{\theta^2} - \frac{nr - \sum x_i}{(1-\theta)^2} \right]_{\theta=\hat{\theta}}$$

$$= -\frac{nr^3}{\overline{x}(r - \overline{x})} < 0 \text{ whenever } 0 < \overline{x} < r$$

Therefore, $\hat{\theta} = \overline{x}/r$ is maxima whenever $0 < \overline{x} < r$. Further, if $\overline{x} = 0$, we have

$$L = (1 - \theta)^{nr}$$

which attains maximum at $\hat{\theta} = 0 = \overline{x}/r$. This shows that irrespective of the value of $\overline{x}$, the MLE of θ is

$$\hat{\theta} = \frac{\overline{x}}{r}$$

Further, the MLE of $\theta(1 - \theta)$, by Theorem 6.6.4, is

$$\hat{\theta}(1 - \hat{\theta}) = \frac{\overline{x}}{r}\left(1 - \frac{\overline{x}}{r}\right)$$

***Example* 6.5** Let $X_1, X_2,\ldots,X_n$ be a random sample drawn from the Bernoulli distribution $b(1, \theta)$ $\theta \in (0, 1)$. Show that the expressions of MoMCS and ML estimators are the same. Comment on the MLE in this case. Discuss the case when $\theta \in [0, 1]$. Also, using Berk (1967), find the MLE of $\theta(1 - \theta)$.

Solution. Let there be n_1 successes in the sample denoted by '1' and n_2 failures denoted by '0' so that $p(1) = \theta$ and $p(0) = 1 - \theta$, $n_1 + n_2 = n$. Given the observations **x**, the likelihood function for θ is given by

$$L(\theta; \mathbf{x}) = \theta^{\sum x_i}(1-\theta)^{n-\sum x_i} = \theta^{n_1}(1-\theta)^{n_2}$$

$$\log L = n_1 \log\theta + n_2 \log(1-\theta)$$

The likelihood equation is given by

$$\frac{\partial}{\partial\theta}\log L = \frac{n_1}{\theta} - \frac{n_2}{1-\theta} = 0$$

$$\hat{\theta} = \frac{n_1}{n}$$

Note that $\hat{\theta}$ may take a value 0 or 1, which is not an admissible value of θ, since $0 < \theta < 1$. Hence, $\hat{\theta}$ is the MLE of θ provided $n_1 > 0$.

Further, the MoMCS equation is given by

$$\sum_{i=1}^{2} \frac{f_i^2}{p_i^2}\frac{dp_i}{d\theta} = 0 = \frac{n_1^2}{\theta^2} + \frac{n_2^2}{(1-\theta)^2}(-1)$$

$$(1-\theta)^2 n_1^2 - \theta^2 n_2^2 = 0$$

$$(n_1^2 - n_2^2)\theta^2 - 2n_1^2\theta + n_1^2 = 0$$

$$\theta = \frac{2n_1^2 \pm \sqrt{4n_1^4 - 4(n_1^2 - n_2^2)n_1^2}}{2(n_1^2 - n_2^2)}$$

$$= \frac{2n_1^2 \pm 2n_1 2n_2}{2(n_1^2 - n_2^2)} = \frac{n_1}{n_1 - n_2} \text{ or } \frac{n_1}{n_1 + n_2}$$

If $n_1 < n_2$, the solution $\hat{\theta} = n_1/(n_1 - n_2)$ is negative, which is not possible since $\hat{\theta}$ is the probability. Hence, the only admissible solution for θ is $n_1/(n_1 + n_2) = n_1/n = \hat{\theta}$. Therefore, the MoMCS estimator coincides with the ML estimator, particularly in this example.

In case $\theta \in [0,1]$, the solution of likelihood equation gives $\bar{X}$ as the MLE of θ.

Next, consider

$$f(x; \theta) = \theta^x(1-\theta)^x$$

$$\log f = x\log\theta + (1-x)\log(1-\theta)$$

$$\frac{\partial}{\partial\theta}\log f = \frac{x}{\theta} - \frac{1-x}{1-\theta}$$

$$\frac{\partial^2}{\partial\theta^2}\log L = -\frac{x}{\theta^2} - \frac{1-x}{(1-\theta)^2}$$

Fisher information in a single observation is, thus, given by

$$I_{X_1}(\theta) = E_\theta\left\{-\frac{\partial^2}{\partial\theta^2}\log L\right\} = \frac{1}{\theta(1-\theta)}$$

Using Theorem 6.6.9, the MLE $\bar{X}$ of θ is asymptotically normal

$$\sqrt{n}(\bar{X}-\theta) \xrightarrow{d} N[0, \theta(1-\theta)]$$

Following Berk (1967) and taking $g(\theta) = \theta(1 - \theta)$, $l(\theta) = \theta$, and $h(\theta) = [g(\theta), l(\theta)] = [\theta(1 - \theta), \theta] = (\eta, \nu) = \xi$, the MLE of $h(\theta)$ is $h(\hat{\theta}) = [g(\hat{\theta}), l(\hat{\theta})] = [\hat{\theta}(1 - \hat{\theta}), \hat{\theta}]$. Thus the MLE of $\eta = g(\theta) = \theta(1 - \theta)$ is $\hat{\theta}(1 - \hat{\theta}) = \bar{X}(1 - \bar{X})$. However, the same result may also be obtained by using the invariance property of MLE due to Zehna (1966), see Sub-section 6.6.5.

***Example* 6.6** Consider a situation of crime in some state of India where sometimes the cases are not reported. Given n information $X_1, X_2, \ldots, X_n$, on such crimes, estimate the true reporting rate of crimes and the total number of their occurrences, k, by the MoM.

Solution. The MoM equations are

$$m_1' = \bar{X} = kp = \mu_1'$$

$$m_2' = \frac{1}{n}\sum X_i^2 = kp(1-p) + k^2p^2 = \mu_2'$$

By solving them for k and p, we get the MoM estimator for k

$$m_1' - \frac{m_1'^2}{k} + m_1'^2 = m_2'$$

$$\hat{k} = \frac{m_1'^2}{m_1' + m_1'^2 - m_2'}$$

$$= \frac{\bar{X}^2}{\bar{X} + \bar{X}^2 - \overline{X^2}}$$

$$= \frac{\bar{X}^2}{\bar{X} - n^{-1}\sum(X_i - \bar{X})^2}$$

Further, the MoM for p is given by

$$\hat{p} = \frac{m_1'}{\hat{k}} = \frac{\bar{X}}{\hat{k}}$$

Since $\bar{X} \xrightarrow{P} kp$ and $(1/n)\Sigma X_i^2 \xrightarrow{P} kp(1 - p) + n^2p^2$, $\hat{k}$ and $\hat{p}$ are consistent estimators for k and p. In case of large variability in the data, the sample mean may be smaller than the sample variance. This may cause MoMEs $\hat{k}$ and $\hat{p}$ to take negative values, which are not the admissible values. Generally, MoMEs are not good estimators. However, MoMEs are often considered as starting trial estimators in some procedures to yield efficient estimators. Satterthwaite (1946) has used MoMEs for getting an approximate estimator of unknown distribution of a statistic.

***Example* 6.7** Let $X_1, X_2, \ldots, X_n$ be a random sample from $P(\theta)$, $\theta > 0$. Find

(i) MoME of θ.

(ii) MoMCSE of θ.

(iii) MLE of θ when $\Theta = (0, \infty)$ and when $\Theta = [0, \infty)$, Is it consistent? Is it asymptotically normal? Discuss variance stabilization in this case. Also find the MLE of θ^2 and $\sin\theta$.

(iv) Find the MLE of $\exp(-\theta)$. Is it asymptotically normal?

Solution.

(i) We have $\mu_1' = E(X) = \theta$ and $\mu_2' = E(X^2) = V(X) + [E(X)]^2 = \theta + \theta^2$. MoM estimators are

$$m_1' = \bar{X} = \hat{\theta}$$

$$\overline{X^2} - \bar{X}^2 = \hat{\theta} = \frac{1}{n}\sum_{1}^{n}(X_i - \bar{X})^2$$

where $\overline{X^2} = n^{-1}\Sigma X_i^2$. We see that there are two MoMEs of θ. In practice, such an estimator of θ is considered as a reasonable MoME that involves lower-order sample moments.

(ii) Here, $p_i(\theta) = [\exp(-\theta)\theta^i]/i!$, $i = 0, 1, \ldots$

$$\log p_i(\theta) = -\theta + i\log\theta - \log(i!)$$

$$\frac{1}{p_i}\frac{\partial}{\partial\theta}p_i = -1 + \frac{i}{\theta}$$

$$\frac{\partial}{\partial\theta}p_i = p_i\left(\frac{i}{\theta} - 1\right)$$

The MoMCS equation is given by

$$\Sigma\frac{f_i^2}{p_i^2}p_i\left[\frac{i}{\theta} - 1\right] = 0$$

or

$$\Sigma\frac{f_i^2}{p_i}\left[\frac{i}{\theta} - 1\right] = 0 \qquad (6.7.1)$$

Note that equation Eq. (6.7.1) is difficult to solve explicitly. Therefore, some numerical procedure is needed to solve this. One numerical procedure is demonstrated here by setting

$$\phi_i(\theta) = \frac{f_i^2}{p_i}\left(\frac{i}{\theta} - 1\right)$$

in Eq. (6.7.1) so that $\Sigma\phi_i(\theta) = 0$. Let θ_1 be a trial value and $\theta = \theta_1 + h$ be the true value that minimizes χ^2, i.e., we, then, have $\Sigma\phi_i(\theta_1 + h) = 0$. Using Taylor's expansion assuming h is small, we have

$$\phi_i(\theta_1 + h) \cong \phi_i(\theta_1) + h\phi_i'(\theta)\big|_{\theta=\theta_1}$$

$$= \frac{f_i^2}{p_i}\left(\frac{i}{\theta} - 1\right) + h\left[-\frac{f_i^2}{p_i^2}\left(\frac{i}{\theta} - 1\right)\cdot\frac{\partial}{\partial\theta}p_i + \frac{f_i^2}{p_i}\left(-\frac{i}{\theta^2}\right)\right]_{\theta=\theta_1}$$

$$= \frac{f_i^2}{p_i}\left(\frac{i}{\theta_1} - 1\right) - h\left[\frac{f_i^2}{p_i}\left(\frac{i}{\theta_1} - 1\right)^2 + \frac{f_i^2}{p_i}\frac{i}{\theta_1^2}\right]$$

$\Sigma\phi_i(\theta_1 + h) = 0$ gives

$$0 = \sum\frac{f_i^2}{p_i}\left(\frac{i}{\theta_1} - 1\right) - h\left\{\frac{f_i^2}{p_i}\left[\left(\frac{i}{\theta_1} - 1\right)^2 + \frac{i}{\theta_1^2}\right]\right\}$$

$$h(\theta) = \theta\frac{\sum(f_i^2/p_i)(i-\theta)}{\sum(f_i^2/p_i)[i + (i-\theta)^2]}$$

At the second iteration, we get better approximation as

$$\theta_2 = \theta_1 + h(\theta)|_{\theta_1}$$

Similarly, at the third iteration,

$$\theta_3 = \theta_2 + h(\theta)|_{\theta_2}$$

We keep repeating the iterations till the subsequent values of θ converge. Such a convergent value is the one that minimizes χ^2.

(iii) The likelihood function of θ is given by

$$L(\theta, \mathbf{x}) = \frac{\exp(-n\theta)\,\theta^{\sum_1^n x_i}}{\prod_{i=1}^{n} x_i}$$

$$\log L = -n\theta + \sum x_i \log\theta - \sum \log x_i$$

$$\frac{\partial}{\partial\theta}\log L = -n + \frac{1}{\theta}\sum x_i - 0$$

$$\hat{\theta}_n = \bar{X}$$

Notice that this solution is unique and the second derivative

$$\frac{\partial^2}{\partial\theta^2}\log L = -\frac{\sum x_i}{\theta^2} < 0$$

unless $\Sigma x_i = 0$. Hence, for $\Sigma x_i > 0$, log-likelihood is strictly concave. Therefore, the solution of likelihood equation gives global maxima and, thus, $\bar{X}$ is the MLE of θ. In case $\Sigma x_i = 0$, the MLE does not exist, since $\bar{X} = 0$ is not in the parameter space Θ.

The given family is not an exponential family, since the *pdf* or *pmf* of an exponential family must hold for every θ in Θ; but in the present case, the *pmf* at $\theta = 0$ is not defined, since it involves θ, which is not defined. In this case, the *pmf* and log-likelihood are expressed by

$$f(x;\theta) = \frac{\exp(-\theta)\theta^x}{x!} I(\theta > 0) + 1 \cdot I(\theta = 0, x = 0)$$

$$L(\theta) = \frac{\exp\left(-n\theta + \log\theta \sum_{i=1}^{n} x_i\right)}{\prod_{i=1}^{n} x_i!} I(\theta > 0) + 1 \cdot I\left(\theta = 0, \sum_{i=1}^{n} x_i = 0\right)$$

If $\Sigma x_i > 0$, there is nothing to prove, since for this case, we have already shown that $\bar{X}$ is MLE. If $\Sigma x_i = 0$, the likelihood reduces to

$$L(\theta) = \exp(-n\theta) I(\theta > 0) + 1 \cdot I(\theta = 0) = \exp(-n\theta) I(\theta \geq 0)$$

$L(\theta)$ is maximized at $\theta = \bar{X} = 0$. Thus $\bar{X}$ is the MLE of θ. Further, by the principle of invariance, $\bar{X}^2$ and $\sin\bar{X}$ are the MLE of θ^2 and $\sin\theta$, respectively.

Further, the regularity conditions A1 to A6 are satisfied. Therefore, the solution of likelihood equation $\hat{\theta}_n = \bar{X}$ is consistent and asymptotically normal, that is

$$\sqrt{n}(\hat{\theta}_n - \theta) \xrightarrow{d} N(0, \sigma^2)$$

where $$\sigma^2 = \frac{1}{E_\theta[(\partial/\partial\theta)\log f(X;\theta)]^2} = \frac{1}{E_\theta[(X/\theta) - 1]^2} = \frac{1}{(1/\theta^2)E_\theta[X - \theta]^2} = \frac{\theta^2}{\theta} = \theta$$

We have $\sigma/\sqrt{n} = \sqrt{\theta/n}$. Thus, we have

$$\sqrt{n}(\hat{\theta}_n - \theta) \xrightarrow{d} N(0, \theta)$$

If we choose a function g such that $g'(\theta) = 1/\sqrt{\theta}$, we get $g(\theta) = 2\sqrt{\theta}$. Thus, the variance gets stabilized

$$\sqrt{n}(\sqrt{\bar{X}} - \sqrt{\theta}) \xrightarrow{d} N(0, 1)$$

(iv) Consider the estimation of $g(\theta) = \exp(-\theta) = P(X = 0)$. Then, the MLE of $\exp(-\theta)$ is $\exp(-\hat{\theta}) = \exp(-\bar{X})$ whenever $\bar{X}$ is an MLE of θ. Further, $g(\hat{\theta}) = \exp(-\bar{X})$ is

$$AN\left[g(\theta), [g'(\theta)]^2 V(\hat{\theta})\right] = AN\left[\exp(-\theta), \frac{\theta\exp(-2\theta)}{n}\right]$$

by delta theorem.

***Example* 6.8** Consider the joint density of $X_1, X_2, \ldots, X_k$

$$f(\mathbf{x};\theta) = \frac{n!}{(n-k)!\,\theta^k} \exp\left(-\frac{\sum_{1}^{k} x_i + (n-k)x_k}{\theta}\right)$$

where $0 \leq x_1 \leq x_2 \leq \cdots \leq x_k$ and $\theta > 0$. Find the MLE of θ and θ^2.

Solution. Assume $u(\mathbf{x}) = \Sigma_1^k x_i + (n - k)x_k$. The log-likelihood is given by

$$\log L(\theta) = \text{const.} - k\log\theta - \frac{u}{\theta}$$

The log likelihood equation $(\partial/\partial\theta)\log L(\theta)$ gives

$$-\frac{k}{\theta} + \frac{u}{\theta^2} = 0;\ \hat{\theta} = \frac{U}{k}$$

which is a unique solution and

$$\left.\frac{\partial^2}{\partial\theta^2}\log L(\theta)\right|_{\hat{\theta}} = \frac{k^3}{u^2} - \frac{2k^3u}{u^3} = -\frac{k^3}{u^2} < 0$$

Hence, $\hat{\theta} = u/k$ is the global maximum of log-likelihood and, therefore, is the MLE of θ. Further, $\hat{\theta}^2$ is the MLE of θ^2 by invariance principle.

***Example* 6.9** Following data is given which relates to a genetical experiment:

Class	*Probability*	*Observed frequency*
AB	$(2 + \theta)/4$	102
Ab	$(1 - \theta)/4$	25
aB	$(1 - \theta)/4$	28
Ab	$\theta/4$	5

Obtain the MoMMCS estimate of θ.

Solution. Modified minimum Chi-square equations are given by

$$\sum \frac{p_i}{f_i}\frac{\partial}{\partial\theta}p_i = 0$$

$$\frac{2+\theta}{4\times 102}\frac{1}{4} - \frac{1-\theta}{4\times 25}\frac{1}{4} - \frac{1-\theta}{4\times 28}\frac{1}{4} + \frac{\theta}{4\times 5}\frac{1}{4} = 0$$

$$\frac{2+\theta}{102} - \frac{1-\theta}{25} - \frac{1-\theta}{28} + \frac{\theta}{5} = 0$$

$$\left(\frac{1}{102} + \frac{1}{25} + \frac{1}{28} + \frac{1}{5}\right)\theta = \frac{1}{25} + \frac{1}{28} - \frac{2}{102}$$

$$\hat{\theta} = 0.1965$$

The method does not provide the variance of the estimator $\hat{\theta}$. However, for large n, the MoMMCS estimates tends to MLE. Hence,

$$V(\hat{\theta}) = \frac{1}{I(\theta)}$$

Here,
$$I(\theta) = E_\theta\left(-\frac{\partial^2}{\partial\theta^2}\log f(\mathbf{X};\theta)\right) = n\sum_{i=1}^{4}\frac{1}{p_i}\left(\frac{\partial p_i}{\partial\theta}\right)^2$$

$$= 160\left[\frac{4}{2+\theta}\cdot\frac{1}{16} + \frac{4}{1-\theta}\cdot\frac{1}{16} + \frac{4}{1-\theta} + \frac{4}{\theta}\cdot\frac{1}{16}\right]$$

$$= 40\left[\frac{1}{2+\theta} + \frac{2}{1-\theta} + \frac{1}{\theta}\right]$$

***Example* 6.10** Let $X_1, X_2, \ldots, X_n$ be *iid* with the *pdf*

$$f(x;\theta) = \begin{cases} \dfrac{\theta}{1-\theta}x^{(2\theta-1)/(1-\theta)}, & 0 < x \le 1, \dfrac{1}{2} < \theta < 1 \\ 0, & \text{otherwise} \end{cases}$$

Find the MLE of θ.

Solution. Consider

$$E_\theta(X) = \frac{\theta}{1-\theta}\int_0^1 x^{[(2\theta-1)/(1-\theta)]+1}\,dx = \frac{\theta}{1-\theta}\int_0^1 x^{[\theta/(1-\theta)]}\,dx = \theta$$

Now, the likelihood function is given by

$$L(\theta;\mathbf{x}) = \left(\frac{\theta}{1-\theta}\right)^n \left(\prod_{i=1}^{n} x_i\right)^{(2\theta-1)/(1-\theta)}$$

$$\log L = n\log\left(\frac{\theta}{1-\theta}\right) + \left(\frac{2\theta-1}{1-\theta}\right)\sum_1^n \log X_i$$

$$\frac{\partial}{\partial\theta}\log L = n\frac{1-\theta}{\theta}\frac{1}{(1-\theta)^2} + \frac{2(1-\theta)+(2\theta-1)}{(1-\theta)^2}\sum_1^n \log X_i$$

$$= \frac{n}{\theta(1-\theta)} + \frac{1}{(1-\theta)^2}\sum_1^n \log X_i$$

$(\partial/\partial\theta)\log L = 0$ implies

$$n(1-\theta) = -\theta\sum_1^n \log X_i$$

$$\hat{\theta} = \frac{n}{n - \sum_1^n \log X_i}$$

We can see that though θ is population mean, $\hat{\theta}$ is not sample mean.

***Example* 6.11** Let $X \sim C(\theta,0)$. Find the MLE of θ.

Solution. We have

$$f(x;\theta)=\frac{\theta}{\pi}\frac{1}{\theta^2+x^2},\quad -\infty<x<\infty, \theta>0$$

$$\log f=\log\theta-\log\pi-\log(\theta^2+x^2)$$

$$\frac{\partial}{\partial\theta}\log f=\frac{1}{\theta}-\frac{2\theta}{\theta^2+x^2}=0$$

$$\theta^2+x^2=2\theta^2 \text{ or } \theta^2=x^2 \text{ or } \hat{\theta}=x$$

***Example* 6.12** Cite an example of several MLEs. Comment on how many of them are consistent when certain regularity conditions are satisfied.

Solution. Consider the *pdf*

$$f(x;\theta)=1+\frac{2x-1}{\sigma}\cos\theta+\frac{3x^2-1}{\sigma}\sin\theta \;\; (0\le x\le 1,\; 0<\theta\le 100\pi)$$

There are at least 50 solutions, at least one between the periods $2\pi r$ and $2\pi(r+1)$, $r = 0, 1,\ldots,49$, since the likelihood function is continuous in θ, periodic with period length 2π. Further, there exists a consistent solution of likelihood equation, since the regularity conditions (6.6.6, A1 to A6) are satisfied. If one of these solutions θ' is consistent for θ_0 (the true value of θ), we have $\theta' \xrightarrow{P} \theta_0$. This implies $\theta' 2\pi r \xrightarrow{P} \theta_0 2\pi r$ for $r = 1,\ldots,49$. Therefore, all other solutions of the likelihood equation corresponding to $r = 1,\ldots,49$ are not consistent; and we conclude that among these solutions, only one is consistent.

***Example* 6.13** A lake contains an unknown number N of fishes from which M are marked with red spots and then released back into the lake. After a while, n fishes are caught again. Of them, there are, suppose, x fishes with red spot. Find the MLE of N.

Solution. The *pmf* of x is given by

$$p(x;N)=\frac{\binom{M}{x}\binom{N-M}{n-x}}{\binom{N}{n}}$$

$$x=0,1,\ldots,\min(n,M),\; N\ge\max(n,M)$$

Thus, the likelihood function is given by

$$L(N,x)=\frac{\binom{M}{x}\binom{N-M}{n-x}}{\binom{N}{n}},\quad N\ge\max(n,M)$$

Since $L(N)$ is a discrete function, we consider the ratios

$$R(N) = \frac{L(N)}{L(N-1)} = \frac{N-n}{N}\frac{N-M}{N-M-n+x}$$

and
$$R(N+1) = \frac{L(N+1)}{L(N)} = \frac{N+1-n}{N+1}\frac{N+1-M}{N+1-M-n+x}$$

$R(N) > 1$ implies that L increases with N and $R(N + 1) < 1$ implies that L decreases as N increases. This gives $R(N) > 1$ if and only if $(N - n)(N - M) > N(N - M - n + x)$ or $nM > Nx$ or $N < nM/x$; and $R(N + 1) < 1$ if and only if $N > (nM/x) - 1$. Therefore, the MLE of N is given by

$$\hat{N} = \left[\frac{nM}{X}\right]$$

where $[Z]$ is the largest integer less than or equal to Z.

***Example* 6.14** Let $X_1, X_2,\ldots,X_n$ be *iid* with common *pdf*

$$f(x;\theta) = \begin{cases} \exp\{-(x-\theta)\}, & x \ge \theta, -\infty < \theta < \infty \\ 0, & \text{otherwise} \end{cases}$$

Find the MLE of θ.

Solution. The likelihood function is given by

$$L(\theta; x) = \prod_{i=1}^{n} f(x_i;\theta) = \begin{cases} \exp\left[-\sum(x_i-\theta)\right], & \text{if } x_i \ge \theta, i = 1,\ldots,n \\ 0, & \text{otherwise} \end{cases}$$

$$= \begin{cases} \exp\left[-\sum(x_i-\theta)\right], & \text{if } x_{(1)} \ge \theta, i = 1,\ldots,n \\ 0, & \text{otherwise} \end{cases}$$

$$\log L = -\sum_{1}^{n}(x_i - \theta)$$

Here, setting $(\partial/\partial\theta)\log L = 0$ does not give MLE, since the likelihood equation gives $n = 0$, which is absurd. Such a value of θ will be an MLE that would maximize $\log L$ or minimize $\Sigma(x_i - \theta)$ subject to $\theta \le x_{(1)}$. $\Sigma_1^n(X_i - \theta)$ will be minimum when each of its term is minimum. This is possible when θ is chosen to be $\hat{\theta} = X_{(1)}$. Therefore, $\hat{\theta} = X_{(1)}$ is an MLE of θ.

***Example* 6.15** Let $X_1, X_2,\ldots,X_n$ be *iid rvs* having Laplace density

$$f(x;\theta) = \frac{1}{2}\exp\left(-|x-\theta|\right),\quad x, \theta \in \mathbb{R}^1$$

Find the MLE of θ.

Solution. Consider the likelihood function

$$L(\theta; \mathbf{x}) = \left(\frac{1}{2}\right)^n \exp\left\{-\sum|x_i - \theta|\right\}$$

$$\log L = -n\log 2 - \sum|x_i - \theta|$$

$$\text{Max } L \equiv \text{Min}\sum|x_i - \theta| = \text{Min } A$$

Let $x_{(1)} < \ldots < x_{(k)} < \cdots < x_{(n)}$ and let $x_{(k)} < \theta x_{(k+1)} > \theta$. Then,

$$A = \sum|x_i - \theta| = -(x_{(1)} - \theta) - \cdots - (x_{(k)} - \theta) + (x_{(k+1)} - \theta) + \cdots + (x_{(n)} - \theta)$$

$$\frac{\partial}{\partial\theta}A = k - (n-k) = 0$$

$\Rightarrow$ $$k = \frac{n}{2}$$

Thus, the MLE of θ is given by $\hat{\theta} = \tilde{X} = \text{median}\{X_1,\ldots,X_n\} = X_{(n/2)}$. For large sample $V(\tilde{X}) = 1/(4nf_1^2)$, $f_1 = 1/2$ or $V(\tilde{X}) = 1/n$. We are not sure whether $\hat{\theta}$ would be consistent, asymptotically normal, and most efficient since $(\partial/\partial\theta)\log f(\mathbf{x};\theta)$ does not exist at 0, therefore, is not differentiable, and the regularity conditions for Theorem 6.6.9 to hold are not satisfied. However,

$$I(\theta) = nE_\theta\left(\frac{\partial}{\partial\theta}\log f(X;\theta)\right)^2$$

We have $$\frac{\partial}{\partial\theta}\log f(x;\theta) = \left(\frac{f'}{f}\right) = \frac{\partial}{\partial\theta}(-|x-\theta|) = \begin{cases}-1, & \text{if } x < \theta \\ +1, & \text{if } x > \theta\end{cases}$$

This gives $$\left(\frac{\partial}{\partial\theta}\log f(x;\theta)\right)^2 = \left(\frac{f'}{f}\right)^2 = 1$$

$\therefore$ $$I_{\mathbf{X}}(\theta) = nI_X(\theta) = n$$

The minimum variance bound of an unbiased estimator is

$$V_\theta(\hat{\theta}) = \frac{1}{I_X(\theta)} = \frac{1}{n}$$

This shows that $\tilde{X}$ is the most efficient.

***Example* 6.16** Let $X_1, X_2,\ldots,X_n$ be *iid* $E(\alpha, \beta)$ with the *pdf*

$$f(x;\alpha,\beta) = \frac{1}{\beta}\exp\left\{-\frac{x-\alpha}{\beta}\right\}I_{(\alpha,\infty)}(x)$$

Find the MLE of α and β.

Solution. The likelihood function is

$$L(\alpha, \beta, \mathbf{x}) = \left(\frac{1}{\beta}\right)^n \exp\left\{-\frac{1}{\beta}\sum(x_i - \alpha)\right\} I_{(\alpha,\infty)}(x_{(1)})$$

For fixed β, L is maximum when α is maximum. This happens when $\hat{\alpha} = X_{(1)}$, which is of course independent of β. Hence, the MLE of α is $\hat{\alpha} = X_{(1)}$. Now,

$$L(\alpha, \beta, \mathbf{x})\big|_{\hat{\alpha}} = \left(\frac{1}{\beta}\right)^n \exp\left\{-\frac{n\bar{x}}{\beta}\right\}\exp\left\{\frac{nx_{(1)}}{\beta}\right\};\ \beta > 0$$

We have $$\frac{\partial}{\partial\beta}\log L = -\frac{n}{\beta} + \frac{n\bar{x}}{\beta^2} - \frac{nx_{(1)}}{\beta^2} = 0$$

$$\hat{\beta} = \bar{X} - X_{(1)}$$

Therefore, the MLEs of α and β are

$$\hat{\alpha} = X_{(1)} \quad \text{and} \quad \hat{\beta} = \bar{X} - X_{(1)}$$

***Example* 6.17** Let $X_1, X_2,\ldots,X_n$ be *iid rvs* with the *pdf*

$$f(x;\theta) = \frac{1}{2}\exp\{-|x-\theta|\}, \quad x, \theta \in \mathbb{R}^1$$

Find the MoM and ML estimators of θ. Compare their efficiency.

Solution. Consider

$$\mu_1' = E_\theta(X) = \frac{1}{2}\int_{-\infty}^{\infty} x\exp\{-|x-\theta|\}\,dx$$

$$= \frac{1}{2}\left\{\int_{-\infty}^{\theta} x\exp(x-\theta)dx + \int_{\theta}^{\infty} x\exp[-(x-\theta)]dx\right\}$$

$$= \frac{1}{2}(\theta - 1 + \theta + 1) = \theta$$

Therefore, the MoM estimator of θ is given by

$$\hat{\theta}_{\text{MoME}} = m_1' = \bar{x}$$

$\hat{\theta}_{\text{MoME}}$ is unbiased since $E(X) = \theta$ and $V(X) = 2$. Therefore, $E(\hat{\theta}_{\text{MoME}}) = \theta$ and $V(\hat{\theta}_{\text{MoME}}) = 2/n$.

In Example 6.7.15, it is shown that the MLE of θ is $\tilde{X}$ = median$\{X_1,\ldots,X_n\}$ = $X_{(n/2)}$. Further, $\tilde{X} \sim N[\theta, 1/(4nf_1^2)]$, where f_1 is the median ordinate. Note that $f_1 = 1/2$ when n is large. Therefore, the asymptotic efficiency of $\tilde{X}$ relative to $\bar{X}$ is

$$\text{ARE}(\tilde{X}, \bar{X}) = \frac{V(\tilde{X})}{V(\bar{X})} = \frac{1/n}{2/n} = \frac{1}{2} < 1$$

Hence, for large n, MLE $\tilde{X}$ is superior to MoME $\bar{X}$.

***Example* 6.18** Let X_1 and X_2 be a random sample from a distribution with the *pdf*

$$f(x;\theta) = \frac{2}{\theta^2}(\theta - x),\ 0 < x < \theta$$

Find the MLE of θ. Based on single observation, show that $2X$ is the MLE of θ but not unbiased.

Solution. Likelihood equation is given by

$$L(\theta; \mathbf{x}) = \left(\frac{2}{\theta^2}\right)^2 (\theta - x_1)(\theta - x_2)\prod_{i=1}^{2} I_{(0,\,\theta)}(x_i)$$

$$= \frac{4}{\theta^4}(\theta - x_1)(\theta - x_2) I_{(0,\theta)}(x_{(n)})$$

$$\log L = \log 4 - 4\log\theta + \log(\theta - x_1) + \log(\theta - x_2)$$

We have likelihood equations

$$\frac{\partial}{\partial\theta}\log L = -\frac{4}{\theta} + \frac{1}{(\theta - x_1)} + \frac{1}{(\theta - x_2)} = 0$$

$$2\theta^2 - 3\theta(x_1 + x_2) + 4x_1x_2 = 0$$

$$\hat{\theta} = \frac{3(x_1 + x_2) + \sqrt{9(x_1 + x_2)^2 - 32x_1x_2}}{4}$$

Based on single observation, the likelihood equation is given by

$$L(\theta; x) = \frac{2}{\theta^2}(\theta - x)\, I_{(0,\,\theta)}(x)$$

$$\log L = \log 2 - 2\log\theta + \log(\theta - x)$$

$$\frac{\partial}{\partial\theta}\log L = -\frac{2}{\theta} + \frac{1}{(\theta - x)} = 0$$

$$\hat{\theta} = 2x$$

Hence, the MLE of θ is $2X$. Further,

$$E_\theta(\hat{\theta}) = E_\theta(2X) = \frac{4}{\theta^2}\int_0^\theta x(\theta - x)dx = \frac{2}{3}\theta \neq \theta$$

Therefore, $2X$ is not unbiased for θ.

***Example* 6.19** Let $X_1, X_2,\ldots,X_n$ be a random sample drawn from $N(\theta,\sigma^2)$, σ^2 is known. Find the MLE of θ and compare it with the sample median $\tilde{X}$, the robust estimator of θ, in terms of ARE.

Solution. The log-likelihood function for θ is

$$\log L = -n\log\left(\sqrt{2\pi}\sigma\right) - \frac{1}{2\sigma^2}\sum(x_i-\theta)^2$$

The likelihood equation

$$\frac{\partial}{\partial\theta}\log L = \frac{1}{\sigma^2}\sum(x_i-\theta) = 0$$

gives $\bar{X} = (1/n)\Sigma X_i$ as the MLE of θ. Using $(\partial^2/\partial\theta^2)\log L = -n/\sigma^2$, $I_X(\theta) = E_\theta\{-(\partial^2/\partial\theta^2)\log L\} = n/\sigma^2$ in Theorem 6.6.9 gives

$$\sqrt{n}(\bar{X}-\theta) \xrightarrow{d} N(0,\sigma^2)$$

Further, a robust estimator of θ is the sample median

$$\tilde{X} = \begin{cases} x_{\left(\frac{n+1}{2}\right)}, & \text{if } n \text{ is odd} \\ \dfrac{x_{\left(\frac{n}{2}\right)} + x_{\left(\frac{n}{2}+1\right)}}{2}, & \text{if } n \text{ is even} \end{cases}$$

Following is the standard result:

$$\sqrt{n}(\tilde{X}-\theta) \xrightarrow{d} N\left(0, \frac{1}{4f_1^2(\theta)}\right)$$

The asymptotic distribution of $\tilde{X}$ in the present example

$$\sqrt{n}(\tilde{X}-\theta) \xrightarrow{d} N\left(0, \frac{\pi\sigma^2}{2}\right)$$

Thus, the ARE of $\bar{X}$ relative to $\tilde{X}$

$$\text{ARE}(\bar{X},\tilde{X}) = \frac{\sigma^2/n}{\pi\sigma^2/2n} = \frac{2}{\pi} < 1$$

implies that $\bar{X}$ is asymptotically superior to $\tilde{X}$. We pay the cost in terms of loosing efficiency by using sample median $\tilde{X}$ as a robust estimator of θ. On the contrary, in the Laplacian model, sample median is superior to sample mean in terms of ARE (see Exercise 6.65).

***Example* 6.20 (Kulldorf).** Let $X_1, X_2,\ldots,X_n$ be a random sample drawn from $N(0,\theta)$, $\theta > 0$. Find the MLE of θ. Is it consistent? Is it asymptotically normal? Find the MLE of $g(\theta) = \theta^2$. Show that the MLE of $g(\theta)$ is asymptotically normal, $AN(\theta^2, 8\theta^4/n)$.

Solution. The likelihood function for θ is

$$L(\theta; \mathbf{x}) = \left(\frac{1}{2\theta\pi}\right)^{n/2} \exp\left\{-\frac{1}{2\theta}\sum x_i^2\right\}$$

$$\log L = c - \frac{n}{2}\log\theta - \frac{1}{2\theta}\sum x_i^2$$

$$\frac{\partial}{\partial\theta}\log L = -\frac{n}{2\theta} + \frac{1}{2\theta^2}\sum x_i^2 = 0$$

$$\hat{\theta} = \frac{1}{n}\sum x_i^2$$

Further, we have $V(X) = E(X^2)$, $U = \Sigma X_i^2/\theta \sim \chi_n^2$, $V(U) = 2n$, $E(U) = n$, $E[(1/n)\Sigma X_i^2] = \theta$, $V[(1/n)\Sigma X_i^2)] = 2\theta^2/n$.

Since $E(\hat{\theta}) = \theta$ and $V(\hat{\theta}) \to 0$ as $n \to \infty$, by Theorem 5.3.1, $\hat{\theta}$ is consistent for θ

$$\hat{\theta} \xrightarrow{P} \theta$$

This shows that the solution of likelihood equation, $\hat{\theta}$, is consistent. By the Kulldorf theorem, $\hat{\theta}$ is asymptotically normal

$$\sqrt{n}(\hat{\theta} - \theta) \xrightarrow{d} N(0, 2\theta^2)$$

Further, for estimating $g(\theta) = \theta^2$, the invariance theorem 6.6.4 gives that the MLE of θ^2 is $g(\hat{\theta}) = \hat{\theta}^2$, where $\hat{\theta} = (1/n)\Sigma X_i^2$ is the MLE of θ. Since $V(\hat{\theta}) = 2\theta^2/n \to 0$ and $\hat{\theta}$ is $AN(\theta, 2\theta^2/n)$, by delta method, $g(\hat{\theta}) = \hat{\theta}^2$ is $AN(g(\theta), [g'(\theta)]^2(2\theta^2/n)) = AN(\theta^2, 8\theta^4/n)$.

***Example* 6.21 (Kulldorf).** Let $X_1, X_2, \ldots, X_n$ be a random sample drawn from $N(\theta, \sigma^2)$, $\theta > 0$. Find (i) the MoME of (θ, σ^2), (ii) the MLE of θ when σ^2 is unknown, (iii) the MLE of σ^2 when θ is known, (iv) the MLE of (θ, σ^2) when both are unknown and (v) the MLE of θ/σ.

Solution.

(i) We have

$$\mu_1' = E(X) = \theta$$
$$\mu_2' = E(X^2) = \theta^2 + \sigma^2$$

Sample moments are $m_1' = (1/n)\Sigma_1^n X_i$ and $m_2' = (1/n)\Sigma_1^n X_i^2$. The moments equations are given by

$$\mu_1' = m_1' \quad \text{or} \quad \theta = m_1'$$

and

$$\mu_2' = m_2' \quad \text{or} \quad \theta^2 + \sigma^2 = m_2'$$

Solving them for θ and σ^2, we have the MoMEs of θ and σ^2 as

$$\hat{\theta} = m_1' = \overline{X}$$

$$\hat{\sigma}^2 = m_2' - (m_1')^2 = \frac{1}{n}\sum_1^n X_i^2 - \overline{X}^2 = \frac{1}{n}\sum(X_i - \overline{X}) = S_n^2$$

(ii) Given σ^2, the likelihood function for θ is given by

$$L(\theta; \mathbf{x}) = \prod_{i=1}^{n} f(x_i, \theta) = \left(\frac{1}{2\pi\sigma^2}\right)^{n/2} \exp\left\{-\frac{1}{2\sigma^2}\sum(x_i - \theta)^2\right\}$$

$$\log L = -\frac{n}{2}\log(2\pi\sigma^2) - \frac{1}{2\sigma^2}\sum(x_i - \theta)^2$$

The likelihood equation for θ is given by

$$\frac{\partial}{\partial\theta}\log L = \frac{1}{2\sigma^2}\sum(x_i - \theta)^2 = 0$$

$$\hat{\theta} = \frac{1}{n}\sum x_i = \overline{x}$$

(iii) The likelihood equation for σ^2 is given by

$$\frac{\partial}{\partial\sigma^2}\log L = \frac{1}{2\sigma^4}\sum(x_i - \theta)^2 - \frac{n}{4\pi\sigma^2}2\pi = 0$$

$$\hat{\sigma}^2 = \frac{1}{n}\sum(x_i - \theta)^2$$

(iv) The likelihood equations for θ and σ^2 are given by

$$\frac{\partial}{\partial\theta}\log L = \frac{1}{\sigma^2}\sum(x_i - \theta) = 0$$

$$\hat{\theta} = \overline{x}$$

$$\frac{\partial}{\partial\sigma^2}\log L = \frac{1}{2\sigma^4}\sum(x_i - \theta)^2 - \frac{n}{2\sigma^2} = 0$$

$$\hat{\sigma}^2 = \frac{1}{n}\sum(x_i - \overline{x})^2 = s_n^2$$

We next verify the second-order condition

$$\frac{\partial}{\partial\theta}\log L = \frac{1}{\sigma^2}\sum(x_i - \theta)$$

$$\frac{\partial^2}{\partial\theta^2}\log L = -\frac{n}{\sigma^2}$$

$$\left(\frac{\partial^2}{\partial\theta\partial\sigma^2}\log L\right) = -\frac{1}{\sigma^4}\sum(x_i - \theta)$$

$$\frac{\partial^2}{\partial(\sigma^2)^2}\log L = -\frac{1}{\sigma^6}\sum(x_i - \theta)^2 + \frac{n}{2\sigma^4}$$

Now, the Hessian matrix is given by

$$\mathbf{H} = \begin{bmatrix} -\dfrac{n}{\sigma^2} & -\dfrac{1}{\sigma^4}\sum(x_i-\theta) \\ -\dfrac{1}{\sigma^4}\sum(x_i-\theta) & \dfrac{n}{2}\left(\dfrac{1}{\sigma^4}-\dfrac{2\sum(x_i-\theta)^2}{n\sigma^6}\right) \end{bmatrix}$$

The value of H at $(\hat{\theta},\ \hat{\sigma}^2)$ is

$$\mathbf{H} = \begin{bmatrix} -\dfrac{n}{s_n^2} & 0 \\ 0 & \dfrac{n}{2}\left(\dfrac{1}{s_n^4}-\dfrac{2}{s_n^4}\right) \end{bmatrix} = \begin{bmatrix} -\dfrac{n}{s_n^2} & 0 \\ 0 & -\dfrac{n}{2s_n^4} \end{bmatrix}$$

The Jacobian of **H** is clearly positive. Therefore, $\bar{X}$ and S_n^2 are the MLEs of θ and σ^2, respectively.

One may proceed differently for calculating the MLEs of θ and σ^2 by using the method of profile likelihood. Consider the minimization of

$$\sum_{i=1}^{n}(x_i-\theta)^2$$

The solution of the equation

$$\frac{\partial}{\partial\theta}\sum_{i=1}^{n}(x_i-\theta)^2 = -\sum_{i=1}^{n}2(x_i-\theta) = 0$$

$\hat{\theta} = \bar{x}$ is unique and it minimizes the above quantity, since

$$\frac{\partial^2}{\partial\theta^2}\sum_{i=1}^{n}(x_i-\theta)^2 = 2n > 0$$

This shows that $\bar{x}$ maximizes

$$\exp\left\{-\frac{1}{2\sigma^2}\sum_{i=1}^{n}(x_i-\theta)^2\right\}$$

irrespective of the value of σ^2. Therefore, $\hat{\theta} = \bar{X}$ is the MLE of θ.

Given that $\bar{X}$ is the MLE of θ, the MLE of σ^2 can be obtained by maximizing the profile likelihood

$$L_p(\sigma^2;\hat{\theta},\mathbf{x}) = \left(\frac{1}{2\pi\sigma^2}\right)^{n/2}\exp\left\{-\frac{1}{2\sigma^2}\sum_{i=1}^{n}(x_i-\bar{x})^2\right\}$$

Maximization of log-profile likelihood

$$\log L_p(\sigma^2; \hat{\theta}, \mathbf{x}) = -\frac{n}{2}\log(2\pi\sigma^2) - \frac{1}{2\sigma^2}\sum_{i=1}^{n}(x_i - \bar{x})^2$$

gives
$$\frac{\partial}{\partial\sigma^2}\log L_p = -\frac{n}{2}\frac{1}{\sigma^2} + \frac{1}{2(\sigma^2)^2}\sum_{i=1}^{n}(x_i - \bar{x})^2 = 0$$

The unique solution of the above equation is

$$\hat{\sigma}^2 = \frac{1}{n}\sum_{i=1}^{n}(x_i - \bar{x})^2$$

This solution maximizes the profile-likelihood since

$$\frac{\partial}{\partial\sigma^2}\log L_p\bigg|_{\hat{\sigma}^2} = \frac{n}{2}\frac{1}{(\hat{\sigma}^2)^2} - \frac{1}{(\hat{\sigma}^2)^3}\sum_{i=1}^{n}(x_i - \bar{x})^2 = \frac{n}{2}\frac{1}{(\hat{\sigma}^2)^2} - \frac{2n\hat{\sigma}^2}{2(\hat{\sigma}^2)^3}$$

$$= -\frac{n}{2(\hat{\sigma}^2)^2} < 0$$

Hence, $\hat{\sigma}^2 = (1/n)\Sigma_{i=1}^{n}(X_i - \bar{X})^2$ is the MLE of σ^2. Thus, $[\bar{X}, (1/n)\Sigma_{i=1}^{n}(X_i - \bar{X})^2] = (\bar{X}, S_n^2)$ is the MLE of (θ,σ^2).

Note that the MLE of σ^2, namely S_n^2 is not the most efficient since its variance $V(\hat{\sigma}^2) = 2(n - 1)\,\sigma^4/n^2$ is not equal to the CR lower bound $2\sigma^4/n$ for the variance of an unbiased estimator of σ^2. Note that $\hat{\sigma}^2$ is not an unbiased estimator of σ^2.

(v) Following Berk (1967), $\boldsymbol{\theta} = (\mu, \sigma^2)'$, $g(\boldsymbol{\theta}) = \mu/\sigma$, $l(\boldsymbol{\theta}) = \mu$, $h(\boldsymbol{\theta}) = [g(\boldsymbol{\theta}), l(\boldsymbol{\theta})] = (\mu/\sigma, \mu) = (\eta, \nu) = \boldsymbol{\xi}$. Note that $l(\boldsymbol{\theta})$ is so chosen that the dimensions of $\boldsymbol{\theta}$ and $\boldsymbol{\xi}$ are the same. The MLE of $h(\boldsymbol{\theta})$ is $h(\hat{\boldsymbol{\theta}}) = (g(\hat{\boldsymbol{\theta}}), l(\hat{\boldsymbol{\theta}})) = (\hat{\mu}/\hat{\sigma}, \hat{\mu})$. Thus, the MLE of $\eta = g(\theta) = \mu/\sigma$ is $\hat{\mu}/\hat{\sigma} = \bar{X}/S$. Same results can, alternatively, be obtained by the invariance principle of MLE due to Zehna.

***Example* 6.22** Let $X_1, X_2,\ldots,X_n$ be a random sample from $N(\theta,\theta^2)$, $\theta \in \mathbb{R}^1$. Find the MLE of θ.

Solution. The likelihood function of θ is given by

$$L(\theta; \mathbf{x}) = \left(\frac{1}{2\pi}\right)^{n/2}\frac{1}{\theta^n}\exp\left\{-\frac{1}{2\theta^2}\sum(x_i - \theta)^2\right\}$$

$$\log L = -\frac{n}{2}\log 2\pi - n\log\theta - \frac{1}{2\theta^2}\left\{\sum x_i^2 - 2\theta\sum x_i + \theta^2\right\}$$

$$= \text{const.} - n\log\theta - \frac{1}{2\theta^2}\sum x_i^2 + \frac{1}{\theta}\sum x_i$$

$$\frac{\partial}{\partial\theta}\log L = -\frac{n}{\theta} + \frac{1}{\theta^3}\sum x_i^2 - \frac{1}{\theta^2}\sum x_i = 0$$

We get $$\theta^2 + \bar{x}\theta - \overline{x^2} = 0$$

$$\hat{\theta} = \frac{-\bar{x} \pm \sqrt{\bar{x}^2 + 4\overline{x^2}}}{2} = -\frac{\bar{x}}{2} \pm \sqrt{\overline{x^2} + \left(\frac{\bar{x}}{2}\right)^2}$$

where $\overline{x^2} = n^{-1}\Sigma x_i^2$. Therefore, the MLE of θ is

$$\hat{\theta} = \begin{cases} -\dfrac{\bar{x}}{2} - \sqrt{\overline{x^2} + \left(\dfrac{\bar{x}}{2}\right)^2}, & \text{if } \bar{x} \le 0 \\ -\dfrac{\bar{x}}{2} + \sqrt{\overline{x^2} + \left(\dfrac{\bar{x}}{2}\right)^2}, & \text{if } \bar{x} \ge 0 \end{cases}$$

***Example* 6.23** Let $X_1, X_2, \ldots, X_n$ be a random sample from a half normal distribution HN(θ,σ^2) with density

$$f(x; \theta, \sigma^2) = \frac{2}{\sqrt{2\pi}\sigma} \exp\left[-\frac{1}{2\sigma^2}(x-\theta)^2\right]$$

where $\sigma > 0$, $x \ge \theta$, $\theta \in \mathbb{R}^1$, and θ and σ^2 are both unknown. Find the MLE of (θ,σ^2).

Solution. The likelihood function of θ is given by

$$L(\theta; \sigma^2) = \text{const.} \prod_{i=1}^{n} I(x_i \ge \theta) \frac{1}{\sigma^n} \exp\left[-\frac{1}{2\sigma^2}\sum_{i=1}^{n}(x_i - \theta)^2\right]$$

$$= \text{const.}\, I(x_{(1)} \ge \theta) \frac{1}{\sigma^n} \exp\left[-\frac{1}{2\sigma^2}\sum_{i=1}^{n}(x_i - \theta)^2\right]$$

where $x_{(1)}$ is the first-order statistic of sample observations $x_1, x_2, \ldots, x_n$.

The maximization of likelihood function $L(\theta)$ for fixed $\sigma^2 > 0$ amounts to the minimization of $\Sigma_{i=1}^n (x_i - \theta)^2$, subject to the constraint $x_{(1)} \ge \theta$. Clearly, this quantity is the minimum when $\hat{\theta} = x_{(1)}$. Hence,

$$\hat{\theta} = X_{(1)}$$

is the MLE of θ. Further, given $\hat{\theta} = X_{(1)}$, the MLE of σ^2 is obtained by maximizing log-profile-likelihood

$$\log L_p(\sigma^2; \hat{\theta}, \mathbf{x}) = \text{const.} - \frac{n}{2}\log\sigma^2 - \frac{1}{2\sigma^2}\sum(x_i - x_{(1)})^2$$

The solution of the corresponding log-profile-likelihood equation for σ^2

$$\frac{\partial}{\partial\sigma^2}\log L_p(\sigma^2; x_{(1)}, \mathbf{x}) = -\frac{n}{2\sigma^2} + \frac{1}{2(\sigma^2)^2}\sum(x_i - x_{(1)})^2$$

$$\hat{\sigma}^2 = \frac{1}{n}\sum(x_i - x_{(1)})^2$$

is the MLE of σ^2 and it is unique as

$$\left.\frac{\partial^2}{\partial(\sigma^2)^2}\log L_p\right|_{\hat{\sigma}^2} = \frac{n}{2(\hat{\sigma}^2)^2} - \frac{1}{(\hat{\sigma}^2)^3}\sum(x_i - x_{(1)})^2$$

$$= \frac{n}{2(\hat{\sigma}^2)^2} - \frac{2n\hat{\sigma}^2}{2(\hat{\sigma}^2)^3} = -\frac{n}{2(\hat{\sigma}^2)^2} < 0$$

Therefore, the MLE of (θ, σ^2) is given by $(X_{(1)}, (1/n)\Sigma(X_i - X_{(1)})^2)$.

***Example* 6.24**

(i) Let $\mathbf{X} \sim N_p(\theta\mathbf{1}, \Sigma)$, if Σ is known. Find the MLE of θ.

(ii) Let $X_1, X_2, \ldots, X_n$ be a random sample from p-variate normal population $N_p(\boldsymbol{\theta}, \Sigma)$. Find the MLEs of $\boldsymbol{\theta}$ and Σ.

Solution.

(i) For single observation, the likelihood function is

$$L(\theta; x, \Sigma) = \frac{1}{(2\pi)^{p/2}|\Sigma|^{1/2}}\exp\left[-\frac{1}{2}(\mathbf{x} - \theta\mathbf{1})'\Sigma^{-1}(\mathbf{x} - \theta\mathbf{1})\right]$$

$$\log L = \text{const.} - \frac{1}{2}(\mathbf{x} - \theta\mathbf{1})'\Sigma^{-1}(\mathbf{x} - \theta\mathbf{1})$$

$$\frac{\partial}{\partial\theta}\log L = (\mathbf{x} - \theta\mathbf{1})'\Sigma^{-1}\mathbf{1} = 0$$

$$\theta\mathbf{1}'\Sigma^{-1}\mathbf{1} = \mathbf{x}'\Sigma^{-1}\mathbf{1}$$

$$\hat{\theta} = \frac{\mathbf{X}'\Sigma^{-1}\mathbf{1}}{\mathbf{1}'\Sigma^{-1}\mathbf{1}}$$

In case X_is are independent with $N(\theta, \sigma_i^2)$, then

$$\Sigma^{-1} = \text{diag}\left(\frac{1}{\sigma_1^2}, \ldots, \frac{1}{\sigma_p^2}\right)$$

(ii) The p-variate normal density is given by

$$f(\mathbf{x}; \boldsymbol{\theta}, \Sigma) = \frac{1}{(2\pi)^{p/2}|\Sigma|^{1/2}}\exp\left[-\frac{1}{2}(\mathbf{x} - \boldsymbol{\theta})'\Sigma^{-1}(\mathbf{x} - \boldsymbol{\theta})\right]$$

The likelihood function is

$$f(\boldsymbol{\theta}, \Sigma; \mathbf{x}_1,\ldots,\mathbf{x}_n) = \frac{1}{(2\pi)^{np/2}|\Sigma|^{n/2}}\exp\left[-\frac{1}{2}\sum_1^n(\mathbf{x}_i-\boldsymbol{\theta})'\Sigma^{-1}(\mathbf{x}_i-\boldsymbol{\theta})\right]$$

$$\log L = \text{const.} - \frac{n}{2}\log|\Sigma| - \frac{1}{2}\sum(\mathbf{x}_i-\boldsymbol{\theta})'\Sigma^{-1}(\mathbf{x}_i-\boldsymbol{\theta})$$

$$\frac{\partial}{\partial\boldsymbol{\theta}}\log L = \sum_1^n \Sigma^{-1}(\mathbf{x}_i-\boldsymbol{\theta}) = 0$$

$$\Sigma^{-1}\boldsymbol{\theta} = \frac{1}{n}\sum_1^n\Sigma^{-1}\mathbf{x}_i = \frac{1}{n}\Sigma^{-1}\mathbf{1}'\mathbf{x}$$

where $\mathbf{x} = (\mathbf{x}_1,\ldots,\mathbf{x}_n)'$. Since Σ^{-1} is nonsingular, multiplying both the sides by Σ, we get

$$\hat{\boldsymbol{\theta}} = \frac{1}{n}\mathbf{1}'\mathbf{X} = \frac{1}{n}\sum_{i=1}^n\mathbf{X}_i$$

Let us write

$$\log L = \text{const.} + \frac{n}{2}\log\left|\Sigma^{-1}\right| - \frac{1}{2}\Sigma(\mathbf{x}_i-\boldsymbol{\theta})'\Sigma^{-1}(\mathbf{x}_i-\boldsymbol{\theta})$$

(Since $|\Sigma||\Sigma^{-1}| = 1$)

$$\frac{\partial}{\partial\Sigma^{-1}}\log L = \frac{n}{2|\Sigma^{-1}|}\text{adj}(\Sigma^{-1}) - \frac{1}{2}\sum_1^n(\mathbf{x}_i-\boldsymbol{\theta})(\mathbf{x}_i-\boldsymbol{\theta})' = \mathbf{0}$$

This implies $\quad \frac{n}{2}(\Sigma^{-1})^{-1} - \frac{1}{2}\sum_1^n(\mathbf{x}_i-\boldsymbol{\theta})(\mathbf{x}_i-\boldsymbol{\theta})' = 0$

$$\hat{\Sigma} = \frac{1}{n}\sum_1^n(\mathbf{x}_i-\hat{\boldsymbol{\theta}})(\mathbf{x}_i-\hat{\boldsymbol{\theta}})'$$

An alternative way of finding the MLE of $(\boldsymbol{\theta},\Sigma)$ is to write log likelihood

$$\log L(\boldsymbol{\theta},\Sigma) = \text{const.} - \frac{n}{2}\log|\Sigma| - \frac{1}{2}\Sigma(\mathbf{x}_i-\boldsymbol{\theta})'\Sigma^{-1}(\mathbf{x}_i-\boldsymbol{\theta})$$

By using the results of multivariate analysis, we can write

$$\Sigma(\mathbf{x}_i-\boldsymbol{\theta})'\Sigma^{-1}(\mathbf{x}_i-\boldsymbol{\theta}) = \text{trace}(\Sigma^{-1}\mathbf{B}) + n(\bar{\mathbf{x}}-\boldsymbol{\theta})'\Sigma^{-1}(\bar{\mathbf{x}}-\boldsymbol{\theta})$$

where $\quad \mathbf{B} = \Sigma(\mathbf{x}_i-\bar{\mathbf{x}})(\mathbf{x}_i-\bar{\mathbf{x}})'$

Thus, the log likelihood is given by

$$\log L(\boldsymbol{\theta},\Sigma) = \text{const.} - \frac{n}{2}\log|\Sigma| - \frac{1}{2}\text{trace}(\Sigma^{-1}\mathbf{B}) - \frac{n}{2}(\bar{\mathbf{x}}-\boldsymbol{\theta})'\Sigma^{-1}(\bar{\mathbf{x}}-\boldsymbol{\theta})$$

A function $g(\mathbf{C})$ with two positive definite matrices $\mathbf{C}$ and $\mathbf{D}$

$$g(\mathbf{C}) = -n\log|\mathbf{C}| - \frac{1}{2}\text{trace}(\mathbf{C}^{-1}\mathbf{D})$$

using the multivariate results can be maximized if

$$\mathbf{C} = \frac{1}{n}\mathbf{D}$$

In the log likelihood, both the matrices Σ and $\mathbf{B}$ are positive definite. Therefore, the term $-(n/2)\log|\Sigma| - (1/2)\text{trace}(\Sigma^{-1}\mathbf{B})$ is maximized when $\Sigma = (1/n)\mathbf{B} = (1/n)\Sigma(\mathbf{x}_i - \bar{\mathbf{x}})(\mathbf{x}_i - \bar{\mathbf{x}})'$. Further, the function $l(\boldsymbol{\theta}) = (\bar{\mathbf{x}} - \boldsymbol{\theta})'\Sigma^{-1}(\bar{\mathbf{x}} - \boldsymbol{\theta}) \geq 0$ since Σ^{-1} is positive definite. $l(\boldsymbol{\theta})$ attains minima or equality if $\boldsymbol{\theta} = \bar{\mathbf{x}}$. This gives the last term in the log likelihood, i.e., $-(n/2)(\bar{\mathbf{x}} - \boldsymbol{\theta})'\Sigma^{-1}(\bar{\mathbf{x}} - \boldsymbol{\theta})$ attains maximum when $\boldsymbol{\theta} = \bar{\mathbf{x}}$. Therefore, the log likelihood is maximized over $(\boldsymbol{\theta},\Sigma)$ when

$$(\hat{\boldsymbol{\theta}}, \hat{\Sigma}) = \left[\bar{X}, \frac{1}{n}\sum_{1}^{n}(X_i - \bar{X})(X_i - \bar{X})'\right]$$

Hence, $(\hat{\boldsymbol{\theta}}, \hat{\Sigma})$ is the MLE of $(\boldsymbol{\theta},\Sigma)$.

***Example* 6.25** Let X be a random variable corresponding to the individual incomes of those who depend on self-employment in some state of India. It is believed that X follows Pareto distribution with the *pdf*

$$f(x;\theta) = \frac{\theta}{x^{\theta+1}}$$

$x \geq 1$, $\theta > 0$. On the basis of n incomes $X_1, X_2, \ldots, X_n$, find a consistent estimator of θ by the method of moments.

Solution. Since $\mu_1' = E(X) = \theta$ does not exist for $0 < \theta < 1$, the method of moments fails in this case. However, if we assume $\theta > 1[\Theta = (1, \infty)]$, then

$$E(X) = \int_1^\infty \frac{x\theta}{x^{\theta+1}}dx = \int_1^\infty \theta x^{-\theta}dx = \mu_1'(\theta) = \frac{\theta}{\theta - 1}$$

Therefore, the moment equation

$$m_1' = \bar{X} = \frac{\theta}{\theta - 1}$$

gives the MoM estimator

$$\hat{\theta} = \frac{\bar{X}}{\bar{X} - 1}$$

which is consistent for θ.

We have seen that if $\Theta = (0, \infty)$, the method of moments fails. In this case, consider the transformation $Y = \log X$. Then,

$$f(y;\theta) = \theta\exp(-\theta y),\ \theta > 0,\ y \geq 0$$

Therefore, by using the results of Example 6.27, the consistent MoME of θ is given by

$$\hat{\theta} = \frac{1}{\bar{Y}} = \frac{n}{\sum \log X_i}$$

***Example* 6.26** Let $X_1, X_2, \ldots, X_n$ be *iid rvs* with the *pdf*

$$f(x;\theta) = \frac{1}{\theta} x^{(1/\theta)-1}$$

$0 < x \le 1$, $\theta > 0$. On the basis of this data, find the MLEs of θ and θ^p for $p \ge 2$.

Solution. The log likelihood gives

$$\log L(\theta) = -n \log \theta + \left(\frac{1}{\theta} - 1\right) \sum \log x_i$$

and the log likelihood equation

$$\frac{\partial}{\partial \theta} \log L(\theta) = -\frac{n}{\theta} - \frac{\sum \log x_i}{\theta^2} = 0$$

gives a unique solution $$\hat{\theta} = -\frac{1}{n} \sum \log X_i$$

so that $$\left.\frac{\partial^2}{\partial \theta^2} \log L(\theta)\right|_{\hat{\theta}} = +\frac{n}{\hat{\theta}^2} - \frac{2n\hat{\theta}}{\hat{\theta}^3} = -\frac{n}{\hat{\theta}^2} < 0$$

Hence, $\hat{\theta}$ is the MLE of θ, and by invariance principle, $\hat{\theta}^p$ is the MLE of θ^p.

***Example* 6.27** Let $X_1, X_2, \ldots, X_n$ be the lifetimes of n bulbs manufactured by some company. We believe that the lifetime follows exponential distribution

(i) $$f(x;\theta) = \theta \exp(-\theta x), \; \theta > 0, \; x > 0$$

Find a consistent estimator of the failure rate (reciprocal of mean lifetime) of bulbs by the MoM.

(ii) $$f(x;\theta) = \frac{\theta^p}{\Gamma(p)} \exp(-\theta x) x^{p-1}, \; x > 0, \theta > 0$$

where p is known. Find the MLE of θ.

(iii) Find the MoM and ML estimators of (θ, p) in (ii), when both θ and p are unknown.

Solution.

(i) Here, $E(X) = \mu_1'(\theta) = 1/\theta$ so that a consistent estimator θ is given by

$$\hat{\theta} = \frac{1}{\bar{X}}$$

(ii) The likelihood function is given by

$$L = \frac{\theta^{np}}{[\Gamma(p)]^n} \exp\left(-\theta \sum x_i\right) \left(\prod x_i\right)^{p-1}$$

$$\log L = np \log \theta - n \log \Gamma(p) - \theta \sum x_i + (p-1) \sum \log x_i$$

The likelihood equation becomes

$$\frac{\partial}{\partial\theta}\log L = \frac{np}{\theta} - \sum x_i = 0$$

$$\hat{\theta} = \frac{p}{\overline{X}}$$

Further, $(\partial^2/\partial\theta^2)\log L = -np/\theta^2$. At $\hat{\theta} = p/\overline{X}$, $(\partial^2/\partial\theta^2)\log L\big|_{\hat{\theta}=p/\hat{x}} = -n\overline{X}^2/p$. This is always less than zero. Hence, $p/\overline{X}$ is an MLE of θ.

(iii) Here, $E(X) = p/\theta$ and $E(X^2) = p(p+1)/\theta^2$. Therefore, moment equations are

$$\mu_1' = m_1'$$

$$\mu_2' = m_2'$$

Now, $p/\theta = m_1'$ and $p(p+1)/\theta^2 = m_2'$. Thus,

$$\frac{p(p+1)/\theta^2}{(p/\theta)^2} = \frac{m_2'}{m_1'^2}$$

$$\frac{(p+1)}{p} = 1 + \frac{1}{p} = \frac{m_2'}{m_1'^2} \quad \text{or} \quad \frac{1}{p} = \frac{m_2'}{m_1'^2} - 1$$

So, the MoM estimators of p and θ are

$$\hat{p} = \frac{m_1'^2}{m_2' - m_1'^2} = \frac{\overline{X}^2}{S^2}$$

and

$$\hat{\theta} = \frac{\hat{p}}{m_1'} = \frac{\overline{X}^2}{S^2}\frac{1}{\overline{X}} = \frac{\overline{X}}{S^2}$$

The likelihood equation is

$$\log L = np\log\theta - n\log\Gamma(p) - \theta\sum x_i + (p-1)\sum\log x_i$$

$$\frac{\partial}{\partial\theta}\log L = \frac{np}{\theta} - \sum x_i = 0$$

$$\frac{\partial}{\partial p}\log L = n\log\theta - n\frac{\partial}{\partial\theta}\log\Gamma(p) + \sum\log x_i = 0$$

Note that for large p, as shown in Example 6.28

$$\frac{\partial}{\partial p}\log\Gamma(p) = \log p - \frac{1}{2p}$$

and let $G = (\Pi x_i)^{1/n}$, We have

$$\log G = \frac{1}{n}\sum\log x_i$$

or

$$n\log G = \sum\log x_i$$

We have $$\frac{\partial}{\partial p}\log L = n\log\theta - n\left[\log p - \frac{1}{2p}\right] + n\log G = 0$$

$$\log\left(\frac{\theta}{p}\right) + \frac{1}{2p} + \log G = 0$$

$$\frac{1}{2p} = -\log G + \log\left(\frac{p}{\theta}\right) = -\log G + \log\overline{x} = \log\left(\frac{\overline{x}}{G}\right)$$

$$\hat{p} = \frac{1}{2\log\left(\frac{\overline{X}}{G}\right)} \tag{6.7.2}$$

This gives $$\hat{\theta} = \frac{1}{2\overline{X}\log\left(\frac{\overline{X}}{G}\right)} \tag{6.7.3}$$

Therefore, the MLEs of θ and p are given by Eqs. (6.7.2) and (6.7.3), respectively.

***Example* 6.28** Let $X_1, X_2, \ldots, X_n$ be a random sample from the population $G_2(\alpha, \beta)$, $\alpha > 0$, $\beta > 0$, both α and β are unknown.

(i) Find the MLE of β assuming α is known.
(ii) Find the MoME and the MLE of α and β assuming both α and β are unknown.

Solution.

(i) The Likelihood function for β, when α is known, is

$$L(\beta; \mathbf{x}) = \frac{\beta^{n\alpha}}{[\Gamma(\alpha)]^n}\prod x_i^{\alpha-1}\exp\{-\beta\sum x_i\};\ x_i > 0, i = 1,\ldots, n$$

$$\log L(\beta; \mathbf{x}) = n\alpha\log\beta - n\log\Gamma(\alpha) + (\alpha - 1)\sum\log x_i - \beta\sum x_i$$

The likelihood equation gives the MLE of β

$$\frac{\partial}{\partial\beta}\log L = 0 = \frac{n\alpha}{\beta} - \sum x_i$$

$$\hat{\beta} = \frac{\alpha}{\overline{X}}$$

We calculate$(\partial^2/\partial^2)\log L = -\,n\alpha/\beta^2 < 0$, since both n and α are positive. Note that $\hat{\beta}$ is the unique point where the partial derivative is 0 and it is a local maximum. Therefore, it is a global maximum. Hence, $\hat{\beta} = \alpha/\overline{X}$ is the MLE of β.

(ii) The moment equations are given by

$$\mu_1' = E(X) = \frac{\alpha}{\beta} = \overline{X}$$

and $$\mu_2' = E(X^2) = \frac{(\alpha+1)\alpha}{\beta^2} = \overline{X^2}$$

On solving these for α and β, we get the MoMEs

$$\hat{\beta} = \frac{\overline{X}}{\overline{X^2} - \overline{X}^2} = \frac{\overline{X}}{S^2}$$

and $$\hat{\alpha} = \overline{X}\hat{\beta} = \frac{\overline{X}^2}{S^2}$$

Therefore, $(\overline{X}^2/S^2, \overline{X}/S^2)$ is the MoME of (α, β).

Now, consider the likelihood function for α and β

$$\log L(\alpha; \beta; \mathbf{x}) = n\alpha \log \beta - n \log \Gamma(\alpha) + (\alpha - 1)\sum \log x_i - \beta \sum x_i$$

It is easy to obtain the likelihood equations for α and β as

$$\frac{\partial}{\partial \beta} \log L = 0 \quad \text{gives} \quad \frac{\alpha}{\beta} - \overline{x} = 0 \tag{6.7.4}$$

and $$\frac{\partial}{\partial \alpha} \log L = 0 \quad \text{gives} \quad \log \alpha - \frac{\Gamma'(\alpha)}{\Gamma(\alpha)} = \log \overline{x} - \frac{1}{n}\sum \log x_i \tag{6.7.5}$$

Equations (6.7.4) and (6.7.5) are not easily solvable for α and β. Therefore, the MLEs of α and β are not obtainable in analytical form. Indeed, some numerical procedure that maximizes L in the two arguments or the use of tables for $\Gamma'(\alpha)/\Gamma(\alpha)$ for solving Eqs. (6.7.4) and (6.7.5) simultaneously may be helpful for calculating the numerical values of $\hat{\alpha}$ and $\hat{\beta}$.

However, it is best to recommend the following: For fixed α, substitute $\hat{\beta}$ by $\alpha/\overline{x}$ in the likelihood function, converting the likelihood into a function of one variable, i.e., of α. Now, maximize this likelihood with respect to α, namely

$$\log L(\alpha; \mathbf{x}) = n\alpha \log\left(\frac{\alpha}{\overline{x}}\right) - n \log \Gamma(\alpha) + (\alpha - 1)\sum \log x_i - \frac{n\alpha \overline{x}}{\overline{x}}$$

$$\frac{\partial}{\partial \alpha} \log L(\alpha; \mathbf{x}) = n \log\left(\frac{\alpha}{\overline{x}}\right) + n\alpha \frac{\overline{x}}{\alpha}\frac{1}{\overline{x}} - n\frac{\Gamma'(\alpha)}{\Gamma(\alpha)} + \sum \log x_i - n = 0$$

$$\log\left(\frac{\alpha}{\overline{x}}\right) - \frac{\Gamma'(\alpha)}{\Gamma(\alpha)} + \frac{1}{n}\sum \log x_i = 0 \tag{6.7.6}$$

Let $$R = \frac{\left(\prod_1^n x_i\right)^{1/n}}{\overline{x}}$$

We have $$\log R = \frac{1}{n}\sum \log x_i - \log \overline{x} \tag{6.7.7}$$

On combining Eq. (6.7.6) with Eq. (6.7.7), we have

$$\log\alpha + \log R - \frac{\Gamma'(\alpha)}{\Gamma(\alpha)} = 0$$

or $$\log\alpha + \log R - \frac{\partial}{\partial\alpha}\log\Gamma(\alpha) = 0 \tag{6.7.8}$$

By Stirling's approximation, for large α, we have

$$\alpha! \cong \sqrt{2\pi}\exp(-\alpha)\alpha^{\alpha+(1/2)}$$

$$\frac{\alpha!}{\alpha} \cong \sqrt{2\pi}\exp(-\alpha)\alpha^{\alpha-(1/2)}$$

$$\log\Gamma(\alpha) \cong \frac{1}{2}\log 2\pi - \alpha + \left(\alpha - \frac{1}{2}\right)\log\alpha$$

$$\frac{\partial}{\partial\alpha}\log\Gamma(\alpha) \cong \log\alpha + \frac{\left(\alpha - \frac{1}{2}\right)}{\alpha} - 1 = \log\alpha - \frac{1}{2\alpha} \tag{6.7.9}$$

Equations (6.7.8) and (6.7.9) give

$$\log R + \frac{1}{2\alpha} = 0$$

$$\hat{\alpha} = \frac{1}{2\log(1/R)} \tag{6.7.10}$$

This gives $$\hat{\beta} = \frac{\hat{\alpha}}{\overline{X}} = \frac{1}{2\overline{X}\log(1/R)}$$

Note. We can easily see that MLEs are not always unique, and are not necessarily unbiased even if they are unique. In terms of mean square error, an MLE may perform poorly.

***Example* 6.29** Consider $X_1, X_2, \ldots, X_n$ as a random sample from $Be(\alpha, \beta)$. Find the MoM estimates for (α, β).

Solution. The *pdf* of $Be(\alpha, \beta)$ is given by

$$f(x; \alpha, \beta) = \frac{1}{B(\alpha, \beta)}x^{\alpha-1}(1-x)^{\beta-1}, \quad \alpha, \beta > 0$$

Moment equations are given by

$$\mu_1' = \frac{\alpha}{\alpha+\beta} = m_1' = \bar{x} \tag{6.7.11}$$

and $$\mu_2' = \frac{\alpha(\alpha+1)}{(\alpha+\beta)(\alpha+\beta+1)} = m_2' = \frac{1}{n}\sum_1^n x_i^2 \tag{6.7.12}$$

Equation (6.7.11) gives

$$\alpha = \frac{m_1'\beta}{1-m_1'}$$

On putting this value of α in Eq. (6.7.12), we get

$$\frac{\left\{\dfrac{m_1'\beta}{(1-m_1')}\right\}\left\{\dfrac{m_1'\beta}{1-m_1'}+1\right\}}{\left\{\dfrac{m_1'\beta}{(1-m_1')}+\beta\right\}\left\{\dfrac{m_1'\beta}{(1-m_1')}+\beta+1\right\}} = m_2'$$

$$\frac{m_1'\beta(m_1'\beta+1-m_1')}{\beta(\beta+1-m_1')} = m_2'$$

$$\beta(m_1'^2 - m_2') = m_1'^2 - m_1' + m_2' - m_1'm_2'$$

The MoME of β is

$$\hat{\beta} = \frac{-m_1'(m_1'-1)+m_2'(m_1'-1)}{(m_2'-m_1'^2)}$$

$$= \frac{(m_2'-m_1')(m_1'-1)}{(m_2'-m_1'^2)}$$

The MoME of α is
$$\hat{\alpha} = \frac{m_1'(m_1'-m_2')}{(m_2'-m_1'^2)}$$

***Example* 6.30** Let $X_1, X_2, \ldots, X_n$ be a random sample from the distribution with the *pdf*

$$f(x;\theta) = \frac{1}{\theta}\exp\left\{-\frac{x}{\theta}\right\},\quad 0 < x < \infty,\ \theta > 0 \tag{6.7.13}$$

(i) Find the MoME and MLE of θ.
(ii) Find the MoME based on the second moment.
(iii) Compare the estimators in (i) and (ii) on the basis of asymptotic variance.
(iv) However, if it is known that k of these observations are less than or equal to M, where M is a fixed positive quantity and $0 \le k \le n$, obtain the MLE of θ.
(v) Let exact values of observations, if they are less than a quantity M, have been recorded as $x_1, x_2, \ldots, x_n$. Moreover, it is also recorded that $N - n$ observations are greater than or equal to M. Using this data, estimate θ by the method of maximum likelihood and also find the standard error of the estimate.

Solution.

(i) The rth order moment is given by

$$\mu_r' = E(X^r) = \frac{1}{\theta}\int_0^\infty x^r \exp\left\{-\frac{x}{\theta}\right\}dx = \frac{\Gamma(r+1)\theta^{r+1}}{\theta} = \Gamma(r+1)\theta^r$$

Thus, we have

$$\mu_1' = E(X) = \Gamma(2)\theta = \theta,\ \mu_2' = E(X^2) = \Gamma(3)\theta^2 = 2\theta^2,\ V(X) = \theta^2$$

The MoME of θ is given by $\hat{\theta} = \bar{X}$. The asymptotic variance of $\bar{X}$ is given by $AV(\bar{X}) = V(X)\left(\frac{\partial}{\partial\theta}\mu(\theta)\right)^{-2} = \theta^2.$

Alternatively, check that the density given in Eq. (6.7.13) is of the form Eq. (6.6.20). Therefore, the MoM estimation would yield the same estimator as given by the ML estimation. We, then, have

$$L(\theta; \mathbf{x}) = \frac{1}{\theta^n}\exp\left\{-\frac{1}{\theta}\sum x_i\right\}$$

$$\log L = -n\log\theta - \frac{1}{\theta}\sum x_i$$

$$\frac{\partial}{\partial\theta}\log L = -\frac{n}{\theta} + \frac{1}{\theta^2}\sum x_i$$

The likelihood equation is

$$\frac{\partial}{\partial\theta}\log L = -\frac{n}{\theta} + \frac{1}{\theta^2}\sum x_i = 0$$

$$\hat{\theta} = \frac{1}{n}\sum x_i = \bar{x}$$

Hence, we see that MoME and MLE are identical. This is the case when the density is of the form Eq. (6.6.20).

(ii) Consider $\mu_2' = 2\theta^2$

This gives that the MoME of θ, $\hat{\theta} = \sqrt{\overline{X^2}/2}$. The asymptotic variance of $\hat{\theta}$ is given by

$$AV\left(\sqrt{\overline{X^2}/2}\right) = V(X^2)\left(\frac{\partial}{\partial\theta}\mu(\theta)\right)^{-2} = V(X^2)(4\theta)^{-2}$$

Note that $g(X) = X^2$, $\mu(\theta) = E[g(X)] = 2\theta^2$, $E(X^4) = \Gamma(5)\theta^4 = 24\theta^4$, and $V(X^2) = 24\theta^4 - 4\theta^4 = 20\theta^4$. Therefore,

$$AV\left(\sqrt{\overline{X^2}/2}\right) = V(X^2)(4\theta)^{-2} = 20\theta^4(4\theta)^{-2} = \frac{5}{4}\theta^2$$

(iii) On comparing the asymptotic variances of MoM estimators in (i) and (ii), we conclude that the MoME in (i) is better.

(iv) Let

$$p = P(X_i \le M) = \int_0^M \frac{1}{\theta} \exp\left\{-\frac{x}{\theta}\right\} dx = 1 - \exp\left\{-\frac{M}{\theta}\right\}$$

$$\theta = \frac{M}{-\log(1-p)}$$

Note that if $\hat{p}$ is the MLE of p, by Theorem 6.6.4, the MLE of θ is given by

$$\hat{\theta} = \frac{M}{-\log(1-\hat{p})}$$

The likelihood function of p is

$$L(p, k) = \binom{n}{k} p^k (1-p)^{n-k}; 0 < p < 1$$

$$\log L = \log\binom{n}{k} + k \log p + (n-k)\log(1-p)$$

$$\frac{\partial}{\partial p} \log L = \frac{k}{p} - \frac{n-k}{1-p} = 0$$

to get $$\hat{p} = \frac{k}{n}$$

Thus, the MLE of θ is given by

$$\hat{\theta} = \frac{M}{\log\left[\dfrac{1}{1-\hat{p}}\right]} = \frac{M}{\log\left[\dfrac{n}{n-k}\right]}$$

(v) The likelihood function in this case is given by

$$L(\theta; \mathbf{x}, N-n) = \prod_{i=1}^{n} \frac{1}{\theta} \exp\left(-\frac{x_i}{\theta}\right) [P(X \ge M)]^{N-n}$$

$$= \left(\frac{1}{\theta}\right)^n \exp\left(-\frac{1}{\theta}\sum x_i\right) [P(X \ge M)]^{N-n}$$

where $$P(X \ge M) = \int_M^\infty \frac{1}{\theta} \exp\left(-\frac{x}{\theta}\right) dx = \exp\left(-\frac{M}{\theta}\right)$$

Hence, the log likelihood function is given by

$$\log L = -n \log\theta - \frac{\sum x_i}{\theta} - \frac{M(N-n)}{\theta}$$

$$\therefore \qquad \frac{\partial}{\partial \theta}\log L = 0$$

gives

$$-\frac{n}{\theta}+\frac{\sum x_i}{\theta^2}+\frac{M(N-n)}{\theta^2}=0$$

or

$$\hat{\theta}=\overline{X}+\frac{M(N-n)}{n}$$

Consider

$$\frac{\partial^2}{\partial \theta^2}\log L=\frac{n}{\theta^2}-\frac{2\sum x_i}{\theta^3}-\frac{2M(N-n)}{\theta^3}$$

$$E_\theta\left[\frac{\partial^2}{\partial \theta^2}\log L\right]=\frac{n}{\theta^2}-\frac{2n\theta}{\theta^3}-\frac{2M(N-n)}{\theta^3}$$

$$=-\frac{n}{\theta^2}-\frac{2M(N-n)}{\theta^3}$$

$$I(\theta)=E_\theta\left[-\frac{\partial^2}{\partial \theta^2}\log L\right]=\frac{n}{\theta^2}+\frac{2M(N-n)}{\theta^3}$$

The asymptotic variance of the MLE is $AV(\hat{\theta}) = 1/I(\theta)$ and the standard error of $\hat{\theta}$ is $1/\sqrt{I(\theta)}$. An estimate of standard error is

$$\widehat{se}(\hat{\theta})=\frac{1}{\sqrt{I(\hat{\theta})}}=\frac{1}{\sqrt{\dfrac{n}{\hat{\theta}^2}+\dfrac{2M(N-n)}{\hat{\theta}^3}}}$$

***Example* 6.31** Let a random sample $X_1, X_2,\ldots,X_n$ be drawn from the lognormal distribution

$$f(x;\mu,\sigma)=\frac{1}{x\sigma\sqrt{2\pi}}\exp\left[-\frac{1}{2\sigma^2}(\log x-\mu)^2\right]$$

where $x > 0$. Find the MoME and MLE of μ and σ^2. Find the MLE of μ^3 when σ^2 is known.

Solution. The kth order population moment of lognormal distribution is given by

$$\mu_k' = E(X^k)=\exp\left(k\mu+\frac{k^2}{2}\sigma^2\right)$$

For $k = 1, 2$, MoM equations

$$m_1'=\exp\left(\mu+\frac{1}{2}\sigma^2\right)$$

and

$$m_2'=\exp(2\mu+2\sigma^2)$$

give

$$\log m_1'-\frac{1}{2}\sigma^2=\mu$$

and $$\frac{1}{2}\log m_2' - \mu = \sigma^2$$

On solving these equations, we get

$$\frac{1}{2}\log m_2' - \log m_1' + \frac{1}{2}\hat{\sigma}^2 = \hat{\sigma}^2$$

$$\log m_2' - 2\log m_1' = \hat{\sigma}^2 = \log \overline{X^2} - \log \overline{X}^2 = \log\left(\frac{\overline{X^2}}{\overline{X}^2}\right)$$

or $$\hat{\sigma}^2 = \log\left(\frac{\overline{X^2}}{\overline{X}^2}\right)$$

On using these estimators, we get

$$\hat{\mu} = \log \overline{X} - \frac{1}{2}(\log \overline{X^2} - \log \overline{X}^2) = 2\log \overline{X} - \frac{1}{2}\log \overline{X^2} = \log\left(\frac{\overline{X}^2}{\sqrt{\overline{X^2}}}\right)$$

Thus, the MoMEs of μ and σ^2 are $\log(\overline{X}^2/\sqrt{\overline{X^2}})$ and $\log(\overline{X^2}/\overline{X}^2)$, respectively.

Consider next the joint density of sample observations

$$f(\mathbf{x};\mu,\sigma^2) = \left(\prod \frac{1}{x_i^2\sigma^2 2\pi}\right)^{1/2} \exp\left\{-\frac{1}{2\sigma^2}\sum(\log x_i - \mu)^2\right\}$$

The log likelihood is given by

$$\log L = -\frac{1}{2}\sum \log(x_i^2\sigma^2 2\pi) - \frac{1}{2\sigma^2}\sum(\log x_i - \mu)^2$$

$$\frac{\partial}{\partial\mu}\log L = \frac{1}{\sigma^2}\sum(\log x_i - \mu) = 0$$

From this, we get a unique solution

$$\hat{\mu} = \frac{1}{n}\sum \log X_i$$

so that $$\left.\frac{\partial^2}{\partial\mu^2}\log L\right|_{\hat{\mu}} = -\frac{n}{\sigma^2} < 0$$

Therefore, $\hat{\mu}$ is the global maximum. Note that the same calculations work when σ^2 is known.

Again, $$\frac{\partial}{\partial\sigma^2}\log L = -\frac{1}{2}\sum \frac{1}{x_i^2\sigma^2 2\pi}x_i^2 2\pi + \frac{1}{2(\sigma^2)^2}\sum(\log x_i - \mu)^2 = 0$$

$$\hat{\sigma}^2 = \frac{1}{n}\sum(\log x_i - \mu)^2$$

Further,

$$\frac{\partial^2}{\partial \mu^2}\log L = -\frac{n}{\sigma^2}$$

$$\frac{\partial^2}{\partial \mu \partial \sigma^2}\log L = -\frac{1}{\sigma^4}\sum \log(x_i - \mu) = -\frac{1}{\sigma^4}\left(\sum \log x_i - n\hat{\mu}\right) = 0$$

$$\frac{\partial^2}{\partial (\sigma^2)^2}\log L = -\frac{n}{2\sigma^4} - \frac{1}{\sigma^6}\sum (\log x_i - \mu)^2$$

Thus, the value of the Hessian matrix **H** at $(\hat{\mu}, \hat{\sigma}^2)$ is

$$\mathbf{H} = \begin{bmatrix} -\dfrac{n}{\hat{\sigma}^2} & 0 \\ 0 & \dfrac{n}{2\hat{\sigma}^4} - \dfrac{n\hat{\sigma}^2}{\hat{\sigma}^6} \end{bmatrix} = \begin{bmatrix} -\dfrac{n}{\hat{\sigma}^2} & 0 \\ 0 & -\dfrac{n}{2\hat{\sigma}^4} \end{bmatrix}$$

$$\mathbf{x}'\mathbf{H}\mathbf{x} = \left(-x_1\frac{n}{\hat{\sigma}^2}, -x_2\frac{n}{2\hat{\sigma}^4}\right)\begin{pmatrix} x_1 \\ x_2 \end{pmatrix} = -x_1^2\frac{n}{\hat{\sigma}^2} - x_2^2\frac{n}{\hat{\sigma}^4} = -n\left[\frac{x_1^2}{\hat{\sigma}^2} + \frac{x_2^2}{\hat{\sigma}^4}\right]$$

$$\leq 0 \ \forall\, \mathbf{x}$$

Therefore,

$$\hat{\mu} = \frac{1}{n}\sum \log X_i$$

and

$$\hat{\sigma}^2 = \frac{1}{n}\sum (\log X_i - \hat{\mu})^2$$

are the likelihood estimators of μ and σ^2, respectively. Further, by invariance principle, $\hat{\mu}^3$ is the MLE of μ^3 when σ^2 is known.

Also, if $X_1, X_2, \ldots, X_n$ are *iid* lognormal (μ, σ^2), then $Y_1, Y_2, \ldots, Y_n = \log X_1, \ldots, \log X_n$ are *iid* $N(\mu, \sigma^2)$. The MLE of (μ, σ^2) can be obtained by using the result on MLE under the transformation in Section 6.6.5.

$$(\hat{\mu}, \hat{\sigma}^2) = \left(\bar{Y}, \frac{1}{n}\sum (Y_i - \bar{Y})^2\right) = \left(\frac{1}{n}\sum \log X_i, \frac{1}{n}\sum\left(\log X_i - \frac{1}{n}\sum \log X_i\right)^2\right)$$

***Example* 6.32** Let $X_1, X_2, \ldots, X_n$ be a random sample drawn from $U(0, \theta)$, $\theta > 0$. Obtain the MLE of θ.

Solution. The likelihood function for θ is

$$L(\theta; \mathbf{x}) = \frac{1}{\theta^n}\prod_{i=1}^{n} I_{(0,\theta)}(x_i) = \frac{1}{\theta^n} I_{(0,\theta)}(x_{(n)})$$

where $x_{(n)}$ is the maximum order statistic. L attains a maximum at $\hat{\theta} = X_{(n)}$. Therefore, the MLE of θ is $\hat{\theta} = X_{(n)}$.

***Example* 6.33** Let $X_1, X_2, \ldots, X_n$ be a random sample drawn from $U(\theta - 1, \theta + 1)$, $\theta > 0$. Obtain the MLE of θ. Are MLEs unique?

Solution. We have

$$f(x;\theta) = \begin{cases} \dfrac{1}{2}, & \theta - 1 < x < \theta + 1 \\ 0, & \text{otherwise} \end{cases}$$

The likelihood function of θ is given by

$$L(\theta;\mathbf{x}) = \begin{cases} \dfrac{1}{2^n}, & \theta - 1 < x_{(1)} < x_{(n)} < \theta + 1 \\ 0, & \text{otherwise} \end{cases}$$

$$= \begin{cases} \dfrac{1}{2^n}, & x_{(n)} - 1 < \theta < x_{(1)} + 1 \\ 0, & \text{otherwise} \end{cases}$$

This shows that likelihood function attains maximum $1/2^n$ for any value of θ lying between $x_{(n)} - 1$ and $x_{(1)} + 1$. Therefore, the MLEs are

$$\hat{\theta} = \alpha(x_{(n)} - 1) + (1 - \alpha)(x_{(1)} + 1) \text{ for } 0 < \alpha < 1$$

This example shows that MLEs need not be unique.

***Example* 6.34** Let $X_1, X_2, \ldots, X_n$ be a random sample drawn from $U(-\theta, \theta)$, $\theta > 0$. Obtain the MLE of θ.

Solution. The *pdf* of X is given by

$$f(x;\theta) = \begin{cases} \dfrac{1}{2\theta}, & -\theta < x < \theta \\ 0, & \text{otherwise} \end{cases}$$

The likelihood function of θ is given by

$$L(\theta;\mathbf{x}) = \begin{cases} \dfrac{1}{(2\theta)^n}, & \text{if } \theta > \max_i |x_i| \\ 0, & \text{otherwise} \end{cases}$$

Note that L is maximum when θ is minimum or $\hat{\theta} = \max_i |x_i|$.

***Example* 6.35** Let $X_1, X_2, \ldots, X_n$ be *iid* $U(-3\theta, 4\theta)$, $\theta > 0$. Obtain the MLE of θ.

Solution. The likelihood function of θ is given by

$$L(\theta;\mathbf{x}) = \begin{cases} \dfrac{1}{(7\theta)^n}, & \dfrac{1}{3}\max_i |x_i| \le \theta \\ 0, & \text{otherwise} \end{cases}$$

Note the L is maximum whenever θ is minimum. Therefore, the MLE of θ is

$$\hat{\theta} = \frac{1}{3}\max_i |x_i|$$

***Example* 6.36** Let $X_1, X_2, \ldots, X_n$ be a random sample from $U(\theta_1, \theta_2)$, $\theta_1 < \theta_2$. Find the MoME and MLE of θ_1 and θ_2.

Solution. MoM equations are

$$\mu_1' = E(X) = \frac{\theta_1 + \theta_2}{2} = m_1'$$

and

$$V(X) = \mu_2' - \mu_1'^2 = \frac{(\theta_2 - \theta_1)^2}{12} = m_2' - m_1'^2$$

We have

$$\theta_2 + \theta_1 = 2m_1'$$

$$\theta_2 - \theta_1 = 2\sqrt{3(m_2' - m_1'^2)}$$

Thus, the required MoMEs are given by

$$\hat{\theta}_1 = m_1' - \{3(m_2' - m_1'^2)\}^{1/2}$$

and

$$\hat{\theta}_2 = m_1' + \{3(m_2' - m_1'^2)\}^{1/2}$$

or

$$\hat{\theta}_1 = \bar{X} - \left\{3\frac{1}{n}\sum(X_i - \bar{X})^2\right\}^{1/2}$$

and

$$\hat{\theta}_2 = \bar{X} + \left\{3\frac{1}{n}\sum(X_i - \bar{X})^2\right\}^{1/2}$$

The likelihood equation is given by

$$L(\theta_1, \theta_2; \mathbf{x}) = \frac{1}{\theta_2 - \theta_1}, \theta_1 < x_{(1)} < x_{(n)} < \theta_2$$

Note that the likelihood is maximum when $(\theta_2 - \theta_1)$ is minimum. Thus, the MLEs of θ_1 and θ_2 are given by $\hat{\theta}_1 = X_{(1)}$ and $\hat{\theta}_2 = X_{(n)}$, respectively.

***Example* 6.37** Consider a random sample $X_1, X_2, \ldots, X_n$ from the population with *pdf*

$$f(x;\theta) = \begin{cases} \dfrac{\alpha}{\theta^\alpha} x^{\alpha-1}, & \text{if } 0 \le x \le \theta, \alpha > 0, 0 \le \theta \le 1 \\ 0, & \text{if } x > \theta \end{cases}$$

Find the MLE of θ and show that it is consistent.

Solution. Note that the support of the distribution depends on the parameter θ. Therefore, the regularity conditions, for Theorem 6.6.9 to hold, are not satisfied. We, thus, cannot use

Theorem 6.6.9 in this case to show that the MLE is asymptotically normal and, therefore, consistent.

The likelihood function is given by

$$L(\theta; \mathbf{x}) = \begin{cases} \dfrac{\alpha^n}{\theta^{n\alpha}} \left(\prod x_i\right)^{\alpha-1}, & \text{if } 0 \le x_{(1)} < x_{(n)} \le \theta \\ & \alpha > 0, 0 \le \theta \le 1 \\ 0, & \text{if } x_i > \theta \end{cases}$$

L will be maximum whenever θ is minimum, which implies $\hat{\theta} = X_{(n)}$. To show that $X_{(n)}$ is consistent consider

$$F_X(x; \theta) = \frac{x^\alpha}{\theta^\alpha}$$

$$f(x_{(n)}; \theta) = n[F(x)]^{n-1} f(x) = \frac{n\alpha}{\theta^{n\alpha}} x^{n\alpha-1}, \quad 0 \le x \le \theta$$

We have

$$E(X_{(n)}) = \frac{n\alpha}{n\alpha + 1}\theta$$

and

$$V\left(\frac{n\alpha + 1}{n\alpha} X_{(n)}\right) = \frac{\theta^2}{(n\alpha + 2)n\alpha}$$

We see that

$$\left[\frac{(n\alpha + 1)}{n\alpha}\right] X_{(n)} \xrightarrow{P} \theta \text{ and } V\left\{\left[\frac{(n\alpha + 1)}{n\alpha}\right] X_{(n)}\right\} \xrightarrow{P} 0 \quad \text{as} \quad n \to \infty$$

Therefore, $[(n\alpha + 1)/n\alpha]X_{(n)}$ or $X_{(n)}$ is consistent estimator for θ.

***Example* 6.38** Let $X_1, X_2, \ldots, X_n$ be a random sample drawn from the logistic model with the *pdf*

$$f(x; \theta) = \frac{\exp\{-(x - \theta)\}}{\{1 + \exp[-(x - \theta)]\}^2}$$

Find the MLE of the location parameter θ.

Solution. Logistic model is a location model. It is difficult to compute MLE for θ in a closed form, since likelihood equation is not easy to solve in this case. Alternatively, we may use Fishers' scoring method to approximate MLE as the most efficient estimator for θ. Note that the Fisher information in location model is constant; in this case, it is $I(\theta) = 1/3$. We can start with any of the two $\sqrt{n}$-consistent estimators, namely sample mean $\bar{X}$ or sample median $\tilde{X}$ in Fishers' scoring method to get asymptotically most efficient estimator

$$\hat{\theta}^{(1)} = \tilde{X} + 3S(\tilde{X})$$

$$= \tilde{X} + 3\nabla L(\theta)\big|_{\theta = \tilde{X}}$$

***Example* 6.39** Let $X_1, X_2, \ldots, X_n$ be a random sample drawn from an extreme-value distribution with the *pdf*

$$f(x; \alpha, \beta) = \alpha \exp\{-\alpha(x-\beta) - \exp[-\alpha(x-\beta)]\}, -\infty < x < \infty$$

This distribution is known as extreme–value distribution. Find the MLEs of α and β.

Solution. The joint density of sample observations is

$$f(\mathbf{x}; \alpha, \beta) = \alpha^n \exp\left\{-\alpha\sum[(x_i-\beta) - \exp[-\alpha(x_i-\beta)]]\right\}$$

$$\log L = n\log\alpha - \alpha\sum(x_i-\beta) - \sum\exp[-\alpha(x_i-\beta)]$$

$$\frac{\partial}{\partial\alpha}\log L = \frac{n}{\alpha} - \sum(x_i-\beta) + \sum(x_i-\beta)\exp[-\alpha(x_i-\beta)]$$

$$\frac{\partial}{\partial\beta}\log L = n\alpha - \sum\alpha\exp[-\alpha(x_i-\beta)]$$

Likelihood equation $(\partial/\partial\beta)\log L = 0$ gives

$$n\alpha - \alpha\sum\exp(-\alpha x_i)\exp(\alpha\beta) = 0$$

$$\exp(-\alpha\beta) = \frac{1}{n}\sum\exp(-\alpha x_i) \tag{6.7.14}$$

Further, $(\partial/\partial\alpha)\log L = 0$ gives

$$\frac{1}{\hat{\alpha}} = \overline{x} - \frac{\sum x_i\exp(-\hat{\alpha}x_i)}{\sum\exp(-\hat{\alpha}x_i)} \tag{6.7.15}$$

Thus, the MLEs of α and β are obtained by solving Eqs. (6.7.14) and (6.7.15).

***Example* 6.40** Let (X, Y) follows bivariate normal distribution with five unknown parameters $\mu_1, \mu_2, \sigma_1^2, \sigma_2^2$, and ρ. Suppose n observations $(X_1, Y_1), \ldots, (X_n, Y_n)$ have been recorded on (X, Y).

(i) Find the MoMEs of $(\mu_1, \mu_2, \sigma_1^2, \sigma_2^2, \rho)$.

(ii) Find the MLEs of $(\mu_1, \mu_2, \sigma_1^2, \sigma_2^2, \rho)$ and show that they are identical as in (i).

Solution.

(i) Bivariate normal *pdf* is given by

$$f[(x,y); \mu_1, \mu_2, \sigma_1^2, \sigma_2^2, \rho] = \frac{1}{2\pi\sigma_1\sigma_2\sqrt{1-\rho^2}}\exp\left\{-\frac{1}{2(1-\rho^2)}\left[\left(\frac{x-\mu_1}{\sigma_1}\right)^2 - 2\rho\left(\frac{x-\mu_1}{\sigma_1}\right)\left(\frac{y-\mu_2}{\sigma_2}\right) + \left(\frac{x-\mu_2}{\sigma_2}\right)^2\right]\right\}$$

$$x, y \in \mathbb{R}^1, \mu_1, \mu_2 \in \mathbb{R}^1, \sigma_1, \sigma_2 > 0, |\rho| < 1$$

Five method of moments equations for X and Y are given by

$$m_1'(X) = \mu_1, \qquad m_2'(X) = \mu_1^2 + \sigma_1^2$$

$$m_1'(Y) = \mu_2, \qquad m_2'(Y) = \mu_2^2 + \sigma_2^2$$

$$m_1'(XY) = \text{cov}(X, Y) + \mu_1 \mu_2 = \rho\sigma_1 \sigma_2 + \mu_1 \mu_2$$
$$\hat{\mu}_1 = m_1'(X)$$
$$\hat{\mu}_2 = m_1'(Y)$$
$$\hat{\sigma}_1^2 = m_1'(X) - m_1'^2(X) = S_X^2$$
$$\hat{\sigma}_2^2 = m_2'(Y) - m_1'^2(Y) = S_Y^2$$
$$\frac{m_1'(XY) - m_1'(X)m_1'(Y)}{S_X S_Y} = \frac{S_{XY}}{S_X S_Y} = \hat{\rho}$$

(ii) The likelihood function is written as
$L[\mu_1, \mu_2, \sigma_1^2, \sigma_2^2, \rho; (x_1, y_1),\ldots,(x_n, y_n)]$

$$= \left(\frac{1}{2\pi\sigma_1\sigma_2\sqrt{1-\rho^2}}\right)^n \exp\left\{-\frac{1}{2(1-\rho^2)\sigma_1^2}\sum x_i^2 + \left[\frac{-1}{2(1-\rho^2)\sigma_2^2}\right]\sum y_i^2\right.$$
$$+\left[\frac{\mu_1}{(1-\rho^2)\sigma_1^2} - \frac{\rho\mu_2}{(1-\rho^2)\sigma_1\sigma_2}\right]\sum x_i + \left[\frac{\mu_2}{(1-\rho^2)\sigma_2^2} - \frac{\rho\mu_1}{(1-\rho^2)\sigma_1\sigma_2}\right]\sum y_i$$
$$\left.+\frac{\rho}{(1-\rho^2)\sigma_1\sigma_2}\sum x_i y_i - \frac{1}{2(1-\rho^2)}\left(\frac{\mu_1^2}{\sigma_1^2} - \frac{2\rho\mu_1\mu_2}{\sigma_1\sigma_2} + \frac{\mu_2^2}{\sigma_2^2}\right)\right\}$$

If we compare this likelihood with the equation given in Eq. (6.6.21), we get $k = 5$, $t_1(x_j, y_j) = x_j$, $t_2(x_j, y_j) = y_j$, $t_3(x_j, y_j) = x_j^2$, $t_4(x_j, y_j) = y_j^2$, and $t_5(x_j, y_j) = x_j y_j$. By the result of Section 6.6.3, MLEs and MoMEs, in this case, are identical. Therefore, setting up the moment equations by

$$\sum_{j=1}^{n} t_i(x_j, y_j) = E_\theta\left[\sum_{j=1}^{n} t_i(X_j, Y_j)\right] = nE_\theta[t_i(X_1, Y_1)]$$

we get the same equations as in part (i). Therefore, the MLEs of these five parameters are the same as the MoMEs obtained in part (i).

Alternatively, the likelihood function may be written as

$$L[\mu_1, \mu_2, \sigma_1^2, \sigma_2^2, \rho; (x_1, y_1),\ldots,(x_n, y_n)] = L_1(\mu_1, \sigma_1^2; \mathbf{x}) \cdot L_2(\boldsymbol{\beta}_\mathbf{x}, \sigma_2^2(1-\rho^2), \mathbf{y}|\mathbf{x})$$

where
$$\beta_{x_i} = E(Y|x_i) = \mu_2 + \rho\frac{\sigma_2}{\sigma_1}(x_i - \mu_1)$$
$$V(Y|x) = \sigma_2^2(1-\rho^2)$$
$$\boldsymbol{\beta}_\mathbf{x} = (\beta_{x_1},\ldots,\beta_{x_n})'$$

$$L_1 = L(\mu_1, \sigma_1^2; \mathbf{x}) = \left(\frac{1}{2\pi\sigma_1^2}\right)^{n/2} \exp\left[-\frac{1}{2\sigma_1^2}\sum(x_i - \mu_1)^2\right]$$

$$L_2 = L(\boldsymbol{\beta}_{\mathbf{x}}, \sigma_2^2(1-\rho^2), \mathbf{y}|\mathbf{x})$$

$$= \left(\frac{1}{2\pi\sigma_2(1-\rho^2)}\right)^{n/2} \exp\left\{-\frac{1}{2\sigma_2^2(1-\rho^2)}\sum\left[y_i - \left(\mu_2 + \rho\frac{\sigma_2}{\sigma_1}(x_i - \mu_1)\right)\right]^2\right\}$$

We know that the estimators of μ_1 and σ_1^2, namely $\bar{X}$ and $S_X^2 = (1/n)\Sigma(X_i - \bar{X})^2$ maximize L_1. We will now check whether these estimators also maximize L_2 for fixed values of μ_2, σ_2^2, and ρ. Let us assume that σ_1^2 is fixed. We have

$$\log L_2 = \text{const.} - \frac{1}{2\sigma_2^2(1-\rho^2)}\sum(y_i - \beta_{x_i})^2$$

$$\frac{\partial}{\partial\mu_1}\log L_2 = \frac{1}{2\sigma_2^2(1-\rho^2)}\sum 2(y_i - \beta_{x_i})\rho\frac{\sigma_2}{\sigma_1} = 0$$

gives

$$\sum(y_i - \mu_2) - \rho\frac{\sigma_2}{\sigma_1}\sum(x_i - \mu_1) = 0$$

or

$$\sum(y_i - \mu_2) = \rho\frac{\sigma_2}{\sigma_1}\sum(x_i - \mu_1) \tag{6.7.16}$$

Similarly, if we write the likelihood by

$$L = L(\mu_2, \sigma_2^2; y)\cdot L[\boldsymbol{\beta}_{\mathbf{y}}, \sigma_1^2(1-\rho^2), x|y]$$

and proceeding similarly, we have

$$\sum(x_i - \mu_1) = \rho\frac{\sigma_1}{\sigma_2}\sum(y_i - \mu_2) \tag{6.7.17}$$

The solutions $\hat{\mu}_1$ and $\hat{\mu}_2$ must satisfy both Eqs. (6.7.16) and (6.7.17). If we assume $\Sigma(x_i - \mu_1) \neq 0$ and $\Sigma(y_i - \mu_2) \neq 0$, we get $\rho = 1$. Therefore, $\Sigma(x_i - \mu_1)$ and $\Sigma(y_i - \mu_2)$ must be 0. Clearly, $\hat{\mu}_1 = \bar{X}$ and $\hat{\mu}_2 = \bar{Y}$. One can easily check that $(\partial^2/\partial\mu_1^2)\log L_2 < 0$. Therefore, $\hat{\mu}_1 = \bar{x}$ maximizes L_2. Further, consider

$$\frac{\partial^2}{\partial\sigma_1^2}\log L_2 = \text{const.}\sum(y_i - \beta_{x_i})\rho\frac{\sigma_2}{\sigma_1}(x_i - \hat{\mu}_1) = 0$$

$$\sum(x_i - \hat{\mu}_1)(y_i - \hat{\mu}_2) - \rho\frac{\sigma_2}{\sigma_1}\sum(x_i - \hat{\mu}_1)^2 = 0$$

$$\sum(x_i - \hat{\mu}_1)(y_i - \hat{\mu}_2) = \rho\frac{\sigma_2}{\sigma_1}\sum(x_i - \hat{\mu}_1)^2 \tag{6.7.18}$$

Similarly, we have

$$\sum(x_i - \hat{\mu}_1)(y_i - \hat{\mu}_2) = \rho\frac{\sigma_1}{\sigma_2}\sum(y_i - \hat{\mu}_2)^2 \tag{(6.7.19)}$$

where $\hat{\mu}_1 = \bar{X}$ and $\hat{\mu}_2 = \bar{Y}$. The estimators $\hat{\sigma}_1^2$ and $\hat{\sigma}_2^2$ are obtained so that they satisfy Eqs. (6.7.18) and (6.7.19). This implies

$$\rho\frac{\sigma_2}{\sigma_1}\sum(x_i - \bar{x})^2 = \rho\frac{\sigma_1}{\sigma_2}\sum(y_i - \bar{y})^2$$

or

$$\frac{\sum(x_i - \bar{x})^2}{\sigma_1^2} = \frac{\sum(y_i - \bar{y})^2}{\sigma_2^2}$$

This implies $\hat{\sigma}_1^2 = c\Sigma(X_i - \bar{X})^2$ and $\hat{\sigma}_2^2 = c\Sigma(Y_i - \bar{Y})^2$. From the fact that $(\bar{x}, S_x^2)$ maximizes L_1, we have $c = 1/n$. This shows that $\bar{x}$ and S_x^2 maximize not only L_1 but also L_2. To find the MLE of ρ, we proceed as follows:

Given the estimators of μ_1, μ_2, σ_1, and σ_2, the likelihood function of ρ is

$$\log L(\rho; \bar{x}, \bar{y}, S_x^2, S_y^2, \mathbf{x}, \mathbf{y})$$

$$= -\frac{n}{2}\log(1-\rho^2)$$

$$-\frac{1}{2(1-\rho^2)}\sum\left\{\frac{(x_i - \bar{x})^2}{S_x^2} - 2\rho\left(\frac{x_i - \bar{x}}{S_x}\right)\left(\frac{y_i - \bar{y}}{S_y}\right) + \left(\frac{y_i - \bar{y}}{S_y^2}\right)^2\right\}$$

$$= -\frac{n}{2}\log(1-\rho^2) - \frac{1}{2(1-\rho^2)}\left\{2n - 2\rho\sum\frac{(x_i - \bar{x})(y_i - \bar{y})}{S_x S_y}\right\}$$

Since $S_x^2 = (1/n)\Sigma(x_i - \bar{x})^2$ and $S_y^2 = (1/n)\Sigma(y_i - \bar{y})^2$, we get

$$\log L = -\frac{n}{2}\log(1-\rho^2) - \frac{n}{(1-\rho^2)} + \frac{\rho}{(1-\rho^2)}B$$

where $B = \Sigma(x_i - \bar{x})(y_i - \bar{y})/S_x S_y$. Further, differentiating with respect to ρ gives

$$\frac{\partial}{\partial\rho}\log L = n\frac{\rho}{(1-\rho^2)} - \frac{2n\rho}{(1-\rho^2)^2} + (1-\rho^2) + \frac{2\rho^2}{(1-\rho^2)^2}B = 0$$

$$\frac{(1+\rho^2)B - 2n\rho + n\rho(1-\rho^2)}{(1-\rho^2)^2} = 0$$

$$\frac{(1+\rho^2)B - n\rho - n\rho^3}{(1-\rho^2)^2} = 0$$

$$(1+\rho^2)B - n\rho(1+\rho^2) = 0$$

$$\hat{\rho} = \frac{B}{n} = \frac{1}{n}\sum \frac{(x_i - \bar{x})(y_i - \bar{y})}{S_x S_y}$$

***Example* 6.41** Let $X_1, \ldots, X_n$ be a random sample from a distribution with the *pdf*

$$f(x;\theta) = \theta(\theta + 1)\, x^{\theta-1}\,(1 - x),\ 0 < x < 1,\ \theta > 0$$

Find the (i) MoME of θ and (ii) the MLE of θ.

Solution.

(i) The estimating equation is

$$\mu_1' = E(X) = \theta(\theta+1)\int_0^1 x^{\theta+1-1}(1-x)^{2-1}\,dx$$

$$= \theta(\theta+1)\frac{\Gamma(\theta+1)\Gamma(2)}{\Gamma(\theta+3)} = \frac{\theta}{\theta+2}$$

Therefore, the MoME of θ is given by

$$\hat{\theta} = \frac{2m_1'}{1-m_1'},\ \text{where}\ m_1' = \frac{1}{n}\sum_1^n x_i$$

(ii) The likelihood function is

$$L(\theta, \mathbf{x}) = \{\theta(\theta+1)\}^n \left\{\prod x_i\right\}^{\theta-1} \prod(1-x_i)$$

$$\log L = n\log\theta + n\log(\theta+1) + (\theta-1)\sum\log x_i + \sum\log(1-x_i)$$

$$\frac{\partial}{\partial\theta}\log L = \frac{n}{\theta} + \frac{n}{\theta+1} + \sum\log x_i$$

The likelihood equation

$$\frac{\partial}{\partial\theta}\log L = 0$$

gives
$$\frac{2\theta+1}{\theta(\theta+1)} = -\log G \text{ where } G = \left(\prod x_i\right)^{1/n}$$

$$(\log G)\theta^2 + (2+\log G)\theta + 1 = 0$$

$$\theta = \frac{-(2+\log G) \pm \sqrt{(\log G)^2 + 4}}{2\log G}$$

The MLE of θ is

$$\hat{\theta} = \frac{-(2+\log G) + \sqrt{(\log G)^2 + 4}}{2\log G}$$

since $\theta > 0$.

***Example* 6.42** Let $X_1, X_2, \ldots, X_n$ be a random sample from $C(1, \theta)$. Find the MLE of θ.

Solution. Note that $C(1, \theta)$ is a location model since $f(x;\theta) = g(x - \theta)$, where $g(x) = (1/\pi)\{1/[1 + (x - \theta)^2]\}$. The likelihood function is given by

$$L(\theta; \mathbf{x}) = \frac{1}{\pi^n} \prod_{i=1}^{n} \frac{1}{[1 + (x_i - \theta)^2]}$$

$$\log L = -n \log \pi - \sum_{1}^{n} \log\{1 + (x_i - \theta)\}^2$$

$$\frac{\partial}{\partial \theta} \log L = \sum_{1}^{n} \frac{2(x_i - \theta)}{\{1 + (x_i - \theta)^2\}} = 0 \Rightarrow \sum_{1}^{n} \frac{(x_i - \theta)}{\{1 + (x_i - \theta)^2\}} = 0$$

This happens if and only if

$$\sum_{1}^{n} (x_i - \theta) \prod_{j \neq i} \{1 + (x_j - \theta)^2\} = 0$$

The above equation has $2n - 1$ roots, since the left-hand side is the polynomial in θ of degree $2(n - 1) + 1 = 2n - 1$. Indeed, the likelihood equation is difficult to solve. Therefore, we use Fisher's scoring method of obtaining MLE. The Fisher information in a single observation is given by

$$I_{X_1}(\theta) = \int_{-\infty}^{\infty} \frac{4(x - \theta)^2}{\pi\{1 + (x - \theta)^2\}^3} dx$$

Let $x - \theta = \tan y$. We have

$$I_{X_1}(\theta) = \frac{4}{\pi} \int_{-\pi/2}^{\pi/2} \frac{\tan^2 y \sec^2 y}{\{1 + \tan^2 y\}^3} dy = \frac{4}{\pi} \int_{-\pi/2}^{\pi/2} \cos^2 y \sin^2 y \, dy$$

$$= \frac{8}{\pi} \int_{0}^{\pi/2} \cos^2 y \sin^2 y \, dy = \frac{1}{2}$$

using the result

$$\int 2 \cos^{2m-1} x \sin^{2n-1} x \, dx = B(m, n)$$

In fact, the Fisher information for location models is always constant. Thus, the Fisher information in the sample is $I_{\mathbf{X}}(\theta) = n/2$. The score function is $S(\mathbf{X};\theta) = 2\Sigma_1^n [x_i - \theta)/[1 + (X_i - \theta)^2]]$. One can use θ_0, the trial value, as median of sample observations $x_1, \ldots, x_n$, $\tilde{\theta} = F_n(1/2)$; and apply the iteration procedure given in Eq. (6.6.93) to obtain the MLE. We get 1-step estimator [Eq. (6.6.95)]

$$\hat{\theta} = \tilde{\theta} + \hat{I}_{\mathbf{X}}(\tilde{\theta})^{-1} S(\mathbf{X}, \theta)$$

$$= \tilde{\theta} + \frac{4}{n} \sum_{1}^{n} \frac{X_i - \tilde{\theta}}{\{1 + (X_i - \tilde{\theta})^2\}}$$

Moreover, for large n, $V(\tilde{\theta}) \cong 2/n$.

***Example* 6.43** In a genetical experiment, there are four phenotype-classes, namely *AB*, *Ab*, *aB*, *ab*. The following data gives the probabilities and observed frequencies of these classes. Find the MLE of θ and its standard error.

Phenotype classes	*Probabilities* (π_j)	*Observed frequencies* (n_j)
AB	$(2 + \theta)/2$	102
Ab	$(1 - \theta)/4$	25
aB	$(1 - \theta)/4$	28
ab	$\theta/4$	5

Solution. The likelihood function for θ given the observations is

$$L = c\left(\frac{2+\theta}{4}\right)^{102} \cdot \left(\frac{1-\theta}{4}\right)^{25} \cdot \left(\frac{1-\theta}{4}\right)^{28} \cdot \left(\frac{\theta}{4}\right)^{5}$$

$$= c \cdot (2+\theta)^{102} (1-\theta)^{53} \theta^5$$

$$\log L = \log c + 102 \log (2 + \theta) + 53 \log (1 - \theta) + 5 \log \theta$$

$$\frac{\partial}{\partial\theta} \log L = \frac{102}{2+\theta} - \frac{53}{1-\theta} + \frac{5}{\theta} = 0$$

$$102\theta(1 - \theta) - 53(2 + \theta)\theta + 5(2 + \theta)(1 - \theta) = 0$$

$$160\theta^2 + 9\theta - 10 = 0$$

Solve for θ to get

$$\hat{\theta} = \frac{-9 \pm \sqrt{81 + 4 \times 160 \times 10}}{2 \times 160}$$

$$= \frac{-9 \pm \sqrt{6481}}{320} = 0.2235$$

Here, $I(\theta) = n\Sigma_{j=1}^{k}(1/\pi_j)[(\partial/\partial\theta)\pi_j]^2$.

π_j	$\frac{\partial}{\partial\theta}\pi_j$	$\left.\frac{\partial}{\partial\theta}\pi_j\right\|_{\theta=0.2235}$
$(2 + \theta)/4$	1/4	1/4
$(1 - \theta)/4$	−1/4	−1/4
$(1 - \theta)/4$	−1/4	−1/4
$\theta/4$	1/4	1/4

$$I(\theta) = 160\left(\frac{1}{\pi_1}\cdot\frac{1}{16}+\frac{1}{\pi_2}\cdot\frac{1}{16}+\frac{1}{\pi_3}\cdot\frac{1}{16}+\frac{1}{\pi_4}\cdot\frac{1}{16}\right)$$

$$= 10\left(\frac{4}{2+\theta}+\frac{4}{1-\theta}+\frac{4}{1-\theta}+\frac{4}{\theta}\right)$$

$$= 40\left(\frac{1}{2+\theta}+\frac{2}{1-\theta}+\frac{4}{\theta}\right)$$

$$I(\theta)|_{\theta=0.2235} = 40(0.4497 + 2.5757 + 17.8971)$$

$$= 836.9$$

Standard error of $\hat{\theta} = \sqrt{1/I(\hat{\theta})} = \sqrt{1/836.9} = 0.0346$

***Example* 6.44**

(i) Assume that $X_1,\ldots,X_n$ is a random sample drawn from a truncated Poisson population truncated from below. Obtain the MLE of θ.

(ii) Consider the following data:

No. of gall cells x in a flower	1	2	3	4	5	6	7	8	9	10
No. of flower heads: f_x	287	272	196	79	29	20	2	0	1	0

Assuming a truncated Poisson distribution, estimate the parameter by using Fisher's scoring method and estimate its standard error.

Solution.

(i) Here, the *pdf* is given by

$$f(x;\theta) = \frac{(\exp(-\theta)\theta^x)}{(1-\exp(-\theta))x!}, \quad x = 1, 2, \ldots, \theta > 0$$

$$= \frac{1}{\exp(\theta)-1}\frac{\theta^{x_i}}{x_i!}$$

The likelihood function is

$$L(\theta;\mathbf{x}) = \frac{1}{(\exp(\theta)-1)^n}\cdot\frac{\prod\theta^{x_i}}{\prod x_i!}$$

$$\log L = -n\log(\exp(\theta)-1) + \sum x_i \log\theta - \sum \log x_i!$$

$$\frac{\partial}{\partial\theta}\log L = -\frac{n\exp(\theta)}{\exp(\theta)-1} + \frac{1}{\theta}\sum x_i \tag{6.7.20}$$

$$\frac{\partial^2}{\partial\theta^2}\log L = -\frac{1}{\theta^2}\sum x_i - \frac{[n\exp(\theta)(\exp(\theta)-1) - n\exp(\theta)\cdot\exp(\theta)]}{(\exp(\theta)-1)^2}$$

$$= -\frac{1}{\theta^2}\sum x_i + \frac{n\exp(\theta)}{(\exp(\theta)-1)^2}$$

$$E\left[\frac{\partial^2}{\partial\theta^2}\log L\right] = -\frac{1}{\theta^2}\sum E(X_i) + \frac{n\exp(\theta)}{(\exp(\theta)-1)^2}$$

Here, $$E(X) = \sum_{i=1}^{\infty}\frac{x\theta^x}{(\exp(\theta)-1)x!} = \frac{\theta}{\exp(\theta)-1}\sum_{i=1}^{\infty}\left(\frac{x\theta^{x-1}}{x!}\right)$$

$$\frac{\theta}{(\exp(\theta)-1)}\sum_{i=1}^{\infty}\frac{\theta^{(x-1)}}{(x-1)!} = \frac{\theta}{\exp(\theta)-1}\sum_{i=0}^{\infty}\frac{\theta^x}{x!} = \frac{\theta\exp(\theta)}{\exp(\theta)-1}$$

We then have

$$E\left[\frac{\partial^2}{\partial\theta^2}\log L\right] = -\frac{n\exp(\theta)}{\theta(\exp(\theta)-1)} + \frac{n\exp(\theta)}{(\exp(\theta)-1)^2}$$

$$= \frac{n\exp(\theta)}{(\exp(\theta)-1)}\left[\frac{1}{(\exp(\theta)-1)} - \frac{1}{\theta}\right]$$

$$\therefore \quad I(\theta) = E\left[-\frac{\partial^2}{\partial\theta^2}\log L\right] = \frac{n\exp(\theta)}{\exp(\theta)-1}\left(\frac{1}{\theta} - \frac{1}{\exp(\theta)-1}\right) \tag{6.7.21}$$

By taking θ_0 as a trial value, correction factor $h^{(1)} = S(\theta_0)/I(\theta_0)$ is calculated to get a further improved estimator $\hat{\theta}^{(1)} = \theta_0 + h^{(1)}$. This process is repeated till the convergence of desired level is achieved.

(ii) We consider the trial value

$$\theta_0 = \frac{\sum_1^{10} x f_x}{\sum_1^{10} f_x} = \frac{2023}{886} = 2.2833$$

$$h = \frac{S(\theta)}{I(\theta)} = \frac{\frac{1}{\theta}\sum x f_x - \frac{n\exp(\theta)}{(\exp(\theta)-1)}}{\frac{n\exp(\theta)}{(\exp(\theta)-1)}\left[\frac{1}{\theta} - \frac{1}{(\exp(\theta)-1)}\right]} = \frac{\frac{\exp(\theta)-1}{\theta\exp(\theta)}\left(\frac{\sum x f_x}{n}\right) - 1}{\frac{1}{\theta} - \frac{1}{(\exp(\theta)-1)}}$$

$$h^{(1)} = \frac{\dfrac{9.809-1}{2.2833\times 9.809}\times 2.2833-1}{\dfrac{1}{2.2833}-\dfrac{1}{9.809-1}} = \frac{\dfrac{8.809}{9.809}-1}{0.438-0.114} = \frac{0.898-1}{0.324}$$

$$= -\frac{0.102}{0.324} = -0.315$$

$$\hat{\theta}(1) = \theta_0 + h^{(1)} = 2.2833 - 0.315 = 1.968$$

$$h^{(2)} = \frac{\dfrac{7.156-1}{1.968\times 7.156}\times 2.2883-1}{\dfrac{1}{1.968}-\dfrac{1}{6.156}}$$

$$= \frac{0.998-1}{0.346} = -0.006$$

$$\hat{\theta}^{(2)} = \hat{\theta}^{(1)} + h^{(2)} = 1.968 - 0.006$$

$$= 1.962$$

Further, the estimate of $V(\hat{\theta}) = 1/I(\hat{\theta}) = \{[(886\times 7.114)/6.114][(1/1.962) - (1/6.114)]\}^{-1}$ $= 0.003$, and the estimate of *s.d.* is $\sqrt{0.003} = 0.055$.

***Example* 6.45** In a human population, the probability distribution of blood groups *O*, *A*, *B*, *AB* and their observed frequencies in a sample of 615 Hindus from Punjab in India is given in the following table:

Blood group	*Theoretical frequency*	*Observed frequency*
O	r^2	199
A	$p^2 + 2pr$	137
B	$q^2 + 2qr$	213
AB	$2pq$	66
		Total: 615

Obtain the MLE of the parameters and estimate their standard error.

Solution. Here, we have two independent parameters, p and q, so that $r = 1 - p - q$; $n = 615$; $\pi_1 = r^2$, $\pi_2 = p^2 + 2pr$, $\pi_3 = q^2 + 2qr$, $\pi_4 = 2pq$, $n_1 = 199$, $n_2 = 137$, $n_3 = 213$, $n_4 = 66$.

We calculate $$S(p) = \sum_{j=1}^{k} \frac{n_j}{\pi_j}\frac{\partial}{\partial p}\pi_j;\quad S(q) = \sum_{j=1}^{k} \frac{n_j}{\pi_j}\frac{\partial}{\partial q}\pi_j$$

$$I_{pp} = n\sum_{j=1}^{k} \frac{1}{\pi_j}\left(\frac{\partial}{\partial p}\pi_j\right)^2;\quad I_{qq} = n\sum_{j=1}^{k} \frac{1}{\pi_j}\left(\frac{\partial}{\partial q}\pi_j\right)^2$$

$$I_{pq} = \sum_{j=1}^{k} \frac{1}{\pi_j}\left(\frac{\partial}{\partial p}\pi_j \frac{\partial}{\partial q}\pi_j\right)$$

These quantities can be calculated from the following table:

Blood groups	π_j	$(\partial/\partial p)\pi_j$	$(\partial/\partial q)\pi_j$
O	$(1 - p - q)^2$	$-2(1 - p - q) = -2r$	$-2(1 - p - q) = -2r$
A	$p^2 + 2p(1 - p - q)$	$2p + 2 - 4p - 2q = 2r$	$-2p$
B	$q^2 + 2q(1 - p - q)$	$-2q$	$2q + 2 - 4q - 2p = 2r$
AB	$2pq$	$2q$	$2p$

The first trial values of p and q can be obtained from the equations

$$n_i = n\pi_i$$
$$n_1 = nr^2$$
$$n_2 = n(p^2 + 2pr)$$
$$n_3 = n(q^2 + 2qr)$$
$$n_4 = 2npq$$
$$n_1 + n_3 = n(r + q)^2 = n(1 - p)^2$$
$$(1 - p) = \sqrt{\frac{n_1 + n_3}{n}};\ p = 1 - \sqrt{\frac{n_1 + n_3}{n}}$$
$$n_1 + n_2 = n(r + p)^2 = n(1 - q)^2$$
$$q = 1 - \sqrt{\frac{n_1 + n_2}{n}}$$

Therefore, the initial values of p and q can be taken as

$$p^{(0)} = 1 - \sqrt{\frac{n_1 + n_3}{n}} = 1 - \sqrt{\frac{199 + 213}{615}} = 0.1815$$

and

$$q^{(0)} = 1 - \sqrt{\frac{n_1 + n_2}{n}} = 1 - \sqrt{\frac{199 + 137}{615}} = 0.2608$$

$$r^{(0)} = 1 - p_0 - q_0 = 0.5577$$

We construct the following table:

Blood groups	n_j	$\pi_j(0)$	$(\partial/\partial p)\pi_j$	$(\partial/\partial q)\pi_j$
O	199	0.3110	–1.1154	–1.1154
A	137	0.2354	1.1154	–0.363
B	213	0.3589	– 0.5216	1.1154
AB	66	0.0947	0.5216	0.3630

$$S(p) = -10.3726;\ S(q) = -9.9071$$
$$I_{pp} = 7944.116;\ I_{qq} = 5792.079;\ I_{pq} = 1635.231$$

$$[\mathbf{I}(\boldsymbol{\theta}^{(0)})]^{-1} = \begin{bmatrix} 0.000133646 & -3.7731E-05 \\ -3.7731E-05 & 0.000183302 \end{bmatrix}$$

i.e.,
$$[\mathbf{I}(\boldsymbol{\theta}^{(0)})]^{-1}\mathbf{S}(\boldsymbol{\theta}^{(0)}) = \begin{bmatrix} -0.0010 \\ -0.0014 \end{bmatrix}$$

$$\boldsymbol{\theta}^{(1)} = \boldsymbol{\theta}^{(0)} + [I(\boldsymbol{\theta}^{(0)})]^{-1}\, S(\boldsymbol{\theta}^{(0)})$$
$$= \begin{bmatrix} 0.1815 \\ 0.2608 \end{bmatrix} + \begin{bmatrix} -0.0010 \\ -0.0014 \end{bmatrix} = \begin{bmatrix} 0.1805 \\ 0.2594 \end{bmatrix}$$
$$r^{(1)} = 1 - p^{(1)} - q^{(1)} = 0.5601$$

This completes the first iteration. Proceeding on to the second iteration, we construct the following table:

Blood groups	n_j	$\pi_j^{(1)}$	$(\partial/\partial p)\pi_j$	$(\partial/\partial q)\pi_j$
O	199	0.3138	–1.1203	–1.12027
A	137	0.2348	1.1203	–0.36098
B	213	0.3578	–0.5188	1.12027
AB	66	0.0936	0.5188	0.36098

$$S(p) = 0.09397;\quad S(q) = 0.08786;\quad I_{pp} = 7977.69;\quad I_{qq} = 5814.12;\quad I_{pq} = 1631.91$$

$$[\mathbf{I}(\boldsymbol{\theta}^{(1)})]^{-1} = \begin{bmatrix} 0.00013299 & -3.73263E-05 \\ -3.7326E-05 & 0.000182472 \end{bmatrix}$$

$$[\mathbf{I}(\boldsymbol{\theta}^{(1)})]^{-1}\mathbf{S}(\boldsymbol{\theta}^{(1)}) = \begin{bmatrix} 0.0000 \\ 0.0000 \end{bmatrix}$$

$\Rightarrow$
$$\boldsymbol{\theta}^{(2)} = \boldsymbol{\theta}^{(1)} + [\mathbf{I}(\boldsymbol{\theta}^{(1)}]^{-1}\,\mathbf{S}(\boldsymbol{\theta}^{(1)})$$
$$= \begin{bmatrix} 0.1805 \\ 0.2594 \end{bmatrix}$$

Finally, we obtain a consistent value of $\boldsymbol{\theta}$ in two iterations. Further,

$$V(\hat{p}) = I^{pp} = 0.00013299$$
$$V(\hat{q}) = I^{qq} = 0.000182472$$

and
$$V(\hat{r}) = V(1 - \hat{p} - \hat{q}) = V(\hat{p}) + V(\hat{q}) + \text{cov}(\hat{p}, \hat{q})$$
$$= 0.0003$$

EXERCISES

1. When would you use minimum chi-square method of estimation? Describe the methods of minimum chi-square and modified minimum chi-square and compare them with maximum likelihood method.
2. Prove that if $\hat{\theta}$ is MLE of θ and $\varphi(\theta)$ is one-to-one function of θ, then $\varphi(\hat{\theta})$ is the maximum likelihood estimator for $\varphi(\theta)$.
3. State the condition under which maximum likelihood estimators (MLEs) are identical with those given by method of moments.
4. Discuss the consistency of the method of moments estimator of population mean when it is finite.
5. Let $X_1, X_2, \ldots, X_n$ be *iid* random variables with common *cdf* given by F, where F is defined on $\mathbb{R}^1$. Assuming that the observations $x_1, x_2, \ldots, x_n$ are all different, find the MLE of F.

   Hint. Sample *df* S_n is the MLE of F.
6. Find the MLE of θ for the exponential family of distributions

$$f(x;\theta) = \exp\{c(\theta) + Q(\theta)T(x) + H(x)\},\ \theta \in \Theta \subseteq \mathbb{R}^1$$

   and prove that there exists a solution of the likelihood function if and only if $T(x)$ and $-c'(\theta)/Q'(\theta)$ have the same range of values.
7. If $f(x;\theta) = \exp\{(1/x)[\theta A'(\theta) - A(\theta)] - A'(\theta) + S(x)\}$ is a distribution indexed by parameter θ, then obtain the MLE of θ.
8. Obtain the least square estimate of the parameters in a simple linear regression model, where the errors are *iid* normal variates. Check whether they are unbiased.
9. Let y_i, $i = 1, 2, \ldots, n$, be independent *rvs* with $E(y_i) = \alpha + \beta x_i$; $V(y_i) = \sigma^2$, where $x_1, x_2, \ldots, x_n$ are given. Obtain the estimators of α and β and the variances of the estimators.
10. Considering that the sum of the angles is 180°, find the best estimates of the three angles A, B, and C for the following data by the method of least squares:

A:	35°	40°	45°
B:	60°	62°	45°
C:	83°	80°	77°

11. $X = x$ is a single observation from the following *pmf*:

x	1	2	3	4
$f(x;\theta)$	1/4	1/4	$(1 + \theta)/4$	$(1 - \theta)/4$

   Find the MLE of θ, $\Theta = [0, 1]$.
12. A discrete random variable X has a distribution dependent on the specification of the value of a parameter θ as indicated in the following table:

X / $P(X)$	0	1	2	3	4
$\theta = 0$	0.1	0.2	0.3	0	0.4
$\theta = 1$	0.2	0.3	0.2	0.1	0.2
$\theta = 2$	0.4	0.3	0.2	0.1	0

Find the MLE of θ when $X = 2$ is an observation of X.

13. Let $X_1, X_2, \ldots, X_n$ be *iid* random variable with common *pmf*

$$P\{X = y_1\} = \frac{1-\theta}{2}, P\{X = y_2\} = \frac{1}{2}, P\{X = y_3\} = \frac{\theta}{2}, \quad (0 < \theta < 1)$$

Find the MLE of θ.

14. Consider a population made up of four different types of individuals occurring in the population with probabilities θ^3, $3\theta^2(1-\theta)$, $3\theta(1-\theta)^2$, and $(1-\theta)^3$, respectively, where $0 < \theta < 1$. Let $X_1, X_2, \ldots, X_n$ be a random sample drawn from this population and n_1, n_2, n_3, and n_4 be the observed frequencies of the above four types of individuals. Determine the maximum likelihood estimator of θ.

15. Let $X_1, X_2, \ldots, X_n$ be *iid* from a distribution with $b(1, \theta)$ *pmf*

$$f(x;\theta) = \theta^x (1-\theta)^{1-x}, x = 0, 1 \text{ and } \theta \in \Theta = (0, 1)$$

(i) Find the MoME and MLE of θ. Are these estimators unbiased? Show that the variance of the MLE of θ attains CR lower bound.

(ii) Find an unbiased estimator of variance of the MLE in (i). Justify why is it not possible to find an unbiased estimator of θ^{-1}.

(iii) Find the MoME and MLE of θ when $\Theta = [0, 1/2]$. Find the mean square error of these estimators and justify which estimator would you prefer.

(iv) Show that $\hat{\eta} = \bar{X}(1-\bar{X})$ is the MLE of $\eta = \theta(1-\theta)$ and $\sqrt{n}(\hat{\eta} - \eta) \xrightarrow{d} N[0, (1-2\theta)^2\eta]$ which is degenerate at $\theta = 1/2$. By appropriate normalization, show that

$$4n\left(\frac{1}{4} - \hat{\eta}\right) = 4n\left(\bar{X} - \frac{1}{2}\right)^2 \xrightarrow{d} \chi_1^2$$

when $\theta = 1/2$.

(v) Explain the jackknife technique of bias reduction of an estimator. Show that the MLE of θ^2 is $T_n = (\Sigma X_i/n)^2$ which is a biased estimator of θ^2. Show that $j_k(T_n) = [X(X-1)]/[n(n-1)]$ where $X = \Sigma X_i$, is a one-step jackknife estimator based on the MLE T_n of θ and show that it not merely reduces the bias but removes it completely. Is the jackknife estimator a best estimator of θ^2? If so, prove it.

16. Consider the two experiments, one of drawing a random sample $X_1, X_2, \ldots, X_n$ *iid* $b(1, \theta^2)$ and second of drawing $Y_1, Y_2, \ldots, Y_n$ *iid* $b(1, \theta)$ for estimating θ^2. Find the

MLE of θ^2 in both the experiments and examine which experiment along with its corresponding MLE is better?

Hint. Calculate the ARE of the MLE of θ^2 in the first experiment with respect to MLE in the second experiment. It calculates to $(1 + \theta)/4\theta$. For $\theta < 1/3$, the second experiment is superior and for $\theta > 1/3$, the first experiment is superior.

17. Let $X \sim b(n, \theta)$, $0 < \theta < 1$.
(i) Find the MLE of θ^2 and its asymptotic distribution.
(ii) Find the MLE of $\theta(1 - \theta)$. Show that the MLE is not unbiased. Construct an unbiased estimator for $\theta(1 - \theta)$ using this MLE.

18. State the invariance property of MLE. Use this to obtain the MLE of $1/\theta$ in sampling from

$$f(x;\theta) = \begin{cases} \binom{m}{x}\theta^x(1-\theta)^{m-x}, & x = 0, 1, \ldots, m \\ 0, & \text{otherwise} \end{cases}$$

19. Let a random sample of size n has been drawn from a multinomial population with parameters $p_1, p_2, \ldots, p_k$, $\Sigma_1^k p_i = 1$. Show that the MLEs of $p_1, p_2, \ldots, p_k$ are $\bar{X}_1, \ldots, \bar{X}_k$ where $\bar{X}_i = n^{-1}\Sigma_{j=1}^n X_{ij}$. Also, show that the variance–covariance matrix of these estimators is

$$V(\bar{X}_1, \ldots, \bar{X}_k) = [n^{-1}(I(ij)\, p_i - p_i p_j)]$$

where
$$I(ij) = \begin{cases} -1, & \text{if } i = j \\ 0, & \text{if } i \neq j \end{cases}$$

20. (i) Discuss the method of scoring for simultaneous estimation of k parameters by the method of maximum likelihood.
(ii) Let X follows a multinomial distribution with probabilities

$$\left\{\frac{1}{4}(2+\theta^2), \frac{1}{4}(1-\theta^2), \frac{1}{4}(1-\theta^2), \frac{1}{4}\theta^2\right\}$$

Obtain the 1-step method of scoring estimator for θ by taking θ_0 as an initial estimator.

21. Let $X_1, X_2, \ldots, X_n$ be a random sample of n observations from a Poisson distribution with the *pmf*

$$f(x;\theta) = \frac{\exp(-\theta)\theta^x}{x!}; \ x = 0, 1, 2, \ldots$$

(i) Find the MLE of θ and $(1 - \theta)^2$, $\theta \geq 0$.
(ii) Is the MLE of θ unbiased?
(iii) Find the MLE of $\exp(-\theta)$. Show by using delta method that the large sample variance of MLE is the same as the CR lower bound for unbiased estimator of $\exp(-\theta)$.

22. Consider an experiment of estimating the total number of fishes, N, in a lake. M fishes are captured from the lake and released in the water after tagging. A few days later, by allowing the tagged fishes to mix with untagged fishes thoroughly, n fishes are recaptured at random and the tagged fishes X are counted. Assuming that the size of the fish population has not changed during this experiment, find the MLE of N, the total number of fishes in the lake.

23. Let $X_1, X_2,\ldots,X_n$ be *iid* from the geometric distribution with the *pmf*

$$f(x;\theta) = \theta(1-\theta)^x,\ x = 0, 1, \ldots \text{ and } 0 < \theta < 1$$

Find the MLE of θ by assuming $\Sigma x_i \in (0, n)$. Show that it is the global maximizer. Find the MLE of $1/\theta^2$. Compute the MLE of $P(X \geq 4)$ based on the data $\mathbf{X} = \mathbf{x} = (2, 7, 6, 5, 9)$.

24. Let $X_1, X_2,\ldots,X_n$ be *iid* with one of the following *pdfs*. If $\theta = 0$, then

$$f(x;\theta) = \begin{cases} 1, & 0 < x < 1 \\ 0, & \text{otherwise} \end{cases}$$

while if $\theta = 1$, then

$$f(x;\theta) = \begin{cases} \dfrac{1}{2\sqrt{x}}, & 0 < x < 1 \\ 0, & \text{otherwise} \end{cases}$$

Find the MLE of θ.

25. Let $X \sim C(\theta,0)$. Find the MLE of θ.

26. Let $X_1, X_2,\ldots,X_n$ be *iid* from the distribution with Lebesgue density

$$f(x;\theta) = \theta f(\theta x)$$

where f is density on $(0, \infty)$ or is symmetric about 0, and $\theta > 0$ is an unknown parameter. Show that the likelihood equation

$$\frac{1}{n}\sum_{i=1}^{n}\left[1 + \frac{\theta X_i f'(\theta X_i)}{f(\theta X_i)}\right] = 0$$

has a unique root if $f'(x)/f(x)$ is continuous and decreasing for $x > 0$. Verify whether the above condition is satisfied when

$$f(x) = \frac{1}{\pi}\frac{1}{(1+x^2)}$$

Show that the nondegenerate asymptotic distribution of RLE $\hat{\theta}$ of θ is

$$\sqrt{n}(\hat{\theta} - \theta) \xrightarrow{d} N\left(0, \frac{1}{I_X(\theta)}\right)$$

where
$$I_X(\theta) = \frac{1}{2\theta^2}$$

Use the identity $$\int_{-\infty}^{\infty} \frac{1}{(1+x^2)^k}\,dx = \frac{\sqrt{\pi}\,\Gamma\left(k-\frac{1}{2}\right)}{\Gamma(k)}$$

27. Let $X_1, X_2, \ldots, X_n$ be two parameters exponential $E(\theta, \lambda)$. Show that the MLE of θ is $X_{(1)}$ for any fixed $\lambda > 0$. Also, find the MLEs of θ and λ by maximizing profile likelihood.

28. Let $X_1, X_2, \ldots, X_n$ be a random sample from an exponential distribution $G_2(1, \theta)$ with the *pdf*

$$f(x;\theta) = \theta \exp(-\theta x),\ x > 0$$

Show that $1/\overline{X}$ is the maximum likelihood estimator of θ, it is biased, consistent, asymptotically normal, and asymptotically efficient. Show that the variance of an unbiased estimator $(n-1)/(n\overline{X})$ of θ is smaller than that of the corresponding MLE. Using the MLE of θ, find the MLE of $E(X) = \theta^{-1}$ and also its asymptotic distribution. Also discuss the case when $x \geq 0$.

29. Let high stress failure time (in hours) of Kevlar/Epoxy spherical vessels used in a sustained pressure environment on the space shuttle are modelled with exponential distribution $G_1(1, \theta)$ with the *pdf*

$$f(x;\theta) = \frac{1}{\theta}\exp\left(-\frac{x}{\theta}\right),\ x > 0$$

(i) Find the MoME and MLE of mean failure time. Examine if the MLE is identical with the MoME of θ. Are these estimators unbiased? Show that the MLE is consistent and asymptotically normal.

(ii) Compare these estimators in terms of their asymptotic variances.

(iii) Compare these estimators with unbiased estimators $nX_{(1)}$ and $E(nX_{(1)} | \Sigma X_i) = \Sigma X_i/n$, where ΣX_i is the complete sufficient statistic for θ.

(iv) Find the MLE of θ^{-1} and its asymptotic distribution.

(v) Suppose there are k observations among n which are less than or equal to M, where M is a fixed positive number. Using this information, find the MLE of θ. Compute the above estimators for the data $\mathbf{X} = \mathbf{x} = (50.1, 70.1, 137.0, 166.9, 170.5, 152.8, 80.5, 123.5, 112.6, 148.5, 160.0, 125.4)$ and comment on the suitability of the model assumed.

Hint. Define $\eta = P(X \leq M) = 1 - \exp(-M/\theta)$, $\theta = M/\log[1/(1-\eta)]$. Use the likelihood of η, $L(\eta; x) = \eta^k (1-\eta)^{n-k}$, to find the MLE $\hat{\eta} = k/n$ of η. Use Zehna's result to get $\hat{\theta} = M/\log[n/(n-k)]$.

30. Let X and Y be two independent exponential random variables with the *pdf*

$$f(x;\theta_1) = \frac{1}{\theta_1}\exp\left(-\frac{x}{\theta_1}\right),\ x > 0$$

$$f(x;\theta_2) = \frac{1}{\theta_2}\exp\left(-\frac{y}{\theta_2}\right),\ y > 0$$

Based on X and Y, define two new random variables

$$S = \min(X, Y)$$

$$U = \begin{cases} 1, & \text{if } S = X \\ 0, & \text{if } S = Y \end{cases}$$

Let $(S_1, U_1),\ldots,(S_n, U_n)$ be *iid* observations on (S, U). Find the MLEs of θ_1 and θ_2.

Hint. Calculate the joint distribution of S and U.

31. Let $X_1, X_2,\ldots,X_n$, $n \geq 2$, be a random sample drawn from the Lebesgue densities $f(\theta,j)$, $j = 1, 2$

$$f(x;\theta,1) = \frac{1}{\sqrt{2\pi\theta}}\exp\left(-\frac{1}{2}\frac{x^2}{\theta^2}\right) \sim N(0,\theta^2),\ x \in \mathbb{R}^1,\ \theta > 0$$

$$f(x;\theta,2) = \frac{1}{2\theta}\exp\left(-\frac{|x|}{\theta}\right),\ \theta > 0$$

Find the MLE $(\hat{\theta}, \hat{j})$ of (θ, j) and show that the MLE $\hat{j}$ is consistent for j. Also show that the MLE of θ is consistent too, and derive its nondegenerate asymptotic distribution.

32. Let $X_1, X_2,\ldots,X_n$, $n \geq 2$, be a random sample drawn from a distribution with the *pmf* $f(x;\theta, j)$, where $\theta \in (0, 1)$, $j = 1, 2$

$$f(x;\theta, 1) = P(\theta)$$

$$f(x;\theta, 2) = b(1, \theta)$$

Find the MLE $(\hat{\theta}, \hat{j})$ of (θ, j). Show that $\hat{\theta}$, and $\hat{j}$ are consistent for θ and j, respectively. Also, derive the asymptotic distribution of $\hat{\theta}$.

33. Let $X_1, X_2,\ldots,X_k$ be *iid* from a population with the *pmf*

$$f(x;\theta) = \frac{1}{x![1-\exp(-\theta)]}\theta^x \exp(-\theta),\ x = 1, 2,\ldots;\theta > 0$$

Find the estimate of θ by the method of moments. Show that the ML equation has a unique root when the sample mean $\bar{X} > 1$. Show whether this root is an MLE of θ.

34. Let $X_1, X_2,\ldots,X_k$ be a random sample drawn from some distribution so that its joint density function is given by

$$f(\mathbf{x};\theta) = \frac{n!}{(n-k)!\,\theta^k}\exp\left[-\frac{\left(\sum_{1}^{k} x_i\right) + (n-k)x_k}{\theta}\right]$$

where $0 \leq x_1 \leq x_2 \leq \cdots \leq x_k$ and $\theta > 0$. Find the MLE of θ and θ^2.

35. Let $X_1, X_2, \ldots, X_n$ be *iid rvs* with the *pmf*

$$f(x;\theta) = \exp(-2\theta)\frac{1}{x!}\exp(x\log 2\theta)$$

for $x = 0, 1, \ldots$, where $\theta > 0$. Find the MLE of θ and θ^3 by assuming at least one of the X_is as positive.

36. Let $X_1, X_2, \ldots, X_n$ be a random sample of n observations on an *rv* X with the *pdf*

$$f(x; 2, \beta) = \frac{1}{\beta^2} x \exp\left(-\frac{x}{\beta}\right), \quad x > 0$$

(i) Find the MLE of β. Find $E(\hat{\beta})$ and $V(\hat{\beta})$.
(ii) Find the MoME of β. Find $E(\hat{\hat{\beta}})$ and $V(\hat{\hat{\beta}})$.

37. Consider the population having density function

$$f(x;\theta) = \frac{2(\theta - x)}{\theta^2}; \ 0 \le x \le \theta$$

Find the MoME and MLE for θ and examine whether these estimators are unbiased.

38. Let X has the *pdf*

$$f(x;\theta) = \frac{\theta}{x^{1+\theta}}, \ x \ge 1$$

where $\theta > 0$. Calculate the distribution of X^2. Find the MoMEs of θ and θ^2.

39. Let $X_1, X_2, \ldots, X_n$ be a random sample of n observations on an *rv* X with the *pdf*

$$f(x;\theta) = \frac{3x^2}{\theta^3} I_{(0,\,\theta)}(x)$$

where $\theta > 0$ and is unknown. Find the UMVUE and MLE of θ.

40. Let X be a random variable with the *pdf*

$$f(x;\theta) = \frac{(x - \theta)^2}{3}, \ \theta - 1 \le x \le \theta + 2$$

Find the MLE of θ and $2\theta - \exp(-\theta^2)$ when $X = 7$ was observed.

Hint. Likelihood function is $L(\theta) = (7 - \theta)^2/3$, $5 \le \theta \le 8$. Likelihood equation gives solution $\hat{\theta} = x = 7$. Values of likelihood function at critical point $\hat{\theta} = 7$ and at end points 5 and 8 are 0, 4/3, 1/3. $\hat{\theta} = 5$ is the MLE of θ.

41. Let X be an observation with the *pdf*

$$f(x;\theta) = \left(\frac{\theta}{2}\right)^{|x|}(1-\theta)^{1-|x|}, \ x = -1, 0, 1; \ 0 \le \theta \le 1$$

Find the MoME and MLE of θ. Find an estimator which is better than the unbiased estimator

$$S(X) = \begin{cases} 2, & \text{if } X = 1 \\ 0, & \text{otherwise} \end{cases}$$

Compare these estimators.

42. Let $X_1, X_2, \ldots, X_n$ be the lengths (in millimeter) of n poultry-farm eggs where lengths may be modelled with the following distribution:

$$P(X \le x; \alpha, \beta) = \begin{cases} 0, & \text{if } x < 0 \\ \left(\dfrac{x}{\beta}\right)^{\alpha}, & \text{if } 0 \le x \le \beta \\ 1, & \text{if } x > \beta \end{cases}$$

where $\alpha > 0$, $\beta > 0$.

(i) Find a two-dimensional sufficient statistic for (α, β) and use it to find the MLEs of α and β.

(ii) Compute these MLEs based on the observations $\mathbf{X} = \mathbf{x} = (58.2, 56.2, 56.1, 53.3, 56.9, 57.8, 55.7, 55.9, 53.5, 54.3, 52.9, 54.3, 57.0, 55.5)$.

43. Let $X_1, X_2, \ldots, X_n$ be a sample from a distribution with density function.

$$f(x;\theta) = \theta(\theta + 1)x^{\theta-1}(1 - x), \quad 0 < x < 1;\ \theta > 0$$

Discuss when are MoME and MoMLE of θ identical.

44. Let a random sample $X_1, X_2, \ldots, X_n$ of size n be drawn from a population with the *pdf*

$$f(x;\theta) = cx^{\theta}(1 - x),\ 0 \le x \le 1$$

where c is some constant. Show that the MLE of θ is

$$\hat{\theta} = \frac{-(3A + 2) + \sqrt{A^2 + 4}}{2A}$$

where A = $(1/n)\log(\Pi_{i=1}^{n} x_i)$.

Hint. $c = (\theta + 1)(\theta + 2)$ when $\theta > -1$.

45. Let the time X until failure from fatigue cracks for underground cable follows an approximate gamma distribution with parameters α and β

$$f(x; \alpha, \beta) = \frac{1}{\Gamma(\alpha)\beta^{\alpha}} \exp\left(-\frac{x}{\beta}\right) x^{\alpha-1},\ x > 0, \alpha > 0, \beta > 0$$

where α is the shape parameter and β is the scale parameter. Let $X_1, X_2, \ldots, X_n$ be n observations taken on the *rv* X. Show that almost surely, the likelihood equation has a unique solution and that it is the MLE of (α, β). Find the MoMEs and MLEs of α

and β. Find the MoME and MLEs of α and β when $(1/n)\Sigma_{i=1}^{n}x_i = 7.29$ and $(1/n)\Sigma_{i=1}^{n}x_i^2 = 85.59$. Show that the Newton–Raphson and the Fisher-scoring iteration equations are

$$\hat{\boldsymbol{\theta}}^{(i+1)} = \hat{\boldsymbol{\theta}}^{(i)} - [R(\hat{\boldsymbol{\theta}}^{(i)})]^{-1} s(\hat{\boldsymbol{\theta}}^{(i)}),\ i = 0, 1, 2, \ldots$$

and
$$\hat{\boldsymbol{\theta}}^{(i+1)} = \hat{\boldsymbol{\theta}}^{(i)} - [F(\hat{\boldsymbol{\theta}}^{(i)})]^{-1} s(\hat{\boldsymbol{\theta}}^{(i)}),\ i = 0, 1, 2, \ldots$$

where $\boldsymbol{\theta} = (\alpha, \beta)', \boldsymbol{S}(\boldsymbol{\theta}) = \dfrac{\partial}{\partial \boldsymbol{\theta}} \log L(\boldsymbol{\theta}) = n\left[-\log\beta + \dfrac{\Gamma'(\alpha)}{\Gamma(\alpha)} + Y, -\dfrac{\alpha}{\beta} + \dfrac{\overline{X}}{\beta^2}\right]$

$\overline{X}$ is sample mean and $Y = n^{-1}\Sigma_1^n \log X_i$.

$$\mathbf{R}(\boldsymbol{\theta}) = \frac{\partial^2}{\partial\boldsymbol{\theta}\partial\boldsymbol{\theta}'}\log L(\boldsymbol{\theta}) = n\begin{pmatrix} \left[\dfrac{\Gamma'(\alpha)}{\Gamma(\alpha)}\right]^2 - \dfrac{\Gamma''(\alpha)}{\Gamma(\alpha)} & -\dfrac{1}{\beta} \\ -\dfrac{1}{\beta} & \dfrac{\alpha}{\beta^2} - \dfrac{2\overline{X}}{\beta^3} \end{pmatrix}$$

and
$$\mathbf{F}(\boldsymbol{\theta}) = E[\mathbf{R}(\boldsymbol{\theta})] = n\begin{pmatrix} \left[\dfrac{\Gamma'(\alpha)}{\Gamma(\alpha)}\right]^2 - \dfrac{\Gamma''(\alpha)}{\Gamma(\alpha)} & -\dfrac{1}{\beta} \\ \dfrac{1}{\beta} & \dfrac{\alpha}{\beta^2} \end{pmatrix}$$

46. Let $X_1, X_2, \ldots, X_n$ be *iid* from a distribution with the *pdf*

$$f(x; \alpha, \beta) = \frac{1}{\Gamma(\beta)}\left(\frac{\beta}{\alpha}\right)^{\beta} \exp\left(-\frac{\beta x}{\alpha}\right) x^{\beta-1}, 0 < x < \alpha$$

Find the MoMEs and MLEs of α and β. Also, find the asymptotic variances of these estimators.

Hint. $\hat{\alpha} = \overline{X}$ and $\hat{\beta} = 1/\Sigma_{i=1}^{n}[2\log\overline{X} - (1/n)\Sigma_{i=1}^{n}\log X_i]$. Use the result that for large n, $(\partial/\partial\beta)\log\Gamma(\beta) = \log\beta - (1/2\beta)$.

47. Based on the random sample $X_1, X_2, \ldots, X_n$ from each of the following distributions, find the MoMEs of their corresponding parameters:

(i) $b(n, \theta)$, both n and p are unknown.
(ii) $U(\theta_1, \theta_2)$, $\theta_1 < \theta_2$, both θ_1 and θ_2 are unknown.

48. Let $X_1, X_2, \ldots, X_n$ be *iid* sample from $N(0, \sigma^2)$, $\sigma > 0$ is unknown.

(i) Find the MLE of σ^2, asymptotic distribution of the estimator, MLE of σ^{-2}, and the asymptotic distribution of the estimator.
(ii) Find the MoME of σ and σ^2.

49. Find the MLE of the parameter in each of the following distributions based on a random sample $X_1, X_2, \ldots, X_n$. Also, discuss the asymptotic distribution of MLEs so obtained.

(i) $N(\theta,1)$. Define

$$Y_i = \begin{cases} 1, & \text{if } X_i < 0 \\ 0, & \text{if } X_i \geq 0 \end{cases}$$

for $i = 1,\ldots,n$. Denote $p = P(X_i < 0) = P(Y_i = 0)$ by $-\Phi(-\theta)$. Find the MLE of θ by using the likelihood function in p.

(ii) $P(\theta)$, $\theta \geq 0$

(iii) $E(\theta)$

Hint. $f(x;\theta) = \theta \exp(-\theta x)$, $x \geq 0$, $\theta > 0$, $(\partial/\partial\theta) \log L = (n/\theta) - \Sigma X_i = 0$ gives $\hat{\theta} = \bar{X}$; $(\partial^2/\partial\theta^2)\log L = -1/\theta^2$, $I(\theta) = E\{-(\partial^2/\partial\theta^2)\log L\} = 1/\theta^2$, $\sqrt{n}\,(\bar{X}^{-1} - \theta) \xrightarrow{d} N(0, \theta^2)$.

50. Let $X_1, X_2,\ldots,X_n$ be a random sample drawn from a distribution with the *pdf*

$$f(x;\theta) = \exp[-(x-\theta)],\ \theta \leq x < \infty$$

Find the MLE of θ. Show that it is unique and consistent for θ. Is it asymptotically normal?

51. Let $X_1, X_2,\ldots,X_n$ be *iid rvs* from each of the following distributions. Find the MoME of vector parameter $\boldsymbol{\theta}$. Also find the MLE of $\boldsymbol{\theta}$ by one or more of the following methods: (i) profile likelihood method, (ii) conditional likelihood method, and (iii) Cassella method:

(a) $N(\theta,\sigma^2)$; $\boldsymbol{\theta} = (\theta,\sigma^2)'$

(b) Location–scale exponential family with *pdf*

$$f(x;\boldsymbol{\theta},\boldsymbol{\sigma}) = \frac{1}{\sigma}\exp\left\{-\frac{(x-\boldsymbol{\theta})}{\boldsymbol{\sigma}}\right\} \text{ if } x \geq \theta, -\infty < \theta < \infty, \sigma > 0, \boldsymbol{\theta} = (\theta,\sigma)'$$

where θ is the location parameter and σ is the scale parameter. Find the MLE of $P_{\theta,\sigma}\{X_1 \geq 1\}$. Also, calculate the ARE of the MLE of θ and σ with respect to the UMVUE of θ and σ.

Hint. UMVUEs of θ and σ are

$$\hat{\theta}_0 = X_{(1)} - \frac{1}{n(n-1)}\sum_{i=1}^{n}(X_i - X_{(1)})$$

and

$$\hat{\sigma}_0 = \frac{1}{(n-1)}\sum_{i=1}^{n}(X_i - X_{(1)})$$

where $X_{(1)}$ is the smallest order statistic. MLEs of θ and σ are

$$\hat{\theta} = X_{(1)} \text{ and } \hat{\sigma} = \frac{1}{n}\sum_{i=1}^{n}(X_i - X_{(1)})$$

$$2(n-1)\frac{\hat{\sigma}_0}{\sigma} \sim \chi^2_{2(n-1)}$$

$$\sqrt{2(n-1)}\left(\frac{\hat{\sigma}_0}{\sigma} - 1\right) \xrightarrow{d} N(0,1)$$

$$\sqrt{n}(\hat{\sigma}_0 - \sigma) \xrightarrow{d} N(0, 2\sigma^2)$$

$\hat{\sigma}$ has the same asymptotic distribution as of $\hat{\sigma}$ since $\hat{\sigma} = [(n-1)/n]\hat{\sigma}_0$. ARE($\hat{\sigma}$, $\hat{\sigma}_0$) = 1. Further,

$$n(\hat{\theta} - \theta) = n(X_{(1)} - \theta) \sim \exp(0, \theta)\ [Z_{0,\theta} \sim \exp(0, \theta)]$$

Then, $$n(\hat{\theta} - \theta) = n\left(X_{(1)} - \theta\right) - \frac{1}{n-1}\sum(X_i - X_{(1)}) \xrightarrow{d} (Z_{0,\theta} - \theta)$$

since $[1/(n-1)]\Sigma(X_i - X_{(1)}) \xrightarrow{P} \theta$.

$$\text{ARE}(\hat{\theta}, \hat{\theta}_0) = \frac{E(Z_{0,\theta} - \theta)^2}{EZ_{0,\theta}^2} = \frac{1}{2}$$

52. Let $X_1, X_2, \ldots, X_n$ be *iid* from the population with *pdf*

$$f(x; \mu, \sigma) = \frac{1}{\sigma}\exp\left(-\frac{x-\mu}{\sigma}\right)$$

where μ and σ are both unknown. Let $X_{(1)}, X_{(2)}, \ldots, X_{(n)}$ be the order statistics corresponding to the above sample. Let $L(\mu, \sigma)$ be the likelihood function. Show that

$$L(X_{(1)}, S) \geq L(\mu, \sigma)\ \forall\ (\mu, \sigma) \in \Theta$$

where $$S = \frac{1}{n}\sum_{i=1}^{n}(X_{(i)} - X_{(1)})$$

Further, show that $X_{(1)}$ is a consistent estimator of μ using the first principle.

53. Let $X_1, X_2, \ldots, X_n$ be *iid* from each of the following distributions. Find the restricted MLE of θ. (Care must be taken to ensure that MLE must belong to the parameter space Θ.)

(i) $N(\theta, 1)$, $\theta \geq 0$

Hint. If $\bar{X}$ is negative, $L(\theta, \mathbf{x})$ is a decreasing function in θ for $\theta > 0$. Therefore, $L(\theta)$ is maximum at $\hat{\theta} = 0$. The MLE of θ is $\hat{\theta} = \begin{cases} \bar{X}, & \text{if } \bar{X} \geq 0 \\ 0, & \text{if } \bar{X} < 0 \end{cases}$.

(ii) $b(1, \theta)$, $0 \leq \theta \leq 1$

(iii) $N(\theta, \sigma^2)$, $\theta \geq 0$

54. Let $X_1, X_2, \ldots, X_n$ be a random sample of size n from a normal distribution with mean μ and variance σ^2. Obtain the MoME and MLE of μ and σ^2. Show that no MLE of μ and σ^2 exists for $n = 1$.

55. Let $X_1, X_2, \ldots, X_n$ be *iid* from normal distribution $N(\mu, 1)$, $\mu \in \mathbb{R}^1$, μ is unknown. Define

$$\theta = P(X_1 \leq c)$$

where c is a known constant. Find the MLE $\hat{\theta}$ of θ and compare it with UMVUE

$$\hat{\theta}_1 = \Phi\left[\frac{c-\bar{X}}{\sqrt{1-(1/n)}}\right]$$

of θ and with another estimator

$$\hat{\theta}_2 = \frac{1}{n}\sum_{i=1}^{n} I_{(-\infty,\, c]}(X_i)$$

in terms of ARE.

56. Let $X_1, X_2,\ldots,X_n$ be *iid* from HN(μ, σ^2) with the *pdf*

$$f(x;\mu,\sigma^2) = \frac{2}{\sqrt{2\pi}\sigma}\exp\left[-\frac{(x-\mu)^2}{2\sigma^2}\right]$$

where $\sigma > 0$ and $x > \mu$, $\mu \in \mathbb{R}^1$. Find the MLE of σ^2 and σ assuming that μ is known.

57. Let $X_1, X_2,\ldots,X_n$ be a random sample drawn from truncated normal $N(\mu, 1)$, truncated at two unknown points α and β, $\alpha < \beta$, with the *pdf*

$$f(x;\mu,\alpha,\beta) = \frac{1}{\sqrt{2\pi}[\Phi(\beta-\mu)-\Phi(\alpha-\mu)]}\exp\left[-\frac{(x-\mu)^2}{2}\right], \alpha < x < \beta$$

where Φ is the *cdf* of $N(0, 1)$. Show that the sample mean $\bar{X}$ is the unique MLE of $\theta = E(X_1)$ and is asymptotically efficient.

58. Let $X_1, X_2,\ldots,X_n$ be *iid* from an $N(a\theta, \theta)$ population, $\theta > 0$. Find the MLE of θ. Show that for $a = 1$,

$$\hat{\theta} = \frac{\sqrt{1+4\left(\frac{1}{n}\right)\sum_{i=1}^{n}(\log X_i)^2} - 1}{2}$$

is the MLE of θ. Asymptotic distribution is given by

$$\sqrt{n}(\hat{\theta}-\theta) \xrightarrow{d} N\left(0, \frac{1}{I_n(\theta)}\right)$$

where
$$I_n(\theta) = \frac{(2\theta+1)n}{2\theta^2}$$

59. Let $X_1, X_2,\ldots,X_n$ be *iid* $N(\theta, a^2\theta^2)$ where $a^2 > 0$ is known and $\theta > 0$. Find the MLE of θ and θ^2.

60. Let $X_1, X_2,\ldots,X_n$ be *iid* from a lognormal distribution lognormal(μ, σ^2) with the *pdf*

$$f(x;\mu,\sigma^2) = \frac{1}{x\sqrt{2\pi}\sigma}\exp\left\{-\frac{(\log x-\mu)^2}{2\sigma^2}\right\}$$

where $\sigma > 0$, $x > 0$, and $\mu \in \mathbb{R}^1$.

(i) Find the method of moments estimators and MLE for σ and σ^2 when $\mu = 0$.

(ii) Find the MLE of μ, σ^2, $E(X)$, and $V(X)$.

Hint. $\hat{\mu} = (1/n)\Sigma\log X_i$, $\hat{\sigma}^2 = (1/n)\ \Sigma_{i=1}^n[\log X_i - (1/n)\Sigma_{i=1}^n \log X_i]^2$, $E(X) = \exp[\mu + (\sigma^2/2)]$, $V(X) = \exp(2\mu + \sigma^2)\ [\exp(\sigma^2) - 1]$

(iii) Show that the MoMEs of μ and σ^2 are $\hat{\mu} = (1/2)\log[\bar{X}^4/(S^2 + \bar{X}^2)]$ and $\hat{\sigma}^2 = \log[(\bar{X}^2 + S^2)/\bar{X}^2]$

Hint. Moment equations are $E(X) = \exp[\mu + (\sigma^2/2)] = (1/n)\Sigma_{i=1}^n X_i$ and $E(X^2) = \{\exp[\mu + (\sigma^2/2)]\}^2 \exp(\sigma^2) = (1/n)\Sigma_{i=1}^n X_i^2$.

61. Let $X_1, X_2, \ldots, X_n$ be *iid rvs* from a distribution with the *pdf*

$$f(x;\sigma) = \frac{2}{\sigma\sqrt{2\pi}} \exp(x) \exp\left\{-\frac{[\exp(x) - 1]^2}{2\sigma^2}\right\}$$

where, $x > 0$ and $\sigma > 0$. Find the MLE of σ and σ^2.

62. Let $X_1, X_2, \ldots, X_n$ be *iid rvs* from a distribution with the *pdf*

$$f(x;\theta) = \frac{2\theta^{3/2}}{\sqrt{\pi}} x^2 \exp(-\theta x^2)$$

where $\theta > 0$, $x \in \mathbb{R}^1$. Find the MLE of θ and θ^{-1}.

63. Suppose that the random variable X follows the distribution with density

$$f(x;\theta) = \frac{1}{\sqrt{2\pi x^3}} \exp\left\{-\frac{(x-\theta)^2}{2x\theta^2}\right\}, x > 0$$

where $\theta > 0$ is an unknown parameter. Show that this distribution is a member of exponential family. Suppose $X_1, X_2, \ldots, X_n$ are *iid* $f(x;\theta)$. Write down complete sufficient statistic. Show that the MLE of θ is $T_n = \bar{X}$ and show that it is an unbiased estimator of θ. Show that the CR lower bound of an unbiased estimator of θ is given by $\text{CRLB}(\theta) = \theta^3/n$. Consider the problem of estimating $g(\theta) = 1/\theta$. Given that $E(1/T_n) = (1/\theta) + (1/n)$ and $V(1/T_n) = 1/(n\theta) + 2/n^2$, show that $1/T_n$ is an efficient estimator of $1/\theta$. Find the UMVUE of $1/\theta$.

64. Let $X_1, X_2, \ldots, X_n$ be *iid rvs* with the *pdf*

$$f(x;\theta,\sigma) = \frac{\sigma^{1/\theta}}{\theta} \exp\left[-\left(1 + \frac{1}{\theta}\right)\log x\right]$$

where $x \geq \sigma$, $\sigma > 0$, and $\theta > 0$. Find the MLE of (σ, θ). Also find the MLE of θ and θ^7 when $\sigma = 1$.

Hint. The MLE of σ is $\hat{\sigma} = X_{(1)}$ irrespective of the values of $\theta > 0$. Find the MLE of θ treating if σ were known.

65. Let $X_1, X_2,\ldots,X_n$ be a random sample from the Laplace location model with the *pdf*

$$f(x;\theta) \propto \exp\left(-\frac{|x-\theta|}{\sigma}\right), \sigma \text{ is known}$$

Show that the $\text{med}(X_1, X_2,\ldots,X_n) = \tilde{X}$ is MLE of θ and find its asymptotic distribution. Show sample mean is worse than sample median, $\text{ARE}(\bar{X}, \tilde{X}) > 1$, where $\tilde{X}$ is the sample median.

66. Let $X_1, X_2,\ldots,X_n$ be a random sample from a distribution with the *pdf*

$$f(x;\theta) = \frac{1}{2\sigma}\exp\left(-\frac{|x|}{\sigma}\right), \sigma > 0$$

Find MoME and MLE of σ.

Hint. The moment equation $E(X) = 0 = (1/n)\Sigma_{i=1}^{n} X_i$ does not give any solution. The second moment equation $E(X^2) = 2\sigma^2 = (1/n)\Sigma_{i=1}^{n} X_i^2$ gives $\hat{\sigma}^2 = \sqrt{(1/2n)\,\Sigma_{i=1}^{n} X_i^2}$. The MLE of σ is $\hat{\sigma} = (1/n)\Sigma_{i=1}^{n}|X_i|$.

67. Let $X_1, X_2,\ldots,X_n$ be *iid* observations drawn from a truncated Laplace distribution with the *pdf*

$$f(x;\theta) = \frac{1}{2[1-\exp(-\theta)]}\exp(-|x|), |x| < \theta, \theta > 0$$

Find the MLE of θ. Comment whether it is unbiased and consistent.

68. Let $X_1, X_2,\ldots,X_n$ be *iid* from each of the following distributions. Find the MoMEs and MLEs of the parameters indexing these distributions. Investigate whether MLEs are also UMVUEs of their respective parameters.

(i) $P(X = 1) = \theta$, $P(X = 0) = 1 - \theta$, $\theta \in [1/4, 3/4]$

(ii) $f(x;\theta) = [\theta/(1-\theta)]x^{(2\theta-1)/(1-\theta)}$, $0 < x \le 1$, $1/2 < \theta < 1$

(iii) $f(x;\theta) = (\theta/x^2)I_{(\theta,\infty)}(x)$, $\theta > 0$ (Find the MLE based on the minimal sufficient statistic of θ)

(iv) Burr (θ,ϕ), ϕ is known

(v) Double exponential, DE(λ, θ), λ is known

(vi) Laplace distribution

$$f(x;\theta) = \frac{1}{2\theta}\exp\left(-\frac{|x|}{\theta}\right), x \in \mathbb{R}^1, \theta > 0$$

Hint. $E(X) = (1/n)\Sigma X_i$ does not yield an estimator of θ since $E(X) = 0$. $E(X^2) = 2\theta^2 = (1/n)\Sigma X_i^2$ yields $\hat{\theta} = \sqrt{(1/2n)\Sigma_1^n X_i^2}$. MLE of θ is given by $(1/n)\Sigma|X_i|$.

(vii) Uniform distribution with the *pmf* $f(x;\theta) = (1/\theta)I_{\{1,\ldots,\theta\}}(x)$, x are discrete values between 0 and θ

(viii) $f(x;\theta) = \exp\{-(x-\theta)\}\ I_{(\theta,\infty)}(x)$, $\theta > 0$

(ix) Gamma $G_1(2, \theta)$. Calculate the variance of MLE $\hat{\theta}$. Also show that $\Sigma X_i^2/3n$ is an unbiased estimator of $V(X)$ and $\bar{X}/2$ is a biased estimator.

(x) Gamma $G_1(\alpha, \beta)$, $\alpha > 0$, $\beta > 0$, both unknown

(xi) $f(x;\theta) = \theta\alpha x^{\alpha-1}\exp(-\theta x^{\alpha})$, $x > 0$ and α known

(xii) Geometric GEO(θ).

(xiii) Inverse Gaussian IG(θ,λ), λ is known

(xiv) Inverse Gaussian IG(λ, θ), λ is known

(xv) Largest extreme value LEV(θ,σ)

$$f(x;\theta,\sigma) = \frac{1}{\sigma}\exp\left[\frac{x-\theta}{\sigma} - \exp\left(\frac{x-\theta}{\sigma}\right)\right]$$

where x and θ are in $\mathbb{R}^1$ and $\sigma > 0$.

(xvi) Logistic distribution $L(\mu, \sigma)$

$$f(x;\mu,\sigma) = \frac{\exp[-(x-\mu)/\sigma]}{\sigma\{1+\exp[-(x-\mu)/\sigma]\}^2}$$

where x and μ are in $\mathbb{R}^1$ and $\sigma > 0$. Also show that the nondegenerate asymptotic distribution of the RLE $\hat{\theta}$ of $\theta = (\mu, \sigma)$ is $\sqrt{n}(\hat{\theta} - \theta) \xrightarrow{d} N[0, 1/I_1(\theta)]$, where $I_1(\theta) = (1/\sigma^2)\begin{pmatrix} 1/3 & 0 \\ 0 & (1/3)+(\pi^2/9) \end{pmatrix}$. Further, using the Newton–Raphson and Fisher scoring methods, find the 1-step MLE of μ when σ is known, the 1-step MLE of σ when μ is known, and the 1-step MLE of (μ, σ). Use $\sqrt{n}$-consistent estimators as initial estimators in the above methods.

(xvii) Double exponential distribution DE(θ,1)

$$f(x;\theta) = \frac{1}{2}\exp(-|x-\theta|);\ x,\ \theta \in \mathbb{R}^1$$

Also find the MLE of θ in terms of order statistics for cases when n is odd and when n is even separately.

(xviii) Double exponential distribution DE(θ,σ)

$$f(x;\theta,\sigma) = \frac{1}{2\sigma}\exp\left(-\frac{|x-\theta|}{\sigma}\right)$$

where x, $\theta \in \mathbb{R}^1$, $\sigma > 0$.

(xix) Half-logistic distribution HL(μ, σ)

$$f(x;\mu,\sigma) = \frac{2\exp[-(x-\mu)/\sigma]}{\sigma\{1+\exp[-(x-\mu)/\sigma]\}^2}$$

where $x \geq \mu$, $\mu \in \mathbb{R}^1$, and $\sigma > 0$.

(xx) Half-normal distribution HN(μ, σ^2)

$$f(x;\mu,\sigma) = \frac{2}{\sqrt{2\pi}\sigma}\exp\left[-\frac{(x-\mu)^2}{2\sigma^2}\right]$$

where $x \geq \mu$, $\mu \in \mathbb{R}^1$, and $\sigma > 0$.

(xxi) $N(\theta,\theta^2)$, $\theta \in \mathbb{R}^1$, $\theta \neq 0$

(xxii) Negative binomial NB(r, θ), r is known

(xxiii) Rayleigh $R(\mu, \sigma)$, μ is known

$$f(x;\mu,\sigma) = \frac{x-\mu}{\sigma^2}\exp\left[-\frac{1}{2}\left(\frac{x-\mu}{\sigma}\right)^2\right], \sigma > 0, \mu \in \mathbb{R}^1, x \geq \mu$$

(xxiv) Pareto distribution PAR(α, β)

$$f(x;\alpha,\beta) = \frac{\alpha\beta^\alpha}{x^{\alpha+1}}, \alpha > 0, 0 < \beta \leq x$$

(xxv) Weibull distribution $W(\alpha, \beta)$

$$f(x;\alpha,\beta) = \frac{\alpha}{\beta}x^{\alpha-1}\exp\left(-\frac{x^\alpha}{\beta}\right), \alpha > 0, x > 0$$

Show that the likelihood equations are equivalent to

$$f(\alpha) = \frac{1}{n}\sum_{i=1}^{n}\log X_i$$

$$\beta = \frac{1}{n}\sum_{i=1}^{n}X_i^\alpha$$

where
$$f(\alpha) = \frac{1}{\left(\sum_{i=1}^{n}X_i^\alpha\right)}\sum_{i=1}^{n}X_i^\alpha\log X_i - \frac{1}{\alpha}$$

and likelihood equations have a unique solution.

(xxvi) Beta distribution Be(α, β)

$$f(x;\alpha,\beta) = \frac{1}{B(\alpha,\beta)}x^{\alpha-1}(1-x)^{\beta-1}, 0 \leq x \leq 1, \alpha > 0, \beta > 0$$

where $B(\alpha, \beta) = [\Gamma(\alpha)\Gamma(\beta)]/\Gamma(\alpha+\beta)$. Also, find the MLE of α when β is known, $\beta = 1$.

(xxvii) $f(x;\alpha,\beta) = (1/\beta^\alpha)\alpha x^{\alpha-1}$, $0 < x < \beta$, $\alpha > 0$, $\beta > 0$

(xxviii) Cauchy distribution $C(\sigma, \mu)$

$$f(x;\alpha,\beta) = \frac{\sigma}{\pi}\frac{1}{\sigma^2 + (x-\mu)^2}, x, \mu \in \mathbb{R}^1 \text{ and } \sigma > 0$$

(xxix) Half-Cauchy distribution HC(μ, σ)

$$f(x;\alpha,\beta) = \frac{2}{\pi\sigma\{1 + [(x-\mu)/\sigma]^2\}}, x \geq \mu, \mu \in \mathbb{R}^1, \text{ and } \sigma > 0$$

(xxx) $f(x;\theta) = \theta(1 - x)^{\theta-1}$, $0 \le x \le 1$, $\theta > 1$. Also, find the asymptotic distribution of the MLE of θ.

(xxxi) $f(x;\theta) = [\theta/(1 - \theta)]x^{(2\theta-1)/(1-\theta)}$, $0 < x < 1$, $\theta \in (1/2, 1)$

(xxxii) $f(x;\theta) = (1/2)(1 - \theta^2)\exp(\theta x - |x|)$, $\theta \in (-1, 1)$

(xxxiii) $f(x;\alpha, \beta) = \theta^x(1 - \theta)^{1-x}$, $x = 0, 1$, $\theta \in [1/2, 3/4]$

(xxxiv) Binomial $b(k, \theta)$, k is known. Show also that the MoMCS estimator of θ is identical with the MLE for any n. Also, show that if $X \neq 0$ or n, the MoMCS estimator is also identical with the MLE of θ.

(xxxv) Binomial $b(k, \theta)$, k is known

(xxxvi) Binomial $b(k, \theta)$, k and θ are both unknown

Hint. MoMEs of k and θ are $\hat{k} = \bar{X}^2/[\bar{X} - (1/n)\Sigma(X_i - \bar{X})^2]$, $\hat{\theta} = \bar{X}/\hat{k}$. Show that $\bar{X} \xrightarrow{P} k\theta$ but $\hat{k} \not\xrightarrow{P} k$.

(xxxvii) Truncated extreme value TEV(θ).

69. Let $X_1, X_2,\ldots,X_n$ be *iid* with *pdf*

$$f(x;\theta) = \theta x^{\theta-1}, \quad 0 \le x \le 1, \quad 0 < \theta < \infty$$

Find the MoME and MLE of θ. Show that the variance of MLE of θ approaches zero as n approaches infinity.

70. Let $X_1, X_2,\ldots,X_n$ be *iid* with *pdf*

$$f(x;\theta) = (\theta + 1)x^{\theta}, \quad 0 \le x \le 1$$

Find the MoME and MLE of θ and show that the MLE is a function of sufficient statistic $\Pi_{i=1}^n X_i$ for θ.

Hint. $-(n/\Sigma \log X_i)-1$ is the MLE of θ and it is a one-to-one function of the sufficient statistic $\Pi_{i=1}^n X_i$ for θ.

71. Let $X_1, X_2,\ldots,X_n$ be *iid* from each of the following distributions. Find the MoME and MLE of parameter(s) as indicated. Calculate the means and variances of the two estimators. Which estimator would you prefer and why?

(i) $U(0, \theta)$, $\theta > 0$. Discuss the cases when x takes values from $(0, \theta)$ and when from $[0, \theta]$. Let $\hat{\theta}$ be the MLE and $\hat{\theta}_0$ be the UMVUE of θ. Obtain MSE($\hat{\theta}_0$)/MSE($\hat{\theta}$) and show that MLE is inadmissible when $n \ge 2$.

Hint. Denote the random variable having exponential distribution on (a, ∞) with scale parameter θ by $Z_{a,\theta}$. Show that $n(\theta - \hat{\theta}) \xrightarrow{d} Z_{0,\theta}$ and $n(\theta - \hat{\theta}_0) \xrightarrow{d} Z_{-\theta,\theta}$. Obtain the ARE of $\hat{\theta}$ with respect to $\hat{\theta}_0$. Use the sufficient statistic $T(\mathbf{X}) = X_{(n)}$.

(ii) $E(\theta,1)$ with density

$$f(x;\theta) = \exp[-(x - \theta)], \; x \ge \theta, \; -\infty < \theta < \infty$$

Hint. $X_{(n)} - (1/2) + \alpha(1 + X_{(1)} - X_{(n)})$ for each α, $0 < \alpha < 1$, is the MLE of θ.

(iii) $U(\theta_1, \theta_2)$, $x \in [\theta_1, \theta_2]$

Hint. MoMEs of θ_1 and θ_2 are given by $\hat{\theta}_1 = \bar{X} - \sqrt{3(n-1)/n\,S^2}$; $\theta_2 = \bar{X} + \sqrt{3(n-1)/n\,S^2}$, respectively, where $S^2 = [1/(n-1)]\Sigma_{i=1}^{n}(X_i - \bar{X})^2$.

(iv) $U(-\theta, \theta)$. Show that the MoME of θ is a biased estimator. Calculate these estimators when the random observations are –2.1, 6.4, –3.5, 2.8, 0, 4.5, –5.2, 5.8.

(v) $U(\theta - 1/2, \theta + 1/2)$

Hint. Any estimator satisfying $X_{(n)} - (1/2) < \hat{\theta} < X_{(1)} + (1/2)$ is an MLE of θ.

(vi) $U(\theta, \theta + 1)$

Hint. $L(\theta) = I(x_{(1)} \geq \theta)I(x_{(n)} \leq \theta + 1) = I(x_{(n)} - 1 \leq \theta \leq x_{(1)})$

(vii) $U(\theta, 2\theta)$. Show that the estimator $\hat{\theta}_1 = [2X_{(n)} + X_{(1)}]/5$ is more efficient than the MLE $\hat{\theta}$. Find a constant c such that $E_\theta(c\hat{\theta}) = \theta$. Find the MoME and MLE of θ when $\mathbf{X} = \mathbf{x} = (2.26, 1.92, 2.01, 2.61, 1.98)$. Which of these estimators can be improved by sufficiency?

(viii) $U(\theta, \theta + |\theta|)$, $\theta \in \Theta$. Calculate these estimators for cases (a) $\Theta = (0, \infty)$, (b) $\Theta = (-\infty, 0)$, $\Theta = \mathbb{R}^1 - \{0\}$.

72. Let $X_1, X_2, \ldots, X_n$ be a random sample from $U(a\theta, b\theta)$, where $a < b$ and are positive constants and $\theta > 0$ is an unknown parameter. Find the MLE of θ and population median.

73. (On invariance property of MLE) Let $X_1, X_2, \ldots, X_n$ be *iid* from

(a) $b(1, \theta)$. Find the MLE of $\sqrt{\theta(1-\theta)}$.

(b) $P(\theta)$. Find the MLE of $P(X \leq 1)$.

(c) $N(\theta, \sigma^2)$. Find MLE of θ/σ, population median, $P(\bar{X} > z) = 0.05$, where z is the 95% percentile of the distribution of $\bar{X}$.

74. Show that if $\hat{\theta}_n$ is maximum likelihood estimate of the real parameter θ of a univariate *pdf*, $f(x;\theta)$, based on a sample $X_1, X_2, \ldots, X_n$ from f, then under some regularity condition to be stated by you the following type of relationship holds:

$$\hat{\theta}_n - \theta = \frac{1}{n}\sum_{\alpha=1}^{n} h(x_\alpha) + \varepsilon_n$$

where (i) $E_\theta\{h(X_\alpha)\}$ exists for all α and for all θ, (ii) the variance of $\{h(x_\alpha)\}$ is finite for all α and for all θ, and (iii) $\varepsilon_n \xrightarrow{P} 0$. Indicate the form of the function h for which the above equality holds.

75. Let $X_1, X_2, \ldots, X_n$ be a random sample drawn from $N(\theta, \sigma_1^2)$ and independently. $Y_1, Y_2, \ldots, Y_m$ be a random sample drawn from $N(\theta, \sigma_2^2)$. Find the MLE of θ, σ_1^2, and σ_2^2 and the variance of these estimators.

76. Let X_{ij}, $j = 1, \ldots, r$, $(r > 1)$, $i = 1, \ldots, n$, be independently distributed as $N(\mu_i, \sigma^2)$. Find the MLE of $\boldsymbol{\theta} = (\mu_1, \mu_2, \ldots, \mu_n, \sigma^2)$. Show that the MLE of σ^2 is not a consistent estimator for σ^2 as $n \to \infty$.

Hint. $\hat{\mu}_i = \bar{X}_{i\cdot} = (1/r)\Sigma_{j=1}^{r} X_{ij}$, $i = 1,\ldots,n$ and $\hat{\sigma}^2 = (1/nr)\Sigma_{i=1}^{n}\Sigma_{j=1}^{r}(X_{ij} - \bar{X}_{i\cdot})^2$ are the MLEs of μ and σ^2. $\hat{\sigma}^2$ is not consistent for σ^2 since $\hat{\sigma}^2 \xrightarrow{P} [(r-1)/r]\sigma^2$ as $n \to \infty$.

77. Suppose m observations are available on (X, Y) while the other $(n - m)$ observations are available only on X where (X, Y) follows bivariate normal distribution $N_2(\theta_1, \theta_2, \sigma_1^2, \sigma_2^2, \rho)$. Find the MLEs of θ_1, θ_2, σ_1^2, σ_2^2, and ρ.

78. Let (X_i, Y_i), $i = 1,\ldots,n$, be *iid* from a bivariate normal distribution $N_2(0, 0, 1, 1, \rho)$, $\rho \in (-1, 1)$. Show that the likelihood equation is a cubic in ρ and the probability that it has a unique root tends to 1 as $n \to \infty$. Show that relative likelihood estimator (RLE) $\hat{\rho}$ of ρ satisfies

$$\sqrt{n}(\hat{\rho} - \rho) \xrightarrow{d} N\left(0, \frac{(1-\rho^2)^2}{1+\rho^2}\right)$$

79. Let n observations are available on (X, Y) where (X, Y) follows bivariate normal distribution $N_2(\mu_X, \mu_Y, \sigma_X^2, \sigma_Y^2, \rho)$. Find the MoMEs and MLEs of all the five parameters and show both the methods yield the same estimators. Show that the MLE of ρ is a root of the equation

$$n\rho^3 - \rho^2\sum xy + \rho\left(\sum x^2 + \sum y^2 - n\right) - \sum xy = 0$$

Also, show that the asymptotic variance of $\hat{\rho}$ is

$$AV(\hat{\rho}) = \frac{(1-\rho^2)^2}{1+(1+\rho^2)}$$

by computing $E[\partial^2/(\partial\rho^2)\log L]$.

Hint. $\hat{\mu}_X = \bar{X}$, $\hat{\mu}_Y = \bar{Y}$, $\hat{\sigma}_X^2 = (1/n)\Sigma(X_i - \bar{X})^2$, $\hat{\sigma}_Y^2 = (1/n)\Sigma(Y_i - \bar{Y})^2$, $\hat{\rho} = (1/n)\Sigma(X_i - \bar{X})(Y_i - \bar{Y})/\hat{\sigma}_X\hat{\sigma}_Y$.

80. Let $Y_1, Y_2,\ldots,Y_n$ be n independent observations from the linear model

$$Y = \beta x + e$$

where x is known and $e \sim N(0, \sigma^2)$, σ^2 is unknown. Find a two-dimensional sufficient statistic for (β, σ^2) and the MLE $\hat{\beta}$ of β. Show that it is unbiased of β. Also, find the distribution of $\hat{\beta}$. Calculate the variance of $\hat{\beta}$, the exact variances of unbiased estimators of β,

$$\hat{\beta}_1 = \frac{\sum Y_i}{\sum x_i}, \quad \hat{\beta}_2 = \frac{1}{n}\sum\frac{Y_i}{x_i}$$

Compare these estimators in terms of their variances.

81. Problem of Nile (Cassela and Berger 2000). Consider *iid* sample (X_i, Y_i), $i = 1,\ldots,n$, on the *rv* (X, Y) with the joint density

$$f(x, y; \theta) = \exp-\left\{-\left(\theta x + \frac{y}{\theta}\right)\right\}, x > 0, y > 0$$

Show that the Fisher information in the sample on θ is $I_X(\theta) = 2n/\theta^2$. Consider the estimators $T = \sqrt{\Sigma Y_i / \Sigma X_i}$ and $U = \sqrt{\Sigma Y_i \, \Sigma X_i}$.

Show that (T, U) is jointly sufficient but complete

$$I_T(\theta) = \left[\frac{2n}{(2n+1)}\right] I_X(\theta) \quad \text{and} \quad I_{(T,U)}(\theta) = I_{\mathbf{X}}(\theta)$$

and show that T is the MLE of θ

$$E(T) = c_1\theta \quad \text{and} \quad E(T^2) = c_2\theta^2$$

where
$$c_1 = \frac{\Gamma\left(n+\frac{1}{2}\right)\Gamma\left(n-\frac{1}{2}\right)}{[\Gamma(n)]^2}, \quad c_2 = \frac{\Gamma(n+1)\Gamma(n-1)}{[\Gamma(n)]^2}$$

Show that U is ancillary. Show that the estimators $\hat{\theta}_1 = (n-1)/(\Sigma X_i)$ and $\hat{\theta}_2 = (\Sigma Y_i)/n$ are both unbiased for θ with variances $V(\hat{\theta}_1) = \theta^2/(n-2)$ and $V(\hat{\theta}_2) = \theta^2/n$. Consider a class of all unbiased estimators of θ

$$C(\theta) = \{\hat{\theta} : \hat{\theta} = \alpha\hat{\theta}_1 + (1-\alpha)\hat{\theta}_2, \; 0 \le \alpha \le 1\}$$

Find the best estimator of θ in the class $C(\theta)$ and compare it with the bias-corrected MLE of θ.

82. Consider one-way random effects model

$$Y_{ij} = \mu + B_i + e_{ij}, \; j = 1,\ldots,n; \; i = 1,\ldots,m$$

$\mu \in \mathbb{R}^1$, B_is are *iid* $N(0, \sigma_b^2)$ and e_{ij}s are *iid* $N(0, \sigma^2)$ and B_is and e_{ij}s are themselves independent, μ, σ_b^2, σ^2 are unknown. Find the MLE of (σ_b^2, σ^2) when $\mu = 0$ and the MLE of $(\mu, \sigma_b^2, \sigma^2)$.

83. Let $\mathbf{X}_1, \mathbf{X}_2, \ldots, \mathbf{X}_n$ be *iid* $N_p(\boldsymbol{\theta}, \Sigma)$. Show that the joint MLE of $\boldsymbol{\theta}$ and Σ is $\{\bar{\mathbf{X}}, [(n-1)/n]\,\mathbf{S}\}$ where $\bar{\mathbf{X}} = (1/n)\Sigma_1^n \mathbf{X}_i$ and $\mathbf{S} = [1/(n-1)]\Sigma_1^n(\mathbf{X}_i - \bar{\mathbf{X}})(\mathbf{X}_i - \bar{\mathbf{X}})'$. Show that $\bar{\mathbf{X}}$ and $\mathbf{S}$ are unbiased for $\boldsymbol{\theta}$ and Σ, respectively. Also, for given $\boldsymbol{\theta} = \boldsymbol{\theta}_0$, find the MLE of Σ.

84. Find the MLE of parameters μ and $\Sigma = \mathbf{M}^{-1}$ of p-dimensional multivariate normal distribution

$$f(\mathbf{x}; \boldsymbol{\mu}, \Sigma) = \frac{1}{(2\pi)^{p/2}|\mathbf{M}^{-1}|}\exp\left\{-\frac{1}{2}(\mathbf{x}-\boldsymbol{\mu})'\mathbf{M}(\mathbf{x}-\boldsymbol{\mu})\right\}, \quad \mathbf{x} \in \mathbb{R}^p$$

on the basis of a random sample $\mathbf{x}_1, \ldots, \mathbf{x}_n$, where, $\mathbf{x} = (x_1, \ldots, x_p)'$, $\boldsymbol{\mu} = (\mu_1, \ldots, \mu_p)'$, $\mathbf{x}$ and $\boldsymbol{\mu}$ take values in $\mathbb{R}^p$; $\mathbf{M}^{-1} = \Sigma$ is a $p \times p$ dimensional variance–covariance matrix of $\mathbf{X}$ which is real symmetric, positive definite.

Hint. The likelihood function is

$$L(\boldsymbol{\mu}, \Sigma; \mathbf{x}) = \frac{|\mathbf{M}|^{n/2}}{(2\pi)^{\pi p/2}} \exp\left[-\frac{1}{2}\sum_{i=1}^{n}(\mathbf{x}_i - \boldsymbol{\mu})'\mathbf{M}(\mathbf{x}_i - \boldsymbol{\mu})\right]$$

$$\log L = \text{const.} + \frac{n}{2}\log|\mathbf{M}| - \frac{1}{2}\sum_{1}^{n}(\mathbf{x}_i - \boldsymbol{\mu})'\mathbf{M}(\mathbf{x}_i - \boldsymbol{\mu})$$

$$\frac{\partial}{\partial \boldsymbol{\mu}}\log L = -\sum_{1}^{n}\mathbf{M}(\mathbf{x}_i - \boldsymbol{\mu}) = \mathbf{0}$$

We have $\hat{\boldsymbol{\mu}} = ((1/n)\Sigma_{j=1}^{n}X_{1j},\ldots,(1/n)\Sigma_{j=1}^{n}X_{pj})' = (\bar{X}_1,\ldots,\bar{X}_p)' = \bar{\mathbf{X}}$, since $\mathbf{M}$ is nonsingular. Therefore, the MLE of $\boldsymbol{\mu}$ is $\hat{\boldsymbol{\mu}} = \bar{\mathbf{X}}$. Further,

$$\frac{\partial}{\partial \mathbf{M}}\log L = \frac{n}{2|\mathbf{M}|}\mathbf{M}_{(\text{cofactor})} - \frac{1}{2}\sum_{1}^{n}(\mathbf{x}_i - \hat{\boldsymbol{\mu}})(\mathbf{x}_i - \hat{\boldsymbol{\mu}}) = 0$$

$$n \cdot \mathbf{M}^{-1} - \sum_{1}^{n}(\mathbf{x}_i - \hat{\boldsymbol{\mu}})(\mathbf{x}_i - \hat{\boldsymbol{\mu}})' = 0$$

$$\hat{\Sigma} = \hat{\mathbf{M}}^{-1} = n^{-1}\sum_{1}^{n}(\mathbf{x}_i - \hat{\boldsymbol{\mu}})(\mathbf{x}_i - \hat{\boldsymbol{\mu}})'$$

85. Consider the linear model

$$\mathbf{Y}_{n\times 1} = \mathbf{X}_{n\times p}\,\boldsymbol{\beta}_{p\times 1} + \mathbf{e}_{n\times 1}$$

where $\mathbf{Y} = (\mathbf{Y}_1',\ldots,\mathbf{Y}_s')'$, $(\mathbf{Y}_i)_{n_i\times 1}$, $i = 1,\ldots,s$, $\Sigma_{i=1}^{s}n_i = n$; $\mathbf{X} = (\mathbf{X}_1',\ldots,\mathbf{X}_s')'$, $(\mathbf{X}_i)_{n_i\times p}$, $i = 1,\ldots,s$, $\boldsymbol{\beta}_{p\times 1} = (\beta_1,\ldots,\beta_p)'$; $\mathbf{e} = (\mathbf{e}_1',\ldots,\mathbf{e}_s')'$, $(\mathbf{e}_i)_{n_i\times 1}$, $i = 1,\ldots,s$; $\boldsymbol{\beta}$ is an unknown vector, $\mathbf{X}$ is a matrix of known constants, $\mathbf{e}$ is a random vector with $\mathbf{e} \sim N_n(\mathbf{0}, \Sigma)$

$$\Sigma = \begin{bmatrix} \sigma_1^2\mathbf{I}_{n_1} & \cdots & \mathbf{0} \\ \vdots & \ddots & \vdots \\ \mathbf{0} & \cdots & \sigma_s^2\mathbf{I}_{n_s} \end{bmatrix}$$

where $\mathbf{I}_{n_i}$ is the unit matrix of order n_i, and $n_i \geq (p + 2)$, $i = 1,\ldots,s$. Show that the MLEs of $\boldsymbol{\beta}$ and σ_i^2, $i = 1,\ldots,s$, are

$$\hat{\boldsymbol{\beta}} = \left[\sum_{i=1}^{s}\hat{\sigma}_i^{-2}\mathbf{X}_i'\mathbf{X}_i\right]^{-1}\left[\sum_{i=1}^{s}\hat{\sigma}_i^{-2}\mathbf{X}_i'\mathbf{Y}_i\right]$$

$$\hat{\sigma}_i^2 = \frac{1}{n_i}(\mathbf{Y}_i - \mathbf{X}_i\hat{\boldsymbol{\beta}})'(\mathbf{Y}_i - \mathbf{X}_i\hat{\boldsymbol{\beta}})$$

86. Generalized version of Hodges' superefficiency. Let $\hat{\theta}_1$ be an estimator of θ, $\theta \in \mathbb{R}^1$, so that

$$\sqrt{n}(\hat{\theta}_1 - \theta) \xrightarrow{d} N[0, v(\theta)]$$

Construct an estimator $\hat{\theta}_2$ such that

$$\sqrt{n}(\hat{\theta}_2 - \theta) \xrightarrow{d} N[0, w(\theta)]$$

with
$$w(\theta) = \begin{cases} v(\theta), & \text{when } \theta \neq \theta_0 \\ t^2 v(\theta_0), & \text{when } \theta = \theta_0 \end{cases}$$

where $t \in \mathbb{R}^1$ and θ_0 is some point in the parameter space.

7

Principle of Equivariance

7.1 INTRODUCTION

In Chapter 2, we have discussed the data reduction principle, namely sufficiency, and have concluded that such consideration results in the reduction of estimation problem. We can restrict only to such class of estimators which are the functions of sufficient statistics. Similarly, we introduced the concept of unbiasedness, which reduces the class of all estimators by eliminating those estimators which give low risks at some parameter values at the cost of high risk at all other parameter values. These principles lead us in reducing the dimensionality of such classes and results in narrowing down the search for optimal estimators. This reduction generally results in the existence of a UMVUE in case of convex loss functions. Typically, UMVUEs exist mainly for exponential families and for convex loss functions and they do not exist for location, scale, and location-scale families and for bounded loss functions. Moreover, very frequently, UMVUEs are inadmissible. The equivariance considerations address these problems and lead to the existence of minimum risk equivariant estimators (MREs) for location, scale, and location-scale families not only for convex loss functions but also for those which are not so restrictive. Moreover, MREs are admissible under mild conditions.

Thus, in the present chapter, we will discuss the *principle of equivariance* that reduces the class of all estimators so that the statistician can make his choice of an optimal estimator more conveniently. If the estimation problem is symmetric (or invariant) under a group of transformations, then the principle of equivariance also requires the estimators to be symmetric under the same transformations.

The principle of equivariance is introduced in Section 7.2 along with the required notations and related concepts. Section 7.3 discusses the minimum risk equivariant estimator (Pitman estimator) for the location family of distributions. The problem of estimation of mean in multivariate normal distribution is also discussed in this section. In Section 7.4, the Pitman estimator for scale family is discussed. In Section 7.5, we have discussed the Pitman estimator for location and scale family of distributions. In Section 7.6, the examples of finding minimum risk equivariant estimators for different location and scale family of distributions are discussed with the purpose of illustrating the applications of theorems and results of the present chapter to different problems. In the last section 7.7, exercises are given.

7.2 PRINCIPLE OF EQUIVARIANCE

7.2.1 Basic Concepts and Definitions

Similar to the impartiality principles discussed in the previous chapters, which reduce the problem of estimation to a great extent, we introduce an alternative impartiality principle in the present chapter that is known as *equivariance principle*. This states that if we observe two points $\mathbf{x}$ and $\mathbf{y}$ so that $\mathbf{y} = g(\mathbf{x})$, for some Borel measurable function g, the inference if we observe $\mathbf{x}$ should have a certain type of relationship to the inference if $\mathbf{y}$ is observed; however, these inferences may be different. This kind of symmetry restrictions usually reduce the estimation problem into a much simpler problem as do all other impartiality principles discussed in the previous chapters.

In fact, the principle of equivariance, as described above, is the combination of two types of equivariance considerations: one is called the *measurement equivariance* and the other is called *formal invariance*. They are defined as follows.

Definition 7.2.1 (Measurement equivariance) This impartiality principle says that the inference should not depend on the unit of measurements used. This gives

$$\delta^*(y) = \tilde{g}(\delta(x))$$

where there are two estimation problems X and $Y = g(X)$; in the first, the estimator δ is used, and in the second, the estimator δ^* is used. $\tilde{g}$ is the change corresponding to the change in the unit of measurement.

Definition 7.2.2 (Formal invariance) If the two estimation problems have the same formal structure in terms of the parameter space Θ, the family of distributions $\mathcal{F} = \{f(x;\ \theta),\ \theta \in \Theta\}$, the sample space χ, allowable action space $\mathcal{A}$, and loss function, then same inferences are used in both the problems. Formally, we can write this as

$$\delta^*(y) = \delta(y) = \delta(g(x))$$

On combining these impartiality (or symmetry) principles, we state that the equivariance principle requires that the inferences must satisfy some structural relationship corresponding to two different estimation problems related by some group relationship

$$\delta(g(x)) = \tilde{g}(\delta(x)) \tag{7.2.1}$$

The equivariance principle restricts us to consider only such estimators which satisfy the above requirement given in Eq. (7.2.1). This principle reduces the class of estimators and makes the search for an optimal estimator into a much simpler problem as it is shown in Sections 7.3, 7.4, and 7.5.

We use the word *equivariance* for the type of estimators satisfying Eq. (7.2.1) which change in a prescribed manner with change in data. Whereas, we use the word *invariance* for all such estimators which remain unchanged under the transformation of data.

7.2.2 Examples on Principle of Equivariance

Before we proceed to the formulation of the problem and establish optimality results, we discuss here some examples that would illustrate the equivariance principle.

Consider the model $X \sim b(n, p)$ and the estimation of p when the number of successes $X = x$ is observed. Based on x, let $\delta(x)$ be the estimator of p so that $1 - \delta(x)$ is the estimate of $1 - p$. Consider, next, the transformed problem as changed in the measurement by $Y = g(X) = n - X \sim b(n, 1 - p)$. Let $\delta^*(y)$ be the estimate of $1 - p$ when $Y = y$ is observed. The principle of measurement equivariance requires that the two estimates of $q = 1 - p$ in the original and the transformed problems must be equal, i.e.,

$$\delta^*(y) = 1 - \delta(x)$$

Further, the formal structures of the original and the transformed problems are the same, e.g., the sample space $\chi = \{0,\ldots,n\}$ is the same, the parameter space $\Theta = \{p: 0 \le p \le 1\}$ is the same, the density is the same, i.e., binomial. So, formal invariance requires that the estimators in the two problems must satisfy

$$\delta^*(y) = \delta(y) \ \forall \ y = 0,\ldots,n$$

Thus, on combining the two principles, we get

$$\delta(y) = \delta[g(x)]$$

$$\delta(n - x) = 1 - \delta(x) \tag{7.2.2}$$

The estimators which satisfy Eq. (7.2.2) are known as equivariant estimators. Equivariance principle, in the present example, eliminates all such estimators from the class of all estimators for which the following does not hold:

$$\delta(n) = 1 - \delta(0)$$

$$\delta(n-1) = 1 - \delta(1)$$

$$\vdots \qquad \vdots$$

$$\delta\left[n - \left(\frac{n}{2}\right)\right] = 1 - \delta\left(\frac{n}{2}\right)$$

where $[n/2]$ is the largest integer not greater than $n/2$.

Note, here, that the reduction in the problem has been achieved due to symmetry. This symmetry says that the inference made at some observed value of X determines the inference to be made at other observed value. Note that the sample proportion of successes, x/n, and the shrinkage estimator $0.9(x/n) + 0.1(0.5)$, which shrinks the sample proportion towards 0.5, are the equivariant estimators since they satisfy Eq. (7.2.2). Note that

$$\delta(y) = \frac{y}{n} = \frac{n-x}{n} = 1 - \frac{x}{n} = 1 - \delta(x)$$

and
$$\delta(y) = 0.9\left(\frac{y}{n}\right) + 0.1(0.5)$$

$$= 0.9\left(\frac{n-x}{n}\right) + 0.1(0.5) = 0.95 - 0.9\left(\frac{x}{n}\right)$$

$$= 1 - \left[0.9\left(\frac{x}{n}\right) + 0.1(0.5)\right] = 1 - \delta(x)$$

However, the estimator $0.8(x/n) + 0.2(0.5)$ is not an equivariance estimator.

We can consider a second example for explaining how the equivariance principle reduces an estimation problem to some simple analytical form. Let the decay time X of some atomic particle has been modelled by an exponential distribution

$$f(x;\theta) = \frac{1}{\theta}\exp\left\{-\frac{x}{\theta}\right\}, \quad x > 0, \theta > 0$$

where the mean decay time θ is to be estimated. Suppose $X = x$ is observed and measured in seconds. Based on this observation x, suppose, we decide to estimate θ by the estimate $\delta(x)$. We will estimate the mean decay time measured in minutes, $\theta' = \theta/60$, by $\delta(x)/60$. Now, consider the transformed problem where the decay time $Y = g(X) = X/60$ is measured in minutes. The density of Y is

$$f(y;\theta') = \frac{1}{\theta'}\exp\left\{-\frac{y}{\theta'}\right\}$$

$y > 0$, $\theta' > 0$. Let the estimator of θ' be given by $\delta^*(y)$. Following the measurement equivariance principle, we must have

$$\delta^*(y) = \frac{\delta(x)}{60}$$

It is easy to see that the sample space, the parameter space, and the density, that is, formal structures in pre-transformed and transformed problems are the same. The invariance principle then prescribes that the same estimator must be used in each of the estimation problems.

$$\delta^*(y) = \delta(y)$$

On combining these two principles, we require the estimator to satisfy

$$\delta(y) = \delta(g(x)) = \delta\left(\frac{x}{60}\right) = \frac{\delta(x)}{60}$$

If the transformed problem, in general, is given by $Y = aX$, $a > 0$, then the above condition for an estimator to be equivariant changes to

$$\delta(x) = a\delta(x)$$
$$= a^{-1}\delta(ax)$$

Consider $a = x^{-1}$

$$\delta(x) = x\delta(1) = cx,\ c > 0$$

Here, we easily see that equivariance considerations prescribe the consideration of only such estimators which are of the type cX. The problem is, thus, greatly reduced.

The equivariance principle and examples that explain this principle depend on the transformation chosen on the sample space so that the formal structure of the problem remains unchanged. For the formal study of this principle, we define a group of transformations.

Consider a sample space $\chi \subseteq \mathbb{R}^1$ and a group of transformations g on χ into $\mathbb{R}^1$ which are one-to-one and onto. A transformation g is said to be one-to-one if $g(x_1) = g(x_2)$ implies $x_1 = x_2$ and is said to be onto if the range of g, i.e., $R(g) = g(\chi) = \chi$.

Definition 7.2.3 A set of measurable one-to-one and onto transformations $\mathcal{G} = \{g(\mathbf{x}): g$ is some transformation on $\chi\}$ on χ into itself is called a *group of transformations* if

(i) (Composition) if $g_1 \in \mathcal{G}$ and $g_2 \in \mathcal{G}$, then $g_2\, g_1 \in \mathcal{G}$

(ii) (Inverse) If $g \in \mathcal{G}$, then there is a $g^{-1} \in \mathcal{G}$ so that $g^{-1}(g(\mathbf{x})) = \mathbf{x}\ \forall\ \mathbf{x}$. The conditions (i) and (ii) give another condition, which is known as *identity*.

(iii) (Identity) The identity transformation $e \in \mathcal{G}$ so that $e(\mathbf{x}) = \mathbf{x}\ \forall\ \mathbf{x}$

7.2.3 Formal Structure

Consider a statistical decision problem $(\Theta, \mathcal{A}, L)$, an *rv* X on the probability space $(\Theta, \mathcal{B}, P)$, so that $X \sim F_\theta$, $\theta \in \Theta$. Let $\mathbf{X} = (X_1,\ldots,X_n)$ be a random sample according to F_θ and the distribution of $\mathbf{X}$ belongs to a family of distributions $\mathcal{F} = \{F_\theta, \theta \in \Theta\}$. The random variable $\mathbf{X}$ takes its values from $\chi \subseteq \mathbb{R}^n$, known as sample space. The principle of equivariance revolves around three groups of transformations: first, $\mathcal{G}$ from sample space χ onto itself; second, $\bar{\mathcal{G}}$ from Θ onto itself; and third, $\tilde{\mathcal{G}}$ from $\mathcal{A}$ onto itself. Readers may refer to Chapter 1 for the definition of group.

Definition 7.2.4 (Invariance of family of distributions $\mathcal{F}$) A family of distributions $\mathcal{F}$ is said to be invariant under a group of transformations $\mathcal{G}$ if for every $g \in \mathcal{G}$ and every $\theta \in \Theta$, there exists a unique $\theta' = \bar{g}\theta \in \Theta$ such that the distribution of $g(\mathbf{X})$ is given by $F_{\bar{g}\theta}$ whenever the distribution of $\mathbf{X}$ is F_θ.

Two useful conditions for a family of distributions $\mathcal{F}$ to be invariant under $\mathcal{G}$ are given as follows:

$$\left.\begin{array}{l} 1.\ P_\theta^{\mathbf{X}}\{g(\mathbf{X}) \in A\} = P_{\bar{g}\theta}^{g(\mathbf{X})}\{\mathbf{X} \in A\} \text{ for every Borel set } A \text{ in } \chi. \\ 2.\ E_\theta\phi[g(\mathbf{X})] = E_{\bar{g}\theta}[\phi(\mathbf{X})] \text{ for every integrable real valued function } \phi. \end{array}\right\} \quad (7.2.3)$$

The advantage of a family of distributions $\mathcal{F}$ to be invariant under a group of transformations $\mathcal{G}$ is that under $\mathcal{G}$, the original family of distributions $\mathcal{F}$, does not change; it changes only the indexing parameter from θ to $\bar{g}\theta$.

The basic transformations g, from χ onto itself, induce two groups of transformations; one, over Θ into itself denoted by $\bar{\mathcal{G}}$ and the other, over $\mathcal{A}$ into itself denoted by $\tilde{\mathcal{G}}$. Thus, we state the following result.

Result 1. If the family of distributions $\mathcal{F}$ is invariant under $\mathcal{G}$, then the group of transformations, $\bar{\mathcal{G}}$, on Θ into itself, is a group.

Definition 7.2.5 (Invariance of decision problem) A decision problem $(\Theta, \mathcal{A}, L)$ along with $\mathcal{F}$, the family of distributions F_θ over χ, is said to be invariant under a group of transformations $\mathcal{G}$ if

(i) the family of distributions is invariant under $\mathcal{G}$ and

(ii) the loss function is invariant under $\mathcal{G}$, i.e., for every $g \in \mathcal{G}$ and $a \in \mathcal{A}$, there exists a unique a' so that

$$L(\theta, a) = L(\bar{g}(\theta), a') \; \forall \; \theta \in \Theta$$

Note that a' is uniquely determined by g and a; so, denote it by $\tilde{g}(a)$.

Result 2. If a decision problem is invariant under a group of transformations $\mathcal{G}$, then $\tilde{\mathcal{G}} = \{\tilde{g}: g \in \mathcal{G}\}$ is a group of transformations on $\mathcal{A}$ into itself.

If the decision problem is invariant, the principle of invariance requires that the decision rules d in the original problem (involving $\mathbf{X}$) and the decision rules d^* in the transformed problem [involving $\mathbf{Y} = g(\mathbf{X})$] must correspond

$$\tilde{g}[d(\mathbf{x})] = d^*(\mathbf{y}) = d(g(\mathbf{x}))$$

Such rules are said to be *equivariant decision rules* and the following definition follows:

Definition 7.2.6 (Equivariant decision rule) If a decision problem is invariant under a group of transformations $\mathcal{G}$, a nonrandomized decision rule d is said to be equivariant under $\mathcal{G}$ if

$$d(g(\mathbf{x})) = \tilde{g}(d(\mathbf{x})) \; \forall \; g \in \mathcal{G} \text{ and } \forall \; \mathbf{x} \in \chi$$

Definition 7.2.7 (Orbits over Θ) Two points θ_1 and θ_2 are said to be equivalent if for some $\bar{g}$ in $\bar{\mathcal{G}}$, $\theta_2 = \bar{g}(\theta_1)$. This gives rise to an equivalence relation over Θ and divides Θ into several orbits of equivalent classes of such points. An orbit containing point θ' is the set

$$\Theta(\theta') = \{\bar{g}(\theta'): \bar{g} \in \bar{\mathcal{G}}\}$$

Definition 7.2.8 (Transitive group of transformations) A group of transformations from Θ into itself, $\bar{\mathcal{G}}$, is said to be transitive if Θ has a single orbit, i.e., for any two points θ_1 and θ_2 in Θ, there is some $\bar{g}$ in $\bar{\mathcal{G}}$ so that $\theta_2 = \bar{g}(\theta_1)$.

An important property of equivalent estimators is that their risk functions are constant over orbits of Θ. This result is stated in the following theorem.

Theorem 7.2.1 If a problem is equivariant under a group of transformations $\mathcal{G}$, then the risk of an equivariant decision rule is constant on the orbits of Θ, i.e.,

$$R(\theta, d) = R(\bar{g}\theta, d) \; \forall \; \theta \in \Theta \; \forall \; g \in \mathcal{G}$$

If $\bar{\mathcal{G}}$ is transitive over Θ, then $R(\theta, d)$ is independent of θ.

Proof. For some $\theta \in \Theta$ and some $g \in \mathcal{G}$, we have

$$\begin{aligned} R(\theta, d) &= E_\theta L[\theta, d(\mathbf{X})] \\ &= E_\theta L[\bar{g}\theta, \tilde{g}d(\mathbf{X})] \end{aligned}$$

(since the loss function is invariant)

$$= E_\theta L[\bar{g}\theta, d(g(\mathbf{X}))]$$

(since d is equivariant)

$$= E_{\bar{g}\theta} L[\bar{g}\theta, d(\mathbf{X})]$$

(by condition 7.2.3, since the family of distribution $\mathcal{F}$ is invariant)

$$R(\theta, d) = R(\bar{g}\theta, d)$$

This shows that the risk function of an equivariant estimator is constant on orbits. Further, if $\bar{\mathcal{G}}$ is transitive over Θ, the risk risk function is independent of θ since there is only one orbit, i.e., $\bar{\mathcal{G}}$ itself. ▣

Definition 7.2.9 An equivariant estimator in the class of equivariant estimators is said to be a *minimum risk equivariant estimator* (MRE) if it minimizes the constant risk.

We introduce a type of statistic which is known as *maximal invariant statistic.*

Definition 7.2.10 A statistic S is said to be maximal invariant if

(i) S is invariant and
(ii) S is maximal, that is, $S(\mathbf{x}_1) = S(\mathbf{x}_2)$ implies

$$\mathbf{x}_1 = g(\mathbf{x}_2) \text{ for some } g \in \mathcal{G}$$

Condition (i) implies that S is constant on the orbits of χ and condition (ii) implies that S distinguishes these orbits.

7.3 MINIMUM RISK EQUIVARIANT ESTIMATOR FOR LOCATION FAMILY

Let X be an *rv* with probability density

$$f(x; \theta) = f(x - \theta), \theta \in \mathbb{R}^1 \tag{7.3.1}$$

where $f(x - \theta)$ does not depend on the parameter θ. This density is known as *location density* and parameter θ is known as *location parameter*. Consider a random sample $X_1, X_2, \ldots, X_n$ drawn from a distribution with location density given in Eq. (7.3.1). Clearly, this family of distributions is invariant under a group of transformations

$$\mathcal{G} = \{g: g(\mathbf{x}) = \mathbf{x} + b, b \in \mathbb{R}^1\}$$

An estimator δ is an invariant estimator of θ if

$$\delta[g(\mathbf{X})] = \delta(\mathbf{X}) + b, b \in \mathbb{R}^1$$

The following result constructs a class of all equivariant estimators for θ for the present problem.

Theorem 7.3.1 An estimator δ is equivariant for estimating θ if and only if

$$\delta(\mathbf{X}) = X_1 + l(X_2 - X_1, \ldots, X_n - X_1) \tag{7.3.2}$$

for some function l.

Proof. Consider an estimator δ evaluated at $g(\mathbf{X})$ and that it satisfies Eq. (7.3.2)

$$\begin{aligned}\delta[g(\mathbf{X})] &= \delta(X_1 + b,\ldots,X_n + b)\\ &= X_1 + b + l(X_2 - X_1,\ldots,X_n - X_1) = \delta(\mathbf{X}) + b\end{aligned}$$

Thus, δ is equivariant.
Conversely, if δ is equivariant, then

$$\begin{aligned}\delta(\mathbf{X}) &= \delta(X_1, X_2,\ldots,X_n)\\ &= \delta(X_1 + X_1 - X_1, X_1 + X_2 - X_1, X_1 + X_3 - X_1,\ldots,X_1 + X_n - X_1)\end{aligned}$$

Assuming $b = X_1$, we have

$$\delta(\mathbf{X}) = X_1 + \delta(0, X_2 - X_1,\ldots,X_n - X_1) \qquad \text{(since } \delta \text{ is equivariant)}$$

The second expression on the right-hand side is a function of $(X_2 - X_1,\ldots,X_n - X_1)$. We may express this by $l(X_2 - X_1,\ldots,X_n - X_1) = \delta(0, X_2 - X_1,\ldots,X_n - X_1)$. This shows that Eq. (7.3.2) holds. ▣

7.3.1 Pitman Estimator

The objective of the present chapter is to find a minimum risk equivariant estimator in the class of all equivariant estimators given in Theorem 7.3.1 for the location family of distributions. The estimator δ_0 is an MRE estimator of θ if

$$R(\theta, \delta_0) = \inf_{\delta} R(\theta, \delta)\ \forall\ \theta \tag{7.3.3}$$

We have already shown that the risk function of an equivariant estimator δ is constant for all θ and that $\bar{\mathcal{G}}$ is transitive. Therefore, by Theorem 7.2.1,

$$R(\theta, \delta) = R(0,\delta) = E_0[\delta(\mathbf{X})]^2\ \forall\ \theta$$

where E_0 corresponds to the *pdf* $f(x; 0)$.
Thus, from Eq. (7.3.3), we have

$$R(0, \delta_0) = \inf_{\delta} R(0, \delta) \tag{7.3.4}$$

For an estimation problem to be invariant, it is essential that the loss function must also be invariant. Thus, a function L in the present problem, for every θ and for every $\bar{g}$,

$$\begin{aligned}L(\theta, \delta) &= L(\bar{g}\theta, \tilde{g}\delta) = L(\theta + b, \delta + b) \quad \forall\ b \in \mathbb{R}^1\\ &= \rho(\delta - \theta)\end{aligned}$$

is invariant if and only if it is a function of $(\delta - \theta)$.

Therefore, the risk corresponding to the MRE estimator δ_0 (in 7.3.4) then simplifies to

$$\begin{aligned}R(0, \delta_0) &= \inf_{\delta} E_0[\rho(\delta)]\\ &= \inf_{q} E_0\{\rho[X_1 - q(\mathbf{Y})]\} \qquad \text{(by Theorem 7.3.1)}\end{aligned}$$

where $\mathbf{Y} = (Y_2,\ldots,Y_n)$, $Y_i = X_i - X_1$, $i = 2,\ldots,n$. The expectation E_0 in the above expression involves the joint *pdf* $h(\mathbf{Y})$ of $\mathbf{Y} = (Y_2,\ldots,Y_n)$ under $\theta = 0$. Denoting the *pdf* of the conditional distribution of X_1 given $\mathbf{Y} = \mathbf{y}$ under $\theta = 0$ by $g(x_1|\mathbf{y})$, the expression for $R(0,\delta_0)$ may be written as

$$R(0,\delta_0) = \inf_q \int_{\mathbf{y}} \left[\int_{x_1|\mathbf{y}} \rho[x_1 - q(\mathbf{y})] g(x_1|\mathbf{y}) dx_1 \right] h(\mathbf{y}) d\mathbf{y}$$

The above integral can be minimized for q if the integrand is minimized for q for each fixed $\mathbf{y}$. Therefore, a minimum risk equivariant estimator δ_0 is the one for which q is chosen so that it minimizes

$$\int \rho(x_1 - q(\mathbf{y}))\, g(x_1|\mathbf{y}) dx_1$$

for every fixed value of $\mathbf{y}$. The q that minimizes the above integral for squared error loss function $\rho(\delta - \theta) = (\delta - \theta)^2$ is

$$q(\mathbf{y}) = \int x_1 g(x_1|\mathbf{y}) dx_1 = E_0\{X_1|\mathbf{Y} = \mathbf{y}\}$$

Thus, the MRE estimator of θ is given by

$$\delta_0(\mathbf{x}) = \mathrm{x}_1 - E_0\{X_1|\mathbf{Y} = \mathbf{y}\} = E\{x_1 - X_1|\mathbf{Y} = \mathbf{y}\} \qquad (7.3.5)$$

This estimator is known as the *Pitman estimator* of θ. To calculate the conditional distribution of X_1 given $\mathbf{Y} = \mathbf{y}$, involved in the Pitman estimator, we calculate the joint *pdf* $X_1, Y_2,\ldots,Y_n$ by

$$f(x_1)f(x_2),\ldots,f(x_n) = f(x_1)f(x_1 + y_2),\ldots,f(x_1 + y_n)$$

Thus, the joint *pdf* of $Y_2,\ldots,Y_n$ is given by

$$\int_{-\infty}^{\infty} f(x_1)f(x_1 + y_2)\ldots f(x_1 + y_n) dx_1$$

This gives the conditional distribution of X_1 given $Y = y$ under $\theta = 0$ as

$$g(x_1|\mathbf{y}) = \frac{f(x_1)f(x_1 + y_2)\ldots f(x_1 + y_n)}{\int_{-\infty}^{\infty} f(x_1)f(x_1 + y_2)\ldots f(x_1 + y_n) dx_1}$$

Define $U = x_1 - X_1$. The conditional distribution of U given $\mathbf{y}$ is then given by $g(x_1 - U|\mathbf{y})$. The Pitman estimator in Eq. (7.3.5) of θ then simplifies to

$$\delta_0(\mathbf{x}) = E_0\{x_1 - X_1|\mathbf{y}\} = E_0\{U|\mathbf{y}\}$$

$$= \frac{\int_{-\infty}^{\infty} u \prod_{j=1}^{n} f(x_j - u) du}{\int_{-\infty}^{\infty} \prod_{j=1}^{n} f(x_j - u) du} \qquad (7.3.6)$$

Pitman and Bayes Estimator

Let $X_1, X_2,\ldots,X_n$ be *iid* with *pdf* $f(x;\theta)$. The joint density of sample observations is then given by $f(\mathbf{x};\theta) = \Pi_{i=1}^n f(x_i;\theta) = \Pi_{i=1}^n f(x_i - \theta)$. Let the prior distribution of θ be given by $\pi(\theta)$. Then the joint density of $\mathbf{X}$ and θ is given by

$$h(\mathbf{x}, \theta) = \pi(\theta)\prod_{i=1}^{n} f(x_i - \theta)$$

and the marginal density of $\mathbf{X}$ is

$$m(\mathbf{x}) = \int_{\Theta} h(\mathbf{x}, \theta)d\theta$$

$$= \int_{-\infty}^{\infty} \pi(\theta)\prod_{i=1}^{n} f(x_i - \theta)d\theta$$

Then the posterior density is given by

$$g(\theta|\mathbf{x}) = \frac{\pi(\theta)\prod_{i=1}^{n} f(x_i - \theta)}{\int_{-\infty}^{\infty} \pi(\theta)\prod_{i=1}^{n} f(x_i - \theta)d\theta}$$

The Bayes estimate of θ with respect to a noninformative improper prior, $\pi(\theta) = 1$ on Θ, under the squared error loss is given by

$$E(\theta|\mathbf{x}) = \frac{\int_{-\infty}^{\infty} \theta\prod_{i=1}^{n} f(x_i - \theta)d\theta}{\int_{-\infty}^{\infty} \prod_{i=1}^{n} f(x_i - \theta)d\theta} \tag{7.3.7}$$

and we immediately conclude that the Pitman estimator of θ in Eq. (7.3.6) and the Bayes estimator of θ in Eq. (7.3.7), in case of noninformative prior and squared error loss function, are the same. Further, the Pitman estimator δ_0, in this case, is also minimax since the risk of δ_0 is constant over all θ in Θ.

Equivariant Estimator and Sufficient Statistic

Let a sampling population admits a sufficient statistic, say, T. On using factorization theorem, the Pitman estimator of θ reduces to

$$\delta_0(\mathbf{x}) = \frac{\int_{-\infty}^{\infty} \theta g(t, \theta)h(\mathbf{x})d\theta}{\int_{-\infty}^{\infty} g(t, \theta)h(\mathbf{x})d\theta}$$

$$= \frac{\int_{-\infty}^{\infty} \theta g(t, \theta) d\theta}{\int_{-\infty}^{\infty} g(t, \theta) d\theta}$$

This shows that the Pitman estimator δ_0 is a function of a sufficient statistic T.

Pitman Estimator is Equivariant (Invariant) with Respect to the Location Group

Consider δ as a Pitman estimator with respect to a location group

$$\mathcal{G} = \{g: g(\mathbf{x} + a) = \mathbf{x} + a, a \in \mathbb{R}^1\}$$

$$\delta(g(\mathbf{x})) = \delta(\mathbf{x} + a)$$

$$= \frac{\int_{-\infty}^{\infty} \theta \prod_{i=1}^{n} f(x_i + a - \theta) d\theta}{\int_{-\infty}^{\infty} \prod_{i=1}^{n} f(x_i + a - \theta) d\theta}$$

Let $$\theta - a = u$$

$$\delta(g(\mathbf{x})) = \frac{\int_{-\infty}^{\infty} (a + u) \prod_{i=1}^{n} f(x_i - u) du}{\int_{-\infty}^{\infty} \prod_{i=1}^{n} f(x_i - u) du}$$

$$= a + \frac{\int_{-\infty}^{\infty} u \prod_{i=1}^{n} f(x_i - u) du}{\int_{-\infty}^{\infty} \prod_{i=1}^{n} f(x_i - u) du}$$

$$= a + \delta(\mathbf{x})$$

This shows that the Pitman estimator is also equivariant.

Estimation of Mean of Multivariate Normal Distribution

Consider the estimation of mean vector $\boldsymbol{\theta}$ in the p-variate normal distribution $N_p(\boldsymbol{\theta}, \mathbf{I})$, $\Theta = \mathcal{A} = \mathbb{R}^p$, under squared error loss function

$$L(\boldsymbol{\theta}, \mathbf{a}) = (\boldsymbol{\theta} - \mathbf{a})'(\boldsymbol{\theta} - \mathbf{a})$$

$$= \sum_{i=1}^{p} (\theta_i - a_i)^2 \tag{7.3.8}$$

Let $\mathbf{X}_1, \ldots, \mathbf{X}_n$ be a random sample taken from the population $N_p(\boldsymbol{\theta}, \mathbf{I})$. In case of single parameter $\theta (p = 1)$, the sample mean $\bar{X}$ is minimax and admissible estimator of θ. Stein has shown that $\bar{\mathbf{X}} = (\bar{\mathbf{X}}_1,\ldots,\bar{\mathbf{X}}_p)'$ is both minimax and admissible in case of $p = 2$ but it is not admissible though remains minimax in case of $p \geq 3$. In addition to it, James and Stein have considered the estimator of $\boldsymbol{\theta}$

$$\boldsymbol{q}_0(\mathbf{X}) = \bar{\mathbf{X}}\left(1 - \frac{p-2}{n\bar{\mathbf{X}}'\bar{\mathbf{X}}}\right) \tag{7.3.9}$$

as an improvement over $\bar{\mathbf{X}}$ for all values of $\boldsymbol{\theta}$. The estimator $\boldsymbol{\theta}_0$ takes $\bar{\mathbf{X}}$ and shifts it towards the origin provided $\bar{\mathbf{X}}'\bar{\mathbf{X}} > p - 2$. This estimator is not unbiased, not MLE, and not equivariant. Clearly, it does not share with the nice statistical properties of the sample mean $\bar{\mathbf{X}}$.

7.4 PITMAN ESTIMATOR FOR SCALE FAMILIES

Let $X_1,\ldots,X_n$ be a random sample from a distribution in the scale family of distributions with common *pdf*

$$f(x;\sigma) = \frac{1}{\sigma} f\left(\frac{x}{\sigma}\right) \tag{7.4.1}$$

where f is known and $\sigma > 0$ is an unknown scale parameter. The joint density of sample observations is given by

$$\begin{aligned} f(\mathbf{x};\sigma) &= \frac{1}{\sigma^n} f\left(\frac{x_1}{\sigma},\ldots,\frac{x_n}{\sigma}\right) \\ &= \frac{1}{\sigma^n}\prod_{i=1}^{n} f\left(\frac{x_i}{\sigma}\right) \end{aligned} \tag{7.4.2}$$

Consider a group of scale transformations on the sample space χ

$$\mathcal{G} = \{g_b: g_b(\mathbf{X}) = (bX_1,\ldots,bX_n),\ b > 0\}$$

Clearly, $\mathcal{G}$ leaves the family of joint distributions of $\mathbf{X}$, $\{f(\mathbf{x};\sigma),\ \sigma > 0\}$, invariant. Subsequently, the induced family of transformations $\bar{\mathcal{G}}$ on Θ is the same as $\mathcal{G}$. We will consider the problem of estimation of σ^r in the present section. A loss function $L(\sigma, a)$ in the present problem is invariant

$$\begin{aligned} L(\bar{g}(\sigma), \tilde{g}(a)) &= L(b\sigma^r, ba) \\ &= L(\sigma^r, a) \end{aligned} \tag{7.4.3}$$

if and only if it is a function of a/σ^r, say $\rho(a/\sigma^r)$. Consider a new set of variables $Y_1,\ldots,Y_{n-1}$, Y_n where

$$Y_1 = \frac{X_1}{X_n},\ldots, Y_{n-1} = \frac{X_{n-1}}{X_n}, Y_n = \frac{X_n}{|X_n|}, X_n \neq 0 \tag{7.4.4}$$

Consider an estimator δ of σ^r. This estimator is equivariant under $\mathcal{G}$ if

$$\delta(g_b(\mathbf{X})) = b^r\delta(\mathbf{X})\ \forall\ b > 0 \tag{7.4.5}$$

which is satisfied if and only if

$$\delta(\mathbf{X}) = \frac{\delta_0(\mathbf{X})}{q(\mathbf{Y})} \tag{7.4.6}$$

where δ_0 is some equivariant estimator of σ^r, and q is some function of $(Y_1,\ldots,Y_n)$. The risk of this estimator δ is defined by

$$R(\sigma, \delta) = E_\sigma L[\sigma, \delta(\mathbf{X})]$$

$$= E_\sigma\left[\rho\left(\frac{\delta}{\sigma^r}\right)\right] = E_\sigma\left[\rho\left(\frac{\delta_0(\mathbf{X})}{q(\mathbf{Y})\sigma^r}\right)\right] \tag{7.4.7}$$

This risk is constant over Θ by Theorem 7.2.1 and $\bar{\mathcal{G}}$ is transitive. Therefore, we may write Eq. (7.4.7) as

$$R(1, \delta) = E_1\left[\rho\left(\frac{\delta_0(\mathbf{X})}{q(\mathbf{Y})}\right)\right] = \int E_1\left\{\rho\left(\frac{\delta_0(\mathbf{X})}{q(\mathbf{y})}\right)\middle|\,\mathbf{y}\right\}g(\mathbf{y})d\mathbf{y} \tag{7.4.8}$$

An MRE or a Pitman estimator of σ^r is δ in Eq. (7.4.6) with such q that minimizes the integrand in Eq. (7.4.8), $E_1\{\rho(\delta_0(\mathbf{X})/q(\mathbf{y}))|\mathbf{y}\}$, over q for each $\mathbf{y}$. Consider a loss function satisfying Eq. (7.4.3)

$$\rho\left(\frac{\delta_0(\mathbf{X})}{q(\mathbf{y})\sigma^r}\right) = \left(\frac{\delta_0(\mathbf{X})}{q(\mathbf{y})\sigma^r} - 1\right)^2$$

This gives
$$E_1\left\{\rho\left(\frac{\delta_0(\mathbf{X})}{q(\mathbf{y})}\right)\middle|\,\mathbf{y}\right\} = E_1\left\{\left(\frac{\delta_0(\mathbf{X})}{q(\mathbf{y})} - 1\right)^2\middle|\,\mathbf{y}\right\}$$

$$= \frac{1}{q^2(\mathbf{y})}E_1\left\{[\delta_0(\mathbf{X}) - q(\mathbf{y})]^2\middle|\,\mathbf{y}\right\}$$

$$= \left\{\left(\frac{1}{q^2(\mathbf{y})}E_1[\delta_0^2(\mathbf{X})] - \frac{2}{q}E_1\delta_0(\mathbf{X}) + 1\right)\middle|\,\mathbf{y}\right\}$$

Thus, the q in Eq. (7.4.6) that minimizes $R(1, \delta)$ in Eq. (7.4.8) is given by

$$\frac{\partial}{\partial q}E_1\left\{\rho\left(\frac{\delta_0(\mathbf{X})}{q(\mathbf{y})}\right)\middle|\,\mathbf{y}\right\} = -\frac{2}{q^3}E_1\delta_0^2 + \frac{2}{q^2}E_1\delta_0 = 0$$

$$\frac{1}{q}E_1\delta_0^2 = E_1\delta_0$$

$$q = \frac{E_1\{\delta_0^2|\mathbf{y}\}}{E_1\{\delta_0|\mathbf{y}\}} \tag{7.4.9}$$

Therefore, from Eqs. (7.4.9) and (7.4.6), the MRE [or the Pitman estimator (1939)] of σ^r is given by

$$\delta(\mathbf{X}) = \frac{\delta_0(\mathbf{X}) E_1\{\delta_0(\mathbf{X}) | \mathbf{y}\}}{E_1\{\delta_0^2(\mathbf{X}) | \mathbf{y}\}} \tag{7.4.10}$$

Let the joint density of Ys be given by $f_{\mathbf{Y}}(y_1,\ldots,y_n)$. Thus, the *Pitman estimator* in Eq. (7.4.10) simplifies to

$$\delta(\mathbf{x}) = \frac{\int_0^\infty v^{n+r-1} f(vx_1,\ldots,vx_n)\, dv}{\int_0^\infty v^{n+2r-1} f(vx_1,\ldots,vx_n)\, dv} \tag{7.4.11}$$

One may easily see that the Pitman estimator in Eq. (7.4.11) is a function of a sufficient statistic if the scale family admits a sufficient statistic for σ.

Next, we will show that the MRE of σ^r and its Bayes estimate with respect to an improper prior $\pi(\sigma) = 1/\sigma$ are the same.

The joint *pdf* of sample observations

$$f(\mathbf{x}; \sigma) = \frac{1}{\sigma^n} f\left(\frac{x_1}{\sigma},\ldots,\frac{x_n}{\sigma}\right)$$

and the improper prior $\pi(\sigma) = 1/\sigma$ gives the joint density of $\mathbf{X}$ and σ by

$$h(\mathbf{x}, \sigma) = \frac{1}{\sigma^{n+1}} f\left(\frac{x_1}{\sigma},\ldots,\frac{x_n}{\sigma}\right)$$

Using the joint density of $\mathbf{X}$ and σ, the marginal density of $\mathbf{X}$ is given by

$$m(\mathbf{x}) = \int_0^\infty \frac{1}{\sigma^{n+1}} f\left(\frac{x_1}{\sigma},\ldots,\frac{x_n}{\sigma}\right) d\sigma$$

Thus, the conditional density of σ given $\mathbf{X} = \mathbf{x}$ is given by

$$g(\sigma|\mathbf{x}) = \frac{\frac{1}{\sigma^{n+1}} f\left(\frac{x_1}{\sigma},\ldots,\frac{x_n}{\sigma}\right)}{\int_0^\infty \frac{1}{\sigma^{n+1}} f\left(\frac{x_1}{\sigma},\ldots,\frac{x_n}{\sigma}\right) d\sigma} \tag{7.4.12}$$

Using the loss function of the type Eq. (7.4.3) and the posterior density given in Eq. (7.4.12), the posterior conditional expected loss is defined by

$$E\{L(\sigma, a) | \mathbf{X} = \mathbf{x}\} = E\left\{\left(\frac{a}{\sigma^r} - 1\right)^2 \middle| \mathbf{x}\right\} \tag{7.4.13}$$

By minimizing the posterior conditional expected loss given in Eq. (7.4.13) with respect to a, we get the Bayes estimate of σ^r

$$\frac{\partial}{\partial a} E\left\{\left(\frac{a}{\sigma^r} - 1\right)^2 \Big| \mathbf{x}\right\} = 0$$

$$E\left\{\frac{2}{\sigma^r}\left(\frac{a}{\sigma^r} - 1\right) \Big| \mathbf{x}\right\} = 0$$

Thus, the Bayes estimator of σ^r is given by

$$\delta^B = \frac{E\{(1/\sigma^r) | \mathbf{x}\}}{E\{(1/\sigma^{2r}) | \mathbf{x}\}} \tag{7.4.14}$$

$$\delta^B = \frac{\int_0^\infty \left(\frac{1}{\sigma^{n+r+1}}\right) f\left(\frac{x_1}{\sigma}, \ldots, \frac{x_n}{\sigma}\right) d\sigma}{\int_0^\infty \left(\frac{1}{\sigma^{n+2r+1}}\right) f\left(\frac{x_1}{\sigma}, \ldots, \frac{x_n}{\sigma}\right) d\sigma}$$

By letting
$$v = \frac{1}{\sigma}, \qquad dv = -\frac{1}{\sigma^2} d\sigma$$

$$\sigma^2 dv = -d\sigma, \quad \frac{1}{v^2} dv = -d\sigma$$

Expression for Bayes estimate of σ^2 reduces to

$$\delta^B = \frac{\int_0^\infty v^{n+r+1} f(vx_1, \ldots, vx_n) \frac{1}{v^2} dv}{\int_0^\infty v^{n+2r+1} f(vx_1, \ldots, vx_n) \frac{1}{v^2} dv}$$

$$= \frac{\int_0^\infty v^{n+r-1} f(vx_1, \ldots, vx_n)\, dv}{\int_0^\infty v^{n+2r-1} f(vx_1, \ldots, vx_n)\, dv} \tag{7.4.15}$$

Therefore, we conclude that the MRE estimator of σ^r given in Eq. (7.4.11) is the same as the Bayes estimator of σ^r with respect to improper prior $\pi(\sigma) = \sigma^{-1}$ given in Eq. (7.4.15). Note that MRE estimator is also minimax since the equivariant estimator has a constant risk over Θ.

7.5 PITMAN ESTIMATOR FOR LOCATION-SCALE FAMILIES

Let the joint density of $\mathbf{X} = (X_1,\ldots,X_n)$ be given by

$$f(\mathbf{x};\mu,\sigma) = \frac{1}{\sigma^n} f\left(\frac{x-\mu}{\sigma},\ldots,\frac{x_n-\mu}{\sigma}\right) \tag{7.5.1}$$

where f is known and both the parameters $\boldsymbol{\theta} = (\mu, \sigma)$ are unknown, $\boldsymbol{\theta} \in \Theta$ and $\Theta = \{(\mu, \sigma):\mu \in \mathbb{R}^1, \sigma > 0\}$. The family of *pdfs* $\mathcal{F} = \{f(x;\boldsymbol{\theta}): \boldsymbol{\theta} \in \Theta\}$ is called *location-scale family*. Consider the problem of estimating σ^r, $r \neq 0$, under the loss function

$$L(\sigma, a) = \rho\left(\frac{a}{\sigma^r}\right) \tag{7.5.2}$$

Consider the group of location and scale transformations

$$\mathcal{G} = \{g_{a,b}(\mathbf{X}): g_{a,b}(\mathbf{X}) = a + b\mathbf{X} = (a + bX_1,\ldots,a + bX_n), a \in \mathbb{R}^1, b > 0\} \tag{7.5.3}$$

These transformations, then, induce similar transformations on Θ by

$$\bar{\mathcal{G}} = \{g_{a,b}(\theta): g_{a,b}(\theta) = (a + b\mu, b\sigma), a \in \mathbb{R}^1, \text{and } b > 0\} \tag{7.5.4}$$

and on the class of estimators D

$$\tilde{D} = \{g_b(d): g_b(d) = b^r d, b > 0, r \neq 0, d \in D\} \tag{7.5.5}$$

where D is the class of decision rules for the above decision problem. Clearly, the location-scale family is invariant under a group of transformations $\mathcal{G}$ and, also, the loss function Eq. (7.5.2) is invariant under the location-scale transformations in Eq. (7.5.3) since

$$L(b\sigma, b^r d) = \rho\left(\frac{b^r d}{b^r \sigma^r}\right) = \rho\left(\frac{d}{\sigma^r}\right) = L(\sigma, d) \tag{7.5.6}$$

Therefore, the decision problem is invariant. An estimator δ of σ^r is location-scale invariant if and only if

$$\delta(a + bX_1,\ldots,a + bX_n) = b^r \delta(X_1,\ldots,X_n), \ \forall \ a \in \mathbb{R}^1, b > 0$$

Therefore, by Theorem (7.2.1), the risk function of any location-scale invariant estimator δ is constant, since the family $\mathcal{F}$ is a location-scale family.

We will now discuss the Pitman estimator of σ^r. For this problem, we consider $b = 1$. The corresponding transformations and spaces are given by

$$\begin{aligned} \mathcal{G} &= \{g_{a,1}(\mathbf{X}): g_{a,1}(\mathbf{X}) = a + \mathbf{X}\} \\ \bar{\mathcal{G}} &= \{\bar{g}_{a,1}(\boldsymbol{\theta}): \bar{g}_{a,1}(\boldsymbol{\theta}) = (a + \mu, \sigma), a \in \mathbb{R}^1\} \\ \tilde{\mathcal{G}} &= \{\tilde{g}(d): \tilde{g}(d) = d\} \end{aligned} \tag{7.5.7}$$

and the equivariant estimator must satisfy

$$\delta(a + X_1,\ldots,a + X_n) = \delta(X_1,\ldots,X_n) \ \forall \ a \in \mathbb{R}^1, b > 0 \tag{7.5.8}$$

The equivariant estimator in Eq. (7.5.8), by Theorem 7.3.1, is a function of differences $Y_i = X_i - X_n$, $i = 1,\ldots,n - 1$; and we write

$$f(x_1,\ldots,x_n; \mu, \sigma) = f(y_1 + x_n,\ldots,y_{n-1} + x_n, x_n; \mu, \sigma)$$

The joint density of Ys can be written as

$$f(y_1,\ldots, y_{n-1}; \mu, \sigma) = \int_{\mathbb{R}^1} (y_1 + u,\ldots, y_{n-1} + u, u)\, du$$

$$= \frac{1}{\sigma^n} \int_{\mathbb{R}^1} f\left(\frac{y_1 + u}{\sigma},\ldots, \frac{y_{n-1} + u}{\sigma}, \frac{u}{\sigma}\right) du$$

Let $u/\sigma = t$. We get

$$f(y_1,\ldots, y_{n-1}; \mu, \sigma) = \frac{1}{\sigma^{n-1}} \int_{\mathbb{R}^1} f\left(\frac{y_1}{\sigma} + t,\ldots, \frac{y_{n-1}}{\sigma} + t, t\right) dt \tag{7.5.9}$$

which is the form of the scale family Eq. (7.4.2) when n is replaced by $n - 1$ and x_is are replaced by y_is. By taking $\delta_0(\mathbf{Y})$ as any finite risk scale equivariant estimator of σ^r based on $\mathbf{Y} = (Y_1,\ldots,Y_{n-1})$, the MRE estimator of σ^r among the estimators Eq. (7.4.5), $[\delta(a + b\mathbf{X}) = b^r\delta(\mathbf{X})]$, by applying the results of Section 7.4, is given by

$$\delta^*(\mathbf{X}) = \left(\frac{\delta_0(\mathbf{Y})}{q^*(\mathbf{Z})}\right) \tag{7.5.10}$$

where $z_i = Y_i/Y_{n-1}$, $i = 1,\ldots,n - 2$, $z_{n-1} = Y_{n-1}/|Y_{n-1}|$, and $q^*(\mathbf{z})$ is any number that minimizes

$$E_{\sigma=1}\left\{L\left(1, \frac{\delta_0(\mathbf{Y})}{q(\mathbf{z})}\right)\middle| \mathbf{Z} = \mathbf{z}\right\} = E_1\left\{\rho\left(\frac{\delta_0(\mathbf{Y})}{q(\mathbf{z})}\right)\middle| \mathbf{z}\right\} \tag{7.5.11}$$

over all function $q(\mathbf{z}) > 0$. Here, E_1 is the conditional distribution of $\mathbf{Y}$ given $\mathbf{Z} = \mathbf{z}$ under the assumption that the unconditional distribution of $\mathbf{Y}$ is given by Eq. (7.5.9) with $\sigma = 1$. Consider the form of the loss function

$$L(\sigma, d) = \rho\left(\frac{d}{\sigma^r}\right) = \left(\frac{d}{\sigma^r} - 1\right)^2 \tag{7.5.12}$$

for the estimation of σ^r. The *MRE estimator* or the *Pitman estimator* $\delta^*(\mathbf{X}) = \delta_0(\mathbf{Y})/q^*(\mathbf{Z})$ in Eq. (7.5.10) may alternatively be expressed by

$$\delta^*(\mathbf{x}) = \frac{\int_0^\infty \sigma^{n+r-1} f(\sigma y_1,\ldots, \sigma y_{n-1})\, d\sigma}{\int_0^\infty \sigma^{n+2r-1} f(\sigma y_1,\ldots, \sigma y_{n-1})\, d\sigma} \tag{7.5.13}$$

wherein the integral involves the joint density of $\mathbf{Y} = (Y_1,\ldots,Y_{n-1})$ under $\sigma = 1$, which is given by

$$f(y_1,\ldots, y_{n-1}) = \int_{-\infty}^{\infty} f(y_1 + t,\ldots, y_{n-1} + t, t)\, dt$$

We, next, consider the estimation of the location parameter μ in the location–scale density given in Eq. (7.5.1). The transformations on the sample space $\mathcal{G}$ and on the parameter space $\bar{\mathcal{G}}$, remain the same as are given in Eqs. (7.5.3) and (7.5.4), but on the decision space, $\tilde{\mathcal{G}}$ changes to

$$\tilde{\mathcal{G}} = \{\tilde{g}: \tilde{g}(d) = a + bd,\ a \in \mathbb{R}^1,\ b > 0\} \tag{7.5.14}$$

Under these transformations, the loss function is invariant if and only if it is some function of $[(d - \mu)/\sigma]$

$$L[(\mu, \sigma), d] = \rho\left(\frac{d-\mu}{\sigma}\right) \tag{7.5.15}$$

An estimator δ of μ is location-scale equivariant if and only if

$$\delta(a + bx_1,\ldots,a + bx_n) = a + b\,\delta(x_1,\ldots,x_n),\ a \in \mathbb{R}^1,\ b > 0 \tag{7.5.16}$$

The risk of such an equivariant estimator δ is constant since $\bar{\mathcal{G}}$ is transitive. Therefore, an MRE estimator of μ exists and it can easily be calculated as follows:
Consider a location-scale equivariant estimator δ_0 of μ as in Eq. (7.5.16)

$$\delta_0(a + b\mathbf{x}) = a + b\delta_0(\mathbf{x}),\ a \in \mathbb{R}^1,\ b > 0 \tag{7.5.17}$$

and an estimator δ_1 of σ taking on only positive values

$$\delta_1(a + b\mathbf{x}) = b\delta_1(\mathbf{x}),\ a \in \mathbb{R}^1,\ b > 0 \tag{7.5.18}$$

Define a function u

$$u(a + b\mathbf{x}) = u(\mathbf{x}),\ a \in \mathbb{R}^1,\ b > 0 \tag{7.5.19}$$

Then, by using Eqs. (7.5.17), (7.5.18), and (7.5.19), the totality of all location-scale equivariant estimators can be expressed by

$$\delta(\mathbf{x}) = \delta_0(\mathbf{x}) - u(\mathbf{x})\, \delta_1(\mathbf{x}) \tag{7.5.20}$$

This estimator is equivariant since

$$\begin{aligned}\delta(a + b\mathbf{x}) &= \delta_0(a + b\mathbf{x}) - u(a + b\mathbf{x})\, \delta_1(\mathbf{a} + b\mathbf{x})\\ &= a + b\delta_0(\mathbf{x}) - u(\mathbf{x})b\delta_1(\mathbf{x})\\ &= a + b[\delta_0(\mathbf{x}) - u(\mathbf{x})\delta_1(\mathbf{x})]\\ &= a + b\delta(\mathbf{x})\end{aligned}$$

Conversely, suppose δ is equivariant. Define u by

$$u(\mathbf{x}) = \left[\frac{\delta(\mathbf{x}) - \delta_0(\mathbf{x})}{\delta_1(\mathbf{x})}\right]$$

then, u at $a + b\mathbf{x}$ is

$$u(a+b\mathbf{x}) = \frac{\delta(a+b\mathbf{x})-\delta_0(a+b\mathbf{x})}{\delta_1(a+b\mathbf{x})}$$

$$= \frac{a+b\delta(\mathbf{x})-a-b\delta_0(\mathbf{x})}{b\delta_1(\mathbf{x})}$$

$$= \frac{\delta(\mathbf{x})-\delta_0(\mathbf{x})}{\delta_1(\mathbf{x})} = u(\mathbf{x})$$

so that Eqs. (7.5.18) and (7.5.19) hold. Therefore, δ in Eq. (7.5.20) satisfies Eq. (7.5.17) if and only if it is of the form Eq. (7.5.20).

Consider, next, a function of $\mathbf{z}$, $u(\mathbf{z})$, say, at $a + b\mathbf{x}$

$$u(\mathbf{z})\big|_{a+b\mathbf{x}} = u\left(\frac{by_1}{by_{n-1}},\ldots,\frac{by_{n-2}}{by_{n-1}},\frac{by_{n-1}}{|by_{n-1}|}\right)$$

Since $y_i = x_i - x_n$, $i = 1,\ldots,n - 1$, we have y_i at $a + b\mathbf{x}$

$$y_i|_{a+b\mathbf{x}} = a + bx_i - a - bx_n$$

$$= b(x_i - x_n) = by_i$$

$$u(\mathbf{z})\big|_{a+b\mathbf{x}} = u\left(\frac{y_1}{y_{n-1}},\ldots,\frac{y_{n-2}}{y_{n-1}},\frac{y_{n-1}}{|y_{n-1}|}\right) \qquad (\text{since } b > 0)$$

$$= u(\mathbf{z})\big|_{\mathbf{x}} \tag{7.5.21}$$

It satisfies Eq. (7.5.19). Conversely, if Eq. (7.5.19) holds, then

$$u(x_1,\ldots,x_n) = u(x_1 - x_n, x_2 - x_n,\ldots,x_{n-1} - x_n) = u(y_1, y_2,\ldots,y_{n-1})$$

$$= u\left(\frac{y_1}{y_{n-1}},\ldots,\frac{y_{n-2}}{y_{n-1}},\frac{y_{n-1}}{|y_{n-1}|}\right) = u(z_1, z_2,\ldots, z_{n-1})$$

u is a function of $\mathbf{z}$. This shows that u depends only on $\mathbf{z}$. Thus, Eq. (7.5.20) holds if and only if u depends on $\mathbf{x}$ only through $\mathbf{z}$. Therefore, the totality of all the equivariant estimators is given by

$$\delta(\mathbf{x}) = \delta_0(\mathbf{x}) - u(\mathbf{z})\,\delta_1(\mathbf{x}) \tag{7.5.22}$$

where $u(\mathbf{z})$ is a Borel function u on $\mathbb{R}^{n-1}$. By Theorem 7.2.1, the risk of the equivariant estimator δ is independent of μ and σ or that the risk is constant. The risk of δ for loss function in Eq. (7.5.15) is given by

$$R[(\mu, \sigma), \delta] = E_{\mu,\sigma}L[(\mu, \sigma), \delta]$$

$$= E_{\mu,\sigma}\rho\left(\frac{\delta-\mu}{\sigma}\right) \tag{7.5.23}$$

Without the loss of generality, we may assume $\mu = 0$ and $\sigma = 1$. Then, we have

$$R[(0,1), \delta] = E_{0,1}\rho[\delta_0(\mathbf{X}) - u(\mathbf{Z})\delta_1(\mathbf{X})]$$
$$= \int E_{0,1}^{\mathbf{X}|\mathbf{z}}\{[\rho(\delta_0(\mathbf{X}) - u(\mathbf{z})\delta_1(\mathbf{X}))]\big|\mathbf{z}\}\,dP_{0,1}(\mathbf{z}) \quad (7.5.24)$$

The MRE estimator δ of μ is the one in Eq. (7.5.22) with $u^*(\mathbf{z})$ that minimizes $R[(0,1), \delta]$ or $E_{0,1}\{\rho[\delta_0(\mathbf{X}) - u(\mathbf{z})\delta_1(\mathbf{X})]|\mathbf{z}\}$ for each $\mathbf{z}$.

If we consider

$$\rho\left(\frac{\delta-\mu}{\sigma}\right) = \left(\frac{\delta-\mu}{\sigma}\right)^2 \quad (7.5.25)$$

then $u^*(\mathbf{z})$ is

$$\frac{\partial}{\partial u(\mathbf{z})} E_{0,1}\{[\delta_0(\mathbf{X}) - u(\mathbf{z})\delta_1(\mathbf{X})]^2|\mathbf{z}\} = 0$$

$$E_{0,1}\{2[\delta_0(\mathbf{X}) - u(\mathbf{z})\,\delta_1(\mathbf{X})]\,\delta_1(\mathbf{X})|\mathbf{z}\} = 0$$

$$u^*(\mathbf{z}) = \frac{E_{0,1}\{\delta_0(\mathbf{X})\,\delta_1(\mathbf{X})\big|\mathbf{z}\}}{E_{0,1}\{\delta_1^2(\mathbf{X})\big|\mathbf{z}\}} \quad (7.5.26)$$

and the MRE estimator of μ, in this case, is given by

$$\delta(\mathbf{X}) = \delta_0(\mathbf{X}) - u^*(\mathbf{z})\delta_1(\mathbf{X}) \quad (7.5.27)$$

7.6 SOLVED EXAMPLES

***Example* 7.1 (A general representation of location equivariant estimators).** Show that an estimator δ is location equivariant if and only if

$$\delta(\mathbf{x}) = \delta_0(\mathbf{x}) + \phi(\mathbf{x})$$

where δ_0 is any equivariant estimator and ϕ is an invariant function.

Solution. Assume that the estimator δ is expressable in the above form. Consider now the value of δ at $g(\mathbf{x}) = a + \mathbf{x}$

$$\delta(a + \mathbf{x}) = \delta_0(a + \mathbf{x}) + \phi(a + \mathbf{x})$$
$$= a + \delta_0(\mathbf{x}) + \phi(\mathbf{x})$$
$$= a + \delta(\mathbf{x})$$

since δ is equivariant estimator and ϕ is invariant function. This shows that δ is an equivariant estimator.

Conversely, suppose δ is an equivariant estimator. Define a function

$$\phi(\mathbf{x}) = \delta(\mathbf{x}) - \delta_0(\mathbf{x})$$

and consider its value at $g(\mathbf{x}) = a + \mathbf{x}$

$$\begin{aligned}\phi(a+\mathbf{x}) &= \delta(a+\mathbf{x}) - \delta_0(a+\mathbf{x})\\ &= a + \delta(\mathbf{x}) - a - \delta_0(\mathbf{x})\\ &\qquad \text{(since } \delta \text{ and } \delta_0 \text{ are both equivariant estimators)}\\ &= \delta(\mathbf{x}) - \delta_0(\mathbf{x}) = \phi(\mathbf{x})\end{aligned}$$

so that the function ϕ is invariant. This proves the assertion that any equivariant estimator δ is necessarily represented by

$$\delta(\mathbf{x}) = \delta_0(\mathbf{x}) + \phi(\mathbf{x})$$

It is important to note that the totality of all equivariant estimators is achieved by characterizing the function ϕ so that it is invariant under location transformations. It can be easily shown that the function ϕ is invariant if and only if it is a function of differences $x_1 - x_n = y_1,\ldots,x_{n-1} - x_n = y_{n-1}$. Therefore, the totality of all equivariant estimators is expressed by

$$\delta(\mathbf{x}) = \delta_0(\mathbf{x}) - q(\mathbf{y}) \ \forall\ \mathbf{x}$$

***Example* 7.2** Let $X_1,\ldots,X_n$ be *iid* $U(\theta - 1/2,\ \theta + 1/2)$, $\theta > 0$. Find the MRE estimator of θ under the squared error loss function $L(\theta, a) = (\theta - a)^2$.

Solution. The statistic $(X_{(1)}, X_{(n)})$ is sufficient for θ. We have

$$\begin{aligned}f(x_1-\theta,\ldots,x_n-\theta) &= \prod_{i=1}^{n} f(x_i-\theta) = 1, \text{ if } \theta - \frac{1}{2} \le x_1,\ldots,x_n \le \theta + \frac{1}{2}\\ &= \prod_{i=1}^{n} I\left(\theta - \frac{1}{2} \le x_i \le \theta + \frac{1}{2}\right)\\ &= I\left(x_{(n)} - \frac{1}{2} \le \theta \le x_{(1)} + \frac{1}{2}\right)\end{aligned}$$

We note that the above family is a location family. Therefore, the Pitman estimator of θ is given by

$$\delta_0(\mathbf{x}) = \frac{\int_{x_{(n)}-(1/2)}^{x_{(1)}+(1/2)} \theta \prod_{i=1}^{n} f(x_i-\theta)\,d\theta}{\int_{x_{(n)}-(1/2)}^{x_{(1)}+(1/2)} \prod_{i=1}^{n} f(x_i-\theta)\,d\theta} = \frac{\int_{x_{(n)}-(1/2)}^{x_{(1)}+(1/2)} \theta\,d\theta}{\int_{x_{(n)}-(1/2)}^{x_{(1)}+(1/2)} 1\,d\theta}$$

$$= \frac{x_{(1)} + x_{(n)}}{2}$$

***Example* 7.3** Let $X_1,\ldots,X_n$ be *iid* $U(0,\theta)$. Find the Pitman estimator of θ.

Solution. $X_{(n)}$ is the complete sufficient statistic for θ. The joint density of sample observations is given by

$$f(x_1,\ldots,x_n;\theta) = \frac{1}{\theta^n} I_{(0,1)}\left(\frac{x_{(n)}}{\theta}\right)$$

Therefore, this family is a scale family with scale parameter θ. The Pitman estimator of θ is, thus, given by

$$\delta_0(\mathbf{X}) = \frac{\int_0^{1/x_{(n)}} v^n dv}{\int_0^{1/x_{(n)}} v^{n+1} dv} = \frac{n+2}{n+1} X_{(n)}$$

We can find the Pitman estimator of θ, alternatively, by noting that $X_{(n)}$ is a complete sufficient statistic and that $\mathbf{Y} = (X_1/X_n,\ldots,X_{n-1}/X_n)'$ is ancillary. By Basu's theorem, $X_{(n)}$ and $\mathbf{Y}$ are independent. Now, consider $\delta_0(\mathbf{X}) = X_{(n)}$ as an equivariant estimator of σ. This gives $E_1(X_{(n)}|\mathbf{Y}) = E_1(X_{(n)})$. Denote the observed values of $X_{(n)}$ by x_n

$$E_1(X_{(n)}) = \int_0^{\theta} x_n \frac{n}{\theta^n} x_n^{n-1} dx_n \bigg|_{\theta=1}$$

$$= n\int_0^1 x_n^n dx_n = \frac{n}{n+1}$$

and

$$E_1(X_{(n)}^2) = n\int_0^1 x_n^{n+1} dx_n = \frac{n}{n+2}$$

Therefore, the Pitman estimator of σ is given by

$$\delta^*(\mathbf{X}) = \frac{\delta_0(\mathbf{X})E_1[\delta_0(\mathbf{X})|\mathbf{Y}]}{E_1[\delta_0^2(\mathbf{X})|\mathbf{Y}]}$$

$$= \frac{X_{(n)}[n/(n+1)]}{[n/(n+2)]} = \frac{n+2}{n+1} X_{(n)}$$

***Example* 7.4** Let $X_1,\ldots,X_n$ be *iid* from exponential family $E(\theta)$ with *pdf*

$$f(x;\theta) = \frac{1}{\theta}\exp\left(-\frac{x}{\theta}\right), x > 0$$

Find the Pitman estimator of θ^r.

Solution. $E(\theta)$ is a scale family with scale parameter θ. The statistic ΣX_i is a complete sufficient statistic for θ and $\mathbf{Y} = (X_1/X_n,\ldots,X_{n-1}/X_n)'$ is an ancillary statistic. Now, consider $\delta_0(\mathbf{X}) = (\Sigma X_i)^r$ as an equivariant estimator of θ^r. Using δ_0, the Pitman estimator of θ^r is given by

$$\delta^*(\mathbf{X}) = \frac{\left(\sum X_i\right)^r E_1\left(\sum X_i\right)^r}{E_1\left(\sum X_i\right)^{2r}}$$

where $\Sigma X_i \sim G_1(n,\theta)$ since $X_i \sim G_1(1,\theta)$. Under $\theta = 1$, $\Sigma X_i|_{\theta=1} \sim G_1(n,1)$. This gives $E_1(\Sigma X_i)^r = \Gamma(n + r)/\Gamma(n)$, $E_1(\Sigma X_i)^{2r} = \Gamma(n + 2r)/\Gamma(n)$. The Pitman estimator, then, simplifies to

$$\delta^*(\mathbf{X}) = \frac{\Gamma(n+r)}{\Gamma(n+2r)}\left(\sum X_i\right)^r$$

***Example* 7.5** Let $X_1,\ldots,X_n$ be *iid* from the distribution

$$f(x;\,\theta) = \exp\{-(x-\theta)\},\ x \geq \theta$$

(i) Under the squared error loss function, find the Pitman estimator of θ.
(ii) Under the location group and squared error loss function, find the MRE estimator of θ.

Solution.

(i) The given family is a location family and the density of sample observations is given by

$$f(x_1,\ldots,x_n;\theta) = \exp\left[-\sum(x_i-\theta)\right],\ x_{(1)} \geq \theta$$

The Pitman estimator of θ is

$$\delta^*(\mathbf{X}) = \frac{\int_{-\infty}^{X_{(1)}} \theta\exp\left[-\sum(x_i-\theta)\right]d\theta}{\int_{-\infty}^{X_{(1)}} \exp\left[-\sum(x_i-\theta)\right]d\theta}$$

$$= \frac{\int_{-\infty}^{X_{(1)}} \theta\exp(n\theta)d\theta}{\int_{-\infty}^{X_{(1)}} \exp(n\theta)d\theta}$$

$$= \frac{(X_{(1)}/n)\exp(nX_{(1)}) - (1/n^2)\exp(nX_{(1)})}{(1/n)\exp(nX_{(1)})} = X_{(1)} - \frac{1}{n}$$

(ii) The location group is given by

$$\mathcal{G} = \{g\colon g(\mathbf{x}) = (x_1 + a,\ldots,x_n + a),\ a \in \mathbb{R}^1\}$$

The group $\mathcal{G}$ induces $\bar{\mathcal{G}}$ on Θ so that $\bar{\mathcal{G}} = \mathcal{G}$. This implies that $\bar{\mathcal{G}}$ is transitive and the problem of estimation of θ is invariant. Thus, by Theorem 7.2.1, the risk of every equivariant estimator is constant.

The problem may be reduced by sufficiency considerations. The statistic $X_{(1)}$ is a minimal sufficient statistic, and an equivariant estimator of θ based on $X_{(1)}$ must be of the form

$$\delta_c(\mathbf{X}) = X_{(1)} + c,\quad c \in \mathbb{R}^1$$

Under the squared error loss function, the risk of this equivariant estimator δ_c is given by

$$R(\theta, \delta_c) = E_\theta(X_{(1)} + c - \theta)^2 = E_0(X_{(1)} + c)^2 = R(0,\delta_c) \ \forall\ \theta$$

(since the risk is constant)

The distribution of the first-order statistic is given by

$$f_{X_{(1)}}(x_1;\ \theta) = n \exp[-n(x_1 - \theta)]$$

The minimization of $R(0,\delta_c)$ with respect to c,

$$\frac{\partial}{\partial c} E_0(X_{(1)} + c)^2 = 0$$

$$c = -E_0(X_{(1)}) = -n\int_0^\infty x \exp(-nx)\, dx$$

$$= -\frac{1}{n}$$

Therefore, the MRE estimator of θ is given by

$$\delta^*(\mathbf{X}) = X_{(1)} - \frac{1}{n}$$

***Example* 7.6** Let $X_1,\ldots,X_n$ be *iid* $N(\theta,1)$ random variables.

(i) Find the Pitman estimator of θ.
(ii) Under the location group and squared error loss function, find the MRE estimator of θ.

Solution.

(i) $N(\theta, 1)$ is a location family and it gives

$$\prod_{i=1}^{n} f(x_i - \theta) = \frac{1}{(2\pi)^{n/2}} \exp\left\{-\frac{\sum (x_i - \theta)^2}{2}\right\}$$

Rewriting the term $\exp\{-\Sigma(x_i - \theta)^2/2\}$ as

$$\exp\left\{-\frac{\sum (x_i - \theta)^2}{2}\right\} = \exp\left\{-\frac{n}{2}(\theta - \bar{x})^2\right\} \exp\left\{-\frac{1}{2}(n-1)s_{n-1}^2\right\}$$

where $$s_{n-1}^2 = \frac{1}{n-1}\sum (x_i - \bar{x})^2$$

This gives

$$\int_{-\infty}^{\infty} \theta \prod_{i=1}^{n} f(x_i - \theta)d\theta = \frac{1}{(2\pi)^{n/2}} \exp\left\{-\frac{1}{2}(n-1)s_{n-1}^2\right\} \int_{-\infty}^{\infty} \theta \exp\left\{-\frac{n}{2}(\theta - \bar{x})^2\right\} d\theta$$

$$= \frac{1}{(2\pi)^{n/2}} \exp\left\{-\frac{1}{2}(n-1)s_{n-1}^2\right\}\left(\frac{2\pi}{n}\right)^{1/2} \bar{x}$$

and $$\int_{-\infty}^{\infty} \prod_{i=1}^{n} f(x_i - \theta)d\theta = \frac{1}{(2\pi)^{n/2}} \exp\left\{-\frac{1}{2}(n-1)s_{n-1}^2\right\}\left(\frac{2\pi}{n}\right)^{1/2}$$

Therefore, the Pitman estimator of location parameter θ is given by

$$\delta^*(\mathbf{X}) = \frac{\int_{-\infty}^{\infty} \theta \prod_{i=1}^{n} f(X_i - \theta)\,d\theta}{\int_{-\infty}^{\infty} \prod_{i=1}^{n} f(X_i - \theta)\,d\theta}$$

$$= \frac{\int_{-\infty}^{\infty} \theta \exp\left[-\frac{n}{2}(\theta - \bar{X})^2\right] d\theta}{\int_{-\infty}^{\infty} \exp\left[-\frac{n}{2}(\theta - \bar{X})^2\right] d\theta}$$

$$= \frac{\int_{-\infty}^{\infty} (\theta - \bar{X} + \bar{X}) \exp\left[-\frac{n}{2}(\theta - \bar{X})^2\right] d\theta}{\int_{-\infty}^{\infty} \exp\left[-\frac{n}{2}(\theta - \bar{X})^2\right] d\theta} = \bar{X}$$

(ii) Under the location group

$$\mathcal{G} = \{g \colon g(x) = (x_1 + a, \ldots, x_n + a),\ a \in \mathbb{R}^1\}$$

the induced group $\bar{\mathcal{G}}$ on Θ is $\bar{\mathcal{G}} = \mathcal{G}$. Clearly, it is transitive and the problem of estimation of θ is invariant. Thus, by Theorem 7.2.1, the risk of every equivariant estimator of θ is constant over Θ. This problem may be reduced by sufficiency considerations. ΣX_i is the minimal and complete sufficient statistic. An equivariant estimator based on this must satisfy

$$\delta_c(\mathbf{X}) = \bar{X} + c,\ c \in \mathbb{R}^1$$

The risk of δ_c is written as

$$R(\theta, \delta_c) = R(0, \delta_c) = E_0(\bar{X} + c - 0)^2$$

This attains minimum

$$\frac{\partial}{\partial c}R(\theta,\delta_c)=E_0\{2(\bar{X}+c)\}=0$$

at

$$c=-E_0(\bar{X})=0$$

Therefore, the MRE estimator of θ is given by

$$\delta^*(\mathbf{X})=\bar{X}$$

***Example* 7.7** Let $X_1,\ldots,X_n$ be *iid* with double exponential distribution DE(θ,1), $\theta \in \mathbb{R}^1$, with density

$$f(x;\theta)=\frac{1}{2}\exp\{-|x-\theta|\}$$

Find the Pitman estimator of θ under the squared error loss.

Solution. Double exponential distribution is not a member of exponential family. However, it belongs to a location family of distributions. The joint density of sample observations is given by

$$f(x_1,\ldots,x_n;\theta)=\frac{1}{2^n}\exp\left(-\sum|x_i-\theta|\right)$$

The values of integrals in the Pitman estimator are unaffected if we rearrange the observations $x_1,\ldots,x_n$ so that $x_1 < x_2 < \cdots < x_n$. If we have some k so that $x_k < \theta < x_{k+1}$, we may write

$$\sum_{i=1}^{n}|x_i-\theta|=-\sum_{1}^{k}(x_i-\theta)+\sum_{k+1}^{n}(x_i-\theta)$$

$$=\sum_{k+1}^{n}x_i-\sum_{1}^{k}x_i+(2k-n)\theta$$

Thus, the joint *pdf* can be written as

$$f(x_1,\ldots,x_n;\theta)=\frac{1}{2^n}\exp\left\{-\left[\sum_{k+1}^{n}x_i-\sum_{1}^{k}x_i+(2k-n)\theta\right]\right\}$$

$$=\frac{1}{2^n}\exp\left\{-\left[\sum_{k+1}^{n}x_i-\sum_{1}^{k}x_i+A\theta\right]\right\}$$

where

$$A=(2k-n)$$

Therefore, the Pitman estimator of θ is given by

$$\delta^*=\frac{\int_{-\infty}^{\infty}\theta\prod_{i=1}^{n}f(x_i-\theta)d\theta}{\int_{-\infty}^{\infty}\prod_{i=1}^{n}f(x_i-\theta)d\theta}$$

$$= \frac{\int_{x_k}^{x_{k+1}} \theta \exp(-A\theta)\, d\theta}{\int_{x_k}^{x_{k+1}} \exp(-A\theta)\, d\theta} = \frac{x_{k+1}\exp(-Ax_{k+1}) - x_k \exp(-Ax_k)}{\exp(-Ax_{k+1}) - \exp(-Ax_k)} + \frac{1}{A}$$

***Example* 7.8** Let $X_1,\ldots,X_n$ be *iid* $N(0,\sigma^2)$. Find the Pitman estimator of σ^2.

Solution. $N(0,\sigma^2)$ is a scale family with scale parameter σ. The joint density of sample observations is given by

$$f(x_1,\ldots,x_n;\sigma^2) = \frac{1}{(2\pi\sigma^2)^{n/2}} \exp\left\{-\frac{1}{2\sigma^2}\sum x_i^2\right\}$$

Consider a scale invariant estimator as

$$\delta_0(\mathbf{X}) = \sum_{i=1}^{n} X_i^2$$

The statistic $\Sigma_{i=1}^{n} X_i^2$ is a complete sufficient statistics for σ^2 and the statistic $Y = (X_1/X_n,\ldots,X_{n-1}/X_n)'$ is ancillary. By Basu's theorem, ΣX_i^2 and $\mathbf{Y}$ are independent. Therefore, the minimization of

$$E_1\left\{\rho\left(\frac{\delta_0(\mathbf{X})}{q(\mathbf{y})}\right)\middle|\mathbf{y}\right\}$$

for q reduces into the minimization of $E_1\{\rho[\delta_0(\mathbf{X})/q(\mathbf{y})]\}$ where $\rho(\delta_0/q) = ((\delta_0/q) - 1)^2$. We then have

$$\frac{\partial}{\partial q} E_1\left(\frac{\sum X_i^2}{q} - 1\right)^2 = 0$$

$$E_1\left\{\left(\frac{\sum X_i^2}{q} - 1\right)\left(-\frac{\sum X_i^2}{q^2}\right)\right\} = 0$$

$$E_1\left\{\frac{\left(\sum X_i^2\right)^2}{q} - \sum X_i^2\right\} = 0$$

$$q^* = \frac{E_1\left(\sum X_i^2\right)^2}{E_1\left(\sum X_i^2\right)}$$

Therefore, the Pitman estimator is given by

$$\delta^*(\mathbf{X}) = \frac{\delta_0(\mathbf{X})}{q^*(\mathbf{Y})} = \frac{\sum X_i^2 E_1(X_i^2)}{E_1\left(\sum X_i^2\right)^2}$$

Under $\sigma = 1$, $\Sigma X_i^2 \sim \chi_n^2 = G_1(n/2, 2)$. This gives $E_1(\Sigma X_i^2) = n$ and $E_1(\Sigma X_i^2)^2 = (n + 2)n$. Thus, the Pitman estimator simplifies to

$$\delta^*(\mathbf{X}) = \frac{\sum X_i^2}{(n+2)}$$

Further, the estimator $\delta_0 = (\Sigma X_i^2)^{r/2}$ is an equivariant estimator for estimating σ^r. The Pitman estimator for σ^r is then given by

$$\delta^*(\mathbf{X}) = \frac{\left(\sum X_i^2\right)^{r/2} E_1\left(\sum X_i^2\right)^{r/2}}{E_1\left(\sum X_i^2\right)^r}$$

On calculating the terms involved in the above expression

$$E_1\left(\sum X_i^2\right)^{r/2} = \frac{2^{r/2}\Gamma[(n+r)/2]}{\Gamma(n/2)}$$

and

$$E_1\left(\sum X_i^2\right)^r = \frac{2^r\Gamma[(n+2r)/2]}{\Gamma(n/2)}$$

gives the Pitman estimator of σ^r

$$\delta^*(\mathbf{X}) = \frac{\left(\sum X_i^2\right)^{r/2} \Gamma[(n+r)/2]}{2^{r/2}\Gamma[(n+2r)/2]}$$

***Example* 7.9** Consider a random sample $X_1,\ldots,X_n$ from the exponential distribution $E(\mu, \sigma)$ with *pdf*

$$f(x;\mu,\sigma) = \frac{1}{\sigma}\exp\left\{-\left(\frac{x-\mu}{\sigma}\right)\right\} I_{(\mu,\infty)}(x),\ \sigma > 0$$

(i) Assume that the location parameter $\mu \in \mathbb{R}^1$ is unknown and the scale parameter $\sigma > 0$ is known. Find the MRE estimator of μ under the loss functions

$$L(\mu,\delta) = |\mu - \delta|$$

and

$$L(\mu,\delta) = I_{(t,\infty)}(|\mu - \delta|)$$

(ii) Assume that $\mu \in \mathbb{R}^1$ and $\sigma > 0$ are both unknown. Determine the MRE estimator of σ for the loss function

$$L(\sigma,\delta) = \left|\frac{\delta}{\sigma} - 1\right|^p, \quad p = 1, 2$$

Consider the estimation of σ^r with the loss function

$$L(\mu, \sigma, \delta) = \rho\left(\frac{\delta}{\sigma^r}\right) = \left(\frac{\delta}{\sigma^r} - 1\right)^2 = \frac{(\delta - \sigma^r)^2}{\sigma^{2r}}$$

and find the MRE estimator of σ. Also, determine the MRE estimator of μ for the loss function

$$L(\mu, \sigma, \delta) = \rho\left(\frac{\delta - \mu}{\sigma}\right) = \frac{(\delta - \mu)^2}{\sigma^2}$$

and compute its bias.

Solution.

(i) The statistic, in this case, $\mathbf{Y} = (Y_1, \ldots, Y_{n-1}) = (X_1 - X_n, \ldots, X_{n-1} - X_n)$ is an ancillary statistic and $X_{(1)}$ is complete sufficient for μ. Thus, by Basu's theorem, $X_{(1)}$ and $\mathbf{Y}$ are independent. Consider $\delta_0 = X_{(1)}$ as a location equivariant estimator of μ. By the results of Section 7.3, the MRE estimator of μ is given by

$$\delta(\mathbf{X}) = \delta_0(\mathbf{X}) - u^*(\mathbf{Y}) = X_{(1)} - u^*(\mathbf{Y})$$

where u^* is the number that minimizes

$$E_0[|\delta_0(\mathbf{X}) - u(\mathbf{y})||\mathbf{y}]$$

for every $\mathbf{y}$ in case of absolute error loss function. Thus, u^* is the number that minimizes

$$E_0\left|X_{(1)} - u\right| = \int_0^\infty |x - u| \frac{n}{\sigma} \exp\left(-\frac{n}{\sigma}x\right) dx$$

where E_0 is the expectation when $\mu = 0$, and the density of $X_{(1)}$ under $\mu = 0$ is

$$f_{X_{(1)}}(x; \theta) = \frac{n}{\sigma} \exp\left(-\frac{n}{\sigma}x\right) I_{(0,\infty)}(x)$$

This gives that u^* is the median of the distribution of $X_{(1)}$ under $\mu = 0$, i.e.,

$$\int_0^{u^*} \frac{n}{\sigma} \exp\left(-\frac{n}{\sigma}x\right) dx = \frac{1}{2}$$

$$1 - \exp\left(-\frac{n}{\sigma}u^*\right) = \frac{1}{2}$$

or

$$u^* = \frac{\sigma}{n}\log 2$$

and the MRE estimator of μ is given by

$$\delta = X_{(1)} - \frac{\sigma}{n}\log 2$$

In case of the loss function $L(\mu, \delta) = I_{(t, \infty)}(|\mu - \delta|)$, the number u^* is obtained by minimizing

$$E_0[|X_{(1)} - u|\, I_{(t,\infty)}(X_{(1)})] = P_0(|X_{(1)} - u| > t)$$

with respect to u. For negative values of u, we have

$$P_0(|X_{(1)} - u| > t) \geq P_0(X_{(1)} > \mathrm{t})$$

Therefore, the negative values of u cannot be the ones which would minimize $P_0(|X_{(1)} - u| > t)$; as opposed to it, it would be some positive values of u. For $u > 0$,

$$P_0(|X_{(1)} - u| > t) = P_0(X_{(1)} < u - t) + P_0(X_{(1)} > u + t)$$

$$= 1 - \exp\left\{-\frac{n}{\sigma}\min[(u-t), 0]\right\} + \exp\left\{-\frac{n}{\sigma}(u+t)\right\}$$

P_0 attains minimum at $u^* = t$. Thus, the MRE estimator of μ is given by $\delta = X_{(1)} - t$.

(ii) The statistics $[X_{(1)}, \Sigma_{i=1}^n(X_i - X_{(1)})]$ are jointly complete sufficient statistics for (μ, σ), and $X_{(1)}$ and $\Sigma_{i=1}^n(X_i - X_{(1)})$ are mutually independent. The density of $X_{(1)}$ is given by

$$f_{X_{(1)}}(x; \mu, \sigma) = \frac{n}{\sigma}\exp\left[-\frac{n(x-\mu)}{\sigma}\right] I_{(\mu,\infty)}(x)$$

We have $$\frac{n(X_{(1)} - \mu)}{\sigma} \sim G_1(1,1)$$

$$\frac{\sum(X_i - \mu)}{\sigma} = \frac{\sum(X_i - X_{(1)} + X_{(1)} - \mu)}{\sigma} = \frac{\sum(X_i - X_{(1)})}{\sigma} + \frac{n(X_{(1)} - \mu)}{\sigma}$$

or $$Q = Q_1 + Q_2$$

$\Rightarrow$ $$Q \sim G_1(n, 1), \quad Q_2 \sim G_1(1, 1)$$

$$Q_1 = \frac{1}{\sigma}\sum_{i=1}^{n}(X_i - X_{(1)}) \sim G_1((n-1), 1)$$

The statistic $\mathbf{W} = (W_1,\ldots,W_{n-1})$, $W_i = (X_i - X_n)/(X_{n-1} - X_n) = Y_i/Y_{n-1}$, $i = 1,\ldots,n - 2$, $W_{n-1} = (X_{n-1} - X_n)/(|X_{n-1} - X_n|) = Y_n/|Y_{n-1}|$ is clearly ancillary. Therefore, by Basu's theorem, $X_{(1)}$, S, and $\mathbf{W}$ are independent. Consider the estimator $S(\mathbf{X})$ of σ

$$S(\mathbf{X}) = \sum_{i=1}^{n}(X_i - X_{(1)})$$

$$= \sum_{i=1}^{n}[X_i - X_n - (X_{(1)} - X_n)]$$

$$= \sum_{i=1}^{n-1} Y_i - n\min\{X_1 - X_n, \ldots, X_{n-1} - X_n\} = S(\mathbf{Y})$$

Note that $S(\mathbf{Y}) = \delta_0(\mathbf{Y})$ is a scale-equivariant estimator of σ based on $\mathbf{Y}$. Therefore, the MRE estimator of σ is given by

$$\delta(\mathbf{Y}) = \frac{\delta_0(\mathbf{Y})}{u^*} = \frac{S(\mathbf{Y})}{u^*}$$

where u^* is any number that minimizes

$$h(u) = E_1\left[L\left(\frac{\delta_0(\mathbf{Y})}{u}\right)\right] = E_1\left|\frac{\delta_0}{u} - 1\right|^p, \ p = 1, 2$$

over $u > 0$. Consider the case for $p = 1$

$$h(u) = E_1\left|\frac{S}{u} - 1\right| = \frac{1}{u}\left[\int_0^u (u-s) f_n(s)ds + \int_u^\infty (s-u) f_n(s)ds\right]$$

where $f_n(s)$ is the density of $G_1(n - 1,1)$. Differentiating $h(u)$ with respect to u, we get

$$h'(u) = \frac{1}{u^2}\left(\int_0^u s f_n(s)ds - \int_u^\infty s f_n(s)ds\right)$$

Putting $h'(u) = 0$ gives

$$\int_0^u s f_n(s)ds = \int_u^\infty s f_n(s)ds$$

$$\int_0^u \frac{1}{\Gamma(n)} \exp(-s) s^{n-1} ds = \int_u^\infty \frac{1}{\Gamma(n)} \exp(-s) s^{n-1} ds$$

Therefore, u^* is the median of $G_1(n, 1)$. Now, consider the case $p = 2$. We have

$$E_1\left(\frac{S}{u} - 1\right)^2 = \frac{1}{u^2}\{E_1 S^2 - 2uE_1 S + u^2\}$$

$$= \frac{n(n-1)}{u^2} - \frac{2(n-1)}{u} + 1$$

On setting

$$\frac{\partial E_1}{\partial u} = 0$$

we get

$$u^* = n$$

Thus, the MRE estimator of σ under the loss function $L(\sigma, a) = |(a/\sigma) - 1|^p$, $p = 2$, is given by

$$\delta(\mathbf{X}) = \frac{1}{n}\sum_{i=1}^{n}(X_i - X_{(1)})$$

Next, consider the loss function for the estimation of σ^r

$$L(\mu, \sigma, \delta) = \rho\left(\frac{\delta}{\sigma^r}\right) = \left(\frac{\delta}{\sigma^r} - 1\right)^2 = \frac{(\delta - \sigma^r)^2}{\sigma^{2r}}$$

We have already seen that $X_{(1)}$ and $\Sigma(X_i - X_{(1)})$ are independently distributed as $E(\mu, \sigma/n)$ and $(1/2)\sigma\chi^2_{2(n-1)}$, respectively. Also, they are jointly sufficient and complete for (μ, σ). Define $Y_i = X_i - X_n$, $i = 1,\ldots,n-1$, and $\mathbf{Z} = (Z_1,\ldots,Z_{n-1})' = (Y_1/Y_{n-1},\ldots,Y_{n-2}/Y_{n-1}, Y_{n-1}/|Y_{n-1}|)'$. $\mathbf{Z}$ is an ancillary statistic. Clearly, by Basu's theorem, $[X_{(1)}, \Sigma(X_i - X_{(1)})]$ are jointly independent of $\mathbf{Z}$. For the loss function

$$L(\mu, \sigma, \delta) = \rho\left(\frac{\delta}{\sigma^r}\right) = \frac{(\delta - \sigma^r)^2}{\sigma^{2r}} \tag{7.6.1}$$

the MRE estimator δ of σ^r is given by

$$d(\mathbf{X}) = \frac{\delta_0(\mathbf{Y})}{q^*(\mathbf{Z})}$$

where $\delta_0(\mathbf{Y})$ is any scale-equivariant estimator of σ^r based on $\mathbf{Y} = (Y_1,\ldots,Y_{n-1})'$ so that

$$\delta_0(a + b\mathbf{Y}) = b^r\delta_0(\mathbf{Y})$$

and, where $q^*(\mathbf{z})$ is any number that minimizes

$$E_{\sigma=1}\left\{\rho\left(\frac{\delta_0(\mathbf{Y})}{q(\mathbf{z})}\right)\middle|\mathbf{z}\right\}$$

For the loss function in Eq. (7.7.1), we have

$$q^*(\mathbf{z}) = \frac{E_{\sigma=1}[\delta_0^2(\mathbf{Y})|\mathbf{z}]}{E_{\sigma=1}[\delta_0(\mathbf{Y})|\mathbf{z}]}$$

We consider $$\delta_0(\mathbf{Y}) = \sum(X_i - X_{(1)})$$

as a scale-equivariant estimator. We have

$$E_1[\delta_0^2(\mathbf{Y})|\mathbf{z}] = E_1[\delta_0^2(\mathbf{Y})]$$

since $\Sigma(X_i - X_{(1)})$ and $\mathbf{Z}$ are independent.

$$E_1[\delta_0^2(\mathbf{Y})] = E_1[\Sigma(X_i - X_{(1)})]^2$$

Note that $\Sigma(X_i - X_{(1)})/\sigma \sim G_1(n-1, 1)$ or $U = 2\Sigma(X_i - X_{(1)})/\sigma \sim \chi^2_{2(n-1)}$. This gives $E(U) = 2(n-1)$ and $E(U^2) = V(U) + \{E(U)\}^2 = 4(n-1) + 4(n-1)^2 = 4n(n-1)$, $E[\Sigma(X_i - X_{(1)})] = (n-1)\sigma$, $E[\Sigma(X_i - X_{(i)})]^2 = n(n-1)\sigma^2$. This gives

$$E_1[\delta_0^2(\mathbf{Y})] = n(n-1)$$

and $$E_1[\delta_0(\mathbf{Y})] = n - 1 \tag{7.6.2}$$

From Eq. (7.6.2), we get

$$q^*(\mathbf{z}) = \frac{n(n-1)}{(n-1)} = n$$

So, the MRE estimator of σ is given by

$$\delta(\mathbf{X}) = \frac{1}{n}\sum(X_i - X_{(1)})$$

Note that this estimator is a biased estimator of σ since $E(\delta) = [(n-1)/n]\sigma$ but is unbiased for large n.
In Chapter 3, we have shown the estimator

$$\delta^*(\mathbf{X}) = \frac{1}{n-1}\sum(X_i - X_{(1)})$$

as UMVUE for σ. MRE estimator and UMVUE of σ are not much different for large n. However, they are optimal for different loss functions.

Consider, now, the location-scale equivariant estimator δ_0 so that $\delta_0(a + b\mathbf{X}) = a + b\delta_0(\mathbf{X})$, $a \in \mathbb{R}^1$, $b > 0$, and the scale equivariant estimator δ_1 so that $\delta_1(a + b\mathbf{X}) = b\delta_1(\mathbf{X})$, $b > 0$. These estimators are $\delta_0 = X_{(1)}$ and $\delta_1 = \Sigma(X_i - X_{(1)})$ for estimating μ and σ. The MRE estimator of μ is given by

$$\delta(\mathbf{X}) = \delta_0(\mathbf{X}) - u^*(\mathbf{z})\delta_1(\mathbf{X})$$

under the loss function

$$L(\mu, \sigma, \delta) = \rho\left(\frac{\delta - \mu}{\sigma}\right) = \left(\frac{\delta - \mu}{\sigma}\right)^2$$

where $u^*(\mathbf{z})$ is the number that minimizes

$$E\left[\frac{\delta_0(X) - u\delta_1(X) - \mu}{\sigma}\middle|\mathbf{z}\right]^2_{\mu=0,\,\sigma=1} = E[\delta_0(X) - u\delta_1(X)|\mathbf{z}]^2$$

with respect to u. This gives

$$u^*(\mathbf{z}) = \frac{E_{0,1}[\delta_0(\mathbf{X})\delta_1(\mathbf{X})|\mathbf{z}]}{E_{0,1}[\delta_1^2(\mathbf{X})|\mathbf{z}]}$$

$$= \frac{E_{0,1}[\delta_0(\mathbf{X})\delta_1(\mathbf{X})]}{E_{0,1}[\delta_1^2(\mathbf{X})]}$$

since $(X_{(1)}, \Sigma(X_i - X_{(1)}))$ is independent of $\mathbf{Z}$. We have

$$u^*(\mathbf{z}) = \frac{E_{0,1}[\delta_0(\mathbf{X})]E_{0,1}[\delta_1(\mathbf{X})]}{E_{0,1}[\delta_1^2(\mathbf{X})]}$$

since $X_{(1)}$ and $\Sigma(X_i - X_{(1)})$ are independent. We have

$$E_{0,1}[\delta_0(\mathbf{X})] = E_{0,1}(X_{(1)}) = \frac{1}{n}$$

$$E_{0,1}[\delta_1^2(\mathbf{X})] = E_{0,1}\left[\sum(X_i - X_{(1)})\right]^2 = n(n-1)$$

$$E_{0,1}[\delta_1(\mathbf{X})] = E_{0,1}\left[\sum(X_i - X_{(1)})\right] = (n-1)$$

$$u^*(\mathbf{z}) = \frac{(1/n)(n-1)}{n(n-1)} = \frac{1}{n^2}$$

Alternatively, from the independence of $X_{(1)}$ and S

$$h(u) = E_{0,1}(X_{(1)} - uS)^2 = E_{0,1}X_{(1)}^2 - 2uE_{0,1}X_{(1)}E_{0,1}S + u^2E_{0,1}S^2$$

$$= \frac{2}{n^2} - \frac{2(n-1)u}{n} + n(n-1)u^2$$

$h'(u) = 0$ gives

$$u^*(\mathbf{z}) = \frac{1}{n^2}$$

Therefore, the MRE estimator of μ is given by

$$\delta(\mathbf{X}) = \delta_0(\mathbf{X}) - \frac{1}{n^2}\delta_1(\mathbf{X})$$

$$= X_{(1)} - \frac{1}{n^2}\sum(X_i - X_{(1)})$$

and the bias in the estimator is given by

$$b[\delta(\mathbf{X})] = E(\delta) - \mu = E\left[X_{(1)} - \left(\frac{1}{n^2}\right)\sum(X_i - X_{(1)}) - \mu\right]$$

$$= \frac{\sigma}{n} - \frac{(n-1)\sigma}{n^2} = \frac{\sigma}{n^2}$$

***Example* 7.10** Let $X_1, \ldots, X_n$ be *iid* from $N(\mu, \sigma^2)$ where μ and $\sigma^2 > 0$ are both unknown. Find the MRE estimator of σ^2 under the loss function

$$L((\mu, \sigma), \delta) = \rho\left(\frac{\delta}{\sigma^2}\right) = \left(\frac{\delta}{\sigma^2} - 1\right)^2$$

Also, find the MRE estimator of μ under the loss function

$$L((\mu, \sigma), \delta) = \rho\left(\frac{\delta - \mu}{\sigma}\right) = \left(\frac{\delta - \mu}{\sigma}\right)^2$$

Solution. The sample variance $S_{n-1}^2 = [1/(n-1)]\Sigma(X_i - \bar{X})^2$ is scale equivariant since it satisfies

$$S_{n-1}^2(a + b\mathbf{X}) = b^2S^2(\mathbf{X})$$

So, we can consider $\delta_0(\mathbf{X}) = S^2$. Further, $[\bar{X}, \Sigma(X_i - \bar{X})^2]$ is jointly sufficient for (μ, σ^2) and also complete and $\mathbf{Z} = ((Y_1/Y_{n-1}),\ldots,(Y_{n-2}/Y_{n-1}), (Y_{n-1}/|Y_{n-1}|))'$, $Y_i = X_i - X_n$, $i = 1,\ldots,n-1$. is ancillary. Therefore, by Basu's theorem, $(\bar{X}, \Sigma(X_i - \bar{X})^2)$ are independent of $\mathbf{Z}$. The MRE estimator of σ^2 under the given loss function is

$$\delta(\mathbf{X}) = \frac{\delta_0(\mathbf{X})}{q^*(\mathbf{Z})}$$

where
$$\delta_0(\mathbf{X}) = \frac{1}{n-1}\sum(X_i - \bar{X})^2 = \frac{1}{n-1}\sum_{i=1}^{n}\left[X_i - X_n - \frac{1}{n}\sum_{i=1}^{n}(X_i - X_n)\right]^2$$
$$= \frac{1}{n-1}\sum(Y_i - \bar{Y})^2 = \delta_0(\mathbf{Y})$$

and $q^*(\mathbf{z})$, for the given loss function, is calculated by

$$q^*(\mathbf{z}) = \frac{E_1(\delta_0^2(\mathbf{Y})|\mathbf{z})}{E_1(\delta_0(\mathbf{Y})|\mathbf{z})} = \frac{E_1(S_{n-1}^4)}{E_1(S_{n-1}^2)}$$

We have $E_1(S_{n-1}^4) = (n + 1)/(n - 1)$ and $E_1(S_{n-1}^2) = 1$ since $(n-1)S_{n-1}^2/\sigma^2|_{\sigma^2=1} = (n-1)\,S_{n-1}^2 \sim \chi_{n-1}^2$. This gives $q^*(z) = (n + 1)/(n - 1)$. Therefore, the MRE estimator of μ is given by

$$\delta(\mathbf{X}) = \frac{1}{n+1}\sum(X_i - \bar{X})^2$$

This is a biased estimator since

$$E[\delta(\mathbf{X})] = \frac{1}{n+1}E\left[\sum(X_i - \bar{X})^2\right]$$
$$= \frac{n-1}{n+1}\sigma^2 \neq \sigma^2$$

Consider, now, the estimation of μ under the given loss function. The statistic $\delta_0(\mathbf{X}) = \bar{X}$ is a location-scale equivariant estimator of μ since it satisfies

$$\delta_0\,(a + b\mathbf{X}) = \delta_0(\mathbf{X})\ \forall\ a \text{ and } b > 0$$

and $\delta_1(\mathbf{X}) = S_{n-1}^2$ is the scale–equivariant estimator of σ^2 since it satisfies

$$\delta_1(a + b\mathbf{X}) = b^2\delta_1(\mathbf{X})\ \forall\ b > 0\ \forall\ a$$

Using the same arguments as before and the use of Basu's theorem gives that $\bar{X}$ and S_{n-1}^2 are independent of $\mathbf{Z}$ and they are independent themselves. Therefore, for the convex and even loss function, and in particular,

$$L((\mu, \sigma^2), \delta) = \rho\left(\frac{\delta - \mu}{\sigma^2}\right) = \frac{(\delta - \mu)^2}{\sigma^4}\ \text{(say)}$$

the minimizing quantity $u^*(\mathbf{z})$,

$$u^*(\mathbf{z}) = \frac{E_{0,1}[\bar{X} S_{n-1}^2 | \mathbf{z}]}{E_{0,1}[S_{n-1}^4 | \mathbf{z}]}$$
$$= \frac{E_{0,1}(\bar{X})\, E_{0,1}(S_{n-1}^2)}{E_{0,1}(S_{n-1}^4)} = 0$$

gives the MRE estimator of μ by

$$\delta(\mathbf{X}) = \delta_0(\mathbf{X}) - u^*(\mathbf{z})\,\delta_1(\mathbf{X}) = \bar{X}$$

***Example* 7.11** Let $X_1,\ldots,X_n$ be *iid* from $U(\mu - (1/2)\sigma, \mu + (1/2)\sigma)$ where both $\mu \in \mathbb{R}^1$ and $\sigma > 0$ are unknown. Find the MRE estimatior of σ under the loss function

$$L((\mu, \sigma), \delta) = \left(\frac{\delta - \sigma}{\sigma}\right)^2$$

and the MRE estimator of μ under the loss function

$$L((\mu, \sigma), \delta) = \rho\left(\frac{\delta - \mu}{\sigma}\right)$$

Solution. The sample range

$$\delta_0(\mathbf{X}) = X_{(n)} - X_{(1)}$$

is the scale equivariant estimator of σ since it satisfies

$$\delta(a + b\mathbf{X}) = b^r \delta(\mathbf{X})$$

It can easily be seen that $\delta_0(\mathbf{X})$ is a function of $Y_1 = X_1 - X_n,\ldots,Y_{n-1} = X_{n-1} - X_n$. Therefore, we can write

$$\delta_0(\mathbf{Y}) = X_n - X_{(1)}$$

The statistic

$$Z_1 = \frac{Y_1}{Y_{n-1}},\ldots, Z_{n-2} = \frac{Y_{n-2}}{Y_{n-1}}, Z_{n-1} = \frac{Y_{n-1}}{|Y_{n-1}|}$$

is ancillary. Therefore, by Basu's theorem, $X_{(n)} - X_{(1)}$ and $\mathbf{Z}$ are independent. The MRE estimator of scale parameter σ is

$$\delta(\mathbf{X}) = \frac{\delta_0(\mathbf{Y})}{q^*(\mathbf{z})}$$

where

$$q^*(\mathbf{z}) = \frac{E_{\sigma=1}[\delta_0^2(\mathbf{Y}) | \mathbf{z}]}{E_{\sigma=1}[\delta_0(\mathbf{Y}) | \mathbf{z}]}$$

The *cdf* of $U(\mu - (1/2)\sigma, \mu + (1/2)\sigma)$ is given by

$$F(x) = \frac{1}{\sigma}\left\{x - \left(\mu - \frac{1}{2}\sigma\right)\right\}$$

The joint density of $X_{(1)}, X_{(n)}$ is

$$f_{X_{(1)}, X_{(n)}}(x_1, x_n; \mu, \sigma) = \frac{n(n-1)}{\sigma^n}(x_n - x_1)^{n-2}, \mu - \frac{1}{2}\sigma < x_1 < x_n < \mu + \frac{1}{2}\sigma$$

Let $R = X_{(n)} - X_{(1)}$, $S = X_{(n)}$, $|J| = 1$.
The joint density of R, S is

$$f_{R,S}(r, s; \sigma) = \frac{n(n-1)}{\sigma^n} r^{n-2}, \;\; \mu - \frac{\sigma}{2} + r < s < \mu + \frac{\sigma}{2}, \;\; 0 < r < \sigma$$

The density of R then calculates to

$$f_R(r; \sigma) = \frac{n(n-1)}{\sigma^n} r^{n-2} \int_r^{\sigma} ds$$

$$= \frac{n(n-1)}{\sigma^n} r^{n-2} (\sigma - r), 0 < r < \sigma$$

This follows $$E_1[\delta_0(\mathbf{Y})] = E_1(R) = \frac{n-1}{n+1}$$

$$E_1[\delta_0^2(\mathbf{Y})] = \frac{n(n-1)}{(n+1)(n+2)}$$

This gives $$q^*(\mathbf{z}) = \frac{n(n-1)}{(n+1)(n+2)} \frac{(n+1)}{(n-1)} = \frac{n}{n+2}$$

Finally, the MRE estimator of σ is given by

$$\delta = \frac{(X_{(n)} - X_{(1)})(n+2)}{n}$$

Consider, now, the estimation of location parameter μ under the given loss function. Consider $\delta_0(\mathbf{X}) = (X_{(1)} + X_{(n)})/2$ which is a location-scale equivariant estimator of μ since it satisfies

$$\delta(a + b\mathbf{X}) = a + b\delta(\mathbf{X}) \;\forall\; a \text{ and } b > 0$$

and $\delta_1(\mathbf{X}) = X_{(n)} - X_{(1)}$ is scale equivariant since it satisfies

$$\delta_1(a + b\mathbf{x}) = b\delta_1(\mathbf{x}) \;\forall\; a \text{ and } \forall\; b > 0$$

The MRE estimator of μ is then given by

$$\delta(\mathbf{X}) = \delta_0(\mathbf{X}) - u^*(\mathbf{z})\delta_1(\mathbf{X})$$

where $$u^*(\mathbf{z}) = \frac{E_{0,1}[\delta_0(\mathbf{X})\delta_1(\mathbf{X})|\mathbf{z}]}{E_{0,1}[\delta_1^2(\mathbf{X})|\mathbf{z}]}$$

$(X_{(1)} + X_{(n)})/2$ and $X_{(n)} - X_{(1)}$ are independent of $\mathbf{Z}$ by Basu's theorem. If the loss function given by $L((\delta - \mu)/\sigma)$ is convex and even, then $u^*(\mathbf{z}) = 0$. This gives the MRE estimator of μ

$$\delta(\mathbf{X}) = \frac{X_{(1)} + X_{(n)}}{2}$$

***Example* 7.12** Let $X_1,\ldots,X_n$ be a random sample drawn from $G_1(\alpha, \sigma)$, where $\alpha > 0$ is known and σ is unknown. Find the MRE estimator of σ under the loss function

$$L(\sigma, \delta) = \left(1 - \frac{\delta}{\sigma}\right)^2$$

Solution. The joint distribution of sample observations is given by

$$f(\mathbf{x}; \sigma) = \frac{1}{[\Gamma(\alpha)]^n \sigma^{n\alpha}} \prod x_i^{\alpha-1} \exp\left\{-\sum\left(\frac{x_i}{\sigma}\right)\right\}, \sigma > 0, \alpha > 0$$

where σ is the scale parameter. The Pitman (MRE) estimator of σ is given by

$$\delta_0(\mathbf{x}) = \frac{\int_0^\infty \sigma^n \exp\left(-\sum \sigma x_i\right) d\sigma}{\int_0^\infty \sigma^{n+1} \exp\left(-\sum \sigma x_i\right) d\sigma}$$

$$= \frac{\Gamma(n+1)\left(\sum x_i\right)^{-(n+1)}}{\Gamma(n+2)\left(\sum x_i\right)^{-(n+2)}} = \frac{1}{(n+1)}\left(\sum X_i\right)$$

***Example* 7.13** Let $X_1,\ldots,X_n$ be a random sample drawn from the folded normal distribution with *pdf*

$$f(x; \mu) = \frac{2}{\sqrt{2\pi}} \exp\left[-\frac{1}{2}(x - \mu)^2\right] I_{[\mu, \infty)}(x)$$

Show that the MRE estimator of the location parameter μ under the squared error loss function is given by

$$\hat{\mu} = \bar{X} - \frac{\exp\left[-\left(\frac{n}{2}\right)(X_{(1)} - \bar{X})^2\right]}{\sqrt{2n\pi} \int_{-\infty}^{\sqrt{n}(X_{(1)} - \bar{X})} \left(\frac{1}{\sqrt{2\pi}}\right) \exp\left(-\frac{z^2}{2}\right) dz}$$

Solution. The joint density of $X_1,\ldots,X_n$ is given by

$$f(\mathbf{x};\mu)=\left(\frac{2}{\pi}\right)^{n/2}\exp\left[-\frac{1}{2}\sum(x_i-\mu)^2\right]I_{[\mu,\infty)}(x_{(1)})$$

$$=\left(\frac{2}{\pi}\right)^{n/2}\exp\left(-\frac{(n-1)S_{n-1}^2}{2}\right)\exp\left[-\frac{n}{2}(\mu-\bar{x})^2\right]I_{[\mu,\infty)}(x_{(1)})$$

where $\bar{x}$ and S_{n-1}^2 are sample mean and sample variance, respectively. The MRE estimator of μ, under the squared error loss function, is given by

$$\delta_0(\mathbf{x})=\frac{\int\mu f(x_1-\mu,\ldots,x_n-\mu)d\mu}{\int f(x_1-\mu,\ldots,x_n-\mu)d\mu}$$

$$=\frac{\int_{-\infty}^{x_{(1)}}\mu\exp[-(n/2)(\mu-\bar{x})^2]d\mu}{\int_{-\infty}^{x_{(1)}}\exp[-(n/2)(\mu-\bar{x})^2]d\mu}$$

$$=\bar{x}+\frac{\int_{-\infty}^{x_{(1)}}(\mu-\bar{x})\exp[-(n/2)(\mu-\bar{x})^2]d\mu}{\int_{-\infty}^{x_{(1)}}\exp[-(n/2)(\mu-\bar{x})^2]d\mu}$$

Let $\sqrt{n}(\mu-\bar{x})=z$. We get

$$\delta_0(\mathbf{x})=\bar{x}+\frac{\int_{-\infty}^{\sqrt{n}(x_{(1)}-\bar{x})}(z/\sqrt{n})\exp[-(z^2/2)](dz/\sqrt{n})}{\sqrt{2\pi}\int_{-\infty}^{\sqrt{n}(x_{(1)}-\bar{x})}(1/\sqrt{2\pi})\exp[-(z^2/2)](dz/\sqrt{n})}$$

Consider the numerator of the second term in the above expression

$$\frac{1}{n}\int_{-\infty}^{\sqrt{n}(x_{(1)}-\bar{x})}z\exp\left(-\frac{z^2}{2}\right)dz$$

and letting $z^2/2 = t$, we get

$$\frac{1}{n}\int_{\infty}^{n(x_{(1)}-\bar{x})^2/2}\exp(-t)dt=-\frac{1}{n}\exp\left[-\left(\frac{n}{2}\right)(x_{(1)}-\bar{x})^2\right]$$

Thus, the MRE estimator of μ is given by

$$\delta_0(\mathbf{X}) = \bar{X} - \frac{\exp[-(n/2)(X_{(1)} - \bar{X})^2]}{\sqrt{2n\pi} \int_{\infty}^{\sqrt{n}(X_{(1)}-\bar{X})} (1/\sqrt{2\pi})\exp[-(z^2/2)]\, dz}$$

***Example* 7.14** Let $X_1,\ldots,X_n$ be *iid* from $U(\theta, 2\theta)$.

(i) Show that $\delta(\mathbf{X}) = (X_{(1)}, X_{(n)})$ is jointly sufficient statistic for θ.
(ii) Is $X_{(n)} - X_{(1)}$ an unbiased estimator of θ? Find an ancillary statistic.
(iii) Determine MRE (Pitman) estimator of the scale parameter θ under the loss function $L(\theta, \delta) = (1 - (\delta/\theta))^2$.

Solution.

(i) Refer Chapter 2.
The joint density of $(X_{(1)}, X_{(n)})$ is given by

$$f_{X_{(1)},X_{(n)}}(x_{(1)}, x_{(n)}) = \frac{n(n-1)}{\theta^n}(x_{(n)} - x_{(1)})^{n-2}$$

Let $R = X_{(n)} - X_{(1)}$ and $S = X_{(n)}$. We get the joint density of R, S as

$$f_{R,S}(r, s) = \frac{n(n-1)}{\theta^n} r^{n-2},\ \theta + r < s < 2\theta;\ 0 < r < \theta$$

This gives the marginal density of R

$$f_R(r) = \frac{n(n-1)}{\theta^n} r^{n-2}(\theta - r), 0 < r < \theta$$

It follows that $\quad E_\theta(R) = E_\theta(X_{(n)} - X_{(1)}) = \dfrac{n-1}{n+1}\theta$

Therefore, the estimator $(X_{(n)} - X_{(1)})$ is not an unbiased estimator of θ.

(iii) The joint density of sample observations is given by

$$f(\mathbf{x}; \theta) = \frac{1}{\theta^n} \text{ if } \theta < x_1,\ldots, x_n < 2\theta$$

$$= \frac{1}{\theta^n} \text{ if } \theta < x_{(1)} < x_{(n)} < 2\theta$$

$$= \frac{1}{\theta^n} \text{ if } 1 < \frac{x_{(1)}}{\theta} < \frac{x_{(n)}}{\theta} < 2$$

$$= \frac{1}{\theta^n}\, \phi\left(1, \frac{x_{(1)}}{\theta}\right) \phi\left(\frac{x_{(1)}}{\theta}, 2\right)$$

so that $$f\left(\frac{x_1}{\theta},\ldots,\frac{x_n}{\theta}\right)=\phi\left(1,\frac{x_{(1)}}{\theta}\right)\phi\left(\frac{x_{(n)}}{\theta},2\right)$$

We then have $$f(\theta x_1,\ldots,\theta x_n) = \phi(1, \theta x_{(1)})\, \phi(\theta x_{(n)}, 2)$$

This gives $$1 < \theta x_{(1)} < \theta x_{(n)} < 2$$
$$x_{(1)}^{-1} < \theta < 2x_{(n)}^{-1}$$

Thus, the Pitman estimator of θ under the loss function $L(\theta, \delta) = (1 - (\delta/\theta))^2$ is given by

$$\delta_0(\mathbf{X})=\frac{\int_{X_{(1)}^{-1}}^{2X_{(n)}^{-1}}\theta^n d\theta}{\int_{X_{(1)}^{-1}}^{2X_{(n)}^{-1}}\theta^{n+1} d\theta}$$

$$=\frac{(n+2)}{(n+1)}\left[\frac{(X_{(n)}/2)^{-(n+1)}-X_{(1)}^{-(n+1)}}{(X_{(n)}/2)^{-(n+2)}-X_{(1)}^{-(n+2)}}\right]$$

***Example* 7.15** Let X_1, X_2 be *iid* with *pdf*

$$f(x;\sigma)=\frac{2}{\sigma}\left(1-\frac{x}{\sigma}\right)I_{(0,\sigma)}(x)$$

Find the MRE estimator of σ^r.

Solution. The joint density of sample observations is given by

$$f(\mathbf{x};\sigma)=\left(\frac{2}{\sigma}\right)^2\prod_{i=1}^{2}\left(1-\frac{x_i}{\sigma}\right),\quad 0<x_{(1)}<x_{(2)}<\sigma$$

$$=\left(\frac{2}{\sigma}\right)^2\prod_{i=1}^{2}\left(1-\frac{x_i}{\sigma}\right)\phi(0,x_{(1)})\phi(x_{(2)},\sigma)$$

where $$f\left(\frac{x_1}{\sigma},\frac{x_2}{\sigma}\right)=\prod_{i=1}^{2}\left(1-\frac{x_i}{\sigma}\right)\phi\left(0,\frac{x_{(1)}}{\sigma}\right)\phi\left(\frac{x_{(2)}}{\sigma},1\right)$$

This gives $$f(\sigma x_1,\sigma x_2)=\prod_{i=1}^{2}(1-\sigma x_i)\phi(0,\sigma x_{(1)})(\sigma x_{(2)},1)$$

This gives $$0<\sigma x_{(1)}<\sigma x_{(2)}<1$$
$$0<\sigma<x_{(2)}^{-1}$$

Therefore, the Pitman (MRE) estimator of σ^r is given by

$$\delta(\mathbf{x}) = \frac{\int_0^{x_{(2)}^{-1}} \sigma^{2+r-1} \prod_{i=1}^{2}(1-\sigma x_i)d\sigma}{\int_0^{x_{(2)}^{-1}} \sigma^{2+2r-1} \prod_{i=1}^{2}(1-\sigma x_i)d\sigma}$$

$$= \frac{\int_0^{x_{(2)}^{-1}} \sigma^{r+1}d\sigma - \left(\sum_1^2 x_i\right)\int_0^{x_{(2)}^{-1}} \sigma^{r+2}d\sigma + \left(\prod_1^2 x_i\right)\int_0^{x_{(2)}^{-1}} \sigma^{r+3}d\sigma}{\int_0^{x_{(2)}^{-1}} \sigma^{2r+1}d\sigma - \left(\sum_1^2 x_i\right)\int_0^{x_{(2)}^{-1}} \sigma^{2r+2}d\sigma + \left(\prod_1^2 x_i\right)\int_0^{x_{(2)}^{-1}} \sigma^{2r+3}d\sigma} \, t$$

$$= \frac{[x_{(2)}^{-(r+2)}/(r+2)] - \left[\left(\sum_1^2 x_i\right)x_{(2)}^{-(r+3)}/(r+3)\right] + \left[\prod_1^2 x_i[x_{(2)}^{-(r+4)}/(r+4)]\right]}{[x_{(2)}^{-(2r+2)}/(2r+2)] - \left(\sum_1^2 x_i\right)x_{(2)}^{-(2r+3)}/(2r+3) + \prod_1^2 x_i[x_{(2)}^{-(2r+4)}/(2r+4)]}$$

$$= x_{(2)}^r \frac{\left\{[1/(r+2)] - \left[x_{(2)}^{-1}\left(\sum_1^2 x_{(i)}\right)/(r+3)\right] + x_{(2)}^{-2}\left[\prod_1^2 x_{(i)}/(r+4)\right]\right\}}{\left\{[1/(2r+2)] - \left[x_{(2)}^{-1}\left(\sum_1^2 x_{(i)}\right)/(2r+3)\right] + \left[x_{(2)}^{-2}\left(\prod_1^2 x_{(i)}\right)/(2r+4)\right]\right\}}$$

$$= x_{(2)}^r \frac{\{[1/(r+2)(r+3)] - [1/(r+3)(r+4)](x_{(1)}x_{(2)}^{-1})\}}{\{[1/(2r+2)(2r+3)] - [1/(2r+3)(2r+4)](x_{(1)}x_{(2)}^{-1})\}}$$

$$= x_{(2)}^r \frac{(2r+3)}{(r+3)} \frac{\{[1/(r+2)] - [x_{(1)}x_{(2)}^{-1}/(r+4)]\}}{\{[1/(2r+2)] - [x_{(1)}x_{(2)}^{-1}/(2r+4)]\}}$$

Thus, the Pitman estimator of σ^r is given by

$$\frac{(2r+2)(2r+3)(2r+4)}{(r+2)(r+3)(r+4)} X_{(2)}^r \frac{[(r+4) - (r+2)X_{(1)}X_{(2)}^{-1}]}{[(2r+4) - (2r+2)X_{(1)}X_{(2)}^{-1}]}$$

***Example* 7.16** Let $X_1,\dots,X_n$ be a random sample from $U(0,\theta)$, with scale parameter $\theta > 0$. Find the MRE (Pitman) estimator of θ under the loss functions.

(i) $L(\theta,\delta) = (\delta/\theta) - 1)^2$ and

(ii) $L(\theta,\delta) = |(\delta/\theta) - 1|$.

Solution. The statistics $Y_1 = X_1/X_n$, $Y_2 = X_2/X_n, \ldots, Y_{n-1} = X_{n-1}/X_n$, $Y_n = X_n/|X_n|$ are ancillary and $\delta_0(\mathbf{X}) = X_{(n)}$ is a complete sufficient statistic of σ. By Basu's theorem, $\mathbf{Y}$ and $X_{(n)}$ are independent. Hence, the MRE estimator of θ is given by $\delta = X_{(n)}/u^*$, where u^* is the number that minimizes $E_1[\rho(X_{(n)}/u)]$ over $u > 0$. The density of $X_{(n)}$ under $\theta = 1$ is given by

$$f_{X_{(n)}}(x;\, \theta = 1) = nx^{n-1},\ 0 < x < 1$$

(i) The Pitman estimator in this case is given by

$$\delta = \frac{X_{(n)}E_1 X_{(n)}}{E_1 X_{(n)}^2} = \frac{(n+2)X_{(n)}}{n+1}$$

(ii) We determine in this case the value of u (≥ 0) that minimizes

$$E_1\left|\frac{X_{(n)}}{u} - 1\right|$$

Therefore, the value of u that minimizes $E_1|(X_{(n)}/u) - 1|$ is between 0 and 1. Consider

$$E_1\left|\frac{X_{(n)}}{u} - 1\right| = \int_0^1 \left|\frac{x}{u} - 1\right| nx^{n-1}dx$$

$$= \int_0^u \left(1 - \frac{x}{u}\right)nx^{n-1}dx + \int_u^1 \left(\frac{x}{u} - 1\right)nx^{n-1}dx$$

$$= \frac{n}{u}\left(\frac{u^{n+1}}{n} - \frac{u^{n+1}}{n+1}\right) + \frac{n}{u}\left[\frac{1-u^{n+1}}{n+1} - \frac{u(1-u^n)}{n}\right]$$

$$= \frac{2}{n+1}u^n + \frac{n}{n+1}u^{-1} - 1$$

Differentiating E_1 with respect to u, we get

$$\frac{\partial}{\partial u}E_1 = 2nu^{n-1} - n\frac{1}{u^2} = 0$$

The gives $u^* = 2^{-(n+1)^{-1}}$. Therefore, the MRE estimator of θ is given by

$$\delta = 2^{(n+1)^{-1}}X_{(n)}$$

***Example* 7.17** Let $X_1, \ldots, X_n$ be a random sample from the Pareto distribution with *pdf*

$$f(x;\alpha,\sigma) = \frac{\alpha\sigma^\alpha}{x^{\alpha+1}}I_{(\sigma,\,\infty)}(x)$$

where $\alpha > 2$ is known and $\sigma > 0$ is an unknown parameter. Find the MRE estimator of σ under the loss function

$$L(\sigma, a) = \left(\frac{a}{\sigma} - 1\right)^2$$

Solution. Consider $Y_1,\ldots,Y_n$, where $Y_i = X_i/X_n$, $i = 1,\ldots,n-1$, $Y_n = X_n/|X_n|$ are ancillary statistics and the statistic $X_{(1)}$ is complete and sufficient. Therefore, by Basu's theorem, **Y** and $X_{(1)}$ are independent. The estimator $\delta_0(\mathbf{X}) = X_{(1)}$ is a scale equivariant estimator of σ. Therefore, the MRE estimator of σ is given by

$$\delta(\mathbf{X}) = \frac{X_{(1)}}{q}$$

where q is the number that minimizes

$$E_1\rho\left(\frac{X_1}{q}\right) = E_1\left(\frac{X_{(1)}}{q} - 1\right)^2, \; q > 0$$

Here, E_1 stands for expectation for the distribution of $X_{(1)}$ under $\sigma = 1$. This corresponding distribution of $X_{(1)}$ under $\sigma = 1$ is given by

$$f_{X_{(1)}}(x; \alpha) = n\alpha \frac{1}{x^{n\alpha+1}} I_{(1,\infty)}(x)$$

$$E_1\left(\frac{X_{(1)}}{q} - 1\right)^2 = \frac{1}{q^2}E_1(X_{(1)} - q)^2 = \frac{1}{q^2}[E_1(X_{(1)}^2) - 2qE_1(X_{(1)}) + q^2]$$

$$= \frac{1}{q^2}\frac{n\alpha}{(n\alpha - 2)} - \frac{2n\alpha}{q(n\alpha - 1)} + 1$$

$$\frac{\partial}{\partial q}E_1 = 0$$

gives

$$q = \frac{n\alpha - 1}{n\alpha - 2}$$

Thus, the MRE estimator of σ is given by

$$\delta(\mathbf{X}) = \frac{n\alpha - 2}{n\alpha - 1}X_{(1)}$$

EXERCISES

1. Consider a random sample $X_1,\ldots,X_n$ from $N(\theta, 1)$. Show that it is a location family and also show that $\delta(\mathbf{X}) = \bar{X}$ is the Pitman estimator of θ.
2. Let $X_1,\ldots,X_n$ be *iid* $N(0,\sigma^2)$. Find the Pitman scale-equivariant estimator of σ^2.
3. Consider a random sample $X_1,\ldots,X_n$ from $E(\theta)$. Find the Pitman scale-equivariant estimator of θ.
4. Let $X_1,\ldots,X_n$ be *iid* $U(0,\theta)$. Find the Pitman scale-equivariant estimator of θ.

5. Let $X_1,\ldots,X_n$ be a random sample from $b(n, p)$, $p \in \Theta = (0,1)$. Consider a group of transformations

$$\mathcal{G} = \{g: g(x) = n - \mathbf{x} = (n - x_1,\ldots,n - x_n)\}$$

(i) Find the class of equivariant estimators with respect to the above group.

(ii) Consider a prior distribution over Θ, Be(α, β). The corresponding posterior distribution is given by Be($\Sigma x_i + \alpha$, $n + \Sigma x_i + \beta$) and the Bayes estimate p is given by

$$p^B = \left(\frac{n}{\alpha+\beta+n}\right)\left(\frac{\sum x_i}{n}\right) + \left(\frac{\alpha+\beta}{\alpha+\beta+n}\right)\left(\frac{\alpha}{\alpha+\beta}\right)$$

Define a class of estimators within this class of Bayes estimators which are equivariant with respect to the above group of transformations.

(iii) Find the MRE estimator of p in the class of equivariant estimators defined in (ii).

6. [Berk (1972), Lehmann (1988)] Let $X_1,\ldots,X_n$ be a random sample from $N(\theta, a\theta^2)$, $\theta > 0$, where $a > 0$ is known.

(i) Find the MLE of θ so that

$$\sqrt{n}(\hat{\theta}_{\text{ML}} - \theta) \xrightarrow{D} N\left(0, \frac{1}{I(\theta)}\right)$$

(ii) Find a minimum risk estimator of θ which is equivariant with respect to a suitable group of transformations.

(iii) For $a = 1$, show that the family of distribution $N(\theta, \theta^2)$ is closed under scale transformation.

(iv) Show that the MLE is equivariant when $a = 1$.

7. Let $X \sim f(x; \theta)$, where the *pdf* $f(x; \theta)$ is given by

$$f(x; \theta) = \exp[-(x - \theta)][1 + \exp(-x - \theta)]^{-2}, x \in \mathbb{R}^1, \theta \in \mathbb{R}^1$$

Find the Pitman estimator of the location parameter θ.

8. (Rohatgi, 2006) Consider a random sample from

(i) a location density $f(x; \theta)$. Consider the group of transformations $\mathcal{G} = \{g: g(\mathbf{x}) = \mathbf{x} + a, a \in \mathbb{R}^1\}$. Show that the following estimators are equivariant under the above group of transformations:

(a) $X_{[np]+1}$, the pth quantile of order p, $0 < p < 1$

(b) $(X_{(1)} + X_{(n)})/2$

(c) $\bar{X} + \bar{Y}$, where $\bar{Y}$ is the mean of a sample of size m, $m \neq n$.

(ii) a location, scale, or location-scale density. Consider $\mathcal{G}$ as location-scale group of transformations. Show that the following statistics are invariant under one of these groups of transformations:

(a) $X_{(n+1-k)} - X_{(k)}$

(b) $\bar{X} - \tilde{X}$, where $\tilde{X}$ is the median of the observations

(c) $\Sigma_{i=1}^n |X_i - \bar{X}|/n$

(d) $\Sigma_{i=1}^{n}(X_i - \bar{X})(Y_i - \bar{Y}) / [\Sigma_{i=1}^{n}(X_i - \bar{X})^2 \Sigma_{i=1}^{n}(Y_i - \bar{Y})^2]^{1/2}$
where $(X_1, Y_1),\ldots,(X_n, Y_n)$ is a random sample from a bivariate distribution.

9. Let $X_1,\ldots,X_n$ be a random sample drawn from the *pdf*

$$f(x; \theta) = f(x - \theta),\ \theta \in \mathbb{R}^1$$

where $f(x - \theta)$ belongs to one-parameter location family of distributions. Using these observations, we want to estimate the location parameter θ. Consider a group of transformations

$$\mathcal{G} = [g_b: g_b(\mathbf{x}) = (x_1 + b,\ldots,x_n + b),\ b \in \mathbb{R}^1]$$

and the loss function satisfy

$$L(\theta, \delta) = \rho(\delta - \theta)$$

where ρ is a real-valued function. Define ancillary statistics by $D_i = X_i - X_n$, $i = 1, 2,\ldots,n - 1$

(i) Show that the estimation problem is invariant under the translation group of transformations defined above.

(ii) Given a δ_0 equivariant estimator of θ, the class of all equivariant estimators is given by

$$\xi = \{\delta: \delta(\mathbf{x}) = \delta_0(\mathbf{x}) - u(\mathbf{d})\ \forall\ \mathbf{x} \in \mathbb{R}^n \text{ for some Borel functions on } \mathbb{R}^n\}$$

Show that the risks of δ in ξ are constant, i.e., they do not depend on θ.

(iii) Show that the MRE estimator of θ is

$$\delta^*(\mathbf{X}) = \delta_0(\mathbf{X}) - u_0(\mathbf{d})$$

where $u_0(\mathbf{d})$ is the number that minimizes the quantity

$$l(\mathbf{d}) = E_0[\rho[\delta_0(\mathbf{X}) - u(\mathbf{d})] | \mathbf{D} = \mathbf{d}]\ \forall\ \mathbf{d}$$

with respect to u, where E_0 involves the distribution of $\mathbf{X}$ with joint *pdf*

$$f(x_1 - \theta,\ldots,x_n - \theta)|_{\theta=0} = f(x_1,\ldots,x_n)$$

Also, show that u_0 exists which minimizes the quantity $l(\mathbf{d})$ for each $\mathbf{d}$ if $L(\cdot)$ is convex and non-monotone; further u_0 is unique if $L(\cdot)$ is strictly convex.

(iv) If δ_0 and $\mathbf{D}$ are independent, then show that u_0 is a constant that minimizes $\mathbf{l} = E_0\{\rho[\delta_0(\mathbf{X}) - u]\}$. Further, if the distribution of δ_0 is symmetric about θ and $L(\cdot)$ is convex and even, then show that $u_0 = 0$.

(v) If the loss function is absolute error loss function

$$L(\theta, \delta) = |\theta - \delta|$$

show that the MRE estimator of θ is

$$\delta^*(\mathbf{X}) = \delta_0(\mathbf{X}) - u_0(\mathbf{D})$$

where $u_0(\mathbf{d})$ is the median of $\delta_0(\mathbf{X})|\mathbf{D} = \mathbf{d}$, $\theta = 0$.

(vi) If the loss function is squared error loss function, i.e.,

$$L(\theta, \delta) = (\delta - \theta)^2$$

Show that the MRE estimator of θ is

$$\delta^*(\mathbf{X}) = \delta(\mathbf{X}) - u_0(\mathbf{D})$$

where $\quad u_0(\mathbf{d}) = E_0[\delta_0(\mathbf{X})|\mathbf{D} = \mathbf{d}]$

Show that $\delta^*(\mathbf{X})$ is the same as given in Eq. (7.3.11), i.e.,

$$\delta^*(\mathbf{x}) = \frac{\int_{-\infty}^{\infty} u \prod_{i=1}^{n} f(x_i - u)du}{\int_{-\infty}^{\infty} \prod_{i=1}^{n} f(x_i - u)du}$$

(vii) Show that the MRE estimator δ^* of θ in (v) is unbiased.

10. Let $X_1,\ldots,X_n$ be a random sample drawn from $N(\mu, \sigma^2)$ where $\mu \in \mathbb{R}^1$ is unknown and $\sigma^2 > 0$ is known.

(i) Use the results of Exercise 7.9 to show that $\bar{X}$ is an MRE estimator of μ under the squared error loss function.

(ii) Show that the estimator $\bar{X}$ remains MRE estimator if the loss function L is even and convex.

(iii) Find the MRE estimator of μ under the loss function

$$L(\mu, \delta) = \begin{cases} -a(\mu - \delta), & \text{if } \mu < \delta \\ b(\mu - \delta), & \text{if } \mu \geq \delta \end{cases}$$

where $a > 0$ and $b > 0$.

11. Let $X_1,\ldots,X_n$ be a random sample drawn from the population having its *pdf*

$$f(x; \theta) = \sqrt{\frac{2}{\pi}} \exp\left[-\frac{(x-\theta)^2}{2}\right], \quad \theta < x < \infty, \theta \in \mathbb{R}^1$$

where θ is unknown. Find the MRE estimator of θ under the squared error loss function.

12. Consider a random sample $X_1,\ldots,X_n$ from $U(0,\sigma)$. Suppose we are interested in estimating σ under the loss function

$$L(\sigma, \delta) = \rho\left(\frac{\delta}{\sigma^r}\right) = \left|\frac{\delta}{\sigma^r} - 1\right|^p = \frac{|\delta - \sigma^r|^p}{\sigma^{pr}}$$

where $p \geq 1$. This loss function is given when one is interested in estimating σ^r. However, in the present example, $r = 1$.

(i) Show that the MRE estimator of σ under the squared error loss function is

$$\delta^*(\mathbf{X}) = \frac{(n+2)(X_{(n)})}{n+1}$$

(ii) Show that the MRE estimator of σ under the loss function

$$L(\sigma, \delta) = \rho\left(\frac{\delta}{\sigma}\right) = \left|\frac{\delta}{\sigma} - 1\right| = \frac{|\delta - \sigma|}{\sigma}$$

is $$\delta^*(\mathbf{X}) = 2^{(n+1)^{-1}} X_{(n)}$$

13. Consider a random sample $X_1,\ldots,X_n$ from an exponential distribution $E(0,\theta)$ with *pdf*

$$f(x;\theta) = \frac{1}{\theta}\exp\left(-\frac{x}{\theta}\right) I_{(0,\infty)}(x)$$

Find the MRE estimator of θ and θ^2 under the loss functions

$$L(\theta,\delta) = \rho\left(\frac{\delta}{\theta}\right) = \left(\frac{\delta}{\theta} - 1\right)^2 = \frac{(\delta-\theta)^2}{\theta^2}$$

and $$L(\theta,\delta) = \rho\left(\frac{\delta}{\theta^2}\right) = \left(\frac{\delta}{\theta^2} - 1\right)^2 = \frac{(\delta-\theta^2)^2}{\theta^4}$$

respectively.

14. Consider a random sample $X_1,\ldots,X_n$ drawn from a population having *pdf*

$$f(x;\sigma) = \frac{2}{\sigma}\left(1 - \frac{x}{\sigma}\right), 0 < x < \sigma$$

where $\sigma > 0$ is a scale parameter and is unknown. Find the Pitman estimator of σ under the loss function

$$L(\sigma,\delta) = \rho\left(\frac{\delta}{\sigma}\right) = \left(\frac{\delta}{\sigma} - 1\right)^2$$

15. Let $X_1,\ldots,X_n$ be a random sample from the Pareto distribution $Pa(\alpha,\sigma)$ with *pdf*

$$f(x;\sigma) = \frac{\sigma\alpha^{\sigma}}{x^{\sigma+1}}, \alpha < x < \infty$$

where $\alpha > 2$ is known and $\sigma > 0$ is an unknown parameter. Find the MRE estimator of σ under the loss function

$$L(\sigma,\delta) = \rho\left(\frac{\delta}{\sigma}\right) = \left(\frac{\delta}{\sigma} - 1\right)^2$$

16. Let $X_1,\ldots,X_n$ be a random sample from the exponential distribution $E(\mu,\sigma)$ with *pdf*

$$f(x;\mu,\sigma) = \frac{1}{\sigma}\exp\left(-\frac{x-\mu}{\sigma}\right), \mu < x < \infty$$

where $\mu \in \mathbb{R}^1$ and $\sigma > 0$ are unknown.

(i) Find the MRE estimators of μ under the loss function

$$L(\mu,\delta) = \left(\frac{\delta-\mu}{\sigma}\right)^2$$

and show that it is not unbiased.

(ii) Find the MRE estimator of σ under the loss function

$$L(\sigma, \delta) = \left|\frac{\delta}{\sigma} - 1\right|^p \text{ with } p = 1 \text{ and } 2$$

17. Let $X_1,\ldots,X_n$ be *iid* $N(0,\sigma^2)$ with unknown scale parameter $\sigma^2 > 0$. Consider the estimation of σ^2 under the loss function

$$L(\sigma, \delta) = \left|\frac{\delta}{\sigma} - 1\right|^2$$

(i) Show that the estimator $\delta_0(\mathbf{X}) = \Sigma_{i=1}^n X_i^2$ is scale-equivariant.

(ii) Show that the MRE estimator of σ^2 is $T^*(\mathbf{X}) = (n + 2)^{-1} \Sigma_{i=1}^n X_i^2$. Note that the UMVUE of σ^2 is $n^{-1}\Sigma_{i=1}^n X_i^2$.

18. Consider a random sample $X_1,\ldots,X_n$ from the normal distribution $N(\mu,\sigma^2)$ where $\mu \in \mathbb{R}^1$ and $\sigma^2 > 0$ are unknown.

(i) Show that $\bar{X}$ is the location-scale invariant estimator and is the MRE estimator of μ under the loss function

$$L((\mu, \delta), \delta) = \rho\left(\frac{\delta - \mu}{\sigma}\right)$$

where $L(\cdot)$ is even and convex.

(ii) Show that the sample variance $\delta^2 = (n - 1)^{-1}\Sigma(X_i - \bar{X})^2$ is location-scale invariant. Also, show that $(n + 1)^{-1} \Sigma_{i=1}^n (X_i - \bar{X})^2$ is the MRE estimator of σ^2 under the loss function

$$L(\sigma, \delta) = \rho\left(\frac{\delta}{\sigma^2}\right) = \left|\frac{\delta}{\sigma} - 1\right|^2$$

19. Consider a random sample $X_1,\ldots,X_n$ from the uniform distribution $U(\mu - (\sigma/2), \mu + (\sigma/2))$ where $\mu \in \mathbb{R}^1$ and $\sigma > 0$ are unknown.

(i) Show that $\delta_0(\mathbf{X}) = X_{(n)} - X_{(1)}$ is a location-scale equivariant estimator of σ. Also, show that

$$\delta^*(\mathbf{X}) = \frac{n+2}{n}(X_{(n)} - X_{(1)})$$

is the MRE estimator of σ under the loss function

$$L(\sigma, \delta) = \rho\left(\frac{\delta}{\sigma}\right) = \left(\frac{\delta}{\sigma} - 1\right)^2$$

(ii) Show that $\delta_0(X) = (X_{(n)} + X_{(1)})/2$ is a location-scale equivariant estimator of μ. Also, show that this estimator is an MRE estimator of μ under the loss function

$$L((\mu, \sigma), \delta) = \rho\left(\frac{\delta - \mu}{\sigma}\right)$$

where $L(\cdot)$ is even and convex.

8 Bayes and Minimax Estimation

8.1 INTRODUCTION

In the previous chapters, we have seen that the impartiality principles, such as unbiasedness and equivariance, when followed, reduce the class of all estimators into such classes of estimators in which an optimal estimator such as uniformly minimum variance estimator exists. However, the drawback of these requirements is their limited applicability. In the present chapter, an alternative approach of minimizing the risk in some overall sense is considered to yield an optimal estimator. Under this approach, there are two risk measures of an estimator δ. The first measure is

$$r(\pi, \delta) = \int R(\theta, \delta)\, \pi(\theta) d\theta \tag{8.1.1}$$

for some weight function $\pi(\theta)$, which is known as *prior distribution*, and $R(\theta, \delta)$ is the *risk function* of the estimator δ which is defined in (1.1.2). The second measure is

$$\sup_{\theta} R(\theta, \delta) \tag{8.1.2}$$

the maximum of the risk function over the parameter space Θ.

The approach that deals with characterizing estimators which minimizes Eq. (8.1.1) is known as the *Bayes estimation* and the one which minimizes Eq. (8.1.2) is known as the *minimax estimation*.

In the present section, some elementary concepts of decision theory are discussed before we discuss the principles of Bayes and minimax estimation in the following sections. Section 8.2 introduces the formal study of Bayes estimation and examines its various aspects. Section 8.3 introduces the concept and importance of natural conjugate prior distributions and list them for various sampling distributions. Section 8.4 shows the duality between loss function and prior distribution. In Section 8.5, noninformative prior distributions are discussed under the situation when no prior information is available. In Section 8.6, the theory and methods of finding minimax estimators are discussed.

8.1.1 Elements of Decision Theory

We will discuss some elementary concepts of decision theory before we move on to discuss Bayes and minimax estimation.

Consider a family of distributions $\mathcal{F} = \{f(x;\theta);\ \theta \in \Theta\}$ where Θ is called the parameter space and $f(x;\theta)$ is the probability density of some random variable X. The value of the parameter θ is unknown and in the decision theoretic framework, it is referred to as a choice of nature that she makes from Θ. As against this choice of nature, the statistician is expected to take an appropriate action from the space of actions available to him; denote it by $\mathcal{A}$. As a result of it, the statistician incurs a loss of amount L so that L: $\Theta \times \mathcal{A} \to \mathbb{R}^1$, where $\Theta \times \mathcal{A} = \{(\theta, a): \theta \in \Theta,\ a \in \mathcal{A}\}$. For choosing a reasonable action against θ, the statistician conducts a random experiment and draws a random sample $X_1, X_2, \ldots, X_n$ on the choices of the nature. The corresponding sample space is denoted by $\chi \subseteq \mathbb{R}^n$. Based on this sample and the loss that he incurs, the statistician makes a choice from the action space by way of defining a function δ on the sample space χ into action space $\mathcal{A}$. The space of all such decision functions δ is called decision space and it is denoted by D. In the previous chapters, we have discussed different choices of estimators for estimating the unknown parameter θ. This is like taking a suitable action $\delta(\mathbf{x})$ against the unknown choice of the nature θ.

We will discuss some commonly used loss functions. A loss function is a non-negative quantity by which a statistician incurs a loss as a result of his choice δ and the nature's choice θ. If this estimator δ is close to θ, then δ is a reasonable estimator and the loss incurred is small; otherwise, if δ is far away from θ, the estimator δ is not a reasonable estimator and it results into a large loss. Thus, small difference between estimator δ and parameter θ results in small loss whereas large difference results in large loss. We list, here, some commonly used loss functions:

1. Squared error loss function. Legendre (1805) first proposed this loss function in 1805 while Gauss in 1810 when he developed the least squares theory. Since then this loss function is being used for estimating θ. The risk of an estimator δ is defined by $R(\theta, \delta) = E(\theta - \delta)^2$, which is nothing but the mean squared error of the estimator δ. The squared error loss function increases with the increase in distance between δ and θ and is zero if the estimator is correct.

A loss function that penalizes over estimation as compared to under estimation can be defined as

$$L(\theta, \delta) = \begin{cases} [\theta - \delta(\mathbf{x})]^2, & \text{if } \delta < \theta \\ k[\theta - \delta(\mathbf{x})]^2, & \text{if } \delta > \theta \end{cases}$$

where $k > 1$.
If $\boldsymbol{\theta} = (\theta_1, \ldots, \theta_p)$ is a vector, then, loss function can be defined as

(i) $L(\boldsymbol{\theta}, \boldsymbol{\delta}) = \|\boldsymbol{\theta} - \boldsymbol{\delta}\|^2 = \sum_{i=1}^{p} (\theta_i - \delta_i)^2$

(ii) $L(\boldsymbol{\theta}, \boldsymbol{\delta}) = \|\boldsymbol{\theta} - \boldsymbol{\delta}\|_p = \left(\sum_{i=1}^{n} |\theta_i - \delta_i|^p \right)^{1/p}$

One advantage of using a squared error loss function is that it is convex. Therefore, one can restrict the class of estimators by excluding all the randomized estimators. Bansal (2007) shows that if loss function is of the form

$$L(\theta, \delta) = \rho(\theta - \delta)$$

for some smooth function ρ, then it can be approximated by squared error loss function. He used Taylor's expansion to prove this result. Thus, if $L(\theta, \delta) = 1 - \exp[-(\theta - \delta)^2/2)]$, then $L(\theta, \delta)$ can be approximated by $(1/2)(\theta - \delta)^2$. The loss function that one prefers to use for estimating θ must satisfy the following conditions:

(i) $L(\theta, \delta) = \rho(\theta - \delta)$, where ρ is continuous and symmetric.

(ii) $L(\theta, \delta)$ is an increasing function of $|\theta - \delta|$ and $0 \le L(\theta, \delta) \le c$ bounded above by some c, $0 < c < \infty$, and bounded below by zero.

However, one problem with the use of squared error loss function is that it grows drastically fast with the increase in error and, consequently, it punishes the high errors very severely. LinEx (linear exponential) loss function is one alternative to this problem, which is defined as 11th type of loss function discussed in this section.

2. Whittle and Lane (1967) squared error loss function. Whittle and Lane worked on to view the Bayes estimate as a UMVUE of θ. They suggested that this is possible if one chooses $w(\theta) = I_X(\theta)$ in the weighted squared error loss function. Let $X \sim b(1, \theta)$

$$I_X(\theta) = \frac{1}{\theta(1-\theta)}$$

The Whittle and Lanes weighted squared error loss function is given by

$$L(\theta, \delta) = w(\theta)\,(\theta - \delta)^2$$

$$= \frac{(\theta - \delta)^2}{\theta(1-\theta)}$$

For example, consider $X \sim N(0, \sigma^2)$. We have $I_X(\sigma^2) = 1/(2\sigma^4)$. The weighted squared error loss function is given by

$$L(\sigma^2, \delta) = w(\sigma^2)\,(\sigma^2 - \delta)^2 = I(\sigma^2)(\sigma^2 - \delta)^2$$

$$= \frac{1}{2\sigma^4}(\sigma^2 - \delta)^2 = \frac{1}{2}\left(1 - \frac{\delta}{\sigma^2}\right)^2$$

3. Absolute error loss function. The absolute error loss function is defined as

$$L(\theta, \delta) = |\theta - \delta(\mathbf{x})|$$

Here too, the loss increases with the increase in distance between δ and θ, and is zero, if the estimator is correct. The statistician is less penalized for heavy errors and more penalized for small errors as opposed to squared error loss functions.

4. $L(\theta, \delta) = |\theta - \delta|^p$ is called L_p loss function.

5. Relative squared error loss function. It is defined as

$$L(\theta, \delta) = \frac{|\theta - \delta(\mathbf{x})|^2}{|\theta|}$$

This penalizes relatively more if θ is close to zero than if $|\theta|$ is large. Some of its variants are $[|\theta - \delta(\mathbf{x})|/|\theta|]^{1/2}$ and $|\theta - \delta(\mathbf{x})|^4$.

6. Bilinear loss function. There are some practical situations where large over-estimation and under-estimations are occasional and if they do occur, we penalize the errors differently for over and under estimations while keeping the penalties low as compared to the squared error loss function. In these situations, the preferred loss function is the bilinear loss function defined as follows:

$$L(\theta, \delta) = \begin{cases} k_1 |\theta - \delta|, & \text{if } \delta \le \theta \\ k_2 |\theta - \delta|, & \text{if } \delta > \theta \end{cases} \tag{8.1.3}$$

where $k_1 > 0$ and $k_2 > 0$. This loss function grows much slower than the squared error loss function and, thus, penalizes much smaller for large over and under estimations as do the squared error loss function. In case $k_1 = k_2$, the bilinear loss function converts into the absolute error loss function. Bilinear loss function is a convex loss function and is asymmetric.

7. $L(\theta, \delta) = \rho(\theta - \delta)$, $\Theta = \mathcal{A} = (0, \infty)$

8. $L(\theta, \delta) = \begin{cases} 0, & \text{if } |\theta - \delta| \le c \\ 1, & \text{if } |\theta - \delta| \ge c \end{cases}, \quad \Theta = \mathcal{A}$

9. $L(\theta, \delta) = \dfrac{(\theta - \delta)^2}{\theta(1 - \theta)}$

10. Stein loss function. (James and Stein 1961; Casella and Berger 2001)

$$L(\sigma^2, \delta) = \frac{\delta}{\sigma^2} - 1 - \log \frac{\delta}{\sigma^2} \tag{8.1.4}$$

This is a reasonable loss function for estimation of such parameters which take positive values, e.g., in case of estimation of population variance σ^2 loss function, $L(\sigma^2, \delta)$ takes value 0 if $\delta = \sigma^2$ and $L(\sigma^2, \delta) \to \infty$ as $\delta \to 0$ or $\delta \to \infty$. In other words, the loss function penalizes equally and highly the under and over-estimation of σ^2.

The loss in Eq. (8.1.4) matches with the likelihood function for σ^2 in case of sampling from normal population. Therefore, this loss function shares all decision theoretic properties which are attached with likelihood estimation.

For the estimation of variance, the squared error loss function

$$L(\sigma^2, \delta) = (\sigma^2 - \delta)^2$$

is not a reasonable loss function since over and under estimations are penalized differently, and under estimation has only a finite penalty, whereas over estimation has infinite penalty. Note that $L(\sigma^2, \delta) \to \sigma^4$, a finite quantity, in case of under estimation, and $L(\sigma^2, \delta) \to \infty$, an infinite quantity, in case of over estimation.

11. LinEx loss function. In real estate assessment, over estimation is treated more seriously than under estimation. The state government may incur heavy losses due to wrong assessment of properties. In these and similar types of situations, Klebanov (1972), Varian (1975), and Zellner (1986) used LinEx (linear exponential), a suitable loss function, which is defined as

$$L(\theta, \delta) = \exp[c(\delta - \theta)] - c(\delta - \theta) - 1 \tag{8.1.5}$$

where $c \neq 0$. It determines the shape of the loss function. If $c > 0$, the loss function increases almost linearly as the under estimation grows, while it increases almost exponentially as the over estimation grows. This is the reason why this loss function is called LinEx. Clearly, this loss function penalizes over estimation much severely as compared to under estimation. If $c < 0$, under estimation penalizes exponentially and over estimation linearly. This shows that if $c < 0$, under estimation is much serious than over estimation.

This loss function is an asymmetric loss function and it grows asymmetrically with increase in $|c|$. However, for small $|c|$, it can be approximated by a squared error loss function

$$L(\theta, \delta) \cong \frac{c^2}{2}(\delta - \theta)^2$$

LinEx loss function and squared error loss function are almost the same. Thus, for small $|c|$, the estimations based on LinEx loss function and that based on squared error loss function are nearly same.

12. Modified LinEx loss function. Basu and Ebrahimi (1991) have defined modified LinEx loss function in case θ is a scale parameter. If θ is estimated by δ, it is defined as

$$L(\theta,\delta) = b\left[\exp c\left(\frac{\delta}{\theta} - 1\right) - c\left(\frac{\delta}{\theta} - 1\right) - 1\right] \tag{8.1.6}$$

where, $b > 0$ and $c \neq 0$.

Having discussed some commonly used loss functions, we will define the risk of a decision function (estimator) δ. The risk in taking a decision function δ is defined as the weighted average of losses over all values of **X** where weights are the probabilities corresponding to the probability mass function or probability density function as is the case $f(x;\theta)$, i.e.,

$$R(\theta, \delta) = E_\theta L[\theta, \delta(\mathbf{X})]$$

Here, for the purpose of brevity, we denote *pmf* or *pdf* by the same notation $f(x;\theta)$ as the case may be. Thus, the risk function of an estimator δ is

$$R(\cdot, \delta)\colon \Theta \to \mathbb{R}^1$$

Thus, the statistician chooses a decision function δ in D on the basis of sample information collected on the choices of the nature to estimate an unknown parameter θ. The statistician chooses such a δ which satisfies certain optimality criterion. In decision theoretic setup, this problem of estimation is denoted by a triplet (Θ, D, R).

We will discuss two methods of estimation: first, the minimax method of estimation, and second, the Bayes method of estimation, in the above decision theoretic framework.

8.1.2 Bayes and Minimax Estimation—Elementary Concepts

So far, in the problem of estimating parameter θ on the basis of the sample observations $x_1, x_2, \ldots, x_n$ drawn from the population with probability density function $f(x;\theta)$ indexing the parameter θ, the parameter θ is treated as fixed, but unknown. This approach of inference is referred to as *classical inference*.

However, in several situations, the past information on the parameter θ are frequently available in the form of a distribution of θ known as *prior information*, and the corresponding probability density is known as *prior density*. A sample $x_1, x_2, \ldots, x_n$ is then taken from the population with probability density function $f(\cdot|\theta)$, indexing the parameter θ. Subsequently, the prior distribution of θ is updated in the light of the sample information (present information) to get an updated prior in the form of the conditional distribution of θ given $\mathbf{x}$, which is known as *posterior distribution*. Any statistical inference drawn on parameter θ by way of using posterior distribution is known as the *Bayesian inference* and, more particularly, the problem of estimating θ is known as *Bayes estimation*. We will discuss Bayes estimation in the present chapter.

In Bayes estimation, before we observe the data, the prior distribution $\pi(\theta)$ describes the stochastic behavior of the parameter, generally with high uncertainty about θ, since we know very little about the true value of parameter θ. Once we collect the data, we use it through the Bayes theorem to find the posterior distribution $\pi(\theta|x)$, so that it sharpens the posterior distribution in terms of cutting down this uncertainty about the parameter θ. Note that the Bayes theorem gives a *pdf* for θ, not a value (point estimate).

Thus, in the Bayesian inference, both X and θ are random variables in contrast to the classical inference, where X is the only random variable while θ is fixed.

Next, consider the estimation problem which is a statistical decision problem in decision theory settings and where nature is treated as opponent; the nature selects the worst possible prior that sets maximum loss to the statistician against his choice of an estimator. The nature sets maximum loss to the statistician by $\sup_{\theta\in\Theta} R(\theta, \delta)$, where δ is an estimator, which is a choice made by the statistician. The statistician prefers one estimator over the other for which this maximum loss is the least. Such an estimator is called *minimax estimator of* θ and the related theory of estimation is called *minimax estimation*.

We will now discuss two types of ordering of estimators over the space of estimators, Bayes and minimax ordering. The most preferred estimators according to these orderings are known as Bayes and minimax estimators.

8.1.3 Bayes and Minimax Ordering

Consider, next, that the past information on θ is available. We assume that θ is a random variable so that its *pdf*, $\pi(\theta)$, is given by utilizing this past information. Therefore, $\pi(\theta)$ is called prior distribution. Denote the density of sample observations by $f(\mathbf{x}|\theta)$ at the place $f(\mathbf{x};\theta)$ for the purpose of clarity that it represents the conditional density of $\mathbf{X}$ indexing θ, that is, when θ is held fixed. So, the joint density of $\mathbf{X}$ and θ is

$$h(\mathbf{x},\theta) = \pi(\theta) f(\mathbf{x}|\theta)$$

The marginal density of **X** is

$$m(\mathbf{x}) = \int_{\Theta} h(\mathbf{x}, \theta) d\theta$$

The posterior density, therefore, is given by

$$\pi(\theta | \mathbf{x}) = \frac{h(\mathbf{x}, \theta)}{m(\mathbf{x})} \tag{8.1.7}$$

provided the marginal density $m(\mathbf{x})$ exists and is positive.

The Bayes risk of an estimator δ with respect to some prior distribution π is defined by

$$r(\pi, \delta) = E^Z R(Z, \delta)$$

where Z is a random variable taking on its values from Θ, the distribution of which is given by π.

The Bayes principle suggests an ordering on the decision space D. According to this principle, an estimator δ_1 is said to be superior to some other estimator δ_2 with respect to some prior distribution π if

$$r(\pi, \delta_1) < r(\pi, \delta_2)$$

This ordering is known as *Bayes ordering*. Note that the Bayes ordering is complete order relationship. A decision function δ_0 for which the Bayes risk is minimum over the Bayes risks of all other decision functions is called Bayes decision function.

Definition 8.1.1 (Bayes estimator) A rule δ_0 is Bayes with respect to some prior distribution π if

$$r(\pi, \delta_0) = \inf_{\delta \in D} r(\pi, \delta)$$

Next, the preference of one estimator over the other based on $\sup_{\theta \in \Theta} R(\theta, \delta)$ induces an order relationship over a class of estimators D; this is known as *minimax ordering*. Under this ordering, an estimator δ_1 is said to be superior to some other estimator δ_2 if

$$\sup_{\theta \in \Theta} R(\theta, \delta_1) \le \sup_{\theta \in \Theta} R(\theta, \delta_2)$$

Minimax principle gives a principle of ordering the estimators on the basis of best possible ("mini" among the "max").

Nature acts as an *adversary* which sets maximum loss to the statistician $\sup_{\theta \in \Theta} R(\theta,\delta)$ for its choice δ as an estimator. The statistician, then, chooses such an estimator for which $\sup_{\theta} R(\theta,\delta)$ is minimum. Such an estimator is called *minimax estimator*.

Definition 8.1.2 (Minimax estimator) An estimator δ_0 in D is called minimax if it is the superior-most estimator in D, that is,

$$\sup_{\theta \in \Theta} R(\theta, \delta_0) = \inf_{\delta \in D} \sup_{\theta \in \Theta} R(\theta, \delta) = \overline{V} \tag{8.1.8}$$

The expression on the right-hand side of Eq. (8.1.8) is called the minimax risk and is known as the *upper value of the game*, $\overline{V}$. Here, the estimation problem (Θ, D, R) is referred to as a game.

There may not exist an estimator that achieves the minimax risk even if it is finite. In such cases, we may find a ε-minimax estimator δ_0 so that it satisfies

$$\sup_{\theta\in\Theta} R(\theta, \delta_0) \le \sup_{\theta\in\Theta} R(\theta, \delta) + \varepsilon \;\forall\, \varepsilon > 0$$

Notes.

1. The prior is a technical or mathematical way of expressing the available information into a decision-making process, leading to an inference known as the Bayesian inference. Sometimes prior distribution may not be normalized and

$$\int_{\Theta} \pi(\theta)d\theta = \infty$$

   Such priors are called *improper priors.* These priors are allowed in the Bayesian estimation as long as the marginal distribution of **X** in the denominator of Eq. (8.1.7) is well-defined. Moreover, many improper priors may be obtained as a limiting proper priors. Consider the example of $N(\eta, \tau^2)$ priors; it becomes flat over its range if we let $\tau^2 \to \infty$. The resulting prior is improper. Similarly, the proper prior $U(a, b)$ becomes flat over its range if we allow $a \to -\infty$, $b \to \infty$, and the limiting prior is improper. The lack of probabilistic interpretation of these improper priors raises controversy on their use.
2. **Posterior distribution depends only on a sufficient statistic.** Assume that $T(\mathbf{X})$ is a sufficient statistic for the family of distributions $\{f(x|\theta), \theta \in \Theta\}$. Using factorization theorem, the conditional distribution of θ given **X** is

$$\pi(\theta|\mathbf{X}) = \frac{g(T(\mathbf{X}), \theta)h(\mathbf{X})\pi(\theta)}{\int g(T(\mathbf{X}),\theta)h(\mathbf{X})\pi(\theta)d\theta}$$

$$= \frac{g(T(\mathbf{X}), \theta)\pi(\theta)}{\int g(T(\mathbf{X}),\theta)\pi(\theta)d\theta} = \pi\left(\theta|T(\mathbf{X})\right)$$

   which is a function of $T(\mathbf{X})$.

Definition 8.1.3 (Least favourable distribution) A prior distribution π_0 in the class Θ^* of all prior probability distributions defined on Θ is said to be *least favourable* if

$$\inf_{\delta} r(\pi_0, \delta) \ge \inf_{\delta} r(\pi, \delta) \;\forall\; \pi \in \Theta^*$$

or

$$\inf_{\delta} r(\pi_0, \delta) \ge \sup_{\pi} \inf_{\delta} r(\pi, \delta) = \underline{V} \tag{8.1.9}$$

The prior distribution π_0 maximizes losses that the statistician cuts by way of choosing an appropriate estimator. It is for this reason that the statistician likes least to be told about what prior nature was choosing; therefore, the name least favorable. The expression on the right-hand side of Eq. (8.1.9) is called maxmin risk or *lower value of the game*, i.e., $\underline{V}$. Further, note that

$$\underline{V} \le \bar{V}$$

Thus, there are two types of optimality criterion of finding optimal estimators: one that minimizes the Bayes risk, known as Bayes estimator, and second that minimizes the maximum risk over all θ, known as minimax estimation.

Definition 8.1.4 (Predictive distribution of *Y*) Suppose that Y follows some distribution $f(y|\theta)$. The posterior predictive distribution of Y

$$f(y|\mathbf{x}) = \int_{\Theta} f(y|\theta)\pi(\theta|\mathbf{x})\,d\theta$$

is called the predictive distribution of Y given the data $(X_1,\ldots,X_n)$.

8.2 BAYES ESTIMATION

We will discuss here the method of Bayes estimation. Assuming $\mathbf{X}$ and θ are continuous random variables, the Bayes risk of an estimator δ is expressed by

$$\begin{aligned} r(\pi, \delta) &= E^{\theta} R(\theta, \delta(\mathbf{x})) = E^{\theta} E_{\theta}^{\mathbf{X}} L(\theta, \delta(\mathbf{x})) \\ &= \int_{\Theta} \left\{ \int_{\chi} L(\theta, \delta(\mathbf{x})) f(\mathbf{x}|\theta)\,d\mathbf{x} \right\} \pi(\theta)\,d\theta \\ &= \int_{\Theta}\int_{\chi} L(\theta, \delta(\mathbf{x}))\, h(\mathbf{x}, \theta)\,d\mathbf{x}\,d\theta \\ &= \int_{\Theta}\int_{\chi} L(\theta, \delta(\mathbf{x}))\, m(\mathbf{x})\pi(\theta|\mathbf{x})\,d\mathbf{x}\,d\theta \end{aligned}$$

Further, assuming that the integration signs are interchangeable, we have

$$r(\pi, \delta) = \int_{\chi} \left\{ \int_{\Theta} L(\theta, \delta(\mathbf{x}))\pi(\theta|\mathbf{x})\,d\theta \right\} m(\mathbf{x})\,d\mathbf{x} \tag{8.2.1}$$

By the expression of Bayes risk in Eq. (8.2.1), it is clear that the rule that minimizes $r(\pi, \delta)$ is the one that minimizes the integral

$$\int_{\Theta} L(\theta, \delta(\mathbf{x}))\pi(\theta|\mathbf{x})\,d\theta$$

for every value of $\mathbf{x}$. The quantity in Eq. (8.2.1) is called *posterior conditional expected loss*, which can also be expressed as

$$E[L(\theta, \delta(\mathbf{x}))|\mathbf{X} = \mathbf{x}]$$

where expectation is taken with respect to the posterior distribution.

Bayes estimator in case of two-dimensional vector $\boldsymbol{\theta} = (\theta_1, \theta_2)'$. Let $X \sim F_{\boldsymbol{\theta}}$, $\boldsymbol{\theta} = (\theta_1, \theta_2)'$, $\theta_j \in \Theta_j$. Let $\pi(\theta_2)$ be the prior density over Θ_2 and for any given θ_2, $\pi(\theta_1|\theta_2)$ be the prior density over Θ_1.

Suppose, for given θ_2, the Bayes estimator of $g'(\theta_1) = g(\theta_1, \theta_2)$ under the squared error loss function is $\delta(\mathbf{X}, \theta_2)$. The Bayes estimator of $g(\theta_1, \theta_2)$ is then given by

$$\delta^B(\mathbf{x}) = \int_{\Theta_2} \delta(\mathbf{x}, \theta_2)\pi(\theta_2|\mathbf{x})d\theta_2$$

where $\pi(\theta_2|\mathbf{x})$ is the posterior distribution of θ_2 given $\mathbf{X} = \mathbf{x}$.

Analogous expressions and results can similarly be found when $\mathbf{X}$ and $\boldsymbol{\theta}$ are discrete random variables.

Note. The behaviour of the posterior distribution for large n is

$$\pi(\theta|\mathbf{x}) \sim N\left(\theta_{ML}, \frac{1}{I_{\mathbf{X}}(\theta_{ML})}\right)$$

Moreover, for Bayes estimators, we may restrict only to a class of nonrandomized estimators. We will now show the existence of a Bayes estimator.

Theorem 8.2.1 Suppose there exists some estimator so that its Bayes risk is finite for a given prior $\pi(\theta)$ and there exists an estimator $\delta_0(x)$ that minimizes the posterior conditional expected loss $E\{L(\theta, \delta(X))|X = x\}$ for almost all x with respect to the marginal distribution of X, $m(x)$. Then δ_0 is a Bayes estimator with respect to π.

Proof. For any estimator δ, we have

$$E[L(\theta, \delta_0)|X = x] \le E[L(\theta, \delta)|X = x] \tag{8.2.2}$$

for almost all x. The expectation of inequality (8.2.2) with respect to $m(x)$ gives

$$E^X E^{\theta|X}[L(\theta, \delta_0)|X] \le E^X E^{(\theta|X)}[L(\theta, \delta)|X]$$

or

$$r(\pi, \delta_0) \le r(\pi, \delta)$$

Hence, δ_0 is Bayes as asserted. ▣

Bayes Estimation under Squared Error Loss Function

We will now discuss the Bayes estimation under the squared error loss function in the following theorem.

Theorem 8.2.2 The Bayes estimator of θ under the squared error loss function is the mean of the posterior distribution, that is,

$$\delta^B(\mathbf{x}) = E\{\theta|\mathbf{X} = \mathbf{x}\}$$

Proof. Assume $\mathbf{X}$ and θ are continuous random variables. The Bayes risk of an estimator δ with respect to some prior distribution π is the average of $\text{MSE}_\theta(\delta)$ over θ

$$r(\pi, \delta) = E^\theta R(\theta, \delta) = E^\theta E_\theta^{\mathbf{X}|\theta} L(\theta, \delta) = E^\theta E_\theta^{\mathbf{X}|\theta}(\delta - \theta)^2 = E^\theta\, \text{MSE}_\theta(\delta)$$

$$= \int_\Theta \int_\chi [\theta - \delta(\mathbf{x})]^2 f(\mathbf{x}|\theta)\pi(\theta) d\mathbf{x}\, d\theta$$

$$= \int_\chi \left[\int_\Theta [\theta - \delta(\mathbf{x})]^2 \pi(\theta|\mathbf{x}) d\theta\right] m(\mathbf{x}) d\mathbf{x}$$

This Bayes risk is minimum if and only if the posterior conditional expected loss is minimum for every **x**.

Therefore, the Bayes estimator is the solution of the equation

$$\frac{\partial}{\partial \delta} E\left\{[\theta - \delta(\mathbf{x})]^2 \middle| \mathbf{x}\right\} = 0$$

or

$$\frac{\partial}{\partial \delta} \int_{\Theta} [\theta - \delta(\mathbf{x})]^2 \pi(\theta|\mathbf{x}) d\theta = 0$$

$$\int_{\Theta} 2[\theta - \delta(\mathbf{x})] \pi(\theta|\mathbf{x}) d\theta = 0$$

$$\delta(\mathbf{x}) = \int \theta \pi(\theta|\mathbf{x}) d\mathbf{x} = E\{\theta|\mathbf{x}\}$$

which is the mean of the posterior distribution. $E(\theta|\mathbf{x})$ minimized the Bayes risk $r(\pi, \delta)$ since

$$\frac{\partial^2}{\partial \delta^2} r(\pi, \delta) = \frac{\partial^2}{\partial \delta^2} E[(\theta - \delta)^2 | \mathbf{x}]$$

$$= \frac{\partial}{\partial \delta} E[-2(\theta - \delta)|\mathbf{x}]$$

$$= E[2] = 2 > 0$$ ■

The proof of Theorem (8.2.2) can also be given by using the following lemma.

Lemma 8.2.1 Let X and Y be two random variables defined on $\mathbb{R}^1$ having same probability space and H be a set of all functions h: $\mathbb{R}^1 \to \mathbb{R}^1$. Then,

$$\min_{h \in H} E[Y - h(X)]^2 = E[Y - E(Y|X)]^2$$

That is, $E(Y|X)$ minimizes $E[Y - h(X)]^2$ over h.

Proof. Let us express

$$E[Y - h(X)]^2 = E[Y - E(Y|X) + E(Y|X) - h(X)]^2$$

$$= E[Y - E(Y|X)]^2 + 2E\{[Y - E(Y|X)][E(Y|X) - h(X)]\} + E\{[E(Y|X) - h(X)]^2\}$$

The cross product term, conditioning over X, gives

$$E\{[Y - E(Y|X)][E(Y|X) - h(X)]\}$$

$$= E^X E^{Y|X}\{YE(Y|X) - Yh(X) - [E(Y|X)]^2 + E(Y|X)h(X)\}$$

$$= E^X\{[E(Y|X)]^2 - E(Y|X)h(X) - [E(Y|X)]^2 + E(Y|X)h(X)\} = 0$$

We have

$$E[Y - h(X)]^2 = E[Y - E(Y|X)]^2 + E[E(Y|X) - h(X)]^2 \tag{8.2.3}$$

The first term in Eq. (8.2.3) is not the function of h; only the second term is the function of h. The second term is minimum, that is, zero, when $h(X) = E(Y|X)$. Therefore, $E[Y - h(X)]^2$ is minimum over h, when $h = E(Y|X)$.

Thus, applying this lemma with $X = X$, $Y = \theta$, and $h = T$ in the theorem, we get the Bayes estimator of θ as

$$\delta^B(\mathbf{x}) = \int \theta\pi(\theta|\mathbf{x})d\mathbf{x} = E\{\theta|\mathbf{x}\} \tag{8.2.4}$$

δ^B is the Bayes estimator of θ and Eq. (8.2.4) is the posterior mean.

Considering the weighted squared error loss function

$$L(\theta, \delta) = w(\theta)\,(\theta - \delta)^2,\ w(\theta) > 0\ \forall\ \theta \in \Theta$$

the Bayes estimate of θ by Theorem 8.2.2 is given by

$$\delta = \frac{E[\theta w(\theta)|\mathbf{x})}{E[w(\theta)|\mathbf{x})} \tag{8.2.5}$$

The estimator δ in Eq. (8.2.5) minimizes the Bayes risk, $r(\pi, \delta)$, since

$$\frac{\partial^2}{\partial\delta^2} r(\pi, \delta) = \frac{\partial^2}{\partial\delta^2} E[w(\theta)(\theta - \delta)^2|\mathbf{x}]$$

$$= 2E[w(\theta)|\mathbf{x}] > 0$$ ▣

The extension of Theorem 8.2.2 to the case when $\boldsymbol{\theta}$ is a vector is given as follows:

***Corollary* 8.2.1** Let $\mathbf{X} \sim f(\mathbf{x}|\boldsymbol{\theta})$, $\boldsymbol{\theta} \in \Theta \subseteq \mathbb{R}^k$, and $\mathbf{x} \in \mathbb{R}^m$. Let $\pi(\boldsymbol{\theta})$ be a prior distribution. The Bayes risk of an estimator $\boldsymbol{\delta}$ of $\boldsymbol{\theta}$ is

$$E\|\boldsymbol{\delta}(\mathbf{X}) - \boldsymbol{\theta}\|^2 = \iint \|\boldsymbol{\delta}(\mathbf{X}) - \boldsymbol{\theta}\|^2 f(\mathbf{x}|\boldsymbol{\theta})\pi(\boldsymbol{\theta})d\mathbf{x}\,d\boldsymbol{\theta}$$

Then, the Bayes estimator of $\boldsymbol{\theta}$ is

$$\boldsymbol{\theta}^B = E(\boldsymbol{\theta}|\mathbf{X}) = (E(\theta_1|\mathbf{X}),\ldots,E(\theta_k|\mathbf{X}))'$$

Quadratic loss function. Let the parameter $\boldsymbol{\theta}$ be a k-dimensional vector estimated by $\boldsymbol{\delta} \in \mathbb{R}^k$. Then the loss function

$$L(\boldsymbol{\theta}, \boldsymbol{\delta}) = (\boldsymbol{\theta} - \boldsymbol{\delta})'\,\mathbf{C}(\boldsymbol{\theta} - \boldsymbol{\delta})$$

is called the *quadratic loss function* where $\mathbf{C}$ is a positive definite real symmetric matrix of order $k \times k$. Further, if $\mathbf{C} = \text{diag}(C_1,\ldots,C_k)$, then

$$L(\boldsymbol{\theta}, \boldsymbol{\delta}) = \sum C_i(\theta_i - \delta_i)^2$$

Bayes Estimator under Absolute Error Loss Function

We will now discuss the Bayes estimation under the absolute error loss function in the following theorem.

Theorem 8.2.3 The Bayes estimator in case of absolute error loss function is the median of the posterior distribution.

Proof. Consider

$$\varphi(\delta) = E(|\theta - \delta||\mathbf{x}) = \int_{\Theta} |\theta - \delta|\pi(\theta|\mathbf{x})d\theta$$

$$= \int_{-\infty}^{\delta} (\delta - \theta)\pi(\theta|\mathbf{x})d\theta + \int_{\delta}^{\infty} (\theta - \delta)\pi(\theta|\mathbf{x})d\theta$$

In this case, the Bayes estimator of θ is

$$\delta^B(\mathbf{x}) = \min_{\delta} \varphi(\delta) = \min_{\delta} \int_{\Theta} |\theta - \delta| \pi(\theta|\mathbf{x}) d\theta$$

$$= \min_{\delta} \left(\int_{-\infty}^{\delta} (\delta - \theta)\pi(\theta|\mathbf{x})d\theta + \int_{\delta}^{\infty} (\theta - \delta)\pi(\theta|\mathbf{x})d\theta \right)$$

Differentiating $\varphi(\delta)$ with respect to δ [using Leibniz's rule (1.2.5)], we have

$$\frac{\partial}{\partial \delta} \int_{-\infty}^{\delta} (\delta - \theta)\pi(\theta|\mathbf{x})d\theta = \int_{-\infty}^{\delta} \pi(\theta|\mathbf{x})\,d\theta$$

$$\frac{\partial}{\partial \delta} \int_{\delta}^{\infty} (\theta - \delta)\pi(\theta|\mathbf{x})d\theta = \int_{\delta}^{\infty} (-1)\pi(\theta|\mathbf{x})\,d\theta$$

The Bayes estimator of θ is the solution of

$$\frac{\partial}{\partial \delta}\varphi(\delta) = -\int_{\delta}^{\infty} \pi(\theta|\mathbf{x})d\theta + \int_{-\infty}^{\delta} \pi(\theta|\mathbf{x})\,d\theta = 0$$

$$\int_{-\infty}^{\delta} \pi(\theta|\mathbf{x})d\theta = \int_{\delta}^{\infty} \pi(\theta|\mathbf{x})\,d\theta$$

$$\therefore \qquad \delta^B(x) = \text{median}[\pi(\theta|\mathbf{x})]$$

The median minimizes $\varphi(\delta)$ since

$$\varphi''(\delta) = \frac{\partial^2}{\partial \delta^2}\varphi(\delta) = 2\pi(\delta|\mathbf{x}) > 0$$

Therefore, the Bayes estimator is the median of the posterior distribution. ▣

Note. If $T(X)$ is a sufficient statistic for the *pdf* $f(x|\theta)$, then $E(\theta|X) = E[\theta|T(X)]$. Consider

$$E(\theta|X) = c\int \theta g[T(X), \theta]\pi(\theta)d\theta = h[T(X)]$$

$$E^{T(X)}E(\theta|X) = E^{T(X)}h[T(X)]$$

$$E^{T(X)}E[\theta|T(X)] = E^{T(X)}\,h[T(X)]$$

[Since $\sigma[T(X)] \subset \sigma(X)$, $E^{T(X)}$ is with respect to the algebra $\sigma(T(X))$]
Thus, $h[T(X)] = E[\theta|T(X)]$.

Characterization of Bayes Estimators

Usually, a Bayes estimator is not unbiased but approximately unbiased; however, it is usually consistent. One can easily check this in different examples and especially when the Bayes estimators are in closed form.

Sometimes, the bias of a Bayes estimator is used to check whether a given estimator can be a Bayes estimator with respect to some prior distribution. The result is summarized in the following theorem.

Theorem 8.2.4 Let $\delta(\mathbf{X})$ be a Bayes estimator of $g(\theta)$ with respect to some prior distribution $\pi(\theta)$ under the squared error loss function. Assume that $V[\delta(\mathbf{X})|\theta] < \infty$ and $V[g(\theta)] < \infty$. Then, $\delta(\mathbf{X})$ is not unbiased except the case when its Bayes risk is zero. i.e.,

$$V(\delta|\theta) = 0 \quad \{r(\pi, \delta) = E^{\mathbf{X},\theta}[\delta(\mathbf{X}) - \theta]^2 = 0\}$$

Proof. It is given that $\delta(\mathbf{x})$ is the Bayes estimate of $g(\theta)$ under the squared error loss function. We have

$$\delta(\mathbf{X}) = E[g(\theta)|\mathbf{X}]$$

where the expectation is taken with respect to the posterior distribution. Suppose $\delta(\mathbf{x})$ is an unbiased estimator of $g(\theta)$

$$E(\delta(\mathbf{X})|\theta) = g(\theta)$$

Consider the variance of $\delta(\mathbf{X})$

$$\begin{aligned} V[\delta(\mathbf{X})] &= E^{\theta} V^{\mathbf{X}|\theta}[\delta(\mathbf{X})] + V^{\theta} E^{\mathbf{X}|\theta}[\delta(\mathbf{X})] \\ &= E^{\theta} V^{\mathbf{X}|\theta}[\delta(\mathbf{X})] + V^{\theta}[g(\theta)] \end{aligned} \tag{8.2.6}$$

Consider, now

$$\begin{aligned} V^{\theta}[g(\theta)] &= E^{\mathbf{X}} V^{\theta|\mathbf{X}}[g(\theta)] + V^{\mathbf{X}} E^{\theta|\mathbf{X}}[g(\theta)] \\ &= E^{\mathbf{X}} V^{\theta|\mathbf{X}}[g(\theta)] + V^{\mathbf{X}}[\delta(\mathbf{X})] \end{aligned} \tag{8.2.7}$$

Equations (8.2.6) and (8.2.7) give

$$E^{\theta} V^{\mathbf{X}|\theta}[\delta(\mathbf{X})] + E^{\mathbf{X}} V^{\theta|\mathbf{X}}[g(\theta)] = 0$$

Note that terms on the left-hand side are non-negative. Therefore, the sum of the two non-negative terms implies

$$E^{\theta} V^{\mathbf{X}|\theta}[\delta(\mathbf{X})] = r(\pi, \delta) = 0$$

This implies

$$V^{\mathbf{X}|\theta}[\delta(\mathbf{X})] = 0$$

since

$$V^{\mathbf{X}|\theta}[\delta(\mathbf{X})] \geq 0$$

Thus, either $V[\delta(\mathbf{x})|\theta] = 0$ or the Bayes estimator $\delta(\mathbf{X})$ is not an unbiased estimator of $g(\theta)$. In other words, the Bayes estimator is not unbiased unless $r(\pi, \delta) = 0$. ▣

However, in Example 8.4.3 a generalized Bayes estimator is shown to be unbiased. Therefore, generalized Bayes estimators may be unbiased. The generalized Bayes estimators are defined in subsection 8.2.2.

Note. If T is unbiased and Bayes, then its Bayes risk must vanish. This result helps in characterizing whether an unbiased estimator is Bayes.

Definition 8.2.1 (Maximum aposteriori (MAP) estimator) An estimator δ_0 is called *maximum aposteriori estimator* if for infinitesimally small ε,

$$\pi(\delta_0|\mathbf{x}) = \max_{\delta} \pi(\delta|\mathbf{x}) = \lim_{\varepsilon \to 0} \max_{\delta} \int_{\delta-\varepsilon}^{\delta+\varepsilon} \pi(\theta|\mathbf{x})d\theta$$

In the limit as $\varepsilon \to 0$ (ε tends to zero), the highest mode of the posterior distribution is chosen as the MAP estimate. This is simple to obtain, since one need not obtain the posterior distribution; rather, it is only to find the maximum of the posterior distribution.

Notice that if the posterior distribution $\pi(\theta|\mathbf{x})$ is unimodal and symmetric, then the Bayes and MAP estimators both coincide.

We will now discuss the relationship between ML and Bayes estimate. Consider

$$\pi(\theta|x) = \frac{f(x|\theta)\pi(\theta)}{\int_{\Theta} f(x|\theta)\pi(\theta)d\theta} \tag{8.2.8}$$

By definition, $f(x|\theta)$ peaks at the ML estimate. If $f(x|\theta)$ peaks sharp (the peak of $f(x|\theta)$ is relatively sharp) and the prior is broad (flat over a reasonable interval of θ), then by Eq. (8.2.8),

$$\pi(\theta|x) \cong f(x|\theta)$$

and the Bayes estimate is approximately the same as the ML estimate. Otherwise, a slanted or peaked prior may push away the MAP from the ML estimate. However, if the uncertainly about θ in the likelihood function is high and the data does not determine the parameter well, then the posterior depends strongly on this prior. The Bayes estimate, in this case, will be close to the prior estimate. However, if likelihood is sharp, i.e., data can strongly determine the parameter θ, then the prior has little influence on the Bayes estimate of θ.

Consider the estimation problem with its likelihood function $f(x|\theta)$ and the prior distribution as uniform $\pi(\theta) = c \neq 0$. The posterior is

$$\pi(\theta|x) = \frac{f(x|\theta)\pi(\theta)}{\int_{\mathbb{R}^1} f(x|\theta)\pi(\theta)d\theta}$$

$$= \frac{cf(x|\theta)}{c\int_{\mathbb{R}^1} f(x|\theta)d\theta} \propto f(x|\theta)$$

It is proportional to the likelihood as long as marginal $m(x)$ is well-defined, i.e., $m(x) < \infty$. Then the MAP estimator of θ is

$$\delta^{\text{MAP}}(x) = \max_{\theta} \pi(\theta|x)$$

$$= \max_{\theta} f(x|\theta) = \delta^{\text{ML}}(x)$$

or the solution of the equation

$$\left.\frac{\partial \log \pi(\theta|x)}{\partial \theta}\right|_{\theta=\theta^{\text{MAP}}} = 0 \tag{8.2.9}$$

On taking logarithm of the terms on both sides of Eq. (8.2.8), we get

$$\log \pi(\theta|x) = \log f(x|\theta) + \log \pi(\theta) - \log m(x)$$

MAP equation takes the form

$$\frac{\partial}{\partial \theta}\log \pi(\theta|x) = \frac{\partial}{\partial \theta}\log f(\theta|x) + \frac{\partial}{\partial \theta}\log \pi(\theta) \tag{8.2.10}$$

since $\log m(x)$ is independent of θ. We know that ML estimate is the solution of the likelihood equation

$$\left.\frac{\partial}{\partial \theta} f(\theta|x)\right|_{\theta=\theta^{\text{ML}}} = 0 \tag{8.2.11}$$

Comparing Eqs. (8.2.9), (8.2.10), and (8.2.11), we observe that if the prior density is sufficiently flat (prior variance approaching high and high), $\log \pi(\theta)$ will also be flat; $(\partial/\partial\theta)\log \pi(\theta)$ is nearly zero. In this case, the posterior density maximizes at the ML estimator, and the MAP and the ML estimates coincide in the limiting sense (prior variance approaching large and large).

Thus, MAP estimator is nothing but ML estimator when the prior is uniform (noninformative) and improper. The MAP estimator is that value of θ at which most likely, the density $f(x|\theta)$ has generated the observed data **x**.

We will discuss, next, the large sample behavior of the estimates δ^B and δ^{MAP}, i.e., when the amount of data grows. Let $\{\delta_n\}$ be a sequence of estimators for estimating the parameter θ and θ_0 be the true value of θ. The estimate δ_n is said to be consistent for θ if

$$\delta_n \xrightarrow{P} \theta_0$$

or for arbitrarily small $\varepsilon > 0$,

$$\lim_{n\to\infty} P\{|\delta_n - \theta_0| > \varepsilon\} = 0$$

We will now discuss the following results without proof, for brevity, since they are beyond the technical level of the book.

Theorem 8.2.5 Assuming that the prior distribution is continuous and positive at the location of the ML estimate, the MAP estimate converges to the ML estimate.

Theorem 8.2.6 Consider certain regularity conditions under which the ML estimate is consistent (see Chapter 6). Accordingly, the MAP estimate is also consistent.

Theorem 8.2.7 In addition to the conditions in Theorems 8.2.5 and 8.2.6, if support $\{x: f(x|\theta) > 0\}$ of observations which are coming from the model $f(x|\theta)$ is independent of θ, then the Bayes estimate converges to the ML estimate, the ML estimate is consistent, and, accordingly, the Bayes estimate is consistent.

Note that the condition in Theorem 8.2.3 that the support is independent of θ is a necessary condition to prove the consistency of an ML estimate.

Moving ahead, we will now show that the posterior distribution is asymptotically normal. A heuristic approach is adopted for brevity by avoiding the formal technical details.

Consider the likelihood function $f(\mathbf{x}|\theta)$ where θ is a real-valued parameter and let $\pi(\theta)$ be the prior distribution on $\Theta \subseteq \mathbb{R}^1$ so that θ_0 is the unique maximum of $\pi(\theta)$, assuming that it exists. Let $X_1,\ldots,X_n$ be n independent and identically distributed observations. The corresponding log likelihood is given by

$$\log f(\mathbf{x}|\theta) = \sum_{i=1}^{n} \log f(x_i|\theta)$$

Let θ^{ML} be the maximum likelihood estimator of θ based on n *iid* observations and is obtained by maximizing log likelihood with respect to θ. The posterior density accordingly is given by

$$\pi(\theta|\mathbf{x}) \propto f(\mathbf{x}|\theta)\ \pi(\theta)$$
$$= \exp\{\log f(\mathbf{x}|\theta) + \log \pi(\theta)\} \tag{8.2.12}$$

On writing Taylor's expansion terms for $\log f(\mathbf{x}|\theta)$ about θ^{ML} and $\log \pi(\theta)$ about θ_0, we get

$$\log f(\mathbf{x}|\theta) = \log f(\mathbf{x}|\theta^{ML}) - \frac{1}{2}(\theta - \theta^{ML})^2 \left.\frac{\partial^2 \log f(\mathbf{x}|\theta)}{\partial \theta^2}\right|_{\theta=\theta^{ML}} + R_n \tag{8.2.13}$$

where R_n is the remainder term and

$$\log \pi(\theta) = \log \pi(\theta_0) - \frac{1}{2}(\theta - \theta_0)^2 \left.\frac{\partial^2 \log \pi(\theta)}{\partial \theta^2}\right|_{\theta=\theta_0} + R_1 \tag{8.2.14}$$

where R_1 is the remainder term. Assume that the terms R_1 and R_n are constant.

We know that the Fisher information about θ based on n *iid* observations is

$$I_{\mathbf{X}}(\theta) = n I_{X_1}(\theta)$$

We may, now, express the term

$$\left.\frac{\partial^2 \log f(\mathbf{x}|\theta)}{\partial \theta^2}\right|_{\theta=\theta^{ML}} = I_{\mathbf{X}}(\theta^{ML}) = n I_{X_1}(\theta^{ML}) \tag{8.2.15}$$

Equations (8.2.13) and (8.2.15) give

$$\log f(\mathbf{x}|\theta) = \log f(\mathbf{x}|\theta^{ML}) - \frac{1}{2}(\theta - \theta^{ML})^2 n I_{X_1}(\theta^{ML}) \tag{8.2.16}$$

Equations (8.2.14) and (8.2.16) give

$$\log f(\mathbf{x}|\theta) + \log \pi(\theta) = \log f(\mathbf{x}|\theta^{ML}) - \frac{1}{2}(\theta - \theta^{ML})^2 n I_{X_1}(\theta^{ML}) + \log \pi(\theta_0)$$
$$- \frac{1}{2}(\theta - \theta_0)^2 \left.\frac{\partial^2 \log \pi(\theta)}{\partial \theta^2}\right|_{\theta=\theta_0}$$

After some algebraic simplifications and putting its value in Eq. (8.2.12), we get

$$\pi(\theta|\mathbf{x}) \propto \exp\left\{-\frac{1}{2}(\theta - \theta^{COM})^2 \left(n I_{X_1}(\theta^{ML}) + \left.\frac{\partial^2 \log \pi(\theta)}{\partial \theta^2}\right|_{\theta=\theta_0} \right)\right\}$$

where $$\theta^{COM} = \frac{[n I_{X_1}(\theta^{ML})]\theta^{ML} + \left\{[(\partial^2/\partial\theta^2)\log \pi(\theta)]_{\theta=\theta_0} \cdot \theta_0\right\}}{n I_{X_1}(\theta^{ML}) + [(\partial^2/\partial\theta^2)\log \pi(\theta)]_{\theta=\theta_0}}$$

$$= \frac{\left[\dfrac{1}{V(\theta^{ML})}\right]\theta^{ML} + \dfrac{1}{\left[\left(\dfrac{\partial^2}{\partial\theta^2}\right)\log\pi(\theta)\right]^{-1}_{\theta=\theta_0}}\theta_0}{\left[\dfrac{1}{V(\theta^{ML})}\right] + \dfrac{1}{\left[\left(\dfrac{\partial^2}{\partial\theta^2}\right)\log\pi(\theta)\right]^{-1}_{\theta=\theta_0}}}$$

The estimate θ^{COM} is treated as a combined estimator, which is a weighted average of ML estimator θ^{ML} and prior mode θ_0 where weights are the inverse of their variances. Further, if the amount of data increases, $n \to \infty$, the weight $[V(\theta^{ML})]^{-1}/\{[V(\theta^{ML})]^{-1} + [(\partial^2/\partial\theta^2)\log\pi(\theta)]^{-1}_{\theta=\theta_0}\}$, attached to θ^{ML} approaches 1 and the weight $\{[(\partial^2/\partial\theta^2)\log\pi(\theta)]^{-1}_{\theta=\theta_0}\}/\{[V(\theta^{ML})]^{-1} + [(\partial^2/\partial\theta^2)\log\pi(\theta)]^{-1}_{\theta=\theta_0}\}$ attached to θ_0 approaches 0. Therefore, the ML estimate θ^{ML}, as $n \to \infty$, in θ^{COM}, dominates completely over the prior estimate θ_0 and the prior becomes irrelevant. Therefore,

$$\lim_{n\to\infty} \pi(\theta|\mathbf{x}) = N(\theta^{ML}, [nI_{X_1}(\theta)]^{-1})$$

Notes.

1. Sometimes, in the state of ignorance about θ, we use noninformative prior (vague prior) such as

$$\pi(\theta) = 1 \text{ (a uniform prior)} \tag{8.2.17}$$

on the whole space Θ, for example, it may be a real time. Clearly, this prior is an improper prior. Attempting to calculate the posterior distribution,

$$\pi(\theta|x) = \frac{f(x|\theta)\pi(\theta)}{\int_\Theta f(x|\theta)\pi(\theta)d\theta} = \frac{f(x|\theta)}{\int_\Theta f(x|\theta)d\theta}$$

is a proper distribution provided $\int_\Theta f(x|\theta)d\theta < \infty$. This shows that if prior is improper, even then we can compute the Bayes estimate as long as posterior distribution is proper. In this case, both MAP and ML estimates coincide and they further coincide with the Bayes estimate if the posterior distribution is unimodal and symmetric. In fact, if the prior knowledge is very weak, i.e., prior variance tends to infinity so that we use prior as in Eq. (8.2.17), a MAP estimate approaches ML estimate.
2. A Bayes estimator with constant risk is minimax.
3. A unique Bayes estimator is admissible.
4. An admissible estimate is either Bayes or limit of Bayes.

8.2.1 Limit of Bayes Estimators

Definition 8.2.2 An estimator δ is said to be a *limit of Bayes estimators* if there exists a sequence of proper prior distributions $\{\pi_\alpha\}$ and a corresponding sequence of Bayes estimators $\{\delta_\alpha(x)\}$ such that

$$\lim_{\alpha\to\infty} \delta_\alpha(x) \to \delta(x)$$

for almost all x in the sense of distribution. The phrase "for almost all x" means that the set B on which the above convergence does not hold has probability zero for all θ, i.e., $P(B) = 0\ \forall\ \theta$.

8.2.2 Generalized Bayes Estimator

In many estimation problems, the statistician has no prior knowledge about the parameter θ and, therefore, he chooses such a prior that treates all parameter values Θ equally. In other words, he chooses such a prior that assigns equal probability mass to all values of θ in Θ. This prior is known as *noninformative prior*, which is often improper, i.e., $\int \pi(\theta)d\theta \neq 1$. These noninformative priors, therefore, are not probability measures though these are σ-finite measures. In such cases, when prior distributions are not proper, their probabilistic interpretations are not possible. Further, in these cases, the marginal distribution of **X** may not be finite and, therefore, one cannot talk of posterior conditional expected loss to find the Bayes estimator, since posterior distribution cannot then be calculated. Therefore, in these cases, we consider the problem of minimization of the quantity

$$\int_\Theta L(\theta,\delta) f(\mathbf{x}|\theta)\pi(\theta)d\theta \tag{8.2.18}$$

with respect to the estimator δ as an alternative to the minimization of posterior conditional expected loss.

An estimator δ that minimizes Eq. (8.2.18) with respect to an improper prior $\pi(\theta)$ is called a *generalized Bayes estimator* of θ.

Note. The Bayes estimates for convex loss functions are unique and admissible. However, generalized Bayes estimates are not.

8.2.3 Empirical Bayes Estimator

We have seen that Bayes estimators depend upon the decision as to what prior distribution is being chosen. The parameters specifying the prior distribution are known as *hyperparameters*. So far, it is assumed that hyperparameters are known. However, in several situations, the hyperparameters are not known and they need to be estimated from the data $x_1,\ldots,x_n$. One way is to view the sample as it is drawn from the marginal distribution

$$m(\mathbf{x}|\xi) = \int_\Theta f(\mathbf{x}|\theta)\pi(\theta|\xi)d\theta$$

and use some statistical method of estimation, e.g., the method of moments or maximum likelihood estimation for estimating the hyperparameter, $\hat{\xi}$ say. Carrying out the regular Bayesian calculations by taking $\pi(\theta|\hat{\xi})$ as prior, the Bayes estimator $\delta^{EB}(\mathbf{x})$, so obtained, is called the *empirical Bayes estimator* of θ.

8.2.4 Hierarchical Bayes Estimators

The method of empirical Bayes estimation is not a pure Bayesian estimation technique in strict sense, since it involves the estimation of parameters by classical method. A pure Bayesian approach in such a situation is the hierarchical Bayes estimation, where some appropriate prior is chosen rather than estimating the hyperparameters. Subsequently, a hierarchical Bayes model is developed to get the Bayes estimate. We use the iterative conditional simulated sampling approach named as the *Markov chain Monte Carlo* (more particularly, in this case, *Gibbs sampling*) to compute the estimates. The Gibbs sampling is used with full conditionals.

Hierarchical Bayes estimation is purely a Bayes way of estimating the hyperparameters. In this, we select a completely specified second-stage prior $\pi_2(\xi)$ in additon to a first-stage prior $\pi_1(\theta|\xi)$ with unknown vector of hyperparameter ξ. If the second-stage prior also involves some other unknown parameters, one may consider a third-stage prior and so on. Usually, one does not require to proceed to third- and further-stage priors since Berger (1985) has shown that the Bayesian inference is robust against misspecifying a second-stage prior. Generally, improper noninformative priors are considered as second-stage priors.

Under the hierarchical approach, one obtains a completely specified prior

$$\pi(\theta) = \int_{\Lambda} \pi_1(\theta|\xi)\pi_2(\xi)d\xi \tag{8.2.19}$$

where ξ is taking values in Λ.

This prior is known as hierarchical prior and the Bayes estimator $T^{HB}(\mathbf{X})$ with respect to the prior $\pi(\theta)$ in Eq. (8.2.19) is known as the *hierarchical Bayes estimator* of θ. Generally, the hierarchical Bayes estimators are better than the empirical Bayes estimators.

The hierarchical Bayes estimator can be calculated by using the Bayes estimator $\delta^B(\mathbf{X})$ with respect to prior $\pi(\theta|\xi)$ when ξ is known, and the posterior distribution of ξ given $\mathbf{x}$ in the following manner:

The marginal distribution of $\mathbf{X}$ is

$$m(\mathbf{x}) = \int_{\Theta} f(\mathbf{x}|\theta)\pi(\theta)d\theta$$

Using Eq. (8.2.19), we get

$$m(\mathbf{x}) = \int_{\Theta}\int_{\Lambda} f(\mathbf{x}|\theta)\pi_1(\theta|\xi)\pi_2(\xi)d\xi\, d\theta$$

The marginal distribution of $\mathbf{X}$ given θ and ξ (ξ is known) is

$$m(\mathbf{x}|\xi) = \int_{\Theta} f(\mathbf{x}|\theta)\pi_1(\theta|\xi)\, d\theta$$

Then the posterior distribution of θ given $\mathbf{x}$ is expressed by

$$\pi(\theta|\mathbf{x}) = \frac{h(\mathbf{x},\theta)}{m(\mathbf{x})} = \frac{f(\mathbf{x}|\theta)\pi(\theta)}{m(\mathbf{x})}$$

From Eq. (8.2.19),

$$\begin{aligned}\pi(\theta|\mathbf{x}) &= \int_\Lambda \frac{f(\mathbf{x}|\theta)\pi_1(\theta|\xi)}{m(\mathbf{x})}\pi_2(\xi)d\xi \\ &= \int_\Lambda \frac{f(\mathbf{x}|\theta)\pi_1(\theta|\xi)}{m(\mathbf{x}|\xi)}\frac{m(\mathbf{x}|\xi)}{m(\mathbf{x})}\pi_2(\xi)d\xi \\ &= \int_\Lambda \underbrace{\pi(\theta|\mathbf{x},\xi)}_{\substack{\text{Posterior distribution}\\ \text{of }\theta\text{ given }\mathbf{x}\text{ and }\xi}} \underbrace{\pi(\xi|\mathbf{x})}_{\substack{\text{Posterior distribution}\\ \text{of }\xi\text{ given }\mathbf{x}}} d\xi\end{aligned}$$

Therefore, given a Bayes estimator $\delta^B(\mathbf{X}, \xi)$ of θ with respect to prior $\pi_1(\theta|\xi)$ when ξ is known, the hierarchical Bayes estimate is

$$\delta(\mathbf{x}) = \int_\Lambda \delta(\mathbf{x},\xi)\pi(\xi|\mathbf{X})d\xi$$

8.2.5 Bayes Estimate and Admissibility

We will prove the result that answers the question "When are Bayes estimators admissible?" The following theorem shows that Bayes estimators are reasonable in the sense that they are admissible under certain conditions.

Theorem 8.2.8 Consider an estimation problem where $\delta(\mathbf{X})$ is Bayes with respect to a prior distribution $\pi(\theta)$. Then, $\delta(\mathbf{X})$ is admissible

(i) if $\delta(\mathbf{X})$ is unique up to equivalence, or
(ii) if Θ is a countable set, the Bayes risk of δ is finite, $r(\pi, \delta) < \infty$, π assigns positive probability to each $\theta \in \Theta$, i.e., $\pi(\theta) > 0\ \forall\ \theta \in \Theta$, or
(iii) if $\Theta = \mathbb{R}^1$, $\delta(\mathbf{X}) \in D$ and $r(\pi, \delta) < \infty$, where D is the class of all estimators of θ so that their risk functions $R(\theta, \delta)$ are continuous in θ, and if the support of the prior distribution π is the entire real line.

Proof.

(i) The meaning of Bayes rules unique up to equivalence is that they yield the same risk functions if there are two Bayes rules with respect to a given prior distribution π. Suppose that the estimator δ is Bayes but not admissible, that is, there exists another estimator δ' better than δ. Thus, the Bayes risk of δ' is smaller than of δ, which is a contradiction. Therefore, the Bayes estimator δ, if unique up to equivalence, is admissible.

(ii) Suppose that δ is not admissible. Then there exists another estimator δ' which is better than δ. We then have

$$R(\theta, \delta') \le R(\theta, \delta) \;\forall\; \theta \in \Theta$$

and $$R(\theta, \delta') < R(\theta, \delta) \text{ for some } \theta \in \Theta$$

On comparing their Bayes risks, we get

$$r(\pi, \delta') = \sum_{\theta\in\Theta} R(\theta, \delta')\pi(\theta) < \sum_{\theta\in\Theta} R(\theta, \delta)\pi(\theta) = r(\pi, \delta)$$

since Θ is countable and $\pi(\theta) > 0 \;\forall\; \theta \in \Theta$. This strict inequality shows that the estimator δ is not Bayes with respect to π. This contradiction proves that the Bayes estimator δ is admissible.

(iii) Assume that δ is not admissible. So, there exists at least one δ' in D such that

$$R(\theta, \delta') \le R(\theta, \delta) \;\forall\; \theta \in \Theta$$

and $$R(\theta_0, \delta') < R(\theta_0, \delta) \text{ for at least one } \theta = \theta_0$$

Define $$\eta = R(\theta_0, \delta) - R(\theta_0, \delta') > 0$$

Then, from the continuity of risk functions of δ and δ', there exists a small $\varepsilon > 0$ such that

$$R(\theta, \delta') \le R(\theta, \delta) - \frac{\eta}{2} \;\forall\; \theta \in (\theta_0 - \varepsilon, \theta_0 + \varepsilon), (|\theta - \theta_0| < \varepsilon)$$

Now, on taking expectation with respect to the prior distribution $\pi(\theta)$,

$$r(\delta, \pi) - r(\delta', \pi) = E\{R(\theta, \delta) - R(\theta, \delta')\}$$

$$\ge \frac{\eta}{2} \int_{\theta_0-\varepsilon}^{\theta_0+\varepsilon} \pi(\theta)\,d\theta$$

$$= \frac{\eta}{2} \pi(\theta_0 - \varepsilon, \theta_0 + \varepsilon)$$

We have $\eta > 0$ and $\pi(\theta_0 - \varepsilon, \theta_0 + \varepsilon) > 0$ since θ_0 is in the support of π. It gives $(\eta/n)\pi(\theta_0 - \varepsilon, \theta_0 + \varepsilon) > 0$. This shows that δ is not Bayes, which is a contradiction to the assumption that δ is Bayes with respect to a given prior distribution π. Hence, if δ is Bayes, it is also admissible under the conditions given in the theorem. This completes the proof. ▣

The converse that an admissible estimator is also Bayes with respect to some prior distribution is shown in the following theorem.

Theorem 8.2.9 If $\Theta = \{\theta_1, \ldots, \theta_k\}$ is finite and δ is admissible, then δ is Bayes with respect to some prior distribution.

Proof. Readers may refer Ferguson (1967) for the complete proof of the theorem.

Note that an admissible estimator is also Bayes with respect to some prior distribution only when the parameter space Θ is finite and no restriction on the prior distribution is needed.

Frequently, the generalized Bayes estimators are the limits of Bayes estimators. Farrell (1968) had shown that the limits of the Bayes estimators are often admissible. However, the generalized Bayes estimators are not necessarily admissible. We will show in the following theorem that an estimator is admissible under limiting conditions on the generalized Bayes risks corresponding to a sequence of improper priors $\{\pi_j\}$.

Theorem 8.2.10 Let $\Theta \subseteq \mathbb{R}^1$ and D be the class of all estimators of θ having continuous risk functions in θ. An estimator δ in D is admissible if there exists a sequence of priors (not necessarily proper) $\{\pi_j\}$ such that
(i) the generalized Bayes risk of δ,

$$r(\pi_j, \delta) < \infty \quad \forall\, j$$

and, (ii) for any $\theta_0 \in \Theta$ and $\eta > 0$,

$$\lim_{j\to\infty} \frac{r(\pi_j, \delta) - r(\pi_j, \delta_0)}{\pi_j(N_\varepsilon(\theta_0))} = 0$$

where $r(\pi_j, \delta_0) = \inf_{\delta\in D(\theta)} r(\pi_j, \delta)$ and $N_\varepsilon(\theta_0)$ is the η-neighbourhood of θ_0, $N_\varepsilon(\theta_0) = \{\theta \in \Theta: |\theta - \theta_0| < \varepsilon\}$ with $\pi_j(N_\varepsilon(\theta_0)) < \infty\ \forall\, j$.

Proof. Assume that δ is not admissible. Then there exists an estimator δ' so that

$$R(\theta, \delta') \le R(\theta, \delta) \quad \forall\, \theta \in \Theta$$

and $$R(\theta, \delta') < R(\theta, \delta) \text{ for some } \theta = \theta_0 \in \Theta$$

For some small $\eta > 0$, there exists an $\varepsilon > 0$, from the continuity of risk functions of δ' and δ, so that

$$R(\theta, \delta') < R(\theta, \delta) - \eta$$

wherever $|\theta - \theta_0| < \varepsilon$ or $\theta \in N_\varepsilon(\theta_0)$.
Consider for any j,

$$\begin{aligned} r(\pi_j, \delta) - r(\pi_j, \delta_0) &\ge r(\pi_j, \delta) - r(\pi_j, \delta') \\ &= \int_\Theta [R(\theta,\delta) - R(\theta,\delta')]\pi_j(\theta)d\theta \\ &\ge \int_{\theta\in N_\varepsilon(\theta_0)} [R(\theta,\delta) - R(\theta,\delta')]\pi_j(\theta)d\theta \\ &\ge \eta\pi_j[N_\varepsilon(\theta_0)] \end{aligned}$$

or $$\frac{r(\pi_j, \delta) - r(\pi_j, \delta_0)}{\pi_j[N_\varepsilon(\theta_0)]} \ge \eta > 0$$

which is a contradiction to the condition (ii) in the theorem. This proves that the estimator δ is admissible if the conditions in the theorem are met. ▣

8.2.6 Bayesian Inference Agrees to the Likelihood Principle

Definition 8.2.3 (Likelihood function) The density $f(x|\theta)$ written as a function of unknown parameter θ depending upon the observed value x of the *rv* X, $L(\theta|x)$, is called *likelihood function*. The meaning of this representation is that the information brought by the observation x on parameter θ is completely contained in the likelihood function.

Let x_1 and x_2 be two observations drawn from two different random experiments, which depend on the same parameter θ, and if

$$L_1(\theta|x_1) = cL_2(\theta|x_2)\ \forall\ \theta$$

then the observations x_1 and x_2 carry the same information on θ and yield identical inferences based on this likelihood function.

The posterior distribution is written as

$$\pi(\theta|x) = \pi(\theta)L(\theta|x) \propto L(\theta|x)$$

We can easily see that the posterior distribution depends on x only through $L(\theta|x)$. Thus, the Bayesian inference works in agreement with the likelihood principle.

We will demonstrate through an example that the Bayesian inference agrees to the likelihood principle.

Binomial Experiment

Let $X \sim b(n, \theta)$ and X be the number of successes in n independent trials. The *pmf* is given by

$$f_1(x|\theta) = \binom{n}{x}\theta^x(1-\theta)^{n-x}$$

Let the prior distribution be given by $\pi(\theta)$. The corresponding posterior distribution is

$$\pi_1(\theta|x) \propto \binom{n}{x}\theta^x(1-\theta)^{n-x}\pi(\theta) \propto \theta^x(1-\theta)^{n-x}\pi(\theta) \tag{8.2.20}$$

Negative Binomial Experiment

Consider the number of trials, N, until we get x successes with the same probability of successes. In this case, $N \sim \text{NB}(x, \theta)$. The *pmf* is given by

$$f_2(n|\theta) = \binom{n-1}{x-1}\theta^x(1-\theta)^{n-x}$$

The corresponding posterior distribution is

$$\pi_2(\theta|x) \propto \theta^x(1-\theta)^{n-x}\pi(\theta) \propto \pi_1(\theta|x) \qquad \text{[from Eq. (8.2.20)]}$$

Thus, we see that the Bayesian inference about θ does not depend whether binomial or negative binomial sampling experiment is used; thus it agrees with the likelihood principle.

However, we may show by the following example that a classical unbiased estimator does not honour likelihood principle.

Consider $X_1,\ldots,X_n$ *iid* $b(1, \theta)$. The unbiased estimator of θ in this case is

$$\hat{\theta} = \frac{\sum X_i}{n}$$

For the observations $\mathbf{x} = (x_1,\ldots,x_n) = (0, 0,\ldots,1)$, the estimator takes the value

$$\hat{\theta} = \frac{1}{n}$$

Consider the experiment that $N \sim$ Geometric (θ). The *pmf* is given by

$$P\{N = n\} = \theta(1 - \theta)^{n-1},\ n = 1, 2,\ldots$$

The only unbiased estimator of θ is

$$\hat{\theta} = \begin{cases} 1, & \text{if } n = 1 \\ 0, & \text{if } n = 2,\ldots \end{cases}$$

Thus, the classical procedure named as unbiasedness violates the likelihood principle. The reason why it violates this principle is that the expectation of the estimator

$$\int_{\chi} \delta(x) f(x|\theta) dx$$

involves integral over the sample space (including $X = x$ which has been observed) and that $f(x|\theta)$ takes on different values for different values of x.

Next, we will show that the conditionality principle is honoured in the Bayesian estimation.

Definition 8.2.4 (Conditionality principle) Suppose two experiments on the parameter θ are available with probability 1/2 each. Then, the resulting inference on θ should depend only on the selected and performed experiment.

Consider the example due to Cox (1958). Let θ be some physical quantity to be measured by some machine. Machine 1 is more reliable since it gives measurements X with $N(\theta, 1)$ compared to machine 2 which gives measurements X with $N(\theta, 100)$. It is for this reason that machine 1 is more busy and is beyond the experimenter's control to use it. Assume that on a given occasion, machine 1 is available to the experimenter with probability 1/2 and it is selected. The conditionality principle says that 95% confidence interval for θ is about $(x - 1.64, x + 1.64)$. However, the frequentist 95% confidence interval for θ is about $(x - 16.4, x + 16.4)$ because of the possibility that machine 2 was selected.

Clearly, we see that conditionality principle is satisfied in the Bayesian estimation.

8.3 NATURAL CONJUGATE PRIOR DISTRIBUTIONS

Any inference in the Bayesian estimation depends on the posterior distribution. In many cases, the calculations of posterior distribution and other functions derived from it are intractable particularly for priors, which are not easy to work with. To circumvent this problem, given a likelihood function, one looks for such priors which capture the available information on θ and lead to mathematically tractable posterior distribution for computational convenience. Hence, this motivates to consider a family of conjugate priors defined as follows.

Definition 8.3.1 (Conjugate priors) Let $\mathcal{F} = \{f(x|\theta),\ \theta \in \Theta\}$ be a family of likelihood functions and $\Pi = \{\pi(\theta)$: π is some *pdf* on $\Theta\}$ be a family of prior distributions on Θ. If for any x, any $f(x|\theta)$ in $\mathcal{F}$, and any $\pi(\theta)$ in Π, the posterior

$$\pi(\theta|x) \propto f(x|\theta)\pi(\theta)$$

belongs to Π, then Π is called a *family of conjugate priors* for $\mathcal{F}$.

Statisticians have solved this problem by considering a pair of families of distributions. One is the family of sampling distributions $\mathcal{F}$ and the other is the family of prior distributions Π, so that if prior distribution is chosen from Π, the posterior distribution also belongs to the same family Π, but with different parameters. Such priors pinned with sampling distributions, i.e., $(\Pi, \mathcal{F})$ are called *conjugate prior distributions.*

We will choose Π as small as possible. The family of conjugate priors leads us only to updating the prior parameters through likelihood function. In fact, the product of the prior and the likelihood function returns the distribution in the same family.

Most often, by using conjugate prior, we not only get posterior distribution easily in exponential family setting, but also the Bayes estimator explicitly in the closed form of X, but not necessarily when we consider the estimation of some parametric function of θ, $g(\theta)$.

In this connection, Raiffa and Schlaiffer (1961) have listed and defined a class of prior densities with the following properties:

(i) If the prior is a member of this family, the posterior is also a member of the same family.

(ii) The class is rich enough to include all such densities, which could reflect the statistician's belief.

(iii) The posterior distribution and expectation of commonly used loss functions can be easily calculated.

Such prior distributions are known as *natural conjugate priors*. We outline a procedure of constructing natural conjugate density as follows:
Consider the likelihood

$$f(\mathbf{x}|\theta) = g(\mathbf{x}|\theta)h(\mathbf{x})$$

where $g(\mathbf{x}|\theta)$ is called the *kernel* of the likelihood and $h(\mathbf{x})$ is the *residual* of the likelihood. If T is a sufficient statistic for θ, we can write

$$f(\mathbf{x}|\theta) = g(t|\theta)h(\mathbf{x})$$

If we view $g(t|\theta)$ as a function of θ indexing the parameter t, we can construct the prior so that

$$\pi(\theta|t) \propto g(\theta|t)\ \forall\ \theta$$

This is called the natural conjugate prior of the likelihood $f(\mathbf{x}|\theta)$.

Having calculated the posterior density $\pi(\theta|\mathbf{x})$, one can easily calculate the marginal density

$$m(\mathbf{x}) = \frac{\pi(\theta)f(\mathbf{x}|\theta)}{\pi(\theta|\mathbf{x})}$$

We list some natural conjugate priors corresponding to some commonly used classes of sampling distributions which are members of exponential family of distributions:

Sampling distribution *pdf (pmf)* $[f(\mathbf{x}\mid\theta)]$	*Conjugate prior* $\pi(\theta)$	*Posterior* $\pi(\theta\mid\mathbf{x})$
$X_1, X_2,\ldots,X_n$ *iid* $b(1,\theta)$	$Be(\alpha,\beta)$ (Beta-binomial)	$Be(\alpha + \Sigma x_i,\ \beta + n - \Sigma x_i)$
$X \sim b(n,\theta)$	$Be(\alpha,\beta)$ (Beta-binomial)	$Be(\alpha + x,\ \beta + n - x)$
$X_1, X_2,\ldots,X_n$ *iid* $P(\theta)$	$G_2(\alpha,\beta)$ (Gamma-Poisson)	$G_2(\alpha + \Sigma x_i,\ \beta + n)$
$X_1, X_2,\ldots,X_n$ *iid* $E(\theta)$ $f(x\mid\theta) = \theta e^{-\theta x},\ x > 0$	$G_2(\alpha,\beta)$ (Gamma-exponential)	$G_2(\alpha + n,\ \beta + \Sigma x_i)$
$X_1, X_2,\ldots,X_n$ *iid* NB(r,p)	$Be(\alpha,\beta)$ (Beta-negative binomial)	$Be(\alpha + nr,\ \beta + \Sigma x_i)$
$X_1, X_2,\ldots,X_n$ *iid* $N(\mu,\sigma^2)$ μ is unknown and σ^2 is known	$N(\eta,\tau^2)$ (Normal-normal)	$N\left(\dfrac{\dfrac{\eta}{\tau^2}+\dfrac{\Sigma x_i}{\sigma^2/n}}{\dfrac{1}{\tau^2}+\dfrac{1}{\sigma^2/n}},\ \dfrac{\dfrac{\sigma^2}{n}\tau^2}{\dfrac{\sigma^2}{n}+\tau^2}\right)$
$X \sim N(\mu,\sigma^2)$, σ^2 is unknown, μ is known	IG$_2(\alpha,\beta)$ (Normal-inverse gamma)	$\text{IG}_2\left[\alpha+\left(\dfrac{1}{2}\right),\ \beta+\dfrac{(\mu-x)^2}{2}\right]$
$X \sim G_1\left(\dfrac{n}{2}, 2\theta\right)$	IG$_1(\alpha,\beta)$	$\text{IG}_1\left(\dfrac{n}{2}+\alpha,\ \left(\dfrac{x}{2}+\dfrac{1}{\beta}\right)^{-1}\right)$
$X_1, X_2,\ldots,X_n$ *iid* $G_2(\nu,\theta)$	$G_2(\alpha,\beta)$ (Gamma-gamma)	$G_2(\alpha + \nu,\ \beta + \Sigma x_i)$
$X_1, X_2,\ldots,X_n$ *iid* $U(0,\theta)$	Pa(α,β)	Pa$(x_{(n)},\ \alpha + n)$

It is important to mention that conjugacy yields simple form of Bayes estimators in exponential settings also.

8.4 DUALITY BETWEEN LOSS FUNCTION AND PRIOR DISTRIBUTION

Let the likelihood function be given by $L(\theta\mid\mathbf{x})$ and the prior distribution $\pi(\theta)$ be modified by multiplying it by a weight function $w(\theta)$

$$\pi_w(\theta) \propto w(\theta)\pi(\theta), \quad w(\theta) > 0$$

Accordingly, the posterior distribution is given by the density

$$\pi_w(\theta|\mathbf{x}) \propto w(\theta)\pi(\theta)L(\theta|\mathbf{x})$$

The Bayes estimate of θ, under the squared error loss function, is obtained by minimizing

$$\int (\theta - \delta)^2 \pi_w(\theta|\mathbf{x})d\theta$$

with respect to δ that yields the mean of the posterior distribution as Bayes estimate.

Consider, now, the weighted squared error loss function defined by

$$L_w(\theta, \delta) = w(\theta)\,(\theta - \delta)^2, \quad w(\theta) \geq 0$$

L_w is called the modified loss function. The Bayes estimate under the loss function L_w and, also, with respect to $\pi(\theta)$, the prior, is such an δ that minimizes

$$\int w(\theta)(\theta - \delta)^2 \pi(\theta) f(\mathbf{x}|\theta)d\theta$$

If we assume that the prior $\pi(\theta)$ is a conjugate prior indexed by some hyperparameter α, we can write this integral as

$$\int w(\theta)(\theta - \delta)^2 \pi(\theta|\alpha) f(\mathbf{x}|\theta)d\theta \tag{8.4.1}$$

Let the posterior density be $\pi(\theta|\mathbf{x},\alpha) \propto \pi(\theta|\alpha)f(\mathbf{x}|\theta)$. The integral may now be written as

$$\int w(\theta)(\theta - \delta)^2 \pi(\theta|\mathbf{x}, \alpha)d\theta \tag{8.4.2}$$

Further, if the weight function $w(\theta)$ and the prior $\pi(\theta|\alpha)$ are such that

$$w(\theta)\pi(\theta|\alpha) = \pi(\theta|\alpha') \tag{8.4.3}$$

the conjugate prior is updated in order to get updated hyper parameter α'. Thus, the integral Eq. (8.4.1) is simplified to

$$\int (\theta - \delta)^2 \pi(\theta|\alpha') f(\mathbf{x}|\theta)d\theta \tag{8.4.4}$$

Let the posterior density be

$$\pi(\theta|\mathbf{x}, \alpha') \propto \pi(\theta|\alpha')f(\mathbf{x}|\theta)$$

The integral in Eq. (8.4.4) may be written as

$$\int (\theta - \delta)^2 \pi(\theta|\mathbf{x}, \alpha')d\theta \tag{8.4.5}$$

Thus, the Bayes estimate of θ under the modified loss function, by minimizing Eq. (8.4.2) for δ, is

$$\frac{E[\theta w(\theta)|\mathbf{x}, \alpha]}{E[w(\theta)|\mathbf{x}, \alpha]} \tag{8.4.6}$$

and the Bayes estimate of θ under the simple loss function, by minimizing Eq. (8.4.5) for δ, is

$$E(\theta|\mathbf{x}, \alpha')$$

which is the mean of the posterior distribution. Thus, we have seen that the problem of estimating θ with modified (weighted) loss function is identical to the problem with simple loss but with modified hyperparameters of the prior distribution while the form of the prior distribution does not change.

Working with the weighted squared error loss function and finding Bayes estimator, as in Eq. (8.4.6), may be difficult. In such a situation, it is rather good to separate out weight $w(\theta)$ from the loss function and use it to modify the conjugate prior distribution as in Eq. (8.4.3) and accordingly the posterior distribution. It simplifies the problem of estimating θ under simple squared error loss function and leads to simple mean of posterior distribution calculated accordingly as Bayes estimate.

This process of attaching the weight of weighted loss function to the prior distribution to get the modified prior distribution and vice-versa is called *duality* between the prior and the loss function, and has been discussed by Robert (2001) and Bansal (2007).

8.5 NONINFORMATIVE PRIORS

We can categorize prior distributions into two types: informative priors and noninformative priors. *Informative priors* describe apriori belief while *noninformative priors* do not describe apriori belief, when no prior information is available.

We will discuss noninformative prior in the present section, that shows that the statistician has no information on the parameter θ. This ignorance about θ is presented by a prior distribution where one has no reason to prefer one value of θ over other. Such a prior distribution is known as *noninformative prior distribution*. If one can choose a uniform distribution over the range of θ as noninformative, formalizing the ignorance about θ, we can easily see that the posterior distribution is proportional to likelihood function and the Bayesian inference leads to the inference that the data speaks for itself, that is the inference remains same as is obtained in classical inference, while staying in the Bayesian settings.

For example, if one is interested in estimating the mean of the normal distribution θ and has no information on possible values of θ, he defines a noninformative prior as a uniform distribution over the entire range of θ, which is $-\infty$ to $+\infty$, by

$$\pi(\theta) = c, -\infty < \theta < \infty$$

This prior density is an improper density since

$$\int_{-\infty}^{+\infty} \pi(\theta) d\theta = c \int_{-\infty}^{+\infty} d\theta = \infty \quad \forall\, c > 0$$

In case of such uniform priors that are defined on the entire real line, we, instead, consider the uniform distribution on the finite interval $(-M, +M)$, $\pi_M(\theta) = 1$, $\theta \in (-M, +M)$, and consider the estimation of parameter θ based on the posterior distribution

$$\pi(\theta|x) = c\pi_M(\theta)L(\theta) \quad \text{as } M \to \infty$$

In the case of $N(\mu, 1)$, the prior showing the ignorance about μ can be considered as $\pi(\mu) = c \neq 0$ (clearly improper). We have

$$\pi(\mu|\mathbf{x}) = f(\mathbf{x}|\mu)$$

and, thus, the *posterior distribution* leads to the likelihood function.

The term *noninfomative prior* for $\pi(\mu) = c$ was given by Jeffreys (1946, 1961). However, Bayes and Laplace contrived this concept much earlier capturing *ignorance* about θ by prior distribution. Generally, these noninformative priors are improper except when the parameter space, Θ, is finite.

Jeffreys (1961) has prescribed a general rule for expressing ignorance about θ by choosing a noninformative prior as a flat prior for some location parameter μ and an inverse prior for scale parameter σ. For example, the Jeffreys prior for mean μ for the model $N(\mu,1)$ is the flat prior $\pi(\mu) = 1$ which has intuitive justification, that if one is ignorant about a location parameter μ, then it can take any value on the real time with equal prior probabilities. Note that this prior is improper. Next, for the standard deviation σ for the model $N(0,\sigma^2)$, the noninformative prior is the inverse $\pi(\sigma) = 1/\sigma$. This prior may equivalently be expressed by $\pi^*(\log\sigma)$ = constant which means that the prior for $\log\sigma$ is flat. Intuitively, if one is totally ignorant about the scale parameter σ, then it is likely to lie in the interval 1 to 10 as it is to lie in the interval 10 to 100. This expression of ignorance justifies flat prior on the log scale. This prior is generally improper since it does not integrate to 1. We must be careful with such models where these improper priors lead to improper posteriors.

Jeffreys rule says that the prior should be so constructed that the inference should not depend on how a model is parameterized. If the child mortality is μ, then the annual survival rate is $\eta = \exp(-\mu)$. Some model makers may use the parameterization μ and some might use η. The Jeffreys rule says that the noninformative prior shall be so constructed that the inference must not depend on the choice of parameterization.

Formalizing the ignorance about θ imbedded in the prior distribution gives an impression that it is a uniform distribution as it was considered by Laplace (1812). However, this is not correct and the prior showing this ignorance is due to what parameterization is chosen in the likelihood function. In fact, the prior distribution showing the ignorance about θ depends on the parameterization chosen and it should be so constructed that it must be invariant under transformation.

Consider, for example, the likelihood function $f(x|\mu)$ as $N(\mu, 1)$ and the uniform prior $\pi(\mu) = c \neq 0$ showing the "complete ignorance" about θ. Note that we have already seen, in this case, that $\pi(\mu|x) = f(x|\mu)$, i.e., posterior distribution and likelihood function both coincide. Suppose, instead of estimating μ, we are interested in estimating the reparameterization $\eta = \exp(\mu)$ which is a one-to-one transformation. If we are ignorant about μ and choose a uniform prior, then so are we about η and intend to choose a uniform prior for this reparameterization problem too, hence, showing our "complete ignorance" for η. By doing so, we expect to get the same answers in either parameterization. By the transformation rule, the probability density function for a transformed random variable η is given by

$$\pi^*(\eta) = \pi(\mu)|_{\mu = \log\eta}\,|J|$$

where
$$|J| = \left|\frac{\partial \mu}{\partial \eta}\right| = \left|\frac{\partial \log \eta}{\partial \eta}\right| = \left|\frac{1}{\eta}\right| = \frac{1}{\eta}$$

is the Jacobian of transformation. Since $\pi(\mu)$ is uniform, we get

$$\pi^*(\eta) = \pi(\log \eta)\frac{1}{\eta} = \frac{c}{\eta} \propto \frac{1}{\eta}$$

This shows that the noninformative prior for η resulting from the uniform prior for μ is not uniform and, thus, it shows further that "complete ignorance" is not always expressed by uniform prior distribution. Clearly, uniform prior distribution is not invariant under exponential transformation. Had it been invariant, we would have inferred that answers in either parameterizations were the same and noninformative prior had maintained consistency. Thus, if we want to maintain consistency, that is, if we want to get the same answers in either parameterizations, in the present example, we must choose such a noninformative prior for η which is proportional to η^{-1} to get the same estimates in either parameterizations.

We will see that the uniform noninformative prior expressing "complete ignorance" is invariant for a the location parameter in the location family of likelihoods. However, the appropriate noninformative prior that shows "complete ignorance" about the scale parameter σ in the scale family of likelihoods and that is scale invariant is not a uniform distribution but is $\pi(\sigma) = 1/\sigma$ which is an improper distribution. In fact, Laplace (1812) introduced the uniform prior showing complete ignorance, but it was later criticized, since it lacked invariance under transformation of the type in the above example.

We will discuss here the noninformative priors for location and scale parameters.

Location parameter: Consider the problem of estimating θ where the likelihood function is $f_X(x|\theta) = f(x - \theta)$, $x \in \chi \subseteq \mathbb{R}^1$, $\theta \in \Theta \subseteq \mathbb{R}^1$, θ is called the *location parameter*, e.g., $N(\mu, \sigma^2)$, μ is unknown, and σ^2 is known. Suppose instead of observing $X = x$, we observe $y = x + c$, where c is known $(x + c \in \chi)$. This is the case when we observe X with changed unit of measurement in which the origin is not zero but c. Define a new parameter by $\eta = \theta + c$, $\eta \in \Theta$. The likelihood function for this new problem is

$$f_Y(y|\eta) = f_X(y - c)|J|,$$

$|J| = 1$, we have

$$f_Y(y|\eta) = f(y - c - \theta) = f(y - (\theta + c)) = f(y - \eta)$$

$y \in \chi \subseteq \mathbb{R}^1$, $\eta \in \Theta \subseteq \mathbb{R}^1$. We, thus, note here that the sample space and parameter space in estimation problems (X, θ) and (Y, η) are the same and we may say that both of these problems are identical in structure. Therefore, estimating η from y is an equivalent problem of estimating θ from x. It is, thus, reasonable to maintain that the noninformative prior in either problems showing *complete ignorance* must be the same and that must be independent of the choice of the origin for deciding the unit of measurements.

Denote the noninformative prior by $\pi(\theta)$ in the problem (X, θ) and by $\pi^*(\theta)$ in the problem (Y, η). Then, for any interval (a, b), we must have

$$\int_a^b \pi(\theta)d\theta = \int_a^b \pi^*(\eta)d\eta \tag{8.5.1}$$

Since $\eta = \theta + c$, we have

$$\pi^*(\eta) = \pi(\eta - c)$$

We get
$$\int_a^b \pi^*(\eta)d\eta = \int_a^b \pi(\eta - c)d\eta = \int_{a-c}^{b-c} \pi(\theta)d\theta \tag{8.5.2}$$

On combining Eqs. (8.5.1) and (8.5.2), we get

$$\int_a^b \pi(\theta)d\theta = \int_{a-c}^{b-c} \pi(\theta)d\theta = \int_a^b \pi(\theta - c)d\theta \quad \forall\, c \in \mathbb{R}^1 \tag{8.5.3}$$

Any prior $\pi(\theta)$ that satisfies Eq. (8.5.3) for any interval (a, b), i.e., $\pi(\theta) = \pi(\theta - c)$, $\forall$ θ and $c \in \mathbb{R}^1$, is said to be a *location invariant prior* for θ. If we set $\theta = c$, then condition (8.5.3) yields

$$\pi(c) = \pi(0) \quad \forall\, c \in \mathbb{R}^1$$

This gives
$$\pi(\theta) = \text{constant} \quad \forall\, \theta \in \mathbb{R}^1$$

Note that
$$\pi^*(\eta) = \text{constant} \quad \forall\, \eta \in \mathbb{R}^1$$

Thus, the noninformative prior for location parameter is a uniform distribution, which is an improper density and is invariant under location transformations.

Scale parameter: Consider an estimation problem where the likelihood function is of the form $f(x|\sigma) = (1/\sigma)f(x/\sigma)$, $x \in \chi \subseteq \mathbb{R}^1$, $\sigma \in \Theta = (0, \infty)$. The density of this model is called the *scale density* and σ is called the *scale parameter*. $N(0,\sigma^2)$, $G(\alpha,\beta)$(α fixed) are the examples of scale densities. Let this estimation problem be denoted by (X,σ) where one estimates σ from x. In this case, suppose instead of observing X in inches (say) we observe $Y = cX$, $c > 0$, $y \in \chi \equiv \mathbb{R}^1$ in feet (say), a changed unit of measurement. Define a parameterization $\eta = c\sigma$, $\eta \in \Theta = (0, \infty)$. The objective in the new problem is to estimate η from y. The likelihood function in the changed problem is

$$f^*(y|\eta) = f\left(\frac{y}{c}\middle|\sigma\right)|J|$$

where
$$|J| = \frac{1}{c}$$

We have
$$f^*(y|\eta) = \frac{1}{\sigma} f\left(\frac{y}{c\sigma}\right)\frac{1}{c}$$

$$= \frac{1}{c\sigma} f\left(\frac{y}{c\sigma}\right) = \frac{1}{\eta} f\left(\frac{y}{\eta}\right)$$

Note that the sample space $\chi = \mathbb{R}^1$ and the parameter space $\Theta = (0, \infty)$ in the estimation problems (X, σ) and (Y, η) are the same. Therefore, we can say that both the problems are equivalent problems, in terms of structure. It is, thus, reasonable to expect that the noninformative prior expressing "total ignorance" about σ in the problem (X, σ) should be the same for expressing *total ignorance* about η in the problem (Y, η). Such a prior which does not change when the unit of measurement is changed is called a *scale invariant noninformative prior*.

Denote the noninformative priors for (X, σ) and (Y, η) problems by π and π^* respectively. They must coincide. For any interval (a, b),

$$\int_a^b \pi(\sigma)d\sigma = \int_a^b \pi^*(\eta)d\eta \tag{8.5.4}$$

Consider the parameterization $\eta = c\sigma$, $c > 0$

$$\pi^*(\eta) = \pi\left(\frac{\eta}{c}\right)|J|$$

$$|J| = \frac{1}{c} \quad \text{since } c > 0$$

We have
$$\pi^*(\eta) = \frac{1}{c}\pi\left(\frac{\eta}{c}\right)$$

Consider the integral on the right-hand side in Eq. (8.5.4)

$$\int_a^b \pi^*(\eta)d\eta = \int_a^b \frac{1}{c}\pi\left(\frac{\eta}{c}\right)d\eta$$

$$= \int_{a/c}^{b/c} \pi(\sigma)d\sigma \tag{8.5.5}$$

On combining Eqs. (8.5.4) and (8.5.5), we get

$$\int_a^b \pi(\sigma)d\sigma = \int_{a/c}^{b/c} \pi(\sigma)d\sigma = \int_a^b \frac{1}{c}\pi\left(\frac{\sigma}{c}\right)d\sigma \;\; \forall\, (a, b) \tag{8.5.6}$$

Any prior satisfying Eq. (8.5.6) is called *scale invariant*, i.e.,

$$\pi(\sigma) = \frac{1}{c}\pi\left(\frac{\sigma}{c}\right) \;\; \forall\, \sigma > 0 \text{ and } c > 0$$

Choosing $\sigma = c$, we have

$$\pi(c) = \frac{1}{c}\pi(1)$$

By setting $\pi(1) = 1$ without loss of generality, we get

$$\pi(c) = \frac{1}{c}$$

This clearly shows that the appropriate noninformative prior that shows "complete ignorance" about the scale parameter σ and the one that is scale invariant is not a uniform distribution but is $\pi(\sigma) = 1/\sigma$ which is an improper distribution.

Thus, the argument that whatever reasonable parameterization one may choose, uniform distribution as a noninformative prior expressing "total ignorance" about θ is not maintainable. Thus, it leads to a search for such a noninformative prior, which is invariant under a chosen transformation.

Notes.

1. Sometimes, more particularly, while working on software such as WinBUGS, these noninformative priors may be approximated by probability densities with a high spread. For example, $\pi(\mu) = 1$ may be approximated by normal density with a high variance, i.e., $N(0, 104)$. Similarly, the inverse prior $\pi(\sigma) = 1/\sigma$ can be approximated by $G(0.001, 0.001)$, where the shape and rate parameters are set very small.
2. It is a usual practice to consider the noninformative prior for the logarithm of scale parameter as uniform distribution, on the real line, in case of estimating the scale of parameter which takes on positive values.

The noninformative priors obtained earlier are subject to criticism, since these depend on the invariance structure. In case of location and scale problems, fortunately, no such controversy arises, since almost all the popular approaches of constructing noninformative priors lead to the same prior. However, this nice thing does not happen for other parameterizations. To get around this problem, Jeffreys (1961) recommended the use of prior as noninformative prior, so that the rule that determines $\pi(\theta)$ should equivalently determine the prior in the same way when applied to reparameterization $\eta = g(\theta)$.

8.5.1 Jeffreys Invariance Principle

Suppose we have a rule that determines prior $\pi(\theta)$. Consider the reparameterization $\eta = g(\theta)$. Using the change of variable method, the corresponding prior for η is given by

$$\pi(\eta) = \pi(\theta)\left|\frac{\partial \theta}{\partial \eta}\right| \tag{8.5.7}$$

Jeffreys says that the rule for determining $\pi(\theta)$ should be such that if applied directly on η, it should give the same prior as obtained in Eq. (8.5.7). This is called *Jeffreys invariance principle* and such priors are called invariant priors under reparameterization. It is seen that the rule actually does not work well if it does not apply to reparameterization.

Harold Jeffreys (1961) proposed a technique of constructing noninformative priors as follows:

$$\pi(\theta) = [I_X(\theta)]^{1/2}$$

where $I_X(\theta)$ is the Fisher information defined by

$$I_X(\theta) = E_\theta\left[-\frac{\partial^2}{\partial \theta^2}\log f(X|\theta)\right]$$

under regularity conditions (i) to (iv) given in Section 4.2 of Chapter 4. Note that the Fisher information measures the sensitivity of an estimator in the neighbourhood of the MLE since it is proportional to the expected curvature of the likelihood at the MLE.

If $\boldsymbol{\theta}$ is a p-dimensional vector, then Jeffreys prior is defined by

$$\pi(\boldsymbol{\theta}) \propto [\det \mathbf{I_X}(\boldsymbol{\theta})]^{1/2}$$

where $\mathbf{I_X}(\boldsymbol{\theta})$ is known as Fisher information matrix defined by

$$\mathbf{I_X}(\boldsymbol{\theta}) = [I_{ij}(\boldsymbol{\theta})]$$

$$I_{ij}(\boldsymbol{\theta}) = E_\theta\left[-\frac{\partial^2}{\partial\theta_i\,\partial\theta_j}\log f(\mathbf{X}|\boldsymbol{\theta})\right]$$

The key importance of this prior is that it is invariant under the reparameterization of θ. It carries special importance when used for scale parameters. Jeffreys rule is well accepted for single-parameter models; however, it is controversial for multi-parameter models.

8.5.2 Invariance of Jeffreys Prior under Reparameterization

Consider the likelihood function $f(\mathbf{x}|\theta)$ and reparameterization $\eta = g(\theta)$, where θ is scalar, and g is some one-to-one continuous transformation. We examine here the relationship between the Jeffreys prior for θ and the reparameterization η. We first compute the Jeffreys prior for θ, and by random variable transformation rule using the Jacobian of transformation to η, subsequently, compute the Jeffreys prior for η. The likelihood function for this parameterization is

$$f^*(\mathbf{x}|\eta) = f(\mathbf{x}|g^{-1}(\eta))$$

g^{-1} exists since g is a one-to-one transformation. Taking log on both the sides,

$$\log f^*(\mathbf{x}|\eta) = \log f(\mathbf{x}|\theta)|_{\theta = g^{-1}(\eta)}$$

By the chain rule of derivatives,

$$\frac{\partial}{\partial\eta}\log f^*(\mathbf{x}|\eta) = \frac{\partial}{\partial\theta}\log f(\mathbf{x}|\theta)|_{\theta=g^{-1}(\eta)}\ \frac{\partial g^{-1}(\eta)}{\partial\eta}$$

$$= \frac{\partial}{\partial\theta}\log f(\mathbf{x}|\theta)|_{\theta=g^{-1}(\eta)}\ g'[g^{-1}(\eta)]$$

where $g'(\cdot)$ is the derivative of $g^{-1}(\eta)$. Squaring both the sides and taking expectation, we get

$$I_{\mathbf{X}}(\eta) = I_{\mathbf{X}}(\theta)\{g'[g^{-1}(\eta)]\}^2$$

or

$$[I_{\mathbf{X}}(\eta)]^{1/2} = [I_{\mathbf{X}}(\theta)]^{1/2}|g'[g^{-1}(\eta)]|$$

Thus, if Jeffreys prior for θ is

$$\pi(\theta) = [I_{\mathbf{X}}(\theta)]^{1/2}$$

then for any one-to-one and continuous reparameterization $\eta = g(\theta)$, the Jeffreys prior for η is

$$\pi^*(\eta) = [I_{\mathbf{X}}(\theta)]^{1/2}\,|g'[g^{-1}(\eta)]| = \pi(\theta)|g'[g^{-1}(\eta)]| \quad (8.5.8)$$

Instead, if the Fisher information is directly computed for reparameterization η, by the results on Fisher information,

$$I_{\mathbf{X}}(\eta) = I_{\mathbf{X}}(\theta)\,\{g'[g^{-1}(\eta)]\}^2$$
$$[I_{\mathbf{X}}(\eta)]^{1/2} = [I_{\mathbf{X}}(\theta)]^{1/2}\,|g'[g^{-1}(\eta)]|$$

By definition, it yields Jeffreys prior for η

$$\pi^*(\eta) = [I_{\mathbf{X}}(\eta)]^{1/2} = [I_{\mathbf{X}}(\theta)]^{1/2}\,|g'[g^{-1}(\eta)]| = \pi(\theta)|g'[g^{-1}(\eta)]| \tag{8.5.9}$$

Thus, on comparing Eqs. (8.5.8) and (8.5.9), we see that the prior for reparameterization $\eta = g(\theta)$ which is computed by computing the prior for θ and, subsequently, by obtaining Jeffreys prior for reparameterization η by variable transformation rule, in Eq. (8.5.8), is the same as obtained directly for reparameterization η in Eq. (8.5.9).

This shows that the prior of η, by applying the rule to θ and then transforming, is the same as the one obtained by directly applying the rule to η.

Therefore, reparameterization η also follows Jeffreys prior, showing that Jeffreys prior is invariant under reparameterization. Alternatively, we can proceed more simply by noting that $I(\theta)$ is invariant for reparameterization and, hence, Jeffreys prior is invariant under reparametrization.

Consider one example where $X \sim N(\mu,1)$. Jeffreys prior for μ is $\pi(\mu) \propto 1$. Consider the replacement $\eta = \exp(\mu)$. Jeffreys prior for η is

$$\pi^*(\eta) \propto \frac{1}{\eta}$$

Further, Eq. (8.5.8) gives

$$\pi^*(\eta) = \pi(\theta)\,|g'[g^{-1}(\eta)]|$$

and we observe that the Jeffreys prior for reparameterization η can directly be obtained by the random variable transformation rule. Since this works well for all one-to-one and continuous reparameterization, Jeffreys noninformative prior apart from it does not rely on any invariance structure; it is treated as a more general way of obtaining noninformative priors.

When $\boldsymbol{\theta}$ is a vector, the probability density of $\boldsymbol{\eta}$ is

$$\pi(\boldsymbol{\eta}) = \pi(\boldsymbol{\theta})\left|\det\frac{\partial\theta_i}{\partial\eta_j}\right| \qquad \text{(by change of variable theorem)}$$

$$\propto \sqrt{\det I_{\mathbf{X}}(\boldsymbol{\theta})\left(\det\frac{\partial\theta_i}{\partial\eta_j}\right)^2} \qquad \text{[since } \pi(\boldsymbol{\theta}) \text{ is the Jeffreys prior]}$$

$$= \sqrt{\det\frac{\partial\theta_k}{\partial\eta_i}\det E\left[\frac{\partial}{\partial\theta_k}\log f(\mathbf{X}|\theta)\frac{\partial}{\partial\theta_l}\log f(\mathbf{X}|\theta)\right]\det\frac{\partial\theta_l}{\partial\eta_j}}$$

(by the definition of Fisher information)

$$= \sqrt{\det E\left[\sum_k \sum_l \frac{\partial \theta_k}{\partial \eta_i} \frac{\partial}{\partial \theta_k} \log f(\mathbf{X}|\theta) \frac{\partial}{\partial \theta_l} \log f(\mathbf{X}|\theta) \frac{\partial \theta_l}{\partial \eta_j}\right]}$$

(since the product of determinants is the determinant of the product of matrices)

$$= \sqrt{\det E\left[\sum_k \sum_l \frac{\partial \theta_k}{\partial \eta_i} \frac{\partial}{\partial \theta_k} \log f(\mathbf{X}|\theta) \frac{\partial}{\partial \theta_l} \log f(\mathbf{X}|\theta) \frac{\partial \theta_l}{\partial \eta_j}\right]}$$

$$= \sqrt{\det E\left[\frac{\partial}{\partial \eta_i} \log f(\mathbf{X}|\theta) \frac{\partial}{\partial \eta_j} \log f(\mathbf{X}|\theta)\right]} = \sqrt{\det I_{\mathbf{X}}(\eta)}$$

This shows that Jeffreys prior is invariant under reparameterization.

It is common to use a uniform prior for $\eta = g(\theta)$, where $g(\theta)$ is some function of θ. It is easy to see that the Fisher information about η, $I_{\mathbf{X}}(\eta)$, is constant. Thus,

$$\pi(\theta) \propto [I_{\mathbf{X}}(\eta)]^{1/2}$$

Such priors are known as Jeffreys priors. The choice for such re-parameterization $\eta = g(\theta)$ for the regular family of distributions $\{f(x;\theta);\ \theta \in \Theta\}$ is given for an arbitrary choice θ_0

$$g(\theta) = \int_{\theta_0}^{\theta} \sqrt{I_X(y)}\,dy$$

where $I_X(\theta)$ is defined as the Fisher information in a single observation about θ

$$I_X(\theta) = E_\theta\left\{-\frac{\partial^2}{\partial \theta^2} \log f(X|\theta)\right\}$$

In this case, the Fisher information in a single observation about η is given by

$$I_X(\eta) = 1$$

Therefore, Jeffreys prior in this case is given by

$$\pi(\theta) = \sqrt{I_X(\eta)} = 1$$

for the overall value of η. If $\hat{\eta}$ is the maximum likelihood estimator, its asymptotic variance is given by $(1/n)$ since $I_X(\eta) = 1$, which is independent of the parameter η. Therefore, the above transformation is variance stabilization.

8.5.3 Jeffreys Prior in Exponential Distribution

Let the *pdf* of a random variable X that belongs to the family of one-parameter exponential distributions be

$$f(x|\theta) = \exp[Q(\theta)T(x) + S(x) + D(\theta)], \quad \theta \in \Theta \subset \mathbb{R}^1$$

The Fisher information is

$$I_X(\theta) = E\left[-\frac{\partial^2}{\partial\theta^2}\log L(\theta|X)\right]$$
$$= -D''(\theta) - Q''(\theta)[E_\theta T(X)]$$
$$= -D''(\theta) + Q''(\theta)\frac{D'(\theta)}{Q'(\theta)} \tag{8.5.10}$$

Thus, Jeffreys prior for θ is given by

$$\pi(\theta) = [I_X(\theta)]^{1/2}$$
$$= \sqrt{Q''(\theta)\frac{D'(\theta)}{Q'(\theta)} - D''(\theta)}$$

For illustration, consider the binomial model, $X \sim b(n, \theta)$ with the *pdf*

$$f(x|\theta) = \binom{n}{x}\theta^x(1-\theta)^{n-x}$$

The Fisher information for θ

$$I_X(\theta) = \frac{n}{\theta(1-\theta)}$$

gives Jeffreys noninformative prior for θ

$$\pi(\theta) = \sqrt{\frac{n}{\theta(1-\theta)}} \propto \frac{1}{\sqrt{\theta(1-\theta)}}$$

Similarly, for the Poisson model, one can show that

$$\pi(\theta) \propto \frac{1}{\sqrt{\theta}}$$

Remarks on Jeffreys Prior

1. This prior may not be proper.
2. It is invariant under reparameterization $\eta = g(\theta)$ where g is some monotone and differentiable transformation.
3. Jeffreys prior depends on the statistical model $f(x|\theta)$.
4. It is generally an improper prior for different models $f(x|\theta)$ since it cannot be normalized.
5. It does not matter what parameterization the Jeffreys prior distribution is derived for as long as it is 1:1 and it naturally contains the same information about θ. Note that Jeffreys prior is invariant under 1:1 transformation.
6. Jeffreys priors favours more only those θ for which $I_{\mathbf{X}}(\theta)$ is large, i.e., the data has more information about these θ, thus, minimizing the effect of prior distribution on

the Bayesian estimation. It is in this sense it expresses noninformative prior and also what data says about θ;

7. Jeffreys priors are sometimes involved for non-conjugate distributions.
8. Box and Tiao (1973) have noted that Jeffreys prior gives strange results in case of $\boldsymbol{\theta}$ as a vector. They have also proposed some necessary modifications.
9. According to the likelihood principle, the inference made about θ by using the likelihood function must be the same given two different statistical experiments involving the same parameter θ and having the same likelihood function. The Fisher information about θ not only depends on the probability of the experimental outcome as a function of θ but in totality, on all the experimental outcomes under the distribution F since, the Fisher information is obtained by averaging over totality of outcomes of the experiment governed by F. Thus, the Fisher information about θ in the two experiments involving the same parameter θ and having the same likelihood function may be different. Hence, using Jeffreys prior, the inferences made about θ may be different since Jeffreys prior for a parameter depends on the statistical model F. This is a violation of the likelihood principle.
10. In the modern statistical inference, the Fisher information and Jeffreys priors are the modern tools used in differential geometric theory of statistical inference. Apart from that, it establishes a link between statistics and information theory.
11. Bernardo (1994) and Robert (1994) observed that the Jeffreys noninformative prior results into the poor Bayesian estimates specially in the case $\boldsymbol{\theta}$ is a vector, e.g., $\boldsymbol{\theta} = (\mu,\sigma^2)'$ in case of estimating μ and σ^2 in $N(\mu,\sigma^2)$.

8.5.4 Jeffreys Prior Violates the Likelihood Principle

Consider first an experiment of drawing n observations $X_1,\ldots,X_n$ *iid* from Bernoulli distribution $b(1, \theta)$ for estimating the parameter θ. Suppose we observe x_0 successes, i.e., $\Sigma X_i = x_0$. The likelihood function for θ,

$$L_1(\theta|\mathbf{x}) = \theta^{x_0}(1-\theta)^{n-x_0} \tag{8.5.11}$$

Jeffreys noninformative prior for θ,

$$\pi_1(\theta) = [I_X(\theta)]^{1/2} \propto \frac{1}{\sqrt{\theta(1-\theta)}}$$

This gives the posterior distribution

$$\pi_1(\theta|\mathbf{x}) \propto \theta^{x_0-(1/2)}(1-\theta)^{n-x_0-(1/2)}$$

$$\pi_1(\theta|\mathbf{x}) = \text{Be}\left(x_0+\frac{1}{2},\, n-x_0+\frac{1}{2}\right) \tag{8.5.12}$$

Next, we consider the second experiment where we report the number X of Bernoulli trials needed to produce x_0 successes

$$f(x|\theta) = \binom{x-1}{x_0-1}\theta^{x_0}(1-\theta)^{x-x_0}, x = x_0, x_0+1,\ldots$$

Suppose in the second experiment, we require $x = n$ Bernoulli trials to produce x_0 successes. The likelihood function is

$$L_2(\theta|n) = \binom{n-1}{x_0-1}\theta^{x_0}(1-\theta)^{n-x_0} \tag{8.5.13}$$

The Jeffreys prior for θ,

$$\pi_2(\theta) \propto \frac{1}{\theta(1-\theta)^{1/2}}$$

This gives the posterior distribution

$$\pi_2(\theta|n) \propto \theta^{x_0-1}(1-\theta)^{n-x_0-(1/2)}$$

$$\Rightarrow \qquad \pi_2(\theta|n) \propto \text{Be}\left(x_0, n-x_0+\frac{1}{2}\right) \tag{8.5.14}$$

One can easily see that if $x_0 = 1$, the likelihood functions in the two experiments, Eqs. (8.5.11) and (8.5.13), are the same, and two different experiments will bring identical inferences based on the likelihood principle. However, the Bayes or posterior inferences are different, since different Jeffreys priors have yielded different posterior distributions in Eqs. (8.5.12) and (8.5.14) respectively. Note that the Jeffreys prior depends on the particular model chosen: binomial in the first experiment and negative binomial in the second experiment. Thus, Jeffreys prior and, subsequently, posterior inferences violate the likelihood principle.

Notes.

1. Hierarchical priors are more flexible than non-hierarchical priors and make the posterior loss sensitive to the prior.
2. If the noninformative prior is improper, we can still find the Bayes estimator provided the posterior density is proper. The sufficient condition for this is that the marginal distribution

$$m(x) = \int_{\Theta} f(x|\theta)\pi(\theta)d\theta$$

is finite for every x. However, if the posterior density is not proper, we find a generalized Bayes estimator defined in Section 8.2.2.

8.6 MINIMAX ESTIMATION

8.6.1 Minimax Estimator

Minimax theory or minimax estimation is discussed in the decision theoretic set up where nature is one player that chooses a prior distribution and the statistician is the second player who chooses an estimator. As a result of these choices of nature and statistician, respectively, the statistician pays the amount $r(\pi, \delta) = \int R(\theta, \delta)\pi(\theta)d\theta$ to the nature which is considered to be the gain of the nature. Therefore, nature's gain is the statistician's loss. Thus, this game is called *zero-sum-two-person game*.

Now, consider the two quantities

$$\underline{V} = \sup_{\theta} \inf_{\delta} R(\theta, \delta)\text{: Lower value of the game} \tag{8.6.1}$$

and

$$\overline{V} = \inf_{\delta} \sup_{\theta} R(\theta, \delta)\text{: Upper value of the game} \tag{8.6.2}$$

The right-hand side of Eq. (8.6.2) is called an *upper value of the game*. This principle is based on the principle of cutting short the losses. Knowing that nature intended to set out a loss to the statistician, the statistician, for some of his choice of estimator, watches out as to what nature would choose; then, he chooses such an estimator that would minimize this damage caused by the nature. In other words, the statistician guards himself against the expected loss and assures himself that the loss will not be greater than $\overline{V}$ by choosing an estimator, no matter what prior distribution nature decides to choose. Such an estimator is called the *minimax estimator*.

On the other hand, nature, from the family of prior distributions, chooses such a prior that does maximum damage to the statistician, no matter what statistician chooses as an estimator (action). He may even choose a minimax estimator. This prior distribution is known as the *least favorable distribution*. A prior distribution π_0 in Θ^* is the least favorable prior if

$$\min_{\delta} r(\pi_0, \delta) = \max_{\pi} \min_{\delta} r(\pi, \delta) = \underline{V} \tag{8.6.3}$$

The value on the right side of Eq. (8.6.3) is called the *lower value of the game*. Nature chooses such a prior which guarantees that whatever be the estimator chosen by the statistician, this prior sets the expected loss of the statistician to at least this lower value. Such a prior distribution is called the *least favorable*, since it favors the least to the statistician.

Interpretation

$\underline{V}$: The amount the statistician pays when he is told what nature chose θ before his choice δ

$\overline{V}$: The amount the statistician pays when nature is told what the statistician chose δ before he chooses θ

There may arise a situation where statistician's expected loss incurred by choosing a best estimator against the worst choice of the nature is the same as the expected loss caused by the nature's choice in response to the statistician's best choice. This expected loss is known as the *value of the game*. Related results are summarized in the following theorem.

Theorem 8.6.1 (Minimax theorem) Suppose that the parameter space $\Theta = \{\theta_1, \ldots, \theta_k\}$ is finite, and the risk set S of the estimation problem (Θ, D, R) is closed and bounded from below. Then, the game has a value

$$\begin{aligned} V = \underline{V} &= \sup_{\pi} \inf_{\delta} r(\pi, \delta) = \inf_{\delta} \sup_{\pi} r(\pi, \delta) \\ &= \inf_{\delta} \sup_{\theta \in \Theta} R(\theta, \delta) = \overline{V} \end{aligned} \tag{8.6.4}$$

and there exists a prior π that is the least favourable.

Proof. Beyond the scope of the present discussion. The interested readers may refer Ferguson (1967) for its proof.

In order to attempt at finding a minimax estimator, we explore the conditions on the estimation problem (Θ, *D*, *R*) under which Eq. (8.6.4) holds.

Note that $\inf_{\delta} r(\pi, \delta) \le r(\pi, \delta) \le \sup_{\pi} r(\pi, \delta)\ \forall\ \pi$ and $\forall\ \delta$

Therefore, for all π and δ,

$$\inf_{\delta} r(\pi, \delta) \le \sup_{\pi} r(\pi, \delta)$$

or

$$\sup_{\pi} \inf_{\delta} r(\pi, \delta) \le \inf_{\delta} \sup_{\pi} r(\pi, \delta)$$

∴

$$\underline{V} \le \bar{V}$$

If the equality holds, $\underline{V} = \bar{V} = V$, we say that the minimax theorem holds. δ_0 is Bayes with respect to the least favorable distribution π_0. δ_0 is minimax. The Bayes and minimax risks both are the same and are equal to the value of the game, i.e., V.

The minimax theorem has been discussed here since it is helpful in finding the minimax estimator in a given estimation problem as shown in the following results.

We will discuss here two commonly used methods of finding minimax estimators. In the first method, we make use of the fact that the statistician generally has an idea of the prior distribution that nature is using about which he would like least to know. Such a prior is known as least favorable distribution. Further, characterizing such a least favorable distribution is considered because the space of priors is easier to handle than the space of estimators. Having guessed about such a least favorable distribution, the statistician finds a Bayes estimator with respect to this prior and checks by some procedure whether it is minimax. This procedure is stated in the following two theorems.

Theorem 8.6.2 If δ_π is Bayes, estimator with respect a prior π such that

$$r(\pi, \delta_\pi) = \int R(\theta, \delta_\pi)\pi(\theta)d\theta = \sup_{\theta\in\Theta} R(\theta, \delta_\pi) \qquad (8.6.5)$$

[equalized risk property, or $R(\theta, \delta_\pi) \le r(\pi, \delta_\pi)\ \forall\ \theta$], then

(i) game (estimation problem) has a value;
(ii) δ_π is minimax;
(iii) the prior π is least favorable; and
(iv) if δ_π is unique Bayes, then it is also unique minimax.

Proof.

(i) Consider

$$\bar{V} = \inf_{\delta} \sup_{\theta} R(\theta, \delta) \le \sup_{\theta} R(\theta, \delta_\pi)$$

$$= r(\pi, \delta_\pi) \le \inf_{\delta} r(\pi, \delta)$$

$$\le \sup_{\pi} \inf_{\delta} r(\pi, \delta) = \underline{V} \qquad (8.6.6)$$

On combining the inequality in Eq. (8.6.6) with the inequality $\underline{V} \le \bar{V}$, we get $\underline{V} = \bar{V} = V$. Thus, we conclude that the game (estimation problem) has a value. Therefore, π is the least favorable and δ_π is minimax.

(ii) Although it is shown in (i) that δ_π is minimax and π is the least favorable, but we will prove these results separately. δ_π is Bayes with respect to the prior π and by the assumptions in the theorem

$$\sup_{\theta\in\Theta} R(\theta, \delta_\pi) = \int R(\theta, \delta_\pi)\pi(\theta)d\theta = r(\pi, \delta_\pi) \le r(\pi, \delta)\ \forall\ \delta$$

$$= \int R(\theta, \delta)\pi(\theta)d\theta \le \sup_{\theta} R(\theta, \delta)$$

On taking infimum over δ on both the sides, we get

$$\sup_{\theta} R(\theta, \delta_\pi) = \inf_{\delta} \sup_{\theta} R(\theta, \delta)$$

This proves that δ_π is minimax.

(iii) For any prior π', consider the corresponding Bayes estimator $\delta_{\pi'}$ and its minimum Bayes risk

$$r(\pi') = r(\pi', \delta_{\pi'}) = \int R(\theta, \delta_{\pi'})\pi'(\theta)d\theta$$

$$\le \int R(\theta, \delta_\pi)\pi'(\theta)d\theta$$

$$\le \sup_{\theta} R(\theta, \delta_\pi) = r(\pi, \delta_\pi) = r(\pi) \tag{8.6.7}$$

where the last step follows from equalized risk property in Eq. (8.6.5). Thus, the inequality in Eq. (8.6.7) shows that π is the least favorable.

(iv) This follows by replacing the inequalities in (ii) by strict inequalities. ▣

Sometimes, the guessed least favorable distribution is not a proper distribution, and then, the following theorem gives the minimax estimator.

Theorem 8.6.3 If $\{\delta_j\}$ is a sequence of Bayes estimators corresponding to the sequence of priors $\{\pi_j\}$ so that

$$r(\pi_j, \delta_j) \to c \quad \text{as } j \to \infty$$

and

$$R(\theta, \delta_0) \le c\ \forall\ \theta$$

then the game (estimation problem) has a value and δ_0 is a minimax estimator.

This theorem is proved as Theorem 8.6.6.

The second method of finding minimax estimator is based on the fact that a minimax estimator is an equalizer estimator. An estimator δ is said to be an equalizer estimator if $R(\theta, \delta) = c\ \forall\ \theta \in \Theta$. ▣

Theorem 8.6.4 Let $\Theta \subseteq \mathbb{R}^1$, $R(\theta, \delta)$ be a continuous function in θ for every δ, the game (estimation problem) has a value V, and δ_0 be a minimax estimator. Then,

$$R(\theta, \delta_0) = V \tag{8.6.8}$$

for all θ in the support of any least favourable distribution π_0.

Proof. It is given that δ_0 is minimax

$$\sup_{\theta} R(\theta, \delta_0) = \overline{V} = V$$

or
$$R(\theta, \delta_0) \le V \ \forall \ \theta \in \Theta \tag{8.6.9}$$

On taking the expectation with respect to some least favorable distribution π_0, we have

$$r(\pi_0, \delta_0) = \int_{\Theta} R(\theta, \delta_0)\pi_0(\theta)d\theta = V$$

Suppose we have a point θ' in the support of π_0 at which the inequality in Eq. (8.6.9) is strict. This gives

$$V = \int_{\Theta} R(\theta, \delta_0)\pi_0(\theta)d\theta < V$$

This contradiction proves that the minimax estimator is an equalizer estimator, i.e.,

$$R(\theta, \delta_0) = V$$

for every θ in the support of π_0. This proves what is asserted in the theorem. ▣

That is why we suspect that an equalizer estimator is minimax. The following theorem states that if an equalizer rule is Bayes, then it is minimax.

Theorem 8.6.5 Let δ_0 be a Bayes estimator of θ with respect to some proper prior π on Θ so that

$$R(\theta, \delta_0) = c$$

for all θ in the support of the prior distribution π. Then δ_0 is minimax. Further, if δ_0 is the unique Bayes estimator with respect to π, then it is the unique minimax estimator.

Proof. Consider any other estimator δ. Then,

$$\begin{aligned}\sup_{\theta} R(\theta, \delta) &\ge \int_{\Theta} R(\theta, \delta)\pi(\theta)d\theta \\ &= r(\pi, \delta) \ge r(\pi, \delta_0) = \int_{\Theta} R(\theta, \delta_0)\pi(\theta)d\theta = c \\ &= \sup_{\theta} R(\theta, \delta_0)\end{aligned} \tag{8.6.10}$$

Since δ is arbitrarily chosen, δ_0 is a minimax estimator of θ. Further, if δ_0 is the unique Bayes estimator, then the second inequality in Eq. (8.6.10) is strict. Thus, it gives δ_0 as a unique minimax estimator. ▣

Clearly, an MRE (minimum risk equivariant estimator, see Chapter 7) is minimax if it is Bayes with respect to some prior distribution since it is an equalizer estimator.

Very often, an equalizer estimator is not Bayes with respect to any prior distribution. However, it is the limit of the Bayes estimators with respect to a sequence of prior distributions. This happens in case of estimation under squared error loss function where an unbiased estimator is not Bayes with respect to any prior distribution. If such an estimator satisfies the condition given in the following theorem, then it is minimax.

Theorem 8.6.6 Let $\{\delta_j\}$ be a sequence of Bayes estimators of θ with respect to a sequence of priors $\{\pi_j\}$ and let the corresponding sequence of Bayes risks be given by $\{r(\pi_j, \delta_j)\}$. Let δ_0 be an equalizer estimator of θ. If

$$r(\pi_j, \delta_j) \to c \quad \text{as } j \to \infty \tag{8.6.11}$$

and there exists an estimator δ_0 for which

$$R(\theta, \delta_0) \le c \;\forall\; \theta \in \Theta$$

then δ_0 is minimax.

Proof. Consider δ as an arbitrarily chosen estimator of θ

$$\sup_\theta R(\theta, \delta) \ge \int_\Theta R(\theta, \delta)\pi_j(\theta)d\theta \quad \text{(for some prior } \pi_j)$$

$$\ge \int_\Theta R(\theta, \delta_j)\pi_j(\theta)d\theta = r(\pi_j, \delta_j)$$

(since δ_j is Bayes with respect to π_j)

By the limiting condition (8.6.11) and given that δ_0 is an equalizer, we get

$$\lim_{j\to\infty} r(\pi_j, \delta_j) \ge \sup_\theta R(\theta, \delta_0)$$

$$\therefore \quad \sup_\theta R(\theta, \delta) \ge \sup_\theta R(\theta, \delta_0)$$

which proves that δ_0 is minimax. ▣

Note that in the above theorem, the minimax estimator may not be unique even if the Bayes estimators δ_js are unique with respect to the prior distributions π_j as contrary to the result in Theorem 8.6.5 which provides the unique minimax estimator if it is a unique Bayes estimator with respect to the given prior distribution. However, Theorem 8.6.6 is more general as compared to Theorem 8.6.5.

Theorem 8.6.7 The minimax estimator of $a\theta + b$, $a \ne 0$, $b \in \mathbb{R}^1$ is $a\delta(\mathbf{X}) + b$ whenever $\delta(\mathbf{X})$ is the minimax estimator of θ under the squared error loss function.

Proof. Assume that some other estimator $U(\mathbf{X}) \not\equiv a\delta(\mathbf{X}) + b$ (say) is the minimax estimator of $a\theta + b$. Then,

$$\sup_\theta R(a\theta + b, U(\mathbf{X})) \le \sup_\theta R(a\theta + b, a\delta(\mathbf{X}) + b)$$

$$\sup_\theta E_\theta[U(\mathbf{X}) - a\theta - b]^2 \le \sup_\theta E_\theta[a\delta(\mathbf{X}) + b - a\theta - b]^2$$

$$a^2 \sup_\theta E_\theta\left(\frac{U(\mathbf{X}) - b}{a} - \theta\right)^2 \le a^2 \sup_\theta E_\theta[\delta(\mathbf{X}) - \theta]^2$$

$$\sup_\theta E_\theta\left(\frac{U(\mathbf{X}) - b}{a} - \theta\right)^2 \le \sup_\theta E_\theta[\delta(\mathbf{X}) - \theta]^2$$

which is a contradiction to the fact that $\delta(\mathbf{X})$ is a minimax estimator of θ. ▣

If $\boldsymbol{\theta}$ is a vector, then the following theorem may be used to find a minimax estimator.

Theorem 8.6.8 Let δ_0 be a minimax estimator of $\boldsymbol{\theta}$ in $\Theta_0 \subset \Theta$. Then δ_0 is a minimax estimator of $\boldsymbol{\theta}$ in Θ if

$$\sup_{\boldsymbol{\theta}\in\Theta} R(\boldsymbol{\theta}, \delta_0) = \sup_{\boldsymbol{\theta}\in\Theta_0} R(\boldsymbol{\theta}, \delta_0)$$

Proof. Assume that δ_0 is not a minimax estimator of $\boldsymbol{\theta}$ in Θ. We then have an estimator δ of $\boldsymbol{\theta}$ so that

$$\sup_{\boldsymbol{\theta}\in\Theta} R(\boldsymbol{\theta}, \delta) < \sup_{\boldsymbol{\theta}\in\Theta} R(\boldsymbol{\theta}, \delta_0)$$

Examine, now, the relationship on the restricted parameter space Θ_0

$$\sup_{\boldsymbol{\theta}\in\Theta_0} R(\boldsymbol{\theta}, \delta) \le \sup_{\boldsymbol{\theta}\in\Theta} R(\boldsymbol{\theta}, \delta)$$

$$< \sup_{\boldsymbol{\theta}\in\Theta} R(\boldsymbol{\theta}, \delta_0) = \sup_{\boldsymbol{\theta}\in\Theta_0} R(\boldsymbol{\theta}, \delta_0)$$

which is a contradiction to the assumption in the theorem that δ_0 is a minimax estimator of $\boldsymbol{\theta}$ when the parametric space is Θ_0. This proves the theorem that δ_0 is a minimax estimator of $\boldsymbol{\theta}$ when the parameter space is Θ. ▣

The following theorem states that if an equalizer estimator of θ is admissible, it is minimax.

Theorem 8.6.9 Suppose δ_0 is an equalizer and admissible estimator of θ. Then, δ_0 is minimax. Further, if the loss function is strictly convex, then δ_0 is the unique minimax estimator of θ.

Proof. If we have an estimator δ, then

$$R(\theta, \delta) \le R(\theta, \delta_0) = c \quad \forall\, \theta$$

since δ_0 is equalizer.
Then by admissibility of δ_0, we must have

$$R(\theta, \delta) = c \quad \forall\, \theta$$

or

$$\sup_{\theta} R(\theta, \delta) = c = \sup_{\theta} R(\theta, \delta_0)$$

This proves that δ_0 is a minimax estimator of θ.

Let δ_1 be a minimax estimator of θ other than δ_0. Then by Jensen's inequality and strict convexity of loss function,

$$R\left(\theta, \frac{\delta_0 + \delta_1}{2}\right) = E_\theta L\left(\theta, \frac{\delta_0 + \delta_1}{2}\right)$$

$$< \frac{1}{2}[E_\theta L(\theta, \delta_0) + E_\theta L(\theta, \delta_1)] = \frac{R(\theta, \delta_0) + R(\theta, \delta_1)}{2}$$

$$= R(\theta, \delta_0)\ \forall\, \theta$$

which is a contradiction to the assumption that δ_0 is admissible. This proves that δ_0 is unique if the loss function is strictly convex. ▣

Notes.

1. Thus, by Theorem 8.6.9, an admissible MRE estimator is minimax. This is usually the case when the family is one-parameter location family.
2. *A unique minimax estimator is always admissible.* Let δ_0 be the unique minimax estimator of θ and let δ be some estimator better than δ_0.

$$R(\theta, \delta) \le R(\theta, \delta_0) \; \forall \; \theta$$

and $$R(\theta, \delta) < R(\theta, \delta_0) \text{ for at least one } \theta$$

This implies $$\sup_\theta R(\theta, \delta) \le \sup_\theta R(\theta, \delta_0)$$

If the inequality is strict, then it is a contradiction to the assumption that δ_0 is minimax; and if it is an equality, then δ is minimax, but it is a contradiction to the assumption that δ_0 is a unique minimax estimator of θ. This implies that we cannot have an estimator δ better than δ_0. Therefore, a unique minimax estimator is always admissible.

We may summarize the methods of finding a minimax estimator by considering Bayes estimator that has constant risk.

***Method* 1.** If an equalizer estimator is Bayes with respect to some prior distribution, then it is minimax. It is stated in Theorem 8.6.5.

***Method* 2.** If an equalizer estimator is a limit of Bayes with respect to a sequence of prior distributions as are the conditions in Theorem 8.6.6, then it is minimax.

The ML Estimator is Approximately Minimax for Large *n*

The risk of an estimator is

$$R(\theta, \delta) = V_\theta(\delta) + \{\text{bias}_\theta(\delta)\}^2$$

For large n, squared bias is roughly of order $O(n^{-2})$ and variance is of order $O(n^{-1})$. Thus,

$$R(\theta, \delta) \simeq V_\theta(\delta)$$

The variance of ML estimator under certain regularity conditions is

$$V(\delta_{\text{ML}}) = \frac{1}{nI_X(\theta)}$$

$$\therefore \quad nR(\theta, \delta_{\text{ML}}) = \frac{1}{I_X(\theta)} \; \forall \; \theta$$

For any estimator δ', it can be shown that

$$R(\theta, \delta') \ge R(\theta, \delta_{\text{ML}}) \quad \text{if } n \text{ is large}$$

Thus, the ML estimator is approximately minimax. ▣

8.7 SOLVED EXAMPLES

***Example* 8.1** Let $\Theta = \mathcal{A}$ be a subset of the real line and let $L(p, \delta) = w(p)(p - \delta)^2$. Show that if a Bayes estimate δ with respect to a prior distribution π is an unbiased estimate of p and $Ew(p) < \infty$, then $r(\pi, \delta) = 0$.

Solution. Consider the Bayes risk of the estimator δ

$$\begin{aligned} r(\pi, \delta) &= E^p E^{X|p} w(p)[p - \delta(X)]^2 \\ &= E^p w(p)\, p^2 - 2E^p w(p)p\, E^{X|p}\delta(X) + E^p\, w(p)E^{X|p}\delta^2(X) \end{aligned}$$

The estimator δ is unbiased and Bayes with respect a given prior distribution π. This gives

$$E^p w(p)p\, E^{X|p}\delta(X) = E^p w(p)p^2$$

and

$$\delta(x) = \frac{E^p w(p)p}{E^p w(p)}$$

Using these results, the Bayes risk simplifies to

$$\begin{aligned} r(\pi, \delta) &= E^p w(p)p^2 - 2E^p w(p)pE^{X|p}\delta(X) + E^p w(p)E^{X|p}\delta^2(X) \\ &= E^p w(p)p^2 - E^p w(p)pE^{X|p}\delta(X) - E^p w(p)pE^{X|p}\delta(X) + E^p w(p)E^{X|p}\delta^2(X) \\ &= E^p w(p)p^2 - E^p w(p)p^2 - E^p w(p)pE^{X|p}\frac{E^p w(p)p}{E^p w(p)} + E^p w(p)E^{X|p}\left(\frac{E^p w(p)p}{E^p w(p)}\right)^2 = 0 \end{aligned}$$

***Example* 8.2** (Dirichlet prior distribution) Suppose we have a frequency data so that $\mathbf{X} \sim$ Multinomial $(n, \boldsymbol{\theta})$ where $\mathbf{X} = (X_1, \ldots, X_k)$, $\boldsymbol{\theta} = (\theta_1, \theta_2, \ldots, \theta_k)$, $0 < \theta_i < 1$, $\Sigma\theta_i = 1$, $\Sigma x_i = n$ with the *pmf*

$$f(\mathbf{x}|\boldsymbol{\theta}) = \frac{n!}{x_1!\, x_2! \ldots \left(n - \sum_{i=1}^{k-1} x_i\right)!} \prod_{i=1}^{k-1} \theta_i^{x_i} \left(1 - \sum_{i=1}^{k-1} \theta_i\right)^{n - \sum_{i=1}^{k-1} x_i}$$

Consider the Dirichlet prior for $\boldsymbol{\theta}$ as

$$\pi(\boldsymbol{\theta}|\boldsymbol{\alpha}) \propto \prod_{i=1}^{k-1} \theta_i^{\alpha_i - 1} \left(1 - \sum_{i=1}^{k-1} \theta_i\right)^{\alpha_k - 1}$$

Find the Bayes estimate of $\boldsymbol{\theta}$ under squared error loss function.

Solution. The posterior distribution $\boldsymbol{\theta}$ given $\mathbf{x}$ is

$$\begin{aligned} \pi(\boldsymbol{\theta}|\mathbf{x}) &\propto \prod_{i=1}^{k-1} \theta_i^{x_i + \alpha_i - 1} \left(1 - \sum_{i=1}^{k-1} \theta_i\right)^{n - \sum_{i=1}^{k-1} x_i + \alpha_k - 1} \\ &= \frac{1}{D(\mathbf{x} + \boldsymbol{\alpha})} \prod_{i=1}^{k-1} \theta_i^{x_i + \alpha_i - 1} \left(1 - \sum_{i=1}^{k-1} \theta_i\right)^{n - \sum_{i=1}^{k-1} x_i + \alpha_k - 1} \end{aligned}$$

where $$D(\mathbf{x}+\boldsymbol{\alpha}) = \frac{\Gamma(x_1+\alpha_1)\ldots\Gamma\left(n-\sum_{i=1}^{k-1} x_i+\alpha_k-1\right)}{\Gamma(n+\alpha_0)}$$

$\Sigma\alpha_i = \alpha_0$ and $\Sigma x_i = n$. Thus, the posterior distribution is $D(\boldsymbol{\theta}|x_1 + \alpha_1,\ldots,x_k + \alpha_k)$. Therefore, the Bayes estimate of $\boldsymbol{\theta}$ with respect to the Dirichlet prior distribution and under the squared error loss function is

$$\theta^B = E(\boldsymbol{\theta}|\mathbf{x}) = \frac{1}{n+\alpha_0}(\mathbf{x}+\boldsymbol{\alpha})$$

$$= \frac{1}{n+\alpha_0}(x_1+\alpha_1,\ldots, x_k+\alpha_k)$$

***Example* 8.3** Find the Bayes estimate of θ under the 0 – 1 loss function with parameter $\varepsilon > 0$ (ε small)

$$L_\varepsilon(\theta,\delta) = \begin{cases} 1, & \text{if } |\theta-\delta| \geq \varepsilon \\ 0, & \text{if } |\theta-\delta| < \varepsilon \end{cases}$$

Solution. The Bayes estimator of θ is given by

$$\delta_\varepsilon(\mathbf{x}) = \min_\delta \int_\theta L_\varepsilon(\theta,\delta)\pi(\theta|\mathbf{x})d\theta$$

$$= \min_\delta\left(1 - \int_{\theta:|\theta-\delta|<\varepsilon} \pi(\theta|\mathbf{x})d\theta\right)$$

$$= \max_\delta \int_{\delta-\varepsilon}^{\delta+\varepsilon} \pi(\theta|\mathbf{x})d\theta$$

Thus, the Bayes estimator under this loss function is that θ around which the area under the posterior distribution $\pi(\theta|\mathbf{x})$ over the interval $(\theta - \varepsilon, \theta + \varepsilon)$ of length 2ε is maximal.

***Example* 8.4 (Furguson, 1967).** Consider a general problem of estimation where the loss functions are

(i) $$L(\theta,\delta) = \begin{cases} k_1|\theta-\delta|, & \text{if } a \leq \theta \\ k_2|\theta-\delta|, & \text{if } a > \theta \end{cases}$$

where $k_1 > 0$ and $k_2 > 0$.

(ii) $$L(\theta,\delta) = \begin{cases} 0, & \text{if} |\theta-\delta| \leq c \\ 1, & \text{if} |\theta-\delta| > c \end{cases}$$

where $c > 0$. Show that for (i), the function $w(\delta) = E^{\theta|x}L(\theta, \delta)$ is minimized when δ is the pth quantile of the distribution of θ given x assuming that its first moment is finite. State the Bayes estimator in this case as a function of k_1 and k_2. Show that for (ii), the function $w(\delta) = E^{\theta|x}L(\theta, \delta)$ is minimized when δ is the midpoint of the modal interval of length $2c$. Find the Bayes estimator by defining "the modal interval of length $2c$" so that it makes sense.

Solution. Consider the loss function in (i), and

$$w(\delta) = EL(\theta, \delta)$$

$$= k_2 \int_{-\infty}^{a} (\delta - \theta)\pi(\theta|x)d\theta + k_1 \int_{a}^{\infty} (\theta - \delta)\pi(\theta|x)d\theta$$

Differentiating $w(\delta)$ with respect to δ and using the result

$$\frac{\partial}{\partial \alpha} \int_{a(\alpha)}^{b(\alpha)} f(x, \alpha)dx = f(b, \alpha)\frac{\partial b}{\partial \alpha} - f(a, \alpha)\frac{\partial a}{\partial \alpha} + \int_{a(\alpha)}^{b(\alpha)} \frac{\partial}{\partial \alpha} f(x, \alpha)dx$$

$$\frac{\partial}{\partial \delta} w(\delta) = k_2 \int_{-\infty}^{\delta} \pi(\theta|x)d\theta - k_1 \int_{\delta}^{\infty} \pi(\theta|x)d\theta = 0$$

we get

$$k_2 \int_{-\infty}^{\delta} \pi(\theta|x)d\theta = k_1 \int_{\delta}^{\infty} \pi(\theta|x)d\theta$$

or

$$k_2 \int_{-\infty}^{\delta} \pi(\theta|x)d\theta + k_2 \int_{\delta}^{\infty} \pi(\theta|x)d\theta = k_1 \int_{\delta}^{\infty} \pi(\theta|x)d\theta + k_2 \int_{\delta}^{\infty} \pi(\theta|x)d\theta$$

we have

$$k_2 = (k_1 + k_2)\int_{\infty}^{\delta} \pi(\theta|x)d\theta$$

$$\int_{\delta}^{\infty} \pi(\theta|x)d\theta = \frac{k_2}{k_1 + k_2} \text{ or } \int_{-\infty}^{\delta} \pi(\theta|x)d\theta = \frac{k_1}{k_1 + k_2}$$

Therefore, the Bayes estimator in a single data problem of θ is the upper $k_1/(k_1 + k_2)$th quantile of the distribution of θ. Consider next the loss function given in (ii)

$$w(\delta) = EL(\theta, \delta) = P(|\theta - \delta| > c|x)$$

$$= 1 - P(|\theta - \delta| \le c|x)$$

$$= 1 - P(\delta - c \le Z \le \delta + c|x)$$

$$= 1 - \int_{\delta-c}^{\delta+c} \pi(\theta|x)d\theta$$

It is clear from the above equation that the value of δ is the value of θ in an interval of fixed length $2c$ for which $w(\delta)$ is the minimum. The probability under the distribution of $\theta|x$ is maximum. Therefore, the Bayes estimator of θ is the mode of the distribution of $\theta|x$.

***Example* 8.5** $X \sim b(1, \theta)$, single observation, Θ has two points $\Theta = \{\theta_1, \theta_2\}$, $L(\theta, \delta) = (\theta - \delta)^2$. Find the minimax estimator of θ and the prior distribution with respect to which this is Bayes. Find them for $\Theta = \{1/3, 2/3\}$.

Solution. Consider an estimator δ as point (x, y) in the plane, with the interpretation that we estimate θ by x if tails is observed and y if heads is observed. The risk of $\delta \equiv (x, y)$ is given by

$$R(\theta, \delta) = E(\delta - \theta)^2 = (x - \theta)^2(1 - \theta) + (y - \theta)^2\theta \tag{8.7.1}$$

Let the prior distribution be denoted by π on $\Theta = \{\theta_1, \theta_2\}$ so that $\pi(\theta_1) = \pi_1$ and $\pi(\theta_2) = 1 - \pi_1$. It is an example of finding minimax estimator when the parameter space is finite. The Bayes risk of δ with respect to π is given by

$$r(\pi, \delta) = R(\theta_1, \delta)\, \pi_1 + R(\theta_2, \delta)(1 - \pi_1) \tag{8.7.2}$$

The estimator δ is Bayes with respect to the prior distribution π if we choose x and y that minimize $r(\pi, \delta)$ in Eq. (8.7.2). Thus, on differentiating $r(\pi, \delta)$ with respect to x and y and equating it to zero, the Bayes estimator is given by

$$x = \frac{\theta_1\pi_1(1-\theta_1) + \theta_2(1-\theta_2)(1-\pi_1)}{(1-\theta_1)\pi_1 + (1-\theta_2)(1-\pi_1)} \tag{8.7.3}$$

and

$$y = \frac{\theta_1^2\pi_1 + \theta_2^2(1-\pi_1)}{\theta_1\pi_1 + \theta_2(1-\pi_1)}$$

If we want the minimax Theorem 8.6.2 to hold, the risk of δ in Eq. (8.7.1) must be constant over all $\theta \in \Theta$ or δ must be an equalizer estimator. This requirement simplifies to

$$(\theta_1 + \theta_2)[1 + 2(x - y)] + (-2x - x^2 + y^2) = 0 \tag{8.7.4}$$

By Theorem 8.6.2, the equalizer estimator δ will be minimax if it is Bayes with respect to the prior distribution π. This prior distribution will be the least favourable distribution. Thus, by putting the values of x and y from Eq. (8.7.3) in Eq. (8.7.4) and solving for π_1, one gets the least favorable prior distribution $\pi = (\pi_1, 1 - \pi_1)$ with respect to which $\delta = (x, y)$ is Bayes and minimax.

If $\Theta = \{1/3, 2/3\}$, we get $\pi = \{1/2, 1/2\}$ as the least favorable distribution and $\delta = \{4/9, 5/9\}$ as the corresponding Bayes and minimax estimator of θ with minimax risk of 2/81. Thus, the minimax estimator of θ is to estimate θ by 4/9 if tail is observed and 5/9 if head is observed.

***Example* 8.6** Let $X \sim b(1, \theta)$, $\theta \in \Theta = [0, 1]$ and the loss function be

$$L(\theta, \delta) = (\theta - \delta)^2$$

Show that the minimax estimator of θ is $\delta = (1/4, 3/4)$ which is also Bayes with respect to the least favorable distribution Be(1/2, 1/2) and the game (estimation problem) has a value 1/16. In case of n observations, plot the risk of minimax estimator and $\bar{X}$ and compare.

Solution. This is an example of finding the minimax estimator of θ when the parameter space is an interval. An estimator δ may be represented by a point (x, y) in the unit square $\{\{x, y\}: 0 \le x \le 1, 0 \le y \le 1\}$ with the meaning that we estimate θ by x if tails is observed and y if heads is observed. The risk of an estimator $\delta \equiv (x, y)$ is given by

$$R(\theta, \delta) = [1 + 2(x - y)]\theta^2 + (y^2 - x^2 - 2x)\theta + x^2 \tag{8.7.5}$$

Consider a prior π on Θ such that

$$E(\theta) = \mu_1'$$
$$E(\theta^2) = \mu_2'$$

The Bayes risk of δ with respect to this prior π is

$$r(\pi, \delta) = [1 + 2(x - y)]\mu_2' + (y^2 - x^2 - 2x)\mu_1' + x^2 \tag{8.7.6}$$

One obtains the Bayes estimator by minimizing $r(\pi, \delta)$ in Eq. (8.7.6) over the variations for x and y. By differentiating $r(\pi, \delta)$ in Eq. (8.7.6) with respect to x and y and setting it to zero, we get the Bayes estimate

$$\left.\begin{aligned} x &= \frac{\mu_1' - \mu_2'}{1 - \mu_1'} \\ y &= \frac{\mu_2'}{\mu_1'} \end{aligned}\right\} \tag{8.7.7}$$

For minimax Theorem 8.6.2 to hold, δ should be an equalizer estimator, i.e., $R(\theta, \delta) =$ constant $\forall\ \theta \in \Theta$. Thus, we must have the coefficients of θ^2 and θ in Eq. (8.7.5) equal to zero, i.e.,

$$1 + 2(x - y) = 0$$

and
$$y^2 - x^2 - 2x = 0$$

This gives
$$x = \frac{1}{4} \quad \text{and} \quad y = \frac{3}{4}$$

On putting these values in Eq. (8.7.7), we get the least favorable distribution π such that $\mu_1' = 1/2$ and $\mu_2' = 3/8$. Thus, $\delta = (x, y) = (1/4, 3/4)$ is Bayes with respect to the least favorable distribution π and, hence, minimax with Bayes risk = value of the game = 1/16. Note that the prior Be(1/2, 1/2) satisfies $\mu_1' = 1/2$ and $\mu_2' = 3/8$. In case of n observations, the least favorable distribution is Be(n/2, n/2) and the minimax risk is $4(\sqrt{n} + 1)^{-2}$.

Consider the sample mean $\bar{X}$ as an estimator of θ. Its risk is given by

$$R(\theta, \bar{X}) = \frac{\theta(1-\theta)}{n}$$

The risk of the minimax estimator is

$$R(\theta, \delta_0) = \frac{1}{4(\sqrt{n}+1)^2}$$

Plotting the risk functions of $\bar{X}$ and δ_0 as a function of θ for different values of n, we get Figure 8.7.1.

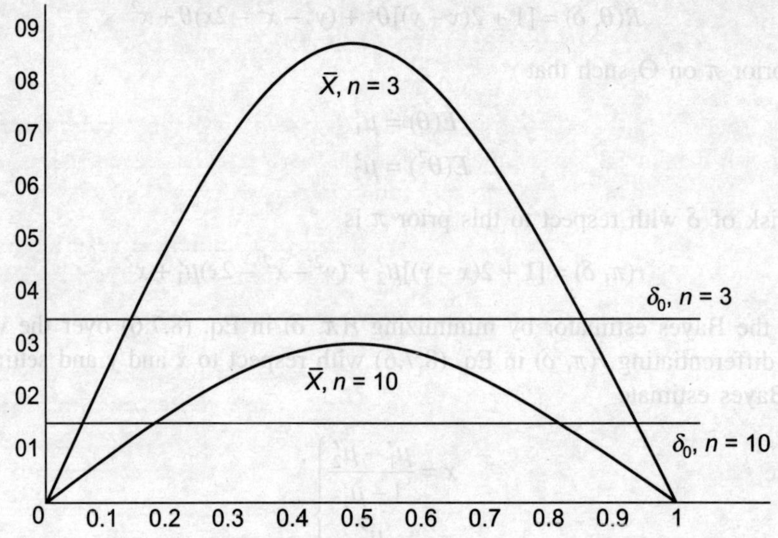

Figure 8.7.1

***Example* 8.7** Let $X_1,\ldots,X_n$ be a random sample drawn from $b(1, p)$, $p \in (0, 1)$. Consider the prior distribution for p on (0, 1) by the uniform distribution $U(0, 1)$. Show that the sample mean $\bar{X}$ is an admissible estimator of p both under the loss functions $L(p, \delta) = w(p)\ (\delta - p)^2$, $w(p) = [p(1-p)]^{-1}$, and the squared error loss function $L(p, \delta) = (\delta - p)^2$.

Solution. We have $T = \Sigma X_i \sim b(n, p)$. The joint distribution of T and p is given by

$$h(t, p) = \binom{n}{t} p^t (1-p)^{n-t}$$

The posterior distribution of p given t is

$$\pi(p|t) = [B(t+1, n-t+1)]^{-1} p^t (1-p)^{n-t}$$

Therefore, the Bayes estimator of p under the weighted squared error loss function is

$$\delta^B(t) = \frac{\int_0^1 p w(p)[B(t+1, n-t+1)]^{-1} p^t (1-p)^{n-t}\, dp}{\int_0^1 w(p)[B(t+1, n-t+1)]^{-1} p^t (1-p)^{n-t}\, dp}$$

$$= \frac{\int_0^1 p^t (1-p)^{n-t-1}\, dp}{\int_0^1 p^{t-1} (1-p)^{n-t-1}\, dp} = \frac{t}{n} = \bar{x}$$

Since δ^B is unique, by Theorem 8.2.10, it is admissible under the weighted squared error loss function $w(p)(\delta - p)^2$. Further, the Bayes estimator δ^B is an equalizer estimator since

$$R(\theta, \delta^B) = \frac{1}{p(1-p)} E(\bar{X} - p)^2 = \frac{1}{n} \ \forall \ p$$

Therefore, by Theorem 8.6.2, δ^B is minimax also. Since δ^B is Bayes, under the squared error loss function, we cannot get an estimator δ which has a smaller Bayes risk as compared to that of δ^B. Therefore, $\delta^B = \bar{X}$ is Bayes and admissible.

***Example* 8.8** Let $X_1,\ldots,X_n$ be *iid* $b(1, \theta)$. Find Jeffreys noninformative prior for θ and find the Bayes and maximum aposteriori (MAP) estimates with respect to this noninformative prior

$$\pi(\theta) = 1 \ \forall \ \theta \in (0, 1)$$

Do Bayes and ML estimates coincide? Show that these estimates, irrespective of noninformative prior distributions, converge to $\lim_{n\to\infty} \Sigma x_i/n$ for large n. Convert the belief that θ is around 1/2 into a proper prior distribution and find the Bayes and MAP estimate in this case. Investigate whether these estimates are now pulled towards 1/2. Note that the noninformative prior, in this case, is not the limiting case of beta conjugate family; it is a particular case with $\alpha = \beta = 1$. Further, this noninformative prior is proper since the parameter space $\Theta = (0, 1)$ is bounded.

Solution. The Fisher information about θ,

$$I_{\mathbf{X}}(\theta) = \frac{n}{\theta(1-\theta)}$$

gives the Jeffreys noninformative prior for θ,

$$\pi(\theta) = [I_{\mathbf{X}}(\theta)]^{1/2} = \sqrt{\frac{n}{\theta(1-\theta)}} \propto \frac{1}{\sqrt{\theta(1-\theta)}} \tag{8.7.8}$$

Beta distribution is a conjugate prior for the Bernoulli likelihood function and the Jeffreys prior computed in Eq. (8.7.8) is $Be(1/2, 1/2)$. Note that if $\theta \sim Be(\alpha, \beta)$, then $E(\theta) = \alpha/(\alpha + \beta)$ and $\theta^{\text{MAP}} \max_\theta \pi(\theta|\mathbf{x}) = (\alpha - 1)/(\alpha + \beta - 2)$ if $\alpha > 1$. Therefore, Jeffreys prior is a conjugate prior. Corresponding to this prior, the posterior distribution is $\text{Be}(\alpha + \Sigma x_i, \beta + n - \Sigma x_i)|_{\alpha=(1/2), \beta=(1/2)} = Be[(1/2) + \Sigma x_i, (1/2) + n - \Sigma x_i]$. Bayes and MAP estimates are

$$\theta^B = \frac{(1/2) + \sum x_i}{n+1}$$

$$\theta^{\text{MAP}} = \frac{-(1/2) + \sum x_i}{n-1}$$

As the amount of data increases, $n \to \infty$, both the estimates converge to $\lim_{n\to\infty} \Sigma x_i/n$. Next, if we consider the uniform prior

$$\pi(\theta) = 1 \ \forall \ \theta \in (0, 1) \tag{8.7.9}$$

the posterior distribution is

$$\pi(\theta|\mathbf{x}) \propto \theta^{\Sigma x_i}(1-\theta)^{n-\Sigma x_i} \propto f(\mathbf{x}|\theta)$$

This shows that the posterior distribution is Be(Σx_i + 1, n – Σx_i + 1). The Bayes and MAP estimates are given by

$$\theta^B = E(\theta|\mathbf{x}) = \frac{\alpha}{\alpha+\beta} = \frac{\sum x_i + 1}{n+2}$$

$$\theta^{\text{MAP}} = \frac{\alpha - 1}{\alpha+\beta-2} \text{ (mode if } \alpha > 1)$$

$$= \frac{\sum x_i + 1 - 1}{n+2-2} = \frac{\sum x_i}{n} = T^{\text{ML}}(\mathbf{x})$$

Thus, in this case, both ML and MAP estimates coincide. These estimates, as data grows to infinity, $n \to \infty$, converge to $\lim_{n\to\infty} \Sigma x_i/n$. So, we conclude that, if data grows to infinity, it does not matter which prior is chosen between the priors given in Eqs. (8.7.8) and (8.7.9), both Bayes and MAP estimates converge to $\lim_{n\to\infty} \Sigma x_i/n$.

If we toss a coin 6 times and get 6 heads, we have $\delta^{\text{MAP}}(\mathbf{x}) = \delta^{\text{ML}}(\mathbf{x}) = 1$ and $\delta^B(\mathbf{x}) = 7/8 = 0.85$. We generally expect θ, the probability of head, to be around 1/2. The values of these estimates do not support this belief. Now, converting this belief into a prior distribution, we expect the prior distribution to be concentrated and symmetric about 1/2, and it goes to 0 as θ goes to 0 or 1. Many priors can satisfy this condition, but several of them yield intractable posterior distribution. However, beta density is one such density that confirms to this belief. Beta prior is a conjugate prior and it yields a simplified posterior distribution, i.e., beta distribution. Let $\theta \sim$ Be(α, β)

$$\pi(\theta|\alpha,\beta) \sim \text{Be}(\alpha,\beta)$$

$$E(\theta|\alpha,\beta) = \frac{\alpha}{\alpha+\beta},$$

$$V(\theta|\alpha,\beta) = \frac{\alpha\beta}{(\alpha+\beta)^2(\alpha+\beta+1)}$$

$$\max_{\theta} \pi(\theta|\alpha,\beta) = \frac{\alpha-1}{\alpha+\beta-2}(\alpha > 1)$$

If we want this prior to imbibe the belief that the estimate must be close to 1/2, we must have $\alpha = \beta$. If one plots beta density for different values of $\alpha = \beta$, the problem that one notices with $\alpha = \beta \le 1$ is that its mode is not at 1/2. However, for higher values of $\alpha = \beta$, this density is concentrated about 1/2 much strongly.

The posterior density corresponding to this conjugate prior, Be(α, β), is

$$\pi(\theta|\mathbf{x}) = Be\left(\alpha + \sum x_i, \beta + n - \sum x_i\right)$$

to yield the estimates

$$\delta^B(\mathbf{x}) = \frac{\alpha + \sum x_i}{\alpha + \beta + n}$$

$$\delta^{\text{MAP}}(\mathbf{x}) = \frac{\alpha + \sum x_i - 1}{\alpha + \beta + n - 2}$$

Consider $\alpha = \beta = 2$ and $n = \Sigma x_i = 6$. We get $\delta^B = (2 + 6)/(2 + 2 + 6) = 8/10 = 0.8$

$$\delta^{\text{MAP}}(\mathbf{x}) = \frac{2+6-1}{2+2+6-2} = \frac{7}{8} = 0.88$$

Now, consider $\alpha = \beta = 3$ and $n = \Sigma x_i = 6$. We get

$$\delta^B(\mathbf{x}) = \frac{3+6}{12} = \frac{9}{12} = 0.75$$

$$\delta^{\text{MAP}}(\mathbf{x}) = \frac{3+6-1}{3+3+6-2} = \frac{8}{10} = 0.80$$

For $\alpha = \beta = 12$, $n = \Sigma x_i = 6$, we get

$$\delta^B(\mathbf{x}) = \frac{12+6}{24+6} = \frac{18}{30} = 0.60$$

$$\delta^{\text{MAP}}(\mathbf{x}) = \frac{12+6-1}{30-2} = \frac{17}{28} \cong 0.61$$

This shows that as $\alpha = \beta$ grows, the prior becomes stronger, the estimates δ^B and δ^{MAP} are pulled towards 1/2 confirming the prior belief.

If the amount of data increases, i.e., n grows to infinity, the estimates δ^B and δ^{MAP} converge to $\lim_{n\to\infty} \Sigma x_i/n$, i.e., true θ, since by the law of large numbers, sample mean converges to population mean.

***Example* 8.9** Let $X_1, \ldots, X_n$ be a random sample from the Bernoulli distribution $b(1, p)$, $p \in (0, 1)$. Consider the statistic $T = \Sigma X_i$.

(i) Can an unbiased estimator $\bar{X}$ of p with $\Theta = [0, 1]$ be Bayes with respect to some prior distribution? Comment. Is $\bar{X}$ a limit of Bayes estimators corresponding to the sequence of beta priors $\{Be(\alpha, \beta)\}$?

(ii) Let the prior distribution of p over (0, 1) be $\pi(p) \sim Be(\alpha, \beta)$, $\alpha > 0, \beta > 0$ are known. Find the Bayes estimator of p under the squared error loss function. Can this estimator be unbiased. Show that prior information is discounted in favour of data, and Bayes estimator approaches sample mean when sample size is large and Bayes estimator approaches prior mean in the absence of data.

(iii) Consider the same prior distribution as in (i). Find the Bayes estimator of $p(1 - p)$ under the squared error loss function and compare it with the UMVUE of $p(1 - p)$. Show that the bias of this Bayes estimator is of order $O(1/n)$ and that it is consistent and admissible.

(iv) Consider the improper prior for p by

$$\pi(p) = \frac{1}{p(1-p)} I_{(0,1)}(p)$$

Show that the posterior density of p given the data, $\pi(p|\mathbf{x})$, is

$$\pi(p|\mathbf{x}) \sim Be(t, n-t)$$

provided that the sample mean $\bar{x} = t/n \in (0, 1)$, where $t = \Sigma x_i$. Also, show that the generalized Bayes estimator of $p(1-p)$ with respect to the above improper prior $\pi(p)$ under the squared error loss function is $T(n - T)/n(n + 1)$.

Solution.

(i) The estimator $\bar{X}$ is unbiased for estimating p, and its risk is given by

$$R(p, \bar{X}) = E(\bar{X} - p)^2 = \frac{1}{n} p(1-p)$$

The Bayes risk of $\bar{X}$ with respect to some prior distribution π is

$$r(\pi, \bar{X}) = \frac{1}{n}\int_0^1 p(1-p)\pi(p)dp$$

This Bayes risk vanishes only if the prior π assigns positive mass to end points $p = 0$ or $p = 1$ and zero to all $p \in (0, 1)$. Thus, $\bar{X}$ is Bayes with respect to any such prior π, whose support is (0, 1), and its Bayes risk is zero. This is a case where there are many Bayes estimators.

We have already shown, for prior $\pi(p) \sim Be(\alpha, \beta)$, that

$$p^B(\mathbf{x}) = \frac{\alpha + \sum x_i}{\alpha + \beta + n} \to \bar{x} \text{ as } \alpha \to 0, \beta \to 0 \qquad (8.7.10)$$

By letting $\alpha \to 0$, $\beta \to 0$, the prior takes the form

$$\pi(p) \propto \frac{1}{p(1-p)}$$

Note that $\pi(p)$ is an improper prior since it cannot be normalized since $\int_0^1 \{1/[p(1-p)]\}dp$ diverges.

In this example, we have a sequence of Bayes estimators $\{p^B_{(\alpha,\beta)}(\mathbf{X})\}$, in Eq. (8.7.10), corresponding to the sequence of priors $\{\pi_{\alpha,\beta}\} = \{Be(\alpha, \beta)\}$ so that

$$\lim_{\substack{\alpha\to 0\\ \beta\to 0}} p^B_{(\alpha,\beta)} = \bar{X}$$

Therefore, $\bar{X}$ is the limit of Bayes estimators with respect to beta–prior, $Be(\alpha, \beta)$.

(ii) The joint density of $T = \Sigma X_i$ and p is

$$h(t, p) \propto p^t(1-p)^{n-t}p^{\alpha-1}(1-p)^{\beta-1}$$
$$= p^{t+\alpha-1}(1-p)^{n-t+\beta-1}$$

Therefore, the posterior density of p given t is

$$\pi(p|t) \propto p^{t+\alpha-1}(1-p)^{n-t+\beta-1}$$

The Bayes estimate of $g(p) = p$ is

$$\delta^B(\mathbf{x}) = \frac{\bar{x}+(\alpha/n)}{1+(\alpha/n)+(\beta/n)}$$

$$= \frac{\sum x_i + \alpha}{\alpha+\beta+n} = (1-w)p_{\text{ML}} + w\,(\text{prior mean } \bar{p})$$

where $\quad w = \dfrac{\alpha+\beta}{\alpha+\beta+n}$; prior mean $= \bar{p} = \dfrac{\alpha}{\alpha+\beta}$; $p_{\text{ML}} = \bar{x}$

The Bayes estimator δ^B is unbiased if $E(\delta^B) = (np + \alpha)/(n + \alpha + \beta) = p$. T^B is not unbiased unless $\alpha = 0 = \beta$, which is not possible since $\theta \sim Be(\alpha, \beta)$ and $\alpha > 0$ and $\beta > 0$. Further, if the sample size is large, that is, if one has plenty of information in favour of X, the prior information becomes irrelevant and the Bayes estimator δ^B approaches $\bar{X}$, which is the usual UMVUE of p. We may note here that prior information is discounted in favor of data.

In case we have no data, that is, $n = 0$, the Bayes estimator δ^B of p becomes

$$p^B = \frac{\alpha}{\alpha+\beta}$$

which is mean of the prior distribution. In other words, if we have no data, we depend on our intuitive guess and suggest Bayes estimator by mean of the prior distribution.

(iii) The Bayes estimator of $g(p) = p(1 - p)$ under the squared error loss function

$$\delta^B(\mathbf{x}) = \frac{\int g(p)\pi(p|t)dp}{\int \pi(p|t)dp}$$

$$= \frac{\int_0^1 p^{t+\alpha}(1-p)^{n-t+\beta}dp}{\int_0^1 p^{t+\alpha-1}(1-p)^{n-t+\beta-1}dp}$$

$$= \frac{(t+\alpha)(n-t+\beta)}{(n+\alpha+\beta)(n+\alpha+\beta+1)}$$

The UMVUE estimate of $p(1 - p)$ is

$$\delta'(\mathbf{x}) = \frac{t(n-t)}{n(n-1)}$$

The bias of the Bayes estimator δ^B is

$$\begin{aligned}\text{bias}(\delta^B) &= E[\delta^B(\mathbf{X})] - p(1-p)\\ &= \frac{\{E[T(n-T)] + (\beta-\alpha)E(T) + \alpha(n+\beta)\}}{(n+\alpha+\beta)(n+\alpha+\beta+1)} - p(1-p)\\ &= \frac{[n(n-1)p(1-p) + (\beta-\alpha)np + \alpha(n+\beta)]}{(n+\alpha+\beta)(n+\alpha+\beta+1)} - p(1-p)\\ &= \left[\frac{n(n-1)}{(n+\alpha+\beta)(n+\alpha+\beta+1)} - 1\right]p(1-p) + \frac{(\beta-\alpha)np + \alpha(n+\beta)}{(n+\alpha+\beta)(n+\alpha+\beta+1)}\\ &= O\left(\frac{1}{n}\right)\end{aligned}$$

Therefore, the Bayes estimator δ^B is not unbiased but approximately unbiased. By the strong law of large numbers, $(1/n)\Sigma_{i=1}^{n} X_i = T/n \xrightarrow{a.s.} E(T/n) = p$.

Therefore, $$\delta^B(\mathbf{X}) = \frac{\left[\left(\frac{T}{n}\right)+\left(\frac{\alpha}{n}\right)+\left(\frac{1}{n}\right)\right]\left[1-\left(\frac{T}{n}\right)+\left(\frac{\beta}{n}\right)\right]}{\left[1+\frac{(\alpha+\beta+1)}{n}\right]\left[1+\frac{(\alpha+\beta)}{n}\right]}$$

$$\delta^B(\mathbf{X}) \xrightarrow{a.s.} p(1-p)$$

This shows that the Bayes estimator $\delta^B(\mathbf{x})$ is consistent. Further, the estimator $\delta^B(\mathbf{x})$ is admissible since it is a unique Bayes estimator.

(iv) The joint distribution of $\mathbf{X}$ and p, in this case, is

$$\begin{aligned}h(\mathbf{x}, p) &\propto p^t(1-p)^{n-t}\frac{1}{p(1-p)}I_{(0,1)}(p)\\ &= p^{t-1}(1-p)^{n-t-1}I_{(0,1)}(p)\end{aligned}$$

Therefore, the posterior distribution of p given $T = t$,

$$\pi(p|t) \propto p^{t-1}(1-p)^{n-t-1}I_{(0,1)}(p)$$

$\pi(p|t) \sim Be(t, n-t)$ if and only if $t \in (0, n)$. Therefore, the generalized Bayes estimate is

$$\delta^{GB}(\mathbf{x}) = \frac{\int_0^1 p(1-p)p^t(1-p)^{n-t}[p(1-p)]^{-1}\,dp}{\int_0^1 p^t(1-p)^{n-t}[p(1-p)]^{-1}\,dp}$$

$$= \frac{\int_0^1 p^t(1-p)^{n-t}\,dp}{\int_0^1 p^{t-1}(1-p)^{n-t-1}\,dp} = \frac{t(n-t)}{n(n+1)}$$

***Example* 8.10** Let $X_1,\ldots,X_n$ be *iid* observations from $b(1,\theta)$, and the conjugate prior distribution of θ be Be(α, β). Consider the weighted loss function

$$L_w(\theta,\delta)=\left(1-\frac{\delta}{\theta}\right)^2=\frac{1}{\theta^2}(\theta-\delta)^2$$

Find the Bayes estimate of θ with respect to the beta prior and the modified beta prior distribution.

Solution. The modified prior is defined as

$$\begin{aligned}\pi_w(\theta)&=w(\theta)\mathrm{Be}(\alpha,\beta)\\&\propto\frac{1}{\theta^2}\,\theta^{\alpha-1}\,(1-\alpha)^{\beta-1},\ \theta\in(0,1)\\&\propto\theta^{\alpha-3}(1-\alpha)^{\beta-1}\end{aligned}$$

Thus, $\pi_w(\theta)$ is $Be(\alpha-2,\beta)$. The posterior distribution then calculates to

$$\pi_w(\theta|\mathbf{x})\propto\theta^{\Sigma x_i+\alpha-3}\,(1-\theta)^{n-\Sigma x_i+\beta-1}$$

Thus, $\pi_w(\theta|x)$ is $Be(\Sigma x_i+\alpha-2,\ n-\Sigma x_i+\beta)$. The Bayes estimator of θ under the squared error loss function is the mean of the posterior distribution

$$\theta^B=\frac{\sum x_i+\alpha-2}{n+\alpha+\beta-2}$$

The Bayes estimate of θ with respect to $Be(\alpha,\beta)$ under the weighted squared error loss function is

$$\theta^B=\frac{E[w(\theta)\theta|\,\mathbf{x}]}{E[w(\theta)|\,\mathbf{x}]}$$

where the posterior distribution is

$$\begin{aligned}\pi(\theta|\mathbf{x})&\propto\theta^{\Sigma x_i}(1-\theta)^{n-\Sigma x_i}\theta^{\alpha-1}(1-\theta)^{\beta-1}\\&=\theta^{\Sigma x_i+\alpha-1}(1-\theta)^{n-\Sigma x_i+\beta-1}\end{aligned}$$

Thus, $\pi(\theta|\mathbf{x})$ is $Be(\Sigma x_i+\alpha,\ n-\Sigma x_i+\beta)$. Thus, Bayes estimate of θ is given by

$$\theta^B=\frac{\int_0^1\theta^{\Sigma x_i+\alpha-2}(1-\theta)^{n+\Sigma x_i+\beta-1}\,d\theta}{\int_0^1\theta^{\Sigma x_i+\alpha-3}(1-\theta)^{n+\Sigma x_i+\beta-1}\,d\theta}$$

$$=\frac{B\left(\sum x_i+\alpha-1,\,n-\sum x_i+\beta\right)}{B\left(\sum x_i+\alpha-2,\,n-\sum x_i+\beta\right)}=\frac{\left(\sum x_i+\alpha-2\right)}{(n+\alpha+\beta-2)}$$

We, therefore, see that the Bayes estimate of θ with respect to the weighted prior distribution under simple squared error loss function is same as the Bayes estimate of θ obtained for unweighted prior and weighted squared error loss function. Notice that we have simple calculations when we proceeded with the weighted prior with modified hyperparameters.

***Example* 8.11** Let $X_1,\ldots,X_n$ be *iid* according to $b(1, p)$.

(i) Show that $\delta(\mathbf{X}) = (\Sigma X_i + \sqrt{n}/2)/(n + \sqrt{n})$ is a minimax estimator of p under the squared error loss function.

(ii) Show that the estimator $\delta(\mathbf{X}) = \bar{X}$ is minimax under the weighted squared loss function $L(p, \delta) = (p - \delta)^2/[p(1 - p)]$.

Solution.

(i) The estimator δ is Bayes for p with respect to the prior distribution Be(α, β), α, $\beta > 0$

$$\delta(\mathbf{X}) = \frac{\sum X_i + \alpha}{\alpha + \beta + n}$$

We have
$$R(p, \delta) = E\left(\frac{\sum X_i + \alpha}{\alpha + \beta + n} - p\right)^2 = E\left[\frac{\left(\sum X_i - np\right) + \alpha - (\alpha + \beta)p}{\alpha + \beta + n}\right]^2$$

$$= \frac{np(1-p) + [\alpha - (\alpha + \beta)p]^2}{(\alpha + \beta + n)^2}$$

However, if we choose $\alpha = \beta = \sqrt{n/4} = \sqrt{n}/2$, the estimator and its risk function take the form

$$\delta(\mathbf{X}) = \frac{\sum X_i + \frac{\sqrt{n}}{2}}{n + \sqrt{n}}$$

$$R(p, \delta) = \frac{n}{4(n + \sqrt{n})^2} = \text{constant}, \ \forall \ p$$

Since δ is Bayes with respect to $Be(\sqrt{n}/2, \sqrt{n}/2)$ and has constant risk, by Theorem 8.6.5, δ is minimax.

(ii) Consider the estimator of p under the weighted squared loss function

$$L(p, \delta) = \frac{(p - \delta)^2}{p(1 - p)}$$

and the estimator
$$\delta(\mathbf{X}) = \frac{1}{n}\sum X_i$$

Its risk is given by

$$R(p, \delta) = E\frac{[\delta(\mathbf{X}) - p]^2}{p(1 - p)}$$

$$= E\left[\frac{\left(\sum X_i - np\right)^2}{n^2 p(1-p)}\right] = \frac{np(1-p)}{n^2 p(1-p)}$$

$$= \frac{1}{n} = \text{constant}, \quad \forall\, p$$

We can show that δ is Bayes with respect to the prior distribution $\pi(\theta) = 1$. Thus, $\delta = \bar{X}$ is minimax.

***Example* 8.12** Let $X_1,\ldots,X_n$ be a random sample drawn from $b(1, \theta)$. Consider the estimation of θ with the following class of estimators:

$$C(\theta) = \{\delta(\mathbf{x}) : \delta(\mathbf{x}) = \bar{x} + \alpha,\ \alpha \in \mathbb{R}^1\}$$

Find a minimax estimator of θ in $C(\theta)$ under the squared error loss function.

Solution. The risk of some estimator δ in $C(\theta)$ is

$$R(\theta, \delta) = E_\theta(\delta - \theta)^2 = E_\theta(\bar{X} + \alpha - \theta)^2$$

$$= \frac{1}{n^2} E\left(\sum X_i - n\theta\right)^2 + \alpha^2 = \frac{1}{n}\theta(1-\theta) + \alpha^2$$

Note, that $\quad \sup\limits_\theta R(\theta, \delta) = \dfrac{1}{n}\dfrac{1}{4} + \alpha^2$

since $R(\theta, \delta)$ attains maximum at $\theta = 1/2$. The infimum of $\sup_\theta R(\theta, \delta)$ over δ is obtained by

$$\frac{\partial}{\partial \alpha} \sup_\theta R(\theta, \delta) = 2\alpha = 0$$

We get $\alpha = 0$. Therefore, $\delta = \bar{X} + \alpha = \bar{X}$ is the minimax estimator.

***Example* 8.13** Let $X \sim b(n, p)$, $p \in (0, 1)$, and the loss function be squared error, i.e., $L(p, \delta) = (p - \delta)^2$.

(i) Find the Bayes estimator of p and p^2 and their risks with respect to the noninformative prior $\pi(p) = 1$ and the Bayes estimator of p with respect to natural conjugate prior $Be(\alpha, \beta)$, known as binomial with beta prior. Show, in case of binomial with beta prior, that the Bayes estimator approaches the sample mean when n grows to infinity or $\alpha \to 0$ and $\beta \to 0$. Show that the Bayes estimate approaches the mean of the prior distribution as $\alpha \to \infty$. Also, find the predictive distribution of Y given the data X. Find the generalized Bayes estimate of p with respect to the prior $\pi(p) \propto [p(1-p)]^{-1}$; and show that the limit of Bayes by letting $\alpha \to 0$ and $\beta \to 0$ in $Be(\alpha, \beta)$ prior is the same as the generalized Bayes estimate. Find the Bayes estimate of p with respect to the Jeffreys noninformative prior. Find the empirical Bayes estimate of p when hyperparameters α and β in prior $Be(\alpha, \beta)$ are unknown. Analyze when is the Bayes estimate of p with respect to $Be(\alpha, \beta)$ prior also minimax. Define a loss function and a prior distribution with respect to which the Bayes estimator is unbiased. Is this estimator also minimax?

(ii) Consider a class of estimators $C = \{\delta\colon \delta = aX + b\}$. Find the Bayes and the minimax estimators of p in C with respect to the uniform distribution $U(0, 1)$ as a prior for p, under the weighted squared error loss function

$$L(p, \delta) = \frac{(p-\delta)^2}{p(1-p)}$$

(iii) Is the maximum likelihood estimator $\delta(X) = X/n$ Bayes for p with respect to some prior distribution under squared error loss function? Discuss the conditions when this unbiased ML estimator is also Bayes.

(iv) Show that the estimation $\delta(X) = X/n$ is generalized Bayes with respect to the measure

$$d\pi(p) = \frac{1}{p(1-p)}dp$$

(v) Consider a group of transformations defined as

$$\mathcal{G} = \{g(x) : g(x) = n - x\}$$

(a) Find the class of estimators $\mathcal{E}(p)$ which are equivariant with respect to the group $\mathcal{G}$.

(b) Consider the class of Bayes estimators of p with respect to Be(α, β) priors

$$\mathcal{B}(p) = \left\{\delta(x) : \delta(x) = \frac{1}{n+\alpha+\beta}x + \frac{\alpha}{n+\alpha+\beta}, \alpha > 0, \beta > 0\right\}$$

Find the estimators in $\mathcal{B}(p)$ which are also equivariant with respect to $\mathcal{G}$.

(c) Find the smallest MLE equivariant Bayes estimators in the class in (b).

Solution.

(i) The posterior density when $\pi(p) = 1$

$$\pi(p|x) = \frac{h(x, p)}{m(x)} = \frac{\Gamma(n+2)}{\Gamma(x+1)\Gamma(n-x+1)} p^x (1-p)^{n-x}$$

is $Be(x + 1, n - x + 1)$. Therefore, under the squared error loss function, the Bayes estimate of p is

$$\delta^B(x) = \frac{x+1}{n+2}$$

and its Bayes risk is given by

$$r(\pi, \delta^B) = E^p E_p^X (\delta^B - p)^2 = E^p E_p^X \left(\frac{X+1}{n+2} - p\right)^2$$

$$= \frac{1}{(n+2)^2} E^p E_p^X [(X - np) + (1 - 2p)]^2$$

$$= \frac{1}{(n+2)^2} E^p[np(1-p) + (1-2p)^2]$$

$$= \frac{1}{(n+2)^2} \int_0^1 \{-(n-4)p^2 + (n-4)p + 1\} dp = \frac{1}{6(n+2)}$$

Next, the Bayes estimate of p^2 is

$$\delta^B(x) = \frac{\int_0^1 p^{x+2}(1-p)^{n-x}\, dp}{\int_0^1 p^x (1-p)^{n-x}\, dp} = \frac{B(x+3, n-x+1)}{B(x+1, n-x+1)} = \frac{(x+1)(x+2)}{(x+2)(x+3)}$$

Consider now the prior $Be(\alpha, \beta)$. The posterior distribution calculates to

$$\pi(p|x) = \frac{\Gamma(n+\alpha+\beta)}{\Gamma(x+\alpha)\Gamma(n-x+\beta)} p^{x+\alpha-1}(1-p)^{n-x+\beta-1}$$

which is $Be(x + \alpha, n - x + \beta)$. The Bayes estimator of p under the squared error loss function is

$$\delta^B(x) = \frac{x+\alpha}{n+\alpha+\beta}$$

$$= \frac{n}{n+\alpha+\beta}\left(\frac{x}{n}\right) + \frac{\alpha+\beta}{n+\alpha+\beta}\left(\frac{\alpha}{\alpha+\beta}\right)$$

By letting

$$w(n, \alpha, \beta) = \frac{n}{n+\alpha+\beta}$$

we have $$\delta^B(x) = w\left(\frac{x}{n}\right) + (1-w)\left(\frac{\alpha}{\alpha+\beta}\right)$$

$$= w(\text{UMVUE of } \theta) + (1-w)(\text{prior mean})$$

Thus, the Bayes estimate of the binomial parameter p with respect to beta prior is the weighted mean of sample mean, the MLE of p, (x/n), and the prior mean $[\alpha/(\alpha + \beta)]$, where weights depend on n, α, and β. If one has large amount of data, $n \to \infty$ (as n grows), $w \to 1$; the prior distribution is discounted. Thus, $\delta^B(\mathbf{x}) \to x/n$. Further, if $\alpha \to 0$ and $\beta \to 0$, even then

$$\delta^B(\mathbf{x}) \to \frac{x}{n}$$

If $\alpha \to \infty$, then $w \to 0$, and $\delta^B(\mathbf{x}) \to [\alpha/(\alpha + \beta)]$, which is the prior mean. Note that $V(p) = \{(\alpha\beta)/[(\alpha + \beta)^2(\alpha + \beta + 1)]\} = O(1/\alpha^2)$ for fixed β. Thus, $V(p) \to 0$ as $\alpha \to \infty$. This justifies $\delta^B(\mathbf{x}) \to [\alpha/(\alpha + \beta)]$ when $\alpha \to \infty$. Further,

$$f(y|x)=\int_{\Theta} f(y|p)\pi(p|x)dp$$

$$=\int_0^1 \binom{n}{y} p^y(1-p)^{n-y}\frac{p^{\alpha+x-1}(1-p)^{n-x+\beta-1}}{B(\alpha+x,\, n-x+\beta)}dp$$

$$=\frac{\binom{n}{y}}{B(x+\alpha,\, n-x+\beta)}\int_0^1 p^{\alpha+x+y-1}(1-p)^{2n-x-y-\beta-1}dp$$

$$=\binom{n}{y}\frac{B(\alpha+x+y,\, 2n-x-y+\beta)}{B(\alpha+x,\, n-x+\beta)}$$

$$y=0,1,2,\ldots,n$$

Gentle (2008) observed that the Bayes estimate $\delta^B(\alpha, \beta) = (x + \alpha)/(\alpha + \beta + n)$ is unbiased if $\alpha = 0 = \beta$ in the beta prior. The resulting prior $\pi(p) \propto [p(1 - p)]^{-1}$ is an improper prior. For computing the generalized Bayes estimate of p, consider the minimization of

$$\int_0^1 (p-\delta)^2 f(x|p)\pi(p)dp$$

$$=\int_0^1 (p-\delta)^2\binom{n}{x}p^x(1-p)^{n-x}\frac{1}{p(1-p)}dp$$

We solve the equation

$$\frac{\partial}{\partial\delta}\int_0^1 (p-\delta)^2 p^{x-1}(1-p)^{n-x-1}dp=0$$

$$\delta^{\text{GB}}=\frac{\int_0^1 p^x(1-p)^{n-x-1}dp}{\int_0^1 p^{x-1}(1-p)^{n-x-1}dp}=\frac{B(x+1,\, n-x)}{B(x,\, n-x)}$$

$$=\frac{x}{n}$$

Thus, the estimate $\delta^{\text{GB}} = x/n$ is the generalized Bayes. Note that it is UMVUE and is the limit of Bayes since

$$\lim_{\substack{\alpha\to 0\\ \beta\to 0}}\delta^B(\alpha,\beta)=\lim_{\substack{\alpha\to 0\\ \beta\to 0}}\frac{x+\alpha}{\alpha+\beta+n}=\frac{x}{n}$$

Next consider the estimation of p with respect to Jeffreys noninformative prior. The Fisher information

$$I_X(p) = \frac{n}{p(1-p)}$$

gives the Jeffreys noninformative prior

$$\pi(p) \propto \sqrt{I_X(p)} \propto \frac{1}{p^{1/2}(1-p)^{1/2}}$$

which is $Be(1/2, 1/2)$. One may alternatively consider the flat non-informative prior as $Be(1,1)$. The distribution $Be(1/2,1/2)$ is an arcsine distribution. Consider the reparametrization $\theta = \sin^2(\eta)$. The Jeffreys prior for η is uniform distribution on the interval $[0, \pi/2]$. As equivalent to it, η is uniformly distributed on the circle $[0, 2\pi]$. Thus, the corresponding posterior distribution is given by

$$\begin{aligned}\pi(p|x) &\propto p^{x-(1/2)}(1-p)^{n-x-(1/2)} \\ &= p^{x+(1/2)-1}(1-p)^{n-x-(1/2)-1} \\ &\sim Be\left(x+\frac{1}{2}, n-x+\frac{1}{2}\right)\end{aligned}$$

Thus, the Bayes estimate of p is

$$\delta^B = \frac{x+(1/2)}{n+1}$$

This Bayes estimate is used when it is rare to get $X = 1$.

For the case when α and β in $Be(\alpha,\beta)$ prior are not known, consider the joint distribution of X and p

$$h(x, p|\alpha, \beta) = \binom{n}{x} p^x(1-p)^{n-x} \frac{\Gamma(\alpha+\beta)}{\Gamma(\alpha)\Gamma(\beta)} p^{\alpha-1}(1-p)^{\beta-1}$$

$$x = 0, 1, \ldots, n;\ 0 < p < 1;\ \alpha, \beta > 0$$

$$= \binom{n}{x} \frac{\Gamma(\alpha+\beta)}{\Gamma(\alpha)\Gamma(\beta)} p^{x+\alpha-1}(1-p)^{n-x+\beta-1}$$

The marginal density of x conditioned on α and β is

$$m(x|\alpha, \beta) = \binom{n}{x} \frac{\Gamma(\alpha+\beta)}{\Gamma(\alpha)\Gamma(\beta)} \frac{\Gamma(x+\alpha)\Gamma(n-x+\beta)}{\Gamma(x+\alpha+\beta)} \tag{8.7.11}$$

which is beta-binomial density. Viewing Eq. (8.7.11) as conditional likelihood of x given α and β, and maximizing this with respect to α and β by adopting some numerical procedure, we get the maximum conditional likelihood estimates of α and β; denote them by $\hat{\alpha}$ and $\hat{\beta}$. Now plugging-in these estimates in the prior $Be(\alpha, \beta)$, we get $Be(\hat{\alpha}, \hat{\beta})$. The corresponding Bayes estimate

$$\delta^B = \frac{\hat{\alpha} + x}{\hat{\alpha} + \hat{\beta} + n}$$

is called the empirical Bayes estimator of p. Note that the estimators $\hat{\alpha}$ and $\hat{\beta}$, in this case, cannot be obtained in the closed form. Therefore, they are obtained by some numerical procedure.

Further, the Bayes estimate

$$\delta^B(\alpha, \beta) = \frac{x + \alpha}{\alpha + \beta + n}$$

is also minimax if it is the equalizer estimate, i.e., its risk $R(p, \delta^B)$ is constant for all p. The risk is

$$R(p, \delta^B) = \frac{1}{(\alpha + \beta + n)^2}\{np(1-p) + [\alpha(1-p) - \beta p]^2\}$$

$$= \frac{1}{(\alpha + \beta + n)^2}\{p^2[(\alpha + \beta)^2 - n] + p[n - 2\alpha(\alpha + \beta)] + \alpha^2\} \qquad (8.7.12)$$

It is constant over all p if and only if $\alpha = \sqrt{n}/2 = \beta$. Thus, the Bayes estimate

$$\delta^B = \frac{n}{\sqrt{n} + n}\frac{x}{n} + \frac{\sqrt{n}}{\sqrt{n} + n}\left(\frac{1}{2}\right)$$

with respect to $Be(\sqrt{n}/2, \sqrt{n}/2)$ prior is also minimax. In fact, it pushes the estimate towards 1/2. Its minimum of the maximum expected loss follows from Eq. (8.7.12)

$$R(p, \delta^B) = \frac{n}{4(\sqrt{n} + n)^2} = \frac{1}{4(1 + \sqrt{n})^2}$$

(ii) Refer Example 8.7. When the prior distribution is uniform prior (proper), the posterior distribution is given by

$$\pi(p|x) \propto p^x(1-p)^{n-x}$$

which is $Be(x + 1, n - x + 1)$. Then, under the weighted squared error loss function, the Bayes estimate in the class C is obtained by minimizing the posterior conditional expected loss

$$\int \frac{(ax + b - p)^2}{p(1-p)}\pi(p|x)dp$$

with respect to a and b. This gives the choice of a and b such that

$$(ax + b)\int \frac{1}{p(1-p)}\pi(p|x)dp - \int \frac{p}{p(1-p)}\pi(p|x)dp = 0$$

or $$(ax+b)B(x, n-x) = B(x+1, n-x)$$

$$ax+b = \frac{x}{n}$$

$$\Rightarrow \qquad a = 1 \quad \text{and} \quad b = 0$$

Thus, the required Bayes estimator in the class C with respect to the uniform prior is given by

$$\delta^B = \frac{X}{n}$$

Further, the risk of an estimator δ in C is given by

$$R(p, \delta) = E_p \frac{(p-\delta)^2}{p(1-p)} = E_p \frac{(aX+b-p)^2}{p(1-p)}$$

$$= E_\theta \frac{[a(X-np)+b+(an-1)p]^2}{p(1-p)}$$

$$= [(an-1)^2 - a^2 n]\frac{p}{1-p} + [a^2 n + 2b(an-1)]\frac{1}{1-p} + \frac{b^2}{p(1-p)}$$

$R(p, \delta)$ is constant if and only if $b = 0$ and $an - 1 = 0$ or $a = 1/n$. Thus, the estimator $\delta(X) = X/n$ is equalizer, which is the same as δ^B. Thus, by Theorem 8.6.2, δ^B is minimax. Note that δ^B is also unbiased.

(iii) $\delta(X) = X/n$ is the MLE of p

$$E^X \delta(X) = p$$

Let $\delta(X)$ is Bayes with respect to some prior $\pi(p)$

$$\delta(X) = E(p|X)$$

The Bayes risk of $\delta(\mathbf{X})$ is

$$r(\pi, \delta(X)) = E\{p - \delta(X)\}^2 = E(p^2 - 2\delta(X)p + \delta^2(X))$$

$$= Ep^2 - 2E^X\delta(X)E^{p|X}p + E^X\delta^2(X) = Ep^2 - E^X\delta^2(X)$$

$$= Ep^2 - E^X E^{p|X} p^2 = 0$$

We also have

$$r(\pi, \delta(X)) = E^p E^{X|p}\left(\frac{X}{n} - p\right)^2$$

$$= \frac{1}{n^2} E^p E^{X|p}(X - np)^2$$

$$= \frac{1}{n} E^p p(1-p) \neq 0$$

which is a contradiction. Therefore, MLE $\delta(X) = X/n$ cannot be Bayes with respect to any prior distribution π.

An alternative solution to this problem may be given by considering the variance of the maximum likelihood estimator X/n

$$V\left(\frac{X}{n}\bigg|\, p\right) = E\left(\frac{X}{n} - p\right)^2 = \frac{p(1-p)}{n}$$

The unbiased estimator X/n is Bayes if $EV(X/n|p)$ is 0, i.e., if and only if $p = 0$ or 1. Note that these points of p are not the points of the parameter space since $\Theta = \{p: 0 < p < 1\}$. If we include these points in Θ, we have $\Theta = [0, 1]$. Define a prior $\pi(p)$ which assigns weight 1 either to $p = 0$ or $p = 1$. Thus, by Theorem 8.2.4, the unbiased ML estimator is also Bayes with respect to this extreme prior when boundary points $p = 0$ and 1 are included in Θ.

(iv) Consider the minimization of the quantity

$$\int L(p, \delta) f(x|\, p) d\pi(p)$$

$$\int_0^1 (\delta - p)^2 \binom{n}{x} p^x (1-p)^{n-x} \frac{1}{p(1-p)} dp$$

We have

$$\delta^{\text{GB}}(x) = \frac{\int_0^1 p^x (1-p)^{n-x-1} dp}{\int_0^1 p^{x-1} (1-p)^{n-x-1} dp}$$

$$\delta^{\text{GB}}(X) = \frac{X}{n}$$

$\delta^{\text{GB}}(X) = \lim_{\alpha\to 0,\, \beta\to 0}(\alpha + X)/(\alpha + \beta + n) = X/n$. Thus, the estimator $\delta^{\text{GB}}(X)$ is a limit of Bayes estimators as $\alpha \to 0, \beta \to 0$.

(v) (a) Consider the group of transformations

$$\mathcal{G} = \{g: g(x) = n - x\}$$

which induces $\bar{\mathcal{G}}$ on Θ

$$\bar{\mathcal{G}} = \{\bar{g}: \bar{g}(\theta) = 1 - \theta\}$$

The class of estimators are equivariant with respect to $\mathcal{G}$ if

$$\xi(p) = \{\delta: \delta(x) = 1 - \delta(n - x)\}$$

(b) The Bayes estimators in $\mathcal{B}(p)$ satisfy the condition in $\xi(p)$, if

$$\frac{1}{n+\alpha+\beta}x + \frac{\alpha}{n+\alpha+\beta} = 1 - \frac{1}{n+\alpha+\beta}(n-x) - \frac{\alpha}{n+\alpha+\beta}$$

$$= \frac{1}{n+\alpha+\beta} x + \frac{\beta}{n+\alpha+\beta}$$

We see that a Bayes estimate is also equivariant if the prior is symmetric, i.e., $\alpha = \beta$. Therefore, the class of all Bayes equivariant estimators are given by [prior Be(α, α)]

$$\mathcal{B}\xi(p) = \left\{ \delta : \delta(x) = \frac{1}{n+2\alpha} x + \frac{\alpha}{n+2\alpha}, \alpha > 0 \right\}$$

(c) The group of transformations $\mathcal{G}$ induces a group of transformations over Θ, that is $\bar{\mathcal{G}}$, which is transitive. Therefore, the risk of any Bayes-equivariant estimator in $\mathcal{B}\xi(p)$ is constant $\forall\ p \in (0, 1)$.

$$R(p, \delta) = E_p \left(\frac{1}{n+2\alpha} X + \frac{\alpha}{n+2\alpha} - p \right)^2 = \text{constant}$$

gives $\quad \alpha = \dfrac{\sqrt{n}}{2}$

and $\quad R(p, \delta) = \left(\dfrac{\alpha}{n+2\alpha} \right)^2 \Bigg|_{\alpha = \sqrt{n}/2} = \dfrac{1}{4(1+\sqrt{n})^2}$

Therefore, $\quad \min\limits_p R(p, \delta) = \dfrac{1}{4(1+\sqrt{n})^2}$

This gives the minimum risk Bayes equivariant estimator

$$\delta(X) = \frac{1}{\sqrt{n}(1+\sqrt{n})} X + \frac{1}{2(1+\sqrt{n})}$$

which is Bayes with respect to Be($\sqrt{n}/2$, $\sqrt{n}/2$); also, it is minimax.

***Example* 8.14** Let X follows $b(n, p)$, $0 \le p \le 1$. Find the minimax estimator of the form $aX + b$ of p under the squared error loss function. Also, comment on the risk of the MLE X/n of p (which is also UMVUE) as compared to the minimax estimator.

Solution. The given estimator $\delta = aX + b$ is an equalizer estimator if its risk is constant for all $p \in \Theta$, i.e.,

$$\begin{aligned} R(p, \delta) &= E_p(aX + b - p)^2 \\ &= \{(an-1)^2 - a^2 n\} p^2 + \{a^2 n + 2b(an-1)\} p + b^2 \\ &= \text{const. } \forall\ p \in (0, 1) \end{aligned}$$

This implies that the coefficients of p^2 and p must be zero. This gives

$$a = \frac{1}{\sqrt{n}(1+\sqrt{n})} \text{ or } \frac{1}{\sqrt{n}(\sqrt{n}-1)}$$

and

$$b = \frac{1}{2(1+\sqrt{n})} \text{ or } -\frac{1}{2(\sqrt{n}-1)}$$

The first set of values are accepted since $0 \le p \le 1$. Therefore, the equalizer estimator of p is given by

$$\delta(X) = \frac{1}{\sqrt{n}(1+\sqrt{n})} X + \frac{1}{2(1+\sqrt{n})} \tag{8.7.13}$$

We will now find a Bayes estimate of p with respect to a natural conjugate prior for p, Be(α, β), under the squared error loss function. The posterior density of p given x is

$$\pi(p|x) \propto f(x|p)\pi(p)$$
$$= p^{x+\alpha-1}(1-p)^{n-x+\beta-1}$$

This gives $$\pi(p|x) = \{\beta(x+\alpha, n-x+\beta)\}^{-1} p^{x+\alpha-1}(1-p)^{n-x+\beta-1}$$

Thus, the Bayes estimate of p under the squared error loss function is expressed by

$$\delta^B = E\{p|x\} = \frac{x+\alpha}{n+\alpha+\beta} = \frac{1}{n+\alpha+\beta} x + \frac{\alpha}{n+\alpha+\beta}$$

If we want $\delta^B(x)$ to be of the form of $\delta(x)$ in Eq. (8.7.13), we must have

$$\frac{1}{\sqrt{n}(1+\sqrt{n})} = \frac{1}{n+\alpha+\beta} \quad \text{and} \quad \frac{1}{2(1+\sqrt{n})} = \frac{\alpha}{n+\alpha+\beta}$$

On solving these equations, we get $\alpha = \sqrt{n}/2 = \beta$. We, therefore, conclude that the equalizer estimator δ in Eq. (8.7.13) is also Bayes with respect to the prior Be($\sqrt{n}/2$, $\sqrt{n}/2$), and, therefore, by Theorem 8.6.5, it is the minimax estimator of p.

The constant risk of this minimax estimator $\delta(x)$ is expressed by

$$R(p, \delta) = \frac{1}{4(1+\sqrt{n})^2} \quad \forall\, p \in \Theta$$

The risk of the MLE and UMVUE, $\delta^* = X/n$, of p is given by

$$R(p, \delta^*) = \frac{p(1-p)}{n}$$

The minimax estimator δ is preferred over MLE δ^* if and only if

$$R(\theta, \delta) < R(\theta, \delta^*)$$
$$\frac{1}{4(1+\sqrt{n})^2} < \frac{p(1-p)}{n}$$
$$\left| p - \frac{1}{2} \right| \le c_n$$
$$\frac{1}{2} - c_n \le p \le \frac{1}{2} + c_n$$

where $c_n = \left(\sqrt{1+2\sqrt{n}}\right)/2(1+\sqrt{n})$. This shows that for some fixed n, the minimax estimator is a better choice in an interval $[(1/2) - c_n, (1/2) + c_n]$ unless one has strong evidence that p is very close to either 0 or 1. We further note that

(i) $$c_n \to 0 \quad \text{as} \quad n \to \infty$$

and

(ii) $$\frac{\sup_p R(p, \delta^*)}{\sup_p R(p, \delta)} = \frac{\sup_p p(1-p)/n}{1/[4(1+\sqrt{n})^2]}$$

$$= \frac{4(1+\sqrt{n})^2}{4n} = \frac{1+2\sqrt{n}+n}{n}$$

$$= \frac{1}{n} + 2\frac{1}{\sqrt{n}} + 1 \to 1 \quad \text{as} \quad n \to \infty$$

From (i), we conclude that for large n, we do not get a point p at which we can claim that minimax estimator δ is better, and from (ii), we conclude that minimax and ML estimators are both asymptotically equivalent in terms of MSE. However, for large n, one may prefer minimax estimator since it is simple to calculate.

***Example* 8.15** Let $X_1,\ldots,X_n$ be a random sample from $b(k, p)$, $p \in (0, 1)$. The prior distribution is given by $Be(\alpha, \beta)$. Show that this prior is a conjugate prior. Find the Bayes estimator of p under the squared loss function. Use the method of moments to find an empirical Bayes estimator of p.

Solution. We have $T = \Sigma X_i \sim b(nk, p)$. The posterior distribution of p given t is $Be(t + \alpha, nk - t + \beta)$. Hence, the prior $Be(\alpha, \beta)$ is a natural conjugate prior and the corresponding Bayes estimate of p is given by

$$\delta^B(\mathbf{x}) = \frac{t+\alpha}{nk+\alpha+\beta}$$

The empirical Bayes estimator of p is obtained by estimating the hyperparameters (α, β) on the basis of data $x_1,\ldots,x_n$. We have

$$E(X_1) = E^p E^{X_1|p}(X_1) = E^p(kp)$$

$$= k\frac{\alpha}{\alpha+\beta} \tag{8.7.14}$$

$$E(X_1^2) = E^p\, E^{X_1|p}(X_1^2)$$

$$= E^p[kp(1-p) + k^2p^2]$$

$$= kE^p p + (k^2 - k)\, E^p p^2$$

$$= k\frac{\alpha}{\alpha+\beta} + (k^2-k)\left[\frac{\alpha(\alpha+1)}{(\alpha+\beta)(\alpha+\beta+1)}\right] \tag{8.7.15}$$

Using Eqs. (8.7.14) and (8.7.15), the moment equations for α and β are

$$m_1' = \frac{1}{n}\sum_{i=1}^{n} X_i = k\frac{\alpha}{\alpha+\beta} \tag{8.7.16}$$

and

$$m_2' = \frac{1}{n}\sum_{i=1}^{n} X_i^2 = k\frac{\alpha}{\alpha+\beta} + (k^2-k)\left[\frac{\alpha(\alpha+1)}{(\alpha+\beta)(\alpha+\beta+1)}\right] \tag{8.7.17}$$

On solving moment Eqs. (8.7.16) and (8.7.17), we get

$$\hat{\alpha} = \frac{m_2' - m_1' - m_1'(k-1)}{m_1'(k-1) + k(1-(m_2'/m_1'))}$$

and

$$\hat{\beta} = \frac{k\hat{\alpha}}{m_1'} - \hat{\alpha}$$

Therefore, the empirical Bayes estimator of p is given by

$$\delta^{\text{EB}}(\mathbf{X}) = \frac{T+\hat{\alpha}}{nk+\hat{\alpha}+\hat{\beta}}$$

***Example* 8.16** Consider a random experiment of counting the Bernoulli trials X, which are necessary to be conducted to get a preassigned number of successes r. The probability mean function is given by

$$f(x|p) = \binom{x-1}{r-1} p^r(1-p)^{x-r}, \; x = r, r+1, \ldots$$

This probability model is known as negative binomial distribution and is denoted by $X \sim \text{NB}(r, p)$, $p \in (0, 1)$ is unknown. Let the prior distribution of p on (0, 1) be $\pi(p) \sim Be(\alpha, \beta)$, $\alpha > 0$, $\beta > 0$ both known. Consider the estimation of $g(p) = p^{-1}$ under the weighted squared error loss function

$$L(p, \delta) = p^2(\delta - p^{-1})^2$$

and show that $(X + 1)/(r + 1)$ is an admissible estimator of p^{-1}. Also, find Jeffreys prior for p.

Solution. The joint distribution of X and p is

$$h(x, p) \propto p^r(1-p)^{x-r}p^{\alpha-1}(1-p)^{\beta-1} = p^{r+\alpha-1}(1-p)^{x-r+\beta-1}$$

Therefore, the posterior distribution of p given x is

$$\pi(p|x) \propto p^{r+\alpha-1}(1-p)^{x-r+\beta-1}$$

Thus, $p|x \sim \text{Be}(r+\alpha, x-r+\beta)$. The loss function for the estimator of p^{-1} is

$$L(p, \delta) = p^2(\delta - p^{-1})^2$$

Consider the minimization of the posterior conditional expected loss

$$\frac{\partial}{\partial \delta} E\{p^2 (\delta - p^{-1})^2 | x\} = 0$$

$$E\{p^2 2(a - p^{-1}) | x\} = 0$$

to get the Bayes estimate of p

$$\delta^B(x) = \frac{E(p|x)}{E(p^2|x)} = \frac{\dfrac{(r+\alpha)}{(x+\alpha+\beta)}}{\dfrac{(r+\alpha+1)(r+\alpha)}{(x+\alpha+\beta+1)(x+\alpha+\beta)}}$$

$$= \frac{(x+\alpha+\beta+1)}{(r+\alpha+1)}$$

The risk of the Bayes estimator $\delta^B(X)$ is given by

$$R(p, \delta^B) = E_p^X L(p, \delta^B)$$

$$= p^2 E\left(\frac{X+\alpha+\beta+1}{r+\alpha+1} - \frac{1}{p}\right)^2$$

$$= \frac{1}{(r+\alpha+1)^2}\left\{p^2 E\left(X - \frac{r}{p}\right)^2 + [p(\alpha+\beta+1) - (\alpha+1)]^2\right\}$$

$$= \frac{1}{(r+\alpha+1)^2}\{r(1-p) + [p(\alpha+\beta+1) - (\alpha+1)]^2\}$$

since

$$E(X) = \frac{r}{p}$$

$$V(X) = \frac{r(1-p)}{p^2}$$

The Bayes risk of δ^B is

$$r(\pi, \delta^B) = E^p R(p, \delta^B)$$

$$= \frac{1}{(r+\alpha+1)^2}\{r[1 - E(p)]\} + \frac{1}{(r+\alpha+1)^2} E[p(\alpha+\beta+1) - (\alpha+1)]^2$$

$$= \frac{r\beta}{(\alpha+\beta)(r+\alpha+1)^2} + \frac{\beta^2}{(\alpha+\beta)^2 (r+\alpha+1)^2} + \frac{a\beta(\alpha+\beta+1)}{(\alpha+\beta)^2 (r+\alpha+1)^2}$$

Consider, now, the risk of the given estimator

$$\delta_0(X) = \frac{X+1}{r+1}$$

and
$$R(p, \delta_0) = p^2 E\left(\frac{X+1}{r+1} - \frac{1}{p}\right)^2 = E\left(\frac{p(X+1)-(r+1)}{(r+1)}\right)^2$$

$$= \frac{1}{(r+1)^2} E(pX - r + p - 1)^2$$

$$= \frac{1}{(r+1)^2}\left[p^2 E\left(X - \frac{r}{p}\right)^2 + (p-1)^2\right]$$

$$= \frac{r(1-p) + (1-p)^2}{(r+1)^2}$$

The Bayes risk of δ_0 is, then, given by

$$r(\pi, \delta_0) = \frac{1}{(r+1)^2} E^p[r(1-p) + (1-p)^2]$$

$$= \frac{r\beta}{(\alpha+\beta)(r+1)^2} + \frac{\beta^2}{(\alpha+\beta)^2(r+1)^2} + \frac{\alpha\beta}{(\alpha+\beta+1)(\alpha+\beta)^2(r+1)^2}$$

Consider the quantity

$$r(\pi, \delta_0) - r(\pi, \delta^B)$$

$$= \frac{\alpha\beta}{\alpha+\beta}\left\{\frac{r}{\alpha}\left[\frac{1}{(r+1)^2} - \frac{1}{(r+\alpha+1)^2}\right] + \frac{\beta}{\alpha(\alpha+\beta)}\left[\frac{1}{(r+1)^2} - \frac{1}{(r+\alpha+1)^2}\right]\right.$$

$$\left. + \frac{1}{(\alpha+\beta)}\left[\frac{1}{(r+1)^2(\alpha+\beta+1)} - \frac{\alpha+\beta+1}{(r+\alpha+1)^2}\right]\right\}$$

If $\alpha \to 0$ and $\beta \to 0$, the first term in the bracket becomes

$$\frac{r}{\alpha}\left[\frac{1}{(r+1)^2} - \frac{1}{(r+\alpha+1)^2}\right] = \frac{r[\alpha + 2(r+1)]}{(r+1)^2(r+\alpha+1)^2} \to \frac{2r(r+1)}{(r+1)^4}$$

$$= \frac{2r}{(r+1)^3}$$

and if $\alpha \to 0$, $\beta \to 0$, and $\alpha/\beta \to 0$, the second term in the bracket becomes

$$\frac{\beta}{\alpha(\alpha+\beta)}\left[\frac{1}{(r+1)^2} - \frac{1}{(r+\alpha+1)^2}\right] = \frac{\beta}{\alpha+\beta}\frac{[\alpha+2(r+1)]}{(r+1)^2(r+\alpha+1)^2}$$

$$= \frac{1}{1+(\alpha/\beta)} \frac{[\alpha + 2(r+1)]}{(r+1)^2 (r+\alpha+1)^2} \to \frac{2}{(r+1)^3}$$

and if $\alpha \to 0$, $\beta \to 0$, and $(\alpha/\beta) \to 0$, the last term becomes

$$\frac{1}{(\alpha+\beta)}\left[\frac{1}{(r+1)^2(\alpha+\beta+1)} - \frac{(\alpha+\beta+1)}{(r+\alpha+1)^2}\right] \to -\frac{2}{(r+1)^2}$$

$$\therefore \quad \frac{\alpha+\beta}{\alpha\beta}[r(\pi, \delta_0) - r(\pi, \delta^B)] \to 0 \quad \text{as } \alpha \to 0, \beta \to 0, \text{and } \frac{\alpha}{\beta} \to 0$$

Consider, next, the probability of $N_\varepsilon(p_0)$ under the prior distribution $\pi(p) \equiv Be(\alpha, \beta)$

$$\pi[N_\varepsilon(p_0)] = \frac{1}{\text{Be}(\alpha, \beta)} \int_{p_0-\varepsilon}^{p_0+\varepsilon} p^{\alpha-1}(1-p)^{\beta-1}\, dp$$

As $\alpha \to 0$ and $\beta \to 0$,

$$\frac{1}{\text{Be}(\alpha, \beta)} = \frac{\Gamma(\alpha+\beta)}{\Gamma(\alpha)\Gamma(\beta)} = \frac{\alpha\beta}{\alpha+\beta} \frac{\Gamma(\alpha+\beta+1)}{\Gamma(\alpha+1)\Gamma(\beta+1)}$$

$$\therefore \quad \frac{(\alpha+\beta)}{\alpha\beta}\pi[N_\varepsilon(p_0)] \to \int_{p_0-\varepsilon}^{p_0+\varepsilon} p^{-1}(1-p)^{-1}\, dp$$

as $\alpha \to 0$ and $\beta \to 0$. This gives

$$\frac{r(\pi, \delta_0) - r(\pi, \delta^B)}{\pi[N_\varepsilon(p_0)]} \to 0 \quad \text{as } \alpha \to 0, \beta \to 0, \text{and} \left(\frac{\alpha}{\beta}\right) \to 0$$

Therefore, by Blyth's theorem, the estimator $\delta_0(X) = (X + 1)/(r + 1)$ is admissible. Consider, now, the Fisher information about p is

$$I_X(p) = E\left[-\frac{\partial^2}{\partial p^2} \log f(X|p)\right]$$

$$= E\left\{-\frac{\partial^2}{\partial p^2}\left[\log\binom{x-1}{r-1} + r\log p + (X-r)\log(1-p)\right]\right\}$$

$$= E\left\{-\frac{\partial}{\partial p}\left[\frac{r}{p} - \frac{X-r}{1-p}\right]\right\} = E\left[\frac{r}{p^2} + \frac{X-r}{(1-p)^2}\right]$$

$$= \sum_{x=r}^{\infty} \binom{x-1}{r-1} p^r (1-p)^{x-r} \left[\frac{r}{p^2} + \frac{x-r}{(1-p)^2}\right] = \frac{r}{p^2(1-p)}$$

This yields Jeffreys prior (improper)

$$\pi(\theta) \propto \frac{1}{p(1-p)^{1/2}}$$

***Example* 8.17** Let X and Y be independent with binomial distributions

$$f_{X,Y}(x, y|p_1, p_2) = \binom{n}{x}\binom{n}{y} p_1^x (1-p_1)^{n-x} p_2^y (1-p_2)^{n-y}$$

$$x = 0, 1, \ldots, n;\ y = 0, 1, \ldots, n$$

Consider the problem of estimation of $p_1 - p_2$ under the squared error loss function $L((p_1, p_2), \delta) = (p_1 - p_2 - \delta)^2$, where $|\delta| \le 1$, $p_1 \in [0, 1]$, $p_2 \in [0, 1]$.

(i) Find the Bayes estimator of $p_1 - p_2$ with respect to the prior which assigns independent $U(0, 1)$ on p_1 and p_2 respectively.

(ii) Find the minimax estimator of $p_1 - p_2$.

Solution.

(i) The joint density of X, Y and p_1, p_2 is

$$h(x, y, p_1, p_2) = \binom{n}{x}\binom{n}{y} p_1^x (1-p_1)^{n-x}\ p_2^y (1-p_2)^{n-y};\quad 0 < p_1 < 1, 0 < p_2 < 1$$

The posterior density

$$\pi(p_1, p_2|x, y) = \frac{p_1^x (1-p_1)^{n-x}\ p_2^y (1-p_2)^{n-y}}{B(x+1, n-x+1)B(y+1, n-y+1)}$$

Thus, the Bayes estimator of $p_1 - p_2$ under the squared error loss function is

$$\delta^B(x, y) = \frac{\int_0^1\int_0^1 (p_1 - p_2) p_1^x (1-p_1)^{n-x}\ p_2^y (1-p_2)^{n-y}\, dp_1 dp_2}{\int_0^1\int_0^1 p_1^x (1-p_1)^{n-x}\ p_2^y (1-p_2)^{n-y}\, dp_1 dp_2}$$

$$= \frac{B(x+2, n-x+1)B(y+1, n-y+1) - B(x+1, n-x+1)B(y+2, n-y+1)}{B(x+1, n-x+1)B(y+1, n-x+1)}$$

$$= \frac{x+1}{n+2} - \frac{y+1}{n+2} = \frac{x-y}{n+2}$$

(ii) Let us assume $p_1 = 1 - p_2 = p$(say). This gives $p_1 - p_2 = 2p - 1$. $X \sim b(n, p)$, $Y \sim b(n, 1-p)$, $n - Y \sim b(n, p)$ gives $S = X + (n - Y) = n + (X - Y) \sim b(2n, p)$. On using the result of Example 8.11, by replacing ΣX_i by S and n by $2n$, the minimax estimator of p is

$$\delta' = \frac{1}{\sqrt{2n}+1}\left(\frac{S}{\sqrt{2n}} + \frac{1}{2}\right)$$

and the minimax estimator of $p_1 - p_2 = 2p - 1$ is

$$\delta^* = \frac{\sqrt{2n}}{1+\sqrt{2n}}\frac{1}{n}(X-Y)$$

The risk of the estimator δ^* of $p_1 - p_2$ is

$$R(p_1-p_2,\delta^*) = E\left[\frac{\sqrt{2n}}{1+\sqrt{2n}}\frac{1}{n}(X-Y)-(p_1-p_2)\right]^2$$

$$= E\left\{\frac{\sqrt{2n}}{1+\sqrt{2n}}\left[\left(\frac{X}{n}-\frac{Y}{n}\right)-(p_1-p_2)\right]-\frac{1}{1+\sqrt{2n}}(p_1-p_2)\right\}^2$$

$$= \frac{2n}{(1+\sqrt{2n})^2}\frac{1}{n^2}E[(X-np_1)-(Y-np_2)]^2 + \frac{1}{(1+\sqrt{2n})^2}(p_1-p_2)^2$$

$$= \frac{1}{(1+\sqrt{2n})^2}[2p_1(1-p_1)+2p_2(1-p_2)+(p_1-p_2)^2]$$

$$\frac{\partial}{\partial p_1}R(p_1-p_2,\delta^*) = \frac{2}{(1+\sqrt{2n})^2}(1-p_1-p_2)$$

$$= \frac{\partial}{\partial p_2}R(p_1-p_2,\delta^*) = 0$$

gives $p_1 + p_2 = 1$. This shows that $R(p_1 - p_2, \delta^*)$ is maximum at p_1, p_2 so that $p_1 + p_2 = 1$.

Assume that $\delta(X, Y)$ is an estimator of $p_1 - p_2$

$$\sup_{p_1,p_2} R(p_1-p_2,\delta) \ge \sup_{\substack{p_1,p_2\\ p_1+p_2=1}} R(p_1-p_2,\delta)$$

Since the restricted space of p_1, p_2 with restriction $p_1 + p_2 = 1$ is the subspace of the unrestricted space, therefore, the maximization over the subspace may be less than or equal to the maximization over the unrestricted space. Further,

$$\sup_{\substack{p_1,p_2\\ p_1+p_2=1}} R(p_1-p_2,\delta) \ge \sup_{\substack{p_1,p_2\\ p_1+p_2=1}} R(p_1-p_2,\delta^*) = \sup_{p_1,p_2} R(p_1-p_2,\delta^*)$$

since δ^* is minimax and its risk maximizes at p_1, p_2 so that $p_1 + p_2 = 1$. This gives

$$\sup_{p_1,p_2} R(p_1-p_2,\delta) \ge \sup_{p_1,p_2} R(p_1-p_2,\delta^*)\ \forall\ \delta$$

since δ was arbitrarily chosen. Therefore,

$$\sup_{p_1,p_2} R(p_1-p_2,\delta^*) = \inf_{\delta}\sup_{p_1,p_2} R(p_1-p_2,\delta)$$

This shows that δ^* is minimax irrespective of the relationship between p_1 and p_2.

***Example* 8.18** Let $X_1,\ldots,X_n$ be a random sample from $P(\theta)$, $\theta > 0$, and a prior distribution for θ is $G_2(1, 1)$. Use the squared error loss function to find the Bayes estimator of θ and of $g(\theta) = \exp(-\theta)$ with respect to this prior distribution.

Solution. The posterior distribution of θ given $\bar{x}$ is proportional to

$$\pi(\theta|\bar{x}) \propto \exp[-(n+1)\theta]\theta^{\Sigma x_i}$$

Therefore, the Bayes estimate of θ is

$$\delta^B(\bar{x}) = \frac{\int_0^\infty \theta \exp[-(n+1)\theta]\theta^{\Sigma x_i}\, d\theta}{\int_0^\infty \exp[-(n+1)\theta]\theta^{\Sigma x_i}\, d\theta}$$

$$= \frac{1}{n+1}\left(\sum x_i + 1\right)$$

Further, the Bayes estimate of $g(\theta) = \exp(-\theta)$ is

$$\delta^B(\bar{x}) = \frac{\int_0^\infty \exp(-\theta)\exp[-(n+1)\theta]\theta^{\Sigma x_i}\, d\theta}{\int_0^\infty \exp[-(n+1)\theta]\theta^{\Sigma x_i}\, d\theta}$$

$$= \frac{\int_0^\infty \exp[-(n+2)\theta]\theta^{\Sigma x_i}\, d\theta}{\int_0^\infty \exp[-(n+1)\theta]\theta^{\Sigma x}\, d\theta}$$

$$= \left(\frac{n+1}{n+2}\right)^{n\bar{x}+1}$$

***Example* 8.19** Assume that radioactive emissions X follow a Poisson process with θ ($\theta > 0$) rate of emission per record which is unknown. Let the prior distribution for θ be $G_1(\alpha, \beta)$, $\alpha > 0$, $\beta > 0$, and the loss function be squared error.

(i) Find the Bayes estimate of θ^k, $k \geq 1$. Based on $x_1,\ldots,x_n$, the independent emissions each on unit time interval, find the Bayes and MAP estimates of θ, and show that these converge to $(1/n)\Sigma x_i$ and they all in turn converge to θ.

(ii) Show that the MLE of θ namely $\delta(X) = X$ is not Bayes.

(iii) Show that the estimator $\delta(X) = X$ is a limit of Bayes.

(iv) Show that the estimator $\delta(X) = X$ is generalized Bayes with respect to the measure $\pi(\theta) = \log\theta$.

Solution.

(i) The posterior distribution of θ given $x_1,\ldots,x_n$ is

$$\pi(\theta|\mathbf{x}) \propto \theta^{\alpha+\Sigma x_i-1}\exp\left[-\left(n+\frac{1}{\beta}\right)\theta\right] \sim G_1\left[\alpha+\sum x_i, \left(n+\frac{1}{\beta}\right)^{-1}\right]$$

The Bayes and MAP estimates of θ are

$$\delta^B(\mathbf{x}) = \left(\alpha+\sum x_i\right)\left(n+\frac{1}{\beta}\right)^{-1}$$

and

$$\delta^{\text{MAP}}(\mathbf{x}) = \left(\alpha+\sum x_i - 1\right)\left(n+\frac{1}{\beta}\right)^{-1}$$

If n tends to infinity, then δ^B and δ^{MAP} approach $\delta^{\text{ML}} = (1/n)\Sigma x_i$ and all these estimates in turn converge to θ.

(ii) Let the MLE of θ, $\delta(X) = X$, is Bayes with respect to some prior distribution π. Then,

$$\delta(X) = X = E(\theta|X)$$

and

$$E[\delta(X)] = E(X) = \theta$$

Further,

$$E[\theta\delta(X)] = E^X E[\theta\delta(X)|X] = E^X[\delta(X)E(\theta|X)] = E[\delta(X)]^2$$
$$= E^\theta E[\theta\delta(X)|\theta] = E^\theta\{\theta E[\delta(X)|\theta]\} = E^\theta\theta^2$$

Consider the Bayes risk of δ,

$$r(\pi, \delta) = E^\theta R(\theta, \delta) = E^\theta E^{X|\theta}(\theta-\delta)^2$$
$$= E^\theta(\theta^2) - 2E^\theta \theta E^{X|\theta}(\delta) + E^{X|\theta}(\delta)^2 = 0$$

Further,

$$r(\pi, \delta) = E^\theta E^{X|\theta}(\theta-\delta)^2$$
$$= E^\theta E^{X|\theta}(X-\theta)^2 = E^\theta(\theta) = \alpha\beta \neq 0$$

$r(\pi, \delta) = 0 = \alpha\beta$ is a contradiction. Therefore, if we include the point zero in the parameter space Θ, then the MLE $\delta(X) = X$ is also Bayes with respect to π; otherwise, the MLE $\delta(X) = X$ cannot be Bayes with respect to any prior distribution π.

(iii) Consider the Bayes estimator

$$\delta^B(X) = \frac{\beta}{\beta+1}(x+\alpha)$$

If $\beta \to \infty$ and $\alpha \to 0$, then $\delta^B \to \delta = X$. Therefore, the estimator $\delta(X) = X$ is a limit of the Bayes estimators.

(iv) Consider the minimization of the quantity

$$\int L(\theta, \delta) f(x|\theta)\pi(\theta)d\theta$$

where $\pi(\theta)$ is a measure so that $\pi(\theta) = 1/\theta$. We have

$$\frac{\partial}{\partial a}\int_0^\infty \frac{1}{x!}(\theta-\delta)^2\theta^{x-1}\exp(-\theta)d\theta = 0$$

gives
$$\delta^{\text{GB}}(x) = \frac{\int_0^\infty \theta^x \exp(-\theta)d\theta}{\int_0^\infty \theta^{x-1}\exp(-\theta)d\theta} = \frac{\Gamma(x+1)}{\Gamma(x)} = x$$

This show that $\delta^{\text{GB}}(x) = x$ is a generalized Bayes estimate of θ with respect to the given measure $\pi(\theta)$.

***Example* 8.20** The emission process of a radioactive substance follows Poisson distribution. Let $x_1,\ldots,x_n$ be the number of emissions counted during periods $T_1,\ldots,T_n$ (measured in seconds). These intervals T_is may be different from each other. The emission rate θ = emission counts/seconds is the parameter of the Poisson model. In general, in the Poisson model, the time interval may be length, area, or volume.

(i) The conjugate prior for the Poisson likelihoods is a family of gamma distributions. Obtain noninformative prior, expressing "complete ignorance" about θ by Jeffreys rule of thumb and show that it is the same as the scale invariant limiting prior of the conjugate prior $G_2(\alpha, \beta)$ by treating θ as a scale parameter. Find Bayes and MAP estimates under this noninformative prior and compare them with the ML estimate. Also obtain the regular Jeffreys noninformative prior for θ. Calculate Bayes estimate of θ with respect to these priors respectively and comment on them for large n.

(ii) Consider $G_1(\alpha, \beta)$ as the conjugate family for $P(\theta)$. Find the posterior distribution of θ and the Bayes estimator of θ under the squared error loss function and also its Bayes risk.

(iii) Consider $G_2(\alpha, \beta)$ as the conjugate family for $P(\theta)$. Find the Bayes estimator of θ under LinEx loss function. Also, find the MAP (maximum/a posteriori) estimate of θ.

Solution.

(i) The noninformative prior for θ by Jeffreys rule of thumb is given by

$$\pi(\theta) \propto \frac{1}{\theta} \tag{8.7.18}$$

since $\Theta = (0, \infty)$. Consider the conjugate prior $G_2(\alpha, \beta)$

$$\pi'(\theta) \propto \theta^{\alpha-1}\exp(-\beta\theta)$$

The limiting conjugate prior by letting $\alpha \to 0$ and $\beta \to 0$,

$$\pi(\theta) \propto \lim_{\substack{\alpha\to 0\\ \beta\to 0}} \theta^{\alpha-1} \exp(-\beta\theta) = \frac{1}{\theta} \tag{8.7.19}$$

which is the same as the scale invariant conjugate noninformative prior in Eq. (8.7.18) as prescribed by Jeffreys as a rule of thumb based on invariance argument. The unit of measurement of the rate parameter θ is counts per unit time-interval. We can, clearly, see that the unit of measurement of time should have no bearing on the estimation results on θ, so, that the present parameterization, that is, the rate parameter θ, can be treated as a scale parameter. The noninformative prior shown in Eq. (8.7.18) is scale invariant by treating θ as a scale parameter.

The Bayes and MAP estimates under noninformative conjugate prior (limiting) in Eq. (8.7.18) are

$$\lim_{\substack{\alpha\to 0\\ \beta\to 0}} \theta^{B} = \lim_{\substack{\alpha\to 0\\ \beta\to 0}} \left(\frac{\alpha + \sum x_i}{n+\beta}\right) = \frac{\sum x_i}{n} = \theta^{\text{ML}}$$

and

$$\lim_{\substack{\alpha\to 0\\ \beta\to 0}} \theta^{\text{MAP}} = \lim_{\substack{\alpha\to 0\\ \beta\to 0}} \left(\frac{\alpha - 1 + \sum x_i}{n+\beta}\right) = \frac{\sum x_i - 1}{n} < \theta^{\text{ML}}$$

We see that in the absence of apriori knowledge about θ, the Bayes estimate coincides with the ML estimate. However, the MAP estimate records slightly less because the noninformative prior slightly pulls-down the estimate. However, this difference disappears when the amount of data is large, i.e., n tends to infinity.

The Fisher information about θ is

$$I_{\mathbf{X}}(\theta) = \frac{n}{\theta}$$

The Jeffreys prior is, thus, given by

$$\pi(\theta) = [I_{\mathbf{X}}(\theta)]^{1/2} \propto \frac{1}{\theta^{1/2}} \tag{8.7.20}$$

It can be observed from Eq. (8.7.20) that the Jeffreys prior for $\theta^{-1/2}$ is non-normalized uniform distribution over $(0, \infty)$. Notice that the Jeffreys priors obtained in Eq. (8.7.18) and Eq. (8.7.20) are different. However, Jeffreys noninformative prior in Eq. (8.7.20) can be seen as a boundary conjugate prior since

$$\lim_{\beta\to 0} G_2\left(\frac{1}{2}, \beta\right) \propto \lim_{\beta\to 0} \theta^{-1/2} \exp(-\beta\theta) \propto \frac{1}{\theta^{1/2}}$$

(ignoring the normalizing constant) yields the Jeffreys prior as obtained in Eq. (8.7.20).

The posterior distributions $[\pi(\theta|\mathbf{x}) = \pi(\theta)L(\theta|\mathbf{x})]$ with respect to these priors are given by $G_2(\Sigma x_i, n)$ and $G_2(\Sigma x_i + 1/2, n)$ and they, in turn, yield the Bayes estimates $\Sigma x_i/n$ and $\Sigma x_i/(n + 1)$, respectively. Thus, one can use Jeffreys versions of non-informative priors interchangeably since they lead to nearly the same posterior distributions for large n and subsequently they yield the same Bayes estimates.

(ii) The Bayes estimate of θ is the function of the sufficient statistic $T = \Sigma X_i \sim P(n\theta)$. The posterior density is given by

$$\pi(\theta|t) = \frac{\theta^{t+\alpha-1}\exp\left(-\dfrac{\theta}{\beta/(n\beta+1)}\right)}{\Gamma(t+\alpha)[\beta/(n\beta+1)]^{t+\alpha}}$$

$$\sim G_1\left(t+\alpha, \frac{\beta}{n\beta+1}\right)$$

Therefore, the Bayes estimate under the squared loss function is

$$\theta^B(\mathbf{x}) = E(\theta|t) = \frac{n\beta}{n\beta+1}\left(\frac{t}{n}\right) + \frac{1}{n\beta+1}(\alpha\beta) = w\left(\frac{t}{n}\right) + (1-w)(\alpha\beta)$$

where $w = n\beta/(n\beta + 1)$. The Bayes risk is given by

$$r(\pi, \theta^B) = E^\theta E^{T|\theta}L[\theta, \theta^B]$$

$$= E^\theta E^{T|\theta}\left[w\left(\frac{T}{n}\right) + (1-w)\alpha\beta - \theta\right]^2$$

$$= E^\theta\left\{\left(\frac{w}{n}\right)^2 E^{T|\theta}(T-n\theta)^2 + [(1-w)(\theta-\alpha\beta)]^2\right\}$$

$$= \left(\frac{w}{n}\right)^2 E(n\theta) + (1-w)^2 E(\theta-\alpha\beta)^2$$

$$= \left(\frac{\beta}{n\beta+1}\right)^2 n\alpha\beta + \left(\frac{1}{n\beta+1}\right)^2 \alpha\beta^2$$

$$= \left(\frac{1}{n\beta+1}\right)^2 \alpha\beta^2(n\beta+1)$$

(iii) The prior $\pi(\theta) \sim G_2(\alpha, \beta)$ gives the posterior distribution

$$\pi(\theta|\mathbf{x}) \sim G_2(\alpha + \Sigma x_i, \beta + n)$$

We know that if $X \sim G_2(\alpha, \beta)$, the moment generating function (mgf) of $G_2(\alpha, \beta)$ is $M_X(t) = [1 - (t/\beta)]^{-\alpha/2}|t/\beta| < 1$. Using this, the mgf of the posterior distribution is given by

$$M_{\theta|\mathbf{x}}(t) = \left(1 - \frac{t}{\beta + n}\right)^{-(\alpha + \Sigma x_i)/2}$$

$$\log M_{\theta|\mathbf{x}}(-c) = -\frac{\alpha + \Sigma x_i}{2} \log\left(1 + \frac{c}{\beta + n}\right)$$

Thus, the Bayes estimate of θ under the LinEx loss function is

$$\theta^B = -\frac{1}{c} \log M_{\theta|\mathbf{x}}(-c) = \frac{\alpha + \Sigma x_i}{2c} \log\left(1 + \frac{c}{\beta + n}\right)$$

The MAP estimate of θ is given by

$$\theta^{\text{MAP}} = \frac{\alpha + \Sigma x_i - 1}{\beta + n}, \quad \alpha + \Sigma x_i \geq 1$$

Notes.

1. If $\pi(\theta) \propto 1$,
$$\pi(\theta|\mathbf{x}) \propto L(\theta|\mathbf{x})$$
then, the classical MLE of θ and MAP coincide.
2. If $\pi(\theta|\mathbf{x})$ is unimodal and symmetric, then the Bayes estimate and the MAP estimate coincide.

***Example* 8.21** Let $X_1,\ldots,X_n$ be a random sample from the Poisson model $P(\theta)$ and the conjugate prior be $G_1(\alpha, \beta)$.

(i) Find the Bayes estimate of θ^r, $r \geq 1$, under the squared error loss function. Also, find the Bayes estimate of the Poisson density at the point x_0, i.e., $P(X = x_0) = \theta^{x_0}\exp(-\theta)/x_0!$ under the squared error loss function. Show that the Bayes estimate of $P(X = x_0)$ under the Jeffreys noninformative prior, which is the limiting distribution of gamma conjugate prior distributions, is negative binomial distribution, NB(Σx_i, $n/(n + 1)$).

(ii) Treating θ as a scale parameter, find the Bayes estimate of θ under squared logarithmic loss function
$$L(\theta, \delta) = (\log \delta - \log \theta)^2$$

Solution.

(i) The posterior distribution is given by $G_1(\alpha + \Sigma x_i, \beta/(n\beta + 1))$. The Bayes estimation of $g(\theta) = \theta^r$ is

$$(\theta^r)^B = E(\theta^r|\mathbf{x})$$

$$= \frac{1}{\Gamma(\alpha + \Sigma x_i)[\beta/(n\beta + 1)]^{\alpha + \Sigma x_i}} \int_0^\infty \theta^{\alpha + r + \Sigma x_i - 1} \exp\left[-\frac{\theta}{\beta/(n\beta + 1)}\right] d\theta$$

$$= \frac{\Gamma(\alpha + r + \Sigma x_i)}{\Gamma(\alpha + \Sigma x_i)} \left(\frac{\beta}{n\beta + 1}\right)^r$$

provided that $\alpha + r + \Sigma x_i > 0$ and $n\beta + 1 > 0$. By letting $\alpha \to 0$ and $\beta \to \infty$, we may see that the Jeffreys noninformative prior, $\pi(\theta) \propto 1/\theta$, is the limiting distribution of gamma conjugate prior distributions. The corresponding Bayes estimate of θ^r in this case takes the form

$$(\theta^r)^B = \frac{1}{n^r}\frac{\Gamma(r+\Sigma x_i)}{\Gamma(\Sigma x_i)}$$

provided $\Sigma x_i > 0$. The Bayes estimate, in this case, of θ is $\Sigma x_i/n$ which is an unbiased estimate of θ, and of θ^2 is $\Sigma x_i(\Sigma x_i + 1)/n^2$, however, it is not an unbiased estimate of θ^2.

Consider, now, the estimation of the Poisson density $P(X = x_0) = \theta^{x_0}\exp(-\theta)/x_0!$. The Bayes estimate of the density at x_0 under the squared error loss function is given by

$$P_{x_0}^B = E\left(\frac{\theta^{x_0}\exp(-\theta)}{x_0!}\middle|\mathbf{x}\right)$$

$$= \frac{1}{\Gamma(\alpha+\Sigma x_i)[\beta/(n\beta+1)]^{\alpha+\Sigma x_i}x_0!}\int_0^\infty \theta^{\alpha+x_0+\Sigma x_i}\exp\left[-\frac{(n+1)\beta+1}{\beta}\theta\right]d\theta$$

$$= \frac{[n+(1/\beta)]^{\alpha+\Sigma x_i}}{x_0![(n+1)+(1/\beta)]^{\alpha+x_0+\Sigma x_i}}\frac{\Gamma(a+x_0+\Sigma x_i)}{\Gamma(\alpha+\Sigma x_i)}$$

We can obtain Jeffreys noninformative prior $\pi(\theta) \propto 1/\theta$ by limiting $\alpha \to 0$ and $\beta \to \infty$. In this case, the Bayes estimate takes the form

$$P_{x_0}^B = \frac{(x_0+\Sigma x_i-1)!}{(\Sigma x_i-1)!x_0!}\left(\frac{n}{n+1}\right)^{\Sigma x_i}\left(\frac{1}{n+1}\right)^{x_0}$$

$$= \binom{\Sigma x_i + x_0 - 1}{x_0}\left(\frac{n}{n+1}\right)^{\Sigma x_i}\left(\frac{1}{n+1}\right)^{x_0}$$

which follows negative binomial distribution, i.e., NB[Σx_i, $n/(n + 1)$]. Note that the UMVUE of the Poisson density at x_0, P is the probability mass function of $b(\Sigma x_i, 1/n)$.

(ii) The posterior distribution is given by

$$\pi(\theta|\mathbf{x}) \propto \exp(-n\theta)\theta^{\Sigma x_i}\theta^{\alpha-1}\exp\left(-\frac{\theta}{\beta}\right)$$

$$= \theta^{\Sigma x_i+\alpha-1}\exp\left[-\left(n+\frac{1}{\beta}\right)\theta\right]$$

which is $G_2[\Sigma x_i + \alpha,\ n + (1/\beta)]$. The Bayes estimate of $\log\theta$ under the given loss function is

$$(\log\theta)^B = E(\log\theta|\mathbf{x})$$

$$= \frac{[n+(1/\beta)]^{\Sigma x_i+\alpha}}{\Gamma\left(\sum x_i+\alpha\right)} \int_0^\infty \log\theta\, \theta^{\Sigma x_i+\alpha-1} \exp\left\{-\left[n+\frac{1}{\beta}\right]\theta\right\} d\theta$$

$$= D\left(\sum x_i+\alpha\right) - \log\left(n+\frac{1}{\beta}\right)$$

by using the result

$$\int_0^\infty \log\theta\, \theta^{\alpha-1}\exp(-\beta\theta)d\theta = \frac{\Gamma(\alpha)}{\beta^\alpha}[D(\alpha)-\log\beta]$$

where $$D(\alpha) = \frac{\partial}{\partial\alpha}\log\Gamma(\alpha)$$

Thus, the Bayes estimate of θ is

$$\theta^B = \exp\left\{D\left(\sum x_i+\alpha\right) - \log\left[n+\frac{1}{\beta}\right]\right\}$$

***Example* 8.22** Let X follows the geometric distribution

$$f(x|p) = p(1-p)^{x-1},\ x = 1, 2, \ldots$$

and $p \in (0, 1]$. The loss function for estimating p is given by

$$L(p, \delta) = \frac{(p-\delta)^2}{p}$$

Find the Bayes estimate of p with respect to a prior distribution $\pi(p)$. Define an estimator

$$\delta_0(x) = \begin{cases} \frac{1}{2}, & \text{if } x = 1 \\ 0, & \text{if } x > 1 \end{cases}$$

Show that δ_0 is a limit of the Bayes estimators. Define an estimator

$$\delta^*(x) = \begin{cases} \text{Arbitrary value}, & \text{if } x = 1 \\ 0, & \text{if } x > 1 \end{cases}$$

Show that δ^* is a generalized Bayes estimate.

Solution. The Bayes estimate of $g(p) = p$ with respect to some prior distribution $\pi(p)$ under the loss function $L(p, \delta) = w(p)[g(p) - \delta]^2$ with $w(p) = p^{-1}$ is given by

$$\delta^B(x) = \frac{\int_\Theta g(p)w(p)f(x|p)\pi(p)dp}{\int_\Theta w(p)f(x|p)\pi(p)dp}$$

$$= \frac{\int_0^1 p(1-p)^{x-1}\pi(p)dp}{\int_0^1 (1-p)^{x-1}\pi(p)dp} \tag{8.7.21}$$

If we consider the prior $Be(\alpha,\alpha)$, $\alpha > 0$, the Bayes estimate given by Eq. (8.7.21) becomes

$$\delta^B(x) = \frac{\int_0^1 p^\alpha(1-p)^{\alpha+x-2}dp}{\int_0^1 p^{\alpha-1}(1-p)^{\alpha+x-2}dp} = \frac{B(\alpha+1, \alpha+x-1)}{B(\alpha, \alpha+x-1)} = \frac{\alpha}{x+2\alpha-1}$$

We have

$$\lim_{\alpha\to 0}\delta^B(x) = \begin{cases}\dfrac{1}{2}, & \text{if } x = 1\\ 0, & \text{if } x > 1\end{cases}$$

$$= \delta_0(x)$$

Therefore, δ_0 is a limit of the Bayes estimates.

Next, consider the posterior risk of an estimator δ with respect to an improper prior $\pi(p) = [p^2(1-p)]^{-1}$

$$\int_0^1 L(p,\delta)f(x|p)\pi(p)dp = \int_0^1 \frac{(p-\delta)^2}{p}p(1-p)^{x-1}\frac{1}{p^2(1-p)}dp$$

$$= \int_0^1 (p-\delta)^2 p^{-2}(1-p)^{x-2}dp$$

The above integral is divergent when $x = 1$ and, consequently, the Bayes estimate is arbitrary in this case. However, the integral is convergent when $x > 1$ if and only if $\delta = 0$. Therefore, the generalized Bayes estimate, in this case, is

$$\delta^*(x) = \begin{cases}\text{Arbitrary value}, & \text{if } x = 1\\ 0, & \text{if } x > 1\end{cases}$$

***Example* 8.23** A lot contains N items out of which there are D defective items. Here, D is unknown. To estimate D, a random sample of n items is taken, out of which X defective items are found. Find a minimax estimator of D under the squared error loss function.

Solution. We have

$$P(X=k|D)=\frac{\binom{D}{k}\binom{N-D}{n-k}}{\binom{N}{n}}$$

with $$E_D(X)=n\frac{D}{N};\ V_D(X)=\frac{nD(N-n)(N-D)}{N^2(N-1)}$$

We expect that the estimator of the form $\delta(X) = aX + b$ is a reasonable estimator of D. This estimator is an equalizer estimator if

$$R(D, \delta) = \text{const.}$$

$$E_D(aX+b-D)^2 = \text{const.}$$

We see that $$E_D(aX+b-D)^2=\beta^2$$

when $$a=\frac{N}{n+\sqrt{n(N-n)/(N-1)}} \quad \text{and} \quad b=\frac{N}{2}\left(1-\frac{an}{N}\right)$$

Consider $b(N, p)$ as a prior for D; denote it by $P(D = d|p)$. Consider Be(α, β) as a prior of p; denote it by $P(p|\alpha, \beta)$. The prior *pmf* of D is

$$P(D=d)=\int_0^1 P(D=d|p)P(p|\alpha,\beta)dp$$

$$=[B(\alpha,\beta)]^{-1}\int_0^1\binom{N}{d}p^d(1-p)^{N-d}p^{\alpha-1}(1-p)^{\beta-1}dp$$

$$d=0,\ldots,N;\ \alpha>0,\ \beta>0$$

$$=[B(\alpha,\beta)]^{-1}\binom{N}{d}B(d+\alpha,N-d+\beta) \tag{8.7.22}$$

The marginal density of X is expressed by

$$P(X=k)=\frac{1}{\binom{N}{n}}[B(\alpha,\beta)]^{-1}\sum_{d=k}^{N-n+k}\frac{\binom{d}{k}\binom{N-d}{n-k}}{\binom{N}{d}}\frac{\Gamma(d+\alpha)\Gamma(N-d+\beta)}{\Gamma(N+\alpha+\beta)}$$

Thus, the posterior distribution is

$$P(D=d|k)=\frac{\binom{d}{k}\binom{N-d}{n-k}\binom{N}{d}\Gamma(d+\alpha)\Gamma(N-d+\beta)}{\sum_{d=k}^{N-n+k}\binom{d}{k}\binom{N-d}{n-k}\binom{N}{d}\Gamma(d+\alpha)\Gamma(N-d+\beta)}$$

The Bayes estimator of D under the squared error loss function is

$$\delta^B(k) = \frac{\sum_{d=k}^{N-n+k} d \binom{d}{k}\binom{N-d}{n-k}\binom{N}{d}\Gamma(d+\alpha)\Gamma(N-d+\beta)}{\sum_{d=k}^{N-n+k} \binom{d}{k}\binom{N-d}{n-k}\binom{N}{d}\Gamma(d+\alpha)\Gamma(N-d+\beta)}$$

On writing $d = (d + \alpha) - \alpha$ and using the identity

$$\binom{d}{k}\binom{N-d}{n-k}\binom{N}{d} = \binom{n}{k}\binom{N-n}{d-k}\binom{N}{n}$$

we get
$$\delta^B(k) = \frac{\sum_{d=k}^{N-n+k} (d+\alpha) \binom{n}{k}\binom{N-n}{d-k}\binom{N}{n}\Gamma(d+\alpha)\Gamma(N-d+\beta)}{\sum_{d=k}^{N-n+k} \binom{n}{k}\binom{N-n}{d-k}\binom{N}{n}\Gamma(d+\alpha)\Gamma(N-d+\beta)} - \alpha$$

Let $d - k = i$

$$\delta^B(k) = \frac{\sum_{i=0}^{N-n} \binom{N-n}{i}\Gamma(d+\alpha+1)\Gamma(N-d+\beta)}{\sum_{i=0}^{N-n} \binom{N-n}{i}\Gamma(d+\alpha)\Gamma(N-d+\beta)} - \alpha$$

$$= k\frac{\alpha+\beta+N}{\alpha+\beta+n} + \frac{\alpha(N-n)}{\alpha+\beta+n}$$

On comparing it with the equalizer estimator $\delta = aX + b$, we get

$$a = \frac{\alpha+\beta+N}{\alpha+\beta+n} \quad \text{and} \quad b = \frac{\alpha(N-n)}{\alpha+\beta+n}$$

Solving it for α and β, we get

$$\alpha = \frac{b}{a-1}, \quad \beta = \frac{N-an-b}{a-1} \tag{8.7.23}$$

Therefore, the equalizer estimator $\delta = aX + b$ is Bayes with respect to the beta–binomial distribution in Eq. (8.7.22) with α and β given in Eq. (8.7.23). Thus, by Theorem 8.3.5, δ is minimax.

$\alpha > 0$ implies $b > 0$ and $\beta > 0$ implies $N > an + b$. If $\alpha > 1$, then $N - n > \sqrt{n(N-1)/(N-1)}$ or $(N - n)(N - 1) > n$, $N^2 - nN - N > 0$ or $N - n - 1 > 0$ or $N > n + 1$. If $N = n + 1$, then $a = (n + 1)/\{n+\sqrt{[n(n+1-n)]/(n+1-1)}\} = 1$ and $b = (n + 1)/2\{1 - [n/(n + 1)]\} = 1/2$. In this case, $aX + b = X + 1/2$ is minimax if $D \sim b(N, 1/2)$. If $N = n$, $a = [n/(n + 0)] = 1$ and $b = (n/2)[1 - (n/n)] = 0$; $aX + b = X$ is minimax.

***Example* 8.24** Let $X_1,\ldots,X_n$ be *iid* according to $N(\mu, \sigma^2)$.

(i) Find the conjugate prior for σ^2 when μ is known.
(ii) Discuss prior for μ and σ^2 when both are unknown.

Solution.

(i) The likelihood function, in this case, is given by

$$L(\sigma^2|\mathbf{x}, \mu) \propto (\sigma^2)^{-n/2} \exp\left[-\frac{1}{2\sigma^2}\sum(x_i-\mu)^2\right]$$
$$= (\sigma^2)^{-n/2} \exp\left(-\frac{1}{2\sigma^2}ns_n^2\right) \tag{8.7.24}$$

where $ns_n^2 = \Sigma_{i=1}^n (x_i - \mu)^2$.

The likelihood function in Eq. (8.7.24) is proportional to inverted gamma distribution $IG_2(\alpha, \beta)$ with the *pdf*

$$f(x:\alpha,\beta) = \frac{\beta^\alpha}{\Gamma(\alpha)} x^{-(\alpha+1)} \exp\left(-\frac{\beta}{x}\right); x > 0, \alpha, \beta > 0$$

This suggests to choose inverted gamma prior for σ^2, $\text{IG}_2(n_0, \sigma_0{}^2)$, as a conjugate prior to get the posterior distribution

$$\pi(\sigma^2|\mathbf{x}, \mu, n_0, \sigma_0^2) \propto (\sigma^2)^{-n/2} \exp\left(-\frac{1}{2\sigma^2}nS_n^2\right)(\sigma^2)^{-(n_0+1)} \exp\left(-\frac{\sigma_0^2}{\sigma^2}\right)$$
$$= (\sigma^2)^{-[(n/2)+n_0+1]} \exp\left\{-\frac{(n/2)S_n^2 + \sigma_0^2}{\sigma^2}\right\}$$

which is $\text{IG}_2((n/2) + n_0, (n/2)S_n^2 + \sigma_0^2)$. Further,

$$\pi((nS_n^2 + \sigma_0^2)\sigma^2|\mathbf{x}, \mu, n_0, \sigma_0^2) = \chi^{-2}_{(n+n_0)}$$

(ii) (Walsh (2002), Introduction to Bayesian Analysis Lecturer Notes) Consider the case when μ and σ^2 are both unknown. We may define a joint prior for μ and σ^2 by a product of two priors, one for μ conditioned on variance and the other for variance

$$\pi(\mu, \sigma^2) = \pi_1(\mu|\sigma^2)\pi_2(\sigma^2)$$

Consider the prior distribution for variance $\pi_2(\sigma^2)$ as an inverted gamma distribution $\text{IG}_2(\nu_0, \sigma_0^2)$ and conditioned on the variance (i.e., assuming that the variance σ^2 is unknown) the prior distribution of μ, as a normal distribution with mean μ_0 and variance σ^2/k_0. Note that the prior variance σ^2/k_0 is scaled by the parameter k_0. We have

$$\pi_2(\sigma^2) = \text{IG}_2(\nu_0, \sigma_0^2) \quad \text{and} \quad \pi_1(\mu|\sigma^2) = N\left(\mu_0, \frac{\sigma^2}{k_0}\right)$$

The posterior distribution of (μ,σ^2) is given by

$$\pi\left(\mu,\sigma^2|\mathbf{x}\right)=c\cdot\frac{1}{(\sigma^2)^{n/2}}\exp\left(-\frac{1}{2\sigma^2}\sum(x_i-\mu)^2\right)\cdot\frac{1}{(\sigma^2)^{1/2}}$$

$$\exp\left(-\frac{k_0}{2\sigma^2}(\mu-\mu_0)^2\right)\cdot\frac{1}{(\sigma^2)^{\nu_0+1}}\exp\left(-\frac{\sigma_0^2}{\sigma_2}\right)$$

$$=\frac{c}{(\sigma^2)^{\frac{n+1}{2}+\nu_0+1}}\exp\left[-\frac{1}{2\sigma^2}\left\{\sum(x_i-\mu)^2+k_0(\mu-\mu_0)^2+2\sigma_0^2\right\}\right]$$

$$=\frac{c}{(\sigma^2)^{\frac{n+1}{2}+\nu_0+1}}\exp\left[-\frac{1}{2\sigma^2}\left\{(n+k_0)\left(\mu-\frac{n\bar{x}+k\mu_0}{nk_0}\right)^2\right.\right.$$

$$\left.\left.+\sum(x_i-\bar{x})^2+\frac{nk_0}{n+k_0}(\bar{x}-\mu_0)^2+2\sigma_0^2\right\}\right]$$

Let $\qquad \mu_n=\dfrac{n\bar{x}+k\mu_0}{nk_0}$

and $\qquad \sigma_n^2=\sum(x_i-\bar{x})^2+\dfrac{nk_0}{n+k_0}(\bar{x}-\mu_0)^2+2\sigma^2$

We have $\qquad \pi\left(\mu,\sigma^2|\mathbf{x}\right)=\dfrac{c}{(\sigma^2)^{\frac{n+1}{2}+\nu_0+1}}\exp\left[-\dfrac{1}{2\sigma^2}\{(n+k_0)(\mu-\mu_n)^2+\sigma_n^2\}\right]$

The posterior marginal for σ^2 is given by

$$\pi_2\left(\sigma^2|\mathbf{x}\right)=\frac{c_1}{(\sigma^2)^{\frac{n}{2}+\nu_0+1}}e^{-\frac{\sigma_n^2}{2\sigma^2}}\int_{-\infty}^{\infty}\frac{\sqrt{n+k_0}}{\sqrt{2\pi}\sqrt{\sigma^2}}\exp\left\{-\frac{(n+k_0)}{2\sigma^2}(\mu-\mu_n)^2\right\}d\mu$$

$$=\frac{c_1}{(\sigma^2)^{\frac{n}{2}+\nu_0+1}}e^{-\frac{\sigma_n^2}{2\sigma^2}}\sim \mathrm{IG}_2\left(\frac{n}{2}+\nu_0,\frac{\sigma_n^2}{2}\right)$$

The Bayes estimate of σ^2 is given by

$$(\sigma^2)^{\beta}=E\left(\sigma^2|\mathbf{x}\right)=\frac{\sigma_n^2/2}{(n/2)+\nu_0-1}=\frac{\sigma_n^2}{n+2\nu_0-1}$$

and $\quad V\left(\sigma^2|\mathbf{x}\right)=\dfrac{(\sigma_n^2/2)^2}{\{(n/2)+\nu_0-1\}\{(n/2)+\nu_0-2\}}=\dfrac{\sigma_n^2}{\{n+2(\nu_0-1)\}\{n+2(\nu_0-2)\}}$

The posterior marginal for μ is given by

$$\pi_1(\mu|\mathbf{x}) = c\int_0^\infty \frac{1}{(\sigma^2)^{\frac{n+1}{2}+\nu_0+1}} \exp\left[-\frac{1}{2\sigma^2}\{(n+k_0)(\mu-\mu_n)^2+\sigma_n^2\}\right] d\sigma^2$$

$$= \frac{c_2}{[(n+k_0)(\mu-\mu_n)^2+\sigma_n^2]^{\frac{n+1}{2}+\nu_0}}$$

$$= \frac{c_3}{\left[1+\frac{(n+k_0)}{\sigma_n^2}(\mu-\mu_n)^2\right]^{\frac{n+2\nu_0+1}{2}}}$$

Let $\sqrt{\frac{n+k_0}{\sigma_n^2}}(\mu-\mu_n) = \frac{t}{\sqrt{n+2\nu_0}}$

We have $\pi_1(\mu|\mathbf{x}) = \frac{c_3}{\left(1+\frac{t^2}{n+2\nu_0}\right)^{\frac{n+2\nu_0+1}{2}}}$

The *pdf* of *t*-distribution with *r* degrees of freedom is given by

$$f(t,r) = \frac{1}{\beta\left(\frac{1}{2},\frac{r}{2}\right)} \cdot \frac{1}{\sqrt{r}} \cdot \frac{1}{\left(1+\frac{t^2}{r}\right)^{\frac{r+1}{2}}}$$

with $E(T) = 0$, $V(T) = r/(r-2)$. Thus

$$\sqrt{n+2\nu_0}\sqrt{\frac{n+k_0}{\sigma_n^2}}(\mu-\mu_n) \sim t_{n+2\nu_0}$$

$E(T) = 0$ implies that the Bayes estimate of μ is given by

$$\mu^\beta = \mu_n$$

and $\quad V(T) = E(T^2) = (n+2\nu_0)\frac{n+k_0}{\sigma_n^2}E(\mu-\mu_n)^2 = \frac{n+2\nu_0}{n+2(\nu_0-1)}$

$$E(\mu-\mu_n)^2 = \frac{\sigma_n^2}{(n+k_0)(n+2(\nu_0-1))}$$

***Example* 8.25** Let $X \sim N(\mu,\sigma^2)$ and prior be $\pi(\mu) = c$, $c \neq 0$ (improper). Show that $\delta(x) = x$ is Bayes, MAP, and ML estimate of μ. Consider a sequence of priors $\{N(\eta, \tau^2)\}$ and the corresponding sequences of Bayes estimators $[\delta^B_{\tau^2}(X)]$. Show that the limit of the Bayes estimates is $\delta(x) = x$ as $\tau \to \infty$. Note that noninformative prior $\pi(\mu) = 1$ is a limiting case of the conjugate family of $N(\eta, \tau^2)$ densities as $\tau^2 \to \infty$.

Solution. The posterior distribution is

$$\pi(\mu|x) = \frac{1}{\sqrt{2\pi}\sigma}\exp\left[-\frac{(\mu - x)^2}{2\sigma^2}\right]$$
$$= f(x|\mu)$$

This gives that $\delta(x) = x$ is the Bayes, maximum a priori (MAP), and ML estimate. Consider, a sequence of priors $\{N(\eta, \tau^2)\}$ so that

$$N(\eta, \tau^2) \to \pi(\mu) = c \neq 0 \text{ as } \tau^2 \to \infty$$

and
$$\delta^B_{\tau^2} \to x \text{ as } \tau^2 \to \infty$$

We, therefore, see that uniform (improper) prior $\pi(\mu) = c \neq 0$ is a limiting distribution of a sequence of priors $\{N(\eta, \tau^2)\}$; and in this case, the Bayes estimate $\delta(x) = x$ coincides with the limit of the Bayes estimates as $\tau \to \infty$.

***Example* 8.26** Let $X \sim N(\mu,1)$, μ is unknown. Can an ML estimator $\delta(X) = X$, which is also unbiased, be Bayes with respect to some proper prior distribution under squared error loss function? Is it the limit of Bayes estimates? Examine.

Solution. Assume that the ML estimator of μ, $\delta(X) = X$, which is also unbiased, is Bayes with respect to some prior distribution $\pi(\mu)$ under the squared error loss function. This gives

$$E[\delta(X)|\mu] = \mu$$

and
$$E(\mu|X) = \delta(X) = X$$

The Bayes risk of δ is

$$r(\pi, \delta) = EE\{[\delta(X) - \mu]^2|\mu\} = EV[\delta(X)|\mu] = E(\delta^2 - 2\delta\mu + \mu^2) = E(\delta^2) - 2E(\delta\mu) + E\mu^2$$

Consider the quantities

$$E(\delta\mu) = E\mu E[\delta(X)|\mu] = E\mu^2$$
$$= E\delta(X)E(\mu|X) = E[\delta(X)]^2$$

This gives
$$r(\pi, \delta) = E[\mu\delta(X)] - 2E[\mu\delta(X)] + E[\mu\delta(X)]$$
$$= 0$$

Further,
$$r(\pi, \delta) = E[\theta - \delta(X)]^2$$
$$= E[E(X - \theta)^2|\theta]$$
$$= E(1) = 1$$

Clearly, we cannot have two values, namely 0 and 1 for the Bayes risk $r(\pi, \delta)$. This contradiction shows that the assumption that the ML estimator $\delta(X) = X$ is Bayes with respect to some prior distribution and which is also unbiased is not true. Thus, $\delta(X) = X$ is not the Bayes estimator with respect to any prior distribution.

Consider the prior

$$\mu \sim N(0, \tau^2)$$

The posterior distribution is given by

$$\pi(\mu|x) = N\left(\frac{x\tau^2}{1+\tau^2}, \frac{\tau^2}{1+\tau^2}\right)$$

The Bayes estimate of μ under the squared error loss function is

$$\delta(x) = \frac{x\tau^2}{1+\tau^2}$$

and its Bayes risk is given by

$$r(\pi, \delta) = \frac{\tau^2}{1+\tau^2}$$

By letting $\tau^2 \to \infty$, one can easily see that the given prior distribution approaches an improper flat distribution which is known as noninformative prior. Under this limiting condition, the sequence of Bayes estimators $\delta_\tau(X)$ approaches the ML estimator $\delta(X) = X$. Thus, we can conclude that though the ML estimator $\delta(X) = X$ is not Bayes with respect to any prior distribution but it is the limit of Bayes.

***Example* 8.27** Let $X \sim N(\mu, 1)$, loss function be squared error, i.e. $L(\mu,\delta) = (\mu - \delta)^2$. We know that $\delta(X) = X$ cannot be Bayes with respect to any prior (proper) distribution. Show that δ is a generalized Bayes estimate of μ.

Solution. Consider a (improper) noninformative prior

$$\pi(\mu) = 1 \ \forall \ \mu \in (-\infty, \infty)$$

The posterior distribution is

$$\pi(\mu|x) \propto \exp\left[-\frac{1}{2}(\mu - x)^2\right]$$

which is $N(x, 1)$. The Bayes estimate is

$$\delta(x) = E(\mu|x) = x$$

where the expectation is with respect to the posterior distribution $N(x, 1)$. Thus, $\delta(x) = x$ is a generalized Bayes estimate of μ.

***Example* 8.28** Let $X_1,\ldots,X_n$ be *iid* according to $N(\mu, \sigma^2)$ (a common model for measurement error), where $\sigma^2 > 0$ is known, and the prior distribution for μ be $N(\eta, \tau^2)$.

(i) Show that $\delta(\mathbf{X}) = \bar{X}$, an ML unbiased estimator of θ, is not Bayes with respect to any prior distribution of θ;

(ii) Find the Bayes estimator of μ under the squared error loss function. Is this Bayes estimator also the ML estimator. Calculate the corresponding Bayes risk. Can it be said that this Bayes estimator is trustworthy as compared to sample mean or expected prior mean η. Comment on the asymptotic behaviour of the Bayes estimator as n grows to infinity. Comment on the sensitivity of the Bayes estimate if one uses different prior by changing its mean.

(iii) Find the Bayes estimate of μ with respect to noninformative prior for μ and normal conjugate prior, $N(\eta, \tau^2)$, for μ, respectively, under LinEx loss function. Discuss conditions under which the Bayes estimator with respect to normal conjugate prior yields a shrinkage estimator that shrinks to prior mean η.

(iv) Find the Bayes estimate of μ under absolute error loss function.

(v) Show that for some constants a and b, the risk of the estimator $\delta(X) = a\bar{X} + b$ is

$$R(\mu, \delta) = a^2 \frac{\sigma^2}{n} + [b - (1-a)\mu]^2$$

(vi) Denote the Bayes estimator in (ii) by δ^B

$$\delta^B = \frac{\tau^2}{\tau^2 + (\sigma^2/n)}\bar{x} + \frac{\sigma^2/n}{\tau^2 + \sigma^2/n}\eta$$

$$= w\bar{x} + (1-w)\eta \quad \text{where } w = \frac{\tau^2}{\tau^2 + (\sigma^2/n)}$$

Show that the risk function of δ^B is

$$R(\mu, \delta^B) = w^2 \frac{\sigma^2}{n} + (1-w)^2(\mu - \eta)^2$$

(vii) Show that the Bayes risk of the estimator δ^B is

$$r(\pi, \delta^B) = (1-w)\tau^2$$

Solution.

(i) The variance of $\delta(\mathbf{X}) = \bar{X}$ is given by

$$V(\bar{X}) = E[E(\bar{X} - \mu)^2 | \mu] = \frac{\sigma^2}{n} > 0$$

Notice that the last expectation is with respect to the prior distribution $\pi(\mu)$ and that σ^2/n is independent of μ. $V(\bar{X})$ can never be zero; therefore, by Theorem 8.2.4, $\delta(\mathbf{X}) = \bar{X}$ is not a Bayes estimator of μ with respect to any prior distribution for μ.

(ii) The posterior distribution depends on the sufficient statistic $T = \bar{X}$ for μ. Therefore, we can write

$$\pi(\mu|\mathbf{x}) = \pi(\mu|\bar{x})$$

The joint density of $\bar{X}$ and μ is

$$h(\bar{x}, \mu) = c\exp\left[-\frac{1}{2\sigma^2/n}(\bar{x}-\mu)^2 - \frac{1}{2\tau^2}(\mu-\eta)^2\right]$$

$$= c\exp\left[-\frac{1}{2}\left(\mu - \frac{\frac{n\bar{x}}{\sigma^2}+\frac{\eta}{\tau^2}}{\frac{n}{\sigma^2}+\frac{1}{\tau^2}}\right)^2\left(\frac{n}{\sigma^2}+\frac{1}{\tau^2}\right)\right]$$

$$= \frac{1}{\sqrt{2\pi}\,\sigma^*}\exp\left[-\frac{1}{2}\left(\frac{\mu-\mu^*}{\sigma^*}\right)^2\right]$$

where $$\mu^* = \frac{\frac{n\bar{x}}{\sigma^2}+\frac{\eta}{\tau^2}}{\frac{n}{\sigma^2}+\frac{1}{\tau^2}} = \left(\frac{n\bar{x}}{\sigma^2}+\frac{\eta}{\tau^2}\right)\sigma^{*2}$$

$$\frac{1}{\sigma^{*2}} = \left(\frac{n}{\sigma^2}+\frac{1}{\tau^2}\right)$$

Therefore, the posterior distribution is $N(\mu^*, \sigma^{*2})$. We can adopt a different procedure for computing this posterior distribution. We have

$$E(\mu) = \eta, \quad E(\bar{X}) = E(\mu) = \eta$$

$$V(\mu) = E^{\mu}E^{\bar{X}|\mu}(\mu-\eta)^2 = E^{\mu}(\mu-\eta)^2 = \tau^2$$

$$V(\bar{X}) = E^{\mu}E^{\bar{X}|\mu}(\bar{X}-\mu+\mu-\eta)^2 = E^{\mu}\left[\frac{\sigma^2}{n}+(\mu-\eta)^2\right] = \frac{\sigma^2}{n}+\tau^2$$

$$\operatorname{cov}(\bar{X}, \mu) = E^{\mu}E^{\bar{X}|\mu}[(\bar{X}-\eta)(\mu-\eta)]$$
$$= E^{\mu}(\mu-\eta)^2 = \tau^2$$

We now calculate the joint distribution of $(\mu, \bar{X})$

$$\begin{pmatrix}\mu\\ \bar{X}\end{pmatrix} \sim N\left[\begin{pmatrix}\eta\\ \eta\end{pmatrix}, \begin{pmatrix}\tau^2 & \tau^2\\ \tau^2 & \frac{\sigma^2}{n}+\tau^2\end{pmatrix}\right]$$

From multivariate theory, if

$$\begin{pmatrix}X_1\\ X_2\end{pmatrix} \sim N\left[\begin{pmatrix}\mu_1\\ \mu_2\end{pmatrix}, \begin{pmatrix}\Sigma_{11} & \Sigma_{12}\\ \Sigma_{21} & \Sigma_{22}\end{pmatrix}\right]$$

then $$X_1|X_2 \sim N(\mu_1 + \Sigma_{12}\Sigma_{22}^{-1}(X_2-\mu_2), \Sigma_{11.2})$$

where $\Sigma_{11.2} = \Sigma_{11} - \Sigma_{12}\Sigma_{22}^{-1}\Sigma_{21}$. On using this result in the present problem, we get

$$\mu_1 + \Sigma_{12}\Sigma_{22}^{-1}(X_2 - \mu_2) = \eta + \tau^2\left(\frac{\sigma^2}{n} + \tau^2\right)^{-1}(\bar{X} - \eta)$$

$$= \eta + \frac{\tau^2}{(\sigma^2/n) + \tau^2}(\bar{X} - \eta) = \frac{1}{\rho}\left(\frac{n\bar{X}}{\sigma^2} + \frac{\eta}{\tau^2}\right)$$

On letting $\rho = (n\tau^2 + \sigma^2)/(\sigma^2\tau^2)$, and

$$\Sigma_{11.2} = \Sigma_{11} - \Sigma_{12}\Sigma_{22}^{-1}\Sigma_{21} = \tau^2 - \tau^2\left(\frac{\sigma^2}{n} + \tau^2\right)^{-1}\tau^2$$

$$= \tau^2 - \frac{\tau^4}{\left(\frac{\sigma^2}{n}\right) + \tau^2} = \frac{1}{\left(\frac{1}{\tau^2}\right) + \left[\frac{1}{(\sigma^2/n)}\right]} = \sigma^{*2}$$

and $$\mu^* = \frac{\frac{n\bar{x}}{\sigma^2} + \frac{\eta}{\tau^2}}{\frac{n}{\sigma^2} + \frac{1}{\tau^2}} = \left(\frac{n\bar{x}}{\sigma^2} + \frac{\eta}{\tau^2}\right)\sigma^{*2}$$

$$\frac{1}{\sigma^{*2}} = \left(\frac{n}{\sigma^2} + \frac{1}{\tau^2}\right)$$

Therefore, the conditional distribution of μ given $\bar{X} = \bar{x}$ is

$$\mu|\bar{x} \sim N\left[\sigma^{*2}\left(\frac{n\bar{x}}{\sigma^2} + \frac{\eta}{\tau^2}\right), \sigma^{*2}\right]$$

Thus, for squared error loss function, the Bayes estimate of μ is

$$\delta^B(\mathbf{x}) = \sigma^{*2}\left(\frac{n\bar{x}}{\sigma^2} + \frac{\eta}{\tau^2}\right) = \frac{\frac{n\bar{x}}{\sigma^2} + \frac{\eta}{\tau^2}}{\frac{\eta}{\sigma^2} + \frac{1}{\tau^2}}$$

$$= \frac{\tau^2\bar{x} + (\sigma^2/n)\eta}{\tau^2 + (\sigma^2/n)}$$

$$= \frac{\tau^2}{\tau^2 + (\sigma^2/n)}\bar{x} + \frac{(\sigma^2/n)}{\tau^2 + (\sigma^2/n)}\eta$$

$$= w(\text{MLE}) + (1 - w)(\text{prior mean})$$

where

$$w = \frac{\tau^2}{(\sigma^2/n) + \tau^2}$$

Thus, the Bayes estimate δ^B of μ is the convex linear combination of usual estimators $\bar{x}$ and the expected prior mean η where the weights are proportional to the inverse of their respective variances or the degree of confidence placed on these estimates (a measure of how much to trust these estimates). So, the Bayes estimate is a compromise between the sample mean and the expected prior mean. Note that the Bayes estimate of μ is the same even when the loss function is an absolute error loss function.

We can easily see that if we choose noninformative prior or diffuse prior ($\tau^2 \to \infty$), the prior information gets worse and worse, and the posterior variance goes to (σ^2/n) and, consequently, $w \to 1$, and $\delta^B \to \bar{x}$, the MLE of μ. Further, if we have enough data, i.e., $n \to \infty$, then also $\delta^B \to \bar{x}$ and the posterior variance goes to σ^2/n. If $\tau \to 0$, i.e., prior information is very precise, then $\delta^B \to \eta$, the expected prior mean. The Bayes risk is given by

$$r(\pi, \delta) = \sigma^{*2}$$

Note that the prior $N(\eta, \tau^2)$ is a conjugate prior since posterior is also normal. The Bayes risk of this Bayes estimator or a posterior variance

$$r(\pi, \delta^B) = E[(\delta^B - \theta)^2 | \mathbf{x}]$$

$$= \frac{1}{(1/\tau^2) + (n/\sigma^2)} < \min\left\{\frac{\sigma^2}{n}, \tau^2\right\}$$

The interpretation of this property is that Bayes estimator is trustworthy as compared to the sample mean $\bar{X}$ or expected prior mean η.

Further, note that in case of large amount of information (data) as n approaches ∞, $\delta^B(\mathbf{x}) \to \bar{x}$, δ^B moves from the prior mean to the ML estimate $\bar{x}$, and the role of the prior distribution becomes unimportant.

$[1/(\sigma^2/n)]$ is interpreted as degree of trust grows linearly with n. In other words, when we have large amount of data (as n tends to infinity), trust in $\bar{X}$ grows, we place no weight to the prior information, $w \to 1$, Bayes estimator $\delta^B \to \bar{X}$ (ML estimator), Bayes risk of δ^B or a posterior variance goes to zero, and the prior becomes irrelevant.

In order to study the sensitivity of Bayes estimate with respect to the change in prior mean, we differentiate the posterior mean with respect to the prior mean

$$\frac{\partial}{\partial \eta} E(\mu | x) = \frac{\sigma^{*2}}{\tau^2}$$

$$= \frac{\text{Posterior variance}}{\text{Prior variance}}$$

We can easily show that the predictive distribution of Y given $X_1,\ldots,X_n$ is

$$Y|\bar{x} \sim N\left[\frac{\tau^2}{(\sigma^2/n)+\tau^2}\bar{x}+\frac{(\sigma^2/n)}{(\sigma^2/n)+\tau^2}\eta,\ \sigma^2+\frac{(\sigma^2/n)\tau^2}{(\sigma^2/n)+\tau^2}\right]$$

Note. $X \sim N(\mu,\sigma^2)$, $\mu \sim N(\eta, \tau^2)$. Note that the Bayes estimate of μ, μ^B, is a weighted mean of sample and prior means. Further, $\mu^B \to \bar{x}$ as $\tau \to \infty$, that is, Bayes estimate tends towards the sample means as the prior information becomes vague and further vague. On the other hand, if the prior information is rich, i.e., $\tau^2 << \sigma^2$, Bayes estimator μ^B gives more weight to the prior mean.

(iii) The *mgf* of posterior distribution $\pi(\mu|\mathbf{x}) \equiv N(\bar{x}, (\sigma^2/n))$ under noninformative prior, $\pi(\mu) \propto$ const. is

$$M_{\mu|\mathbf{x}}(t) = \exp\left(\bar{x}t + \frac{1}{2}\frac{t^2\sigma^2}{n}\right)$$

The value of *mgf* at $t = -c$

$$M_{\mu|\mathbf{x}}(-c) = \exp\left(-c\bar{x} + \frac{1}{2}\frac{c^2\sigma^2}{n}\right)$$

or

$$\log M_{\mu|\mathbf{x}}(-c) = -c\bar{x} + \frac{1}{2}c^2\frac{\sigma^2}{n}$$

yields the Bayes estimate of μ under LinEx loss function

$$\delta^{B(1)}_{\text{LinEx}} = \bar{x} - \frac{1}{2}c\frac{\sigma^2}{n}$$

This Bayes estimate when σ^2 is not known is given by

$$\delta^{B(1)}_{\text{LinEx}} = \bar{x} - \frac{1}{2}c\frac{S_n^2}{n}$$

where $S_n^2 = n^{-1}\Sigma(x_i - \bar{x})^2$. [See Zellner (1986), Bansal (2007)].

In case of normal conjugate prior $N(\eta, \tau^2)$, the mgf of the posterior distribution $N\{[\eta(\sigma^2/n) + \bar{x}\tau^2)]/[\tau^2 + (\sigma^2/n)], [\tau^2(\sigma^2/n)]/[\tau^2 + (\sigma^2/n)]\}$ is

$$M_{\mu|\mathbf{x}}(t) = \exp\left[\sigma^{*2}\left(\frac{n\bar{x}}{\sigma^2}+\frac{\eta}{\tau^2}\right)t + \frac{1}{2}t^2\sigma^{*2}\right]$$

and its value at $t = -c$ is

$$M_{\mu|\mathbf{x}}(-c) = \exp\left[-\sigma^{*2}\left(\frac{n\bar{x}}{\sigma^2}+\frac{\eta}{\tau^2}\right)c+\frac{1}{2}c^2\sigma^{*2}\right]$$

$$\log M_{\mu|\mathbf{x}}(-c) = -c\sigma^{*2}\left(\frac{n\bar{x}}{\sigma^2}+\frac{\eta}{\tau^2}\right)+\frac{1}{2}c^2\sigma^{*2}$$

This yields the Bayes estimate of μ under the LinEx loss function $L(\mu, \delta) = \exp[c(\delta - \mu)] - c(\delta - \mu) - 1$, i.e.

$$\min_{\delta} \int L(\mu,\delta)\pi(\mu|\mathbf{x})d\mu$$

$$\int \{c\exp[c(\delta-\mu)]-c\}\pi(\mu|\mathbf{x})d\mu = 0$$

$$\exp(c\delta)\int \exp(-c\mu)\pi(\mu|\mathbf{x})d\mu = 1$$

$$\exp(c\delta)M_{\mu}(-c) = 1$$

$$c\delta + \log M_{\mu}(-c) = \log 1$$

$$\delta + \frac{1}{c}\log M_{\mu}(-c) = 0$$

$$\delta^{B(2)}_{\text{LinEx}} = -\frac{1}{c}\log M_{\mu|x}(-c)$$

$$= \frac{1}{\rho}\left(\frac{n\bar{x}}{\sigma^2}+\frac{\eta}{\tau^2}\right)-\frac{1}{2}c\frac{1}{\rho}$$

$$= \frac{1}{\rho(\sigma^2/n)}\left(\bar{x}-\frac{1}{2}c\frac{\sigma^2}{n}\right)+\frac{\eta}{\rho\tau^2}$$

$$= \frac{1}{\rho(\sigma^2/n)}\delta^{B(1)}_{\text{LinEx}}+\frac{\eta}{\rho\tau^2}$$

As $\tau^2 \to 0$, $(\sigma^2/n)/\tau^2 \to \infty$, $\rho\tau^2 = [1 + (\sigma^2/n)/\tau^2]/[(\sigma^2/n)/\tau^2] \to 1$ and $\rho(\sigma^2/n) = 1 + (\sigma^2/n)/\tau^2 \to \infty$. Under these conditions, $\delta^{B(2)}_{\text{LinEx}} \to \eta$ and we conclude that instead of using a flat noninformative, if one uses conjugate prior for μ, the Bayes estimate under LinEx loss function yields a shrinkage estimator that shrinks to prior mean η.

(iv) The Bayes estimate in absolute error loss function is the median of the posterior distribution. In this case, the estimate obtained in (i) is Bayes since posterior distribution is symmetric about its mean.

(v) We have

$$R(\mu, \delta) = E(a\bar{X} + b - \mu)^2$$
$$= E\{a(\bar{X} - \mu) + [b - (1 - a)\mu]\}^2$$
$$= a^2 \frac{\sigma^2}{n} + [b - (1 - a)\mu]^2$$

(vi) We have

$$R(\mu, \delta^B) = E[w\bar{X} + (1 - w)\eta - \mu]^2$$
$$= E\{w(\bar{X} - \mu) - (1 - w)(\mu - \eta)\}^2$$
$$= w^2 E(\bar{X} - \mu)^2 + (1 - w)^2(\mu - \eta)^2$$
$$= w^2 \frac{\sigma^2}{n} + (1 - w)^2 (\mu - \eta)^2$$

where $$w = \frac{\tau^2}{(\sigma^2/n) + \tau^2}$$

(vii) The Bayes risk of δ^B is

$$r(\pi, \delta^B) = E^{\mu} R(\mu, \delta^B)$$
$$= w^2 \frac{\sigma^2}{n} + (1 - w)^2 \tau^2$$
$$= (1 - w)\tau^2$$

***Example* 8.29** Let $X_1,\ldots,X_n$ be *iid* according to $N(\mu,1)$ and let the prior *pdf* of μ be $N(0,\tau^2)$. Let the loss function be squared error.

(i) Find the minimax estimator of μ.
(ii) Is the minimax estimator obtained in (i) also admissible?

Solution.

(i) The Bayes estimator of μ under the squared error loss function can be obtained from the results in Example 8.28(ii) by putting $\sigma^2 = 1$ and $\eta = 0$

$$\delta^B(\mathbf{x}) = \frac{n\bar{x}}{n + (1/\tau^2)} = \bar{x} \frac{n\tau^2}{n\tau^2 + 1}$$

The Bayes risk of δ is

$$r(\pi, \delta^B) = \frac{1}{n}\left(\frac{n\tau^2}{1 + n\tau^2}\right)$$

Consider a sequence of priors $\{N(0, \tau^2), \tau^2 > 0\}$ and the corresponding sequence of Bayes estimators $\{\delta_\tau(\mathbf{X})\}$, $\delta_\tau(\mathbf{X}) = \bar{X}\{(n\tau^2)/[(1 + n\tau^2)]\}$ with Bayes risk

$$r(\pi, \delta_\tau) = \frac{1}{n}\left(\frac{n\tau^2}{1+n\tau^2}\right)$$

$r(\pi, \delta_\tau) \to 1/n = r^*$ as $\tau^2 \to \infty$. Consider $\bar{X}$, the MLE of μ. Its risk is

$$R(\mu, \bar{X}) = \frac{1}{n} \ \forall \ \mu$$

or
$$\sup_\mu R(\mu, \bar{X}) = \frac{1}{n} = r^*$$

Therefore, by Theorem 8.6.2, $\bar{X}$ is minimax.

(ii) Suppose that $\bar{X}$ is not admissible. Then there exists some estimator δ^* so that

$$R(\mu, \delta^*) \le R(\mu, \bar{X}) \ \forall \ \mu$$

and
$$R(\mu, \delta^*) < R(\mu, \bar{X}) \text{ for at least one } \mu = \mu_0$$

Using the fact that the risk function $R(\mu, \delta)$ is a continuous function of μ, the second condition may be written as

$$R(\mu, \delta^*) - R(\mu, \bar{X}) < -\varepsilon \ \forall \ \mu \in (\mu_0 - \varepsilon, \mu_0 + \varepsilon)$$

Taking expectation over μ, we have

$$E^\mu R(\mu, \delta^*) - E^\mu R(\mu, \bar{X}) < -\varepsilon E^\mu(1)$$

$$r(\pi, \delta^*) - r(\pi, \bar{X}) < -\frac{\varepsilon}{\sqrt{2\pi\tau}} \int_{\mu_0-\varepsilon}^{\mu_0+\varepsilon} \exp\left(-\frac{1}{2\tau^2}\mu^2\right) d\mu \tag{8.7.25}$$

Consider now

$$r(\pi, \bar{X}) - r(\pi, \delta_{\tau^2}) = \frac{1}{n} - \frac{1}{n}\left(\frac{n\tau^2}{1+n\tau^2}\right)$$

$$= \frac{1}{n}\left(\frac{1}{1+n\tau^2}\right) \tag{8.7.26}$$

On comparing Eqs. (8.7.25) and (8.7.26), we get

$$r(\pi, \delta^*) - r(\pi, \delta_{\tau^2}) \le \frac{\varepsilon}{\sqrt{2\pi\tau}} \int_{\mu_0-\varepsilon}^{\mu_0+\varepsilon} \exp\left(-\frac{1}{2\tau^2}\mu^2\right) d\mu - \frac{1}{n}\left(\frac{1}{1+n\tau^2}\right)$$

or
$$\tau[r(\pi, \delta^*) - r(\pi, \delta_{\tau^2})] \le -\frac{\varepsilon}{\sqrt{2\pi}} \int_{\mu_0-\varepsilon}^{\mu_0+\varepsilon} \exp\left(-\frac{1}{2\tau^2}\mu^2\right) d\mu - \frac{\tau}{n}\left(\frac{1}{1+n\tau^2}\right)$$

As $\tau \to \infty$, $-(\varepsilon/\sqrt{2\pi})\int_{\mu_0-\varepsilon}^{\mu_0+\varepsilon} \exp[-(1/2\tau^2)\mu^2]d\mu \to -(\varepsilon^2/\sqrt{2\pi})$ and $(\tau/n)[1/(1+n\tau^2)] \to 0$. Therefore, as $\tau \to \infty$,

$$r(\pi, \delta^*) - r(\pi, \delta_{\tau^2}) \to -\frac{\varepsilon}{\sqrt{2\pi}\tau}$$

which shows that δ^* cannot be admissible. Therefore, the minimax $\bar{X}$ obtained in (i) is admissible.

***Example* 8.30** Let $X_1,\ldots,X_n$ be *iid* from the normal distribution $N(\mu,1)$, and let a priori *pdf* of μ be $N(0,1)$.

(i) Find the Bayes estimate of μ under the weighted square error loss function

$$L_w(\mu, \delta) = \mu(\mu - \delta)^2$$

(ii) Use the weight, $w(\mu) = \mu$ in L_w to define the modified prior distribution

$$\pi_w(\mu) = \mu\, N(0, 1)$$

Find the Bayes estimate of μ with respect to the modified prior π_w under the squared error loss function.

Also, show that the Bayes estimates of μ in (i) and (ii) are identical. Comment.

Solution.

(i) The posterior distribution is given by

$$\pi(\mu|\mathbf{x}) = N\left(\frac{n\bar{x}}{n+1}, \frac{1}{n+1}\right)$$

Thus, the Bayes estimates of μ under the weighted squared error loss function is

$$\mu^B = \frac{E[w(\mu)\cdot\mu|\mathbf{x}]}{E[w(\mu)|\mathbf{x}]} = \frac{E(\mu^2|\mathbf{x})}{E(\mu|\mathbf{x})} = \frac{[1/(n+1)] + [n\bar{x}/(n+1)]^2}{n\bar{x}/(n+1)}$$

$$= \frac{n+1+(n\bar{x})^2}{n(n+1)\bar{x}}$$

(ii) The prior $\pi(\mu)$ is $N(0, 1)$

$$\pi(\mu)\frac{1}{\sqrt{2\pi}}\exp^{(-\mu^2/2)}$$

The modified prior is, then, given by

$$\pi_w(\mu) = w(\mu)\pi(\mu)$$

$$= \frac{1}{\sqrt{2\pi}}\mu \exp^{(-\mu^2/2)}$$

Thus, the modified posterior distribution is given by

$$\pi_w(\mu|\mathbf{x}) = \frac{1}{(2\pi)^{n/2}}\exp\left[-\frac{1}{2}\sum(x_i-\mu)^2\right]\frac{1}{\sqrt{2\pi}}\mu\exp\left(-\frac{\mu^2}{2}\right)$$

$$= c\mu\exp\left[-\frac{1}{2}(n\mu^2 - 2n\bar{x}\mu + \mu^2)\right]$$

$$= c\mu\exp\left[-\frac{(n+1)}{2}\left(\mu^2 - \frac{2n\bar{x}}{n+1}\mu\right)\right]$$

$$= c_1\mu\left[-\frac{(n+1)}{2}\left(\mu - \frac{n\bar{x}}{n+1}\right)^2\right]$$

where

$$\frac{1}{c_1} = \left\{\frac{\sqrt{n+1}}{\sqrt{2\pi}}\int\mu\exp\left[-\frac{n+1}{2}\left(\mu - \frac{n\bar{x}}{n+1}\right)^2\right]d\mu\right\}\frac{\sqrt{2\pi}}{\sqrt{n+1}}$$

$$= \frac{\sqrt{2\pi}}{\sqrt{n+1}}\frac{n\bar{x}}{n+1}$$

i.e.,

$$c_1 = \frac{\sqrt{n+1}}{\sqrt{2\pi}}\frac{n+1}{n\bar{x}}$$

Thus, the Bayes estimate with respect to the modified prior π_w under the squared error loss function is given by

$$\mu^B = E(\mu|\mathbf{x}) = \int\mu\,\pi_w(\mu|\mathbf{x})d\mu$$

$$= \frac{n+1}{n\bar{x}}\frac{\sqrt{n+1}}{\sqrt{2\pi}}\int_{-\infty}^{\infty}\mu^2\exp\left[-\frac{n+1}{2}\left(\mu - \frac{n\bar{x}}{n+1}\right)^2\right]d\mu$$

$$= \left[\frac{1}{n\bar{x}} + \frac{n\bar{x}}{n+1}\right] = \frac{(n+1)+(n\bar{x})^2}{n(n+1)\bar{x}}$$

We notice that the calculation for the Bayes estimate when working with weighted prior distribution under simple squared error loss function is tedious. However, it is simple and convenient when the loss function is weighted squared error. This observation is limited only to this example. More particularly, when weight and conjugate prior are such that the prior hyperparameter can be modified through the relationship

$$w(\theta)\pi(\theta|\alpha) = \pi(\theta|\alpha')$$

in such cases, the calculation of the Bayes estimate simplifies to the calculation of the mean of the posterior distribution when the loss function is squared error since posterior distribution belongs to the same family as do the prior.

***Example* 8.31** Let $X_1,\ldots,X_n$ be *iid* random variables according to $N(\mu, \sigma^2)$. Find the Jeffreys prior for (i) μ when σ^2 is known, (ii) σ^2 when μ is known, and, (iii) μ when coefficient of variation when σ/μ is known and is equal to 1, and (iv) $\boldsymbol{\theta} = (\mu, \sigma^2)'$ when both μ and σ^2 are unknown. Also, find the posterior distribution.

Solution.

(i) Consider the case when σ^2 is known.

$$f(\mathbf{x}; \mu) \propto f(\bar{x}; \mu)$$
$$\propto \exp\left[-\frac{n(\bar{x}-\mu)^2}{2\sigma^2}\right]$$

This gives
$$\log f(\mathbf{x}; \mu) \propto -\frac{n(\bar{x}-\mu)^2}{2\sigma^2}$$

or
$$-\frac{\partial^2}{\partial\mu^2}\log f(\mathbf{x}; \mu) \propto \frac{n}{\sigma^2}$$

On taking expectation, we get

$$I_{\mathbf{X}}(\mu) \propto \frac{n}{\sigma^2}$$

Thus, the Jeffreys prior for μ is given by

$$\pi(\mu) \propto \sqrt{I_{\mathbf{X}}(\mu)} \propto \left(\frac{n}{\sigma^2}\right)^{1/2} = c$$

since σ^2 is known and constant. This prior is an improper prior. Note that the Jeffreys noninformative prior is the same as the one obtained by requiring the noninformative prior to be invariant under location transformation. Therefore, in case of μ as a location parameter, the Jeffreys prior is an un-normalized uniform distribution over the entire real line and unique translation invariant distribution on $\mathbb{R}^1$. Thus, it is a Haar measure with respect to the translation group on $\mathbb{R}^1$. The translation invariance of Jeffreys prior for a location parameter maintains that no information about the location parameter is available which is actually shown by Jeffreys prior.

The corresponding posterior distribution is given by

$$\pi(\mu|\mathbf{x}) \equiv N\left(\bar{x}, \frac{\sigma^2}{n}\right)$$

Notice, in this case, that we can accept Jeffreys prior since it yields proper posterior. The Bayes estimate of μ is given by

$$\mu^B = \bar{x} = \mu_{\text{ML}}$$

Note that complete ignorance about μ results in identical Bayes and ML estimates, $V(\mu_{ML}) = \sigma^2/n$. Further, $(1 - \alpha)$ level confidence interval and credible interval for μ are also the same, and it is $(\bar{x} \mp \sigma(z_{\alpha/2}/\sqrt{n}))$.

(ii) This is the case when μ is known. The likelihood function is given by

$$f(\mathbf{x}; \sigma^2) \propto \frac{1}{(\sigma^2)^{n/2}} \exp\left\{-\frac{\Sigma(x_i - \mu)^2}{2\sigma^2}\right\}$$

$$\log f(\mathbf{x}; \sigma^2) \propto -\frac{n}{2}\log\sigma^2 - \frac{\Sigma(x_i - \mu)^2}{2\sigma^2}$$

$$-\frac{\partial^2}{\partial(\sigma^2)^2}\log f(\mathbf{x}; \sigma^2) \propto -\frac{n}{2(\sigma^2)^2} + \frac{\Sigma(x_i - \mu)^2}{(\sigma^2)^3}$$

We know that $\Sigma(X_i - \mu)^2/\sigma^2 \sim \chi_n^2 \Rightarrow E[\Sigma(X_i - \mu)^2/\sigma^2] = n$, $E[\Sigma(X_i - \mu)^2] = n\sigma^2$. Thus, the Fisher information for σ^2 is given by

$$I_{\mathbf{X}}(\sigma^2) \propto \frac{n\sigma^2}{(\sigma^2)^3} \propto \frac{1}{(\sigma^2)^2}$$

Hence, Jeffreys prior in this case is

$$\pi(\sigma^2) \propto \sqrt{I(\sigma^2)} = \frac{1}{\sigma^2}$$

This prior for the scale parameter is unnormalized (improper) prior on $(0, \infty)$ and unique (up to a multiplicative constant) scale-invariant. Therefore, this prior is the Haar measure with respect to a multiplicative group. Scale-invariance of Jeffreys prior maintains that no information about scale is available that Jeffreys prior shows.

Note that the Jeffreys prior for $\log\sigma^2$ is uniform distribution over $\mathbb{R}^1$ and normalized.

(iii) The Fisher information is

$$I_{\mathbf{X}}(\mu) = E\left[-\frac{\partial^2}{\partial\mu^2}\log L(\mu|\mathbf{X})\right] \propto \frac{1}{\mu^2}$$

as $\sigma = \mu$. This yields Jeffreys noninformative prior for μ

$$\pi(\mu) \propto \sqrt{I_{\mathbf{X}}(\mu)} \propto \frac{1}{|\mu|},\ \mu \in \mathbb{R}^1$$

(iv) Consider the case when $\boldsymbol{\theta} = (\mu, \sigma^2)'$ is unknown. We have

$$\log f(\mathbf{x}|\theta) = \left[-\frac{n}{2}\log 2\pi\sigma^2 - \frac{1}{2\sigma^2}\sum(x_i - \mu)^2\right]$$

$$= \left[-\frac{n}{2}\log 2\pi\sigma^2 - \frac{(n-1)s_{n-1}^2 + n(\bar{x} - \mu)^2}{2\sigma^2}\right]$$

where $$s_{n-1}^2 = (n-1)^{-1}\sum(x_i - \bar{x})^2$$

Using this log-likelihood, we have

$$\frac{\partial}{\partial\mu}\log f(\mathbf{x}|\boldsymbol{\theta}) = \frac{n}{\sigma^2}(\bar{x}-\mu)$$

$$\frac{\partial^2}{\partial\mu^2}\log f(\mathbf{x}|\boldsymbol{\theta}) = -\frac{n}{\sigma^2}$$

$$\frac{\partial^2}{\partial\mu\partial\sigma^2}\log f(\mathbf{x}|\boldsymbol{\theta}) = -\frac{n(\bar{x}-\mu)}{\sigma^4}$$

$$\frac{\partial}{\partial\sigma^2}\log f(\mathbf{x}|\boldsymbol{\theta}) = -\frac{n}{2\sigma^2} + \frac{1}{2\sigma^4}[(n-1)s_{n-1}^2 + n(\bar{x}-\mu)^2]$$

$$\frac{\partial^2}{\partial(\sigma^2)^2}\log f(\mathbf{x}|\boldsymbol{\theta}) = \frac{n}{2\sigma^4} - \frac{1}{\sigma^6}[(n-1)s_{n-1}^2 + n(\bar{x}-\mu)^2]$$

Thus, the Fisher information matrix is given by

$$\mathbf{I}_{\mathbf{X}}(\mu, \sigma^2) = E\begin{bmatrix} \dfrac{n}{\sigma^2} & 0 \\ 0 & \dfrac{n}{2\sigma^4} \end{bmatrix}$$

$$[\det \mathbf{I}_{\mathbf{X}}(\boldsymbol{\theta})] = \frac{n^2}{2(\sigma^2)^3}$$

The Jeffreys noninformative prior for $\boldsymbol{\theta} = (\mu, \sigma^2)'$ is

$$\pi(\boldsymbol{\theta}) \propto [\det \mathbf{I}_{\mathbf{X}}(\boldsymbol{\theta})]^{1/2} \propto \left(\frac{1}{\sigma^2}\right)^{3/2} \tag{8.7.27}$$

Next, the noninformative prior for location parameter μ, which is invariant under translation transformation, is

$$\pi(\mu) \propto c$$

and for scale parameter σ^2, which is scale invariant under scale transformation, is

$$\pi(\sigma^2) \propto \frac{1}{\sigma^2}$$

Assuming the independence of μ and σ, we have

$$\pi(\boldsymbol{\theta}) \propto \frac{1}{\sigma^2} \tag{8.7.28}$$

Jeffreys (1961) has recommended to use Eq. (8.7.28) as a noninformative prior for a location-scale problem since it is the right invariant Haar density. Jeffreys (1961) observed that in some cases, the noninformative prior in Eq. (8.7.28) is superior to that in Eq. (8.7.27).

Therefore, the posterior distribution in case of Jeffreys prior is given by

$$\pi(\mu, \sigma^2|\mathbf{x}) \propto (\sigma^2)^{-(3/2)}(\sigma^2)^{-n/2} \exp\left[-\frac{(n-1)s_{n-1}^2 + n(\bar{x}-\mu)^2}{2\sigma^2}\right]$$

$$\propto \left(\frac{1}{\sigma^2/n}\right)^{1/2} \exp\left[-\frac{(\mu-\bar{x})^2}{2\sigma^2/n}\right]\frac{1}{(\sigma^2)^{(n/2)+1}}\exp\left[-\frac{(n-1)s_{n-1}^2}{2\sigma^2}\right]$$

$$= N\left(\bar{x}, \frac{\sigma^2}{n}\right)\text{IG}_2\left[\frac{n}{2}, \frac{(n-1)s_{n-1}^2}{2}\right]$$

$$= \pi(\mu|\sigma^2, \mathbf{x})\,\pi(\sigma^2|\mathbf{x})$$

Sometimes, ignorance is expressed by some other priors, which are known as default or reference priors. These priors too satisfy the invariance principle as do the Jeffreys prior. If one chooses one such prior as

$$\pi(\mu, \sigma^2) = \frac{1}{\sigma^2} \text{(a default choice)}$$

then the corresponding posterior distribution is given by

$$\pi(\mu, \sigma^2|\mathbf{x}) = \pi(\mu|\sigma^2, \mathbf{x})\,\pi(\sigma^2|\mathbf{x})$$

$$= N\left(\bar{x}, \frac{\sigma^2}{n}\right)\text{IG}_2\left(\frac{n-1}{2}, \frac{(n-1)s_{n-1}^2}{2}\right)$$

The Bayes estimate of μ using this posterior is identical to the ML estimate. Also, $(1 - \alpha)$ level credible and confidence intervals for μ are identical, that is, intervals are shown by $(\bar{x} \mp s_{n-1}\, t_{n-1,\alpha}/\sqrt{(n)})$.

***Example* 8.32** Let $X_1,\ldots,X_n$ be a random sample from $N(\mu,\sigma^2)$, where $\mu \in \mathbb{R}^1$ be unknown and $\sigma^2 > 0$ be known. Can the unbiased estimator $\bar{X}$ be Bayes with respect to some prior distribution. The prior distribution for μ, $\pi_{\tau^2}(\mu)$, is $N(\eta, \tau^2)$ where η and τ^2 are known. Show that $\bar{X}$ is a minimax estimator of μ and remains minimax even when $\sigma^2 > 0$ is unknown.

Solution. The estimator $\bar{X}$ is unbiased for μ and

$$R(\mu, \bar{X}) = E_\mu^{\bar{X}}(\bar{X}-\mu)^2 = \frac{\sigma^2}{n}$$

The Bayes risk of $\bar{X}$ with respect to some prior distribution π is

$$r(\pi, \bar{X}) = E^{\mu} R(\mu, \bar{X}) = E^{\mu}\left(\frac{\sigma^2}{n}\right) = \frac{\sigma^2}{n} > 0$$

Thus, by Theorem 8.2.4, $\bar{X}$ cannot be Bayes with respect to any prior distribution π.

Next, consider the sequence of the Bayes estimators of μ as $\{\delta_{\tau^2}(\mathbf{X})\}$ corresponding to the sequence of priors $\{\pi_{\tau^2}(\mu)\}$, where

$$\delta_{\tau^2}(\mathbf{X}) = \frac{\sigma^2}{n\tau^2 + \sigma^2}\eta + \frac{n\tau^2}{n\tau^2 + \sigma^2}\bar{X}$$

with the Bayes risk

$$r_{\tau^2} = r(\pi_{\tau^2}, \delta_{\tau^2}) = \frac{\tau^2\sigma^2}{n\tau^2 + \sigma^2}, \tau^2 > 0$$

Consider $$\lim_{\tau^2 \to \infty} \inf r(\pi_{\tau^2}, \delta_{\tau^2}) = \frac{\sigma^2}{n}$$

Consider the estimator $\bar{X}$

$$R(\mu, \bar{X}) = EL(\mu, \bar{X}) = E(\mu - \bar{X})^2 = \frac{\sigma^2}{n} \;\; \forall\, \mu$$

It is an equalizer estimator of μ and satisfies the condition

$$\lim_{\tau^2 \to \infty} \inf r(\pi_{\tau^2}, \delta_{\tau^2}) \ge R(\mu, \bar{X}) = \frac{\sigma^2}{n} \quad \forall \mu \in \mathbb{R}^1$$

Therefore, by Theorem 8.6.6, $\bar{X}$ is minimax.

Consider an alternative way of showing that $\bar{X}$ is minimax. Let δ be an estimator of μ

$$\begin{aligned}
\sup_{\mu \in \mathbb{R}^1} R(\mu, \delta) &\ge \int_{\mu \in \mathbb{R}^1} R(\mu, \delta)\pi_{\tau^2}(\mu)d\mu \\
&\ge \int_{\mathbb{R}^1} R(\mu, \delta_{\tau^2})\pi_{\tau^2}(\mu)d\mu = r(\pi_{\tau^2}, \delta_{\tau^2}) \\
&= E^{\mu} R(\mu, \delta_{\tau^2}) = E^{\mu} E^{\mathbf{X}|\mu}(\mu - \delta_{\tau^2})^2 \\
&= E^{\mathbf{X}} E^{\mu|\mathbf{X}}(\mu - \delta_{\tau^2})^2 \\
&= E^{\mathbf{X}} \frac{\tau^2\sigma^2}{n\tau^2 + \sigma^2} = \frac{\tau^2\sigma^2}{n\tau^2 + \sigma^2}
\end{aligned}$$

$$\left\{\text{since } \mu|\mathbf{x} \text{ is } N\left(\delta_{\tau^2}(\mathbf{x}), \frac{(\tau^2\sigma^2)}{(n\tau^2 + \sigma^2)}\right)\right\}$$

$$\to \frac{\sigma^2}{n} \quad \text{as } \tau^2 \to \infty$$

$$= \sup_{\mu \in \mathbb{R}^1} R(\mu, \bar{X}) \qquad \left[\text{since } \bar{X} \sim N(\mu, (\sigma^2/n))\right]$$

This shows that $\bar{X}$ is minimax.

If σ^2 is fixed at σ_0^2(say), we have $\Theta_0 = \{(\mu, \sigma_0^2): \mu \in \mathbb{R}^1\}$ and as in the above calculations, we have already shown that $\bar{X}$ is a minimax estimator of μ. Now consider

$$\Theta = \{(\mu, \sigma^2): \mu \in \mathbb{R}^1, 0 < \sigma^2 \le \sigma_0^2, \sigma_0^2 > 0\}$$

We have

$$\sup_{\theta \in \Theta_0} R(\theta, \bar{X}) = \frac{\sigma^2}{n} = \sup_{\theta \in \Theta} R(\theta, \bar{X})$$

where $\boldsymbol{\theta} = (\mu, \sigma^2)'$. Therefore, by Theorem 8.6.8, $\bar{X}$ is a minimax estimator of μ when the parameter space is Θ.

***Example* 8.33** Let $X_1,\ldots,X_n$ be not independent observations drawn from normal distribution. They have a common unknown mean μ but known variance-covariance matrix Σ. The joint distribution of $X_1,\ldots,X_n$ is given by

$$f(\mathbf{x}|\mu) = \left(\frac{1}{(2\pi)^n \det(\Sigma)}\right)^{1/2} \exp\left[-\frac{1}{2}(\mathbf{x} - \mu\mathbf{1})'\Sigma^{-1}(\mathbf{x} - \mu\mathbf{1})\right]$$

where $\mathbf{1} = (1,\ldots,1)'$. Consider the prior distribution $\pi(\mu) \sim N(\eta, \tau^2)$. Find the Bayes estimate of μ if

(i) $\Sigma = \sigma^2 I$.

(ii) the observations are independent but each have different variances, i.e.,

$$\Sigma = \text{diag}\{\sigma_1^2, \sigma_2^2,\ldots,\sigma_n^2\}$$

(iii) the observations are not independent and consider $n = 2$ (bivariate case) for the simplicity of results.

Solution. After regular calculations, we get the Bayes estimate

$$\delta^B(\mathbf{x}) = \left(\mathbf{1}'\Sigma^{-1}\mathbf{1} + \frac{1}{\tau^2}\right)^{-1}\left(\mathbf{x}'\Sigma^{-1}\mathbf{1} + \frac{\eta}{\tau^2}\right)$$

Assume $\mathbf{w} = \Sigma^{-1}\mathbf{1}$. The ith element in $\mathbf{w}$ is the sum of the ith row of Σ^{-1}. The Bayes estimate in the changed notations is given by

$$\delta^B(\mathbf{x}) = \frac{\mathbf{x}'\mathbf{w} + (\eta/\tau^2)}{\mathbf{1}'\mathbf{w} + (1/\tau^2)} = \frac{\sum_{i=1}^{n} w_i x_i + (\eta/\tau^2)}{\sum_{i=1}^{n} w_i + (1/\tau^2)}$$

This shows that the Bayes estimate is a weighted average of observations and prior expected mean η.

(i) We have $\Sigma^{-1} = (1/\sigma^2)I$ and $w_i = (1/\sigma^2)$

$$\delta^B = \left(\frac{\bar{x}}{\sigma^2/n} + \frac{\eta}{\tau^2}\right) \Big/ \left(\frac{1}{\sigma^2/n} + \frac{1}{\tau^2}\right)$$

We have studied this case in Example 8.4.28.

(ii) In this case,

$$\Sigma^{-1} = \text{diag}\left(\frac{1}{\sigma_1^2}, \frac{1}{\sigma_2^2}, \ldots, \frac{1}{\sigma_n^2}\right)$$

$w_i = 1/\sigma_i^2$. The Bayes estimate of μ reduces to

$$\delta^B = \frac{\sum_{i=1}^{n}(x_i/\sigma_i^2) + (\eta/\tau^2)}{\sum_{i=1}^{n}(1/\sigma_i^2) + (1/\tau^2)}$$

Note that the weights are inversely proportional to their corresponding observations, which is intuitively justified.

(iii) The variance-covariance matrix can be written as

$$\Sigma^{-1} = \frac{1}{1-\rho^2}\begin{bmatrix} \frac{1}{\sigma_1^2} & -\frac{\rho}{\sigma_1\sigma_2} \\ -\frac{\rho}{\sigma_1\sigma_2} & \frac{1}{\sigma_2^2} \end{bmatrix}$$

where ρ is the correlation coefficient between X_1 and X_2. This gives

$$w_i = \frac{1}{1-\rho^2}\left(\frac{1}{\sigma_i^2} - \frac{\rho}{\sigma_1\sigma_2}\right), i = 1, 2$$

Note that if $\rho = 0$, the weights are $w_i = 1/\sigma_i^2$ as in (ii). Further, if $\rho \neq 0$, let us assume that $\sigma_1^2 = \sigma_2^2 = \sigma^2$. The Bayes estimate simplifies to

$$\mathbf{x}'\mathbf{w} = \sum \frac{1}{1-\rho^2}\left(\frac{1}{\sigma^2} - \frac{\rho}{\sigma^2}\right)x_i$$

$$= \sum \frac{1}{(1+\rho)}\frac{1}{\sigma^2}x_i$$

$$= \frac{x_1 + x_2}{\sigma^2(1+\rho)}$$

We get
$$\delta^B(\mathbf{x}) = \left[\frac{x_1 + x_2}{\sigma^2(1+\rho)} + \frac{\eta}{\tau^2}\right] \Big/ \left[\frac{2}{\sigma^2(1+\rho)} + \frac{1}{\tau^2}\right]$$

***Example* 8.34** (Normal with normal prior) Consider the linear model

$$X_i = \theta + e_i,\ i = 1, 2, \ldots, n$$

where
$$\theta \sim N(\mu, \sigma^2)$$

and $e_i \sim N(0, 1)$ are *iid* and e_is and θ are independent. Then,

$$X_i|\theta \sim N(\theta, 1)$$

Find the Bayes estimate of θ. Comment on this estimator when n is large.

Solution. The posterior distribution is given by

$$\theta|\mathbf{x} \sim N\left(\frac{n\sigma^2}{n\sigma^2+1}\bar{x}, \frac{\sigma^2}{n\sigma^2+1}\right)$$

The Bayes estimate of θ is

$$\theta^B = \frac{n\sigma^2}{n\sigma^2+1}\bar{x} \to \bar{x} \quad \text{as } n \to \infty$$

If one has large data, i.e., n large, the prior information is discredited and sample average is the Bayes estimate.

***Example* 8.35** Let $X_1, \ldots, X_n$ be *iid* according to a random sample from $N(0, \sigma^2)$ with scale parameter σ^2 which is unknown. Suppose that the prior distribution of σ^2 is Jeffreys noninformative prior $\pi(\sigma^2) \propto 1/\sigma^2$. Find the Bayes estimate of σ^2 with respect to this prior under the following loss functions:

(i) Squared error loss function

$$L(\sigma^2, \delta) = (\delta - \sigma^2)^2$$

(ii) Squared logarithmic loss function

$$L(\sigma^2, \delta) = (\log\delta - \log\sigma^2)^2$$

(iii) Scale invariant loss function

$$L(\sigma^2, \delta) = \left(1 - \frac{\delta}{\sigma^2}\right)^2$$

(iv) Modified LinEx loss function

Solution. The posterior distribution of σ^2 given $\mathbf{x}$ is

$$\pi(\sigma^2|\mathbf{x}) \propto \frac{1}{(\sigma^2)^{(n/2)+1}} \exp\left(-\frac{\Sigma x_i^2}{2\sigma^2}\right)$$

which is inverted gamma $IG_2(n/2, \Sigma x_i^2/2)$. If $X \sim IG_2(a, b)$, then $E(X) = b/(a-1)$, $a > 1$, and $V(X) = b^2/[(a-1)^2(a-2)]$, $a > 2$.

(i) The Bayes estimate of σ^2 with respect to Jeffreys prior under the squared error loss function is

$$(\sigma^2)^B = E(\sigma^2|\mathbf{x}) = \frac{\Sigma x_i^2/2}{(n/2)-1} = \frac{\Sigma x_i^2}{(n-2)}$$

(ii) Under the squared logarithmic loss function, the Bayes estimate of $\log \sigma^2$ is

$$(\log\sigma^2)^B = E(\log\sigma^2)$$

where the expectation is taken with respect to the posterior distribution $IG_2(n/2, \Sigma x_i^2/2)$. This gives

$$(\log\sigma^2)^B = \frac{\left(\sum x_i^2/2\right)^{n/2}}{\Gamma(n/2)} \int_0^\infty \log\sigma^2 \frac{1}{(\sigma^2)^{(n/2)+1}} \exp\left(-\frac{\Sigma x_i^2}{2\sigma^2}\right) d\sigma^2$$

After some simplifications, we get

$$(\log\sigma^2)^B = \log\left(\frac{\Sigma x_i^2}{2}\right) - D\left(\frac{n}{2}\right)$$

where $D(u) = (\partial/\partial u)\log\Gamma(u)$. Jahnke and Emde (1994) have approximated

$$2\exp\left[D\left(\frac{n}{2}\right)\right] \simeq n-1 \quad \text{(for large } n\text{)}$$

This gives

$$(\log\sigma^2)^B = \log\left(\frac{\Sigma x_i^2}{2}\right) - \log\left(\frac{n-1}{2}\right)$$

$$= \log\left(\frac{\Sigma x_i^2}{n-1}\right)$$

Thus, the Bayes estimate of σ^2 is

$$(\sigma^2)^B = \frac{\Sigma x_i^2}{n-1}$$

(iii) The loss function may be expressed by

$$L(\sigma^2, \delta) = \frac{1}{\sigma^4}(\sigma^2 - \delta)^2 = w(\sigma^2)(\sigma^2 - \delta)^2$$

Thus, the Bayes estimate of σ^2 under the scale invariant loss function is

$$(\theta^2)^B = \frac{E[w(\sigma^2)\sigma^2|\mathbf{x}]}{E[w(\sigma^2)|\mathbf{x}]} = \frac{E(1/\sigma^2|\mathbf{x})}{E(1/\sigma^4|\mathbf{x})}$$

$$= \frac{\left[\frac{\left(\Sigma x_i^2/2\right)^{n/2}}{\Gamma(n/2)}\right]\int_0^\infty \frac{1}{(\sigma^2)^{(n/2)+2}}\exp\left(-\frac{\Sigma x_i^2}{2\sigma^2}\right)d\sigma^2}{\left[\frac{\left(\Sigma x_i^2/2\right)^{n/2}}{\Gamma(n/2)}\right]\int_0^\infty \frac{1}{(\sigma^2)^{(n/2)+3}}\exp\left(-\frac{\Sigma x_i^2}{2\sigma^2}\right)d\sigma^2}$$

$$= \frac{\Gamma[(n/2)+1]}{\left(\Sigma x_i^2/2\right)^{(n/2)+1}}\,\frac{\left(\Sigma x_i^2/2\right)^{(n/2)+2}}{\Gamma[(n/2)+2]} = \frac{\Sigma x_i^2}{n+2}$$

(iv) We have computed the posterior distribution of σ^2 given **x** as $IG_2(n/2, \Sigma x_i^2/2)$. This gives

$$E\left(\frac{1}{\sigma^2}\middle|\mathbf{x}\right) = \frac{\left(\Sigma x_i^2/2\right)^{n/2}}{\Gamma(n/2)}\int_0^\infty \frac{1}{(\sigma^2)^{(n/2)+2}}\exp\left(-\frac{1}{2\sigma^2}\Sigma x_i^2\right)d\sigma^2$$

$$= \frac{\left(\Sigma x_i^2/2\right)^{n/2}}{\Gamma(n/2)}\,\frac{\Gamma[(n/2)+1]}{\left(\Sigma x_i^2/2\right)^{(n/2)+1}} = \frac{n}{\sum x_i^2}$$

and $$E\left\{\frac{1}{\sigma^2}\exp\left[c\left(\frac{\delta}{\sigma^2}\right)\right]\middle|\mathbf{x}\right\} = \frac{\left(\Sigma x_i^2/2\right)^{n/2}}{\Gamma(n/2)}\int_0^\infty \frac{1}{(\sigma^2)^{(n/2)+2}}\exp\left[-\left(\frac{\Sigma x_i^2}{2} - c\delta\right)\frac{1}{\sigma^2}\right]d\sigma^2$$

$$= \frac{\left(\Sigma x_i^2/2\right)^{n/2}}{\Gamma(n/2)}\,\frac{\Gamma[(n/2)+1]}{\left[\left(\Sigma x_i^2/2\right) - c\delta\right]^{(n/2)+1}}$$

$$= \frac{n\left(\Sigma x_i^2\right)^{n/2}}{\left(\Sigma x_i^2 - 2c\delta\right)^{(n/2)+1}}$$

The Bayes estimate of σ^2 under modified LinEx loss function, given in Eq. (8.1.6), is obtained by minimizing the posterior conditional expected loss. This is given by solving the equation

$$E\left[\frac{1}{\sigma^2}\exp\left(\frac{c\delta}{\sigma^2}\right)\middle|\mathbf{x}\right] = \exp(c)E\left(\frac{1}{\sigma^2}\middle|\mathbf{x}\right)$$

by replacing θ by σ^2 in Eq. (8.1.6). Thus, we have

$$\frac{n\left(\Sigma x_i^2\right)^{n/2}}{\left(\Sigma x_i^2 - 2c\delta\right)^{(n/2)+1}} = \exp(c)\frac{n}{\Sigma x_i^2}$$

$$\left(\Sigma x_i^2 - 2c\delta\right)^{(n/2)+1} = \exp(-c)\left(\Sigma x_i^2\right)^{(n/2)+1}$$

$$\Sigma x_i^2 - 2c\delta = \exp\left[-\frac{c}{(n/2)+1}\right]\Sigma x_i^2$$

$$(\sigma^2)^B = \left[1-\exp\left(-\frac{2c}{n+2}\right)\right]\frac{\Sigma x_i^2}{2c}$$

Notice that the Bayes estimates of σ^2 under different loss functions considered here, for large n, are same.

***Example* 8.36** Let $X_1,\ldots,X_n$ be *iid* observations from $N(0, \sigma^2)$, σ^2 is unknown. Consider the estimation of σ^2 and a convenient parameterization $\theta = 1/\sigma^2$. Consider the conjugate prior distribution for θ as $G_1(\alpha, \beta)$.

(i) Find Bayes, MAP, and ML estimates of θ and show, for large n and $\alpha = 1/\beta$, that Bayes and MAP estimates converge to ML estimate. Show that Bayes and MAP estimates converge to θ more rapidly when the above prior becomes stronger, i.e., $\alpha = 1/\beta$ increases.

(ii) Show that the improper noninformative prior $\pi(\theta) = \theta^{-1}$ is a limiting distribution of gamma conjugate family of prior distributions $G_1(\alpha, \beta)$. Show that this noninformative prior is invariant under parameterization $\theta = 1/\sigma^2$.

(iii) Show that the scale invariant noninformative prior for σ^2 using invariance considerations and Jefferys prior for σ^2 are both identical. Find Bayes and MAP estimates of σ^2.

(iv) Show that the MAP estimate in (iii) is a minimum risk estimate of σ^2 of the form $c\Sigma x_i^2$. Show that the minimax estimate of σ^2 does not exist.

(v) Show that the noninformative prior is uniform distribution when scale parameter is expressed in logarithmic units. Show that the Bayes estimate of θ with respect to this noninformative prior is identical to the ML estimate, and the MAP estimate is less than the ML estimate since noninformative prior $1/\theta$ pulls down the estimate.

Solution.

(i) The joint density of the sample observations is given by

$$f(\mathbf{x}|\theta) = \left(\frac{\theta}{2\pi}\right)^{n/2}\exp\left(-\frac{\theta}{2}\Sigma x_i^2\right)$$

and the conjugate prior for θ is $G_1(\alpha, \beta)$. We have

$$E(\theta|\alpha, \beta) = \alpha\beta;\ V(\theta|\alpha, \beta) = \alpha\beta^2$$

This leads to the posterior density

$$\pi(\theta|\mathbf{x}) \propto \theta^{\alpha+(n/2)-1}\exp\left[-\left(\frac{1}{\beta}+\frac{1}{2}\Sigma x_i^2\right)\theta\right]$$

$$= G_1\left[\alpha+\frac{n}{2}, \left(\frac{1}{\beta}+\frac{1}{2}\Sigma x_i^2\right)^{-1}\right]$$

This yields the following estimates

$$\theta^B = \left(\frac{2\alpha}{n}+1\right)\left(\frac{2}{n\beta}+\frac{1}{n}\Sigma x_i^2\right)^{-1}$$

$$\theta^{\text{MAP}} = \left(\frac{2\alpha}{n}+1-\frac{2}{n}\right)\left(\frac{2}{n\beta}+\frac{1}{n}\Sigma x_i^2\right)^{-1}$$

The ML estimator, in this case, is given by

$$\theta^{\text{ML}} = \frac{1}{(1/n)\Sigma x_i^2}$$

Notice that as n grows to infinity, $\alpha = 1/\beta$, the estimates θ^B and θ^{MAP} converge to θ^{ML}, the role of prior becomes redundant, and these estimates converge to true $\theta = 1/\sigma^2$.

If we take $\alpha = 1/\beta$ and increase the magnitude of α, though the prior mean remains the same, the prior variance cuts down by β times, and the prior becomes stronger. In this case, the estimates θ^B and θ^{MAP} converge to true θ more rapidly as compared to the case when α is small.

(ii) Consider $G_1(\alpha, \beta)$ as a conjugate prior for the parameterization $\theta = 1/\sigma^2$. By letting $\alpha \to 0$ and $\beta \to 0$, the mean $E(\theta|\alpha, \beta)$ and mode $\max_\theta \pi(\theta|\alpha, \beta)$ are indeterminate and variance $V(\theta|\alpha, \beta) \to +\infty$. If we ignore the normalizing constant, the limiting prior is given by

$$\lim_{\substack{\alpha\to 0\\ \beta\to 0}} G_1(\alpha, \beta) = \lim_{\substack{\alpha\to 0\\ \beta\to 0}} \theta^{\alpha-1}\exp\left(-\frac{\theta}{\beta}\right)$$

$$\pi(\theta) = \frac{1}{\theta}, \theta \in (0, \infty)$$

Thus, we get an improper noninformative prior as a limiting distribution of gamma conjugate family of prior distributions. Note that this prior resembles $\exp(-\theta)$. The density for $\sigma = \theta^{-(1/2)}$ is

$$\pi^*(\sigma) = \pi\left(\frac{1}{\sigma^2}\right)|J|;\ |J| = \frac{2}{\sigma^3}$$

We have $$\pi^*(\sigma) = \sigma^2 \cdot \frac{2}{\sigma^3} \propto \frac{1}{\sigma}$$

Note that this prior for the parameterization σ is the same as the noninformative scale invariant prior obtained for scale parameter σ. Hence, it is invariant under the above parameterization.

(iii) Consider the parameterization $\lambda = g(\sigma) = \sigma^2$. By random variable transformation rule,

$$\pi(\lambda) = \pi^*(\sqrt{\lambda})\left|\frac{\partial\sigma}{\partial\lambda}\right| \propto \frac{1}{\lambda}$$

or
$$\pi(\sigma^2) \propto \frac{1}{\sigma^2}$$

Further, the Fisher information for σ^2,

$$I_{\mathbf{X}}(\sigma^2) = \frac{n}{\sigma^4}$$

gives the Jeffreys prior for σ^2 as

$$\pi(\sigma^2) = [I_{\mathbf{X}}(\sigma^2)]^{1/2} = \frac{\sqrt{n}}{\sigma^2} \propto \frac{1}{\sigma^2}$$

This shows that the scale invariant noninformative prior for σ^2 using invariance considerations and Jeffreys prior for σ^2 are both identical. Under this Jeffreys prior, the posterior density is

$$\pi(\sigma^2|\mathbf{x}) \propto \left(\frac{1}{\sigma^2}\right)^{(n/2)+1} \exp\left[-\frac{1}{2\sigma^2}\sum_{i=1}^{n} x_i^2\right]$$

It is a proper density. We assume $n > 2$ since $E(\sigma^2|\mathbf{x})$ exists only when $n > 2$; otherwise the integral $\int_0^\infty \sigma^2\pi(\sigma^2|\mathbf{x})d\sigma^2 = \infty$. The Bayes and MAP estimates are, respectively, given by

$$(\sigma^2)^B = E(\sigma^2|\mathbf{x}) = \frac{1}{n-2}\sum_{i=1}^{n} x_i^2, \ n > 2$$

and
$$(\sigma^2)^{\text{MAP}} = \frac{1}{n+2}\sum_{i=1}^{n} x_i^2$$

(iv) Consider the estimator of σ^2 of the following form

$$\hat{\sigma}^2 = c\sum x_i^2$$

The risk of such estimators for the squared error loss function is

$$R(\sigma^2, \hat{\sigma}^2) = \text{MSE}(\hat{\sigma}^2)$$
$$= E^{\mathbf{X}|\sigma^2}\left(c\sum X_i^2 - \sigma^2\right)^2 \quad [\text{since } E(X^4) = 3(\sigma^2)^2 \text{ as } \Sigma X_i^2/\sigma^2 \sim \chi_n^2]$$
$$= [c^2(n^2 + 2n) - 2cn + 1](\sigma^2)^2$$

Now, solving the equation

$$\frac{\partial}{\partial c}R(\sigma^2, \hat{\sigma}^2) = 0$$

we get
$$c = \frac{1}{n+2}$$

Thus the minimum risk estimate

$$\hat{\sigma}^2 = \frac{1}{n+2}\sum_{i=1}^{n} x_i^2$$

is identical to the MAP estimate obtained under Jeffreys prior.
The minimax estimator, in this case, for some $c > 0$,

$$\sup_{\sigma^2} R(\sigma^2, \hat{\sigma}^2) = [c^2(n^2 + 2n) - 2cn + 1] \sup_{\sigma^2} (\sigma^2)^2 = \infty$$

since $[c^2(n^2 + 2n) - 2cn + 1] > 0$ and $\sup_{\sigma^2}(\sigma^2)^2 = \infty$. We cannot talk of minimizing this supremum for an estimator of the form $c\Sigma X_i^2$ by choosing an appropriate c. Thus, in this case, the minimax estimator of σ^2 does not exist.

(v) Usually, it is natural to express the scale parameter σ in logarithmic units and, thus, we may think of choosing the transformation $\lambda = \log(\sigma)$ when $\sigma \in (0, \infty)$ and $\lambda \in (-\infty, \infty)$. The probability density of λ is

$$\pi'(\lambda) = \pi(e^{\lambda})|J|$$

where $|J| = e^{\lambda}$. We have

$$\pi'(\lambda) \propto \frac{1}{e^{\lambda}} e^{\lambda} = 1$$

Thus, the complete ignorance about σ can be expressed by a uniform noninformative prior for the parameterization $\lambda = \log(\sigma)$. Finally, the Bayes and MAP estimates of θ for this noninformative prior, respectively, are

$$\lim_{\substack{\alpha\to 0\\ \beta\to\infty}} \theta^B = \lim_{\substack{\alpha\to 0\\ \beta\to\infty}} \left(\frac{2\alpha}{n}+1\right)\left(\frac{2}{n\beta}+\frac{1}{n}\Sigma x_i^2\right)^{-1} = \frac{1}{(1/n)\Sigma x_i^2} = \theta^{\text{ML}}$$

$$\lim_{\substack{\alpha\to 0\\ \beta\to\infty}} \theta^{\text{MAP}} = \lim_{\substack{\alpha\to 0\\ \beta\to\infty}} \left(\frac{2\alpha}{n}+1-\frac{2}{n}\right)\left(\frac{2}{n\beta}+\frac{1}{n}\sum x_i^2\right)^{-1} = \frac{1}{[1/(n-2)]\Sigma x_i^2} < \theta^{\text{ML}}$$

Thus, in case of complete ignorance about θ, the Bayes estimate and the ML estimate are identical to the MAP estimate. However, θ^{MAP} is less than the ML estimate since the prior $1/\theta$ pulls down the estimate.

***Example* 8.37** Let $X_1,\ldots,X_n$ be *iid* observations from $N(0, \sigma^2)$. Consider the (improper) prior distribution $\pi(\sigma^2) \propto (\sigma^2)^k$. Find the generalized Bayes estimate of σ^2.

Solution. The posterior distribution is

$$\pi(\sigma^2|\mathbf{x}) = f(\mathbf{x}|\sigma^2)\pi(\sigma^2) \propto \frac{1}{(\sigma^2)^{n/2}}\exp\left(-\frac{1}{2}\frac{\sum x_i^2}{\sigma^2}\right)(\sigma^2)^k$$

$$= \frac{1}{(\sigma^2)^{[(n/2)-k-1]+1}}\exp\left(-\frac{\sum x_i^2}{2\sigma^2}\right)$$

which is a proper distribution $IG_2((n/2) - k - 1, \Sigma x_i^2/2)$ provided

$$\frac{n}{2} - k - 1 > 0 \quad \text{or} \quad k < \frac{n-2}{2}$$

Under the squared error loss function, the generalized Bayes estimate, since the prior distribution of σ^2 is improper, is

$$(\sigma^2)^{\text{GB}} = E(\sigma^2|\mathbf{x}) = \frac{\sum x_i^2/2}{(n/2)-k-1-1} = \frac{\sum x_i^2}{n-2k-4}$$

provided $(n/2) - k - 1 > 1$ or $k < (n - 4)/2$.

***Example* 8.38** Consider a linear model

$$Y_{ij} = \sum_{l=1}^{p} \beta_l X_{il} + \varepsilon_{ij}$$

$$= \boldsymbol{\beta}'\mathbf{X}_i + \varepsilon_{ij},\ j = 1,\ldots, n_i,\ i = 1,\ldots, k$$

where $\mathbf{X}_i$s are known vectors; ε_{ij}s are independent with $\varepsilon_{ij} \sim N(0, \sigma_i^2)$, $j = 1,\ldots,n_i$ and $i = 1,\ldots,k$; $\sigma_i^2 > 0$, $i = 1,\ldots,k$, are unknown; $\boldsymbol{\beta} \in \mathbb{R}^p$ is unknown; $\mathbf{Y} = (Y_{ij})$, $j = 1,\ldots,n_i$ and $i = 1,\ldots,k$. Assume $\sigma_i' = (2\sigma_i^{2})^{-1}$ and the parameter vector $\boldsymbol{\theta} = (\boldsymbol{\beta}, \boldsymbol{\sigma}')$ is unknown where $\boldsymbol{\beta} = (\beta_1,\ldots,\beta_p)'$ and $\boldsymbol{\sigma}' = (\sigma_1',\ldots,\sigma_p')'$. Assume that priors are $\pi(\boldsymbol{\beta})$ for $\boldsymbol{\beta}$ and $\pi(\sigma_i') = G_1(\alpha, \beta)$, (α, β) is known, $\alpha > 0$, $\beta > 0$. Find the Bayes estimators of σ_i^2, $i = 1,\ldots,k$, and of $\mathbf{l}'\boldsymbol{\beta}$, $\mathbf{l} \in \mathbb{R}^p$, under the squared error loss function.

Solution. The model

$$Y_{ij} \sim N(\boldsymbol{\beta}'\mathbf{X}_i, \sigma_i^2)\ \forall\ j = 1,\ldots,n_i \text{ and } i = 1,\ldots,k$$

and the prior

$$\pi(\boldsymbol{\theta}) \propto \pi(\boldsymbol{\beta})\prod_{i=1}^{k} \sigma_i'^{\alpha-1}\exp\left(-\frac{\sigma_i'}{\beta}\right)$$

give the joint distribution of $\mathbf{Y}$ and $\boldsymbol{\theta}$ by

$$f(\mathbf{y}, \boldsymbol{\theta}) \propto \pi(\boldsymbol{\beta})\prod_{i=1}^{k} \sigma_i'^{(n_i/2)+\alpha-1}\exp\left[-\left(\frac{1}{\beta} + \sum_{j=1}^{n_i}(Y_{ij} - \boldsymbol{\beta}'\mathbf{X}_i)\right)\sigma_i'\right]$$

The posterior distribution of $\boldsymbol{\theta}$ given $\mathbf{y}$ is then proportional to

$$\pi(\boldsymbol{\theta}|\mathbf{y}) \propto \pi(\boldsymbol{\beta})\prod_{i=1}^{k}\sigma_i'^{(n_i/2)+\alpha-1}\exp\left\{-\left[\left(\frac{1}{\beta}\right)+v_i(\boldsymbol{\beta})\right]\sigma_i'\right\}$$

where $v_i(\boldsymbol{\beta}) = \Sigma_{j=1}^{n_i}(Y_{ij} - \boldsymbol{\beta}'\mathbf{X}_i)$. The Bayes estimator of $\sigma_i^2 = (2\sigma_i')^{-1}$ given $\boldsymbol{\beta}$ under the squared error loss function is given by

$$(\sigma_i^2)^B(\mathbf{y}, \boldsymbol{\beta}) = \frac{\int(2\sigma_i')^{-1}\pi(\boldsymbol{\theta}|\mathbf{y})d\sigma'}{\int\pi(\boldsymbol{\theta}|\mathbf{y})d\sigma'}$$

$$= \frac{\int(2\sigma_i')^{-1}\sigma_i'^{(n_i/2)+\alpha-1}\exp\{-[(1/\beta)+v_i(\boldsymbol{\beta})]\sigma_i'\}d\sigma_i'}{\int\sigma_i'^{(n_i/2)+\alpha-1}\exp\{-[(1/\beta)+v_i(\boldsymbol{\beta})]\sigma_i'\}d\sigma_i'}$$

$$= \frac{(1/2)\Gamma[(n_i/2)+\alpha-1][(1/\beta)+v_i(\boldsymbol{\beta})]^{-[(n_i/2)+\alpha-1]}}{\Gamma[(n_i/2)+\alpha][(1/\beta)+v_i(\boldsymbol{\beta})]^{-[(n_i/2)+\alpha+1]}} = \frac{(1/\beta)+v_i(\boldsymbol{\beta})}{(n_i+2\alpha-2)}$$

The posterior distribution of $\boldsymbol{\beta}$ given $\mathbf{y}$ is

$$\pi(\boldsymbol{\beta}|\mathbf{y}) = \int\pi(\boldsymbol{\theta}|\mathbf{y})d\sigma_i'$$

$$\propto \pi(\boldsymbol{\beta})\prod_{i=1}^{k}\int\sigma_i'^{(n_i/2)+\alpha-1}\exp\left[-\left(\frac{1}{\beta}+v_i(\boldsymbol{\beta})\right)\sigma_i'\right]d\sigma_i'$$

$$\propto \pi(\boldsymbol{\beta})\prod_{i=1}^{k}\left(\frac{1}{\beta}+v_i(\boldsymbol{\beta})\right)^{-[(n_i/2)+\alpha]}$$

Therefore, the Bayes estimate of σ_i^2 is

$$(\sigma_i^2)^B(\mathbf{y}) = c\int(\sigma_i^2)^B(\mathbf{y}, \boldsymbol{\beta})\pi(\boldsymbol{\beta}|\mathbf{y})d\boldsymbol{\beta}$$

$$= c\int\frac{(1/\beta)+v_i(\boldsymbol{\beta})}{n_i+2\alpha-2}\pi(\boldsymbol{\beta}|\mathbf{y})d\boldsymbol{\beta}$$

Further, the Bayes estimate of $\mathbf{l}'\boldsymbol{\beta}$ is

$$(\mathbf{l}'\boldsymbol{\beta})^B(\mathbf{y}) = \mathbf{l}'E(\boldsymbol{\beta}|\mathbf{y})$$

where the above expectation involves the *pdf* $\pi(\boldsymbol{\beta}|\mathbf{y})$. Note that these Bayes estimators cannot be obtained in closed form. Therefore, some numerical method can be applied to evaluate them.

***Example* 8.39** Let $X_1,\ldots,X_n$ be a random sample from $N(\mu, \sigma^2)$ with unknown $\mu \in \mathbb{R}^1$ and $\sigma^2 > 0$. Let the priors for μ and σ^2 are

$$\sigma' = \frac{1}{2\sigma^2} \sim G_1(\alpha, \beta) \text{ with } \alpha \text{ and } \beta \text{ are known}$$

$$\mu|\sigma' \sim N\left(\theta, \frac{\tau^2}{\sigma'}\right)$$

Find the Bayes estimates of μ and σ^2 under the squared error loss function. Show that the biases of these estimators are of the order $O(1/n)$ and that they are consistent.

Solution. The posterior distribution of μ, σ' given $\mathbf{x}$ is given by

$$\pi(\mu, \sigma'|\mathbf{x}) \propto \sigma'^{[(n+1)/2]+\alpha-1} \exp\left\{-\left[nS_n^2 + n(\bar{x}-\mu)^2 + \frac{1}{\beta} + \frac{1}{2\tau^2}(\mu-\theta)^2\right]\sigma'\right\}$$

where

$$\sum(x_i-\mu)^2 = \sum(x_i-\bar{x})^2 + n(\bar{x}-\mu)^2 = nS_n^2 + n(\bar{x}-\mu)^2$$

Expressing the term

$$n(\bar{x}-\mu)^2 + \frac{1}{2\tau^2}(\mu-\theta)^2 = \left(n + \frac{1}{2\tau^2}\right)\mu^2 - 2\left(n\bar{x} + \frac{\theta}{2\tau^2}\right)\mu + \left(n\bar{x}^2 + \frac{\theta^2}{2\tau^2}\right)$$

$$= \left(n+\frac{1}{2\tau^2}\right)[\mu-\mu(\mathbf{x})]^2 + \left(n\bar{x}^2 + \frac{\theta^2}{2\tau^2}\right) - \left(n+\frac{1}{2\tau^2}\right)\mu(\mathbf{x})^2$$

where

$$\mu(\mathbf{x}) = \frac{n\bar{x} + (\theta/2\tau^2)}{n + (1/2\tau^2)}$$

The posterior distribution simplifies to

$$\pi(\mu, \sigma'|\mathbf{x}) \propto \sigma'^{[(n+1)/2]+\alpha-1} \exp\left\{-\left[nS_n^2 + \frac{1}{\beta} + n\bar{x}^2 + \frac{\theta^2}{2\tau^2} - \left(n+\frac{1}{2\tau^2}\right)\{\mu(\mathbf{x})\}^2 + \left(n+\frac{1}{2\tau^2}\right)[\mu-\mu(\mathbf{x})]^2\right]\sigma'\right\}$$

By integrating out μ, we get the posterior distribution of $\sigma'|\mathbf{x}$,

$$\pi(\sigma'|\mathbf{x}) \propto \sigma'^{[(n/2)+\alpha]-1}\exp(-\sigma' y)$$

that is, $\sigma'|\mathbf{x} \sim G_1((n/2)+\alpha, y^{-1})$ where $y = nS_n^2 + (1/\beta) + n\bar{x}^2 + [\theta^2/(2\tau^2)] - [n + (1/2\tau^2)]\mu^2(\mathbf{x})$.
Further, the posterior distribution of $\mu|\sigma', \mathbf{x}$

$$\pi(\mu|\sigma', \mathbf{x}) \propto \sigma'^{(1/2)} \exp\left\{-\sigma'\left(n+\frac{1}{2\tau^2}\right)[\mu-\mu(\mathbf{x})]^2\right\}$$

which is $\pi(\mu|\sigma',\mathbf{x}) \sim N(\mu(\mathbf{x}), [(2n+\tau^{-2})\sigma']^{-1})$.
Therefore, under the squared error loss function, the Bayes estimator of μ is

$$\mu^B(\mathbf{x}) = \mu(\mathbf{x})$$

Now, consider the bias in $\mu^B(\mathbf{x})$

$$\text{bias}[\mu^B(\mathbf{X})] = E(\mu(\mathbf{X}) - \mu)$$

$$= \frac{1}{n + (1/2\tau^2)}\left[nE(\bar{X}) + \frac{\theta}{2\tau^2}\right] - \mu$$

$$= \frac{n\mu + (\theta/2\tau^2)}{n + (1/2\tau^2)} - \mu = \frac{\theta - \mu}{2n\tau^2 + 1} = O\left(\frac{1}{n}\right)$$

Though $\mu^B(\mathbf{x})$ is not an unbiased estimator of μ, but for large n, it is unbiased. Now, calculating the variance of $\mu^B(\mathbf{x})$,

$$V[\mu^B(\mathbf{X})] = \frac{n^2}{[n + (1/2\tau^2)]^2} V(\bar{X})$$

$$= \frac{n^2}{[n + (1/2\tau^2)]^2} \frac{\sigma^2}{n} = O\left(\frac{1}{n}\right)$$

shows that $V(\mu^B(\mathbf{X})) \to 0$ as $n \to \infty$. Therefore, the Bayes estimator $\mu^B(\mathbf{X})$ is consistent.

Consider, next, the Bayes estimate of σ^2, i.e., of $(2\sigma')^{-1}$

$$(\sigma^2)^B(\mathbf{x}) = \frac{y^{(n/2)+\alpha}}{\Gamma[(n/2)+\alpha]} \int_0^\infty (2\sigma')^{-1} \sigma'^{\{(n/2)+\alpha\}-1} e^{-\sigma' y} d\sigma'$$

$$= \frac{y^{(n/2)+\alpha}}{\Gamma[(n/2)+\alpha]} \frac{1}{2} \frac{\Gamma[(n/2)+\alpha-1]}{y^{[(n/2)+\alpha-1]}}$$

$$= \frac{y}{n + 2\alpha - 2}$$

provided that $n + 2\alpha - 2 > 0$. Consider the bias in $(\sigma^2)^B(\mathbf{x})$

$$\text{bias}[(\sigma^2)^B(\mathbf{x})] = \frac{E(Y)}{n + 2\alpha - 2} - \sigma^2$$

$$= \frac{1}{n + 2(\alpha - 1)}\left\{\sigma^2(n-1) + \beta^{-1} + \sigma^2 + n\mu^2 + \frac{\theta^2}{2\tau^2}\right.$$

$$\left. - \frac{1}{[n + (1/2\tau^2)]}\left[n\sigma^2 + \left(n\mu + \frac{\theta}{2\tau^2}\right)^2\right]\right\} - \sigma^2$$

The convergence of the terms in this bias expression are as follows:

(a) $[n + 2(\alpha - 1)]^{-1}[n\sigma^2 + \beta^{-1} + n\mu^2 + (\theta^2/2\tau^2)] \to \sigma^2 + \mu^2$ as $n \to \infty$

(b) $[n + 2(\alpha - 1)]^{-1}\ [n + (1/2\tau^2)]^{-1} n\sigma^2 \to 0$ as $n \to \infty$

(c) $[n + 2(\alpha - 1)]^{-1}[n + (1/2\tau^2)]^{-1}[n\mu + (\theta/2\tau^2)]^2 \to \mu^2$ as $n \to \infty$; it gives bias $[(\sigma^2)^B(\mathbf{X})] \to 0$ as $n \to \infty$

Therefore, though $(\sigma^2)^B(\mathbf{X})$ is not an unbiased estimator of σ^2, however, it is asymptotically unbiased since bias $[(\sigma^2)^B(\mathbf{X})] = O(1/n)$. One can also see that $(\sigma^2)^B(\mathbf{X})$ is consistent.

***Example* 8.40** Let $X_1,\ldots,X_n$ be a random sample from $N(\mu, \sigma^2)$ where $\mu \in \mathbb{R}^1$ is unknown and $\sigma^2 > 0$ is known. Consider the first-stage prior

$$\pi_1(\mu|\boldsymbol{\eta}) = N(\theta, \tau^2),\ \boldsymbol{\eta} = (\theta, \tau^2)'$$

and the second-stage prior

$$\pi_2(\boldsymbol{\eta}|a, v) = N(a, v^2) \times \text{ Lebesgue measure on } (0, \infty)$$

where a and v are known. Find the hierarchical Bayes estimate of μ under the squared error loss function.

Solution. We have already shown that the Bayes estimator of μ when the hyperparameters $\boldsymbol{\eta} = (\theta, \tau^2)'$ were known,

$$\delta(\mathbf{x}, \boldsymbol{\eta}) = \frac{\sigma^2}{n\tau^2 + \sigma^2}\theta + \frac{n\tau^2}{n\tau^2 + \sigma^2}\bar{x}$$

The hierarchical Bayes estimate of μ is

$$\delta^{\text{HB}}(\mathbf{x}) = \int \delta(\mathbf{x}, \boldsymbol{\eta})\pi(\boldsymbol{\eta}|\mathbf{x})d\boldsymbol{\eta}$$

where $\pi(\boldsymbol{\eta}|\mathbf{x})$ is the posterior distribution of $\boldsymbol{\eta}$ given $\mathbf{x}$

$$\pi(\boldsymbol{\eta}|\mathbf{x}) = \frac{m(\mathbf{x}|\boldsymbol{\eta})\pi_2(\boldsymbol{\eta}|a, v)}{m(\mathbf{x})}$$

Here, $m(\mathbf{x}|\boldsymbol{\eta})$ is the marginal distribution of $\mathbf{X}$ given θ and τ^2 assuming $\boldsymbol{\eta} = (\theta, \tau^2)'$ is known

$$m(\mathbf{x}|\boldsymbol{\eta}) = m(\mathbf{x}|\theta, \tau^2) = \int_{\mu \in \mathbb{R}^1} f(\mathbf{x}|\mu)\pi_1(\mu|\theta, \tau^2)d\mu$$

It is clear that

$$m(\mathbf{x}|\theta, \tau^2)\cdot\pi_2(\theta, \tau^2|a, v)$$

is the joint density of $\mathbf{X}$, θ, τ^2, i.e., $f(\mathbf{x}, \theta, \tau^2)$ (say). This density may be obtained by considering the joint density of $\mathbf{X}$, μ, θ, τ^2

$$\left(\frac{1}{2\pi\sigma^2}\right)^{n/2} \exp\left[-\frac{1}{2\sigma^2}\sum(x_i - \mu)^2\right]\frac{1}{\sqrt{2\pi\tau}}\exp\left[-\frac{1}{2\tau^2}(\mu - \theta)^2\right]\frac{1}{\sqrt{2\pi v}}\exp\left[-\frac{1}{2v^2}(\theta - a)^2\right]$$

and integrating out μ, we get

$$f(\mathbf{x}|\theta,\tau^2)=\left(\frac{1}{2\pi\sigma^2}\right)^{n/2}\frac{1}{2\pi\tau v}\exp\left[-\frac{1}{2v^2}(\theta-a)^2\right]$$

$$\int_{\mu\in\mathbb{R}^1}\exp\left[-\frac{1}{2\sigma^2}\sum(x_i-\mu)^2-\frac{1}{2\tau^2}(\mu-\theta)^2\right]d\mu$$

On expressing

$$\int_{\mu\in\mathbb{R}^1}\exp\left\{-\frac{1}{2}\left[\frac{1}{\sigma^2}\sum(x_i-\mu)^2+\frac{1}{\tau^2}(\mu-\theta)^2\right]\right\}d\mu$$

$$=\int_{\mu\in\mathbb{R}^1}\exp\left\{-\frac{1}{2}\left[\left(\frac{n}{\sigma^2}+\frac{1}{\tau^2}\right)\mu^2-2\left(\frac{n\bar{x}}{\sigma^2}+\frac{\theta}{\tau^2}\right)\mu+\left(\frac{\sum x_i^2}{\sigma^2}+\frac{\theta^2}{\tau^2}\right)\right]\right\}d\mu$$

and using the result

$$\int_{\mu\in\mathbb{R}^1}\exp\left\{-\frac{1}{2}[at^2-2bt+c]\right\}dt=\left(\frac{2\pi}{a}\right)^{1/2}\exp\left(\frac{b^2}{2a}-\frac{c}{2}\right)$$

The above integral simplifies to

$$\frac{(2\pi)^{1/2}}{[(n/\sigma^2)+(1/\tau^2)]^{1/2}}\exp\left\{\frac{[(n\bar{x}/\sigma^2)+(\theta/\tau^2)]^2}{2[(n/\sigma^2)+(1/\tau^2)]}-\frac{1}{2}\left(\frac{\sum x_i^2}{\sigma^2}+\frac{\theta^2}{\tau^2}\right)\right\}$$

$$\therefore\qquad f(\mathbf{x}|\theta,\tau^2)=\frac{1}{(2\pi)^{(n+1)/2}\sigma^n v\tau[(n/\sigma^2)+(1/\tau^2)]^{1/2}}\exp\left\{-\frac{1}{2\sigma^2}\sum x_i^2-\frac{1}{2\tau^2}\theta^2\right.$$

$$\left.-\frac{1}{2v^2}(\theta-a)^2+\frac{[(n\bar{x}/\sigma^2)+(\theta/\tau^2)]^2}{2[(n/\sigma^2)+(1/\tau^2)]}\right\}$$

Integrating out θ from the above density yields the joint density of $\mathbf{X}$ and τ^2,

$$f(\mathbf{x}|\tau^2)=\frac{\exp\left[-(1/2\sigma^2)\sum x_i^2\right]}{(2\pi)^{(n+1)/2}\sigma^n v\tau[(n/\sigma^2)+(1/\tau^2)]^{1/2}}\int_{\theta\in\mathbb{R}^1}\left\{-\frac{1}{2}\left[\left(\frac{1}{\tau^2}+\frac{1}{v^2}-\frac{1}{[(n/\sigma^2)+(1/\tau^2)]\tau^4}\right)\theta^2\right.\right.$$

$$\left.\left.-2\left(\frac{a}{v^2}+\frac{n\bar{x}}{[(n/\sigma^2)+(1/\tau^2)]\sigma^2\tau^2}\right)\theta-\frac{a^2}{v^2}+\frac{(n\bar{x}/\sigma^2)^2}{[(n/\sigma^2)+(1/\tau^2)]}\right]\right\}d\theta$$

$$=\frac{\exp\left[-(1/2\sigma^2)\sum x_i^2\right]}{(2\pi)^{(n+1)/2}\sigma^n v\tau[(n/\sigma^2)+(1/\tau^2)]^{1/2}}\frac{(2\pi)^{1/2}}{\left\{(1/\tau^2)+(1/v^2)-\dfrac{1}{[(n/\sigma^2)+(1/\tau^2)]\tau^2}\right\}^{1/2}}$$

$$\exp\left\{\frac{\left\{\left(\frac{a}{v^2}\right)+\frac{n\bar{x}}{\{[(n/\sigma^2)+(1/\tau^2)]\sigma^2\tau^2\}}\right\}^2}{2\left[\left(\frac{1}{\tau^2}\right)+\left(\frac{1}{v^2}\right)-\frac{1}{[(n/\sigma^2)+(1/\tau^2)]\tau^4}\right]}-\frac{1}{2}\left(\frac{(n\bar{x}/\sigma^2)^2}{[(n/\sigma^2)+(1/\tau^2)]}-\frac{a^2}{v^2}\right)\right\}$$

Further, on integrating out τ^2 from $f(\mathbf{x}|\tau^2)$, we get the marginal density of $\mathbf{X}$,

$$m(\mathbf{x}) = \int_{\tau^2>0} f(\mathbf{x}|\tau^2)d\tau^2$$

This gives the hierarchical Bayes estimator of μ by

$$\begin{aligned}\delta^{\text{HB}}(\mathbf{x}) &= \frac{1}{m(\mathbf{x})}\int_{\tau^2>0}\int_{\theta\in\mathbb{R}^1}\delta(\mathbf{x},\theta,\tau^2)f(\mathbf{x}|\theta,\tau^2)d\theta d\tau^2\\ &= \frac{1}{m(\mathbf{x})}\int_{\tau^2>0}\int_{\theta\in\mathbb{R}^1}\delta(\mathbf{x},\theta,\tau^2)f(\theta|\mathbf{x},\tau^2)f(\mathbf{x}|\tau^2)d\theta d\tau^2\\ &= \frac{1}{m(\mathbf{x})}\int_{\tau^2>0}[E^{\theta|\mathbf{x},\tau^2}\delta(\mathbf{x},\theta,\tau^2)]f(\mathbf{x}|\tau^2)d\tau^2\end{aligned}$$

We have $$E^{\theta|\mathbf{x},\tau^2}\delta(\mathbf{x},\theta,\tau^2) = \frac{\sigma^2}{n\tau^2+\sigma^2}E(\theta|\mathbf{x},\tau^2)+\frac{n\tau^2}{n\tau^2+\sigma^2}\bar{x}$$

where $$E(\theta|\mathbf{x},\tau^2) = \frac{a\tau^2(n\tau^2+\sigma^2)-n\tau^2v^2\bar{x}}{\tau^2(n\tau^2+\sigma^2)+n\tau^2v^2}$$

We have $$\delta^{\text{HB}}(\mathbf{x}) = \frac{1}{m(\mathbf{x})}\int_0^\infty\left(\frac{\sigma^2}{n\tau^2+\sigma^2}E(\theta|\mathbf{x},\tau^2)+\frac{n\tau^2}{n\tau^2+\sigma^2}\bar{x}\right)f(\mathbf{x}|\tau^2)d\tau^2$$

***Example* 8.41** Let $X_1,\ldots,X_n$ be a random sample from $N(\mu, \sigma^2)$, where $\mu \in \mathbb{R}^1$ is unknown and $\sigma^2 > 0$ is known. Suppose that the first-stage prior distribution of μ is

$$\pi_1(\mu|\theta) \sim N(\theta, \tau^2)$$

where θ is unknown and τ^2 is known. Let the second-stage prior for θ, $\pi_2(\theta)$, be the Lebesgue measure on $\mathbb{R}^1$ which is an improper prior. Find the hierarchical Bayes estimate for μ under the squared error loss function.

Solution. The Bayes estimator of μ when θ is known,

$$\delta(\mathbf{x},\theta) = \frac{\sigma^2}{n\tau^2+\sigma^2}\theta+\frac{n\tau^2}{n\tau^2+\sigma^2}\bar{x}$$

The hierarchical Bayes estimator for μ is then given by

$$\delta^{\text{HB}}(\mathbf{x}) = E^{\theta|\mathbf{x}}\delta(\mathbf{x},\theta)$$

The joint distribution of $\mathbf{X}$, μ, θ is

$$f(\mathbf{x}, \mu, \theta) \propto \exp\left[-\frac{n}{2\sigma^2}(\bar{x}-\mu)^2\right]\exp\left[-\frac{1}{2\tau^2}(\mu-\theta)^2\right]$$

Integrating out μ in $f(\bar{x}, \mu, \theta)$ yields the joint *pdf* of $\mathbf{X}$ and θ,

$$f(\mathbf{x}, \theta) \propto \int_{\mathbb{R}^1} \exp\left[-\frac{n}{2\sigma^2}(\bar{x}-\mu)^2 - \frac{1}{2\tau^2}(\mu-\theta)^2\right]d\mu$$

$$\propto \int_{\mathbb{R}^1} \exp\left\{-\frac{1}{2}\left[\left(\frac{n}{\sigma^2}+\frac{1}{\tau^2}\right)\mu^2 - 2\left(\frac{n\bar{x}}{\sigma^2}+\frac{\theta}{\tau^2}\right)\mu + \frac{\theta^2}{\tau^2}\right]\right\}d\mu$$

By ignoring the term $\exp(n\bar{x}/2\sigma^2)$, we get

$$f(\mathbf{x}, \theta) = \exp\left\{\frac{[(n\bar{x}/\sigma^2)+(\theta/\tau^2)]^2}{2[(n/\sigma^2)+(1/\tau^2)]} - \frac{\theta^2}{2\tau^2}\right\}$$

by using the result $\int_{\mathbb{R}^1}\exp[-(1/2)(at^2 - 2bt + c)]dt = (2\pi/a)^{1/2}\exp[(b^2/2a) - (c/2)]$. Therefore, the distribution of $\theta|\mathbf{x}$ is

$$f(\theta|\mathbf{x}) \propto \exp\left\{-\frac{1}{2}\left[-\frac{(n\tau^2\bar{x}+\theta\sigma^2)^2}{\tau^2\sigma^2(n\tau^2+\sigma^2)} - \frac{\theta^2}{\tau^2}\right]\right\}$$

$$= \exp\left\{-\frac{1}{2}\left[-\frac{(n\tau^2\bar{x}+\theta\sigma^2)^2+\theta^2\sigma^2(n\tau^2+\sigma^2)}{\tau^2\sigma^2(n\tau^2+\sigma^2)}\right]\right\}$$

$$= \exp\left\{-\frac{1}{2}\frac{n[(\theta-\bar{x})^2-(n\tau^2+\sigma^2)(\bar{x}^2/\sigma^2)]}{(n\tau^2+\sigma^2)}\right\}$$

$$\propto \exp\left[-\frac{1}{2}\frac{n(\theta-\bar{x})^2}{n\tau^2+\sigma^2}\right]$$

This gives $\theta|\mathbf{x} \sim N(\bar{x}, \tau^2 + (\sigma^2/n))$. Therefore, the hierarchical Bayes estimate of μ is

$$\delta^{\text{HB}}(\mathbf{x}) = \frac{\sigma^2}{n\tau^2+\sigma^2}E(\theta|\mathbf{x}) + \frac{n\tau^2}{n\tau^2+\sigma^2}\bar{x}$$

$$= \frac{\sigma^2}{n\tau^2+\sigma^2}\bar{x} + \frac{n\tau^2}{n\tau^2+\sigma^2}\bar{x} = \bar{x}$$

***Example* 8.42** Let $X_1,\ldots,X_n$ be a random sample drawn from $N(\mu, \sigma^2)$, where $\mu \in \mathbb{R}^1$ is unknown and $\sigma^2 > 0$ is known. Let $\pi(\mu)$ be a given prior density and the loss function be squared error. Show that the Bayes estimate of μ is

$$\delta^B(x) = x + \frac{\sigma^2}{n}\frac{\partial}{\partial x}\log m(x)$$

where $\bar{X} = x$, the sample mean, and $m(x)$ is the marginal density of $\bar{X}$. Find the posterior variance of μ given $\bar{X} = x$ and express it as a function of the first two derivatives of $\log m(x)$. Find the Bayes estimate $\delta^B(x)$ when prior is a mixture of distributions

$$\pi(\mu) = \varepsilon\, I(\theta_1) + (1 - \varepsilon)N(\theta, \tau^2)$$

where $I(\theta_1)$ indicates that the prior assigns its full mass at the point θ_1.

Solution. The joint density of $\bar{X}$ and μ is

$$h(x, \mu) = \frac{\sqrt{n}}{\sqrt{2\pi}\sigma}\exp\left[-\frac{n}{2\sigma^2}(x-\mu)^2\right]\pi(\mu) \quad \left[\text{Since } \bar{X} \sim N\left(\mu, \frac{\sigma^2}{n}\right)\right]$$

This gives the marginal density of $\bar{X}$ (unconditional on μ),

$$m(x) = \frac{\sqrt{n}}{\sqrt{2\pi}\sigma}\int \exp\left[-\frac{n}{2\sigma^2}(x-\mu)^2\right]\pi(\mu)d\mu$$

The first and the second derivatives of $m(x)$ are

$$m'(x) = \frac{\sqrt{n}}{\sqrt{2\pi}\sigma}\frac{n}{\sigma^2}\int (\mu - x)\exp\left[-\frac{n}{2\sigma^2}(x-\mu)^2\right]\pi(\mu)d\mu$$

and
$$m''(x) = \frac{\sqrt{n}}{\sqrt{2\pi}\sigma}\frac{n^2}{\sigma^4}\int (\mu - x)^2\exp\left[-\frac{n}{2\sigma^2}(x-\mu)^2\right]\pi(\mu)d\mu - \frac{n}{\sigma^2}m(x)$$

The Bayes estimate of μ is

$$\delta^B(x) = \frac{\sqrt{n}}{m(x)\sqrt{2\pi}\sigma}\int \mu\exp\left[-\frac{n}{2\sigma^2}(x-\mu)^2\right]\pi(\mu)d\mu$$

$$= x + \frac{\sqrt{n}}{m(x)\sqrt{2\pi}\sigma}\int (\mu - x)\exp\left[-\frac{n}{2\sigma^2}(x-\mu)^2\right]\pi(\mu)d\mu$$

$$= x + \frac{\sigma^2}{n}\frac{m'(x)}{m(x)} = x + \frac{\sigma^2}{n}\frac{\partial}{\partial x}\log m(x)$$

Consider

$$E[(\mu - x)^2 | \bar{X} = x] = \frac{\sqrt{n}}{m(x)\sqrt{2\pi}\sigma}\int (\mu - x)^2\exp\left[-\frac{n}{2\sigma^2}(\mu - x)^2\right]\pi(\mu)d\mu$$

$$= \frac{[m''(x) + (n/\sigma^2)m(x)](\sigma^4/n^2)}{m(x)}$$

$$= \frac{\sigma^4}{n^2}\frac{m''(x)}{m(x)} + \frac{\sigma^2}{n}$$

Therefore, the posterior variance is

$$V(\mu|\bar{X} = x) = E\{(\mu - x)^2|\bar{X} = x\} - \{E[(\mu - x)|\bar{X} = x]\}^2$$

$$= \frac{\sigma^4}{n^2}\frac{m''(x)}{m(x)} + \frac{\sigma^2}{n} - \left[\frac{\sigma^2}{n}\frac{m'(x)}{m(x)}\right]^2$$

$$= \frac{\sigma^4}{n^2}\left[\frac{m''(x)}{m(x)} - \left(\frac{m'(x)}{m(x)}\right)^2\right] + \frac{\sigma^2}{n}$$

$$= \frac{\sigma^4}{n^2}\frac{\partial^2}{\partial x^2}\log m(x) + \frac{\sigma^2}{n}$$

Consider, now, the joint density $h(x, \mu)$ when $N(\theta, \tau^2)$ is the prior for μ. Note that $\bar{X} \sim N(\mu, \sigma^2/n)$. This gives that the marginal of $\bar{X}$, $m(x)$, as normal with mean

$$E(\bar{X}) = E^{\mu}E^{\bar{X}|\mu}(\bar{X}) = E^{\mu}\mu = \theta$$

and variance

$$V(\bar{X}) = E^{\mu}V^{\bar{X}|\mu}(\bar{X}) + V^{\mu}E^{\bar{X}|\mu}(\bar{X})$$

$$= E^{\mu}\left(\frac{\sigma^2}{n}\right) + V^{\mu}(\mu) = \frac{\sigma^2}{n} + \tau^2$$

Thus, $m(x)$ is $N[\theta, (\sigma^2/n) + \tau^2]$. In case of prior that assigns probability one at some point θ_1, the marginal of $\bar{X}$ is

$$m(x) = \frac{\sqrt{n}}{\sqrt{2\pi}\sigma}\exp\left[-\frac{n}{2\sigma^2}(x - \theta_1)^2\right]$$

which is $N(\theta_1, \sigma^2/n)$. Consider, now, the prior for μ as the mixture of distributions $N(\theta, \sigma^2/n)$ and $N(\theta, \tau^2)$

$$\pi(\mu) = \varepsilon I(\theta = \theta_1) + (1 - \varepsilon)N(\theta, \tau^2)$$

The marginal is also a mixture of distributions

$$m(x) = \varepsilon\phi\left(\frac{x - \theta_1}{\sqrt{\sigma^2/n}}\right) + (1 - \varepsilon)\phi\left(\frac{x - \theta}{\sqrt{(\sigma^2/n) + \tau^2}}\right) \quad (8.7.29)$$

where $\phi(x) = (1/\sqrt{2\pi})\exp(-x^2/2)$. We have

$$m'(x) = \varepsilon\frac{(\theta_1 - x)}{\sqrt{\sigma^2/n}}\phi\left(\frac{x - \theta}{\sqrt{\sigma^2/n}}\right) + (1 - \varepsilon)\frac{(\theta - x)}{\sqrt{(\sigma^2/n) + \tau^2}}\phi\left(\frac{x - \theta}{\sqrt{(\sigma^2/n) + \tau^2}}\right) \quad (8.7.30)$$

The Bayes estimate of μ in this case is given by

$$\delta^B(x) = x + \frac{\sigma^2}{n}\frac{m'(x)}{m(x)}$$

with values of $m(x)$ and $m'(x)$ calculated in Eqs. (8.7.29) and (8.7.30), respectively.

***Example* 8.43** Let $X_1,\ldots,X_n$ be a random sample from $N(\mu, \sigma^2)$, where $\mu \in \mathbb{R}^1$ is unknown and $\sigma^2 > 0$ is known. Let $\Theta = \mathbb{R}^1$, $\pi(\mu)$ be an improper prior defined as a Lebesgue measure on $\mathbb{R}^1$, and the loss function be squared error loss function. Find the generalized Bayes estimate of μ. Consider, next, the sequence of prior distributions $\{N(0, j)\}$. Find the limit of the Bayes estimates and show that it is the generalized Bayes estimate. Also, show that $\delta^{\text{GB}} = \bar{X}$ is admissible.

Solution. Consider the minimization of the quantity

$$\int_{\Theta} L(\mu, \delta) f(\mathbf{x}|\mu)\pi(\mu)d\mu$$

$$\frac{\partial}{\partial \delta}\int_{-\infty}^{\infty}(\mu-\delta)^2\left(\frac{1}{2\pi\sigma^2}\right)^{n/2}\exp\left[-\frac{1}{2}\Sigma(x_i-\mu)^2\right]d\mu = 0$$

The generalized Bayes estimate with respect to the Lebesgue measure $\pi(\mu)$ is

$$\mu^{\text{GB}}(\mathbf{x}) = \frac{\int_{-\infty}^{\infty}\mu\exp\left[-\left(\frac{1}{2}\right)\Sigma(x_i-\mu)^2\right]d\mu}{\int_{-\infty}^{\infty}\exp\left[-\left(\frac{1}{2}\right)\Sigma(x_i-\mu)^2\right]d\mu}$$

Expressing $\Sigma(x_i - \mu)^2 = \Sigma(x_i - \bar{x})^2 + n(\bar{x} - \mu)^2$, we have

$$\mu^{\text{GB}} = \frac{\int_{-\infty}^{\infty}\mu\exp\left[-\left(\frac{n}{2}\right)(\mu-\bar{x})^2\right]d\mu}{\int_{-\infty}^{\infty}\exp\left[-\left(\frac{n}{2}\right)(\mu-\bar{x})^2\right]d\mu} = \bar{x}$$

We have already seen that the Bayes estimate of μ with respect to the prior distribution $\pi(\mu) \sim N(\theta, \tau^2)$,

$$\mu^B_{\tau^2}(\mathbf{x}) = \frac{\sigma^2}{n\tau^2+\sigma^2}\theta + \frac{n\tau^2}{n\tau^2+\sigma^2}\bar{x}$$

$\mu^B_{\tau^2}(\mathbf{x}) \to \mu^{\text{GB}} = \bar{x}$ as $\tau^2 \to \infty$, shows that μ^{GB} is the limit of the Bayes estimates $\mu^B_{\tau^2}(\mathbf{x})$. Note that the Bayes estimator is not unbiased, but a generalized Bayes estimate is unbiased.

Consider, now, the sequence of prior distributions $\{\pi_j\} = \{N(0, j)\}$. The risk function of any estimator δ, $R(\mu, \delta) = E^{\mathbf{X}}_{\mu}L[\mu, \delta(\mathbf{X})]$ is continuous in μ since the distribution of $\mathbf{X}$ is one-parameter exponential family. Consider the estimator $\bar{X}$. Its risk is given by

$$R(\mu, \bar{X}) = \frac{\sigma^2}{n}$$

$$r(\pi_j, \bar{X}) = \frac{\sigma^2}{n} < \infty \; \forall \, j$$

This satisfies the first condition of Theorem 8.2.10. The Bayes estimator with respect to the prior distribution $\pi_j \equiv N(0, j)$ is

$$\delta_j^0(\mathbf{x}) = \frac{nj}{nj + \sigma^2} \bar{x}$$

$$R(\mu, \delta_j^0) = E^{\bar{X}} \left(\frac{nj}{nj + \sigma^2} \bar{X} - \mu \right)^2 = \frac{nj^2 \sigma^2 + \sigma^4 \mu^2}{(nj + \sigma^2)^2}$$

since $\bar{X} \sim N(\mu, \sigma^2/n)$. The Bayes risk of δ_j^0 is

$$r(\pi_j, \delta_j^0) = \frac{nj^2 \sigma^2 + j\sigma^4}{(nj + \sigma^2)^2} = \frac{\sigma^2 j(nj + \sigma^2)}{(nj + \sigma^2)^2} = \frac{\sigma^2 j}{nj + \sigma^2}$$

Consider that for any $\varepsilon > 0$, $N_\varepsilon(\mu_0) = \{\mu \colon \mu_0 - \varepsilon \le \mu \le \mu_0 + \varepsilon\}$

$$\pi_j[N_\varepsilon(\mu_0)] = P_{\pi_j}[\mu_0 - \varepsilon \le \mu \le \mu_0 + \varepsilon]$$

$$= P_{\pi_j}\left(\frac{\mu_0 - \varepsilon}{\sqrt{j}} \le \frac{\mu}{\sqrt{j}} \le \frac{\mu_0 + \varepsilon}{\sqrt{j}} \right)$$

$$= \Phi\left(\frac{\mu_0 + \varepsilon}{\sqrt{j}} \right) - \Phi\left(\frac{\mu_0 - \varepsilon}{\sqrt{j}} \right) = \frac{2\varepsilon}{\sqrt{j}} \Phi'(\mu_j)$$

where Φ is the *cdf* of $N(0, 1)$ distribution, Φ' is its first derivative, and μ_j is some value of μ

$$\frac{\mu_0 - \varepsilon}{\sqrt{j}} \le \mu_j \le \frac{\mu_0 + \varepsilon}{\sqrt{j}}$$

Consider the quantity

$$\frac{r(\pi_j, \bar{X}) - r(\pi_j, \delta_j^0)}{\pi_j[N_\varepsilon(\theta_0)]} = \frac{\sigma^4/[n(nj + \sigma^2)]}{(2\varepsilon/\sqrt{j})\Phi'(\mu_j)}$$

$$= \frac{\sigma^4 \sqrt{j}}{2\varepsilon n(nj + \sigma^2)\Phi'(\mu_j)}$$

Since $j \to \infty$, $\Phi'(\mu_j) \to \Phi'(0) = (\sqrt{2\pi})^{-1}$ and

$$\lim_{j \to \infty} \frac{r(\pi_j, \bar{X}) - r(\pi_j, \delta_j^{(0)})}{\pi_j[N_\varepsilon(\theta_0)]} = 0$$

Thus, the second condition of Theorem 8.2.10 is also satisfied. Therefore, $\bar{X}$ is admissible.

***Example* 8.44** Consider a random sample $X_1,\ldots,X_n$ from $N(\mu, \sigma^2)$ and an improper prior for (μ, σ^2) with respect to the Lebesgue measure on $\mathbb{R}^2$

$$\pi(\mu, \sigma^2) = \frac{1}{\sigma^2}, \sigma^2 > 0$$

Show that the posterior density of (μ, σ^2) given $\mathbf{x} = (x_1,\ldots,x_n)$ is

$$\pi(\mu, \sigma^2|\mathbf{x}) = \pi_1(\mu|\sigma^2, \mathbf{x})\, \pi_2(\sigma^2|\mathbf{x})$$

where π_1 is $N(\overline{x}, \sigma^2/n)$ and π_2 is the density of $1/\sigma^2$, where σ^2 is $G_1((n-1)/2, 2/\Sigma(x_i-\overline{x})^2)$. Find the marginal posterior density of μ given $\mathbf{x}$ which is a t-distribution with $(n-1)$ degree of freedom, t_{n-1}. Find the generalized Bayes estimator of μ/σ under the squared error loss function.

Solution. The joint distribution of $\mathbf{X}$ and (μ, σ^2) is

$$h(\mathbf{x}, (\mu, \sigma^2)) = \left(\frac{1}{2\pi}\right)^{n/2} \frac{1}{\sigma^n} \exp\left(-\frac{1}{2\sigma^2}\sum(x_i-\mu)^2\right)\frac{1}{\sigma^2}$$

$$= \left(\frac{1}{2\pi}\right)^{n/2} \frac{1}{\sigma^{n+2}} \exp\left[-\frac{n}{2\sigma^2}(\mu-\overline{x})^2\right] \exp\left(-\frac{1}{2\sigma^2}\sum(x_i-\overline{x})^2\right)$$

The posterior distribution of (μ, σ^2) given $\mathbf{x}$ is

$$\pi(\mu, \sigma^2|\mathbf{x}) \propto \frac{1}{(\sigma^2)^{1/2}} \exp\left[-\frac{n}{2\sigma^2}(\mu-\overline{x})^2\right] \frac{1}{(\sigma^2)^{[(n-1)/2]+1}} \exp\left[-\frac{1}{2\sigma^2}\sum(x_i-\overline{x})^2\right] \quad (8.7.31)$$

The first term on the right gives $\pi_1(\mu|\mathbf{x})$, so that $\pi_1(\mu|\sigma^2, \mathbf{x}) \sim N(\overline{x}, \sigma^2/n)$ and $\pi_2(\sigma^2|\mathbf{x}) \equiv \text{IG}_2[(n-1)/2, \Sigma(x_i-\overline{x})^2/2]$. This gives $\pi_1(1/\sigma^2|\mathbf{x}) \equiv G_2[(n-1)/2, \Sigma(x_i-\overline{x})^2/2]$. [Note that if $Y \sim \text{IG}_2(\alpha, \beta)$, $X = 1/Y$, then $X \sim G_2(\alpha, \beta)$]. Thus the second and the last term in Eq. (8.7.31) is $\pi_2(\sigma^2|\mathbf{x})$ which is the density of $1/\sigma^2$ where $\sigma^2|\mathbf{x} \sim \text{IG}_2((n-1)/2, \Sigma(x_i-\overline{x})^2/2)$. This proves the required result

$$\pi(\mu, \sigma^2|\mathbf{x}) = \pi_1(\mu|\sigma^2, \mathbf{x})\, \pi_2(\sigma^2|\mathbf{x})$$

The marginal density of μ given $\mathbf{x}$ is

$$\pi(\mu|\mathbf{x}) = \int_0^\infty \pi(\mu, \sigma^2|\mathbf{x})\,d\sigma^2$$

$$\propto \int_0^\infty \frac{1}{\sigma^{n+2}} \exp\left\{-\frac{1}{2\sigma^2}[n(\mu-\overline{x})^2 + \sum(x_i-\overline{x})^2]\right\} d\sigma^2 \quad (8.7.32)$$

$$= c\left[n(\mu-\overline{x})^2 + \sum(x_i-\overline{x})^2\right]^{-n/2}$$

$$\propto \left[1 + \frac{1}{n-1}u\right]^{-n/2} \sim t_{n-1}$$

where $u = n(\overline{x} - \mu)^2/s_{n-1}^2$, $s_{n-1}^2 = [1/(n-1)]\Sigma(x_i - \overline{x})^2$, $u/(n-1) \sim F_{1,n-1}$. This shows that the marginal distribution of μ given $\mathbf{x}$, $\pi(\mu|\mathbf{x})$, is $f(u)$ where f is the density of t-distribution, t_{n-1}. The generalized Bayes estimator of $g(\mu, \sigma^2) = \mu/\sigma$ is

$$\delta^{GB}(\mathbf{x}) = \int \frac{\mu}{\sigma}\pi(\mu, \sigma^2|\mathbf{x})d\mu d\sigma^2$$

$$= \iint \frac{\mu}{\sigma}\pi_1(\mu|\sigma^2, \mathbf{x})\pi_2(\sigma^2|\mathbf{x})d\mu d\sigma^2$$

$$= \int \frac{1}{\sigma}\left[\int \mu\pi_1(\mu|\sigma^2, \mathbf{x})d\mu\right]\pi_2(\sigma^2|\mathbf{x})d\sigma^2$$

$$= \overline{x}\int_0^\infty \frac{1}{\sigma}\pi_2(\sigma^2|\mathbf{x})d\sigma^2$$

$$= \overline{x}\frac{\left[\sum(x_i - x)^2/2\right]^{(n-1)/2}}{\Gamma[(n-1)/2]}\int_0^\infty \frac{1}{(\sigma^2)^{(n/2)+1}}\exp\left[-\frac{\sum(x_i-\overline{x})^2}{2\sigma^2}\right]d\sigma^2$$

$$= \overline{x}\frac{\left[\sum(x_i - \overline{x})^2/2\right]^{(n-1)/2}}{\Gamma[(n-1)/2]}\frac{\Gamma(n/2)}{\left[\sum(x_i - \overline{x})^2/2\right]^{n/2}}$$

$$= \frac{\Gamma(n/2)}{\Gamma[(n-1)/2]}\left[\frac{2}{\sum(x_i-\overline{x})^2}\right]^{1/2}\overline{x}$$

***Example* 8.45** Let $\mathbf{X} \sim N_p(\boldsymbol{\theta}, I_p)$ and the prior distribution of $\boldsymbol{\theta} = (\theta_1,\ldots,\theta_p)'$, $\pi(\boldsymbol{\theta})$, is $N_p(\mathbf{0}, \tau I_p)$, $\tau > 0$. Consider the loss function $L(\boldsymbol{\theta}, \boldsymbol{\delta}) = \|\boldsymbol{\theta} - \boldsymbol{\delta}\|^2$.

(i) Find the posterior distribution of $\boldsymbol{\theta}|\mathbf{X}$ and Bayes estimator of $\boldsymbol{\theta}$.
(ii) Compare the estimators $\boldsymbol{\theta}^B$ and UMVUE $\boldsymbol{\theta}_1(\mathbf{X}) = \mathbf{X}$ in terms of MSE.
(iii) Show that the UMVUE of $\boldsymbol{\theta}$, $\boldsymbol{\theta}(\mathbf{X}) = \mathbf{X}$, is minimax for any $p \geq 1$. Is it admissible?

Solution.

(i) From the given prior, we have

$$E(\boldsymbol{\theta}) = \mathbf{0} \text{ and } V(\boldsymbol{\theta}) = D(\boldsymbol{\theta}) = \tau I_p$$

Further, $E(\mathbf{X}) = E^{\boldsymbol{\theta}}E^{\mathbf{X}|\boldsymbol{\theta}}(\mathbf{X}) = E^{\boldsymbol{\theta}}(\boldsymbol{\theta}) = \mathbf{0}$ and $V(\mathbf{X}) = E^{\boldsymbol{\theta}}\, E^{\mathbf{X}|\boldsymbol{\theta}}\|\mathbf{X} - \boldsymbol{\theta}\|^2$
We can verify

$$E(\mathbf{X}\boldsymbol{\theta}') = \tau I_p \text{ and } E(\mathbf{X}\mathbf{X}') = (1+\tau)I_p$$

So, the joint distribution of $\mathbf{X}$ and $\boldsymbol{\theta}$ is given by bivariate normal distribution

$$\begin{pmatrix}\boldsymbol{\theta}\\ \mathbf{X}\end{pmatrix} \sim \left[\begin{pmatrix}\mathbf{0}\\ \mathbf{0}\end{pmatrix}\begin{pmatrix}\tau I_p & \tau I_p\\ \tau I_p & (1+\tau)I_p\end{pmatrix}\right]$$

Therefore, by multivariate standard result,

$$\mu_1 + \Sigma_{12}\Sigma_{22}^{-1}(X_2 - \mu_2) = \mathbf{0} + \tau I_p(1+\tau)^{-1} I_p(\mathbf{X} - \mathbf{0}) = \frac{\tau}{1+\tau}\mathbf{X}$$

and
$$\Sigma_{11.2} = \Sigma_{11} - \Sigma_{12}\Sigma_{22}^{-1}\Sigma_{21} = \tau I_p - \tau I_p(1+\tau)^{-1} I_p^{-1} \tau I_p$$

$$= \left(\tau - \frac{\tau^2}{1+\tau}\right) I_p = \tau\left(1 - \frac{\tau}{1+\tau}\right) I_p$$

$$= \frac{\tau}{1+\tau} I_p$$

This gives
$$\boldsymbol{\theta} | \mathbf{x} \sim N\left(\frac{\tau}{1+\tau}\mathbf{x}, \frac{\tau}{1+\tau} I_p\right)$$

Therefore, the Bayes estimator of $\boldsymbol{\theta}$ is

$$\boldsymbol{\theta}^B = \frac{\tau}{1+\tau}\mathbf{x}$$

(ii) We have

$$\text{MSE}_{\boldsymbol{\theta}}[\boldsymbol{\theta}_1(X)] = E_{\boldsymbol{\theta}}\|\mathbf{X} - \boldsymbol{\theta}\|^2 = p \tag{8.7.33}$$

which is independent of $\boldsymbol{\theta}$, and

$$\text{MSE}_{\theta}[\boldsymbol{\theta}^B(\mathbf{X})] = E_{\boldsymbol{\theta}}\left\|\frac{\tau}{1+\tau}\mathbf{X} - \boldsymbol{\theta}\right\|^2$$

$$= E_{\boldsymbol{\theta}}\left\|\frac{\tau}{1+\tau}(\mathbf{X} - \boldsymbol{\theta}) - \frac{1}{1+\tau}\boldsymbol{\theta}\right\|^2$$

$$= \left(\frac{\tau}{1+\tau}\right)^2 E_{\boldsymbol{\theta}}\|\mathbf{X} - \boldsymbol{\theta}\|^2 + \left(\frac{1}{1+\tau}\right)^2 \|\boldsymbol{\theta}\|^2$$

$$= \left(\frac{\tau}{1+\tau}\right)^2 p + \left(\frac{1}{1+\tau}\right)^2 \|\boldsymbol{\theta}\|^2 \to p \text{ as } \tau \to \infty \tag{8.7.34}$$

In this case, prior is a uniform distribution over $\mathbb{R}^1$. Thus, the UMVUE $\boldsymbol{\theta}_1(\mathbf{X})$ and Bayes estimate of $\boldsymbol{\theta}$ are identical in terms of MSE when prior has no information about $\boldsymbol{\theta}$.

(iii) Any estimator $\boldsymbol{\delta}_0(\mathbf{X})$ that satisfies

$$\|\boldsymbol{\delta\theta}\|^2 \inf_{\boldsymbol{\delta}} \sup_{\boldsymbol{\theta}} E_{\boldsymbol{\theta}} \|\boldsymbol{\delta}(\mathbf{X}) - \boldsymbol{\theta}\|^2 = V$$

is called the minimax estimator of $\boldsymbol{\theta}$. Consider

$$V = \inf_{\boldsymbol{\delta}} \sup_{\boldsymbol{\theta}} E_{\boldsymbol{\theta}} \|\boldsymbol{\delta}(\mathbf{X}) - \boldsymbol{\theta}\|^2 \le \sup_{\boldsymbol{\theta}} E_{\boldsymbol{\theta}} \|\mathbf{X} - \boldsymbol{\theta}\|^2 = p \tag{8.7.35}$$

Further,
$$V = \inf_{\boldsymbol{\delta}} \sup_{\boldsymbol{\theta}} E_{\boldsymbol{\theta}} \|\boldsymbol{\delta}(\mathbf{X}) - \boldsymbol{\theta}\|^2 \ge \inf_{\boldsymbol{\delta}} \int E_{\boldsymbol{\theta}} \|\boldsymbol{\delta}(\mathbf{X}) - \boldsymbol{\theta}\|^2 \pi_\tau(\boldsymbol{\theta}) d\boldsymbol{\theta}$$

where $\pi_\tau(\boldsymbol{\theta}) \equiv N_p(\mathbf{0}, \tau I_p)$. We have

$$V \geq \inf_{\boldsymbol{\delta}} r(\pi_\tau, \boldsymbol{\delta}) = r(\pi_\tau, \boldsymbol{\theta}^B(\mathbf{X}))$$

$$= \left(\frac{\tau}{1+\tau}\right)^2 p + \left(\frac{1}{1+\tau}\right)^2 E^\theta \|\boldsymbol{\theta}\|^2 \qquad \text{[by Eq. (8.7.34)]}$$

$$= \left(\frac{\tau}{1+\tau}\right)^2 p + \left(\frac{1}{1+\tau}\right)^2 p\tau = \left(\frac{\tau}{1+\tau}\right) p \qquad [\text{since } \tau > 0]$$

This shows $\qquad V \geq p \text{ as } \tau \to \infty$

On combining Eqs. (8.7.34) and (8.7.35), we get

$$V = p \text{ as } \tau \to \infty$$

Since $\qquad \sup_{\boldsymbol{\theta}} E_{\boldsymbol{\theta}} \|\mathbf{X} - \boldsymbol{\theta}\|^2 = V = p$

from Eq. (8.7.33), $\theta_1(\mathbf{X}) = \mathbf{X}$ is minimax.

Blyth (1951), for $p = 1$, has shown that $\boldsymbol{\theta}(\mathbf{X}) = \mathbf{X}$ is admissible, and Stein (1961), for $p = 2$, has shown that it is an admissible estimator of $\boldsymbol{\theta}$. However, Stein (1956), for $p \geq 3$, has shown that $\boldsymbol{\theta}(\mathbf{X}) = \mathbf{X}$ is not admissible and proposed an estimator

$$\boldsymbol{\delta}_c(\mathbf{X}) = \left(1 - c\frac{p-2}{\|\mathbf{X}\|^2}\right)\mathbf{X}$$

so that $\qquad E\left\|\boldsymbol{\delta}_c(\mathbf{X}) - \boldsymbol{\theta}\right\|^2 = p - (p-2)^2 E\left[\frac{c(2-c)}{\|\mathbf{X}\|^2}\right]$

Note that $c(2 - c) \geq 0$ whenever $0 \leq c \leq 2$ and $c(2 - c)$ attains maximum at $c = 1$. The estimator $\boldsymbol{\delta}_c(\mathbf{X})$ outperforms $\boldsymbol{\theta}(\mathbf{X}) = \mathbf{X}$. The estimator $\boldsymbol{\delta}_1(\mathbf{X})$ is popularly known as the *James-Stein estimator* and it outperforms all $\boldsymbol{\delta}_c(\mathbf{X})$, $c \neq 1$.

***Example* 8.46** Let $\mathbf{X} = (\mathbf{X}_1,\ldots,\mathbf{X}_n)$ be a random sample from $N_k(\boldsymbol{\mu}, \mathbf{I})$, $\boldsymbol{\mu} \in \mathbb{R}^k$, and the prior distribution of $\boldsymbol{\mu}$ be $N_k(\boldsymbol{\theta}_0, \Sigma_0)$. Show that it is a conjugate prior. Find the Bayes estimator of $\boldsymbol{\mu}$ and its Bayes risk.

Solution. We know that the statistic $\mathbf{T}(\mathbf{X}) = \Sigma_{i=1}^n \mathbf{X}_i = \mathbf{1}'\mathbf{X}$ is sufficient for $\boldsymbol{\mu}$ and is distributed as $N_k(n\boldsymbol{\mu}, n\mathbf{I})$. The joint density of $\mathbf{T}$ and $\boldsymbol{\mu}$ is given by

$$h(\mathbf{t}, \boldsymbol{\mu}) = \frac{1}{[(2\pi)^k |n\mathbf{I}|]^{1/2}} \exp\left[-(\mathbf{t} - n\boldsymbol{\mu})'(n\mathbf{I})^{-1}\frac{(\mathbf{t} - n\boldsymbol{\mu})}{2}\right] \frac{1}{[(2\pi)^k |\Sigma_0|]^{1/2}} \exp\left[-(\boldsymbol{\mu} - \boldsymbol{\theta}_0)'\Sigma_0^{-1}\frac{(\boldsymbol{\mu} - \boldsymbol{\theta}_0)}{2}\right]$$

$$\propto \exp\left[-\frac{(1/n)(\mathbf{t} - n\boldsymbol{\mu})'(\mathbf{t} - n\boldsymbol{\mu})}{2} - \frac{(\boldsymbol{\mu} - \boldsymbol{\theta}_0)'\Sigma_0^{-1}(\boldsymbol{\mu} - \boldsymbol{\theta}_0)}{2}\right]$$

Assuming $\mathbf{C} = n\mathbf{I} + \Sigma_0^{-1}$,

$$h(\mathbf{t}, \boldsymbol{\mu}) \propto \exp\left\{-\frac{\left[\boldsymbol{\mu} - \mathbf{C}^{-1}(\Sigma_0^{-1}\boldsymbol{\theta}_0 + \mathbf{t})\right]' \mathbf{C}\left[\boldsymbol{\mu} - \mathbf{C}^{-1}(\Sigma_0^{-1}\boldsymbol{\theta}_0 + \mathbf{t})\right]}{2}\right\}$$

Thus, the posterior distribution of $\boldsymbol{\mu}$ given $\mathbf{X}$ is

$$\pi(\boldsymbol{\mu}|\mathbf{t}) \propto \exp\left\{-\frac{1}{2}\left[\boldsymbol{\mu} - \mathbf{C}^{-1}(\Sigma_0^{-1}\boldsymbol{\theta}_0 + \mathbf{t})\right]' \mathbf{C}\left[\boldsymbol{\mu} - \mathbf{C}^{-1}(\Sigma_0^{-1}\boldsymbol{\theta}_0 + \mathbf{t})\right]\right\}$$

i.e., $$N\left[\mathbf{C}^{-1}(\Sigma_0^{-1}\boldsymbol{\theta}_0 + \mathbf{t}), \mathbf{C}^{-1}\right]$$

Thus, the Bayes estimator of $\boldsymbol{\mu}$ is given by

$$\boldsymbol{\mu}^B = E(\boldsymbol{\mu}|\mathbf{X}) = \mathbf{C}^{-1}(\Sigma_0^{-1}\boldsymbol{\theta}_0 + \mathbf{t})$$

and the Bayes risk of $\boldsymbol{\mu}^B$ is

$$r(\pi, \boldsymbol{\mu}^B) = V(\boldsymbol{\mu}|\mathbf{X}) = \mathbf{C}^{-1}$$

***Example* 8.47** Though normal distribution is a natural conjugate family of priors, in case of sampling from normal distribution, sometimes it does not precisely reflect the prior belief of a statistician. In this case, a different prior is sought. Let $X_1,\ldots,X_n$ be *iid* from $N(\mu, \sigma^2)$ and $\mu \sim \text{DE}(\tau)$ with the *pdf*

$$\pi(\mu) = \frac{1}{2\tau}\exp\left(-\frac{|\mu|}{\tau}\right)$$

where τ is known. Find the Bayes estimate of μ.

Solution. The joint *pdf* of $\bar{x}$ and μ is

$$h(\bar{x}, \mu|\tau) \propto \exp\left[-\frac{n}{2\sigma^2}(\bar{x} - \mu)^2 - \frac{|\mu|}{\tau}\right]$$

Consider the simplification of the term in the exponent

$$-\frac{n}{2\sigma^2}(\bar{x} - \mu)^2 - \frac{|\mu|}{\tau} = -\frac{n}{2\sigma^2}\left(\mu^2 - 2\mu\bar{x} + \frac{2\sigma^2|\mu|}{n\tau} + \bar{x}^2\right)$$

$$= \begin{cases} -\dfrac{n}{2\sigma^2}\left[\mu - \left(\bar{x} + \dfrac{\sigma^2}{n\tau}\right)\right]^2 + \left(\dfrac{\bar{x}}{\tau} + \dfrac{\sigma^2}{2n\tau^2}\right), & \text{if } \mu < 0 \\ -\dfrac{n}{2\sigma^2}\left[\mu - \left(\bar{x} - \dfrac{\sigma^2}{n\tau}\right)\right]^2 + \left(-\dfrac{\bar{x}}{\tau} + \dfrac{\sigma^2}{2n\tau^2}\right), & \text{if } \mu > 0 \end{cases}$$

Therefore, the posterior distribution is given by

$$\pi(\mu|\bar{x},\tau) \propto \begin{cases} \exp\left\{-\dfrac{n}{2\sigma^2}(\mu-\mu(\mathbf{x})_-)^2\right\}, & \text{if } \mu<0 \\ \exp\left\{-\dfrac{n}{2\sigma^2}(\mu-\mu(\mathbf{x})_+)^2\right\}, & \text{if } \mu>0 \end{cases}$$

where $\mu(\mathbf{x})_- = \bar{x} + [\sigma^2/(n\tau)]$ when $\mu < 0$ and $\mu(\mathbf{x})_+ = \bar{x} - [\sigma^2/(n\tau)]$ when $\mu > 0$. Therefore, the Bayes estimate of μ is given by

$$\mu^B = E(\mu|\mathbf{x}) = \frac{\int_{-\infty}^{\infty} \mu \exp\left\{-\left(\dfrac{n}{2\sigma^2}\right)(\mu-\mu(\mathbf{x})_\pm)^2\right\}d\mu}{\int_{-\infty}^{\infty} \exp\left\{-\left(\dfrac{n}{2\sigma^2}\right)(\mu-\mu(\mathbf{x})_\pm)^2\right\}d\mu}$$

Consider the denominator, we have

$$\int_{-\infty}^{\infty} \exp\left\{-\frac{n}{2\sigma^2}(\mu-\mu(\mathbf{x})_\pm)^2\right\}d\mu$$

$$= \int_{-\infty}^{0} \exp\left[-\frac{n}{2\sigma^2}(\mu-\mu(\mathbf{x})_-)^2\right]d\mu + \int_{0}^{\infty} \exp\left[-\frac{n}{2\sigma^2}(\mu-\mu(\mathbf{x})_+)^2\right]d\mu$$

$$= 2\frac{1}{2}\sqrt{2\pi}\frac{\sigma}{\sqrt{n}} = \sqrt{\frac{2\pi\sigma^2}{n}}$$

Consider, now, the term in the numerator

$$\int_{-\infty}^{\infty} \mu \exp\left\{-\frac{n}{2\sigma^2}(\mu-\mu(\mathbf{x})_\pm)^2\right\}d\mu$$

$$= \int_{-\infty}^{0} \mu \exp\left\{-\frac{n}{2\sigma^2}(\mu-\mu(\mathbf{x})_-)^2\right\}d\mu + \int_{0}^{\infty} \mu \exp\left\{-\frac{n}{2\sigma^2}(\mu-\mu(\mathbf{x})_+)^2\right\}d\mu$$

Simplifying the second term, we get

$$\int_{0}^{\infty} \mu \exp\left\{-\frac{n}{2\sigma^2}(\mu-\mu(\mathbf{x})_+)^2\right\}d\mu$$

$$= \int_{0}^{\infty} (\mu-\mu(\mathbf{x})_+) \exp\left\{-\frac{n}{2\sigma^2}(\mu-\mu(\mathbf{x})_+)^2\right\}d\mu + \int_{0}^{\infty} \mu(\mathbf{x})_+ \exp\left\{-\frac{n}{2\sigma^2}(\mu-\mu(\mathbf{x})_+)^2\right\}d\mu$$

Let $(n/2\sigma^2)(\mu - \mu(\mathbf{x})_+)^2 = t$; $(\mu - \mu(\mathbf{x})_+)\,d\mu = (\sigma^2/n)dt$

$$\int_0^{\infty} \mu \exp\left\{-\frac{n}{2\sigma^2}(\mu - \mu(\mathbf{x})_+)^2\right\} d\mu$$

$$= \int_{(n/2\sigma^2)\mu^2(\mathbf{x})_+}^{\infty} \exp(-t)\frac{\sigma^2}{n}dt + \mu(\mathbf{x})_+ \frac{1}{2}\sqrt{2\pi}\frac{\sigma}{\sqrt{n}}$$

$$= \frac{\sigma^2}{n}\exp\left[-\frac{n}{2\sigma^2}\mu^2(\mathbf{x})_+\right] + \mu(\mathbf{x})_+\sqrt{\frac{\pi\sigma^2}{2n}}$$

Similarly, we obtain for the first term in the numerator,

$$\int_{-\infty}^{0} \mu \exp\left[-\frac{n}{2\sigma^2}(\mu - \mu(\mathbf{x})_-)^2\right] d\mu$$

$$= \frac{\sigma^2}{n}\exp\left[-\frac{n}{2\sigma^2}\mu^2(\mathbf{x})_-\right] + \mu(\mathbf{x})_-\sqrt{\frac{\pi\sigma^2}{2n}}$$

Therefore, the term in the numerator simplifies to

$$\int_{-\infty}^{\infty} \mu \exp\left[-\frac{n}{2\sigma^2}(\mu - \mu(\mathbf{x})_\pm)^2\right] d\mu$$

$$= \frac{\sigma^2}{n}\left[\exp\left(-\frac{n\mu^2(\mathbf{x})_+}{2\sigma^2}\right) + \exp\left(-\frac{n\mu^2(\mathbf{x})_-}{2\sigma^2}\right)\right] + \sqrt{\frac{\pi\sigma^2}{2n}}(\mu(x)_+ + \mu(x)_-)$$

Therefore, the Bayes estimate of μ is given by

$$\mu^B = E(\theta|\mathbf{x}) = \left(\sqrt{\frac{2\pi\sigma^2}{n}}\right)^{-1}\left\{\frac{\sigma^2}{n}\left[\exp\left(-\frac{n\mu^2(\mathbf{x})_+}{2\sigma^2}\right) + \exp\left(-\frac{n\mu^2(\mathbf{x})_-}{2\sigma^2}\right)\right] + \sqrt{\frac{\pi\sigma^2}{2n}}(\mu(\mathbf{x})_+ + \mu(\mathbf{x})_-)\right\}$$

***Example* 8.48** Let $X_1,\ldots,X_n$ be *iid* according to $N(\mu, \sigma^2)$ where both μ and σ^2 are unknown. Define the conjugate prior for σ^2 by inverted gamma distribution $\mathrm{IG}_2(\alpha, \beta)$ with the *pdf*

$$\pi(\sigma^2) = \frac{\beta^\alpha}{\Gamma(\alpha)}\left(\frac{1}{\sigma^2}\right)^{\alpha+1}\exp\left(-\frac{\beta}{\sigma^2}\right), \sigma^2 > 0$$

where α and β are both positive. Show that the posterior distribution of σ^2 is $\mathrm{IG}_2\{(n-1)/2) + \alpha, [(n-1)S_{n-1}^2/2] + \beta\}$. Find the Bayes estimator of σ^2 under squared error loss function.

Solution. For $N(\mu, \sigma^2)$, S_{n-1}^2 is a sufficient statistic for σ^2. Therefore, the Bayes estimator of σ^2 must be a function of S_{n-1}^2. We know $U = (n-1)S_{n-1}^2/\sigma^2 \sim \chi_{n-1}^2 = G_1[(n-1)/2, 2]$

$$f(u; n) = \frac{1}{\Gamma[(n-1)/2]2^{(n-1)/2}}u^{[(n-1)/2]-1}\exp\left(-\frac{u}{2}\right), u > 0$$

The distribution of $s_{n-1}^2 = \sigma^2 U/(n-1)$ is given by

$$f(s_{n-1}^2|\sigma^2) = \frac{1}{\Gamma[(n-1)/2]2^{(n-1)/2}}\left[\frac{(n-1)s_{n-1}^2}{\sigma^2}\right]^{[(n-1)/2]-1} \exp\left[-\frac{(n-1)s_{n-1}^2}{2\sigma^2}\right]|J|$$

where $|J| = [(n-1)/\sigma^2]$. We have

$$f(s_{n-1}^2|\sigma^2) = \frac{1}{\Gamma[(n-1)/2]2^{(n-1)/2}}\left[\frac{(n-1)s_{n-1}^2}{\sigma^2}\right]^{(n-1)/2} \exp\left[-\frac{(n-1)s_{n-1}^2}{2\sigma^2}\right]\frac{1}{s_{n-1}^2}$$

The joint density of s_{n-1}^2 and σ^2 is given by

$$h(s_{n-1}^2, \sigma^2) = f(s_{n-1}^2|\sigma^2)\pi(\sigma^2)$$

$$= \frac{1}{\Gamma[(n-1)/2]\,2^{(n-1)/2}}\left[\frac{(n-1)s_{n-1}^2}{\sigma^2}\right]^{(n-1)/2} \exp\left[-\frac{(n-1)s_{n-1}^2}{2\sigma^2}\right]\frac{1}{S_{n-1}^2}$$

$$\frac{\beta^\alpha}{\Gamma(\alpha)}\left(\frac{1}{\sigma^2}\right)^{\alpha+1} \exp\left(-\frac{\beta}{\sigma^2}\right); \sigma^2 > 0$$

Thus, the posterior density is given by

$$\pi(\sigma^2|s_{n-1}^2) \propto h(s_{n-1}^2, \sigma^2)$$

$$\propto \left(\frac{1}{\sigma^2}\right)^{[(n-1)/2]+\alpha+1} \exp\left\{-\frac{1}{\sigma^2}\left[\frac{(n-1)s_{n-1}^2}{2}+\beta\right]\right\}$$

which is $IG_2\{[(n-1)/2] + \alpha, [(n-1)s_{n-1}^2/2] + \beta\}$. We know that if $X \sim IG_2(\alpha, \beta)$, then $E(X) = \beta/(\alpha - 1)$, provided $\alpha > 1$. Therefore, the Bayes estimator of σ^2 is given by

$$(\sigma^2)^B = E(\sigma^2|s_{n-1}^2) = \frac{[(n-1)s_{n-1}^2/2] + \beta}{[(n-1)/2] + \alpha - 1}$$

***Example* 8.49** Let $X_1,\ldots,X_n$ be a random sample from $G_2(1, \theta)$ and the prior distribution be $G_2(1, 1)$. Find the Bayes estimator of θ under the squared error loss function.

Solution. ΣX_i is a sufficient statistic for θ and $\Sigma X_i \sim G_2(n, \theta)$. The joint density of ΣX_i and θ is

$$h(\Sigma x_i, \theta) = \frac{\theta^n}{\Gamma(n)}(\Sigma x_i)^{n-1} \exp[-(\Sigma x_i + 1)\theta]$$

This gives the posterior distribution as

$$\pi(\theta|\Sigma x_i) \propto \exp[-(\Sigma x_i + 1)\theta]\theta^n$$

Therefore, the Bayes estimate of θ is

$$\delta^B(\Sigma x_i) = \frac{\int_0^\infty \theta^{n+1}\exp[-(\Sigma x_i + 1)\theta]\,d\theta}{\int_0^\infty \theta^n \exp[-(\Sigma x_i + 1)\theta]\,d\theta}$$

$$= \frac{\Gamma(n+2)[1/(\Sigma x_i + 1)]^{n+2}}{\Gamma(n+1)[1/(\Sigma x_i + 1)]^{n+1}}$$

$$= \frac{(n+1)}{(\Sigma x_i + 1)}$$

Example **8.50** Let $X_1,\ldots,X_n$ be *iid* $E(\theta)$ with the *pdf*

$$f(x|\theta) = \theta\exp(-\theta x),\ x \ge 0$$

Let the prior distribution of θ be $G_2(\alpha, \beta)$. Show that the Bayes estimator of θ approaches $\overline{X}^{-1}$ as $n \to \infty$.

Solution. The posterior distribution is

$$\theta|\mathbf{x} \sim G_2(\alpha + n, \beta + \Sigma x_i)$$

The Bayes estimate of θ is

$$\theta^B = E(\theta|\mathbf{x}) = \frac{\alpha + n}{\beta + \Sigma x_i} = \frac{(\alpha/n) + 1}{(\beta/n) + \overline{x}} \to \frac{1}{\overline{x}} \text{ as } n \to \infty$$

Example **8.51** Let $X_1,\ldots,X_n$ is a random sample from exponential distribution

$$f(x|\theta) = \frac{1}{\theta}\exp\left(-\frac{x}{\theta}\right)I_{(0,\,\infty)}(x),\ \theta > 0$$

and the prior distribution of θ be the inverted gamma distribution $\mathrm{IG}_1(\alpha,\beta)$ with shape parameter α and scale parameter β, $\alpha > 0$, $\beta > 0$.

(i) Show that this prior is a conjugate prior and find the Bayes estimate of θ. Show that this Bayes estimator is not unbiased but consistent.

(ii) Show that the UMVUE of θ, $\overline{X}$, is not Bayes with respect to any prior distribution but the limit of Bayes estimate corresponding to priors $\{G_1(\alpha,\beta)\}$ for θ^{-1} and also is generalized Bayes with respect to an improper prior

$$\pi(\theta^{-1}) = I_{(0,\,\infty)}(\theta^{-1})$$

Also, find the Bayes estimate of $\exp(-c/\theta)$ under the squared error loss function.

(iii) Find the empirical Bayes estimate of θ by using the method of moments for estimating hyperparameters.

Solution.

(i) Since $X \sim G_1(1,\theta)$, let $T = \Sigma X_i \sim G_1(n, \theta)$. The joint density of t and θ is given by

$$h(t,\theta) \propto \left(\frac{1}{\theta}\right)^{n+\alpha+1} \exp\left[-\left(t+\frac{1}{\beta}\right)\frac{1}{\theta}\right]$$

This gives that the posterior distribution of θ given t is $IG_1[n + \alpha, (t + (1/\beta))^{-1}]$. Thus, the Bayes estimate of θ is

$$\delta^B(\mathbf{x}) = \frac{t+(1/\beta)}{n+\alpha-1}$$

since $E(X) = \beta(\alpha - 1)$ if $X \sim IG_1(\alpha, \beta)$. The bias in $\delta^B(\mathbf{X})$ is given by

$$\text{bias}[\delta^B(\mathbf{X})] = E[\delta^B(\mathbf{X})] - \theta$$

$$= \frac{n\beta\theta+1}{\beta(n+\alpha-1)} - \theta = \frac{1-(\alpha-1)\beta\theta}{\beta(n+\alpha-1)} = O\left(\frac{1}{n}\right)$$

The Bayes estimator $\delta^B(\mathbf{X})$ is not unbiased although it is approximately unbiased since bias$[\delta^B(\mathbf{X})] \to 0$ as $n \to \infty$.
Further, the variance of $\delta^B(\mathbf{X})$ is

$$V[\delta^B(\mathbf{X})] = \frac{1}{(n+\alpha-1)^2}V(T) = \frac{n\theta^2}{(n+\alpha-1)^2} = O\left(\frac{1}{n}\right)$$

Thus, the Bayes estimator $\delta^B(\mathbf{X})$ of θ is consistent since $V[T^B(\mathbf{X})] \to 0$ and $E[T^B(\mathbf{X})] \to \theta$ as $n \to \infty$.

(ii) The sample mean $\bar{X}$ is a UMVUE of θ. This estimator is not Bayes with respect to any prior distribution by Theorem 8.2.4 since

$$r(\pi, \bar{x}) = E^\theta E^{\bar{X}|\theta}(\bar{X}-\theta)^2 = E^\theta \frac{1}{n^2} E^{T|\theta}(T-n\theta)^2$$

$$= E^\theta\left(\frac{n\theta^2}{n^2}\right) = E^\theta\left(\frac{\theta^2}{n}\right) > 0, \text{ for any } \pi$$

Next, consider the prior (improper) for $\theta' = 1/\theta$ over $(0, \infty)$

$$\pi(\theta') = I_{(0,\infty)}(\theta')$$

The solution of the following equation gives the generalized Bayes estimate of θ

$$\frac{\partial}{\partial\delta}\int_0^\infty (\theta'^{-1}-\delta)^2 f(t|\theta')\pi(\theta')d\theta' = 0$$

$$\delta^{GB}(\mathbf{x}) = \frac{\int_0^\infty \theta'^{-1}\theta'^n \exp(-t\theta')d\theta'}{\int_0^\infty \theta'^n \exp(-t\theta')d\theta'}$$

$$= \frac{\Gamma(n)}{t^n} \frac{t^{n+1}}{\Gamma(n+1)} = \frac{t}{n} = \bar{x}$$

Consider the limit of the Bayes estimates with respect to the priors of θ', $\{G_1(\alpha, \beta)\}$

$$\delta^B(\mathbf{x}) = \frac{t + (1/\beta)}{n + \alpha - 1} \to \bar{x}$$

as $\alpha \to 1$ and $\beta \to \infty$.

Therefore, we have seen that $\bar{x}$ is generalized Bayes and the limit of Bayes estimates, though it is not Bayes with respect to any prior distribution. The Bayes estimate under the squared error loss function for estimating the parametric function $\exp(-c\theta')$ is

$$\delta_c^B(\mathbf{x}) = \frac{[t + (1/\beta)]^{n+\alpha}}{\Gamma(n+\alpha)} \int_0^\infty \theta'^{n+\alpha-1} \exp\left\{-\theta'\left[t + c + \left(\frac{1}{\beta}\right)\right]\right\} d\theta'$$

$$= \frac{[t + (1/\beta)]^{n+\alpha}}{\Gamma(n+\alpha)} \Gamma(n+\alpha)\left[t + c + \left(\frac{1}{\beta}\right)\right]^{-(n+\alpha)}$$

$$= \left[1 + \frac{c}{t + (1/\beta)}\right]^{-(n+\alpha)}$$

(iii) Further the empirical Bayes estimate of θ is obtained by estimating hyperparameters α and β. We consider the moments

$$E(X_1) = E^\theta E^{X|\theta}(X_1) = E^\theta(\theta)$$

$$= E^{\theta'}(\theta')^{-1} = \int_0^\infty (\theta')^{-1} \frac{1}{\Gamma(\alpha)\beta^\alpha} (\theta')^{\alpha-1} \exp\left(-\frac{\theta'}{\beta}\right) d\theta'$$

$$= \frac{\Gamma(\alpha-1)\beta^{\alpha-1}}{\Gamma(\alpha)\beta^\alpha} = \frac{1}{\beta(\alpha-1)}$$

and $$E(X_1^2) = E^\theta E^{X|\theta}(X_1^2) = E^\theta(2\theta^2)$$

$$= 2\int_0^\infty (\theta')^{-2} \frac{1}{\Gamma(\alpha)\beta^\alpha} (\theta')^{\alpha-1} \exp\left(-\frac{\theta'}{\beta}\right) d\theta'$$

$$= 2\frac{\Gamma(\alpha-2)\beta^{\alpha-2}}{\Gamma(\alpha)\beta^\alpha} = \frac{2}{\beta^2(\alpha-1)(\alpha-2)}$$

to get the following moment equations

$$\frac{1}{n}\sum_{i=1}^n x_i = m_1' = E(X_1) = \frac{1}{\beta(\alpha-1)}$$

and $$\frac{1}{n}\sum_{i=1}^{n} x_i^2 = m_2' = E(X_1^2) = \frac{2}{\beta^2(\alpha-1)(\alpha-2)}$$

Solving these equations, we get the estimates of α and β as

$$\hat{\alpha} = \frac{2m_2' - m_1'^2}{m_2' - 2m_1'^2}$$

and $$\hat{\beta} = \frac{1}{(\hat{\alpha}-1)m_1'}$$

Thus, the empirical Bayes estimate is given by

$$\delta^{\text{EB}}(\mathbf{x}) = \frac{\hat{\beta}t+1}{\hat{\beta}(n+\hat{\alpha}-1)}$$

***Example* 8.52** Let $X_1,\ldots,X_n$ be *iid* from

(i) $G_1(\alpha_0, \theta)$ with the *pdf*

$$f(x|\alpha_0,\theta) = \frac{1}{\theta^{\alpha_0}\Gamma(\alpha_0)} x^{\alpha_0-1}\exp\left(-\frac{x}{\theta}\right),\quad 0<x<\infty,\ \alpha_0,\theta>0$$

where α_0 is known and θ is unknown. Show that the inverted gamma $IG_2(a, b)$ is the conjugate prior for θ. Find the Bayes estimate of θ^{-1} under the squared error loss function.

(ii) $G_2(\alpha_0, \theta)$ with the *pdf*

$$f(x|\alpha_0,\theta) = \frac{\theta^{\alpha_0}}{\Gamma(\alpha_0)} x^{\alpha_0-1}\exp(-\theta x),\ 0<x<\infty;\ \alpha_0,\theta>0$$

where α_0 is known and θ is unknown. Show that the gamma $G_2(a, b)$ is the conjugate prior for θ. Find the Bayes estimate of θ^{-1} under the scale invariant loss function

$$L(\theta,\delta) = (1-\delta\theta)^2$$
$$= \theta^2(\theta^{-1}-\delta)^2 = w(\theta)\,[g(\theta)-\delta]^2$$

where, $w(\theta)$ and $g(\theta)$ are the weight function attached to the loss function and $g(\theta)$ is the parametric function respectively.

Solution.

(i) The likelihood function is

$$L(\theta|\mathbf{x}) = \frac{1}{\theta^{n\alpha_0}(\Gamma\alpha_0)^n}\prod_{i=1}^{n} x_i^{\alpha_0-1}\exp\left(-\frac{\Sigma x_i}{\theta}\right)$$

We know that Σx_i is the sufficient statistic. Therefore, the kernel $k(\Sigma x_i|\theta)$ of the likelihood function $L(\theta|\mathbf{x})$ is

$$k\left(\sum x_i|\theta\right) = \frac{1}{\theta^{n\alpha_0}}\exp\left(-\frac{\sum x_i}{\theta}\right)$$

On replacing $n\alpha_0$ and Σx_i by a and b, we get the conjugate prior density

$$\pi(\theta|a, b) = c\frac{1}{\theta^a}\exp\left(-\frac{b}{\theta}\right)$$

The normalizing constant is obtained by

$$\frac{1}{c} = \int_0^\infty \frac{1}{\theta^{(a-1)+1}}\exp\left(-\frac{b}{\theta}\right)d\theta = \frac{\Gamma(a-1)}{b^{a-1}}$$

Thus, the conjugate prior is given by

$$\pi(\theta|a, b) = \frac{b^{a-1}}{\Gamma(a-1)}\frac{1}{\theta^{(a-1)+1}}\exp\left(-\frac{b}{\theta}\right)$$

which is inverted gamma $IG_2(a - 1, b)$, where a and b are known hyperparameters. Thus, $IG_2(a, b)$ is the conjugate prior for θ. Accordingly, the posterior density is

$$\pi(\theta|\mathbf{x}) \propto \frac{1}{\theta^{n\alpha_0}}\exp\left(-\frac{\sum x_i}{\theta}\right)\frac{1}{\theta^{a+1}}\exp\left(-\frac{b}{\theta}\right)$$

$$= \frac{1}{\theta^{n\alpha_0+a+1}}\exp\left(-\frac{\sum x_i + b}{\theta}\right)$$

which is $IG_2(n\alpha_0 + a, \Sigma x_i + b)$.

Thus, the Bayes estimate of θ^{-1} under the squared error loss function is

$$(\theta^{-1})^B = E[g(\theta)|\mathbf{x}] = E\left[\frac{1}{\theta}\middle|\mathbf{x}\right] = \frac{(\sum x_i + b)}{n\alpha_0 + a - 1} \text{ if } n\alpha_0 + a > 1$$

(ii) The likelihood function, in this case, is

$$L(\theta|\mathbf{x}) = \frac{\theta^{n\alpha_0}}{[\Gamma(\alpha_0)]^n}\prod_{i=1}^n x_i^{\alpha_0-1}\exp\left(-\theta\sum x_i\right)$$

Σx_i is the sufficient statistic for θ. The kernel of the likelihood function is given by

$$k\left(\sum x_i|\theta\right) = \theta^{n\alpha_0}\exp\left(-\theta\sum x_i\right)$$

Replacing $n\alpha_0 + 1$ and Σx_i by the hyperparameters a and b, we get the conjugate prior density

$$\pi(\theta) \propto \theta^{a-1}\exp(-b\theta)$$

Thus, the conjugate prior is

$$\pi(\theta) = \frac{b^a}{\Gamma(a)}\theta^{a-1}\exp(-b\theta)$$

which is a gamma density $G_2(a, b)$. Accordingly, the posterior density is

$$\pi(\theta|\mathbf{x}) \propto \theta^{n\alpha_0+a-1} \exp\left[-\left(\sum x_i + b\right)\theta\right]$$

which is $G_2(n\alpha_0 + a, \Sigma x_i + b)$.

The scale invariant loss function is mostly used for estimating the reciprocal of the scale parameter θ. The Bayes estimate of $g(\theta) = \theta^{-1}$ under the given scale invariant loss function is

$$(\theta^{-1})^B = \frac{E[w(\theta)\theta^{-1}]}{E[w(\theta)]} = \frac{\sum x_i + b}{n\alpha_0 + a + 1}$$

***Example* 8.53** Let $X_1,\ldots,X_n$ be *iid* from $G_2(\alpha_0, \theta)$, gamma density of second type, with the *pdf*

$$f(x|\alpha_0, \theta) = \frac{\theta^{\alpha_0}}{\Gamma(\alpha_0)} x^{\alpha_0-1} \exp(-\theta x)$$

(i) Consider the prior distribution for θ by Jeffreys noninformative prior

$$\pi(\theta) \propto 1/\theta$$

which is indeed not proper. Find the generalized Bayes estimate of θ.

(ii) Consider $G_2(\alpha, \beta)$ as a conjugate prior for θ. Find the Bayes estimate of θ^{-1} under the Whittle and Lane weighted squared error loss function.

Solution.

(i) The posterior distribution is

$$\pi(\theta|\mathbf{x}) \propto \theta^{n\alpha_0} \exp\left(-\theta\sum x_i\right)\frac{1}{\theta}$$

$$= \theta^{n\alpha_0-1} \exp\left(-\theta\sum x_i\right)$$

which is $G_2(n\alpha_0, \Sigma x_i)$. The generalized Bayes estimate of θ under squared error loss function is

$$\theta^B = E(\theta|\mathbf{x}) = \frac{n\alpha_0}{\sum x_i}$$

(ii) The Fisher information in single observation about θ is given by

$$I_{X_1}(\theta) = \frac{\alpha_0}{\theta^2}$$

The Whittle and Lane weighted squared error loss function for estimating parameterization $1/\theta$ is given by

$$L(\theta, \delta) = I(\theta)\left(\delta - \frac{1}{\theta}\right)^2 = \frac{\alpha_0}{\theta^2}\left(\delta - \frac{1}{\theta}\right)^2$$

The prior $G_2(\alpha, \beta)$ modifies to

$$\pi_w(\theta) = w(\theta)\, G_2(\alpha, \beta) \propto \theta^{\alpha-3} \exp(-\beta\theta)$$

Thus, the modified prior is given by

$$\pi_w(\theta) \equiv G_2(\alpha - 2, \beta)$$

The corresponding posterior distribution is given by

$$\pi_w(\theta|\mathbf{x}) \propto \theta^{n\alpha_0+\alpha-3} \exp\left[-(\beta + \textstyle\sum x_i)\theta\right]$$

which is $G_2(n\alpha_0 + \alpha - 2, \beta + \Sigma x_i)$. Thus, the Bayes estimate of $1/\theta$ with respect to the modified prior distribution $G_2(\alpha - 2, \beta)$ under the simple squared error loss function is

$$\left(\frac{1}{\theta}\right)^B = E(\theta^{-1}|\mathbf{x})$$

where the expectation is taken with respect to the posterior distribution $G_2(n\alpha_0 + \alpha - 2, \beta + \Sigma x_i)$. This gives

$$\left(\frac{1}{\theta}\right)^B = \int_0^\infty \frac{(\beta + \sum x_i)^{n\alpha_0+\alpha-2}}{\Gamma(n\alpha_0 + \alpha - 2)} \theta^{n\alpha_0+\alpha-4} \exp\left[-\left(\beta + \textstyle\sum x_i\right)\theta\right] d\theta$$

$$= \frac{(\beta + \sum x_i)^{n\alpha_0+\alpha-2}}{\Gamma(n\alpha_0 + \alpha - 2)} \frac{\Gamma(n\alpha_0 + \alpha - 3)}{(\beta + \sum x_i)^{n\alpha_0+\alpha-3}} = \frac{\beta + \sum x_i}{n\alpha_0 + \alpha - 3}$$

***Example* 8.54** Let $X_1,\ldots,X_n$ be a random sample from $U(\theta, \theta + 1)$, $\theta \in \mathbb{R}^1$. Consider the prior density $\pi(\theta)$ as some continuous density over $\mathbb{R}^1$. Find the Bayes estimator of θ under the squared error loss function and show that it is consistent.

Solution. The joint density of $\mathbf{X}$ and θ is

$$f(\mathbf{x}; \theta) = I_{(\theta,\,\theta+1)}(x_{(1)}) I_{(\theta,\theta+1)}(x_{(n)}) \pi(\theta)$$

The posterior density of θ given $\mathbf{x}$ is

$$\pi(\theta|\mathbf{x}) \propto \pi(\theta) I_{(X_{(n)}-1,\, X_{(1)})}(\theta)$$

The Bayes estimate of θ is

$$\delta^B(x) = \frac{\int_{x_{(n)}-1}^{x_{(1)}} \theta\pi(\theta)d\theta}{\int_{x_{(n)}-1}^{x_{(1)}} \pi(\theta)d\theta}$$

For large n, $X_{(1)} \to \theta$ and $X_{(n)} \to \theta + 1$ or $X_{(n)} - 1 \to \theta$ almost surely. Therefore, for large n, the interval $(X_{(n)}-1, X_{(1)})$ converges to a single point θ with probability one. Since $\pi(\theta)$ is continuous

$$\int_{x_{(n)}-1}^{x_{(1)}} \theta\pi(\theta)d\theta \to \theta\pi(\theta)$$

and

$$\int_{x_{(n)}-1}^{x_{(1)}} \pi(\theta)d\theta \to \pi(\theta)$$

as $n \to \infty$. Thus, $\delta^B(\mathbf{x}) \to \theta$ as $n \to \infty$ with probability one. This shows that δ^B is consistent.

***Example* 8.55** Let $X_1, X_2, \ldots, X_n$ be *iid* $U(0, \theta)$.

(i) Find the Jeffreys noninformative prior for θ and the Bayes estimate of θ with respect to this prior. Show that this Jeffreys prior is a limiting distribution of the Pareto distribution which is conjugate prior.

(ii) Consider the prior distribution $\pi(\theta)$ as $\text{IG}_2(\alpha, \beta)$. Find the Bayes estimate of θ under squared error loss function.

(iii) Consider $n = 1$ and the prior distribution on θ as $G_2(2, 1)$ having the *pdf*

$$\pi(\theta) = \theta \exp(-\theta),\ \theta > 0$$

Find the Bayes estimate of θ when the loss function is squared error and absolute error.

Solution.

(i) The Fisher information

$$I_X(\theta) = -\frac{1}{\theta^2}$$

gives the Jeffreys prior for θ by

$$\pi(\theta) \propto \frac{1}{\theta}$$

Consider the Pareto distribution which is a conjugate prior in this case

$$\pi'(\theta) = \alpha\theta_0^\alpha \theta^{-\alpha-1} I_{(\theta_0,\infty)}(\theta) \propto \theta^{-\alpha-1} I_{(\theta_0,\infty)}(\theta)$$

Under the limiting conditions $\alpha \to 0$ and $\theta_0 \to 0$

$$\pi'(\theta) \to \frac{1}{\theta}, \theta > 0$$

i.e., conjugate prior distribution, under the above limiting conditions, approaches Jeffreys noninformative prior.

The posterior distribution is given by

$$\pi(\theta|\mathbf{x}) = \frac{nx_{(n)}^n}{\theta^{n+1}}, \quad \theta > x_{(n)}$$

Thus, the Bayes estimate of θ with respect to the Jeffreys prior is

$$\theta^B = E(\theta|\mathbf{x}) = nx_{(n)}^n \int_{x_{(n)}}^{\infty} \theta^{-n} d\theta = \frac{n}{n-1} x_{(n)}$$

(ii) The posterior distribution is given by

$$\pi(\theta|\mathbf{x}) \propto \frac{1}{\theta^{n+\alpha+1}} \exp\left(-\frac{\beta}{\theta}\right), \quad \theta > x_{(n)}$$

$$= c \frac{1}{\theta^{n+\alpha+1}} \exp\left(-\frac{\beta}{\theta}\right)$$

where c is obtained by

$$c \int_{x_{(n)}}^{\infty} \frac{1}{\theta^{n+\alpha+1}} \exp\left(-\frac{\beta}{\theta}\right) d\theta = 1$$

Letting $\beta/\theta = y$, we have

$$\frac{c}{\beta^{n+\alpha}} \int_0^{\beta/x_{(n)}} y^{n+\alpha-1} \exp(-y) dy = 1$$

By using the relation between Poisson and incomplete gamma function Eq. (1.2.17), we have

$$\frac{c\Gamma(n+\alpha-1)}{\beta^{n+\alpha}} (1 - P(X < n+\alpha)) = 1$$

where $X \sim P(\beta/x_{(n)})$. Thus

$$\frac{c\Gamma(n+\alpha-1)}{\beta^{n+\alpha}} \sum_{i=n+\alpha}^{\infty} \frac{\exp(-\beta/x_{(n)})(\beta/x_{(n)})^i}{i!} = 1$$

$$c = \frac{\beta^{n+\alpha}}{s_0 \Gamma(n+\alpha-1)}, \text{ where } \sum_{i=n+\alpha}^{\infty} \frac{\exp(-\beta/x_{(n)})(\beta/x_{(n)})^i}{i!} = s_0$$

Thus, the Bayes estimate of θ is given by

$$\theta^B(\mathbf{x}) = \frac{\beta^{n+\alpha}}{s_0 \Gamma(n+\alpha-1)} \int_{x_{(n)}}^{\infty} \frac{1}{\theta^{n+\alpha}} \exp\left(-\frac{\beta}{\theta}\right) d\theta$$

$$= \frac{\beta}{(n+\alpha-1)} \left[\frac{1}{s_0} \cdot \frac{\exp(-\beta/x_{(n)})(\beta/x_{(n)})^{n+\alpha-1}}{(n+\alpha-1)!} + 1 \right]$$

(iii) The posterior distribution is given by

$$\pi(\theta|x) = \frac{h(x,\theta)}{m(x)} \propto \frac{1}{\theta}\theta e^{-\theta} = e^{-(\theta - x)}, \theta > x_{(n)}$$

The Bayes estimate of θ is

$$\theta^B = E(\theta|x) = x + 1$$

since $$E(\theta - x) = \int_x^\infty (\theta - x)\exp[-(\theta - x)]d\theta = \int_0^\infty y\exp(-y)dy = 1$$

The posterior expected loss of the Bayes estimate θ^B is

$$E[(\theta - \theta^B)^2 | x] = \int_x^\infty [\theta - (x+1)]^2 \exp[-(\theta - x)]d\theta$$

Let $\theta - x = u$

$$E[(\theta - \theta^B)^2 | x] = \int_0^\infty (u-1)^2 e^{-u} du$$

$$= \int_0^\infty (u^2 - 2u + 1)e^{-u} du$$

$$= \Gamma(3) - 2\Gamma(2) + \Gamma(1) = 1$$

In the case of absolute error loss function,

$$L(\theta, \delta) = |\theta - \delta|$$

The median of the posterior distribution

$$\pi(\theta|x) = e^{x-\theta}, \theta > x$$

is the Bayes estimate of θ, i.e.,

$$\int_x^\delta e^{x-\theta}d\theta = \int_\delta^\infty e^{x-\theta}d\theta = \frac{1}{2}$$

This gives $$\theta^B(x) = x + \log 2$$

***Example* 8.56** Let $X_1,\ldots,X_n$ be *iid* uniform $U(0,\theta)$ random variables, where $\theta > 0$ is unknown and is to be estimated. Consider the Pareto distribution, Pa(α, β), as conjugate prior to $U(0, \theta)$

$$\pi(\theta|\alpha, \beta) = \frac{\alpha\beta^\alpha}{\theta^{\alpha+1}}, \quad \alpha > 0, \beta > 0, \theta > \beta$$

where $\alpha > 0$, $\beta > 0$, β is the scale parameter, and α is the shape parameter.

(i) Show that $\pi(\theta|\alpha, \beta)$ is a conjugate prior. Find the Bayes, MAP, and ML estimates of θ under the squared error loss function. Find the Jeffreys prior and show that it is the boundary noninformative prior of the Pareto conjugate family of priors for θ.

(ii) Find the empirical Bayes estimate of θ:
  (a) when only one hyperparameter β is unknown, and
  (b) when both the hyperparameters α, β are unknown.

(iii) Assume that the hyperparameter α is known and β is unknown, and consider the uniform noninformative prior for β. Find the hierarchical Bayes estimate of θ.

Solution.

(i) The maximum order statistic $X_{(n)}$ is a complete sufficient statistic for θ. Its *pdf* is given by

$$f_{X_{(n)}}(x|\theta) = n\left(\frac{x}{\theta}\right)^{n-1}\frac{1}{\theta} = \frac{n}{\theta^n}x^{n-1},\ 0 < x < \theta$$

The joint density of $X_{(n)}$ and θ is

$$h(x, \theta) = \frac{n}{\theta^n}x^{n-1}I_{(0,\,\theta)}(x)\frac{\alpha\beta^\alpha}{\theta^{\alpha+1}}I_{(\beta,\,\infty)}(\theta)$$

The posterior distribution of θ given $X_{(n)} = x$

$$\pi(\theta|x) \propto \frac{1}{\theta^{n+\alpha+1}}I_{(\max(x,\beta),\ \infty)}(\theta)$$

$$\sim \text{Pa}(n+\alpha, \max\{x_{(n)}, \beta\})$$

This shows that the Pareto distribution is a conjugate prior with respect to the uniform observation model. Using the results for the Pareto distribution, Pa(α, β),

$$E(\theta|\alpha, \beta) = \begin{cases} \dfrac{\alpha\beta}{\alpha-1}, & \alpha > 1 \\ \text{Mean does not exist,} & \text{otherwise} \end{cases}$$

$$V(\theta|\alpha, \beta) = \frac{\alpha\beta^2}{(\alpha-1)^2(\alpha-2)} \text{ provided } \alpha > 2$$

and $$\theta^{\text{MAP}} = \beta$$

The Bayes estimate of θ is given by

$$\theta^B(x) = \frac{\int_{\max(\beta,x)}^{\infty}\theta^{-(n+\alpha)}d\theta}{\int_{\max(\beta,x)}^{\infty}\theta^{-(n+\alpha+1)}d\theta} = \frac{n+\alpha}{n+\alpha-1}\max\{\beta, x\}$$

and the Bayes risk is

$$r(\pi, \theta^B) = V(\theta|x) = \frac{n+\alpha}{(n+\alpha-1)^2(n+\alpha-2)}[\max\{x_{(n)}, \beta\}]^2$$

$$\theta^{\text{MAP}} = \max\{x_{(n)}, \beta\}$$

$$\theta^{\text{ML}} = x_{(n)}$$

Note that θ^B and θ^{MAP} do not converge to θ^{ML} with the increase in data. The Fisher information in **X** about θ is

$$I_{\mathbf{X}}(\theta) = \frac{n}{\theta^2}$$

This yields the Jeffreys prior

$$\pi(\theta) = [I_{\mathbf{X}}(\theta)]^{1/2} \propto \frac{1}{\theta} \tag{8.7.36}$$

We have already seen that the Pareto distribution is a conjugate prior

$$\pi(\theta) \propto \frac{1}{\theta^{\alpha+1}}, \theta > \beta$$

As $\alpha \to 0$, $\beta \to 0$

$$\lim_{\substack{\alpha\to 0\\ \beta\to 0}} \frac{1}{\theta^{\alpha+1}} I_{(\beta,\infty)}(\theta) = \frac{1}{\theta}$$

which is Jeffreys prior in Eq. (8.7.36). Thus, the Jeffreys prior can be seen as the boundary noninformative prior of the Pareto conjugate family of priors for θ.

(ii) Consider now the first- and second-order moments

$$E(X_1) = E^{\theta} E^{X_1|\theta}(X_1) = E^{\theta}\left(\frac{\theta}{2}\right)$$

$$= \frac{1}{2}\int_{\beta}^{\infty} \theta \frac{\alpha\beta^{\alpha}}{\theta^{\alpha+1}} d\theta = \frac{\beta\alpha}{2(\alpha-1)}$$

and

$$E(X_1^2) = E^{\theta} E^{X_1|\theta}(X_1^2)$$

$$= E^{\theta}\left(\frac{\theta^2}{3}\right) = \frac{\beta^2\alpha}{3(\alpha-2)}$$

On solving the moment equations

$$m_1' = \frac{\beta\alpha}{2(\alpha-1)} = \frac{1}{n}\sum_{i=1}^{n} x_i$$

$$m_2' = \frac{\beta^2 \alpha}{3(\alpha-2)} = \frac{1}{n}\sum_{i=1}^{n} x_i^2$$

(a) We get moment estimator of β, when α is known

$$\hat{\beta} = \frac{2(\alpha-1)}{\alpha} m_1'$$

Thus, the empirical Bayes estimate of θ is

$$\theta^{\text{EB}}(x) = \frac{n+\alpha}{n+\alpha-1}\max\{x, \hat{\beta}\}$$

(b) If α and β are both unknown, their moment estimators are

$$\hat{\alpha} = 1 + \sqrt{\frac{3m_2'}{(3m_2' - 4m_1'^2)}}$$

and $$\hat{\beta} = \frac{2m_1'(\hat{\alpha}-1)}{\hat{\alpha}}$$

In this case, the empirical Bayes estimate of θ is

$$\theta^{\text{EB}}(x) = \frac{n+\hat{\alpha}}{n+\hat{\alpha}-1}\max\{x, \hat{\beta}\}$$

(iii) Let us calculate the joint density of $\mathbf{X}$, θ, β

$$f(\mathbf{x}, \theta, \beta) = f(\mathbf{x}|\theta)\, \pi_1(\theta|\beta)\pi_2(\beta)$$

$$= \frac{1}{\theta^n} I_{(0,\theta)}(X_{(n)} = x)\frac{\alpha\beta^\alpha}{\theta^{\alpha+1}} I_{(t,\infty)}(\theta)\frac{1}{\beta} I_{(0,\theta)}(\beta)$$

$$= \frac{\alpha\beta^{\alpha-1}}{\theta^{n+\alpha+1}} I_{(t,\beta)}(\theta) I_{(0,\theta)}(\beta)$$

where $t = \max\{x, \beta\}$

Integrating out β from the above density yields the joint density of $\mathbf{X}$, θ

$$f(\mathbf{x}, \theta) = \frac{\alpha}{\theta^{n+\alpha+1}} I_{(t,\infty)}(\theta)\int_0^\theta \beta^{\alpha-1} d\beta$$

$$= \frac{1}{\theta^{n+1}} I_{(t,\infty)}(\theta)$$

The posterior distribution of θ given $\mathbf{x}$ is then

$$\pi(\theta|\mathbf{x}) \propto \frac{1}{\theta^{n+1}} I_{(t,\infty)}(\theta)$$

Thus, the hierarchical Bayes estimate of θ is

$$\theta^B(\mathbf{x}) = \frac{\int_t^\infty \theta\theta^{-(n+1)}d\theta}{\int_t^\infty \theta^{-(n+1)}d\theta} = \frac{n}{n-1}t$$

***Example* 8.57** Let $X_1,\ldots,X_n$ be *iid* according to $U(0, \theta)$. Consider an arbitrary continuous prior $\pi(\theta)$. Find the Bayes and MAP estimates of θ. Show that the MAP estimate converges to the ML estimate as the amount of data grows. Can this prior be the conjugate Pareto distribution?

Solution. The posterior density, in this case, is given by

$$\pi(\theta|\mathbf{x}) \propto f(\mathbf{x}|\theta)\pi(\theta) = \frac{\pi(\theta)}{\theta^n},\ \theta \geq x_{(n)}$$

The MAP estimate, δ^{MAP}, is the value of $\theta \geq x_{(n)}$ which maximizes $\pi(\theta)/\theta^n$ and the Bayes estimate of θ is given by

$$\delta^B = \frac{\int_{\theta \geq x_{(n)}} \left(\frac{\pi(\theta)}{\theta^{n-1}}\right) d\theta}{\int_{\theta \geq x_{(n)}} \left(\frac{\pi(\theta)}{\theta^{n}}\right) d\theta}$$

The estimate δ^{MAP} converges to the ML estimate $x_{(n)}$ as the amount of data grows or as n approaches infinity, since

$$\begin{aligned}\lim_{n\to\infty} \delta^{\text{MAP}} &= \lim_{n\to\infty} \max_{\theta \geq x_{(n)}} \left[\frac{\pi(\theta)}{\theta^n}\right],\ \theta \geq x_{(n)} \\ &= \lim_{n\to\infty} \max_{\theta \geq x_{(n)}} [\log \pi(\theta) - n\log\theta] \\ &= \lim_{n\to\infty} \min_{\theta \geq x_{(n)}} [n\log\theta] = x_{(n)}\end{aligned}$$

This convergence does not take place when the Pareto distribution is considered as conjugate prior since it is not continuous.

***Example* 8.58** Let $X_1,\ldots,X_n$ be n observations drawn independently and identically from $U(0, \theta)$ distribution. Let the noninformative prior distribution be given by

$$\pi(\theta) = 1,\ \theta \in [0, 1]$$

Find the Bayes estimate of θ under the weighted squared error loss function

$$L(\theta, \delta) = \frac{1}{\theta^2}(\delta - \theta)^2$$

Solution. The likelihood function is written as

$$L(\theta|\mathbf{x}) = \left(\frac{1}{\theta}\right)^n, \ 0 < x_{(n)} < \theta$$

The marginal distribution is

$$m(\mathbf{x}) = \int_{x_{(n)}}^{1} \left(\frac{1}{\theta}\right)^n d\theta, 0 < x_{(n)} < \theta$$

$$= \frac{1}{n-1}\left(\frac{1}{x_n^{n-1}} - 1\right)$$

Thus, the posterior distribution is

$$\pi(\theta|\mathbf{x}) = \frac{(1/\theta)^n I_{(x_{(n)},1)}(\theta)}{[1/(n-1)][(1/x_n^{n-1}) - 1]}$$

Here, $w(\theta) = 1/\theta^2$. Thus,

$$E[\theta w(\theta)|\mathbf{x}] = E\left(\frac{1}{\theta}\bigg|\mathbf{x}\right)$$

$$= \int_{x_{(n)}}^{1} \frac{1}{\theta^{n+1}} d\theta = -\frac{1}{n}\left(1 - \frac{1}{x_{(n)}^n}\right)$$

$$E[w(\theta)|\mathbf{x}] = \int_{x_{(n)}}^{1} \frac{1}{\theta^{n+2}} d\theta = -\frac{1}{n+1}\left(1 - \frac{1}{x_n^{n+1}}\right)$$

Thus, the Bayes estimate of θ is given by

$$\delta = \frac{E[\theta w(\theta)|\mathbf{x}]}{E[w(\theta)|\mathbf{x}]} = \frac{n+1}{n}\left(\frac{x_{(n)}^n - 1}{x_{(n)}^{n+1} - 1}\right) x_{(n)}$$

***Example* 8.59** Let $X_1,\ldots,X_n$ be a random sample from $U(0, \theta)$, $\theta > 0$, and the prior of θ be given by log-normal distribution with parameters (μ, σ^2), where $\mu \in \mathbb{R}^1$ and $\sigma^2 > 0$ are known constants. Find the posterior density of θ given $\mathbf{X}$ and of $\log \theta$ given $\mathbf{X}$. Using this, find the rth posterior moment. Also, find the value of θ at which the posterior density is maximum.

Solution. The joint density of $\mathbf{X}$ and θ is

$$h(\mathbf{x}, \theta) = \frac{1}{\theta^n} I_{(0,\theta)}(x_{(n)}) \frac{1}{\theta\sigma\sqrt{2\pi}} \exp\left[-\frac{(\log\theta - \mu)^2}{2\sigma^2}\right] I_{(0,\infty)}(\theta)$$

or the posterior density of θ given $\mathbf{x}$ is

$$\pi(\theta|\mathbf{x}) = c\frac{1}{\theta^{n+1}} \exp\left[-\frac{(\log\theta - \mu)^2}{2\sigma^2}\right] I_{(x_{(n)},\infty)}(\theta)$$

where c is determined by

$$c\int_{x_{(n)}}^{\infty}\frac{1}{\theta^{n+1}}\exp\left[-\frac{(\log\theta-\mu)^2}{2\sigma^2}\right]d\theta=1$$

Let $\eta = \log\theta$; $d\eta = (1/\theta)d\theta$

$$c\int_{\log x_{(n)}}^{\infty}\exp\left[-n\eta-\frac{1}{2\sigma^2}(\eta-\mu)^2\right]d\eta=1$$

The exponent may be written as

$$-n\eta-\frac{1}{2\sigma^2}(\eta-\mu)^2=-\frac{1}{2\sigma^2}[\eta^2-2\eta(\mu-n\sigma^2)+(\mu-\eta\sigma^2)^2-(\mu-\eta\sigma^2)^2+\mu^2]$$

$$=-\frac{1}{2\sigma^2}[\eta-(\mu-n\sigma^2)]^2-\left(n\mu-\frac{n^2\sigma^2}{2}\right)$$

We have $$c\exp\left[-\left(n\mu-\frac{n^2\sigma^2}{2}\right)\right]\int_{\log x_{(n)}}^{\infty}\exp\left\{-\frac{1}{2\sigma^2}[\eta-(\mu-n\sigma^2)]^2\right\}d\eta=1$$

Let $z = [\eta - (\mu - n\sigma^2)]/\sigma$

$$c\sigma\sqrt{2\pi}\exp\left[-\left(n\mu-\frac{n^2\sigma^2}{2}\right)\right]\frac{1}{\sqrt{2\pi}}\int_{[\log x_{(n)}-(\mu-n\sigma^2)]/\sigma}^{\infty}\exp\left(-\frac{1}{2}z^2\right)dz=1$$

$$c\sigma\sqrt{2\pi}\exp\left[-\left(n\mu-\frac{n^2\sigma^2}{2}\right)\right]\left[1-\Phi\left(\frac{\log x_{(n)}-(\mu-n\sigma^2)}{\sigma}\right)\right]=1$$

$$c=\frac{\exp[n\mu-(n^2\sigma^2)/2]}{\sigma\sqrt{2\pi}\Phi[(\mu-n\sigma^2-\log x_{(n)})/\sigma]}\qquad[\text{Since } 1-\Phi(x)=\Phi(-x)]$$

This gives the posterior distribution of θ given $\mathbf{x}$

$$\pi(\theta|\mathbf{x})=\frac{\exp[n\mu-(n^2\sigma^2/2)]}{\sigma\sqrt{2\pi}\,c(x)}\frac{1}{\theta^{n+1}}\exp\left[-\frac{(\log\theta-\mu)^2}{2\sigma^2}\right]I_{(x_{(n)},\infty)}(\theta)$$

where $c(x) = \Phi[(\mu - n\sigma^2 - \log x_{(n)})/\sigma]$. After simplifications,

$$\pi(\theta|\mathbf{x})=\frac{\exp(-\log\theta)}{\sigma\sqrt{2\pi}c(x)\theta}\exp\left[-\frac{(\log\theta-\mu+n\sigma^2)^2}{2\sigma^2}\right]I_{(x_{(n)},\,\infty)}(\theta)$$

The function

$$\exp\left[-\log\theta-\frac{1}{2\sigma^2}(\log\theta-\mu+n\sigma^2)^2\right]$$

$$= \exp\left\{-\frac{1}{2\sigma^2}[\log\theta - (\mu-(n+1)\sigma^2)]^2\right\}\exp\left[\left(n+\frac{1}{2}\right)\sigma^2 - \mu\right]$$

is maximum at $\theta = \exp[\mu - (n+1)\sigma^2]$. Therefore, the posterior distribution of θ given $\mathbf{X}$ is maximum at

$$\theta = \max\{\exp[\mu - (n+1)\sigma^2], X_{(n)}\}$$

Consider the reparameterization

$$\eta = \log\theta$$

The Jacobian of transformation is

$$|J| = \exp(\eta)$$

The posterior distribution of η given $\mathbf{x}$ is

$$\pi(\eta|\mathbf{x}) = c\exp\left[-(n+1)\eta + \eta - \frac{1}{2\sigma^2}(\eta-\mu)^2\right]I_{(\log x_{(n)},\infty)}(\eta)$$

The exponentiation simplifies to

$$-\frac{1}{2\sigma^2}[(\eta-\mu)^2 + n\eta] = -\frac{1}{2\sigma^2}[\eta - (\mu - n\sigma^2)]^2 - \left(n\mu - \frac{n^2\sigma^2}{n}\right)$$

This simplifies to

$$\pi(\eta|\mathbf{x}) = \frac{1}{\sigma\sqrt{2\pi}c(\mathbf{x})}\exp\left\{-\frac{1}{2\sigma^2}[\eta-(\mu-n\sigma^2)]^2\right\}I_{(\log x_{(n)},\infty)}(\eta)$$

which is a truncated normal distribution. The rth posterior moment of θ is

$$E(\theta^r|\mathbf{x}) = E[\exp(r\eta)|\mathbf{x}]$$

$$= \frac{1}{\sigma\sqrt{2\pi}c(\mathbf{x})}\int_{\log x_{(n)}}^{\infty}\exp\left\{r\eta - \frac{1}{2\sigma^2}[\eta-(\mu-n\sigma^2)]^2\right\}d\eta$$

$$= \frac{\exp\left\{\frac{r}{2}[2\mu-(2n-r)\sigma^2]\right\}}{\sigma\sqrt{2\pi}c(\mathbf{x})}\int_{\log x_{(n)}}^{\infty}\exp\left\{-\frac{1}{2\sigma^2}[\eta-(\mu-(n-r)\sigma^2)]^2\right\}d\eta$$

$$= c^{-1}(\mathbf{x})\exp\left[\frac{r}{2}[2\mu-(2n-r)\sigma^2]\right]\left\{1-\Phi\left(\frac{\log x_{(n)} - [\mu-(n-r)\sigma^2]}{\sigma}\right)\right\}$$

$$= c^{-1}(\mathbf{x})\exp\left\{\left(\frac{r}{2}\right)\left[2\mu-(2n-r)\sigma^2\right]\right\}\Phi\left\{\frac{[\mu-(n-r)\sigma^2 - \log x_{(n)}]}{\sigma}\right\}$$

***Example* 8.60** Let $X \sim f(x|\theta)$ be *iid* with the *pdf* $f(x|\theta) = \exp[-(x-\theta)]$, $x > \theta$. Take $\pi(\theta) = \exp(-\theta)$, $\theta > 0$. Find the Bayes estimate of θ under quadratic loss function.

Solution. The posterior distribution is given by

$$\pi(\theta|x) = c\exp[-(x-\theta)]\exp(-\theta) = c\exp(-x),\ \theta < x$$

where c is
$$\frac{1}{c} = \int_0^x \exp(-x)d\theta = x\exp(-x)$$

This gives
$$\pi(\theta|x) = \frac{1}{x},\ \theta < x$$

Therefore, the Bayes estimator of θ under squared error loss function is

$$\delta(x) = E(\theta|x) = \frac{1}{2}x$$

***Example* 8.61** Let $X_1,\ldots,X_n$ be a random sample from $f(x|\theta) = \exp[-(x-\theta)]$, $x > \theta$, and the prior distribution of θ is given by $G_1(1, 1)$. Find the Bayes estimate of θ under the quadratic loss function. Show that the minimax estimator in the class of estimators

$$C = \left\{\delta : \delta = a\left[X_{(1)} - (1/n)\right], a > 0\right\}$$

is given by $X_{(1)} - (1/n)$.

Solution. The posterior distribution is given by

$$\pi(\theta|\mathbf{x}) = c\exp((n-1)\theta),\ \theta < x_{(1)}$$

where,
$$\frac{1}{c} = \int_0^{x_{(1)}} \exp((n-1)\theta)\,d\theta = \frac{1}{(n-1)}[\exp((n-1)x_{(1)}) - 1]$$

Therefore, the Bayes estimate of θ is given by

$$\begin{aligned}
\theta^B &= E(\theta|\mathbf{x}) \\
&= \frac{(n-1)}{\exp((n-1)x_{(1)}) - 1}\int_0^{x_{(1)}} \theta\exp((n-1)\theta)\,d\theta \\
&= \frac{(n-1)}{\exp((n-1)x_{(1)}) - 1}\left[\frac{x_{(1)}\exp((n-1)x_{(1)})}{n-1} - \frac{\exp((n-1)x_{(1)}) - 1}{(n-1)^2}\right] \\
&= \left[\frac{x_{(1)}\exp((n-1)x_{(1)})}{\exp((n-1)x_{(1)}) - 1} - \frac{1}{n-1}\right]
\end{aligned}$$

Consider, next, the risk of some estimator $\delta = a[X_{(1)} - (1/n)]$, $a > 0$ in C

$$\begin{aligned}
R(\theta, \delta) &= \left\{a\left[X_{(1)} - (1/n)\right] - \theta\right\}^2 \\
&= E\left[aX_{(1)} - ((a/n) + \theta)\right]^2
\end{aligned}$$

$$= a^2 E(X_{(1)}^2) - 2[(a/n) + \theta] E(X_1) + [(a/n) + \theta]^2$$

$E(X_{(1)})$ and $E(X_{(1)}^2)$ are computed as follows

$$E(X_{(1)}) = n\exp(n\theta)\int_0^\infty x\exp(-nx)dx = \theta + \frac{1}{n}$$

and
$$E(X_{(1)}^2) = n\exp(n\theta)\int_0^\infty x^2\exp(-nx)dx = \theta^2 + \frac{2\theta}{n} + \frac{2}{n^2}$$

Thus, $R(\theta, \delta)$ simplifies to

$$R(\theta, \delta) = a^2\left(\theta^2 + \frac{2\theta}{n} + \frac{2}{n^2}\right) - 2a\left(\frac{a}{n} + \theta\right)\left(\theta + \frac{1}{n}\right) + \left(\frac{a}{n} + \theta\right)^2$$

$$= (a-1)^2\theta^2 + \frac{a^2}{n^2}$$

and the estimator $\delta = a(X_{(1)} - (1/n))$ is minimax at $a = 1$.

***Example* 8.62** Let X follows the Weibull distribution with the *pdf* (Bansal, 2007)

$$f(x|\alpha, \beta) = \frac{\alpha}{\beta}x^{\alpha-1}\exp\left(-\frac{x^\alpha}{\beta}\right), x > 0, \alpha > 0, \beta > 0$$

Find Jeffreys prior for (α, β).

Solution. If we assume α and β are apriori independent, then the joint prior for (α, β) is defined by

$$\pi(\alpha, \beta) = \pi_1(\alpha)\ \pi_2(\beta)$$

By Jeffreys rule of thumb for constructing noninformative priors,

$$\pi_1(\alpha) \propto \frac{1}{\alpha} \quad \text{and} \quad \pi_2(\beta) \propto \frac{1}{\beta}$$

Since the range of α and β is $(0, \infty)$, we have

$$\pi(\alpha, \beta) \propto \frac{1}{\alpha\beta}$$

Next, if α and β are not apriori independent, the Jeffreys prior for (α, β) is given by

$$\pi(\alpha, \beta) \propto \sqrt{\det[I_{\mathbf{X}}(\alpha, \beta)]}$$

The Fisher information matrix for (α, β) is given by

$$\mathbf{I}_{\mathbf{X}}(\alpha, \beta) = \begin{bmatrix} E\left\{-\frac{\partial^2}{\partial\alpha^2}\log L(\alpha, \beta|X)\right\} & E\left\{-\frac{\partial^2}{\partial\alpha\partial\beta}\log L(\alpha, \beta|X)\right\} \\ E\left\{-\frac{\partial^2}{\partial\alpha\partial\beta}\log L(\alpha, \beta|X)\right\} & E\left\{-\frac{\partial^2}{\partial\beta^2}\log L(\alpha, \beta|X)\right\} \end{bmatrix}$$

After some simplifications,

$$\mathbf{I_X}(\alpha, \beta) = \begin{bmatrix} \frac{1}{\alpha^2}[1+G'(2)+2G(2)\log\beta+(\log\beta)^2] & \frac{1}{\alpha\beta}[G(2)+\log\beta] \\ \frac{1}{\alpha\beta}[G(2)+\log\beta] & \frac{1}{\beta^2} \end{bmatrix}$$

where $$G(x) = \int_0^\infty t^{x-1}\exp(-t)\log t\, dt$$

and $$G'(x) = \frac{\partial}{\partial x}G(x) = \int_0^\infty t^{x-1}\exp(-t)(\log t)^2\, dt$$

are called digamma and trigamma functions, respectively.
Thus, Jeffreys prior for (α, β) is given by

$$\pi(\alpha, \beta) \propto \sqrt{\det[\mathbf{I_X}(\alpha, \beta)]} = \frac{1}{\alpha\beta}[1+G'(2)-(G(2))^2]^{1/2} = \frac{1}{\alpha\beta}$$

We see that irrespective of α and β are apriori independent or they are not, the appropriate noninformative prior for (α, β) is

$$\pi(\alpha, \beta) \propto \frac{1}{\alpha\beta}$$

***Example* 8.63** Let $X_1,\ldots,X_n$ be a random sample drawn from the distribution with the *pdf*

$$f(x|\theta) = \sqrt{\frac{2}{\pi}}\exp\left[-\frac{1}{2}(x-\theta)^2\right]I_{(\theta,\infty)}(x)$$

where $\theta \in \mathbb{R}^1$ is unknown. Let the prior distribution of θ be a Lebesgue measure on $\mathbb{R}^1$, i.e. $\pi(\theta)\,d\theta = d\theta$ which is an improper prior. Find the generalized Bayes estimate of θ under the squared error loss function.

Solution. The joint density of $\mathbf{x}$ and θ is

$$h(\mathbf{x}, \theta) = \left(\frac{2}{\pi}\right)^{n/2}\exp\left[-\frac{1}{2}\sum(x_i-\theta)^2\right]I_{(\theta,\infty)}(x_{(1)})$$

$\delta(\mathbf{x})$ is a generalized Bayes estimate if it minimizes the integral

$$\int L(\theta, \delta)f(\mathbf{X}|\theta)\pi(\theta)d\theta = \int_{-\infty}^{x_{(1)}}(\theta-\delta)^2\left(\frac{2}{\pi}\right)^{n/2}\exp\left[-\frac{1}{2}\sum(x_i-\theta)^2\right]d\theta$$

This gives $$\delta^{\text{GB}} = \frac{\int_0^{x_{(1)}}\theta\exp\left[-\left(\frac{1}{2}\right)\sum(x_i-\theta)^2\right]d\theta}{\int_0^{x_{(1)}}\exp\left[-\left(\frac{1}{2}\right)\sum(x_i-\theta)^2\right]d\theta}$$

$$= \frac{\int_0^{x_{(1)}} \theta \exp\left[-\left(\frac{n}{2}\right)(\bar{x}-\theta)^2\right] d\theta}{\int_0^{x_{(1)}} \exp\left[-\left(\frac{n}{2}\right)(\bar{x}-\theta)^2\right] d\theta}$$

$$= \bar{x} + \frac{\int_0^{x_{(1)}} (\theta-\bar{x}) \exp\left[-\left(\frac{n}{2}\right)(\theta-\bar{x})^2\right] d\theta}{\int_0^{x_{(1)}} \exp\left[-\left(\frac{n}{2}\right)(\theta-\bar{x})^2\right] d\theta} \tag{8.7.37}$$

By letting $\sqrt{n}(\theta - \bar{x}) = z$ in the integral, the above expression simplifies to

$$\int_{-\sqrt{n}\bar{x}}^{\sqrt{n}(x_{(1)}-\bar{x})} \exp\left\{-\frac{z^2}{2}\right\} dz = \frac{\sqrt{2\pi}}{\sqrt{n}}\{\Phi[\sqrt{n}(x_{(1)}-\bar{x})] - \Phi(-\sqrt{n}\bar{x})\} \tag{8.7.38}$$

Differentiating Eq. (8.7.38) with respect to θ, we get

$$-\int_0^{x_{(1)}} n(\theta-\bar{x}) \exp\left[-\frac{n}{2}(\theta-\bar{x})^2\right] d\theta = \sqrt{2\pi}\{\Phi'(\sqrt{n}(x_{(1)}-\bar{x})) - \Phi'(-\sqrt{n}\bar{x})\}$$

or

$$\int_0^{x_{(1)}} (\theta-\bar{x}) \exp\left[-\frac{n}{2}(\theta-\bar{x})^2\right] d\theta = \frac{\sqrt{2\pi}}{n}\{\Phi'(-\sqrt{n}\bar{x}) - \Phi'(\sqrt{n}(x_{(1)}-\bar{x}))\} \tag{8.7.39}$$

Therefore, on using Eqs. (8.7.38) and (8.7.39) in Eq. (8.7.37), we get the generalized Bayes estimate of θ

$$\delta^{\text{GB}} = \bar{x} + \frac{1}{\sqrt{n}} \frac{\Phi'(-\sqrt{n}\bar{x}) - \Phi'[\sqrt{n}(x_{(1)}-\bar{x})]}{\Phi[\sqrt{n}(x_{(1)}-\bar{x})] + \Phi(-\sqrt{n}\bar{x})}$$

***Example* 8.64** Let $X_1,\ldots,X_n$ be a random sample from the exponential distribution on $(0, \infty)$ with scale parameter $(0, \infty)$. Define

$$T = X_1 + \cdots + X_\theta$$

where θ is some positive integer. Consider the prior distribution for θ by geometric distribution

$$\pi(\theta) = p(1-p)^{\theta-1},\ \theta = 1, 2, \ldots;\ p \in (0, 1)$$

so that $E(\theta) = 1/p$, p is known. Consider the problem of estimation of θ under the loss function

$$L(\theta, \delta) = \frac{(\theta-\delta)^2}{\theta}$$

Show that the posterior conditional expected loss of some estimator δ of θ is

$$E[L(\theta, \delta)|T = t] = 1 + \eta - 2\delta + \frac{[1 - \exp(-\eta)]\delta^2}{\eta}$$

where $\eta = (1 - p)t$. Find the Bayes estimate of θ and calculate its posterior conditional expected loss. Also, find the explicit expression for the Bayes risk of the above Bayes estimator of θ.

Solution. For a given θ, the distribution of T_θ is $G_1(\theta, 1)$. Therefore, the joint density of T_θ and θ is

$$f(t, \theta) = f(t|\theta)\pi(\theta)$$

$$= \frac{1}{(\theta - 1)!} t^{\theta-1} \exp(-t) p(1 - p)^{\theta-1}, t > 0, \theta = 1, 2, \ldots$$

The marginal density of T is

$$m(t) = \sum_{\theta=1}^{\infty} f(t, \theta) = p\exp(-t) \sum_{\theta=1}^{\infty} \frac{1}{(\theta - 1)!} [t(1 - p)]^{\theta-1}$$

$$= p\exp(-t)\exp[t(1 - p)] = p\exp(-pt)$$

This shows that $T \sim G_2(1, p)$, $T > 0$. This gives the posterior distribution of θ given t

$$\pi(\theta|t) = \frac{1}{(\theta - 1)!} [t(1 - p)]^{\theta-1} \exp[-t(1 - p)]$$

$$= \frac{1}{(\theta - 1)!} \eta^{\theta-1} \exp(-\eta), \theta = 1, 2, \ldots$$

assuming $\eta = t(1 - p)$. Note that the posterior distribution is the truncated Poisson distribution.

The posterior conditional expected loss is given by

$$E[L(\theta, \delta)|T = t] = \sum_{\theta=1}^{\infty} \frac{(\theta - \delta)^2}{\theta} \frac{1}{(\theta - 1)!} \eta^{\theta-1} \exp(-\eta)$$

The Bayes estimate δ of θ is obtained by minimizing the posterior conditional expected loss with respect to δ, that is, by solving the equation

$$\frac{1}{\theta^B} = \sum_{\theta=1}^{\infty} \frac{1}{\theta} \cdot \frac{1}{(\theta - 1)!} \eta^{\theta-1} \exp(-\eta) = \frac{\exp(-\eta)}{\eta} \sum_{\theta=1}^{\infty} \frac{\eta^\theta}{\theta!}$$

$$= \frac{\exp(-\eta)}{\eta} (\exp(\eta) - 1) = \frac{1 - \exp(-\eta)}{\eta}$$

Thus, we get the Bayes estimate of θ under the given loss function by

$$\theta^B(t) = \frac{\eta}{1 - \exp(-\eta)} = \frac{(1 - p)t}{1 - \exp[-(1 - p)t]}$$

The posterior conditional expected loss of the Bayes estimate $\theta^B(t)$ is

$$E\{L[\theta,\theta^B(t)]\big|T=t\}=1+\eta-2\frac{\eta}{1-\exp(-\eta)}+\frac{\eta}{1-\exp(-\eta)}$$

$$=1-\frac{\eta\exp(-\eta)}{1-\exp(-\eta)}$$

$$=1-\eta\sum_{i=1}^{\infty}\exp(-i\eta)$$

Therefore, the Bayes risk of θ^B is given by

$$r(\pi,\theta^B)=E^T E[L(\theta,\theta^B(T))|T]$$

$$=1-E^T\left\{(1-p)T\sum_{i=1}^{\infty}\exp[-i(1-p)T]\right\}$$

$$=1-(1-p)\sum_{i=1}^{\infty}\int_0^{\infty}t\exp[-i(1-p)t]p\exp(-pt)dt$$

$$=1-p(1-p)\sum_{i=1}^{\infty}\int_0^{\infty}t\exp[-i(1-p)t+pt]\,dt$$

$$=1-p(1-p)\sum_{i=1}^{\infty}\frac{1}{[i(1-p)+p]^2}$$

***Example* 8.65 (Ferguson, 1967; Bansal, 2007).** Let $X_1,\ldots,X_n$ be a random sample from a Pareto distribution with density

$$f(x|\sigma)=\frac{\alpha_0\sigma^{\alpha_0}}{x^{\alpha_0+1}},\ \sigma<x<\infty$$

where $\alpha_0>0$ is known and $\sigma>0$ is unknown. The conjugate prior for σ is

$$\pi(\sigma)=\frac{\beta}{k^{\beta}}\sigma^{\beta-1},\ 0<\sigma<k$$

Find the Bayes estimate of σ under the following loss functions:

(i) $L(\sigma, a) = [(a/\sigma) - 1]^2,\ n \ge 3$
(ii) $L(\sigma, a) = |\log a - \log\sigma|$
(iii) $L(\sigma, a) = |(a/\sigma) - 1|,\ n \ge 2.$

Solution.

(i) The posterior distribution is given by

$$\pi(\sigma|\mathbf{x})=c\sigma^{n\alpha_0+\beta-1},\ 0<\sigma<k_1=\min\{k,x_{(1)}\}$$

where $$\frac{1}{c} = \int_0^{k_1} \sigma^{n\alpha_0+\beta-1} d\sigma$$

$$= \frac{k_1^{n\alpha_0+\beta}}{n\alpha_0+\beta}$$

Thus, we have

$$\pi(\sigma|\mathbf{x}) = \frac{n\alpha_0+\beta}{k_1^{n\alpha_0+\beta}} \sigma^{n\alpha_0+\beta-1}, 0<\sigma<k_1$$

We can clearly see that prior $\pi(\sigma)$ is a conjugate prior. The Bayes estimate of σ under the weighted squared error loss function is given by

$$\sigma^B = \frac{E(\sigma w(\sigma)|\mathbf{x})}{E(w(\sigma)|\mathbf{x})}$$

$$= \frac{\int_0^{k_1} \sigma^{n\alpha_0+\beta-2} d\sigma}{\int_0^{k_1} \sigma^{n\alpha_0+\beta-3} d\sigma} = \frac{k_1^{n\alpha_0+\beta-1}}{n\alpha_0+\beta-1} \frac{n\alpha_0+\beta-2}{k_1^{n\alpha_0+\beta-2}}$$

$$= \frac{n\alpha_0+\beta-2}{n\alpha_0+\beta-1} k_1$$

(ii) The Bayes estimate of σ under the loss function $L(\sigma, \delta) = |\log\delta - \log\sigma|$ is such an δ that minimizes the posterior conditional expected loss

$$E(|\log\delta - \log\sigma||\mathbf{x})$$

We set $$\frac{\partial}{\partial\delta} E(|\log\delta - \log\sigma||\mathbf{x}) = 0$$

$$\frac{\partial}{\partial\delta}\left[\int_0^{\delta} (\log\delta - \log\sigma)\pi(\sigma|\mathbf{x})d\sigma - \int_{\delta}^{k_1} (\log\delta - \log\sigma)\pi(\sigma|\mathbf{x})d\sigma\right] = 0$$

$$\frac{1}{\delta}\int_0^{\delta} \pi(\sigma|\mathbf{x})d\sigma - \frac{1}{\delta}\int_{\delta}^{k_1} \pi(\sigma|\mathbf{x})d\sigma = 0$$

Thus, the Bayes estimate of σ is given by choosing such an δ that satisfies

$$\int_0^{\delta} \pi(\sigma|\mathbf{x})d\sigma = \int_{\delta}^{k_1} \pi(\sigma|\mathbf{x})d\sigma$$

or $$\int_0^{\delta} \frac{n\alpha_0+\beta}{k_1^{n\alpha_0+\beta}} \sigma^{n\alpha_0+\beta-1} d\sigma = \frac{1}{2}$$

$$\frac{n\alpha_0+\beta}{k_1^{n\alpha_0+\beta}}\frac{\sigma^{n\alpha_0+\beta}}{n\alpha_0+\beta}\bigg|_0^{\delta}=\frac{1}{2}$$

$$2\delta^{n\alpha_0+\beta}=k_1^{n\alpha_0+\beta}$$

$$\sigma^B=k_1\left(\frac{1}{2}\right)^{1/(n\alpha_0+\beta)}$$

(iii) The Bayes estimate of θ under the loss function $L(\sigma, \delta) = |(\delta/\sigma) - 1|$ is such an δ that minimizes the expected conditional expected loss

$$\frac{\partial}{\partial\delta}E[(\sigma,\delta)|\mathbf{x}]=0$$

$$\frac{\partial}{\partial\delta}E\left[\left|\frac{\delta}{\sigma}-1\right|\mathbf{x}\right]=0$$

$$\frac{\partial}{\partial\delta}\left[\int_0^{\delta}\left(\frac{\delta}{\sigma}-1\right)\pi(\sigma|\mathbf{x})d\sigma+\int_{\delta}^{k_1}\left(1-\frac{\delta}{\sigma}\right)\pi(\sigma|\mathbf{x})d\sigma\right]=0$$

$$\int_0^{\delta}\frac{1}{\sigma}\pi(\sigma|\mathbf{x})d\sigma=\int_{\delta}^{k_1}\frac{1}{\sigma}\pi(\sigma|\mathbf{x})d\sigma$$

The left-hand side expression

$$\int_0^{\delta}\frac{n\alpha_0+\beta}{k_1^{n\alpha_0+\beta}}\sigma^{n\alpha_0+\beta-2}d\sigma=\frac{n\alpha_0+\beta}{k_1^{n\alpha_0+\beta}}\frac{\delta^{n\alpha_0+\beta-1}}{n\alpha_0+\beta-1}$$

and the right-hand side expression

$$\int_{\delta}^{k_1}\frac{1}{\sigma}\pi(\sigma|\mathbf{x})d\sigma=\frac{n\alpha_0+\beta}{k_1^{n\alpha_0+\beta}}\frac{k_1^{n\alpha_0+\beta-1}-\delta^{n\alpha_0+\beta-1}}{n\alpha_0+\beta-1}$$

gives

$$\delta^{n\alpha_0+\beta-1}=k_1^{n\alpha_0+\beta-1}-\delta^{n\alpha_0+\beta-1}$$

$$2\delta^{n\alpha_0+\beta-1}=k_1^{n\alpha_0+\beta-1}$$

Thus, the Bayes estimate of σ is given by

$$\sigma^B=k_1\left(\frac{1}{2}\right)^{n\alpha_0+\beta-1}$$

***Example* 8.66** Consider the LinEx loss function (1986)

$$L(\theta,\delta)=\exp[c(\delta-\theta)]-c(\delta-\theta)-1$$

where $c > 0$. The use of this loss function is prescribed in the case where asymmetries are to be converted into symmetries in a smooth way by changing c.

(i) Plot $L(\theta, \delta) = f(\theta - \delta)$ as a function of $(\theta - \delta)$, for $c = 0.2$, 0.5, and 1.

(ii) Let $X \sim f(x|\theta)$, $\theta \sim \pi(\theta)$. Show that the Bayes estimate $\delta^B(x) = -(1/c)\log[\exp(-c\theta)|x]$

(iii) Let $X_1,\ldots,X_n$ be *iid* $N(\theta, \sigma^2)$, $\sigma^2 > 0$ is known. Consider the prior distribution of θ, $\pi(\theta)$, as $\pi(\theta) = 1$, a non–informative prior. Show that the Bayes estimate under LinEx loss is given by

$$\delta^B(\overline{x}) = \overline{x} - \frac{c\sigma^2}{2n}$$

(iv) Calculate the posterior conditional expected loss of $\delta^B(\overline{X})$ and of $\overline{X}$ using LinEx loss.

(v) Calculate the posterior conditional expected loss of $\delta^B(\overline{X})$ and of $\overline{X}$ using squared error loss.

Solution.

(i) Exercise for readers.

(ii) We know that the Bayes estimator is the one that minimizes the posterior conditional expected loss, i.e.,

$$E^{\theta|x} L(\theta, \delta) = E\{\exp[c(\delta - \theta)]|x\} - cE[(\delta - \theta)|x] - 1$$
$$= \exp(c\delta)E[\exp(-c\theta|x)] - c[\delta - E(\theta|x)] - 1$$

Differentiating with respect to δ, we get

$$\frac{\partial}{\partial \delta}E[L(\theta, \delta)|x] = c\exp(c\delta)E[\exp(-c\theta)|x] - c = 0$$
$$\exp(-c\delta) = E[\exp(-c\theta)|x]$$
$$\delta^B(x) = -\frac{1}{c}\log E[\exp(-c\theta)|x]$$

Note that the second derivative is positive. So, $\delta^B(x)$ minimizes the posterior conditional expected loss.

(iii) The posterior distribution, in this case, is given by

$$\pi(\theta|\overline{x}) \sim N\left(\overline{x}, \frac{\sigma^2}{n}\right)$$

The moment generating function of $N(\mu, \sigma^2)$ is

$$M_X(t) = \exp\left(\mu t + \frac{1}{2}t^2\sigma^2\right)$$

This gives $\quad M_\theta(-c) = E[\exp(-c\theta)|\overline{x}] = \exp\left(-\overline{x}c + \frac{1}{2}c^2\frac{\sigma^2}{n}\right)$

Therefore, the Bayes estimate is

$$\delta^B(\overline{x}) = -\frac{1}{c}\log\exp\left(-\overline{x}c + \frac{1}{2}c^2\frac{\sigma^2}{n}\right) = \overline{x} - \frac{1}{2}c\frac{\sigma^2}{n}$$

(iv) Consider the LinEx posterior expected loss of the estimator of the form $\bar{X} + d$ when $\pi(\theta|\bar{x}) \sim N(\bar{x}, \sigma^2/n)$

$$E[L(\theta, \bar{x} + d)|\mathbf{x}] = E\exp[c(\bar{x} + d - \theta)] - cE(\bar{x} + d - \theta) - 1$$

$$= \exp[c(\bar{x}+d)]\exp\left(-\bar{x}c + \frac{c^2\sigma^2}{2n}\right) - cd - 1$$

$$= \exp\left(cd + \frac{c^2\sigma^2}{2n}\right) - cd - 1$$

When $d = 0$, the LINEX expected loss for $\bar{X}$ is

$$E[L(\theta, \bar{x})|\mathbf{x}] = \exp\left(\frac{c^2\sigma^2}{2n}\right) - 1$$

and when $d = -c\sigma^2/2n$, the LinEx expected loss for the Bayes estimate $\delta^B = \bar{x} - (c\sigma^2/2n)$ is

$$E\left[L\left(\theta, \bar{x} - \frac{c\sigma^2}{2n}\right)\bigg|\mathbf{x}\right] = \frac{c^2\sigma^2}{2n}$$

Note that the Bayes risk of any estimator of the form $\bar{X} + d$ [$\bar{X}$ and $\bar{X} - (c\sigma^2/2n)$] is infinite, since the marginal distribution of $\bar{X}$, $m(\bar{x})$, is one.

(v) The posterior conditional expected risk of any estimator of the form $\bar{X} + d$ for squared error loss function is

$$E[L(\theta, \bar{x} + d|\mathbf{x}) = E[(\theta - \bar{x} - d)^2|\bar{x})$$

$$= E[(\theta - \bar{x})^2|\bar{x}) + d^2$$

$$= \frac{\sigma^2}{n} + d^2$$

since $\pi(\theta|\bar{x})$ is $N(\bar{x}, \sigma^2/n)$. This shows that in terms of posterior, the conditional expected loss of the estimator $\bar{X}$ is less than that of the Bayes estimate $\delta^B(\bar{x})$. Therefore, the former is superior. However, the Bayes risk of both the estimators is infinite.

EXERCISES

1. Explain the Bayesian point and interval estimation based on the posterior distribution. Show that a unique Bayes estimator δ of θ with respect to some prior distribution π is admissible.

2. In a decision problem $D = \{d_1, d_2, d_3\}$, $\Theta = \{\theta_1, \theta_2, \theta_3\}$, and the risk function $R(\theta, d)$ is as given below:

D \ Θ	d_1	d_2	d_3
θ_1	97	103	102
θ_2	105	100	101
θ_3	101	101	98

Assume the prior distribution:

$$P(\theta = \theta_1) = 2P(\theta = \theta_2) = 2P(\theta = \theta_3)$$

Which are the best and the worst decision rules, respectively, (use Bayes principle)?

(i) d_1 and d_2
(ii) d_1 and d_3
(iii) d_2 and d_1
(iv) These cannot be computed with present data.

3. A coin is tossed for which the probability θ of getting a head is either 1/3 or 2/3. We are asked to decide which value of θ is true. The coin is tossed once and the loss function is considered as $L(\theta, \delta) = (\theta - \delta)^2$. Find

(i) All possible decision rules
(ii) Risk for all decision rules
(iii) Minimax decision rule

4. Let $X \sim f(x|\theta)$, $\pi(\theta)$ be the prior distribution, and the uniform cost loss function

$$L(\theta, \delta) = \begin{cases} 0, & \text{if } |\theta - \delta| \le \dfrac{\varepsilon}{2} \\ 1, & \text{if } |\theta - \delta| > \dfrac{\varepsilon}{2} \end{cases}$$

This loss function shows no loss as long as error is below $\varepsilon/2$. However, it assigns uniform cost when it exceeds $\varepsilon/2$. Find the Bayes estimate of θ.

Hint. $E[L(\theta, \delta)|x] = 1 - \int_{\delta-\varepsilon/2}^{\delta+\varepsilon/2} \pi(\theta|x)d\theta$. The integral $\int_{\delta-\varepsilon/2}^{\delta+\varepsilon/2} \pi(\theta|x)d\theta$ is maximized when θ is the midpoint of the modal interval of length ε. Therefore, the mode of the posterior distribution is the Bayes estimator of θ.

5. Show that in an estimation problem having the loss function

$$L(\theta, \delta) = \left|1 - \frac{\delta}{\theta}\right|$$

the Bayes estimate of θ is the equation

$$E(\theta^{-1}|\mathbf{x}) = 2\int_0^{\infty} \theta^{-1}\pi(\theta|\mathbf{x})d\theta$$

6. Let $X_1,\ldots,X_n$ be a random sample from the distribution whose probability mass function is $f(x; \theta) = \theta^x(1-\theta)^{1-x}$, $x = 0, 1$.

(i) Assume that the prior distribution of the unknown parameter θ is the uniform distribution over the interval (0, 1). Find the Bayes estimator of θ and $\theta(1-\theta)$. Find the $(1-\alpha)$ level credible interval for θ.

Hint. The posterior distribution of θ given Σx_i is $Be(\Sigma x_i + 1, n - \Sigma x_i + 1)$; $\theta^B = (\Sigma x_i + 1)/(n + 2)$; $[\theta(1-\theta)]^B = (\Sigma x_i + 1)(n - \Sigma x_i + 1)/(n + 3)(n + 2)$.

(ii) Consider a class of estimators δ of θ

$$C = \{\delta\colon \delta(x) = n^{-1}\sum x_i + a, \ a \text{ is some constant}\}$$

and the loss function

$$L(\theta, \delta) = (\theta - \delta)^2$$

Find the minimax estimator of θ.

7. Let $X_1,\ldots,X_n$ be *iid* $b(1, \theta)$ and (i) $\theta \sim Be(\alpha, \beta)$, α and β are known, (ii) $\theta \sim U(0, 1)$. Find the Bayes estimate of θ. Show that the Bayes estimate of θ in (i) is a weighted average of the MLE of θ and the mean $\alpha/(\alpha + \beta)$ of the prior distribution. Compare these estimators with the MLE of θ, i.e., $\bar{X}$ for $\alpha = \sqrt{n/4} = \beta$ in case (i). Show that the generalized maximum likelihood estimator of θ is $\theta^{\text{GMLE}} = (\Sigma x_i + \alpha + 1)/(n + \alpha + \beta - 2)$ when $1 - \alpha < \Sigma x_i < n + \beta - 1$.

Hint. Bayes estimator for θ when $\alpha = \sqrt{n/4} = \beta$ is $\theta^B = (\Sigma_{i=1}^n x_i + \sqrt{n/4})/(n + \sqrt{n})$.

8. Let $X_1, X_2,\ldots,X_n$ be a random sample from the distribution

$$f(x, \theta) = \theta^x(1-\theta)^{1-x};\ x = 0, 1;\ 0 \le \theta \le 1$$

If the loss function is squared error and apriori distribution of θ is

$$g(\theta) = \frac{1}{B(a, b)}\theta^{a-1}(1-\theta)^{b-1},\ 0 \le \theta \le 1$$

obtain the Bayes estimate for θ.

9. (Bernoulli experiment with Θ finite) Let θ be the probability of head of a coin and the parameter space Θ has two points {1/3, 2/3}. By noting that the parameter space Θ is discrete, a coin is tossed and a mapping δ on the event space $\{H, T\}$ is defined to consider $\delta(H)$ as an estimate of θ if head comes up and $\delta(T)$ if tail comes up. The loss function is considered as squared error loss function $L(\theta, \delta) = (\theta - \delta)^2$.

(i) Find the Bayes estimate with respect to the prior distribution that gives probability π to $\theta = 1/3$ and probability $1 - \pi$ to $\theta = 2/3$.

(ii) Show that the estimate

$$\delta_0(H) = \frac{5}{9},\ \delta_0(T) = \frac{4}{9}$$

is a minimax estimate of θ; $\pi = 1/2$ to $\theta = 1/3$ and $\pi = 1/2$ to $\theta = 2/3$ is the least favorable and the value of the game, V, is 2/81.

10. (Bernoulli experiment with Θ continuous) Let θ be the probability of coming up of heads of a coin, $\Theta = [0, 1]$. It is required to estimate the parameter θ on the basis of toss of the coin. An estimate is defined as mapping δ on $\{H, T\}$ into [0, 1]. Consider the squared error loss function, namely $L(\theta, \delta) = (\theta - \delta)^2$.

(i) Find the Bayes estimate of θ with respect to given prior distribution π. Show that any other prior distribution, with the same first two moments as for π, gives the same Bayes estimate of θ.

(ii) Show that the equalizer rule

$$\delta_0(H) = \frac{3}{4}, \quad \delta_0(T) = \frac{1}{4}$$

is a minimax estimate of θ when it is Bayes with respect to a prior distribution $\pi(\theta)$ having their moments

$$E(\theta) = \mu_1' = \frac{1}{2}, \quad E(\theta^2) = \mu_2' = \frac{3}{8}$$

Is such a prior distribution least favorable? Show that the Bayes (and hence minimax) risk, i.e., the value of the game is $V = 1/16$. Show that in case of n observations, $Be(n/2, n/2)$ is the least favorable and the minimax risk is $[4(\sqrt{n} + 1)^2]^{-1}$.

11. Consider a Bernoulli experiment, perform n independent trials with the probability of success θ, and count the number of successes; denote it by x. In the second experiment, independent trials are repeated until one gets x successes. Suppose both the experiments are conducted and that the number of trials and the number of successes are the same, i.e., n and x are same in both the experiments. Show that the likelihood functions are the same and that they carry equal information on θ in both the experiments. Show that the Jeffreys priors, in these cases, are not the same. Thus, Jeffreys prior violates the likelihood principle.

Hint. Let $X|\theta \sim b(n, \theta)$, n is known. The Jeffreys prior for θ is

$$\pi_1(\theta) \propto \frac{1}{\sqrt{\theta(1-\theta)}}$$

Consider $n \sim \text{NB}(x, \theta)$. The Jeffreys prior for θ is

$$\pi_2(\theta) \propto \frac{1}{\theta\sqrt{(1-\theta)}}$$

$\pi_1(\theta) \neq \pi_2(\theta)$

12. Let $X_1, \ldots, X_n$ be a random sample drawn from $b(1, p)$; and let the prior distribution of p on (0, 1), $\pi(p)$, be given by $Be(\alpha, \beta)$ with $\alpha = \beta = \sqrt{n/4}$. Show that under the squared error loss function, the Bayes estimate of p is

$$\delta_p^B(\mathbf{x}) = \frac{\alpha + \sum x_i}{\alpha + \beta + n} = \rho\bar{x} + (1-\rho)\frac{\alpha}{\alpha+\beta} = \frac{\sum x_i + \sqrt{n/4}}{n + \sqrt{n}}$$

where $\rho = 1 - (\alpha + \beta)/(\alpha + \beta + n)$. Show that if the sample size n is large and that the prior information becomes irrelevant

$$\delta_p^B(\mathbf{X}) \to \bar{X}$$

the Bayes estimator becomes the usual likelihood estimator and if $n = 0$

$$\delta_p^B(\mathbf{X}) \to \frac{\alpha}{\alpha + \beta}$$

the Bayes estimator becomes the prior mean, our intuitive belief. Show that $\bar{X}$ is a generalized Bayes with respect to an improper prior

$$\pi(p) \propto \frac{1}{p(1-p)} [\alpha = 0, \beta = 0 \text{ in Be}(\alpha, \beta)]$$

Also, show that $\bar{X}$ is a limit of the Bayes estimators with respect to $Be(\alpha, \beta)$ priors.

13. Let $X \sim b(n, p)$,

(i) the prior distribution be noninformative, $\pi(p) = 1$ for $0 < p < 1$. The Bayes estimator of p is

(a) X
(b) $(X + 1)/(n + 2)$
(c) $X/(n + 1)$
(d) $X/(n + 2)$

(ii) Consider the estimator $\delta(X) = X/n$ and loss function as absolute error $L(\theta, \delta) = |\delta - \theta|$. Show that the risk of δ is

$$R(\theta, \delta) = 2\binom{n-1}{k-1} p^k (1-p)^{n-k+1} \text{ for } \frac{k}{n}\frac{1}{n} \le \theta \le \frac{k}{n}$$

Plot this risk function for $n = 3$, 4, and 6.

Hint. Use the identity $\binom{n}{x}(x - n\theta) = n\left[\binom{n-1}{x-1}(1-\theta) - \binom{n-1}{x}\theta\right]$, $1 \le x \le n$

14. Let X has $b(5, p)$ distribution, on which an observation $x = 4$ is obtained. Under the Bayesian approach of estimation of p, a $Be(2, 3)$ prior distribution [with density proportional to $p(1-p)^2$] is considered. Then the posterior

(i) distribution of p is uniform on (0,1), (ii) the mean of p is 6/10, and (iii) the distribution of p is $Be(6, 4)$.

15. Let $X \sim b(n, p)$, $0 < p < 1$, and prior of p be $Be(\alpha, \beta)$, where $\alpha > 0$, $\beta > 0$ are known. Consider the loss function $L(p, \delta) = (\delta - p)^2/p^r(1-p)^s$, where r and s are known integers.

(i) Find the Bayes estimate of p with respect to the given loss function.

(ii) Show the Bayes estimate of p, for $r = 0$, $s = 0$, is

$$\delta_p^B(x) = \frac{\alpha + x}{\alpha + \beta + n}$$

Show that the classical estimate of p, $\delta(x) = x/n$, is not Bayes but the limit of Bayes and extended Bayes. Find a measure τ on (0, 1): with respect to which $\delta(x) = x/n$ is generalized Bayes. Also, find the Bayes estimate of $p(1 - p)$ when $r = 0$, $s = 0$.

(iii) Find the condition under which the Bayes estimate δ_p^B is an equalizer estimate. Show that the minimax estimate of p is

$$\delta_p^0(x) = \frac{x + \sqrt{n}}{n + \sqrt{n}}$$

Show that δ_p^0 is admissible.

(iv) Find the minimax estimate of p when $r = 1 = s$.

(v) Show that the classical estimate $\delta(x) = x/n$ of p under the loss function with $r = 1 = s$ is Bayes with respect to the uniform prior on (0, 1). Show that $\delta(x) = x/n$ is a minimax estimate of p with constant risk $1/n$.

16. Let X be a binomial variate with parameters n and p $(0 < p < 1)$. Using apriori distribution of p as $\pi(p) = 1$, for $0 < p < 1$, find the Bayes and minimax estimates for p and p^2 and $p(1 - p)$. Find the posterior conditional expected loss of the estimate of $p(1 - p)$ with respect to the Jeffreys noninformative prior for p

(i) under a quadratic error loss function, and

(ii) under the loss function of the type $L(p, \delta) = (p - \delta)^2/[p(1 - p)]$.

Hint. The Bayes estimate of p under the squared error loss function is $\delta_p^B = (x + 1)/(n + 2)$ and its Bayes risk is $r(\pi, \delta_p^B) = [6(n + 2)]^{-1}$.

17. Let $X \sim b(n, \theta)$ and the prior density for θ be $Be(\alpha, \beta)$. Consider the squared error loss function.

(i) Show that the Bayes estimate of θ is

$$\theta^B = \frac{x + \alpha}{n + \alpha + \beta}$$

(ii) Show that

$$R(\theta, \theta^B) = \frac{1}{(n + \alpha + \beta)^2}\{[(\alpha + \beta)^2 - n]\theta^2 + [n - 2\alpha(\alpha + \beta)\theta + \alpha^2]\}$$

(iii) If $\alpha = \beta = \sqrt{n/2}$, then show that $R(\theta, \theta^B)$ is a constant and independent of n.

(iv) Show that the Bayes estimate in (i) with $\alpha = \sqrt{n/2} = \beta$ is minimax.

(v) Consider the Haldane's (noninformative) prior

$$\pi(\theta) = \frac{1}{\theta(1 - \theta)}$$

Note that it is an improper prior. Show that the posterior distribution is

$$\pi(\theta|x) = Be(x, n-x)$$

and the Bayes estimate of θ is

$$\theta^B = \frac{x}{n}$$

provided $x \neq 0$ or n. Note that if $x = 0$ or n, the marginal density

$$m(x) \propto \int_0^1 \frac{1}{\theta(1-\theta)} \binom{n}{x} \theta^x (1-\theta)^{n-x}\, d\theta$$

$$= \binom{n}{x} \int_0^1 \theta^{x-1} (1-\theta)^{n-x-1}\, d\theta$$

is not defined for $x = 0$ or n.

(vi) For $x = 0$ or n, define the prior by heuristic considerations

$$\pi(\theta) \equiv Be(\alpha, \beta),\ \alpha, \beta > 0$$

which converges the Haldane's prior as in (v) as $\alpha, \beta \to 0$. Show that the posterior distribution is Be($\alpha + x$, $\beta + n - x$). Show that the Bayes estimate of θ is

$$\theta^B(\alpha, \beta) = \frac{\alpha + x}{\alpha + \beta + n} \to \frac{x}{n} = \theta^B$$

as in (v) by letting $\alpha, \beta \to 0$.

18. If $X \sim b(n, \theta)$ and $\pi(\theta) = Be(\alpha, \beta)$, find the Bayes estimate under the weighted squared error loss function

$$L(\theta, \delta) = \frac{(\theta - \delta)^2}{\theta(1-\theta)}$$

What treatment would you suggest in case of $x = 0$ and $x = n$?

19. Let X and Y be independent with bivariate distribution

$$f(x, y|p_1, p_2) = \binom{n}{x}\binom{n}{y} \theta_1^x \theta_1^y (1 - \theta_1)^{n-x} (1 - \theta_2)^{n-y};\ x = 0, 1, \ldots, n;\ y = 0, 1, \ldots, n;$$

$0 < (\theta_1, \theta_2) < 1$

Suppose that θ_1 and θ_2 are apriori independent with each other and having $U(0, 1)$. Find the Bayes estimate of $(\theta_1 - \theta_2)$ under the squared error loss function

$$L(\theta_1 - \theta_2, \delta) = (\theta_1 - \theta_2 - \delta)^2$$

Also, find the Bayes estimate of $\eta = \theta_1/\theta_2$ under the relative squared error loss function

$$L(\eta, \hat{\eta}) = \theta_2^2 (\eta - \hat{\eta})^2$$

20. Let $X_1,\ldots,X_n$ be a random sample from the probability density function

$$f(x; \lambda) = \lambda \exp(-\lambda x); x > 0, \lambda > 0$$

(i) Let apriori distribution over $\Theta = \{\lambda: \lambda > 0\}$ be given by the *pdf*

$$g(\lambda) = 3\lambda \exp(-3\lambda); \lambda > 0$$

The Bayes estimate of $P[1 < X_1 < 2]$ based on squared error loss function is

(a) $[(3 + \Sigma x_i)/(4 + \Sigma x_i)]^n - [(3 + \Sigma x_i)/(5 + \Sigma x_i)]^n$
(b) $[(3 + \Sigma x_i)/(4 + \Sigma x_i)]^{n+1} - [(3 + \Sigma x_i)/(5 + \Sigma x_i)]^{n+1}$
(c) $[(3 + \Sigma x_i)/(4 + \Sigma x_i)]^{n+2} - [(3 + \Sigma x_i)/(5 + \Sigma x_i)]^{n+2}$
(d) $[(3 + \Sigma x_i)/(4 + \Sigma x_i)]^{n+3} - [(3 + \Sigma x_i)/(5 + \Sigma x_i)]^{n+3}$

(ii) Let apriori distribution over $\Theta = \{\lambda: \lambda > 0\}$ be given by the *pdf*

$$g(\lambda) = \exp(-\lambda); \lambda > 0$$

Find the Bayes estimate of λ and $\exp(-\lambda)$ under the squared error loss function.

21. Let $X \sim P(\theta)$, $\theta > 0$ $[\Theta = (0, \infty)]$, and the prior for θ be $G_1(\alpha, \beta)$, $\alpha > 0$, $\beta > 0$ are known.

(i) Show that under the squared error loss function, the Bayes estimate of θ is

$$\delta_\theta^B(x) = \frac{\beta(\alpha + x)}{\beta + 1}$$

(ii) Show that under the squared error loss function, the Bayes estimate of θ is

$$\delta_\theta^B(x) = \frac{1}{a}(\alpha + x)\log\left(1 + \frac{\alpha}{\beta + 1}\right)$$

(iii) Show that the classical (maximum likelihood) estimate $\delta_\theta(x) = x$ is not Bayes with respect to any prior distribution. Show that if zero is included in Θ, $\delta_\theta(x) = x$ is Bayes.

(iv) Show that $\delta(x) = x$ is a limit of the Bayes estimates.

(v) Show that $\delta(x) = x$ is generalized Bayes with respect to the measure $\pi(\theta) = \log(\theta)$ $[d\pi(\theta) = (1/\theta)d\theta]$.

(vi) Show that the estimator $\delta(x) = x$ is an extended Bayes estimate.

(vii) Let the number of fire cases reported in Delhi be modelled as Poisson with parameter θ. In the absence of prior information on the average fire cases, a noninformative prior

$$\pi(\theta) = \frac{1}{\theta}, 0 < \theta < \infty$$

has been considered. The number of fire cases reported per week are 0, 1, 0, 1, 0. Construct 90% highest posterior density (HPD) credible interval for θ. (see Section 9.3.3).

22. Let $X \sim P(\theta)$, $\theta \in (0, \infty)$, and let $L(\theta, \delta) = (\theta - \delta)^2/\theta$.

(i) Show that the usual estimator (maximum likelihood estimator) of θ, $\delta_0(X) = X$ is generalized Bayes with respect to the Lebesgue measure on $\Theta = (0, \infty)$ [$\pi(\theta) \propto \theta^{-1}$, $\theta > 0$] and $A = [0, \infty)$. Show that δ_0 is an equalizer estimator.

(ii) Find a Bayes estimator of θ with respect to a prior distribution $\pi(\theta)$ on $(0, \infty)$ given by $G_2(\alpha, \beta)$, where hyperparameters α and β are known.

(iii) Show that δ_0 is extended Bayes and, therefore, minimax.

23. Let $X_1,\ldots,X_n$ be a random sample from the Poisson distribution $P(\theta)$ and let θ has the conjugate prior distribution given by

$$\pi(\theta) = \frac{\beta^\alpha}{\Gamma(\alpha)}\exp(-\beta\theta)\theta^{\alpha-1}d\theta, \alpha > 0, \beta > 0$$

(i) Find the Bayes estimate of θ and its risk under the squared error loss function and comment when $n \to \infty$

$$\left[\delta_\theta^B(\mathbf{x}) = \frac{\sum x_i + \alpha}{n + \beta} \to \bar{x} \text{ if } n \to \infty\right]$$

(ii) Show that the generalized maximum likelihood estimate of θ is $(\Sigma x_i + \alpha - 1)/(n + \beta)$ provided $\alpha + \Sigma x_i > 1$.

(iii) Find $(1 - \alpha)$-level credible interval for θ.

24. Malkov (1994).

(i) Consider $X_1,\ldots,X_n$ a random sample from $P(\theta)$ and a prior distribution on θ as $G_2(\alpha, \beta)$. Find the Bayes estimate of θ under the following loss functions:

(a) $L_1(\theta, \delta) = (\theta - \delta)^2$

(b) $L_2(\theta, \delta) = (1/\theta)(\theta - \delta)^2$

(c) $L_3(\theta, \delta) = (1/\theta^2)(\theta - \delta)^2$

(d) $L_4(\theta, \delta) = (\log\delta - \log\theta)^2$

(e) $L_5(\theta, \delta) = (\sqrt{\theta} - \sqrt{\delta})^2$

Further, show that the Bayes estimates of θ when $n = 1$, $\alpha = 0 = \beta$ are x, $x - 1$, $x - 2$, $\exp[D(x)]$, and $[\Gamma(x + 0.5)/\Gamma(x)]^2$ under the loss functions (a) to (e) where $D(x) = (\partial/\partial x)\log\Gamma(x)$.

(ii) [El Sayyad and Freeman (1973); Bansal (2007)] Let $X \sim P(\theta)$ and the prior distribution of θ be given by $\pi(\theta) = 1/\theta$, $\theta > 0$.

(a) Show that the Bayes estimate of θ under the squared error loss function is $\delta(x) = x$ and its posterior conditional loss is x.

(b) Show that the Bayes estimate of θ under the loss function $L(\theta, \delta) = (\theta - \delta)^2/\theta$ is $\delta(x) = x - 1$ and its posterior conditional expected loss is 1.

(c) Show that the Bayes estimate of θ under the loss function $L(\theta, \delta) = (\theta - \delta)^2/\theta^2$ is $\delta(x) = x - 2$ and its posterior conditional expected loss is $1/(x - 1)$.

(d) Show that the Bayes estimate of θ under the loss function $L(\theta, \delta) = [\log(\delta/\theta)]^2$ is $\delta(x) = \exp\Psi(x)$ and its posterior conditional expected loss is $\Psi'(x) \cong 1/x$.

(e) Show that the Bayes estimate of θ under the loss function $L(\theta, \delta) = (\sqrt{\theta} - \sqrt{\delta})^2$ is $\delta(x) = [\Gamma(x + 0.5)/\Gamma(x)]^2$ and its posterior conditional expected loss is $x - [\Gamma(x + 0.5)/\Gamma(x)]^2$.

[$\Psi(x) = (\partial/\partial x)\log\Gamma(x)$ and $\Psi'(x)$ are digamma and trigamma functions, respectively.)

25. Let X have geometric distribution

$$f(x|\theta) = (1 - \theta)\theta^x,\ x = 0, 1, 2,\ldots;\ \theta \in [0, 1)$$

(i) Let the loss function be given by

$$L(\theta, \delta) = \frac{(\theta - \delta)^2}{(1 - \theta)}$$

Write the risk of some estimator $\delta(X)$, $R(\theta, \delta)$, as a power series in θ. Show that

$$\delta(0) = \frac{1}{2},\ \delta(1) = \delta(2) = \cdots = 1 \tag{8.8.1}$$

is the only nonrandomized equalizer estimate. Show that an estimate is Bayes with respect to a prior distribution π if and only if

$$\delta(i) = \frac{\mu'_{i+1}}{\mu'_i},\ i = 0, 1, 2,\ldots$$

where μ'_i is the ith moment of the prior π density. Show that the estimate in Eq. (8.8.1) is extended Bayes, and therefore, minimax.

(ii) Let the loss function be given by

$$L(\theta, \delta) = (\delta - \theta)^2$$

Find the minimax estimate of θ.

(iii) Let θ has a prior distribution Be(α, β). Find the (1 – α)-level credible interval for θ.

26. Let X follows hypergeometric distribution with the *pmf*

$$f(x, \theta) = \frac{\binom{\theta}{x}\binom{M - \theta}{n - x}}{\binom{M}{n}};\ x = \max(0, \theta + n - M), \min(\theta, n)$$

Suppose the prior distribution of θ is beta-binomial distribution

$$\pi(\theta|\alpha, \beta) = \binom{n}{\theta}\frac{B(\theta + \alpha, n + \beta - \theta)}{B(\alpha, \beta)};\ \theta = 0, 1,\ldots, n$$

Find the Bayes estimate of θ under squared error loss function.

Hint. $\theta^B = [(M + \alpha + \beta)x + \alpha(M - n)]/(n + \alpha + \beta)$

27. Let $X \sim \text{NB}(r, p)$, $0 < p < 1$, r is known positive constant. Let us reparameterize the distribution so that

$$E(X) = r\frac{p}{1-p} = r\theta$$

and the *pmf* be given by

$$f(x|\theta) = \binom{r+x-1}{x}\theta^x(\theta+1)(r+x), x = 0, 1, 2, \ldots; \theta \in \Theta = (0, \infty)$$

We are now interested in estimating the parameter θ under the loss function

$$L(\theta, \delta) = \frac{(\theta-\delta)^2}{\theta(\theta+1)}$$

(i) Consider the classical estimator $\delta_0(X) = X/r$. Show that δ_0 is an equalizer and generalized Bayes with respect to the Lebsegue measure on $(0, \infty) = \Theta$, $r > 1$. Comment on the case when $r = 1$.

(ii) Consider the prior distribution on Θ, $\pi(\theta)$, with the *pdf*

$$\pi(\theta|\alpha, \beta) = \frac{\Gamma(\alpha+\beta)}{\Gamma(\alpha)\Gamma(\beta)}\theta^{\alpha-1}(\theta+1)^{-(\alpha+\beta)}, 0 < \theta < \infty$$

This distribution is derived from

$$\theta = \frac{p}{1-p} \text{ where } p \sim \text{Be}(\alpha, \beta)$$

Show that the Bayes estimate of θ with respect to this prior $\pi(\theta)$ is

$$\delta_0^B(x) = \frac{\alpha+x-1}{\beta+r+1}$$

(iii) Show that

$$\delta_\theta^M(x) = \frac{x}{r+1}$$

is minimax.

(iv) Comment on whether δ_0 is admissible?
(δ_0 is not minimax, therefore, not admissible.)

28. Let $X_1,\ldots,X_n$ be a random sample from negative binomial distribution $\text{NB}(r, \theta)$ and the prior on θ be $Be(\alpha, \beta)$. Show that the posterior distribution is

$$\text{Be}\left(\alpha + nr, \beta + \sum_{i=1}^{n} x_i\right)$$

29. Consider that (X_1, X_2, X_3) follows a trinomial distribution with the *pmf*

$$f(x_1, x_2, x_3; \theta_1, \theta_2, \theta_3) = \frac{(x_1 + x_2 + x_3)!}{x_1!x_2!x_3}\theta_1^{x_1}\theta_1^{x_2}\theta_1^{x_3}$$

with parameters θ_1, θ_2, θ_3 so that $\theta_1 + \theta_2 + \theta_3 = 1$. Find the Jeffreys prior for $(\theta_1, \theta_2, \theta_3)$.

30. Suppose $X \sim N(\theta, \sigma^2)$ and suppose the prior distribution of θ is also

$$N\left(\frac{\tau^2}{\tau^2 + \sigma^2}x + \frac{\sigma^2}{\tau^2 + \sigma^2}\mu, \frac{\sigma^2\tau^2}{\tau^2 + \sigma^2}\right)$$

Find the Bayes estimate of θ under the squared error loss function and calculate the corresponding Bayes risk.

31. Let $X_1,\ldots,X_n$ be independent observations according to cdfs $F(x|\theta_i)$. Let $T_i^{\pi_i}(X_i)$ be a Bayes estimator of θ_i under the loss function $L(\theta_i, a_i)$ and prior $\pi_i(\theta_i)$. Then show that

$$\mathbf{T}^{\pi}(\mathbf{X}) = (T^{\pi_1}(X_1),\ldots,T^{\pi_n}(X_n))$$

is Bayes for estimating $\boldsymbol{\theta} = (\theta_1,\ldots,\theta_n)$ under the loss function $\Sigma_{i=1}^n L(\theta_i, a_i)$ and prior $\pi(\theta) = \Pi_{i=1}^n \pi_i(\theta_i)$.

32. Let $X_1,\ldots,X_n$ be a random sample from $N(\mu, \sigma^2)$, where μ and σ^2 are both unknown. For the purpose of estimating σ^2, use the joint density function $f(x|\mu, \sigma^2)$ as $f(s_{n-1}^2|\sigma^2)$ where $(n-1)s_{n-1}^2/\sigma^2 \sim \chi_{n-1}^2$. Let the conjugate prior distribution for σ^2 be given by $\text{IG}_1(\alpha, \beta)$ with the *pdf*

$$\pi(\sigma^2) = \frac{1}{\Gamma(\alpha)\beta^{\alpha}}\frac{1}{(\sigma^2)^{\alpha+1}}\exp\left(-\frac{1}{\beta\sigma^2}\right), 0 < \sigma^2 < \infty$$

where $\alpha > 0$, $\beta > 0$ are known. Show that the posterior distribution of $\pi(\sigma^2|\mathbf{x})$ is $\text{IG}_1\{\alpha + (n-1)/2, [((n-1)^2/2) + (1/\beta)]^{-1}\}$. Show that the Bayes estimator of σ^2 under the squared error loss function, is

$$\delta_{\sigma^2}^B(S_{n-1}^2) = \frac{[(n-1)/2]S_{n-1}^2 + (1/\beta)}{[(n-1)/2] + \alpha - 1}$$

{***Note.*** $X \sim \text{IG}_1(\alpha, \beta)$ $E(X) = [(\alpha - 1)\beta]^{-1}$}

33. (Casella & Berger) Consider the observation $X_1,\ldots,X_n$ so that

$$X_i|\mu_i \sim N(\mu_i, \sigma^2), i = 1,\ldots,n, \text{ independent}$$
$$\mu_i \sim N(\eta, \tau^2), i = 1,\ldots,n, \text{ independent}$$

(i) Show that the marginal distribution of X_i is $N(\eta, \sigma^2 + \tau^2)$ and that marginally, $X_1,\ldots,X_n$ are *iid*. The joint density of $X_1,\ldots,X_n$ is used as a likelihood function to estimate the prior parameter in the empirical Bayes estimation.

(ii) Extending the result in (i), show that if

$$X_i|\mu_i \sim f(x|\mu_i), i = 1,\ldots,n, \text{ independent}$$
$$\mu_i \sim \pi(\eta|\tau), i = 1,\ldots,n, \text{ independent}$$

then marginally, $X_1,\ldots,X_n$ are *iid*.

34. Let $X_1,\ldots,X_n$ be a random sample drawn from $N(\mu, \sigma^2)$ for estimating μ where σ^2 is known. Assume the prior distribution for μ is double exponential DE(0 , τ) with the *pdf*

$$\pi(\mu|\tau) = \frac{1}{2\tau}\exp\left(-\frac{|\mu|}{\tau}\right)$$

where $\tau > 0$ is known. Find the Bayes estimate of μ under the squared error loss function.

35. (i) Let $X \sim N(\mu, 1)$ and consider the estimate of μ under the squared error loss function. The prior distribution for μ is

$$\pi(\mu) \sim N(0, \tau^2)$$

Show that the Bayes estimate of μ with respect to the prior distribution is

$$\delta_\mu^B(x) = \frac{x\tau^2}{1+\tau^2}$$

with its Bayes risk

$$r(\pi, \delta_\mu^B) = \frac{\tau^2}{1+\tau^2}$$

Show that the classical estimate $\delta(x) = x$ is the limit of Bayes and extended Bayes. Show that $\delta(x)$ is the equalizer and minimax estimate of μ.

(a) Assume $\tau^2 = 1$. Show that the Bayes estimate of μ under the loss function $L(\mu, \delta) = (\mu - \delta)^2 \exp(3\mu^2/4)$ is $\delta_\mu^B(x) = 2x$ which has uniformly higher risk as compared to the estimate $\delta(x) = x$. Also, calculate the expected risk and comment. Further, assume $\tau^2 = 1$ and consider the loss function $L(\mu, \delta) = 1$ if $\delta < \mu$ and zero otherwise. Show that no Bayes estimate of μ exists.

(b) Let μ be a measure of some positive quantity ($\mu > 0$). Consider the prior distributions for μ

$$\pi_1(\mu) = 1 \text{ for } \mu > 0,\ \pi_2(\mu) = G_1(1, 1).$$

Find the Bayes estimate of μ under the squared error loss function with respect to these prior distributions.

$$\left[\delta_{\pi_1}^B = x + \frac{\exp(-x^2/2)}{\sqrt{2\pi}P(Z > -x)} \text{ where } z \sim N(0,1);\ \delta_{\pi_2}^B = \bar{x} - \left(\frac{1}{n}\right)\right]$$

(c) Find the Bayes risk of the estimator of the type

$$\delta_c = cX$$

and, thus, find the Bayes estimator of μ with respect to the above prior distribution and the under squared error loss function by using the Bayes principle. Show that the Bayes risk of δ^B is $\tau^2/(1 + \tau^2)$. Also, show that the estimator δ_1 is superior to δ_c for $c > 1$.

(ii) Construct a 95% HPD credible interval for the mean μ when precision is 1 and the observed sample is 1.6, 2.4, 1.3, –1.9, 0.6, 2.2. Compare it with 95% confidence interval for μ. Highest posterior density (HPD) credible interval is defined in 9.3.3.

(iii) Consider the interval estimation of μ on the basis of n *iid* observations $X_1,\ldots,X_n$ from $N(\mu, 1)$. Show $(1 - \alpha)$-level credible interval for μ is

$$\left(\frac{n\bar{x}\tau^2}{1+n\tau^2} - z_{\alpha/2}\frac{\tau^2}{1+n\tau^2}, \frac{n\bar{x}\tau^2}{1+n\tau^2} + z_{\alpha/2}\frac{\tau^2}{1+n\tau^2}\right)$$

What happens to this credible interval if the prior variance τ^2 grows to infinity? Comment. Is this credible interval shorter in length as compared to the length of classical confidence interval estimate

$$\left(\bar{x} - \frac{z_{\alpha/2}}{n}, \bar{x} + \frac{z_{\alpha/2}}{n}\right)$$

for μ at level $(1 - \alpha)$? Comment.

(iv) Consider the interval estimation of μ on the basis of n *iid* observations $X_1,\ldots,X_n$ from $N(\mu, 1)$, and with respect to the prior on μ by $U(-1, 1)$. Find the $(1 - \alpha)$-level credible interval for μ.

(v) Consider the estimation of μ on the basis of n *iid* observations $X_1,\ldots,X_n$ from $N(\mu, 1)$ under the absolute error loss function $L(\theta, \delta) = |\delta - \theta|$. Show that $\bar{X}_n$ is an admissible estimator of θ.

36. (Bansal, 2007) Let $X \sim N(\mu, 1)$ and the prior distribution on θ be such that its density is unimodal with median 0 and quartiles ±1. The data has been observed $X = 6$.

(i) Find the 90% HPD credible interval for μ with respect to the prior distribution $N(0, 2.19)$.

(ii) Find 90% HPD credible interval for μ with respect to the prior distribution $C(0, 1)$.

(iii) Based on a random sample $X_1,\ldots,X_n$ from $N(\theta, 1)$ and prior on θ

$$\pi(\theta) = 1 \ \forall \ \theta \in \mathbb{R}^1$$

Find the Bayes estimate of θ under the LinEx loss function.

37. (Berger 1985, Bansal 2007) Let $X \sim N(\theta, 100)$, the prior distribution be given by $N(100, 225)$, and the loss function be

$$L_w(\theta, \delta) = \exp\left[-\left(\frac{\theta - 100}{30}\right)^2\right](\theta - \delta)^2$$

$X = 115$ has been observed on X to estimate θ. Calculate the modified prior distribution and show that the corresponding posterior distribution of $\theta|\mathbf{x}$ is $N(112.28, 81.81)$. Show that the Bayes estimate of θ is 112.28. Report, also, its corresponding Bayes risk.

38. (Normal model) Let $X_1,\ldots,X_n$ be *iid* random variables according to $N(\mu, \sigma^2)$ where both μ and σ^2 are unknown. Use the Jeffreys prior for $\boldsymbol{\theta} = (\mu, \sigma^2)$ to find the posterior predictive distribution. Consider normal-inverse gamma prior. Show that it is a conjugate prior and find the posterior distribution. Jeffreys (1961) recommended noninformative independence prior [not the Jeffreys prior $\pi(\boldsymbol{\theta}) \propto \sqrt{I_{\mathbf{X}}(\boldsymbol{\theta})}$]

$$\pi(\mu, \sigma^2) = \frac{1}{\sigma}$$

Find the posterior distribution and also the Bayes estimates of μ and σ.

39. (i) Consider the estimation problem of $\theta > 0$ under two types of loss functions

$$L_1(\theta, \delta) = \frac{(\theta - \delta)^2}{\theta}$$

$$L_2(\theta, \delta) = \frac{(\theta - \delta)^2}{\delta}$$

Find the Bayes estimate of θ, respectively, under these loss functions.

Hint. $\theta_{L_1}^B = [E(\theta^{-1}|x)]^{-1}$, $\theta_{L_2}^B = [E(\theta^2|x)]^{-1/2}$

(ii) Let $X \sim N(0, \sigma^2)$ where σ^2 is unknown. Find the Bayes estimate of σ^2 with respect to Jeffreys prior for σ^2 under the loss functions in (i), respectively.

40. (i) Let X follows lognormal distribution with parameters (θ, σ^2), where θ and σ^2 are both unknown. Let the loss function be squared error and the prior distribution on (θ, σ^2) be given by

$$\pi(\theta, \sigma^2) \propto \frac{1}{\sigma^c}, c > 0$$

Find the Bayes estimates of θ and σ^2.

(ii) Let $X_1,\ldots,X_n$ be a random sample from lognormal distribution with parameters (θ, σ^2), where σ^2 is known. Consider the prior distribution on θ which is given by the *pdf*

$$\pi(\theta) = \exp(-\theta), 0 < \theta < \infty$$

Show that the Bayes estimate of θ under the squared error loss function is

$$\exp\left[\bar{z} + \frac{\sigma^2}{2}\left(1 - \frac{1}{n}\right)\right]$$

where
$$\bar{z} = \frac{1}{n}\sum \log x_i$$

41. Let $X_1,\ldots,X_n$ as a random sample from $G_2(1, \theta)$ and the prior distribution on θ be $\pi(\theta) = \exp(-\theta)$, $\theta > 0$. Find the Bayes estimate of θ under the squared error loss function.

42. Consider $X_1,\ldots,X_n$ as a random sample from $G_2(2, \theta)$ and the prior on θ is given by $G_2(\alpha, \beta)$. Find the Bayes estimate of the reciprocal of scale parameter θ, i.e., θ^{-1} under the modified LinEx loss function.

43. Let X follows a distribution which has the following *pdf*

$$f(x|\mu,\sigma)=\frac{1}{\sigma}\exp\left(-\frac{x-\mu}{\sigma}\right);0<\mu<x<\infty$$

Consider the joint noninformative prior for μ and σ with the *pdf*

$$\pi(\mu,\sigma)\propto\sigma^{-1},\ \sigma>0$$

Find the Bayes estimate of μ under the squared error loss function.

44. Suppose a random sample of lifetimes $X_1,\ldots,X_n$ has been drawn from a population having the exponential distribution density

$$f(x|\theta)=\frac{1}{\theta}\exp\left(-\frac{x}{\theta}\right);x>0$$

Consider the loss function of the type

$$L(\theta,\delta)=\left(1-\frac{\delta}{\theta}\right)^2$$

Construct and identify the conjugate prior for θ and find the Bayes estimate of θ. Let the prior for expected lifetime θ be modelled by the density

$$\pi(\theta)=\frac{\alpha^\beta}{\Gamma(\beta)}\theta^{-(\beta+1)}e^{(-\alpha/\theta)},\theta>0$$

Show that the Bayes estimate of survival function [Basu and Thompson (1996), Bansal (2007)]

$$S(x_0|\theta)=P(X>x_0|\theta)=\exp\left(-\frac{x_0}{\theta}\right),x_0>0$$

is $\quad [S(t)]^B=\left(1+\frac{x_0}{\alpha+t}\right)^{-(\beta+n)}$

and its posterior variance is given by

$$V[S(x_0|\theta)|x]=\left(1+\frac{2x_0}{\alpha+t}\right)^{-(\beta+n)}-\left(1+\frac{x_0}{\alpha+t}\right)^{-(\beta+n)}$$

45. Let X follows the exponential distribution with the *pdf*

$$f(x|\theta)=\exp[-(x-\theta)],\ \theta\le x<\infty$$

The prior on θ is a Cauchy distribution, $\pi(\theta)\equiv C(0,1)$, with the *pdf*

$$\pi(\theta)=\frac{1}{\pi(1+\theta^2)}$$

Show that x is the MAP estimate of θ. Show that it is also the generalized maximum likelihood estimate of θ.

46. Let $X_1,\ldots,X_n$ be *iid* with the *pdf*

$$f(x|\theta)=\frac{1}{2}\exp(-|x-\theta|),\ x\in\mathbb{R}^1$$

Consider the prior distribution of θ by *pdf*

$$\pi(\theta)=1;\ \theta\in\Theta=\mathbb{R}^1$$

Find the Bayes estimate of θ under the squared and absolute error loss functions. Discuss the cases for $n = 2$ and 3.

47. Let $X_1,\ldots,X_n$ be *iid* according to

$$f(\mathrm{x}|\theta)=\exp[-(x-\theta)]\ ,\ x>\theta$$

(i) Suppose that the prior distribution of θ is

$$\pi(\theta)=\exp(-\theta),\ \theta>0$$

Find the Bayes estimate of θ under the squared error loss function. Show that the minimax estimator of θ, under the squared error loss function, is $X_{(1)}-(1/n)$ in the class of estimators

$$C=\left[\delta:\delta=a\left(X_{(1)}-\frac{1}{n}\right)\ \forall\, a>0\right]$$

(ii) Let the prior defined on θ be $\pi(\theta) = C(0, 1)$

$$\pi(\theta)=\frac{1}{\pi}\frac{1}{1+\theta^2}$$

Find the HPD estimate of θ.

Hint. $\pi(\theta|\mathbf{x})$ is an increasing function of θ, $\theta\le x_{(1)}$, since $(\partial/\partial\theta)\pi(\theta|x)>0$, $\theta^{\text{HPD}}=x_{(1)}$.

48. Let X follows such a gamma distribution that its mean and variance are equal to some positive quantity θ. Assume that the prior distribution of θ is given by the geometric distribution with the *pdf*

$$\pi(\theta|a)=(1-a)a^{\theta-1};\ \theta=1,2,\ldots;\ 0<a<1$$

Show that the posterior distribution of $\theta-1$ given $X = x$ is the Poisson distribution $P(ax)$. Find the Bayes estimate of θ under the loss function

$$L(\theta,\delta)=\frac{(\theta-\delta)^2}{\theta},\ \theta=1,2,\ldots$$

and show that the posterior conditional expected loss is

$$E(\theta|x)-[E(\theta^{-1}|x)]^{-1}=1-ax\sum_{\theta=1}^{\infty}\exp(-ax\theta)$$

49. Let $X \sim G_1(n/2, 2\theta)$ $(X/\theta \sim \chi^2_n)$ and prior on θ be given $IG_1(\alpha, \beta)$. Show that the posterior distribution is $IG_1((n/2) + \alpha, [(x/2) + \beta^{-1}]^{-1})$. Find the Bayes estimate of θ for the loss function

$$L(\theta, \delta) = \frac{(\theta - \delta)^2}{\theta^2}$$

50. Suppose that $X_1,\ldots,X_n$ is a random sample from an exponential distribution with parameter $\theta > 0$ with density

$$f(x|\theta) = \theta \exp(-\theta x), x > 0$$

Suppose that the prior density $\pi(\theta)$ is $G_2(\alpha, \beta)$. Show that the posterior distribution of θ given **x** is $G_2(\alpha + n, \beta + \Sigma x_i)$.

51. Let X follows the Maxwell distribution with probability density function

$$f(x; \theta) = \sqrt{\frac{2}{\pi}} \theta^{3/2} x^2 \exp\left(-\frac{\theta x}{2}\right), x \ge 0, \theta > 0$$

(i) Find the Jeffreys prior for θ.
(ii) Suggest the parameterization for which the corresponding prior is uniform.

52. Suppose $X_1,\ldots,X_n$ is a random sample drawn from $U(0, \theta)$.

(i) Suppose the prior on θ is a Pareto distribution $Pa(\theta_0, \alpha)$

$$\pi(\theta) = \frac{\alpha\theta_0^\alpha}{\theta^{\alpha+1}} \text{ for } \theta \ge \theta_0$$

Show that the posterior distribution is

$$Pa(\max\{\theta_0, x_1,\ldots,x_n\}, \alpha + n)$$

Find the Bayes estimate of θ under the squared loss function.

Hint. $\theta^B = (\alpha + n)\max(\theta_0, x_{(n)})/(\alpha + n - 1)$

(ii) (Bansal, 2007) Consider single observation case, i.e., $n = 1$. Let the loss function be squared error and the two prior distributions on θ be

$$\pi_1(\theta) = G_1(2, \beta), \quad \pi_2(\theta) = 6\alpha^{-2}\theta\left(1 + \frac{\theta}{\alpha}\right)^4, \theta > 0$$

The median of the prior is felt to be 6. Find the Bayes estimate of θ with respect to the above prior distributions, respectively, by way of determining α and β, and their corresponding risks $R(\theta, \delta)$. Which estimate would you prefer for large θ?

53. Suppose $X_1,\ldots,X_n$ is a random sample drawn from the Pareto distribution $P(\alpha_0, \theta)$ with the *pdf*

$$f(x; \theta) = \frac{\theta\alpha_0^\theta}{x^{\theta+1}}, x \ge \alpha_0 > 0$$

where α_0 is given. Consider the prior distribution of θ,

$$\pi(\theta) \propto \frac{1}{\theta}, 0 < \theta < \infty$$

Show that the Bayes estimate of θ under the squared error loss function is $[\log(G/\alpha_0)]^{-1}$, where $G = (\Pi x_i)^{1/n}$ is geometric mean.

54. Let $X_1,\ldots,X_n$ be *iid* from the Rayleigh *pdf*

$$f(x;\theta) = \frac{x}{\theta^2}\exp\left(-\frac{x^2}{2\theta^2}\right); x, \theta > 0$$

Derive the natural conjugate family of prior distributions for θ. Find the Bayes estimate of θ under the squared error loss function.

9 Confidence Interval Estimation

9.1 INTRODUCTION

So far, we have discussed the estimation of unknown population parameter θ in F_θ, whenever $X \sim F_\theta$, $\theta \in \Theta$. In that, the point estimator is just a single number based on the sample observations as a best guess for θ. How credible is this estimate or close to the true value of the parameter is measured by its standard error. This is why the variance or the standard error of the estimator is also reported along with this estimate. Point estimate gives only a single value which may be close to the true value. However, it may not be equal to the true value for continuous distributions, since for such distributions, the point probabilities are zero. Under these and other practical considerations, it is sometimes reasonable to report an interval as a function of sample observations by combining the estimated value and its standard error with a high probability or credibility, that it will contain the true value of θ. This interval is formally known as *confidence interval* and the probability is known as *confidence level*. The shorter the interval, the better is the precision or credibility that it captures the true value of the parameter. For example, let $X_1, X_2, \ldots, X_n$ be *iid* $N(\mu,1)$. We are not interested here in estimating μ by $\bar{X}$ but by the interval estimator $(\bar{X} - 1, \bar{X} + 1)$.

Consider another example where one requires the estimation of confidence interval. If X measures the nicotine percentage in a certain brand of cigarettes, the quality assurance manager may be interested in estimating the upper and lower bounds for the average nicotine percentage in the cigarettes to ensure whether it is in the limits as prescribed by some regulatory body. In another a bulb manufacturing company measures the life of bulbs and is interested in estimating the lower bound for the mean lifetime of bulbs.

Clearly, the above examples show that the experimenter is not interested in a point estimation of the parameter θ which is the mean of the random variable X, $E(X)$, rather he is interested in knowing the lower or upper or both bounds for θ.

In Section 9.2, some basic definitions are stated which are required to discuss confidence interval estimation in subsequent sections. Section 9.3 deals with constructing confidence interval estimators using different methods. Section 9.4 discusses the optimality of confidence interval estimators by using different reasonable criteria of evaluating interval estimators. In Section 9.5, equivariant confidence interval estimators are discussed.

9.2 BASIC NOTATIONS AND DEFINITIONS

Definition 9.2.1 (Random set) Let the data $\mathbf{X} \sim f(\mathbf{x}; \boldsymbol{\theta})$, $\boldsymbol{\theta} \in \Theta \subseteq \mathbb{R}^k$. The family of subsets $S(\mathbf{x})$ of Θ is called a family of random sets, where $S(\mathbf{x})$ is a function of data $x_1, x_2, \ldots, x_n$ and not of θ.

Definition 9.2.2 (Random interval) If $\Theta \subseteq \mathbb{R}^1$, $S(\mathbf{x}) = (\underline{\theta}(\mathbf{x}), \overline{\theta}(\mathbf{x}))$ is called the family of random intervals, where $\underline{\theta}(\mathbf{x})$ and $\overline{\theta}(\mathbf{x})$ are called the lower and upper bounds of θ, respectively. These bounds are only the functions of $\mathbf{x}$, not of θ. In particular, random intervals may be $S(\mathbf{x}) = [-\infty, \overline{\theta}(\mathbf{x})]$ if $\underline{\theta}(\mathbf{x}) = -\infty$ and $S(\mathbf{x}) = (\underline{\theta}(\mathbf{x}), \infty)$ if $\overline{\theta}(\mathbf{x}) = +\infty$.

Definition 9.2.3 (Family of (1 – α)-level confidence sets) A family of random sets $S(\mathbf{x})$ of $\Theta \subseteq \mathbb{R}^k$ for the parameter $\boldsymbol{\theta}$ is called the family of $(1 - \alpha)$-level confidence sets, if

$$P_{\boldsymbol{\theta}}[S(\mathbf{X}) \ni \boldsymbol{\theta}] \geq 1 - \alpha \;\forall\; \boldsymbol{\theta} \in \Theta \qquad (9.2.1)$$

If θ is a scalar, $\Theta \subseteq \mathbb{R}^1$, the confidence interval $S(\mathbf{x}) = (\underline{\theta}(\mathbf{x}), \overline{\theta}(\mathbf{x}))$ in Eq. (9.2.1) is called an interval estimator for θ and the observed interval $(\underline{\theta}(\mathbf{x}), \overline{\theta}(\mathbf{x}))$ is called an *interval estimate*. The probability in Eq. (9.2.1) is called the *coverage probability*.

Condition (9.2.1) states that the confidence set $S(\mathbf{X})$ covers the unknown $\boldsymbol{\theta}$ with a high probability not less than $(1 - \alpha)$; that is, the probability of true coverage is at least $(1 - \alpha)$. The confidence set $S(\mathbf{X})$ may cover $\boldsymbol{\theta}$ or may not when data $\mathbf{X} = \mathbf{x}$ is observed. The interpretation of the confidence set $S(\mathbf{X})$ satisfying Eq. (9.2.1) is that if $\alpha = 0.05$, $1 - \alpha = 0.95$ and that the statistical experiment is repeated 100 times and $S(\mathbf{x})$ is computed each time, 95 of these confidence sets among a total of 100 contain the true unknown value of $\boldsymbol{\theta}$.

Definition 9.2.4 (Level of significance and confidence coefficient) If

$$\inf_{\theta \in \Theta} P_{\theta}[S(\mathbf{X}) \ni \theta] \geq 1 - \alpha \qquad (9.2.2)$$

the family of such confidence sets $\{S(\mathbf{x})\}$ is called a *family of confidence sets* for $\boldsymbol{\theta}$ at level of significance $(1 - \alpha)$ and the quantity on the left-hand side of Eq. (9.2.2) is called the *confidence coefficient* of $S(\mathbf{X})$. The confidence coefficient indicates the highest possible level of significance for $S(\mathbf{X})$.

The confidence set when $\Theta \subseteq \mathbb{R}^1$, $S(\mathbf{x}) = (-\infty, \overline{\theta}(\mathbf{x}))$, is called the *upper confidence bound*; $S(\mathbf{x}) = (\underline{\theta}(\mathbf{x}), \infty)$ is called the *lower confidence bound*; and $S(\mathbf{x}) = (\underline{\theta}(\mathbf{x}), \overline{\theta}(\mathbf{x}))$ is called the *confidence interval* for θ.

Definition 9.2.5 (Uniformly most accurate bound) A family of $(1 - \alpha)$-level confidence set $[S(\mathbf{x})]$ is said to be a uniformly most accurate (UMA) family of confidence sets for $\boldsymbol{\theta}$ if

$$P_{\boldsymbol{\theta}}[S(\mathbf{X}) \ni \boldsymbol{\theta}'] \leq P_{\boldsymbol{\theta}}[S'(\mathbf{X}) \ni \boldsymbol{\theta}'] \;\forall\; \boldsymbol{\theta} \text{ and } \boldsymbol{\theta}' \qquad (9.2.3)$$

where $S'(\mathbf{X})$ is any other family of confidence sets for θ at level $(1 - \alpha)$.

Thus, the family of confidence sets $S(\mathbf{X})$, in the class of all $(1 - \alpha)$-level confidence sets, if it satisfies Eq. (9.2.3), it has the smallest probability of false or incorrect coverage of the parameter θ.

Definition 9.2.6 (Expected length) Given an interval $(\underline{\theta}(\mathbf{x}), \overline{\theta}(\mathbf{x}))$ of the parameter θ, the quantity

$$E_\theta[\overline{\theta}(\mathbf{X}) - \underline{\theta}(\mathbf{X})]$$

is called its *expected length.*

In the present chapter, we will be interested in constructing such a family of confidence sets that have at least a prespecified (high) value of probability of correct coverage of the parameter θ and have certain other optimal properties.

9.3 METHODS OF CONSTRUCTING CONFIDENCE INTERVALS

We will discuss here a few methods of constructing confidence sets at a given confidence level.

9.3.1 Confidence Interval Based on a Pivotal Quantity

This method of construpting confidence sets or intervals is based on pivots defined in the following definition.

Definition 9.3.1 (Pivot) A function of data $X_1, X_2, \ldots, X_n$ and $\boldsymbol{\theta}$, $T(\mathbf{X}, \boldsymbol{\theta})$, is called a pivotal quantity if its distribution does not depend on $\boldsymbol{\theta}$ whenever $\mathbf{X} \sim F_\theta$, $\boldsymbol{\theta} \in \Theta$.

Note that a pivot is not an ancillary statistic since it depends on θ.

Definition 9.3.2 (Asymptotic pivotal quantity) A function of data $X_1, X_2, \ldots, X_n$ and $\boldsymbol{\theta}$, $T_n(\mathbf{X}, \boldsymbol{\theta})$, is called asymptotic pivotal quantity if the limiting distribution of T_n as $n \to \infty$ is independent of $\boldsymbol{\theta}$.

Note that in the above definition, the distribution of T_n for some finite n may depend on $\boldsymbol{\theta}$. Generally, central limit theorem and asymptotic normality of maximum likelihood estimators help us in identifying the asymptotic pivotal quantity.

We give some examples on the choices of pivotal quantities for different distributions.

Let $X_1, X_2, \ldots, X_n$ be a random sample from $E(\theta)$ [or $G_2(1, \theta)$]. Then, $T(\mathbf{X}, \theta) = \theta\Sigma X_i$ is a pivotal quantity since its distribution does not depend on θ. Let $X_1, X_2, \ldots, X_n$ be a random sample from $U(0, \theta)$. Then, $T(\mathbf{X}, \theta) = X_{(n)}/\theta$ is a pivotal quantity since its distribution $f_T(t) = nt^{n-1}$, $0 \le t \le 1$, does not depend on θ. Let $X_1, X_2, \ldots, X_n$ be a random sample from $N(\mu, \sigma^2)$, where μ is unknown and σ is known. Then, $T(\mathbf{X}, \theta) = \sqrt{n}(\overline{X} - \mu)/\sigma \sim N(0, 1)$ is a pivotal quantity since its distribution does not depend on the parameter μ. In case μ and σ^2 are both unknown, $T(\mathbf{X}, \theta) = \sqrt{n}(\overline{X} - \mu)/S_{n-1}^2 \sim t_{n-1}$, where $S_{n-1}^2 = (n-1)^{-1}\Sigma(X_i - \overline{X})^2$, and $T(\mathbf{X}, \theta) = (n-1)S_{n-1}^2/\sigma^2 \sim \chi_{n-1}^2$ are the pivotal quantities used for constructing confidence intervals for μ and σ^2, respectively.

We will now give some examples to illustrate the use of pivots for constructing confidence intervals.

Confidence interval for μ in a normal population. Consider a random sample $X_1, X_2, \ldots, X_n$ drawn from $N(\mu, \sigma^2)$, where σ^2 is known. The quantity

$$T(\mathbf{X}, \mu) = \frac{\sqrt{n}(\bar{X} - \mu)}{\sigma} \sim N(0, 1)$$

is a pivotal quantity. Using this

$$P\left(-z_{\alpha/2} \le \frac{\sqrt{n}(\bar{X} - \mu)}{\sigma} \le z_{\alpha/2}\right) = 1 - \alpha$$

or

$$P\left(\bar{X} - z_{\alpha/2}\frac{\sigma}{\sqrt{n}} \le \mu \le \bar{X} + z_{\alpha/2}\frac{\sigma}{\sqrt{n}}\right) = 1 - \alpha$$

We get $(\bar{X} - z_{\alpha/2}(\sigma/\sqrt{n}), \bar{X} + z_{\alpha/2}(\sigma/\sqrt{n}))$ as $(1 - \alpha)$-level confidence interval for μ. The length of this class interval is $2z_{\alpha/2}\ (\sigma/\sqrt{n})$.

The sample size required for estimating population mean μ correct to within d units with probability $(1 - \alpha)$ is

$$n_0 = \left(2z_{\alpha/2}\frac{\sigma}{d}\right)^2$$

It is important to note that the above confidence interval for μ still holds even if the assumption on sampled population being normal is dropped. In case sample size is large n (>30), central limit theorem guarantees that the approximate distribution of $\sqrt{n}(\bar{X} - \mu)/\sigma$ is normal, irrespective of sampled population. Consider, now, the construction of $(1 - \alpha)$-level confidence interval for μ when μ and σ^2 are both unknown. The pivotal quantity

$$T(\mathbf{X}, \mu) = \frac{\sqrt{n}(\bar{X} - \mu)}{S_{n-1}} \sim t_{n-1}$$

is used in case of small n ($\le$30) to construct $(1 - \alpha)$-level confidence interval for μ

$$P\left(-t_{n-1,\alpha/2} \le \frac{\sqrt{n}(\bar{X} - \mu)}{S_{n-1}} \le t_{n-1,\alpha/2}\right) = 1 - \alpha$$

On simplification, we get

$$P\left(\bar{X} - t_{n-1,\alpha/2}\frac{S_{n-1}}{\sqrt{n}} \le \mu \le \bar{X} + t_{n-1,\alpha/2}\frac{S_{n-1}}{\sqrt{n}}\right) = 1 - \alpha$$

The construction of the above confidence interval for μ requires normality assumption of the sampled population. However, this is not required in case n ($\ge$30) is large. Then, the $(1 - \alpha)$-level confidence interval for μ is

$$P\left(\bar{X} - z_{\alpha/2}\frac{S_{n-1}}{\sqrt{n}} \le \mu \le \bar{X} + z_{\alpha/2}\frac{S_{n-1}}{\sqrt{n}}\right) = 1 - \alpha$$

Confidence interval for σ^2 in a normal population. Further, $(1 - \alpha)$-level confidence interval for σ^2 is constructed by considering the pivot

$$T(\mathbf{X}, \sigma^2) = \frac{(n-1)S_{n-1}^2}{\sigma^2} = \frac{\sum (X_i - \bar{X})^2}{\sigma^2} \sim \chi_{n-1}^2$$

The range of χ^2 is $(0, \infty)$. It is not symmetric, positively skewed with long upper tail and the density curve becomes symmetric for large degrees of freedom. The required class interval is

$$P\left(\chi_{n-1,\,1-\alpha/2}^2 \le \frac{(n-1)S_{n-1}^2}{\sigma^2} \le \chi_{n-1,\alpha/2}^2\right) = 1 - \alpha$$

or
$$P\left(\frac{(n-1)S_{n-1}^2}{\chi_{n-1,\,\alpha/2}^2} \le \sigma^2 \le \frac{(n-1)S_{n-1}^2}{\chi_{n-1,\,1-\alpha/2}^2}\right) = 1 - \alpha$$

and the $(1 - \alpha)$-level confidence interval for σ is

$$P\left(\sqrt{\frac{(n-1)S_{n-1}^2}{\chi_{n-1,\,\alpha/2}^2}} \le \sigma \le \sqrt{\frac{(n-1)S_{n-1}^2}{\chi_{n-1,\,1-\alpha/2}^2}}\right) = 1 - \alpha$$

Simultaneous confidence interval for (μ, σ^2) in a normal population. Consider a random sample $X_1, X_2, \ldots, X_n$ drawn from $N(\mu, \sigma^2)$, where μ and σ^2 are both unknown. The quantities

$$T_1(\mathbf{X}, \mu) = \frac{\sqrt{n}(\bar{X} - \mu)}{S_{n-1}} \sim t_{n-1}$$

and
$$T_2(\mathbf{X}, \sigma^2) = \frac{(n-1)S_{n-1}^2}{\sigma^2} = \frac{\sum (X_i - \bar{X})^2}{\sigma^2} \sim \chi_{n-1}^2$$

are considered as pivotal quantities. For events A and B, Boole's inequality is given by

$$P(AB) \ge 1 - P(\bar{A}) - P(\bar{B})$$

Using Boole's inequality, we can construct $(1 - \alpha_1 - \alpha_2)$-level confidence interval for (μ, σ^2)

$$P\left(-t_{\alpha_1/2} < \frac{\sqrt{n}(\bar{X} - \mu)}{S_{n-1}} < t_{\alpha_1/2},\ \chi_{n-1,\alpha_2/2}^2 < \frac{\sum (X_i - \bar{X})^2}{\sigma^2} < \chi_{n-1,1-(\alpha_2/2)}^2\right)$$

$$\ge 1 - P\left(\frac{\sqrt{n}(\bar{X} - \mu)}{S_{n-1}} < -t_{n-1,\alpha_1/2} \text{ or } \frac{\sqrt{n}(\bar{X} - \mu)}{S_{n-1}} \ge t_{n-1,\alpha_1/2}\right)$$

$$- P\left(\frac{\sum (X_i - \bar{X})^2}{\sigma^2} \le \chi_{n-1,\alpha_2/2}^2 \text{ or } \frac{\sum (X_i - \bar{X})^2}{\sigma^2} \ge \chi_{n-1,1-(\alpha_2/2)}^2\right)$$

$$= 1 - \alpha_1 - \alpha_2$$

Thus, $(1 - \alpha_1 - \alpha_2)$-level simultaneous confidence interval for (μ, σ^2) is given by

$$\left(\bar{X} - \frac{S_{n-1}}{\sqrt{n}} t_{n-1,\,\alpha_1/2},\ \bar{X} + \frac{S_{n-1}}{\sqrt{n}} t_{n-1,\alpha_1/2}\right) \times \left(\frac{(n-1)S_{n-1}^2}{\chi^2_{n-1,\alpha_2/2}},\ \frac{(n-1)S_{n-1}^2}{\chi^2_{n-1,1-(\alpha_2/2)}}\right)$$

Confidence interval for $\mu_1 - \mu_2$ in two normal populations. Consider two independent samples of sizes n_1 and n_2 from normal populations $N(\mu_1, \sigma_1^2)$ and $N(\mu_2, \sigma_2^2)$, respectively. We will discuss three cases for constructing $(1 - \alpha)$-level confidence interval for $(\mu_1 - \mu_2)$.

(i) If $\sigma_1^2 = \sigma_2^2 = \sigma^2$, σ^2 is known. Let the means and variances of the samples be given by $\bar{X}_1$, S_1^2 and $\bar{X}_2$, S_2^2. We consider the pivot

$$T(\mathbf{X}_1, \mathbf{X}_2, \mu_1, \mu_2) = \frac{(\bar{X}_1 - \bar{X}_2) - (\mu_1 - \mu_2)}{\sigma\sqrt{(1/n_1) + (1/n_2)}} \sim N(0, 1)$$

Based on this pivot, we can construct $(1 - \alpha)$-level confidence interval for $(\mu_1 - \mu_2)$ by choosing a and b such that

$$P\{a < T < b\} = 1 - \alpha$$

Let the probability α be equally distributed in the tails of the distribution of T

$$P\{T < a\} = \frac{\alpha}{2} = P\{T > b\}$$

Since the distribution of T is symmetric about zero

$$b = z_{\alpha/2},\ a = -z_{\alpha/2}$$

Therefore, the $(1 - \alpha)$-level confidence interval for $\mu_1 - \mu_2$ is given by

$$\left((\bar{X}_1 - \bar{X}_2) - z_{\alpha/2}\sigma\sqrt{\frac{1}{n_1} + \frac{1}{n_2}},\ (\bar{X}_1 - \bar{X}_2) + z_{\alpha/2}\sigma\sqrt{\frac{1}{n_1} + \frac{1}{n_2}}\right)$$

However, if σ_1^2 and σ_2^2 are known but different, we have the pivot

$$T(\mathbf{X}_1, \mathbf{X}_2, \mu_1, \mu_2) = \frac{(\bar{X} - \bar{Y}) - (\mu_1 - \mu_2)}{\sqrt{(\sigma_1^2/n_1) + (\sigma_2^2/n_2)}} \sim N(0, 1)$$

Therefore, the required $(1 - \alpha)$-level confidence interval for $\mu_1 - \mu_2$ is given by

$$\left((\bar{X}_1 - \bar{X}_2) - z_{\alpha/2}\sqrt{\frac{\sigma_1^2}{n_1} + \frac{\sigma_2^2}{n_2}},\ (\bar{X}_1 - \bar{X}_2) + z_{\alpha/2}\sqrt{\frac{\sigma_1^2}{n_1} + \frac{\sigma_2^2}{n_2}}\right)$$

(ii) When $\sigma_1^2 = \sigma_1^2 = \sigma^2$ is not known. The independence of $(n_1 - 1)S_1^2/\sigma^2 \sim \chi^2_{n_1-1}$ and $(n_2 - 1)S_2^2/\sigma^2 \sim \chi^2_{n_2-1}$ gives

$$\frac{(n_1 + n_2 - 2)}{\sigma^2} S_p^2 = \frac{(n_1 - 1)S_1^2}{\sigma^2} + \frac{(n_1 - 1)S_2^2}{\sigma^2} \sim \chi^2_{n_1+n_2-2}$$

where S_1^2 and S_2^2 are the sample variances with divisors $(n_1 - 1)$ and $(n_2 - 1)$, respectively, and is called the pooled variance. The pivotal quantity is defined by

$$\frac{\dfrac{\bar{X}_1 - \bar{X}_2 - (\mu_1 - \mu_2)}{\sigma\sqrt{\left(\dfrac{1}{n_1} + \dfrac{1}{n_2}\right)}}}{\sqrt{\dfrac{\dfrac{(n_1 - 1)S_1^2}{\sigma^2} + \dfrac{(n_2 - 1)S_2^2}{\sigma^2}}{(n_1 + n_2 - 2)}}} = \frac{\bar{X}_1 - \bar{X}_2 - (\mu_1 - \mu_2)}{\sqrt{\dfrac{(n_1 - 1)S_1^2 + (n_2 - 1)S_2^2}{(n_1 + n_2 - 2)}\left(\dfrac{1}{n_1} + \dfrac{1}{n_2}\right)}}$$

$$= \frac{\bar{X}_1 - \bar{X}_2 - (\mu_1 - \mu_2)}{S_p\sqrt{\left(\dfrac{1}{n_1} + \dfrac{1}{n_2}\right)}} \sim t_{n_1 + n_2 - 2}$$

Note that the statistics in numerator and denominator are independent. Thus, the pivot follows t-distribution with $(n_1 + n_2 - 2)$ degrees of freedom. Based on this pivot, we can now construct $(1 - \alpha)$-level confidence interval for $(\mu_1 - \mu_2)$ by choosing a and b such that

$$P\{a < T < b\} = 1 - \alpha$$

If the probability α is distributed in the tails of the distribution of T equally, we choose a and b so that

$$P\{T < a\} = \frac{\alpha}{2} = P\{T > b\}$$

Since the distribution of T is symmetric about zero, we have

$$b = t_{\alpha/2},\ a = -t_{\alpha/2}$$

Thus, the required confidence interval is given by

$$P\left[(\bar{X}_1 - \bar{X}_2) - t_{n_1 + n_2 - 2,\, \alpha/2} S_p \sqrt{\frac{1}{n_1} + \frac{1}{n_2}} \le (\mu_1 - \mu_2)\right.$$

$$\left.\le (\bar{X}_1 - \bar{X}_2) + t_{n_1 + n_2 - 2,\, \alpha/2} S_p \sqrt{\frac{1}{n_1} + \frac{1}{n_2}}\right] = 1 - \alpha$$

(iii) (Behren–Fisher problem) When σ_1^2 and σ_2^2 are not known and $\sigma_1^2 \neq \sigma_2^2$. We define two variables U and V by

$$U = \frac{(\bar{X}_1 - \bar{X}_2) - (\mu_1 - \mu_2)}{\sqrt{(\sigma_1^2/n_1) + (\sigma_2^2/n_2)}} \sim N(0, 1) \qquad (9.3.1)$$

since

$$\bar{X}_1 - \bar{X}_2 \sim N\left(\mu_1 - \mu_2, \frac{\sigma_1^2}{n_1} + \frac{\sigma_2^2}{n_2}\right)$$

and

$$cV = \frac{S_1^2}{n_1} + \frac{S_2^2}{n_2};\ V \sim \chi_k^2 \qquad (9.3.2)$$

where $$\frac{(n_1-1)S_1^2}{\sigma_1^2} = \frac{\sum(X_{1i}-\bar{X}_1)^2}{\sigma_1^2} \sim \chi^2_{n_1-1}$$

$$\frac{(n_2-1)S_2^2}{\sigma_2^2} = \frac{\sum(X_{2i}-\bar{X}_2)^2}{\sigma_2^2} \sim \chi^2_{n_2-1} \tag{9.3.3}$$

$$E(S_1^2) = \sigma_1^2;\ V(S_1^2) = \frac{2}{(n_1-1)}\sigma_1^4$$

$$E(S_2^2) = \sigma_2^2;\ V(S_2^2) = \frac{2}{(n_2-1)}\sigma_2^4$$

From Eqs. (9.3.2) and (9.3.3), we have

$$E(cV) = ck = \frac{\sigma_1^2}{n_1} + \frac{\sigma_2^2}{n_2}$$

and $$V(cV) = c^2(2k) = 2c(ck)$$

$$= \frac{1}{n_1^2}\frac{2}{n_1-1}\sigma_1^4 + \frac{1}{n_2^2}\frac{2}{n_2-1}\sigma_2^4 \tag{9.3.4}$$

From Eq. (9.3.4), we have

$$c = \frac{(1/n_1^2)[1/(n_1-1)]\sigma_1^4 + (1/n_2^2)[1/(n_2-1)]\sigma_2^4}{[(\sigma_1^2/n_1) + (\sigma_2^2/n_2)]}$$

$$k = \frac{[(\sigma_1^2/n_1) + (\sigma_2^2/n_2)]^2}{(1/n_1^2)[1/(n_1-1)]\sigma_1^4 + (1/n_2^2)[1/(n_2-1)]\sigma_2^4}$$

$$= \frac{(\sigma_2^2/n_2)^2[(\sigma_1^2/n_1)(n_2/\sigma_2^2)+1]^2}{(\sigma_2^4/n_2^2)\{(\sigma_1^2/n_1^2)(n_2^2/\sigma_2^2)[1/(n_1-1)] + [1/(n_2-1)]\}}$$

Denoting $$R = \frac{\sigma_1^2/n_1}{\sigma_2^2/n_2}$$

k can be rewritten as

$$k = \frac{(1+R)^2}{\left[\dfrac{R}{(n_1-1)} + \dfrac{1}{n_2-1}\right]}$$

or $$\frac{1}{k} = \left[\left(\frac{R}{1+R}\right)^2 \frac{1}{n_1-1} + \left(\frac{1}{1+R}\right)^2 \frac{1}{n_2-1}\right] \tag{9.3.5}$$

Thus, on using Eqs. (9.3.1), (9.3.2), and (9.3.5), we get the pivot

$$T = \frac{U}{\sqrt{V/k}} = \frac{\dfrac{(\bar{X}_1 - \bar{X}_2) - (\mu_1 - \mu_2)}{\sqrt{(\sigma_1^2/n_1) + (\sigma_2^2/n_2)}}}{\sqrt{\dfrac{(S_1^2/n_1) + (S_2^2/n_2)}{ck}}}$$

$$= \frac{\dfrac{(\bar{X}_1 - \bar{X}_2) - (\mu_1 - \mu_2)}{\sqrt{(\sigma_1^2/n_1) + (\sigma_2^2/n_2)}}}{\sqrt{\dfrac{(S_1^2/n_1) + (S_2^2/n_2)}{(\sigma_1^2/n_1) + (\sigma_2^2/n_2)}}} = \frac{(\bar{X}_1 - \bar{X}_2) - (\mu_1 - \mu_2)}{\sqrt{(S_1^2/n_1) + (S_2^2/n_2)}} \sim t_k$$

where k is given by Eq. (9.3.5) with R replaced by $\hat{R}$

$$\hat{R} = \frac{(S_1^2/n_1)}{(S_2^2/n_2)}$$

Finally, the pivot simplifies to

$$\frac{(\bar{X}_1 - \bar{X}_2) - (\mu_1 - \mu_2)}{\sqrt{(S_1^2/n_1) + (S_2^2/n_2)}} \sim t_k \tag{9.3.6}$$

Using the pivot in Eq. (9.3.6), the $(1 - \alpha)$-level confidence interval for $(\mu_1 - \mu_2)$ is given by

$$P\left[(\bar{X}_1 - \bar{X}_2) - t_{k,\alpha/2}\sqrt{\frac{S_1^2}{n_1} + \frac{S_2^2}{n_2}} \le (\mu_1 - \mu_2) \le (\bar{X}_1 - \bar{X}_2) + t_{k,\alpha/2}\sqrt{\frac{S_1^2}{n_1} + \frac{S_2^2}{n_2}}\right] = 1 - \alpha$$

Confidence interval for ratio of variances in two normal populations. Consider the pivotal quantity

$$T(\mathbf{X}, \sigma_1^2, \sigma_2^2) = \frac{S_1^2/\sigma_1^2}{S_2^2/\sigma_2^2} \sim F_{n_1-1,\, n_2-1}$$

for constructing $(1 - \alpha)$-level confidence interval for the ratio of the variances σ_1^2/σ_2^2, when μ_1, μ_2 are not known

$$P\left(F_{n_1-1,\, n_2-1,\, \alpha_1} \le \frac{S_1^2}{S_2^2}\frac{\sigma_2^2}{\sigma_1^2} \le F_{n_1-1,\, n_2-1,\, \alpha_2}\right) = 1 - \alpha_1 - \alpha_2$$

or

$$P\left(\frac{S_1^2}{S_2^2}\frac{1}{F_{n_1-1,\, n_2-1,\, \alpha_2}} \le \frac{\sigma_1^2}{\sigma_2^2} \le \frac{S_1^2}{S_2^2}\frac{1}{F_{n_1-1,\, n_2-1,\, \alpha_1}}\right) = 1 - \alpha_1 - \alpha_2$$

so that $\alpha = \alpha_1 + \alpha_2$.

Confidence interval for μ_d in a bivariate normal population. Let a random sample (X_i, Y_i), $i = 1, 2, \ldots, n$ (n small, $n \leq 30$), be selected from a bivariate normal population, such that the population of paired differences $d_i = Y_i - X_i$ is approximately normal. Then, $(1 - \alpha)$-level confidence interval for $\mu_d = \mu_Y - \mu_X$ is

$$P\left(\bar{d} - t_{n-1,\,\alpha/2}\frac{S_d}{\sqrt{n}} \leq \mu_d \leq \bar{d} + t_{n-1,\,\alpha/2}\frac{S_d}{\sqrt{n}}\right) = 1 - \alpha$$

where $\bar{d}$ and S_d^2 are the mean and sample variance of n sample differences, respectively.

Confidence interval for probability of success in a binomial population. Consider a large random sample of size n (> 30) from a binomial distribution $b(n, p)$, to construct a confidence interval for probability of success p. Suppose the proportion of successes in the sample is $\hat{p}$. For infinite population or for finite population sampling with replacement, we consider the pivot

$$T(\mathbf{X}, p) = \frac{\hat{p} - p}{\sqrt{[\hat{p}(1-\hat{p})/n]}} \sim N(0, 1)$$

to get $(1 - \alpha)$-level approximate confidence interval for p

$$P\left[\hat{p} - z_{\alpha/2}\sqrt{\frac{\hat{p}(1-\hat{p})}{n}} \leq p \leq \hat{p} + z_{\alpha/2}\sqrt{\frac{\hat{p}(1-\hat{p})}{n}}\right] = 1 - \alpha$$

If one wishes to estimate population proportion p, to be within d units with probability $(1 - \alpha)$, the sample size required is

$$n = \left(\frac{2z_{\alpha/2}}{d}\right)^2 \hat{p}(1-\hat{p})$$

In case of finite population without replacement, $(1 - \alpha)$-level confidence interval for p is

$$P\left[\hat{p} - z_{\alpha/2}\sqrt{\frac{\hat{p}(1-\hat{p})}{n}}\sqrt{\frac{N-n}{N-1}} \leq p \leq \hat{p} + z_{\alpha/2}\sqrt{\frac{\hat{p}(1-\hat{p})}{n}}\sqrt{\frac{N-n}{N-1}}\right] = 1 - \alpha$$

Confidence interval for difference of probability of successes in two binomial populations. Consider two independent samples from binomial populations $b(n_1, p_1)$ and $b(n_2, p_2)$ with $\hat{p}_1$ and $\hat{p}_2$ as estimates for proportions of successes, respectively. The $(1 - \alpha)$-level confidence interval for the difference of probability of successes is given as

$$P\left[(\hat{p}_1 - \hat{p}_2) - z_{\alpha/2}\sqrt{\frac{\hat{p}(1-\hat{p}_1)}{n_1} + \frac{\hat{p}_2(1-\hat{p}_2)}{n_2}} \leq p_1 - p_2\right.$$

$$\left. \leq (\hat{p}_1 - \hat{p}_2) + z_{\alpha/2}\sqrt{\frac{\hat{p}(1-\hat{p}_1)}{n_1} + \frac{\hat{p}_2(1-\hat{p}_2)}{n_2}}\right] = 1 - \alpha$$

Further, to estimate the difference $p_1 - p_2$ to be within d units with probability $(1 - \alpha)$, the sample sizes required are

$$n_1 = n_2 = \left(\frac{2z_{\alpha/2}}{d}\right)^2 [\hat{p}_1(1-\hat{p}_1) + \hat{p}_2(1-\hat{p}_2)]$$

Estimation of $(1 - \alpha)$-level confidence interval for p based on maximum likelihood estimate (MLE) of p for large n. If $\hat{\theta}$ is the MLE of θ, then under certain regularity conditions in subsection (6.6.6), A1 to A6,

$$\sqrt{nI_{\mathbf{X}}(\hat{\theta})}(\hat{\theta} - \theta_0) \xrightarrow{D} N(0, 1)$$

where θ_0 is the true value of the parameter.
Based on this pivotal quantity, we can, therefore, construct $(1 - \alpha)$-level confidence interval for θ

$$P\left[-z_{\alpha/2} \le \sqrt{nI_{\mathbf{X}}(\hat{\theta})}(\hat{\theta} - \theta_0) \le z_{\alpha/2}\right] = 1 - \alpha$$

$$S(\mathbf{X}) = \left(\hat{\theta} - z_{\alpha/2}\frac{1}{\sqrt{nI_{\mathbf{X}}(\hat{\theta})}}, \hat{\theta} + z_{\alpha/2}\frac{1}{\sqrt{nI_{\mathbf{X}}(\hat{\theta})}}\right)$$

Here, we see that the estimate and the estimated error, both combine to construct a $(1 - \alpha)$-level confidence interval. The length of the interval decreases with increase in the sample size n and for smaller confidence level $(1 - \alpha)$. We will see later in this chapter that the length of the class interval becomes meaningful when it is considered as one of the criterion of judging its precision.

By increasing the level of confidence to 100%, we get a meaningless class interval $(-\infty, +\infty)$ of infinite length. Therefore, the confidence level is fixed at $(1 - \alpha)$ and the length of the interval at d to determine the sample size by solving

$$\overline{\theta}(\mathbf{x}) - \underline{\theta}(\mathbf{x}) = d$$

for n. Consider, for example, a $(1 - \alpha)$-level confidence interval for μ based on a random sample $X_1, X_2, \ldots, X_n$ of size n drawn from $N(\mu, 28^2)$

$$P\left(\overline{X} - z_{\alpha/2}\frac{\sigma}{\sqrt{n}} \le \mu \le \overline{X} + z_{\alpha/2}\frac{\sigma}{\sqrt{n}}\right) = 1 - \alpha$$

To get the sample size necessary to get the above interval at 95% confidence level with the width $d = 12$, we solve

$$2z_{\alpha/2}\frac{\sigma}{\sqrt{n}} = d$$

for n, to get

$$n = \left(2z_{\alpha/2}\frac{\sigma}{d}\right)^2 = \left(2 \times 1.96\frac{28}{12}\right)^2 \cong 84$$

Having illustrated the use of pivots in constructing confidence intervals of desired level, we now theorize these steps in the form of a theorem.

Theorem 9.3.1 Let $T(\mathbf{X}, \theta)$ be a pivot taking the values in $\Lambda \subseteq \mathbb{R}^1$. If

(i) for each fixed θ, $T(\mathbf{X}, \theta)$ is a statistic;
(ii) for each fixed $\mathbf{x} \in \mathbb{R}^n$, T is monotone in θ; and
(iii) for each fixed $\mathbf{x}$ and t in Λ, the equation

$$T(\mathbf{x}, \theta) = t$$

is solvable for θ, then the confidence interval for θ can be constructed at any level.

Proof. For any α, $0 < \alpha < 1$, we get a set $(t_1(\alpha), t_2(\alpha))$ not necessarily unique in Λ with t_1, $t_2 \in \Lambda$

$$P_\theta[t_1(\alpha) < T(\mathbf{X}, \theta) < t_2(\alpha)] \geq 1 - \alpha \ \forall \ \theta$$

Note that t_1 and t_2 do not depend on θ since T is a pivot and its distribution does not depend on θ. For every $\mathbf{x} \in \mathbb{R}^n$, we can solve the following equations for θ

$$T(\mathbf{x}, \theta) = t_1(\alpha)$$

and
$$T(\mathbf{x}, \theta) = t_2(\alpha) \tag{9.3.7}$$

since T is monotone in θ. This gives the $(1 - \alpha)$-level confidence interval $S(\mathbf{X}) = (\underline{\theta}(\mathbf{X}), \overline{\theta}(\mathbf{X}))$ so that

$$P_\theta(\underline{\theta}(\mathbf{X}) < \theta < \overline{\theta}(\mathbf{X})) \geq 1 - \alpha \ \forall \ \theta$$

This completes the proof. ▣

The multivariate version of the theorem can similarly be stated.

Notes.

1. Note that the two equations in (9.3.7) can be solved for θ to yield a confidence interval of level which is exactly equal to $(1 - \alpha)$ if the distribution of T is continuous and monotone in θ. However, if the distribution of T is not continuous, the solutions of the inequalities $T < t_2$ and $T > t_1$ when solved for θ may not give a single confidence interval rather a set of confidence intervals in Θ. The level of confidence of these intervals may be a bit higher than $(1 - \alpha)$. As a result, the length of the interval also increases.
2. If the pivot T is monotone in θ, then by the application of Theorem 9.3.1, one obtains $(1 - \alpha)$-level confidence interval for θ, $P_\theta(a \leq T(\mathbf{X}, \theta) \leq b) = 1 - \alpha$, by $S(\mathbf{X}) = (\theta$: $\underline{\theta}(\mathbf{X}, a) \leq \theta \leq \overline{\theta}(\mathbf{X}, b))$, if T is increasing in θ and by $S(\mathbf{X}) = (\theta$: $\underline{\theta}(\mathbf{X}, b) \leq \theta \leq \overline{\theta}(\mathbf{X}, a))$, if T is decreasing in θ.
3. Pivotal quantities do not always exist in every case. However, for location and scale families, they always exist.
4. Let $Z_1, Z_2, \ldots, Z_n$ be *iid* according to $f(z)$ where $f(z)$ is the *pdf* of the *rv* Z, so that $f(z)$ is independent of θ. Let $X_1, X_2, \ldots, X_n$ be *iid* with $f(x - \theta)$. So, $(X_1, X_2, \ldots, X_n)$ and $(Z_1 + \theta, Z_2 + \theta, \ldots, Z_n + \theta)$ are equivalent. The distribution of $\bar{X} - \theta \sim (\bar{Z} + \theta) - \theta = \bar{Z}$ does not depend on θ. Therefore, $T(\mathbf{X}, \theta) = \bar{X} - \theta$ is the pivot.

5. Let $X_1, X_2, \ldots, X_n$ be *iid* with $(1/\sigma)f(x/\sigma)$; $(X_1, X_2, \ldots, X_n)$ and $(\sigma Z_1, \sigma Z_2, \ldots, \sigma Z_n)$ are equivalent. The distribution of $\bar{X}/\sigma \sim \sigma\bar{Z}/\sigma = \bar{Z}$ is independent of σ. Therefore, $T(\mathbf{X}, \theta) = \bar{X}/\sigma$ is the pivot.

6. Let $X_1, X_2, \ldots, X_n$ be *iid* with $(1/\sigma)f((x - \theta)/\sigma))$; $(X_1, X_2, \ldots, X_n)$ and $(\theta + \sigma Z_1, \theta + \sigma Z_2, \ldots, \theta + \sigma Z_n)$ are equivalent. The distribution of $(\bar{X} - \theta)/S_X \sim (\overline{\theta + \sigma Z} - \theta)/S_{\theta+\sigma Z} = \sigma\bar{Z}/\sigma S_Z = \bar{Z}/S_Z$ does not depend on θ and σ. Therefore, $T(\mathbf{X}, \theta) = (\bar{X} - \theta)/S_X$ is the pivot.

We state another result without proof that helps in identifying a pivot in the probability density of some statistic S of sample observations $X_1, X_2, \ldots, X_n$.

Theorem 9.3.2 Let S be some real-valued statistic so that its *pdf* is expressed by

$$f(s;\theta) = h[T(s,\theta)]\left|\frac{\partial}{\partial s}T(s,\theta)\right| \tag{9.3.8}$$

where T is monotone in s for each fixed $\theta \in \Theta$. Then $T(S, \theta)$ is a pivot. Also, condition (9.3.8) is satisfied by taking $h \equiv 1$ and $T(s, \theta) = F(s, \theta)$, i.e., the *cdf* of T.

We will demonstrate the result of Theorem 9.3.2 in identifying a pivot for a given family of exponential distributions.

Consider a random sample $X_1, X_2, \ldots, X_n$ from $G_1(1, \theta)$, $f(x; \theta) = (1/\theta)\exp(-x/\theta)$, $x > 0$. We have $S = \Sigma X_i \sim G_1(n, \theta)$, $f(s, \theta) = [1/(\Gamma(n)\theta^n)] \exp(-s/\theta)\, s^{n-1}$, $s > 0$. We can easily identify the quantity s/θ in this density as $T(s, \theta)$ in Eq. (9.3.8) and the above theorem applies to give $T(S, \theta) = S/\theta = \Sigma X_i/\theta$ as pivot with its distribution $G_1(n, 1)$.

Distribution function as pivot for constructing (1 – α)-level confidence intervals. Consider $X_1, X_2, \ldots, X_n$ as *iid* according to some continuous distribution function $F(X; \theta)$. We have a corresponding random sample $Y_1, Y_2, \ldots, Y_n$ where $Y_i = F(X_i; \theta)$

$$P(Y \le y) = P[F(X,\theta) \le y]$$
$$= P[X \le F^{-1}(y)] = y$$

so that each $Y_i = F(X_i; \theta) \sim U(0, 1)$. We know that $-\log Y_i \sim G_1(1, 1)$. Therefore, $-\Sigma_1^n \log Y_i \sim G_1(n, 1)$. We have

$$T(\mathbf{X},\theta) = \prod Y_i$$

$$-\log T(\mathbf{X},\theta) = -\sum_1^n \log Y_i = -\sum_1^n \log F(X_i;\theta) \sim G_1(n,1)$$

$-\log T(\mathbf{X}, \theta)$ is the pivot. (1 – α)-level confidence interval based on this pivot is constructed by choosing a and b such that

$$P(a < -\log T(\mathbf{X},\theta) < b) = 1 - \alpha$$

or

$$\int_a^b f(x,n)dx = 1 - \alpha$$

where $f(x, n)$ is the *pdf* of $G_1(n, 1)$ distribution.

Note that if F is continuous, it does not guarantee that the pivot $-\Sigma_1^n \log F(X_i;\, \theta)$ can be inverted to get $(1 - \alpha)$-level confidence interval for θ. However, if $-\Sigma_1^n \log F(X_i;\, \theta)$ is monotone in θ for given $\mathbf{x}$, it can be inverted to give a $(1 - \alpha)$-level confidence interval for θ.

Another way of pivoting the *cdf* is to consider some statistic U of sample observations $X_1, X_2, \ldots, X_n$. Let the *cdf* of U be given by

$$F(u;\, \theta) = P_\theta(U \le u)$$

Define a random variable T

$$T(\mathbf{X},\, \theta) = F(U;\, \theta)$$

Consider that the *cdf* of T is

$$P_\theta(T \le t) = P_\theta\{F^{-1}[F(U;\, \theta)] \le F^{-1}(t)\}$$

(Since F^{-1}, the *cdf* of U, is increasing)

$$= P_\theta[U \le F^{-1}(t)] = F[F^{-1}(t)] = t$$

This shows that $T \sim U(0, 1)$. Therefore, T is a pivot. Based on this pivot, the $(1 - \alpha)$-level confidence interval is given by

$$P_\theta[a < T(\mathbf{X},\, \theta) < b] = 1 - \alpha$$

or

$$\int_a^b dt = b - a$$

If $T(\mathbf{X},\, \theta) = F(U,\, \theta)$ is decreasing in θ for all values of U, then we solve

$$T(\mathbf{x},\, \theta) = F(u;\, \theta) = b$$

for θ to get $\underline{\theta}(U)$, and solve

$$T(\mathbf{x},\, \theta) = F(u;\, \theta) = a$$

for θ to get $\bar{\theta}(U)$. Thus, the $(1 - \alpha)$-level confidence interval is given by $(\underline{\theta}(U),\, \bar{\theta}(U))$. Further, if $F(U;\, \theta)$ is increasing in θ for all values of U, $\underline{\theta}(U)$ is calculated by solving

$$F(U;\, \theta) = a$$

and $\bar{\theta}(U)$ by

$$F(U;\, \theta) = b$$

Consider the testing problem $H_0: \theta = \theta_0$ against $H_1: \theta \ne \theta_0$. We may define a test

$$\phi(u) = \begin{cases} 1, & F_U(u, \theta_0) < a \text{ or } F_U(u, \theta_0) > b \\ 0, & \text{otherwise} \end{cases}$$

such that

$$E_{\theta_0}\, \phi(U) = \alpha$$

$$P(a < F_U(U,\, \theta_0) < b) = 1 - \alpha$$

$$P(T \le b) - P(T \le a) = 1 - \alpha$$

$$F_T(b) - F_T(a) = 1 - \alpha$$
$$b - a = 1 - \alpha$$

Therefore, we can select $b = 1 - \alpha_2$ and $a = \alpha_1$ so that

$$1 - \alpha_1 - \alpha_2 = 1 - \alpha$$

i.e., $\alpha_1 + \alpha_2 = \alpha$. The acceptance region of this test, which is of size α, is given by

$$A(\theta_0) = \{u: \alpha_1 \leq F_U(u, \theta_0) \leq 1 - \alpha_2\}$$

and the corresponding confidence set at level $(1 - \alpha)$ is

$$S(\mathbf{X}) = \{\theta: U \in A(\theta)\}$$
$$= \{\theta: \alpha_1 \leq F_U(u, \theta) \leq 1 - \alpha_2\} \quad (9.3.9)$$

This confidence set is not necessarily an interval. The following theorem states the condition on F that if F is continuous and monotone in θ, then the confidence set in Eq. (9.3.9) is an interval.

Theorem 9.3.3 Let U be a statistic of sample observations $X_1,\ldots,X_n$ with the continuous *cdf* $F_U(u, \theta)$. Then, for each u in $\mathcal{U}$, the sample space of U, $(\underline{\theta}(u), \overline{\theta}(u))$, is a $(1 - \alpha)$, $\alpha_1 + \alpha_2 = \alpha$, $0 < \alpha < 1$, level confidence interval for θ where

(i) $\overline{\theta}$ and $\underline{\theta}$ are obtained by solving

$$F_U(u, \theta) = \alpha_1 \quad \text{and} \quad F_U(u, \theta) = 1 - \alpha_2$$

respectively, for θ if $F_U(u, \theta)$ is a decreasing function of θ.

(ii) $\overline{\theta}$ and $\underline{\theta}$ are obtained by solving

$$F_U(u, \theta) = 1 - \alpha_2 \quad \text{and} \quad F_U(u, \theta) = \alpha_1$$

respectively, for θ if $F_U(u, \theta)$ is an increasing function of θ.

Proof. Assume that we can construct an acceptance region of size α for testing $H_0:\theta = \theta_0$ against $H_1:\theta \neq \theta_0$

$$A(\theta_0) = \{u: \alpha_1 \leq F_U(u, \theta) \leq 1 - \alpha_2\}$$

The corresponding $(1 - \alpha)$-level confidence set is

$$S(u) = \{\theta: \alpha_1 \leq F_U(u, \theta) \leq 1 - \alpha_2\}$$

(i) If for each fixed $u \in \mathcal{U}$, $F_U(u, \theta)$ is an decreasing function of θ, and F is continuous in θ, then the corresponding confidence interval is a decreasing function of θ. Thus, the corresponding confidence interval is given by $(\underline{\theta}, \overline{\theta})$ so that $\underline{\theta}$ is the solution of

$$F_U(u; \theta) = \alpha_1$$

or

$$\int_{-\infty}^{u} f_U(u; \theta)du = \alpha_1 \quad (9.3.10)$$

and $\bar{\theta}$ is the solution of

$$F_U(u;\,\theta) = 1 - \alpha_2$$

or

$$\int_u^{\infty} f_U(u;\,\theta)du = \alpha_2 \tag{9.3.11}$$

where $\underline{\theta}$ and $\bar{\theta}$ are unique by assuming $\alpha_1 < 1 - \alpha_2$. Moreover,

$$F_U(u;\,\theta) < \alpha_1 \;\forall\; \theta > \bar{\theta}$$

and

$$F_U(u;\,\theta) > 1 - \alpha_2 \;\forall\; \theta < \underline{\theta}$$

The proof of part (ii) is similar. This completes the proof. ▣

Theorem 9.3.3 can be used to construct one-sided intervals if the family of *pdf*s $\{f(x;\,\theta),\ \theta \in \Theta\}$ has an MLR. Consider the cdfs at points θ_1, θ_2 in Θ, $\theta_1 > \theta_2$,

$$\frac{\partial}{\partial x}[F(x;\,\theta_1) - F(x;\,\theta_2)] = f(x;\,\theta_1) - f(x;\,\theta_2)$$

$$= f(x;\,\theta_2)\left[\frac{f(x;\,\theta_1)}{f(x;\,\theta_2)} - 1\right]$$

The ratio $f(x;\,\theta_1)/f(x;\,\theta_2)$ is increasing since f has MLR. Therefore, the derivative can only change sign from negative to positive showing that any interior extremum is a minimum. Thus, the function is maximum at $+\infty$ or $-\infty$, which is zero. This shows that

$$F(x;\,\theta_1) \le F(x;\,\theta_2)$$

whenever $\theta_1 > \theta_2$. Therefore, if the family $\{f(x;\,\theta),\ \theta \in \Theta\}$ has an MLR, then the corresponding family of pdfs is stochastically increasing in θ. Theorem 9.3.3 holds in this case, since the condition that $F_U(u;\,\theta)$ is monotonic is satisfied. The usual practice is to consider $\alpha_1 = \alpha_2 = \alpha/2$. However, it is not always reasonable. One-sided intervals can be constructed by taking either $\alpha_1 = 0$ or $\alpha_2 = 0$.

The $(1 - \alpha)$-level confidence interval in Theorem 9.3.3 is not necessarily obtained for analytic solutions of Eqs. (9.3.10) and (9.3.11). If the analytic solutions to Eqs. (9.3.10) and (9.3.11) cannot be obtained, then numerical solutions are obtained. Note that the proof of Theorem 9.3.3 does not necessary require analytic solutions.

We, now, consider the construction of confidence interval at level $(1 - \alpha)$ when the *cdf* is discrete.

Theorem 9.3.4 Consider a statistic U with discrete *cdf* $F_U(u;\,\theta) = P(U \le u;\,\theta)$. Then, $(1 - \alpha)$-level confidence interval $(\underline{\theta}(U), \bar{\theta}(U))$ for θ can be constructed based on $F_U(u;\,\theta)$ by solving

$$P(U \le u;\,\theta) \le \alpha_1$$

to get $\bar{\theta}(U)$; and by solving

$$P(U \ge u;\,\theta) \le \alpha_2$$

to get $\underline{\theta}(U)$, whenever $F_U(u;\ \theta)$ is decreasing function of θ for each fixed $u \in \mathcal{U}$, where $\alpha = \alpha_1 + \alpha_2$ with $\alpha < 1/2$. Similarly, $(1 - \alpha)$-level confidence interval $(\underline{\theta}(U),\ \overline{\theta}(U))$ can be constructed by solving

$$P(U \geq u;\ \theta) \leq \alpha_2$$

to get $\overline{\theta}$; and by solving

$$P(U \leq u;\ \theta) \leq \alpha_1$$

to get $\underline{\theta}$, whenever $F_U(u;\ \theta)$ is an increasing function of θ for each fixed $u \in \mathcal{U}$.

Proof. Let $T = F_U(u;\ \theta)$ and $S \sim U(0, 1)$. Note that $F_S(t) = P(S \leq t) = t\ \forall\ t \in (0, 1)$. Consider

$$F_T(t) = P(T \leq t) = P(F_U(u;\ \theta) \leq t) \tag{9.3.12}$$

Define a set

$$A_t = \{u\colon F_U(u;\ \theta) \leq t\}$$

Since $F_U(u)$ is discrete and nondecreasing, A_t is either $(-\infty,\ u(t))$ or $(-\infty,\ u(t)]$. We may write Eq. (9.3.12) as

$$F_T(t) = P(T \leq t) = P(U \in A_t) = F_U[u(t);\ \theta] \leq t$$

If $A_t = (-\infty,\ u(t)]$, i.e., A_t is open, we consider

$$F_T(t) = P(U \in A_t) = P\{U \in [-\infty,\ u(t)]\}$$

$$= \lim_{u \uparrow u(t)} P[U \in (-\infty,\ u]] = \lim_{u \uparrow u(t)} F_U(u;\ \theta) \leq y$$

This inequality is true since $F_U(u;\ \theta) \leq y$ for every $u \in A_t$, i.e., for every $u < u(t)$. Therefore,

$$F_T(t) \leq t \text{ for every } t$$

in both the cases when A_t is closed or open.

Let $u(t)$ be the jump-point or the point of discontinuity

$$\lim_{u \uparrow u(t)} F_U(u;\ \theta) < t < F_U(u(t);\ \theta)$$

In this case,

$$A_t = (-\infty,\ u(t))$$

and

$$F_T(t) = \lim_{u \uparrow u(t)} F_U(u;\ \theta) < t$$

$\therefore$

$$P[F_U(u;\ \theta) \leq t] = F_T(t) \leq t = P(S \leq t) = F_s(t)\ \forall\ t$$

and

$$P[F_U(u;\ \theta) \leq t] = F_T(t) < t = F_s(t) \quad \text{for some } t \tag{9.3.13}$$

This shows that $T = F_U(u;\ \theta)$ is stochastically greater than $S \sim U(0, 1)$. One can, similarly, show if $F_U(u;\ \theta)$ is stochastically greater than $U(0, 1)$, then

$$P_U(U \geq u;\ \theta) = 1 - P_U(U < u;\ \theta) = 1 - F_U(u-;\ \theta)$$

is stochastically greater than $S \sim U(0, 1)$

$$P[P(U \geq u;\, \theta) \leq t] \leq t \;\forall\, t$$

or

$$P[1 - P(U < u;\, \theta) \leq t] \leq t \;\forall\, t$$

or

$$P[1 - F_U(u-;\, \theta) \leq t] \leq t \;\forall\, t$$

$$P[F_U(u-;\, \theta) \geq 1 - t] \leq t \;\forall\, t$$

and

$$P[F_U(u-;\, \theta) \geq 1 - t] < t \quad \text{for some } t \tag{9.3.14}$$

Since $F_U(u;\, \theta)$ is decreasing in θ, $\underline{\theta}$ and $\overline{\theta}$, in the interval $(\underline{\theta}, \overline{\theta})$ are obtained by solving

$$P(U < u;\, \theta) = 1 - \alpha_2$$

or

$$F_U(u-;\, \theta) = 1 - \alpha_2$$

for $\underline{\theta}$, and

$$P(U \leq u;\, \theta) = \alpha_1$$

Moreover,

$$F_U(u-;\, \theta) \geq 1 - \alpha_2 \;\forall\, \theta < \underline{\theta}$$

and

$$F_U(u;\, \theta) \leq \alpha_1 \;\forall\, \theta > \overline{\theta}$$

Under these conditions

$$\begin{aligned} P(\underline{\theta} < \theta < \overline{\theta}) &= P(\theta < \overline{\theta}) - P(\theta < \underline{\theta}) \\ &= P[F_U(u;\, \theta) \geq \alpha_1] - P[F_U(u-;\, \theta) \geq 1 - \alpha_2] \\ &= 1 - P[F_U(u;\, \theta) \leq \alpha_1] - P[F_U(u- ;\, \theta) \geq 1 - \alpha_2] \\ &\geq 1 - \alpha_1 - \alpha_2 = 1 - \alpha \end{aligned}$$

[By Eqs. (9.3.13) and (9.3.14)]

This shows that the above interval is a $(1 - \alpha)$-level confidence interval. ▣

9.3.2 Confidence Interval by Inverting Acceptance Region of a Test

Testing of hypotheses and confidence interval estimation, both imply each other but looked upon with different perspective. In hypotheses testing, the experimenter hypothesizes the parameter in the null hypothesis statement by fixing it to some constant value. Using this, he defines an appropriate region of the sample space, namely, acceptance region, with the meaning that the sample points falling in this region are in consistence with the value of the parameter being fixed. While in confidence interval estimation, the experimenter collects a sample and based on it he defines an interval that captures such values of the parameter that make this sample most plausible.

We will now show that the acceptance region, $AR(\theta)$, of a size-α test and a family of $(1 - \alpha)$-level confidence sets $S(\mathbf{X})$ imply each other by restricting only to non-randomized size-α tests.

Let us consider a testing problem $H_0: \theta \in \Theta_0$ versus $H_1: \theta \in \Theta_1$ denoted by $H_0(\Theta_0)$ and $H_1(\Theta_1)$, respectively. Let the class of level-α tests be denoted by

$$C_\alpha = \left\{\phi \in \mathcal{D} : \sup_{\theta \in \Theta_0} E_\theta \phi(\mathbf{X}) \le \alpha\right\}$$

$$= \left\{A \in \mathbb{R}^n : \sup_{\theta \in \Theta_0} P_\theta(A) \ge 1 - \alpha\right\}$$

$$= \left\{S(\mathbf{x}) : \min_{\theta \in \Theta_0} P_\theta[S(\mathbf{X}) \ni \theta] \ge 1 - \alpha\right\}$$

This shows that a class of α-level tests for testing $(\alpha, H_0(\Theta_0), H_1(\Theta_1))$ and of confidence sets $S(\mathbf{x}) \subset \Theta_0$ at level $(1 - \alpha)$ imply each other. The UMP size-α test with its acceptance region $A(\theta_0)$ gives

$$P_\theta(\mathbf{X} \in A(\theta_0)) \le P_\theta(\mathbf{X} \in A^*(\theta_0)) \; \forall \; \theta \in \Theta_1 \; (\theta \in H_1(\Theta_1))$$

or
$$P_\theta(S(\mathbf{X}) \ni \theta_0) \le P_\theta(S^*(\mathbf{X}) \ni \theta_0) \; \forall \; \theta \in \Theta_1$$

This implies that the acceptance region corresponding to a UMP size-α test, $A(\Theta_0)$, minimizes

$$P_\theta[S(\mathbf{X}) \ni \theta'] \; \forall \; \theta \in H_1(\Theta_1) \text{ and for every } \theta' \text{ in } \Theta_0$$

Therefore, the family of confidence sets corresponding to a UMP acceptance region is UMA family at $(1 - \alpha)$-level.

We will now give an important result that establishes a correspondence between UMP size-α acceptance region and UMA $(1 - \alpha)$-level family of confidence sets. Let us consider the null hypothesis $H_0{:}\theta = \theta_0$. Denote it by $H_0(\theta_0)$ against any one- or two-sided alternative hypothesis; denote it by $H_1(\theta_0)$.

Theorem 9.3.5 Let $A(\theta_0)$, $\theta_0 \in \Theta$, denotes the acceptance region of a size-α test of $H_0(\Theta_0)$. For each observation $\mathbf{x} = (x_1,\ldots,x_n)$, let $S(\mathbf{x})$ denotes the set

$$S(\mathbf{x}) = \{\theta : \mathbf{x} \in A(\theta),\, \theta \in \Theta\}$$

Then, $S(\mathbf{x})$ is a $(1 - \alpha)$-level family of confidence sets for θ. Conversely, if $S(\mathbf{x})$ is a $(1 - \alpha)$-level confidence set for θ, for any θ_0, define the acceptance region of a test for testing hypothesis $H_0{:}\,\theta = \theta_0$ by $A(\theta_0) = \{\mathbf{x}{:}\; \theta_0 \in S(\mathbf{x})\}$. This test has level α. Further, if $A(\theta_0)$ is UMP for the above problem, $(\alpha, H_0(\theta_0), H_1(\theta_0))$, then $S(\mathbf{X})$ is UMA $(1 - \alpha)$-level family of confidence sets, i.e., $S(\mathbf{X})$ minimizes

$$P_\theta\{S(X) \ni \theta'\} \; \forall \; \theta \in H_1(\theta')$$

among all $(1 - \alpha)$-level families of confidence sets.

Proof. Let us consider the size-α acceptance region $A(\theta)$ such that

$$P_\theta\{\mathbf{X} \in A(\theta)\} \ge 1 - \alpha \; \forall \; \theta \in H_0(\theta_0)$$

This may be equivalently written as

$$P_\theta\{S(\mathbf{X}) \ni \theta\} \ge 1 - \alpha$$

since $\mathbf{X} \in A(\theta)$ if and only if $S(\mathbf{X}) \ni \theta$. By definition 9.2.1, $S(\mathbf{X})$ is a family of $(1 - \alpha)$-level confidence sets. This establishes the correspondence between a size-α acceptance region and the family of confidence sets for θ at level $(1 - \alpha)$.

Let us consider $A^*(\theta)$ as any other size-α acceptance region or

$$P_\theta[\mathbf{X} \in A^*(\theta)] \geq 1 - \alpha$$

The corresponding $(1 - \alpha)$-level family of confidence sets is given by $S^*(\mathbf{x})$ so that

$$P_\theta[S^*(\mathbf{X}) \ni \theta] \geq 1 - \alpha$$

It is given that $A(\theta_0)$ is a UMP size-α acceptance region or

$$P_\theta[\mathbf{X} \in A(\theta_0)] \leq P_\theta[\mathbf{X} \in A^*(\theta_0)] \ \forall \ \theta \in H_1(\theta_0)$$

Further, $S(\mathbf{X})$ has been given as the family of confidence sets corresponding to $A(\theta_0)$. We then have

$$P_\theta[S(\mathbf{X}) \ni \theta_0] \leq P_\theta[S^*(\mathbf{X}) \ni \theta_0] \ \forall \ \theta \in H_1(\theta_0)$$

Since $A^*(\theta)$ was taken as an arbitrary size-α acceptance region, the $(1 - \alpha)$-level family of confidence sets $S(\mathbf{X})$ corresponding to UMP size-α acceptance region $A(\theta_0)$ is the one that minimizes $P_\theta[S(X) \ni \theta_0]$ [the probability that $S(\mathbf{X})$ captures the false value of the parameter while true value being θ] over $\theta \in H_1(\theta_0)$. This completes the proof. ▣

Notes.

1. The acceptance region corresponding to the test of size α for testing
   (i) $H_0{:}\theta = \theta_0$ against $H_1{:}\theta \neq \theta_0$, when inverted gives $(1 - \alpha)$-level confidence interval $(\underline{\theta}(\mathbf{X}), \bar{\theta}(\mathbf{X}))$.
   (ii) $H_0{:}\theta = \theta_0$ against $H_1{:}\theta > \theta_0$, when inverted gives $(1 - \alpha)$-level confidence interval $[\underline{\theta}(\mathbf{X}), \infty)$.
   (iii) $H_0{:}\theta = \theta_0$ against $H_1{:}\theta < \theta_0$, when inverted gives $(1 - \alpha)$-level confidence interval $(-\infty, \bar{\theta}(X)]$.
2. Let T be such that

$$P_\theta[a(\theta) < T(\mathbf{X}) < b(\theta)] = 1 - \alpha$$

and

$$P_\theta[T(\mathbf{X}) < a(\theta)] = \alpha_1$$

$$P_\theta[T(\mathbf{X}) > b(\theta)] = \alpha_2$$

so that $\alpha_1 + \alpha_2 = \alpha$. Then, $a(\theta)$ and $b(\theta)$ are the increasing functions of θ if T has an MLR.

9.3.3 Confidence Intervals Based on Posterior Distribution

Definition 9.3.3 (Credible region) A region $S(\mathbf{x})$ is called a $(1 - \alpha)$-level credible region for θ with respect to the posterior distribution $\pi(\theta|\mathbf{x})$ if

$$\int_S \pi(\theta|\mathbf{x})d\theta = 1 - \alpha \tag{9.3.15}$$

If $S(\mathbf{x}) \subset \mathbb{R}^1$, $S(\mathbf{x})$ is an interval, it is called a *credible interval*.

The distribution involved in Eq. (9.3.15) is the posterior distribution of θ conditional on $\mathbf{X} = \mathbf{x}$. Let the observable random variable $\mathbf{X}$ has the *pdf (pmf)* $f(\mathbf{x}|\theta)$ and the prior distribution of θ be $\pi(\theta)$ on Θ. Using the joint distribution of $\mathbf{X}$ and θ,

$$h(\mathbf{x}, \theta) = f(\mathbf{x}|\theta)\pi(\theta)$$

and the marginal of $\mathbf{X}$,

$$m(\mathbf{x}) = \int_{\Theta} h(\mathbf{x}, \theta)d\theta$$

The posterior distribution of θ given $\mathbf{x}$ is calculated by

$$\pi(\theta|\mathbf{x}) = \frac{f(x|\theta)\pi(\theta)}{m(\mathbf{x})}, \; m(\mathbf{x}) > 0$$

Using this posterior distribution, the $(1 - \alpha)$-level Bayes interval for θ is constructed by

$$P(\underline{\theta}(X) < \theta < \overline{\theta}(\mathbf{X})|\mathbf{X} = \mathbf{x}) \geq 1 - \alpha$$

or

$$\int_{\underline{\theta}(\mathbf{x})}^{\overline{\theta}(\mathbf{x})} \pi(\theta|\mathbf{x})d\theta \geq 1 - \alpha$$

assuming $\pi(\theta|\mathbf{x})$ is the *pdf.* The above interval $(\underline{\theta}, \overline{\theta})$ is also called *credible interval* in the Bayesian setup.

The set S in Eq. (9.3.15) is not unique. The criterion of optimality of credible intervals is to consider a credible interval as the most optimal which is the smallest in length or has the highest posterior density.

The classical confidence interval with a specified coverage probability is usually obtained so that its length is smallest. Like in classical statistics in the Bayesian set up, we are interested in such a confidence interval $S(\mathbf{x})$ which is of specified coverage probability and has shortest length i.e.

(i) $\int_{S(\mathbf{x})} \pi(\theta|\mathbf{x})d\theta = 1 - \alpha$, and

(ii) $L(S(\mathbf{x})) \leq \text{L}(S'(\mathbf{x}))$

for any other $S'(\mathbf{x})$ satisfying $\int_{S'(\mathbf{x})} \pi(\theta|\mathbf{x})d\theta \geq 1 - \alpha$

Definition 9.3.4 If $S(\mathbf{x})$ is an interval of the type (a, b) and the posterior density $\pi(\theta|\mathbf{x})$ is unimodal, then for given α the smallest length credible interval for θ is obtained by applying Theorem 9.4.1 and it is given by

$$S(\mathbf{x}) = \{\theta \in \Theta\colon \pi(\theta|\mathbf{x}) \geq k(\alpha)\}$$

where $k(\alpha)$ is the largest integer such that

$$P(\theta \in S(\mathbf{x})|\mathbf{x}) \geq 1 - \alpha$$

The credible interval $S(\mathbf{x})$ is called $(1 - \alpha)$-level highest posterior density (HPD) interval with respect to the posterior density $\pi(\theta|\mathbf{x})$.

The HPD credible interval consist of all such values of θ, for which the posterior density is highest, therefore, it is called as *highest posterior density* (HPD) interval.

By the Theorem 9.4.1 HPD intervals are the shortest length credible interval when the posterior distribution $\pi(\theta|\mathbf{x})$ is *unimodal*.

If the posterior distribution is symmetric then the HPD credible interval (a, b) is obtained by solving the equations

$$\int_a^b \pi(\theta|\mathbf{x})\,d\theta = \alpha$$

and

$$\pi(a|\mathbf{x}) = \pi(b|\mathbf{x})$$

However, if the posterior distribution $\pi(\theta|\mathbf{x})$ is not symmetric then HPD credible interval is obtained by solving

$$\int_a^b \pi(\theta|\mathbf{x})\,d\theta = \alpha$$

and

$$\pi(a+h|\mathbf{x}) = \pi(a|\mathbf{x})$$

when $h > 0$ is taken as small as possible.

Usually we expect symmetric posterior distribution for location parameter problems and asymmetric posterior distribution for scale parameter problems.

In case, the posterior density have several modes, then the highest mode is determined and HPD credible interval is constructed around it. Such an interval may not be unique. Sometimes, HPD region is union of two or more intervals in cases mixture of information are considered by taking mixing priors. However, in case of natural conjugate priors, unimodal posteriors are obtained and we get an unique HPD credible interval.

In classical statistics, the confidence interval is a random interval. A realized value of a confidence interval may cover the fixed and unknown value of the parameter θ with probability 0 or 1. When we say that a given confidence interval has $100(1 - \alpha)\%$ chance of coverage, we mean that $100(1 - \alpha)\%$ of random intervals cover the true value of parameter θ under the repetition of the experimental situations. We do not mean that the parameter θ is inside the confidence interval with probability $(1 - \alpha)$ since the parameter is assumed to be fixed, it does not move. Whereas, in Bayesian set up the parameter is assumed to be a random variable with certain probability distribution. Thus for the credible intervals we can state that the parameter with respect to the posterior distribution is inside an interval with some probability. This is not 0 or 1 as in the case of classical confidence interval. It may be noticed that the construction of Baysian credible intervals are computationally straightforward and have clear edge over probabilistic interpretation as compared to that of a classical confidence interval.

However, the case of computations and better-probabilistic interpretation of Bayesian credible interval is achieved at the cost of requiring an additional information which is the prior knowledge about the parameter θ to get the posterior distribution.

Consider a random sample $X_1,\ldots,X_n$ from $N(\theta, 1)$. Let the prior distribution for θ is noninformative prior given by $\pi(\theta) \propto 1$. The posterior distribution of θ given $\mathbf{x}$ is $N(\bar{x}, 1/n)$. Notice that the classical confidence and Bayesian credible intervals are numerically same that is $(\bar{x} - 1.96/\sqrt{n}, \bar{x} + 1.96/\sqrt{n})$ and each have the same length, i.e. $2(1.96)/\sqrt{n}$. Though these intervals are same but they have quiet different interpretations. Note that the non-informative prior reflects the experimenters subjective belief that he has no information about the parameter apriori. When the prior distribution is updated with the data, i.e. with the likelihood function, we get the posterior distribution. Notice, in the present example, posterior distribution and likelihood function both contains the same amount of information on the parameter, thus they have yielded the same intervals. In case of Bayesian credible interval, an assertion of 95% coverage means the experimenter after updating the prior distribution with present data to the posterior distribution is sure of 95% coverage. However, 95% classical coverage probability means that in a long sequence of identical trials, 95% of the observed confidence interval will cover the true parameter. The classical coverage probability reflects the uncertainity in the sampling procedure.

In the above example if the noninformative prior is replaced by an informative prior $N(0,1)$, then the posterior distribution of θ given $\mathbf{x}$ is $N((n\bar{x}/(n + 1)), (1/(n + 1)))$.

Now, the HPD credible interval is given by $([(n\bar{x}/(n + 1)) - 1.96/\sqrt{(n+1)}], [(n\bar{x}/(n + 1)) + 1.96/\sqrt{(n+1)}]\}$. Notice that the length of the HPD interval is reduced to $2(1.96/\sqrt{(n+1)})$. This reduction of length is significant when n is small. Thus the use of informative prior has resulted into reduction of length of the credible interval.

We can not argue the superiority of one type of interval over other. In some statistical problems classical confidence interval is better than Bayesian credible interval and some problems may best be solved with Bayesian credible intervals.

Consider, for example, a random sample $X_1, X_2,\ldots,X_n$ from $N(\theta, \sigma^2)$, σ^2 is known, and prior $\theta \sim N(\eta, \tau^2)$, η and τ^2 are known. Then the posterior distribution of θ given $\mathbf{X} = \mathbf{x}$ is $N[(n\bar{x}/\sigma^2) + (\eta/\tau^2)/(n/\sigma^2) + (1/\tau^2), 1/(n/\sigma^2) + (1/\tau^2)]$. Hence, $(1 - \alpha)$-level Bayesian credible interval for θ is given by

$$P\left(\frac{\frac{n\bar{X}}{\sigma^2}+\frac{\eta}{\tau^2}}{\frac{n}{\sigma^2}+\frac{1}{\tau^2}} - z_{\alpha/2,n-1}\frac{1}{\left(\frac{n}{\sigma^2}\right)+\frac{1}{\tau^2}} \le \theta \le \frac{\frac{n\bar{X}}{\sigma^2}+\frac{\eta}{\tau^2}}{\frac{n}{\sigma^2}+\frac{1}{\tau^2}} + z_{\alpha/2,n-1}\frac{1}{\left(\frac{n}{\sigma^2}\right)+\frac{1}{\tau^2}}\right) = 1-\alpha$$

9.3.4 Confidence Intervals Based on Large Samples

Frequently, confidence intervals cannot be obtained in closed form. In such cases, asymptotic results are useful to obtain much simplified confidence intervals.

Definition 9.3.5 (Family of $(1 - \alpha)$-level confidence intervals for large n) A family of intervals $(\underline{\theta}(\mathbf{X}), \bar{\theta}(\mathbf{X}))$ is called a family of asymptotic confidence intervals at level $1 - \alpha$ if

$$P_\theta(\underline{\theta}, \bar{\theta}) \to 1 - \alpha \;\forall\; \theta \in \Theta$$

as $n \to \infty$.

Definition 9.3.6 (Asymptotic length) If the length of a $(1 - \alpha)$-level confidence interval $(\underline{\theta}, \bar{\theta})$ for θ for large n

$$n^\delta(\bar{\theta} - \underline{\theta}) \xrightarrow{P} L_\alpha$$

as $n \to \infty$, the quantity L_α is called the *scaled asymptotic length* of the class interval. The value of δ is usually $\delta = 1/2$. An interval, at fixed $(1 - \alpha)$ and δ, is said to be better than other at the same level $(1 - \alpha)$ and δ if its asymptotic length is smaller than of other.

Definition 9.3.7 (Asymptotic relative efficiency) The asymptotic relative efficiency of one class interval as compared to other at the same level $(1 - \alpha)$ and δ is defined as

$$\text{ARE}(I_1, I_2) = \left(\frac{L_{2,\alpha}}{L_{1,\alpha}}\right)^{1/\delta}$$

Usually, central limit theorem and asymptotic normality of MLEs under certain regularity conditions are used to construct asymptotic $(1 - \alpha)$-level confidence interval for θ.

Central limit theorem. Let $X \sim F_\theta$, $\theta \in \Theta$, the population F_θ be unknown, and $\{T_n(\mathbf{X})\}$ be a sequence of statistics so that

$$\sqrt{n}(T_n - \theta) \xrightarrow{d} N[0, v(\theta)]$$

The asymptotic variance $v(\theta)$ depends on θ which creates a problem in constructing a confidence interval for θ. Cramer dealt with this problem by choosing such a smooth transformation $g(\cdot)$ and the sequence of transformed statistics $\{g(T_n)\}$ that stabilizes the asymptotic variance. This method is popularly known as delta method.

$$\sqrt{n}[g(T_n) - g(\theta)] \xrightarrow{d} N(0, [g'(\theta)]^2\, v(\theta))$$

The choice of g is such that

$$g'(\theta) = \frac{c}{\sqrt{v(\theta)}}$$

This choice of g then stabilizes the variance, and

$$\sqrt{n}[g(T_n) - g(\theta)] \xrightarrow{d} N(0, c^2)$$

The quantity $\sqrt{n}[g(T_n) - g(\theta)]$ is called an asymptotic pivotal quantity, based on which an asymptotic $(1 - \alpha)$-level confidence interval for θ can, thus, be constructed in the closed form

$$P(-z_{\alpha/2} \le \sqrt{n}[g(T_n) - g(\theta)] \le z_{\alpha/2}) \approx (1 - \alpha)$$

We will illustrate this method of constructing confidence interval by considering one example. Let $X_1, X_2, \ldots, X_n$ be a random sample from $P(\theta)$. Consider $\hat{\theta} = \bar{X}$ the MLE of θ

$$\sqrt{n}(\bar{X} - \theta) \xrightarrow{d} N(0, \theta)$$

If we want $g'(\theta) = c/\sqrt{\theta}$ for variance stabilization, we must choose

$$g(\theta) = 2c\sqrt{\theta},\ c = 1$$

We then have

$$\sqrt{n}\left(2\sqrt{\bar{X}} - 2\sqrt{\theta}\right) \xrightarrow{d} N(0, 1)$$

Thus, an asymptotic (1 – α)-level confidence interval for $\sqrt{\theta}$ in closed form is given by

$$P\left[-z_{\alpha/2} \le \sqrt{n}\left(2\sqrt{\bar{X}} - 2\sqrt{\theta}\right) \le z_{\alpha/2}\right] \approx (1-\alpha)$$

or
$$P\left(\sqrt{\bar{X}} - \frac{1}{2\sqrt{n}} z_{\alpha/2} \le \sqrt{\theta} \le \sqrt{\bar{X}} + \frac{1}{2\sqrt{n}} z_{\alpha/2}\right) \approx (1-\alpha)$$

Another way to handle the dependence of the asymptotic variance $v(\theta)$ on θ is to consider some statistic $U(\mathbf{X})$ such that

$$U(\mathbf{X}) \xrightarrow{P} v(\theta)$$

Then, by Slutsky's theorem,

$$\sqrt{n}\frac{T(\mathbf{X}) - \theta}{\sqrt{U(\mathbf{X})}} \xrightarrow{d} N(0, 1)$$

or, we have
$$P\left[-z_{\alpha/2} \le \sqrt{n}\frac{T(\mathbf{X}) - \theta}{\sqrt{U(\mathbf{X})}} \le z_{\alpha/2}\right] \ge 1-\alpha$$

or
$$P\left[-z_{\alpha/2}\frac{\sqrt{U(\mathbf{X})}}{\sqrt{n}} \le T(\mathbf{X}) - \theta \le z_{\alpha/2}\frac{\sqrt{U(\mathbf{X})}}{\sqrt{n}}\right] = 1-\alpha$$

or
$$P\left[T(\mathbf{X}) - z_{\alpha/2}\frac{\sqrt{U(\mathbf{X})}}{\sqrt{n}} \le \theta \le T(\mathbf{X}) + z_{\alpha/2}\frac{\sqrt{U(\mathbf{X})}}{\sqrt{n}}\right] = 1-\alpha$$

Clearly, in this case, the (1 – α)-level asymptotic confidence interval for θ in closed form is $(T(\mathbf{X}) - z_{\alpha/2}(\sqrt{U(\mathbf{X})}/\sqrt{n}), T(\mathbf{X}) + z_{\alpha/2}(\sqrt{U(\mathbf{X})}/\sqrt{n}))$.

Consider, for example, $X_1, X_2, \ldots, X_n$ to be a random sample from an unknown population with unknown mean μ and variance σ^2. Then, by central limit theorem, the quantity

$$\frac{\bar{X} - \mu}{\sigma/\sqrt{n}} \sim N(0, 1) \quad \text{as } n \to \infty$$

is a pivotal quantity. Thus, we may approximate (1 – α)-level confidence interval by

$$P\left(-z_{\alpha/2} \le \frac{\bar{X} - \mu}{\sigma/\sqrt{n}} \le z_{\alpha/2}\right) \approx 1-\alpha$$

or
$$P\left(\bar{X} - z_{\alpha/2}\frac{\sigma}{\sqrt{n}} \le \mu \le \bar{X} + z_{\alpha/2}\frac{\sigma}{\sqrt{n}}\right) \approx 1-\alpha$$

Using $S_n \xrightarrow{P} \sigma$ for large n, we may replace σ by sample standard deviation S_n to get $(1 - \alpha)$-level confidence interval for μ

$$\left(\bar{X} - z_{\alpha/2}\frac{S_n}{\sqrt{n}}, \bar{X} + z_{\alpha/2}\frac{S_n}{\sqrt{n}}\right)$$

Consider another example where $X \sim b(n, p)$, X is the number of successes in a sample of size n from $b(1, p)$. Assuming $np > 10$ and $n(1 - p) > 10$ and estimating p by $\hat{p} = X/n$, by central limit theorem, the quantity

$$\frac{\hat{p} - p}{\sqrt{[p(1-p)/n]}} \sim N(0, 1)$$

The $(1 - \alpha)$-level confidence interval for p is

$$P\left(-z_{\alpha/2} \le \frac{\hat{p} - p}{\sqrt{[p(1-p)/n]}} \le z_{\alpha/2}\right) \approx 1 - \alpha$$

On solving the following equation for p, we get

$$p = \hat{p} \pm z_{\alpha/2}\sqrt{\frac{p(1-p)}{n}}$$

$$p^2\left(1 + \frac{z_{\alpha/2}^2}{n}\right) - p\left(2\hat{p} + \frac{z_{\alpha/2}^2}{n}\right) + \hat{p}^2 = 0$$

$$p = \frac{\hat{p} + (z_{\alpha/2}^2/2n) \pm z_{\alpha/2}\sqrt{[\hat{p}(1-\hat{p})/n] + (z_{\alpha/2}^2/4n^2)}}{1 + (z_{\alpha/2}^2/n)}$$

Thus, the $(1 - \alpha)$-level confidence interval for p is given by

$$P\left(\frac{\hat{p} + (z_{\alpha/2}^2/2n) - z_{\alpha/2}\sqrt{[\hat{p}(1-\hat{p})/n] + (z_{\alpha/2}^2/4n^2)}}{1 + (z_{\alpha/2}^2/n)} \le p \right.$$

$$\left. \le \frac{\hat{p} + (z_{\alpha/2}^2/2n) + z_{\alpha/2}\sqrt{[\hat{p}(1-\hat{p})/n] + (z_{\alpha/2}^2/4n^2)}}{1 + (z_{\alpha/2}^2/n)}\right) \approx 1 - \alpha$$

For large n, this reduces to

$$P\left(\hat{p} - z_{\alpha/2}\sqrt{\frac{\hat{p}(1-\hat{p})}{n}} \le p \le \hat{p} - z_{\alpha/2}\sqrt{\frac{\hat{p}(1-\hat{p})}{n}}\right) \approx 1 - \alpha$$

Asymptotic normality of MLEs. Yet another way of approximating $(1 - \alpha)$-level confidence interval is when θ is estimated by MLE $\hat{\theta}$ so that under certain regularity conditions as specified in subsection (6.6.6), A1 to A6,

$$\frac{(\hat{\theta} - \theta)}{\sqrt{v(\hat{\theta})}} \xrightarrow{d} N(0, 1) \quad \text{as } n \to \infty$$

where $$v(\hat{\theta}) = \frac{1}{nI_X(\theta)}$$

and $$I_X(\theta) = E_\theta\left[\frac{\partial}{\partial\theta}\log f(X;\theta)\right]^2 = -E_\theta\left[\frac{\partial^2}{\partial\theta^2}\log f(X;\theta)\right]$$

On using this result, the asymptotic $(1 - \alpha)$-level confidence interval for θ is

$$P_\theta\left[-z_{\alpha/2} \le \sqrt{n}\frac{(\hat{\theta} - \theta)}{I_X^{-1/2}(\theta)} \le z_{\alpha/2}\right] \to 1 - \alpha$$

or $$P_\theta\left[\hat{\theta} - \frac{1}{\sqrt{nI_X(\theta)}}z_{\alpha/2} \le \theta \le \hat{\theta} + \frac{1}{\sqrt{nI_X(\theta)}}z_{\alpha/2}\right] \to 1 - \alpha$$

This confidence interval depends on θ since $I_{\mathbf{X}}(\theta)$ depends on θ. To deal with this problem, let us assume that $I_{\mathbf{X}}(\theta)$ is continuous in θ so that

$$\frac{I_{\mathbf{X}}(\theta_n)}{I_{\mathbf{X}}(\theta)} \xrightarrow{P} 1 \quad \text{whenever} \quad \theta_n \xrightarrow{P} \theta$$

θ_n may be any estimator satisfying the above conditions; it may be the MLE of θ, i.e., $\theta_n = \hat{\theta}$. Then by Slutsky's theorem,

$$\sqrt{nI_{\mathbf{X}}(\theta_n)}(\theta_n - \theta) = \sqrt{\frac{I_{\mathbf{X}}(\theta_n)}{I_{\mathbf{X}}(\theta)}}\sqrt{nI_X(\theta)}(\theta_n - \theta) \xrightarrow{d} N(0, 1)$$

since $\sqrt{I_{\mathbf{X}}(\theta_n)/I_{\mathbf{X}}(\theta)} \xrightarrow{P} 1$ and $\sqrt{nI_X(\theta)}(\theta_n - \theta) \xrightarrow{d} N(0, 1)$. Using $\sqrt{nI_X(\theta_n)}(\theta_n - \theta)$ as asymptotic pivotal quantity, an asymptotic $(1 - \alpha)$-level confidence interval for θ can be constructed by

$$P\left[\left|\sqrt{nI_X(\theta_n)}(\theta_n - \theta)\right| \le z_{\alpha/2}\right] \to 1 - \alpha$$

and $$S(\mathbf{X}) = \left(\theta_n - \frac{1}{\sqrt{nI_X(\theta_n)}}z_{\alpha/2}, \theta_n + \frac{1}{\sqrt{nI_X(\theta_n)}}z_{\alpha/2}\right)$$

Another way to deal with the dependence of $I_{\mathbf{X}}(\theta)$ on θ is to approximate it by

$$I_{\mathbf{X}}(\theta)\Big|_{\theta=\theta_n} = -E_\theta\left(\frac{\partial^2}{\partial\theta^2}\log f(\mathbf{X};\theta)\right)_{\theta=\theta_n} > 0$$

$$= \frac{1}{n}\sum_{i=1}^{n} -\frac{\partial^2}{\partial\theta^2}\log f(X_i;\theta)\Big|_{\theta=\theta_n} + O_p(1)$$

$$= I(\mathbf{X},\theta_n) + O_p(1)$$

In this approximation, $I_{\mathbf{X}}(\theta)$ is estimated by the sample Fisher information $I(\mathbf{X}, \theta_n)$ such that

$$\sqrt{nI(X,\theta_n)}(\theta_n - \theta) \xrightarrow{d} N(0,1)$$

$\therefore$
$$P\left(\left|\sqrt{nI(X,\theta_n)}(\theta_n - \theta)\right| \le z_{\alpha/2}\right) \to 1-\alpha$$

and
$$S(\mathbf{X}) = \left(\theta - \frac{1}{\sqrt{nI(X,\theta_n)}} z_{\alpha/2}, \theta_n + \frac{1}{\sqrt{nI(X,\theta_n)}} z_{\alpha/2}\right)$$

is an asymptotic $(1 - \alpha)$-level confidence interval for θ.
If $\boldsymbol{\theta}$ is k-dimensional,

$$\sqrt{n}(\hat{\boldsymbol{\theta}} - \boldsymbol{\theta}) \sim N(\mathbf{0}, \mathbf{I}_{\mathbf{X}}^{-1}(\boldsymbol{\theta}))$$

We use the pivot
$$n(\hat{\boldsymbol{\theta}} - \boldsymbol{\theta})'\, \mathbf{I}_{\mathbf{X}}(\boldsymbol{\theta})(\hat{\boldsymbol{\theta}} - \boldsymbol{\theta}) \sim \chi_k^2$$

The problem is that $\mathbf{I}_{\mathbf{X}}(\boldsymbol{\theta})$ involves unknown parameter $\boldsymbol{\theta}$. We, therefore, use the estimator $I_{\mathbf{X}}^{(n)}(\hat{\boldsymbol{\theta}})$ so that

$$(\hat{\boldsymbol{\theta}} - \boldsymbol{\theta})' I_{\mathbf{X}}^{(n)}(\hat{\boldsymbol{\theta}})(\hat{\boldsymbol{\theta}} - \boldsymbol{\theta}) \to (\hat{\boldsymbol{\theta}} - \boldsymbol{\theta})' I_{\mathbf{X}}(\boldsymbol{\theta})(\hat{\boldsymbol{\theta}} - \boldsymbol{\theta})$$

asymptotically. For example, let $X \sim b(n, p)$, $\hat{p} = X/n$, and $(\hat{p} - p)/\sqrt{\hat{p}(1-\hat{p})/n} \sim N(0, 1)$. We can use the pivot $(\hat{p} - p)/\sqrt{\hat{p}(1-\hat{p})/n}$ for constructing the confidence interval for p.

9.3.5 Confidence Intervals Based on Chebyshev's Inequality

We discuss here the construction of $(1 - \alpha)$-level confidence interval yielded by Chebyshev's inequality for some $\varepsilon > 0$

$$P_\theta\left[\left|\hat{\theta}(\mathbf{X}) - \theta\right| < \varepsilon\sqrt{\text{MSE}(\hat{\theta})}\right] > 1 - \frac{1}{\varepsilon^2}$$

where $\hat{\theta}(\mathbf{X})$ is some estimator of θ not necessarily unbiased, and $\text{MSE}(\hat{\theta}) = E(\hat{\theta} - \theta)^2$. The $(1 - (1/\varepsilon^2))$ level confidence interval for θ in this case is

$$\left(\hat{\theta} - \varepsilon l(\theta), \theta + \varepsilon l(\theta)\right)$$

where $l(\theta) = \sqrt{\text{MSE}(\hat{\theta})}$. The quantity $l(\theta)$ is unknown since it is the function of θ. Under certain mild consistency conditions, we replace $l(\theta)$ by $l(\hat{\theta})$ so that

$$l(\hat{\theta}) \xrightarrow{P} l(\theta)$$

where $\hat{\theta} \xrightarrow{P} \theta$ as $n \to \infty$. It is not necessary that the limiting distribution θ be normal.

$$\frac{\hat{\theta}-\theta}{\sqrt{l(\hat{\theta})}} \xrightarrow{d} N(0,1)$$

One drawback of this method is that the length of the interval constructed by this method is too large.

However, the larger the length of the interval, the higher is the confidence level of trapping the true value of the parameter, and the less meaningful it is. Note that the interval $(-\infty, \infty)$, which completely ignores the data $\mathbf{X}$, captures the true values of θ with confidence level 1. However, such a large interval has no relevance in the inference, since it carries no information about θ.

9.4 OPTIMALITY OF CONFIDENCE INTERVAL ESTIMATORS

We will discuss in this section different criteria of evaluating interval estimators. There are three common criteria of evaluating and constructing confidence intervals. One is the length of an interval estimate, second is the probability of an interval estimate of covering the false value of the parameter, and third is the expected length of a confidence interval. We will discuss each of these in the following subsections.

9.4.1 Shortest-Length Confidence Interval

By the size of a confidence set, we mean length if it is an interval and volume if it is multidimensional set. The smaller the size and higher coverage probability, the better is the confidence interval. By increasing the size of the confidence interval, the probability of coverage may be increased. One extreme example of this is the interval $(-\infty, \infty)$ with coverage probability one. This interval is so wide that it is meaningless, though it has coverage probability one. Therefore, for optimal interval estimation, there must be a balance between its size and coverage probability.

Thus, we construct such an interval estimate that has the uniformly shortest length among all intervals at a given level of confidence. However, in most of the cases, the uniformly shortest length confidence intervals do not exist among the class of all $(1 - \alpha)$-level confidence intervals.

We will discuss here three ways of minimizing the length of the intervals; one, when the *pdf* of the pivot is unimodal, the second by Lagrange's method, and the third due to Guenther (1969).

Following theorem constructs a shortest length confidence interval at $(1 - \alpha)$-level when the *pdf* of the pivot is unimodal.

Definition 9.4.1 The *pdf* $f(t)$ of a pivot $T(\mathbf{X}, \theta)$ is said to be unimodal at $t_0 \in \mathbb{R}^1$ if $f(t)$ is nondecreasing for $t \le t_0$ and nonincreasing for $t \ge t_0$.

Theorem 9.4.1 Let $U(\mathbf{X})$ be a real-valued statistic for a real-valued parameter θ.

(i) (Location case). Let $T(\mathbf{X}, \theta) = (U(\mathbf{X}) - \theta)/S(\mathbf{X})$, $S(\mathbf{X}) > 0$, be a pivot with its unimodal *pdf* $f(t)$. If the interval (a, b) satisfies

$$\int_a^b f(t)dt = 1 - \alpha \tag{9.4.1}$$

and
$$f(a) = f(b) > 0 \tag{9.4.2}$$

and $a \le t_0 \le b$ where t_0 is a mode of $f(t)$, then interval $(U - bS, U - aS)$ is the shortest length among all the families of confidence intervals satisfying Eq. (9.4.1).

(ii) (Scale case). Let $U(\mathbf{X}) > 0$ and $\theta > 0$, $T(\mathbf{X}, \theta) = U/\theta$ be a pivot with its *pdf* $f(t)$, and $t^2 f(t)$ be unimodal at t_0. If the interval (a, b) satisfies

$$\int_a^b f(t)dt = 1 - \alpha \tag{9.4.3}$$

and
$$a^2 f(a) = b^2 f(b) > 0$$

and $a \le t_0 \le b$, where t_0 is a mode of $t^2 f(t)$, then the interval $(U/b, U/a)$ is of the shortest length among all the families of confidence intervals satisfying Eq. (9.4.3).

Proof.

(i) Assume that the interval $(U - bS, U - aS)$ is not the shortest, so that there exists another interval $(U - b'S, U - a'S)$ such that

$$b' - a' < b - a$$

We consider here the case where $a' \le a$, and the case $a' > a$ can be dealt similarly. Further, there are two cases: (i) $a' \le b' \le a \le t_0 \le b$ and (ii) $a' \le a < b' < b$, $b' < b$; since if b' were greater than or equal to b it would had implied $b' - a' > b - a$, violating the assumption.

Consider the case (i) $a' \le b' \le a \le t_0 \le b$. The coverage probability of the interval (a', b') is given by

(a)
$$P[(a', b') \ni \theta] = \int_{a'}^{b'} f(t)dt \le f(b')(b' - a')$$

since $t \le b' \le t_0$ with t_0 as mode implies $f(t) < f(b')$;

(b)
$$P[(a', b') \ni \theta] \le f(a)(b' - a')$$

since $b' \le a \le t_0$ with t_0 as mode implies $f(b') \le f(a)$;

(c)
$$P[(a', b') \ni \theta] < f(a)(b - a)$$

since $b' - a' < b - a$ and $f(a) > 0$;

(d) $$P[(a', b') \ni \theta] \le \int_a^b f(t)dt = 1-\alpha$$

since $f(a) = f(b)$, $a \le t_0 \le b$, and the unimodality of $f(t)$ at t_0 implies $f(t, \theta) \ge f(a)$ for $a \le t \le b$.

This contradicts Eq. (9.4.3).

Next, consider case (ii) $a' \le a < b' < b$. The coverage probability of the interval (a', b') is

$$P[(a', b') \ni \theta] = \int_{a'}^{b'} f(t)dt = \int_{a'}^{a} f(t)dt + \int_{a}^{b'} f(t)dt$$

$$= \int_{a'}^{a} f(t)dt + \int_{a}^{b} f(t)dt - \int_{b'}^{b} f(t)dt$$

$$= \int_{a}^{b} f(t)dt + \left\{ \int_{a'}^{a} f(t)dt - \int_{b'}^{b} f(t)dt \right\}$$

Consider the difference of integrals in the curly brackets. Unimodality and $f(a) = f(b)$ gives

$$\int_{a'}^{a} f(t)dt \le f(a)(a-a')$$

and $$\int_{b'}^{b} f(t)dt \ge f(b)(b-b')$$

This gives $$\int_{a'}^{a} f(t)dt - \int_{b'}^{b} f(t)dt \le f(a)(a-a') - f(b)(b-b')$$

$$= f(a)[(b'-a')-(b-a)] \qquad \text{[since } f(a) = f(b)\text{]}$$

$$< 0 \qquad \text{[since } b' - a' < b - a \text{ and that } f(a) > 0\text{]}$$

This shows that $P[(a', b') \ni \theta] < 1 - \alpha$

which is a contradiction. Thus, we cannot have an interval (a', b') with length smaller than (a, b). Similar argument proves the same for the case $a' > a$. This proves the interval (a, b) as the shortest length confidence interval at level $(1 - \alpha)$.

(ii) We are to minimize the length $((1/a) - (1/b))U$ of the interval $(U/b, U/a)$ under the constraint

$$\int_a^b f(t)dt = 1-\alpha$$

The transformation $V = 1/T$ changes the constraint to

$$\int_{1/b}^{1/a} \frac{1}{v^2} f\left(\frac{1}{v}\right) dv = 1 - \alpha$$

The function $(1/v^2)f(1/v)$ is unimodal at $1/t_0$ since $t^2 f(t)$ is unimodal at t_0. Applying the result (i) to the function $(1/v^2)f(1/v)$ completes the proof. Clearly, Eq. (9.4.2) reduces to

$$a^2 f(a) = b^2 f(b) > 0$$ ▣

Notes.

1. Further, if $f(t)$ in Theorem 9.4.1 is continuous, the proof of the theorem simplifies to give $(1 - \alpha)$-level shortest length confidence interval $(a, a + d)$ by choosing a and d such that

$$\int_a^{a+d} f(t)dt = 1 - \alpha \tag{9.4.4}$$

and

$$f(a + d) = f(a) \tag{9.4.5}$$

For proving this result, one can consider the maximization of the integral $\int_a^{a+d} f(t)dt$ as a function of a for fixed d. Such a value of a is that which satisfies

$$\frac{\partial}{\partial a} \int_a^{a+d} f(t)dt = 0$$

$$f(a + d) - f(a) = 0$$

from the unimodality of f. Therefore, the values of a and d that satisfy Eq. (9.4.5) subject to Eq. (9.4.4) constitute the shortest length confidence interval $(a, a + d)$.

2. If $f(t)$ is symmetric unimodal, the shortest length confidence interval is $(U - bU, U - aS)$ where a and b are chosen such that

$$\int_{-\infty}^{a} f(t)dt = \frac{\alpha}{2} = \int_b^{\infty} f(t)dt \tag{9.4.6}$$

We will show that the interval $(U - bS, U - aS)$ with choices of a and b according to Eq. (9.4.6) is a $(1 - \alpha)$-level shortest length confidence interval by showing that such a and b satisfy the conditions of Theorem 9.4.1.

The value of a such that $\int_{-\infty}^{a} f(t)dt = \alpha/2$, $\alpha > 0$, is unique since $f(t)$ is a unimodal *pdf*. Choose b such that $b = 2\mu - a$ where μ is the point of symmetry of the *pdf* f. We have $\int_b^{\infty} f(t)dt = \alpha/2$ and $f(b) = f(a)$. Further, a is below μ since

$$\int_{-\infty}^{a} f(t)dt = \frac{\alpha}{2} \le \frac{1}{2} = \int_{-\infty}^{\mu} f(t)dt$$

i.e., $a \leq \mu$. Similarly, b is above μ since

$$\int_b^\infty f(t)dt = \frac{\alpha}{2} \leq \frac{1}{2} = \int_\mu^\infty f(t)dt$$

i.e., $b \geq \mu$. Further, the condition

$$\int_{-\infty}^a f(t)dt = \frac{\alpha}{2} > 0$$

implies that $f(t) > 0$ for some $t < a$. The unimodality of f and $a \leq \mu$ imply that

$$f(a) \geq f(t) \ \forall \ t \leq a$$

$$\therefore \qquad f(a) > 0$$

which shows that $f(b) = f(a) > 0$. Thus, the conditions of Theorem 9.4.1 hold to show that the interval $(U - bS, U - aS)$ is $(1 - \alpha)$-level shortest length confidence interval for θ where a and b are such that it satisfies Eq. (9.4.6).

3. Let $f(t)$ be strictly decreasing on $[0, \infty)$. Consider the interval (a, b) and the interval $(a - \varepsilon, b - \varepsilon)$ by shifting (a, b) by ε towards the left. Consider the difference of probabilities under these intervals

$$\int_a^b f(t)dt - \int_{a-\varepsilon}^{b-\varepsilon} f(t)dt = \int_{b-\varepsilon}^b f(t)dt - \int_{a-\varepsilon}^a f(t)dt$$

$$\leq f(b-\varepsilon)[b-(b-\varepsilon)] - f(a)[a-(a-\varepsilon)]$$

[since $f(t)$ is strictly decreasing]

$$= \varepsilon[f(b-\varepsilon) - f(a)] \leq 0$$

We see that the probability under the interval $(a - \varepsilon, b - \varepsilon)$ grows as the interval moves towards zero and is maximum when a is moved and set at zero. Thus, of all the $(1 - \alpha)$-level confidence intervals so that

$$\int_a^b f(t)dt = 1 - \alpha$$

the shortest length confidence interval is obtained by choosing $a = 0$ and b such that

$$\int_0^b f(t)dt = 1 - \alpha$$

and the interval is given by $(U - bS, U)$.

Similarly, let $f(t)$ be an increasing function on $[0, \infty)$, consider a interval (a, b) and interval $(a + \varepsilon, b + \varepsilon)$ obtained by shifting the interval (a, b) towards right by $\varepsilon > 0$.

The difference of probabilities

$$\int_a^b f(t)\,dt - \int_{a+\varepsilon}^{b+\varepsilon} f(t)\,dt = \int_a^{a+\varepsilon} f(t)\,dt - \int_b^{b+\varepsilon} f(t)\,dt$$

$$\leq \varepsilon(f(a+\varepsilon) - f(b+\varepsilon)) \leq 0$$

shows that the probability under the interval $(a + \varepsilon, b + \varepsilon)$ grows as the interval moves towards right and is maximum when b is set at ∞. Thus, of all the $(1 - \alpha)$-level confidence intervals (a, b) so that

$$\int_a^b f(t)\,dt = 1 - \alpha$$

the shortlist length confidence interval is obtained by choosing $b = \infty$ and a by solving

$$\int_a^\infty f(t)\,dt = 1 - \alpha$$

The corresponding $(1 - \alpha)$-level shortest length confidence interval is given by $(-\infty, U - aS)$.

4. [Huola(1993), Shao(2003)] Let $T(\mathbf{X}, \theta)$ be the pivot so that

$$\int_a^b f(t)dt = 1 - \alpha$$

where $f(t)$ is the *pdf* of T. If the length of the interval is of the form

$$L(a, b) = b - a \quad \text{or} \quad \frac{1}{b^2} - \frac{1}{a^2}$$

so that it can be expressed as

$$L(a, b) = \int_a^b g(t)dt$$

then $(1 - \alpha)$-level shortest length confidence interval is the solution of $\min_{\{a,b\}} \int_a^b g(t)dt$ subject to $\int_a^b f(t)dt = 1 - \alpha$, or $\min_C \int_a^b f(t)dt$ subject to $\int_C f(t)dt \geq 1 - \alpha$. Define an indicator function

$$\phi(t) = \begin{cases} 1, & t \in C \\ 0, & \text{otherwise} \end{cases}$$

where $C = \{t: g(t)/f(t) < k\}$ and k is chosen such that

$$\int_C f(t)dt = 1 - \alpha$$

Let C' be some other set with indicator function $\phi'(t)$

$$\phi'(t) = \begin{cases} 1, & t \in C' \\ 0, & \text{otherwise} \end{cases}$$

so that
$$\int_{C'} f(t)dt \geq 1 - \alpha$$

Then, $[\phi(t) - \phi'(t)][g(t) - kf(t)] \leq 0$ implies

$$\begin{aligned} &\int [\phi(t) - \phi'(t)][g(t) - kf(t)]dt \\ &= \int_C g(t)dt - \int_{C'} g(t)dt - k\left[\int_C f(t)dt - \int_{C'} f(t)dt\right] \\ &\geq \int_C g(t)dt - \int_{C'} g(t)dt = L(C) - L(C') \leq 0 \end{aligned}$$

showing that C has the smallest length subject to

$$\int_C f(t)dt = 1 - \alpha$$

A second way to obtain the shortest length confidence interval (a, b) is to obtain a and b so that it minimizes $b - a$ subject to $\int_a^b f(t)dt = 1 - \alpha$ by Lagrange's multiplier method.

The third way is the Guenther (1969) procedure of finding $(1 - \alpha)$-level shortest length confidence interval based on a pivot $T(\mathbf{X}, \theta) = T_\theta$ of θ. Indeed, the shortest length confidence interval under this procedure is not of the shortest length among all intervals at a given level, as one may get another pivot that produces a confidence interval at the same level which is shorter in length.

The procedure due to Guenther is summarized by taking a random sample $X_1, \ldots, X_n$ from $f(x; \theta)$, $\theta \in \Theta$, and a pivot of θ,

$$T(\mathbf{X}, \theta) = T_\theta$$

We may choose (a, b) so that

$$P_\theta(a < T_\theta < b) \geq 1 - \alpha \tag{9.4.7}$$

which, on inverting the inequalities for θ, reduces to

$$P_\theta[\underline{\theta}(\mathbf{X}) < \theta < \overline{\theta}(\mathbf{X})] \geq 1 - \alpha \tag{9.4.8}$$

Equation (9.4.7) may be written as

$$F(b) - F(a) = 1 - \alpha \tag{9.4.9}$$

where F is the *cdf* of T_θ. The length of the interval in Eq. (9.4.8) is

$$L = \overline{\theta}(\mathbf{X}, a, b) - \underline{\theta}(\mathbf{X}, a, b)$$

The numbers a and b are so chosen that it minimizes L subject to Eq. (9.4.9) or

$$f(b)\frac{\partial b}{\partial a} - f(a) = 0 \tag{9.4.10}$$

where $f(\cdot)$ is the density of T_θ. We solve

$$\frac{\partial}{\partial a}L = 0$$

or

$$h\left(\frac{\partial b}{\partial a}\right) = 0 \tag{9.4.11}$$

where h is some function of $\partial b/\partial a$. The function h is obtained by putting $\partial b/\partial a = f(a)/f(b)$ in Eq. (9.4.11). We get

$$h\left[\frac{f(a)}{f(b)}\right] = 0$$

The values a and b satisfying Eq. (9.4.11) give the $(1 - \alpha)$-level shortest length confidence interval (a, b) based on T_θ at level $(1 - \alpha)$.

9.4.2 Minimum Probability of False Coverage Confidence Intervals—Unbiased Confidence Intervals

So far, we have considered the families of confidence sets $\{S(\mathbf{x})\}$ which are at level $(1 - \alpha)$ with the meaning that such sets trap the true value of the parameter θ with probability at least $(1 - \alpha)$. However, in the present subsection, we will consider such confidence sets as optimal of level $(1 - \alpha)$ which minimize the probability of false converge of the parameter. Clearly, one searches for such a family of sets for which the probability of true coverage is at least $(1 - \alpha)$ and the probability of false coverage is at most $(1 - \alpha)$. Such a family of confidence sets $\{S(\mathbf{x})\}$ is said to be unbiased at level $(1 - \alpha)$.

The coverage probability is a function of θ and, thus, has infinite number of values. So, the minimum of the coverage probability over θ is considered as a measure of coverage probability performance and it is denoted by *confidence coefficient*.

Definition 9.4.2 A family of confidence sets $\{S(\mathbf{x})\}$ for a parameter θ is said to be unbiased at level $(1 - \alpha)$ if

$$P_\theta(S(\mathbf{X}) \ni \theta) \geq 1 - \alpha \;\forall\; \theta$$

and

$$P_\theta(S(\mathbf{X}) \ni \theta') \leq 1 - \alpha \;\forall\; \theta, \theta' \in \Theta \quad \text{and} \quad \theta \neq \theta'$$

Definition 9.4.3 A family of confidence sets $\{S_0(\mathbf{x})\}$ among the families of all unbiased confidence sets at level $(1 - \alpha)$ which minimizes

$$P_\theta(S(\mathbf{X}) \ni \theta') \;\forall\; \theta, \theta' \in \Theta \text{ and } \theta \neq \theta'$$

is called the UMA unbiased family of confidence set at level $1 - \alpha$.

The UMA unbiased family of confidence sets at level $(1-\alpha)$ is obtained from the UMP unbiased size-α test by the test inversion procedure discussed in subsection 9.3.2. Following theorem summarizes this result.

Theorem 9.4.2 Consider the testing problem $H_0(\theta_0)$: $\theta = \theta_0$ against $H_1(\theta_0)$: $\theta \neq \theta_0$. Let $A_0(\theta_0)$ be the acceptance region corresponding to the UMP size-α test for testing $H_0(\theta_0)$ against $H_1(\theta_0)$. Then, the confidence set $S_0(\mathbf{x}) = \{\theta: \mathbf{x} \in A_0(\theta)\}$ is a UMA unbiased family of confidence sets at level $1-\alpha$.

Proof. Let the UMP unbiased size-α test be given by $\phi_0(\mathbf{x})$. Unbiasedness of the test ϕ_0 gives

$$E_{\theta'}\phi_0(\mathbf{X}) \geq \alpha \;\forall\; \theta' \in H_1(\theta_0)$$

or

$$E_{\theta'}(1-\phi_0(\mathbf{X})) \leq 1-\alpha \;\forall\; \theta' \in H_1(\theta_0)$$

or

$$P_{\theta'}(\mathbf{X} \in A_0(\theta)) \leq 1-\alpha$$

or

$$P_{\theta'}(S_0(\mathbf{X}) \ni \theta) \leq 1-\alpha$$

This shows that the family of confidence sets $S_0(\mathbf{X})$ is unbiased.

Next, consider any other unbiased size-α test $\phi'(\mathbf{x})$ with acceptance region $A'(\theta)$; we get a corresponding $(1-\alpha)$-level family of unbiased confidence sets $S'(\mathbf{X})$, i.e.,

$$P_{\theta'}(S'(\mathbf{X}) \ni \theta) \leq 1-\alpha \;\;\forall\; \theta' \in H_1(\theta_0)$$

The test $\phi_0(\mathbf{x})$ is UMP. We have

$$E_{\theta'}[\phi_0(\mathbf{X})] \geq E_{\theta'}[\phi'(\mathbf{X})] \;\forall\; \theta' \in H_1(\theta_0)$$

or

$$E_{\theta'}[1-\phi_0(\mathbf{X})] \leq E_{\theta'}[1-\phi'(\mathbf{X})]$$

or

$$P_{\theta'}(\mathbf{X} \in A_0(\theta)) \leq P_{\theta'}(\mathbf{X} \in A'(\theta))$$

or

$$P_{\theta'}(S_0(\mathbf{X}) \ni \theta) \leq P_{\theta'}(S'(\mathbf{X}) \ni \theta) \;\forall\; \theta' \in H_1(\theta_0)$$

This implies that the family of confidence sets $S_0(\mathbf{X})$ at level $(1-\alpha)$ is UMA unbiased. This proves the theorem. ▣

9.4.3 Minimum Expected Length Confidence Interval

Uniformly shortest length confidence intervals usually do not exist among all $(1-\alpha)$-level confidence intervals, even for the most commonly used distributions. Therefore, this cannot be taken as a measure of precision of a confidence interval. In this light, Pratt (1961) had suggested to take expected length of a confidence interval as a measure of its precision. He had shown that the minimum expected length of a confidence interval is attained when one restricts attention only to unbiased confidence intervals. This is met when one restricts attention to a class of unbiased tests in case of two-sided testing problems or in case of testing problems that involve nuisance parameters, where for certain families, UMP unbiased test exists. By Theorem 9.4.2, the corresponding family of confidence sets is UMA unbiased at level $(1-\alpha)$. Pratt (1961) had shown that such a family of UMA unbiased confidence sets at level $(1-\alpha)$ has the minimum expected length among the families of unbiased $(1-\alpha)$-level confidence sets. The result is presented in the following theorem.

Theorem 9.4.3 Let $X \sim f(x; \theta)$, $\theta \in \Theta \subseteq \mathbb{R}^1$. Let $S(\mathbf{X})$ be a family of $(1 - \alpha)$-level confidence intervals of finite length, that is, let $S(\mathbf{X}) = (\underline{\theta}(\mathbf{X}), \overline{\theta}(\mathbf{X}))$ so that $(\overline{\theta}(\mathbf{X}) - \underline{\theta}(\mathbf{X}))$ be a random variable taking on finite values. Then,

$$E_\theta[\overline{\theta}(\mathbf{X}) - \underline{\theta}(\mathbf{X})] = \int_{\theta' \neq \theta} P_\theta(S(\mathbf{X}) \ni \theta')d\theta' \;\; \forall\, \theta \in \Theta$$

or, in simple words, the expected length of the confidence interval is $P_\theta(S(\mathbf{X}) \ni \theta')$, averaged over all false values of $\theta' \neq \theta$.

Proof. Consider the interval

$$\overline{\theta} - \underline{\theta} = \int_{\underline{\theta}}^{\overline{\theta}} d\theta'$$

Taking expectation on both sides, we get

$$E_\theta[\overline{\theta}(\mathbf{X}) - \underline{\theta}(\mathbf{X})] = E_\theta\left\{\int_{\underline{\theta}}^{\overline{\theta}} d\theta'\right\} = \int_{\mathbf{x}}\left\{\int_{\underline{\theta}}^{\overline{\theta}} d\theta'\right\} f(\mathbf{x}; \theta)d\mathbf{x}$$

$$= \int\left\{\int_{\underline{\theta}(\mathbf{x})}^{\overline{\theta}(\mathbf{x})} f(\mathbf{x}; \theta)d\mathbf{x}\right\} d\theta' = \int_{\theta' \neq \theta} P_\theta(S(\mathbf{X}) \ni \theta')d\theta' \;\; \forall\, \theta \in \Theta$$

Hence, the theorem is proved. ▣

If a test ϕ_0 is UMP unbiased of size α in the class of all size-α unbiased tests, then, correspondingly, we have a family of UMAU confidence interval of θ at level $(1 - \alpha)$, $S(\mathbf{X})$, that minimizes $P_\theta(S(\mathbf{X}) \ni \theta')$ for all false parameter values $\theta' \neq \theta$ that it captures. On using Pratt's result, such a family of UMAU $(1 - \alpha)$-level of confidence sets, $S(\mathbf{X})$, in the class of all families of $(1 - \alpha)$-level unbiased confidence sets, minimizes the expected length of the confidence intervals, $E_\theta^{\mathbf{X}} S(\mathbf{X})$, which is taken as a reasonable measure of precision of confidence intervals.

Guenther (1971) has proposed a useful procedure of finding a family of $(1 - \alpha)$-level unbiased confidence intervals. Let $X_1,\ldots,X_n$ be a random sample drawn from a continuous *pdf* $f(x; \theta)$. Let $T(\mathbf{X}, \theta)$ be a pivot so that

$$P_\theta\{\lambda_1 < T(\mathbf{X}, \theta) < \lambda_2\} = 1 - \alpha$$

which can be written in the form of a confidence interval

$$P_\theta\{\underline{\theta}(\mathbf{X}) < \theta < \overline{\theta}(\mathbf{X})\} = 1 - \alpha$$

Further, unbiasedness of $(1 - \alpha)$-level confidence interval implies

$$P_\theta\{\underline{\theta}(\mathbf{X}) < \theta' < \overline{\theta}(\mathbf{X})\} < 1 - \alpha \;\; \forall\, \theta' \neq \theta$$

Let us define

$$P(\theta, \theta') = P_\theta\{\underline{\theta}(\mathbf{X}) < \theta' < \overline{\theta}(\mathbf{X})\}$$
$$= P[\gamma(\theta, \theta')]$$

Assuming this, the probability depends on some function γ of θ and θ'. The above conditions may be restated in terms of γ by the equation

$$P(\gamma) = \begin{cases} = 1-\alpha, & \text{if } \theta' = \theta \\ < 1-\alpha, & \text{if } \theta' \neq \theta \end{cases}$$

i.e., $P(\gamma)$ attains maximum at $\theta' = \theta$. Thus, λ_1 and λ_2 can be finally obtained by solving

$$P_{\theta'=\theta}(\gamma) = 1-\alpha \quad \text{and} \quad \left.\frac{\partial}{\partial\gamma}P(\gamma)\right|_{\theta'=\theta} = 0$$

9.5 EQUIVARIANT CONFIDENCE INTERVALS

Let $X_1,\ldots,X_n$ be a random sample from $f(x;\theta)$, $\theta \in \Theta \subseteq \mathbb{R}^1$, where $f(x;\theta)$ is a distribution in a family of distributions $\mathcal{F} = \{f(x;\theta): \theta \in \Theta\}$. Let $\mathcal{F}$ be invariant under a group of transformations $\mathcal{G}$ on χ. Let $S(\mathbf{x})$ be a family of confidence intervals for θ at level $(1-\alpha)$.

Definition 9.5.1 Let the family of distributions be invariant under a group of transformations $\mathcal{G}$ on χ. A family of confidence intervals $S(\mathbf{x})$ for θ is equivariant under $\mathcal{G}$ if

$$S(\mathbf{x}) \ni \theta \Leftrightarrow S(g(\mathbf{x})) \ni \bar{g}\theta \;\forall\, \mathbf{x} \in \chi,\ \forall\, \theta \in \Theta \quad \text{and} \quad \forall\, g \in \mathcal{G} \tag{9.5.1}$$

Since inversion procedure is one of the most popular methods of constructing confidence intervals at a certain level, one may find the family of equivariant confidence intervals by inverting the acceptance region of an invariant test under certain conditions. However, the relationships between invariant tests and confidence intervals are not simple in general. Therefore, to check whether a confidence interval is equivariant, the easy way is to check condition (9.5.1) in the definition.

9.6 SOLVED EXAMPLES

***Example* 9.1** Let $X_1,\ldots,X_n$ be a random sample from $P(\theta)$. Construct $(1-\alpha)$-level confidence interval for the Poisson mean θ by pivoting the *cdf* of the sufficient statistic $U = \Sigma X_i$. Also, construct an asymptotic confidence interval for θ.

Solution. The statistic $U = \Sigma X_i$ is a complete sufficient statistic for θ and follows the Poisson distribution $P(n\theta)$. Let for given observations $X_1, X_2,\ldots,X_n$, the value of U be u. The *cdf* of U is

$$F_U(u,\theta) = \sum_{i=0}^{u} \frac{\exp(-n\theta)(n\theta)^i}{i!}, \; u = 0, 1, 2, \ldots$$

It is known that the Poisson family has monotone likelihood ratio in the statistic $U = \Sigma X_i$. Therefore, $F_U(u,\theta)$ is strictly decreasing in θ; $F_U(u,\theta) \to 1$ as $\theta \to 0$ and $\to 0$ as $\theta \to \infty$. Therefore, by Theorem 9.3.4, $(1-\alpha)$-level confidence interval $(\underline{\theta}, \overline{\theta})$ is constructed by solving

$F_U(u-;\ \theta) = F_U(u-1;\ \theta) = 1 - \alpha_2$ to get $\underline{\theta}$ when $u > 0$ ($\underline{\theta} = 0$ when $u = 0$) and by solving $F_U(u;\ \theta) = \alpha_1$ to get $\overline{\theta}$. These conditions are written as

$$\sum_{i=0}^{u-1} \frac{\exp(-n\theta)(n\theta)^i}{i!} = 1 - \alpha_2 \tag{9.6.1}$$

and
$$\sum_{i=0}^{u} \frac{\exp(-n\theta)(n\theta)^i}{i!} = \alpha_1 \tag{9.6.2}$$

Consider $X \sim P(\lambda)$ and $Y \sim \chi^2_{2r} \equiv G_1(r,\ 2)$. We have

$$P(X \le r-1) = \sum_{x=0}^{r-1} \frac{\exp(-\lambda)\lambda^x}{(x)!} = \frac{1}{\Gamma(r)} \int_{\lambda}^{\infty} t^{r-1} \exp(-t)dt$$

By assuming $y = 2t$, we have

$$P(X \le r-1) = \frac{1}{\Gamma(r)2^r} \int_{2\lambda}^{\infty} y^{r-1} \exp\left(-\frac{y}{2}\right)dy$$

$$= P(Y \ge 2\lambda)$$

Using this relationship between gamma and Poisson distributions, we may write Eq. (9.6.1) as

$$P(P(n\theta) \le u-1) = P(\chi^2_{2u} > 2n\theta) = 1 - \alpha_2$$

This gives
$$2n\theta = \chi^2_{2u,\alpha_2}$$

$$\underline{\theta} = \frac{1}{2n}\chi^2_{2u,\alpha_2}$$

Similarly, Eq. (9.6.2) may be written as

$$P(P(n\theta) \le u) = P(\chi^2_{2(u+1)} > 2n\theta) = \alpha_1$$

This gives
$$2n\theta = \chi^2_{2(u+1),1-\alpha_1}$$

$$\overline{\theta} = \frac{1}{2n}x^2_{2(u+1),1-\alpha_1}$$

Therefore, $(1-\alpha)$-level confidence interval for θ is given by

$$\left(\frac{1}{2n}\chi^2_{2u,\,\alpha_2},\ \frac{1}{2n}\chi^2_{2(u+1),1-\alpha_1}\right)$$

The MLE of θ is $\hat{\theta} = \bar{X}$ and the Fisher information of $P(\theta)$ is $I_X(\theta) = 1/\theta$. Hence, $(1 - \alpha)$-level asymptotic confidence interval for θ, assuming $\sqrt{n}(\hat{\theta} - \theta) \to N(0, 1/I_X(\theta))$, is

$$P\left(-z_{\alpha/2} \le \sqrt{nI_X(\hat{\theta})}(\hat{\theta}-\theta) \le z_{\alpha/2}\right) = 1-\alpha$$

or

$$P\left(-z_{\alpha/2} \le \sqrt{\frac{n}{\bar{X}}}(\bar{X}-\theta) \le z_{\alpha/2}\right) = 1-\alpha$$

or

$$P\left(\bar{X} - z_{\alpha/2}\sqrt{\frac{\bar{X}}{n}} \le \theta \le \bar{X} + z_{\alpha/2}\sqrt{\frac{\bar{X}}{n}}\right) = 1-\alpha$$

***Example* 9.2** Consider a random sample $X_1,\ldots,X_n$ from $N(\mu, \sigma^2)$.

(i) Construct a $(1 - \alpha)$-level confidence interval for μ by pivot method for the cases when σ^2 is known and σ^2 is not known. Also, find the shortest length confidence interval. Find also the required sample size so that the length of the $(1 - \alpha)$-level confidence interval does not exceed 2d.

(ii) Assuming σ as known, let the prior distribution of μ be $N(0, 1)$. Find the $(1 - \alpha)$-level confidence interval for μ.

(iii) By test inversion method, construct a UMA (unbiased) $(1 - \alpha)$-level confidence interval for μ corresponding to the UMP (unbiased) test for using $H_0(\mu_0){:}\mu = \mu_0$ against $H_1(\mu_0){:}\mu \ne \mu_0$ in both the cases when σ^2 is known and when it is not known. Is it also minimum expected length confidence interval? Also, find the equivariant confidence interval for μ at level $(1 - \alpha)$.

(iv) Construct a UMA $(1 - \alpha)$-level confidence interval for μ by inverting the acceptance region of size-α UMP test for testing $H_0(\mu_0)$: $\mu \ge \mu_0$ against $H_1(\mu_0){:}\mu < \mu_0$ and $H_0(\mu_0)$: $\mu \le \mu_0$ against $H_1(\mu_0)$: $\mu > \mu_0$ when σ^2 is known.

Solution.

(i) $\bar{X}$ is the complete sufficient statistics for μ. Therefore, the desired confidence interval will be the function of $\bar{X}$. We expect the confidence interval to be of the type $(\bar{X} - a, \bar{X} + b)$. The constants a and b are obtained so that the confidence interval is $(1 - \alpha)$-level

$$P_\mu(\bar{X} - a < \mu < \bar{X} + b) \ge 1 - \alpha$$

or

$$P_\mu(\mu - b < \bar{X} < \mu + a) \ge 1 - \alpha$$

Assuming σ is known,

$$\sqrt{n}\frac{(\bar{X}-\mu)}{\sigma} \sim N(0,1)$$

can be used to obtain the constants a and b such that

$$P_\mu\left(-\sqrt{n}\frac{b}{\sigma} < \sqrt{n}\frac{(\bar{X}-\mu)}{\sigma} < \sqrt{n}\frac{a}{\sigma}\right) = 1-\alpha \qquad (9.6.3)$$

There are infinitely many such pairs (a, b) and, thus, intervals $(-\sqrt{n}(b/\sigma), \sqrt{n}(a/\sigma))$ which satisfy condition (9.6.3). As a rule of thumb, we choose such a and b that divide the total probability α equally into two tails. This gives

$$\sqrt{n}\frac{a}{\sigma} = z_{\alpha/2}$$

$$-\sqrt{n}\frac{b}{\sigma} = z_{1-(\alpha/2)} = -z_{\alpha/2}$$

Thus, the $(1 - \alpha)$-level confidence interval for μ is

$$\left(\bar{X} - z_{\alpha/2}\frac{\sigma}{\sqrt{n}}, \bar{X} + z_{\alpha/2}\frac{\sigma}{\sqrt{n}}\right)$$

The length of this interval is

$$L = 2z_{\alpha/2}\frac{\sigma}{\sqrt{n}}$$

Consider, next, the estimation of confidence interval for μ at level $(1 - \alpha)$ based on the pivot

$$T(\mathbf{X}, \mu) = \sqrt{n}\frac{\bar{X} - \mu}{\sigma}$$

since T follows $N(0, 1)$ distribution which is independent of μ. The interval (a, b) is constructed so that

$$P_\mu(a < T(\mathbf{X}, \mu) < b) = 1 - \alpha \tag{9.6.4}$$

and by solving $a = T(\mathbf{X}, \mu)$ and $b = T(\mathbf{X}, \mu)$ for μ. This gives

$$P_\mu\left(\bar{X} - \frac{\sigma}{\sqrt{n}}b < \mu < \bar{X} - \frac{\sigma}{\sqrt{n}}a\right) = 1 - \alpha$$

It is reasonable to consider that

$$a = -b = -z_{\alpha/2}$$

The $(1 - \alpha)$-level confidence interval is now given by

$$\left(\bar{X} - \frac{\sigma}{\sqrt{n}}z_{\alpha/2}, \bar{X} + \frac{\sigma}{\sqrt{n}}z_{\alpha/2}\right)$$

Consider now the problem of constructing such a $(1 - \alpha)$-level confidence interval satisfying Eq. (9.6.4) for which the length $L = (\sigma/\sqrt{n})(b - a)$ is minimum. We may express Eq. (9.6.4) as

$$\Phi(b) - \Phi(a) = \int_a^b \frac{1}{\sqrt{2\pi}}\exp\left(-\frac{x^2}{2}\right)dx = \int_a^b \phi(x)dx = 1 - \alpha \tag{9.6.5}$$

where ϕ and Φ are the *pdf* and the *cdf* of $N(0, 1)$ distribution, respectively. On differentiating Eq. (9.6.5) and the length of the interval with respect to a, we get

$$\phi(b)\frac{\partial b}{\partial a} - \phi(a) = 0$$

and
$$\frac{\partial L}{\partial a} = \frac{\sigma}{\sqrt{n}}\left(\frac{\partial b}{\partial a} - 1\right)$$

Solving
$$\frac{\partial L}{\partial a} = 0$$

gives
$$\frac{\sigma}{\sqrt{n}}\left(\frac{\phi(a)}{\phi(b)} - 1\right) = 0$$

$$\phi(a) = \phi(b)$$

This is satisfied if $a = b$ or $a = -b$. $a = b$ does not satisfy Eq. (9.6.5). Therefore, $a = -b = -z_{\alpha/2}$ is the acceptable solution. Therefore, $(1 - \alpha)$-level shortest length confidence interval is given by

$$\left(\bar{X} - \frac{\sigma}{\sqrt{n}} z_{\alpha/2}, \bar{X} + \frac{\sigma}{\sqrt{n}} z_{\alpha/2}\right)$$

Next, to construct a $(1 - \alpha)$-level confidence interval having its length not more than $2d$,

$$L \le 2d$$

$$2\frac{\sigma}{\sqrt{n}} z_{\alpha/2} \le 2d$$

a random sample of size at least

$$n \ge \left(\frac{\sigma}{d} z_{\alpha/2}\right)^2$$

is required. In this case, the experimenter ensures himself by suggesting an estimator $\bar{X}$ based on a sample of size $n \ge [(\sigma/d)z_{\alpha/2}]^2$ for estimating μ that he is $100(1 - \alpha)\%$ confident that the error in the estimator $\bar{X}$ does not exceed d.

Next, consider the estimation of confidence interval for μ when σ^2 is not known. Assuming the confidence interval is of the form $(\bar{X} - a, \bar{X} + b)$ and proceeding as before, a and b are obtained so that

$$P_\mu(-b < (\bar{X} - \mu) < a) = 1 - \alpha$$

or
$$P_\mu\left[-\frac{\sqrt{n}b}{S_{n-1}} < \frac{\sqrt{n}(\bar{X} - \mu)}{S_{n-1}} < \frac{\sqrt{n}a}{S_{n-1}}\right] = 1 - \alpha \qquad (9.6.6)$$

where $\sqrt{n}(\bar{X} - \mu)/S_{n-1} \sim t_{n-1}$. There may be infinitely many such (a, b) which satisfy Eq. (9.6.6) but as a rule of thumb, we choose such a and b that equally divide α in the tails. Since t-distribution is symmetric about zero, we have

$$\frac{\sqrt{n}a}{S_{n-1}} = t_{n-1,\,\alpha/2} \quad \text{and} \quad -\frac{\sqrt{n}b}{S_{n-1}} = t_{n-1,\,1-(\alpha/2)} = -t_{n-1,\,\alpha/2}$$

since t_{n-1} is symmetric about zero. We have

$$P_\mu\left[-t_{n-1,\,\alpha/2} < \frac{\sqrt{n}(\bar{X}-\mu)}{S_{n-1}} < t_{n-1,\,\alpha/2}\right] = 1-\alpha$$

$$P_\mu\left[\bar{X} - \frac{S_{n-1}}{\sqrt{n}}t_{n-1,\,\alpha/2} < \mu < \bar{X} + \frac{S_{n-1}}{\sqrt{n}}t_{n-1,\alpha/2}\right] = 1-\alpha$$

Therefore, the $(1 - \alpha)$-level confidence interval for μ is given by

$$\left(\bar{X} - \frac{S_{n-1}}{\sqrt{n}}t_{n-1,\alpha/2},\, \bar{X} + \frac{S_{n-1}}{\sqrt{n}}t_{n-1,\alpha/2}\right)$$

Note that the length $L = (2S_{n-1}/\sqrt{n})\,t_{n-1,\,\alpha/2}$ is a random variable. The expected length of this confidence interval is given by

$$E(L) = \frac{2}{\sqrt{n}}t_{n-1,\alpha/2}E(S_{n-1})$$

$$= \frac{2}{\sqrt{n}}t_{n-1,\alpha/2}\sqrt{\frac{2}{n-1}}\,\frac{\sqrt{(n/2)}}{\sqrt{(n-1)/2}}\sigma$$

$$= \frac{2}{\sqrt{n}}t_{n-1,\alpha/2}k_n\sigma$$

which increases with the increase in σ^2 and can be made small by choosing large n. Further, a suitable pivot is chosen as

$$T(\mathbf{X}, \mu) = \sqrt{n}\frac{\bar{X}-\mu}{S_{n-1}}$$

which follows t-distribution with $(n - 1)$ degrees of freedom. The distribution of T is independent of μ. Therefore, by Theorem 9.3.1, we can choose a and b uniquely so that

$$P_\mu[a < T(\mathbf{X}, \mu) < b] = 1 - \alpha \;\forall\; \mu$$

Solving the equation

$$a = T(\mathbf{X}, \mu) = \sqrt{n}\frac{\bar{X}-\mu}{S_{n-1}}$$

or

$$\bar{\mu} = \bar{X} - \frac{S_{n-1}}{\sqrt{n}}a$$

and solving $$b = T(\mathbf{X}, \mu) = \sqrt{n}\,\frac{\overline{X} - \mu}{S_{n-1}}$$

gives $$\underline{\mu} = \overline{X} - \frac{S_{n-1}}{\sqrt{n}}\,b$$

and, thus, the $(1 - \alpha)$-level confidence interval is

$$P_\mu\left(\overline{X} - \frac{S_{n-1}}{\sqrt{n}}b \le \mu \le \overline{X} - \frac{S_{n-1}}{\sqrt{n}}a\right) = 1 - \alpha$$

In the above confidence interval, the constants a and b are unknown. For the choice of a and b in the above equation, let us assume that the probability α is equally divided in tails. Since t-distribution is symmetric about zero,

$$b = -a = t_{n-1,\alpha/2}$$

This gives the $(1 - \alpha)$-level confidence interval for μ

$$\left(\overline{X} - \frac{S_{n-1}}{\sqrt{n}}t_{n-1,\alpha/2}, \overline{X} + \frac{S_{n-1}}{\sqrt{n}}t_{n-1,\alpha/2}\right)$$

We will now find the shortest length confidence interval for μ. This is obtained by solving two equations

$$P_\mu(a \le T(\mathbf{X}, \mu) \le b) = \int_a^b f(t, n-1)dt = 1 - \alpha \tag{9.6.7}$$

and $$\frac{\partial L}{\partial a} = \frac{\partial}{\partial a}(b - a)\frac{S_{n-1}}{\sqrt{n}} = \left(\frac{\partial b}{\partial a} - 1\right)\frac{S_{n-1}}{\sqrt{n}} = 0 \tag{9.6.8}$$

On differentiating Eq. (9.6.7) with respect to a, we get

$$f(b, n-1)\frac{\partial b}{\partial a} - f(a, n-1) = 0$$

$$\frac{\partial b}{\partial a} = \frac{f(a, n-1)}{f(b, n-1)}$$

and putting its value in Eq. (9.6.8), we get

$$\frac{\partial L}{\partial a} = \left[\frac{f(a, n-1)}{f(b, n-1)} - 1\right]\frac{S_{n-1}}{\sqrt{n}} = 0$$

$$f(a, n-1) = f(b, n-1)$$

This implies that either $a = b$ or $a = -b$. The value $a = b$ is rejected, since it does not satisfy Eq. (9.6.7). Therefore, $a = -b$ is the only acceptable value. By Eq. (9.6.7),

$$\int_{-b}^{b} f(t, n-1)dt = 1 - \alpha$$

We have by the symmetry of t-distribution

$$b = t_{n-1,\,\alpha/2}$$

Therefore, $(1-\alpha)$-level shortest length confidence interval based on the pivot $T(\mathbf{X}, \mu)$ is

$$\left(\overline{X} - \frac{S_{n-1}}{\sqrt{n}} t_{n-1,\,\alpha/2},\ \overline{X} + \frac{S_{n-1}}{\sqrt{n}} t_{n-1,\,\alpha/2}\right)$$

and it also minimizes $E(L) = 2\,(\sigma/\sqrt{n})\,t_{n-1,\,\alpha/2}\,k_n$.

(ii) Given a prior $\pi(\mu) \equiv N(0, 1)$, the posterior distribution of μ given $\mathbf{x}$ is

$$\pi(\mu|\mathbf{x}) \sim N\left(\frac{x}{1+(\sigma^2/n)}, \frac{\sigma^2/n}{1+(\sigma^2/n)}\right)$$

Therefore, the $(1-\alpha)$-level credible interval for μ is

$$P_\mu\left(\frac{a - [\overline{x}/(1+\sigma^2/n)]}{\sqrt{\dfrac{\sigma^2/n}{1+(\sigma^2/n)}}} \le N(0,1) \le \frac{b - [\overline{x}/(1+\sigma^2/n)]}{\sqrt{\dfrac{\sigma^2/n}{1+(\sigma^2/n)}}}\right) \ge 1-\alpha$$

Solving

$$\frac{b - \dfrac{\overline{x}}{1+(\sigma^2/n)}}{\sqrt{\dfrac{\sigma^2/n}{1+(\sigma^2/n)}}} = z_{\alpha/2}$$

for b, we get

$$b = \frac{n\overline{x}}{\sigma^2+n} + \sqrt{\frac{\sigma^2/n}{1+(\sigma^2/n)}}\, z_{\alpha/2}$$

and, solving

$$\frac{a - \dfrac{\overline{x}}{1+(\sigma^2/n)}}{\sqrt{\dfrac{\sigma^2/n}{1+(\sigma^2/n)}}} = -z_{\alpha/2}$$

for a, we get

$$a = \frac{n\overline{x}}{\sigma^2+n} - \sqrt{\frac{\sigma^2}{\sigma^2+n}}\, z_{\alpha/2}$$

Therefore, the required credible interval for μ is given by

$$\left(\frac{n\overline{x}}{\sigma^2+n} - \frac{\sigma}{\sqrt{n+\sigma^2}} z_{\alpha/2},\ \frac{n\overline{x}}{\sigma^2+n} + \frac{\sigma}{\sqrt{n+\sigma^2}} z_{\alpha/2}\right)$$

The length of this interval

$$L = 2\frac{\sigma}{\sqrt{n+\sigma^2}} z_{\alpha/2} < 2\frac{\sigma}{\sqrt{n}} z_{\alpha/2}$$

is shorter than the length of the interval obtained by frequentist approach.

(iii) The UMP unbiased size-α test for testing H_0: $\mu = \mu_0$ against H_1: $\mu \neq \mu_0$, assuming that σ^2 is known, is given by

$$\phi_0(\mathbf{x}) = \begin{cases} 1, & \text{if } \dfrac{\sqrt{n}|\bar{x} - \mu_0|}{\sigma} \geq c \\ 0, & \text{otherwise} \end{cases}$$

where c is obtained by

$$E_{\mu_0}[\phi_0(\mathbf{X})] = \alpha$$

$$P_{\mu_0}\left\{\left|\frac{\sqrt{n}(\bar{X} - \mu_0)}{\sigma}\right| \geq c\right\} = \alpha$$

We get $c = z_{\alpha/2}$ since $\sqrt{n}(\bar{X} - \mu_0)/\sigma \sim N(0, 1)$. The acceptance region corresponding to this test is given by

$$A(\mu_0) = \left\{\mathbf{x} : \left|\frac{\sqrt{n}(\bar{x} - \mu_0)}{\sigma}\right| \leq z_{\alpha/2}\right\}$$

The UMA unbiased family of $(1 - \alpha)$-level confidence interval for μ corresponding to this acceptance region, by Theorems 9.3.5 and 9.4.2, is, therefore, given by

$$\begin{aligned} S(\mathbf{x}) &= \{\mu : \mathbf{x} \in A(\mu)\} \\ &= \left\{\bar{x} - \frac{\sigma}{\sqrt{n}} z_{\alpha/2} \leq \mu \leq \bar{x} + \frac{\sigma}{\sqrt{n}} z_{\alpha/2}\right\} \\ &= \left(\bar{x} - \frac{\sigma}{\sqrt{n}} z_{\alpha/2}, \bar{x} + \frac{\sigma}{\sqrt{n}} z_{\alpha/2}\right) \end{aligned}$$

This interval is UMA unbiased confidence interval at level $(1 - \alpha)$ and also by Theorem 9.4.3, is minimal expected length.

The acceptance region of the UMP unbiased test ϕ_0 of size α for testing H_0: $\mu = \mu_0$ against H_1: $\mu \neq \mu_0$, when σ is not known, gives UMA unbiased confidence interval for μ at level $(1 - \alpha)$. This UMP unbiased test is

$$\phi_0(\mathbf{x}) = \begin{cases} 1, & \text{if } \left|\dfrac{\sqrt{n}(\bar{x} - \mu_0)}{s_{n-1}}\right| > c \\ 0, & \text{otherwise} \end{cases}$$

where c is obtained by

$$E_{\mu_0}\phi(\mathbf{X}) = \alpha$$

$$P_{\mu_0}\left\{\left|\frac{\sqrt{n}(\bar{X}-\mu_0)}{S_{n-1}}\right| > c\right\} = \alpha$$

We get $c = t_{n-1,\alpha/2}$ since $\sqrt{n}(\bar{X} - \mu_0)/S_{n-1} \sim t_{n-1}$. The acceptance region corresponding to the UMP unbiased size α test Φ_0 is

$$A_0(\mu_0) = \left\{\mathbf{x}: \left|\frac{\sqrt{n}(\bar{x}-\mu_0)}{s_{n-1}}\right| \le t_{n-1,\alpha/2}\right\}$$

Therefore, by Theorems 9.3.5 and 9.4.2, the UMA unbiased family of confidence intervals at level $(1 - \alpha)$ is

$$S_0(\mathbf{X}) = \{\mu: \mu \in A_0(\mu)\}$$

$$= \left(\bar{X} - \frac{S_{n-1}}{\sqrt{n}} t_{n-1,\,\alpha/2},\, \bar{X} + \frac{S_{n-1}}{\sqrt{n}} t_{n-1,\alpha/2}\right)$$

This confidence interval $S_0(\mathbf{X})$, by Theorem 9.4.3, is the minimal expected length confidence interval.

The test ϕ_0 is also UMP invariant of size α under a group of location transformations

$$\mathcal{G} = \{g : g_b(\mathbf{x}) = \mathbf{x} + b, \text{ for some } b \in \mathbb{R}^1\}$$

Under this group, the family $\mathcal{F} = \{N(\mu, \sigma^2): \mu \in \mathbb{R}^1, \sigma^2 > 0\}$ is invariant so that

$$\bar{g}(\mu) = \mu + b$$

Consider, now

$$S_0(g_b(\mathbf{X})) = \left(\bar{X} - \frac{S_{n-1}}{\sqrt{n}} t_{n-1,\alpha/2} + b, \bar{X} + \frac{S_{n-1}}{\sqrt{n}} t_{n-1,\alpha/2} + b\right)$$

Since the condition

$$S_0(\mathbf{X}) \ni \mu \quad \Leftrightarrow \quad S_0(g_b(\mathbf{X})) \ni \bar{g}(\mu)$$

holds, the confidence interval for μ at level $(1 - \alpha)$ is equivariant under the group of translations.

(iv) The UMP size-α test for testing H_0: $\mu \ge \mu_0$ against H_1: $\mu < \mu_0$, assuming that σ^2 is known, is given by

$$\phi_0(\mathbf{x}) = \begin{cases} 1, & \text{if } \dfrac{\sqrt{n}(\bar{x}-\mu_0)}{\sigma} < c \\ 0, & \text{otherwise} \end{cases}$$

where c is obtained by

$$P_{\mu_0}\left\{\frac{n(\bar{X}-\mu_0)}{\sigma}<c\right\}=\alpha$$

We get $c = -z_\alpha$ since $\sqrt{n}(\bar{X}-\mu_0)/\sigma \sim N(0, 1)$. The acceptance region corresponding to this test is

$$A(\mu_0)=\left[\mathbf{x}: \frac{\sqrt{n}(\bar{x}-\mu_0)}{\sigma}\ge -z_\alpha\right]$$

The UMA family of $(1-\alpha)$-level confidence interval for μ corresponding to this acceptance region, by Theorem 9.3.5, is, therefore, given by

$$S(\mathbf{x})=\{\mu: \mathbf{x}\in A(\mu)\}=\left(-\infty, \bar{x}+\frac{\sigma}{\sqrt{n}}z_{\alpha/2}\right)$$

The UMP size-α test for testing H_0: $\mu \le \mu_0$ against H_1: $\mu > \mu_0$, assuming that σ^2 is known, is given by

$$\phi_0(\mathbf{x})=\begin{cases}1, & \text{if } \dfrac{\sqrt{n}(\bar{x}-\mu_0)}{\sigma}>z_\alpha\\ 0, & \text{otherwise}\end{cases}$$

The acceptance region corresponding to this test is

$$A(\mu_0)=\left\{\mathbf{x}: \frac{\sqrt{n}(\bar{x}-\mu_0)}{\sigma}\le z_\alpha\right\}$$

The UMA family of $(1-\alpha)$-level confidence interval for μ corresponding to this acceptance region, by Theorem 9.3.5, is, therefore, given by

$$S(\mathbf{x})=\left(\bar{x}-\frac{\sigma}{\sqrt{n}}z_{\alpha/2}, \infty\right)$$

***Example* 9.3** Let X follows normal $N(100, 25)$ distribution. Determine how large a sample should be drawn from this population so that the sample mean does not differ from 100 by more than 1, with probability 0.95.

Solution. Given that the sample mean does not differ from 100 by more than 1,

$$P(-1\le \bar{X}-100\le 1)=0.95$$

specifies that the length of interval is $L = 2$. Further, $(1-\alpha)$-level confidence interval for population mean μ is

$$P\left(-z_{\alpha/2} \le \frac{(\bar{X}-\mu)}{(\sigma/\sqrt{n})} \le z_{\alpha/2}\right) = 0.95$$

$$P\left(\bar{X} - z_{\alpha/2}\frac{\sigma}{\sqrt{n}} \le \mu \le \bar{X} + z_{\alpha/2}\frac{\sigma}{\sqrt{n}}\right) = 0.95$$

The length of this interval,

$$L = 2z_{\alpha/2}\frac{\sigma}{\sqrt{n}}$$

must be equal to 2. This gives

$$n = (1.96 \times 5)^2 \cong 96$$

***Example* 9.4** Let $X_1,\ldots,X_n$ be a random sample from $N(\mu, \sigma^2)$ where σ^2 is known. Let the prior for μ be $N(\eta, \tau^2)$ where η and τ^2 are known. Find $(1 - \alpha)$-level credible interval for μ and compare it with frequentist interval and comment.

Solution. The posterior distribution of μ given $\mathbf{x}$ is

$$N\left(\frac{(n\bar{x}/\sigma^2) + (\eta/\tau^2)}{(n/\sigma^2) + (1/\tau^2)}, \frac{1}{(n/\sigma^2) + (1/\tau^2)}\right)$$

The posterior mean and variance are, respectively,

$$E(\mu|\mathbf{x}) = \begin{cases} \bar{x}, & \text{if } \tau^2 \gg \sigma^2 \\ \eta, & \text{if } \dfrac{\sigma^2}{n} \gg \tau^2 \end{cases}$$

This indicates that if the prior variance is very high as compared to the variance of the sample observations i.e., $\tau^2 >> \sigma^2/n$ then the posterior mean converges to mean of the observations $\bar{x}$, and if the sample observations are highly variable as compared to the prior variance then the posterior mean converges to the mean of prior distribution i.e. η. The posterior variance is

$$V(\mu|\mathbf{x}) = \frac{1}{(n/\sigma^2) + (1/\tau^2)} < \min\left\{\frac{\sigma^2}{n}, \tau^2\right\}$$

indicates that the posterior variance is smaller than the original variance.

The credible interval for μ at level $(1 - \alpha)$ is

$$P\left(-z_{\alpha/2} \le \frac{\mu - E(\mu|\mathbf{x})}{\sigma(\mu|\mathbf{x})} \le z_{\alpha/2}\right) = 1 - \alpha$$

$$P\{E(\mu|\mathbf{x}) - z_{\alpha/2}\sigma(\mu|\mathbf{x}) \le \mu \le E(\mu|\mathbf{x}) + z_{\alpha/2}\sigma(\mu|\mathbf{x})\} = 1 - \alpha \qquad (9.6.9)$$

If $\tau^2 \to \infty$, then $E(\mu|\mathbf{x}) \to \bar{x}$ and $V(\mu|\mathbf{x}) \to \sigma^2/n$ and Eq. (9.6.9) reduces to frequentist-like confidence interval at level $(1-\alpha)$

$$P\left(\bar{x} - z_{\alpha/2}\frac{\sigma}{\sqrt{n}} \le \mu \le \bar{x} + z_{\alpha/2}\frac{\sigma}{\sqrt{n}}\right) = 1-\alpha$$

The credible interval $(\bar{x} \mp z_{\alpha/2}(\sigma/\sqrt{n}))$ is not the frequentist interval $(\bar{X} \mp z_{\alpha/2}(\sigma/\sqrt{n}))$ where the randomness is due to the distribution of $\mathbf{X}$ indexing on μ and applies before the sample $\mathbf{X}$ is observed in the credible interval, the randomness is due to the distribution of μ given the sample $\mathbf{X} = \mathbf{x}$ and is, therefore, reported after the sample is actually observed.

***Example* 9.5** Let $X_1,\ldots,X_n$ be a random sample from $N(\mu, \sigma^2)$. Construct shortest length, UMA unbiased, minimum expected length and equivariant confidence interval for σ^2 at $(1-\alpha)$-level for the cases when μ is known and when μ is not known.

Solution. Since $\Sigma(X_i - \mu)^2$ is the sufficient statistic for σ^2 when μ is known, we may consider the confidence interval for σ^2, $(c_1\Sigma(x_i-\mu)^2, c_2\Sigma(x_i-\mu)^2)$ where $c_1, c_2 > 0$ are obtained so that

$$P_{\sigma^2}\left\{c_1\sum(X_i-\mu)^2 < \sigma^2 < c_2\sum(X_i-\mu)^2\right\} = 1-\alpha$$

$$P_{\sigma^2}\left[\frac{1}{c_2} < \frac{\sum(X_i-\mu)^2}{\sigma^2} < \frac{1}{c_1}\right] = 1-\alpha \tag{9.6.10}$$

where $\Sigma(X_i - \mu)^2/\sigma^2 \sim \chi^2_n$. One can choose infinitely many such pairs of (c_1, c_2) satisfying Eq. (9.6.10), however, conventionally one chooses such a pair (c_1, c_2) that each of the two tails carries equal probability $\alpha/2$. This gives

$$P_{\sigma^2}\left[\frac{\sum(X_i-\mu)^2}{\sigma^2} \ge \frac{1}{c_1}\right] = \frac{\alpha}{2} = P_{\sigma^2}\left[\frac{\sum(X_i-\mu)^2}{\sigma^2} \le \frac{1}{c_2}\right]$$

We then have

$$\frac{1}{c_1} = \chi^2_{n,\alpha/2} \quad \text{and} \quad \frac{1}{c_2} = \chi^2_{n,1-\alpha/2}$$

Thus, the $(1-\alpha)$-level confidence interval for σ^2, when μ is known, is

$$\left(\frac{\sum(X_i-\mu)^2}{\chi^2_{n,\alpha/2}}, \frac{\sum(X_i-\mu)^2}{\chi^2_{n,1-\alpha/2}}\right)$$

In case when μ is not known, $S^2_{n-1} = [1/(n-1)]\Sigma(X_i - \bar{X})^2$ is a sufficient statistics for σ^2. Based on S^2_{n-1}, we consider $(c_1S^2_{n-1}, c_2S^2_{n-1})$, $c_1, c_2 > 0$, as the required confidence interval where c_1 and c_2 are obtained by

$$P_{\sigma^2}\{c_1S^2_{n-1} \le \sigma^2 \le c_2S^2_{n-1}\} \ge 1-\alpha$$

Proceeding similarly as before, we choose (c_1, c_2) satisfying

$$P_{\sigma^2}\left\{\frac{1}{c_2} \le \frac{S_{n-1}^2}{\sigma^2} \le \frac{1}{c_1}\right\} = 1-\alpha$$

and $$P_{\sigma^2}\left[\frac{(n-1)S_{n-1}^2}{\sigma^2} \ge \frac{n-1}{c_1}\right] = \frac{\alpha}{2} = P_{\sigma^2}\left[\frac{(n-1)S_{n-1}^2}{\sigma^2} \le \frac{n-1}{c_2}\right]$$

where $(n-1)S_{n-1}^2/\sigma^2 \sim \chi_{n-1}^2$. It gives

$$\frac{n-1}{c_1} = \chi_{n-1,\alpha/2}^2 \quad \text{and} \quad \frac{n-1}{c_2} = \chi_{n-1,1-\alpha}^2$$

Therefore, $(1-\alpha)$-level confidence for σ^2, when μ is not known, is given by

$$\left(\frac{(n-1)S_{n-1}^2}{\chi_{n-1,\,\alpha/2}^2}, \frac{(n-1)S_{n-1}^2}{\chi_{n-1,\,1-\alpha/2}^2}\right)$$

The shortest length confidence interval for σ^2, when μ is known, based on the pivot

$$T(\mathbf{X}, \sigma^2) = \frac{\sum (X_i - \mu)^2}{\sigma^2}$$

which follows χ_n^2, can be obtained as

$$P_{\sigma^2}\left\{a < \frac{\sum (X_i - \mu)^2}{\sigma^2} < b\right\} = 1-\alpha$$

$$P_{\sigma^2}\left\{\frac{\sum (X_i - \mu)^2}{b} < \sigma^2 < \frac{\sum (X_i - \mu)^2}{a}\right\} = 1-\alpha$$

$$\int_a^b f(t, n)\,dt = 1-\alpha \tag{9.6.11}$$

where $f(t, n)$ is the *pdf* of χ_n^2. We obtain the confidence interval $((1/b)\Sigma(X_i - \mu)^2, (1/a)\Sigma(X_i - \mu)^2)$ for σ^2 so that it satisfies condition (9.6.11) and it minimizes the length

$$L = \left(\frac{1}{a} - \frac{1}{b}\right)\sum (X_i - \mu)^2 \tag{9.6.12}$$

On differentiating Eq. (9.6.11) with respect to a, we get

$$f(b, n)\frac{\partial b}{\partial a} - f(a, n) = 0$$

$$\frac{\partial b}{\partial a} = \frac{f(a, n)}{f(b, n)} \tag{9.6.13}$$

and, differentiating Eq. (9.6.12) with respect to a, we get

$$\frac{\partial L}{\partial a} = \left(-\frac{1}{a^2} + \frac{1}{b^2}\frac{\partial b}{\partial a}\right)\sum (X_i - \mu)^2 = 0 \tag{9.6.14}$$

On combining Eqs. (9.6.13) and (9.6.14), we get

$$\left[-\frac{1}{a^2} + \frac{1}{b^2}\frac{f(a,n)}{f(b,n)}\right]\sum (X_i - \mu)^2 = 0$$

which gives

$$\frac{1}{b^2}\frac{f(a,n)}{f(b,n)} = \frac{1}{a^2} \tag{9.6.15}$$

Tate and Klett have solved Eq. (9.6.15) numerically and tabulated the solutions (a, b). For simplicity, equal-tails interval for σ^2 is used

$$\left(\frac{\sum (X_i - \mu)^2}{\chi^2_{n,\alpha/2}}, \frac{\sum (X_i - \mu)^2}{\chi^2_{n,1-\alpha/2}}\right)$$

Proceeding similarly as before, shortest length confidence interval for σ^2 when μ is unknown based on the pivot

$$T(\mathbf{X}, \sigma^2) = \frac{\sum (X_i - \bar{X})^2}{\sigma^2} = \frac{(n-1)S^2_{n-1}}{\sigma^2}$$

which follows χ^2_{n-1}, the required confidence interval is obtained as

$$\left((n-1)\frac{S^2_{n-1}}{\chi^2_{n-1,\alpha/2}}, (n-1)\frac{S^2_{n-1}}{\chi^2_{n-1,1-(\alpha/2)}}\right)$$

To obtain UMA unbiased confidence interval for σ^2 at level $(1 - \alpha)$, we consider the UMP unbiased size-α test, when μ is given, for testing $H_0{:}\sigma^2 = {\sigma_0}^2$ against $H_1{:}\sigma^2 \neq {\sigma_0}^2$

$$\phi_0(\mathbf{x}) = \begin{cases} 1, & \text{if } \dfrac{\sum (x_i - \mu)^2}{\sigma_0^2} \le c_1 \text{ or } \ge c_2 \\ 0, & \text{otherwise} \end{cases}$$

where c_1 and c_2 are obtained by

$$\int_{c_1}^{c_2} f(y; n)dy = 1 - \alpha$$

and $f(y; n)$ is the *pdf* of χ^2_n or, for simplicity, choosing equal-tails test

$$\int_0^{c_1} f(y; n)dy = \frac{\alpha}{2} = \int_{c_2}^{\infty} f(y; n)dy$$

we obtain $c_1 = \chi^2_{n,1-\alpha/2}$, $c_2 = \chi^2_{n,\alpha/2}$.

The acceptance region corresponding to the UMP unbiased test ϕ_0 is

$$A_0(\sigma_0) = \left\{ \mathbf{x}: \chi^2_{n,1-\alpha/2} \le \frac{\sum (x_i - \mu)^2}{\sigma_0^2} \le \chi^2_{n,\alpha/2} \right\}$$

The corresponding UMA unbiased confidence interval for σ^2 at $(1 - \alpha)$, by Theorem 9.3.5, is

$$S_0(\mathbf{x}) = \{\sigma^2: \mathbf{x} \in A_0(\sigma)\}$$

$$= \left(\frac{\sum (x_i - \mu)^2}{\chi^2_{n,\alpha/2}}, \frac{\sum (x_i - \mu)^2}{\chi^2_{n,1-\alpha/2}} \right)$$

Similarly, one can obtain UMA unbiased confidence interval for σ^2 at level $(1 - \alpha)$ when μ is not known by using Theorems 9.3.5 and 9.4.2,

$$S(\mathbf{x}) = \left(\frac{(n-1)S^2_{n-1}}{\chi^2_{n-1,\alpha/2}}, \frac{(n-1)S^2_{n-1}}{\chi^2_{n-1,1-\alpha/2}} \right)$$

Consider, now, Guenther's procedure for constructing $(1 - \alpha)$-level unbiased confidence interval for σ^2 based on the pivot

$$T(\mathbf{X}, \sigma^2) = \frac{(n-1)S^2_{n-1}}{\sigma^2} \tag{9.6.16}$$

assuming μ as unknown. Consider

$$P(\sigma^2, \sigma^2) = P_{\sigma^2}\left(a < \frac{(n-1)S^2_{n-1}}{\sigma^2} < b \right) = 1 - \alpha$$

$$= P_{\sigma^2}\left(\frac{(n-1)S^2_{n-1}}{b} < \sigma^2 < \frac{(n-1)S^2_{n-1}}{a} \right) = P(l)\big|_{l=1}$$

and

$$P(\sigma^2, \sigma'^2) = P_{\sigma^2}\left(\frac{(n-1)S^2_{n-1}}{b} < \sigma'^2 < \frac{(n-1)S^2_{n-1}}{a} \right)$$

$$= P_{\sigma^2}\left(\frac{(n-1)S^2_{n-1}}{b\sigma^2} < \frac{\sigma'^2}{\sigma^2} < \frac{(n-1)S^2_{n-1}}{a\sigma^2} \right)$$

$$= P_{\sigma^2}\left(a\frac{\sigma'^2}{\sigma^2} < \frac{(n-1)S^2_{n-1}}{\sigma^2} < b\frac{\sigma'^2}{\sigma^2} \right)$$

$$= P^2_{\sigma}(al < T(\mathbf{X}, \sigma^2) < bl) = P(l) < 1 - \alpha, \text{ where } l \ne 1$$

Therefore, $P(l)$ maximizes at $\sigma^2 = \sigma'^2$ or at $l = 1$ with $P(l)|_{l=1} = 1 - \alpha$. The numbers a and b are, therefore, obtained such that

$$P(1) = 1 - \alpha \tag{9.6.17}$$

and
$$\left.\begin{array}{r}\dfrac{\partial}{\partial l}P(l)\Big|_{l=1}=0\\ bf(n-1,b)-af(n-1,a)=0\end{array}\right\}\tag{9.6.18}$$

where $f(n-1,\cdot)$ is the density of χ^2_{n-1}. Tate and Klett have tabulated the values of a and b satisfying Eqs. (9.6.17) and (9.6.18). Therefore, $(1-\alpha)$-level unbiased confidence interval for σ^2, when μ is unknown, is given by

$$\left(\frac{(n-1)S^2_{n-1}}{b},\frac{(n-1)S^2_{n-1}}{a}\right)$$

Note that this confidence interval is different from the shortest-length confidence interval, UMA unbiased confidence interval, etc., obtained earlier and may stand poor in terms of length. However, for large n, all these intervals constructed so far are roughly the same.

Example 9.6 Let (X_i, Y_i), $i = 1,\ldots,n$, be a random sample of size n from a bivariate normal population $(\mu_1, \mu_2, \sigma_1^2, \sigma_2^2, \rho)$.

(i) Find a shortest-length $(1-\alpha)$-level confidence interval for $\mu_1-\mu_2$.
(ii) Construct a $(1-\alpha)$-level confidence set for $\theta = \mu_2/\mu_1$, the ratio of normal means. This confidence set is known as Feller's (1954) confidence set.

Solution.

(i) Corresponding to the bivariate sample $(X_1, Y_1),\ldots,(X_n, Y_n)$, we have independent sample $D_1,\ldots,D_n$ from $N(\mu_1-\mu_2, \sigma_1^2+\sigma_2^2-2\rho\sigma_1\sigma_2)$, where

$$D_i = X_i - Y_i,\ i = 1,\ldots,n$$

We may consider the pivot

$$T=\frac{\sqrt{n}[\bar{D}-(\mu_1-\mu_2)]}{S_D}$$

where
$$S_D^2=\frac{1}{n-1}\sum_i(D_i-\bar{D})^2$$

so that $T \sim t_{n-1}$. Based on this pivot T, $(1-\alpha)$-level confidence interval for $(\mu_1-\mu_2)$ is

$$P\left\{a<\frac{\sqrt{n}[\bar{D}-(\mu_1-\mu_2)]}{S_D}<b\right\}=1-\alpha$$

or
$$\int_a^b f(t)dt=1-\alpha\tag{9.6.19}$$

Thus, $(1-\alpha)$-level of confidence interval for $(\mu_1-\mu_2)$ is given by

$$P\left(\bar{D}-\frac{S_D}{\sqrt{n}}b<\mu_1-\mu_2<\bar{D}-\frac{S_D}{\sqrt{n}}a\right)=1-\alpha$$

The length of this confidence interval is

$$L = \frac{S_D}{\sqrt{n}}(b-a)$$

Proceeding similarly as in Example 9.5, a and b that minimize L subject to Eq. (9.6.19), we get

$$b = t_{(n-1),\alpha/2}$$

$$a = -b = -t_{(n-1),\alpha/2}$$

Thus, $(1-\alpha)$-level shortest length confidence for $(\mu_1 - \mu_2)$ is given by

$$\left(\bar{D} - \frac{S_D}{\sqrt{n}} t_{(n-1),\,\alpha/2}, < \bar{D} + \frac{S_D}{\sqrt{n}} t_{(n-1),\,\alpha/2}\right)$$

(ii) Define

$$D(\theta) = Y - \theta X$$

We have $D(\theta) \sim N(0, V_D(\theta))$, where $V_D(\theta) = \sigma_2^2 - 2\theta\sigma_{12} + \theta^2\sigma_1^2$, $\sigma_2^2 = V(Y)$, $\sigma_1^2 = V(X)$, $\sigma_{12} = \text{cov}(X, Y)$. Then, $D_1(\theta),\ldots,D_n(\theta)$ are *iid* $N(0, V_D(\theta))$ and $\bar{D}(\theta) = \bar{Y} - \theta\bar{X} \sim N(0, V_{\bar{D}}(\theta))$ where $V_{\bar{D}}(\theta) = V_D(\theta)/n$. Variance $V_{\bar{D}}(\theta)$ can be estimated by

$$\hat{V}_{\bar{D}}(\theta) = \frac{1}{n(n-1)}\sum_{i=1}^{n}(D_i(\theta) - \bar{D}(\theta))^2$$

$$= \frac{1}{(n-1)}(S_2^2 - 2\theta S_{12} + \theta^2 S_1^2)$$

where $S_1^2 = (1/n)\Sigma_{i=1}^n (X_i - \bar{X})^2$ and $S_2^2 = (1/n)\Sigma_{i=1}^n (Y_i - \bar{Y})^2$ are the sample variances and $S_{12} = (1/n)\Sigma_{i=1}^n (X_i - \bar{X})(Y_i - \bar{Y})$ is sample covariance. When sampling from normal population, it is a standard result that $\hat{V}_{\bar{D}}(\theta)$ is independent of $\bar{D}(\theta)$ and $(n-1)\hat{V}_{\bar{D}}(\theta)/V_{\bar{D}}(\theta) \sim \chi^2_{n-1}$. Thus,

$$\frac{\bar{D}(\theta)/\sqrt{V_{\bar{D}}(\theta)}}{\sqrt{(n-1)\hat{V}_{\bar{D}}(\theta)/(n-1)V_{\bar{D}}(\theta)}} = \frac{\bar{D}(\theta)}{\sqrt{\hat{V}_{\bar{D}}(\theta)}} \sim t_{n-1}$$

is a pivot. Therefore, $(1-\alpha)$-level confidence set for θ is given by

$$S(\mathbf{X}) = \left\{\theta : \left|\frac{\bar{D}(\theta)}{\sqrt{\hat{V}_D(\theta)}}\right| \le t_{n-1,\alpha/2}\right\}$$

The confidence set $S(\mathbf{X})$ is obtained by solving the parabola in θ,

$$h(\theta) = \frac{\bar{D}(\theta)^2}{\hat{V}_{\bar{D}}(\theta)} - t^2_{n-1,\alpha/2} \le 0$$

We may express the parabola by

$$h(\theta) = A\theta^2 - 2B\theta + C^2 \le 0$$

where $A = (\bar{x}^2 - tS_1^2)$, $B = (\bar{x}\bar{y} - tS_{12})$, $C = (\bar{y}^2 - tS_2^2)$ and $t = [1/(n-1)]\, t^2_{n-1,\alpha/2}$. Consider the value of $h(\theta)$ at $\theta = \bar{y}/\bar{x}$

$$h(\theta)\big|_{\theta=\bar{y}/\bar{x}} = (\bar{x}^2 - tS_1^2)\left(\frac{\bar{y}}{\bar{x}}\right)^2 - 2(\bar{x}\bar{y} - tS_{12})\left(\frac{\bar{y}}{\bar{x}}\right) + (\bar{y}^2 - tS_2^2)$$

$$= -t\left[\left(\frac{\bar{y}}{\bar{x}}\right)^2 S_1^2 - 2\left(\frac{\bar{y}}{\bar{x}}\right)S_{12} + S_2^2\right]$$

$$= -\frac{t}{n}\sum_{i=1}^{n}\left[\left(\frac{\bar{y}}{\bar{x}}\right)^2 (x_i - \bar{x})^2 - 2\left(\frac{\bar{y}}{\bar{x}}\right)(x_i - \bar{x})(y_i - \bar{y}) + (y_i - \bar{y})^2\right]$$

$$= -t\sum_{i=1}^{n}\left[\frac{\bar{y}}{\bar{x}}(x_i - \bar{x}) - (y_i - \bar{y})\right]^2 < 0$$

This shows that the parabola opens upwards if $A > 0$ and downwards if $A < 0$ and this ensures that there exists at least one real root. The parabola $h(\theta)$ has no real root if $B^2 - 4AC < 0$ which may occur if $A < 0$.

Note the condition A < 0, i.e.,

$$\left|\frac{\sqrt{n-1}\,\bar{x}}{S_1} < t_{n-1,\alpha/2}\right|$$

that refers to the acceptance region of UMP unbiased test for testing $H_0{:}\mu_1 = 0$ against $H_1{:}\mu_1 \ne 0$ at level α.

***Example* 9.7** Let $X_1,\ldots,X_n$ be a random sample from the Pareto population with the *pdf*

$$f(x;\theta) = \frac{\theta}{x^2}, x \ge \theta$$

Consider the pivot $T(\mathbf{X}, \theta) = \theta/X_{(1)}$.

(i) Show that the $(1-\alpha)$-level shortest length confidence interval for θ based on the pivot T is $(X_{(1)}\alpha^{1/n}, X_{(1)})$.

(ii) Find the shortest length $(1-\alpha)$-level unbiased confidence interval for θ based on the pivot $T(\mathbf{X}, \theta)$.

Solution.

(i) Probability density of $X_{(1)}$ is

$$f_{X_{(1)}}(x;\theta) = \frac{n\theta^n}{x^{n+1}}, x > \theta$$

The probability density of $T(\mathbf{X}, \theta) = \theta/X_{(1)}$ is $f_T(t) = nt^{n-1}$, $0 < t < 1$. The $(1 - \alpha)$-level confidence interval for θ based on $T(\mathbf{X}, \theta)$ is $(aX_{(1)}, bX_{(1)})$ where a and b are such that it minimizes the length of the interval $L = (b - a)X_{(1)}$ subject to

$$P(a < T(\mathbf{X}, \theta) < b) = 1 - \alpha$$

$$P\left(a < \frac{\theta}{X_{(1)}} < b\right) = 1 - \alpha$$

$$\int_a^b nt^{n-1}dt = 1 - \alpha$$

$$b^n - a^n = 1 - \alpha \tag{9.6.20}$$

where $b^n - a^n < b^n$ or $(1 - \alpha)^{1/n} < b \le 1$. We now have

$$\frac{\partial L}{\partial b} = \left(1 - \frac{\partial a}{\partial b}\right) X_{(1)} \tag{9.6.21}$$

and differentiating Eq. (9.6.20) with respect to b, we get

$$nb^{n-1} - n\alpha^{n-1}\frac{\partial a}{\partial b} = 0$$

or

$$\frac{\partial a}{\partial b} = \frac{b^{n-1}}{a^{n-1}} \tag{9.6.22}$$

Equations (9.6.21) and (9.6.22) give

$$\frac{\partial L}{\partial b} = \left(1 - \frac{b^{n-1}}{a^{n-1}}\right) X_{(1)} < 0$$

This implies that the length of the interval L is minimum for the maximum value of b that is 1. On putting $b = 1$ in Eq. (9.6.20), we get $a = \alpha^{1/n}$. Therefore, $(1 - \alpha)$-level shortest length confidence interval for θ is $(X_{(1)}\ \alpha^{1/n}, X_{(1)})$.

Alternatively, we can use note 3 to the Theorem 9.4.1 to construct $(1-\alpha)$-level shortest length confidence interval for θ. The density of the pivot T is increasing since $(\partial/\partial t)\, f_T(t) = n(n - 1)t^{n-2} > 0$ for $0 < t < 1$. Therefore, the shortest length confidence interval is given by $(a < \theta/X_{(1)} < 1)$ or $(aX_{(1)}, X_{(1)})$ where a is obtained by

$$\int_a^1 nt^{n-1}dt = 1 - \alpha$$

$$1 - a^n = 1 - \alpha$$

or

$$a = \alpha^{1/n}$$

Thus, the required shortest length confidence interval at level-$(1 - \alpha)$ is given by $(X_{(1)}\alpha^{1/n}, X_{(1)})$.

(ii) The $(1-\alpha)$-level confidence interval based on this pivot T is

$$P_\theta\left(a<\frac{\theta}{X_{(1)}}<b\right)=P_\theta(aX_{(1)}<\theta<bX_{(1)})=1-\alpha$$

Define $P(\theta, \theta') = P(\gamma)$ with $\gamma = \theta'/\theta$

$$\begin{aligned}P(\gamma) &= P_\theta(aX_{(1)}<\theta'<bX_{(1)})\\ &= P_\theta\left(\frac{aX_{(1)}}{\theta}<\frac{\theta'}{\theta}<\frac{bX_{(1)}}{\theta}\right)\\ &= P_\theta\left(\frac{aX_{(1)}}{\theta}<\gamma<\frac{bX_{(1)}}{\theta}\right)\\ &= P_\theta\left(\frac{a}{\gamma}<\frac{\theta}{X_{(1)}}<\frac{b}{\gamma}\right)\\ &= P_\theta\left(\frac{a}{\gamma}<T<\frac{b}{\gamma}\right)\\ &= F\left(\frac{b}{\gamma}\right)-F\left(\frac{a}{\gamma}\right)\end{aligned}$$

The confidence interval $(aX_{(1)}, bX_{(1)})$ is unbiased at the level $(1-\alpha)$ if a and b are chosen such that

$$\begin{aligned}P(1) &= P_\theta(a<T<b)\\ &= \int_a^b nt^{n-1}\,dt = t^n\Big|_a^b = b^n-a^n = 1-\alpha\end{aligned}$$

and
$$\frac{\partial}{\partial\gamma}P(\gamma)\Bigg|_{\gamma=1} = -\frac{b}{\gamma^2}f\left(\frac{b}{\gamma}\right)+\frac{a}{\gamma^2}f\left(\frac{a}{\gamma}\right)\Bigg|_{\gamma=1} = 0 = n(a^n-b^n)$$

or
$$b^n - a^n = 0$$

This implies
$$b = 1,\ a = \alpha^{1/\alpha}$$

***Example* 9.8** Let $X_1, X_2, \ldots, X_n$ be a random sample from the population having the *pdf*

$$f(x;\theta) = \theta x^{\theta-1},\ 0<x<1$$

Find $(1-\alpha)$-level confidence interval for θ by pivoting the *cdf*. Using the MLE $\hat{\theta} = -n/\log(\Pi_{i=1}^n X_i)$ of θ, find an approximate 90% confidence interval for θ.

Solution. Let $-\theta\log x = y$ or $x = \exp(-y/\theta)$. Then the *pdf* of Y is $G_1(1, 1)$. $-\theta\Sigma_{i=1}^n \log X_i = \Sigma_{i=1}^n Y_i = U \sim G_1(n, 1)$. We choose a and b such that

$$P\{a<U<b\} = (1-\alpha)$$

$$\int_a^b f_n(t)\,dt = (1-\alpha)$$

where $f_n(t)$ is the density of $G_1(n, 1)$ distribution. With these values of a and b, the $(1 - \alpha)$-level confidence interval is given by

$$P\left(a < -\theta \log \prod_{i=1}^{n} X_i < b\right) = (1-\alpha)$$

$$P\left(-\frac{b}{\log \prod_{i=1}^{n} X_i} < \theta < -\frac{a}{\log \prod_{i=1}^{n} X_i}\right) = (1-\alpha)$$

Therefore, the $(1 - \alpha)$-level confidence interval is given by

$$\left(-\frac{b}{\log \prod_{i=1}^{n} X_i}, -\frac{a}{\log \prod_{i=1}^{n} X_i}\right)$$

We will now construct the confidence interval for θ based on the MLE. The MLE of θ is

$$\hat{\theta} = -\frac{n}{\log \prod_{i=1}^{n} X_i}$$

The Fisher information

$$I_{\mathbf{X}}(\theta) = nI_{\mathbf{X}}(\theta) = nE_\theta\left(-\frac{\partial^2}{\partial\theta^2}\log L(\theta; \mathbf{X})\right) = \frac{n}{\theta^2}$$

helps in constructing the asymptotic pivotal quantity

$$T(\mathbf{X}, \theta) = \frac{\hat{\theta} - \theta}{\sqrt{I_{\mathbf{X}}^{-1}(\theta)}} = \frac{\hat{\theta} - \theta}{\sqrt{\theta^2/n}} \sim N(0,1)$$

since MLE is asymptotically normal. We estimate $I_{\mathbf{X}}(\theta)$ by its estimator $I_{\mathbf{X}}(\hat{\theta})$ since it is a function of θ. We then have the asymptotic pivotal quantity

$$T(\mathbf{X}, \theta) = \frac{\hat{\theta} - \theta}{\sqrt{I_{\mathbf{X}}^{-1}(\hat{\theta})}} = \frac{\hat{\theta} - \theta}{\sqrt{\hat{\theta}^2/n}} \sim N(0,1)$$

Therefore, the approximate 90% level confidence interval for θ is

$$\left(\hat{\theta} - z_{\alpha/2}\sqrt{\frac{\hat{\theta}^2}{n}}, \hat{\theta} + z_{\alpha/2}\sqrt{\frac{\hat{\theta}^2}{n}}\right) = \left(\hat{\theta} - 1.64\sqrt{\frac{\hat{\theta}^2}{n}}, \hat{\theta} + 1.64\sqrt{\frac{\hat{\theta}^2}{n}}\right)$$

***Example* 9.9** Suppose $X_1, X_2, \ldots, X_n$ are n observations drawn randomly from the population having the *pdf*

$$f(x; \theta) = \frac{x^2}{2\theta^3}\exp\left(-\frac{x}{\theta}\right), 0 < x < \infty$$

where $\theta > 0$. Construct an approximate $(1-\alpha)$-level confidence interval for θ based on the MLE of θ namely $\hat{\theta} = \bar{X}/3$.

Solution. The Fisher information is

$$I_{\mathbf{X}}(\theta) = E\left(-\frac{\partial^2}{\partial\theta^2}\log L(\theta, \mathbf{X})\right) = \frac{3n}{\theta^2}$$

The normality of MLE gives

$$T(\mathbf{X}, \theta) = \frac{\hat{\theta}-\theta}{\sqrt{I_{\mathbf{X}}^{-1}(\theta)}} = \frac{\hat{\theta}-\theta}{\sqrt{\theta^2/3n}} \sim N(0,1)$$

as pivotal quantity. The approximate pivotal quantity on replacing $I_{\mathbf{X}}(\theta)$ by its estimator $I_{\mathbf{X}}(\hat{\theta})$ becomes

$$T(\mathbf{X}, \theta) = \frac{\hat{\theta}-\theta}{\sqrt{\hat{\theta}^2/3n}} \sim N(0,1)$$

Thus, the $(1-\alpha)$-level confidence interval for θ is

$$\left(\hat{\theta} - z_{\alpha/2}\sqrt{\frac{\hat{\theta}^2}{3n}}, \hat{\theta} + z_{\alpha/2}\sqrt{\frac{\hat{\theta}^2}{3n}}\right)$$

where $\hat{\theta} = \bar{X}/3$.

***Example* 9.10** Let $X_1, X_2, \ldots, X_n$ be *iid* according to the distribution

$$F_X(x; \beta) = \begin{cases} 0, & \text{if } x \le 0 \\ \left(\dfrac{x}{\beta}\right)^a, & \text{if } 0 < x < \beta \\ 1, & \text{if } x \ge \beta \end{cases}$$

If a is known $a = a_0$, construct a $(1-\alpha)$-level upper confidence limit for β by using $X_{(n)}/\beta$ as pivot since β is the scale parameter, where $X_{(n)}$ is the MLE of β.

Solution. The $(1-\alpha)$-level confidence limit for β based on the pivot $T(\mathbf{X}, \beta) = X_{(n)}/\beta$ is $P(X_{(n)}/\beta > c) = P(\beta < X_{(n)}/c) = 1-\alpha$; $P(X_{(n)} \le c\beta) = \alpha$; $P(\text{all } X_i \le c\beta) = \alpha$; $(c\beta/\beta)^{na_0} = \alpha$; $c = \alpha^{1/na_0}$. Therefore, the $(1-\alpha)$-level upper confidence bound for β is

$$\left(0, \frac{X_{(n)}}{\alpha^{1/na_0}}\right)$$

***Example* 9.11** Let $X \sim Be(\theta, 1)$. Find a pivotal quantity and construct a confidence interval for θ at the confidence coefficient same as of the interval $(y/2, y)$ where $y = -(\log x)^{-1}$.

Solution. The *pdf* of Y is

$$f_Y(y; \theta) = \frac{\theta}{y^2}\exp\left(-\frac{\theta}{y}\right), 0 < y < \infty$$

The distribution of Y is $IG_2(1, \theta)$. The confidence coefficient of the given interval calculates to

$$P\left(\frac{Y}{2} < \theta < Y\right) = P(\theta < Y < 2\theta) = \int_{\theta}^{2\theta} \frac{\theta}{y^2} \exp\left(-\frac{\theta}{y}\right) dy$$

$$= \exp\left(-\frac{1}{2}\right) - \exp(-1) = 0.239$$

The density of X is given by

$$f_X(x; \theta) = \theta x^{\theta-1}, \quad 0 < x < 1$$

Consider

$$T(X, \theta) = X^{\theta}$$

as pivot. The density of T is given by

$$f_T(t) = 1, 0 < t < 1$$

Thus, 0.239 level confidence interval for θ is

$$P(a < T(X, \theta) < b) = b - a = 0.239$$

$$P(a < X^b < b) = P\left(\frac{\log a}{\log X} \le \theta \le \frac{\log b}{\log X}\right) = 0.239$$

where choices of a and b are such that $b - a = 0.239$.

***Example* 9.12** Let $X_1, X_2, \ldots, X_n$ be a random sample from the *pdf*

$$f(x; \theta) = \exp[-(x - \theta)], x \ge \theta$$

(i) Find a $(1 - \alpha)$-level confidence interval for θ based on the pivot $T(\mathbf{X}, \theta) = 2n(X_{(1)} - \theta)$ so that the probability α is equally divided in the tails of the distribution of T. Also, obtain the $(1 - \alpha)$-level confidence interval by likelihood ratio method and compare the two.

(ii) Find the unbiased confidence interval based on the pivot $T(\mathbf{X}, \theta)$ in (i).

(iii) Find the shortest-length confidence for θ at level $(1 - \alpha)$ based on the same pivot $T(\mathbf{X}, \theta)$.

(iv) Is the unbiased confidence interval obtained in (ii) also a shortest length interval?

Solution.

(i) Consider the sufficient statistic $X_{(1)}$. Its *pdf* is

$$f_{X_{(1)}}(x; \theta) = n \exp[-n(x - \theta)], \quad x > \theta$$

The distribution of the pivot

$$T(\mathbf{X}, \theta) = 2n(X_{(1)} - \theta)$$

is

$$f_T(t) = \frac{1}{2} \exp\left(-\frac{t}{2}\right), \; t > 0$$

which is a $G_1(1, 2) \equiv \chi_2^2$ distribution. The $(1 - \alpha)$-level confidence interval for θ based on pivot T is

$$P(a < 2n(X_{(1)} - \theta) < b) = 1 - \alpha$$

$$= P\left(\frac{a}{2n} < X_{(1)} - \theta < \frac{b}{2n}\right)$$

$$= P\left(X_{(1)} - \frac{b}{2n} < \theta < X_{(1)} - \frac{a}{2n}\right)$$

Based on T, the $(1 - \alpha)$-level confidence interval for θ is $(X_{(1)} - (b/2n), X_{(1)} - (b/2n))$ so that

$$P_\theta(a < T(\mathbf{X}, \theta) < b) = 1 - \alpha \ \forall\ \theta \tag{9.6.23}$$

There are infinite such pairs (a, b) which satisfy Eq. (9.6.23). For simplicity, one can choose such (a, b) that equally divides α in the tails of the distribution of T

$$P_\theta(T(\mathbf{X}, \theta) < a) = \frac{\alpha}{2} = P_\theta(T(\mathbf{X}, \theta) > b)$$

Solving $$\int_b^\infty \frac{1}{2}\exp\left(-\frac{t}{2}\right)dt = \frac{\alpha}{2}$$

$$b = -2\log\left(\frac{\alpha}{2}\right)$$

and $$\int_0^a \frac{1}{2}\exp\left(-\frac{t}{2}\right)dt = \frac{\alpha}{2}$$

$$a = -2\log\left(1 - \frac{\alpha}{2}\right)$$

Therefore, the desired confidence interval for θ is given by $(X_{(1)} + (1/n)\log(\alpha/2)$, $X_{(1)} + (1/n)\log(1 - \alpha/2))$.

We will now compare the confidence intervals based on the pivotal quantity and likelihood ratio test on the basis of their lengths. Consider, now, the likelihood ratio test for testing $H_0{:}\theta = \theta_0$ against $H_1{:}\theta \neq \theta_0$ based on the sufficient statistic $X_{(1)}$

$$\lambda(x) = \frac{\sup\limits_{\theta=\theta_0} L(\theta, x)}{\sup\limits_{\theta\in\mathbb{R}^1} L(\theta, x)} = \frac{n\exp[-n(x-\theta_0)]I_{[\theta_0,\infty)}(x)}{n\exp[-n(x-x)]I_{[\theta_0,\infty)}(x)}$$

$$= \begin{cases} 0, & \text{if } x < \theta_0 \\ \exp[-n(x-\theta_0)], & \text{if } x \geq \theta_0 \end{cases}$$

The rejection region of the likelihood ratio test is

$$\{x: \lambda(x) < c\} = \{x: \exp[-n(x - \theta_0)] < c \text{ or } x < \theta_0\}$$

where c is obtained by the size condition

$$P\{\exp(-n(X_{(1)} - \theta_0)) < c \quad \text{or} \quad X_{(1)} < \theta_0 | H_0\} = \alpha$$

$$P[-n(X_{(1)} - \theta_0) < \log c | \theta = \theta_0] = \alpha$$

$$P\left(X_{(1)} > \theta_0 - \frac{1}{n}\log c | \theta = \theta_0\right) = \int_{\theta_0 - (1/n)\log c}^{\infty} n\exp[-n(x - \theta_0)]dx$$

$$= -\exp[-n(x - \theta_0)]\Big|_{\theta_0 - (1/n)\log c}^{\infty}$$

$$= c = \alpha$$

Therefore, the rejection region is

$$\left\{\mathbf{x}: x_{(1)} > \theta_0 - \frac{1}{n}\log\alpha, x_{(1)} > \theta_0\right\}$$

and the corresponding acceptance region is

$$A(\theta_0) = \left\{\mathbf{x}: \theta_0 \le x_{(1)} \le \theta_0 - \frac{1}{n}\log\alpha\right\}$$

On inverting this acceptance region, we get $(1 - \alpha)$-level confidence interval for θ,

$$S(x) = \left(\theta: x_{(1)} + \frac{1}{n}\log\alpha \le \theta \le x_{(1)}\right)$$

Computing the lengths of confidence intervals obtained from likelihood ratio test and from pivotal quantity

$$L(\text{LRT}) = x_{(1)} - \left(x_{(1)} + \frac{1}{n}\log\alpha\right)$$

$$= -\frac{1}{n}\log\alpha$$

$$L(\text{Pivot}) = \left[x_{(1)} + \frac{1}{n}\log\left(1 - \frac{\alpha}{2}\right)\right] - \left[x_{(1)} + \frac{1}{n}\log\left(\frac{\alpha}{2}\right)\right]$$

$$= \frac{1}{n}\log\frac{1 - (\alpha/2)}{\alpha/2}$$

We observe that the $(1 - \alpha)$-level confidence interval based on the likelihood ratio test is superior to the one obtained by pivotal quantity since

$$L(\text{LRT}) < L(\text{Pivotal})$$

(ii) The $(1-\alpha)$-level confidence interval for θ based on pivot T is

$$P(a < 2n(X_{(1)} - \theta) < b) = 1 - \alpha$$

$$= P\left(\frac{a}{2n} < X_{(1)} - \theta < \frac{b}{2n}\right)$$

$$= P\left(X_{(1)} - \frac{b}{2n} < \theta < X_{(1)} - \frac{a}{2n}\right)$$

Let us define

$$P(\theta, \theta') = P_\theta\left(X_{(1)} - \frac{b}{2n} < \theta' < X_{(1)} - \frac{a}{2n}\right)$$

$$= P\left(X_{(1)} - \theta - \frac{b}{2n} < \theta' - \theta < X_{(1)} - \theta - \frac{a}{2n}\right)$$

$$= P\left(\frac{T_\theta}{2n} - \frac{b}{2n} < \gamma < \frac{T_\theta}{2n} - \frac{a}{2n}\right)$$

$$= P\left(\gamma + \frac{a}{2n} < \frac{T_\theta}{2n} < \gamma + \frac{b}{2n}\right)$$

$$P(\gamma) = P(2n\gamma + a < T_\theta < 2n\gamma + b) = F(2n\gamma + b) - F(2n\gamma + a)$$

where $\gamma = \theta' - \theta$ and F is the *cdf* of T_θ. The interval $(X_{(1)} - (b/2n), X_{(1)} - (a/2n))$ is $(1-\alpha)$-level unbiased for such values of a and b so that

$$P(0) = \int_a^b \chi^2(2) d\chi^2 = 1 - \alpha$$

and
$$\left.\frac{\partial}{\partial \gamma} P(\gamma)\right|_{\gamma=0} = 2nf(b) - 2nf(a) = 0$$

or
$$f(a) = f(b)$$

(iii) Equation (9.6.23) can be written as

$$\int_a^b f_T(t)dt = 1 - \alpha \tag{9.6.24}$$

The *pdf* of T is decreasing. Therefore, from note 3 of Section 9.4, the $(1-\alpha)$-level shortest-length confidence interval based on T is $(0, b)$ where b is obtained by

$$\int_0^b \frac{1}{2} \exp\left(-\frac{t}{2}\right) dt = 1 - \alpha$$

This gives $b = 2\log(1/\alpha)$ to get

$$P\left(0 < 2n(X_{(1)} - \theta) < 2\log(1/\alpha)\right) = 1 - \alpha$$

$$P\left(X_{(1)} + \frac{1}{n}\log\alpha \le \theta \le X_{(1)}\right) = 1 - \alpha$$

Thus, $(X_{(1)} + (1/n)\log\alpha \le \theta \le X_{(1)})$ is $(1 - \alpha)$-level shortest-length confidence interval for θ.

(iv) The interval (a, b) is a $(1 - \alpha)$-level confidence interval so that

$$P[a < T < b] = 1 - \alpha$$

or

$$\int_a^b f(t)dt = 1 - \alpha$$

This implies

$$f(b)\frac{\partial b}{\partial a} - 1 = 0$$

The length of the interval is $L = (1/2n)(b - a)$. This interval is of shortest length if

$$\frac{\partial L}{\partial a} = \frac{\partial b}{\partial a} - 1 = 0$$

This implies

$$f(b) = f(a)$$

i.e., we choose a and b such that

$$\int_a^b f(t)dt = 1 - \alpha$$

and $f(a) = f(b)$. Thus, unbiased and shortest length confidence intervals in (ii) at level $(1 - \alpha)$ are both identical.

***Example* 9.13** Let $X_1, X_2, \ldots, X_n$ be a random sample from $N(\mu, \sigma^2)$ where μ and σ^2 are both unknown. Consider the pivot

$$T(\mathbf{X}, \mu, \sigma^2) = \frac{\sqrt{n}(\bar{X} - \mu)}{S_{n-1}}$$

Show that the equal-tails interval $(\bar{X} - t_{n-1,\alpha/2}\, S_{n-1}/\sqrt{n},\ \bar{X} + t_{n-1,\alpha/2}\, S_{n-1}/\sqrt{n})$ is the shortest-length unbiased interval for μ at level $(1 - \alpha)$.

Solution. Consider the testing problem

$$H_0\colon \mu = \mu_0 \quad \text{vs.} \quad H_1\colon \mu \ne \mu_0$$

The UMP unbiased test for the above testing problem is

$$\phi(\mathbf{x}) = \begin{cases} 1, & \left|\dfrac{\sqrt{n}(\bar{x} - \mu_0)}{S_{n-1}}\right| > c \\ 0, & \text{otherwise} \end{cases}$$

where $\bar{x} = \Sigma x_i/n$ and $S_{n-1}^2 = (n-1)^{-1}\Sigma(x_i - \bar{x})^2$. The constant c is obtained by the size condition

$$P_{\mu_0}\left[\left|\frac{\sqrt{n}(\bar{X}-\mu_0)}{S_{n-1}}\right| > c\right] = \alpha$$

which gives $c = t_{n-1,\alpha/2}$. The acceptance region corresponding to this test is

$$A(\mu_0) = \left\{\mathbf{x}: \left|\frac{\sqrt{n}(\bar{x}-\mu_0)}{S_{n-1}}\right| < t_{n-1,\alpha/2}\right\}$$

The family of confidence intervals corresponding to this acceptance region

$$S(\mathbf{x}) = \{\mu: \mathbf{x} \in A(\mu)\}$$

$$= \left(\bar{x} - \frac{S_{n-1}}{\sqrt{n}} t_{n-1,\alpha/2}, \bar{x} + \frac{S_{n-1}}{\sqrt{n}} t_{n-1,\alpha/2}\right)$$

is UMA unbiased at level $(1 - \alpha)$. We have already shown that the above confidence interval for μ is the shortest length interval. Therefore, the above interval is the shortest length unbiased at level $(1 - \alpha)$.

***Example* 9.14** Let $X_1, X_2, \ldots, X_n$ be a sample from $G_1(1,\theta)$

$$f(x;\theta) = \frac{1}{\theta}\exp\left(-\frac{x}{\theta}\right), x > 0$$

Find the unbiased confidence interval for θ based on the pivot $T(\mathbf{X}, \theta) = 2\Sigma X_i/\theta$. Also, find the shortest-length confidence interval.

Solution. We define

$$P(\theta,\theta') = P(\gamma) = P\left(\frac{2\sum X_i}{b\theta} < \frac{\theta'}{\theta} < \frac{2\sum X_i}{a\theta}\right)$$

$$= P\left(\frac{2\sum X_i}{b\theta} < \gamma < \frac{2\sum X_i}{a\theta}\right)$$

$$P(\gamma) = P(\gamma a < T < \gamma b) = \int_{\gamma a}^{\gamma b} \chi^2_{2n}(t)dt = F(\gamma b) - F(\gamma a)$$

where F is the *df* of χ^2_{2n} distribution. We have

$$\frac{\partial}{\partial\gamma}P(\gamma) = bf(\gamma b) - af(\gamma a)$$

The $(1-\alpha)$-level confidence interval for θ is given by $(2\Sigma X_i/b,\ 2\Sigma X_i/a)$ where a and b are obtained by the conditions

$$\int_a^b \chi^2_{2n}(t)\,dt = 1-\alpha$$

and
$$\frac{\partial}{\partial \gamma}P(\gamma)\bigg|_{\gamma=1} = bf(b) - af(a) = 0$$

The interval (a, b) based on T is at level $(1-\alpha)$ if

$$P[a < T < b] = 1-\alpha$$

$$\Rightarrow \qquad f(b)\frac{\partial b}{\partial a} - f(a) = 0$$

The length of the interval $L = 2\Sigma X_i[(1/a) - (1/b)]$ is minimized by solving the equation

$$\frac{\partial L}{\partial a} = 2\sum X_i\left(-\frac{1}{a^2} + \frac{1}{b^2}\frac{\partial b}{\partial a}\right) = 0$$

$$\Rightarrow \qquad \frac{\partial b}{\partial a} = \frac{b^2}{a^2}$$

Thus, the $(1-\alpha)$-level confidence interval for θ with the shortest length is given by (a, b). The constants a and b are chosen such that

$$\int_a^b f(t)dt = 1-\alpha$$

and
$$b^2 f(b) = a^2 f(\text{a})$$

***Example* 9.15** Let $X_1,\ldots,X_n$ be a random sample from $U(0, \theta)$, $\theta > 0$. Find the shortest-length $(1-\alpha)$-level for θ based on the pivot $T(\mathbf{X}, \theta) = X_{(n)}/\theta$, UMA confidence interval, MLE, and Chebyshev's inequality based confidence intervals. Also prescribe the smallest sample size such that the $(1-\alpha)$-level confidence interval $(X_{(n)}, \alpha^{-1/n}X_{(n)})$ has length not more than d. Show that the unbiased confidence interval for θ based on the pivot T as defined above coincides with the shortest length confidence interval.

Solution. Maximum statistic $X_{(n)}$ is the sufficient statistic and its distribution is given by

$$f_{X_{(n)}}(x;\theta) = n\frac{x^{n-1}}{\theta^n}, \quad 0 < x < \theta$$

Based on $X_{(n)}$, the pivot may be defined as

$$T(\mathbf{X}, \theta) = \frac{X_{(n)}}{\theta}$$

with its *pdf*
$$f_T(t) = nt^{n-1}, \ 0 < t < 1$$

We construct a $(1 - \alpha)$-level confidence interval

$$P[a < T(\mathbf{X}, \theta) < b] = 1 - \alpha$$

The *pdf* of T is increasing on (0,1). Therefore, by note 3 of Theorem 6.4.1, the shortest length $(1 - \alpha)$-level confidence interval $(X_{(n)}/b, X_{(n)}/a)$ is obtained by setting $b = 1$ and choosing a such that

$$\int_a^1 f_T(t)dt = 1 - \alpha$$

$$1 - a^n = 1 - \alpha$$

or

$$a = \alpha^{1/n}$$

Thus, $(1 - \alpha)$-level shortest-length confidence interval is $(X_{(n)}, \alpha^{-1/n} X_{(n)})$ which converges to a single point $X_{(n)}$ or length converges to zero as $n \to \infty$.

The second way to find the shortest-length confidence interval is to note that the length of the interval can be expressed as

$$L = \left(\frac{1}{a} - \frac{1}{b}\right) X_{(n)} \propto \left(\frac{1}{a} - \frac{1}{b}\right) = \int_a^b \frac{1}{t^2} dt$$

$\Rightarrow$

$$g(t) = \frac{1}{t^2}$$

so that the note 4 of Section 9.4 applies. This gives the best set by

$$C = \left\{t : \frac{g(t)}{f(t)} < k\right\} = \left\{t : \frac{1}{t^2 n t^{n-1}} < k\right\} = \{t : a \le t \le 1\}$$

where a is such that

$$\int_a^1 n t^{n-1} dt = 1 - \alpha \quad \Rightarrow \quad a = \alpha^{1/n}$$

Thus, $(X_{(n)}, X_{(n)}/\alpha^{1/n})$ is the shortest-length confidence interval for θ at level $(1 - \alpha)$.

Consider the testing problem $H_0: \theta = \theta_0$ against $H_1: \theta \ne \theta_0$ for constructing UMA confidence interval at level $(1 - \alpha)$. For this, the test

$$\phi_0(\mathbf{x}) = \begin{cases} 1, & \text{if } X_{(n)} > \theta_0 \text{ or } X_{(n)} \le \theta_0 \alpha^{1/n} \\ 0, & \text{otherwise} \end{cases}$$

is a UMP test of size α for testing problem H_0 against H_1. The corresponding acceptance region (UMP) is

$$A_0(\theta_0) = \{\mathbf{x}: \theta_0 \alpha^{1/n} < X_{(n)} < \theta_0\}$$

where x is the observed value of $X_{(n)}$. Corresponding to this acceptance region, the UMA confidence interval for θ at level $(1 - \alpha)$ is, then, given by

$$S_0(\mathbf{X}) = \{\theta : \mathbf{X} \in A_0(\theta)\} = (X_{(n)}, \alpha^{-1/n} X_{(n)})$$

Consider, next, the construction of $(1 - \alpha)$-level confidence interval based on the MLE $\hat{\theta} = X_{(n)}$. The Fisher's information is

$$I_{\mathbf{X}}(\theta) = nI_{X_1}(\theta) = nE_\theta\left[\frac{\partial}{\partial\theta}\log(X_1;\theta)\right]^2 = \frac{n}{\theta^2}$$

and

$$V(\hat{\theta}) = \frac{1}{I_{\mathbf{X}}(\theta)} = \frac{\theta^2}{n}$$

so that

$$\frac{(\hat{\theta}-\theta)}{\sqrt{V(\hat{\theta})}} \xrightarrow{d} N(0,1) \quad \text{as } n \to \infty$$

The $(1 - \alpha)$-level confidence interval for θ may then be approximated by

$$P_\theta\left(-z_{\alpha/2} < \frac{(\hat{\theta}-\theta)}{\sqrt{V(\hat{\theta})}} < z_{\alpha/2}\right) = 1-\alpha$$

$$P_\theta\left(\hat{\theta} - z_{\alpha/2}\sqrt{V(\hat{\theta})} < \theta < \hat{\theta} + z_{\alpha/2}\sqrt{V(\hat{\theta})}\right) = 1-\alpha$$

and this may be estimated by

$$\left(\left(1-\left(z_{\alpha/2}/\sqrt{n}\right)\right)X_{(n)}, \left(1+\left(z_{\alpha/2}/\sqrt{n}\right)X_{(n)}\right)\right)$$

Next, consider the construction of $(1 - \alpha)$-level confidence interval based on Chebyshev's inequality for every $\varepsilon > 0$

$$P_\theta\left(\left|\hat{\theta}-\theta\right| < \varepsilon\sqrt{\text{MSE}(\hat{\theta})}\right) \geq 1 - \frac{1}{\varepsilon^2}$$

Since $X_{(n)}$ is the sufficient statistic and $E_\theta(X_{(n)}) = [n/(n + 1)]\theta$, we consider $\hat{\theta} = X_{(n)}$. Further, $\text{MSE}(\hat{\theta}) = E_\theta(X_{(n)} - \theta)^2 = 2/[(n + 1)(n + 2)]\theta^2$. Thus, we have

$$P_\theta\left(\left|X_{(n)}-\theta\right| < \varepsilon\sqrt{\frac{2}{(n+1)(n+2)}}\,\theta\right) \geq 1 - \frac{1}{\varepsilon^2}$$

Since θ is unknown in the above inequality, it can be replaced by $X_{(n)}$ since $X_{(n)} \xrightarrow{P} \theta$. Therefore, the above inequality may be approximated by

$$P_\theta\left[\left|X_{(n)}-\theta\right| < \varepsilon\sqrt{\frac{2}{(n+1)(n+2)}}\,X_{(n)}\right] \geq 1 - \frac{1}{\varepsilon^2}$$

Thus, $1 - (1/\varepsilon^2)$ level confidence interval is approximated by

$$\left(\left(1-\varepsilon\sqrt{2/(n+1)(n+2)}\right)X_{(n)},\left(\left(1+\varepsilon\sqrt{2/(n+1)(n+2)}\right)X_{(n)}\right) \tag{9.6.25}$$

On choosing $\varepsilon = 1/\sqrt{\alpha}$, we have $1 - (1/\varepsilon^2) = (1 - \alpha)$ and for large n, we can approximate $\sqrt{2/[(n+1)(n+2)]}$ by $\sqrt{2}/n$.

Further, by $P(X_{(n)} \le \theta) = 1$, Eq. (9.6.25) can be approximated by

$$\left(X_{(n)}, \left(1+\frac{1}{n}\sqrt{\frac{2}{\alpha}}\right)X_{(n)}\right)\ \forall\,\theta$$

Note that the $(1 - \alpha)$-level confidence interval for θ narrows down to a single point $X_{(n)}$ as $n \to \infty$.

The minimum sample size, so that the length of the interval $(X_{(n)}, \alpha^{-1/n} X_{(n)})$ is not more than d, is obtained by

$$(\alpha^{-1/n} - 1)X_{(n)} < d$$

$$\alpha^{-1/n} < \frac{d}{X_{(n)}} + 1$$

$$\frac{1}{n}\log\left(\frac{1}{\alpha}\right) < \log\left(\frac{d}{X_{(n)}} + 1\right)$$

$$n > \frac{\log(1/\alpha)}{\log[(d/X_{(n)}) + 1]}$$

To construct $(1 - \alpha)$-level unbiased confidence interval based on the pivot $T(\mathbf{X}, \theta) = X_{(n)}/\theta$, consider

$$P_\theta\left(a < \frac{X_{(n)}}{\theta} < b\right) = 1 - \alpha$$

$$P_\theta\left(\frac{X_{(n)}}{b} < \theta < \frac{X_{(n)}}{a}\right) = 1 - \alpha$$

We may define for $\theta' \ne \theta$,

$$P(\theta, \theta') = P_\theta\left(\frac{X_{(n)}}{b} < \theta' < \frac{X_{(n)}}{a}\right)$$

$$= P_\theta\left(\frac{X_{(n)}}{b\theta} < \frac{\theta'}{\theta} < \frac{X_{(n)}}{a\theta}\right)$$

Let $\gamma = \theta'/\theta$

$$P(\theta, \theta') = P_\theta\left(\frac{X_{(n)}}{b\theta} < \gamma < \frac{X_{(n)}}{a\theta}\right)$$

$$= P_\theta(\gamma a < T_\theta < \gamma_b)$$

We have
$$P(\theta, \theta')|_{\theta=\theta'} = P(1) = b^n - a^n = 1 - \alpha$$
$$P(\gamma) = (\gamma b)^n - (\gamma a)^n < 1 - \alpha \ \forall\ \theta' \neq \theta \ (\gamma \neq 1)$$

This confirms that the shortest-length confidence interval and unbiased confidence interval at level $(1 - \alpha)$ are the same.

***Example* 9.16** Let X follows

(i) $U(\theta - (1/2), \theta + (1/2))$ and
(ii) $f(x; \theta) = 2x/\theta^2$, $0 < x < \theta$, $\theta > 0$.

Construct a $(1 - \alpha)$-level confidence interval for θ in each case.

Solution.

(i) We may consider the pivot

$$T(X; \theta) = X - \theta \sim U\left(-\frac{1}{2}, \frac{1}{2}\right)$$

to construct a $(1 - \alpha)$-level confidence interval for θ, namely $(X - b, X - a)$ such that

$$P(a < X - \theta < b) = b - a = 1 - \alpha$$

or
$$P(X - b < \theta < X - a) = 1 - \alpha$$

We may choose $a = -1/2 + \alpha/2$, $b = 1/2 - \alpha/2$ to get $(1 - \alpha)$-level interval $(X - 1/2 + \alpha/2, X + 1/2 - \alpha/2)$.

(ii) Consider the pivot

$$T(X, \theta) = \frac{X}{\theta}$$

It is the pivot since

$$f_T(t) = 2t, 0 \le t \le 1$$

The *pdf* of T is an increasing function on (0,1). Therefore, the shortest length $(1 - \alpha)$-level confidence interval $(X/b, X/a)$ is obtained by setting $b = 1$ and choosing a by solving the equation

$$P\left(a \le \frac{X}{\theta} \le 1\right) = \int_a^1 2t\, dt = 1 - a^2 = 1 - \alpha$$

or $a = \sqrt{\alpha}$. Thus, the required interval is given by $(X, X/\sqrt{\alpha})$.

***Example* 9.17** Let $X_1, X_2, \ldots, X_n$ be a random sample drawn from

$$f_X(x;\, \theta) = \exp(i\theta - x), \;\; i\theta \le x < \infty$$

The statistic $S = \min_i (X_i/i)$ is the sufficient statistic for θ. Use the pivot $T(\mathbf{X}, \theta) = S - \theta$ to construct a $(1 - \alpha)$-level shortest-length confidence interval for θ of the form $(S + a, S + b)$.

Solution. The probability

$$P(S > s) = \prod_{i=1}^{n} P(X_i > is) = \prod \int_{is}^{\infty} \exp(i\theta - x)dx = \prod \exp[i(\theta - s)]$$

$$= \exp\left[-\frac{n(n+1)}{2}(s-\theta)\right]$$

gives the *cdf* and *pdf* of S by

$$P(S \le s) = 1 - \exp\left[-\frac{n(n+1)}{2}(s-\theta)\right]$$

and

$$f_s(s) = \frac{n(n+1)}{2}\exp\left[-\frac{n(n+1)}{2}(s-\theta)\right], s \ge \theta$$

respectively, where θ is the location parameter. The quantity $T(\mathbf{X}, \theta) = S - \theta$ is the pivot since its distribution

$$f_T(t) = \frac{n(n+1)}{2}\exp\left\{-\frac{n(n+1)}{2}t\right\}, t \ge 0$$

is independent of θ. Based on this pivot, the $(1 - \alpha)$-level confidence interval for θ is

$$P(a \le T \le b) = 1 - \alpha \tag{9.6.26}$$

$$P(S - b \le \theta \le S - a) = 1 - \alpha$$

We require such values of a and b that minimize the length of the interval $L = b - a$ subject to Eq. (9.6.26). Since the density of T is strictly decreasing, the length is minimum if $a = 0$ and b is chosen such that

$$P(0 \le T \le b) = F(b) - F(0) = 1 - \alpha$$

or

$$\exp\left[-\frac{n(n+1)}{2}b\right] = \alpha$$

$$b = 2\frac{\log(1/\alpha)}{n(n+1)}$$

Thus, the required shortest confidence interval is

$$\big(S - [2\log(1/\alpha)/(n(n+1))],\, S\big)$$

***Example* 9.18** Let $X_1,\ldots,X_m$ and $Y_1,\ldots,Y_n$ be two independent random samples drawn from two normal populations $N(\mu_1, \sigma_1^2)$ and $N(\mu_2, \sigma_2^2)$, respectively. Construct $(1-\alpha)$-level confidence interval for the ratio $\sigma_2{}^2/\sigma_1{}^2$

(i) when μ_1 and μ_2 are known;
(ii) when μ_1 and μ_2 are unknown; and
(iii) when one of μ_1 and μ_2 is known and the other is unknown.

Solution.

(i) μ_1 and μ_2 are known

$$\frac{\sum_1^m (X_i-\mu_1)^2}{\sigma_1^2} \sim \chi_m^2 \tag{9.6.27}$$

and

$$\frac{\sum_1^n (Y_i-\mu_2)^2}{\sigma_2^2} \sim \chi_n^2 \tag{9.6.28}$$

Since the statistics in Eqs. (9.6.27) and (9.6.28) are independent, we may define the pivot by

$$T(\mathbf{X},\mathbf{Y},\sigma_1^2,\sigma_2^2) = \frac{\left[\sum_1^m (X_i-\mu_1)^2\right]\Big/ m\sigma_1^2}{\left[\sum_1^n (Y_i-\mu_2)^2\right]\Big/ n\sigma_2^2} \sim F(m,n)$$

The $(1-\alpha)$ level confidence interval for σ_2^2/σ_1^2 at level $(1-\alpha)$ is

$$P(a < T(\mathbf{X},\mathbf{Y},\sigma_1^2,\sigma_2^2) < b) = 1-\alpha$$

One can choose, for the purpose of simplicity, a and b so that

$$P(T<a) = \frac{\alpha}{2} = P(T>b)$$

giving $b = F(\alpha/2, m, n)$ and $a = F(1-\alpha/2, m, n)$. Thus, the required confidence interval for σ_2^2/σ_1^2 is

$$\left[F(1-(\alpha/2),m,n)\frac{m\sum_1^n (Y_i-\mu_2)^2}{n\sum_1^m (X_i-\mu_1)^2},\, F((\alpha/2),m,n)\frac{m\sum_1^n (Y_i-\mu_2)^2}{n\sum_1^m (X_i-\mu_1)^2}\right]$$

(ii) The pivot, when μ_1 and μ_2 are unknown, is

$$T(\mathbf{X}, \mathbf{Y}, \sigma_1^2, \sigma_2^2) = \frac{(m-1)S_1^2(n-1)\sigma_2^2}{(m-1)\sigma_1^2(n-1)S_2^2}$$

$$= \frac{S_1^2}{S_2^2}\frac{\sigma_2^2}{\sigma_1^2} \sim F(m-1, n-1)$$

Proceeding similarly as before, we get the (1 – α)-level confidence interval for σ_2^2/σ_1^2 for equal probability in tails of the distribution of T

$$\left[F\big(1-(\alpha/2), m-1, n-1\big)\frac{S_2^2}{S_1^2}, F\big((\alpha/2), m-1, n-1\big)\frac{S_2^2}{S_1^2}\right]$$

(iii) In case when either of μ_1 or μ_2 is known and the other unknown, consider μ_1 is known and μ_2 unknown, the (1 – α)-level confidence interval for σ_2^2/σ_1^2 is

$$\left(F\big(1-(\alpha/2), m, n-1\big)\frac{S_2^2}{(1/m)\sum_1^m (X_i-\mu_1)^2}, F\big(1-(\alpha/2), m, n-1\big)\frac{S_2^2}{(1/m)\sum_1^m (X_i-\mu_1)^2}\right)$$

One can, similarly, construct a confidence interval for σ_2^2/σ_1^2 when μ_1 is unknown and μ_2 is known.

***Example* 9.19** Consider the linear model

$$\mathbf{Y} = \mathbf{X}\boldsymbol{\beta} + \boldsymbol{e}, \boldsymbol{e} \sim N_n(\mathbf{0}, \sigma^2\mathbf{I})$$

where $\mathbf{Y} = (Y_1,\ldots,Y_n)$, $\boldsymbol{e} = (e_1,\ldots,e_n)$, $\mathbf{X}$ is an $n \times p$ matrix, β is a p-vector of unknown parameters, $p < n$. Construct a (1 – α)-level confidence set for $\boldsymbol{\beta}$.

Solution. Any solution for $\boldsymbol{\beta}$ in the normal equations

$$\mathbf{X}'\mathbf{X}\boldsymbol{\beta} = \mathbf{X}'\mathbf{Y}$$

is the least square estimator of $\boldsymbol{\beta}$. Assuming $\mathbf{X}$ is the full-rank matrix, i.e., $r(\mathbf{X}) = p$,

$$\hat{\boldsymbol{\beta}} = (\mathbf{X}'\mathbf{X})^{-1}\mathbf{X}'\mathbf{Y}$$

is the unique solution to the above normal equations.
From the results of linear model, we know

$$(\hat{\boldsymbol{\beta}} - \boldsymbol{\beta})'\, \mathbf{X}'\mathbf{X}(\hat{\boldsymbol{\beta}} - \boldsymbol{\beta})/\sigma^2 \sim \chi_p^2$$

and

$$\left\|\mathbf{Y} - \mathbf{X}\hat{\boldsymbol{\beta}}\right\|^2 = \frac{(\mathbf{Y} - \mathbf{X}\hat{\boldsymbol{\beta}})'(\mathbf{Y} - \mathbf{X}\hat{\boldsymbol{\beta}})}{\sigma^2} \sim \chi_{n-p}^2$$

This gives a pivotal quantity

$$h(\mathbf{Y}, \boldsymbol{\beta}) = \frac{(\hat{\boldsymbol{\beta}} - \boldsymbol{\beta})'\mathbf{X}'\mathbf{X}(\hat{\boldsymbol{\beta}} - \boldsymbol{\beta})/p}{(\mathbf{Y} - \mathbf{X}\hat{\boldsymbol{\beta}})'(\mathbf{Y} - \mathbf{X}\hat{\boldsymbol{\beta}})/(n-p)} \sim F(p, (n-p))$$

Based on this pivot, one can construct a $(1 - \alpha)$-level confidence set

$$P(a \le h(\mathbf{Y}, \boldsymbol{\beta}) \le b) = 1 - \alpha$$

The set $\{\boldsymbol{\beta}: h(\mathbf{Y}, \boldsymbol{\beta}) \le \text{const.}\}$ is an ellipsoid in $\mathbb{R}^p$.

***Example* 9.20** Consider n observations $(X_1, Y_1),\ldots,(X_n, Y_n)$ independently drawn from the following linear model

$$Y|x = \beta_0 + \beta_1 x + e$$

where $e \sim N(0, \sigma^2)$. Find a pivot for β_0 and based on this pivot, construct a $(1 - \alpha)$-level confidence interval for β_0 when σ^2 is known and when it is not known.

Solution. We know from linear models that $\hat{\beta}_0 = \bar{Y} - \hat{\beta}_1\bar{x}$ and $\hat{\beta}_1 = S_{xy}/S_{xx}$ is the MLE of β_0 and β_1 where $S_{xy} = \Sigma_{i=1}^n x_iY_i - n\bar{x}\bar{Y}$, $S_{xx} = \Sigma_{i=1}^n x_i^2 - n\bar{x}^2$. The estimators $\hat{\beta}_0$ and $\hat{\beta}_1$ are unbiased for β_0 and β_1, respectively. We have $V(\hat{\beta}_1) = \sigma^2/S_{xx}$ and $V(\hat{\beta}_0) = V(\bar{Y}) + \bar{x}^2V(\hat{\beta}_1) - 2\bar{x}$ $\text{cov}(\bar{Y}, \hat{\beta}_1)$. Let us compute the covariance term

$$\text{cov}(\bar{Y}, \hat{\beta}_1) = \frac{1}{S_{xx}}\left[\text{cov}\left(\bar{Y}, \sum_{i=1}^n x_iY_i\right) - \text{cov}(\bar{Y}, n\bar{x}\bar{Y})\right]$$

$$= \frac{1}{S_{xx}}\left[\sum_{i=1}^n x_i\,\text{cov}(\bar{Y}, Y_i) - n\bar{x}\,\text{var}(\bar{Y})\right]$$

$$= \frac{1}{S_{xx}}\left[\sum_{i=1}^n \frac{x_i\sigma^2}{n} - \frac{n\bar{x}\sigma^2}{n}\right]$$

$$= \frac{\sigma^2}{nS_{xx}}\left[\sum_{i=1}^n (x_i - n\bar{x})\right] = 0$$

This gives
$$V(\hat{\beta}_0) = \frac{\sigma^2}{n} + \bar{x}^2\frac{\sigma^2}{S_{xx}} = \frac{\sigma^2}{nS_{xx}}(S_{xx} + n\bar{x}^2)$$

$$= \frac{\sigma^2}{nS_{xx}}\sum_{i=1}^n x_i^2$$

We may express $\hat{\beta}_1$ and $\hat{\beta}_0$ as

$$\hat{\beta}_1 = \frac{S_{xy}}{S_{xx}} = \frac{1}{S_{xx}}\sum_{i=1}^n (x_i - \bar{x})Y_i$$

$$\hat{\beta}_0 = \overline{Y} - \hat{\beta}_1 \overline{x} = \frac{1}{n}\sum_{i=1}^{n} Y_i - \frac{\overline{x}}{S_{xx}} \sum_{i=1}^{n} (x_i - \overline{x}) Y_i$$

$$= \frac{1}{nS_{xx}} \sum_{i=1}^{n} [S_{xx} - n\overline{x}(x_i - \overline{x})] Y_i$$

We see that $\hat{\beta}_1$ and $\hat{\beta}_0$ are the linear functions of $Y_1, \ldots, Y_n$ where each Y_i is normally distributed. Therefore, the quantities

$$T(\mathbf{Y}, \beta_1) = \frac{\hat{\beta}_1 - \beta_1}{\sigma / \sqrt{S_{xx}}} \sim N(0,1)$$

and

$$T(\mathbf{Y}, \beta_0) = \frac{\hat{\beta}_0 - \beta_0}{\sigma \sqrt{\left(\sum_{i=1}^{n} x_i^2\right) \Big/ nS_{xx}}} \sim N(0,1)$$

are the pivots for β_1 and β_0, respectively. Based on these pivots, the $(1 - \alpha)$-level confidence intervals for β_1 and β_0 are given by

$$\left(\hat{\beta}_1 - z_{\alpha/2} \left(\sigma / \sqrt{S_{xx}}\right), \hat{\beta}_1 + z_{\alpha/2} \left(\sigma / \sqrt{S_{xx}}\right)\right)$$

and

$$\left(\hat{\beta}_0 - z_{\alpha/2} \sigma \sqrt{\left(\sum_{i=1}^{n} x_i^2\right) \Big/ nS_{xx}}, \hat{\beta}_0 + z_{\alpha/2}\, \sigma \sqrt{\left(\sum_{i=1}^{n} x_i^2\right) \Big/ nS_{xx}}\right)$$

respectively. In case the model parameter σ^2 is not known, it is estimated by

$$S^2 = \frac{1}{n-2} \sum_{i=1}^{n} (Y_i - \hat{\beta}_0 - \hat{\beta}_1)^2$$

and the quantities

$$T(\mathbf{Y}, \beta_1) = \frac{\hat{\beta}_1 - \beta_1}{S / \sqrt{S_{xx}}} \sim t_{n-2}$$

and

$$T(\mathbf{Y}, \beta_0) = \frac{\hat{\beta}_0 - \beta_0}{S \sqrt{\left(\sum_{i=1}^{n} x_i^2\right) \Big/ nS_{xx}}} \sim t_{n-2}$$

are the pivots for β_1 and β_0, respectively. Based on these pivots, the $(1 - \alpha)$-level confidence intervals for β_1 and β_0 are given by

$$\left(\hat{\beta}_1 - t_{n-2,\, \alpha/2} \left(S / \sqrt{S_{xx}}\right), \hat{\beta}_1 + t_{n-2,\, \alpha/2} \left(S / \sqrt{S_{xx}}\right)\right)$$

and

$$\left[\hat{\beta}_0 - t_{n-2,\, \alpha/2}\, S \sqrt{\left(\sum_{i=1}^{n} x_i^2\right) \Big/ nS_{xx}}, \hat{\beta}_0 + t_{n-2,\, \alpha/2}\, S \sqrt{\left(\sum_{i=1}^{n} x_i^2\right) \Big/ nS_{xx}}\right]$$

respectively.

***Example* 9.21** Let $X \sim G_1(k, \beta)$ with known shape parameter k. Find the shortest $(1 - \alpha)$-level pivotal confidence interval for scale parameter β.

Solution. The pivot is $T = X/\beta \sim G_1(k, 1)$. The $(1 - \alpha)$-level confidence interval for β based on T is

$$P\left(a < \frac{X}{\beta} < b\right) = 1 - \alpha$$

$$P\left(\frac{X}{b} < \beta < \frac{X}{a}\right) = 1 - \alpha$$

Theorem 9.4.1(ii) applies for constructing $(1 - \alpha)$-level shortest confidence interval for the scale parameter β by choosing a and b such that

$$\int_a^b f(t)dt = 1 - \alpha$$

$$\int_a^b G_1(k, 1)dt = \int_a^b f_k(t)dt = 1 - \alpha$$

and
$$b^2 f(b) = a^2 f(a)$$

$$b^2 \frac{1}{\Gamma(k)} b^{k-1} \exp(-b) = a^2 \frac{1}{\Gamma(k)} a^{k-1} \exp(-a)$$

$$b^{k+1} \exp(-b) = a^{k+1} \exp(-a)$$

Another way to construct a $(1 - \alpha)$-level shortest-length confidence interval for β is to note that the length can be expressed as

$$L = \left(\frac{1}{a} - \frac{1}{b}\right)X \propto \left(\frac{1}{a} - \frac{1}{b}\right) = \int_a^b \frac{1}{t^2} dt \quad \Rightarrow \quad g(t) = \frac{1}{t^2}$$

Therefore, by note 4 of Section 9.4, the best set is

$$\left\{t: \frac{g(t)}{f(t)} < k\right\} = \left\{t: \frac{\Gamma(k)}{t^2 t^{k-1} \exp(-t)} < k\right\} = \left\{t: \frac{t^{k+1} \exp(-t)}{\Gamma(k+2)} > k'\right\}$$

$$= \{t: f_{k+2}(t) \geq k'\}$$

So, the optimal values of a and b are the ones which satisfy

$$\int_a^b f_k(t)dt = 1 - \alpha$$

and
$$f_{k+2}(a) = f_{k+2}(b)$$

or
$$b^{k+1} \exp(-b) = a^{k+1} \exp(-a)$$

***Example* 9.22** Let $X_1,\ldots,X_n$ be *iid* according to the *pdf*

$$f(x;\theta)=\frac{1}{\theta}\exp\left(-\frac{x}{\theta}\right),\ x>0$$

Find the shortest-length confidence interval for θ at level $(1-\alpha)$ based on the pivot $T(\mathbf{X},\theta)=2\Sigma X_i/\theta$.

Solution. Each $X_i \sim G_1(1,\theta)$, $i=1,\ldots,n$; ΣX_i is sufficient for θ and it follows $G_1(n,\theta)$ distribution. Consider now the pivot

$$T(\mathbf{X},\theta)=\frac{2\sum X_i}{\theta}$$

which follows $G_1(n,2)\equiv\chi^2_{2n}$. The $(1-\alpha)$-level confidence interval based on this pivot is

$$P_\theta(a<T(\mathbf{X},\theta)<b)=1-\alpha$$

or

$$\int_a^b \chi^2_{2n}(t)\,dt=1-\alpha \tag{9.6.29}$$

where $T(\mathbf{X},\theta)\sim G_1(n,2)=\chi^2_{2n}$. Further, based on this pivot, the $(1-\alpha)$-level confidence interval is given by

$$P_\theta\left(a<\frac{2\sum X_i}{\theta}<b\right)=P_\theta\left(\frac{2\sum X_i}{b}<\theta<\frac{2\sum X_i}{a}\right)=1-\alpha$$

The length of this interval $L=2[(1/a)-(1/b)]\Sigma X_i$ can be minimized subject to Eq. (9.6.29) by putting the value of

$$\frac{\partial b}{\partial a}=\frac{\chi^2_{2n}(a)}{\chi^2_{2n}(b)}$$

in $\partial L/\partial b$ and setting it to zero

$$\frac{\partial L}{\partial a}=\left(-\frac{1}{a^2}+\frac{1}{b^2}\frac{\partial b}{\partial a}\right)=0$$

$$=\frac{1}{b^2}\frac{\chi^2_{2n}(a)}{\chi^2_{2n}(b)}-\frac{1}{a^2}=0$$

This gives

$$a^2\chi^2_{2n}(a)=b^2\chi^2_{2n}(b) \tag{9.6.30}$$

Therefore, $(1-\alpha)$-level shortest-length confidence interval for θ is $((2\Sigma X_i)/b,\ (2\Sigma X_i)/a)$ where a and b are obtained by solving Eqs. (9.6.29) and (9.6.30).

***Example* 9.23** Let $X_1,\ldots,X_n$ be a random sample from the *pdf*

$$f(x;\theta)=\frac{1}{\theta_2-\theta_1},\ \theta_1\le x\le\theta_2,\theta_1<\theta_2$$

Let $R = X_{(n)} - X_{(1)}$. Using $T(\mathbf{X}, \theta) = R/(\theta_1 - \theta_2)$ as a pivot for estimating $\theta_2 - \theta_1$, show that the shortest-length confidence interval is of the from $(R, R/c)$, where c is determined from the level as a solution of

$$c^{n-1}[(n-1)c - n] + \alpha = 0$$

Solution. The distribution function of the given *pdf* is

$$F(x) = \int_{\theta_1}^{x} \frac{1}{\theta_2 - \theta_1} dx = \frac{x - \theta_1}{(\theta_2 - \theta_1)}$$

The joint density of $X_{(1)}$ and $X_{(n)}$ is

$$f_{X_{(1)}, X_{(n)}}(x_1, x_n; \theta_1, \theta_2) = n(n-1)\frac{(x_n - x_1)^{n-2}}{(\theta_2 - \theta_1)^{n-2}} \frac{1}{(\theta_2 - \theta_1)^2},\ \theta_1 < x_1 < x_2 < \theta_2$$

Let $R = X_{(n)} - X_{(1)}$, $S = X_{(n)}$, $|J| = 1$. The joint distribution of R and S is

$$f_{R,S}(r, s; \theta_1, \theta_2) = \frac{n(n-1)}{(\theta_2 - \theta_1)} r^{n-2}, r + \theta_1 < s < \theta_2$$

This gives

$$f_R(r; \theta_1, \theta_2) = \frac{n(n-1)}{(\theta_2 - \theta_1)^n} r^{n-2}(\theta_2 - \theta_1 - r), 0 < r < \theta_2 - \theta_1$$

Consider, now, $T(\mathbf{X}, \theta_1, \theta_2) = R/(\theta_1 - \theta_2)$ as a pivot, having the *pdf* with $|J| = |\partial R/\partial T| = \theta_1 - \theta_2$

$$\begin{aligned} f_T(t) &= \frac{n(n-1)}{(\theta_2 - \theta_1)^n} [(\theta_2 - \theta_1)t]^{n-2}(\theta_2 - \theta_1)(1-t)(\theta_2 - \theta_1) \\ &= n(n-1)\, t^{n-2}(1-t), 0 < t < 1 \end{aligned}$$

The $(1 - \alpha)$-level confidence interval for θ based on the pivot T is given by

$$P(c < T(\mathbf{X}, \theta_1, \theta_2) < b) = 1 - \alpha$$

$$P\left(c < \frac{R}{\theta_2 - \theta_1} < b\right) = P\left(\frac{R}{b} < \theta_2 - \theta_1 < \frac{R}{c}\right) = 1 - \alpha$$

Since the *pdf* of T is increasing on (0,1), we can find the shortest length $(1 - \alpha)$-level confidence interval $(R/b, R/c)$ by note 3 of Theorem 9.4.1 and setting $b = 1$ and choosing c by solving the equation

$$\int_c^1 f_T(t)\, dt = 1 - \alpha$$

or

$$c^{n-1}[(n-1)c - n] + \alpha = 0$$

***Example* 9.24** Let $X_1,\ldots,X_n$ be a random sample from $U(0, \theta)$. Define $R = X_{(n)} - X_{(1)}$. Find a $(1 - \alpha)$-level confidence interval for θ of the form $(R, R/c)$. Calculate the expected length of this interval and compare it with the expected length of the shortest length confidence interval $(X_{(n)}, \alpha^{-1/n} X_{(n)})$.

Solution. The joint density of $X_{(1)}$ and $X_{(n)}$ is expressed by

$$f_{X_{(1)},X_{(n)}}(x_1, x_n; \theta) = n(n-1)\frac{1}{\theta^n}(x_n - x_1)^{n-2},\ 0 < x_1 < x_n < \theta$$

The joint *pdf* of $R = X_{(n)} - X_{(1)}$ and $S = X_{(n)}$,

$$f_{R,S}(r, s; \theta) = \frac{n(n-1)}{\theta^n} r^{n-2}, r < s < \theta$$

gives the marginal *pdf* of R,

$$f_R(r; \theta) = \frac{n(n-1)}{\theta^n} r^{n-2}(\theta - r), 0 < r < \theta$$

Consider the pivot $$T(\mathbf{X}, \theta) = \frac{R}{\theta}$$

The *pdf* of T is

$$f_T(t) = n(n-1)\, t^{n-2}(1-t), \quad 0 < t < 1$$

The *pdf* of T is increasing on (0, 1). Therefore by the note 3 of Theorem 9.4.1, the $(1 - \alpha)$-level shortest length confidence interval $(c < (R/\theta) < d)$ or $((R/d), (R/c))$ is obtained as

$$\left(R, \frac{R}{c}\right) \tag{9.6.31}$$

by setting $d = 1$, in the above interval, c is chosen by solving the equation

$$n(n-1)\int_c^1 t^{n-2}(1-t)\,dt = 1 - \alpha$$

or $$c^{n-1}[(n-1)c - n] + \alpha = 0 \tag{9.6.32}$$

The expected length of this interval is

$$E\left(\frac{R}{c} - R\right) = \left(\frac{1}{c} - 1\right)\theta E\left(\frac{R}{\theta}\right)$$

$$= \left(\frac{1}{c} - 1\right)\theta n(n-1)\int_0^1 t^{n-1}(1-t)dt$$

$$= \left(\frac{1}{c} - 1\right)\theta n(n-1)B(n, 2)$$

$$= \left(\frac{1}{c} - 1\right)\left(\frac{n-1}{n+1}\theta\right) \tag{9.6.33}$$

The $(1-\alpha)$-level shortest length confidence interval for θ based on the pivot $T(\mathbf{X},\theta) = R/c$ is $(R, R/c)$, where c is obtained from Eq. (9.6.32). The expected length of this confidence interval is given in Eq. (9.6.33) where c is obtained as the solution of the Eq. (9.6.32). Note that the solution of Eq. (9.6.32) is $c = \alpha^{1/n}$ for large n. Thus, the asymptotic expected length of the above confidence interval is given by $(\alpha^{-1/n}-1)\theta$. However, for the same problem in Example, 9.15, the $(1-\alpha)$-level shortest length confidence interval for θ is obtained as $(X_{(n)}, \alpha^{-1/n}X_{(n)})$ which is based on the pivot $T(\mathbf{X},\theta) = X_{(n)}|\theta$. The expected length of this confidence interval is given by

$$E(L) = (\alpha^{-1/n} - 1)E(X_{(n)}) = (\alpha^{-1/n} - 1)\frac{n}{n+1}\theta \tag{9.6.34}$$

Note that, the asymptotic expected length of this confidence interval is given by $(\alpha^{-1/n} - 1)\theta$. Thus, we see that the asymptotic length of the confidence intervals based on the pivots R/θ and $X_{(n)}/\theta$ are same.

***Example* 9.25** Let $X_1,\ldots,X_n$ be a random sample from the population with the *pdf*

$$f(x;\theta) = \frac{1}{2\theta}\exp\left(-\frac{|x|}{\theta}\right), x \in \mathbb{R}, \theta > 0$$

Find the shortest-length confidence interval for θ based on the sufficient statistic $\Sigma_{i=1}^{n}|X_i|$.

Solution. The distribution of the sufficient statistic $\Sigma_{i=1}^{n}|X_i|$ is $G_1(n, \theta)$ since each $|X_i|$ is *iid* $G_1(1, \theta)$. Consider the pivot

$$T(\mathbf{X}, \theta) = \frac{2\sum|X_i|}{\theta}$$

$T(\mathbf{X}, \theta) \sim G_1(n, 2) \equiv \chi^2_{2n}$. The $(1-\alpha)$-level confidence interval based on T is

$$P\left(a < \frac{2\sum|X_i|}{\theta} < b\right) = 1-\alpha$$

$$P\left(\frac{2\sum|X_i|}{b} < \theta < \frac{2\sum|X_i|}{a}\right) = 1-\alpha$$

The $(1-\alpha)$-level shortest-length $L = ((1/a) - (1/b))2\Sigma|X_i|$ confidence interval is $(2\Sigma|X_i|/b, 2\Sigma|X_i|/a)$ with such choices of a and b that minimize L subject to

$$\int_a^b f_T(t)dt = 1-\alpha \tag{9.6.35}$$

where $f_T(t)$ is the *pdf* of T. We then have

$$\frac{\partial L}{\partial a} = \left(-\frac{1}{a^2} + \frac{1}{b^2}\frac{\partial b}{\partial a}\right)2\sum|X_i| = 0$$

which implies
$$b^2 f_T(b) = a^2 f_T(a)$$

or
$$b^{n+1} = a^{n+1}\exp\left(\frac{b-a}{2}\right) \tag{9.6.36}$$

Therefore, the required confidence interval is $(2\Sigma|X_i|/b,\ 2\Sigma|X_i|/a)$ where a and b are the values which satisfy Eqs. (9.6.35) and (9.6.36).

***Example* 9.26** Let $X_1,\ldots,X_n$ be a random sample from $N(\mu, 1)$ and suppose that the prior distribution of μ is a noninformative prior given by $\pi(\mu) = 1$, $\mu \in \mathbb{R}$. Construct $(1-\alpha)$-level credible interval for μ.

Solution. The posterior distribution $\pi(\mu|\mathbf{x})$ is proportional to

$$\pi(\mu|\mathbf{x}) \propto \exp\left[-\frac{n}{2}(\mu-\bar{x})^2\right]$$

i.e., $\mu|\mathbf{x} \sim N(\bar{x}, 1/n)$. Thus, the $(1-\alpha)$-level credible interval for μ is

$$P(a < \mu < b|\mathbf{x}) = 1 - \alpha$$

$$P(-z_{\alpha/2} < \sqrt{n}(\mu - \bar{x}) < z_{\alpha/2}) = 1 - \alpha$$

or
$$P\left(\bar{x} - z_{\alpha/2}/\sqrt{n} < \mu < \bar{x} + z_{\alpha/2}/\sqrt{n}\,|\mathbf{x}\right) = 1 - \alpha$$

Thus, the required credible interval for μ at level $(1-\alpha)$ is $(\bar{x} - z_{\alpha/2}/\sqrt{n},\ \bar{x} + z_{\alpha/2}/\sqrt{n})$.

***Example* 9.27** Let $X_1,\ldots,X_n$ be a random sample from geometric distribution with *pmf*

$$f(x;\theta) = \theta(1-\theta)^{x-1},\ x = 1, 2,\ldots;\ 0 < \theta < 1$$

(i) $\hat{\theta} = \bar{X}^{-1}$ is the MLE of θ. Find MLE-based asymptotic $(1-\alpha)$-level confidence interval for θ.

(ii) Assume that the prior distribution of θ is given by $\text{Be}(\alpha, \beta)$. Find the $(1-\alpha)$-level credible interval for θ.

Solution.

(i) Fisher information in a single observation

$$I_{X_1}(\theta) = E_\theta\left[-\frac{\partial^2}{\partial\theta^2}\log L(\theta, X_1)\right] = \frac{1}{\theta^2(1-\theta)}$$

gives Fisher information in the sample as

$$I_{\mathbf{X}}(\theta) = nI_{X_1}(\theta) = \frac{n}{\theta^2(1-\theta)}$$

$$\therefore \qquad V(\hat{\theta}) = \frac{1}{nI_{X_1}(\theta)} = \frac{\theta^2(1-\theta)}{n}$$

Further, Fisher information $I_{\mathbf{X}}(\theta)$ is estimated by

$$I_{\mathbf{X}}(\hat{\theta}) = \frac{n}{\hat{\theta}^2(1-\hat{\theta})} = \frac{n\overline{X}^3}{\overline{X}-1}$$

The asymptotic normality of the MLE $\hat{\theta}$,

$$\frac{\hat{\theta}-\theta}{\sqrt{\hat{\theta}^2(1-\hat{\theta})/n}} \sim N(0,1)$$

$$\frac{\overline{X}^{-1}-\theta}{\sqrt{(\overline{X}-1)/n\overline{X}^3}} \sim N(0,1)$$

provides us the basis to construct an asymptotic $(1-\alpha)$-level confidence interval for θ as

$$\left(\frac{1}{\overline{X}} - z_{\alpha/2}\sqrt{\frac{\overline{X}-1}{n\overline{X}^3}}, \frac{1}{\overline{X}} + z_{\alpha/2}\sqrt{\frac{\overline{X}-1}{n\overline{X}^3}}\right)$$

(ii) The joint distribution of $\mathbf{X}$ and θ

$$h(\mathbf{x}, \theta) \propto \theta^n (1-\theta)^{\Sigma x_i - n}\, \theta^{\alpha-1}(1-\theta)^{\beta-1}, i = 0, 1, \ldots; 0 < \theta < 1$$

Thus, the posterior distribution,

$$\pi(\theta|\mathbf{x}) \propto \theta^{n+\alpha-1}(1-\theta)^{\Sigma x_i+\beta-n-1}$$

which is $\text{Be}(n+\alpha, \Sigma x_i + \beta - n)$. The $(1-\alpha)$-level credible interval for θ is given by (a, b) satisfying

$$\int_a^b \pi(\theta|\mathbf{x})\, d\theta = 1-\alpha$$

***Example* 9.28** Let $X_1,\ldots,X_n$ be a random sample from $P(\theta)$ and let the prior distribution for θ be $G_1(\alpha, \beta)$. Find a $(1-\alpha)$-level confidence credible interval for θ.

Solution. The posterior distribution

$$\pi(\theta|\mathbf{x}) \propto \theta^{\alpha+\Sigma x_i-1} \exp\left[-\left(n+\frac{1}{\beta}\right)\theta\right]$$

Clearly, the posterior distribution $\pi(\theta|\mathbf{x})$ is $G_1(\alpha + \Sigma x_i, [n + (1/\beta)]^{-1})$. Converting it to a χ^2-distribution, consider the transformation $2[n + (1/\beta)]\theta \sim \chi^2_{2(\alpha+\Sigma x_i)}$, assuming α is some positive integer. Therefore, $(1-\alpha)$-level credible interval is

$$P\left(a < 2\left(n+\frac{1}{\beta}\right)\theta < b\right) = 1-\alpha$$

On assuming the equal probability $\alpha/2$ under each of the two tails of $\chi^2_{2(\alpha + \Sigma x_i)}$ distribution, the above interval simplifies to

$$P\left(\frac{\chi^2_{2(\alpha+\Sigma x_i),1-\alpha/2}}{2[n+(1/\beta)]} < \theta < \frac{\chi^2_{2(\alpha+\Sigma x_i),\alpha/2}}{2[n+(1/\beta)]}\right) = 1-\alpha$$

***Example* 9.29** Let $X \sim f(x;\, \theta) = (2/\theta^2)(\theta - x),\ 0 < x < \theta$. Find the shortest-length confidence interval for θ at level $(1 - \alpha)$ based on the statistic X/θ.

Solution. Consider the pivot

$$T(X, \theta) = 1 - \frac{X}{\theta}$$

The *pdf* of T is $\quad f(t) = 2t,\ 0 < t < 1$

The $(1 - \alpha)$-level confidence interval based on T is

$$P(a < T < b) = 1 - \alpha$$

$$P\left(a < 1 - \frac{X}{\theta} < b\right) = P\left(\frac{X}{1-a} < \theta < \frac{X}{1-b}\right) = 1 - \alpha$$

On minimizing $L = [(1/(1 - b)) - (1/(1 - a))]X$ subject to

$$\int_a^b f(t)dt = b^2 - a^2 = 1 - \alpha \tag{9.6.37}$$

we set

$$\frac{\partial L}{\partial a} = \left[\frac{1}{(1-b)^2}\frac{\partial b}{\partial a} - \frac{1}{(1-a)^2}\right]X = 0$$

or

$$\frac{\partial b}{\partial a} = \frac{(1-b)^2}{(1-a)^2} \tag{9.6.38}$$

On differentiating Eq. (9.6.37) with respect to a, we get

$$\frac{\partial b}{\partial a} = \frac{a}{b} \tag{9.6.39}$$

On comparing Eqs. (9.6.38) and (9.6.39), we get

$$a(1 - a)^2 = b(1 - b)^2 \tag{9.6.40}$$

Therefore, $(1 - \alpha)$-level shortest-length confidence interval for θ based on T is given by $(X/(1 - a),\ X/(1 - b))$ where a and b are chosen such that these satisfy Eqs. (9.6.37) and (9.6.40).

***Example* 9.30** Show that one can construct a $(1 - \alpha)$-level confidence interval based on the pivot of form $T(\mathbf{X}, \theta) = T_1(\mathbf{X}) - \theta$ that maximizes the confidence coefficient for the intervals of fixed length d.

(i) Use the results to construct the interval for μ when sampling is taken from $N(\mu, 1)$ population. Find the smallest sample size for which this confidence interval has a confidence coefficient not less than $1 - \alpha$.

(ii) Do the same exercise when a random sample is drawn from an exponential *pdf*

$$f(x; \theta) = \exp[-(x-\theta)],\ x > \theta$$

Solution. We can construct a confidence interval $(T_1(\mathbf{X}) - (a + d), T_1(\mathbf{X}) - a)$ of length d by choosing such an a so that the confidence coefficient

$$Q(a) = P(a < T(\mathbf{X}, \theta) < a + d)$$
$$= P(a < T_1(\mathbf{X}) - \theta < a + d) = F(a+d) - F(a)$$

is maximum.

(i) Consider the pivot $T(\mathbf{X}, \theta) = \bar{X} - \mu$. Note that $T \sim N(0, 1/n)$. The confidence coefficient

$$Q(a) = P(a < \bar{X} - \mu < a + d)$$
$$= P(\sqrt{n}a < \sqrt{n}(\bar{X} - \mu) < \sqrt{n}(a+d))$$
$$= \Phi(\sqrt{n}(a+d)) - \Phi(\sqrt{n}a)$$

$Q(a)$ is maximum when $a + d = d/2$ or $a = -d/2$. Therefore, the required confidence interval is $(T_1(\mathbf{X}) - (a + d), T_1(\mathbf{X}) - a) = (\bar{X} - (d/2), \bar{X} + (d/2))$. For smallest sample size, so that

$$P\left(-\frac{\sqrt{n}}{2}d < N(0,1) < \frac{\sqrt{n}}{2}d\right) \ge 1 - \alpha$$

we must have

$$\frac{\sqrt{n}}{2}d \ge z_{\alpha/2}$$

or

$$n \ge \left[4\left(\frac{z_{\alpha/2}}{d}\right)^2\right]$$

(ii) Pivot, in this case, is considered as

$$T(\mathbf{X}, \theta) = X_{(1)} - \theta$$

with *pdf*

$$f(t) = n\exp(-nt), \quad t > 0$$

The required confidence interval is given by $(X_{(1)} - (a + d), X_{(1)} - a)$ so that the confidence coefficient

$$Q(a) = P(a < X_{(1)} - \theta < a + d)$$

is maximum. Note that $Q(a)$ is maximum when $a = 0$. So, the required confidence interval is $(X_{(1)} - d, X_{(1)})$. The smallest sample size for which the confidence interval to have a confidence coefficient not less than $1 - \alpha$ is

$$\int_0^d f(t)dt \geq 1-\alpha$$

$$n\int_0^d \exp(-nt)dt \geq 1-\alpha$$

$$n \geq \left(\frac{1}{d}\right)\log\left(\frac{1}{\alpha}\right)$$

***Example* 9.31** Let X and Y be independent random variables with the *pdf* $\lambda\exp(-\lambda x)$, $x > 0$, and $\mu\exp(-\mu y)$, $y > 0$, respectively. Find a $(1-\alpha)$-level confidence region for (λ, μ) of the form $\{(\lambda, \mu): \lambda X + \mu Y \leq k\}$.

Solution. The $(1-\alpha)$-level confidence region for (λ, μ) is determined by prescribing k so that

$$\int_{x=0}^{k/\lambda} f(x;\lambda) \int_{y=0}^{(k-\lambda x)/\mu} f(y;\mu)dy\,dx = 1-\alpha$$

$$\lambda\mu \int_{x=0}^{k/\lambda} \exp(-\lambda x) \int_{y=0}^{(k-\lambda x)/\mu} \exp(-\mu y)dy\,dx = 1-\alpha$$

$$\lambda \int_{x=0}^{k/\lambda} \exp(-\lambda x)\{1-\exp[-(k-\lambda x)]\}dx = 1-\alpha$$

$$\lambda \int_{x=0}^{k/\lambda} [\exp(-\lambda x)-\exp(-k)]dx = 1-\alpha$$

$$1-\exp(-k)-k\exp(-k) = 1-\alpha$$

$$(k+1)\exp(-k) = \alpha$$

***Example* 9.32** Let $X_1,\ldots,X_n$ be a random sample drawn from

(i) $G_2(1, \theta)$. Approximate $(1-\alpha)$-level confidence interval for θ based on the MLE of θ and by using Chebyshev's inequality.

(ii) $P(\theta)$. Approximate $(1-\alpha)$-level confidence interval for θ based on the MLE of θ.

Solution.

(i) We know that $\bar{X}^{-1}$ is the MLE of θ and $I_{\mathbf{X}}(\theta) = n/\theta^2$ or $V(\bar{X}^{-1}) = [I_{\mathbf{X}}(\theta)]^{-1} = \theta^2/n$. Then, under certain regularity conditions and for large n,

$$\frac{\bar{X}^{-1}-\theta}{\sqrt{V(\bar{X}^{-1})}} \xrightarrow{d} N(0,1) \quad \text{as } n\to\infty$$

$$\frac{\sqrt{n}(\bar{X}^{-1}-\theta)}{\theta} \xrightarrow{d} N(0,1) \quad \text{as } n\to\infty$$

The (1 – α)-level confidence interval for θ may be approximated by

$$P_\theta\left[-z_{\alpha/2} < \frac{\sqrt{n}(\bar{X}^{-1}-\theta)}{\theta} < z_{\alpha/2}\right] = 1-\alpha$$

$$P_\theta\left\{\left[\bar{X}\left(1+\frac{z_{\alpha/2}}{\sqrt{n}}\right)\right]^{-1} < \theta < \left[\bar{X}\left(1-\frac{z_{\alpha/2}}{\sqrt{n}}\right)\right]^{-1}\right\} = 1-\alpha$$

(ii) In this case, $\bar{X}$ is the MLE of θ and $I_{\mathbf{X}}(\theta) = n/\theta$ or $V(\bar{X}) = 1/I_{\mathbf{X}}(\theta) = \theta/n$. For large n,

$$\frac{\sqrt{n}(\bar{X}-\theta)}{\theta} \xrightarrow{d} n(0,1) \quad \text{as } n \to \infty$$

For large n, the (1 – α)-level confidence interval for θ may be approximated by

$$P\left[-z_{\alpha/2} < \frac{\sqrt{n}(\bar{X}-\theta)}{\theta} < z_{\alpha/2}\right] = 1-\alpha$$

$$P\left[\left(1-\frac{z_{\alpha/2}}{\sqrt{n}}\right) < \frac{\bar{X}}{\theta} < \left(1+\frac{z_{\alpha/2}}{\sqrt{n}}\right)\right] = 1-\alpha$$

$$P\left[\frac{\bar{X}}{1+(z_{\alpha/2}/\sqrt{n})} < \theta < \frac{\bar{X}}{1-(z_{\alpha/2}/\sqrt{n})}\right] = 1-\alpha$$

***Example* 9.33** Let $X_1,\ldots,X_n$ be a random sample drawn from the uniform distribution on N points. Find an upper (1 – α)-level confidence bound for N based on the maximum statistic $X_{(n)}$.

Solution. The distribution of $X_{(n)}$ is given by

$$\begin{aligned} P[X_{(n)} = x] &= F_n(x) - F_n(x-1) \\ &= P[X_{(n)} \le x] - P[X_{(n)} \le x-1] \\ &= \left(\frac{x}{N}\right)^n - \left(\frac{x-1}{N}\right)^n;\ x = 1,\ldots,N \end{aligned}$$

Consider the random variable

$$T(\mathbf{X}, N) = \frac{X_{(n)}}{N}$$

The distribution of T is

$$P[T = t] = \left[(t)^n - \left(t-\frac{1}{N}\right)^n\right];\ t = \frac{1}{N},\ldots,1$$

Consider $(1 - \alpha)$-level confidence upper bound

$$P(a < T) = 1 - \alpha$$

$$P\left(a < \frac{X_{(n)}}{N}\right) = 1 - \alpha$$

$$P\left(N < \frac{X_{(n)}}{a}\right) = 1 - \alpha$$

Therefore, the $(1-\alpha)$-level confidence upper bound is given by $(0, X_{(n)}/a)$ where a is obtained by solving

$$\sum_{t=a}^{1}\left[t^n - \left(t - \frac{1}{N}\right)^n\right] = 1 - \alpha$$

***Example* 9.34** Let X follows one-parameter exponential family. Consider the testing problem $(\alpha, H_0(\Theta_0), H_1(\Theta_1))$ defined as

$$H_0: \theta = \theta_0 \text{ against } H_1: \theta < \theta_0$$

Find a UMP size-α acceptance region and the corresponding UMA confidence sets at level $1 - \alpha$.

Solution. The density function of the one-parameter exponential family is given by

$$f_X(x|\theta) = c(\theta)h(x)\exp(Q(\theta)T(x))$$

where, $Q(\theta)$ is a nondecreasing function of θ. On applying the Karlin and Rubin theorem, the UMP size-α test for testing H_0: $\theta = \theta_0$ against H_1: $\theta < \theta_0$ is given by

$$\phi_0(T(x)) = \begin{cases} 1, & \text{if } T \le t \\ 0, & \text{if } T > t \end{cases}$$

where t is obtained by

$$E_{\theta_0}\phi(T) = \alpha \quad \text{or} \quad P_{\theta_0}(T \le t) = \alpha$$

This shows that t is a function of θ_0. The corresponding UMP size-α acceptance region is

$$A(\theta_0) = \{x: T > t(\theta_0)\}$$

Since $A(\theta_0)$ is UMP size-α acceptance region,

$$P_{\theta'}(T(X) \le t(\theta_0)) \ge P_{\theta_0}(T(X) \le t(\theta_0)) = \alpha \;\forall\; \theta' < \theta_0, \text{ i.e., } \theta' \in H_1(\Theta_1)$$

But $$P_{\theta'}(T(X) \le t(\theta')) = \alpha$$

This gives $$P_{\theta'}(T(X) \le t(\theta_0)) \ge P_{\theta'}(T(X) \le t(\theta')) \text{ for } \theta' < \theta_0$$

which implies $t(\theta_0) \geq t(\theta')$. This shows that $t(\theta)$ is a nondecreasing function of θ. We have, now,

$$S(x) = \{\theta: x \in A(\theta)\} = \{\theta: T(x) > t(\theta)\}$$

$$= \{\theta: \theta < t^{-1}[T(x)]\} \qquad \text{(since } t \text{ is nondecreasing)}$$

$$= (-\infty, t^{-1}(T(x)))$$

where t^{-1} is defined by

$$t^{-1}(T(x)) = \sup_{\theta} \{\theta: t(\theta) \leq T(x)\}$$

***Example* 9.35** Let X be a random variable with density

$$f(x;\theta) = \begin{cases} \dfrac{1}{\theta}\exp\left(-\dfrac{x}{\theta}\right), & \text{if } x > 0 \\ 0, & \text{otherwise} \end{cases}$$

where $\theta > 0$. Consider the testing problem $H_0\colon \theta = \theta_0$ against $H_1\colon \theta < \theta_0$. Find the UMA $(1-\alpha)$-level confidence interval for θ corresponding to the size-α UMP test.

Solution. The given family belongs to the one-parameter exponential family. The UMP size-α acceptance region is given by

$$A(\theta) = \{x\colon T(x) \geq c(\theta)\}$$

$$= \{x\colon x \geq c(\theta)\}$$

where we choose $c(\theta)$ by

$$P_{\theta_0}(A(\theta_0)) = 1 - \alpha$$

or

$$\int_0^{c(\theta_0)} \frac{1}{\theta_0}\exp\left(-\frac{x}{\theta_0}\right)dx = \alpha$$

or

$$c(\theta_0) = \theta_0 \log\frac{1}{1-\alpha}, \; 0 < \alpha < 1$$

Therefore, the corresponding UMA $(1-\alpha)$-level of confidence interval for θ is given by

$$S(x) = \{\theta: x \in A(\theta)\} = \left\{\theta: x \geq \theta \log\frac{1}{1-\alpha}\right\}$$

$$= \left\{\theta: \theta \leq \frac{x}{\log[1/(1-\alpha)]}\right\} = \left(0, \frac{x}{\log[1/(1-\alpha)]}\right] \qquad \text{(since } \theta > 0\text{)}$$

***Example* 9.36** Let $X_1,\ldots,X_n$ be a sample from $N(\mu, \sigma^2)$ where both μ and σ^2 are unknown. For testing $H_0\colon \mu = \mu_0$ against $H_1\colon \mu \neq \mu_0$, find the UMA unbiased $(1-\alpha)$-level confidence interval for μ.

Solution. The UMP unbiased size-α test for the testing $H_0: \mu = \mu_0$ against $H_1: \mu \neq \mu_0$ is given by

$$\phi(\mathbf{x}) = \begin{cases} 1, & \text{if} \left|\sqrt{n-1}\dfrac{(\bar{x}-\mu_0)}{s_n}\right| > c \\ 0, & \text{otherwise} \end{cases} \tag{9.6.41}$$

where $\bar{x} = (1/n)\Sigma x_i$, $s_n^2 = (1/n)\Sigma(x_i - \bar{x})^2$. The constant c is obtained by the size condition

$$E_{\mu_0}(\phi(\mathbf{X})) = \alpha$$

or
$$E_{\mu_0}(1 - \phi(\mathbf{X})) = 1 - \alpha$$

or
$$P_{\mu_0}\left[\left|\frac{\sqrt{n-1}(\bar{X}-\mu_0)}{S_n}\right| \le c\right] = 1-\alpha$$

where
$$\frac{(\bar{X}-\mu_0)/(\sigma/\sqrt{n})}{\sqrt{nS_n^2/[\sigma^2(n-1)]}} = \frac{\sqrt{(n-1)}(\bar{X}-\mu_0)}{S_n} \sim t_{n-1}$$

This yields $c = t_{n-1,\alpha/2}$. Thus, the acceptance region corresponding to UMP unbiased size-α test for testing $H_0: \mu = \mu_0$ against $H_1: \mu \neq \mu_0$ in Eq. (9.6.41) is

$$A(\mu_0) = \left\{\mathbf{x}: \left|\frac{\sqrt{n-1}(\bar{x}-\mu_0)}{S_n}\right| \le t_{n-1,\alpha/2}\right\}$$

By Theorem 9.4.2, the corresponding UMAU $(1 - \alpha)$-level confidence interval for μ is

$$S(\mathbf{x}) = \{\mu: \mathbf{x} \in A(\mu)\}$$

$$= \left\{\mu: \left|\frac{\sqrt{n-1}(\bar{x}-\mu_0)}{S_n}\right| \le t_{n-1,\alpha/2}\right\}$$

$$= \left(-t_{n-1,\alpha/2} \le \frac{\sqrt{n-1}(\bar{x}-\mu_0)}{S_n} \le t_{n-1,\alpha/2}\right)$$

$$= \left(\bar{x} - \frac{S_n}{\sqrt{n-1}}t_{n-1,\alpha/2} \le \mu \le \bar{x} + \frac{S_n}{\sqrt{n-1}}t_{n-1,\alpha/2}\right)$$

***Example* 9.37** Let $X_1,\ldots,X_n$ be a random sample from $N(\mu, \sigma^2)$ where σ^2 is known for testing $H_0: \mu = \mu_0$ against $H_1: \mu \neq \mu_0$. Find the UMA $(1 - \alpha)$-level confidence interval for μ.

Solution. For testing the hypothesis $H_0: \mu = \mu_0$ against $H_1: \mu \neq \mu_0$, the UMP unbiased size-α test is given by

$$\phi(\mathbf{x}) = \begin{cases} 1, & \text{if} \left|\dfrac{\sqrt{n}(\bar{x}-\mu_0)}{\sigma}\right| > c \\ 0, & \text{otherwise} \end{cases}$$

This test is known as the z-test. The constant c is determined by the size condition

$$E_{\mu_0}[\phi(\mathbf{X})] = \alpha$$

or

$$P_{\mu_0}\left(\frac{\sqrt{n}\,|\bar{X}-\mu_0|}{\sigma} > c\right) = \frac{\alpha}{2}$$

which gives

$$c = z_{\alpha/2}$$

Thus, the acceptance region corresponding to this UMP unbiased size-α test is given by

$$A(\mu_0) = \left\{\mathbf{x}: \left|\frac{\sqrt{n}(\bar{x}-\mu_0)}{\sigma}\right| \le z_{\alpha/2}\right\}$$

By utilizing Theorem 9.4.2, the UMA $(1-\alpha)$-level unbiased confidence interval for μ is given by

$$S(\mathbf{x}) = \{\mu : \mathbf{x} \in A(\mu)\}$$

$$= \left(-z_{\alpha/2} \le \frac{|\sqrt{n}(\bar{x}-\mu)|}{\sigma} \le z_{\alpha/2}\right)$$

$$= \left(-\frac{\sigma}{\sqrt{n}} z_{\alpha/2} \le (\mu - \bar{x}) \le \frac{\sigma}{\sqrt{n}} z_{\alpha/2}\right)$$

$$= \left(\bar{x} - \frac{\sigma}{\sqrt{n}} z_{\alpha/2} \le \mu \le \bar{x} + \frac{\sigma}{\sqrt{n}} z_{\alpha/2}\right)$$

***Example* 9.38** Let $X_1,\ldots,X_n$ be a sample from $U(0, \theta)$. Consider the testing problem $H_0: \mu = \mu_0$ against $H_1: \mu \ne \mu_0$.

Find the $(1-\alpha)$-level UMA confidence intervals for θ.

Solution. The UMP unbiased size-α test for the above testing problem is given by

$$\phi(\mathbf{x}) = \begin{cases} 1, & \text{if } \max(x_1,\ldots,x_n) < c_1 \\ & \text{or } \max(x_1,\ldots,x_n) > c_2 \\ 0, & \text{otherwise} \end{cases}$$

Let us denote $X_{(n)} = \max(X_1,\ldots,X_n)$ and let c_1 and c_2 be obtained by the size condition

$$E_{\theta_0}[\phi(X_{(n)})] = \alpha \quad \text{or} \quad \int_{c_1}^{c_2} g_{X_{(n)}}(y)\Big|_{\theta=\theta_0} dy = 1-\alpha$$

where the density of $X_{(n)}$ is given by $g_{X_{(n)}}(y) = ny^{n-1}\theta_0^{-n}$ under H_0. Let α be divided equally in tails. We get c_1 and c_2 such that

$$\int_0^{c_1} ny^{n-1}\theta_0^{-n}\,dy = \frac{\alpha}{2} = \int_{c_2}^{\theta_0} ny^{n-1}\theta_0^{-n}\,dy$$

We have
$$c_1^n = \theta_0^n\left(\frac{\alpha}{2}\right) \quad \text{or} \quad c_1 = \theta_0\left(\frac{\alpha}{2}\right)^{1/n}$$

and
$$\theta_0^n - c_2^n = \theta_0^n\left(\frac{\alpha}{2}\right)$$

$$c_2^n = \left(1-\frac{\alpha}{2}\right)\theta_0^n \quad \text{or} \quad c_2 = \theta_0\left(1-\frac{\alpha}{2}\right)^{1/n}$$

The corresponding acceptance region is given by

$$A(\theta_0) = \left\{\mathbf{x}: \theta_0\left(\frac{\alpha}{2}\right)^{1/n} \le X_{(n)} \le \theta_0\left(1-\frac{\alpha}{2}\right)^{1/n}\right\}$$

Utilizing the result of Theorem 9.4.2, the UMA (1 – α)-level unbiased confidence interval for θ is given by

$$S(\mathbf{x}) = \{\theta: \mathbf{x} \in A(\theta)\}$$

$$= \left\{\mathbf{x}: \theta\left(\frac{\alpha}{2}\right)^{1/n} \le X_{(n)} \le \theta\left(1-\frac{\alpha}{2}\right)^{1/n}\right\}$$

$$= \left(\left(\frac{\alpha}{2}\right)^{1/n} \le \frac{X_{(n)}}{\theta} \le \left(1-\frac{\alpha}{2}\right)^{1/n}\right)$$

$$= \left(\frac{1}{(1-\alpha/2)^{1/n}} \le \frac{\theta}{X_{(n)}} \le \frac{1}{(\alpha/2)^{1/n}}\right)$$

$$= \left(\frac{X_{(n)}}{(1-\alpha/2)^{1/n}} \le \theta \le \frac{X_{(n)}}{(\alpha/2)^{1/n}}\right)$$

***Example* 9.39** Let $X_1,\ldots,X_n$ be a random sample from $N(\mu_0, \sigma^2)$, where σ^2 is unknown and $\mu = \mu_0$ is known. Consider the hypothesis

$$H_0: \sigma = \sigma_0 \text{ against } H_1: \sigma \ne \sigma_0$$

and find the UMA (1 – α)-level unbiased confidence interval for σ^2.

Solution. The UMP unbiased test for the testing problem $H_0: \sigma = \sigma_0$ against $H_1: \sigma \ne \sigma_0$ is given by

$$\phi(\mathbf{x}) = \begin{cases} 1, & \text{if } \dfrac{\sum(x_i-\mu_0)^2}{\sigma^2} < c_1 \\ & \text{or } \dfrac{\sum(x_i-\mu_0)^2}{\sigma^2} > c_2 \\ 0, & \text{otherwise} \end{cases}$$

where constants c_1 and c_2 are determined by the size condition

$$E_{\sigma_0}[\phi(\mathbf{X})] = \alpha$$

or
$$P_{\sigma_0}\left[c_1 \le \frac{\sum (X_i - \mu_0)^2}{\sigma_0^2} \le c_2\right] = 1 - \alpha$$

or
$$\int_0^{c_1} \chi_n^2(y)dy = \frac{\alpha}{2} = \int_{c_2}^{\infty} \chi_n^2(y)dy$$

This gives
$$c_1 = \chi^2_{n,(1-\alpha/2)};\ c_2 = \chi^2_{n,\,\alpha/2}$$

Thus, the acceptance region corresponding to UMP unbiased size-α test $\phi(\mathbf{x})$ is given by

$$A(\sigma_0) = \left\{\mathbf{x} : \chi^2_{n,(1-\alpha/2)} \le \frac{\sum (x_i - \mu_0)^2}{\sigma^2} \le \chi^2_{n,\alpha/2}\right\}$$

Further, by utilizing the result of Theorem 9.4.2, UMAU $(1 - \alpha)$-level confidence interval for σ is given by

$$S(\mathbf{x}) = \{\sigma : \mathbf{x} \in A(\sigma)\}$$

$$= \left(\chi^2_{n,1-\alpha/2} \le \frac{\sum (x_i - \mu_0)^2}{\sigma^2} \le \chi^2_{n,\alpha/2}\right)$$

$$= \left(\frac{\sum (x_i - \mu_0)^2}{\chi^2_{n,\alpha/2}} \le \sigma^2 \le \frac{\sum (x_i - \mu_0)^2}{\chi^2_{n,\,1-\alpha/2}}\right)$$

***Example* 9.40** Let $X_1,\ldots,X_m$ and $Y_1,\ldots,Y_n$ be independent samples from $N(\mu, \sigma^2)$ and $N(\theta, \sigma^2)$, respectively. Consider the testing problem

$$H_0: \mu - \theta = \delta \text{ against } H_1: \mu - \theta \ne \delta$$

Find UMA $(1 - \alpha)$-level unbiased confidence interval for $(\mu - \theta)$.

Solution. Here, μ, θ, and σ^2 are unknown. We have already shown the following test as the UMP unbiased of size-α

$$\phi(\mathbf{x}, \mathbf{y}) = \begin{cases} 1, & \text{if } \left| \dfrac{(\bar{x} - \bar{y} - \delta)/\sqrt{(1/m) + (1/n)}}{\sqrt{\left(\sum (x_i - \bar{x})^2 + \sum (y_i - \bar{y})^2\right)/(m+n-2)}} \right| > c \\ 0, & \text{otherwise} \end{cases} \qquad (9.6.42)$$

The constant c is determined by the size condition

$$E_{\delta}[\phi(\mathbf{X}, \mathbf{Y})] = \alpha$$

$$P_\delta\left[\frac{|\bar{X}-\bar{Y}-\delta|/\sqrt{(1/m)+(1/n)}}{S_p}>c\right]=\alpha$$

where $$S_p^2=\frac{1}{m+n-2}\left[\sum(X_i-\bar{X})^2+\sum(Y_i-\bar{Y})^2\right]$$

and $$c=t_{m+n-2,\alpha/2}$$

Therefore, the acceptance region corresponding to the UMP unbiased test in Eq. (9.6.42) is given by

$$A(\delta)=\left\{(\mathbf{x},\mathbf{y}):\left|\frac{|\bar{X}-\bar{Y}-\delta|/\sqrt{(1/m)+(1/n)}}{S_p}\right|\le t_{m+n-2,\alpha/2}\right\}$$

By utilizing the result of Theorem 9.4.2, the UMAU $(1-\alpha)$-level family of confidence sets $S(\mathbf{X})$ is given by

$$S(\mathbf{x},\mathbf{y})=\{\delta:(\mathbf{x},\mathbf{y})\in A(\delta)\}$$

$$=\left\{-t_{m+n-2,\alpha/2}\le\frac{(\bar{x}-\bar{y})-\delta}{S_p\sqrt{(1/m)+(1/n)}}\le t_{m+n-2,\alpha/2}\right\}$$

$$=\left((\bar{x}-\bar{y})-S_p\sqrt{\frac{1}{m}+\frac{1}{n}}\;t_{m+n-2,\alpha/2}\le(\mu-\theta)\le(\bar{x}-\bar{y})+S_p\sqrt{\frac{1}{m}+\frac{1}{n}}\;t_{m+n-2,\alpha/2}\right)$$

***Example* 9.41** Let $X_1,\ldots,X_n$ be a random sample from $N(\mu,\sigma^2)$, where, μ and σ^2 are both unknown. Show that

$$S(\mathbf{X})=\left(\bar{X}-t_{n-1,\alpha/2}\frac{S_n}{\sqrt{n-1}},\bar{X}+t_{n-1,\alpha/2}\frac{S_n}{\sqrt{n-1}}\right)$$

is the minimum expected-length confidence interval.

Solution. Let us consider the pivot

$$T(\mathbf{X},\mu)=\frac{\bar{X}-\mu}{S_n/\sqrt{n-1}}\sim t_{n-1}$$

where $$S_n^2=\frac{1}{n}\sum_{i=1}^{n}(X_i-\bar{X})^2$$

We know from the theory of confidence interval estimation that the given confidence interval $S(\mathbf{X})$ is of the shortest length and that this is UMAU confidence interval that corresponds to the UMP unbiased test of size α. Therefore, by utilizing the results of Theorem 9.4.3, $S(\mathbf{X})$ is the confidence interval with minimum expected length, i.e., $E_\theta^{\mathbf{X}}S(\mathbf{X})$.

***Example* 9.42** Let $X_1,\ldots,X_n$ be a random sample drawn from $N(\mu, \sigma^2)$.

(i) Find $(1-\alpha)$-level unbiased confidence intervals for σ^2 by using the Guenther procedure.
(ii) Find $(1-\alpha)$-level shortest length confidence intervals for σ^2.
(iii) Discuss whether the unbiased confidence interval for σ^2 obtained in (i) minimizes the expected length, i.e., $E[S(\mathbf{X})]$.

Solution.

(i) Consider the pivot

$$T(\mathbf{X},\sigma^2)=\frac{nS_n^2}{\sigma^2}\sim\chi^2_{n-1}\quad\text{where}\quad S_n^2=\frac{1}{n}\sum_{i=1}^{n}(X_i-\bar{X})^2$$

We have
$$P_{\sigma^2}\left(\lambda_1<\frac{nS_n^2}{\sigma^2}<\lambda_2\right)=1-\alpha$$

or
$$P_{\sigma^2}\left(n\frac{S_n^2}{\lambda_2}<\sigma^2<n\frac{S_n^2}{\lambda_1}\right)=1-\alpha$$

Consider
$$P(\theta,\theta')=P(\sigma^2,\sigma'^2)=P_{\sigma^2}\left(n\frac{S_n^2}{\lambda_2}<\sigma'^2<n\frac{S_n^2}{\lambda_1}\right)$$

Let
$$\gamma=\frac{\sigma'^2}{\sigma^2};\;P(\sigma^2,\sigma'^2)=P_{\sigma^2}\left(\frac{1}{\lambda_2}\frac{nS_n^2}{\sigma^2}<\frac{\sigma'^2}{\sigma^2}<\frac{1}{\lambda_1}\frac{nS_n^2}{\sigma^2}\right)$$

$$P(\sigma^2,\sigma'^2)=P_{\sigma^2}\left(\frac{T(\mathbf{X},\sigma^2)}{\lambda_2}<\gamma<\frac{T(\mathbf{X},\sigma^2)}{\lambda_1}\right)$$

or
$$P(\gamma)=P(\lambda_1\gamma<T(\mathbf{X},\sigma^2)<\lambda_2\gamma)$$

We have
$$P(1)=1-\alpha\quad\text{if}\quad\sigma'^2=\sigma^2$$

$$\frac{\partial}{\partial\gamma}P(\gamma)\big|_{\gamma=1}=\lambda_2 f_{n-1}(\lambda_2)-\lambda_1 f_{n-1}(\lambda_1)=0\tag{9.6.43}$$

where f_{n-1} is the *pdf* of $T(\mathbf{X},\sigma^2)$ which is χ^2_{n-1}. Tate and Klett (1959) have solved Eq. (9.6.43) for λ_1 and λ_2 numerically.
Therefore, $(1-\alpha)$-level unbiased confidence interval is given by

$$\left(n\frac{S_n^2}{\lambda_2},\,n\frac{S_n^2}{\lambda_1}\right)$$

(ii) The interval $(nS_n^2/\lambda_2,\ nS_n^2/\lambda_1)$ is a $(1-\alpha)$-level confidence interval for σ^2 so that

$$\int_{\lambda_1}^{\lambda_2} f_{n-1}(t)\,dt=1-\alpha\tag{9.6.44}$$

or $$f_{n-1}(\lambda_2)\frac{\partial \lambda_2}{\partial \lambda_1} = f_{n-1}(\lambda_1)$$

The length of this interval $L = n((1/\lambda_1) - (1/\lambda_2))S_n^2$ is minimized for λ_1 and λ_2 so that

$$\frac{\partial L}{\partial \lambda_1} = 0 \quad \Rightarrow \quad -\frac{1}{\lambda_1^2}\frac{\partial \lambda_1}{\partial \lambda_2} + \frac{1}{\lambda_2^2} = 0$$

or $$\frac{\partial \lambda_2}{\partial \lambda_1} = \frac{\lambda_2^2}{\lambda_1^2}$$

This gives $$\lambda_1^2 f_{n-1}(\lambda_1) = \lambda_2^2 f_{n-1}(\lambda_2) \tag{9.6.45}$$

Thus, $(1 - \alpha)$-level confidence interval of the shortest length for σ^2 is $(nS_n^2/\lambda_2, nS_n^2/\lambda_1)$ where λ_1 and λ_2 are the solutions of Eqs. (9.6.43) and (9.6.45).

(iii) The test

$$\phi_0(\mathbf{x}) = \begin{cases} 1, & \text{if } \dfrac{\sum (x_i - \bar{x})^2}{\sigma_0^2} < \lambda_1 \text{ or } > \lambda_2 \\ 0, & \text{otherwise} \end{cases} \tag{9.6.46}$$

is a UMP unbiased size-α test for testing the hypotheses $H_0: \sigma = \sigma_0$, $\mu \in \mathbb{R}^1$, against $H_0: \sigma \neq \sigma_0$, $\mu \in \mathbb{R}^1$. The constants λ_1 and λ_2 involved in the test are obtained by the following two conditions

$$E_{\sigma_0}[\phi_0(\mathbf{X})] = \alpha \tag{9.6.47}$$

and $$E_{\sigma_0}[\mathbf{X}\phi_0(\mathbf{X})] = \alpha E_{\sigma_0}[\mathbf{X}] \tag{9.6.48}$$

We know that $\Sigma(X_i - \bar{X})^2/\sigma^2 \,|_{\sigma^2=\sigma_0^2} \sim \chi_{n-1}^2$. Let f_{n-1} be the *pdf* of χ_{n-1}^2-distribution. Then, Eqs. (9.6.47) and (9.6.48) reduce to

$$\int_{\lambda_1}^{\lambda_2} f_{n-1}(y)dy = 1 - \alpha \tag{9.6.49}$$

$$\frac{1}{(n-1)}\int_{\lambda_1}^{\lambda_2} y f_{n-1}(y)dy = 1 - \alpha \tag{9.6.50}$$

respectively. On using the identity

$$y f_{n-1}(y) = (n-1) f_{n+1}(y)$$

in Eq. (9.6.50) along with Eqs. (9.6.47) and (9.6.50) it reduces to

$$\int_{\lambda_1}^{\lambda_2} f_{n+1}(y)dy = 1 - \alpha = \int_{\lambda_1}^{\lambda_2} f_{n-1}(y)dy$$

In practice, α is divided equally under the tails of χ^2_{n-1} distribution. In this case, we get λ_1 and λ_2 so that

$$\int_0^{\lambda_2} f_{n-1}(y)dy = \int_{\lambda_2}^{\infty} f_{n-1}(y)dy = \frac{\alpha}{2}$$

$$\Rightarrow \qquad \lambda_1 = \chi^2_{n-1,\,1-(\alpha/2)};\ \lambda_2 = \chi^2_{n-1,\,(\alpha/2)}$$

We have the acceptance region corresponding to the UMPU size-α test in Eqs. (9.6.46), (9.6.47), and (9.6.48) given by

$$A(\sigma_0) = \left\{\mathbf{x}: \lambda_1 < \frac{nS_n^2}{\sigma_0^2} < \lambda_2\right\}$$

Therefore, the corresponding UMAU (1 – α)-level confidence interval is given by

$$S(\mathbf{x}) = \left(n\frac{S_n^2}{\lambda_2}, n\frac{S_n^2}{\lambda_1}\right) \tag{9.6.51}$$

Note that λ_1 and λ_2 are obtained by the conditions

$$\int_{\lambda_1}^{\lambda_2} f_{n-1}(y)dy = 1-\alpha \tag{9.6.52}$$

or $$P_{\sigma_0^2}\left(\lambda_1 < n\frac{S_n^2}{\sigma_0^2} < \lambda_2\right) = 1-\alpha \equiv P(1) = 1-\alpha$$

and, on differentiating Eq. (9.6.52) with respect to y, we get

$$\lambda_2 f_{n-1}(\lambda_2) - \lambda_1 f_{n-1}(\lambda_1) = 0 \tag{9.6.53}$$

Note that UMAU (1 – α)-level confidence interval in Eq. (9.6.51) is obtained by choosing such λ_1 and λ_2 which satisfy Eqs. (9.6.52) and (9.6.53). This interval minimizes the expected length, i.e., $E(S(\mathbf{X}))$.

Therefore, we see that the family of (1 – α)-level unbiased confidence interval for σ^2, following Guenther's procedure obtained in (i) and the UMAU (1 – α)-level confidence interval, are both identical. Further, the (1 – α)-level shortest length confidence interval in (ii) is obtained by getting λ_1 and λ_2 so that

$$P(\lambda_1 < \chi^2_{n-1} < \lambda_2) = 1 - \alpha$$

and $$\lambda_1^2 f_{n-1}(\lambda_1) = \lambda_2^2 f_{n-1}(\lambda_2)$$

If we compare the conditions for obtaining the UMAU (1 – α)-level minimum expected length confidence interval and that of the shortest length confidence interval, the length of the former is considerably large as compared to the latter. However, it has been investigated that for large n, both the lengths of confidence intervals are approximately the same.

EXERCISES

1. Explain the problem of interval estimation. Describe a general method of obtaining confidence intervals. Also, mention other methods for finding confidence interval.
2. Explain how the method of confidence intervals is connected with the theory of testing of hypotheses.
3. Describe the procedure for setting distribution-free confidence interval for population quartile.
4. Obtain a $100(1-\alpha)\%$ confidence interval for the population median based on the sign test.
5. In a random sample of size 200 (without replacement) from a population of size 1000, 75 persons happened to be smokers. Estimate the proportion of smokers and the total number of smokers in the population and calculate 95% confidence interval for the proportion of smokers and also for the total number of smokers in the population. (Given that for 95% confidence probability, $t = 1.96$.)
6. Suppose that the coefficient of variation of the income of household in a region with 2000 households is 75%. How large random sample of households is required for a margin of error ±5% with 95% confidence coefficient. (For 95% confidence probability, $t = 1.96$.)
7. Let $X_1,\ldots,X_n$ be *iid* observations from the Bernoulli distribution $b(1, p)$. Show by inverting the acceptance region of the UMP size-α test for testing that

$$H_0: p = p_0 \text{ against } H_1: p > p_0$$

gives $(1-\alpha)$-level confidence set

$$S(t) = \{p_0 : t \le k(p_0)\}$$

where $k(p_0)$ is the integer between 0 and n that simultaneously satisfies the conditions

$$\sum_{y=0}^{k(p_0)} \binom{n}{y} p_0^y (1-p_0)^{n-y} \ge 1-\alpha$$

and

$$\sum_{y=0}^{k(p_0)-1} \binom{n}{y} p_0^y (1-p_0)^{n-y} < 1-\alpha$$

Show that the confidence set $S(t)$ is a $(1-\alpha)$-level lower confidence bound of the form

$$\begin{aligned} S(t) &= \{p_0 : t \le k(p_0)\} \\ &= \{p_0 : p_0 > k^{-1}(t)\} \\ &= (k^{-1}(t), 1) \end{aligned}$$

where $k^{-1}(t)$ is obtained by

$$k^{-1}(t) = \sup\left\{p: \sum_{y=0}^{t-1}\binom{n}{y}p^y(1-p)^{n-y} \ge 1-\alpha\right\}$$

Also, show by inverting the acceptance region of size-α UMP test for testing that $H_0{:}p = p_0$ against $H_1{:}p < p_0$ gives $(1 - \alpha)$-level upper confidence bound on p

$$S(t) = [p_0{:}\, t \ge k(p_0)]$$

where $k(p_0)$ satisfies simultaneously

$$\sum_{y=k(p_0)}^{n}\binom{n}{y}p_0^y(1-p_0)^{n-y} \ge 1-\alpha$$

and $$\sum_{k(p_0)+1}^{n}\binom{n}{y}p_0^y(1-p_0)^{n-y} < 1-\alpha$$

8. Let $X_1,\ldots,X_n$ be *iid* Bernoulli random variable $b(1, p)$, $0 < p < 1$. Use the *pdf* of $T = \Sigma_{i=1}^n X_i$ to construct $(1 - \alpha)$-level confidence interval for p

$$\left(\frac{1}{1+[(n-T+1)/T]F_{2(n-T+1),2T,\alpha_2}}, \frac{[(T+1)/(n-T)]F_{2(T+1),\,2(n-T),\alpha_1}}{1+[(T+1)/(n-T)]F_{2(T+1),2(n-T),\alpha_1}}\right)$$

where $F_{\alpha_1,\alpha_2,\alpha}$ is the $(1 - \alpha)$-th quantile of the F-distribution, F_{α_1,α_2}; and $F_{\alpha_1,0,\alpha}$ is defined as ∞.

Hint. If $Y \sim \text{Be}\,(t, n - t + 1)$, then show that $P(T \ge t) = P(Y \le p)$

9. Given a random sample of n from a Bernoullian population with constant probability p. If we assume that the sample estimate of p, $\bar{X}$, is approximately $N(p, p(1 - p)/n)$, show that the shortest confidence interval for p is $(p_1 < p < p_2)$, where p_1 and p_2 are solutions of the quadratic equation

$$n(\bar{X} - p)^2 = (1.96)^2(p - p^2)$$

10. Let $X_1, X_2,\ldots,X_n$ be n observations from Bernoulli distribution $b(1, p)$. The prior distribution of p is $U(0, 1)$. Find $(1 - \alpha)$-credible confidence interval for p.

11. Let $X_1,\ldots,X_n$ be a random sample from the Bernoulli distribution $b(1, p)$. Assume that p has conjugate $\text{Be}(\alpha, \beta)$ prior. Construct a $(1 - \alpha)$-level HPD credible set for p. Use the relationship between F and beta distribution to express this credible set in the form of the confidence interval for p based on the *cdf* of ΣX_i. Compare these intervals.

12. Let $X_{11},\ldots,X_{1n_1}$ be *iid* $b(1, p_1)$ and $X_{21},\ldots,X_{2n_2}$ be *iid* $b(1, p_2)$. Construct $(1 - \alpha)$-level confidence interval for $(p_1 - p_2)$.

Hint. $X_1 = \Sigma_{i=1}^{n_1} X_{1i} \sim b(n_1, p_1)$ and $X_2 = \Sigma_{i=1}^{n_2} X_{2i} \sim b(n_2, p_2)$. Use $[((X_1/n_1) - (X_2/n_2)) - (p_1 - p_2)]/\sqrt{(p_1(1-p_1)/n_1)+(p_2(1-p_2)/n_2)} \sim N(0,1)$ as pivot. Denominator involves unknown p_1 and p_2 and it creates problem in constructing confidence interval for $p_1 - p_2$ based on this pivot. We replace p_1 by X_1/n_1 and p_2 by X_2/n_2 and use $[((X_1/n_1) - (X_2/n_2)) - (p_1 - p_2)]/\sqrt{[(X_1/n_1)(1-(X_1/n_1))/n_1]+[(X_2/n_2)(1-(X_2/n_2))/n_2]} \sim N(0, 1)$ as pivot to construct $(1 - \alpha)$-level confidence interval for $p_1 - p_2$. Note that $X_1/n_1 \xrightarrow{P} p_1$ and $X_2/n_2 \xrightarrow{P} p_2$.

13. Let $X_1, X_2,\ldots,X_n$ be a random sample from geometric distribution with parameter θ, $\theta \in (0, 1)$. If the prior of θ is $Be(\alpha, \beta)$, α, β are known, find the $(1 - \alpha)$-level credible interval for θ.

14. Let X follows *pmf*

(i) $f(x; \theta) = p(1 - p)^{x-\theta}$, $x = \theta, \theta + 1,\ldots$; p known

(ii) $f(x; \theta) = \binom{n}{x-\theta} p^{x-\theta}(1-p)^{n-x+\theta}$, $x = \theta, \theta + 1,\ldots,n + \theta$, and p is known. Find

$(1 - \alpha)$-level confidence interval for θ.

15. Let $X_1, X_2,\ldots,X_n$ be a random sample from $P(\theta)$. The UMP test of size-α for testing H_0: $\theta \le \theta_0$ against H_1: $\theta > \theta_0$ is given by

$$\phi(\mathbf{x}) = \begin{cases} 1, & \sum_{i=1}^{n} x_i > c \\ 0, & \text{otherwise} \end{cases}$$

where for large n, c is obtained by

$$E_{\theta_0}\phi(\mathbf{X}) = \alpha$$

Find the corresponding one-sided confidence interval for θ at level $(1 - \alpha)$.

Hint. $\alpha = P_\theta(\Sigma X_i > c) \approx 1 - \Phi[(c - n\theta_0)/\sqrt{n\theta_0})]$ gives $c = n\theta_0 + \sqrt{n\theta_0}\, z_\alpha$. Acceptance region is $\bar{X} \le \theta_0 + \sqrt{\theta_0/n}\, z_\alpha$. The corresponding $(1 - \alpha)$-level confidence interval for θ is $\left((1/2)\left[\sqrt{(z_\alpha^2/n) + 4\bar{X}} - z_\alpha/\sqrt{n}\right]^2, \infty\right)$.

16. [Shao (2002)] Let $X_1,\ldots,X_n$ be a random sample from a Poisson distribution $P(\theta)$.

(i) Find $(1 - \alpha)$-level confidence interval for θ based on the *pdf* of the statistic $T(\mathbf{X}) = \Sigma X_i$ and by inverting the likelihood ratio test. Also, compare these intervals.

(ii) Let the potato crop is arranged in rows in a given field and assume the number of aphids as counted in each row follow the Poisson distribution with mean number of aphids per row as parameter θ. The aphids' count per nine randomly selected rows is available: 155, 104, 66, 50, 36, 40, 30, 35, 42. Find a 90% likelihood ratio test and the confidence interval for θ based on the *pdf* of the statistic $T(\mathbf{X})$ as obtained in (i).

17. Let $X_1,\ldots,X_n$ be a random sample from $P(\theta)$. Find $(1 - \alpha)$-level confidence interval for θ by test inversion method for testing H_0: $\theta = \theta_0$ against H_1: $\theta > \theta_0$ so that it is UMA interval.

18. Let $X_1,\ldots,X_n$ be *iid* according to the Poisson distribution $P(\theta)$ and θ has the conjugate prior $G_1(\alpha, \beta)$, α, β are known, assuming α as an integer. Show that $(1 - \alpha)$-level credible interval for θ is

$$\left(\frac{\beta}{2(n\beta+1)}\chi^2_{2(\Sigma x_i+\alpha),(1-\alpha/2)}, \frac{\beta}{2(n\beta+1)}\chi^2_{2(\Sigma x_i+\alpha),\alpha/2}\right) \tag{9.7.1}$$

and further show that the coverage probability of the above credible interval, as $\theta \to \infty$, becomes

$$P\left(\frac{n\beta}{n\beta+1}\chi^2_{2(\Sigma x_i+\alpha),(1-\alpha/2)} \le \chi^2_{2\Sigma X_i} \le \frac{n\beta}{n\beta+1}\chi^2_{2(\Sigma x_i+\alpha),\alpha/2}\right) \tag{9.7.2}$$

The coverage probability in Eq. (9.7.2) approaches zero as $\theta \to \infty$, thus, showing that it does not fare good when evaluated as confidence set. Also, find the $(1 - \alpha)$-level HPD credible set for θ.

19. Let X be a single observation from the negative binomial distribution NB(r, p) with the *pmf*

$$f(x; p, r) = \binom{x-1}{r-1} p^r (1-p)^{x-r}, \; x = r, r+1, \ldots$$

where $p \in (0, 1)$ is unknown and $r = 1, 2, \ldots$ is known. Using the *cdf* of $T = X$, show that the $(1 - \alpha)$-level confidence interval for p is

$$\left[\frac{1}{1+[(T+1)/(r+1)]F_{2(T+1),2(r+1),\alpha_2}}, \frac{[(r+1)/T]F_{2(r+1),2T,\alpha_1}}{1+[(r+1)/T]F_{2(r+1),2T,\alpha_1}}\right]$$

where $F_{\alpha_1,\alpha_2,\alpha}$ is the $(1 - \alpha)$-th quantile of the F-distribution, F_{α_1,α_2}.

20. (Asymptotic confidence interval for the Hardy-Weinberg equilibrium) Construct $(1 - \alpha)$-level asymptotic confidence interval for θ based on the data $(X_1, X_2, X_3) \sim$ Multinomial(n, p_1, p_2, p_3) where $p_1 = (1 - \theta)^2$, $p_2 = 2\theta(1 - \theta)$, $p_3 = \theta^2$ for $0 < \theta < 1$.

21. Let $X_1,\ldots,X_n$ be a random sample from $N(\mu, \sigma^2)$, σ^2 is known. Use pivotal method to find a $100(1 - \alpha)\%$ confidence interval for μ.

22. Suppose both the parameters μ and σ^2 of $N(\mu, \sigma^2)$ distribution are unknown. Obtain confidence interval estimators for μ and σ^2 based on the sample $X_1,\ldots,X_n$.

23. Let $X_1,\ldots,X_n$ be a random sample from $N(\theta, \sigma^2)$.

(i) Show that $(1 - \alpha)$-level upper confidence bound for θ, when σ^2 is known, is $(-\infty, \bar{x} + z_\alpha \sigma/\sqrt{n})$ and is unbiased.

(ii) Show that $(1 - \alpha)$-level upper confidence bound for θ, when σ^2 is unknown, by inverting the acceptance region of likelihood ratio test is $(-\infty, \bar{x} + t_{n-1,\alpha}S_{n-1}/\sqrt{n})$. Also, show that it is an unbiased interval.

(iii) Show that $(1 - \alpha)$-level confidence interval for μ when σ^2 is unknown by inverting the acceptance region of a likelihood ratio test is

$$\left(\overline{x} - t_{n-1,\alpha/2}\left(\frac{s_{n-1}}{\sqrt{n}}\right), \overline{x} + t_{n-1,\alpha/2}\left(\frac{s_{n-1}}{\sqrt{n}}\right)\right)$$

Also, show that it is an unbiased interval.

24. Let $X_1,\ldots,X_n$ be a random sample from $N(\mu, \sigma^2)$.

(i) Find the minimum sample size n required for the 95% confidence interval for μ, when σ^2 is known, to have its length no more than $\sigma/4$. Also, find the minimum n that guarantees, with probability 90%, that 95% confidence interval for μ, when σ^2 is unknown, to have its length no more than $\sigma/4$.

(ii) Find the expected length of $(1 - \alpha)$-level confidence intervals for μ when σ^2 is known and when σ^2 is not known. Also, compare these expected lengths.

25. Let $X_1, X_2,\ldots,X_n$ be a random sample from $N(\mu, \sigma^2)$ where σ^2 is unknown. Find the shortest length confidence interval for μ at level $(1 - \alpha)$ and show that it is equal tailed.

26. Let $X_1, X_2,\ldots,X_n$ be *iid* according to $N(\theta, \theta)$ with $\theta > 0$ unknown. Define a pivot; use it to construct $(1 - \alpha)$-level confidence interval for θ.

27. Let $X_1,\ldots,X_n$ be a sample from $N(\mu, \sigma^2)$ population. Explain how much is the sample size needed to have a prescribed level $(1 - \alpha)$ and the length of the confidence interval at most $2d$ when σ^2 is known and when an upper bound η^2 on σ^2 is known. What is the minimum sample size needed if $\alpha = 0.001$ and $\sigma^2 = 5$ with $d = 0.05$?

28. Let $X_1, X_2,\ldots,X_n$ be a random sample from $N(\mu, \sigma^2)$ when both μ and σ^2 are unknown.

(i) Find $(1 - \alpha)$-level confidence interval for μ.

(ii) Find $(1 - \alpha)$-level upper confidence bound for μ.

(iii) Find $(1 - \alpha)$-lower confidence bound for μ.

29. Let $X_1, X_2,\ldots,X_n$ be *iid* $f(x; \theta)$. Based on an appropriate pivot, find $(1 - \alpha)$-level confidence interval for

(i) θ when $f(x; \theta)$ is the *pdf* of $E(\theta)$.

(ii) θ when $f(x; \theta)$ is the *pdf* of $N(\theta, \sigma^2)$, σ^2 is known.

(iii) θ when $f(x; \theta)$ is the *pdf* of $N(\theta, \sigma^2)$, σ^2 is unknown.

(iv) σ when $f(x; \theta)$ is $N(\mu, \sigma^2)$, μ is known.

(v) σ when $f(x; \theta)$ is $N(\mu, \sigma^2)$, μ is unknown.

30. Let $X_1,\ldots,X_n$ be *iid* according to the normal distribution $N(\mu, \sigma^2)$ where μ and σ^2 are unknown. Consider the prior for (μ, σ^2)

$$\pi(\mu, \sigma^2|\eta, \tau^2, \alpha, \beta) = \pi_1(\mu|\eta, \tau^2\sigma^2)\pi_2(\sigma^2|\alpha, \beta)$$

where π_1 is $N(\eta, \tau^2\sigma^2)$ and π_2 is $\mathrm{IG}_1(\alpha, \beta)$. Note that this prior is a conjugate prior. Find $(1 - \alpha)$-level credible set for μ. Investigate about the limiting sequence of τ^2, α, β under which this credible set approaches the $(1 - \alpha)$-level classical confidence set for μ

$$\left\{\mu : |\mu - \overline{x}|^2 \le F_{1,n-1,\alpha}\left(\frac{s_{n-1}^2}{n}\right)\right\}$$

31. Let $X_1,\ldots,X_n$ be *iid* $N(\mu, \sigma^2)$ and let μ have the conjugate prior $N(\eta, \tau^2)$ where η, σ^2, and τ^2 are all known.

(i) Show that the $(1-\alpha)$-level credible set for μ is

$$\left(\mu^B(\bar{x}) - z_{\alpha/2}\sqrt{V(\mu|\bar{x})},\ \mu^B(\bar{x}) + z_{\alpha/2}V(\mu|\bar{x})\right) \tag{9.7.3}$$

where $$\mu^B(\bar{x}) = \frac{\sigma^2}{\sigma^2 + n\tau^2}\eta + \frac{n\tau^2}{\sigma^2 + n\tau^2}\bar{x}$$

and $$V(\mu|\bar{x}) = \frac{\sigma^2\tau^2}{\sigma^2 + n\tau^2}$$

Is the credible interval for μ in Eq. (9.7.3) also HPD?

(ii) Prove that the classical coverage probability of HPD interval, using the probability model $\bar{X} \sim N(\mu, \sigma^2/n)$, is

$$P_\theta\left(\left|\mu - \mu^B(\bar{X})\right| \le z_{\alpha/2}\sqrt{V(\mu|\bar{X})}\right)$$

$$= P_\theta\left[-\sqrt{1+\nu}\,z_{\alpha/2} + \frac{\nu(\mu-\eta)}{\sigma/\sqrt{n}} \le Z \le \sqrt{1+\nu}\,z_{\alpha/2} + \frac{\nu(\mu-\eta)}{\sigma/\sqrt{n}}\right] \tag{9.7.4}$$

where $\nu = \sigma^2/(n\tau^2)$ and $\sqrt{n}(\bar{X} - \mu)/\sigma = Z \sim N(0, 1)$. Show that the above $(1-\alpha)$-credible set is not a $(1-\alpha)$-confidence set. Next, consider the $(1-\alpha)$-level (classical) confidence set for $\{\mu{:}|\mu - \bar{x}| \le z_{\alpha/2}\sigma/\sqrt{n}\}$. Prove that the credible probability of this set is

$$P\left(|\mu - \bar{x}| \le z_{\alpha/2}\sigma/\sqrt{n}\right)$$

$$= P\left[-\sqrt{1+\nu}\ z_{\alpha/2} + \frac{\nu(\bar{x}-\eta)}{\sqrt{1+\nu}(\sigma/\sqrt{n})} \le Z \le \sqrt{1+\nu}\ z_{\alpha/2} + \frac{\nu(\bar{x}-\eta)}{\sqrt{1+\nu}(\sigma/\sqrt{n})}\right]$$

Also, show that this $(1-\alpha)$-level confidence set is not a $(1-\alpha)$-credible set.

Hint. (i) $\pi(\mu|\bar{x}) \sim N(\mu^B, V(\mu|\bar{x}))$. HPD credible interval is given by $\{\mu{:}\, \pi(\mu|\bar{x}) \ge c\}$ where c is chosen so that $\int_{\{\mu:\pi(\mu|\bar{x}) \ge c\}} \pi(\mu|\bar{x})d\mu = 1 - \alpha$. This condition is met if we choose $(\underline{\mu}, \bar{\mu})$ such that $\pi(\underline{\mu}|\bar{x}) = \pi(\bar{\mu}|\bar{x})$ and $\int_{\underline{\mu}}^{\bar{\mu}} \pi(\mu|\bar{x})d\mu = 1-\alpha$, and we get the same credible interval as the HPD credible interval in Eq. (9.7.3).

(ii) $\mu^B(\bar{X})$ and $V(\mu|\bar{X})$ is given by

$$\mu^B(\bar{X}) = \frac{\nu}{1+\nu}\eta + \frac{1}{1+\nu}\bar{X}$$

and $$V(\mu|\bar{X}) = \frac{\sigma^2}{n(1+\nu)}$$

Fix $\mu \ne \eta$ and let $\tau = \sigma/\sqrt{n}$. It gives $\nu = 1$. Let $\sigma/\sqrt{n} \to 0$ and if $\mu > \eta$, the lower bound goes to ∞, and if $\mu < \eta$ the upper bound goes to $-\infty$. The probability in

Eq. (9.7.4) approaches 0. If $\mu = \eta$, the probability in Eq. (9.7.4) is bounded away from zero.

32. Let $X_1,\ldots,X_n$ be a random sample from $N(\mu, \sigma^2)$, where $\sigma^2 > 0$ is known. Construct $(1 - \alpha)$-level confidence interval for μ by test-inversion procedure for each of the following testing problems:

(i) $H_0{:}\theta = \theta_0$ against $H_1{:}\theta \neq \theta_0$

(ii) $H_0{:}\theta \leq \theta_0$ against $H_1{:}\theta > \theta_0$

(iii) $H_0{:}\theta \geq \theta_0$ against $H_1{:}\theta < \theta_0$

33. Let $X_1,\ldots,X_n$ be *iid* $N(\mu, \sigma^2)$ with μ and σ^2 unknown. Construct $(1 - \alpha)$-level upper confidence bound for μ by inverting the acceptance region of a likelihood ratio test. Show that it is a UMAU upper confidence bound.

Hint. $(-\infty, \theta_U)$, $\theta_U = \bar{X} + t_{n-1,\alpha} S_{n-1}/\sqrt{n}$, where $t_{n-1,\alpha}$ is the $(1 - \alpha)$-th quantile of the t-distribution.

34. Let $X_1,\ldots,X_n$ be a random sample drawn from $N(\mu, \sigma^2)$ where both the parameters μ and σ^2 are unknown.

(i) Use the Bonferroni inequality to construct one simultaneous confidence set at level $(1 - \alpha)$ for (μ,σ) by combining the two confidence sets $(\bar{x} \mp k(S_{n-1}/\sqrt{n}))$ and $((n - 1)S_{n-1}^2/b,\ (n - 1)S_{n-1}^2/a)$ for μ and σ^2 respectively. Discuss how are the constants k, a and b are choosen so that simultaneous confidence set is at level $(1 - \alpha)$.

(ii) Use Bonferroni inequality to construct one confidence set at level $(1 - \alpha)$ for (μ,σ) by combining two confidence sets namely, $(\bar{x} \mp k(\sigma/\sqrt{n}))$ and $((n-1)S_{n-1}^2/b,$ $(n-1)S_{n-1}^2/a)$ for μ and σ^2 respectively.

(iii) Comment on the simultaneous sets obtained in (i) and (ii).

35. In sampling from $N(\mu, \sigma^2)$, μ unknown, derive $100(1 - \alpha)\%$ shortest confidence interval for σ^2.

36. Let $X_1,\ldots,X_n$ be a random sample from $N(\mu, \sigma^2)$, where both μ and σ^2 are unknown. Consider the $(1 - \alpha)$-level confidence interval for σ^2,

$$\left(\frac{(n-1)S_{n-1}^2}{b}, \frac{(n-1)S_{n-1}^2}{a}\right) \tag{9.7.5}$$

so that

$$\int_a^b f_{n-1}(t)dt = 1 - \alpha \tag{9.7.6}$$

where $(n - 1)S_{n-1}^2/\sigma^2 \sim \chi_{n-1}^2$ and f_{n-1} is the *pdf* of χ_{n-1}^2.

(i) Show that the $(1 - \alpha)$-level confidence interval obtained by inverting the likelihood ratio test of $H_0{:}\sigma = \sigma_0$ against $H_1{:}\sigma \neq \sigma_0$ is of the form (9.7.5) where the constants a and b are obtained by Eq. (9.7.6) and

$$f_{n+2}(a) = f_{n+2}(b)$$

(ii) Show that the $(1 - \alpha)$-level shortest length confidence interval for σ^2 is of the form (9.7.5) where the constants a and b are obtained by Eq. (9.7.6) and

$$f_{n+3}(a) = f_{n+3}(b)$$

(iii) Show that the $(1 - \alpha)$-level shortest unbiased confidence interval for σ^2 is of the form (9.7.5) when the constants a and b are obtained by Eq. (9.7.6) and

$$f_{n+1}(a) = f_{n+1}(b)$$

Show that this interval is also obtained by minimizing the ratio of the end points.

(iv) Equal-tail interval for σ^2 is of the form (9.7.5) where a and b are obtained by

$$\int_0^a f_{n-1}(t)dt = \frac{\alpha}{2} = \int_b^\infty f_{n-1}(t)dt$$

(v) Compute 90% confidence intervals for $n = 4$. Compare these intervals in terms of their lengths.

[Tate and Klett (1959) have derived these intervals and calculated some cutoff points.]

37. Obtain uniformly most accurate unbiased and minimum length confidence interval for σ^2 at 95% level based on a random sample $X_1, X_2,\ldots,X_n$ from $N(\mu, \sigma^2)$ with unknown mean μ.

38. Let $X_1,\ldots,X_n$ be *iid* observations from $N(\mu, \sigma^2)$ and let σ^2 has conjugate prior inverse gamma $\text{IG}_1(\alpha, \beta)$ with the *pdf*

$$\pi(\sigma^2; \alpha, \beta) = \frac{1}{\Gamma(\alpha)\beta^\alpha}\left(\frac{1}{\sigma^2}\right)^{\alpha+1} \exp\left(-\frac{1}{\beta\sigma^2}\right), 0 < \sigma^2 < \infty$$

Find $(1 - \alpha)$-level HPD credible set for σ^2 based on the sample variance S_{n-1}^2. Also, find the limiting HPD credible set for σ^2 under the limiting conditions $\alpha \to 0$ and $\beta \to \infty$.

39. Let $X_1, X_2,\ldots,X_m$ be a random sample from $N(0, \sigma_1^2)$ and let $Y_1, Y_2,\ldots,Y_n$ be a random sample from $N(0, \sigma_2^2)$ independent of Xs. Define $\tau = {\sigma_2}^2/{\sigma_1}^2$.

Find a level-α likelihood ratio test of $H_0: \tau = 1$ against $H_1: \tau > 1$. Construct an lower $(1 - \alpha)$-confidence interval of the form (L, ∞) for τ.

40. Suppose $(X_1,Y_1),\ldots,(X_n,Y_n)$ represent a random sample from $N(0, 0, \sigma_1^2, \sigma_2^2, \rho)$. Suppose $\rho = \rho_0$ (known). Then, find a confidence interval of σ_1/σ_2 with confidence coefficient $(1 - \alpha)$ that incorporates the information that $\rho = \rho_0$.

41. Let (X_i, Y_i), $i = 1,\ldots,n$ be a random sample from a bivariate normal distribution $N_2(\boldsymbol{\mu}, \Sigma)$ with $\boldsymbol{\mu}$ and Σ unknown; ρ is the population correlation coefficient. Construct an approximate $(1 - \alpha)$-level confidence interval for ρ.

Hint. Sample correlation coefficient $r = \text{cov}(X, Y)/\sigma_X\sigma_Y$ is the MLE of ρ which under regularity conditions and for large n follows asymptotically normal distribution

$$N(\rho, (1-\rho^2)^2/n)$$

for large n. This asymptotic distribution is good only when n is large; otherwise for small n, the distribution of r is highly skewed. Consider the Fisher z-transformation

$$z = \frac{1}{2}\log\frac{1+r}{1-r}$$

Using delta method $\qquad Z \sim N(\theta, (1/n))$

where $\qquad \theta = \frac{1}{2}\log\frac{1+\rho}{1-\rho}$

Based on the pivot $\sqrt{n}(Z - \theta)$, the $(1 - \alpha)$-level confidence interval for θ is given by

$$P\left(-z_{\alpha/2} \le \frac{Z-\theta}{1/\sqrt{n}} \le z_{\alpha/2}\right) = 1-\alpha$$

$$P\left(Z - \frac{1}{\sqrt{n}} z_{\alpha/2} \le \theta \le Z + \frac{1}{\sqrt{n}} z_{\alpha/2}\right) = 1-\alpha$$

or $$P\left(\frac{\exp\left[2\left(Z-(1/\sqrt{n})z_{\alpha/2}\right)\right]-1}{\exp\left[2\left(Z-(1/\sqrt{n})z_{\alpha/2}\right)\right]+1} \le \rho \le \frac{\exp\left[2\left(Z+(1/\sqrt{n})z_{\alpha/2}\right)\right]-1}{\exp\left[2\left(Z+(1/\sqrt{n})z_{\alpha/2}\right)\right]+1}\right) = 1-\alpha$$

42. Let $X_1, X_2,\ldots,X_n$ be a random sample from uniform distribution on $\{1, 2,\ldots,N\}$. Find an upper $(1 - \alpha)$-level confidence bound for N.

Hint. $\alpha^{-1/n}X_{(n)}$

43. Let $X_1,\ldots,X_n$ be a random sample from the uniform distribution $U(0, \theta)$. Obtain $100(1 - \alpha)\%$ confidence interval for θ and explain how the length of the interval be made shortest.

44. Let $X_1, X_2,\ldots,X_n$ be a random sample from a distribution with the *pdf* $f(x; \theta) = 1/\theta$, $0 < x < \theta$. Find a confidence interval for θ based on the MLE of θ.

45. Let $X_1,\ldots,X_n$ be a random sample from the uniform distribution $U(\theta - (1/2), \theta + (1/2))$ where $\theta \in R$. Find $(1 - \alpha)$-level confidence interval for θ.

46. Let X be a single observation from $U(\theta - (1/2), \theta + (1/2))$, $\theta \in \mathbb{R}^1$.

(i) Show that the quantity $T(X) = X - \theta$ is a pivotal quantity. Show that a confidence interval of the type $(X + c, X + d)$, with c and d such that $-1/2 < c < 1/2$, is $(1 - \alpha)$-level if and only if its length $L = d - c = 1 - \alpha$.

(ii) Show that the *cdf* $F_\theta(x)$ of X is not an increasing function in θ for fixed x. Use this *cdf* to construct a $(1 - \alpha)$-level confidence interval for θ.

47. Let $X_1, X_2,\ldots,X_n$ be a random sample from $U(0, \theta)$ and θ has the prior density

$$\pi(\theta) = \frac{\alpha a^\alpha}{\theta^{\alpha+1}}, \theta > a$$

Find the minimum length $(1 - \alpha)$-level credible interval for θ.

Hint. $(\max(a, x_1,\ldots,x_n), \alpha^{-1/(n+\alpha)} \max(a, x_1,\ldots,x_n))$

48. If X_r and X_s , $r < s$, are order statistics for a random sample of size n from a continuous distribution, then show that (X_r, X_s) is a confidence interval for the population quantile of order p, with confidence coefficient $I_p(r, n - r + 1) - I_p(s, n - s + 1)$, where $I_p(\nu_1, \nu_2)$ is the incomplete beta function.

49. Let $X_1,\ldots,X_n$ be *iid* observations drawn from a location density $f(x;\ \theta) = f(x - \theta)$, where f is known and θ is unknown. Show that the coverage probability of the interval $S(\mathbf{X}) = (\bar{X} - c_1, \bar{X} + c_2)$, $P(S(\mathbf{X}) \ni \theta)$, for some constants c_1 and c_2, is constant.

50. Let $X_1, X_2,\ldots,X_n$ be a random sample from the Laplace distribution

$$f(x;\theta) = \frac{1}{2}\exp(-|x-\theta|),\ x \in \mathbb{R}^1,\ \theta \in \mathbb{R}^1$$

(i) Show that

$$P_\theta(X_{(1)} \le \theta \le X_{(n)}) = 1 - \frac{1}{2^{n-1}}.$$

Find n so that this confidence coefficient is at least $(1 - \alpha)$.

(ii) Prove that $P_\theta(X_{(n)} - X_{(1)} < c) > [1 - \exp(c/2)]^n$

Hint. (i) $P_\theta(X_{(1)} \le \theta \le X_{(n)}) = P_\theta(\min_i(X_i - \theta) \le 0 \le \max_i(X_i - \theta)) = P_0(X_{(1)} \le 0 \le X_{(n)}) = 1 - P_0(X_{(1)} > 0) - P_0(X_{(n)} < 0) = 1 - \Pi_{i=1}^n P_0(X_i > 0) - \Pi_{i=1}^n P_0(X_i < 0) = 1 - (1/2^n) - (1/2^n) = 1 - (1/2^{n-1})$; $1 - (1/2^{n-1}) \ge (1 - \alpha)$ gives $n \ge 1 + [\log(1/\alpha)/\log 2]$

(ii) $P_\theta(X_{(n)} - X_{(1)} < c) = P_0(X_{(n)} - X_{(1)} < c) > P_0(-c/2 < X_1 < c/2,\ldots,-c/2 < X_n < c/2) = \left(\int_{-c/2}^{c/2}(1/2)\exp(-|x|)dx\right)^n = [1 - \exp(-c/2)]^n$.

51. Let $X \sim$ Logistic$(\theta, 1)$ distribution with the location *pdf*

$$f(x;\theta) = \frac{\exp[-(x-\theta)]}{\{1+\exp[-(x-\theta)]\}^2},\ \mu \in \mathbb{R}^1$$

Find $(1 - \alpha)$-level UMA upper confidence bound for θ based on just single observation.

52. Let $X_1, X_2,\ldots,X_n$ be *iid* according to exponential distribution $E(\mu, 1)$, with *pdf*

$$f(x;\mu) = \exp[-(x-\mu)],\ x \ge \mu,\ \mu \in \mathbb{R}^1$$

(i) Find the $(1 - \alpha)$-pivotal based confidence interval for μ. Also find the confidence interval for μ by pivoting the *cdf* of the sufficient statistic $T(\mathbf{X}) = X_{(1)}$. Show that these confidence intervals are the same.

(ii) Compare the intervals obtained in (i) with the intervals obtained by likelihood and pivotal methods.

(iii) Find the shortest length $(1 - \alpha)$-level confidence interval based on the sufficient statistic $X_{(1)}$.

53. If $f(x;\ \theta) = \exp[-(x - \theta)]$, $\theta \le x < \infty$, $-\infty < \theta < \infty$, then show that

$$P\left[X_{(1)} - \frac{1}{n}\log\alpha \le \theta \le X_{(1)}\right] = 1 - \alpha$$

54. Given a random sample of size n from the *pdf*

$$f(x\,;\mu) = \exp[-(x-\mu)],\ x > \mu\ ,\ -\infty < \mu < \infty$$

Derive the UMA confidence interval for μ.

55. Let $X_1, X_2, \ldots, X_n$ be a random sample drawn from a location scale family of distribution with the *pdf*

$$f(x;\mu,\sigma) = \frac{1}{\sigma} f\left(\frac{x-\mu}{\sigma}\right)$$

where $\mu \in \mathbb{R}^1$, $\sigma > 0$; and f is known. Construct $(1-\alpha)$-level confidence set for $\boldsymbol{\theta} = (\mu, \sigma)'$ based on a suitable pivot. Find this set when f is the *pdf* of exponential distribution $E(0, 1)$.

56. Let $X_1, X_2, \ldots, X_n$ be a random sample drawn from $E(0, \sigma)$, $\sigma > 0$ unknown

$$f(x;\sigma) = \frac{1}{\sigma}\exp\left(-\frac{x}{\sigma}\right)$$

where $x > 0$, $\sigma > 0$.

(i) Find $(1-\alpha)$-level confidence set for σ based on the pivot $\bar{X}/\sigma$.

(ii) Find $(1-\alpha)$-level confidence set for σ based on the *cdf* of the statistic $T = \bar{X}$.

(iii) Show that the UMA $(1-\alpha)$-level confidence interval for σ by inverting the acceptance region corresponding to the UMP test for testing $H_0{:}\sigma = \sigma_0$ against $H_1{:}\sigma < \sigma_0$ is

$$S(\mathbf{x}) = (0, 2\Sigma x_i/\chi^2_{2n,\alpha})$$

Also, find the expected length of $S(\mathbf{X})$.

57. Let $X_1, X_2, \ldots, X_n$ be a random sample from an $E(\theta)$, with mean θ. Find $(1-\alpha)$-level upper confidence bound for θ.

Hint. $n\bar{X}/\theta \sim G_1(n, 1)$ is pivot. $P_\theta(n\bar{X}/\theta > c) = 1-\alpha$ gives $c = G_1[(n, 1), 1-\alpha]$ where $G_1[(n, 1), 1-\alpha]$ is upper $(1-\alpha)$ quantile of $G_1(n, 1)$ distribution. Thus, $P_\theta(\theta \le n\bar{X}/G_1[(n, 1), 1-\alpha]) = (1-\alpha)$ gives $(1-\alpha)$-level upper confidence interval for θ by $(0, n\bar{X}/G_1[(n, 1), 1-\alpha])$.

58. Let $X_1, X_2, \ldots, X_n$ be *iid* according to $E(\theta)$. Show that (i) X_1/θ, (ii) $\Sigma X_i/\theta$, and (iii) $2\Sigma X_i/\theta$ are all pivots for θ. Also, find their respective distributions and use them to construct $(1-\alpha)$-level confidence interval for θ.

59. Explain the method of finding confidence interval for large sample. Let $X_1, X_2, \ldots, X_n$ be a random sample from a distribution with the *pdf*

$$f(x;\theta) = \theta e^{-\theta x},\ 0 \le x \le \infty$$

Determine 95% confidence interval of θ where n is a large.

Hint. $(1 \pm (1.19/\sqrt{n})/\bar{X})$.

60. Let $X_1, X_2, \ldots, X_n$ be *iid* $E(\theta)$. Find $(1-\alpha)$-level confidence interval for θ by inverting the acceptance region of size-α test for testing $H_0:\theta = \theta_0$ against $H_1:\theta \neq \theta_0$.

Hint. Use likelihood ratio test.

61. Let $X_1, \ldots, X_n$ be *iid* observations from $E(\theta)$ with the *pdf*

$$f(x;\theta) = \frac{1}{\theta}\exp\left(-\frac{x}{\theta}\right),\quad 0 < x < \infty$$

Using the conjugate prior $IG_2(\alpha, \beta)$ for θ, find the $(1-\alpha)$-level highest posterior density interval (HPD) for θ.

62. Suppose single observation has been drawn from the *pdf*

$$f(x;\theta) = \frac{2}{\theta^2}(\theta - x),\ 0 < x < \theta$$

Let θ have the prior density

$$\pi(\theta) = \frac{1}{16}\theta^2\exp\left(-\frac{\theta}{2}\right), \theta > 0$$

Find $(1-\alpha)$-level credible interval for θ. Compare the obtained Bayesian credible interval with the classical interval obtained in solved Example 9.29.

Hint. $\pi(\theta|x) = [(\theta - x)/4]\exp[(\theta - x)/2],\ 0 < x < \theta$

63. Let $X_1, X_2, \ldots, X_n$ be a random sample drawn from the population having the *pdf*

$$f(x;\theta) = \frac{a}{\theta}\left(\frac{x}{\theta}\right)^{a-1}, 0 < x < \theta$$

where $a \geq 0$ is known and $\theta > 0$ is unknown.

(i) Find $(1-\alpha)$-level confidence interval for θ based on the pivotal quantity.

(ii) Find the confidence interval for θ based on the *cdf* of the statistic $T = X_{(n)}$. Show that the confidence intervals in (i) and (ii) are the same.

Also, compare the results in (i) and (ii) when $a = 1$, and comment.

64. Let $X_1, \ldots, X_n$ be *iid* from the Weibull distribution $W(\alpha, \theta)$ with the *pdf* $f(x;\theta) = (\alpha/\theta) x^{\alpha-1}\exp(-x^{\alpha}/\theta)$, $0 < x < \infty$, where $\alpha > 0$, $\theta > 0$ are unknown. Show that the quantity $T(\mathbf{X}, \alpha, \theta) = \Pi_{i=1}^{n} X_i^{\alpha}/\theta$ is a pivot. Find $(1-\alpha)$-level confidence set for (α, θ) based on this pivot.

65. Let $X_1, \ldots, X_n$ be a random sample drawn from the Pareto distribution $P_a(a, \theta)$, $a > 0$, $\theta > 0$, with the *pdf*

$$f(x; a, \theta) = \theta a^{\theta} x^{-(\theta+1)},\ a < x < \infty,\ a > 0,\ \theta > 0$$

(i) Consider a and θ as unknown. Construct $(1-\alpha)$-level confidence interval for (a, θ) based on the pivotal quantity. Find the confidence interval for θ using the *cdf* of the statistic $T(\mathbf{X}) = \Pi_{i=1}^{n}(X_i/X_{(1)})$ assuming $n \geq 2$.

(ii) Consider θ as known. Construct $(1-\alpha)$-level confidence interval for a based on the pivot and the *pdf* of the statistic $T = X_{(1)}$ separately.

66. Let $X \sim f(x;\theta) = \theta x^{\theta-1}$, $0 < x < 1$, $\theta > 0$. Construct $(1-\alpha)$-level confidence interval for θ based on the pivot X^{θ}.

Hint. $\alpha = \alpha_1 + \alpha_2$, $X^{\theta} \sim U(0, 1)$; $P(\alpha_1 \le X^{\theta} \le \alpha_2) = 1 - \alpha_1 - \alpha_2$; $(\log[1/(1-\alpha_2)]/\log(1/X), \log(1/\alpha_1)/\log(1/X))$.

67. Let $X_1,\ldots,X_n$ be *iid* according to a $Be(\theta, 1)$ and a prior $\pi(\theta)$ for θ be $G_2(\alpha, \beta)$, α, β known. Find $(1-\alpha)$-level credible confidence interval for θ.

68. Consider a single observation X from $Be(\theta, 1)$ with the *pdf*

$$f(x;\theta) = \theta x^{\theta-1}, \quad 0 < x < 1, \quad \theta > 0$$

Consider the transformation $t = -\log(x)$. The *pdf* of T is given by $f(t;\theta) = \theta\exp(-\theta t)$. Find the confidence coefficient of the interval $[t/2, t]$, which is based on the statistic T. Find the pivotal quantity to construct the confidence interval for θ at the level of the interval $[t/2, t]$. Also, compare the two confidence intervals.

69. Let X_{ij}, $j = 1,\ldots,n_i$, $i = 1,\ldots,m$, be independent random variables having distribution $N(\mu_i, \sigma^2)$, $i = 1,\ldots,m$. This is one way ANOVA model expressed by

$$X_{ij} = \mu_i + \varepsilon_{ij}, \quad j = 1,\ldots,n_i,\ i = 1,\ldots,m$$

where ε_{ij} are independently distributed as $N(0, \sigma^2)$. Define $\bar{\mu} = (1/n)\Sigma_{i=1}^{m} n_i\mu_i$ and $\theta = (1/\sigma^2)\Sigma_{i=1}^{m} n_i(\mu_i - \bar{\mu})^2$. Construct an upper confidence bound for θ at level $(1-\alpha)$ based on the statistic

$$T = \frac{n-m}{m-1}\frac{\text{SSA}}{\text{SSR}}$$

where $\text{SSR} = \Sigma_{i=1}^{m}\Sigma_{j=1}^{n_i}(X_{ij} - \bar{X}_{i\cdot})^2$, $\text{SST} = \Sigma_{i=1}^{m}\Sigma_{j=1}^{n_i}(X_{ij} - \bar{X})^2$ and $\text{SSA} = \Sigma_{i=1}^{m} n_i(\bar{X}_{i\cdot} - \bar{X})^2$.

Hint. X_{ij}s are independent, $\Sigma_{j=1}^{n_i}(X_{ij} - \bar{X}_{i\cdot})^2 \sim \chi^2_{n_i-1}$, $i = 1,\ldots,m$. Therefore, $\text{SSR} = \Sigma_{i=1}^{m}\Sigma_{j=1}^{n_i}(X_{ij} - \bar{X}_{i\cdot})^2 \sim \chi^2_{n-m}$, where $n = \Sigma_{i=1}^{m} n_i$. Let $\hat{\mu} = (\bar{X}_{1\cdot},\ldots,\bar{X}_{m\cdot})$, $\bar{X}_{i\cdot} = (1/n_i)\Sigma_{j=1}^{n_i}X_{ij}$, $i = 1,\ldots,m$, and $A = \text{diag}(\sqrt{n_1}/\sigma,\ldots,\sqrt{n_m}/\sigma)$. Then $A\hat{\mu} \sim N_m(\eta, I_m)$, where I_m is the identity matrix of order m and $\eta = (\mu_1\sqrt{n_1}/\sigma,\ldots,\mu_m\sqrt{n_m}/\sigma)'$.
Further, define a column vector $B_m = (\sqrt{n_1},\ldots,\sqrt{n_m})'$. We get $B_m' B_m = n$ and $(I_m - n^{-1} B_m B_m')^2 = I_m - n^{-1} B_m B_m'$, $r(I_m - n^{-1} B_m B_m') = m - 1$. Therefore, by Cochran's theorem, $(A\hat{\mu})'(I_m - n^{-1}B_m B_m')A\hat{\mu} \sim \chi^2_{m-1}(\delta)$ (non-central chi-square distribution) where the non-centrality parameter δ is

$$\delta = [E(A\hat{\mu})]'(I_m - n^{-1}B_m B_m')[E(A\hat{\mu})] = \theta$$

On simplification, we get

$$\hat{\mu}'A(I_m - n^{-1}B_m B_m')A\hat{\mu} = \sum_{i=1}^{m} n_i(\bar{X}_{i\cdot} - \bar{X})^2 = \text{SSA} \sim \chi^2_{m-1}(\theta)$$

We know that SST = SSA + SSR. Also, by Basu's theorem, SSA and SSR are independent. Therefore,

$$\frac{\text{SSA}/(m-1)}{\text{SSR}/(n-m)} = T - \frac{n-m}{m-1} \sim F_{m-1,n-m}(\theta)$$

(non-central F-distribution)

One can show that the *cdf* of T, $F_\theta(t)$ for some fixed t, is a non-increasing function of θ. Therefore, by Theorem 9.3.3(ii), the $(1-\alpha)$-level upper bound bound for θ is

$$\theta_U = \sup[\theta : F_\theta(T) \geq \alpha]$$

70. Let $X_{11},\ldots,X_{1n_1}$ and $X_{21},\ldots,X_{2n_2}$ are two random samples independently drawn from $N(\mu_1, \sigma_1^2)$ and $N(\mu_2, \sigma_2^2)$, respectively. Denote the sample mean and variance by $\bar{X}_i$ and S_i^2 for the ith sample, $i = 1, 2$.

(i) Find $(1-\alpha)$-level confidence bound or interval for $\mu_2 - \mu_1$, by assuming $\sigma_1^2 = \sigma_2^2$, by inverting the acceptance region of the tests $H_0{:}\mu_2 - \mu_1 \leq 0$ against $H_1{:}\mu_2 - \mu_1 > 0$ and $H_0{:}\mu_2 - \mu_1 = 0$ against $H_1{:}\mu_2 - \mu_1 \neq 0$, and by using the pivotal quantity

$$T(\mathbf{X}_1, \mathbf{X}_2, \mu_1, \mu_2) = \frac{[(\bar{X}_2 - \bar{X}_1) - (\mu_2 - \mu_1)]/\sqrt{n_1^{-1} + n_2^{-1}}}{\sqrt{[(n_1-1)S_1^2 + (n_2-1)S_2^2]/(n_1+n_2-2)}}$$

Show that T is asymptotically pivotal if $n_1/n_2 \to 1/2$ or $\sigma_1^2 = \sigma_2^2$.

(ii) Construct $(1-\alpha)$-level confidence interval for σ_2^2/σ_1^2 by inverting the acceptance regions of the test $H_0{:}\,\mu_2 - \mu_1 \leq 0$ against $H_1{:}\,\mu_2 - \mu_1 > 0$ and $H_0{:}\,\mu_2 - \mu_1 = 0$ against $H_1{:}\,\mu_2 - \mu_1 \neq 0$, and by using the pivotal quantity.

$$T(\mathbf{X}_1, \mathbf{X}_2, \sigma_1^2, \sigma_2^2) = \frac{S_2^2/\sigma_2^2}{S_1^2/\sigma_1^2} = \frac{S_2^2}{\theta S_1^2}$$

where $\theta = \sigma_2^2/\sigma_1^2$.

(iii) Show that the quantity

$$U(\mathbf{X}_1, \mathbf{X}_2, \mu_1, \mu_2) = \frac{[(\bar{X}_2 - \bar{X}_1) - (\mu_2 - \mu_1)]}{\sqrt{n_1^{-1}S_1^2 + n_2^{-1}S_2^2}}$$

is asymptotically pivotal by assuming $n_1/n_2 \to 1$. Construct $(1-\alpha)$-level asymptotically correct confidence interval for θ using U.

Hint. (i) The critical region for UMPU size-α test for testing $H_0{:}\,\mu_2 - \mu_1 \leq 0$ against $H_1{:}\,\mu_2 - \mu_1 \geq 0$ is $T(\mathbf{X}_1, \mathbf{X}_2, \mu_1, \mu_2)|_{\mu_2-\mu_1=0} > t_{n_1+n_2-2,\,\alpha}$. The corresponding acceptance region is $T \leq t_{n_1+n_2-2,\,\alpha}$. Define

$$S_p^2 = \frac{(n_1-1)S_1^2 + (n_2-1)S_2^2}{n_1+n_2-2}$$

We have $\dfrac{[(\bar{X}_2 - \bar{X}_1) - (\mu_2 - \mu_1)]/\sqrt{n_1^{-1} + n_2^{-1}}}{S_p} \le t_{n_1+n_2-2,\alpha}$

$$[(\bar{X}_2 - \bar{X}_1) - (\mu_2 - \mu_1)] \le S_p\sqrt{n_1^{-1} + n_2^{-1}}\, t_{n_1+n_2-2,\alpha}$$

$$(\mu_2 - \mu_1)_L = (\bar{X}_2 - \bar{X}_1) - S_p\sqrt{n_1^{-1} + n_2^{-1}}\, t_{n_1+n_2-2,\alpha} \le \mu_2 - \mu_1$$

Similarly, the acceptance region corresponding to the UMPU size-α test for testing H_0: $\mu_2 - \mu_1 = 0$ against H_1: $\mu_2 - \mu_1 \ne 0$ is $|T| \le t_{n_1+n_2-2,\,\alpha/2}$. This gives $(1 - \alpha)$-level confidence interval $((\bar{X}_2 - \bar{X}_1) \mp S_p\sqrt{n_1^{-1} + n_2^{-1}}\, t_{n_1+n_2-2,\,\alpha/2})$.

(ii) Define $\Delta = \sigma_2^2/\sigma_1^2$. Consider the testing problem H_0:$\Delta \le \Delta_0$ against H_1:$\Delta \ge \Delta_0$. The UMPU test rejects H_0 if $F > F_{n_2-1,\,n_1-1,\,\alpha}$ where $F = S_2^2/\Delta S_1^2$ and that $F|_{\Delta=\Delta_0} \sim F_{n_2-1,\,n_1-1}$, $F_{n_2-1,\,n_1-1,\,\alpha}$ is the $(1 - \alpha)$-th quantile of the F-distribution, $F_{n_2-1,\,n_1-1}$. The corresponding acceptance region is

$$F \le F_{n_2-1,\,n_1-1,\,\alpha}$$

$$\frac{S_2^2}{S_1^2} \le F_{n_2-1,\,n_1-1,\,\alpha}$$

$$\Delta_L = \frac{S_2^2}{S_1^2}\, F^{-1}_{n_2-1,\,n_1-1,\,\alpha} \le \Delta$$

So, the testing problem H_0:$\Delta \le \Delta_0$ gives $(1 - \alpha)$-level lower bound of $\Delta = \sigma_2^2/\sigma_1^2$. Consider next the testing problem

$$H_0\!:\Delta = \Delta_0 \text{ against } H_1\!:\Delta \ne \Delta_0$$

Define the statistic

$$V = \frac{(n_2 - 1)F}{(n_1 - 1) + (n_2 - 1)F}$$

$$V|H_0 : \Delta = \Delta_0 \sim \text{Be}\left(\frac{n_2 - 1}{2}, \frac{n_1 - 1}{2}\right)$$

The acceptance region corresponding to the UMPU test for this testing problem is

$$c_1 < V < c_2$$

where c and c_2 are obtained by

$$\int_{c_1}^{c_2} v f_{(n_2-1)/2,(n_1-1)/2}(v)dv = \int_{c_1}^{c_2} f_{(n_2+1)/2,(n_1-1)/2}(v)dv = 1-\alpha$$

$$c_1 < \frac{(n_2-1)(S_2^2/\Delta S_1^2)}{(n_1-1)+(n_2-1)(S_2^2/\Delta S_1^2)} < c_2$$

$$c_1 < \frac{1}{1+\dfrac{(n_1-1)}{(n_2-1)(S_2^2/\Delta S_1^2)}} < c_2$$

$$\frac{1}{c_2}-1 < \frac{(n_1-1)S_1^2}{(n_2-1)S_2^2}\Delta < \frac{1}{c_1}-1$$

The $(1-\alpha)$-level confidence interval for Δ is

$$\left(\frac{1}{c_2}-1\right)\frac{(n_2-1)S_2^2}{(n_1-1)S_1^2} < \Delta < \left(\frac{1}{c_1}-1\right)\frac{(n_2-1)S_2^2}{(n_1-1)S_1^2}$$

Bibliography

Abramovich, F. and Ritov, Y. (2013). *Statistical Theory: A Concise Introduction*. Chapman and Hall/CRC, New York.

Aitkin, M. (2010). *Statistical Inference: An Integrated Bayesian/Likelihood Approach*, 1st ed., Chapman and Hall/CRC, USA.

Anderson, T.W. (1964). *An Introduction to Multivariate Statistical Analysis*, 2nd ed., Wiley, NY.

Apostol, T.M. (1957). *Mathematical Analysis: A Modern Approach to Advanced Calculus*, Addison-Wesley, Reading, MA.

Arnold, S.F. (1990). *Mathematical Statistics*, Prentice Hall, Upper Saddle. NJ.

Ash, R.B. (2011). *Statistical Inference: A Concise Course*, http://www.math.uiuc. edu/~r-ash, Dover Publications, NY.

Azzalini, A. (1996). *Statistical Inference Based on Likelihood*, Chapman and Hall/CRC, Florida.

Bahadur, R.R. (1954). Sufficiency and statistical decision functions, *Ann. Math. Statist.*, 25, 423–462.

Bahadur, R.R. (1957). On unbiased estimates of uniformly minimum variance, *Sankhya*, 18, 211–224.

Bahadur, R.R. (1958). Examples of inconsistency of maximum likelihood estimates, *Sankhya*, 20, 207–210.

Bahadur, R.R. (1966). A note on quantiles in large samples, *Ann. Math. Statist.*, 37, 577–580.

Bahadur, R.R. (1971). *Some Limit Theorems in Statistics*, NSF-CBMS Monograph No. 4, Society for Industrial and Applied Mathematics, Philadelphia.

Bain, L.J. and Engelhardt, M. (1992). *Introduction to Probability and Mathematical Statistics*, Duxbury Press, Boston.

Balakrishnan, N. and Basu, A.P. (1995). *The Exponential Distributions: Theory, Methods and Applications* (edited volume), Gordon and Breach, Amsterdam.

Banks, D.L. (1996). A conversation with I.J. Good, *Statist. Sci.*, 11, 1–19.

Bansal, A.K. (2007). *Bayesian Parametric Inference*, Narosa Publishing House, New Delhi.

Barankin, E.W. and Maitra, A. (1963). Generalization of Fisher-Darmois-Koopman-Pitman theorem on sufficient statistics, *Sankhya*, Ser. A, 25, 217–244.

Barker, L. (2002). A Comparison of nine confidence intervals for a Poisson parameter when the expected number of events is ≤ 5, *The American Statistician*, 56(2), 86–89.

Barnard, G. (1992). Review of "Statistical inference and analysis: selected corresondence of R.A. Fisher (Edited by J.H. Bennett)", *Statist. Sci.*, 7, 5–12.

Barnard, G.A. (1949). Statistical inference (with discussion), *J. Roy. Statist., Ser. B*, 11, 115–139.

Barnard, G.A. (1982). A new approach to the Behrens-Fisher problem, *Utilitas Mathematica*, 21B, 261–271.

Barndorff-Nielsen, O.E. (1978). *Information and Exponential Families in Statistical Theory*, John Wiley and Sons, NY.

Barndorff-Nielsen, O.E. (1982). Exponential Families, in Banks, D.L., Read, C.B. and Kotz, S., Editors, *Encyclopedia of Statistical Sciences (9 vols, plus supplements)*, vol. 2, Wiley, NY, 587–596.

Barndorff-Nielsen, O.E. and Cox, D.R. (1994). *Inference and Asymptotics*, Chapman and Hall, New York.

Barnett, V. (1999). *Comparative Statistical Inference*, 3rd ed., John Wiley and Sons, New York.

Bartoszynski, R. and Bugaj, N.M. (2008). *Probability and Statistical Inference*, John Wiley and Sons, NY.

Basu, A., Shioya, H. and Park, C. (2011). *Statistical Inference: The Minimum Distance Approach*, Chapman and Hall/CRC, NY.

Basu, D. (1955a). On statistics independent of a complete sufficient statistic, *Sankhya*, 15, 377–380.

Basu, D. (1955b). An inconsistnecy of the the method of maximum likelihood, *Ann. Math. Statist.*, 26, 144–145.

Basu, D. (1958). On statistics independent of sufficient statistic, *Sankhya*, 20, 223–226.

Basu, D. (1959). The family of ancillary statistics, *Sankhya,* Ser. A, 21, 247–256.

Basu, D. (1964). Recovery of ancillary information, *Sankhya*, Ser. A, 26, 3–16.

Basu, D. (1965). Recovery of ancillary information. *Contributions to Statistics, the 70th Birthday Festschrift Volume Presented to P.C. Mahalanobis*, Pergamon Press, Oxford.

Basu, S. and Thompson, M.J. (1996). On constructing seismic models of sun, *Astonomy and Astophysics*, vol. 305, 631.

Bayes, T. (1783). An essay towards solving a problem in the doctrine of chances, *Phil Trans. Roy. Soc.*, 53, 370–418.

Behrens, W.V. (1929). Ein betrang zur fehlerbercchnung bei weniger beobachtungen, *Landwirtschaftliche Jahrbiicher*, 68, 807–837.

Bennett, J.H. (1971–1974). *Collected Works of R.A. Fisher, Volumes 1–5* (edited volumes), University of Adelaide, South Australia.

Bennett, J.H. (1990). *Statistical Inference and Analysis: Selected Correspondence of R.A. Fisher* (edited volume), Oxford University, Oxford.

Berger, J.O. and Wolpert, R.W. (1984). *The Likelihood Principle*. Institute of Mathematical Statistics Lecture Notes-Monograph Series. Hayward, CA: IMS.

Berger, J.O. (1985). *Statistical Decision Theory and Bayesian Analysis*, 2nd ed., Springer-Verlag, New York.

Berk, R.H. (1967). Review 1922 of "Invariance of Maximum Likelihood Estimators" by Peter W. Zehna, *Mathematical Reviews,* 33, 342–343.

Berk, R.H. (1972). Consistency and asymptotic normality of MLEs for exponential models, *The Annals of Mathematical Statistics*, 43, 193–204.

Bernardo, J.M. and Smith, A.F.M. (1994). *Bayesian Theory*, Wiley, New York.

Berry, D.A. and Lindgren, B.W. (1995). *Statistics, Theory and Methods,* 2nd ed., Duxbury Press, Belmont, CA.

Bertsekas Dimitri, P. (1999). *Nonlinear Programming,* 2nd ed., Athena Scientific, Belmont, Massachusetts.

Bhattacharyya, A. (1946). On some analogues of the amount of information and their use in satstistical estimation, *Sankhya ser A*, 8, 1–14, 201–218 (1947), 315–328 (1948).

Bickel, P.J. and Doksum, K.A. (1977). *Mathematical Statistics*, Holden Day, San Francisco.

Bickel, P.J. and Doksum, K.A. (2002). *Mathematical Statistics*, 2nd ed., Holden Day, San Francisco.

Bickel, P.J. and Doksum, K.A. (2006). *Mathematical Statistics*, vol I, 2nd ed., Pearson.

Bickel, P.J. and Doksum, K.A. (2007). *Mathematical Statistics: Basic Ideas and Selected Topics*, vol. 1., 2nd ed., Upper Saddle River, NJ.

Bickel, P.J. and Mallows, C.L. (1988). A note on unbiased Bayes esimates, *Amer. Statist.*, 42, 132–134.

Birnbaum, A. (1962). On the foundations of statistical inference (with discussion), *J. Amer. Statist. Assoc.*, 57, 269–326.

Blackwell, D. (1947). Conditional expectation and unbiased sequential estimation, *Ann. Math. Stat.*, 18, 105–110.

Blyth, C.R. (1951). On minimax statistical decision procedures and their admissibility, *Ann. Math. Stat.*, 22(1), 22–42.

Boole, G. (2010). *An investigations of the Laws of thoughts on which are founded the Mathematical Theories of Logic and Probabilities*. Watchmaker Publishing, New York.

Boos, D.D. (1985). A converse to Scheffe's theorem, *Annals of Statistics*, 13, 423–427.

Boos, D.D. and Hughes-Oliver, J.M. (1998). Applications of Basu's Theorem, *The American Statistician*, 52, 218–221.

Boos, D.D. and Stefanski, L.A. (2013). *Essential Statistical Inference: Theory and Methods*, Springer-Verlag, NY.

Bowman, K. and Shenton, L. (1988). *Properties of Estimators for the Gamma Distribution*, Marcel Dekker, NY.

Box, G.E.P. and Tiao, G.C. (1973). *Bayesian Inference in Statistical Analysis*, Addison-Wesley, Reading, MA USA.

Box, G.E.P. and Tiao, G.C. (1992). *Bayesian Inference in Statistical Analysis*, Wiley-Interscience, USA.

Brewster, J.F. and Zidek, J.V. (1974). Improving on equivariant esimators. *Ann. Statist.*, 2, 21–38.

Bromik, T. and Pleszezynska, E. (2011). *Statistical Inference: Theory and Practice*, Springer.

Brown, L.D. (1964). Sufficient statistics in the case of indepencent ramdom variables. *Ann. Math. Statist.*, 35, 1456–1474.

Brown, L.D. (1986). *Fundamentals of Statistical Exponential Families with Applications in Statistical Decision Theory*, Institute of Mathematical Statistics Lecture Notes – Monograph Series, IMS, Haywood, CA.

Brown, L.D., Cai, T.T. and DasGupta, A. (2001). Interval estimation for a binomial proportion, (with discussion), *Statistical Science*, 16, 101–133.

Brown, L.D., Cai, T.T. and DasGupta, A. (2002). Confidence intervals for a binomial proportion and asymptotic expansions, *The Annals of Statistics*, 30, 160–201.

Brown, L.D., Cai, T.T. and DasGupta, A. (2003). Interval estimation in exponential families, *Statistica Sinica*, 13, 19–49.

Brownstein, N. and Pensky, M. (2008). Application of transformations in parametric inference, *Journal of Statistical Education*, vol 16(1).

Buehler, R.J. (1982). Some ancillary statistics and their properties (with discussion), *J. Amer. Statist. Assoc.*, 77, 590–591.

Bhler, W. and Sehr, J. (1987). Some remarks on exponential families, *The American Statistician*, 41, 279–280.

Byrne, J. and Kabaila, P. (2005). Comparison of Poisson confidence intervals, *Communications in Statistics: Theory and Methods*, 34, 545–556.

Carlin, B.P. and Louis, T.A. (1996). *Bayes and Empirical Bayes Methods for Data Analysis*. London: Chapman and Hall.

Cassella, G. and George, E.I. (1992). Explaining the Gibbs Sampler, *Amer. Statist.*, 46, 167–174.

Cassella, G. and Berger, E.L. (2002). *Statistical Inference*, Duxbury, 2nd ed., Duxbury, Belmont, CA.

Chakravarti, I.M. (1980). *Asymptotic Theory of Statistical Tests and Estimation: In Honour of Wassily Hoeffding*, Academic Press.

Chance, B.L. (2014). *An Active Approach to Statistical Inference*, Wiley, New York.

Chanda, K.C. (1954). A note on the consistency and maxima of the roots of likelihood equations. *Biometrika*, 41, 56–61.

Chapman, D.G. and Robbins, H. (1951). Minimum variance estimation without regularity assumptions, *Ann. Math. Statist.*, 22, 581–586.

Chernoff, H. and Moses, L.E. (1986). *Elementary Decision Theory*, Dover Publications, NY.

Cornfield, J. (1969). The Bayesian outlook and its application (with discussions), *Biometrika*, 25, 617–657.

Cox, C. (1984). An elementary introduction to maximum likelihood estimations for multinomial models: Birch's theorem and the Delta method, *The American Statistician*, 38, 283–287.

Cox, D.R. (1958). Some problems connected with statistical inference, *The Ann. of Math. Stats.*, 29(2), 357–372.

Cox, D.R. (1971). The choice between ancillary statistics, *J. Roy. Statist. Soc.*, Ser. B, 33, 251–255.

Cox, D.R. (2006). *Principles of Statistical Inference*, Cambridge University Press, Cambridge, New York.

Cox, D.R. and Hinkley, D.V. (1974). *Theoretical Statistics,* Chapman and Hall, London.

Cox, D.R. and Hinkley, D.V. (1979). *Theoretical Statistics*, CRC Press, USA.

Cramér, H. (1946a), *Mathematical Methods of Statistics*, Princeton University Press, Princeton, NJ.

Cramér, H. (1946b). A contribution to the theory of statistical estimation. *Skand. Aktuarietidskr.*, 29, 85–94.

Darmois, G. (1935). Sur les lois de probabilities a estimation exhaustivd, *C. R. Acad. Sci.*, Paris, 260, 1265–1266.

DasGupta, A. (2008). *Asymptotic Theory of Statistics and Probability*, Springer-Verlag, NY.

Datta, G.S. (2005). An alternative derivation of the distributions of the maximum likelihood estimators of the parameters in an inverse gaussian distribution, *Biometrika*, 92, 975–977.

David, H.A. (1981). *Order Statistics*, 2nd ed., John Wiley and Sons Inc., New York.

David, H.A. and Nagaraja, H.N. (2003) *Order Statistics*, 3rd ed., Wiley Interscience (John Wiley and Sons), Hoboken, NJ.

Davidson, R.R. and Solomon, D.L. (1974). Moment-type estimation in the exponential Family, *Communications in Statistics*, vol. 3, pp. 1101–1108.

DeGroot, M.H. (1986). *Probability and Statistics*, 2nd ed., Addison-Wesley, New York.

DeGroot, M.H. and Schrevish, M.J. (2011). *Probability and Statistics*, 4th ed., Addison-Wesley, Pearson, UK.

Desu, M.M. (1971). Optimal confidence intervals of fixed width. *Am. Stat.,* 25 (2), 27–29.

Dudewicz, E.J. (1976). *Introduction to Statistics and Probability*, Hott, Rinehart and Winston, New York.

Dudewicz, E.J. (Ed.) (1985). *The Frontiers of Modern Statistical Inference Procedures*, American Science Press, Columbus, Ohio.

Dudewicz, E.J. and Mishra, S.N. (1988). *Modern Mathematical Statistics*, Wiley, New York.

Durbin, J. (1970). On Birnbaum's theorem and the relation between sufficiency, conditionality and likelihood, *J. Amer. Statist. Assoc.*, 65, 395–398.

Edwards, A.W.F. (1992). *Likelihood* (2nd ed.), Johns Hopkins University Press, Baltimore.

Efron, B. (1982). The Jackknife, the bootstrap and other resampling plans, *Society for Industrial and Applied Mathematics*, Philadelphia.

Efron, B. (2010). *Large-scale Inference: Empirical Bayes Methods for Estimation Testing and Prediction*, Cambridge University Press, UK.

El-Sayyed, G.M. and Freeman, P.R. (1973). Bayesian sequential estimation of a Poisson process rate, *Biometrika*, 60, 289–296.

Fahmeir, L. and Kaufmann, H. (1985). Consistency and asymptotic normality of maximum likelihood estimator in generalized linear models, *Ann. Statist.*, 13, 342–368.

Feller, W. (1968). *An Introduction to Probability Theory and Its Applications* (3rd ed., vol. 1), Wiley, New York.

Feller, W. (1971). *An Introduction to Probability Theory and Its Applications* (2nd ed., vol. 2), Wiley, New York.

Ferentinos, K.K. (1990). Shortest confidence intervals for families of distributions involving truncated parameters. *Am. Stat.,* 44, 40–41.

Ferguson, T.S. (1967). *Mathematical Statistics: A Decision Theoretic Approach*, Academic Press, New York.

Ferguson, T.S. (1996). *A Course in Large Sample Theory,* Chapman and Hall, NY.

Fisher, R.A. (1912). On an absolute criterion for fitting frequency curves, *Messenger of Mathematics*, 41, 155–160.

Fisher, R.A. (1920). A mathematical examination of the methods of determining the accuracy of an observation by the mean error, and by the mean square error, *Monthly Notices of the Royal Astronomical Society*, 80, 758–770.

Fisher, R.A. (1922). On mathematical foundations of theoretical statistics. *Philos. Trans. R. Soc.* , A222, 309–386.

Fisher, R.A. (1925). Theory of statistical estimation, *Proceedings of the Cambridge Philosophical Society*, 22, 700–725.

Fisher, R.A. (1937). Professor Karl Pearson and method of moments, *Annals of Eugenics*, 7, 303–318.

Fisher, R.A. (1954). *Statistical Methods for Research Workers* (12th ed.), *Biological monographs and manuals*, Oliver and Boyd Ltd., Edinburgh, 299–320.

Fisher, R.A. (1959). *Statistical Methods and Scientific Inference* (2nd ed.), Hafner Publishing Co., New York.

Fisz, M. (1963). *Probability Theory and Mathematical Statistics* (3rd ed.), Wiley, New York.

Foutz, R.V. (1977). On the unique consistent solution of the likelihood equations, *Jour. Amer. Stat. Assoc.*, 72, 147–148.

Fraser, D.A.S. (1968). *The Structure of Inference*, Wiley, New York.

Fréchet, M. (1943). Sur l'extension de certaines evaluations statistiques au cas de petits echantillons. *Revue de l'Institut International de Statistique/Review of the International Statistical Institute* 11(3/4), 182–205.

Freedman, D., Pisani, R., Purves, R. and Adhikari, A. (1991). *Statistics*, 2nd ed., Norton, New York.

Galambos, J. and Kotz, S. (1977). *Characterizations of Probability Distributions*, Lecture Notes in Mathematics, Springer-Verlag, Heidelberg.

Gathwaite, P.H., Jolliffe, I.T. and Jones, B. (2002). *Statistical Inference*, 2nd ed., Oxford University Press, Oxford, UK.

Gauss, C.F. (1809). *Theoria motus corporum coelestium in sectionibus conicis solem ambientium.* Perthles et Besser, Hamburg, *Werke*, 7, 1–280. Translated by C.H. Davis as *Theory of the Motion of the Heavenly Bodies about the Sun in Conic Sections*. Little, Brown, Boston, 1857. Reprint by Dover, New York, 1963.

Gauss, C.F. (1823). Theoria combinationis observationum erroribus minimis obnoxiae: Pars prior. *Commentatines societatis regiae scientarium Gottingensis recentiores*, 5. English translation by G.W. Stewart, 1995.

Geisser, S. and Johnson, W.O. (2006). *Modes of Parametric Statistical Inference*, John Wiley and Sons, New Jersey.

Gelfand, A.E. and Smith, A.F.M. (1990). Sampling-based approaches to calculating marginal densities, *J. Amer. Statist. Assoc.*, 85, 398–409.

Gelfand, A. and Rubin, D.B. (1992). Inference from iterative simulation using multiple sequences (with discussion), *Statist. Sci.*, 7, 457–511.

Gentle, J.E. (2008). *A Companion for Mathematical Statistics*, Faifax Couty, Virginia.

Guenther, W.C. (1969). Shortest confidence intervals, *Am. Stat.*, 23, No. 1, 22–25.

Guenther, W.C. (1971). Unbiased confidence intervals, *Am. Stat.*, 25, No. 1, 51–53.

Guenther, W.C. (1978). Some easily found ninimum variance unbiased estimators, *The American Statistician*, 32, 29–33.

Hahn, G.J. and Meeker, M.Q. (1991). *Statistical Intervals: A Guide for Practitioners*, Wiley, Hoboken, NJ.

Hald, A. (2010). *A History of Parametric Statistical Inference from Bernoulli to Fisher*, 1713–1935, Springer-Verlag, NY.

Halmos, P.R. and Savage, L.J. (1949). Application of Radon-Nikodym theorem to the theory of sufficient statistics, *Ann. Math. Stat.*, 20, 225–241.

Held, L. and Bove, D.S. (2014). *Applied Statistical Inference: Likelihood and Bayes*, Springer-Verlag, New York.

Hinkley, D.V. (1980a). *Fisher's development of conditional inference, R.A. Fisher: An Appreciation* (S.E. Fienberg and D.V. Hinkley, Eds.), 101–108, Springer-Verlag, New York.

Hinkley, D.V. (1980b). Likelihood, *Can. J. Statist.*, 8, 151–163.

Hinkley, D.V. (1980c). *R.A. Fisher: Some introductory remarks, R.A. Fisher: An Appreciation* (S.E. Fienberg and D.V. Hinkley, Eds.), 1–5, Springer-Verlag, New York.

Hipp, C. (1974). Sufficient statistics and exponential families. *Ann. Statist.*, 2, 1283–1292.

Hodges, J.L. and Lehmann, E.L. (1950). Some problems in minimax point estimation, *Ann. Math. Stat.*, 21, 182–197.

Hoeffding, W. (1948). A class of statistics with asymptotically normal distribution, *Ann. Math. Stat.*, 19, 293–325.

Hoff, P.D. (2009). *A First Course in Bayesian Statistical Methods*. Springer-Verleg, New York.

Hogg, R.V. and Tanis, E.A. (2006). *Introduction to Mathematical Statistics,* 7th ed., Pearson, UK.

Hogg, R.V., Mkcean, J. and Craig, A.T. (2006). *Probability and Statistical Inference*, 7th ed., Pearson Education, New Delhi, India.

Huntsberger, D.V. and Billingley, P. (1987). *Elements of Statistical Inference*, Brown (Willam C.) Co, U.S.

Huzurbazar, V.S. (1948). The likelihood equation consistency, and maxima of the likelihood function, *Ann. Eugen.* (London), 14, 185–200.

Huzurbazar, V.S. (1955). Confidence intervals for the parameter of a distribution admitting a sufficient statistic when the range depends on the parameter, *Jour. Roy. Stat. Soc. B*, 17, 86–90.

Ibragimov, I.A. and Khasminskii, R.Z. (1981). *Statistical Estimation, Asymptotic Theory*, Springer, New York, Heidelberg Berlin.

Jahnke, E., Emde, F. and Lsch, F. (1966). *Tables of Higher Functions* (7th ed.), Stuttgart. B.G. Teubner. Reprint of 6th ed. (McGraw Hill, New York, 1960).

James, W. and Stein, C. (1961). Estimation with quadratic loss, *In Proc. 4th Berkeley Symp. Math. Stat. Probab.*, 1, 361–379. University of California Press, CA.

Jeffreys, H. (1957). *Scientific Inference*, Cambridge University Press, London.

Jeffreys, H. (1961). *Theory of Probability*, 3rd ed., Oxford University Press, London.

Joshi, V.M. (1976). On the Attainment of the Cramér-Rao Lower Bound, *The Annals of Statistics,* 4, 998–1002.

Kalbfleisch, J.D. (1975). Sufficiency and conditionality, *Biometrika*, 62, 251–268.

Kalbfleisch, J.G. (1985). *Probability and Statistical Inference*, 2nd ed., Springer-Verlag, New York.

Kale, B.K. (1999). *A First Course on Parametric Inference*, Narosa Publishing House, New Delhi.

Kale, B.K. (2005). *A First Course on Parametric Inference*, 2nd ed., Alpha Science International Ltd.

Kallianpur, G. and Rao, C.R. (1955). On Fisher's lower bound to asymptotic variance of a consistent estimate, *Sankhaya*, 15, 331–342.

Karakostas, K.X. (1985). On minimum variance unbiased estimators, *The American Statistician,* 39, 303–305.

Karlin, S. (1992). R.A. Fisher and evolutionary theory, *Statist. Sci.*, 7(1), 13–33.

Keener, R.W. (2010). *Theoretical Statistics*, Springer-Verlag, NY.

Kendall, M. and Stuart, A. (1994). *The Advanced Theory of Statistics, Vol. I, Distribution Theory,* 6th ed., John Wiley, NY.

Kendall, M. and Stuart, A. (1979). *The Advanced Theory of Statistics, Vol. II, Inference and Relationship*, 4th ed., Macmillan, NY.

Kiefer, J. (1952). On minimum variance estimators, *Ann. Math. Stat.*, 23, 627–629.

Kiefer, J.C. (1987). *Introduction to Statistical Inference*, Springer-Verlag, New York.

Klebanov, L.B. (1972). Unbiased loss functions and unbiased estimation, *Dokl. Akad. Nauk SSSR* 203(6): 1249–1251. (in Russian)

Klebanov, L.B., Link, Yu. V. and Rukhin, A.L. (1971). Unbiased estimation and matrix loss functions, *Soviet Math Dokl.*, 12, 1526–1528.

Knight, K. (2000). *Mathematical Statistics*, Chapman and Hall/CRC, Boca Raton, FL.

Koehn, U. and Thomas, D.L. (1972). On statistics independent of a sufficient statistic: Basu's lemma, *The American Statistician*, 29, 40–42.

Kolmogorov, A.N. (1950a). Unbiased estimation, *Amer. Math. Soc. Translation,*, No. 90, 155–177.

Kolmogorov, A.N. (1950b). *Foundations of the Theory of Probability* (German edition, 1933). Chelsea, New York.

Koopman, B.O. (1936). On distributions admitting a sufficient statistic, *Trans. Amer. Math. Soc.*, 39, 399–409.

Kubacek, L. (1988). *Foundations of Estimation Theory*, Elsevier, Amsterdam-Oxford, New York-Tokyo.

Kulldorf, G. (1957). On the condition for consistency and asymptotic efficiency of maximum likelihood estimates, *Skand. Aktuarietidskr.*, 40, 129–144.

Laplace, P. (1812). *Théorie analytique des probabilités*, Courcier, Paris.

Laha, R.G. and Rohatgi, V.K. (1979). *Probability Theory*, Wiley, New York.

Lane, D.A. (1980). *Fishers, Jeffreys and the Nature of Inference, R.A. Fisher: An Appreciation (S. E. Fienberg and D.V. Hinkley, Eds.)*, 148–160, Springer-Verlag, New York.

Larry, Wasserman (2005). *All of Statistics: A Concise Course in Statistical Inference*, Springer-Verlag, New York.

Larsen, R.J. and Marx, M.L. (2001). *Introduction to Mathematical Statistics and Its Applications*, 3rd ed., Prentice Hall, Upper Saddle River, NJ.

LeCam, L. (1953). On Some asymptotic properties of maximum likelihood estimates and related Bayes' estimates. *Univ. of Calif. Publ. in Statist.*, 1, 277–330.

LeCam, L. (1956). On the asymptotic theory of estimation and testing hypotheses, *Proc. Third Berkley Symp. Math. Statist. Probab.*, 1, 129–156, University of California Press, Berkeley.

LeCam, L. (1986). *Asymptotic Methods in Statistical Decision Theory*, Springer-Velag, New York.

LeCam, L. and Yang, G.C. (2000). *Asymptotics in Statistics: Some Basic Concepts*, 2nd ed., Springer-Verlag, New York.

Legendre, A.M. (1805). *Nouvelles Méthodes pour la Détermination des Orbites des Comètes*, Firmin Didot, Paris; second edition Courcier, Paris, 1806. Pages 72–75 of the appendix reprinted in Stigler (1986, p. 56). English translation of these pages by H.A. Ruger and H.M. Walker in D.E. Smith, *A Source Book of Mathematics*, McGraw-Hill Book Company, New York, 1929, 576–579.

Lehmann, E.L. (1980). Efficient Likelihood Estimators, *The American Statistician*, 34, 233–235.

Lehmann, E.L. (1981). An interpretation of completeness and Basu's theorem, *J. Am. Stat. Assoc.*, 76, 335–340.

Lehmann, E.L. (1983). *Theory of Point Estimation*, John Wiley and Sons., New York.

Lehmann, E.L. (1986). *Testing Statistical Hypotheses*, 2nd ed., John Wiley and Sons, New York.

Lehmann, E.L. and Scheffe, H. (1950). Completeness, similar regions, and unbiased estimation-Part-I, *Sankhya*, Ser. A, 10, 305–340.

Lehmann, E.L. and Scheffe, H. (1955). Completeness, similar regions, and unbiased estimation-Part-II, *Sankhya*, Ser. A, 15, 219–236. Corrections, 17, 250.

Lehmann, E.L. (1993). Mentors and early collaborators: Reminiscences from the years 1940–1956 with an epilogue, *Statist. Sci.*, 8, 331–341.

Lehmann, E.L. (1999). *Elements of Large-Sample Theory*, Springer-Verlag, NY.

Lehmann, E.L. and Casella, G. (2003). *Theory of Point Estimation,* 2nd ed., Springer, NY.

Lehmann, E.L. and Scheffe, H. (1950, 1955, 1956). Completeness, similar regions, and unbiased estimation. *Sankhya*, Ser. A, 10, 305–340; 15, 219–236; correction 17, 250.

Levy, M.S. (1985). A note on nonunique MLE's and sufficient statistics, *The American Statistician,* 39, 66.

Liero, H. and Zwanzig, S. (2011). *Introduction to the Theory of Statistical Inference*, Chapman and Hall/CRC, London.

Lindgren, B.W. (1993). *Statistical Theory,* 4th ed., Chapman and Hall/CRC, Boca Ratan, FL.

Lindley, D.V. (1965). *Introduction to Probability and Statistics from a Bayesian Point of View*, Cambridge University Press, London.

Lindley, D.V. and Smith, A.F.M. (1972). Bayes estimates for the linear model. *J. Roy. Statist. Soc. Ser. B*, 34, 1–41.

Lindsey, J.K. (1996). *Parametric Statistical Inference*, Oxford University Press, Oxford, UK.

Madansky, A. (1962). More on length of confidence intervals. *J. Amer. Statist. Assoc.*, 57, 586–589.

McCulloch, R.E. (1988). Information and the likelihood function in exponential families, *The American Statistician*, 42, 73–75.

McCullagh, P. and Nelder, J.A. (1989). *Generalized Linear Models*, 2nd ed., Chapman and Hall, London.

Miescke, K.J. and Liese, F. (2008). *Statistical Decision Theory: Estimation, Testing, Selection*, Springler-Verlag, New York.

Migon, H.S., Gamerman, D. and Neto, F.L. (1999). *Statistical Inference: An Integrated Approach*, Arnold, London.

Migon, H.S., Gamerman, D. and Neto, F.L. (2014). *Statistical Inference: An Integrated Approach*, 2nd ed., Chapman and Hall/CRC, London.

Miller, R.G. (1974). The Jackknife-A Review, *Biometrka*, 61, 1–15.

Millar, R.B. (2011). *Maximum likelihood estimation and inference: with examples in R, SAS, and ADMB*, John Wiley, New York.

Mood, A.M., Graybill, F.A., and Boes, D.C. (1974). *Introduction to the Theory of Statistics*, 3rd ed., McGraw-Hill, NY.

Moore, D.S. (1971). Maximum likelihood and sufficient statistics, *The American Mathematical Monthly*, 78, 50–52.

Moore, D.S. (2007). *The Basic Practice of Statistics*, 4th ed., W.H. Freeman, NY.

Morris, C.N. (1983). Parametric empirical Bayes inference: Theory and Applications (with discussion). *Amer. Statist. Assoc.*, 78, 47–65.

Mukhopadhyay, N. (2000). *Probability and Statistical Inference*, Marcel Dekker, NY.

Mukhopadhyay, N. (2006). *Introductory Statistical Inference*, Chapman and Hall/CRC, Boca Raton, FL.

Mukhopadhyay, P. (2009). *Mathematical Statistics*, 2nd. ed., Books and Allied (P) Ltd., Kolkata, India.

Neyman, J. (1935a). Sur un teorema concernente le cosidette statistiche sufficienti, *Giorn.1st. Ital. Atti.* 6, 320–334.

Neyman, J. (1935b). On the problem of confidence interval, *Ann. Math. Statist.*, 6(3), 111–116.

Neyman, J. (1937). Outline of a theory of statistical estimation based on the classical theoy of probability, *Phil. Trans. Roy. Soc.*, Ser. A, 236, 333–380.

Neyman, J. and Scott, E.L. (1948). Consistent estimates based on partially consistent observations, *Econometrica*, 16, 1–32.

Nagaraja, H.N. and Nevzorov, V.B. (1997). On characterizations based on record values and order statistics, *J. Statist. Plann. Inference*, 63, 271–284.

Oakes, M.W. (1990). *Statistical Inference*, Epidemiology Resources, Inc. Chestnut Hill, MA.

Olive, D.J. (2004). Does the MLE Maximize the Likelihood?, Unpublished Document, see (www.math.siu.edu/olive/infer.htm).

Olive, D.J. (2007). A Simple Limit Theorem for Exponential Families, Unpublished Document, see (www.math.siu.edu/olive/infer.htm).

Olive, D.J. (2008), Using Exponential Families in an Inference Course, Unpublished Document, see (www.math.siu.edu/olive/infer.htm).

Olive, D.J. (2014). *Statistical Theory and Inference*, Springer-Verlag, New York.

Oliver, D.J. (2008). A Course in Statistical Theory, dolive@math.siu.edu, Maicode 40008.

Oliver, E.H. (1972). A maximum likelihood oddity, *Am. Stat.*, 26, No. 3, 43–44.

Pal, N. and Berry, J. (1992). On invariance and maximum likelihood estimation, *The American Statistician,* 46, 209–212.

Pawitan, Y. (2013). *In All Likelihood: Statistical Modelling and Inference using Likelihood*, Oxford University press, UK/USA.

Pearson, K. (1920). The fundamental problem of practical statistics. *Biometrika*, 13(1): 1–16.

Pearson, K. (1936). Method of moments and method of maximum likelihood, *Biometrika*, 28, 34–59.

Peressini, A.L., Sullivan, F.E. and Uhl, J.J. (1988). *The Mathematics of Nonlinear Programming*, Springer-Verlag, New York.

Perlman, M.D. (1972). On the strong consistency of approximate maximum likelihood estimators–an introduction, *Proceedings of the Sixth Berkeley Symposium on Mathematical Statistics and Probability,* 1, 263–281.

Pfanzagl, J. (1993). Sequences of optimal unbiased estimators need not be asymptotically optimal, *Scandinavian Journal of Statistics,* 20, 73–76.

Pitman, E.J.G. (1936). Sufficient statistics and intrinsic accuracy, *Proc. Camb. Phil. Soc.*, 32, 567–579.

Pitman, E.J.G. (1979). *Some Basic Theory for Statistical Inference*, Chapman and Hall, New York.

Portnoy, S. (1977). Asymptotic efficiency of minimum variance unbiased estimators, *The Annals of Statistics,* 5(3), 522–529.

Pratt, J.W. (1961). Length of confidence intervals, *J. Am. Stat. Assoc.*, 56, 260–272.

Quenouille, M.H. (1956). Notes on bias in estimation, *Biometrika*, 43, 353–360.

Raiffa, H. and Schlaifer, R. (1961, 2000). *Applied Statistical Decision Theory*, John, Wiley, NY.

Rao, B.L.S.P. (1987). Asymptotic theory of statistical inference, Wiley, NY.

Rao, C.R. (1945). Information and accuracy attainable in the estimation of statistical parameters, *Bull. Calcutta Math. Soc*., 37, 81–91.

Rao, C.R. (1947). Large sample tests of statistical hypotheses concerning several parameters with applications to problem of estimation, *Proc. Com. Phil. Soc.*, 44, 50–57.

Rao, C.R. (1949). Sufficient statistics and minimum variance unbiased estimates, *Proc. Cambridge Philos. Soc*., 45, 213–218.

Rao, C.R. (1962). Efficient estimates and optimum inference procedures in large samples (with discussion), *Jour. Roy. Stat. Soc.* B, 24, 46–72.

Rao, C.R. (1963). Criterion of estimation in large samples. *Sankhya* A, 25, 189–206.

Rao, C.R. (2001). *Linear Statistical Inference and Its Applications*, 2nd ed., Wiley, New York.

Rencher, A.C. (1998). *Multivariate Statistical Inference and Applications*, John Wiley and Sons, New York.

Robert, C.P. (1994). *The Bayesian Choice: A Decision Theoretic Motivation*, Springer-Verlag, New York.

Robert, C.P. (2001). *The Bayesian Choice: From Decision Theoretic Foundation to Computational Implementation*, 2nd ed., Springer.

Rohatgi, V.K. (1984). *Statistical Inference*, Wiley, New York.

Rohatgi, V.K., Saleh, A.K. and Md. Ehsanes (2006). *An Introduction to Mathematical Statistics*, John Wiley and Sons, 2nd ed., New York.

Ross, S.M. (1988). *A First Course in Probability Theory,* 3rd ed., Macmillan, New York.

Roussas, G.G. (1997). *A Course in Mathematical Statistics*, 2nd ed., Academic Press, San Diego, CA.

Roussas, G.G. (2003). *An Introduction to Probability and Statistical Inference*, Academic Press.

Roy, J. and Mitra, S. (1957). Unbiased minimum variance estimation in a class of discrete distributions, *Sankhya,* 18, 371–378.

Roy, R., LePage, Y. and Moore, M. (1974). On the power series expansion of the moment generating function, *Am. Stat.*, 28, 58–59.

Samaniego, F.J. (2010). *A Comparision of the Bayesian and Frequentist Approaches to Estimation*, Springer-Verlag, NY.

Sampson, A. and Spencer, B. (1977). Sufficiency, minimal sufficiency and lack thereof, *Am. Stat.*, 30, 34–35. Correction (1977), 31, 54.

Satterthwaite, F.E. (1946). An approximate distribution of estimates of variance components, *Biometrics Bulletin*, vol 2, No. 6, 110–114.

Savage, S.L. (1972). *The Foundations of Statistics*, 2nd (revised) ed., Dover Publication, New York.

Saw, J.G., Yang, M.C.K. and Mo, T.C. (1984). Chebychev's inequality with estimated mean and variance, *Amer. Statist.*, 38, 130–132.

Scheffe, H. (1943). On solutions of the Behrens-Fisher problem based on the *t*-distribution, *Ann. Math. Stat.*, 14, 33–44.

Scheffe, H. (1944). A note on Behrens-Fisher problem, *Ann. Math. Stat.*, 15, 430–432.

Scheffe, H. (1947). A useful convergence theorem for probability distributions, *Ann. Math. Stat.*, 18, 434–438.

Scheffe, H. (1954). *The Foundations of Statistics*, Wiley, New York.

Scheffe, H. (1959). *Analysis of Variance*, Wiley, New York.

Scheffe, H. (1970). Practical solution of the Behrens-Fisher problem, *Jour. Amer. Statist. Assoc.*, 65, 1501–1508.

Schervish, M.J. (1995). *Theory of Statistics*, Springer-Verlag, NY.

Searle, S.R. (1971). *Linear Models*, Wiley, New York.

Sen, P.K. and Singer, J.O. (1993). *Large Sample Methods in Statistics: An Introduction with Applications*, Chapman and Hall, NY.

Serfling, R.J. (1979). *Approximation Theorems of Mathematical Statistics*, John Wiley and Sons, New York.

Severine, T.A. (2001). *Likelihood Methods in Statistics*, Oxford University Press, NY.

Shao, J. and Tu, D. (1995). *The Jackknife and the Bootstrap*. Springer Verlag, New York.

Shao, Jun (2003). *Mathematical Statistics*, 2nd ed., Springer, New York.

Shao, Jun (2005). *Mathematical Statistics: Exercises and Solutions*, Springer, New York.

Silvey, S.D. (1975). *Statistical Inference*, Chapman and Hall/CRC, USA.

Smith, A.F.M. and Gelfand, A.E. (1992). Bayesian statistics without tears: A sampling resampling perspective, *Amer. Statist.*, 46(2), 84–88.

Smith, A.F.M. and Roberts, G.O. (1993). Bayesian computation via the Gibbs sampler and related Markov Chain Monte Carlo methods (with discussion), *J. Roy. Statist. Soc. Ser. B*, 55, 3–24.

Sprott, D.A. (1990). Inferential estimation, likelihood, and linear pivotals, *Canadian Journal of Statistics*, 18, 1–5.

Sprott, D.A. (2000). *Statistical Inference in Science*, Springer-Verlag, New York.

Sprott, D.A. and Kalbfleisch, J.D. (1969). Examples of likelihoods and comparison with point estimates and large sample approximations, *Journal of the American Statistical Association*, 64, 468–484.

Stapleton, J.H. (2008). *Models for Probability and Statistical Inference: Theory and Applications*, John Wiley and Sons, New Jersey.

Stein, C. (1956). Inadmissibility of the usual estimator of the mean of a multivariate normal distribution, *Pro. Third Berkley Symp. Math. Statist. Prob.*, 1, 197–206, University of California Press, Berkley, CA.

Stein, C. (1959). The admissibility of Pitman's estimator of a single location parameter, *Ann. Math. Statist.*, 30, 970–979.

Stephen, M.S., Wing, H.W. and Daming, X. (2002). R.R. Bahadur's Lectures on the theory of estimation, monograph volume 39 series, Institute of Mathematical Statistics, USA.

Stigler, S.M. (1972). Completeness and unbiased estimation, *Am. Stat.*, 26, 28–29.

Stigler, S.M. (2007). The epic journey of maximum likelihood, *Statistical Science*, 22, 598–620.

Stuart, A. and Ord, J.K. (1991). *Kendall's Advanced Theory of Statistics, Volume II,* 5th ed., New York: Oxford University Press.

Stuart, A., Ord, J.K. and Arnold, S. (1999). *Advanced Theory of Statistics, Volume 2A, Classical Inference and the Linear Model,* 6th ed., Oxford University Press, London.

Sundaram, R.K. (1996). *A First Course in Optimization Theory*, Cambridge University Press, UK.

Tanner, M.A. (1996). *Tools for Statistical Inference: Methods for the Exploration of the Posterior Distributions and Likelihood Functions*, 3rd ed., Springer-Verlag, New York.

Tanner, M.A. (1996). *Tools for Statistical Inference: Observed Data and Data Augmentation Methods,* 3rd ed., Springer-Verlag, New York.

Tate, R.F. and Klett, G.W. (1959). Optimal confidence intervals for the variance of a normal distribution, *J. Am. Stat. Assoc.*, 54, 674–682.

Thompson, R.D. and Basu, A.P. (1996). Asymmetric loss functions for estimating system reliability. In D.A. Berry, K.M. Chaloner and J.K. Geweke, Editors, *Bayesian Analysis in Statistics and Econometrics*, pages 471–482, John Wiley and Sons, New York.

Tukey, J.W. (1977). *Exploratory Data Analysis*, Addison-Wesley, Reading, MA.

van der Vaart, A.W. (1998). *Asymptotic Statistics*, Cambridge University Press, Cambridge.

Varian, H.R. (1975). A Bayesian approach to real estate assessment. In *Studies in Bayesian Econometrics and Statistics in Honor of Leonard J. Savage* (S.E. Fienberg and Zellner, Eds.), 195–208, North-Holland Publ., Amsterdam.

Wald, A. (1949). Note on the consistency of the maximum likelihood estimate, *Ann. Math. Stat.*, 20, 595–601.

Wald, A. (1950). *Statistical Decision Functions*, Wiley, New York.

Wasserman, L. (2004). *All of Statistics: A Concise Course in Statistical Inference*, Springer, NY.

Welsh, A.H. (1996). *Aspects of Statistical Inference*, John Wiley and Sons, New York.

Westfall, P.H. and Henning, K.S.S. (2013). *Understanding Advanced Statistical Methods*, Chapman and Hall/CRC Press, New York.

Whittle, P. and Lane, R.O.D. (1967). A class of situations in which a sequential estimation procedure is non-sequential, *Biometrika*, 54, 229–234.

Widder, D.V. (1961). *Advanced Calculus*, Prentice-Hall, 2nd ed., Englewood Cliffs, N.J.

Wijsman, R.A. (1973). On the attainment of the Cramér-Rao lower bound, *The Annals of Statistics,* 1(3), 538–542.

Wilks, S.S. (1962). *Mathematical Statistics*, Wiley, New York.

Wolfowitz, J. (1965). Asymptotic efficiency of maximum likelihood estimators, *Theory of Prob. and its Appl.*, 10, 247–260.

Wu, C.F.J. (1986). Jackknife, bootstrap and other resampling methods in regression analysis (with discussions), *Ann. Statist.*, 14, 1261–1350.

Young, G.A. and Smith, R.L. (2005). *Essentials of Statistical Inference*, Cambridge University Press, UK.

Zacks, S. (1971). *The Theory of Statistical Inference*, Wiley, New York.

Zehna, P.W. (1966). Invariance of maximum likelihood estimators, *Ann. Math. Stat.*, 37, 744.

Zellner, A. (1986). Bayesian methods and prediction using asymmetric loss function, *J. Amer. Statist. Assoc.*, 81, 446–451.

Index

Ancillary statistic, 50
Asymptotic mean, 284
Asymptotic variance, 284

Bayes estimation
  absolute error loss function, 496
  admissibility of Bayes estimators, 514
  Bayes estimator, 5, 500
  Bayes ordering, 500
  Bayes risk, 4
  bilinear loss function, 497
  conditionality principle, 518
  conjugate prior distributions, 519
  elements of decision theory, 495
  empirical Bayes estimator, 513
  generalized Bayes estimator, 512
  hierarchical Bayes estimators, 513
  Jeffreys non informative prior, 527
  Jeffreys prior in exponential distribution, 530
  least favourable distribution, 501
  limit of Bayes estimators, 512
  LinEx loss function, 498
  maximum aposteriori (MAP) estimator, 507
  minimax estimator, 500
  minimax ordering, 500
  minimax theorem, 534
  modified LinEx loss function, 498
  noninformative priors, 522
  noninformative priors for location family of distributions, 524
  noninformative priors for scale family of distributions, 525
  posterior conditional expected loss, 502
  predictive distribution of *Y*, 502
  squared error loss function, 495
  Stein loss function, 497
  value of a game, 534
  Whittle and Lane (1967) squared error loss function, 496
Bernoulli distribution, 13
Beta distribution, 17
Binomial distribution, 13
Bivariate normal distribution, 19
Borel field, 2

Cauchy distribution, 19
Chebyshev's inequality, 111
Chebyshev's WLLN, 28
Chi-square (χ^2) distribution, 21
Completeness of a family of distributions, 51
Complete sufficient statistic, 50
Confidence interval, 6
  ARE for comparing confidence intervals, 700
  asymptotic confidence interval, 699
  asymptotic confidence interval based on MLE, 703

asymptotic confidence interval based on an asymptotic pivot, 700
Behren–Fisher problem, 683
confidence interval by Chebyshev's inequality, 704
confidence intervals based on large samples, 699
credible interval, 697
delta method for constructing asymptotic confidence interval, 700
equivariant confidence intervals, 715
Guenther method for constructing shortest length confidence intervals, 711
Guenther method for constructing unbiased confidence interval, 714
highest posterior density (HPD) interval, 697
minimum expected length confidence interval, 714
shortest length confidence interval, 705
shortest length confidence interval for location parmeter, 706
shortest length confidence interval for scale parameter, 706
UMA confidence interval obtained from UMP acceptance region, 695
UMA unbiased confidence interval, 713
unbiased confidence interval, 712
uniformly most accurate (UMA) confidence interval, 678
unimodal pivot, 706

Confidence interval
$(1 - \alpha)$-level confidence sets, 678
confidence coefficient, 678
coverage probability, 678
pivot, 679
construction based on a pivot, 688

Consistency, 266, 267
asymptotically most efficient estimator, 285
asymptotic efficiency of MLE, 288
asymptotic relative efficiency, 283, 284
best asymptotically normal (BAN) estimator, 284
CAN estimator, 283
CAN estimator by method of moments, 290
CAN estimator by method of percentiles, 297
consistent estimation by method of moments, 278
consistent estimator by method of percentiles, 280
Fisher, 269
invariance principle of, 274
joint consistency, 268
marginal consistency, 267
mean squared error, 269
principle of invariance of CAN estimators, 285
rate of consistency, 276
simple consistency, 267
sufficient condition for consistency, 270

Convex loss function, 58

Data summarization, 41
Delta method, 32
Discrete uniform distribution, 20

Efficiency of an estimator, 112
Estimate, 3
Estimator, 3, 39
Exponential family of distributions, 11

F-Distribution, 22

Game, 38
Gamma distribution, 15
Geometric distribution, 14
Group, 10
Group family of distributions, 8

Hessian matrix, 59
Hypergeometric distribution, 14

Information lower bound
asymptotically most efficient estimator, 193
attainment of CR lower bound, 194
Bhattacharyya lower bound, 202
Chapman, Robbins, and Kiefer lower bound, 206
efficiency of an estimator, 193
Fisher information, 187
Fisher information under reparameterization, 188
most efficient estimator, 193

Rao and Cramer lower bound, 190
regularity conditions, 185
relationship between CRLB and CRKLB, 207
score function, 186
Inverse Chi-square distribution, 16
Inverted-gamma distribution, 16

Jensen's inequality, 5, 59

Kolmogorov's Strong law of large numbers, 28
Koopman–Pitman–Darmois, 11

Laplace or double exponential distribution, 19
Laplace transform
bilateral, 25, 53
unilateral, 25, 52
Lindeberg–Feller central limit theorem, 30
Lindeberg–Levy central limit theorem, 30
Location family, 9
Location-scale family, 9
Logistic distribution, 17
Lognormal distribution, 18
Loss function, 3

Markov's weak law of large numbers, 28
method of moment estimation (MoM), 278, 337
Methods of estimation
estimating equations, 347
for exponential family of distribution, 351
Fisher information matrix, 347
general properties, 354
invariance property of MLE, 356
large sample properties, 358
log-likelihood function, 346
method of least squares (MoLS) estimation, 344
method of maximum likelihood (MoML) estimation, 345
method of minimum chi-square (MoMCS) estimation, 340
method of modified minimum chi-square (MoMMCS) estimation, 343
method of moment (MoM) estimation, 278, 337
MLE, obtained by Fisher's scoring method, 369
for multiparameters, 350
observed information matrix, 347
regularity conditions, 358
superefficient estimator, 367
variance stabilization, 368
Minimal sufficient statistic, 46
Minimax estimation
Bayes estimation is minimax, 535
equalizer estimator, 536
minimax estimator, 5
minimax risk, 4
Multinomial distribution, 14
multivariate normal distribution, 20
Multivariate extension of CR inequality, 199

Negative binomial distribution, 13
Non-central t-distribution, 21
Normal distribution, 15

Parameter, 38
Parameter exponential family, 11
Pareto distribution, 17
Pitman family of distributions, 12
Poisson distribution, 13
Principle of equivariance
Bayes estimate for a scale parameter, 459
class of equivariance estimator, 448
equivariant decision rules, 450
formal invariance, 446
formal structure, 449
invariance of decision problem, 450
invariance of family of distributions, 449
measurement equivariance, 446
minimum risk equivariant estimator, 6
Pitman and Bayes estimator for a location parameter, 454
Pitman estimator for a location parameter (minimum risk equivariant estimator), 452
Pitman estimator for a scale parameter, 458
Pitman or minimum risk equivariant (MRE) estimator for location parameter in location-scale family of distributions, 464
Pitman or minimum risk equivariant (MRE) estimator for scale parameter in location-scale family of distributions, 461
Principle of sufficiency, 42
Prior distribution, 5

Probability measure, 2
Probability space, 2

Random variable, 2
Rao-blackwell theorem, 5, 60
Rayleigh distribution, 18
Risk of an estimator, 4, 39

Sample space, 2
Scale family of distributions, 9
Scaled–inverse chi–square distribution, 17
σ–field, 2
Slutsky's theorem, 30
Squared error loss function, 111
Statistic, 39
Statistical decision problem, 38
Statistical model, 3
Sufficiency, 41
Sufficient statistics, 5, 41

t-Distribution, 21

Unbiased estimation, 111
  bias of an estimator, 111
  Chebyshev's inequality, 112
  efficient estimator, 113
  Jackknife estimator, 145
  Lehmann–Scheffe theorem, 116
  locally minimum variance unbiased estimator, 112
  method of finding UMVU estimator, 112
  sufficiency and UMVU estimation, 116
  uniformly minimum variance unbiased estimator, 112
  zero estimator, 111
Uniform distribution, 13, 14

Weibull distribution, 18